Handbook of Numerical Analysis

General Editors:

P.G. Ciarlet

Analyse Numérique, Tour 55–65
Université Pierre et Marie Curie
4 Place Jussieu
75005 PARIS, France

J.L. Lions

Collège de France
Place Marcelin Berthelot
75005 PARIS, France

ELSEVIER
Amsterdam • Lausanne • New York • Oxford • Shannon • Tokyo

Finite Element Methods (Part 2)
Numerical Methods for Solids (Part 2)

Volume IV

Finite Element Methods (Part 2)

Numerical Methods for Solids (Part 2)

1996
ELSEVIER
Amsterdam • Lausanne • New York • Oxford • Shannon • Tokyo

ELSEVIER SCIENCE B.V.
Sara Burgerhartstraat 25
P.O. Box 211, 1000 AE Amsterdam, The Netherlands

For information on published and forthcoming volumes URL = http://www.elsevier.nl/locate/hna

Library of Congress Catalog Card Number: 89-23314

ISBN: 0 444 81794 8

This book is printed on acid-free paper.

Printed and bound by Antony Rowe Ltd, Eastbourne
Transferred to digital printing 2005

General Preface

During the past decades, giant needs for ever more sophisticated mathematical models and increasingly complex and extensive computer simulations have arisen. In this fashion, two indissociable activities, *mathematical modeling* and *computer simulation*, have gained a major status in all aspects of science, technology, and industry.

In order that these two sciences be established on the safest possible grounds, mathematical rigor is indispensable. For this reason, two companion sciences, *Numerical Analysis* and *Scientific Software*, have emerged as essential steps for validating the mathematical models and the computer simulations that are based on them.

Numerical Analysis is here understood as the part of *Mathematics* that describes and analyzes all the numerical schemes that are used on computers; its objective consists in obtaining a clear, precise, and faithful, representation of all the "information" contained in a mathematical model; as such, it is the natural extension of more classical tools, such as analytic solutions, special transforms, functional analysis, as well as stability and asymptotic analysis.

The various volumes comprising the *Handbook of Numerical Analysis* will thoroughly cover all the major aspects of Numerical Analysis, by presenting accessible and in-depth surveys, which include the most recent trends.

More precisely, the Handbook will cover the *basic methods of Numerical Analysis*, gathered under the following general headings:

- Solution of Equations in $\mathbb{R}^n$,
- Finite Difference Methods,
- Finite Element Methods,
- Techniques of Scientific Computing,
- Optimization Theory and Systems Science.

It will also cover the *numerical solution of actual problems of contemporary interest in Applied Mathematics*, gathered under the following general headings:

- Numerical Methods for Fluids,
- Numerical Methods for Solids,
- Specific Applications.

"Specific Applications" include: Meteorology, Seismology, Petroleum Mechanics, Celestial Mechanics, etc.

Each heading is covered by several *articles*, each of which being devoted to a specialized, but to some extent "independent", topic. Each article contains a thorough description and a mathematical analysis of the various methods in actual use, whose practical performances may be illustrated by significant numerical examples.

Since the Handbook is basically expository in nature, only the most basic results are usually proved in detail, while less important, or technical, results may be only stated or commented upon (in which case specific references for their proofs are systematically provided). In the same spirit, only a "selective" bibliography is appended whenever the roughest counts indicate that the reference list of an article should comprise several thousand items if it were to be exhaustive.

Volumes are numbered by capital Roman numerals (as Vol. I, Vol. II, etc.), according to their *chronological appearance*.

Since all the articles pertaining to a given *heading* may not be simultaneously available at a given time, a given heading usually appears in more than one volume; for instance, if articles devoted to the heading "Solution of Equations in $\mathbb{R}^n$" appear in Volumes I and III, these volumes will include "Solution of Equations in $\mathbb{R}^n$ (Part 1)" and "Solution of Equations in $\mathbb{R}^n$ (Part 2)" in their respective titles. Naturally, all the headings dealt with within a given volume appear in its title; for instance, the complete title of Volume I is "Finite Difference Methods (Part 1)—Solution of Equations in $\mathbb{R}^n$ (Part 1)".

Each article is subdivided into *sections*, which are numbered consecutively throughout the article by *Arabic numerals*, as Section 1, Section 2, . . . , Section 14, etc. Within a given section, *formulas*, *theorems*, *remarks*, *and figures*, have their own independent numberings; for instance, with Section 14, formulas are numbered consecutively as (14.1), (14.2), etc., theorems are numbered consecutively as Theorem 14.1, Theorem 14.2, etc. For the sake of clarity, the article is also subdivided into *chapters*, numbered consecutively throughout the article by *capital Roman numerals*; for instance, Chapter I comprises Sections 1 to 9, Chapter II comprises Sections 10 to 16, etc.

P.G. CIARLET
J.L. LIONS
May 1989

Contents of Volume IV

Contents of the Handbook

Finite Element Methods
(Part 2)

Origins, Milestones and Directions of the Finite Element Method— A Personal View

O.C. Zienkiewicz

Institute for Numerical Methods in Engineering
University College of Swansea, Wales
UK

HANDBOOK OF NUMERICAL ANALYSIS, VOL. IV
Finite Element Methods (Part 2)—Numerical Methods for Solids (Part 2)
Edited by P.G. Ciarlet and J.L. Lions

Contents

CHAPTER I

Origins, Milestones and Directions of the Finite Element Method—A Personal View

1. Introduction

It is now over thirty years since I became involved in "the finite element method" which during most of that period dominated my research activity. The invitation to write this article provides me with a most welcome opportunity to record the story of its origins and of its subsequent development, highlighting the important milestones and directions. Clearly the latter parts are very much a personal view and hence selective. Apologies are given in advance for omission of those who perhaps may view other steps as more important. A full and thorough mathematical analysis of the finite element method as well as many more theoretically oriented references will be found in Volume II of this Handbook.

Since my early introduction to the possibilities offered by numerical approximation by Sir Richard SOUTHWELL [1940, 1946, 1956] viz. his relaxation methods and Allen [1955], my objective has been always that of providing solutions for otherwise intractible problems of interest to applied science and engineering. This objective indeed was shared by others with similar background and led to the development of the finite element method in the late fifties and sixties.

This method was only made possible by the advent of the electronic, digital computer which at the time was making its entry into the field of large arithmetic processing. Indeed the rapid rise and widespread recognition of the methodology of Finite Elements is clearly linked with the development of the computer. This of course led to a rapid development of the method which today, through various commercial and research codes, provides the key for rational design of structures, study of aeronautical fluid dynamics and electromagnetic devices needed by physics.

This article has been published in *Archives of Computational Methods in Engineering (ARCME)*, Vol. 2, No. 1 (1995) 1–48.

It is therefore not surprising that much of the development and direction of the finite element method was provided by applied scientists (engineers) seeking to solve real problems. Though recognising the roots of the methodology and the mathematical basis of the procedures such work frequently omitted the very rigorous proofs of the quality satisfying the pure mathematicians. It was therefore of much value to the field that in the seventies more formal, mathematical, approaches were introduced generally confirming the validity of the previous reasoning and adding a deeper understanding. However, in what follows I shall try to present a view of the discovery process and of the motivation which led to our present knowledge.

2. The origins

The search for the exact origins of the finite element method could be as fruitless as the search for the inventor of the wheel! The roots of the method are deeply embedded in the mathematics of continua and in engineering where assemblies of discrete "elements" were, and continue to be, the only practicable way of dealing with complex design systems.

Indeed in the earliest days of science the "atomistic" or "discrete" concepts introduced by Aristotles has governed the thinking and the mathematical concept of a continuum permitting infinite subdivision was introduced only as a convenient fiction, as late as the end of the sixteenth century by Newton and Leibnitz.

While the concept of a continuum is a useful one and has led to the fuller comprehension of modern mechanics, the engineer and physicist have learned to understand that it must not be pushed beyond certain limits. Thus the infinitesimal "dx" is limited in size by the problem at hand, including for instance atomic dimensions in a study of crystals or dimensions of several metres in a study of rock mass behaviour. The solution of practical problems can thus frequently be approached keeping in mind the "duality" of description in much the same manner as prevails in physics in the description of light phenomena (the quantum and the wave nature).

The choice of the solution procedure is, for this reason, frequently governed by mere convenience. Examples of complex *discrete* structures being modelled as continua (using a process of homogenization) or of *continua* modelled by a *physical* discretization process are, and were for at least a century, a common device. There is here an obvious necessity. The final solution of the problem *on the scale considered* must be (nearly) identical whichever approach is used! Here indeed lies the first important intersection of the engineer's and mathematicians approaches (though the two are by no means very different species).

As an example we can consider the modelling of an elastic continuum as a uniform structural bar assembly by HRENIKOFF [1941] and MCHENRY [1943] who were able to show the equivalence between this and a continuum approach to plane elasticity indicating *convergence of the discretization process* (or simply that as the size of discrete bars decreased the solution of the continuum problem was approached).

Unfortunately the Hrenikoff–McHenry "discretization" procedure was only available for rectangular mesh assemblies. To overcome this difficulty Turner together with Clough working in the aircraft industry showed that a more direct discretisation process was possible by approximating the behaviour of a continuum using "elements" of arbitrary triangular or rectangular forms. In these, simple strain states were assumed, and using such elements, solution of full scale continuum problems could be again achieved in a "convergent" manner.

Although their work was presented in January 1954 at a meeting of the Institution of Aeronautical Sciences in New York, it was not published until 1956 (TURNER, CLOUGH, MARTIN and TOPP [1956]). That paper to many is the start of the engineering finite element method although that name was only first used in 1960 by CLOUGH [1960]. The rapid development of the methodology from those early days is of course linked with the meteoric rise of the computer power permitting realistic calculations on an unprecedented scale. In the Appendix a verbatim extract from a later paper by CLOUGH [1979] is given where he describes this exciting phase of the developments.

Much contribution to this early engineering work was doubtless made by a systematic organisation of the computations using a matrix methodology coupled with energy methods in structural analysis by FALKENHEIMER [1951], LANGEFORS [1952] and ARGYRIS [1955].

Indeed the last author shows in the context of aircraft skin models that a triangular panel behaviour may well be approximated by using energy minimisation procedures—which were shown later to provide a firm basis for finite element formulation.

Priorities are of course, as shown above, difficult to establish due to much independent work containing the germs of the basic ideas. Indeed some of these due to FENG [1965] appeared in China quite early, and were totally unknown in the west.

3. The "variational" approaches via extremum principles

Although the earliest finite element forms were derived by a direct consideration of inter-element "forces" equivalent to the internal stresses in the continuum and the "displacements" of connecting nodes, it soon became evident that a more general approach could be obtained by seeking the minimum of the total potential energy within the constraint of an assumed displacement field. One of the first to adopt this approach was SZMELTER [1958] but others independently arrived at the same conclusions.

Thus if the strain field in an elastic continuum were defined by a suitable operator $\boldsymbol{S}$ acting on the displacement $\boldsymbol{u}$ as

$$\boldsymbol{\varepsilon} = \boldsymbol{S}\boldsymbol{u} \tag{3.1}$$

with the corresponding stresses given as

$$\boldsymbol{\sigma} = \boldsymbol{D}\boldsymbol{\varepsilon}\,, \tag{3.2}$$

where $\boldsymbol{D}$ was a matrix of elastic constants, then the finite element solution sought could be obtained by the minimisation of the potential energy defined as

$$\Pi = \tfrac{1}{2}\int_{\Omega} \boldsymbol{\varepsilon}^{\mathrm{T}}\boldsymbol{D}\boldsymbol{\varepsilon}\,\mathrm{d}\Omega - \int_{\Gamma_t} \bar{\boldsymbol{t}}^{\mathrm{T}}\boldsymbol{u}\,\mathrm{d}\Gamma - \int_{\Omega} \boldsymbol{b}^{\mathrm{T}}\boldsymbol{u}\,\mathrm{d}\Omega \tag{3.3}$$

in which the displacement field is approximated as

$$\boldsymbol{u}^h = \boldsymbol{N}\bar{\boldsymbol{u}}\,. \tag{3.4}$$

In the above, $\bar{\boldsymbol{u}}$ are "nodal" values of $\boldsymbol{u}$ or other parameters satisfying prescribed displacements on the boundary Γ_u, $\bar{\boldsymbol{t}}$ are the given tractions on the boundary Γ_t and $\boldsymbol{b}$ are the body forces. The function $\boldsymbol{N}$ are given in terms of the coordinates and are variously known as "shape" or "basis" functions.

Clearly such a minimisation will lead to the final, "discrete", algebraic equation of the form

$$\boldsymbol{K}\bar{\boldsymbol{u}} = \boldsymbol{f}\,, \tag{3.5}$$

where

$$\begin{aligned} \boldsymbol{K} &= \sum \boldsymbol{K}_e\,, \quad \boldsymbol{K}_e = \int_{\Omega_e} (\boldsymbol{S}\boldsymbol{N})^{\mathrm{T}}\boldsymbol{D}(\boldsymbol{S}\boldsymbol{N})\,\mathrm{d}\Omega\,, \\ \boldsymbol{f} &= \sum \boldsymbol{f}_e\,, \quad \boldsymbol{f}_e = \int_{\Gamma_t^e} \boldsymbol{N}^{\mathrm{T}}\bar{\boldsymbol{t}}\,\mathrm{d}\Gamma + \int_{\Omega_e} \boldsymbol{N}^{\mathrm{T}}\boldsymbol{b}\,\mathrm{d}\Omega \end{aligned} \tag{3.6}$$

and the simple, additive, rule of structural assembly common to standard engineering problems is preserved.

With Ω_e (and Γ^e) corresponding to "elements" into which the whole continuum problem is divided, i.e.,

$$\Omega \cup \Omega_e\,, \qquad \Gamma_e \cup \Gamma_t^e \tag{3.7}$$

Eq. (3.6) provides a convenient means of generating "element" stiffness coefficients and forces providing the approximation shape functions of Eq. (3.4) are defined on a local basis.

Clearly such a derivation of the finite element procedure showed it to be but a particular case of the approaches introduced much earlier by Lord Rayleigh (STRUTT) [1870] and RITZ [1909] which were well known and used in engineering circles. Indeed the main difference appears in the computational advantage of using a local definition of the shape functions $\boldsymbol{N}$ yielding a banded structure of the assembled stiffness matrix K of Eq. (3.5) and in preserving the local assembly structure of matrix equations.

Further, the definition of the process implies immediately certain conditions which are sufficient, but not always necessary, for the convergence of the approximation. These were first given in the early 1960s and are still valid today:

(i) that the displacement $\boldsymbol{u}$ is so defined by the shape functions (Eq. (3.4)) that no discontinuities (leading to infinite strains) develop;

(ii) that, if polynomial shape functions are used, the terms leading to constant strain values in an element can be given any arbitrary values.

The full details of such finite element requirements were presented, together with many corollaries at a meeting held in Swansea in January 1964 and published formally as text in 1965 (editors Zienkiewicz and Holister) where many seminal papers appeared, e.g., CLOUGH [1965], FRAEIJS DE VEUBEKE [1965], ZIENKIEWICZ [1965], MASSONNET [1965]. Of particular importance is the contribution of Fraeijs De Veubeke who was the first to realise that other variational, extremum, principles can be used in addition to the principle of minimum potential energy to derive approximation to problems of structural, elastic continua. Here, Fraeijs De Veubeke introduces for the first time the concept of "equilibrium" finite elements basing these on the maximisation of complementary potential energy. Such new elements are capable of ensuring equilibrium of stresses and thus provide an upper bound on the strain energy of the appropriate solution while the potential energy form provides the appropriate minimum of this. Although the direct formulation of such "equilibrating" elements presents many difficulties of stability (which were overcome much later by the introduction of stress function approximation by FRAEIJS DE VEUBEKE and ZIENKIEWICZ [1967]) the provision of energy bounds was important in bracketing the error of the approximation.

Even more important is the contribution that Fraeijs De Veubeke introduces to the so-called *mixed formulations* based on the REISSNER [1950] variational principle (later referred to as the Hellinger–Reissner principle by WASHIZU [1975]). This work led to many later publications by others and introduced the so-called *principle of limitation* which is of fundamental importance in judging the possible performance of the mixed methods. This principle originally limited to elastic problems can be paraphrased as (viz. ZIENKIEWICZ and TAYLOR [1989]) the observation that* "if mixed and irreducible approximation to the same problem can result in the same approximation they will do so—and both will yield identical results".

Thus for instance as stated by Fraeijs De Veubeke: "it is useless to look for a

* The subdivision of finite element approximations into *mixed* and *irreducible* follows the nomenclature of ZIENKIEWICZ and TAYLOR [1989]. The name is associated with the differential equations from which the approximation starts. In the irreducible form the dependent functions are reduced to the essential minimum (if necessary using penalty functions). The remainder is of course *mixed*.

better solution (of a displacement form) by injecting additional degrees of freedom for the stresses . . .". This fact, not realised widely, led to many false computational claims for mixed approaches.

The fact that finite element forms can be derived for any variational principle and not only those referring to solid mechanics led to the extension of the finite element method beyond the domain of structural mechanics. The first applications of this extension have been made in the context of the variational principle corresponding to the quasi-harmonic Poisson's equations by ZIENKIEWICZ and CHEUNG [1965], ZIENKIEWICZ, MAYHER and CHEUNG [1966] and ZIENKIEWICZ, ARLETT and BAHRANI [1967] allowing problems governed by such equations in both solid, fluid mechanics, and electro-magnetics to be solved.

Indeed at that time the earlier suggestions of Courant for a similar solution of the Laplace equations were discovered. It is interesting to note that the classic paper of COURANT [1943] is indeed not the first one of his contributions in which the use of triangular finite elements is suggested. These are alluded to by him as a possibility as early as 1923 (see COURANT [1923]) (though he appears at that time not to anticipate the importance of the discovery!).

4. Some early applications of the finite element method and alternatives

At this stage it is perhaps of interest to show some very early examples of finite element application which illustrate the versatility of the process. Here I have chosen a civil engineering problem—that of a dam solved circa 1963 and I believe the first *real* application of the method to design. The problem is taken from my first text on the finite element method in 1967 in which I tried to assemble the state of the knowledge available at the time (ZIENKIEWICZ and CHEUNG [1967]).

Incidentally the various editions of this book (ZIENKIEWICZ [1971, 1977] and ZIENKIEWICZ and TAYLOR [1989, 1991]) record the growth of the subject in which today hundreds of texts exist. However, till 1971 the above texts were the sole source—fortunately for my ego!

In Fig. 4.1 the dam, later constructed at Clywedog, Wales, is illustrated together with the subdivision into linear triangular elements. The mesh subdivision shows how easily the finite element method allows a graded mesh to be achieved solving simultaneously the problem of foundation and the dam with largely different finite element sizes. Figure 4.2 shows the stress distribution achieved and the *natural* treatment of non-orthogonal boundaries.

Up to the time of this example the only alternative to the solution of this realistic and important problem was by the use of finite differences. In 1910, L.F. Richardson presented the first analysis of the Aswan Dam using a mesh with some 250 nodes and solving the equations by a laborious Gauss–Seidel process. Figure 4.3 shows this mesh composed of rectangles and in Fig. 4.4 a more elaborate mesh with some 900 nodes is presented from ZIENKIEWICZ

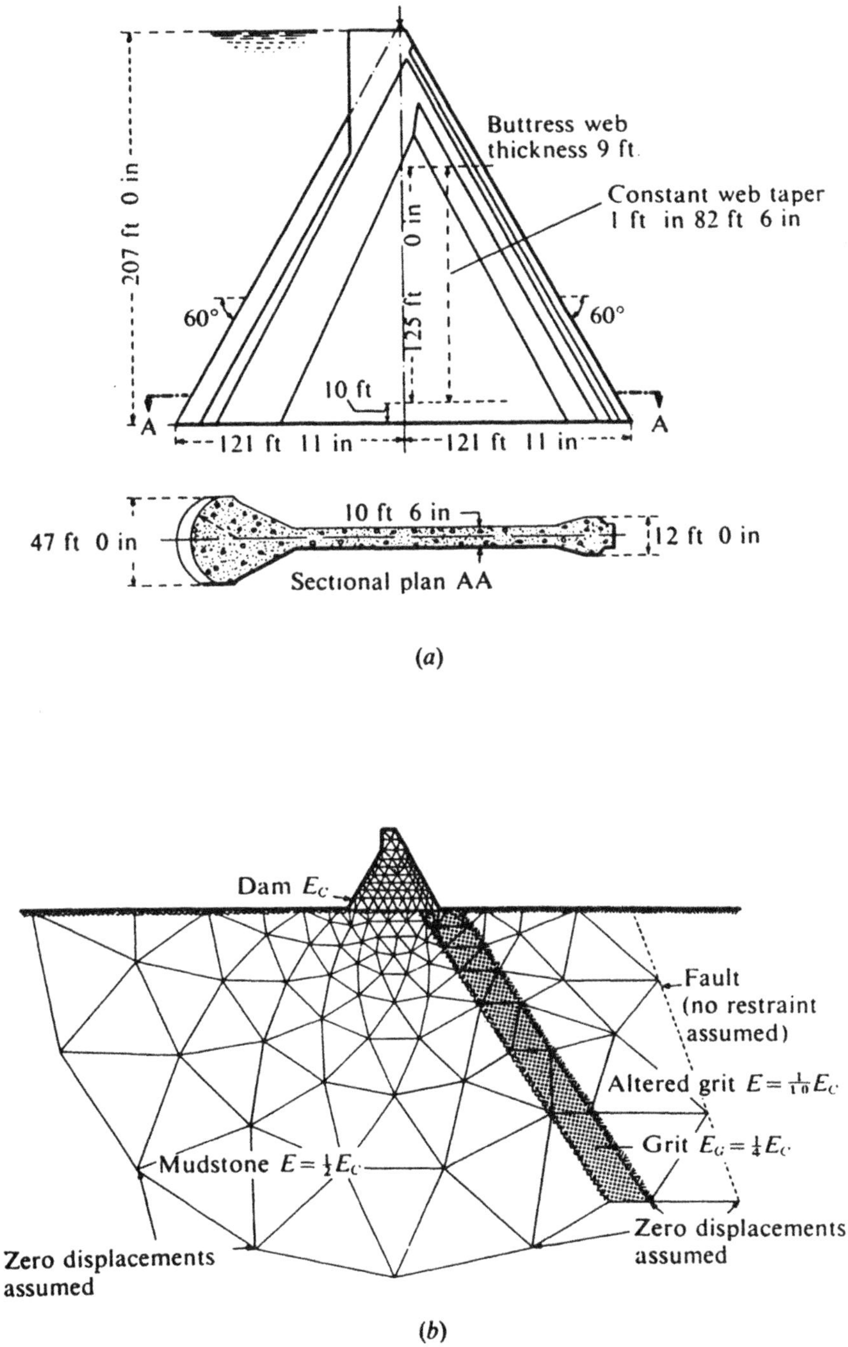

FIG. 4.1. Finite element analysis of the Clywedog buttress dam (1963). Plane stress and plane strain assumptions with linear elasticity. Note ease of simultaneous modelling of small and large detail. Reproduced by permission of McGraw Hill.

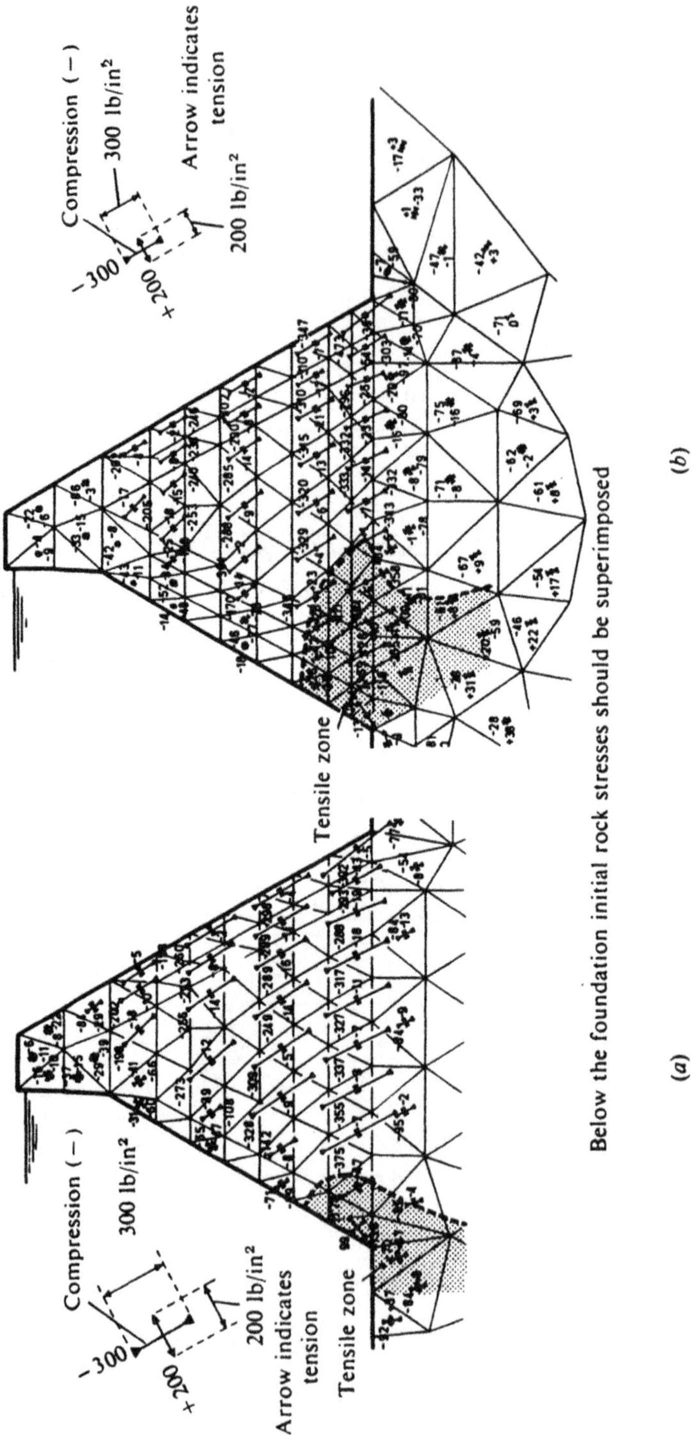

FIG. 4.2. Principal stresses in analysis in Clywedog dam, (a) with external loads only, (b) including pore pressure effects. Reproduced by permission of McGraw Hill.

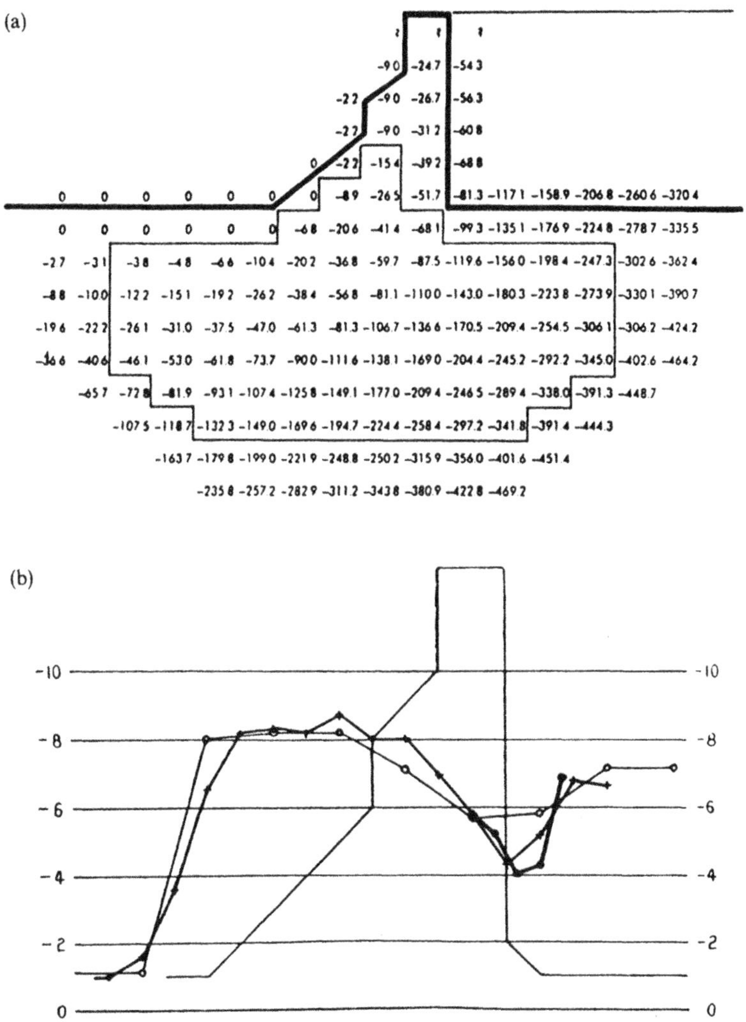

FIG. 4.3. The first "practical" numerical analysis. Assuan Dam by Richardson (1910) using a stress function formulation and finite differences with 250 nodes, (a) mesh, (b) vertical stresses at base.

[1947] for a somewhat similar structure. Here solution was obtained using the Southwell relaxation procedure with some results shown in Fig. 4.5.

Such finite difference procedures illustrate the difficulty of grading the mesh size in rectangular meshes—and the difficulties of dealing with arbitrary boundaries.

These examples show indirectly why the finite element procedures had to wait for the digital computer! In the original Richardson solution and in the later relaxation solution the tedious calculation of residuals could only be

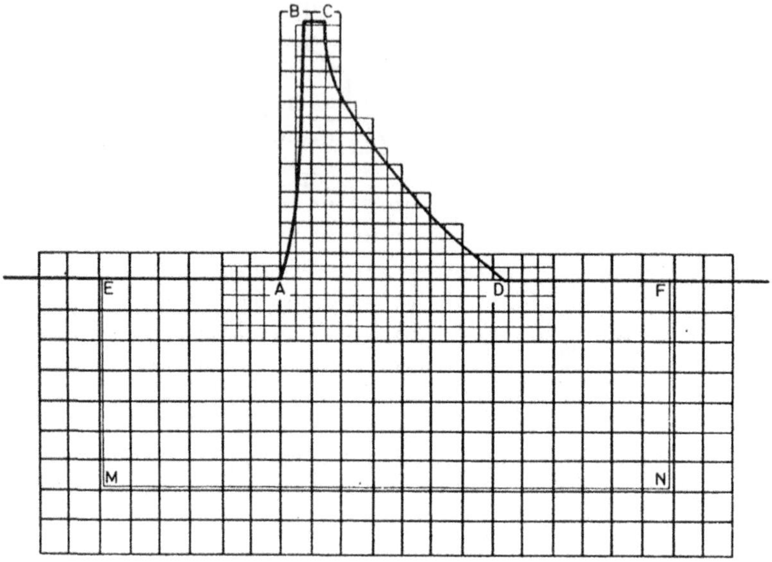

FIG. 4.4. A relaxation solution of finite difference equations for a gravity dam by ZIENKIEWICZ (1947) using circa 900 nodes.

accomplished by the device of memorising simple patterns of the finite "stencil" and using this for the residual distribution.

On the other hand, the arbitrary (though banded) structure of the finite element equations preserves no simple pattern and needs to use the facilities of modern electronic computers to "memorise" the matrices and solve them.

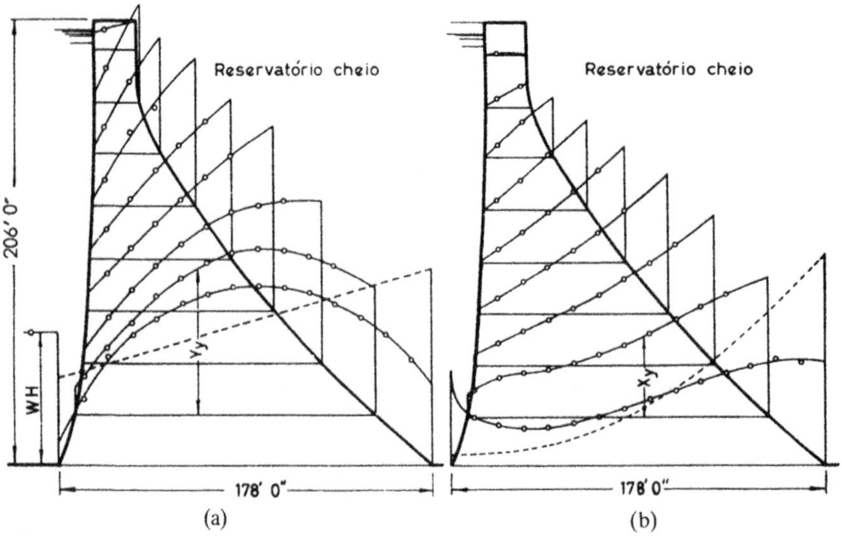

FIG. 4.5. Distribution of vertical (a) and shear (b) stresses in the problem of Fig. 4.4.

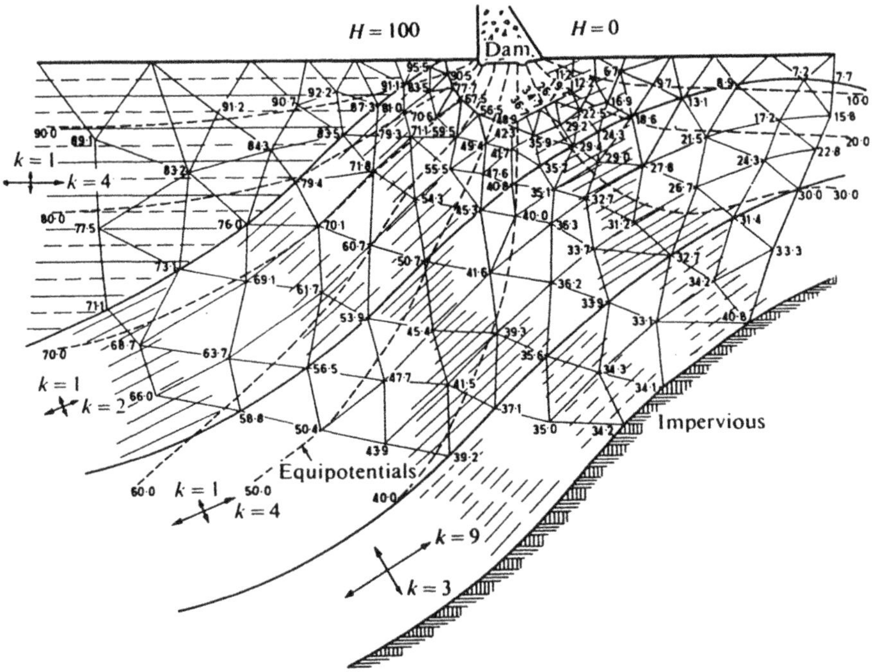

FIG. 4.6. Finite elements used for solution of seepage flow in a highly inhomogenous and anisotropic foundation (ZIENKIEWICZ, ARLETT and BAHRANI [1967]). Reproduced by permission of McGraw Hill.

Little wonder that the formulation inherent in the early work of Courant was not considered by him as a practical possibility.

Another example from this era as illustrated in Fig. 4.6 is one relating to fluid mechanics and shows a complex problem of fluid flow in a hypothetical, anisotropic, foundation (ZIENKIEWICZ, ARLETT and BAHRANI [1967]). Such complex problems presented serious difficulties to the then conventional finite difference processes—especially if abrupt material changes occurred.

Clearly by the mid-sixties the *start* of the finite element method was firmly established.

5. Virtual work and weighted residual approaches. Generalised finite element method and other approximations

An alternative derivation of the finite element approximation to that of using potential energy minimisation was developed in the early sixties (ZIENKIEWICZ [1965], ZIENKIEWICZ and CHEUNG [1964]). This was based on the use of the well-known virtual work principle. This principle simply states that if an equilibrating system of stresses $\boldsymbol{\sigma}$, boundary traction $\bar{t}$ and body forces $\boldsymbol{b}$ is subject to virtual displacement system $\boldsymbol{W}$, then the internal and external work is equal.

Thus if the strains corresponding to the virtual displacement $\boldsymbol{W}$ are given by expression (3.1), i.e.,

$$\boldsymbol{\varepsilon}^{W} = \boldsymbol{S}\boldsymbol{W} \tag{5.1}$$

then the virtual work equality requires that

$$\int_{\Omega} (\boldsymbol{S}\boldsymbol{W})^{\mathrm{T}} \boldsymbol{\sigma} \, \mathrm{d}\Omega = \int_{\Gamma_t} \bar{\boldsymbol{t}}^{J} \boldsymbol{w} \, \mathrm{d}\Gamma + \int_{\Omega} \boldsymbol{b}^{\mathrm{T}} \boldsymbol{W} \, \mathrm{d}\Omega \tag{5.2}$$

providing the system $\boldsymbol{W}$ satisfies

$$\boldsymbol{W} = 0 \quad \text{on } \Gamma_u . \tag{5.3}$$

Of course if the virtual displacement system is made such that

$$\boldsymbol{W} = \boldsymbol{N} \tag{5.4}$$

and the stress system $\boldsymbol{\sigma}$ is that due to a displacement approximation of Eq. (3.4)

$$\boldsymbol{u}^{h} = \boldsymbol{N}\bar{\boldsymbol{u}} \tag{5.5}$$

then, using definitions (3.1) and (3.2) the virtual work Eq. (5.2) will yield again an algebraic equation set identical to Eq. (3.5), i.e.,

$$\boldsymbol{K}\bar{\boldsymbol{u}} = \boldsymbol{f} \tag{5.6}$$

with the same definition of the stiffness matrix and "forces" as given in Eq. (3.6). Now, of course, if the virtual work principle is applied to a single element, equivalent, inter-element forces can be obtained and the physical analogy with the early approximation is available as shown in ZIENKIEWICZ and CHEUNG [1967]. However, more important corollaries follow:

(i) Virtual displacements of a form different to $\boldsymbol{N}$ can be used if desired (though of course the satisfaction of the condition (5.4) is optimal from the point of view of an energy minimisation).

(ii) It is possible to use non-symmetric forms of the $\boldsymbol{D}$ matrix such as arise for instance in non-associative plasticity, viz. ZIENKIEWICZ, VALLIAPPAN and KING [1969], NAYAK and ZIENKIEWICZ [1972], for which the potential energy can not be defined.

However, there is more. The mathematician will recognise in Eq. (5.2) the standard *bilinear form* used later in their approach to the finite element theory—and written for a typical scalar problem as

$$a(w, u) + b(w) = 0 . \tag{5.7}$$

Others will note that on using integration by parts Eq. (5.2) can be written as

$$\int_{\Omega} W^{\mathrm{T}}[S^{\mathrm{T}}\sigma + b]\,\mathrm{d}\Omega - \int_{\Gamma_p} W^{\mathrm{T}}(n\sigma - \bar{t})\,\mathrm{d}\Gamma = 0 \tag{5.8}$$

with $\sigma = DSu$.

This is, of course, a *weighted residual* form of the equilibrium equation

$$S^{\mathrm{T}}\sigma + b = 0 \quad \text{in } \Omega \tag{5.9}$$

together with the boundary conditions

$$n\sigma - \bar{t} = 0 \quad \text{on } \Gamma_t \tag{5.10}$$

if Eq. (3.5) is used for approximating the unknown u.

The name of "weighted residual" appears to be introduced into literature of approximate numerical solutions by CRANDALL [1956] as a general procedure, recognising that GALERKIN [1915] was the first to use it formally with the special case of weighting of Eq. (5.4) i.e.,

$$W = N. \tag{5.11}$$

However, point collocation, method of moments etc. present other possibilities. Today it appears customary to refer to the case of Eq. (5.11) as the Galerkin–Bubnov method while all other possibilities are simply lumped as Petrov–Galerkin methods for which

$$W \neq N. \tag{5.12}$$

The original references, to this classification are not clear—but "what's in a name?". The important matter is that by the late 60s the recognition of the relation between weighted residual forms and the finite element method was well established opening the doors to the solution of most problems cast in terms of differential governing equations and boundary conditions.

Clearly it now became possible to apply the Galerkin (Bubnov) process to non-self-adjoint equation systems such as arise in fluid mechanics, viz. ODEN and SAMOGYI [1969], ODEN and WELLFORD [1972], ODEN [1973], ZIENKIEWICZ and TAYLOR [1973] and finally TAYLOR and HOOD [1973] who were the first to solve successfully the Navier–Stokes problem.

Equally clearly it became possible to interpret most other approximation procedures such as finite differences, finite volumes, boundary methods, etc., as variants of the general process implied in Eq. (5.8). Let us write for such a process the governing equation as

$$L(u) = 0 \quad \text{in } \Omega \tag{5.13}$$

and the boundary condition (not automatically satisfied by the approximation) as

$$B(u) = 0 \quad \text{on } \Gamma. \tag{5.14}$$

Using the approximation

$$u \approx \hat{u} = N\bar{u} = \sum N_i \bar{u}_i\,, \quad i = 1 - u \tag{5.15}$$

the discrete approximation is formed as

$$\int_\Omega W^{\mathrm{T}} L(\hat{u})\, \mathrm{d}\Omega + \int_\Gamma W^{\mathrm{T}} B(\hat{u})\, \mathrm{d}\Gamma = 0\,. \tag{5.16}$$

For any such approximation the summation rule of the finite element method obviously applies though of course the banded, sparse, matrix structure will arise only if locally based shape functions N are used.

With such a general statement it is easy to see that for instance we can now interpret *finite differences* as a point collocation process in which

$$W_i = \delta_i \quad \text{(Dirac)}\,. \tag{5.17}$$

In Fig. 5.1 we show for instance an approximation to a one-dimensional problem governed by

$$\frac{\mathrm{d}^2 u}{\mathrm{d}x^2} + q = 0\,. \tag{5.18}$$

Here the shape functions N_i are simple parabolas determined by u_{i-1}, u_i and

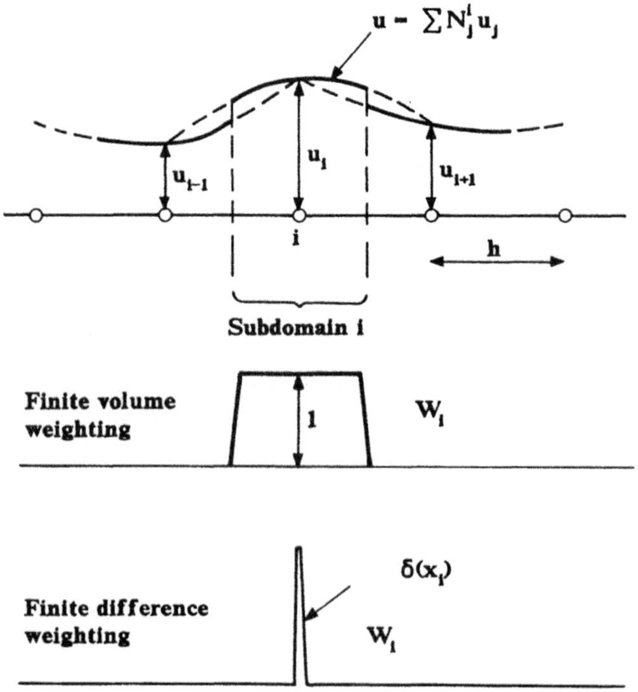

FIG. 5.1. The interpretation of finite difference and volume approximations as special cases of the *generalised finite method*. Note shape functions N_j^i depend on subdomain i considered.

u_{i+1}, which give a discontinuous approximation shown violating the previously stated continuity requirements. However, the contribution of the discontinuities in the weighted form disappears and the reader can verify that application of Eq. (5.16) will result simply in the well-known difference form

$$u_{i+1} - 2u_i + u_{i-1} + h^2 q_i = 0 . \tag{5.19}$$

In a similar manner the so called *finite volume* approximation can be interpreted as a particular kind of a finite element. This is in fact a subdomain collocation (to use the classification of CRANDALL [1955]) with

$$W_i = 1 \tag{5.20}$$

over a "volume" h_i shown again in Fig. 5.1. Now the approximation equation is again of similar form to (5.17) but with the last term modified:

$$u_{i+1} - 2u_i + u_{i-1} + h^2 \tilde{q}_i = 0 , \tag{5.21}$$

where

$$\tilde{q}_i = \frac{1}{h} \int_{i=1/2}^{i+1/2} q \, \mathrm{d}x . \tag{5.22}$$

The above approximation can of course be contrasted with that obtained using simple C_0 continuous linear shape functions N_i^h. Here again the form is identical to that of Eq. (5.21) but with $\tilde{q}_i$ now defined as

$$\tilde{q}_i = \frac{1}{h_i} \int_{i-1}^{i+1} N_i q \, \mathrm{d}x . \tag{5.23}$$

Such a standard (Galerkin) finite element form is in fact optimal delivering exact nodal values in this case as shown by TONG [1969]. However, the finite volume process has a useful physical interpretation and has gained widespread use in fluid mechanics where the problems are generally non-self-adjoint! The above interpretation puts that approximation of course in a form suitable for finite element type computation, viz. ZIENKIEWICZ and OÑATE [1991], and OÑATE, CERVERA and ZIENKIEWICZ [1994].

Boundary methods (TREFFTZ [1926]) are yet another alternative approximate for which many advantages are sometimes claimed. These procedures choose the basis (shape) functions $\boldsymbol{N}$ in such a manner that the differential equations are satisfied identically throughout the domain Ω by these. Now of course, for linear problems, Eq. (5.13) is

$$\boldsymbol{L}\hat{\boldsymbol{u}} \equiv 0 \tag{5.24}$$

and the approximation in Eq. (5.16) becomes simply

$$\int_{\Gamma} WB(\hat{u})\, d\Gamma = 0 \tag{5.25}$$

thus reducing the dimensionality of the problem. The possibility of deriving banded matrix systems is of course not available but the total number of unknown parameters can well be reduced for a given accuracy. The use of such "boundary elements" has been extensive as shown in recent texts, viz. BEER and WATSON [1992].

In the above I have attempted to show that all well-known approximation methods have very similar origins and a basic form of the finite element procedure. Taking a chauvinistic viewpoint we could thus embrace all in the name of a *generalised finite element* procedure though the alternative of *generalised Galerkin method* would be equally applicable, FLETCHER [1984]. The name is not important! What is, is the recognition of the educational aspect of similarity of the processes and more importantly the realisation that practical advantage may occasionally accrue by utilising these in combined computer codes.

A fairly obvious combination is the use of boundary procedures for treatment of semi-infinite domains of linear kind together with non-linear finite elements of a "standard" kind in other subregions. Suggestions for such a *mariage à la mode*, made first by the author and his colleagues (ZIENKIEWICZ, KELLY and BETTESS [1977b]) has found many applications.

While the so-called *spectral methods* have not been specifically examined here these obviously are another general application of the basic formulation of Eq. (5.16). These are of course mirrored in the use of very high-order polynomial based functions in the standard finite element approaches to which we shall refer later. In this context it is important to realise that "nodal values" are not the essential requirement of any finite element approximation. From the earliest days "nodeless" variables have been used in standard codes with such variables generally confined still to single elements. A "landmark" in the use of such finite element interpolation was the introduction of the so-called *hierarchical interpolation* (ZIENKIEWICZ, IRONS, CAMPBELL and SCOTT [1970]).

In the hierarchic form the approximation in each element is still local and of the form (for a scalar variable)

$$u = \boldsymbol{N}\bar{\boldsymbol{u}} = \sum_{i=1}^{n} N_i \bar{u}_i\,, \tag{5.26}$$

but $\bar{u}_i$ parameters are no longer nodal values of u and the usual condition that

$$\sum N_i = 1 \tag{5.27}$$

does not apply. Moreover, the expression is so constructed that the shape (basis) functions N_i are independent of the number of parameters as chosen for the approximation (i.e., the expression of Eq. (5.26) is in a *series* form).

This has the advantage that not only the matrices of lower orders do not change as elements are refined by adding higher-order polynomials (p-refinement)—but also the equation conditioning is vastly improved. The hierarchical form is today widely adopted—and particularly useful in adaptive refinement (ZIENKIEWICZ, GAGO and KELLY [1983], PEANO [1976]).

While the lowest hierarchic form still preserves element-variable "banding" of the matrices an alternative in which global functions are superposed on the local ones is possible. This, first introduced by MOTE [1971] allows known exact solutions to be used as an addition to the basis functions with the true finite element refinement being a perturbation in this solution and thus requiring smaller accuracy (and cost). The solution matrix no longer banded still remains sparse and this use of this type of hierarchical form deserves to be further explored.

6. Non-conforming approximation and the patch test as a necessary and sufficient condition of finite element method convergence

6.1. Plate bending—Conforming and non-conforming variants

We have already stated that one of the early requirements of the finite element approximation was the choice of shape function which did not lead to infinite strains on element interfaces and which therefore preserved a necessary degree of continuity. Satisfaction of this requirement guarantees convergence and in the case of energy minimisation processes, an energy bound.

While in the case of simple elasticity and other self-adjoint problems governed by second-order equations such continuity is easy to satisfy—this requirement in case of thin plate bending using the Kirchhoff–Germain postulates, *where fourth-order equations arise, is much more difficult to achieve.* (The name of Germain is added here to honour the young French lady who, many years before Kirchhoff, formulated a nearly correct plate theory.) Now C_1 continuity has to be introduced (and the continuity of both the function and of its normal gradient assured). This requirement was well known in the early days of the sixties decade, and the development of plate bending elements followed two lines.

(i) the search for elements satisfying rigorously the continuity or

(ii) designing elements where continuity of slopes was imposed only at the nodes with the hope that the interface work would tend to zero as the refinement proceeded, and that convergence would be still achieved.

The first approach was, as mentioned before, difficult. The problem lies in the fact that if only the value of the displacement function w and of its slopes (∇w) are used as parameters, at the nodes it is *impossible* to determine unique polynomial shape functions in the element. The proof of this was established later by IRONS and DRAPER [1965] (viz. ZIENKIEWICZ and TAYLOR [1991], pp. 12–13), who show that for continuous polynomial shape functions it is necessary to specify some of the second displacement derivatives. In general, specification of such nodally continuous second derivatives, though used successfully by some authors (viz. here the simultaneous development of quintic triangular elements with 21 degrees of freedom and their reduction to 18 degrees of freedom by ARGYRIS, FRIED and SCHARPF [1968], COWPER, KOSKO, LINDBERG and OLSEN [1968], BOSSHARD [1968], VISSER [1968], BELL [1969] and IRONS [1969]), does not permit abrupt thickness changes in the plate (or moment discontinuity) such as may occur at irregularities and should in our opinion not be used as if it were a case of *excessive continuity imposition*. The single exception to this is the continuity of the cross derivative, $\partial^2 w/\partial x\, \partial y$, on an orthogonal mesh, and an element of rectangular shape, using this derivative as a nodal variable, was presented by BOGNER, FOX and SCHMIT [1965] at the first Wright-Patterson Conference in Dayton, Ohio.

At the same meeting, however, an alternative derivation of fully conforming triangles was presented by Clough and Tocher [1965] and by the authors group BAZELEY, CHEUNG, IRONS and ZIENKIEWICZ [1965] and much later modified by SPECHT [1988]. In both of these elements the shape functions are so formed that non-unique second derivatives arise at the nodes. In the first element the shape functions are derived by using three separate polynomials in different parts of the triangular element. In the second a non-polynomial expression is used to achieve the desired end.

With much exercise of ingenuity the various "conforming" elements described above proved, as expected, convergent but unfortunately were poor performers.

Much more fortune was however experienced by those who flouted the rules and ignored the interelement continuity requirements. Here, the first element of rectangular shape was produced independently by CLOUGH and ADINI [1961] in an unpublished report and by ZIENKIEWICZ and CHEUNG [1964]. This element performed well and experiments showed it to be convergent.

Similar non-conforming triangular elements were described in BAZELEY, CHEUNG, IRONS and ZIENKIEWICZ [1965] and much later modified by SPECHT [1988]. In Fig. 6.1 we show a typical example comparing the convergence of the conforming and non-conforming triangular elements discussed above. However, the assurance of convergence was not automatic and needed further investigation. The answer to this was proposed by Irons and published with the author in BAZELEY, CHEUNG, IRONS and ZIENKIEWICZ [1965]. This answer was the *patch* test.

6.2. The patch test

The original idea of the patch test stemmed from very physical, intuitive, considerations. It simply stated that a "patch" or any assembly of elements of the form shown in Fig. 6.2 for an elastic continuum must be able to reproduce *exactly* all constant stress and strain states when subject to appropriate (linear) variation of displacements.

The patch test was first applied to the non-conforming plate bending elements in which constant curvature/moment states were modelled. It was

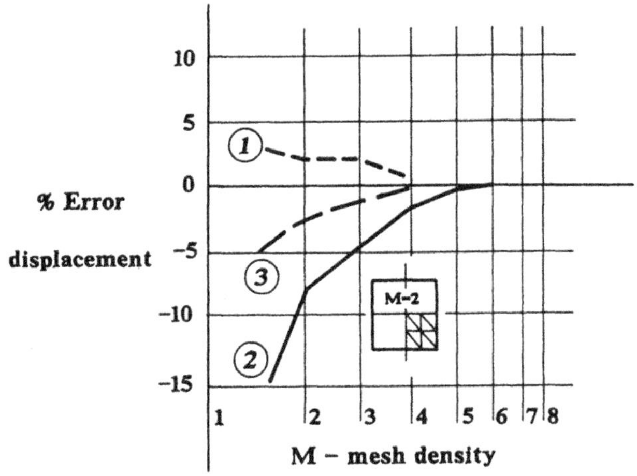

① Nonconforming triangle (Bazeley et al 1965)
② Conforming triangle (Bazeley et al 1965, Clough and Tocher 1965)
③ Nonconforming triangle (Specht 1984)

FIG. 6.1. Performance comparison of some conforming and non-conforming thin plate bending triangles with 9 DOF Error in displacement for a uniformly loaded, simply supported square plate.

easily shown that the non-conforming rectangle was fully convergent but the triangle derived by BAZELEY, CHEUNG, IRONS and ZIENKIEWICZ [1965] was only convergent for regular meshes formed by three sets of intersecting parallel lines (though its performance was roughly acceptable for other meshes!).

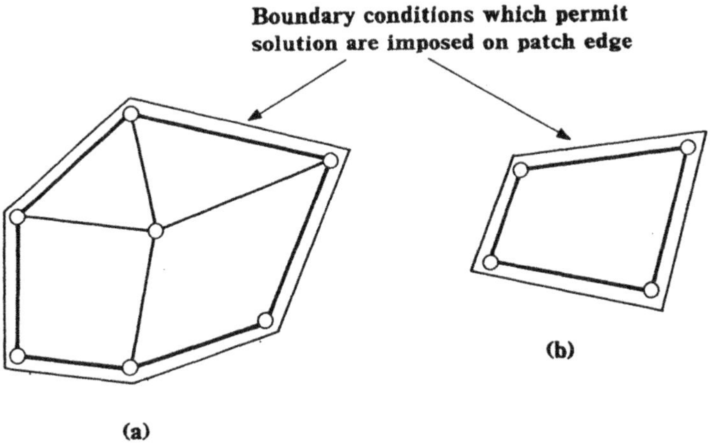

FIG. 6.2. Patch test of (a) multiple element and (b) single element type. Boundary conditions and internal forces imposed on the patch correspond (at least) to any arbitrary lowest order solution necessary for convergence (e.g., constant stress conditions). Higher order solution can be imposed to check convergence order.

Though the application of the patch test was first envisaged for testing non-conforming elements, its application, by the original arguments, was universal. It soon became obvious that it should be universally used for all elements, even those based on fully convergent assumptions, as a test of the correctness of programming, viz. IRONS and RAZZAQUE [1972], FRAEIJS DE VEUBEKE [1974] and OLIVEIRA [1977]. For such purposes, it even got the blessing of some mathematicians (viz. STRANG and FIX [1973]) as a *necessary* condition for convergence and in due course became a standard test required for the verification of all codes.

However, in the form originally given it was only a *necessary condition* of convergence but clearly not *sufficient*. Considerable confusion was engendered in more recent days when a mathematician denied even its necessity (STUMMEL [1980]) and some updating of its validity was needed. This provided some of us with a stimulus for further research. In 1984 the author with his colleagues published a paper which extended the test to provide both necessary and sufficient conditions for convergence of all finite element forms, viz. TAYLOR, ZIENKIEWICZ, SIMO and CHAN [1986].

The essence of the extension was that of testing the stability of patches by ensuring that no zero eigenvalues (singularity) were present under *any acceptable boundary condition* on a patch. Indeed this part of the test extended the usual structural ideas in which non-singularity of individual elements was generally imposed but now the question was posed in a more mathematical form valid for all applications.

A further extension was that permitting the assessment of the convergence order by the imposition of higher-order test solutions.

A very useful corrolary of the test was that made later for mixed formulations where it provided a guide for the "design" and testing of various possible approximations (ZIENKIEWICZ, QU, TAYLOR and NAKAZAWA [1986]). Here it gave a simpler alternative to the well-known BABUŠKA [1971, 1973], BREZZI [1974] conditions which are recognized to be necessary and sufficient for convergence. While generally the patch test has to be applied numerically with randomly selected low-order exact solutions, in the case of mixed forms, which in two field situations frequently give rise to algebraic equations of the type containing a zero on the diagonal

$$\begin{bmatrix} A & Q \\ Q^{\mathrm{T}} & 0 \end{bmatrix} \begin{Bmatrix} \bar{\boldsymbol{u}} \\ \bar{\boldsymbol{p}} \end{Bmatrix} = \begin{Bmatrix} f_1 \\ f_2 \end{Bmatrix}, \tag{6.1}$$

it is possible to determine the existence of singularity "a priori". Here $\bar{\boldsymbol{u}}$ and $\bar{\boldsymbol{p}}$ are the variables describing the two fields and some indication of behaviour can be obtained by *inspection*. It is simple to show that to obtain the solution and avoid zero eigenvalues it is *necessary* that

$$n_u \geqslant n_p\,, \tag{6.2}$$

where n_u and n_p stand for the number of variables in the $\bar{\boldsymbol{u}}$ and $\bar{\boldsymbol{p}}$ sets. The verification that this condition is satisfied for *all* patches of elements (from a single element onwards) helps to eliminate most of the elements which were known to be either unusable or non-robust by virtue of failing the Babuška–Brezzi conditions. Further it indicates immediately as "possible" candidates other elements which have hitherto been derived by more elaborate means. However, it must be stressed that the simple "count" provides a necessary condition only, and that in general further numerical tests on zero eigenvalues are needed.

If $\bar{\boldsymbol{u}}$ and $\bar{\boldsymbol{p}}$ of Eq. (6.1) stand for instance for displacements and pressures in typical incompressible elasticity equations (HERRMANN [1965]) then it is clear from Fig. 6.3(a) that the triangular element with linear, continuous interpolation of $\bar{u}$ and $\bar{p}$ fails in all patch assemblies on which $\boldsymbol{u}$ is prescribed on boundaries. (Note that one value of p in any patch needs to be always prescribed for mathematical reasons.)

In Fig. 6.3(b) a simple "bubble" function with $\boldsymbol{u}$ displacement parameters has been added and immediately the count is satisfied for all patches. This indeed provides now a convergent element (as can be verified by eigenvalue tests), but of course merely reproduces a well-known element of FORTIN and FORTIN [1982]. It is of course difficult to add to the study of such a well-known problem (and the count here is strictly educational) but in the case of some later elements for plate bending problems where three independent fields interact, the integer inequalities similar to Eq. (6.1) have led to novel forms giving amongst others, the first fully robust element based on Reissner–Mindlin assumptions (ZIENKIEWICZ and LEFEBVRE [1987, 1988]).

It should be mentioned here that the robustness of mixed forms of limiting kind such as those given by Eq. (6.1) is of importance even if a non-zero, but small, diagonal term exists permitting the elimination of the variable "p". Now the element is usually identical to the irreducible form—and if robust in the limit will behave well generally. Thus for instance it is well known that the simple linear triangle behaves very erratically as incompressibility is approached as in that limit is not solvable.

The patch test so far discussed is an essentially numerical verification procedure. However, it is often possible to derive from its requirements analytical conditions required for convergence. Here a very useful criterion for the design of incompatible elements has been postulated by TAYLOR, ZIENKIEWICZ, SIMO and CHAN [1986] which can lead to the development of incompatible element displacement functions fully satisfying convergence criteria. SPECHT [1988] develops on this basis an incompatible element superceding that of BAZELEY, CHEUNG, IRONS and ZIENKIEWICZ [1965]. This element is convergent for all types of meshes and performs well as was shown in Fig. 6.1.

Indeed if we accept that the satisfaction of the patch test is the only convergence criterion this could be used as the sole means of deriving element stiffness matrices without specifying displacement shape functions, or indeed the background theory, directly. Such elements have been successfully derived by BERGAN and HANSSEN [1977] and BERGAN and NYGARD [1984].

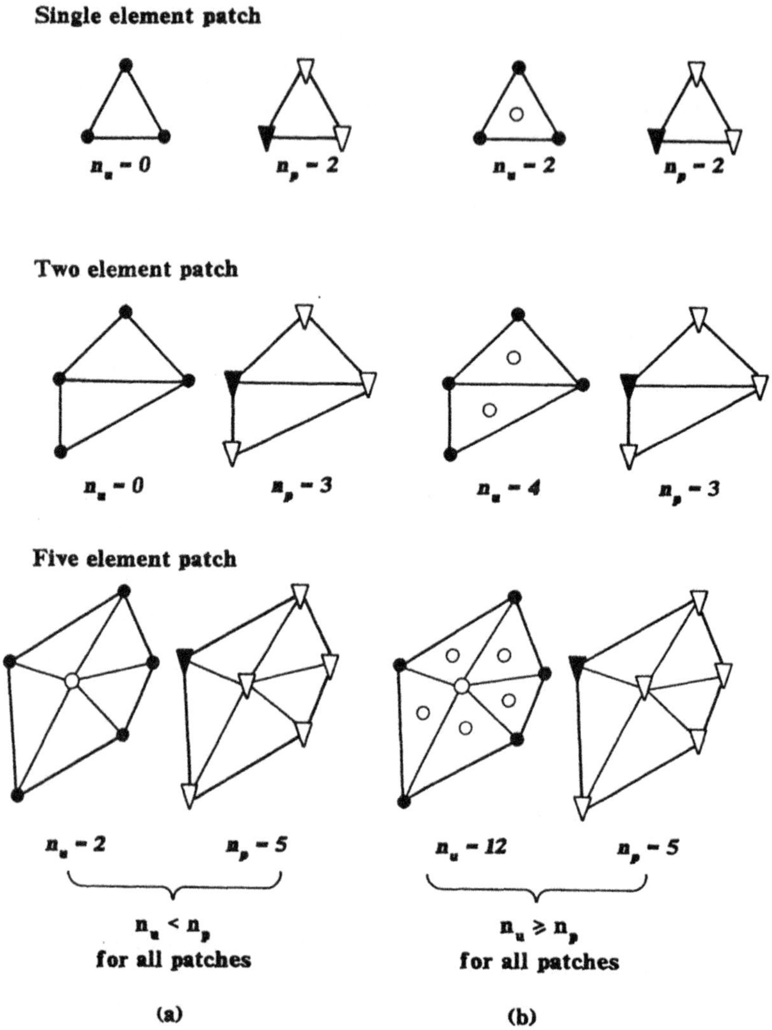

FIG. 6.3. Patch test count for incompressible elasticity (or Stokes flow) with C_0 continuous interpolation of displacements u and pressures p. (a) Simple linear interpolation of both variables fails count on all patches. (b) Addition of bubble function on displacement u ensures count is satisfied on all patches. ○: Node with 2 displacement DOF; ▽: Node with 1 pressure DOF; ●, ▼: Restrained nodes.

Clearly, the establishment of the patch test is a major landmark in providing a basis for finite elements.

6.3. *Diffuse element approximation*

In the previous sections we discussed the use and limitations of "non-conforming", discontinuous, shape functions and showed under what conditions such conformity was restored in the limit $h \to 0$. However, there is a very simple way of ensuring such limiting conformity by use of overlapping shape function. In Fig. 5.1 we showed the genesis of such overlapping functions for the purpose of using a finite difference (or finite volume) approximations. In that figure, quadratic approximations generated by a nodal overlap were used and it is physically evident that in the limit the approximation will ensure the continuity of both the function u and of the first derivative. A formal proof of the above is available in NEYROLLES, TOUZOT and VILLON [1992].

The idea was first put to practical use by NAY and UTKU [1973] who derived shape functions for plate bending elements by using quadratic polynomial approximations obtained by least square fitting of a number of nodes in the vicinity of a particular node, considered in the manner shown in Fig. 6.4. Here the seven nodes together with the node i are used to define the quadratic (six parameters) for the displacement variable and a *tributary area* Ω_i defines the element.

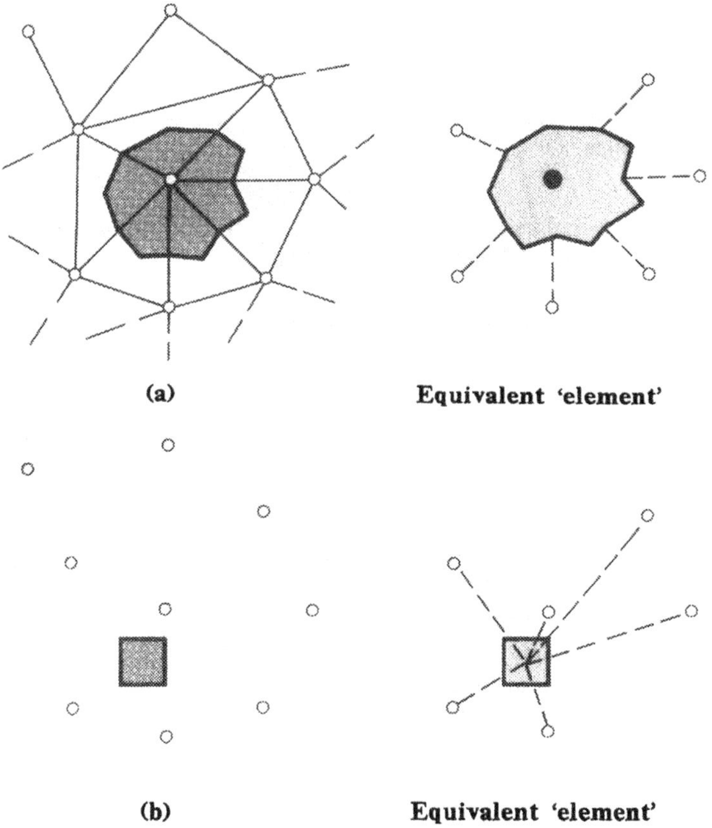

FIG. 6.4. Diffuse approximation. (a) Based on underlying mesh structure (Utku [1971], Orkisz [1980]). (b) Based on nearest 7 nodes to an integration module (element) (NEYROLLES et al. [1991]).

Clearly now we have elements with external nodes but the identical formulation to that previously used can be followed, now indeed achieving limiting compatibility without use of gradients as nodal variables. (This indeed has been of some recent practical interest.) Of course special treatment of boundaries is now necessary (which we shall not discuss here) and it is of interest to note that approximations are convergent and of a form similar to the generalised finite difference approximation. However, standard finite element assembly etc. still apply.

The procedure was extended and widely used by PAVLIN and PERRONE [1979], LISZKA and ORKISZ [1980] and very recently adapted by NEYROLLES, TOUZOT and VILLON [1991, 1992] for precise modelling. The ideas are basically simple and it is surprising that such processes had to wait so long to see practical use. It may well be that in the future their merits will become more evident.

Of course there are many possible ways of achieving the interpolation needed or creating the tributary element volumes as shown in Fig. 6.4 and the interested reader should consult the literature. Here of interest is the recent work of BELYTSCHKO, LU and GU [1994]. Indeed it can be observed that in such formulations similarities exist with the so-called "wavelet" forms as shown by LIU, ADEE and JUN [1993].

Clearly, the patch test again has a dominant role for all these problems to assure convergence.

7. Higher-order elements and isoparametric mapping. Three-dimensional analysis, plates, shells and reduced integration

7.1. The need governs development—Isoparametric mapping

In the early sixties only simplest linear elements were generally used and the solution of realistic problems was possible (with some difficulty) for relatively small 2D problems on the computers available in that era. However, the need for solving three-dimensional problems was pressing and for these the obvious linear, tetrahedral element rapidly overstretched the capability of computation (GALLAGHER, PADLOG and BIJLAARD [1962], MELOSH [1963]). It was clear that higher-order elements would be required to achieve the accuracy desired with a reasonable number of degrees of freedom. Here other problems were immediately encountered. If for instance the quadratic triangle originally suggested by FRAEIJS DE VEUBEKE [1965] were extended to its equivalent three-dimensional, tetrahedral form then purely geometrical difficulties would be encountered in dealing with complex boundary shape as obviously a smaller number of such elements would now be used. Clearly some form of coordinate mapping would be necessary to deal with this situation.

As in general the interpolation can be chosen independently of the integration volumes many opportunities are offered as the computation does not require element specification. It is of interest to note the similarity with "wavelet" theory noted by WING [1993].

It was my good fortune to encounter at the time Bruce Irons who then worked at Rolls Royce, and whom I managed to persuade to join the academia where he would work with people who had more interest in reading his reports and acting upon them. It was clear that he had given the problem much thought and that he had already formulated a possible way of mapping using the essential element shape functions for many higher-order polynomials. Further he suggested that the complex transformation integrals could be evaluated numerically. This work published in 1966 by Irons (IRONS [1966a,b]) generalised an original suggestion of TAIG [1961] who first succeeded in arbitrary mapping a rectangle into a general quadrilateral form. Together we succeeded in elaborating and applying the idea to both two- and three-dimensional problems viz. ERGATOUDIS, IRONS and ZIENKIEWICZ [1968a,b], ZIENKIEWICZ, IRONS, ERGATOUDIS, AHMAD and SCOTT [1969].

One of the first major problems at the time was the analysis of a nuclear pressure vessel shown in Fig. 7.1 where, by practical use of quadratic brick elements, the problem was adequately solved using only 2 121 DOF in 1966. The same figure shows a similar analysis made by RASHID and ROCKHAUSER [1968], using some 18 000 DOF with linear tetrahedra. Obviously both analyses were done almost simultaneously but "secrecy" of commercial nature prevented any comparison of results.

At the same time the writer was engaged in a civil engineering project attempting to devise a method for computing stresses in arch dams. Figure 7.2 shows some typical results of this study using full three-dimensional analysis with quadratic and cubic elements. Clearly the gain of accuracy (per degree of freedom) between quadratic and linear elements was dramatic, though at the time we could not appreciate this fully as accuracy of different solutions could not be fully compared. However, in Fig. 7.3 we show (using adaptivity concepts) an identical problem solved with the same accuracy by linear and quadratic elements and the reduction of degrees of freedom is very substantial (ZIENKIEWICZ and ZHU [1987]). This of course does not necessarily mean that computational cost is proportionally reduced though with conventional solvers this indeed can be the case.

However, I would classify the step of isoparametric mapping as one of the major landmarks which by permitting the practical use of higher-order elements, had a very substantial effect on the finite element scene. Since the early days described above many other alternative mapping processes have been introduced (viz. ZIENKIEWICZ and TAYLOR [1989]) but none have become so universally popular.

7.2. *Physics govern theory—The essential development for plates and shells*

With the easy and readily accessible form of isoparametric mapping in the late sixties, it appeared to the author that perhaps the difficulties initially associated with plates and shells which were inherent in the Kirchhoff–Germain thin plate assumption could be overcome by simply treating such structures using thin

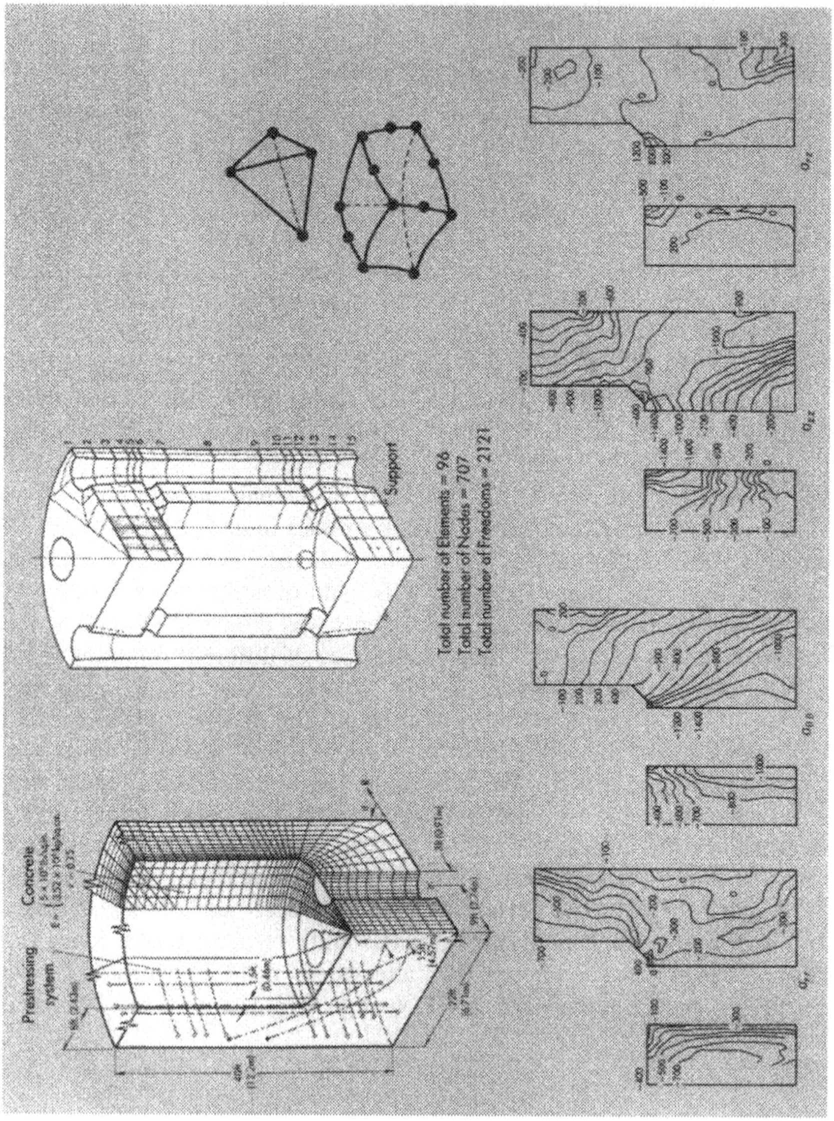

FIG. 7.1. Analysis of a nuclear reactor using linear tetrahedra (10 000 DOF) and quadratic hexahedra (2 000) DOF. Reproduced by permission of McGraw Hill.

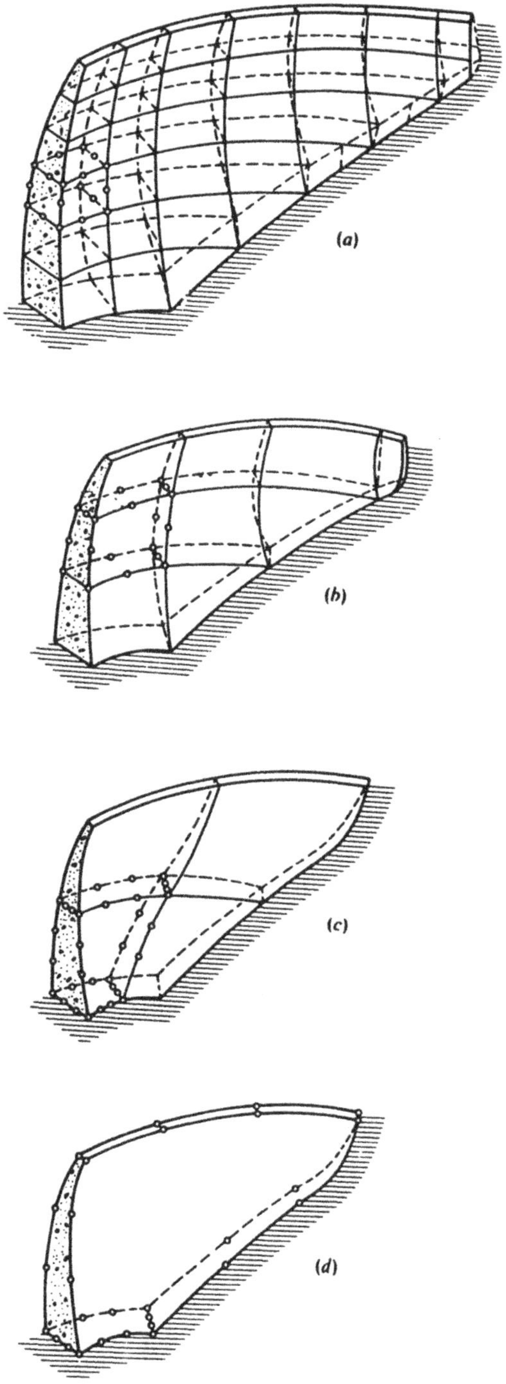

FIG. 7.2. A test analysis of an arch dam (1968). (a) and (b) Two meshes for quadratic elements. (c) and (d) Two meshes for cubic elements.

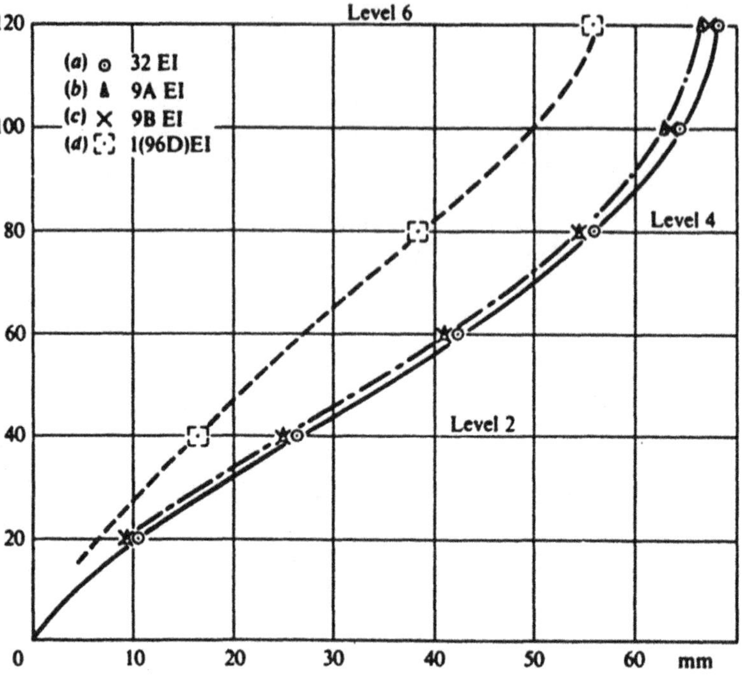

FIG. 7.2. (e) Results giving deflections on centre-line. Reproduced by permission of McGraw Hill.

isoparametric elements of the three-dimensional, solid kind. It was obvious that here a neglect of the stresses in the transverse direction should be made and that linear variation of other stresses in that direction would be sufficient to model the relatively thin behaviour of plates and shells. Thus the idea illustrated in Fig. 7.4 was introduced almost simultaneously with full three-dimensional analysis. Here plates and shells could be treated by identical processes already developed for 3D analysis.

The first work published on this by AHMAD, IRONS and ZIENKIEWICZ [1968, 1970] was a little disappointing. It was found that the performance of the new approach was good only when fairly thick elements were used and deteriorated rapidly as this thickness was reduced (or as the Kirchhoff–Germain assumptions were enforced). However, as the direct approach obviated completely the complex plate and shell theory and did not require the enforcement of slope, continuity methods of improving the poor performance had to be urgently found.

Here intuition and some rather heuristic arguments pointed to an answer which, at least in part, was provided by using *reduced* (or *selectively reduced*) integration (ZIENKIEWICZ, TOO and TAYLOR [1971], PAWSEY and CLOUGH [1971]). It turned out that, simply by using a certain lower order integration, answers could be dramatically improved both for thin and thick plate and shell forms. From this day the formulation based on the three-dimensional degene-

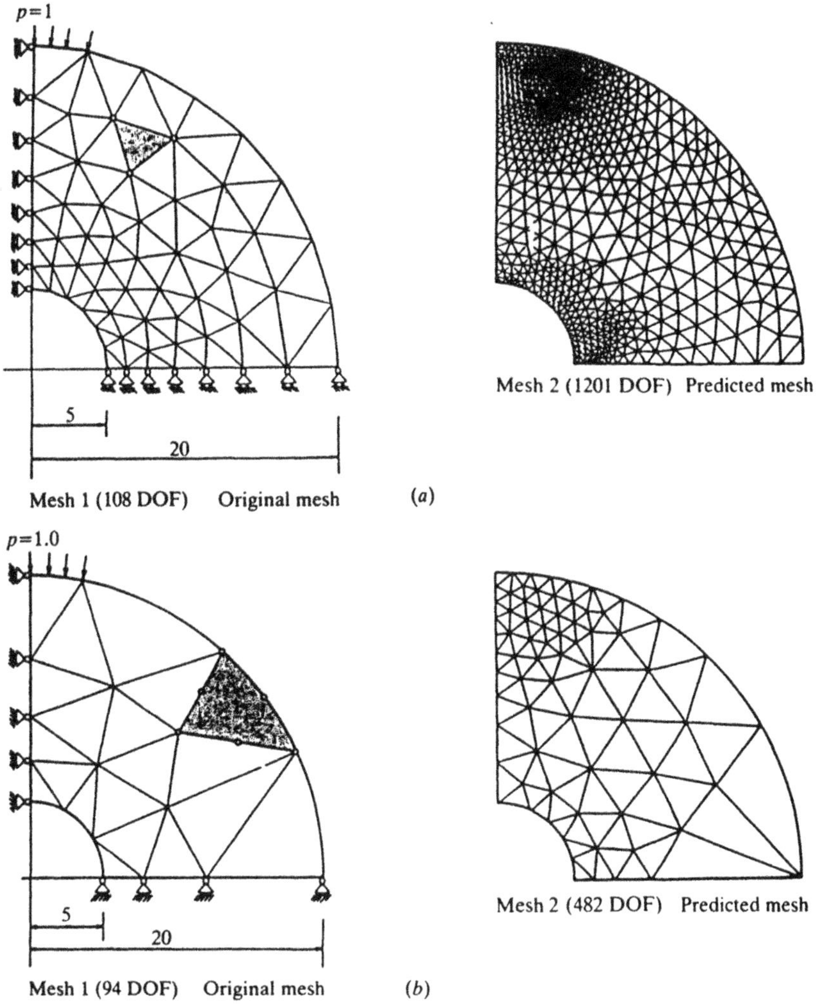

FIG. 7.3. Performance of linear (a) and quadratic (b) triangular elements in an elastic solution. Adaptive refinement of one stage is followed aiming at 5% error in energy norm error. η – energy norm error; θ – effectivity of error estimator. (a) Left: $\eta = 16\%$, $\theta = 0.6$; right: $\eta = 8\%$, $\theta = 0.9$. (b) Left: $\eta = 12\%$, $\theta = 0.7$; right: $\eta = 3.5\%$, $\theta = 0.9$. Reproduced by permission of John Wiley and Sons Limited.

ration (or alternatively by reintroducing the REISSNER [1945] and MINDLIN [1951] assumptions) became the favourite approach to this class of problems with many high-order elements being used, viz. ZIENKIEWICZ and TAYLOR [1991].

From a practical point of view the problem was thus *nearly* solved with the various elements performing reasonably in most situations though occasional malfunctions occurred. From the theoretical aspect, however, a difficulty remained: Why should a mere reduction of the integration labour result in an

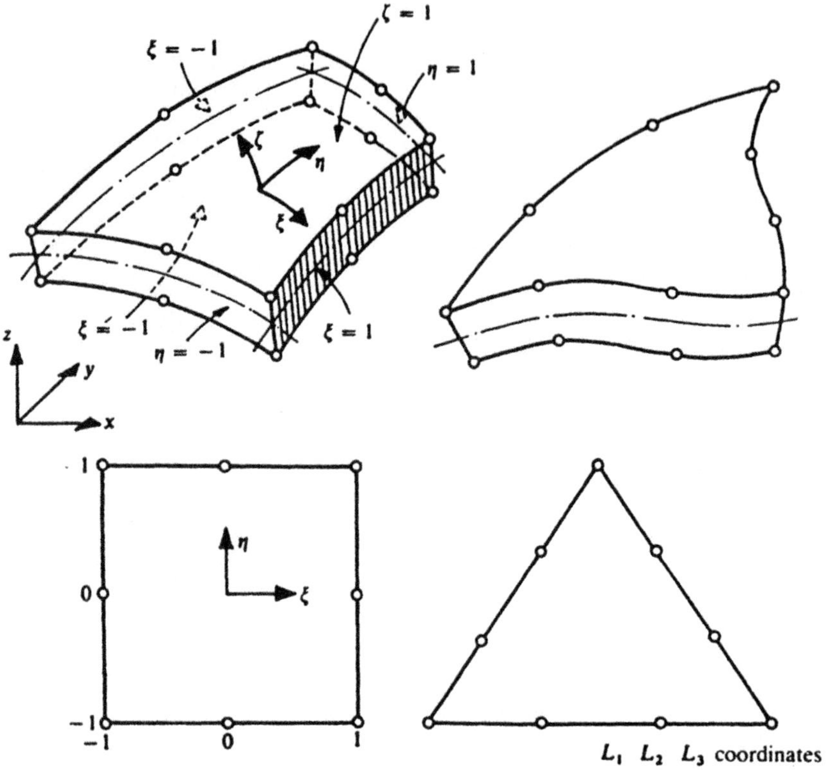

FIG. 7.4. Modelling of plates and shells by using thin, isoparametric element forms. Reproduced by permission of McGraw Hill.

improvement? (This surely went against the "protestant work ethic" as said by some at the time!)

The answers to this problem took a long time in coming. The first important step here was made by MALKUS and HUGHES [1978] who showed that reduced integration was (in many cases) equivalent to the use of a mixed formulation in which in addition to displacement interpolation, the shear forces were independently interpolated from "nodes" associated with the integration points.

The second important step was, we believe, the introduction of the requirements of the patch test which led to the development of first two *robust* elements which could be guaranteed to work in all conditions, ZIENKIEWICZ and LEFEBVRE [1988] and ARNOLD and FALK [1987]. Other approaches were at the same time becoming popular and again could be justified by the same arguments DVORKIN and BATHE [1984], HUANG and HINTON [1986].

Now at last the problem could be considered solved though there are many corollaries yet subject to exploration as discussed by ZIENKIEWICZ and TAYLOR [1991].

8. Adaptivity and error estimation

In the preceding section I have dwelt at some length on development and understanding of matters initially introduced by engineers rather than mathematicians (though the difference of these two "species" is by no means clear in the individuals—though perhaps is more evident in the language!). The serious problem of estimating errors of the finite element discretisation economically was however first addressed by mathematicians. Here the work of BABUŠKA and RHEINBOLDT [1978, 1979] was the major landmark which not only showed the possibilities of economic error estimation but indicated how, by successive, adaptive refinement of meshes a desired accuracy of the numerical solutions could be reached. This work was obviously important as the only practical procedures previously available to judge the accuracy of the solution by were:

(1) Comparison with, occasionally available, exact solutions.

(2) Full, uniform subdivision of the mesh and use of an extrapolation—which of course was too costly for the majority of problems.

To facilitate the knowledge transfer, Babuška and myself "teamed" up to translate and develop the procedures for a wider audience. This resulted in a series of papers (KELLY, NAKAZAWA and ZIENKIEWICZ [1980]) and an international meeting on the subject, BABUŠKA, ZIENKIEWICZ, GAGO and OLIVEIRA [1986]. However, much needed to be done and the subject became a largely popular area of research in which the processes of h-refinement (element size adjustment), p-refinement (uniform or non-uniform increase of polynomial order) or even h–p combinations were investigated.

Nevertheless, practical use of adaptive procedures remained quite small. What was missing, especially in the context of h refinement which could be used in general purpose finite elements codes, centred around three important points.

(a) A robust, simple and economical process of error estimation on an existing mesh.

(b) A means of predicting directly the required mesh density satisfying economically the specified accuracy and thus avoiding the previously used and costly progressive refinement strategies.

(c) A mesh generator capable of deriving a mesh of specified density.

An important step in this direction was taken by ZIENKIEWICZ and ZHU [1987, 1989] who spell out above requirements and make use of the first triangular mesh generator specifically designed for the adaptive process by PERAIRE, VAHDATI, MORGAN and ZIENKIEWICZ [1987].

The error estimator, postulated in this paper for self-adjoint problems, is based on the approximation that the error in fluxes (stresses) can be written as

$$e_\sigma = \boldsymbol{\sigma} - \boldsymbol{\sigma}^h \approx \boldsymbol{\sigma}^* - \boldsymbol{\sigma}^h \tag{8.1}$$

in which $\boldsymbol{\sigma}$ are the exact values of the stresses (or other fluxes), $\boldsymbol{\sigma}^h$ is the finite element approximation to these and $\boldsymbol{\sigma}^*$ are "*recovered*" *values of* σ^h *obtained by some post-processing operation which improves their accuracy.*

Indeed it was shown by ZIENKIEWICZ and ZHU [1992] that the effectivity of such an error estimator, i.e.,

$$\theta = \frac{\text{estimated error}}{\text{actual error}}, \tag{8.2}$$

where the errors are specified in any suitable norm is always bounded by

$$1 - \alpha \leqslant \theta \leqslant 1 + \alpha , \tag{8.3}$$

where

$$\alpha = \frac{\|e_\sigma^*\|}{\|e_\sigma^h\|} . \tag{8.4}$$

In the above the numerator is the error of the recovered solution $\boldsymbol{\sigma}^*$ and the denominator that of the finite element solution $\boldsymbol{\sigma}^h$. The need for a small value of α is obvious. Clearly the effectivity of the estimation process will depend on the accuracy attainable by the recovery process. In the 1987 paper the currently much used process of "nodal averaging" or of the so-called L_2 projection (BRAUCHLI and ODEN [1971], HINTON and CAMPBELL [1974]) were used with some success though with these frequently the value of α was not always very small.

A major step forward was made here only very recently by ZIENKIEWICZ and ZHU [1992a,b] where the well-known properties of "superconvergence" of σ^h at certain sampling points were used. This process, now given the name of SPR (superconvergent patch recovery), is so simple and self-evident that in retrospect I found it hard to believe it was not used before! How did we all miss it?

The "SPR" algorithm simply assumes that in patches of elements surrounding a typical corner node the value of $\boldsymbol{\sigma}$ is approximated by a simple polynomial expansion one order higher than that exactly reproduced by the element in question. This expansion which can always be written as

$$\tilde{\boldsymbol{\sigma}}^* = \boldsymbol{P}\boldsymbol{a} , \tag{8.5}$$

where

$$\boldsymbol{P} = \begin{bmatrix} \boldsymbol{p} & & 0 \\ & \boldsymbol{p} & \\ 0 & & \boldsymbol{p} \end{bmatrix}, \quad \boldsymbol{p} = [1, x, y, \ldots], \quad \boldsymbol{a} = \begin{Bmatrix} a_1 \\ a_2 \\ \vdots \\ a_j \end{Bmatrix}, \tag{8.6}$$

with j being the number of components of $\bar{\boldsymbol{\sigma}}^*$ which we assume to fit in a least square sense to superconvergent values of $\boldsymbol{\sigma}_k^h$ where k is the appropriate sampling point.

Minimisation with respect to $\boldsymbol{a}$ of

$$\sum_k |\boldsymbol{\sigma}_k^h - \tilde{\boldsymbol{\sigma}}^*|^{\mathrm{T}} |\boldsymbol{\sigma}_k^h - \tilde{\boldsymbol{\sigma}}^*| \tag{8.7}$$

gives

$$P^{T}[\sigma_k^h - Pa] = 0 \tag{8.8}$$

or

$$a = [P^{T}P]^{-1}P^{T}\sigma_k^h \tag{8.9}$$

with summation with respect to k implied determines $\tilde{\sigma}^*$ uniquely.

The nodal values $\tilde{\sigma}^*$ of the final "recovered" approximation A which is given below and in which N are the "displacement" shape functions

$$\sigma^* = N\bar{\sigma}^* \tag{8.10}$$

can now be simply determined. Clearly, in the above each component of stress can be solved separately.

Figure 8.1 illustrates some typical two-dimensional patches which can be used and in Fig. 8.2 we show the dramatic improvement in the value of recovered stresses for a simple elastic problem.

Of course the process is equally applicable in three-dimensions or indeed in

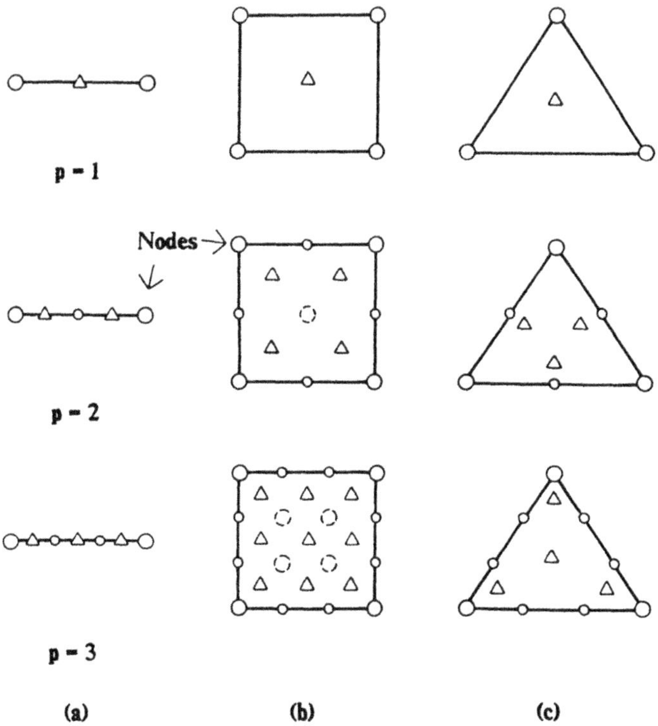

FIG. 8.1. Superconvergent points in one dimension (a), two-dimensional quadrilaterals (b), and triangles (c) in which a continuous polynomial approximates the stress distribution. (Superconvergence in quadrilaterals is only true for simple Laplace equations and for triangles their sampling points are not truly superconvergent.)

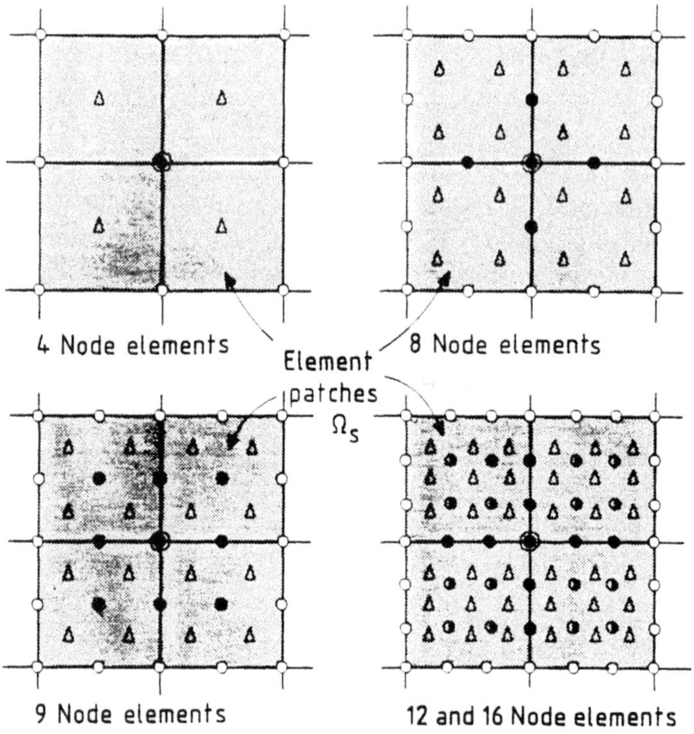

(d)

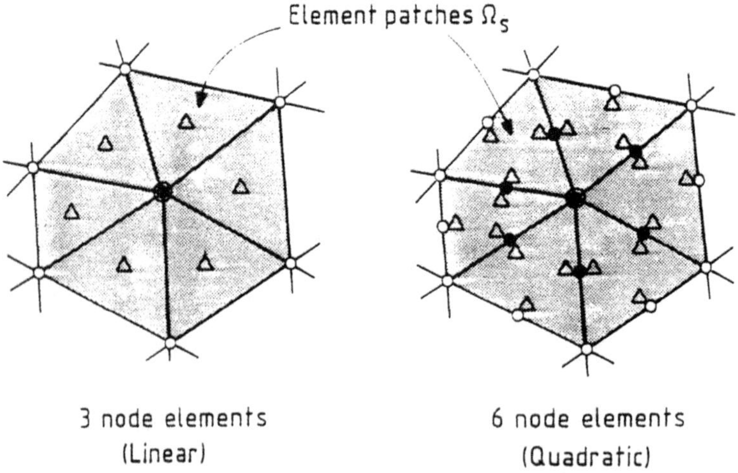

(e)

FIG. 8.1 (continued). Computation of superconvergent nodal values for typical recovery patches of different quadrilateral and triangular elements. Reproduced by permission of John Wiley and Sons Limited.

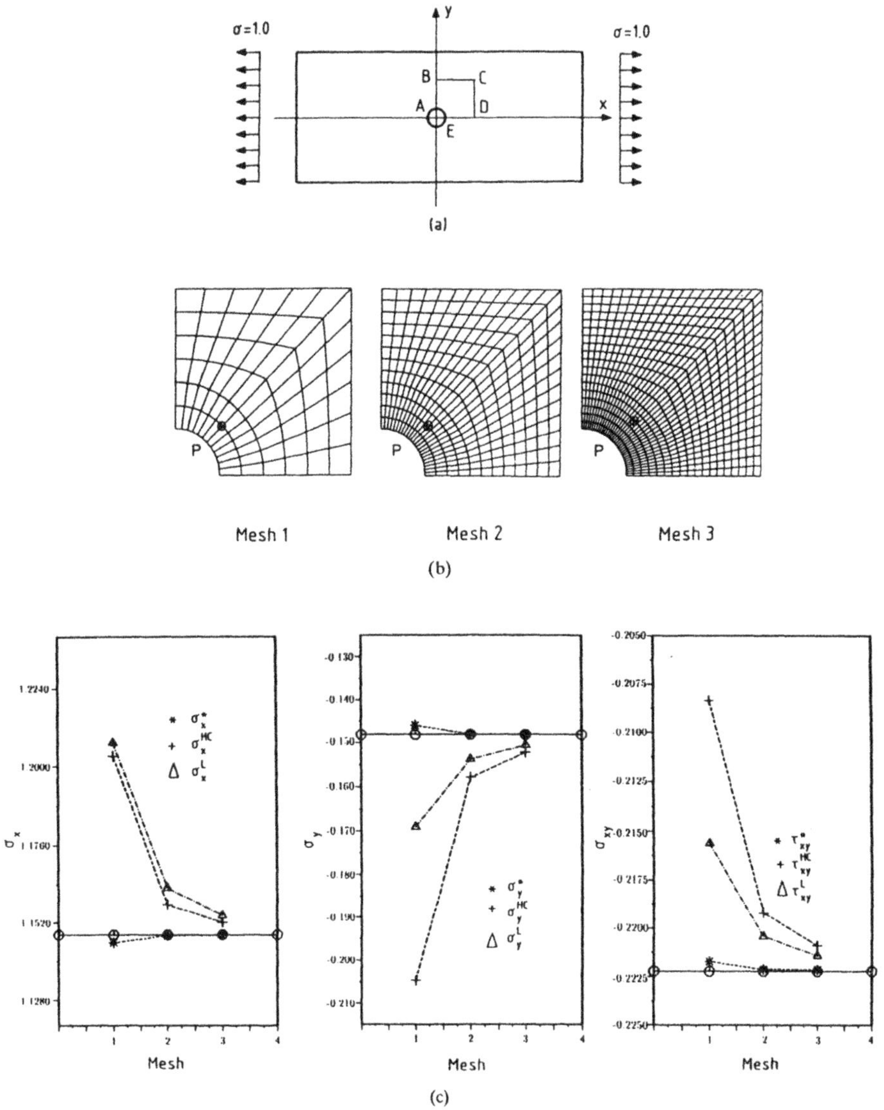

FIG. 8.2. Accuracy of the superconvergent patch recovery (SPR) illustrated on an elastic perforated continuum for which exact solution is available (a) the analysis region, (b) three meshes using nine node, quadratic, quadrilaterals, (c) error of stresses obtained on various meshes using SPR, L^2 recovery (sup. L) and local L^2 recovery with averaging (sup. HC). Reproduced by permission of John Wiley and Sons Limited.

one. For the latter it is easy to see from Fig. 8.3 why all the values recovered are superconvergent, with the linear u element being capable of modelling *exactly* a linear variation of σ, quadratic, a quadratic variation, etc. In Fig. 8.4 indeed we show how typical nodal values show superconvergence in a typical,

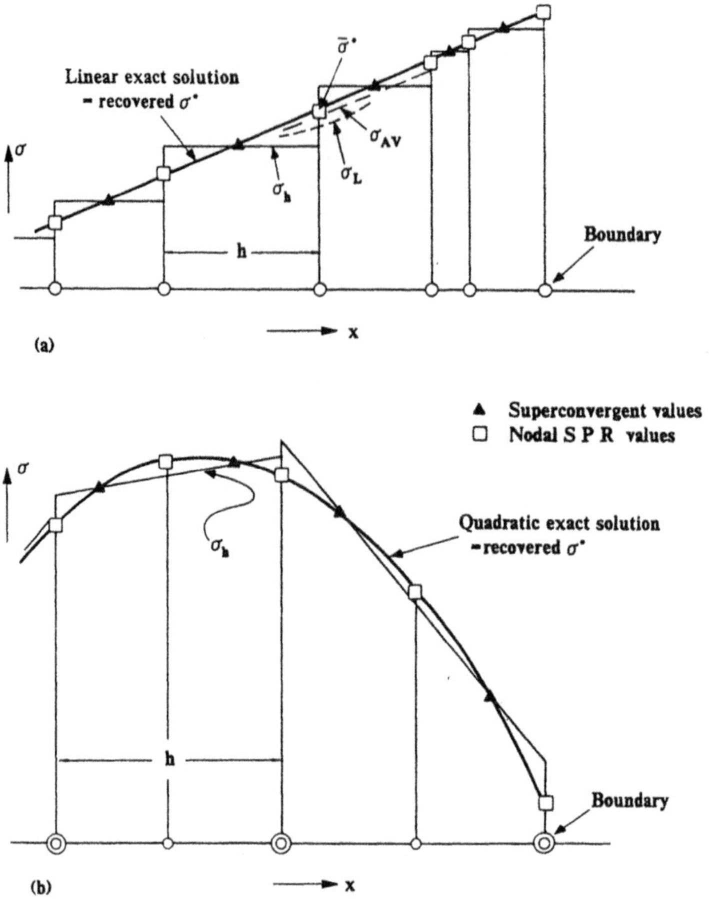

FIG. 8.3. Recovery of exact solution by SPR in a 1D problem. (a) Linear elements recover exact linear stresses. (b) Quadratic elements recover exact quadratic stresses variation, etc. Reproduced by permission of John Wiley and Sons Limited.

one-dimensional, second-order problem for various degrees of polynomial used in the original approximation.

Certainly error estimates based on $\boldsymbol{\sigma}^*$ are now extremely accurate (though a further improvement can be added by modifying the functional of Eq. (8.8) to include overall equilibrium satisfaction—viz. WIBERG and ABDULAHAB [1992]).

The error estimator is now ready to be used with adaptivity and the process of predicting the new mesh density will not be here described as it is much dependent on the objectives of the analysis; reference to the original papers should be made for details. In Fig. 8.5 we show results of adaptive analysis applied to a simple heat conduction problem with distributed sources for which the exact solution is known, ZIENKIEWICZ and ZHU [1992b]. The fast convergence to the required solution accuracy and the efficiency of the estimators should be noted.

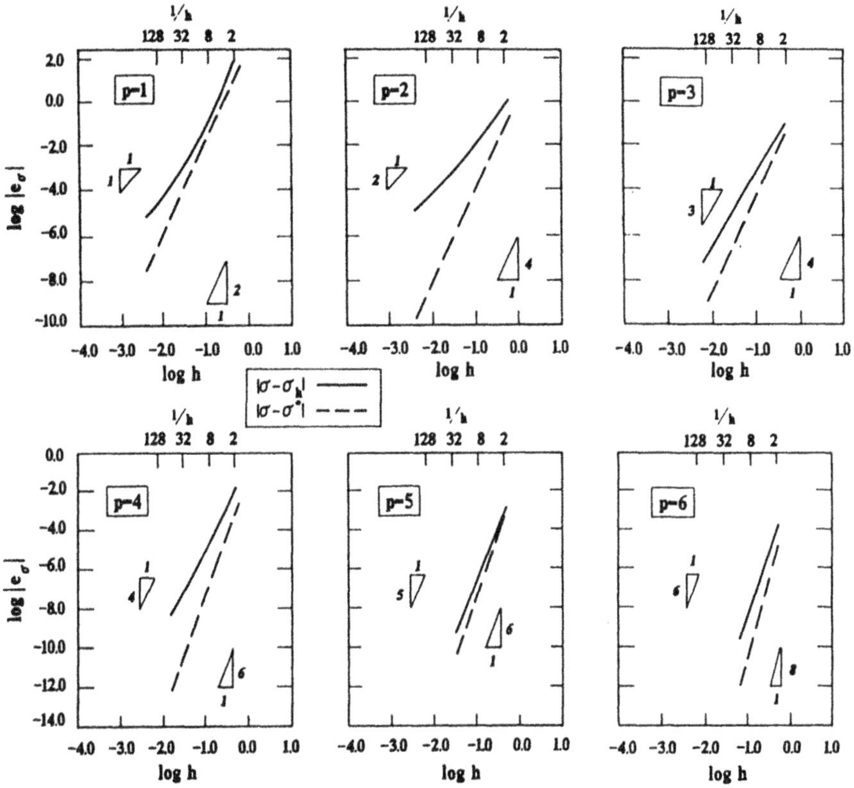

FIG. 8.4. Convergence of stresses recovered by SPR in a 1D model problem (dotted line) and convergence of the original finite element stresses. Various orders of elements $p=1$ to $p=6$ shown. (Note two order higher convergence rate for even order elements—this occurs only for equal size elements.) Reproduced by permission of John Wiley and Sons Limited.

Clearly, in practical applications the degree of effectivity of the estimator will not be tested, but as shown in a typical problem of Fig. 8.6, optimal meshes are readily generated with considerable assurance of quality.

In parallel with the "h" adaptivity procedures much progress has been made in recent years on p and h–p processes, viz. DEMKOWICZ, ODEN and BABUŠKA [1992], however, space does not permit to record here all the major steps achieved.

9. Fluid mechanics and non-self-adjoint problems

I have already referred to the wide application of the finite element method beyond structured mechanics which occurred after 1965 and in particular to fluid mechanics where its application seemed natural. The extension to potential problems of ideal, inviscid flow and indeed to predominantly viscous,

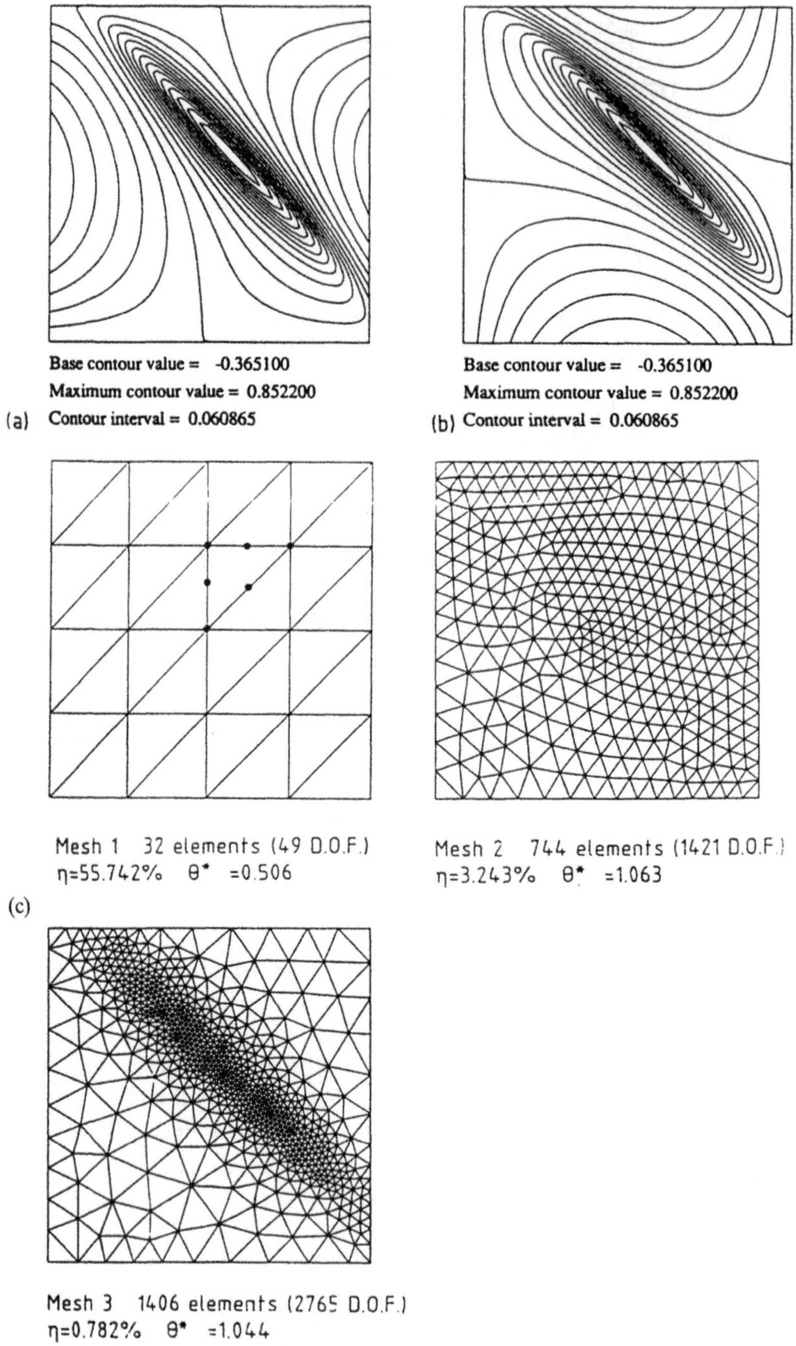

FIG. 8.5. Adaptive solution of a model 2D problem. Heat conduction with source terms leading to exact solution in (a) for $\partial u/\partial x$ and (b) for $\partial u/\partial y$. (c) shows three stages of adaptive solution using quadratic triangles reducing the energy norm error below 1% from 55%.

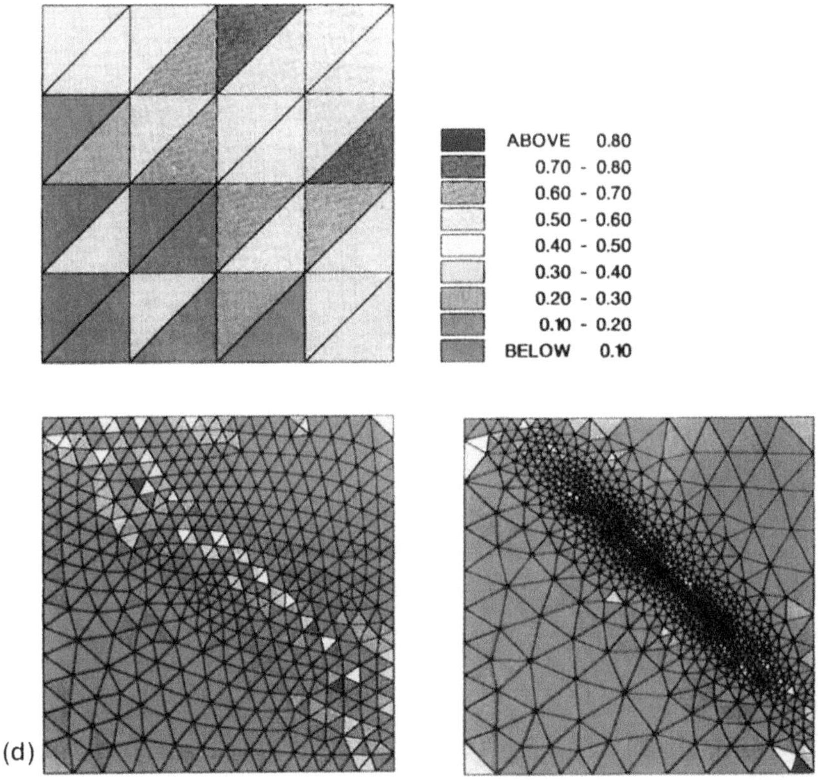

FIG. 8.5. (d) $(1-\theta)$ for local effectivity indices during the adaptive refinement. Reproduced by permission of John Wiley and Sons Limited.

Stokes flow was obvious. The previously developed approaches (and indeed frequently the same computer codes) could be used directly without modification (MARTIN [1968], ATKINSON, CARD and IRONS [1970].

However, the situation where convective accelerations occur which are typical of Navier–Stokes equations presented a difficulty. Here the non-self-adjoint nature of the equations precluded the use of variational (extremum) principles and the only discretisation possibility presented itself via the use of weighted residual, Galerkin-type approaches. This possibility was first outlined by ODEN [1969, 1973] and ZIENKIEWICZ and TAYLOR [1973] with first realistic solutions produced by TAYLOR and HOOD [1973]. Another article in this Handbook series, written by R. Glowinski, on "Numerical simulation of incompressible viscous flows" will give a full account of this phase of development.

However, simple application of the Galerkin weighting was soon found to be inapplicable in problems in which convective terms were dominant. Indeed, here the same difficulties of oscillatory (or divergent) solutions observed earlier by finite difference practitioners using central differences were observed and it

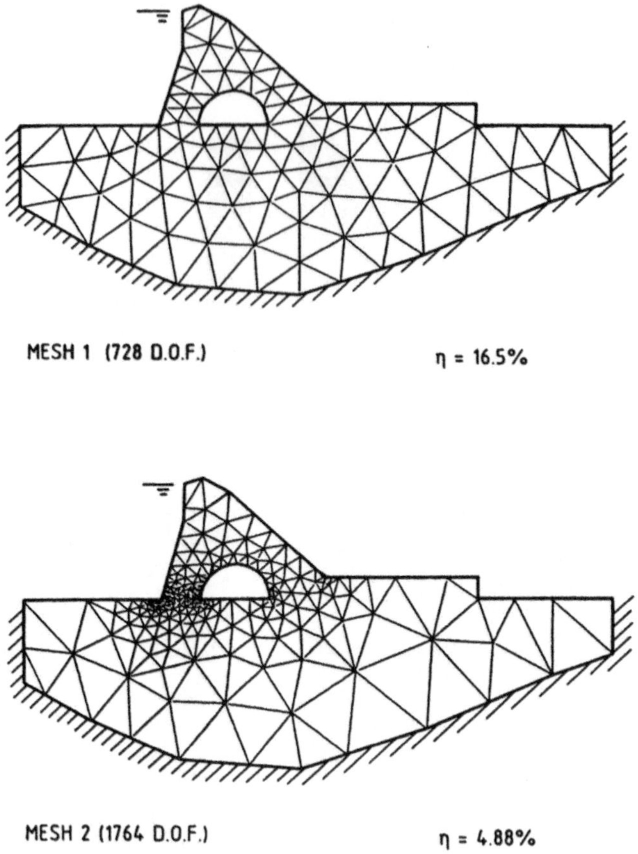

FIG. 8.6. Automatic adaptive elastic stress analysis. A single re-analysis achieves specified 5 per cent energy norm error. Example of a dam. Reproduced by permission of John Wiley and Sons Limited.

was natural to seek remedies from those working in that field. Indeed, a standard "medicine" for such problems in the finite difference fraternity was the use of "upwind differences" (SPALDING [1972]) and the translation of this process to upwind weighting of the Petrov–Galerkin kind in the finite element context was first suggested by ZIENKIEWICZ, GALLAGHER and HOOD [1976]. The detailed elaboration of this by CHRISTIE, GRIFFITHS, MITCHELL and ZIENKIEWICZ [1976] and ZIENKIEWICZ, HEINRICH, HUYAKORN and MITCHELL [1977a] led to the adoption of such upwind Petrov–Galerkin methods for effective solutions of both simple convective problems and of the incompressible Navier–Stokes equation permitting in the latter a larger range of Reynolds numbers to be dealt with effectively.

An important development of the techniques originally suggested was made by HUGHES and BROOKS [1979] and KELLY, NAKAZAWA and ZIENKIEWICZ [1980] and later JOHNSON, NÄVERT and PITKÄRANTA [1984] resulting in schemes which avoid the introduction of any "cross wind" diffusion.

But this first step allowing the treatment of steady state problems was not sufficient as the use of upwind Petrov–Galerkin procedure was not justified when source or transient term existed. Here the major step forward was taken by the introduction of a characteristic Galerkin process making use of the wave propagation features of these together with the optimality of Galerkin methods for the self-adjoint, diffusion part of the problem. This indeed led to a family of new algorithms, BERCOVIER, PIRONNEAU, HARBANI and LIVNE [1982], PIRONNEAU [1982], ZIENKIEWICZ, LÖHNER, MORGAN and NAKAZAWA [1984], ZIENKIEWICZ, LÖHNER, MORGAN and PERAIRE [1986], LÖHNER, MORGAN and ZIENKIEWICZ [1984] which permitted a wide and new range of problems to be solved. The simple algorithm produced in the last two references, in which the characteristic "search" is dealt with by a simple Taylor expansion could be reinterpreted as a finite element version of the LAX and WENDROFF [1960] method (now ensuring a better accuracy due to the Galerkin approximation) and was identified with the so-called Taylor–Galerkin method, DONEA [1984]. The latter, not based on a single characteristic velocity, can deal with a number of characteristic speeds and as such allowed extension to problems of high speed compressible gas flow to be made by the mid 1980s.

This indeed was a landmark and within the last decade this procedure for the solution within both compressible (and incompressible) flows was widely used and revolutionised this field of mechanics. This work of Morgan, Löhner, Peraire, the writer and others in the years of 1985–1990 is reported in the bibliography and has opened the doors to finite element analysis of very complex aeronautical problems. The development of algorithms for compressible flow is still in progress despite the achievements recorded, viz. ZIENKIEWICZ, MORGAN, SATYA SAI, CODINA and VAZQUEZ [1995].

The rapid progress made in this context has thrown up many interesting points:

(1) Adaptive refinement is essential for realistic compressible flow solution if the capture of such features as shock, boundary layers, etc. is to be made. Further, the overall refinement necessary in such complex, three-dimensional problems as the analysis of the whole aircraft, viz. PERAIRE, PEIRO, FORMAGGIA, MORGAN and ZIENKIEWICZ [1988] requires a very large number of elements and degrees of freedom (a million is quite a common mean!) and only by the adaptive process can economy be achieved.

(2) Lowest-order elements have been universally used in all solutions for two reasons; first because it is felt that these are optimal in modelling discontinuities, second because with these explicit dynamic transient problems can be most efficiently computed.

(3) Steady state solutions are always achieved by an iterative process (Gaussian elimination being not applicable to such large systems). Here all the tools available to finite difference practitioners who invariably use such iterations are available, once again have been reinterpreted in terms of

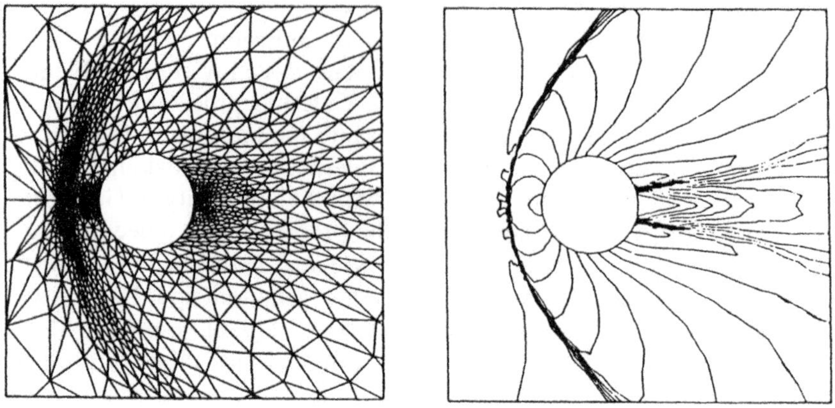

FIG. 9.1. Use of adaptive refinement for shock capture in compressible flow aerodynamics. Mach 2 flow of ideal gas round a cylinder. Note mesh elongation adopted for economy. Reproduced by permission of McGraw Hill.

unstructured finite element forms including such devices as multigrid techniques, etc.

Figures 9.1–9.4 illustrate some typical compressible and incompressible analysis in both steady and unsteady situations.

The procedures shown have been extended to viscous compressible flow and here problems become even more complex as shown in Fig. 9.5.

At this stage it is perhaps worth pondering whether finite element procedures have taken over from finite difference ones or whether simply a merging of the methods has occurred in the field of fluid mechanics. Certainly the unstructured nature of typical finite elements has been duplicated recently by the, so-called, finite volume techniques which can and indeed do on occasions use identical triangular forms (JAMESON, BAKER and WEATHERILL [1986]). Adaptivity can therefore be used on a similar basis. However, the finite volumes here bear a closer resemblance to finite elements than they do to standard finite differences. Computationally both procedures are extremely similar and there is little that can be said to distinguish differences. However, the clear cut finite element approximation is preferable in transient problems where the correct "mass matrices" generated add considerably to the accuracy, in contrast to the lumped forms traditionally used by the "difference" proponents.

The algorithms used for the solution are, as already mentioned, similar but once again the finite element forms allow in principle to use a more rational derivation and generally avoid rather arbitrary "artificial diffusion" operators.

The biggest controversy still raging is that between the use of structured and unstructured meshes. The first of course can lead to simpler numerical

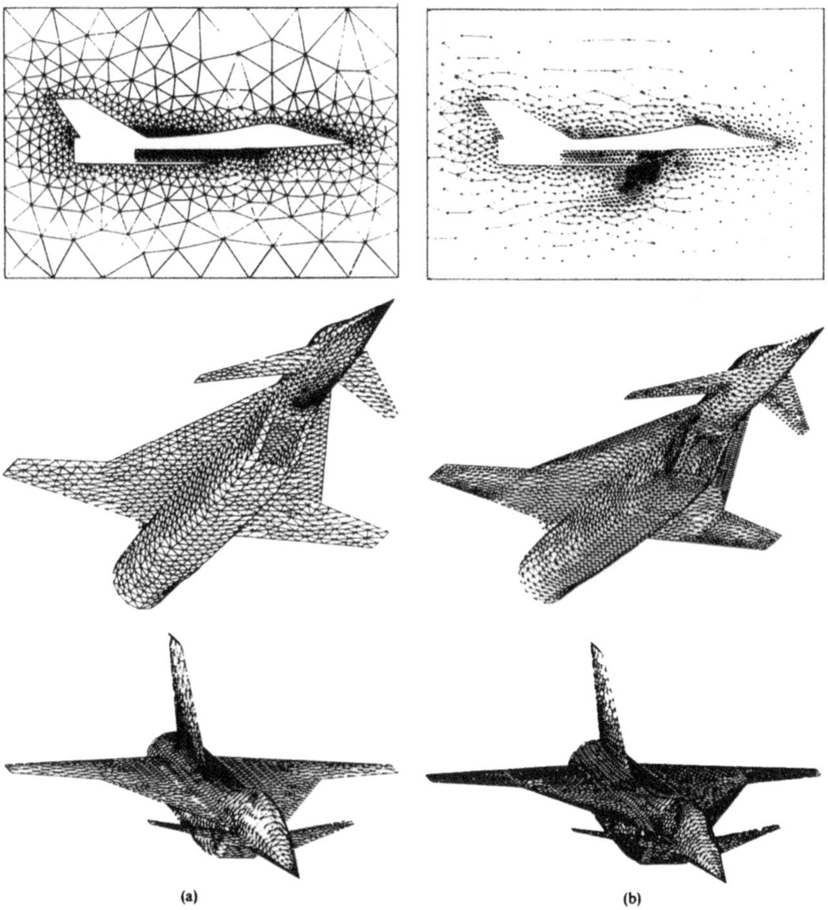

FIG. 9.2. Three-dimensional, adaptive solution of Euler equations at Mach 2 for a whole aircraft (NASA generic fighter configuration): (a) original mesh; (b) adaptively refined mesh. Reproduced by permission of John Wiley and Sons Limited.

algorithms but here the difficulties associated with the mapping of complex shapes and the impossibility of achieving optimal meshes by adaptivity, carry severe penalties of computational time.

It is my opinion that in fact the balance has turned today firmly in favour of finite element approximation in this field—particularly if these are given a liberal interpretation and make use of strictly numerical improvements made by "the other side". I have already remarked that simple elements are much in vogue in the fluid field, reversing the swing of the late sixties and seventies towards higher-order approximation. The same phenomenon has been observed in the development of solid mechanics codes based on explicit dynamic formulation in the early eighties (see, e.g., DYNA [1980]).

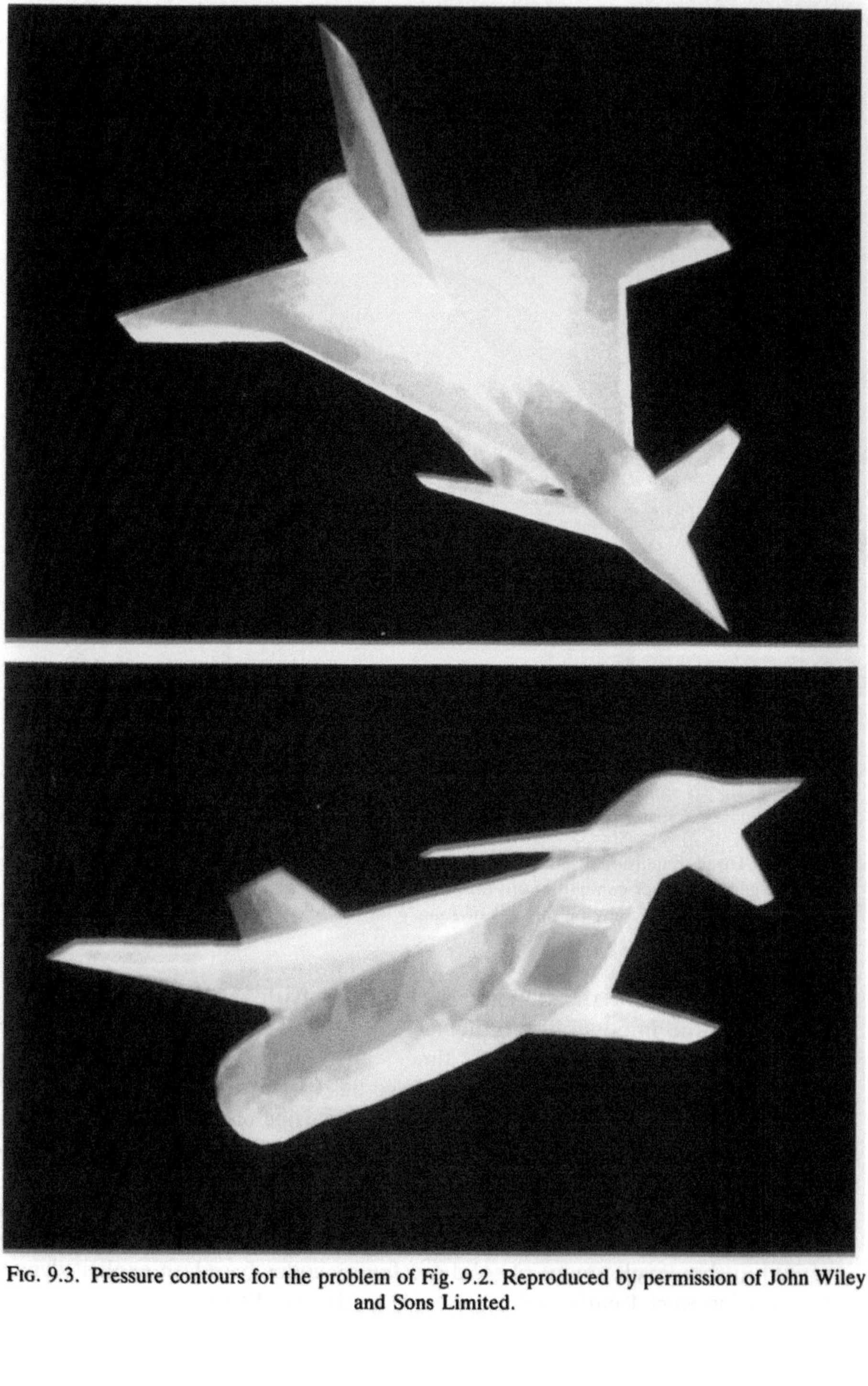

FIG. 9.3. Pressure contours for the problem of Fig. 9.2. Reproduced by permission of John Wiley and Sons Limited.

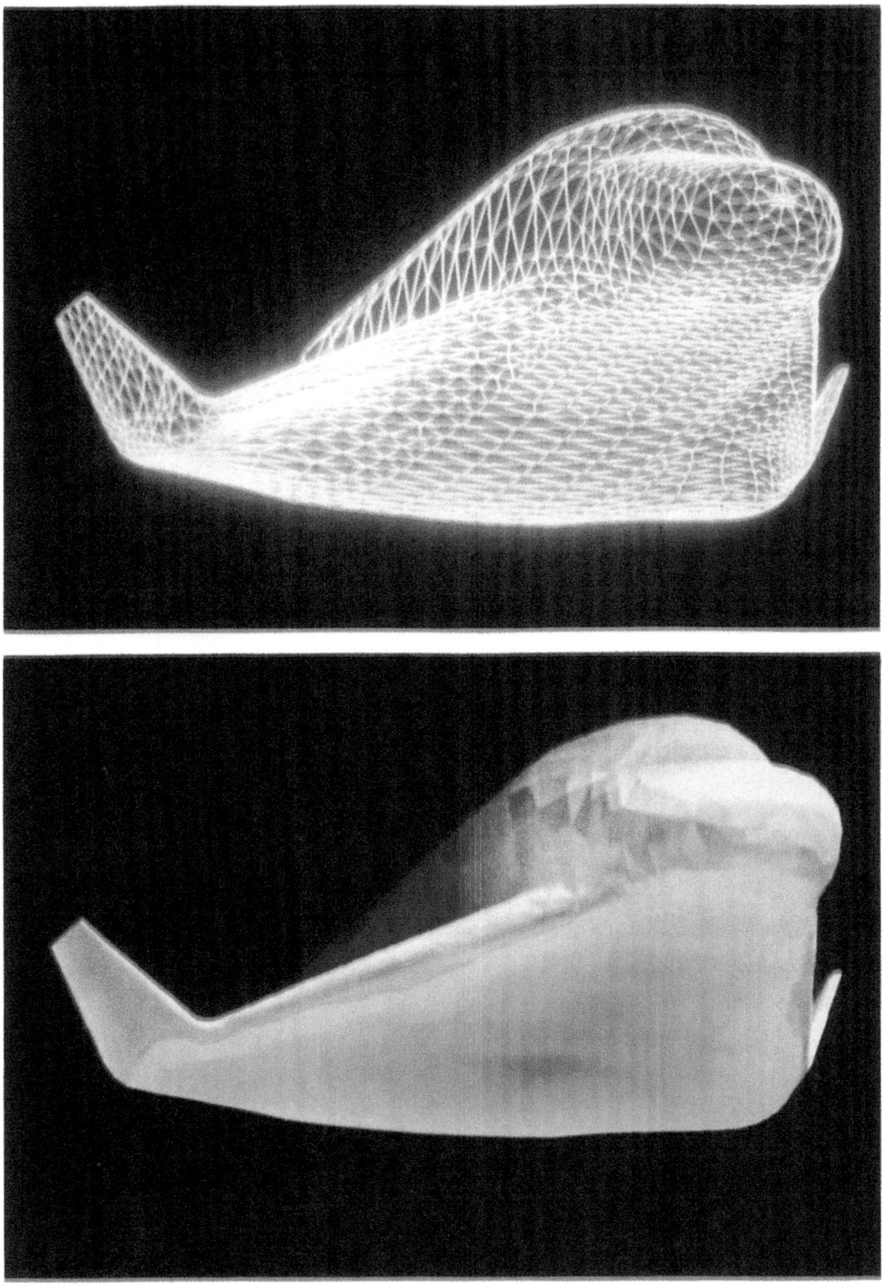

FIG. 9.4. An early study of the European shuttle (Euler equations): (a) Mesh; (b) Pressure contours. Reproduced by permission of John Wiley and Sons Limited.

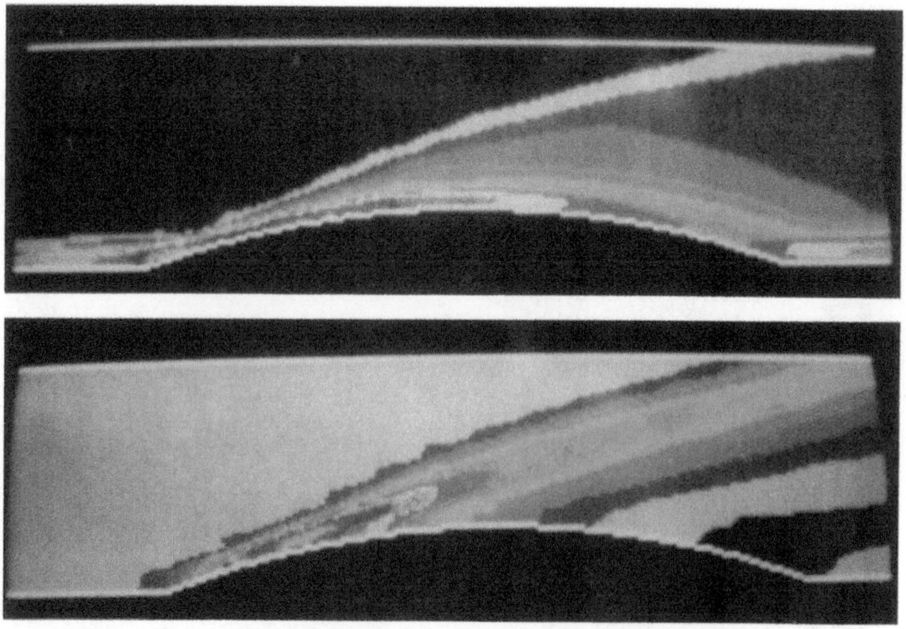

FIG. 9.5. Flow around a surface "bump" at Mach 3 in viscous, compressible flow. Temperature and pressure contours. Reproduced by permission of John Wiley and Sons Limited.

Is it possible that the future will see a possible swing of the pendulum and a return to higher-order elements? Only time will tell.

10. Epilogue

In concluding this recital of the "landmarks" I am very conscious that omission has been made of many achievements which others would choose in this context. In particular I have not talked about the important aspect of non-linear application in the fields of structurally plasticity and large deformation, metal forming and geomechanics (which inevitably occupied time of my own and many others—ZIENKIEWICZ, VALLIAPPAN and KING [1969], NAYAK and ZIENKIEWICZ [1972]). Nor have I discussed the many problems associated with such non-linear computations to which many have contributed. The reasons for this are manifold; what I intended to present are in the main the various features of the *generalised finite element formulation* which can be widely applied and offers many possibilities. The view that "the finite element method is simply a systematic technique for construction of Ritz–Galerkin approximations for irregular domains" is in my opinion too restrictive and I hope that wider possibilities are implied in the name.

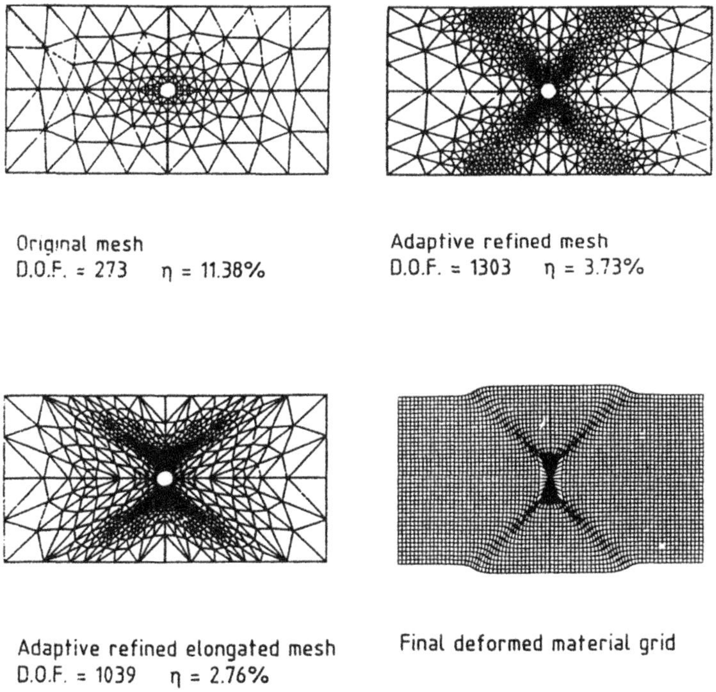

FIG. 10.1. Technology transfer. Adaptive methods developed for compressible flow capture a strain localization in plastic flow. Reproduced by permission of John Wiley and Sons Limited.

However, nothing in the field of finite element method activity is done in isolation. The technology transfer is now rapid between one or another area of activity. In Fig. 10.1 I show how the developments for adaptive shock capture in fluid mechanics have influenced the currently fashionable problems of stress localization in plasticity.

Finally, let me stress that the process is a playground of many, including both engineers and mathematicians. The first, using intuition, frequently act before proof of correctness is made available by the latter. The history shows that in general this has been the path of progress!

Appendix

Extract from R.W. CLOUGH ([1979], pp. 1.6–1.7)

The work which I associate with the beginning of the computerized FEM was done during summer 1953 when I was again employed by Boeing Airplane Company on their summer faculty program. Again, I was assigned to Mr. M.J. Turner's Structural Dynamics Unit, to work on methods of evaluating the stiffness of a delta airplane wing for use in flutter analysis. Because the bar assemblage approach tried during the previous summer had been unsatisfactory, Mr. Turner suggested that we should merely try dividing the wing skin into appropriate triangular segments. The stiffness properties of these segments were to be evaluated by assuming constant normal and shear stress states within the triangles and applying Castigliano's theorem; then the stiffness of the complete wing system (consisting of skin segments, spars, stringers, etc.) could be obtained by appropriate addition of the component stiffnesses (the direct stiffness method). Thus, at the beginning of the summer, 1953, Mr. Turner had completely outlined the FEM concept and those of us working on the project merely had to carry out the details and test the results by numerical experiment.

Our paper describing this initial effort was presented at the New York meeting of the Institute of Aeronautical Sciences in January 1954 (TURNER, CLOUGH, MARTIN and TOPP [1956]). I have never known why the decision was made not to submit the paper for publication until 1955, so the publication date of September 1956 was more than two years after the first presentation and over three years after the work was done. As was mentioned, this is graphic evidence that the FEM did not attain instant recognition. Undoubtedly, a major factor which limited its acceptability was that the original work was done in the Structural Dynamics Unit, where the objective was limited to stiffness and deflection analysis; it was several years before the concept was accepted and put to use by the stress analysis groups at Boeing. Thus, it is possible that the orientation of this initial step toward a specific engineering application tended to obscure the general applicability of the FEM concept, even though the individuals working with the development at Boeing were quite aware of its broader implications.

Although I maintained close contact with several of my Boeing colleagues for many years after 1953, I did not work there again and I had no opportunity for further study of the FEM until 1956–1957, when I spent my first sabbatical leave in Norway (with the Skipsteknisk Forsknings Institutt in Trondheim). This "Norwegian connection" also was a factor in my decision to prepare a historical summary for this Conference; it was this period which made possible my continued contact with the finite element concept. Lack of computer facilities in Norway limited the type of work I could do at this time, but I was intrigued by plane stress application of the method and I carried out some very simple analyses of rectangular and triangular element assemblages using a desk calculator. Although this work was too trivial to warrant publication, it convinced me of the potential of the FEM for the solution of general continuum problems.

About the time I returned to Berkeley from my sabbatical leave, the Engineering College acquired an IBM 701 Computer (replacing the old Card Programmed Calculator) and we began to develop structural analysis capabilities with this machine. For educational purposes, a Matrix Interpretive Program of the type pioneered in England (HUNT [1956]) offered the best means of making the computer capabilities accessible to the students, and most of my early efforts went into developing such a program (CLOUGH [1958]). Then it was possible to continue my work with the FEM which had been undergoing continuing development at Boeing but had attracted only little attention elsewhere.

Early FEM studies at Berkeley were greatly limited by the two thousand word central processor capacity of the IBM 701, but by utilizing the seven 2000 word drum storage units it was possible to carry out some creditable analyses. Our first concentrated effort toward plane stress analysis was in response to a challenge by one of my continuum mechanics colleagues who was skeptical of the validity of the procedure and wanted to see a solution of some classical problem. To me it seemed obvious that the method could solve any plane stress problem to any desired accuracy—limited only by the time and energy one wished to expend on the calculations. But in the hopes of attracting wider interest toward the FEM concept, I allocated part of a small NSF research grant to the solution of a few sample plane stress problems. The results were as good as I had expected, so a paper was prepared. The principal problem that arose in writing the paper was choosing a suitable name for this analytical procedure and I decided finally on the Finite Element Method. This name first appeared in that paper (CLOUGH [1960]), and I can only conclude from subsequent history that it was an apt choice.

In retrospect, the next red letter event in my personal FEM history occurred in December 1960, when Professor O.C. Zienkiewicz invited me to Northwestern University to give a seminar lecture on the new procedure. We were friends from previous meetings, and I knew that he had been brought up in the Southwell finite difference tradition, so it was apparent that his invitation was prompted by skepticism and a desire to discuss the relative merits of finite elements vs. finite differences. Certainly, we did have such discussions during my visit, but Professor Zienkiewicz obviously is a very intelligent person and was quick to recognise the advantages of the FEM. During that short visit an illustrious convert was won to the cause, and I think it is not coincidental that rapid worldwide acceptance of the FEM started almost from that moment.

References

AHMAD, S., B.M. IRONS and O.C. ZIENKIEWICZ (1968), Curved thick shell and membrane elements with particular reference to axi-symmetric problems, in *Proc. 2nd Conf. on Matrix Methods in Structural Mechanics*, Wright-Patterson AF Base, OH.

AHMAD, S., B.M. IRONS and O.C. ZIENKIEWICZ (1970), Analysis of thick and thin shell structures by curved elements, *Internat. J. Numer. Methods Engrg.* **2**, 419–451.

ALLEN, D.N. DE G. (1955), *Relaxation Methods* (McGraw-Hill, New York).

ARGYRIS, J.H. (1955), Energy theorems and structural analysis, *Aircraft Engrg.*; reprinted: Butterworths, London, 1960.

ARGYRIS, J.H., I. FRIED and D.W. SHARPF (1968), The TUBA family of plate elements for the matrix displacement method, *Roy. Aeronaut. Soc.* **72**, 701–709.

ARGYRIS, J.H., G. MARECZEK and D.W. SCHARPF (1969), Two and three dimensional flow using finite elements, *J. Roy. Aero. Soc.* **73**, 961–964.

ARNOLD, D.N. and R.S. FALK (1987), A uniformly accurate finite element method for Mindlin–Reissner plate, IMA Preprint Series No. 307, Institute for Mathematics and its Applications, University of Minnesota.

ATKINSON, B., C.C.M. CARD and B.M. IRONS (1970), Application of the finite element method to creeping flow problems, *Trans. Inst. Chem. Engrg.* **48**, 276–284.

BABUŠKA, I (1971), Error bounds for finite element methods, *Numer. Math.* **16**, 322–333.

BABUŠKA, I. (1973), The finite element method with Lagrange multipliers, *Numer. Math.* **20**, 179–192.

BABUŠKA, I. and W.C. RHEINBOLDT (1978), A-posteriori error estimates for the finite element method, *Internat. J. Numer. Methods Engrg.* **11**, 1597–1615.

BABUŠKA, I. and W.C. RHEINBOLDT (1979), Adaptive approaches and reliability estimates in finite element analysis, *Comput. Methods Appl. Mech. Engrg.* **17/18**, 519–540.

BABUŠKA, I., O.C. ZIENKIEWICZ, J.P. DE S.R. GAGO and E.R. DE ARRANTES OLIVEIRA, eds. (1986), *Accuracy Estimates and Adaptive Refinement in Finite Element*. Wiley, New York.

BAZELEY, G.P., Y.K. CHEUNG, B.M. IRONS and O.C. ZIENKIEWICZ (1965), Triangular elements in bending—conforming and non-conforming solutions, in: *Proc. Conf. Matrix Methods in Structural Mechanics*, Air Force Inst. Tech., Wright-Patterson AF Base, OH.

BEER, G. and J.O. WATSON (1992), *Introduction to Finite and Boundary Element Methods for Engineering* (Wiley, Chichester, UK).

BELL, K. (1969), A refined triangular plate bending element, *Internat. J. Numer. Methods Engrg.* **1**, 101–122.

BELYTSCHKO, T., X.Y. LU and L. GU (1994), Element free Galerkin methods, *Internat. J. Numer. Methods Engrg.* **37**, 229–256.

BERCOVIER, M., O. PIRONNEAU, Y. HARBANI and E. LIVNE (1982), Characteristics and finite element methods applied to equations of fluids, in: J.R. WHITEMAN, ed., *The Mathematics of Finite Elements and Applications*, Vol. V (Academic Press, London) 471–478.

BERGAN, P.G. and L. HANSSEN (1977), A new approach for deriving "good" element stiffness matrices, in: J.R. WHITEMAN, ed., *The Mathematics of Finite Elements and Applications* (Academic Press, London) 483–497.

BERGAN, P.G. and M.K. NYGARD (1984), Finite elements with increased freedom in choosing shape functions, *Internat. J. Numer. Methods Engrg.* **20**, 643–663.

BOGNER, F.K., R.L. FOX and L.A. SCHMIT (1965), The generation of interelement-compatible stiffness and mass matrices by the use of interpolation formulae, in: *Proc. Conf. on Matrix Methods in Structural Mechanics*, Air Force Institute of Technology, Wright-Patterson AF Base, OH.

BOSSHARD, W. (1968), Ein neues vollverträgliches endliches Element für Plattenbiegung, *Mt. Ass. Bridge Struct. Eng. Bull.* **28**, 27–40.

BRAUCHLI, H.J. and J.T. ODEN (1971), On the calculation of consistent stress distribution in finite element applications, *Internat. J. Numer. Methods Engrg.* 3, 317–325.

BREZZI, F. (1974), On the existence, uniqueness and approximation of saddle point problems arising from lagrangian multipliers, *Rairo* **8**-R2, 129–151.

CHRISTIE, I., D.F. GRIFFITHS, A.R. MITCHELL and O.C. ZIENKIEWICZ (1976), Finite element methods for second order differential equations with significant first derivatives, *Internat. J. Numer. Methods Engrg.* **10**, 1389–1396.

CLOUGH, R.W. (1958), Structural analysis by means of a matrix algebra program, in: *Proc. A.S.C.E. Conf. on Electronic Computation*, Kansas City, November, 109–132.

CLOUGH, R.W. (1960), The finite element in plane stress analysis, in: *Proc. 2nd ASCE Conf. on Electronic Computation*, Pittsburgh, PA.

CLOUGH, R.W. (1965), The finite element method in structural mechanics, in: O.C. ZIENKIEWICZ and G.S. HOLISTER, eds., *Stress Analysis*, Chapter 7 (Wiley, New York).

CLOUGH, R.W. (1979), The finite element method after twenty-five years—A personal view, in: A.S. Computas, ed., *Engineering Applications of the Finite Element Method* (Det Norske Veritas, Høvik, Norway) 1.1–1.34.

CLOUGH, R.W. and A. ADINI (1961), Analysis of plate bending by the finite element method, Report to National Science Foundation USA G.7337.

CLOUGH, R.W. and J.L. TOCHER (1965), Finite element stiffness matrices for analysis of plates in bending, in: *Proc. Conf. on Matrix Methods in Structural Mechanics*, Air Force Institute of Technology, Wright-Patterson AF Base, OH.

COURANT, R. (1923), On a convergence principle in calculus of variation (German), Kön Gesellschaft der Wissenschaften zu Götingen Wachrichten, Berlin.

COURANT, R. (1943), Variational methods for the solution of problems of equilibrium and vibration, *Bull. Amer. Math. Soc.* **49**, 1–23.

COWPER, G.R., E. KOSKO, G.M. LINDBERG and M.D. OLSON (1968), Formulation of a new triangular plate bending element, *Trans. Canad. Aero-Space Inst.* **1**, 86–90 (See also NRC Aero report LR514, 1968).

CRANDALL, S.H. (1955), *Engineering Analysis* (McGraw-Hill, New York).

DEMKOWICZ, L., J.T. ODEN and I. BABUŠKA (1992), Reliability in computational mechanics, Special issue *Comput. Methods Appl. Mech. Engrg.* **101**, 1–492.

DONEA, J. (1984), A Taylor–Galerkin method for convective transport problems, *Internat. J. Numer. Methods Engrg.* **20**, 101–119.

DVORKIN, E.N. and K.J. BATHE (1984), A continuum mechanics based four noded shell element for non linear analysis, *Engrg. Comput.* **1**, 77–88.

ERGATOUDIS, J.G., B.M. IRONS and O.C. ZIENKIEWICZ (1968a), Curved isoparametric "quadrilateral" elements for finite element analysis, *Internat. J. Solids and Structures* **4**, 31–42.

ERGATOUDIS, J.G., B.M. IRONS and O.C. ZIENKIEWICZ (1968b), Three dimensional analysis of arch dams and their foundations, in: *Symposium on Arch Dams*, Inst. Civ. Eng., London.

FALKENHEIMER, K. (1951), Systematic calculation of the elastic characteristics of hyperstatic systems, *Rech. Aeronaut.*, 17–23.

FINLAYSON, B.A. (1972), *The Method of Weighted Residuals and Variational Principles* (Academic Press, New York).

FENG, KANG. (1965), Difference schemes based on variational principles (in Chinese), *Appl. Comput. Math.* **2**, 238–262.

FLETCHER, C.A.T. (1984), *Computational Galerkin Methods* (Springer, Berlin).

FORTIN, M. and N. FORTIN (1982), Newer and newer elements for incompressible flow, in: R.H. Gallagher et al., eds., *Finite Elements in Fluids* (J. Wiley, Chichester) Chapter 7, Vol. 6.

GALERKIN, B.G. (1915), Series solution of some problems of elastic equilibrium of rods and plates (Russian), *Vestnik. Inzh. Tech.* **19**, 897–908.

GALLAGHER, R.H., J. PADLOG and P.P. BIJLAARD (1962), Stress analysis of heated complex shapes, *ARS J.*, 700–707.

HASSAN, O., K. MORGAN and J. PERAIRE (1989), An implicit–explicit scheme for compressible viscous high speed flows, *Comput. Methods Appl. Mech. Engrg.* **76**, 245–258.

HERRMANN, L.R. (1965), Elasticity equations for incompressible or nearly incompressible materials by a variational theorem, *J. AIAA* **3**, 1896–1900.

HINTON, E., and J. CAMPBELL (1974), Local and global smoothing of discontinuous finite element function using a least squares method, *Internat. J. Numer. Methods Engrg.* **8**, 461–480.

HRENIKOFF, A. (1941), Solution of problems in elasticity by the framework method, *J. Appl. Mech.* **A8**, 169–175.

HUANG, E.C. and E. HINTON (1986), Elastic, plastic and geometrically non linear analysis of plates and shells using a new, nine noded element, in: P. BERGAN et al., eds., *Finite Elements for Non Linear Problems* (Springer, Berlin) 283–297.

HUNT, C.P. (1956), The electronic digital computer in Aircraft structural analysis, *Aircraft Engrg.* **28**, March p. 40, April p. 11, May p. 155.

HUGHES, T.J.R. and A. BROOKS (1979), A multi-dimensional upwind scheme with no cross wind diffusion, in: T.J.R. HUGHES, ed., *Finite Elements for Convection Dominated Flows*, AMD 34 (ASME, New York).

IRONS, B.M. (1966a), Numerical integration applied to finite element methods, in: *Conf. Use of Digital Computers in Struct. Engrg.* Univ. of Newcastle.

IRONS, B.M. (1966b), Engineering application of numerical integration in stiffness method, *J. AIAA* **14**, 2035–2037.

IRONS, B.M. (1969), A conforming quartic triangular element for plate bending, *Internat. J. Numer. Methods Engrg.* **1**, 29–46.

IRONS, B.M. and J.K. DRAPER (1965), Inadequacy of nodal connections in a stiffness solution for plate bending, *J. AIAA* **3**, 5.

IRONS, B.M. and A. RAZZAQUE (1972), Experience with the patch test for convergence of finite element method, in: A.K. AZIZ, ed., *Mathematical Foundations of the Finite Element Method* (Academic Press, New York) 557–587.

JAMESON, A., T.J. BAKER and N.P. WEATHERILL (1986), Calculation of inviscid transonic flow over a complete aircraft, in: *Proc. AIAA 24th Aerospace Sci. Meeting*, Reno, Nevada, AIAA paper 86-0103.

JOHNSON, C., V. NÄVERT and J. PITKÄRANTA (1984), Finite element methods for linear, hyperbolic problems, *Comput. Methods Appl. Mech. Engrg.* **45**, 285–312.

KELLY, D.W., S. NAKAZAWA and O.C. ZIENKIEWICZ (1980), A note on anisotropic balancing dissipation in the finite element method approximation to convective diffusion problems, *Internat. J. Numer. Methods Engrg.* **15**, 1705–1711.

LANGEFORS, B. (1952), Analysis of elastic structures by matrix transformation with special regard to semi monocaque structures, *J. Aero. Sci.* **19** (17).

LAX, P.D. and B. WENDROFF (1960), Systems of conservative laws, *Comm. Pure Appl. Math.* **13**, 217–237.

LISZKA, T. and J. ORKISZ (1980), The finite difference method at arbitrary irregular grids and its application in applied mechanics, *Comput. & Structures* **11**, 83–95.

LIU, W.R., J. ADEE and S. JUN (1993), Reproducing kernel particle methods for elastic and plastic problems, in: D.J. BENSON et al., eds, *Adv. Comput. Methods Mater. Model., AMD* **180**, 175–190.

LÖHNER, R., K. MORGAN and O.C. ZIENKIEWICZ (1984), The solution of non-linear hyperbolic equation systems by the finite element method, *Internat. J. Numer. Methods Fluids* **4**, 1043–1063.

LÖHNER, R., K. MORGAN and O.C. ZIENKIEWICZ (1985), An adaptive finite element procedure for compressible high speed flows, *Comput. Methods Appl. Mech. Engrg.* **51**, 441–465.

MALKUS, D.S. and T.J.R. HUGHES (1978), Mixed finite element methods in reduced and selective integration techniques: A unification of concepts, *Comput. Methods Appl. Mech. Engrg.* **15**.

Martin, H.C. (1968), Finite element analysis of fluid flows, in: *Proc. 2nd Conf. on Matrix Methods in Struct. Mech.*, Wright-Patterson AF Base, OH.

Massonet, C.E. (1965), Numerical use of integral procedures, in: O.C. Zienkiewicz and G.S. Holister, eds., *Stress Analysis*, Chapter 10 (Wiley, New York) 198–235.

McHenry, D. (1943), A lattice analogy for the solution of plane stress problems, *J. Inst. Civ. Engrg.* 21, 59–82.

Melosh, R.J. (1963), Structural analysis of solids, *Proc. Amer. Soc. Civ. Engrg.* **ST4**, 205–223.

Mindlin, R.D. (1951), Influence of rotatory inertia and shear in flexural motions of isotropic elastic plates, *J. Appl. Mech.* **18**, 31–38.

Mote, C.D. (1971), Global-local finite element, *Internat. J. Numer. Methods Engrg.* **3**, 565–574.

Nay, R.A. and S. Utku (1973), An alternative for the finite element method, in: C.A. Brebbia and H. Tottenham, eds., *Variational Methods in Engineering* (Southampton Univ. Press, Southampton) 3162–3174.

Nayak, G.C. and O.C. Zienkiewicz (1972), Elasto-plastic stress analysis. Generalization for various constitutive relations including strain softening, *Internat. J. Numer. Methods Engrg.* **5**, 113–135.

Neyrolles, B., G. Touzot and P. Villon (1991), Le methode des elements diffus. *C. R. Acad. Sci. Paris Ser. II* **313**, 293–296.

Neyrolles, B., G. Touzot and P. Villon (1992), Generalizing the finite element method. Diffuse approximation and diffuse elements, *Comput. Mech.* **10**, 3–18.

Oden, J.T. (1969), A general theory of finite elements—I: Topological considerations, II: Applications, *Internat. J. Numer. Methods Engrg.* **1**, 205–221, 247–260.

Oden, J.T. (1973), The finite element in fluid mechanics, in: J.T. Oden and E.R.A. Oliveira, eds., *Lectures on Finite Element Method in Continuum Mechanics, 1970, Lisbon* (University of Alabama Press, Huntsville) 151–186.

Oden, J.T. and D. Samogyi (1969), Finite element application in fluid dynamics, *Proc. Amer. Soc. Civ. Engrg.* **95**, 821–826.

Oden, J.T., T. Strouboulis and P. Devloo (1987), Adaptive finite element methods for high speed compressible flows, *Internat. J. Numer. Methods Engrg.* **7**, 1211–1228.

Oden, J.T. and L.C. Wellford (1972), Analysis of viscous flow by the finite element method, *JIAA* **10**, 1590–1599.

Oliveira, E.R. De Arantes (1977), The patch test and the general convergence criteria of the finite element method, *Internat. J. Solids & Structures* **13**, 159–178.

Oñate, E., M. Cervera and O.C. Zienkiewicz (1994), A finite volume format for structural mechanics, *Internat. J. Numer. Methods Engrg.* **37**, 181–201.

Pavlin, V. and N. Perrone (1979), Finite difference energy technique for arbitrary meshes applied to linear plate problems, *Internat. J. Numer. Methods Engrg.* **14**, 647–664.

Pawsey, S.F. and R.W. Clough (1971), Improved numerical integration of thick slab finite elements, *Internat. J. Numer. Methods Engrg.* **3**, 575–586.

Peano, A.G. (1976), Hierarchics of conforming finite elements for elasticity and plate bending, *Comput. Math. Appl.* **2**, 3–4.

Peraire, J., J. Peiro, L. Formaggia, K. Morgan and O.C. Zienkiewicz (1988), Finite element Euler computations in 3-D, *Internat. J. Numer. Methods Engrg.* **26**, 2135–2159.

Peraire, J., M. Vahdati, K. Morgan and O.C. Zienkiewicz (1987), Adaptive remeshing for compressible flow computations, *J. Comput. Phys.* **72**, 449–466.

Pironneau, O. (1982), On the transport diffusion algorithm and its application to the Navier Stokes equation, *Numer. Math.* **38**, 309–332.

Prager, W. and J.L. Synge (1947), Approximation in elasticity based on the concept of function space, *Quart. J. Appl. Math.* **5**, 241–269.

Rashid, Y.R. and W. Rockenhauser (1968), Pressure vessel analysis by finite element techniques, in: *Proc. Conf. On Prestressed Concrete Pressure Vessels*, Inst. Civ. Eng.

Reissner, E. (1945), The effect of transverse shear deformation on the bending of elastic plates, *J. Appl. Mech.* **12**, 69–76.

Reissner, E. (1950), On a variational theorem in Elasticity, *J. Math. Phys.* **29**, 90–95.

RITZ, W. (1909), Über eine neue Methode zur Lösung gewissen Variations—Probleme der mathematischen Physik, *J. Reine Angew. Math.* **135**, 1–61.
SOUTHWELL, R.V. (1940), *Relaxation Methods in Engineering Science* (Oxford University Press, Oxford).
SOUTHWELL, R.V. (1946), *Relaxation Methods in Theoretical Physics*, Vol. I (Clarendon Press, Oxford).
SOUTHWELL, R.V. (1956), *Relaxation Methods in Theoretical Physics*, Vol. II (Clarendon Press, Oxford).
SPALDING, D.B. (1972), A novel finite difference formulation for differential equations involving both first and second derivatives, *Internat. J. Numer. Methods Engrg.* **4**, 551–559.
SPECHT, B. (1988), Modified shape functions for the three node plate bending element passing the patch test, *Internat. J. Numer. Methods Engrg.* **26**, 705–715.
STRANG, G. and G.J. FIX (1973), *An Analysis of the Finite Element Method* (Prentice-Hall, Englewood Cliffs, NJ).
STRUTT, J.W. (Lord Rayleigh) (1870), On the theory of resonance, *Trans. Roy. Soc. London* **A161**, 77–118.
STUMMEL, F. (1980), The limitations of the patch test, *Internat. J. Numer. Methods Engrg.* **15**, 177–188.
SZMELTER, J. (1959), The energy method of networks of arbitrary shape in problems of the theory of elasticity, in: W. OLSZAK, ed., *Proc. IUTAM 1958 Symposium on Non-Homogeneity in Elasticity and Plasticity* (Pergamon Press, Oxford).
TAIG, I.C. (1961), Structural analysis by the matrix displacement method, Engl. Electric Aviation Report No. S017.
TAYLOR, C. and P. HOOD (1973), A numerical solution of the Navier–Stokes equations using the finite element techniques, *Comput. Fluids* **1**, 73–100.
TAYLOR, R.L., O.C. ZIENKIEWICZ, J.C. SIMO and A.H.C. CHAN (1986), The patch test—A condition for assessing f.e.m. convergence, *Internat. J. Numer. Methods Engrg.* **22**, 39–62.
TONG, P. (1969), Exact solution of certain problems by the finite element method, *J. AIAA* **7**, 179–180.
TREFFTZ, E. (1926), Ein gegenstuck zum Ritz'schen Verfahren, in: *Proc. 2nd Int. Congr. Appl. Mech.*, Zürich.
TURNER, M.J., R.W. CLOUGH, H.C. MARTIN and L.J. TOPP (1956), Stiffness and deflection analysis of complex structures, *J. Aero. Sci.* **23**, 805–823.
VEUBEKE, B. FRAEIJS DE (1965), Displacement and equilibrium models in finite element method, in: O.C. ZIENKIEWICZ and G.S. HOLISTER, eds., *Stress Analysis*, Chapter 9 (Wiley, New York) 145–197.
VEUBEKE, B. FRAEIJS DE (1974), Variational principles and the patch test, *Internat. J. Numer. Methods Engrg.* **8**, 783–801.
VEUBEKE, B. FRAEIJS DE and O.C. ZIENKIEWICZ (1967), Strain energy bounds in finite element analysis, *J. Strain Analysis* **2**, 265–271.
VISSER, W. (1968), The finite element method in deformation and heat conduction problems, Dissertation, Tech. University, Delft.
WASHIZU, K. (1975), *Variational Methods in Elasticity and Plasticity* (Pergamon Press, Oxford, 2nd ed.).
WIBERG, N.E. and F. ABDULAHAB (1992), An efficient post-processing technique for stress problems based on superconvergence derivatives and equilibrium, in: Ch. HIRSCH et al., eds., *Numerical Methods in Engineering* (Elsevier, Amsterdam).
ZIENKIEWICZ, O.C. (1947), The stress distribution in gravity dams, *Proc. Inst. Civil. Engrg.* **27**, 244–247.
ZIENKIEWICZ, O.C. (1965), Finite element procedures in the solution of plate and shell problems, in: O.C. ZIENKIEWICZ and G.S. HOLISTER, eds., *Stress Analysis*, Chapter 8 (Wiley, New York) 120–144.
ZIENKIEWICZ, O.C. (1971), *The Finite Element Method in Engineering Science* (McGraw-Hill, New York) 521.

Zienkiewicz, O.C. (1977), *The Finite Element Method* (McGraw-Hill, New York, 3rd ed.) 787.

Zienkiewicz, O.C., P.L. Arlett and A.K. Bahrani (1967), Solution of three-dimensional field problems by the finite element method, *The Engineer*, 27 Oct.

Zienkiewicz, O.C. and Y.K. Cheung (1964), The finite element method for analysis of elastic isotropic and orthotropic slabs, *Proc. Inst. Civ. Engrg.* **28**, 471–488.

Zienkiewicz, O.C. and Y.K. Cheung (1965), Finite elements in the solution of field problems, *The Engineer*, 507–510.

Zienkiewicz, O.C. and Y.K. Cheung (1967), *The Finite Element Method in Structural Mechanics* (McGraw-Hill, New York) 272.

Zienkiewicz, O.C. and R. Codina (1995), A general algorithm for compressible and incompressible flow, *Internat. J. Numer. Math. Fluids*, to appear.

Zienkiewicz, O.C., J.P. De S.R. Gago and D.W. Kelly (1983), The hierarchical concept in finite element analysis, *Comput. & Structures* **16**, 53–65.

Zienkiewicz, O.C., R.H. Gallagher and P. Hood (1976), Newtonian and non-Newtonian viscous incompressible flow. Temperature induced flows and finite element solutions, in: J. Whiteman, ed., *The Mathematics of Finite Elements and Applications*, Vol. II (Academic Press, London).

Zienkiewicz, O.C., J.C. Heinrich, P.S. Huyakorn and A.R. Mitchell (1977), An upwind finite element scheme for two dimensional convective transport equations, *Internat. J. Numer. Methods Engrg.* **11**, 131–144.

Zienkiewicz, O.C. and G.S. Holister, eds. (1965), *Stress Analysis* (Wiley, Chicago).

Zienkiewicz, O.C., B.M. Irons, J. Campbell and F.C. Scott (1970), Three dimensional stress analysis, in: *Proc. Int. Un. Th. Appl. Mech. Symposium on High Speed Computing in Elasticity*, Liege.

Zienkiewicz, O.C., B.M. Irons, I. Ergatoudis, S. Ahmad and F.C. Scott (1969), Isoparametric and associated element families for two and three dimensional analysis, from *Finite Element Methods in Stress Analysis* (Tapir Press, Norway) 38–432.

Zienkiewicz, O.C., D.W. Kelly and P. Bettess (1977), The coupling of the finite element method and boundary solution procedures, *Internat. J. Numer. Methods Engrg.* **11**, 355–375.

Zienkiewicz, O.C. and D. Lefebvre (1987), Three field mixed approximation and the plate bending problem, *Comm. Appl. Numer. Methods* **3**, 301–309.

Zienkiewicz, O.C. and D. Lefebvre (1988), A robust triangular plate bending element of the Reissner–Mindlin type, *Internat. J. Numer. Methods Engrg.* **26**, 1169–1184.

Zienkiewicz, O.C., R. Löhner, K. Morgan and S. Nakazawa (1984), Finite elements in fluid mechanics—a decade of progress, in: R.H. Gallagher et al., eds., *Finite Elements in Fluids*, Vol. 5, Ch. 1 (Wiley, Chichester) 1–26.

Zienkiewicz, O.C., R. Löhner, K. Morgan and J. Peraire (1986), High speed compressible flow and other advection dominated problems of fluid mechanics, in: R.H. Gallagher et al., eds., *Finite Elements in Fluids*, Vol. 6, Ch. 2 (Wiley, Chichester) 41–88.

Zienkiewicz, O.C., P. Mayer and Y.K. Cheung (1966), Solution of anisotropic seepage problems by finite elements, *Proc. Amer. Soc. Civ. Engrg.* **92**, 111–120.

Zienkiewicz, O.C., K. Morgan, B.V.K. Satya Sai, R. Codina and M. Vazquez (1995), A general algorithm for compressible and incompressible flow, Part II—Tests on the explicit form. *Internat. J. Numer. Methods Fluids*, to appear.

Zienkiewicz, O.C. and E. Oñate (1991), Finite volumes vs. finite elements. Is there really a choice?, in: P. Wriggers and W. Wagner, eds., *Nonlinear Computational Mechanics. State of the Art* (Springer, Berlin) 240–254.

Zienkiewicz, O.C., E. Oñate and J.C. Heinrich (1978), *Plastic Flow in Metal Forming* (ASME, San Francisco, CA) 107–120.

Zienkiewicz, O.C., S. Qu, R.L. Taylor and S. Nakazawa (1986), The patch test for mixed formulation, *Internat. J. Numer. Methods Engrg.* **23**, 1873–1883.

Zienkiewicz, O.C. and C. Taylor (1973), Weighted residual processes in finite elements with particular reference to some transient and coupled problems, in: J.T. Oden and E.R.A. Oliveira, eds., *Lectures on Finite Element Method in Continuum Mechanics, 1970, Lisbon* (University of Alabama Press, Huntsville) 415–458.

ZIENKIEWICZ, O.C. and R.L. TAYLOR (1989), *The Finite Element Method*, Vol. I (Basic formulation and Linear problems) (McGraw-Hill, New York) 648.

ZIENKIEWICZ, O.C. and R.L. TAYLOR (1991), *The Finite Element Method*, Vol. II (Solid and Fluid Mechanics, Dynamics and Nonlinearity) (McGraw-Hill, New York) 807.

ZIENKIEWICZ, O.C., J. TOO and R.L. TAYLOR (1971), Reduced integration technique in general analysis of plates and shells, *Internat. J. Numer. Methods Engrg.* **3**, 275–290.

ZIENKIEWICZ, O.C., S. VALLIAPPAN and I.P. KING (1968), Stress analysis of rock as a "no-tension" material, *Geotechnique* **18**, 56–66.

ZIENKIEWICZ, O.C., S. VALLIAPPAN and I.P. KING (1969), Elasto-plastic solutions of engineering problems. Initial stress, finite element approach, *Internat. J. Numer. Methods Engrg.* **1**, 75–100.

ZIENKIEWICZ, O.C. and J.Z. ZHU (1987), A simple error estimator and adaptive procedure for practical engineering analysis, *Internat. J. Numer. Methods Engrg.* **24**, 337–357.

ZIENKIEWICZ, O.C. and J.Z. ZHU (1989), Error estimation and adaptive refinement for plate bending problems, *Internat. J. Numer. Methods Engrg.* **28**, 2839–2853.

ZIENKIEWICZ, O.C. and J.Z. ZHU (1992a), The superconvergent patch recovery (SPR) and adaptive finite element refinement, *Comput. Methods Appl. Mech. Engrg.* **101**, 207–224.

ZIENKIEWICZ, O.C. and J.Z. ZHU (1992b), Superconvergent patch recovery and a-posteriori error estimation in the finite element method, Part I: A general superconvergent recovery technique, *Internat. J. Numer. Methods Engrg.* **33**, 1331–1364.

ZIENKIEWICZ, O.C. and J.Z. ZHU (1992c), Superconvergent patch recovery and a-posteriori error estimation in the finite element method, Part II: The Zienkiewicz–Zhu, a-posteriori error estimator, *Internat. J. Numer. Methods Engrg.* **33**, 1365–1381.

Subject Index

Automatic Mesh Generation and Finite Element Computation

P.L. George

INRIA, Domaine de Voluceau
Rocquencourt, B.P. 105
78153 Le Chesnay Cedex
France

HANDBOOK OF NUMERICAL ANALYSIS, VOL. IV
Finite Element Methods (Part 2)—Numerical Methods for Solids (Part 2)
Edited by P.G. Ciarlet and J.L. Lions

Contents

Preface

A large range of physical phenomena can be formalized in terms of partial differential equations and thus can be solved numerically using the finite element method.

This method was first conceived and used by engineers in the early fifties, in particular for problems with a structural mechanic nature. Due to the constant improvement of computers since this date, the finite element method became more and more popular for numerical simulation of more and more diverse physical problems, including elasticity, heat transfer analysis, electromagnetic, flow computation, ...

During this time, both theoretical foundations and practical aspects of the method were developed and, in this respect, various finite elements created, analyzed and then used in simulations. These latter became more and more sophisticated as applied in complicated problems and both their size (in terms of the number of nodes defined) and their complexity (in terms of the nature of the problem: time dependent problem, non linear problem, coupled problem, ...) becomes greater and greater. On the other hand, their fields of application (in terms of the geometries dealt with) become more and more realistic and as a consequence more and more complex.

Mesh generation represents the first step of any finite element method that engineers have to implement practically after the necessary theoretical analysis of the problem they want to solve.

This task must be carefully done as the mesh is responsible for the accuracy of the solution computed with such a support. In particular, it is of great importance to capture the geometry of the domain where the problem is posed as well as possible, with special attention paid to the way the boundaries of the domain are approximated and to capture the physical behavior of the problem as accurately as possible. It means that both the number of nodes and elements and their nature must be adequate (in terms of shape, size, density, variation from region to region, ...).

However, mesh construction can be very expensive in terms of time and thus is costly. This is the reason why the generation of meshes is a crucial point in the numerical implementation of the finite element method and one can therefore deplore that the techniques, algorithms and tricks related to the creation of meshes have probably not received adequate attention over the last years (at least not by people considering only the theoretical aspects of the finite element method).

The present article aims to make the different notions encountered when considering the mesh generation aspects of the finite element method as clear as possible and would pay special attention to the automatic mesh generation methods. The discussion consists of several chapters divided into sections. Chapter I introduces useful definitions such as conformal mesh, mesh structure, control structure and control space. Finally, a methodology for conceiving a mesh is proposed. Chapter II outlines the most popular mesh generation methods and algorithms to help the reader to be familiar with these. Manual method, product method, algebraic method and method using the solution of partial differential equations are shortly presented as solutions in the case where the domain enjoys some appropriate properties. The multiblock method is briefly discussed as a semi-automatic solution a priori possible in general. Automatic methods, including quadtree–octree techniques, advancing-front and Voronoï type approach are introduced. The two latter are then fully detailed in Chapter III (advancing-front method) and Chapter IV (Voronoï type method) for the general situation (geometrical point of view) and in the case where properties to be satisfied are specified in advance (using a control space).

CHAPTER I

General Definitions and Problem

Introduction

In practice, the finite element method consists of the construction of a finite dimensional space on which suitable approximation properties of the requested functions hold and such that a computer implementation can be developed easily. The first part of this construction relies on the creation of a suitable mesh of the domain under consideration. This mesh must enjoy certain properties which can be classified in terms of general properties, such that the mesh is suitable for the finite element method, and properties connected to the problem under consideration. General properties are essentially of a geometric nature while the second type may include physical considerations.

This chapter firstly recalls the main aspects of the finite element method (Section 1), after which general mesh definitions are given (Section 2). Suitable information that must be contained in a mesh in order to be computationally convenient is then indicated (Section 3) after which an example of mesh structure is given (Section 4). The notion of a control space is introduced (Section 5), which appears to be a possible way of specifying the properties (geometric and physical) the mesh must enjoy. A methodology for mesh conception is then proposed (Section 6). Data structure corresponding to the mesh generation algorithm and resulting from the above methodology is briefly described (Section 7). The chapter ends with some useful formulas concerning triangles and tetrahedra which will be used in the chapters which follow (Section 8).

1. Basic ideas of finite element methods

Numerous physical problems can be formulated as follows:

$$\begin{cases} Au = f & \text{in } \Omega, \\ Bu = g & \text{on } \Gamma, \end{cases} \tag{1.1}$$

where Ω is a subset of R^2 or R^3, Γ the boundary of Ω, A and B partial differential operators, f and g are the prescribed data and u is the solution sought.

This formulation, called the *continuous problem*, presents several disadvantages as excessive regularity is demanded for A, B, f, g and u. On the one hand, the data f and g as well as the unknown u must be defined at every point of the domain or its boundary and, on the other hand, u must be sufficiently differentiable, in particular in order to have Au and Bu well defined. Thus, it is poorly adapted in terms of the practical computation and discontinuities as well as irregular data or solutions are not easy to consider.

This is the reason why a *variational formulation* or *weak formulation* is introduced. It can be summarized as follows:

$$\begin{cases} \text{Find } u \text{ in a space of admissible functions } V \text{ such that} \\ \quad a(u, v) = f(v) \text{ for all } v \in V, \end{cases} \tag{1.2}$$

where $a(\cdot, \cdot)$ is a bilinear form and $f(\cdot)$ a linear form using integrals on Ω and Γ such that this weak formulation is in some sense equivalent to (1.1). While the exact solution of (1.1) or (1.2) is generally impossible to obtain, in terms of computation, the main advantage of formulation (1.2) is related to the two following results: firstly, it is more confident with the physical problem and, secondly, it can be approximated by a discrete formulation which is suitable for computation. The discrete form of the above formulation is formally given by:

$$\begin{cases} \text{Find } u_h \text{ in a space of admissible functions } V_h \text{ such that} \\ \quad a(u_h, v_h) = f(v_h) \text{ for all } v_h \in V_h, \end{cases} \tag{1.3}$$

where V_h is a finite dimensional subspace of V. In terms of *conforming finite element methods*, see CIARLET [1991], which is the solution method retained in the context of this article, V_h is a member of a family of subspaces associated with parameter h. The resulting discrete scheme is said to be convergent if:

$$\lim_{h \to 0} \|u - u_h\| = 0,$$

where $\|\cdot\|$ denotes the norm in space V. A typical error estimate then has the form:

$$\|u - u_h\| \leqslant C(u)h^\beta,$$

i.e. for a sufficiently smooth solution u, there exists a constant $C(u)$, independent of h, such that the above relation holds for β, the order of convergence. In practice, parameter h is defined as the maximum of the diameters of the elements in the *mesh*. The geometrical nature of the mesh discretizing the domain Ω is thus clearly responsible for the quality of the computed solution.

Numerous papers, devoted to the theoretical aspects of the finite element method, discuss the properties of the above formulations, show the way to construct suitable discrete approximations of the problem and give corresponding error estimates (as above or slightly different according to the nature of the problem). A general and comprehensible presentation dealing in details with all these points, can be found in CIARLET [1991].

In practice, the finite element method consists of the construction of a space V_h of finite dimension such that, on the one hand, suitable approximation properties are obtained and, on the other hand, it is convenient from the point of view of computer implementation. This construction is based on the three following basic ideas:

- the creation of a *mesh* or *triangulation*, denoted by $\mathcal{T}_h$, of domain Ω. The domain is written as the finite union of elements K, as will be described later (cf. Definition 2.1);
- the definition of space V_h as the set of functions v_h, whose restriction to each element K in $\mathcal{T}_h$ is a polynomial;
- the existence of a basis for space V_h whose functions have a *small support.*

Thus, the restriction of a function u_h to any element K is written as

$$u_h = \sum_{i=1}^{N} u_i v_i,$$

where N denotes the *number of degrees of freedom* of element K, u_i is the value of the function at the degrees of freedom and v_i is the basis function i of the polynomial space defined previously. A finite element is then characterized by a suitable choice of:

- K the geometrical element (triangle, quadrilateral, etc.);
- Σ_K the set of degrees of freedom defined on K. These degrees are the values at the *nodes* of the function (Lagrange type finite element), or those values and directional derivatives of the function (Hermite type finite element);
- P_K the space of polynomials on K.

The *implementation* of a finite element method can be summarized as follows:

- the mathematical analysis of the problem with, more specifically, its related variational formulation and the investigation of the properties of the latter;
- the construction of a triangulation (the mesh) of the domain under consideration, i.e., the creation of $\mathcal{T}_h$;
- the definition of the different finite elements, i.e. the choice of (K, P_K, Σ_K);
- the generation of element matrices due to the contribution of each element, K, to the matrix and right-hand side of the system;
- the assembly of the global system;
- the consideration of *essential* boundary conditions (depending on the physical problem and the variational formulation);
- the solution of the system, i.e. the computation of the solution field approaching the sought solution;
- the post-processing of results.

Nevertheless, in this article, the reader needs only note that the discrete approximation of the problem leads to a spatial discretization of the domain Ω where the problem is posed. This discretization is nothing else than the mesh, or triangulation as also referred to hereafter.

2. Mesh definition

A mesh of a domain Ω is defined by a set $\mathcal{T}_h$ consisting of a finite number of segments in one dimension, segments, triangles and quadrilaterals in two dimensions and the above elements, tetrahedra, pentahedra and hexahedra in three dimensions. Depending on the space dimension, segments are used to model beam problems, triangles and quadrilaterals are used for plane, plate or shell problems while tetrahedra, pentahedra and hexahedra are used in purely three-dimensional problems. The elements, denoted by K, of such a mesh must satisfy certain properties which will be introduced hereafter.

The above description corresponds to the general finite element situation. In some situations, one may prefer to have only a mesh composed of quadrilaterals (in two dimensions) or hexahedra (in three dimensions) which is a source of simplification at the time of solving the resulting system but, on the other hand, gives less flexibility at the mesh generation step.

In general, finite element solvers require a conformal mesh (also referred to as a conforming mesh) (Fig. 2.1) as support (note that there exist some solvers which do not require this property). Such a mesh is defined by:

DEFINITION 2.1. $\mathcal{T}_h$ is a *conformal* mesh of Ω if the following conditions hold:

(1) $\overline{\Omega} = \bigcup_{K \in \mathcal{T}_h} K$,

(2) all elements K of $\mathcal{T}_h$ have a non-empty interior,

(3) the intersection of two elements in $\mathcal{T}_h$ is such that it is either reduced to the empty set, an element vertex, an element edge or an element face.

This definition implies that $\mathcal{T}_h$ covers Ω in a *conformal* manner, i.e. without overlapping or intersecting elements and with some continuous properties at element interfaces.

In practice, $\mathcal{T}_h$ is a partitioning of Ω covering this domain as accurately as possible. When Ω is not a polygonal (polyhedral) domain, $\mathcal{T}_h$ will only be an approximate partitioning of the domain.

DEFINITION 2.2. The *connectivity* of a mesh defines the connection between its vertices.

Two major types of meshes are currently used in numerical simulations. They correspond to a particular connectivity:

DEFINITION 2.3. A mesh is called *structured* if its connectivity is of the finite difference type.

And on the contrary

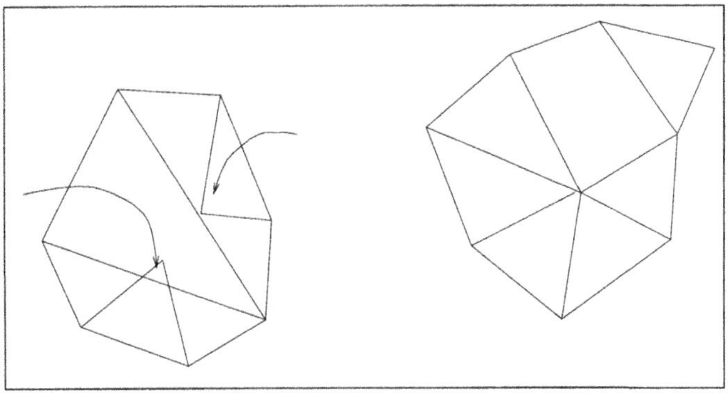

FIG. 2.1. Conformity.

DEFINITION 2.4. A mesh is called *unstructured* if its connectivity is of any other type.

Structured meshes are also referred to as *grids*.

DEFINITION 2.5. The *topology* of an element is its description (including that of its edges and faces) in terms of its vertices.

Thus the definition of a mesh consists of its connectivity, its topology and, of course, the data of the coordinates of its vertices.

A structured mesh consists of a set of coordinates and connectivities which naturally map into the elements of a matrix. Let (i, j) be the indices of a vertex in the mesh, then its neighbors (Fig. 2.2) are implicitly the vertices with indices $(i + 1, j)$, $(i, j+1)$, $(i-1, j)$ and $(i, j-1)$ in two dimensions and are similarly defined in three dimensions using a third index k. For such a mesh, the topology of any

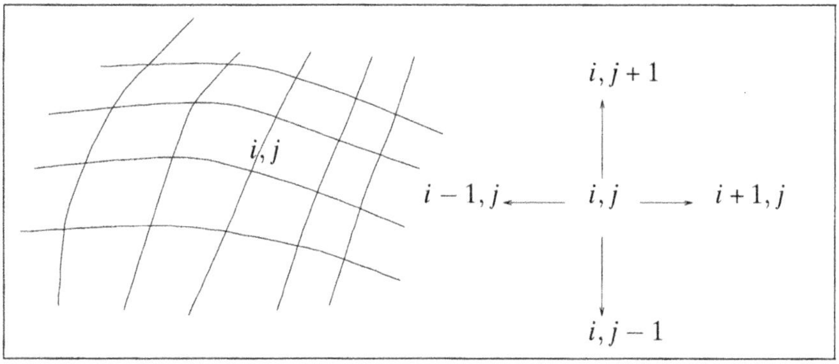

FIG. 2.2. Sample structured mesh.

TABLE 2.1. Connectivities.

	Element								
	1	2	3	4	5	6	7	8	...
Vertex 1	1	9	1	9	9	9	7	7	.
Vertex 2	9	7	2	2	3	4	4	5	.
Vertex 3	8	8	9	3	4	7	5	6	.

element, for example an element (in two dimensions) with vertices $(i, j), (i+1, j), (i+1, j+1), (i, j+1)$, can be defined as follows:

- edge 1 connects vertex (i, j) with vertex $(i+1, j)$,
- edge 2 connects vertex $(i+1, j)$ with vertex $(i+1, j+1)$,
- edge 3 connects vertex $(i+1, j+1)$ with vertex $(i, j+1)$,
- edge 4 connects vertex $(i, j+1)$ with vertex (i, j).

Identical definitions apply in three dimensions where edges and faces must be defined similarly.

An unstructured mesh is defined similarly but, in this case, the connectivities and topologies must be explicitly defined using a connectivity matrix and a topology matrix which must be established at the mesh generation step before being used at the solution step.

The connectivity matrix is then an ordered relation from vertex to vertex. For example, Table 2.1 shows the connectivities associated with the mesh in Fig. 2.3. We should note that the manner in which the connectivities is defined is not unique, while it must basically include the information listed above. Connectiv-

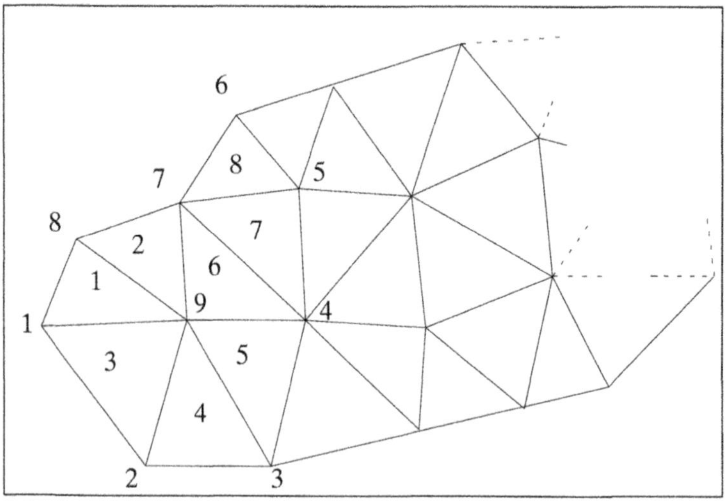

FIG. 2.3. Sample unstructured mesh.

ity is an ordered list that induces several useful properties. Let (a, b, c) be an element in the mesh, then it means that a is its first vertex, b its second and, finally, c its last vertex. This element differs from element (b, a, c), for example, which gives a sense to the sign of the surface of the element. In fact, all circular permutations in the triplet (a, b, c) define exactly the same element, in this respect, triplets $(a, b, c), (b, c, a)$ and (c, a, b) represent different enumerations of the same element.

The topology can be defined, for each element, as follows: the first edge is that connecting vertex 1 (a in the previous example) with vertex 2 (b), the second edge is defined from vertex 2 to vertex 3 and edge number three is that connecting vertex 3 and vertex 1. This gives a sense to the way an edge is defined and permits, without ambiguity, the computation of inward or outward normals, ...

REMARK 2.1. A different definition is also pertinent for a mesh consisting of triangles or tetrahedra. It consists, in two dimensions, of defining edge i as that opposite to vertex i and can be similarly extended to three dimensions.

In practice, the topology must be carefully defined explicitly for all types of elements (segments, triangles, quadrilaterals, tetrahedra, pentahedra and hexahedra) that can be encountered in the mesh. It should be noticed that this definition is given explicitly at the mesh generation step and then becomes implicit at the computational step.

As an example of possible topologies, we give those used in the Modulef software package, see BERNADOU et al. [1988]. It consists of defining the ordered list of vertices once the first is fixed and then the list of edges and faces. According to the element type, it is as follows:

(1) The segment:

- (1) → (2)

(2) The triangle: the vertices are numbered in a counterclockwise direction in two dimensions so that the surface has a direction. In three dimensions, the surface has no sign.

- edge [1]: (1) → (2)
- edge [2]: (2) → (3)
- edge [3]: (3) → (1)

(3) The quadrilateral (or quadrangle): the vertices are numbered following the edges in the same manner (same surface property yields)

- edge [1]: (1) → (2)
- edge [2]: (2) → (3)
- edge [3]: (3) → (4)
- edge [4]: (4) → (1)

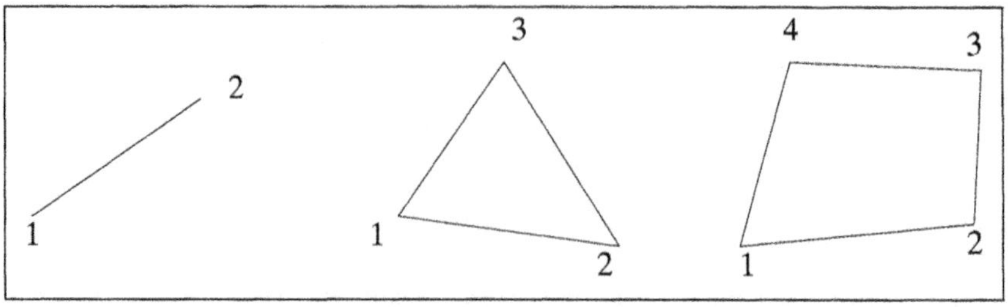

FIG. 2.4. Segment, triangle and quadrilateral.

(4) The tetrahedron: trihedron $(\vec{12}, \vec{13}, \vec{14})$ is assumed positive. ($\vec{ij}$ denotes the vector with origin i and extremity j.)

- edge [1]: (1) → (2)
- edge [2]: (2) → (3)
- edge [3]: (3) → (1)
- edge [4]: (1) → (4)
- edge [5]: (2) → (4)
- edge [6]: (3) → (4)

Any face, seen from the exterior, is in the positive direction:

- face [1]: (1) (3) (2)
- face [2]: (1) (4) (3)
- face [3]: (1) (2) (4)
- face [4]: (2) (3) (4)

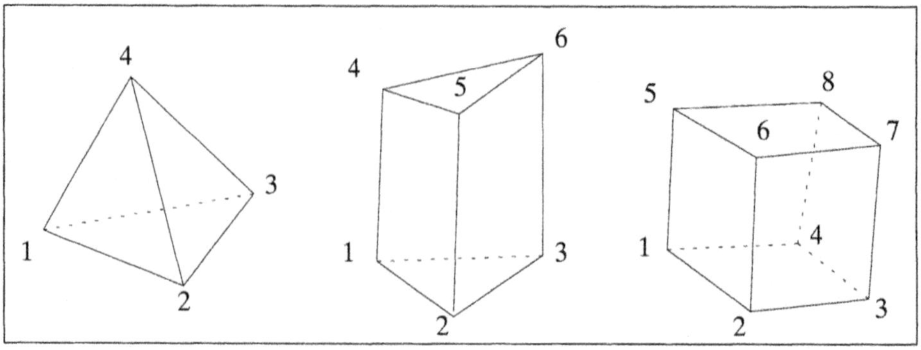

FIG. 2.5. Tetrahedron, pentahedron and hexahedron.

(5) The pentahedron: trihedron $(\vec{12}, \vec{13}, \vec{14})$ is assumed positive.

- edge [1]: (1) → (2)
- edge [2]: (2) → (3)
- edge [3]: (3) → (1)
- edge [4]: (1) → (4)
- edge [5]: (2) → (5)
- edge [6]: (3) → (6)
- edge [7]: (4) → (5)
- edge [8]: (5) → (6)
- edge [9]: (6) → (4)

Any face, seen from the exterior, is in the positive direction:

- face [1]: (1) (3) (2)
- face [2]: (1) (4) (6) (3)
- face [3]: (1) (2) (5) (4)
- face [4]: (4) (5) (6)
- face [5]: (2) (3) (5) (6)

(6) The hexahedron: trihedron $(\vec{12}, \vec{14}, \vec{15})$ is assumed positive.

- edge [1]: (1) → (2)
- edge [2]: (2) → (3)
- edge [3]: (3) → (4)
- edge [4]: (4) → (1)
- edge [5]: (1) → (5)
- edge [6]: (2) → (6)
- edge [7]: (3) → (7)
- edge [8]: (4) → (8)
- edge [9]: (5) → (6)
- edge [10]: (6) → (7)
- edge [11]: (7) → (8)
- edge [12]: (8) → (5)

Any face, seen from the exterior, is in the positive direction:

- face [1]: (1) (4) (3) (2)
- face [2]: (1) (5) (8) (4)
- face [3]: (1) (2) (6) (5)
- face [4]: (5) (6) (7) (8)
- face [5]: (2) (3) (7) (6)
- face [6]: (3) (4) (8) (7)

In the following section, the reader is given a possible structure for a mesh which includes, either explicitly or implicitly, the above information (connectivity, topology, coordinates and additional attributes to make the assignment of relevant physical values and prescriptions possible when computing the solution). The practical mesh structure, designed so that this information is present, will be discussed in Section 4.

3. Mesh contents

The structure chosen for storing a mesh must be convenient in view of computation. It must obviously include, in some sense, all the previous information (coordinates, connectivities and topologies) and in addition (as previously mentioned), it must be such that physical data can be easily applied (boundary conditions, applied loads, prescribed temperature, material characteristics, etc., ... depending on the problem to be solved).

There is no unique manner of storing these values. The choice between one type of mesh structure and another depends strongly on the usage required of the mesh during computation. During this step, different actions must be possible, for example:

- to easily access the vertices, edges, faces of a given element,
- to be able to apply the physical prescriptions,
- to compute the normals, derivatives, ..., volumes, surfaces which are needed,
- to obtain the list of the neighbors of any given vertex,
- etc.

Thus the simplest data structure can be designed so that it contains the following:

- the number of vertices in the mesh,
- their coordinates,
- the number of elements in the mesh,
- the list of the elements including, for each, its geometrical nature (this information is needed when the mesh contains elements which are geometrically different, for example triangles and quadrilaterals), its constituent material, the number and the list of its vertices (connectivity) and their physical characterizations, the logical definition of its items: edges and faces (topology), and the physical characteristics of the latter.

The characterization of any item (point, edge, face, element) with respect to the physical values that must be applied can be determined by assigning an integer value, or attribute, to the item under consideration to make possible the following relation:

integer value $\rightarrow$ given processing

To practically construct and then use such a structure, it is necessary to use arrays, lists, linked lists, trees and other *basic structures* as described, for example, in AHO, HOPCROFT and ULLMAN [1983].

To illustrate that there is no unique manner to construct a mesh structure, we propose a scheme slightly different to the above. A mesh structure can be designed so that it contains the following:

- the number of elements in the mesh,
- the list of these elements including, for each, the number and list of its faces and its constituent material,
- the number of faces in the mesh,
- the list of these faces including, for each, the number and list of its edges and its physical characterization,
- the number of edges in the mesh,
- the list of these edges including, for each, the list of its endpoints and its physical characterization,
- the number of vertices in the mesh,
- the list of these vertices including, for each, its coordinates and its physical characterization.

It is clear that the second manner of structuring a mesh includes exactly the same information as the first, but, in this case, the access to a value is fundamentally different. Note that the construction of such a storage relies on the same basic structures as above.

In addition to these values, it may be necessary for the mesh structure to contain information concerning the way the mesh was conceived. Mesh generation methods (as described in Chapters II, III and IV), generally use the boundaries of the domain used for the simulation as data. These boundaries are specified in terms of points, lines and surfaces, which have been defined in some way (see Section 7). It can be useful to know that such or such point in the mesh belongs to such or such data item and similarly that such or such edge or face belongs to such or such curve or surface, etc. To make this point possible, it is necessary to ensure that the mesh structure includes the information needed to reconstitute the *history* of such items.

The above information is clearly needed in the case where the computation relies, not on P_1 finite elements whose nodes are the vertices (Fig. 3.1, left-hand side), but, for example, on P_2 finite elements (same figure, right-hand side). In this case, the nodes are the vertices of the elements and the midpoints of the edges.

The question is now what is the best strategy for creating nodes which are not vertices. There are a priori two possible answers: the first one consists in designing the mesh generation process in such a way that all the required nodes are defined; the second solution consists in enriching the mesh structure, assumed to be P_1, by an additional processing of the mesh created by the mesh generation process. It appears that this second method is more flexible and permits the design of a simpler mesh creation process as only geometry is involved. For efficiency, the mesh structure must contain all the information needed when the nodes are defined. On the contrary, mesh creation processes in which nodes can be defined in all cases complicate the mesh generation step.

Using the "P_1 mesh generation" approach, the history (defined as above) is

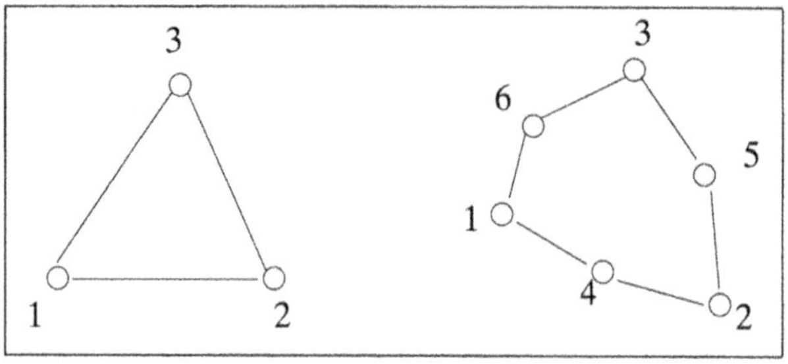

FIG. 3.1. P_1 and P_2 triangles.

clearly needed. To illustrate this point, assume that we want to create a P_2 finite element mesh. To do this we need to know the definition of the edges in the mesh. These can be defined, for example in two dimensions, by:

- a portion of a line which is known by an equation of the type $y = y(x)$, i.e. $f(x, y) = 0$, by equations of the type $x = x(t)$ and $y = y(t)$, by a spline given as a series of control points and a polynomial form, or using a different manner,
- a portion of a curve which can be defined as above,
- an item defined by a constructive process. For example, a point is defined as the intersection of two lines, a line as that of two surfaces, ... ,
- etc.

Assuming that information of these types are accessible, the definition of the P_2 nodes includes the definition of the vertex nodes, already known as vertices of the mesh resulting from the mesh generation method, and the definition of the nodes which are edge midpoints and thus need to be computed using the definition of the edges. To define the edge midpoints, several techniques are possible depending on the way the edges are defined. For example (see Fig. 3.2):

- let AB be the edge under consideration and $f(M) = f(x, y) = 0$ be the definition of the curve that the edge must follow, then we define $P = \frac{1}{2}(A + B)$ and compute $\vec{n}$, the normal to AB. If $f(P) = 0$, P is the solution, if not, define, using $\vec{n}$, point M such that $f(P) \cdot f(M) < 0$. The following algorithm gives the solution:
  (1) define $N = \frac{1}{2}(P + M)$,
  (2) if $f(N) = 0$, N is the solution,
  (3) if $f(N) \cdot f(M) \leqslant 0$ then replace P by N, else replace M by N, and return to (a),
- let AB be the edge under consideration and $(x(t), y(t))$ be the definition of the edge, then the midpoint can be computed as $(x(t), y(t))$ with $t = \frac{1}{2}(t_A + t_B)$ (t_A is the value of the parameter associated with endpoint A) or using the curvilinear abscissa directly,
- etc.

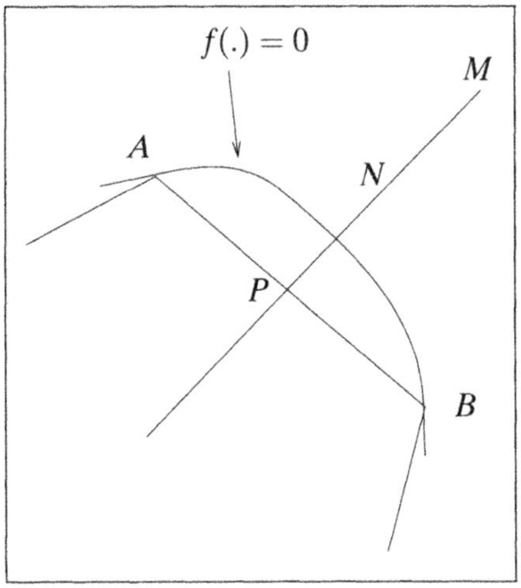

FIG. 3.2. Defining the midpoint.

Thus, for convenience sake, a mesh structure must be enriched in some way to contain the information related to the history which may be required in the sequel. The latter is not so clear and norms are not, at present, specified to do this; at least it is reasonable to think that, in a very near future, an international norm (regarding the norms possible, one could consult norms SET, PDES or IGES) will appear to clarify this point and unify, as best as possible, the ideal contents of a mesh structure. In addition, the language used for implementation clearly interacts with the way the structure may be defined. In this respect, examples are given in the following section for FORTRAN and C^{++}.

4. Mesh structure

The mesh structure is the actual implementation designed so that the mesh contents defined in Section 3 can be organized easily. Thus, the mesh structure must be defined carefully in accordance with its usage. It generally combines two meanings. At the mesh generation algorithm step, it must be designed in such a way that the algorithms can be implemented efficiently while at the computation step it must be convenient in order to perform the desired operations.

The first mesh structure, namely the *internal structure*, may be different from the *external structure*. Algorithms deal with the internal structure and produce the external structure as output. In both cases, structures (internal or external) must be conceived to facilitate the different operations required.

At the mesh generation algorithm step, the internal structure relies on the basic structures already mentioned. It is composed of tables, linked lists and adequate trees so that information is readily accessible and can be manipulated easily. The necessary operations include computations (length, surface, volume, normal, ...), fast researchs of items (using tables, linked lists, ...), construction of stacks, etc.

When the mesh is generated, information necessary while creating it may become irrelevant and need not be stored in the external structure. Therefore, the latter consists of information extracted from the internal structure which will be necessary at the computation step. As an example of an (external) mesh structure, we detailed that used in the Modulef software package, see BERNADOU et al. [1988]. Referred to as the NOPO data structure, it corresponds to an element by element description of the meshes and provides a sequential access to information. This organization is convenient with regard to the computations required during the different steps of the finite element method and, in addition, the amount of memory occupied is relatively little.

The NOPO data structure of the Modulef software package is written in Fortran. It is composed of six tables of a pre-defined organization listed below. The first tables, with a fixed number of variables, contain the general description of the mesh which enable us to dimension the other tables in the structure dynamically. Subroutines, included in the package, make the different operations necessary when dealing with the mesh structure easy. The NOPO structure is given below:

- Table NOP0: General information. This table contains a general description.
- Table NOP1: Supplementary table descriptor, if any.
- Table NOP2: General mesh description. This table contains the values describing the mesh. It includes, for example, the dimension of the space (2 or 3), the number of reference numbers (physical attribute), the number of sub-domain numbers (physical attribute), the number of elements in the mesh, the number of elements reduced to a point, the number of segments, the number of triangles, the number of quadrilaterals, the number of tetrahedra, the number of pentahedra, the number of hexahedra, the number of points, ...
- Table NOP3: Optional pointer if the next table is segmented.
- Table NOP4: Vertex coordinates.
- Table NOP5: Sequential element description. This table describes each element in the mesh sequentially by specifying, for each, its geometric code, its sub-domain number, the number of nodes and their list, the number of points and their list and then the physical attributes associated with its points, edges and faces.

It is important to note that the topology of the elements is not present explicitly in table NOP5; the conventions assumed for this topology are those indicated in Section 2 so that the data of the corresponding list of vertices clearly defines the topology.

Physical attributes are used mainly for physical reasons (assignment of physical values), but can also serve to indicate the history (see Section 3) of a particular item.

Mesh generation algorithms present in the Modulef code produce "P_1 elements" which can be post-processed to add the non-vertex nodes if necessary. The structure contains the number of nodes, but not their location, which is defined at the step the finite element interpolation is established.

It should be noticed that the elements (when the mesh contains only elements of the same type) are described as the list of their vertices included at some index in table NOP5 of the NOPO mesh structure, while they could be defined (see Section 2) in a simple table as a triplet (in the case of triangle) containing their vertices, or as a double triplet (in the same case) including their vertices and their neighbors.

Finally, note that the usage of a different language can strongly modify the notion of a mesh data structure (both from the internal and external point of view). In this respect, object-oriented languages appear to offer great facilities for storing information. The remainder of this section presents the main notions corresponding to such a choice. The example of C^{++} (or a similar language) is assumed to illustrate the discussion.

Classes are defined which correspond to the natural items of the language. Each class is associated with values, pointers, items of different classes, functions and mathematical operators. A class of a lower level inherits the properties of those which are in a upper level. Classes can be illustrated by the following example (we assume that the mesh is two-dimensional):

- Class Coordinates: defined by a couple (x, y) and a series of operations: operation +, operation −, scalar product, scaling, matrix product.
- Class Vertex: this class inherits from class Coordinates and has a physical attribute as value.
- Etc.

5. Mesh prescriptions

The mesh associated with the domain under consideration and the physical problem must be adequate both with regard to the geometry and in view of computation. So, it must take into account properties of geometric and physical nature. Depending on the data provided, two cases are encountered: either the data consist only of the boundaries, in which case, because no other information is provided, only the creation of an isotropic mesh can be reasonably envisaged, or the data includes, in addition, some specifications about the nature of the elements desired.

The first type of data consisting of the boundaries, or more exactly a discrete form of the latter, is included naturally in the data required by the mesh generation algorithm. As no additional data is known, the mesh generator can only profit by this information when creating the mesh. Thus, the mesh creation process can only be governed in this way and the created mesh can only reflect the geometric aspects included in the data while its elements must satisfy the general finite element properties. These are the following:

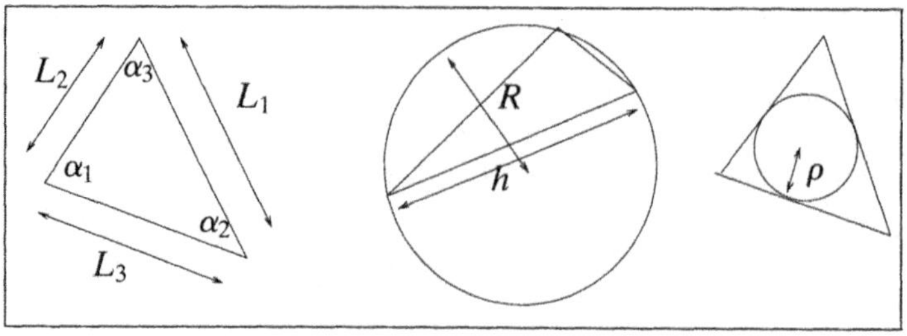

FIG. 5.1. Possible parameters for appreciating a triangle.

An adequate element size must be obtained (it is measured by the diameter, denoted by h, of the elements which is supposed to tend to 0 in the theoretical error analysis).

The element must be well shaped. This property can be measured in different ways. Among these are the aspect ratio defined by R/ρ where R is the circum-radius of the element and ρ its in-radius, the quality denoted by $Q = \alpha(h/\rho)$ (α is a normalization factor), the amplitude of the element angles, the ratio $\max(L_i)/\min(L_i)$ where L_i is the length of edge number i, etc.

Variation must be smooth from one element to another (in terms of size).

The number of elements created, and consequently that of nodes, must be sufficient to obtain a precise simulation and not too large to minimize the computational effort.

Note that implicitly, elements are isotropic in this case.

The second type of data includes, of course, the boundaries of the domain as well as some additional information (provided in some way) to be taken into account when creating the mesh accordingly. An additional problem is now to define a way to specify this information and the associated properties which must be satisfied as best as possible. The latter must in principle be as above (classical error estimate) but, in practice can be slightly different, in particular the aspect ratio and the quality introduced above, entering in the finite element error estimate, can be defined in a different way using some pertinent criteria related to the physical nature of the problem.

In this case, i.e. when the physical behavior of the problem is considered at the mesh generation step, the manner in which these properties are specified is not trivial or unique. A popular method for specifying the desired properties is to use a *background grid* which is nothing else than the current mesh on which a solution is known. Then an analysis of this solution enables us to create a new mesh whose generation is governed by criteria derived from this solution. More generally, what we need in order to create the desired mesh, is the boundary definitions and a *control space*. Such a space can be defined as a couple (Δ, H) using Definitions 5.1 and 5.2.

DEFINITION 5.1. Δ is a *control structure* for mesh T of domain Ω if:
- $\Omega \subseteq \Delta$ where Δ is a covering-up enclosing Ω.

Examples of control structures are:
(a) a partitioning of type "quadtree" in two dimensions, or "octree" in three dimensions;
(b) a regular partitioning (of finite difference type) or grid;
(c) a pre-existing mesh (for example mesh T);
(d) a mesh specified by the user.
Using this notion of a control structure, a control space is as follows:

DEFINITION 5.2. (Δ, H) is a *control space* of mesh T of domain Ω if:
- Δ is a control structure of mesh T of domain Ω, and
- A functional $H(P,d)$ is associated with each $P \in \Delta$, where d is a direction of sphere S^2 (of circle S^1 in dimension 2):

$$H(P,d) : \Delta \times S^2 \to R.$$

In addition to this covering-up feature (via Δ), the control space includes, via the functional H, global information about the physical nature of the problem. The associated values can be used to evaluate if mesh T satisfies the functional at all points P.

Note that the background grid, as briefly discussed above, enters into this framework by setting Δ to be the current mesh and by computing H according to the known solution.

The problems associated with the definition and the efficient use of such a control space are of different natures:

- at the definition step: the proper definition of Δ and H;
- at the mesh generation step: the effective taking into account of the information included in H by the mesh generation algorithm.

To obtain H, one of the following approaches can be used:

- compute the local h, i.e. the desired stepsize, associated with the points specified, and derive H by a generalized interpolation, (this analysis is purely geometrical in the sense that it relies on the geometrical properties of the data: for example the size of the faces of the surface of the domain);
- define, for each element of the covering-up Δ, the value of H manually (for example, specify the desired stepsize in the space—*isotropic* control—or along a direction—*anisotropic* control);
- after an initial computation on a mesh of Ω, define H from the associated solution (or from its gradient or any quantity computed from the solution), or from an error estimate, ...;
- specify H manually by giving its value for all elements of the covering-up constructed for this purpose (this is a control structure of type (d) seen above);

- in case (a) above, the size of the boxes can be used to encode the value of H, meaning that the partitioning is created so that some property holds.

Following its definition, the application of a control space is clearly to govern the mesh generation algorithm in such a way that the created points and elements satisfy the desired properties. Regarding this point, the reader is referred to Chapters III and IV.

Additional difficulties concern the transfer of information from control space to mesh and vice versa which lead to solve intersection, tree search, interpolation and extrapolation problems.

It should be noticed that this mesh generation approach consists in creating the mesh in accordance to the information contained in the control space so that the mesh is adapted in some way. Quite different are the techniques involving local modifications to adapt an existing mesh to a solution. These techniques generally rely on local subdivision of elements, moving of points, ...

6. Methodology

The mesh construction, in the context of P_1 elements as discussed above, must be done with regard to the geometry of the domain under consideration, the result expected and, obviously, the features of available mesh generation algorithms. To achieve these objectives it is convenient to define a suitable methodology. It appears that a simple way to do this is to follow a *top-down* analysis of the problem and a *bottom-up* conception resulting from this analysis.

The purpose of the top-down analysis is to best utilize the geometrical features of the domain in view of minimizing the work required to generate a mesh, as well as making this step more reliable. Such a method leads to splitting the problem under consideration (possibly a complex one from a geometrical point of view) into a series of sub-problems which are easier to solve or which are better adapted with regard to the capabilities of the available algorithms. This step has three objectives:

(1) To minimize the mesh generation operation and make it more reliable by considering the possible repetitive features present in the domain. In this respect, it is often useless to create the mesh of the entire domain if some parts of the domain can easily be obtained from others.

(2) To adapt the region under consideration to the capabilities or robustness of the mesh generator. Any shape can be considered successfully by a "powerful" mesh generator, but it is often advisable to use a specific algorithm to deal with a particular shaped domain, and one has to split the domain into adapted parts when a "poor" mesh generator is used.

(3) To obtain a mesh enjoying some special features, for example: to create quadrilaterals in some regions of the domain, to obtain a varying density of the elements, to impose a point, a line, a surface, ... in a mesh or to obtain symmetry in the resulting mesh, as this symmetry is not produced directly by the mesh

generator used; in this case, only a part of the domain is processed after which the symmetry is ordered explicitly.

In most cases, a domain can be subdivided into geometrically simple sub-domains. Various possibilities exist to obtain this partitioning:

- The geometrical repetitive features (existence of symmetries, translations, rotations, etc.) are recorded to define the sub-sets, called *primal sub-sets*, which will effectively be considered for mesh generation opposed to the sub-sets, called *secondary sub-sets*, which can be derived with the help of usual transformations (symmetries, translations, rotations, etc.).
- If such possibilities do not exist, one can split the domain artificially into geometrically simpler regions which form the primal sub-sets.
- Another way of achieving this simplification is to obtain different meshes which are only slightly different, by creating only one of them, considered as a primal sub-set, and creating the others by simple modification of the latter.
- In addition, the valid gluing together of meshes (to ensure conformal features) can lead to the definition of a priori non-evident primal sub-sets.
- Etc.

The bottom-up construction of the mesh can then be envisaged resulting from the previous analysis and the associated decomposition of the problem. It consists in first dealing with the primal sub-sets and then deduces the mesh of the secondary sub-sets before forming the final mesh.

7. Data structure

In this section data structure refers to the data structure necessary when using the desired mesh generation algorithm. Several points must be considered. The data of the algorithm generally differs from the data the user has to provide. Let us consider as example, an algorithm of the advancing-front or Voronoï class used without a control space (i.e. the problem is only to create a suitable mesh enjoying the classical geometric properties). In this case, algorithms use a discrete form of the boundary of the domain to be meshed as data. The boundary mesh is composed of a collection of the segments which approach the boundary and can therefore be seen as a list. Such a list must possess some properties: It may be oriented to be able to define the internal part of the domain; if the boundary has more than one connected component, information must be given to define these components exactly. Obviously, these types of data items are not so easy to specify. The user would prefer to enter the data in a simpler manner, by:

- defining some points,
- constructing lines using these points and specifying the geometry,
- ordering the discretization of these lines by specifying some parameters,
- defining each component on the contour as a list of these discretized lines,
- etc.,

and to have the data for the algorithm created automatically by a program processing the information given by the user. Thus, this program is in charge of:

- processing the given points;
- splitting the lines by creating intermediary points;
- forming the components of the contour by joining the relevant lines while controlling the correct closure of the resulting contour;
- checking that the data thus defined is consistent with the mesh generator envisaged.

The user data can be captured by different methods. For example, by:

- creating a file containing the relevant values given in a specified format,
- creating a data file automatically by entering the relevant values manually, by capturing the values interactively in the case where graphic facilities are available, or by converting information resulting from a CAD system into that necessary.

The creation of a data file is an easy way of keeping this data in memory. Thus, at a later stage, the *history* (as introduced in Section 3) of any value can be accessed, so that any modification or required computation can be done easily.

As a conclusion of this rapid discussion, it appears that the existence of a data file containing all the pertinent information is clearly a good solution which offers great flexibility when the mesh generation process is involved once only. When it is included in a loop or an iterative process, a different solution must be adopted.

8. Properties and formula about triangles and tetrahedra

As will be seen in Chapters III and IV, automatic mesh generators are, in general, designed to create simplices, i.e. triangles in two dimensions and tetrahedra in three dimensions. This is why it seems to be advisable to recall some basic properties and useful formula related to these elements.

The surface S of a triangle is given by:

$$S = \frac{1}{2}\begin{vmatrix} x_2 - x_1 & x_3 - x_1 \\ y_2 - y_1 & y_3 - y_1 \end{vmatrix},$$

where x_i, y_i are the coordinates of vertex i.

The in-radius of a triangle is given by:

$$r = \frac{2S}{L_1 + L_2 + L_3},$$

where S is the triangle area and L_i are the three edges lengths.

The circum-radius of a triangle is:

$$R = \frac{L_1 L_2 L_3}{4S}.$$

The volume V of a tetrahedron is given by:

$$V = \frac{1}{6}\begin{vmatrix} x_2 - x_1 & x_3 - x_1 & x_4 - x_1 \\ y_2 - y_1 & y_3 - y_1 & y_4 - y_1 \\ z_2 - z_1 & z_3 - z_1 & z_4 - z_1 \end{vmatrix},$$

where x_i, y_i, z_i are the coordinates of vertex i.

The in-radius of a tetrahedron is given by:

$$r = \frac{3V}{S_1 + S_2 + S_3 + S_4},$$

where V is the tetrahedron volume and S_i are the four faces surfaces.

The circum-radius R of a tetrahedron must be explicitly computed (contrary to that of a triangle expressed as above) as the distance between the circumcenter and any vertex of the element.

CHAPTER II

Review of Mesh Generation Methods

Introduction

Mesh generation methods can be classified broadly into two main classes. The first corresponds to the algorithm complexity while the second corresponds to the field of applications in terms of the geometry to which the algorithms apply. In this chapter we would like to outline the most popular mesh generation methods according to the above classification.

The methods which are available for the realization of meshes in dimensions 2, $2^{1/2}$ (this notation is employed for the surfaces in the R^3 space) and 3 (or 2D, $2^{1/2}$D and 3D meshes) can be classified, in terms of algorithms, according to several types. The following enumeration is not unique but indicates the most popular methods. One can find similar, and a little bit different, overviews of the main mesh generation algorithms in numerous papers, for instance in THACKER [1980], HO LE [1988], SHEPHARD, GRICE, LO and SCHROEDER [1988], BAKER [1989a], etc. It appears that there is a shortage of books devoted to these topics, for example THOMPSON, WARSI and MASTIN [1985] and GEORGE [1990]. On the other hand, many international conferences have been held on mesh or grid generation methods, for instance, SENGUPTA, HAUSER, EISEMAN and THOMPSON [1988], ARCILLA, HAUSER, EISEMAN and THOMPSON [1991], CASTILLO [1991b].

There are several types of mesh generation methods, which may be classified as follows:

Type 1:

- M1: "*Manual mesh generation method*". In this case, the user defines the elements by their vertices and provides all useful information. This approach (Section 9) is quite suitable for domains with very simple geometries or for those which can be covered by a limited number of elements. The resulting mesh will often be post-processed by transformations to obtain a more accurate and precise mesh.
- M2: "*Product method*". By using a simple mesh as a basis for a more complex construction which can be seen as a product of the latter. This approach

(Section 10) is suitable for domains with hexahedral or cylindrical geometries: from a two-dimensional mesh of a section, the corresponding volumetric elements are defined layer by layer.

Type 2:

- M3: Generation of a mesh by transformation of a reference mesh using an *algebraic method* (Section 11). This corresponds to a domain with an elementary geometry (triangle, quadrilateral, etc.). The mapping function used in the process is predefined to ensure certain properties (respecting contours, etc.).
- M4: Generation of a mesh by the *explicit solution of partial derivative equations* (Section 11) formulated on a reference mesh. It corresponds to a domain with an elementary geometry in which a structured grid is developed easily. A mapping function is defined by solving a PDE system to ensure some required properties (respecting contours, orthogonality of elements, variable density of elements, etc.).

Type 3:

- M5: Generation of the final mesh by the *structured partitioning* of a coarse mesh composed of blocks (Section 12), assuming a simple geometric shape. Firstly, a coarse partitioning or *decomposition* of the domain into blocks with simple elementary geometry (segments, triangles, quadrilaterals, tetrahedra, etc.) is given. The interfaces between these blocks must be defined carefully, on the one hand, to ensure the validity of the result and, on the other hand, to determine the nature of the partitioning to which they will be subject (number of subdivisions, etc.) precisely. This technique enables us to consider a domain with an arbitrary shape; nevertheless, in the case of a complex shape, one has to define a large number of initial blocks.

Type 4:

- M6: Application of techniques of *overlapping* and *deforming* of simple meshes in such a way that the real domain is covered accurately. The domain is included in a grid which is composed of quadrilateral or hexahedral elements, constructed in order to satisfy some properties (by modifying the number, the size and the repartitioning of these elements according to the problem). This grid is then deformed after which its elements are split, if desired, to obtain the elements of the final mesh. Quadtree and octree techniques (Section 13) enter in this class of methods.

Type 5:

- M7: Generation of a mesh from the boundaries of the domain using an *advancing-front method* (Section 13 and Chapter III). The boundaries are defined either in a global or discrete way as a polygonal approximation of the contour in the form of segments, in two dimensions, and, in three dimensions, as a polyhedral approximation in terms of triangular faces. A front, initialized by the contour, is established. From an edge (a face) of the front, one or several internal elements are created. The front is then updated by suppressing or adding some edges (faces). The process is iterated as long as the front is not empty. This technique is suitable for the creation of triangular or tetrahedral meshes of arbitrary domains.

- M8: Generation of the mesh from the points of the boundaries of the domain by an approach of type *Delaunay–Voronoï* (Section 13 and Chapter IV). The boundaries are defined as for method M7. This type of mesh generators usually carries this name as it relies on the Delaunay–Voronoï construction which leads to the creation of a mesh of a convex hull of a set of points. Nevertheless, the method applies in the case of arbitrary domains.

Type 6:

- M9: Generation of meshes by *combining* previously created meshes. This approach relies on the use of geometrical transformations (symmetry, etc.) or topological transformations (global refinement, combining of two meshes, etc.). In this case, the geometrical properties of the domain are considered fully in order to mesh only the necessary parts by applying the most appropriate technique; the meshes of the various parts, which can be derived from others (by symmetry, rotation, etc.), are then obtained using simple transformations and the final mesh is simply the pasting together of all the different sub-meshes.

The remainder of this chapter is devoted to presenting some examples of the above methods briefly by selecting one in each of the five first types. It also discusses the problem of surface mesh generation (Section 14) shortly. In George [1990], the reader can find some detailed examples of practical applications of Type 6 methods.

9. Manual methods

Such methods simply consist of enumerating, in a proper way, all the elements in the mesh to be constructed. More precisely, it is the user's responsibility to define:

- the number of elements to be created by specifying for each:
  - its geometrical nature,
  - the list of its vertices in a proper way,
  - the physical attributes associated with the element and its constituent items (faces, edges and vertices).
- the number of vertices and their coordinates.

It is clear that only simple meshes, or simple sections of more complex meshes, can be created using this approach. These meshes can be used in more complicated creations via mesh modification methods (geometrical transformations of these initial meshes (for example by translation, rotation, symmetry, ...) or combination methods (creation of a mesh by juxtaposition of two existing meshes sharing a common zone)).

10. Product methods

The aim of the product method consists in creating elements of dimension $d+1$ from the data, on the one hand, of elements of dimension d and, on the other hand, of a meshed line serving as a *generation line*.

Thus, locally, a point (i.e. an element reduced to a point, a 0-dimensional item) defined in the d-dimensional space produces a series of segments (item of one dimension) defined in the $d+1$-dimensional space. A segment (item of one dimension) defined in the d-dimensional space produces quadrilaterals in $d+1$-dimensional space. A triangle serves as support for creating pentahedra, while a quadrilateral produces hexahedra (see Fig. 10.1).

Degeneracies may be encountered for some special positions of the generation line with respect to the given mesh. In this case segments may produce, not only quadrilaterals, but triangles, triangles may produce degenerated pentahedra and quadrilaterals may produce degenerated hexahedra. Depending on the situation, the degenerate elements are valid, or not, in the usual finite element context. In this respect, for example, hexahedra may degenerate into pentahedra, which are admissible elements, or into non-admissible elements (see Fig. 10.2), while pentahedra may lead to the creation of tetrahedra or non-admissible elements.

Product methods (also referred to as "extrusion" methods) can be applied to domains which enjoy the desired properties, i.e. which enter in the following topological types:

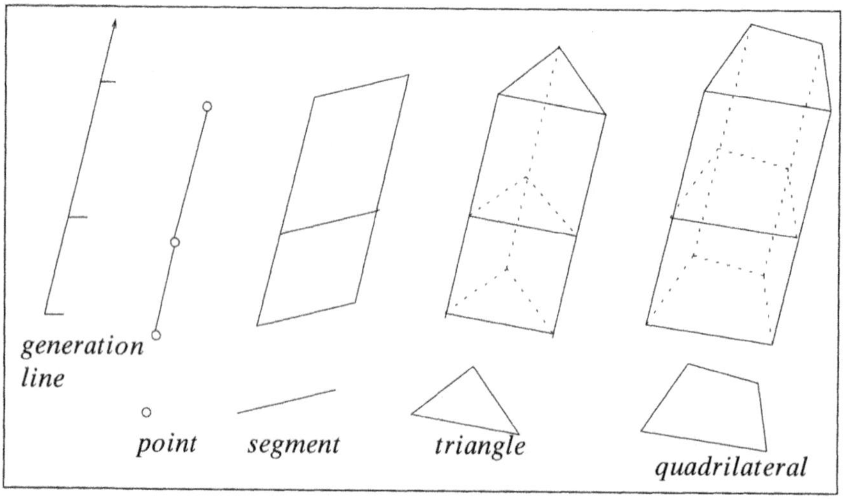

FIG. 10.1. Correspondences.

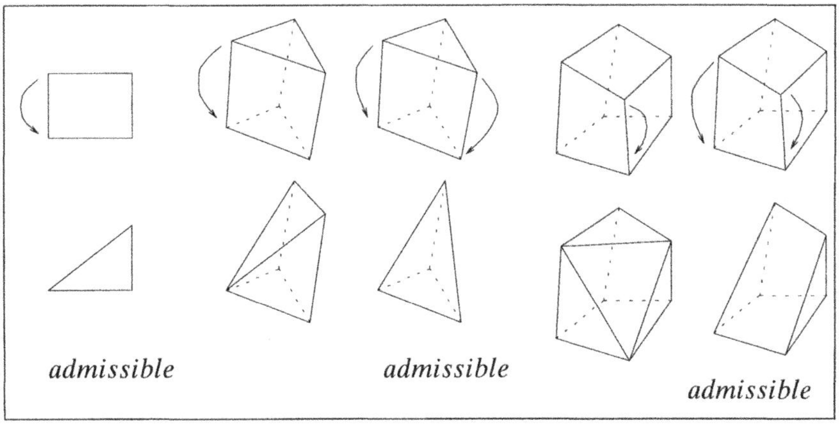

FIG. 10.2. Correspondences with degeneracies.

- cylindrical topology: the domain can be described via the data of a two-dimensional mesh and a generation line defining layers with which the three-dimensional elements are associated,
- hexahedral topology: the domain can be described via the data of a one-dimensional mesh and a generation line which allow for the construction of a two-dimensional mesh which is then coupled with a second generation line and produces the desired three-dimensional elements.

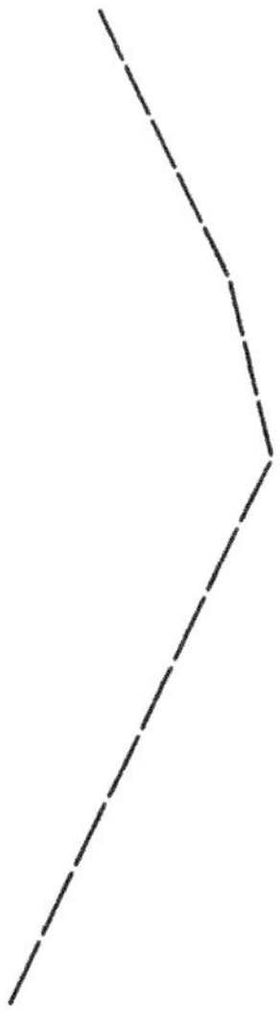

FIG. 10.3. Line serving as data.

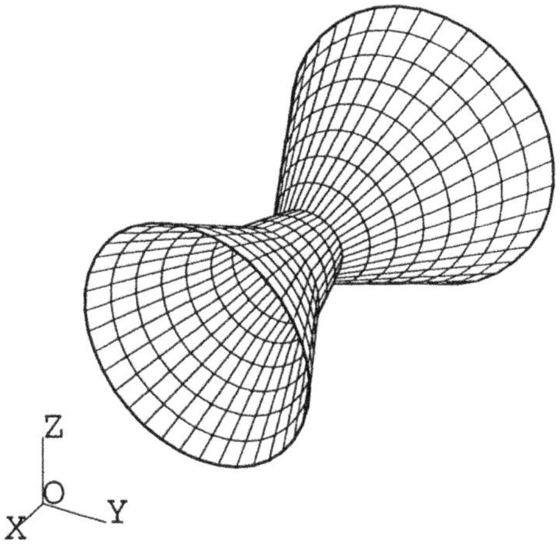

FIG. 10.4. Resulting mesh.

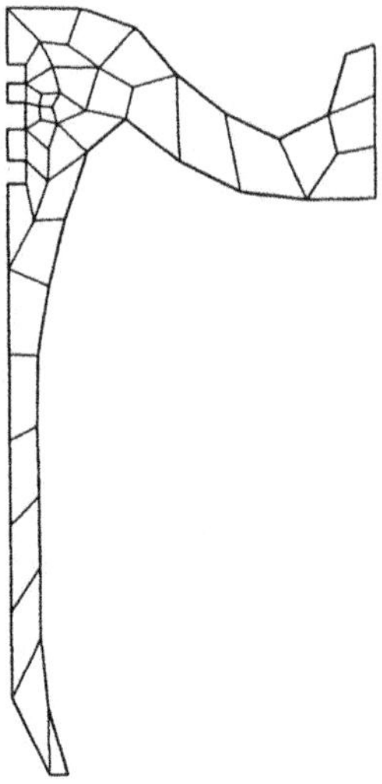

FIG. 10.5. 2D mesh serving as data.

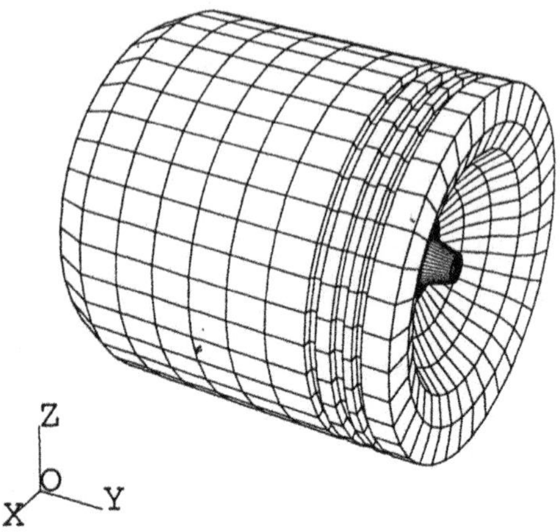

FIG. 10.6. Resulting mesh.

Figures 10.3–10.6 display two examples constructed by a product method applied respectively to a meshed line and a two-dimensional mesh serving as data and a rotation used as generation line. In the first example, the generation line is not connected with the data and classical elements are constructed (quadrilaterals), while in the second example, the generation line is common with an edge of the data and pentahedra (instead of hexahedra) are produced in this region.

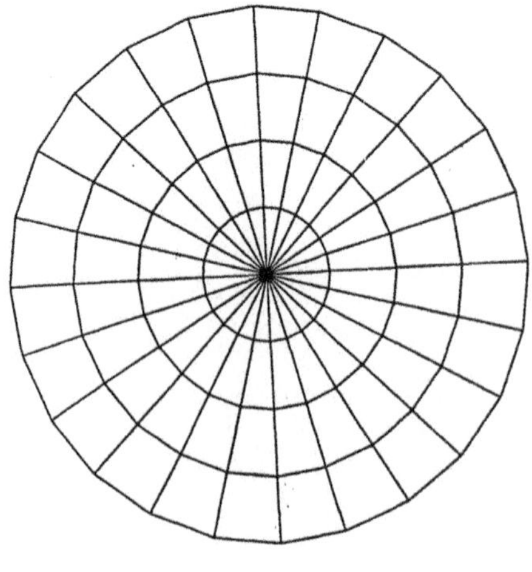

FIG. 10.7. Classical radial mesh.

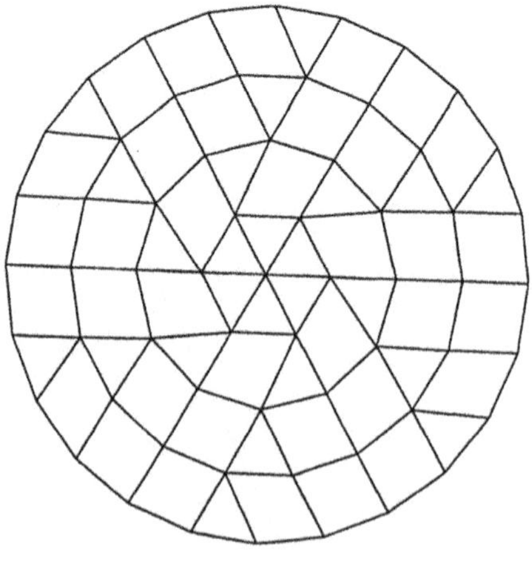

FIG. 10.8. Non-classical radial mesh.

Variations to degenerated cases, in particular to balance the shape of the created elements better, can be envisaged. In Fig. 10.7, one can see the classical result obtained from the data of a mesh composed of segments and a generation line consisting of a rotation and Fig. 10.8 shows the mesh obtained using a variation developed by HECHT [1992].

11. Special application methods

In the event that the domain under consideration cannot be dealt with easily using the previous approaches, one must consider a different solution to the mesh generation problem. Historically, the first tentative for automating the mesh generation process relied on the use of automatic methods developed for simple shapes.

11.1. Algebraic methods

Let us consider a unit square. We suppose that a series of points is given along the four boundary edges in such a way that the number of points present on one edge is identical to that present on the opposite edge. Thus, two parameters, n_1 and n_2, are introduced which allow for some flexibility and enable us to control the process (see Fig. 11.1 where $n_1 = 3$ and $n_2 = 4$). Then, it is obvious to create a mesh of the domain composed of quadrilaterals.

This construction consists of forming the lines joining the points corresponding to one side to the other. Consequently, the points defined by the intersections of these lines are created and enable us to enumerate the elements of the mesh easily.

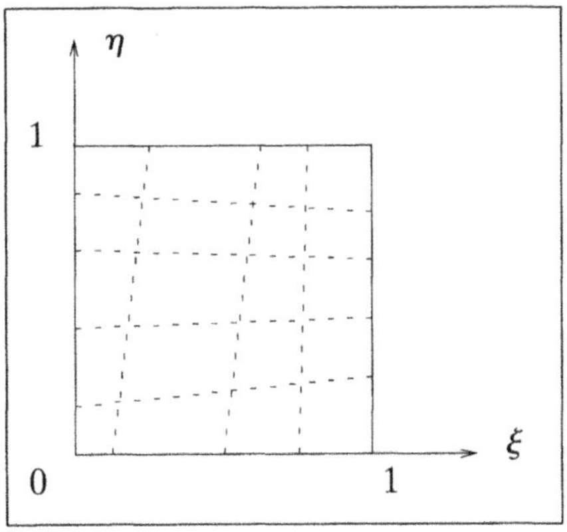

FIG. 11.1. Logical mesh on the unit square.

We suppose therefore that the domain (namely Ω) under consideration is defined by its boundary (Γ) which is approximated by a polygon, the latter being a collection of segments defined in such a way that it is possible to identify four sides clearly. We assume in addition that two opposed sides possess the same number of segments. Then the method can be described as:

- the identification of each side of the domain with the corresponding side of the unit square,
- the creation of the points on the sides of the unit square, corresponding to those of the given contour in such a way that the relative distances between points are respected,
- the creation of the corresponding mesh in the unit square,
- the mapping of this mesh onto the real domain.

To achieve the last step, a mapping function is defined which allows us to transfer any point $\hat{M}(\xi, \eta)$ in the logical mesh onto a point $M(x, y)$ "on" the real domain. An example of such a function is given by the transfinite interpolation (GORDON and HALL [1973], THOMPSON, WARSI and MASTIN [1985]). It is defined by:

$$\begin{aligned} F(\xi,\eta) = {} & (1-\eta)\phi_1(\xi) + \xi\phi_2(\eta) + \eta\phi_3(\xi) + (1-\xi)\phi_4(\eta) \\ & -((1-\xi)(1-\eta)a_1 + \xi(1-\eta)a_2 + \xi\eta a_3 + (1-\xi)\eta a_4). \end{aligned} \tag{11.1}$$

where the ϕ_i are parametrizations of the sides of the domain (in practice, a discrete form of these sides) and the a_i are their endpoints, i.e. the corners of the domain.

It can be proved easily that function F satisfies the "corner" identities, i.e. that $F(0,0) = a_1$, $F(1,0) = a_2$, $F(1,1) = a_3$ and $F(0,1) = a_4$, and matches the boundary (in terms of its discretization), i.e. that $F(\xi, 0) = \phi_1$, $F(1, \eta) = \phi_2$, $F(\xi, 1) = \phi_3$ and $F(0, \eta) = \phi_4$.

Such a function guarantees that the mesh generation method can be applied with success for a convex domain (the image of any point inside the unit square falls inside the domain). For non-convex geometries, the method can produce a mesh which does not cover the domain properly. Nevertheless, it has been proven that for a not too "twisted" geometry, this mesh generation method is suitable.

For a triangular shaped domain (in the same sense), a quite similar method can be employed. It is based on the same principles and depends on only one parameter ($n_1 = 3$ in Fig. 11.2).

In this case the initial process is applied to a unit triangle, all three sides having the same number of points. Three families of lines are constructed by joining the relevant points; the intersection of three associated lines defines a zone which enables us to construct the points inside the unit triangle necessary to create the requested mesh.

The steps of the mesh generation method are then identical. The mapping from the unit triangle to the real domain is defined, in this case, by the following

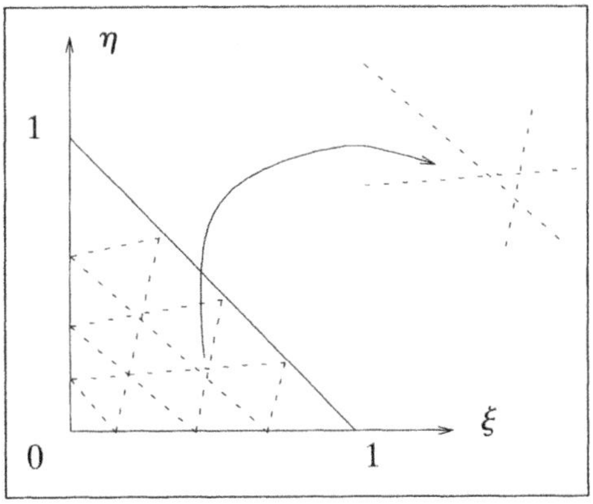

FIG. 11.2. Logical mesh on the unit triangle.

function:

$$F(\xi,\eta) = \frac{1-\xi-\eta}{1-\xi}\phi_1(\xi) + \frac{\xi}{1-\eta}\phi_2(\eta) + \frac{\eta}{\xi+\eta}\phi_3(1-\xi-\eta) - \left(\frac{\eta}{\xi+\eta}(1-\xi-\eta)a_1 + \frac{1-\xi-\eta}{1-\xi}\xi a_2 + \frac{\xi}{1-\eta}\eta a_3\right) \tag{11.2}$$

using the same notation. This function satisfies the "corner" identities and matches the boundary, as can be verified easily.

Algebraic methods, among which the previous mesh generation methods enter, can be employed to construct a mesh for domains having the required shape, or can be used for more complex geometries as local mesh generation processes as will be discussed later (see the section concerning the multiblock methods).

Variations to function F lead to alternative mesh generation methods. The reader is referred to THOMPSON, WARSI and MASTIN [1985], GILDING [1988] and CASTILLO [1991b] for more details.

The previous methods can be employed for surface meshes and similar methods can be used in three dimensions for domains enjoying the same assumptions and defined in a similar way (see COOK [1974] and PERRONNET [1983]).

An example (derived from a method proposed in PERRONNET [1983]) of a function suitable for a hexahedral domain is given by:

$$F(\xi,\eta,\zeta) = \frac{\sum_{i=1}^{6}\beta_i\phi_i(\xi,\eta,\zeta)}{\sum_{i=1}^{6}\alpha_i}, \tag{11.3}$$

where the α_i and the β_i are the following functions of ξ,η,ζ:

$$\alpha_i = (1-x_{i-1})\frac{x_i}{(x_i+x_{i-1})}\frac{1-x_i}{(1-x_i+x_{i-1})}\frac{x_{i+1}}{(x_{i+1}+x_{i-1})}\frac{1-x_{i+1}}{(1-x_{i+1}+x_{i-1})},$$

where $i = 1,3$ with $x_{j+1} = \xi,\eta$ or ζ, $j = 0,2$ (j modulo 3).

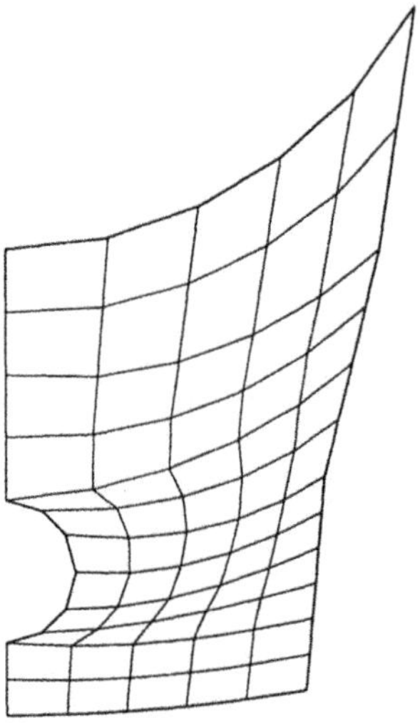

FIG. 11.3. Algebraic method for a "quadrilateral" domain.

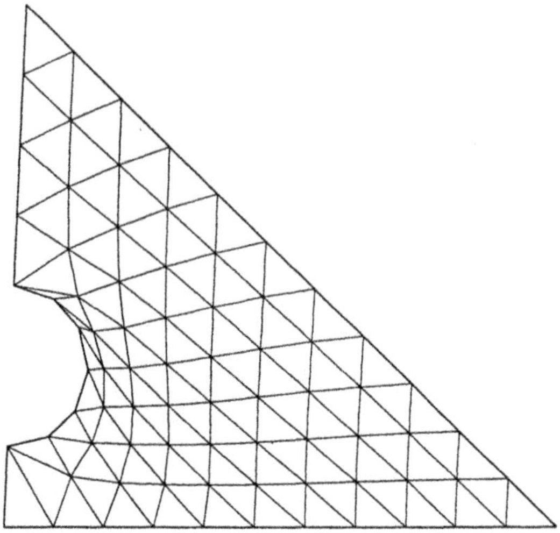

FIG. 11.4. Algebraic method for a "triangular" domain.

$$\alpha_{i+3} = x_{i-1}\frac{x_i}{(x_i+1-x_{i-1})}\frac{1-x_i}{(2-x_{i-1}-x_i)}\frac{x_{i+1}}{(x_{i+1}+1-x_{i-1})}\frac{1-x_{i+1}}{(2-x_{i+1}-x_{i-1})},$$

where $i = 1,3$ and ϕ_i are the parametrizations of the faces of the domain, and

$$\beta_i = (1-x_{i-1})(\alpha_i + \alpha_{i+3}) \quad i = 1,3,$$

$$\beta_{i+3} = x_{i-1}(\alpha_i + \alpha_{i+3}) \quad i = 1,3.$$

As before, this function satisfies the "corner" identities and matches the boundary.

Similar expressions apply to tetrahedral and pentahedral shaped domains.

Figures 11.3–11.6 display examples constructed by an algebraic method. The two first examples correspond to domains which are well shaped with respect to the method (the domain of Fig. 11.3 is not convex but the resulting mesh is valid). Example of Fig. 11.5 shows the mesh obtained by the method when it fails while Fig. 11.6 depicts the mesh resulting from local corrections of the previous one.

11.2. PDE methods

PDE mesh generation methods represent an elegant alternative to algebraic methods used when the domain (Ω with boundary denoted by Γ hereafter) under consideration can be identified by a quadrilateral (in two dimensions), or a cuboïd (in three dimensions). Contrary to algebraic methods, a transformation from the domain to this quadrilateral (cuboïd) called the logical region is sought.

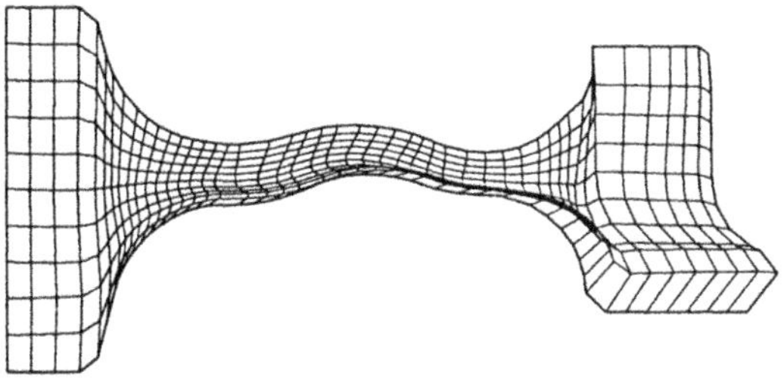

FIG. 11.5. Domain inducing degeneracies.

A *generation system* is associated with such a transformation, which allows us to compute the requested mesh. Variables x, y, (x, y, z) describe the domain (see Fig. 11.7) while the logical region is described using variables ξ, η, (ξ, η, ζ). The problem now becomes: find the functions

$$x = x(\xi, \eta),$$
$$y = y(\xi, \eta)$$

in two dimensions, and

$$x = x(\xi, \eta, \zeta),$$
$$y = y(\xi, \eta, \zeta),$$
$$z = z(\xi, \eta, \zeta)$$

in three dimensions, assuming that the transformation maps the logical region one-to-one onto the domain and that the boundaries are preserved.

The one-to-one property is ensured by requiring that the jacobian of the transformation is non-zero. The transformation (for example in two dimensions) is defined by the matrix

$$M = \begin{pmatrix} x_\xi x_\eta \\ y_\xi y_\eta \end{pmatrix},$$

where x_ξ is equivalent to $\partial x/\partial\xi$, etc.

Its jacobian is

$$J = x_\xi y_\eta - x_\eta y_\xi.$$

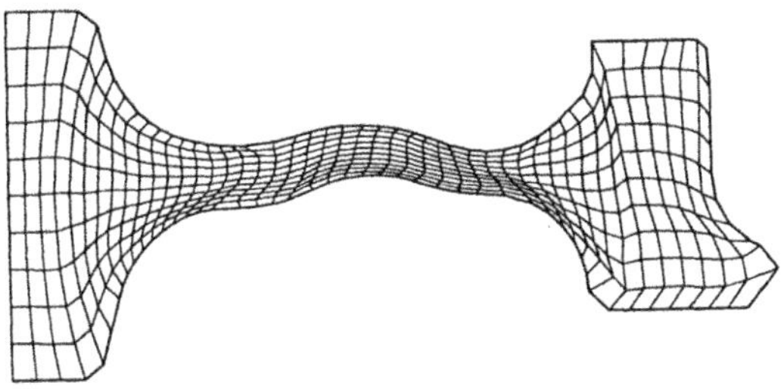

FIG. 11.6. Mesh resulting from local corrections.

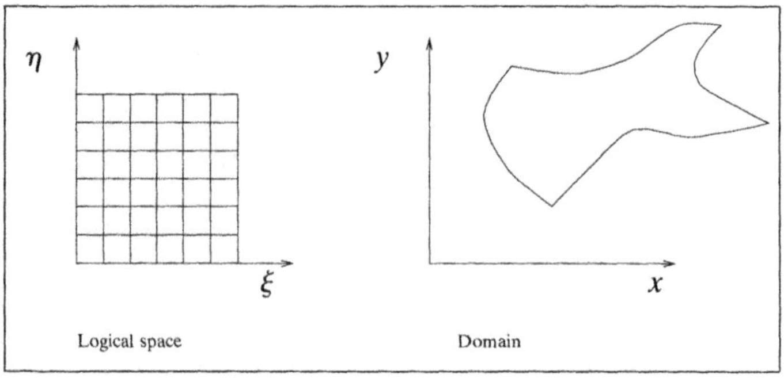

FIG. 11.7. Logical mesh and real domain.

As it is assumed non-zero, the inverse of the transformation exists and variables ξ, η can be expressed in terms of x, y as follows:

$$\xi = \xi(x, y),$$
$$\eta = \eta(x, y).$$

The two ways of expressing the variables are mathematically equivalent and lead to two possibilities for solving the problem. If variables ξ, η are expressed in terms of x, y, the logical mesh can be transformed into a mesh on the domain, and the physical problem is solved in the domain as in the classical way. On the other hand, if variables x, y are expressed in terms of ξ, η, either the physical problem can be written in terms of these variables and then solved in the logical region, or we return to the classical solution above.

As a simple example of PDE methods, we briefly consider the following generation system based on the regularizing properties of the Laplacian operator Δ. We consider the two following systems:

$$\begin{cases} \xi_{xx} + \xi_{yy} = 0 \quad \text{in } \Omega, \\ \text{Boundary conditions on } \Gamma \end{cases} \tag{11.4}$$

and

$$\begin{cases} \eta_{xx} + \eta_{yy} = 0 \quad \text{in } \Omega, \\ \text{Boundary conditions on } \Gamma \end{cases} \tag{11.5}$$

which are then inversed in order to find $x(\xi, \eta)$ and $y(\xi, \eta)$, thus we obtain as generation system:

$$\begin{cases} g_{11}x_{\xi\xi} + g_{22}x_{\eta\eta} + 2g_{12}x_{\xi\eta} = 0, \\ g_{11}y_{\xi\xi} + g_{22}y_{\eta\eta} + 2g_{12}y_{\xi\eta} = 0 \end{cases} \tag{11.6}$$

with

$$g_{ij} = \sum_{m=1}^{2} A_{mi}A_{mj},$$

where $A_{mi} = (-1)^{i+m}(\text{Cofactor}_{m,i} \text{ of } M)$ and M is the above matrix.

The result is a system posed in the logical space (where a mesh exists); it is a non-linear coupled system which can be solved using relaxation techniques or, more generally, iterative methods after an initialization by a solution in which the real boundary conditions, are prescribed.

Variations on the previous generation systems can be experimented with to obtain special properties. For example, adding a non-zero right-hand side and considering:

$$\begin{cases} \xi_{xx} + \xi_{yy} = P \\ \text{Boundary conditions} \end{cases} \tag{11.7}$$

and

$$\begin{cases} \eta_{xx} + \eta_{yy} = Q \\ \text{Boundary conditions} \end{cases} \tag{11.8}$$

enables us to control the distribution of points inside the domain. In this situation, the inverse system is:

$$\begin{cases} g_{11}x_{\xi\xi} + g_{22}x_{\eta\eta} + 2g_{12}x_{\xi\eta} + J^2(Px_\xi + Qx_\eta) = 0, \\ g_{11}y_{\xi\xi} + g_{22}y_{\eta\eta} + 2g_{12}y_{\xi\eta} + J^2(Py_\xi + Qy_\eta) = 0 \end{cases} \tag{11.9}$$

using the same notation and J, the jacobian, defined by $J = \det(M)$.

The right-hand sides P and Q interact as follows: For $P > 0$, the points are attracted to the "right", $P < 0$ induces the inverse effect. For $Q > 0$ the points are attracted to the "top", $Q < 0$ leads to the inverse effect. Close to the boundary, P and Q induce an inclination of lines $\xi =$ constant or $\eta =$ constant.

P (Q) can also be used to concentrate lines $\xi =$ constant or $\eta =$ constant towards a given line, or to attract them towards a given point. To achieve this, one can define the right-hand side as follows:

$$\begin{aligned} P(\xi,\eta) &= -\sum_{i=1}^{n} a_i \operatorname{sign}(\xi - \xi_i)\, e^{-c_i|\xi-\xi_i|} \\ &\quad -\sum_{i=1}^{m} b_i \operatorname{sign}(\xi - \xi_i)\, e^{-d_i\left[(\xi-\xi_i)^2+(\eta-\eta_i)^2\right]^{1/2}}, \end{aligned}$$

where n and m denote the number of lines in ξ and η of the grid. Such a control function induces the following:

for $a_i > 0$, $i = 1, n$, lines ξ are attracted to line ξ_i,

for $b_i > 0$, $i = 1, m$, lines ξ are attracted to point (ξ_i, η_i).

These effects are modulated by the amplitude of a_i (b_i) and by the distance from the attraction line (attraction point), modulated by coefficients c_i and d_i. For $a_i < 0$ ($b_i < 0$), the attraction is transformed into a repulsion. When $a_i = 0$ ($b_i = 0$), no particular action is connected to line ξ_i (or point (ξ_i, η_i)).

A right-hand side Q of the same form produces analogous effects with respect to η by interchanging the roles of ξ and η.

The major difficulty for automating this class of mesh generation systems is how to choose the control functions (P, Q, ...) and the parameters they involve. Anyway, these methods can be extended to three dimensions and, for a complete discussion, the reader is referred to THOMPSON, WARSI and MASTIN [1985], where other forms of right-hand sides P and Q producing other properties are discussed (for example, the concentration of lines ξ or η towards an arbitrary line and not only towards a particular one (ξ = constant or η = constant) or towards a given point to increase the mesh density near this point). In the above mentioned reference, and some others, other types of generation systems, including parabolic and hyperbolic operators, are discussed and numerous examples are provided.

When using a generation method of the present class, it is convenient to find the best analogy from the domain to a logical shape (quadrilateral or cuboïd). Such an analogy is often obtained either by partitioning the domain into several simpler domains, or by identifying the domain with the requested shape, using several methods.

For example, in two dimensions, there are several major classes of decompositions of the domain under consideration from which different kinds of grids will result in order to capture the physics of the problem as well as possible. In this respect, a domain can be discretized following an O-type, C-type or H-type analysis (see Fig. 11.8). To obtain such an analogy, artificial cuts must be introduced. Such analogies extend, more or less, to three dimensions.

PDE methods can also be employed to construct a mesh for surfaces or domains having more complex geometries as local mesh generation processes as will be discussed below (see Section 12 concerning the multiblock methods).

Figures 11.9 and 11.10 display an example that results from an elliptic method constructed as an extension of a purely two-dimensional method. The first figure, corresponding to variables u and v which describe the surface domain, shows the mesh obtained by controlling the cell areas, while the second figure shows the resulting mesh of the surface.

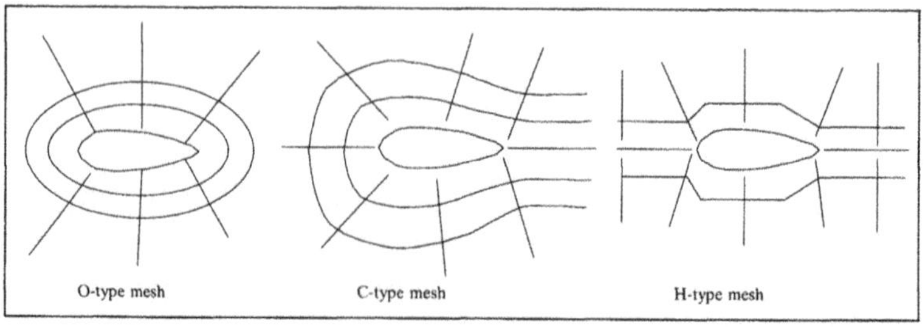

FIG. 11.8. O-type, C-type and H-type decompositions.

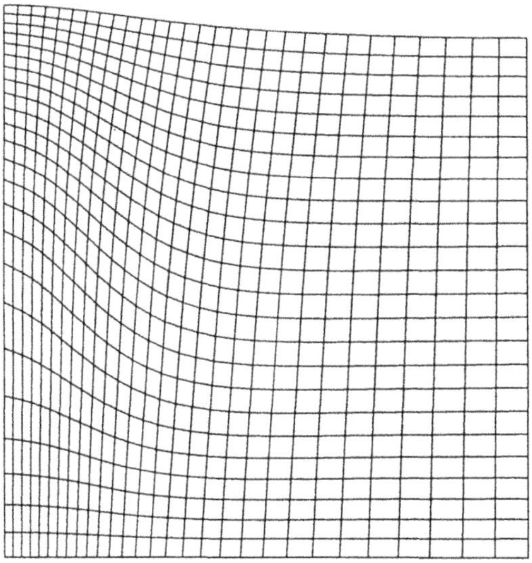

FIG. 11.9. PDE method governed by cell areas.

12. Semi-automatic methods

A method is said semi-automatic if, while possessing automatic features, it requires some interaction of the user. Typical semi-automatic methods are those based on the multiblock approach (ECER, SPYROPOULOS and MAUL [1985], THOMPSON [1987], FRITZ [1987]).

These methods represent a solution for the creation of meshes of domains with complex geometries, while a structured connectivity is maintained locally. First introduced in the context of structured meshes, a multiblock method basically consists in partitioning the domain into several blocks having a quadrilateral shape in two dimensions, or a hexahedral shape in three dimensions. This partitioning is such that an automated method can be applied for each block.

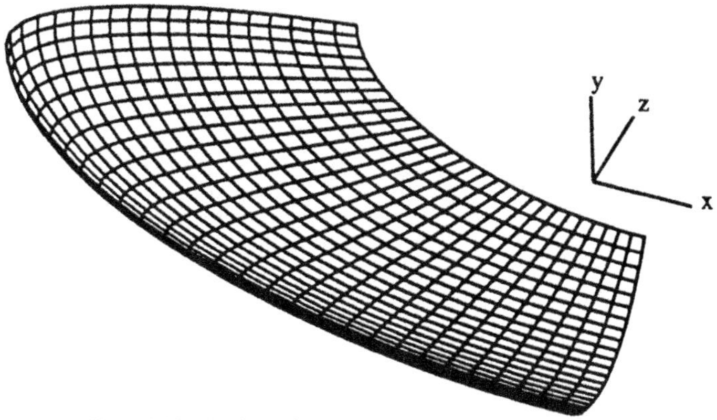

FIG. 11.10. Real mesh resulting from the previous mesh.

This local mesh generation process can be of algebraic type, conformal mapping method, or falls into the PDE class (see above).

Thus, a possible scheme for a multiblock method can be summarized as follows:

- Partition the entire domain into suitable blocks, by:
  - numbering the vertices of the blocks required,
  - defining the edges and faces of those which form the interfaces between blocks.
- Discretize these interfaces with the following objectives:
  - capture the geometry as well as possible,
  - ensure a proper assembly of the different meshes when creating the mesh of the entire domain. There are two ways of achieving this property: firstly, the chosen decomposition is conformal by itself and, secondly, although it is not, it has been carefully defined in such a way that interfaces between blocks are defined in a unique manner,
  - ensure that each block is well defined with respect to the local mesh generation process (for instance, if an algebraic local mesh generation method is elected, the number of points lying on the opposite edges must be equal).
- Apply the mesh generation method locally to create the internal points within each block.
- Construct the final mesh by "adding" the meshes of the blocks. This phase can be done easily if phase 2 of the global process was carried out properly, i.e. the connection from one block to another is implicitly since present.

Such a block subdivision coupled with a local structured mesh generation method into blocks provides flexibility when dealing with complex domains, and the resulting mesh possesses a structured connectivity locally.

The same principles can be applied to non-structured meshes. In this situation, the initial partitioning consists in exhibiting, not only quadrilaterals (hexahedra in three dimensions), but quadrilaterals and triangles (hexahedra, pentahedra and tetrahedra in three dimensions) and then to apply the adequate local mesh generation method to each block (see for example the algebraic method previously described which is suitable both for quadrilateral and triangular shapes in two dimensions). The final mesh then consists of elements of the same nature as the initial blocks, and some structured features are still present in the final mesh.

This last approach offers a better flexibility than the previous in the sense that it is easier to cover a complex domain in two dimensions with quadrilaterals

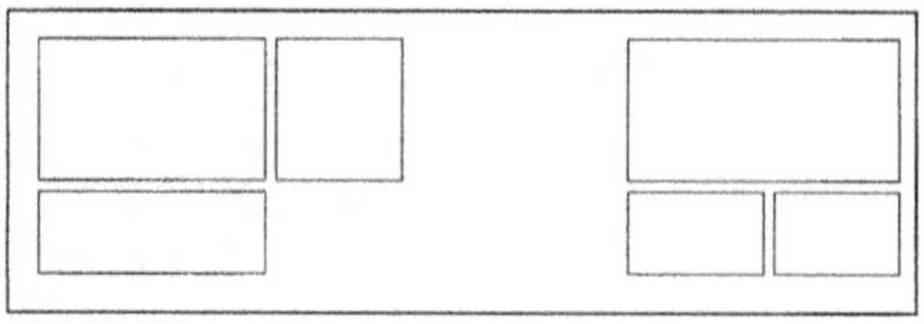

FIG. 12.1. Conformal and non-conformal decomposition.

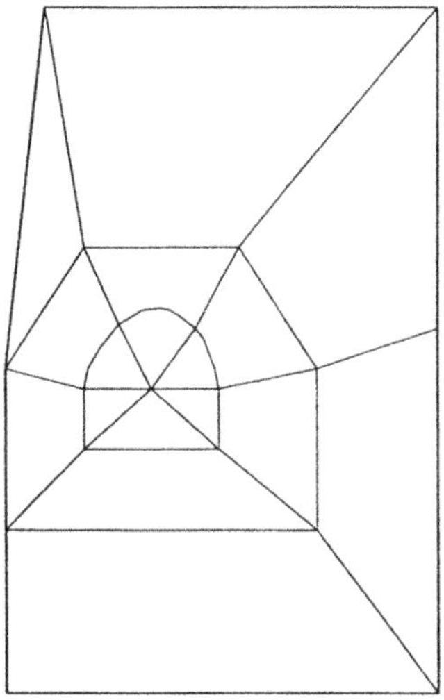

FIG. 12.2. Data.

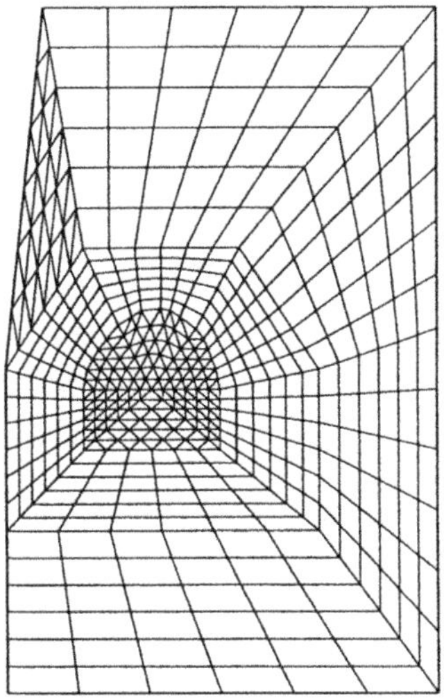

FIG. 12.3. Resulting mesh.

and triangles than with quadrilaterals only, this property is also valid in three dimensions where tetrahedra, pentahedra and hexahedra can be used instead of only hexahedra.

The main difficulty of any multiblock method is the construction of a suitable partitioning, for example, in order to enjoy a O-type, C-type or H-type structure in some regions. This task is not trivial, not unique and time consuming. Furthermore, it is usual to have to define a large number of blocks.

Anyway, multiblock methods can be a good solution to the mesh generation problem in some situations. In particular it is well suited for parallel computation or domain decomposition when solving the system resulting from the mesh created.

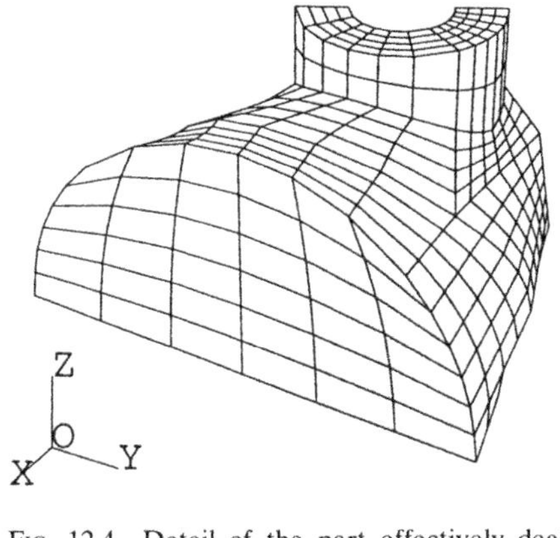

FIG. 12.4. Detail of the part effectively dealt with.

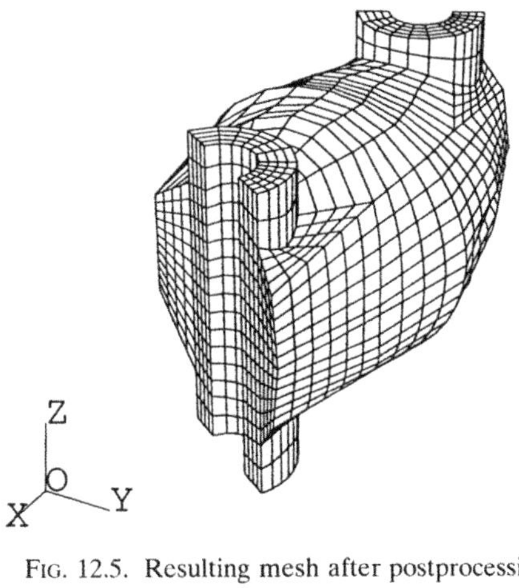

FIG. 12.5. Resulting mesh after postprocessing.

Figures 12.2–12.5 display application examples of a multiblock method in two and three dimensions respectively.

13. Automatic methods

Contrary to all previous methods, a method is said automatic if it requires no intervention from the user who only needs to provide the necessary data. In general, the data corresponds to a discretization of the boundary of the domain under consideration and, in some cases, include the data of all (or some) internal points.

Among the automatic methods, three classes of methods are encountered. The first type of methods relies on an easy to build decomposition covering the domain which is then updated in a proper way to obtain the final mesh. The second type of method relies on the advancing-front technique while the third type is based on the Delaunay–Voronoï principle.

13.1. Quadtree–octree type methods

Assuming the data of a polygonal discretization of the contour of the given domain (the two-dimensional case is considered for a simplified discussion), it is easy to construct a *regular grid* enclosing this polygon. Then a possible mesh can be obtained from this grid by considering its cells. Firstly, the latter are classified as follows:

- those enclosed within the domain,
- those which are outside the domain,
- and those which are intersected by the contour or close to it.

Then the mesh generation process consists in:

- suppressing the cells which are exterior,
- slightly modifying the cells which intersect the given polygon and those close to the boundary,
- identifying as elements those cells which fall inside the domain.

This simple method is strongly dependent of the stepsize of the cells composing the initial grid chosen. This stepsize, which is constant within the grid, must be sufficiently small to capture the details of the given geometry and thus the method can require large memory occupation. In addition, the resulting mesh is nothing else but an approximate covering of the domain as the boundary is more and less accurately present in it.

A more elegant and flexible approach consists of using a quadtree or octree (in three dimensions) grid created so that it covers the domain. Then the mesh generation method follows the same principles as above.

The quadtree or octree structure is a special tree structure which is now introduced briefly.

A tree is a (non-empty) oriented collection of items, namely *nodes*, which are connected using a hierarchical relationship. Therefore, two nodes are connected if there exist one *edge* between them, or if there exists a *path* (i.e. a series of successive edges) from one to the other. Tree structures enables us to access any node by visiting the appropriate edges of the tree. Given a node, the nodes which are connected to it are called the *sons* while the initial node is called *parent*. Several types of tree structures are possible depending on the number of edges resulting from a given node. Special tree structures play an important role in the present context. Among them are the binary tree, the quadtree and the octree. A binary tree is a tree in which a node has exactly zero or two sons; for a quadtree, the structure used in this section, a node has exactly zero or four sons while for an octree, it has exactly zero or eight sons.

The main interest of trees is the fact that they provide an easy access to information. To illustrate this feature, let us consider a quadtree and point P with coordinates (x, y) (Fig. 13.1).

A binary index is associated with every cell in the quadtree. Thus, cell a in Fig. 13.1 is known via its index, $\text{ind}_a = 0011$, or by the couple, ind $x_a = 01$, and ind $y_a = 01$. Similarly cell b in the figure is defined by $\text{ind}_b = 100111$, or by the couple $(101, 011)$. Let n be the granulation of the quadtree, then coordinates of point P are scaled so that the quadtree runs from 0 to 2^n. Let p_x and p_y be the coordinates of P in this system, then p_x and p_y are a sequence of digits 0 and 1. In the example, we have:

$$p_x = 100\ldots,$$

$$p_y = 011\ldots.$$

From these indices, it is clear that P falls within the cell with index 100001. The cell at the right of P is then obtained as the result of the binary operation $100 + 001$ so that its index in terms of x is 101. Similarly, its index in terms of y is $011 = 011 + 000$, meaning that cell b is the cell to the right of P. Similarly,

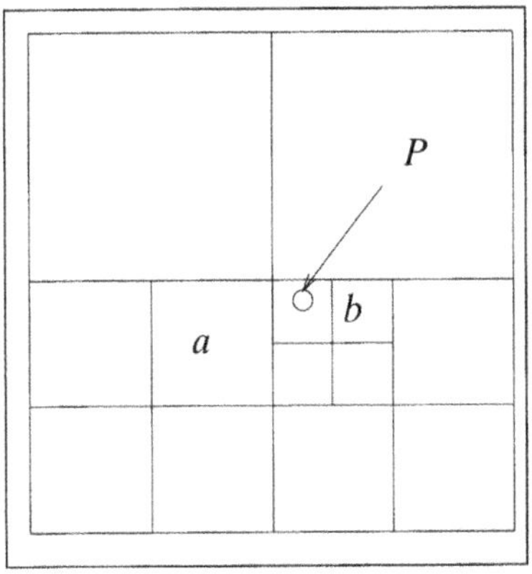

FIG. 13.1. Visiting a quadtree.

cell a is found to be on the left of P as result of the operations $100 - 001$ and $011 - 000$.

This kind of operation applies to all types of trees and is a source of efficiency as it is easy to locate a point in a tree, constructed for this purpose, and to find the information corresponding to a given region.

In the present context, YERRI and SHEPHARD [1983], SCHROEDER and SHEPHARD [1990], SHEPHARD, GUERINONI, FLAHERTY, LUDWIG and BAEHMANN [1988], the quadtree or octree structure is used to generate a mesh as follows (for simplicity, the two-dimensional case is assumed):

- the initial grid is a square or a rectangle such that the domain is enclosed in it and thus, is composed of only one cell or quadrant. This type of item will be referred to as a *quad* henceforth,
- the quads of the current grid are recursively split into 4 sub-quads, using the quadtree structure, until there is no quad containing more than one point of the given polygon (see Fig. 13.2) which approaches the boundaries of the domain.
- resulting quads are then considered as before. Thus, it leads to suppress the quads which are exterior to the domain, retain the quads clearly inside the domain which form quadrilateral elements or are split into two or four triangles (see Fig. 13.3), or finally, modify the quads which intersect the given polygon and those close to the boundary slightly by relocating their nodes in such a way that they lie on the boundary, or by creating the intersection point. The quads are then split into triangles according to the situation (same figure displays some possible solutions according to the number of nodes present on the sides of the considered cell and such that quadrilaterals may or may not exist. Note that variations and other patterns can be found easily).

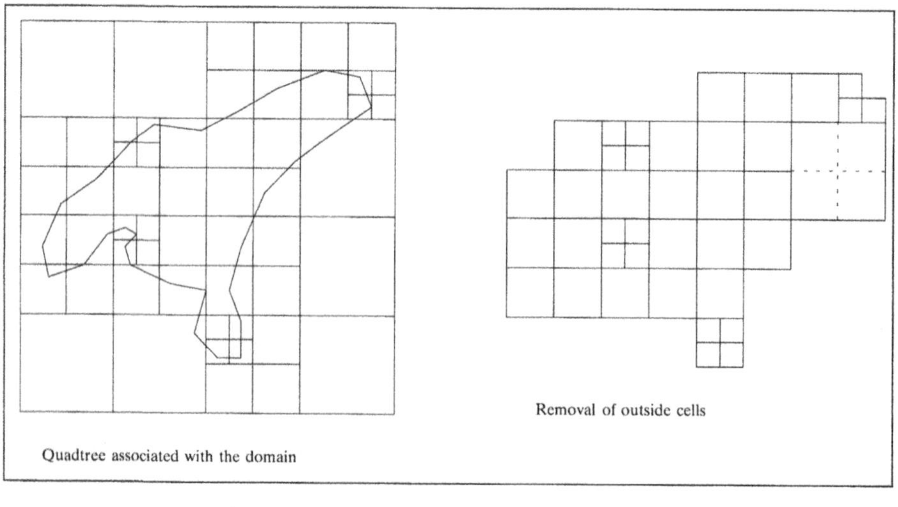

FIG. 13.2. Initial quadtree.

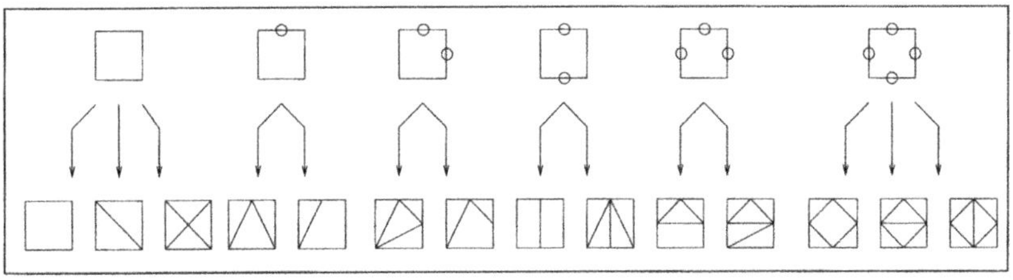

FIG. 13.3. Examples of local treatments of the quads.

From the above method, it is clear that the vertices of the final mesh are corners of quads or points defined by a quad boundary intersection. Thus, in general, the discrete boundary data is not present in the created mesh.

The flexibility of such a method is due to the possibility of varying the size of the quads which enable us to obtain large or small quads according to the details present in the given geometry. Furthermore, the number of quads is reduced and the memory occupation can be minimized.

The same technique applies in three dimensions by replacing the notion of quadtree by that of octree whose members are called octants. In this case, octants are partitioned into eight sub-octants, depending on the details that must be captured.

Typical examples of meshes created using the quadtree (octree) method are given in Figs. 13.4 and 13.5 (courtesy of Control Data). In these meshes, it can be seen that elements are mostly good shaped elements and that the maximal gap in size, from one element to the other, is about 2.

13.2. Advancing-front methods

A detailed description of an advancing-front method is given in Chapter III with a list of relevant references. In this section only an outlined presentation will be given.

A mesh generator of this type constructs the mesh of the domain from its boundaries. The elements created are triangles in two dimensions and tetrahedra in three dimensions.

The data required consists of the boundaries defined in terms of a polygonal discretization (dimension 2), i.e. a list of segments, or a polyhedral discretization (dimension 3), i.e. a list of triangular faces.

The process is iterative: a *front*, initialized by the collection of boundary items, is analyzed to determine a *departure zone*, from which one or several internal elements are created; the front is then updated and the element creation process is pursued as long as the front is not empty. The process can be summarized as follows:

- Initialization of the front;

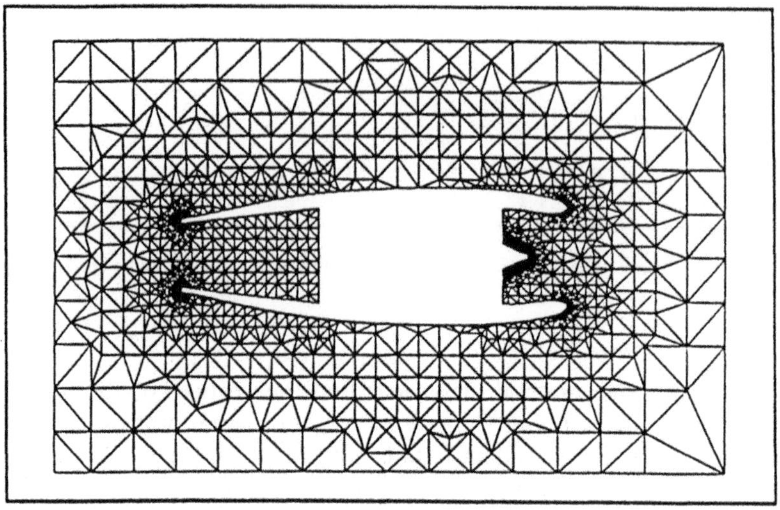

FIG. 13.4. Quadtree method.

- (a) Analysis of the front:
  - Determination of the departure zone;
  - Analysis of this zone:
    - * Creation of internal point(s) and internal element(s);
    - * Update of the front.
- As long as the front is not empty, return to (a).

The determination of the departure zone, as well as the analysis of the front and the way in which the elements are created, can be done in several ways.

The zone of departure can be chosen as:

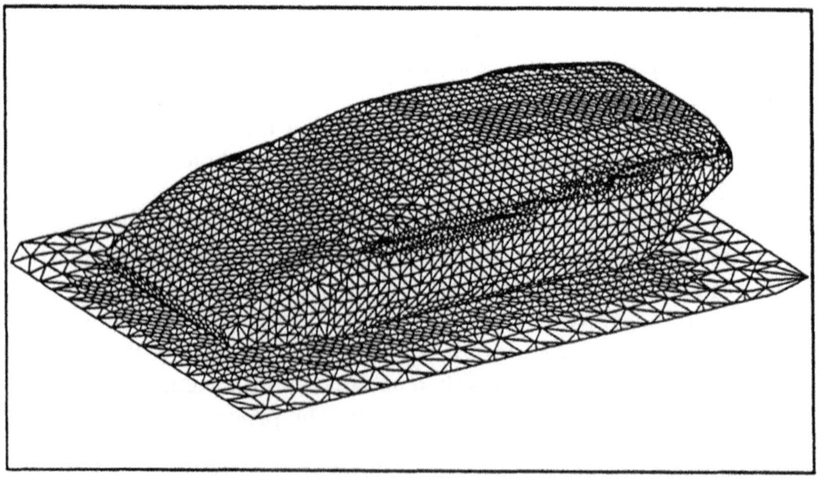

FIG. 13.5. Octree method.

- a portion of the contour where items (segments, faces, angles, ...) satisfy certain conditions;
- the entire front, i.e. its items, are considered in some defined order.

The first approach caters to the treatment, first of all, of particular zones (for example those with "small" angles). The second approach produces an inflation of the initial front or a propagation from an initial zone.

From the current state of the front, the front analysis is based on the examination of the geometrical properties of the segments (faces) that constitute it in terms of lengths, angles, and the present context. This analysis induces the construction of elements by connecting existing points or the creation of points and associated elements.

To illustrate this point, let us consider the 2-dimensional case and let α be the angle formed by two consecutive segments of the front. Then three situations or *patterns* are identified (Fig. 13.6):

- $\alpha < \frac{1}{2}\Pi$, the two segments with angle α are retained and form the two edges of the single triangle created (see same figure);
- $\frac{1}{2}\Pi \leqslant \alpha \leqslant \frac{2}{3}\Pi$, from the two segments with angle α, an internal point and two triangles are generated;
- $\frac{2}{3}\Pi < \alpha$, one segment is retained, a triangle is created with this segment as an edge and an internal point.

Other patterns slightly different in two dimensions, as well as suitable patterns in three dimensions, can be defined to make the mesh generation process efficient.

To update the front, one has to:

- suppress the edges of the former front which are included in the triangle just created,
- add to the front the edges of the triangles just formed which are not common to two elements,

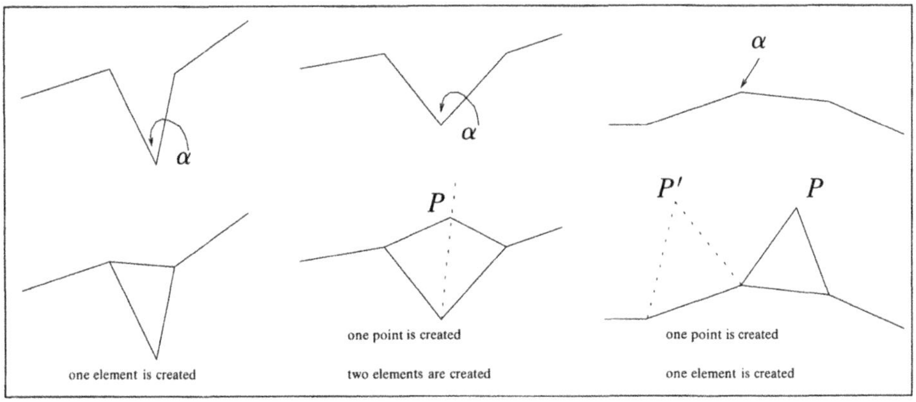

FIG. 13.6. The three retained patterns and associated constructions.

in two dimensions and,

- remove the faces belonging to the tetrahedron now formed,
- add those which are not shared by two elements existing now,

in three dimensions.

The main difficulties when implementing such a method are:

- determining the type of situation encountered when it is not evident,
- finding an adequate location for a point when such a point must be created, or deciding what point must be connected when all points are already known. Two constraints must be then satisfied: firstly the point must be internal and secondly it must be located at a proper place, or selected properly in the collection of existing points.

To ensure that a point is internal to the domain, one has to be sure that:

- all edges including this point do not intersect any edge of the front,
- no elements created intersect any edge of the front,

in two dimensions and,

- all edges including this point do not intersect any face of the front,
- all faces including this point are not intersected by any edge of the front,
- all elements created do not intersect any edge of the front,

in three dimensions. To find or locate a point, local properties (distances, angles, ...) are involved (see Chapter III).

Detailed discussion as well as numerous examples can be found throughout Chapter III.

13.3. Voronoï type methods

Chapter IV gives a detailed description of a Voronoï type method and lists numerous relevant references. Only an outlined presentation will be given in this section.

The aim of any method of this type is an insertion point process. The present discussion follows an incremental approach which consists of inserting one point in an existing mesh. A divide-and-conquer method can also be used which is based on the recursive decomposition of the problem into two problems of the same nature with a smaller size coupled with an algorithm which allow to merge the two meshes created in order to obtain the entire mesh. Possible in the case where all the points are explicitly known this approach does not seem suitable at the step the internal points are defined (see below). Popular in the context of computational geometry, the divide-and-conquer method is rarely used in finite element mesh generation. So, the insertion point process we would like to discuss in this article enters in the incremental approach.

Let $\{P_k\} = \{P_1, P_2, ..., P_n\}$ be a set of n two- or three-dimensional distinct points, and T_{old} a Delaunay triangulation whose elements are denoted by K_j

and whose element vertices are the first i points of $\{P_k\}$, then triangulation T_{new} can be derived from triangulation T_{old} in such a way that point P_{i+1} is one of its element vertices.

Introducing the set $\mathcal{B}_{old} = \bigcup_j B_j$ where B_j is the circumcircle (the circumsphere) of element K_j of T_{old} then point P_{i+1} necessarily falls into one of the three following situations:

Case (a) $P_{i+1} \in T_{old}$, i.e. $\exists K_j \in T_{old}$ such that $P_{i+1} \in K_j$,

Case (b) $P_{i+1} \notin \mathcal{B}_{old}$, i.e. P_{i+1} is not in any circle (sphere) B_j,

Case (c) $P_{i+1} \notin T_{old}$ but $P_{i+1} \in \mathcal{B}_{old}$, i.e. P_{i+1} is not in an element K_j of T_{old} but is in a circle (sphere) B_j.

A constructive method for creating the triangulation T_{new} corresponds to each of these cases. For example, in situation (a) and assuming the 2-dimensional case, the new mesh obtained after insertion of point P_{i+1} is defined by:

$$T_{new} = (T_{old} - \mathcal{S}) \cup \{F_j, P_{i+1}\}_j \quad 1 \leqslant j \leqslant p, \tag{13.1}$$

where $\mathcal{S}$ is the set of elements of T_{old} whose circumcircle contains point P_{i+1} and $F_1, \ldots, F_p$ are the external edges of this set.

T_{new} contains the first $(i+1)$ points of set $\{P_k\}$ as element vertices and is obtained by replacing the elements of $\mathcal{S}$ by new elements created by joining point P_{i+1} to the external edges F_j, enumerated in such a way that the element surfaces are positive.

In Cases (b) and (c), T_{new} is obtained following a similar construction (see Chapter IV). The insertion process is initialized by creating a first simplex by selecting 3 (4) points in set $\{P_k\}$ which are not colinear in two dimensions and not coplanar in three dimensions. Then points of $\{P_k\}$ are inserted one-at-a-time. The final triangulation T_{new} which includes all the P_k is that of the convex hull of set $\{P_k\}$ and is Delaunay.

The application of such an insertion point process for the creation of a mesh of a given domain consists of the following phases:

(1) Creation of the set of points associated with the data, i.e. the boundary points (the endpoints of the segments (triangles) discretizing its boundary) and the field points if they are known.

(2) Calculation of the location of four (eight) supplementary points in such a way that the quadrilateral (cuboïd) so formed contains all the points in the set.

(3) Creation of the mesh of this box (the box is formed by a quadrilateral, in two dimensions, but may be constituted by a single triangle. Variations exist which result in other choices, the only property desired to facilitate the process is that the box is convex) using two triangles (five tetrahedra).

(4) Insertion, one-at-a-time, of the points of the initial set to obtain a mesh that includes these points as element vertices.

(5) Regeneration of the boundaries and definition of outside elements.

(6) Creation of internal points (if necessary) and their insertion.

(7) Removal of outside elements.

(8) Smoothing processes.

Phases (2) and (3) are optional and can be performed to simplify the method. After these operations, only Case (a) of the insertion point process is encountered. Phase (4) relies on the application of the insertion points process as described above. Phase (5) is necessary in most cases as the current mesh contains as element vertices the endpoints of the given segments (triangles) of the boundaries but *does not* contain necessarily these items. It comprises two parts: the first consists in regenerating the boundary items in some sense after which the second phase consists in defining outside and inside elements. Phase (6) is obviously required if the current mesh only has the points of the given boundary as vertices (when the internal points have not been provided initially), the creation of the internal points can be achieved using different techniques after which they are inserted using the insertion point process described previously. Phase (7) consists of the classification (after phase (5)) of elements with respect to the boundary of the domain. Finally phase (8) is generally executed to obtain better shaped elements.

For a complete discussion and application examples, the reader is referred to Chapter IV.

14. Surfaces

Surface mesh generation is a crucial aspect of the mesh generation problem. In the case of three-dimensional problems, the surface mesh constitutes the natural data for the mesh generation algorithms which require a discretization of the boundaries of the domain as data (see Chapters III and IV). Thus, surface meshes influence the meshes which will be created through two aspects:

- the manner in which the surface is approximated by the surface mesh,
- the quality of the elements created close to the boundaries which include one of the items of the surface mesh as a face, as well as that of all the elements created.

The rest of this section attemps to give some keys about surface mesh generation by listing the problems which arise immediately. It does not claim for completeness.

The creation of a surface mesh is a delicate problem which is more and less correctly solved within CAD-CAM software packages which include this option. The main difficulty lies in the fact that surface description and surface mesh generation are two distinct problems. They are strongly connected but they do not have the same constraints and purposes.

The most popular systems for surface description provide a description of a surface consisting of patches lying on different kinds of interpolation (Splines, Bezier, NURBS, etc...). The number of patches can be very large. Patches have been defined solely in terms of surface approximation: it means that for some

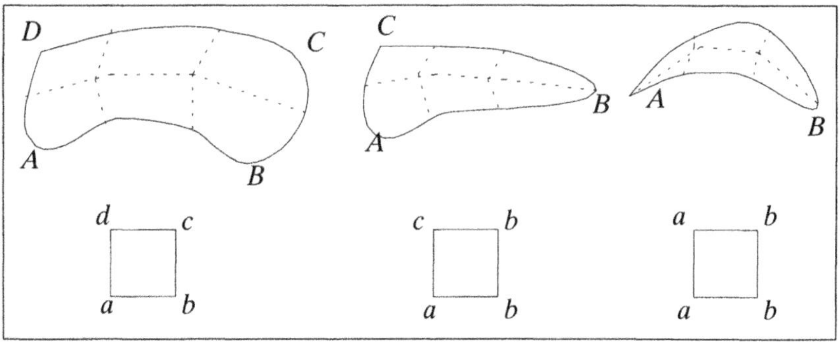

FIG. 14.1. Different types of patches.

geometrical reasons patches can be very different in size, degree of interpolation and, in addition, patches can overlap.

By itself, a patch can be a generalized quadrangle (including identical points) with constrained lines or specified points serving as geometrical support which must be preserved in the surface mesh (see Fig. 14.1).

It can also be part of a patch only (a face or a restricted patch), i.e. the domain with geometrical support the region defined as the part of this patch delimited by some given lines (see Fig. 14.2).

Thus, it is clear that surface mesh generation is a complex problem which cannot be reduced to a loop over the patches describing it. In particular it appears that surface generation algorithms need to use the mesh description by itself (it is not reasonable to develop surface mesh algorithms independently).

Nevertheless, surface mesh generation algorithms exist, such as:

- local algorithms designed to deal with a given patch,
- more global algorithms able to deal with a collection of patches enjoying some particular properties.

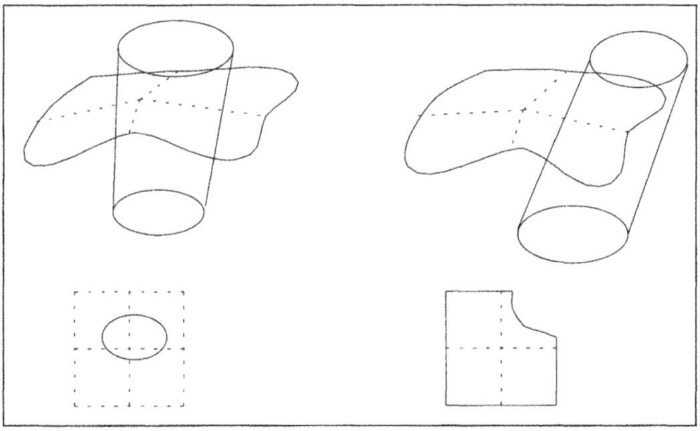

FIG. 14.2. Different types of restricted patches.

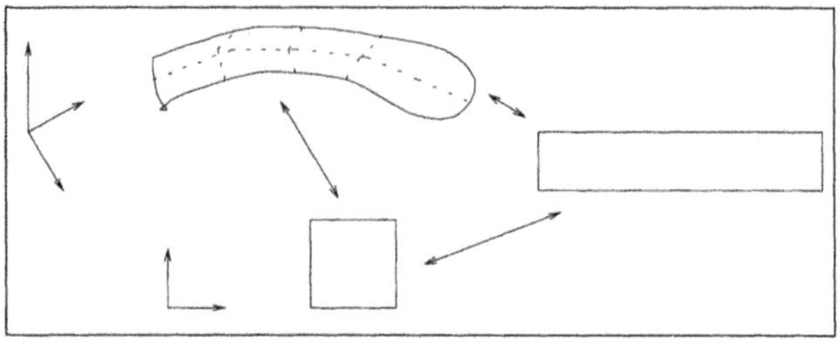

FIG. 14.3. Real patch, u, v patch and "metric" patch.

For a quadrilateral patch, a local algorithm generally consists of using an algebraic method (or an equivalent method) with respect to the variables u, v serving at the description of the patch under consideration. The implementation needs to evaluate the distance from point to point on the real domain to govern the generation method applied in the u, v space. A mesh is created in the u, v space after which it is projected onto the real domain, see Fig. 14.3.

For a generalized patch, a local algorithm can be based on any of the automatic methods valid in the u, v space (see the methods described in the present chapter for the two-dimensional case) with a close attention paid to the effective metric. The thus created mesh is then projected onto the real surface.

Global algorithms can be of the above type. To use this type of approach, it is firstly necessary to group together a certain number of patches according to the geometrical properties they enjoy at their interfaces (see Fig. 14.4). In this case, the initial patch interfaces may or may not exist in the resulting patch.

To be suitable, all of these methods require some properties regarding the surface patch description. In particular, it is necessary that patches form a conformal covering-up of the entire surface. When this property is not achieved, a special treatment must first be applied to attain it.

It seems that CAD-CAM software packages with surface mesh options do not include automatic mesh generators but a large list of algorithms the user must use to generate *and adjust* the mesh interactively until the entire surface mesh is completed correctly. Among these algorithms we find gathering tools to assemble several patches, local and global mesh generation methods, projection processes, topological and geometrical transformations and powerful visualization tools to help the user to check the different steps of the mesh generation process.

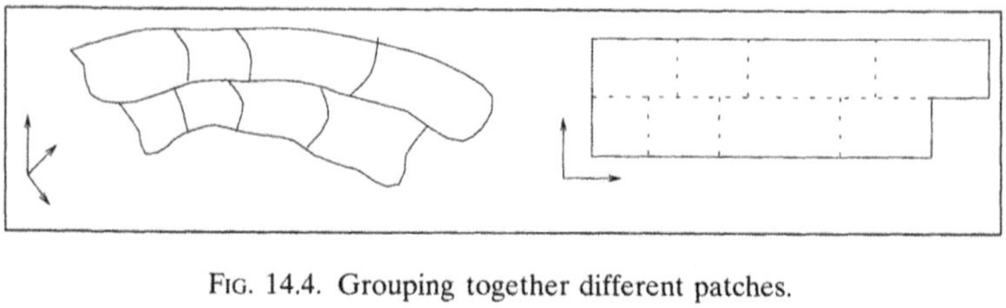

FIG. 14.4. Grouping together different patches.

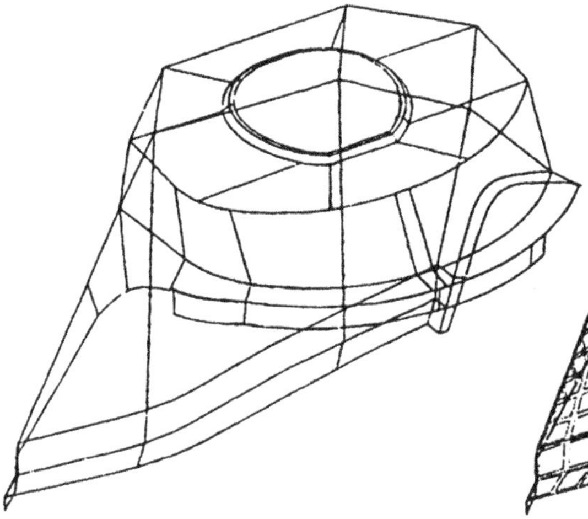

FIG. 14.5. Surface definition.

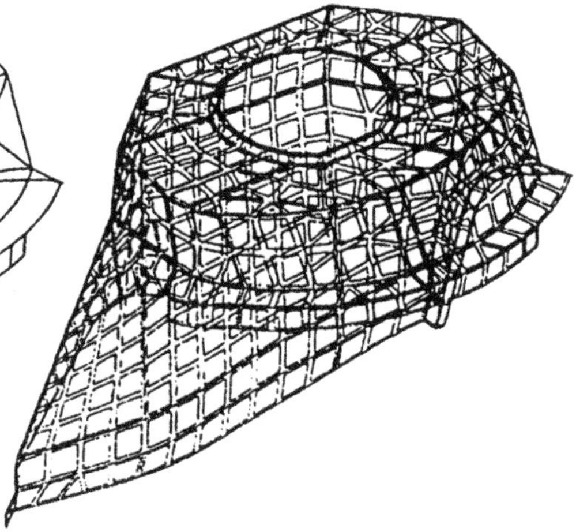

FIG. 14.6. Surface mesh.

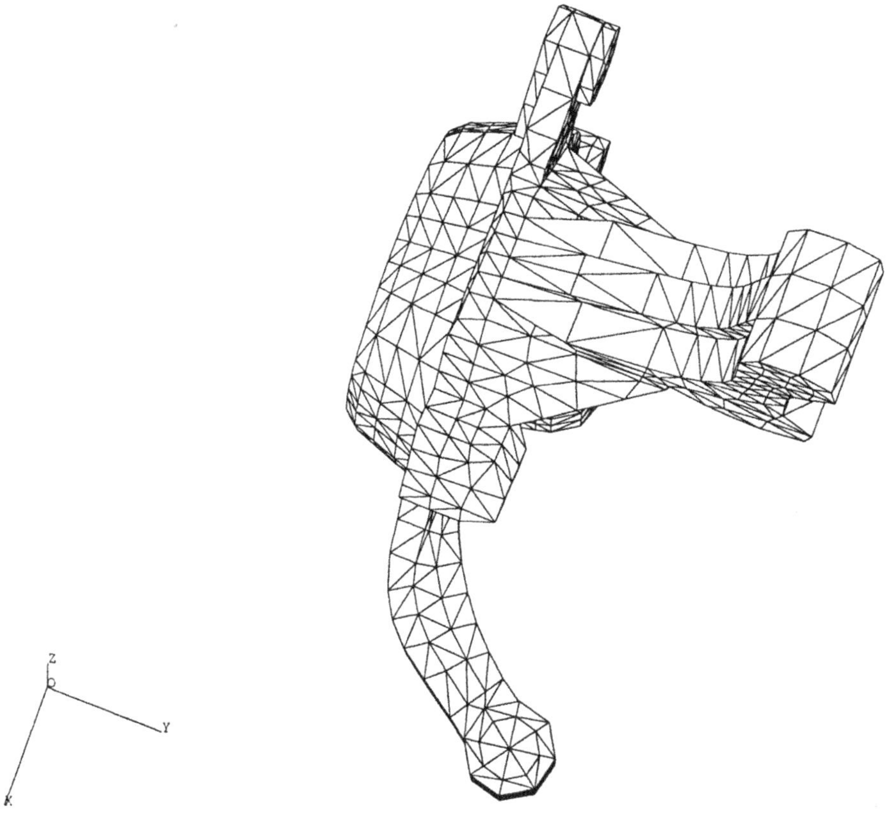

FIG. 14.7. Surface mesh.

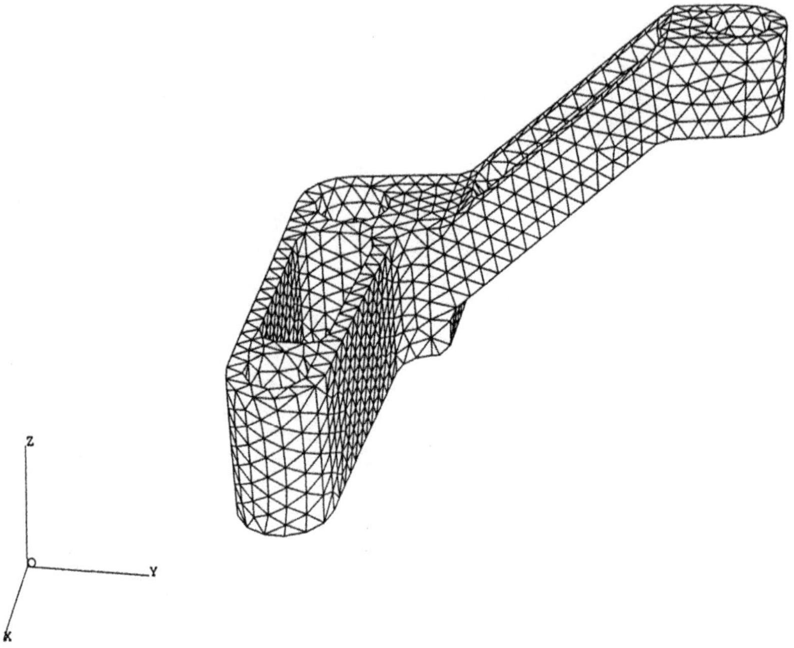

FIG. 14.8. Surface mesh.

Figure 14.5 shows an example of a surface definition including patches with different sizes while Fig. 14.6 illustrates the corresponding surface mesh (from Strim100, the CAD-CAM system by Cisigraph).

Figure 14.7 (courtesy of Cisigraph) and Fig. 14.8 (courtesy of SDRC) illustrate two other examples of surface meshes created by the two mentioned CAD-CAM systems.

CHAPTER III

Automatic Method (1): Advancing-front Type Mesh Generation

Introduction

This class of mesh generators, adapted to arbitrary geometries, has been studied by GEORGE [1971], LÖHNER and PARIKH [1988a,b], PERAIRE, PEIRO, FORMAGGIA, MORGAN and ZIENKIEWICZ [1988], and, more recently, by GOLGOLAB [1989]. This kind of mesh generator constructs the mesh of the domain from its boundary. The elements created are triangles in two dimensions and tetrahedra in three dimensions. Variations exist, in two dimensions, which enable us to create quadrilaterals almost everywhere in the domain, depending on the number of sides forming the given boundary (see LO [1991] and ZHU, ZIENKIEWICZ, HINTON and WU [1991]).

The data required consists of the boundary, or more precisely, a polygonal discretization of it (dimension 2), input as a set of segments, or a polyhedral discretization (dimension 3), input as a set of triangular faces. Additional items (points, edges or faces), which must be considered as prescribed items and thus which must be present in the created triangulation, can be included in the data.

The process is iterative: a *front*, initialized by the set of given items (those describing the boundary and the specified items, if any), is analyzed to determine a *departure zone*, from which one or several internal elements are created; the front is then updated and the element creation process is pursued as long as the front is not empty.

The analysis of the front and the way in which the elements are created can be done in several ways, which will be recalled later. Section 15 attempts to give a survey on the classical meaning of the advancing-front method. Possible variations, necessary steps, as well as difficulties will be mentioned. Section 16 presents a new method combining the Voronoï approach and the advancing-front technique. It will be shown that this method permits us to avoid the classical difficulties of the advancing-front mesh generation algorithms, as well as enabling us to consider the situation in which a given control must be used to govern the mesh generation process (Section 18).

Section 19 depicts some application examples while Section 20 indicates some details and indications concerning the practice of this kind of mesh generators by including information about data structures and effective algorithms.

15. General principles

The process of any advancing-front method can be summarized as follows:

- Initialization of the front;
- Analysis of the front by:
  - Determining the departure zone;
  - Analyzing this region by:
    - * Creating internal point(s) and internal element(s);
    - * Updating the front.
- As long as the front is not empty, go to "Analysis of the front".

The determination of the departure zone, the analysis of the front and the way in which the elements are created can be done in several ways. With each choice a particular advancing-front method is associated. The departure zone (see Fig. 15.1) can be chosen as:

- a portion of the current front where some properties are satisfied (length of consecutive segments, size of faces, angles, ...);
- a front defined iteratively as the translation of a set of items of the initial contour;
- the entire current front.

The first approach caters, first of all, the treatment of particular zones (for example those with "small" angles or "small" edges (faces)). The second choice induces a propagation from an initial zone, while the third produces an inflation from a given contour. It is not clear how to define an optimal strategy optimal in all geometrical cases.

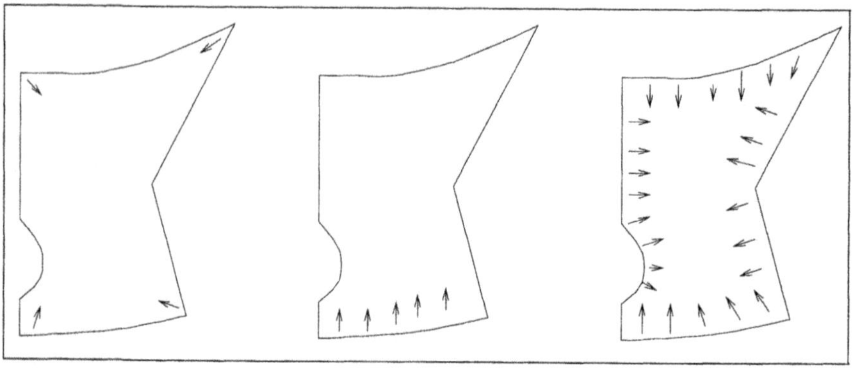

FIG. 15.1. Three types of departure zones.

Whatever the choice, the retained items are analyzed in terms of geometrical properties with regard to the local tools available for the creation of points and elements. For example, GEORGE [1971], the first reference in this domain, lists a certain number of patterns (see Fig. 13.4 in Chapter II) and finds if there exist such configurations in the examined zone. A point and element construction is associated with each of these cataloged patterns. After a construction, the front is updated and the process is repeated as long as the front is not empty.

The difficulties when considering such an algorithm are well identified. They consist of:

- determining the type of situation which governs the type of construction;
- knowing if a point is inside the domain or not;
- defining a suitable position when a point is created. Three constraints must be satisfied: the created point must be inside the domain; the element(s) resulting from this point creation must be well-shaped and it (they) must be such that the remaining zone can be dealt with without difficulties at a later stage.

In practice we have to answer the following two questions: *where are we* and *what do we have to do* in the place we are? The first question means that we need to check, at each state, that every point under consideration (Fig. 15.2, right-hand side) is inside the domain, and need to know if there is any existing point, edge or face that can possibly interfere with the envisaged construction. The second question means that we want to decide if the computed position of the so created point is optimal with regard to some neighborhood. A typical example of an ambiguous situation results from the junction of elements which are very different in size and must be joined. This configuration appears when the initial discretization of the contour includes segments with very different lengths (Fig. 15.2, left-hand side). The traditional advancing-front method leads to merge two fronts of widely differing length scales and, thus, may generate cross-over or, at least, badly shaped elements. Dealing first with "smallest" items (in terms of an adequate definition of the departure zone) and creating non-equilateral elements is a method to avoid this difficulty in some cases. Another

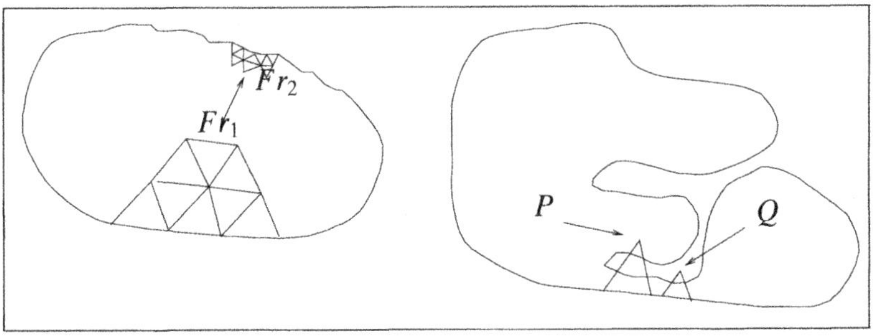

FIG. 15.2. An example of a delicate situation.

solution can be used to solve this problem: the method includes a step that enables to remove some elements created previously when they cause trouble at a later step. Obviously, something is conceptually missing: a control space, i.e. a way to govern the process by indicating, in some way, what the optimality is in the zone under consideration.

The method discussed in the following sections aims to give a new way of considering the advancing-front technique, while avoiding all the trouble mentioned above by introducing a control space and a neighborhood space. It means that we assume that we have a fast way of deciding, in any region of the domain, what to do. As a consequence, while the general scheme remains similar to that of the classical approach, the above difficulties disappear.

16. Creation and insertion of an optimal point

Let us introduce a control space and a neighborhood space. These structures will be described in details later on and are only outlined here. The control space (see also Chapter I) consists of a covering-up of the domain with which, at each point of the domain, a stepsize (denoted by h_{loc} in that follows) is associated. Similarly, the neighborhood space is a covering-up of the domain with which pointers are associated. These pointers are used to know what the items located in some neighborhood of any examined region are. Thus, it is now possible to perform all the steps of the advancing-front method, so revisited, without difficulty. These steps are discussed below.

16.1. Position of a point with respect to the domain

One of the problems arising when implementing any advancing-front is to decide quickly if a given point is inside the domain or not. In practice, to ensure that a point is internal to the domain, one has to verify that:
close up space

- all edges including this point as endpoint do not intersect any edge belonging to the front,
- all created elements do not include any point,

in two dimensions (see Fig. 16.1), and

- all edges including this point as endpoint do not intersect any face belonging to the front,
- all faces including this point as endpoint are not intersected by any edge belonging to the front,
- all created elements do not include any point,

in three dimensions.

The previous properties involve a priori only the items of the current front and not all the existing items. In fact, only part of the front needs to be considered

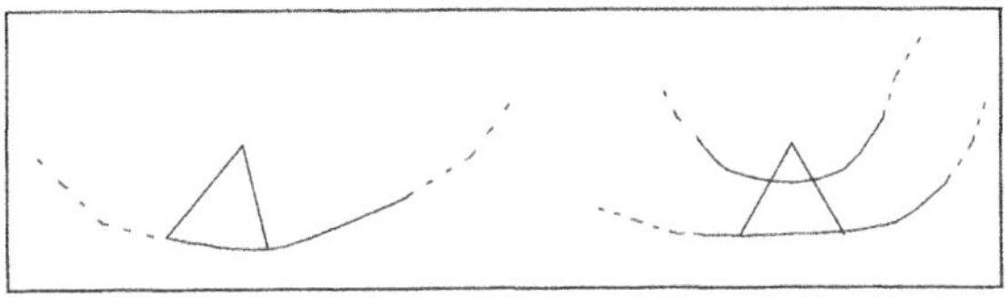

FIG. 16.1. Position of a point with respect to the domain.

and the decision is therefore obtained rapidly. To define the part of the front which must be visited, we use the control and the neighborhood spaces.

16.2. *Finding an optimal location*

With regard to the manner in which the current front is analyzed and the nature of the departure zone, front items are dealt with in a predefined order. Assume that we examine a front item (edge or face according to the space dimension) obtained by one method or another, then we are able to define an optimal point very easily. The process is as follows (for simplicity, we first examine the two-dimensional case and denote by AB the front edge under consideration (Fig. 16.2), after which the three-dimensional case will be discussed briefly):

- point C is constructed so that triangle ABC is equilateral,
- $h_{\mathrm{loc}}(C)$ is computed using the control space: according to h_{loc} and the current distances from C to A and B, point C is relocated on the mediatrice of AB in such a way that these three values become equal,
- (a) points of the front other than C, falling in some neighborhood, are examined in order to decide if they can be connected with AB in such a way that triangles ABP_i, where P_i denote these points, are valid. The retained points form the set $\{P_i\}, i = 1, n$,

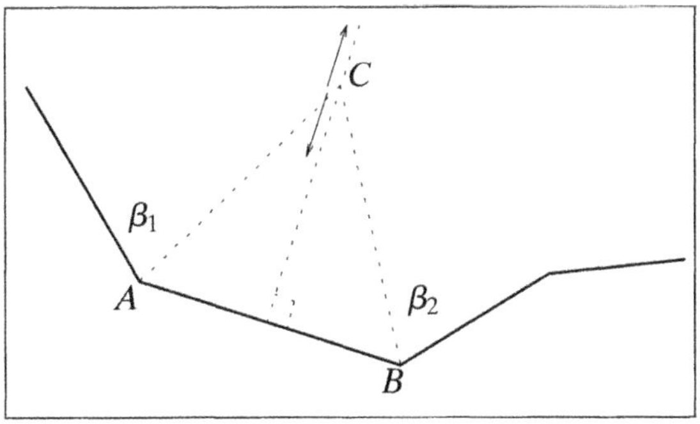

FIG. 16.2. Optimal point in two dimensions.

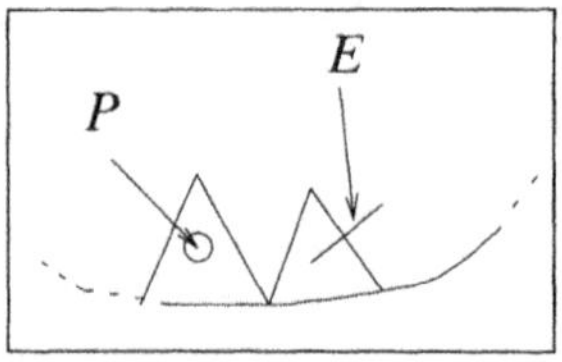

FIG. 16.3. Triangles in test and isolated items.

- this set of points is sorted with respect to the distance from C, after which point C is placed at the end of the so formed set,
- the previous set is now visited: we pick the points in the set (starting with the first) and, considering one of them (namely P_k), we check if P_k is inside the domain (see above discussion), if triangle ABP_k is positive and empty and, in addition, if angles β_1 and β_2 are sufficiently large. For example, in Fig. 16.5, a case is depicted where the optimal point C is the only point in the list but will cause trouble furthermore. In this case, r, the size defining the examined region, is increased until a point P is found which can become a member of the list and can then be retained. As soon as a point verifies all these properties, it is retained. In the case where no point is selected, we return to (a) while the size of the visited neighborhood (r in Fig. 16.4) is increased,
- triangle ABP_k is formed (see below discussion) and the front is updated.

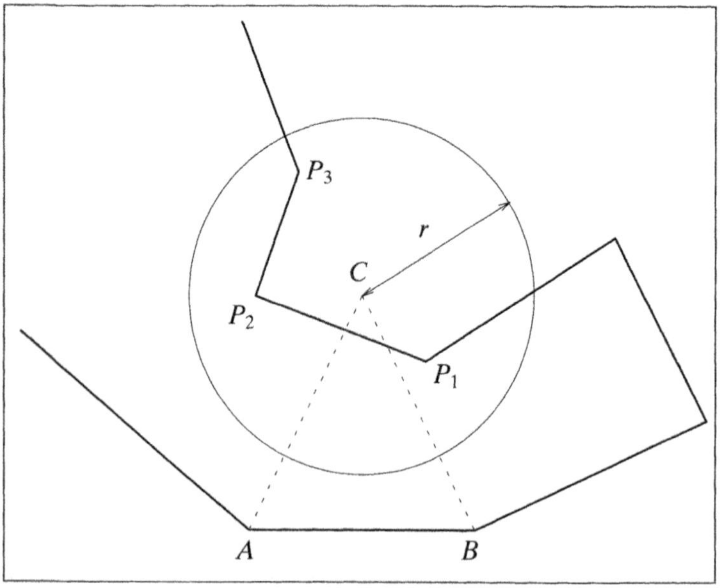

FIG. 16.4. Candidates.

Angles β_1 and β_2 are used to ensure that the element in creation will not cause trouble latter (Fig. 16.5). The created triangle must be empty (see Fig. 16.3), i.e. no specified point must be contained in it and no specified edge must intersect one of the constructed edges. This latter requirement is necessary if the data include some specified items other than that of the contour discretization: while a specified edge may be included in the initial front so that this additional treatment becomes useless, this is not the case for specified points, if any, which must be taken into account in this way.

The above principle formally extends to three dimensions. In this case, the method consists in constructing the optimal point associated with the triangle of the current front under consideration. Such a point is said optimal if the tetrahedron formed by joining this point with the triangle serving as support is well shaped. Thus the method similarly involves the following steps:

- point D is constructed so that tetrahedron $ABCD$ (where ABC is the front triangle under examination) is equilateral: let G be the centroïd of triangle ABC, p the perimeter of this triangle and $\vec{n}$ the inward normal, then (Fig. 16.6) D is constructed on $\vec{n}$ at distance $p/3$ from G,
- $h_{\text{loc}}(D)$ is computed using the control space: according to h_{loc} and the current distances from D to A, B and C, point D is relocated in such a way that these values become equal,
- (a) points of the front other than D, falling in some neighborhood, are examined in order to decide if they can be connected with ABC in such a way that tetrahedra $ABCP_i$, where P_i denote these points, are valid. The retained points form the set $\{P_i\}, i = 1, n$,
- this set of points is sorted with respect to the distance from D, after which point D is placed at the end of the so formed set,
- the previous set is now visited: we pick the points in the set (starting with the first) and considering one of them (namely P_k), we check if P_k is inside

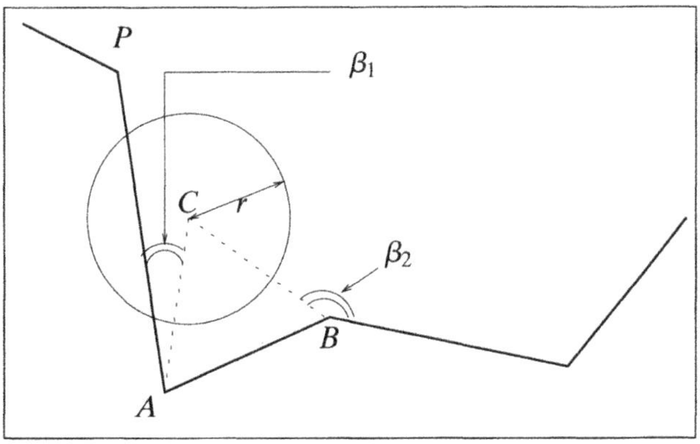

FIG. 16.5. Case where no "natural" candidates other than C exist.

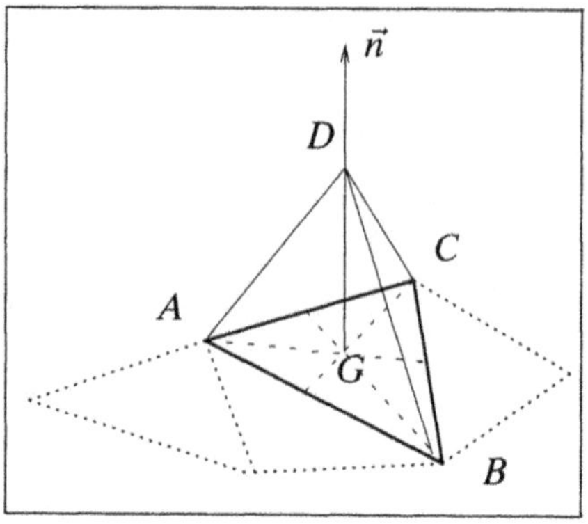

FIG. 16.6. Optimal point in three dimensions.

the domain (see above discussion), if element $ABCP_k$ is positive and empty and, in addition, if the region delimited by this element and the neighboring faces of the front remains well shaped. As soon as a point verifies all these properties, it is retained. When no point is selected, we return to (a) while the size of the visited neighborhood is increased,

- tetrahedron $ABCP_k$ is formed (see below discussion) and the front is updated.

Contrary to the two-dimensional case, an additional difficulty appears in three dimensions in the case where an existing point is selected. The problem lies in the fact that the polyhedron containing the current front triangle is also limited by existing face(s) and, thus, could be impossible to mesh (see Fig. 23.11). This situation is detected at the step the position of the point is examined (see above) as new edges are exactly contained in an existing face. The solution consists in creating a new point, probably not optimal in terms of position, so that the problem disappears.

Due to the method in which the points are created, the method prefers to use an existing point rather than to create a new one. This remark leads to minimize further delicate situations.

16.3. Insertion

The effective insertion of the point defined in the previous step consists in creating *only one element* by connecting the point with the item of the front from which it has been constructed. The only thing to ensure is that the surface (the volume) of the element created is positive.

In some situations, it is clear that several elements, and not only one element, may be created when considering a point. In the present discussion, the way in which a point is defined ensures that the two methods are equivalent and, as a consequence, the scheme, involving only one case, is simplified.

16.4. Updating the front

To update the front in two dimensions, one has to:

- suppress the edge of the former front which was used to define the point just created, in the case where such a point has effectively been constructed, or suppress the edges which are included in the triangles just created in the case where an already existing point has been selected as an optimal point,
- add to the front the edges of the triangle(s) just formed which are not common to two elements,

and, in three dimensions:

- remove the faces of the former front included in a tetrahedron now formed,
- add those which are not shared by two elements existing now.

Once the front is empty, the method has converged and the mesh of the domain is completed.

16.5. Control space

The control space is a way to associate a stepsize all over the domain easily. The reader is referred to Chapter I where this notion is introduced in a general context. The control space consists of a control structure, i.e. a covering-up of the domain with which a function defining the desirable size in each point is associated. Several choices can be made to construct the control structure (Chapter I). An example is given below depicting such a choice for the case where the only data at our disposal is the contour discretization of the domain.

Thus, the control structure is defined as a regular grid enclosing the domain. A cell in this grid has a size Δ_c depending on the size of the initial data items. For example, in two dimensions, a solution is to set Δ_c to be twice the length of the smallest edge in the data. This value may induce a too large number of cells so an alternative solution is to set Δ_c to be the average length of the given edges. When such choices lead to problems, the solution is to define the control structure as a quadtree (octree) structure which offers the desired flexibility (see also Chapter II, the section devoted to the quadtree or octree mesh generation method).

We have now to define the function associated with the control structure and, more precisely, with the cell vertices. The problem is, again, that the only data known consists of the boundary discretization. While it is easy to find the grid cells intersecting one edge of the data and then define a desirable stepsize by means of interpolation based on the size of these edges, it is not easy to define this stepsize for the other cells. This is the reason why we first use the first steps of a Voronoï mesh generation method to obtain the boundary mesh of the domain. This boundary mesh (see Chapter IV) is defined as a mesh resulting from the insertion of the boundary points which are the only points we know at this stage. Once the triangles of this boundary mesh are constructed, we define the stepsizes at the grid cell vertices as follows (Fig. 16.7):

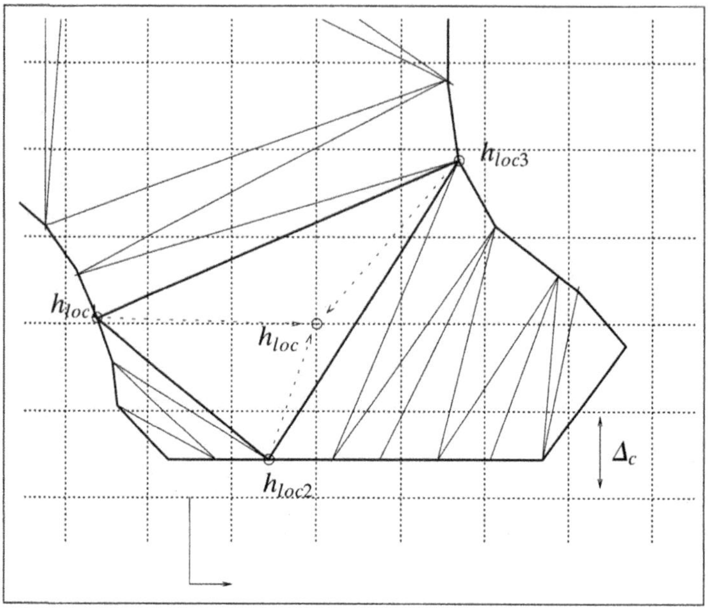

FIG. 16.7. Control space (construction).

For all cell vertices:

- find the triangle in the boundary mesh containing it,
- compute the desirable stepsize , h_{loc}, as the P_1-interpolate of the h_{loc} of the three vertices of the triangle.

On completion of this operation, we have fully defined the control space which can now be used (Fig. 16.8) in the following manner: let P be the point under consideration in the current construction, then $h_{\mathrm{loc}}(P)$ is computed as follows:

- find the grid cell within which point P falls,
- compute the desirable stepsize, $h_{\mathrm{loc}}(P)$, as the Q_1-interpolate of the h_{loc} of the four vertices of the cell.

When the data includes information other than the boundary discretization, similar processes can be used by effectively taking this information into account (see Section 18).

The above method for constructing the control space relies on the boundary mesh and clearly applies to three dimensions. Nevertheless, classical advancing-front methods are not conceived in this way. It means that they cannot take advantage of a boundary mesh to help the control space construction. In this situation, it is more delicate to develop a mesh generation algorithm able to deal with complex shapes. The main problem, which is not easy to solve, deals with the configuration shown in Fig. 15.2 (left-hand side) where elements with very

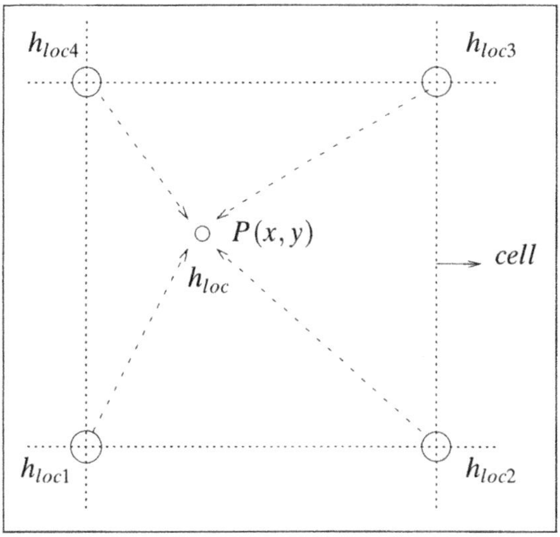

FIG. 16.8. Control space (utilization).

different size must be joined. To overcome this difficulty, the mesh generation algorithm can be modified as follows:

- at the definition of the departure zone step: it seems that the first encounter with small edges (in two dimensions) and small or badly shaped triangles (in three dimensions) leads to avoid the problem of joining two parts of the front including items with different sizes.
- at the element creation step: when only a badly shaped element can be constructed, the method must include a removal procedure able to suppress elements previously created in such a way that the present difficulty disappears (such a procedure must be carefully designed, in particular to avoid recursivity).

16.6. Neighborhood space

The neighborhood space is a way to easily find items (points, edges or faces in three dimensions) located in a given neighborhood of the region under consideration. Similar to the notion of a control space, the neighborhood space is composed of a covering-up of the domain (possibly identical to the previous) with which the useful information is associated. Several choices can be envisaged of which only an example is given.

The control structure is defined as a regular grid enclosing the domain. The cell of this grid has a size Δ_n which can be equal to Δ_c. We then construct a binary tree (see Chapter II, the section discussing the quadtree or octree mesh generation method) governed by the data points (points of the boundary

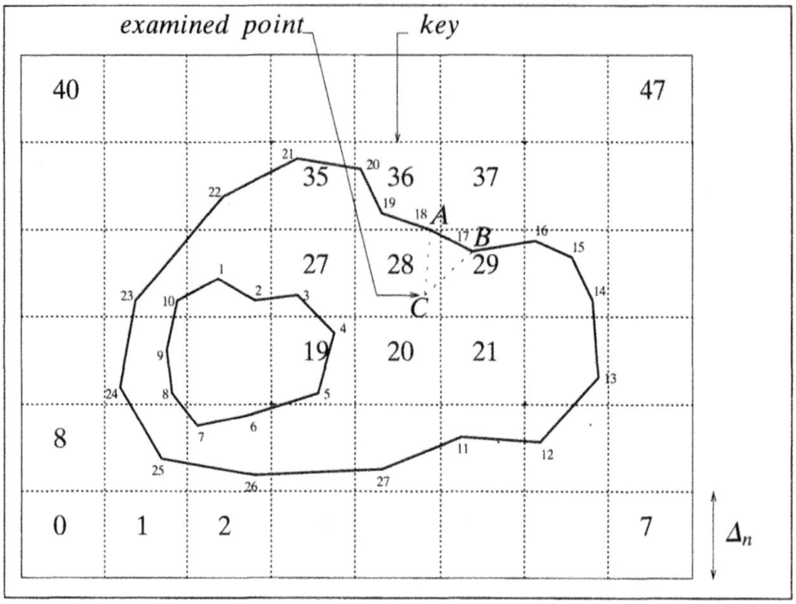

FIG. 16.9. Neighborhood space.

discretization and specified points if any). This construction is done as follows: for each point (see Fig. 16.9 as well as Fig. 16.10):

- find the cell n in which point P_i under consideration falls ($n = 26$ for point number $P_i = 1$);
- (a) consider the next point ($P_{i+1} = 2$) and find the cell containing it (26): if the cell number is n, classify P_{i+1} in cell n, else if the cell number is $p \neq n$, define the right branch ($p > n$) or the left branch (otherwise) attached to node n and classify P_{i+1} in this cell. Then set $i = i + 1$ and return to (a).

In this way, we obtain an ordered tree attached to the points whose nodes point out to the grid cells.

On completion of this operation, we have defined the neighborhood space which can now be used in the following manner: let P be the point under consideration in the current construction, then the points which must be examined are those pointed out in the cells located in the appropriate neighborhood of the cell containing the point used to define point P.

Due to the way the previous tree was created, this operation can be done very rapidly.

Variations exist to construct this neighborhood space but it seems that there are no major difficulties to create a suitable structure containing, in one way or another, the desired information accessible via a binary tree structure or a different basic structure (quadtree, octree, ...).

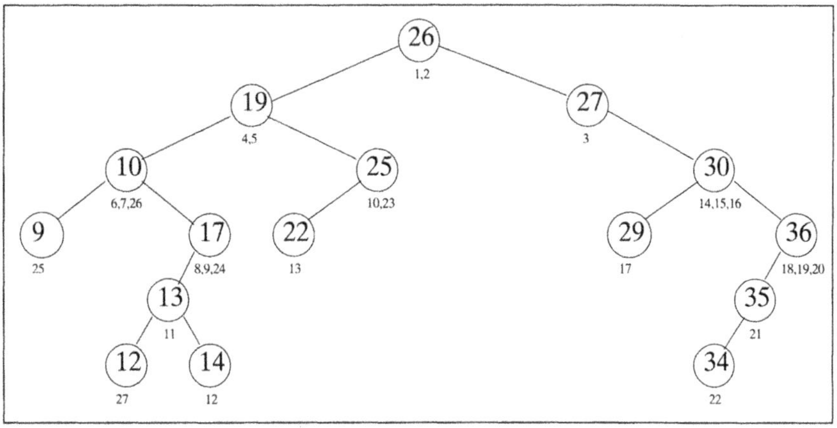

FIG. 16.10. Corresponding tree.

16.7. Mesh point and element smoothing

It can be of interest to smooth the triangulation resulting from the mesh generation method. This process can be based on two different techniques which can be coupled.

- Moving the points: this operation can be performed in three ways:
  Barycentrage: The set of points surrounding a given point P is determined; they correspond to the vertices of the elements sharing vertex P. Point P is then repositioned at the weighted barycenter of its surrounding points:

$$P = \frac{1}{n}\sum_{j=1}^{n} \alpha_j P_{k_j}, \tag{16.1}$$

  where n denotes the number of points connected to point P, P_{k_j} denote these points with numbers k_j, and α_j is the associated weight ($\sum \alpha_j = 1$).
  Relaxation: This variant consists of "relaxing" the above expression:

$$P^{m+1} = (1-\omega)P^m + \frac{\omega}{n}\sum_{j=1}^{n} \alpha_j P_{k_j}^m \tag{16.2}$$

  using identical notation with ω a relaxation parameter.
  Iterations governed by a criterion: Let P be a vertex of the mesh and Q be an associated measure (for example, the maximum of the quality (see Chapter I) of the elements with vertex P), then an iterative process for the displacement of P along an arbitrary direction is defined, where the amplitude is initialized to some given value. The displacement of P is simulated and the associated value Q is computed by successive iterations; the way in which it varies directs the process.

- Local modifications: The transformations described throughout Chapter IV, when discussing the boundary integrity problem, can be used to obtain better shaped elements. Note that such transformations are nowadays currently used in different contexts (see, e.g., FREY and FIELD [1991], GEORGE [1993]) to enhance the triangulation quality.

All of the above smoothing methods must be controlled to guarantee that the triangulation remains valid and that its quality is enhanced with respect to the given measure of element quality.

17. Scheme for the mesh generator

Results of the previous sections are used to propose the following steps as a possible automatic mesh generator based on the advancing-front method:

(1) Create the set of points associated with the data, i.e. the boundary points of the domain (the endpoints of the segments (triangles) discretizing this boundary) and the field points if they are known.
(2) Compute the location of four (eight) supplementary points in such a way that the quadrilateral (cuboïd) so formed contains all the points in the set.
(3) Create the mesh of this box using two triangles (five tetrahedra).
(4) Insert, using the method described in Chapter IV, the points of the initial set to obtain a mesh including these points as element vertices.
(5) Regenerate, using the same method, the boundaries and define the outside elements.

These steps are similar to the first steps of the scheme proposed in the case of a Voronoï type method (see Chapter IV) and enable us to define the useful control and neighborhood spaces we need in the present case easily and rapidly. Thus, the following steps can be envisaged:

(6) Create the control and the neighborhood spaces.
(7) Define the initial front.
(8) Define a departure zone, create an optimal point or find an existing point satisfying the suitable properties, insert the retained point to construct one or several elements and update the front.
(9) If the current front is not empty, return to (8).
(10) Apply smoothing processes.

Obviously Steps (1) to (5) must be suppressed when a background grid (see below) is known which enables us to construct the control and neighborhood spaces needed in the following steps of the algorithm directly. They must also be omitted when a different way to construct equivalent information is possible The steps have been indicated only as an example of construction. In particular, when a background space (i.e. the pair background grid and associated function) is known, it can be employed directly as a control space. The main difference lies in the way in which the information is accessible: in this case the "control structure" is a mesh of the same nature than the mesh in creation and the search problems are more complicated to solve as the members of the structure are not cells but triangles or tetrahedra.

18. Applications in the case of a given control

In this section we assume that we have information concerning the physical problem to be solved. For example, a first computation with a mesh obtained by the above method is done whose solution indicates that some regions present a stiff gradient that must be introduced into the mesh generation process in order to govern it. Another usual situation is that where an a priori knowledge of the behavior of the physical solution can be used to indicate that the mesh must enjoy some properties in some regions (finer or larger elements by specifying the mesh density, stretched elements by giving the stretching ratio and the direction of stretching, ...).

The first and simplest idea is based on the *h-method*, i.e. on local mesh enrichment of the mesh created by the above method. Enrichment is performed in regions with large solution error, large gradient, ... and may conform to some simple rules: a smooth change of element size, the maintaining of conformity, etc. In this respect, simplices are flexible as they are easy to split into sub-elements of the same simplicial nature using different manners. This technique has proved to produce a good adaptivity with respect to the physical behavior of the problem, but generally leads to meshes consisting of too many elements (so that the time required when computing the solution of the problem is too long).

A tentative of response to avoid this trouble is to simultaneously use a de-enrichment procedure in some regions. This task is not so easy to perform: for instance the only thing we can do in a mesh consisting of simplices is to delete a vertex shared by only 3 elements, in two dimensions, and 4 elements, in three dimensions. When it is required to remove a vertex shared by more than 3 (4) elements, local transformation procedures can be used to reduce the number of elements to which the vertex belongs. It is not clear when such reductions succeed or fail, and the quality of the so modified elements can be very poor.

In addition, we should note that the so created meshes are implicitly isotropic in the sense that no special directions are a priori involved.

The above naturally leads to consider the control and adaption problem in a different way and, in particular, not only locally by involving global *remesh* techniques governed by some criteria. In this respect, a mesh generator of the advancing-front type offers great flexibility in the way in which internal points are created by varying their number and location.

Let us first write down the internal point creation process formally. This process uses the control and neighborhood spaces defined previously. The neighborhood space being unchanged, we define the following pair as a control space:

- Δ a control structure enclosing the domain Ω under consideration;
- H a function such that $H(P, \vec{d})$ defined in Δ is the value to be verified at point P for a given direction $\vec{d}$.

In this way we have defined a *control space* (following Definition 5.2) which is more general than that previously defined.

To return to the above method for the creation of internal points, the following configuration can be chosen:

- Δ is nothing but R^2 in two dimensions and R^3 in three dimensions;
- $H(P, \vec{d}) = H(P)$ with

$$H(P) = \min_{M \in \Delta} f(P, M), \tag{18.1}$$

where f is the function defined over all points M, different from P in Δ by:

$$f(P, M) = \frac{h_{\text{loc}P} + h_{\text{loc}M}}{2}, \tag{18.2}$$

where h_{loc} is defined as before and $h_{\text{loc}M}$ is as follows:
– if M falls within Ω, $h_{\text{loc}M}$ is known,
– if M falls within $\Delta - \Omega$, $h_{\text{loc}M}$ is set to ∞.

Thus, any modification concerning the choice of Δ or H leads to the possibility of governing the creation of internal point P. Note that, in relation (18.1), H is isotopric as $\vec{d}$ is not used explicitly.

From a geometrical point of view, Δ is an arbitrary partitioning, for example, of one of the following types (see also Chapter I):

(a) a partitioning of the type "quadtree" in two dimensions, or "octree" in three dimensions;
(b) a regular partitioning (of finite difference type);
(c) a pre-existing mesh (for instance the current mesh);
(d) a pre-existing mesh specified by the user.

With space Δ a function H is associated, which can be:

- derived from the local h_{loc} associated with the points specified, using a generalized interpolation, (such an analysis is purely geometrical in the sense that it depends on the geometrical properties of the data: size of the faces, surfaces, lengths, etc.);
- defined manually, for each element of the covering-up Δ (for example, by specifying the desired stepsize in space—*isotropic* control—or along a direction—*anisotropic* control);
- specified manually by constructing the covering-up for this purpose (this case falls in type (d) seen above);
- defined from the physical solution (or from its gradient, etc.), or from an error estimate, ... evaluated at a previous computation;

- related, as in case (a) above, to the size of the cells which are used to encode the value of H, meaning that the partitioning was created so that some property holds.

Such a formal writing of the mesh generation problem using a control space (the pair "control structure, function") offers great flexibility and can be adapted to the physical problem under consideration without difficulty. This is done by defining an adequate control structure Δ properly, in accordance with domain Ω and such that function H can be constructed. To obtain H, a pertinent method is to select a *key variable*, σ, and to control a quantity related to this variable. PERAIRE, VAHDATI, MORGAN and ZIENKIEWICZ [1987] propose the computation of the interpolation error estimated for this variable with the purpose of equidistributing it. More precisely, a mesh is created in such a way that:

$$H(P, ...) = c_0 g(P, \sigma)$$

for a selected function H with c_0 a given tolerance and g the control function with respect to the variation of the key variable.

As an example of constructing Δ and H we follow VALLET, HECHT and MANTEL [1991] who consider the Navier–Stokes equations to model viscous flows with the velocity as key variable. In this situation H is constructed as follows:

$$H(P, \vec{d}) = \frac{\| X_d \|}{(X^t_{\vec{d}}\,\mathrm{Mat}(P) X_{\vec{d}})^{1/2}},$$

where

- $\mathrm{Mat}(P)$ is a matrix determining the metric associated with the desired control,
- $\| \cdot \|$ denotes the norm associated with the considered metric,
- $X_{\vec{d}}$ is the unit vector in direction $\vec{d}$.

Note that the metric is defined locally for each point P. With such equipment the control function g is given by:

$$g(P, \sigma) = \left| \frac{\partial^2 \sigma(P)}{\partial d^2} \right| .$$

The process of control is then summarized as follows:

- set $i = 0$,
- create mesh T_i of domain Ω using the mesh generation method involving only the geometrical properties of the data,
- (a) compute solution number i of the problem,
- define the control mesh to be T_i,
- compute function g of key variable σ and then function H,
- compare H and c_0 and create the required points,
- set $i = i + 1$,

- create mesh T_i,
- while convergence is not achieved, return to (a).

19. Application examples

The examples depicted in this section rely on geometrical considerations only. It means that the triangulation created is governed by the given boundary of the domain under consideration. The finer the discretization of the boundary is the finer the triangulation is in the corresponding regions. Figures 19.1 (courtesy of TELMA and Simulog) and 19.2 illustrate the triangulations obtained in this case.

Triangulation 1 consists of 4 440 elements, 2 495 vertices and its worst quality (i.e. the ratio $\alpha(h/\rho)$, where h is the diameter of the element, ρ the radius of its inscribed circle and α a normalization factor) is 2.33, while triangulation 2 includes 197 elements and 125 vertices, with a quality equal to 2.36.

A parameter can be introduced to dilute the boundary information in such a way that elements are larger in regions far from it. Figure 19.3 depicts the mesh obtained by the advancing-front method in the same domain than that of Fig. 27.1 (triangulation resulting from a Voronoï method) while Fig. 19.4 is the mesh resulting from the advancing-front method applied with a dilution parameter of 1.2. The first triangulation consists of 514 elements, 314 vertices, its worst quality being 1.76, while the diluted triangulation includes only 304 elements and 209 vertices, with a quality of 2.21.

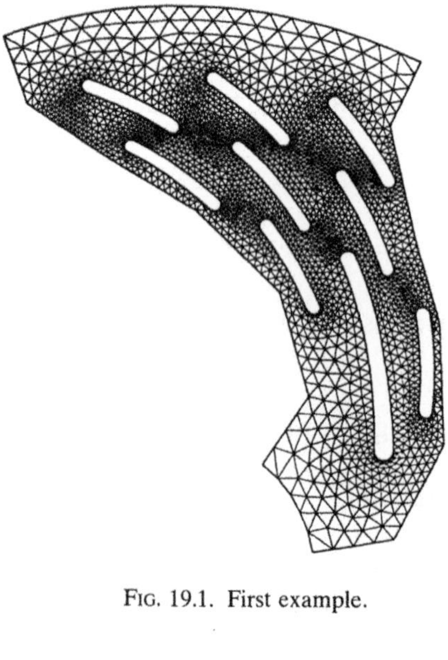

FIG. 19.1. First example.

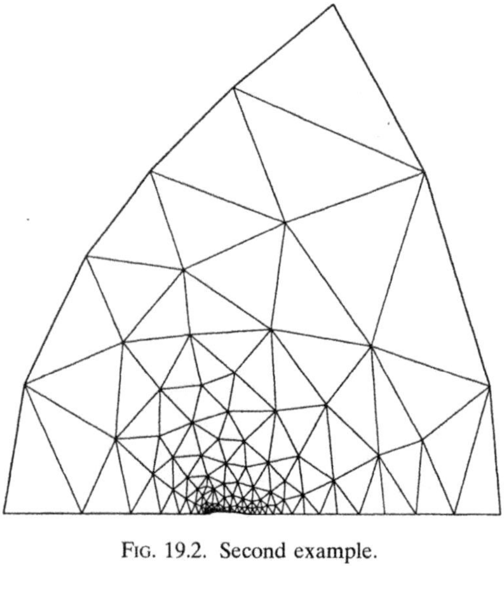

FIG. 19.2. Second example.

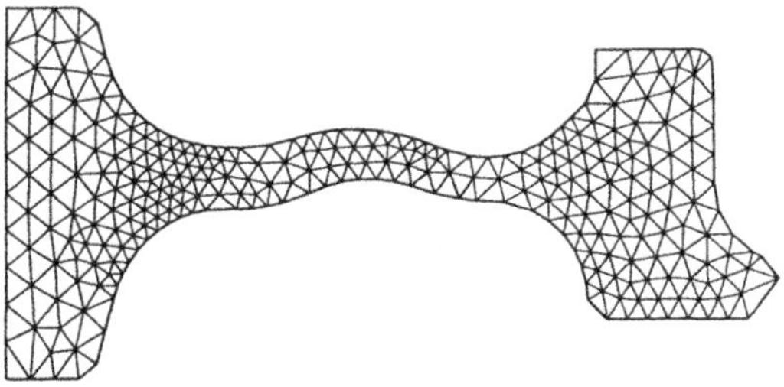

FIG. 19.3. Third example (without dilution).

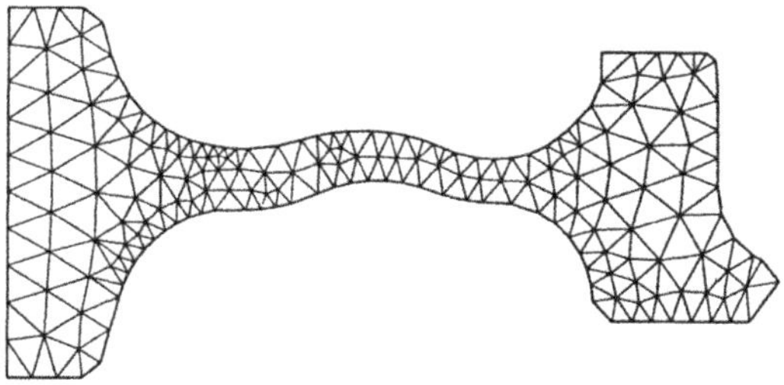

FIG. 19.4. Same example with dilution.

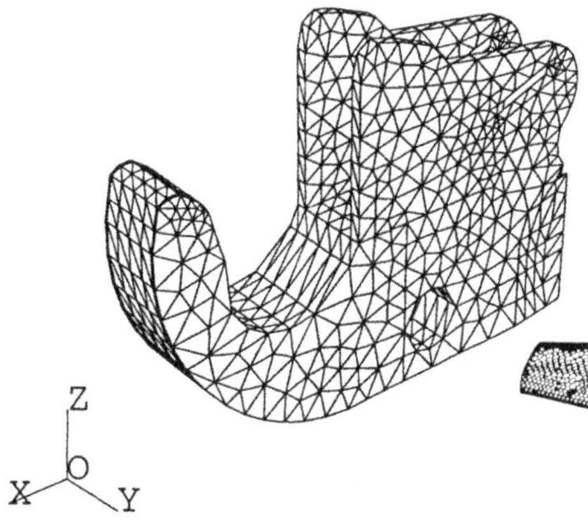

FIG. 19.5. 3-dimensional example.

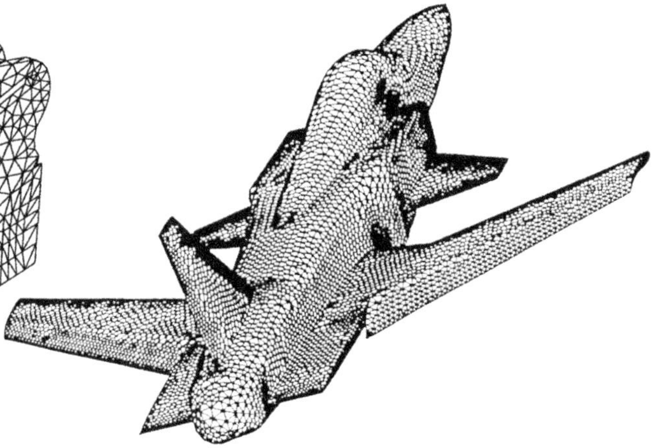

FIG. 19.6. 3-dimensional example.

In three dimensions, we give the following examples: the first example (Fig. 19.5) shows a mesh consisting of 4698 elements and 1502 vertices, its quality being 8.79. Figure 19.6 (courtesy of J. Peraire) depicts a mesh created for a fluid problem in the case of the Euler equations (PERAIRE and PEIRO [1992]).

Mesh qualities must be appreciated in comparison to the quality of the surface meshes serving as data. In the first example (Fig. 19.5) the expected worst quality is 7.10. In general, the advancing-front method creates a mesh such that its elements are mostly distributed in the good quality part of the histograms that can be easily constructed to appreciate the mesh quality (an histogram indicates the number of elements in the mesh whose quality falls in some range).

20. Structures and algorithms

Some structures and algorithms can be used in multiple contexts (and in this respect, when using a Voronoï–Delaunay method). In this section, we mention only some details directly related to the present mesh generation method.

20.1. *Basic structure*

The advancing-front approach leads to a very simple data structure suitable in both two and three dimensions. It is as follows:

- Any element in the mesh is well defined via the oriented list of its 3 (4) vertices. This orientation ensures that elements have positive surfaces (volumes).

20.2. *Background structures*

The advancing-front method clearly involves the correct management of the front, defined at each step, and that of an appropriate neighborhood of the zone where a point is created. The simplest way to store the front is to define the linked list of edges (face) in the current front so that those which have common items can be easily found. In two dimensions, it means, for edge AB of the front, that we need to know the next edge, say BC, and the previous edge, say XA:

$$\cdots \to X \to A \to B \to C \to \cdots$$

Thus, when inserting point P, for example, when dealing with edge AB, the list is updated and becomes:

$$\cdots \to X \to A \to P \to B \to C \to \cdots$$

i.e. the link from A to B is replaced by the links from A to P and P to B. Operations necessary when updating this list are the addition of links (the length of the list increases) and the suppression of links (after which this length decreases).

The method, as discussed throughout this chapter, involves a priori more sophisticated structures: two binary trees associated with the neighborhood structure. The first tree corresponds to the points as described above. The second tree corresponds to the items of the front.

The binary tree associated with the points is initialized by the given points and constructed as described in Section 16. After each internal point creation, this tree is enriched in such a way that the new point is incorporated. This task is very easy to perform as it is sufficient to find the grid cell within which the considered point falls.

The binary tree associated with the items of the current front is initialized by the given items and constructed similarly. After each internal point creation, this tree is enriched or de-enriched in such a way that a new item is incorporated, or a former item is removed. Adding an item in the tree is very easy to perform, while removing an item leads to the redefinition of some connections in the tree from nodes which should be disconnected when removing an item.

In PERAIRE, PEIRO, FORMAGGIA, MORGAN and ZIENKIEWICZ [1988], a different technique is described for constructing a structure suitable for point, edge and face management. Instead of defining one tree for the points and another for the front items, only one tree, the so-called alternative digital tree, is constructed. It contains links to points, edges, faces and elements in the current mesh and naturally offers the possibility of inserting and removing members in order to follow the mesh construction dynamically. This solution also offers a fast access to the only items which must be examined when creating a point and so helps to solve the geometrical intersection problems arising when checking if the current construction is valid or not.

The use of a quadtree (octree) structure and linked lists are discussed in LÖHNER and PARIKH [1988a], as well as in MAVRIPLIS [1992], to serve as support for the necessary searching operations and to help with point location.

20.3. Basic algorithms

There is a series of procedures which is currently used in the mesh generation process. Among them are those concerning the insertion of a point, the search for a cell containing a given point, the search for an element containing a given point, the search for elements intersected by a given segment, etc.

The first operation is very easy to do in the case of a regular grid, one has to compare the coordinates of the given point and that of the cell vertices.

The second task is more delicate as it must be, on the one hand, as cheap as possible and, on the other hand, exact. The last requirement is not obtained obviously as round-off errors may cause inacurate decisions. To avoid such a problem, the retained solution consists in using the following algorithm assuming that the (surface) volume computation is numerically exact: Let K be a simplex in the current mesh and $F_j, j = 1,4$ be its four faces, we consider a point P and define the four *virtual* tetrahedra, denoted as K_j^*, constructed with point P and any of the F_j (it means that P takes the place of the vertex opposite to the considered face). Then the computation of the four volumes determines in what region point P falls (15 regions can be identified. For simplicity, Fig. 20.1 presents the two-dimensional problem where only 7 regions can be identified). More precisely the following holds:

- $\forall j, 1 \leqslant j \leqslant 4, \mathrm{Vol}(K_j^*) > 0 \leftrightarrow P \in K$;
- $\exists j_1, ..., j_q, 1 \leqslant j_1 \neq \cdots \neq j_q \leqslant 4$ such that $\mathrm{Vol}(K_{j_1}^*) = \cdots = \mathrm{Vol}(K_{j_q}^*) = 0 \leftrightarrow \{P$ is in the intersection of the q planes defined by the faces $F_{j_1} ... F_{j_q}\}$.
- $\exists j_1, ..., j_q, 1 \leqslant j_1 \neq \cdots \neq j_q \leqslant 4$ such that $\mathrm{Vol}(K_{j_1}^*) < 0, ..., \mathrm{Vol}(K_{j_q}^*) < 0 \leftrightarrow \{P$ falls in the intersection of the q semi-spaces defined by the faces $F_{j_1} ... F_{j_q}\}$.

Using this algorithm, it is now possible to find the element within which point P falls:

- Initialization: $K = K_0$ where K_0 is an arbitrary element in the mesh;
- (a) If $\forall j$ $\mathrm{Vol}(K_j^*) > 0$ then $P \in K$, end of algorithm.
- If $\exists j$ $\mathrm{Vol}(K_j^*) < 0$, find element K' such that $K \cap K' = F_j$ where F_j is face j of K, so do $K = K'$ and go to (a);
- If $\exists j_1, ..., j_q$ such that $\mathrm{Vol}(K_{j_1}^*) = \cdots = \mathrm{Vol}(K_{j_q}^*) = 0$ then $x \in \mathscr{C}(\bigcap_{k=1,q}(F_{j_k}))$ where $\mathscr{C}$ is the set of elements including $\bigcap F_{j_k}$ as an item, end of algorithm.

In some special configuration, this algorithm may lead to a cyclic path. To avoid this trouble, Step 3 must be modified, for instance, by selecting randomly

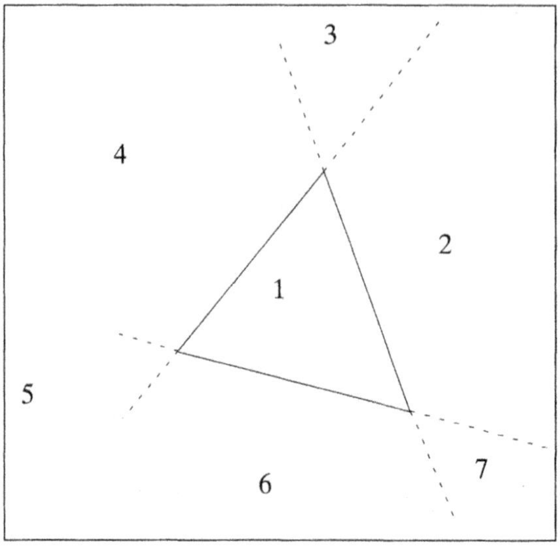

FIG. 20.1. The 7 regions associated with a triangle in two dimensions.

one of the faces serving to find element K in the case where not only one face induces a negative volume.

In three dimensions, the previous algorithm determines if P is inside an element, is located on an edge or a face common to several or two elements, or is identical to another vertex.

The complexity of this method is not optimal but, in practice, element K_0 initializing the process can be chosen sufficiently close to point P so that the process is not expensive. In the present context, this process is used to construct the control space. We consider an element in the boundary mesh and find the cell containing one of its vertices, K_0 is set to be this element and we find the elements containing the cell vertices: obviously they are close to K_0.

The third question can be solved using the same principle and, in this way, relies on surface (volume) computations only.

CHAPTER IV

Automatic Method (2): Voronoï Type Mesh Generation

Introduction

In this chapter we discuss a *Voronoï type method governed by a given control.* Before doing so, we first introduce the outlines of the classical Voronoï method and its application to finite element mesh generation.

Firstly (Sections 21 and 22), we discuss the main framework of any Voronoï method, namely the process to create a Delaunay triangulation which includes a series of specified points as element vertices.

Unfortunately, the mesh generation problem in the context of finite element computation cannot only relate to the application of the Voronoï method. The latter considers only the notion of points and a priori does not consider edges or faces. Thus, as domains are generally described through their boundaries in terms of a list of edges and faces, the latter must be considered as a collection of specified items. So while the Voronoï method is suitable for the creation of a mesh including the points of the boundary, it can be seen that, in general, some edges and faces of this boundary are not present in such a triangulation. As soon as at least one edge (one face) is specified, the desired mesh is called a *constrained* mesh and cannot be created using this method by itself. The second part (Section 23) of this chapter will be devoted to this problem.

In a third part (Section 24), we consider the problem of creating the field points inside the domain and then propose (Section 25) a general scheme for conceiving an automatic mesh generator based on the Voronoï method.

Applications for a given control will then be discussed (Section 26), after which (Section 27) application examples are presented. Finally (Section 28), the chapter ends with some details and indications concerning the implementation of these kinds of mesh generators and includes information about data structures and effective algorithms.

21. General principles

Let $\{P_k\}$ be a set of points in two or three dimensions. The Voronoï method consists in creating a triangulation composed of simplices (triangles in two dimensions and tetrahedra in three dimensions) whose vertices are the members of set $\{P_k\}$.

An initial triangulation can be obtained as the dual of the Voronoï cells V_i constructed as follows:

$$V_i = \{P \text{ such that } \| P - P_i \| \leqslant \| P - P_j \|, \forall j \neq i\}.$$

It is clear that V_i is a closed convex polygon (polyhedron in three dimensions); these cells (Fig. 21.1) cover the space and do not overlap, they are known as the *Dirichlet tesselation* of the entire space which includes all the initial points (DIRICHLET [1850]).

The cell sides are midway between the two points they separate so that the perpendicular bisector of all sides defines a set of segments joining the given points. It is then obvious to obtain the *Delaunay triangulation* associated with the Voronoï cells by constructing the elements formed by the bisector of all sides.

More precisely, the above construction enables us to create the triangulation of the convex hull of the specified points which consists of triangles and convex polygons (other than triangles) in two dimensions (see Fig. 21.2) and tetrahedra and convex polyhedra (different from tetrahedra) in three dimensions (COXETER, FEW and ROGERS [1959]).

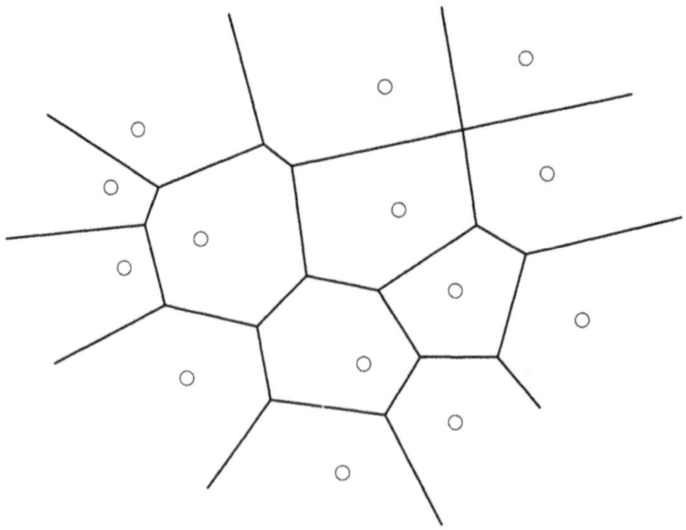

FIG. 21.1. Voronoï cells.

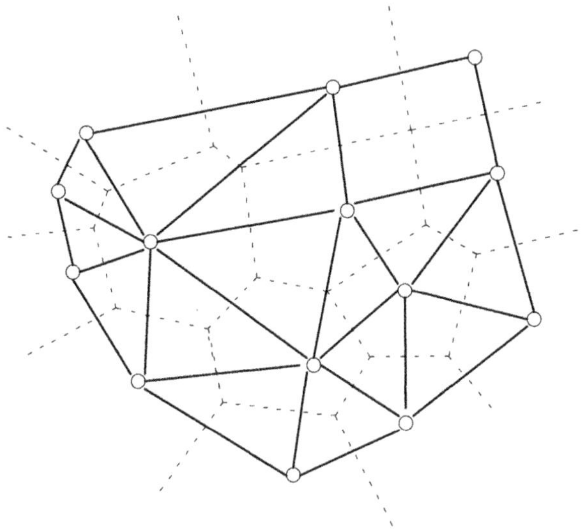

FIG. 21.2. *The* Delaunay triangulation.

In practice, elements different from simplices can be easily split into simplices (see Fig. 21.3). The resulting triangulation is also said to be a Delaunay triangulation of the convex hull of the given set of points and such elements resulting from a partition are said to be *special*. When special elements exist, the Delaunay triangulation is not unique. In the literature, the notion of a special item

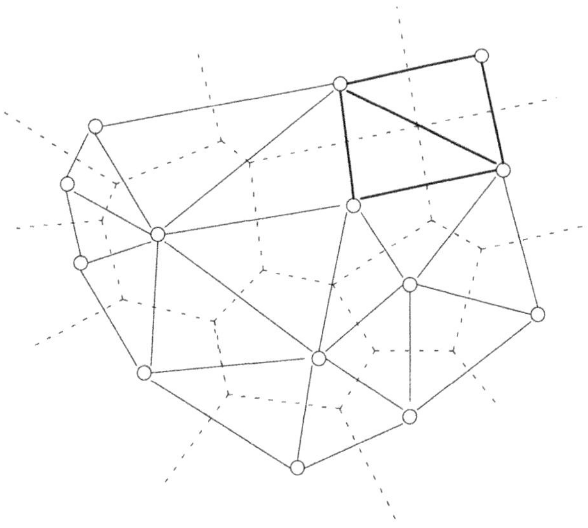

FIG. 21.3. *A* Delaunay triangulation.

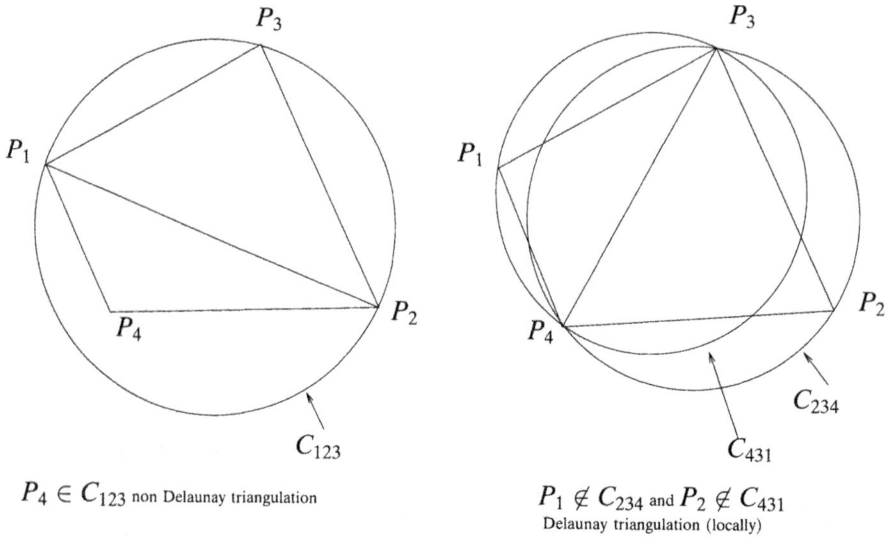

FIG. 21.4. Property of a Delaunay triangulation.

applies frequently to points and not to elements. For simplicity, we extend this notion to elements as there is an intuitive correlation.

This type of triangulation enjoys a series of properties. Among them are:

(1) The *in-circle (sphere)* or *empty circle (sphere)* criterion: The circumcircle (circumsphere) associated with the non-special elements of a Delaunay triangulation does not contain any other point of the triangulation (Fig. 21.4).

(2) The *maximum–minimum angle* criterion: In two dimensions, each pair of triangles sharing an edge and forming a convex quadrilateral is such that the smallest of the angles formed by two consecutive edges is maximal.

(3) The *circumcenter* property: The circumcenter of any simplex is the intersection of the sides of the Voronoï cells which are intersected by the edges of this simplex.

Properties (1) and (2) mean that a Delaunay triangulation is optimal, with respect to the related criterion, for the given set of specified points.

22. Connection from points to points

The above method for constructing a Delaunay triangulation of a given set of points proves to be ineffective in practice in this form. This is why alternative methods have been proposed to deal with this problem. The method depicted below was introduced, almost simultaneously, by HERMELINE [1980] and WATSON [1981] in the early 80s. It is referred to by different names, the most usual being Watson's algorithm.

22.1. Original insertion point method

Let $\{P_k\} = \{P_1, P_2, ..., P_n\}$ be a set of n two- or three-dimensional distinct points, and T_{old} a Delaunay triangulation whose elements are denoted by K_j and whose element vertices are the first i points of $\{P_k\}$, then triangulation T_{new} can be derived from triangulation T_{old} in such a way that point P_{i+1} is one of its element vertices.

Introducing the set $\mathscr{B}_{\text{old}} = \bigcup_j B_j$ where B_j is the circumcircle (the circumsphere) of element K_j of T_{old}, then point P_{i+1} necessarily falls into one of the three following situations (see Fig. 22.1 in two dimensions):

Case (a): $P_{i+1} \in T_{\text{old}}$, i.e. $\exists K_j \in T_{\text{old}}$ such that $P_{i+1} \in K_j$,
Case (b): $P_{i+1} \notin \mathscr{B}_{\text{old}}$, i.e. P_{i+1} is not in any circle (sphere) B_j,
Case (c): $P_{i+1} \notin T_{\text{old}}$ but $P_{i+1} \in \mathscr{B}_{\text{old}}$, i.e. P_{i+1} is not in an element K_j of T_{old} but is in a circle (sphere) B_j.

A constructive method for creating the triangulation T_{new} is associated with each of these cases. In the remainder of this section, the two-dimensional case is described, which can be extended easily to the three-dimensional case by replacing notions of planes, circles and edges by those of spaces, spheres and faces. (In fact, such a construction formally extends to higher dimensions.)

For simplicity we denote P_{i+1} by P in the following.

Case (a): Let $\mathscr{S}$ be the set of elements of T_{old} whose circumcircle contains point P and let $F_1 ... F_p$ be the external edges of this set, then:

$$T_{\text{new}} = (T_{\text{old}} - \mathscr{S}) \cup \{F_j, P\}_j \quad 1 \leqslant j \leqslant p. \tag{22.1}$$

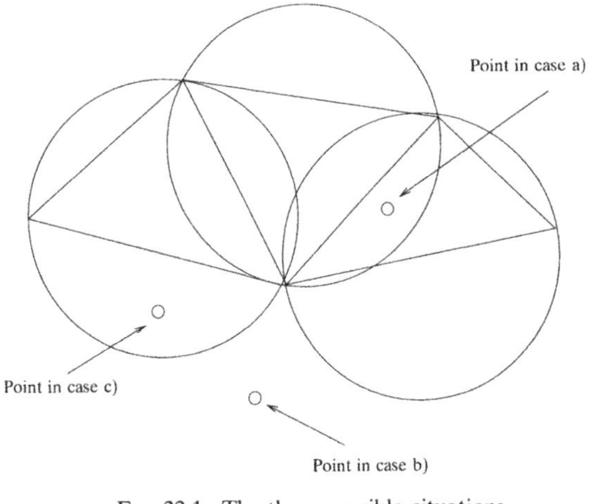

FIG. 22.1. The three possible situations.

Case (b): As $P \notin \mathcal{B}_{\text{old}}$, $P \notin T_{\text{old}}$, we exhibit $F_1, \ldots, F_p$, the external edges of T_{old}, which define a half-plane strictly separating P and T_{old}, then:

$$T_{\text{new}} = T_{\text{old}} \cup \{F_j, P\}_j \quad 1 \leqslant j \leqslant p. \tag{22.2}$$

Case (c): Let $\mathcal{S}$ be the set of elements of T_{old} whose circumcircle contains P and let $F_{i_1}, \ldots, F_{i_q}$ be the edges not common to two elements of $\mathcal{S}$ defining a half-plane not separating P and T_{old}, and, among the external edges of elements of T_{old} not already dealt with, let $F_{i_{q+1}}, \ldots, F_{i_p}$ be those defining a half-plane separating P and T_{old}, then:

$$T_{\text{new}} = T_{\text{old}} - \mathcal{S} \cup \{F_{i_j}, P\}_j \quad 1 \leqslant j \leqslant p. \tag{22.3}$$

In all of these situations, the resulting mesh T_{new} includes the first $(i+1)$ points of set $\{P_k\}$ as element vertices and is obtained as follows:

Case (a): by replacing the elements of $\mathcal{S}$ by new elements created by joining point P to the external edges F_j (Fig. 22.2) numbered in such a way that their surfaces are positive.

Case (b): by connecting the selected edges F_j and point P (Fig. 22.3) while ensuring positive surfaces for the so created elements.

Case (c): as in Case (b) with the edges selected here (Fig. 22.4) in such a way that the same property holds.

This method is initialized by creating an initial simplex by selecting 3 (4) points in set $\{P_k\}$ which are not colinear in two dimensions and not coplanar in three dimensions. Then the points of $\{P_k\}$ are inserted one-at-a-time.

The final triangulation, T_{new}, covers the convex hull of set $\{P_k\}$ and is Delaunay. A proof can be found in different papers, for example HERMELINE [1980], BAKER [1989a], RISLER [1991]. The proof, for example in Case (a), can be outlined as follows: T_{old} is assumed to be a valid Delaunay triangulation, then it suffices to prove that T_{new} enjoys the same property. Thus, first it is shown that $\mathcal{S}$ is star-shaped with respect to P and contains no points other than P and the endpoints of the edges of its elements so that T_{new} is a triangulation including P as vertex; then it is proven that T_{new} is Delaunay by involving the in-circle criterion.

REMARK 22.1. In R^2, a polygon is star-shaped if there exists an internal point M such that the line joining M with any point of the polygon falls entirely within the polygon. This definition applies in R^3 in the case of a polyhedron.

Several variations are popular when implementing such a method:

- A preliminary sorting of the points with respect to their coordinates allows us to construct a first simplex which defines two regions separating the points. The plane defined by the face opposite to the so-called first point separates all remaining points from this point. The nearest point is then inserted and the same method is applied recursively. As a consequence, the construction method relies only on Cases (b) or (c).

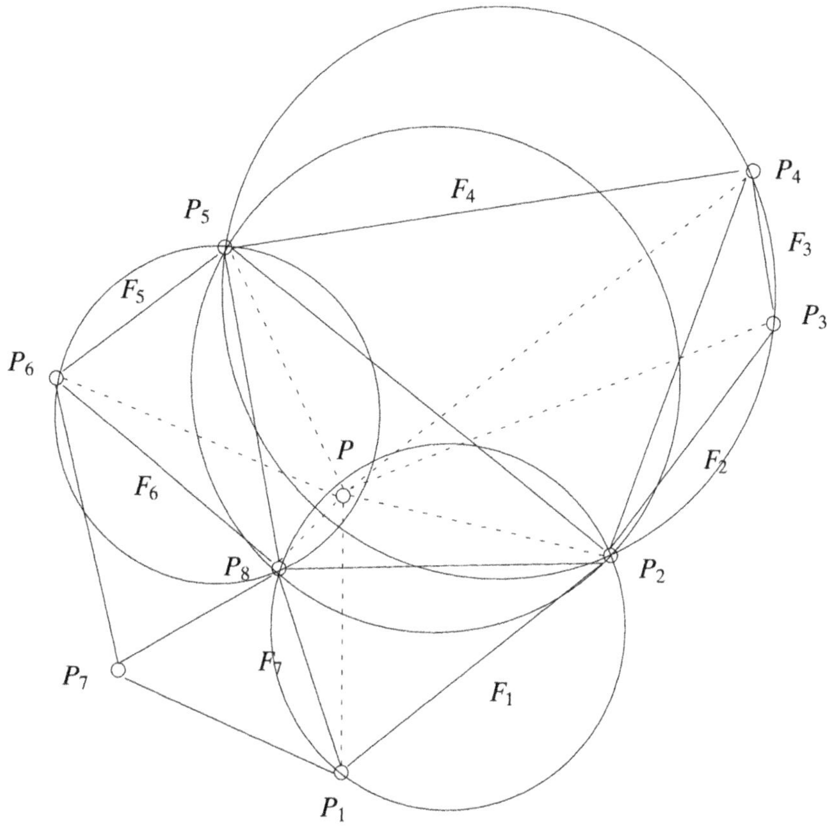

FIG. 22.2. Case (a): Insertion of point P (by Hermeline).

- The introduction of a simplex or a box (easily split into simplices) such that all points of $\{P_k\}$ are included in it, leads to Case (a). In this variation, the resulting triangulation is that of the box introduced and, when deleting all elements which have one of the additional points as vertices, the triangulation obtained is not necessarily that of the convex hull of $\{P_k\}$ (GEORGE and HERMELINE [1992]).
- Using the same introduction of an enclosing simplex or box, but instead of exhibiting the appropriate set $\mathcal{S}$, this variation consists in searching for the element(s) containing the point P under consideration after which this(these) element(s) is(are) split into simplices by joining P with its(their) external edges (faces). A swapping procedure (as described latter) is then used recursively to optimize the created elements and their neighbors. A suitable sequence of such optimizations ensures that the resulting triangulation is Delaunay (see CHERFILS and HERMELINE [1990] and JOE [1991b]).

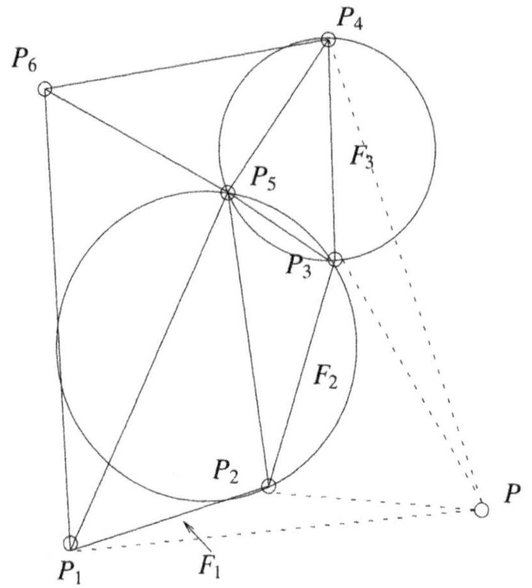

FIG. 22.3. Case (b): Insertion of point P (ibidem).

22.2. *Derived insertion point method*

An alternative way to construct a triangulation, not necessarily strictly of the Delaunay type (this does not matter as constrained items related to the boundaries must be present in the final triangulation), can be derived from the previous method. We first introduce, as in the latter variation, an enclosing simplex or box to fall into Case (a). Then, the only thing to do is to construct a suitable set $\mathcal{S}$ from T_{old}; to do this, we only require the following properties:

- external sides of $\mathcal{S}$ are visible from P,
- $\mathcal{S}$ is a connected set,
- $\mathcal{S}$ does not contain points of $\{P_k\}$ other than P.

As will be seen below, the main interest of this method is that it involves only the computation of surfaces (volumes) and does not necessitate the consideration of circumcircles (circumspheres). The following algorithm is proposed:

(1) Find the element(s) in T_{old} enclosing P;

(2) Stack this(these) element(s) in set $\mathcal{S}$;

(3) Find $F_1, \ldots, F_p$, the external sides of $\mathcal{S}$ (i.e. the sides not shared by two elements in $\mathcal{S}$);

(4) Check if $\mathcal{S}$ is a connected set. In practice, it is sufficient to verify that each element in this set has at least one neighbor in the set (if more than one

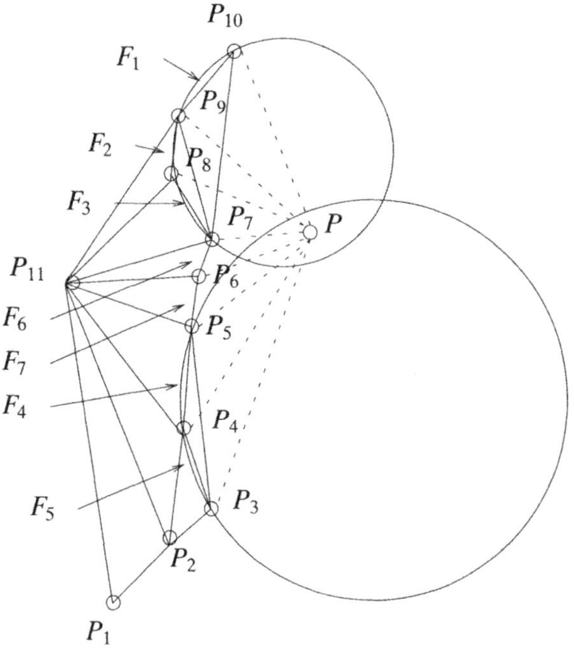

FIG. 22.4. Case (c): Insertion of point P (ibidem).

element is in the set). If one element fails in this property, it must be removed from $\mathscr{S}$; Define the p simplices K_j formed by joining the sides F_j with P and denote by $\mathrm{Det}(K_j)$ the surface (volume) of element K_j; Define n_F to be the number of distinct endpoints of sides F_j; Define n_S to be the number of distinct vertices of elements in $\mathscr{S}$;

(5) Set $\mathscr{S}' = \mathscr{S}$. Loop for $j = 1, p$:

- If $\mathrm{Det}(K_j) > 0$, continue;
- If $\mathrm{Det}(K_j) < 0$, $\mathscr{S}' = \mathscr{S}' - K_j$ where K_j is the simplex of $\mathscr{S}'$ with side F_j;
- If $\mathrm{Det}(K_j) = 0$, $\mathscr{S}' = \mathscr{S}' \cup K_j$ where K_j is the simplex of T_{old}, with side F_j, not in $\mathscr{S}'$.

(6) If $\mathscr{S}'$ is not affected by Step (5), compute n_F and $n_S = n_{S'}$ and then go to (7); if not, update the set of F_j and return to (4);

(7) If n_F and n_S are equal, set $\mathscr{S} = \mathscr{S}'$ and go to (8);
If not, set $\mathscr{S}' = \mathscr{S}' - K$ and return to (4) (K is any possible simplex in T_{old} with a vertex, distinct from P, in $\mathscr{S}'$ and not in F_j);

(8) Then, following (22.1), $T_{\mathrm{new}} = (T_{\mathrm{old}} - \mathscr{S}) \cup \{F_j, P\}_j, 1 \leqslant j \leqslant p$, End.

This process is repeated for all points in $\{P_k\}$ and obviously produces a valid triangulation of the enclosing box after which the boundary integrity phase can be started.

REMARK 22.2. In Step (7) above, a simplex is said to be possible if it is not a member of the set established in Step (1) and if set $\mathcal{S}'$ remains connected after it has been removed.

Note that $\mathrm{Det}(K_j)$ is nothing but:

$$\mathrm{Det}(K_j) = \begin{vmatrix} x_2 - x_1 & x_3 - x_1 \\ y_2 - y_1 & y_3 - y_1 \end{vmatrix},$$

where x_1, x_2, x_3 are the abscissa of the vertices of K_j and y_1, y_2, y_3 their ordinates. A similar expression applies in three dimensions. Note that this value is twice the surface of K_j, in two dimensions, and a similar expression is the volume of the element multiplied by six, in three dimensions (see Chapter I).

23. Boundary integrity

All the above methods can be employed to create a triangulation of either the convex hull of a given set of points, or a box enclosing these points. In the context of finite element simulation, domains are commonly known via their boundaries. The latter consists of a set of edges in two dimensions and a set of faces (assumed to be triangular) in three dimensions. A natural idea is to constitute the set of all the endpoints of these edges (faces) as set $\{P_k\}$ and to apply one of the above methods to the so formed set of specified points.

This results in the creation of a triangulation where the element vertices are all the points of the boundaries. But, as mentioned earlier, such a triangulation does not a priori include the specified items of the given boundaries (see Fig. 23.1 where a simple two-dimensional example is depicted in which boundary edges A_1B_1 and A_2B_2 are not present in the created triangulation). The following sections discuss the way in which these missing items can be, in some sense, regenerated.

23.1. The two-dimensional case

Two methods are usually used to deal with the problem of maintaining the integrity of the boundary of the domain. Both produce the desired solution.

23.1.1. Creation of the midpoint of every missing boundary edge

When all specified points have been inserted, a check of the so created mesh enables us to list the missing boundary edges. The following algorithm,

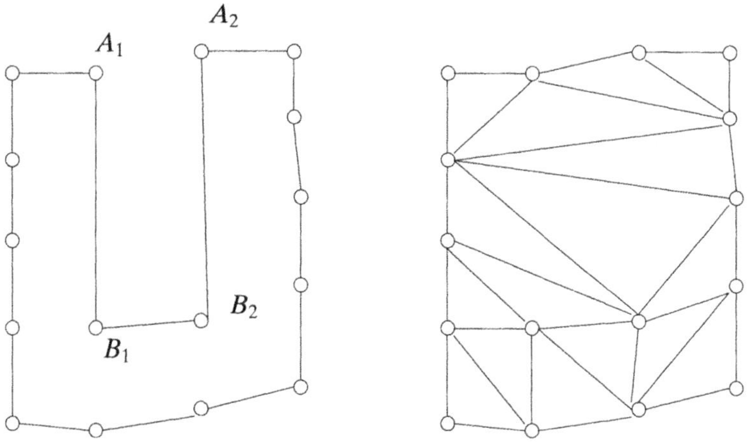

FIG. 23.1. Edges are missing while their endpoints exist.

WEATHERILL [1990a], is then proposed. It consists, for every missing edge P_iP_j, of:

- creating point P_{ij} such that $P_{ij} = \frac{1}{2}(P_i + P_j)$,
- removing edge P_iP_j from the list of specified edges and insert edges P_iP_{ij} and $P_{ij}P_j$ in this list,
- updating the current mesh by inserting point P_{ij}.

It is then checked to see if all new specified edges are present in the mesh and, if not, the same process is applied.

Intuitively, such a method converges as the radius of the circumcircles associated with the elements decrease and finally contains no undesirable points.

A phase can then be added, consisting of removing all the so created points in order to obtain a mesh which contains exactly the initial boundary edges and not a partition of them. It is clear that such a mesh is not Delaunay in the region where points have been created and then deleted.

Figure 23.2 shows an example in which two edges are missing in the triangulation (right-hand side) after the insertion of the endpoints of the specified boundary edges (left-hand side). In Fig. 23.3 one can see the Delaunay triangulation (left-hand side) resulting from the creation of the two associated midpoints and then (right-hand side) the local modification of the triangulation to re create the original edges.

REMARK 23.1. Following the advancing-front approach in its classical form, a solution consisting in creating a point on every boundary edge in such a way that every boundary edge exists in the mesh can be employed. It is clear that such a mesh is not Delaunay in the region close to its boundary. For the remaining

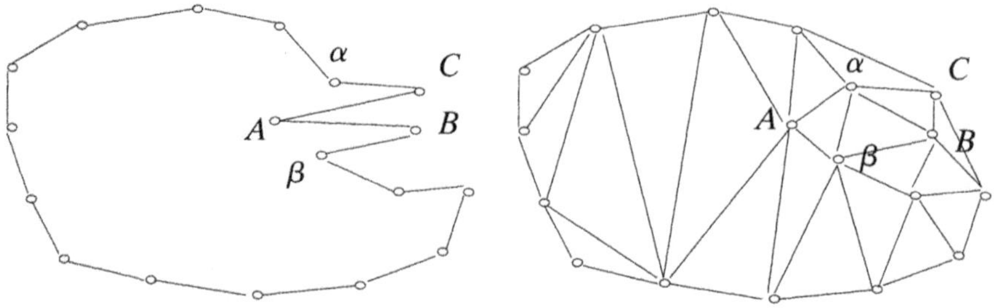

FIG. 23.2. The given boundary and the associated Delaunay triangulation.

domain, almost Delaunay triangulation can easily be obtained as there is no longer a constraint due to specified edges.

23.1.2. Local diagonal swapping procedures

Firstly, all specified points are inserted and a check of this mesh enables us to list the missing boundary edges. After having obtained the list of all of them, we define their corresponding *pipes*.

The *pipe* associated with the missing edge P_iP_j is a collection of the following elements:

- the first contains vertex P_i and P_iP_j intersects the edge opposite to P_i,
- the last contains vertex P_j and the edge opposite to P_j is intersected by P_iP_j,
- the elements possessing two edges intersected by P_iP_j.

We define the diagonal swapping procedure as the local modification of two elements sharing an edge and which form a *convex* quadilateral into two new elements consituted by using as common edge the other diagonal of the initial quadrilateral (Fig. 23.4).

The following algorithm is proposed (the example of one pipe is depicted in Figs. 23.5–23.7 where the evolution of the triangulation can be seen). The method consists in processing every pipe in the mesh and:

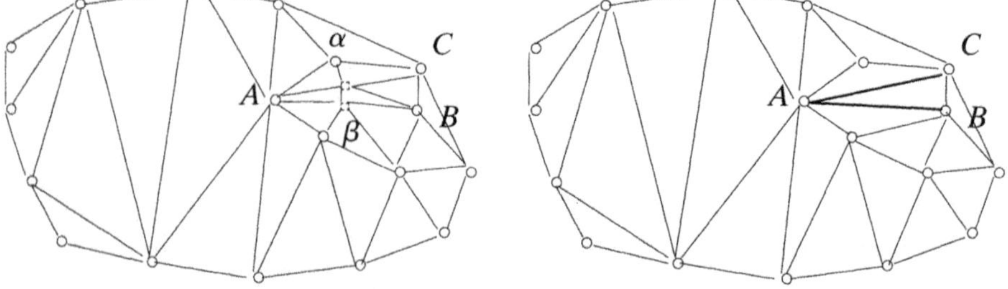

FIG. 23.3. Creation of midpoints, Delaunay triangulation and removal of added points.

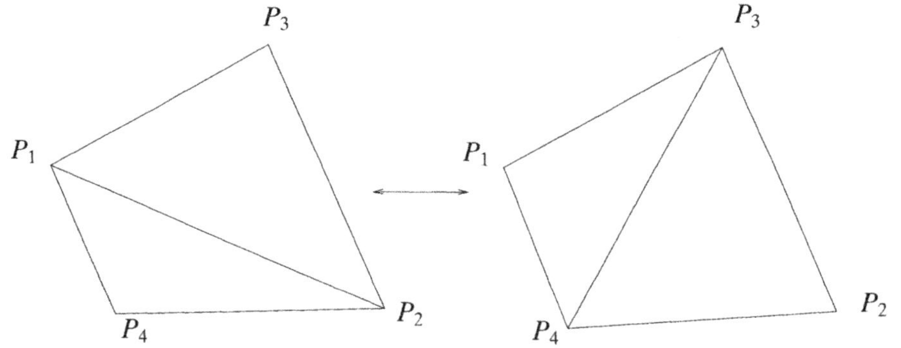

FIG. 23.4. Diagonal swapping.

- if the pipe contains only two members, swap them and go to the next pipe,
- if not, consider its first member and define as a second member the element sharing the edge intersected by the missing edge. If these two elements can be swapped: remove them and create the alternative situation. If the swap is not possible then:
  - if the pipe has not yet been reserved, reverse the pipe and try again
  - if it has been reversed, look for two consecutive members of the pipe which can be swapped
- end of loop for the pipe under consideration.

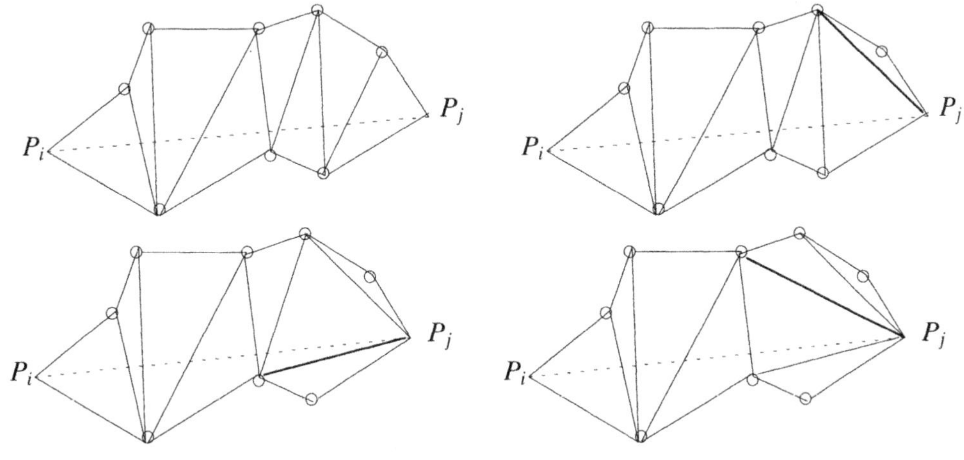

FIG. 23.5. Dealing with the pipe P_iP_j.

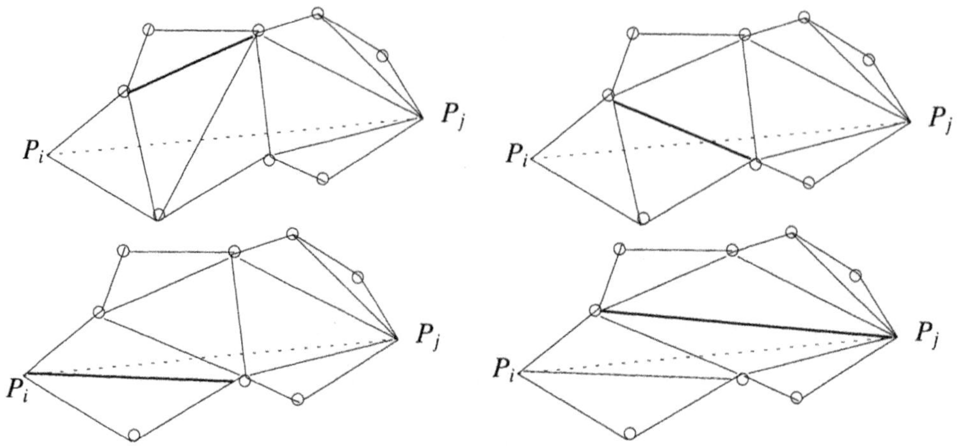

FIG. 23.6. Dealing with the pipe P_iP_j, continued.

This algorithm converges to the desired solution as soon as the boundary of the domain is not crossed (this is trivially true). A proof can be found in GEORGE [1991] in which this method is used, as well as in HECHT and SALTEL [1990].

23.2. *The three-dimensional case*

The two methods discussed in the case of two dimensions extend in this situation and prove to produce the desired solution.

23.2.1. *Creation of the midpoint of all missing edges and creation of a point at the centroïd of all missing faces*

When all specified points have been inserted, the following algorithm (see

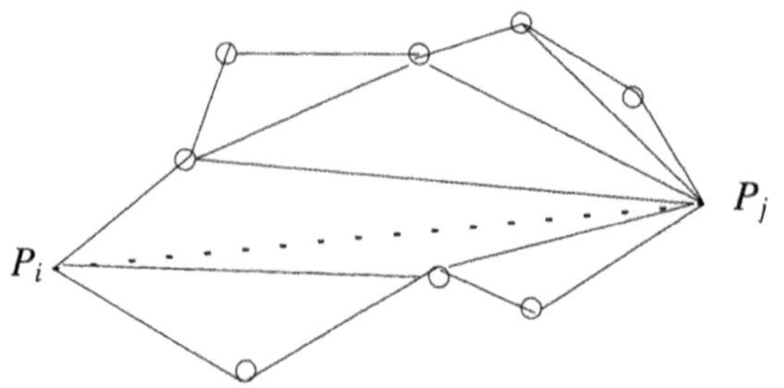

FIG. 23.7. Resulting triangulation, P_iP_j exists.

PERRONNET [1988] and WEATHERILL [1992]) is applied to the current triangulation. It first examines every missing edge, P_iP_j:

- create point P_{ij} such that $P_{ij} = \frac{1}{2}(P_i + P_j)$,
- remove edge P_iP_j from the list of specified edges and insert edges P_iP_{ij} and $P_{ij}P_j$ in this list,
- update the list of specified faces by taking the new point into account,
- update the current mesh by inserting point P_{ij}

and then visit every missing face, $P_iP_jP_k$:

- create point P_{ijk} at the centroïd of face $P_iP_jP_k$,
- remove face $P_iP_jP_k$ from the list of specified faces and insert faces $P_iP_jP_{ijk}$, $P_jP_kP_{ijk}$ and $P_kP_iP_{ijk}$ in this list,
- update the current mesh by inserting point P_{ijk}.

Several variations can be employed to obtain the same result. For example, the two loops can be merged into only one, i.e. the process loops over the faces and, for every face, loops over the three edges.

The convergence of this algorithm relies on the same principle as in two dimensions (see the remark given in this case). Note that in this case, the removal phase for all the created points is not trivial.

23.2.2. Local modifications coupled with possible creation of internal points
Firstly, as in two dimensions, all specified points are inserted and a check of the so created mesh enables us to list the missing boundary edges and faces. The method considers, firstly, the missing edges and then the missing faces. We define the two following sets of elements:

- the *pipes* associated with a missing edge,
- the *shell* associated with an edge.

Similar to two dimensions, the *pipe* associated with missing edge P_iP_j is the collection of the following elements:

- the first contains vertex P_i and P_iP_j intersects the face opposite to P_i,
- the last contains vertex P_j and P_iP_j intersects the face opposite to P_j,
- the elements possessing two faces intersected by P_iP_j.

The *local shell* associated with any edge P_iP_j is composed of all the elements sharing this edge.

Considering one missing edge, namely P_iP_j, we find the list of elements having at least one edge or one face intersected by P_iP_j. Two cases are encountered: either this set of elements is a pipe or there exists at least one element with one edge $\alpha\beta$ intersected by P_iP_j so that the set of interesting elements contains the local shell $\alpha\beta$.

We define some local element modifications which result in the creation or the removal of edge(s) or face(s). Among these we have the following: for two elements sharing a face and forming a *convex* polyhedron: let $\alpha\beta\gamma A$ and $B\alpha\beta\gamma$ be these elements, then the associated polyhedron can be remeshed by elements $AB\alpha\beta$, $AB\beta\gamma$ and $AB\gamma\alpha$. It implies that face $\alpha\beta\gamma$ is deleted and edge AB is created. This process is, for simplicity, denoted by $\mathrm{Tr}_{2\to3}$ (2 elements are replaced by 3 elements, see Fig. 23.8).

For a shell: *Removal of the shell*: any *convex* shell associated with a given edge $\alpha\beta$ can be remeshed in such a way that this edge is deleted. Let us denote by $M_i\alpha\beta M_{i+1}$ for $i = 1, n$ (with $M_{n+1} = M_1$), the n elements of the shell, then it suffices to remesh the polyhedron defined by the M_i using triangles and to connect these triangles to both α and β to obtain the desired solution. As examples we consider the cases where $n = 3$ and $n = 4$. For $n = 3$, the three original elements are $M_1\alpha\beta M_2$, $M_2\alpha\beta M_3$ and $M_3\alpha\beta M_1$, the polyhedron defined by the M_i is a simple triangle $M_1M_2M_3$, so that the solution is simply $M_1M_2M_3\alpha$ and $\beta M_1M_2M_3$ (it can be observed that this transformation is opposite to the transformation $\mathrm{Tr}_{2\to3}$ introduced above (see Fig. 23.9)). For $n = 4$, the four original elements are $M_1\alpha\beta M_2$, $M_2\alpha\beta M_3$, $M_3\alpha\beta M_4$ and $M_4\alpha\beta M_1$, the polyhedron defined by the M_i is a quadrilateral and can be remeshed in two ways, so there are two solutions $M_1M_2M_4$ and $M_2M_3M_4$ connected with, on one hand α and, on the other hand β, or $M_1M_3M_4$ and $M_1M_2M_3$ connected similarly (Fig. 23.10). Note that, in the case of a non-convex shell, the above remeshing process can be applied successfully when the shell is star-shaped.

Reduction of the shell: the application of the above process $\mathrm{Tr}_{2\to3}$ to each pair of elements in a shell can be used in order to reduce the number of members of the shell.

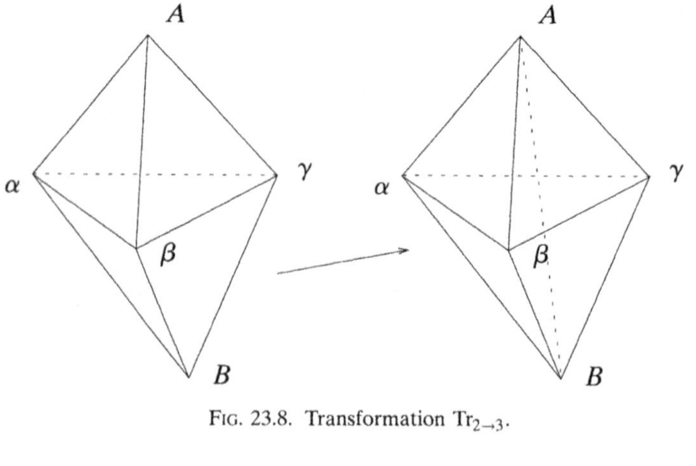

FIG. 23.8. Transformation $\mathrm{Tr}_{2\to3}$.

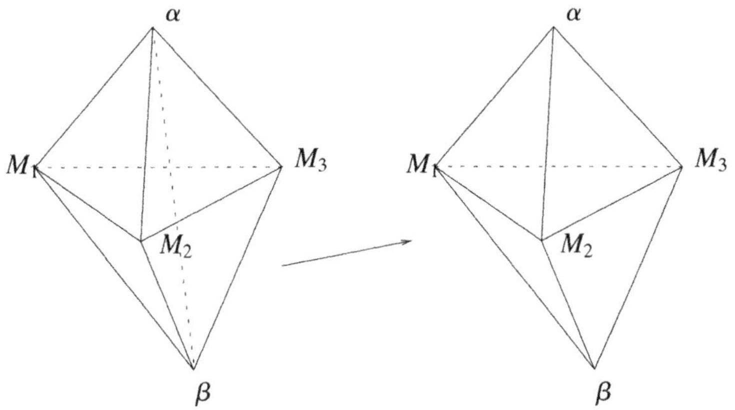

FIG. 23.9. Transformation $Tr_{3\to 2}$.

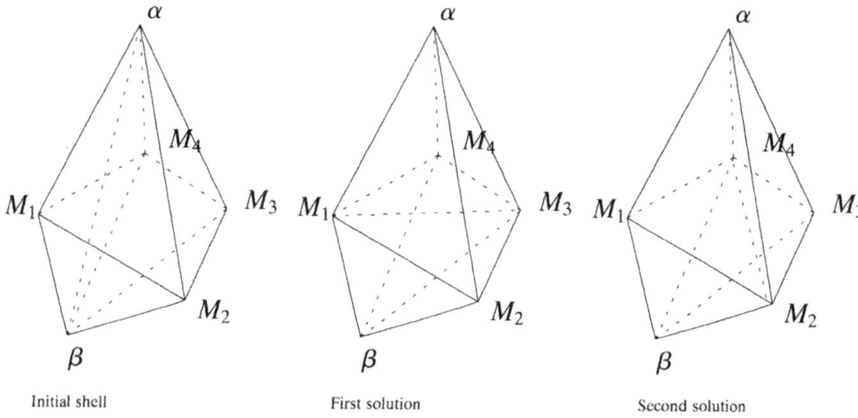

FIG. 23.10. Transformation $Ti_{4\to 4}$.

The proposed algorithm consists of visiting every pipe in the mesh, and:

- (a) if the pipe contains only two members, swap them and go to the next pipe
- if not, consider its first member and as a second member, the element sharing the face intersected by the missing edge, then:
  - if these two elements can be swapped: remove them and create the alternative situation
  - if the swap is not possible and if the pipe has not yet been reversed, reverse it and try again, otherwise, look for two consecutive members in the pipe

which can be swapped, and if a swap is possible leading to a pipe with less elements: perform it or if a swap is not possible: create a point (a Steiner point) on the intersected face common to element 2 and 3 of the pipe and return to (a)
- end of loop for the pipe under consideration

and every local shell in the mesh, and:
- if the shell contains only three members and if the swap is possible, swap them and go to the next shell
- if not, transform the shell in order to decrease its number of elements, if possible, and if not create a point to delete the shell and go to the next shell
- end of the loop for shells
- return to the treatment of pipes.

This algorithm converges to the desired solution as soon as the boundary of the domain is not crossed (this is trivially true). A detailed proof can be found in GEORGE, HECHT and SALTEL [1991].

Now that all specified edges exist in the triangulation, we apply a similar algorithm to all missing faces.

- For each of them, consider the triangle whose sides are the three edges of the face and,
  - find all the edges intersecting this triangle and the associated local shells
  - for every shell, if it contains only three members and is convex, swap them and go to the next shell and if not, try to remove the shell or to reduce it if possible, otherwise create a point to delete the shell and go to the next shell
- end of the loop for faces.

Now all specified faces exist in the triangulation.

A variation of this method consists in locally modifying the triangulation in such a way that missing items are created. This method uses the same local tools concerning an alternative mesh of the polyhedron formed by two elements sharing a face or by elements sharing an edge. In principle this method does not converge (see Fig. 23.11, the case where it is possible to split a prism into three tetrahedra while respecting the given triangular faces and, conversely, the case where the given faces are such that no triangulation is possible without creating at least one internal point), but, in practice it proves to produce the desired result in most cases. When it fails, Steiner point creations can be envisaged as above so that the process converges.

23.3. Definition of the inside of the domain

Once the boundary items have been regenerated using one of the above meth-

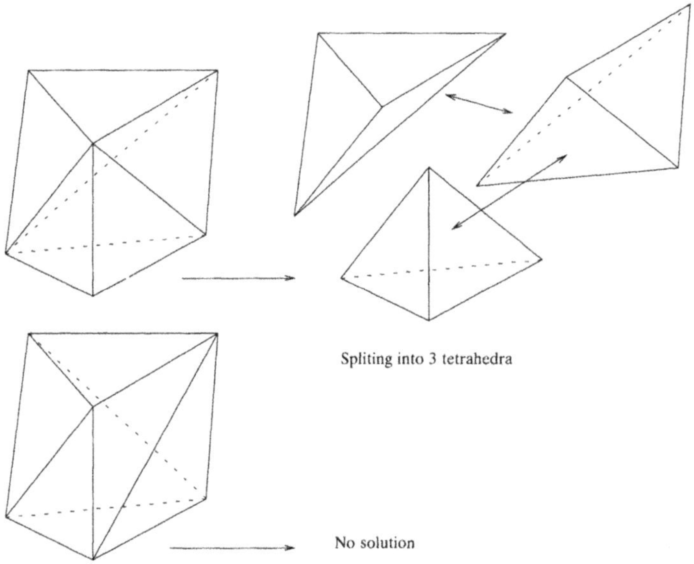

FIG. 23.11. An impossible constrained triangulation for a simple prism.

ods, it is possible to define the elements which are inside the domain (assumed to be non convex) precisely from those which are outside the domain.

The following algorithm is proposed:

- Assign the value $v = -1$ to all elements and set $n = 0$;
- Find an element clearly outside the domain (for example, in the case where a box is employed, an element with one of the extra points as vertex), assign the value $v = n$ to it;
- (a) Visit its neighbors recursively:
  - if the value assigned to this element is not -1 return to (a),
  - else, if the common face is not a boundary face, assign the value $v = n$,
  - else, if the common face is a boundary face, return to (a).
- If an element with value $v = -1$ can be found, set $n = n + 1$ and return to (a).

In this way, elements can be classified easily with respect to the different connected components of the domain and the suitable components can thus be retained (the outside corresponds to $v = 0$).

REMARK 23.2. This algorithm gives the principles to be followed when defining the position of the elements and, can obviously be optimized.

REMARK 23.3. The above algorithm applies to domains with multiple connected components.

Note that the deletion of outside elements is generally not performed at this step but only at the very end of the generation process. This is to simplify the phase corresponding to the creation and insertion of field points.

24. Creation of the field points

If the data consist only of the boundaries of the domain, it is required to create and insert points in the elements of the triangulation obtained at the previous step, called the boundary triangulation, which are inside the domain.

24.1. Creation and insertion of the internal points

One of the advantages of the Voronoï method is that a triangulation of the domain is known since the boundary points have been taken into account and the integrity of the boundaries has been completed. Thus, this triangulation can be used as a background to help in the creation of internal points. Two major classes of methods are currently used: one class consists in creating iteratively, element by element, a certain number of internal points and inserting them as long as some criterion is not reached; the other class involves a different method (advancing-front, algebraic, elliptic, quadtree, octree, ...) to create the internal points and then uses the Voronoï method to update the triangulation by connecting them.

REMARK 24.1 (*Important*). In the case where an edge or a face has been forced, the current triangulation is not strictly Delaunay. Thus, the classical insertion method may *fail* and must be adapted to ensure, similarly as the derived method, that the elements created are valid and do not overlap with any other elements.

In this section, we assume that only data of geometrical nature is known. Thus, the creation of internal points can only involve geometrical considerations (the case where physical information is available to govern this process will be discussed latter).

24.2. First type method

As an example of a method belonging to the first class, we propose one which is purely local and involves only the Voronoï method. First a local (isotropic) stepsize h_{loc} is associated with each data point. This value is computed by taking into account the length, the surface, ... of the boundary edges and faces. Then, the method is as follows: consider the elements in the mesh, and:

- Insert a point if the element is too large with respect to a value related to the h_{loc} associated with its vertices, HERMELINE [1980]. A variation consists of inserting a point inside each element whose circumcircle (circumsphere) is "large", HOLMES and SNYDER [1988]. The location of the internal point to be created is computed by a weighted barycenter.

or, with a more complete analysis:

- Evaluate the element by computing its quality (i.e. $Q = \alpha(h/\rho)$, where h is the diameter of the element, ρ the radius of its inscribed circle (sphere) and α a normalization factor; note that this quality is clearly related to the interpolation error of any finite element method), its surface or volume and the consistency between the length of its edges and the local h_{loc}'s of its vertices. If the element fails in one of these criteria, the optimal point, associated with its smallest edge, is created. If this point is inside the element or is located inside one of its neighbors, the point is inserted.

The loop over the elements is then repeated until no point creation is required.

24.3. *Second type method*

As an example of a method belonging to the second class, we propose the following which is based on a simplified algebraic method (see Chapter II) for the creation of points and uses the Voronoï method to insert them. As above, a local (isotropic) stepsize h_{loc} is associated with each data point. Then, the method is as follows: consider the elements in the mesh, and:

- For every non-boundary edge e_i of the element
  - compare the length of the edge with the local stepsize of its endpoints, $h_{\text{loc}1}$ and $h_{\text{loc}2}$

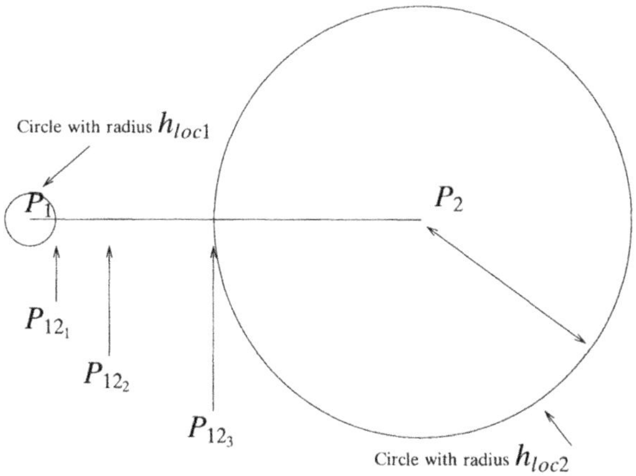

FIG. 24.1. Distribution of points along one edge.

  - determine n_i, the number of points to create on edge e_i, such that they follow a given distribution (arithmetic or geometric) varying smoothly from h_{loc1} to h_{loc2}, see Fig. 24.1
  - stack the so created points and compute their h_{loc}
- End of loop

- Suppress from the stack those points which are too close to another point
- Insert the points in the resulting stack (if it is not empty).

This method has proved to be efficient both in two and three dimensions. A variation (discussed in GEORGE [1992] for the two-dimensional case) consists in creating all the points in only one iteration (and avoid the loop over the entire process). It is as follows, we consider the elements in the mesh and:

- For every edge e_i of the element
  - compare the length of the edge with the local stepsize of its endpoints, h_{loc1} and h_{loc2}
  - determine n_i, the number of points to create on edge e_i, such that they follow a given distribution (arithmetic or geometric) varying smoothly from h_{loc1} to h_{loc2}
- End of loop
  - adjust n_i such that only one or two values are retained in such a way that a catalogued pattern is encountered (see Fig. 24.2 where two patterns are depicted: the left-hand side one corresponds to $n_1 = n_2 = n_3$ and the right-hand side one uses only two splitting parameters $p_1 = n_1 = n_2$ and $p_2 = n_3$ for example)
  - create the n_i desired points and pile them
  - create points inside the element in accordance with the retrieved pattern and stack them

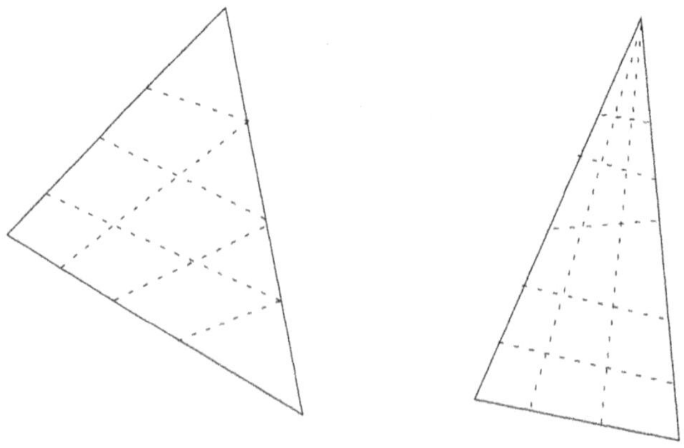

FIG. 24.2. The two patterns for creating the points.

- Suppress from the stack the points which are too close to another point or located on a boundary edge
- Insert the points in the resulting stack.

The main advantage of this method is, as mentioned earlier, that all the points are created in only one iteration, thus they can be easily inserted using a randomization that leads to efficiency. The extension of this method to the three-dimensional situation is easy to conceive: first points on the edges are created, then points on the faces are constructed, after which points inside elements are derived; a priori suitable, this method necessitates a large memory occupation.

24.4. Mesh point and element smoothing

It can be of interest to smooth the triangulation obtained after inserting all the field points. Such a process can be based upon the two different techniques, possibly coupled, introduced in Chapter III.

- Moving the points using a barycentrage technique: when the set of points neighboring a given point P is determined, point P is redefined as:

$$P = \frac{1}{n}\sum_{j=1}^{n} \alpha_j P_{k_j}, \tag{24.1}$$

  where n denotes the number of points connected to point P, P_{k_j} denote these points with numbers k_j, and α_j is the associated weight.
- Using a relaxation method to relocated above point P by "relaxing" the above expression:

$$P^{m+1} = (1-\omega)P^m + \frac{\omega}{n}\sum_{j=1}^{n} \alpha_j P_{k_j}^m \tag{24.2}$$

  using identical notation and where ω is a relaxation parameter.
- Apply some iterations governed by the increase in the element quality: to point P is applied an iterative process which consists of moving P along an arbitrary direction while controlling the effect of this relocation in terms of element quality.
- Local modifications: The transformations firstly developed when dealing with the boundary integrity problem (in the third approach) can be used with the purpose of obtaining better shaped elements.

All of these smoothing methods must be controlled in such a way that it is guaranteed that the triangulation remains valid and, on the other hand, becomes improved with respect to the given element quality measure.

25. Scheme for the mesh generator

Results of the previous sections lead to the following steps as an example of an automatic mesh generator:

(1) Create the set of points associated with the data, i.e. the boundary points of the domain (the endpoints of the segments (triangles) discretizing this boundary) and the field points if they are known.

(2) Compute the location of four (eight) supplementary points in such a way that the quadrilateral (cuboïd) so formed contains all the points in the set.

(3) Create the mesh of this box using two triangles (five tetrahedra).

(4) Insert, one by one, the points of the initial set to obtain a mesh including these points as element vertices.

(5) Regenerate the boundaries and define the outside elements.

(6) Create the internal points (if desired) and insert them.

(7) Remove the outside elements.

(8) Apply the smoothing processes.

Obviously this scheme does not apply when the classical advancing-front method is used for solving the boundary problem. In this case, certain adaptations must be made.

26. Applications in the case of a given control

In this section we assume that we have information concerning the physical problem in solution. For instance, a previous computation with a mesh obtained by the above method is known, from which the solution indicates that some regions present a stiff gradient that must be introduced into the mesh generation process to govern it. Another usual situation is where an a priori knowledge of the behavior of the physical solution can be used to indicate that the mesh must enjoy some properties (finer or larger elements, stretched elements, ...) in some regions.

Possible ideas for updating the mesh created by the method previously described are similar to those discussed in Chapter III, Section 18. These ideas include local mesh enrichment, local mesh de-enrichment in some regions. Such techniques are not easy to implement and implicitly lead to isotropic meshes in the sense that no special directions are a priori involved.

All these reasons lead us to consider the problem of control and adaptation in a different way and, more precisely, not only locally but by involving global *remesh* techniques governed by some criteria. In this respect, a Voronoï type mesh generator offers great flexibility in the different phases and permits a global approach to the desired solution. Such an approach enables us to act, on the one hand, on the way the points are inserted and, on the other hand, on the way the internal points are created by varying the number and the location of these points.

Let us first introduce a more formal writing of the insertion point method described above. To arrive at this, we consider, without loss of generality, Case (a) above and we introduce:

- C_r a criterion (or a set of criteria)
- C_o a constraint (or a set of constraints)

Then the construction (22.1):

$$T_{new} = (T_{old} - \mathcal{S}) \cup \{F_j, P\}_j \quad 1 \leqslant j \leqslant p$$

is just seen as: given T_{old} and P, find an adequate set $\mathcal{S}$. The latter is nothing but the set of elements K_j in T_{old} such that:

(H1) K_j satisfies C_r, and
(H2) K_j satisfies C_o.

To obtain $\mathcal{S}$, and as a consequence the F_j, it is sufficient to find C_r and C_o such that the above construction is effective. For example selecting C_r as:

$$\begin{cases} K_j \text{ satisfies } C_r \text{ iff } P \in B_j \\ \text{no } C_o \end{cases} \tag{26.1}$$

leads to the classical insertion point method and

$$\begin{cases} K_j \text{ satisfies } C_r \text{ iff } \cup K_j \text{ is star-shaped w.r.t. } P \\ \text{and does not contain points other than } P \\ \text{no } C_o \end{cases} \tag{26.2}$$

leads to the derived insertion point method as initially seen.

It is then clear that the modification of C_r and the addition of an adequate C_o such that the process remains valid enable us to define a different way to govern the insertion point process.

Similarly, let us outline the internal point creation process formally. Let us consider the following pair:

- Δ (see Section 18) a control structure enclosing the domain Ω under consideration
- H a function as follows: $H(P, \vec{d})$ (see same section) defined in Δ is the value to be verified at point P for a given direction $\vec{d}$

We have now defined a *control space* (following Definition 5.2).

To return to the above method for the creation of internal points, the following choice is made:

- Δ is nothing but R^2 in two dimensions and R^3 in three dimensions
- $H(P, \vec{d}) = H(P)$ with

$$H(P) = \min_{M \in \Delta} f(P, M), \tag{26.3}$$

where f is the function defined for all points M, different from P in Δ by:

$$f(P, M) = \frac{h_{locP} + h_{locM}}{2}, \tag{26.4}$$

where h_{loc} is defined as above and $h_{\text{loc}M}$ is as follows:
– if M falls within Ω, $h_{\text{loc}M}$ is known,
– if M falls within $\Delta - \Omega$, $h_{\text{loc}M}$ is set to ∞.

Thus any modification concerning the choice of Δ or H leads to a possibility of governing the creation of internal point P. Note that, in the above relation (26.3), H is isotopric as $\vec{d}$ is not explicitly used.

This formal writing of the mesh generation problem using a control space (the pair "control structure, function") offers great flexibility and can be adapted to the physical problem under consideration without difficulty. This is done by defining an adequate control mesh Δ properly in accordance with domain Ω and such that function H can be constructed. To obtain H, a pertinent method is to select a *key variable* σ and control a quantity related to this variable. PERAIRE, VAHDATI, MORGAN and ZIENKIEWICZ [1987] propose to compute the interpolation error estimated for this variable with the purpose of equi-distributing it. More precisely, a mesh is created in such a way:

$$H(P, \ldots) = c_0 g(P, \sigma)$$

for function H selected and c_0 a given tolerance and g the control function with respect to the variation of the key variable.

As an example of constructing Δ and H we follow VALLET, HECHT and MANTEL [1991] who consider the Navier–Stokes equations modeling viscous flows with the velocity as key variable. In this situation, H is constructed as

$$H(P, \vec{d}) = \frac{\| X_d \|}{(X^t_{\vec{d}} \operatorname{Mat}(P) X_{\vec{d}})^{1/2}},$$

where

- $\operatorname{Mat}(P)$ is a matrix determining the metric associated with the desired control,
- $\| \cdot \|$ denotes the norm in the considered metric,
- $X_{\vec{d}}$ is the unit vector in direction $\vec{d}$.

Note that the metric is defined locally at each point P. With this equipment the control function g is as follows:

$$g(P, \sigma) = \left| \frac{\partial^2 \sigma(P)}{\partial d^2} \right|.$$

The control process is then summarized as follows:

- set $i = 0$,
- create mesh T_i of domain Ω using the mesh generation method involving only the geometrical properties of the data,
- (a) compute solution number i of the problem,
- define the control mesh to be T_i,
- compute function g of key variable σ and then function H,
- compare H and c_0 and create needed points,
- set $i = i + 1$,

- create mesh T_i so defined,
- as long as convergence is not achieved, return to (a).

An example of such a process will be depicted in the following section. This example can be easily adapted by modifying the key variable and the metric for a given physical problem.

27. Application examples

The first examples rely on geometrical considerations only. It means that the triangulation created is governed by the boundary of the domain under consideration. The finer the discretization of the boundary is, the finer the triangulation is in the corresponding regions. Figures 27.1 and 27.2 illustrate the triangulations obtained, note that these meshes have not been smoothed.

Triangulation 1 consists of 536 elements, 325 vertices, its worst quality (i.e. the ratio $\alpha(h/\rho)$, where h is the diameter of the element, ρ the radius of it inscribed circle and α a normalization factor) is 1.66 while triangulation 2 includes 6750 elements and 3 527 vertices, with a quality of 2.22.

A parameter can be introduced to dilute the information related to the boundary in such a way that elements are larger in regions far from it. The two previous examples lead to the two triangulations depicted in Figs. 27.3 and 27.4.

Due to a parameter value of 1.2, triangulation 1 now consists of 456 elements, 285 vertices, its worst quality being 1.68 while triangulation 2 now includes 3 968 elements and 2 136 vertices, with an unchanged quality.

In three dimensions, we give the following examples. The first is a rod where the created triangulation consists of 7 054 elements and 1 893 vertices. Element qualities are distributed as shown in Table 27.1, the cpu time is about 7 sec. (HP 735).

The second is the exterior of the aircraft depicted in the figure where the created triangulation consists of 65 115 elements and 11 919 vertices. Element qualities are distributed as shown in Table 27.2, the cpu time is about 30 sec. (same computer).

The qualities of the meshes must be appreciated with respect to the quality

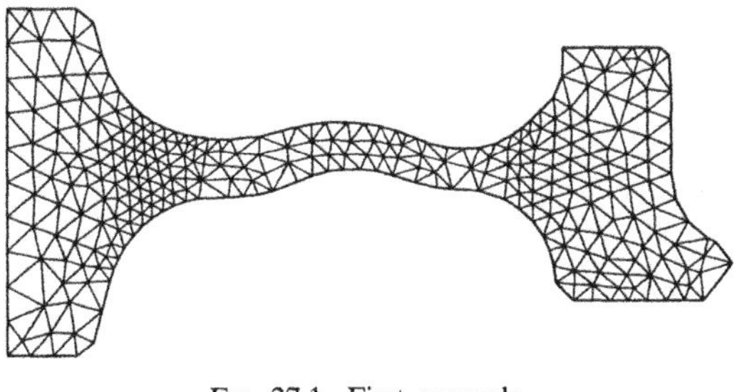

FIG. 27.1. First example.

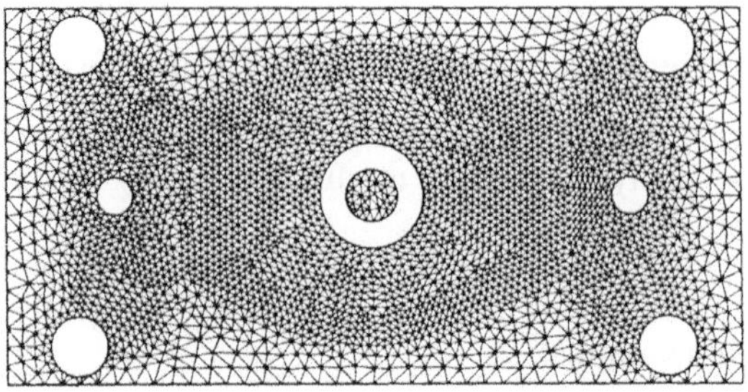

FIG. 27.2. Second example.

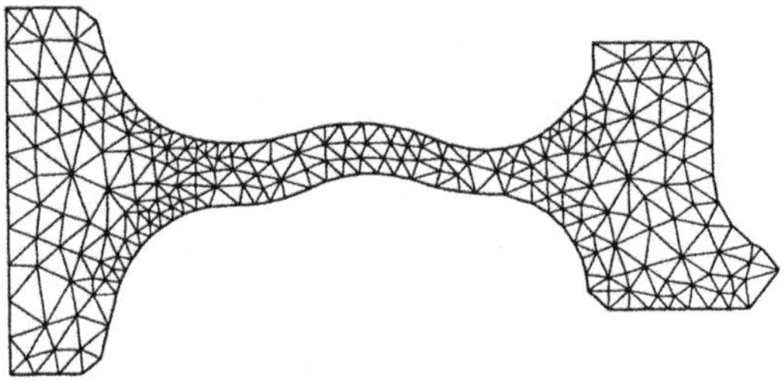

FIG. 27.3. First example.

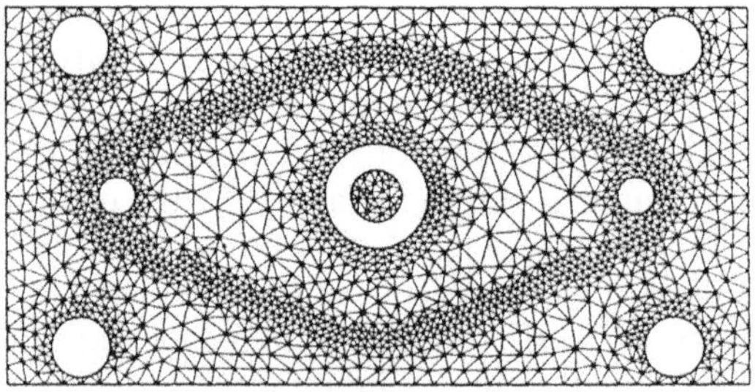

FIG. 27.4. Second example.

TABLE 27.1. Histogram of quality (example 3D number 1).

Quality	Percentage	Number of elements
$1 < Q < 2$	74	5271
$2 < Q < 3$	19	1397
$3 < Q < 4$	3	257
$4 < Q < 5$	.	61
$5 < Q < 6$	.	35
$6 < Q < 7$	.	16
$7 < Q < 8$	.	4
$8 < Q < 9$	.	6
$9 < Q < 10$	.	5
$10 < Q$	.	2

TABLE 27.2. Histogram of quality (example 3D number 2).

Quality	Percentage	Number of elements
$1 < Q < 2$	88	57417
$2 < Q < 3$	9	6176
$3 < Q < 4$	1	804
$4 < Q < 5$	.	281
$5 < Q < 6$		149
$6 < Q < 7$		91
$7 < Q < 8$		50
$8 < Q < 9$		52
$9 < Q < 10$	.	23
$10 < Q$	.	72

of the surface meshes serving as data. In the first example (Fig. 27.5) the worst quality expected is 9.95 and the final mesh has a worst quality of 10.55 and contains only two tetrahedra with a quality greater than the expected value. For the second example (Fig. 27.6), the expected quality is 37.77 while that of the resulting mesh is 39.31 and only one tetrahedron has a quality which is worse than the expected value. Nevertheless, one can see that the elements are distributed mostly in the good quality part of the histograms as can be observed in the above tables.

The examples depicted below illustrate the case where physical information can be used to govern the mesh creation process (see previous section). More precisely, Fig. 27.7 depicts the initial mesh resulting from the automatic method governed only by geometrical considerations while Fig. 27.8 shows the associated velocity. Two iterations of the method in its controlled version are then proceeded and the resulting mesh is that shown in Fig. 27.9 while the obtained solution is shown in Fig. 27.10.

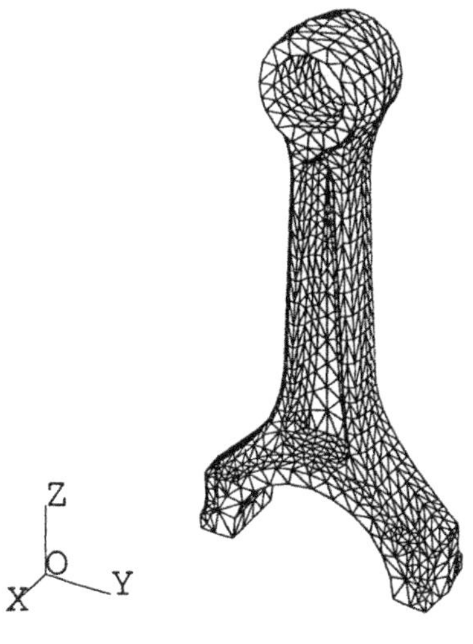

FIG. 27.5. 3-dimensional example.

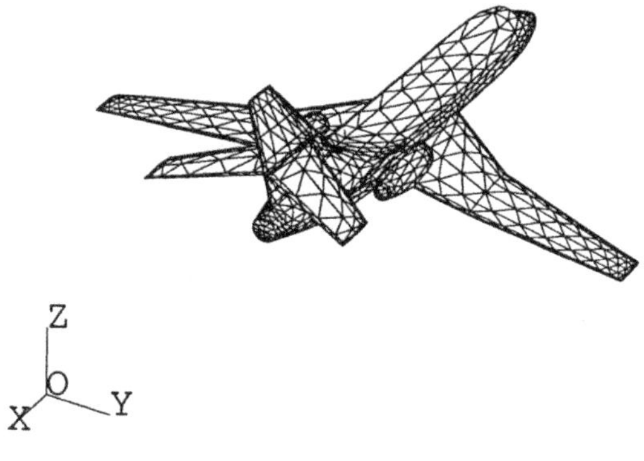

FIG. 27.6. 3-dimensional example.

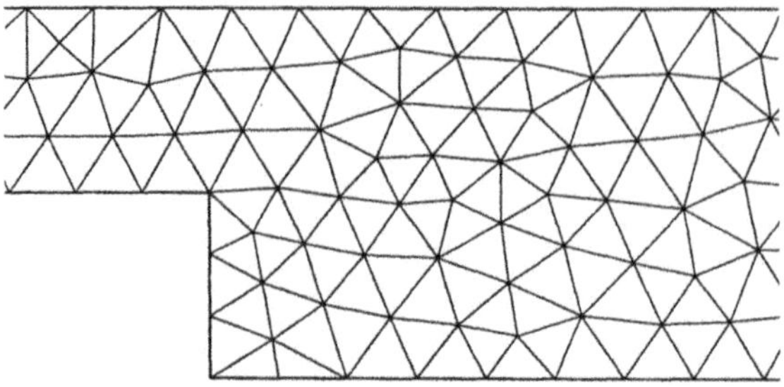

FIG. 27.7. Initial mesh.

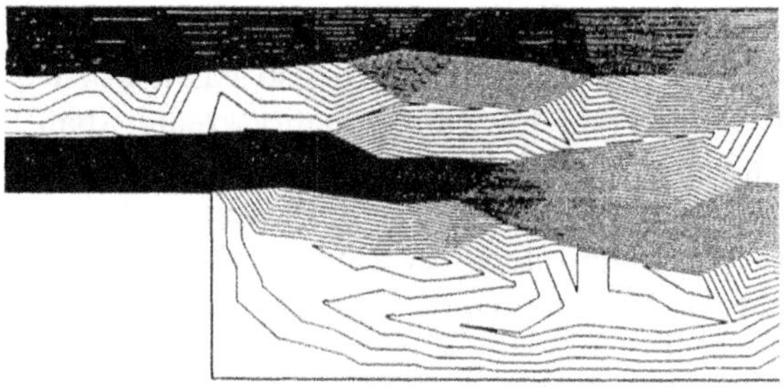

FIG. 27.8. Corresponding solution.

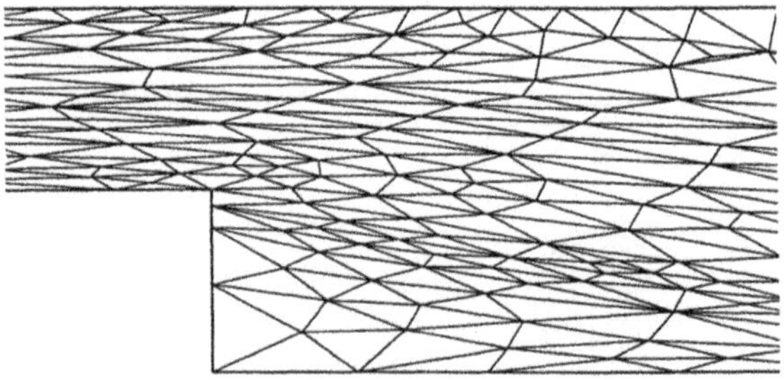

FIG. 27.9. Final mesh.

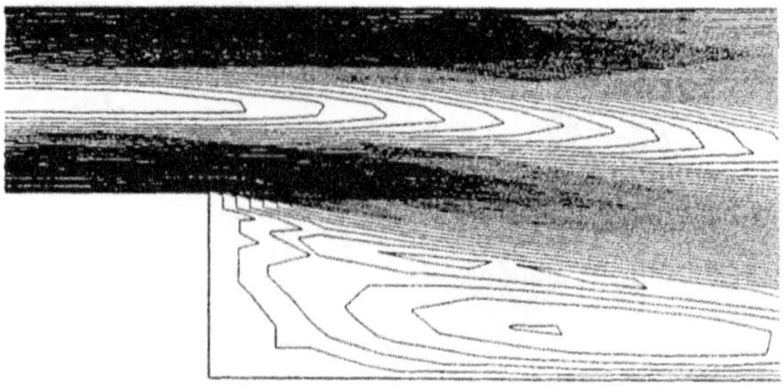

FIG. 27.10. Corresponding solution.

28. Structures and algorithms

Some structures and algorithms can be used in multiple contexts (and in this respect, when using an advancing-front method). In this section, we give some details directly related to the present mesh generation method.

28.1. Basic structure

The Delaunay–Voronoï approach leads to a very simple data structure suitable in both two and three dimensions. It is as follows:

- any element in the mesh is well defined via the oriented list of its 3 (4) vertices. This orientation ensures that elements have positive surfaces (volumes);
- the list of its 3 (4) neighbors is associated with every element, using the following conventions:
  - neighbor number i of element K is opposite to vertex number i of this element;
  - if neighbor i does not exist (i.e. the side (face) opposite to vertex i is a boundary member), it is set to 0.

Table 28.1 shows a part of the basic structure associated with part of the mesh displayed in Fig. 28.1.

When circumcircle (sphere) is used, it is natural to associate the following information with each element:

- the radius of the circumcircle (sphere),
- the coordinates of its centre.

28.2. Background structures

The creation of internal points (see Section 24) uses a stack in which points are stored and then possibly deleted if two points are judged too close. To help with such a decision a regular grid or a quadtree (octree) grid can be employed. First points are encoded, i.e. they are stored in the elements of the grid, then these elements are analyzed to decide which points must be discarted or retained.

TABLE 28.1. Example of basic structure.

	Element							
	1	2	3	4	5	6	7	8
Vertex 1	1	9	1	9	9	9	7	7
Vertex 2	9	7	2	2	3	4	4	5
Vertex 3	8	8	9	3	4	7	5	6
Neighbor 1	2	0	4	0	..	7	..	..
Neighbor 2	0	1	1	5	6	2	8	0
Neighbor 3	3	6	0	3	4	5	6	7

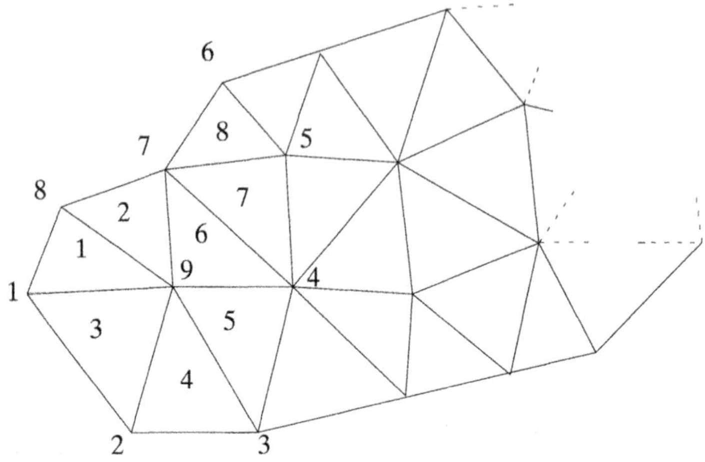

FIG. 28.1. Sample mesh associated with the structure in Table 28.1.

The effective implementation of such a control is detailed hereafter in the case where a regular grid is used.

Firstly, we initialize the control grid. Assuming that the coordinates of the given points lie between 0 and max, the control grid consists of a regular partitioning of a cuboïd composed of *nc* cells in three directions. Depending on their position, points are distributed among the cells. Let x, y and z the coordinates of the point under consideration, then the three indices are:

$$\mathrm{ind}_x = nc\frac{x}{\max + 1}$$

$$\mathrm{ind}_y = nc\frac{y}{\max + 1}$$

$$\mathrm{ind}_z = nc\frac{z}{\max + 1}$$

If the cell with indices ind_x, ind_y and ind_z is empty, the point is encoded in it by setting its number in the cell (at indices ind_x, ind_y, ind_z of the cell table), otherwise the point is linked with the points already present in the cell. This encodage being completed for all the points, the desired decision can be obtained as follows: given a point, the cell in which it falls is found. Then all points already present in the cell and the neighboring cells are analyzed to decide if the given point is too close to one of them. In this case, the given point is removed, otherwise it is added to the control structure (the latter being updated).

Other useful background structures, as quadtree (octree), can be required for some processes. The reader is referred to Chapter II (section discussing the quadtree or octree mesh generation method) for an introduction to these structures.

28.3. Basic algorithms

Regardless of the method selected, there is a series of procedures which are currently used. Among them are those concerning the insertion of a point (already described, for instance in Section 22), the search for an element containing a given point, the search of elements intersected by a given segments, etc.

The second problem has been detailed in Section 20, thus, the only example we will give concerns the third problem which is posed when completing the boundary integrity as it is necessary to examine the context for each missing edge or face. This task implies the knowledge of the elements in the current mesh which are intersected by the missing item. Let us consider the case of a missing edge, say P_iP_j, then the following algorithm can be employed:

- find one element in the mesh with P_i as vertex,
- among all elements sharing vertex P_i, find the one(s) intersected by P_iP_j:
  - *Case* 1: if face F exists, opposite to P_i, intersected by P_iP_j, then element FP_i is the first member of the requested set,
  - *Case* 2: if edge $\alpha\beta$ exists, member of the face opposite to P_i, intersected by P_iP_j, then all elements sharing this edge are member of the requested set,
  - if a vertex exists, different from P_i, identical to P_j, then the element includes edge P_iP_j and this latter now exists: end of algorithm,
- if one of the previous elements contains P_j as vertex, then the requested set is completed,
- otherwise consider:
  - in Case 1, the element with face F opposite to P_i,
  - in Case 2, the elements sharing edge $\alpha\beta$ and not including vertex P_i,
- apply a similar process to this (these) element(s) until P_j is found as vertex.

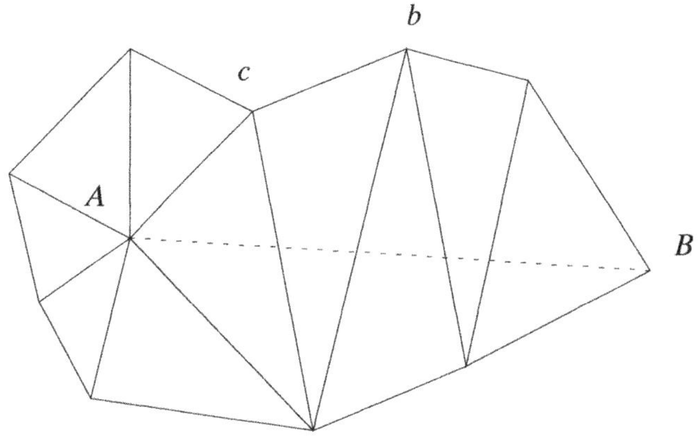

FIG. 28.2. Triangles intersected by a line in two dimensions.

Figure 28.2 corresponds to the same problem in two dimensions (while P_iP_j is replaced by AB).

The above information and programs have been included in this article to illustrate some of the difficulties that occur when implementing a Voronoï method (other methods lead to similar problems). It is clear that the materials provided are not sufficient to write a computer program but aim to help the reader by pointing out some numerical problems. In particular, it seems to be obvious that a pertinent choice of elementary structures leads to facilitate the task and proves to be a solution to obtain some efficiency. As a final remark, the author wishes to indicate that he has experimented with the effective implementation of a Voronoï method in two and three dimensions. The first package consists of about 36 subroutines and 3 900 lines while the second includes about 130 subroutines and 12 658 lines. For comparison, the advancing-front method described in Chapter III has also been experimented in two dimensions. It results in about 30 subroutines and comprises of about 4 000 lines.

References

AHO, A., J. HOPCROFT and J. ULLMAN (1983), *Data Structures and Algorithms* (Addison-Wesley, Reading, MA).

ARCILLA, A.S., J. HAUSER, P.R. EISEMAN and J.F. THOMPSON (1991), Numerical grid generation in computational fluid dynamics and related fields, in: *Proc. Conf.*, Barcelona (North-Holland, Amsterdam) 841–854.

AURENHAMER, F. (1991), Voronoï diagrams. A survey of a fundamental data structure, *ACM Computing Surveys* **23** (3), 345–405.

BAKER, T.J. (1987a), Three dimensional mesh generation by triangulation of arbitrary point sets, AIAA Paper, 87-1124.

BAKER, T.J. (1987b), Mesh generation by a sequence of transformations, in: *Proc. AIAA 8th Comp. Fluid Dynamics Conf.*, Honolulu, HI.

BAKER, T.J. (1988a), Generation of tetrahedral meshes around complete aircraft, in: *Proc. Numerical Grid Generation in Computational Fluid Mechanics '88*, Miami.

BAKER, T.J. (1988b), Unstructured mesh generation and its application to the calculation of flows over complete aircraft, in: *Proc. Tennessee Univ. Space Institute Workshop.*

BAKER, T.J. (1989a), Developments and trends in three-dimensional mesh generation, *Appl. Numer. Math.* **5**, 275–309.

BAKER, T.J. (1989b), Automatic mesh generation for complex three-dimensional regions using a constrained Delaunay triangulation, *Engrg. Comput.* **5**, 161–175.

BAKER, T.J. (1991), Shape reconstruction and volume meshing for complex solids, *Internat. J. Numer. Methods Engrg.* **32**, 665–675.

BATHE, K.J. and S.W. CHAE (1989), On automatic mesh construction and mesh refinement in finite element analysis, *Computers & Structures* **32** (3/4), 911–936.

BERGER, M. (1978), *Géométrie Tome 3: Convexes et Polytopes, Polyèdres Réguliers, Aires et Volumes* (Fernand Nathan, Paris).

BERNADOU, M., P.L. GEORGE, A. HASSIM, P. JOLY, P. LAUG, A. PERRONNET, E. SALTEL, D. STEER, G. VANDERBORCK and M. VIDRASCU (1988), *Modulef: A Modular Library of Finite Elements* (INRIA ed.).

BOWYER, A. (1981), Computing Dirichlet tesselations, *Comput. J.* **24** (2), 162–167.

CABELLO, J., R. LÖHNER and O.P. JACQUOTTE (1991), A variational method for the optimization of directionally stretched elements generated by the advancing front method, in: A.S. ARCILLA, J. HAUSER, P.R. EISEMAN and J.F. THOMPSON, eds., *Numerical Grid Generation in Computational Fluid Dynamics and Related Fields, Conf.*, Barcelona (North-Holland, Amsterdam) 521–532.

CABELLO, J., R. LÖHNER and O.P. JACQUOTTE (1992), A variational method for the optimization of two- and three-dimensional unstructured meshes, *30-th Aerospace Sciences Meeting*, Reno.

CASTILLO, J.E. (1991a), The discrete grid generation on curves and surfaces, in: A.S. ARCILLA, J. HAUSER, P.R. EISEMAN and J.F. THOMPSON, eds., *Numerical Grid Generation in Computational Fluid Dynamics and Related Fields, Conf.*, Barcelona (North-Holland, Amsterdam) 915–924.

CASTILLO, J.E. (1991b), *Mathematical Aspects of Numerical Grid Generation* (SIAM, Philadelphia, PA).

CAUGHEY, D.A. (1978), A systematic procedure for generating useful conformal coordinates mappings, *Internat. J. Numer. Methods Engrg.* **12**, 1651–1657.

CAVENDISH, J.C., D.A. FIELD and W.H. FREY (1985), An approach to automatic three-dimensional finite element mesh generation, *Internat. J. Numer. Methods Engrg.* **21**, 329–347.

CENDES, Z.J., D.N. SHENTON and H. SHAHNASSER (1985), Magnetic field computations using Delaunay triangulations and complementary finite element methods, *IEEE Trans. Magnetics* **21**.

CHERFILS, C. and F. HERMELINE (1990), Diagonal swap procedures and characterizations of 2D-Delaunay triangulations, *RAIRO Modél. Math. Anal. Numér.* **24** (5), 613–626.

CHEW, L.P. (1989), Constrained Delaunay triangulations, *Algorithmica* **4**.

CHEW, L.P. and R.L. DRYSDALE (1985), Voronoï diagrams based on convex distance functions, *ACM 0-89791-163-6*, 235–244.

CIARLET, P.G. (1991), Basic error estimates for elliptic problems, in: P.G. CIARLET and J.L. LIONS, eds., *Finite Element Methods*, Handbook of Numerical Analysis, Vol. II (North-Holland, Amsterdam) 17–351.

COOK, W.A. (1974), Body oriented coordinates for generating 3-dimensional meshes, *Internat. J. Numer. Methods Engrg.* **8**, 27–43.

COULOMB, J.L. (1987), Maillage 2D et 3D. Experimentation de la triangulation de Delaunay, in *Proc. Conf. on Automated Mesh Generation and Adaptation*, Grenoble.

COXETER, H.S.M., L. FEW and C.A. ROGERS (1959), Covering space with equal spheres, *Mathematika* **6**, 147–151.

DELAUNAY, B. (1934), Sur la sphère vide, *Bull. Acad. Sci. URSS Class. Sci. Nat.*, 793–800.

DIRICHLET, G.L. (1850), Uber die Reduction der positiven quadratischen formen mit drei understimmten ganzen Zahlen, *Z. Angew. Math. Mech.* **40** (3), 209–227.

DWYER, R.A. (1987), A faster divide-and-conquer algorithm for constructing Delaunay triangulations, *Algorithmica* **2**, 137–151.

ECER, A., J.T. SPYROPOULOS and J.D. MAUL (1985), A three-dimensional block-structured finite element grid generation scheme, *AIAA J.* **23**, 1483–1490.

FIELD, D.A. (1988), Laplacian smoothing and Delaunay triangulations, *Comm. Appl. Numer. Methods* **4**, 709–712.

FIELD, D.A. and W.D. SMITH (1991), Graded tetrahedral finite element meshes, *Internat. J. Numer. Methods Engrg.* **31**, 413–425.

FIELD, D.A. and K. YARNALL (1989), Three dimensional Delaunay triangulations on a Cray X-MP, in: *Proc. Supercomputing 88*, Vol. 2, Science et Applications (IEEE Computer Soc. Press, Silver Spring MD and ACM, New York).

FREY, W.H. and D.A. FIELD (1991), Mesh relaxation: A new technique for improving triangulations, *Internat. J. Numer. Methods Engrg.* **31**, 1121–1133.

FRITZ, W. (1987), Two-dimensional and three-dimensional block structured grid generation techniques, in: *Conf. on Automated Mesh Generation and Adaption*, Grenoble.

GAREY, D.M.R., S. JOHNSON, F.P. PREPARATA and R.E. TARJAN (1978), Triangulating a simple polygon, *Inform. Process Lett.* **7**.

GEORGE, J.A. (1971), Computer implementation of the finite element method, Stan-CS, Ph.D.

GEORGE, P.L. (1990), *Génération Automatique de Maillages, Applications aux Méthodes d'Elements Finis* (Masson, Paris) RMA no. 16;
also as *Automatic Mesh Generation, Applications to Finite Element Methods* (Wiley, New York, 1991).

GEORGE, P.L. (1991), Génération de maillages par une méthode de type Voronoï, *Rapport de Recherche INRIA*, no. 1398.

GEORGE, P.L. (1992), Génération de maillages par une méthode de type Voronoï, Partie 2: Le cas tridimensionnel, *Rapport de Recherche INRIA*, no. 1664.

GEORGE, P.L. (1993), Transformation d'un maillage en un maillage aigu, *J. Phys.* **3**, 55–68.

GEORGE, P.L., F. HECHT and E. SALTEL (1988), Constraint of the boundary and automatic mesh generation, in: *Numerical Grid Generation in Computational Fluid Mechanics* (Pineridge Press, Swansea).

GEORGE, P.L., F. HECHT and E. SALTEL (1989), Automatic triangulations by a pseudo-Voronoï technique in 3D, in: *Proc. Workshop on Computational Mathematics and Applications*, Pavie, Italie.

GEORGE, P.L., F. HECHT and E. SALTEL (1990a), Fully automatic mesh generator for 3D domains of any shape, *Impact Comput. Sci. Engrg.* **2** (3), 187–218.

GEORGE, P.L., F. HECHT and E. SALTEL (1990b), Automatic 3D mesh generation with prescribed meshed boundaries, *IEEE Trans. Magnetics* **26** (2), 771–774.

GEORGE, P.L., F. HECHT and E. SALTEL (1991), Automatic mesh generation with specified boundary, *Comput. Methods Appl. Mech. Engrg.* **92**, 269–288.

GEORGE, P.L., F. HECHT and M.G. VALLET (1991), Creation of internal points in Voronoï type method, control adaption, *Adv. Engrg. Software* **13** (5/6) 303–313.

GEORGE, P.L. and F. HERMELINE (1992), Delaunay's mesh of a convex polyhedron in dimension d. Application for arbitrary polyhedra, *Internat. J. Numer. Methods Engrg.* **33**, 975–995.

GEORGE, P.L. and E. SEVENO (1992), Génération de maillages par une méthode de type frontal, *Rapport de Recherche INRIA*, no. 1725.

GILDING, B. (1988), A numerical grid generation technique, *Comput. & Fluids* **16**, 47–58.

GOLGOLAB, A. (1989), Mailleur tridimensionnel automatique pour des géométries complexes, *Rapport de Recherche INRIA*, no. 1004.

GORDON, W.J. and C.A. HALL (1973), Construction of curvilinear co-ordinate systems and applications to mesh generation, *Internat. J. Numer. Methods Engrg.* **7**, 461–477.

HALL, C.A. (1976), Transfinite interpolation and applications to engineering problems, in: A.G. Law and B.N. Sahney, *Theory of Approximation* (Academic Press, New York) 308–331.

HAUSER, J. and C. TAYLOR (1986), *Numerical Grid Generation in Computational Fluid Mechanics CFD 86* (Pineridge, Swansea).

HECHT, F. (1992), Thèse d'habilitation, Université Pierre et Marie Curie, Paris.

HECHT, F. and E. SALTEL (1990), Emc2: Un logiciel d'édition de maillages et de contours bidimensionnels, *Rapport Technique INRIA* no. 118.

HERMELINE, F. (1980), Une méthode automatique de maillage en dimension n, Thèse, Université Paris 6, Paris.

HERMELINE, F. (1982), Triangulation automatique d'un polyèdre en dimension N, *RAIRO Anal. Numér.* **16** (3), 211–242.

HO LE, K. (1988), Finite element mesh generation methods: A review and classification, *Comput. Aided Design* **20**, 27–38.

HOLMES, D.G. and D.D. SNYDER (1988), The generation of unstructured triangular meshes using Delaunay triangulation, in: *Numerical Grid Generation in Computational Fluid Mechanics* (Pineridge, Swansea).

JOE, B. (1991a), Delaunay versus max-min solid angle triangulations for three-dimensional mesh generation, *Internat. J. Numer. Methods Engrg.* **31**, 987–997.

JOE, B. (1991b), Construction of three-dimensional Delaunay triangulations using local transformations, *Comput. Aided Geom. Design* **8**, 123–142.

JOHNSTON, B.P. and M. SULLIVAN (1992), Fully automatic two dimensional mesh generation using normal offsetting, *Internat. J. Numer. Methods Engrg.* **33**, 425–442.

LAWSON, C.L. (1986), Properties of n-dimensional triangulations, *Comput. Aided Design* **3**, 231–246.

LEE, D.T. (1980), Two-dimensional Voronoï diagrams in Lp-Metric, *J. ACM* **27** (4), 604–618.

LO, S.H. (1985), A new mesh generation scheme for arbitrary planar domains, *Internat. J. Numer. Methods Engrg.* **21**, 1403–1426.

LO, S.H. (1989), Delaunay triangulation of non-convex planar domains, *Internat. J. Numer. Methods Engrg.* **28**, 2695–2707.

LO, S.H. (1991a), Automatic mesh generation and adaptation by using contours, *Internat. J. Numer. Methods Engrg.* **31**, 689–707.

LO, S.H. (1991b), Volume discretization into tetrahedra – I: Verification and orientation of boundary surfaces, *Comput. Struct.* **39**, 493–500.

LO, S.H. (1991c), Volume discretization into tetrahedra – II: 3D triangulation by advancing front approach, *Comput. Struct.* **39**, 501–511.

LÖHNER, R. (1988), Some useful data structures for the generation of unstructured grids, *Comm. Appl. Numer. Methods* **4**, 123–135.

LÖHNER, R., J. CAMBEROS and M. MERRIAM (1992), Parallel unstructured grid generation, *Comput. Meth. Appl. Mech. Engrg.* **95**, 343–357.

LÖHNER, R. and P. PARIKH (1988a), Three-dimensional grid generation by the advancing-front method, *Internat. J. Numer. Methods Fluids* **8**, 1135–1149.

LÖHNER, R. and P. PARIKH (1988b), Generation of 3D unstructured grids by the advancing front method, in: *Proc. AIAA 88 0515, 26th Aerospace Sciences Meeting*, Reno, NV.

MAVRIPLIS, D.J. (1991), Unstructured and adaptive mesh generation method, in: A.S. ARCILLA, J. HAUSER, P.R. EISEMAN and J.F. THOMPSON, eds., *Numerical Grid Generation in Computational Fluid Dynamics and Related Fields, Conf.*, Barcelona (North-Holland, Amsterdam) 79–92.

MAVRIPLIS, D.J. (1992), An advancing front Delaunay triangulation algorithm designed for robustness, *Icase Report* no 92–49.

MESHKAT, S., J. RUPPERT and H. LI (1991), Three-dimensional automatic unstructured grid generation based on Delaunay tetrahedrization, in: A.S. ARCILLA, J. HAUSER, P.R. EISEMAN and J.F. THOMPSON, eds., *Numerical Grid Generation in Computational Fluid Dynamics and Related Fields, Conf.*, Barcelona (North-Holland, Amsterdam) 841–854.

PERAIRE, J. and J. PEIRO (1992), A 3D Finite Element Multigrid Solver for the Euler Equations, *AIAA*, Reno, NV.

PERAIRE, J., J. PEIRO, L. FORMAGGIA, K. MORGAN and O.C. ZIENKIEWICZ (1988), Finite element Euler computations in three dimensions, *Internat. J. Numer. Methods Engrg.* **26**, 2135–2159.

PERAIRE, J., J. PEIRO and K. MORGAN (1992), Adaptive remeshing for three-dimensional compressible flow computations, *J. of Comp. Phys.* **103**, 269–285.

PERAIRE, J., M. VAHDATI, K. MORGAN and O.C. ZIENKIEWICZ (1987), Adaptive remeshing for compressible flow computations, *J. Comput. Phys.* **72** (2), 449–466.

PERRONNET, A. (1983), Logical and physical representation of an object, modularity for the programming of finite element methods, in: *Proc. IFIP Working Group 2.5 on Numerical Software*, Soderkoping, Sweden.

PERRONNET, A. (1988), A generator of tetrahedral finite elements for multi-material object and fluids, in: *Numerical Grid Generation in Computational Fluid Mechanics* (Pineridge, Swansea).

PREPARATA, F.P. and M.I. SHAMOS (1985), *Computational Geometry, an Introduction* (Springer, Berlin).

RISLER, J.J. (1991), *Méthodes Mathématiques pour la C.A.O.* (Masson, Paris) RMA no. 18.

SAMET, H. (1984), The quadtree and related hierarchical data structures, *Comput. Surveys* **16** (2), 187–285.

SCHROEDER, W.J. and M.S. SHEPHARD (1990), A combined Octree–Delaunay method for fully automatic 3D mesh generation, *Internat. J. Numer. Methods Engrg.* **29**, 37–55.

SENGUPTA, S., J. HAUSER, P.R. EISEMAN and J.F. THOMPSON (1988), *Numerical Grid Generation in Computational Fluid Mechanics 88* (Pineridge, Swansea).

SHAW, J.A. and N.P. WEATHERILL (1992), Automatic topology generation for multiblock grids, *Appl. Math. Comput.* **52**, 355–388.

SHEPHARD, M.S. (1988), Approaches to the automatic generation and control of finite element meshes, *Appl. Mech. Rev.* **41**, 169–185.

SHEPHARD, M.S. and M.K. GEORGES (1992), Reliability of automatic 3D mesh generation, *Comput. Methods Appl. Mech. Engrg.* **101**, 443–462.

SHEPHARD, M.S., K.R. GRICE, J.A. LO and W.J. SCHROEDER (1988), Trends in automatic three-dimensional mesh generation, *Comput. & Structures* **30** (1/2), 421–429.

SHEPHARD, M.S., F. GUERINONI, J.E. FLAHERTY, R.A. LUDWIG and P.L. BAEHMANN (1988), Adaptive solutions of the Euler equations using finite quadtree and octree grids, *Comput. Structures* **30**, 327–336.

SHEPHARD, M.S. and M.A. YERRI (1982), An approach to automatic finite element mesh generation, *Comput. Engrg.* **3**, 21–28.

SLOAN, S.W. (1987), A fast algorithm for constructing Delaunay triangulations in the plane, *Adv. Engrg. Software* **9** (1), 34–55.

SLOAN, S.W. and G.T. HOULSBY (1984), An implementation of Watson's algorithm for computing 2-dimensional Delaunay triangulations, *Adv. Engrg. Software* **6** (4), 192–197.

THACKER, W.C. (1980), A brief review of techniques for generating irregular computational grids, *Internat. J. Numer. Methods Engrg.* **15**, 1335–1341.

THOMPSON, J.F. (1987), A general three dimensional elliptic grid generation system on a composite block-structure, *Comput. Methods Appl. Mech. Engrg.* **64**, 377–411.

THOMPSON, J.F., Z.U.A. WARSI and C.W. MASTIN (1985), *Numerical Grids Generation, Foundations and Applications* (North-Holland, Amsterdam).

VALLET, M.G. (1990), Génération de maillages anisotropes adaptés – Application à la capture de couches limites, *Rapport de Recherche INRIA* no. 1360.

VALLET, M.G., F. HECHT and B. MANTEL (1991), Anisotropic control of mesh generation based upon a Voronoï type method, in: *Numerical Grid Generation in Computational Fluid Dynamics and Related Fields* (North-Holland, Amsterdam).

VORONOÏ, G. (1908), Nouvelles applications des paramètres continus à la théorie des formes quadratiques. Recherches sur les parallélloedres primitifs, *J. Angew. Math.* **134**.

WATSON, D.F. (1981), Computing the n-dimensional Delaunay tesselation with applications to Voronoï polytopes, *Comput. J.* **24** (2), 167–172.

WEATHERILL, N.P. (1985), The generation of unstructured grids using Dirichlet tesselation, MAE report no. 1715, Princeton Univ.

WEATHERILL, N.P. (1988), A method for generating irregular computational grids in multiply connected planar domains, *Internat. J. Numer. Methods Fluids* **8**, 181–197.

WEATHERILL, N.P. (1990a), The integrity of geometrical boundaries in the 2-dimensional Delaunay triangulation, *Comm. Appl. Numer. Methods* **6**, 101–109.

WEATHERILL, N.P. (1990b), Numerical grid generation, *Lecture Series 1990-06*, Von Karman Institute for Fluid Dynamics, Brussels.

WEATHERILL, N.P. (1992), Delaunay triangulation in computational fluid dynamics, *Comput. Math. Appl.* **24** (5/6), 129–150.

WORDENWEBER, B. (1984), Finite element mesh generation, *Comput. Aided Design* **16**, 285–291.

YERRI, M.A. and M.S. SHEPHARD (1983), A modified quadtree approach to finite element mesh generation, *IEEE Comput. Graphics Appl.* **3**, 36–46.

YERRI, M.A. and M.S. SHEPHARD (1985), Automatic mesh generation for three-dimensional solids, *Comput. & Structures* **20**, 211–223.

ZHU, J.Z., O.C. ZIENKIEWICZ, E. HINTON and J. WU (1991), A new approach to the development of automatic quadrilateral mesh generation, *Internat. J. Numer. Methods Engrg.* **32**, 849–866.

ZIENKIEWICZ, O.C. and D.V. PHILLIPS (1971), An automatic mesh generation scheme for plane and curved surfaces by isoparametric co-ordinates, *Internat. J. Numer. Methods Engrg.* **3**, 519–528.

ZIENKIEWICZ, O.C. and J.Z. ZHU (1991), Adaptivity and mesh generation, *Internat. J. Numer. Methods Engrg.* **32**, 783–810.

Subject Index

Numerical Methods for Solids (Part 2)

Limit Analysis of Collapse States

Edmund Christiansen

Department of Mathematics and Computer Science
Odense University
Campusvej 55
DK-5230 Odense
Denmark

HANDBOOK OF NUMERICAL ANALYSIS, VOL. IV
Finite Element Methods (Part 2)—Numerical Methods for Solids (Part 2)
Edited by P.G. Ciarlet and J.L. Lions

Contents

Preface

Limit analysis is the analysis of the simplest possible model for the collapse of a material subject to a static load distribution. The model involves a duality problem in infinite dimensional mathematical programming, formally similar to the min–max problem of game theory. It involves:

(a) Mathematical analysis within the framework of Sobolev spaces.

(b) Numerical analysis based on the finite element method.

(c) Nonlinear optimization methods to solve the discrete problem formulated in (b).

The object of limit analysis is to improve the understanding of plastic materials, or rather, our models for plastic material behaviour, specifically equilibrium equation and yield condition. From a numerical point of view the ambition is to make limit analysis useful for applications as a supplement to the elasticity analysis, which is now a standard tool in engineering. The model of linear elasticity results in linear elliptic boundary value problems, which are so well behaved, both mathematically and numerically, that realistic problems can be handled in great detail. However, even small forces may cause unbounded elastic stresses, clearly in conflict with the physical reality, and the very question of collapse is beyond the model of elasticity.

As is the case in elasticity theory, the numerical analysis is closely related to the mathematical analysis of the model: Both employ the theory for partial differential equations using Sobolev spaces. However, since one of the fundamental concepts for plastic materials, the yield condition, is given as a pointwise condition, i.e., in terms of L^∞, the spaces involved are different from those in elasticity theory. The minimum energy principle in elasticity is replaced by a saddle point problem for a bilinear form in stresses and plastic flow. Uniqueness and regularity, known from the theory of elliptic boundary value problems, do not hold. For the numerical analysis this means that mixed finite elements must be used, and there is a delicate balance between the discrete spaces for stresses and plastic flow. Also the convergence results for the numerical solution are weaker than in the elliptic case.

The solution methods for large nonlinear optimization problems are based on a sequence of steps, each of which involves the sort of sparse numerical linear algebra that is used in elasticity problems. Also in this respect limit analysis stands on the shoulders of the elastic model.

The conclusion of this presentation is that limit analysis is ready for applications. Realistic problems can be solved. However, we need more computational experience on such problems to help us choose between the possible combinations of finite element spaces and several possible optimization methods for the discrete problem.

CHAPTER I

Introduction

1. The physical problem

The plastic analysis of solids must be seen as a supplement to the elastic analysis. The linear elastic model results in elliptic boundary value problems, which are essentially well behaved from a mathematical and numerical viewpoint, whereas the models of plasticity are intrinsically harder. The plastic analysis will probably always lag behind the elastic analysis. However, in view of the quality which practical structural analysis has reached, it is apparent that the linear elastic model is insufficient: In standard applications even small forces may cause unbounded elastic stresses and thus require special treatment and, by definition, the problem of collapse is beyond the model of elasticity. It is a safe, and not at all new, prediction that nonlinear analysis of solids and structures will move from the research table to practical engineering applications, but since the survey by COHN [1979] most progress has still been at the research level. One goal still to be achieved is that the finite element programs, used every day by the working engineer, will be able to perform a plastic collapse analysis in addition to the linear elastic analysis, which is now standard.

Another immediate goal for computations in limit analysis is to supply a feedback to the model. The development of good yield criteria for different materials has been impeded by the lack of consequential computations to compare with experiment. It is impossible to refine a model without the ability to compute the behaviour, which the model predicts. With the recent progress in mathematical optimization and computing power such computations have become possible.

Limit analysis is the simplest way to perform plastic analysis of a solid. The material is assumed to be *rigid, perfectly plastic*, which means that small forces will be neutralized by stresses in the material without any deformation taking place. These stresses are not uniquely determined by the equilibrium equation, which in the rigid plastic case is non-elliptic. If the forces are too large, so that the material cannot sustain stresses to neutralize them, a plastic, i.e., permanent, deformation will occur, and the material will flow for as long as the forces remain and geometric changes can be ignored. Hence the deformation of the material is described by flow or displacement rates, and not by displacements. We shall not try to justify the rigid, perfectly plastic model here, but refer to, e.g., SAVE [1979].

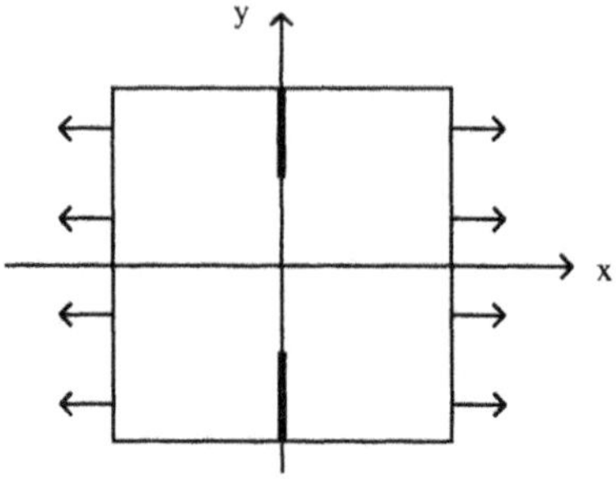

FIG. 1.1. Test problem in plane strain.

Within the model of rigid, perfectly plastic materials the problem of limit analysis can be formulated as follows: Consider a fixed force distribution on the continuum, consisting of volume and surface forces. What is the maximum multiple of this force distribution, which the solid can sustain without collapsing, i.e., without plastic flow occurring? And, at the moment of collapse, what are the fields of stresses and flow in the material? In particular, what is the *plastified region*, i.e., the region of the continuum where the stresses are at the limit, and where local deformation may occur?

Clearly the model of limit analysis is based on a very idealized situation corresponding to a rigid, perfectly plastic material and static or slowly increasing loads. It is the simplest model, which makes it possible to analyse quantitatively the plastic collapse of a material. For a justification of the model of limit analysis we refer to, e.g., KALISZKY [1975] or SAVE [1979].

An alternative approach is the incremental analysis in an elastic, perfectly plastic model, but as pointed out by SAVE [1979, Section 3.1], the validity of both models requires that deformations for loads smaller than the limit load do not alter the initial geometry of the material, and in that case the incremental method offers no improvement. The model of limit analysis describes only the fields for stresses and flow at the collapse moment, i.e., when plastic deformation first occurs. It does not give information about these fields, when deformation has changed the initial geometry. It is, in a way, a flash photo of the collapse moment.

In order to visualize the difference between the elastic and the plastic analysis of a solid we shall consider a specific example. It will be used as a reference test problem later.

EXAMPLE 1.1 (*A test problem in plane strain*). Consider a homogeneous and isotropic block of material, e.g., a metal, as shown in Fig. 1.1. The block is assumed very wide in the z-direction, and quadratic in the $x-y$ plane. All forces and deformations take place in the $x-y$ plane, reducing the problem to 2 space dimensions. This is the plane strain model. The external load consists of a uniform tensile force in the x-direction applied at the ends $x = \pm 1$. To make it more in-

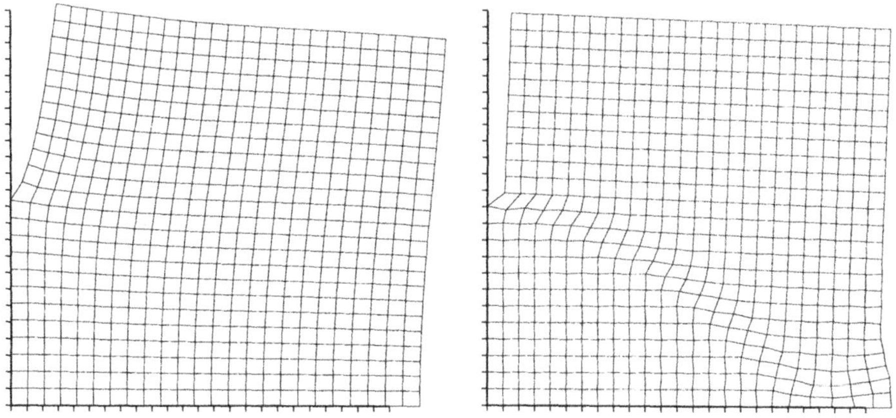

FIG. 1.2. Elastic and plastic solutions for the test problem in Fig. 1.1.

teresting we make symmetric thin cuts at $x = 0$ half way to the mid-plane $y = 0$ (see Fig. 1.1). Due to symmetry the solution need only be computed in the first quadrant, and only this part is shown in Fig. 1.2.

Figure 1.2 shows on the left side the elastic deformation with the tensile force suitably scaled and, on the right side, the plastic limit solution, both computed using the finite element method. The plastic solution needs explanation: The collapse flow field $\boldsymbol{u}$ has been computed at the grid points and multiplied by a time constant. The resulting displacement field is then plotted. Since the collapse velocity is determined only at the instant of collapse, this displacement may never be seen, but there is general agreement that this is the best way to visualize the flow field. It gives a better idea of the local deformations at collapse than the velocity field itself does. We see that the elastic deformation is smooth and distributed over the solid, while the plastic flow field is of a different nature. We can identify two rigid regions separated by a relatively narrow interface, which seems to contain the entire collapse mechanism. The stresses are not shown in the figure, although they have been computed as part of the solution. The elastic stress field may be computed from the deformation. The exact stress field is known to have a singularity at the tip of the cuts and to be smooth everywhere else. The plastic collapse stress tensor $\boldsymbol{\sigma}(\boldsymbol{x})$ is much harder to find. It is not uniquely determined outside the plastified region, i.e., the part of the material, where there is deformation taking place. In the chapter on computation we shall see several ways to visualize the collapse stresses, also for the problem in this example.

The qualitative difference between the elastic and the plastic behaviour of a solid is reflected in the mathematics and the numerics of the problem. The plastic model does not have the nice mathematical properties of the elastic model, and numerically the solution is much harder to compute. All useful numerical solution methods in limit analysis for a continuous medium known to this author are based on discretizations of a variational principle using the finite element method. As is the case for elliptic problems the numerical analysis of the finite element method applied to limit analysis is closely related to the mathematical analysis of the continuous problem, so we have to start there. After formulating the continuous problem we shall attempt to point out some of the steps, mainly analysis of corresponding formulations for discrete problems and structures, that have led the way.

2. The mathematical model

We first formulate the problem in its most basic form. Subsequently it will be indicated how the analysis applies also to more general boundary conditions and forces. We concentrate on the mathematical formulation to be analysed later. For a closer link to the mechanics we refer to, e.g., KACHANOV [1971], MARTIN [1975, Chapter 9] or ZAVELANI-ROSSI [1979].

Let Ω denote the bounded domain in space occupied by a rigid, perfectly plastic continuum. The boundary $\partial\Omega$ consists of two parts, S and T: $\partial\Omega = S \cup T$. On S the boundary is fixed, i.e., $\boldsymbol{u} = \boldsymbol{0}$ on S, where $\boldsymbol{u} = \boldsymbol{u}(\boldsymbol{x})$ denotes the plastic displacement rate at each point $\boldsymbol{x}$ in $\overline{\Omega}$, the closure of Ω. On T there are given surface forces $\boldsymbol{g}(\boldsymbol{x}), \boldsymbol{x} \in T$, (possibly zero), and inside the material there is given a field of volume forces $\boldsymbol{f}(\boldsymbol{x})$, $\boldsymbol{x} \in \Omega$ (e.g., zero or gravity forces). This pair of force fields $(\boldsymbol{f}, \boldsymbol{g})$ is kept fixed. The problem of limit analysis is to find the *limit multiplier*, i.e., the upper limit λ^* of the set of real multipliers λ, such that the body of material can "carry" the force distribution $(\lambda\boldsymbol{f}, \lambda\boldsymbol{g})$.

Associated with $\boldsymbol{u}$ is the *strain rate tensor* $\boldsymbol{\varepsilon} = \boldsymbol{\varepsilon}(\boldsymbol{u})$ defined by

$$\varepsilon_{ij}(\boldsymbol{u}) = \frac{1}{2}\left(\frac{\partial u_i}{\partial x_j} + \frac{\partial u_j}{\partial x_i}\right). \tag{2.1}$$

This is the notation normally used for the strains themselves, and not their time derivatives. Since displacements and strains do not occur in limit analysis (only the corresponding rates), we ask the reader to accept this notation. It can also be justified by the fact that the time scale does not occur in the model.

Given a force distribution $(\boldsymbol{f}, \boldsymbol{g})$ a virtual displacement rate $\boldsymbol{u}$ in the continuum will be associated with the *external work rate*

$$F(\boldsymbol{u}) = \int_{\Omega} \boldsymbol{f}\cdot\boldsymbol{u} + \int_{T} \boldsymbol{g}\cdot\boldsymbol{u}. \tag{2.2}$$

The stresses inside the material are given by the symmetric *stress tensor* $\boldsymbol{\sigma} = \boldsymbol{\sigma}(\boldsymbol{x})$. The classical form of the equation of equilibrium between $\boldsymbol{\sigma}$ and the force distribution $(\boldsymbol{f}, \boldsymbol{g})$ defined above is

$$\begin{aligned} -\boldsymbol{\nabla}\cdot\boldsymbol{\sigma} &= \boldsymbol{f} \quad \text{in } \Omega, \\ \boldsymbol{\nu}\cdot\boldsymbol{\sigma} &= \boldsymbol{g} \quad \text{on } T, \end{aligned} \tag{2.3}$$

where $\boldsymbol{\nabla}$ denotes the divergence operator, here applied to the columns of $\boldsymbol{\sigma}$, and $\boldsymbol{\nu}$ is the outward normal to Ω. The coordinate form of Eq. (2.3) is

$$-\sum_{j=1}^{3} \frac{\partial \sigma_{ij}}{\partial x_j} = f_i \quad \text{in } \Omega \quad \text{for } i = 1, 2, 3,$$

$$\sum_{j=3}^{3} \nu_j \sigma_{ij} = g_i \quad \text{on } T \quad \text{for } i = 1, 2, 3.$$

In rigid plasticity there is a priori no relation given between the strains and the stresses. In the collapse state, however, such a relation will be deduced.

We shall always interpret Eq. (2.3) as being equivalent to the more precise variational form Eq. (2.6), which we now derive: To a given pair of stresses $\boldsymbol{\sigma}$ and displacement rates $\boldsymbol{u}$ we associate the *internal work rate* or *energy dissipation rate*

$$a(\boldsymbol{\sigma}, \boldsymbol{u}) = \int_{\Omega} \langle \boldsymbol{\sigma}, \boldsymbol{\varepsilon}(\boldsymbol{u}) \rangle$$

$$= \int_{\Omega} \sum_{i,j} \sigma_{ij} \varepsilon_{ij}(\boldsymbol{u}) \tag{2.4}$$

$$= -\int_{\Omega} (\boldsymbol{\nabla} \cdot \boldsymbol{\sigma}) \cdot \boldsymbol{u} + \int_{T} (\boldsymbol{\nu} \cdot \boldsymbol{\sigma}) \cdot \boldsymbol{u}. \tag{2.5}$$

For the moment we ignore technical details and assume that $\boldsymbol{\sigma}$ and $\boldsymbol{u}$ are sufficiently smooth. The second equality then follows from Green's formula. The equation of virtual work (rate) states that the external work rate defined by Eq. (2.2) and the internal work rate defined by Eq. (2.4) and Eq. (2.5) must be equal for all admissible $\boldsymbol{u}$:

$$a(\boldsymbol{\sigma}, \boldsymbol{u}) = F(\boldsymbol{u}) \quad \text{for all } \boldsymbol{u} \text{ satisfying the boundary condition on } S. \tag{2.6}$$

Equating Eq. (2.2) and Eq. (2.5) we see that the classical form of the equilibrium equation Eq. (2.3) is equivalent to the generalized form Eq. (2.6), at least for sufficiently smooth fields $\boldsymbol{\sigma}$ and $\boldsymbol{u}$. Note that the boundary condition on the stresses (the second part of Eq. (2.3)) is only implicit. In the terminology of elliptic boundary value problems the boundary condition on the displacement field is an *essential boundary condition*, while the condition on the stress field is a *natural boundary condition* as in the Neumann problem.

There is a limit to the strength of a plastic material. In this model it means that the stress tensor must satisfy a *yield condition* in addition to the equilibrium equation Eq. (2.6). The yield condition is a pointwise constraint of the form

$$\boldsymbol{\sigma}(\boldsymbol{x}) \in B(\boldsymbol{x}) \quad \text{for all } \boldsymbol{x} \in \Omega. \tag{2.7}$$

If the material is assumed to be homogeneous $B(\boldsymbol{x})$ is independent of $\boldsymbol{x}$ and will be denoted B. For the mathematical theory it is no complication to allow the yield condition to depend on $\boldsymbol{x}$. About the set of admissible stresses $B(\boldsymbol{x})$ at each point we make the following assumptions:

(a) There exists $\varepsilon > 0$, independent of $\boldsymbol{x}$, such that $\sum_{ij} |\sigma_{ij}| \leqslant \varepsilon$ implies $\boldsymbol{\sigma} \in B(\boldsymbol{x})$.

(b) $B(\boldsymbol{x})$ is a convex subset of the space of stress tensors.

(c) $B(\boldsymbol{x})$ is a closed subset of $\mathbb{R}^9$.

These three conditions are crucial for the mathematical and numerical analysis of the model. The theory of convex optimization can be applied to prove that the concept of a collapse state is well defined. (a) states that the material can sustain any stress field, if only the stresses are sufficiently small at every point, i.e., there are no points or directions of "zero-strength" for the solid. (b) corresponds to adequate physical assumptions, while condition (c) is purely mathematical without physical significance. It states that a limit of admissible stresses is itself admissible.

It is important to note that $B(\boldsymbol{x})$ is not assumed to be bounded. In three-dimensional problems and in the two-dimensional plane strain model B is unbounded for many materials. It is bounded in plane stress and for plate bending problems. An unbounded yield set means that certain types of stresses may be arbitrarily large. In order to understand this, let us consider a specific classical yield condition:

The Von Mises yield condition in three space dimensions for a homogeneous material can be written as follows

$$(\sigma_{11} - \sigma_{22})^2 + (\sigma_{22} - \sigma_{33})^2 + (\sigma_{33} - \sigma_{11})^2 + 6(\sigma_{12}^2 + \sigma_{23}^2 + \sigma_{31}^2) \leqslant 2\sigma_0^2, \qquad (2.8)$$

σ_0 being the yield stress in simple one-dimensional tension. Define the trace of a tensor $\boldsymbol{\sigma}$ as

$$\operatorname{tr}(\boldsymbol{\sigma}) = \sum \sigma_{ii}$$

and let $\boldsymbol{I}$ denote the identity tensor (δ_{ij}). Then we define the *deviator* of $\boldsymbol{\sigma}$ as the tensor

$$\boldsymbol{\sigma}^{\mathrm{D}} = \boldsymbol{\sigma} - \frac{1}{N} \operatorname{tr}(\boldsymbol{\sigma}) \cdot \boldsymbol{I}, \qquad (2.9)$$

where N is the spatial dimension (2 or 3) of the problem. Clearly $\boldsymbol{\sigma}$ satisfies the yield condition Eq. (2.8), if and only if $\boldsymbol{\sigma}^{\mathrm{D}}$ does, but since $\boldsymbol{\sigma}^{\mathrm{D}}$ has trace zero, Eq. (2.8) puts a bound on its single components. Typically the yield condition will be of the form

$$\boldsymbol{\sigma}^{\mathrm{D}}(\boldsymbol{x}) \in B(\boldsymbol{x}) \quad \text{for all } \boldsymbol{x} \in \Omega, \qquad (2.10)$$

where $B(\boldsymbol{x})$ is a bounded subset of $\mathbb{R}^9$. Such yield conditions, which can be formulated exclusively in terms of the deviator $\boldsymbol{\sigma}^{\mathrm{D}}$ of the stresses, are insensitive to the addition of tensors of the form $\boldsymbol{\sigma} = \phi \boldsymbol{I}$, where ϕ is any scalar function defined on Ω. Hence the material can sustain any such tensor field, no matter how large ϕ is. These tensors correspond to hydrostatic pressure in the material, so using a yield condition of this form is equivalent to assuming that the material does not yield under hydrostatic pressure. We shall see later that there is a corresponding dual property of the plastic flow $\boldsymbol{u}$: It is divergence free, or incompressible: $\nabla \cdot \boldsymbol{u} = 0$. We do not have to require this property. It follows automatically, but it is important to be aware of this dual relationship in the mathematical and numerical analysis of the problem.

We can now formulate the *static principle* of limit analysis: Find the limit multiplier λ^* of a given force distribution $(\boldsymbol{f}, \boldsymbol{g})$ that the solid can "carry". In other words, we want to maximize λ subject to the constraints that there exists a stress field $\boldsymbol{\sigma}$, which is in equilibrium with $(\lambda \boldsymbol{f}, \lambda \boldsymbol{g})$ and which satisfies the yield condition Eq. (2.7). With the above notation this may be written

$$\lambda^* = \max\{\lambda \mid \exists \boldsymbol{\sigma} \in K\colon a(\boldsymbol{\sigma}, \boldsymbol{u}) = \lambda F(\boldsymbol{u}) \text{ for all } \boldsymbol{u} \text{ with } \boldsymbol{u} = \boldsymbol{0} \text{ on } S\}, \tag{2.11}$$

where K denotes the set of stress fields that satisfy the pointwise yield condition at every point. Details will be made precise later.

Since GREENBERG and PRAGER [1952] it is customary to term a stress field $\boldsymbol{\sigma}$ *statically admissible* with a given force distribution $(\boldsymbol{f}, \boldsymbol{g})$, if it satisfies the equilibrium equation Eq. (2.3) or equivalently Eq. (2.6). If, in addition, the stress field satisfies the yield condition, it is called *safe*. Similarly a velocity field $\boldsymbol{u}$ is termed *kinematically admissible*, if it satisfies the boundary condition $\boldsymbol{u} = \boldsymbol{0}$ on S.

The *kinematic principle* of limit analysis is formulated in terms of the plastic flow: We define the *energy dissipation rate* associated with $\boldsymbol{u}$ as follows

$$D(\boldsymbol{u}) = \max_{\sigma \in K} a(\boldsymbol{\sigma}, \boldsymbol{u}). \tag{2.12}$$

The kinematic principle states that the limit multiplier λ^* is the minimum of $D(\boldsymbol{u})$ taken over all $\boldsymbol{u}$ with external work rate normalized to unity:

$$\lambda^* = \min_{F(u)=1} D(\boldsymbol{u}). \tag{2.13}$$

The physical reasoning behind the kinematic principle is as follows: If $\lambda > \lambda^*$, then there is a $\boldsymbol{u}$ with $F(\boldsymbol{u}) = 1$, such that $\lambda > D(\boldsymbol{u})$. This implies that for all stress tensors satisfying the yield condition we have that $\lambda = \lambda F(\boldsymbol{u}) > a(\boldsymbol{\sigma}, \boldsymbol{u})$. Hence, for this particular flow field $\boldsymbol{u}$, no admissible stress tensor can neutralize the work rate of the external forces $(\lambda \boldsymbol{f}, \lambda \boldsymbol{g})$, and the material will yield. Conversely, if $\lambda < \lambda^*$, no such $\boldsymbol{u}$ exists.

The static and kinematic principles are formulated independently of each other. However, we shall see that they are dual mathematical programming problems. In particular, Eq. (2.11) and Eq. (2.13) give the same value for λ^*, and the extremal values are attained for some stress field $\boldsymbol{\sigma}^* \in K$ for the problem Eq. (2.11) and some admissible velocity field $\boldsymbol{u}^*$ for the problem Eq. (2.13). The pair $(\boldsymbol{\sigma}^*, \boldsymbol{u}^*)$ is a saddle point for the bilinear form $a(\boldsymbol{\sigma}, \boldsymbol{u})$ and can be interpreted as the fields for stress and flow in the collapse state. More precisely, the following holds for all $\boldsymbol{\sigma}$ satisfying the yield condition and all $\boldsymbol{u}$ satisfying the boundary conditions and normalized to $F(\boldsymbol{u}) = 1$:

$$a(\boldsymbol{\sigma}, \boldsymbol{u}^*) \leqslant \lambda^* = a(\boldsymbol{\sigma}^*, \boldsymbol{u}^*) \leqslant a(\boldsymbol{\sigma}^*, \boldsymbol{u}). \tag{2.14}$$

This variational principle, formulated for discrete structures or continuous solids, is as fundamental to limit analysis as the minimum energy principle is to the elliptic problems of elasticity.

In order to simplify the notation we have made a couple of assumptions in the above formulation: The boundary condition on $\boldsymbol{u}$ either prescribes $\boldsymbol{u}$ completely, or leaves it free. A more general boundary condition may allow the surface to move only in certain directions, e.g., tangential to the surface, and be subject to surface forces in these directions. Also, there may be a fixed force distribution such as gravity, sometimes referred to as a pre-load or a dead load, on the continuum in addition to the load $(\boldsymbol{f}, \boldsymbol{g})$, for which we want to find the maximal multiplier. The analysis of the next chapter is readily modified to cover these cases as well.

3. Remarks on the development of limit analysis

Like so many other numerical methods limit analysis has its origin in clever hand calculations. Suppose that for some value λ we are able to guess or construct a stress field, which satisfies the yield condition and is in equilibrium with the forces $(\lambda\boldsymbol{f}, \lambda\boldsymbol{g})$. Then, by the static principle, λ is a lower bound for the limit multiplier λ^*. Similarly, by the kinematic principle, for any flow field $\boldsymbol{u}$, which satisfies the boundary condition and has the work rate normalized to $F(\boldsymbol{u}) = 1$, $D(\boldsymbol{u})$ will provide an upper bound of λ^*. Thus the two principles can be used to establish bounds for the collapse multiplier, simply by guessing stress or flow fields. This approach is frequently referred to as *lower bound methods*, respectively *upper bound methods*. In the lower bound method it may be difficult to verify the equilibrium equation for the stresses, while the calculation of $D(\boldsymbol{u})$ is the hard part when using an upper bound method. If the yield condition is of the form Eq. (2.10), then $D(\boldsymbol{u})$ is equal to $+\infty$ unless $\boldsymbol{u}$ is divergence free. Only in rare cases is it possible to determine the exact limit multiplier this way by getting the same value for the upper and lower bound.

The static principle was formulated in a special case already by KAZINCZY [1914] and KIST [1917]. Precise physical arguments for both the static and kinematic principles (often referred to as "theorems" in the literature of mechanics) based on the convexity of the yield set were given in GVOZDEV [1938]. More details about the early history of limit analysis are given in NEAL [1956, Chapter 3]. It must be emphasized that all early formulations and analysis are given for simple structures and not for a general continuous medium.

In 1951–52 the theory expanded in two directions: In DRUCKER, PRAGER and GREENBERG [1952] the static and kinematic principles were established for continuous media, and in CHARNES and GREENBERG [1951] the connection between the static and kinematic principles for trusses on one side and linear programming duality on the other side was pointed out. It was later proved for frames in CHARNES, LEMKE and ZIENKIEWICZ [1959]. The realization of the connection to duality theory seemed to surface simultaneously in several places and immediately made a major impact. The application of linear programming and the simplex method to limit analysis made it possible in general to find the exact limit multiplier for finitely determined structures. This started an explosive development in the plastic analysis of structures as reflected in a multitude of papers. Many problems for simple structures can adequately be formulated with piecewise linear yield conditions, and nonlinear optimization algorithms was not ready for use in this context yet.

Limit analysis for continuous media needed more time to develop. Computations had to await the success of the finite element method. It appears that the first descriptions of the finite element method applied to limit analysis was ARGYRIS [1967, Chapter IV] (bound methods for discrete structures), HAYES and MARCAL [1967] (upper bound method in plane stress), and HODGE and BELYTSCHKO [1968] (upper and lower bound methods for plates). Finite element function spaces were used to represent stresses or flow field, and the associated computed bound was optimized over the discrete space. Thus the exact limit multiplier for the *discretized structure* was determined, and this value was then a bound (lower or upper) for the limit multiplier for the continuous problem. In both cases nonlinear optimization methods were used, but of course only for rather coarse elements. Convergence, as the element size tends to zero, was not discussed. At this time the development again exploded with papers using the finite element method to discretize continuum problems combined with convex and linear programming. At first only bound methods were used with finite element representations of stresses or flow, but not both. In CAPURSO [1971] and ANDERHEGGEN and KNÖPFEL [1972] mixed finite elements were introduced to discretize stresses and flow simultaneously, in both cases using linear programming, where the duality between the static and the kinematic principle was established. Without a duality theorem the mathematical foundation of the mixed method is missing, a point emphasized by CASCIARO and DICARLO [1974, p. 172].

Mathematical respectability, in form of a proof within a rigorous mathematical frame of the duality between the static and kinematic principles, took even longer. NAYROLES [1970] gave a formulation of the problem as a duality problem in infinite-dimensional mathematical programming. The duality was proved, and also the existence of a collapse mechanism in the duality between $\boldsymbol{\sigma}$ with components in L^∞ and $\boldsymbol{\varepsilon}$ with components in the dual space $(L^\infty)^*$. In order to recover $\boldsymbol{u}$ from $\boldsymbol{\varepsilon}$ in this formulation it was necessary to work within the quotient space modulo rigid displacements. The spaces and boundary conditions do not quite match the physical problem. In the fundamental work DUVAUT and LIONS [1972] the rigid perfectly plastic problem is introduced as a limit case of the elasto-plastic model.

MOREAU [1976] and CHRISTIANSEN [1976] deduced from analysis of the duality that the strain rates $\boldsymbol{\varepsilon}(\boldsymbol{u})$ must be measures. The space of velocities $\boldsymbol{u}$ with precisely this property was introduced in MATTHIES, STRANG and CHRISTIANSEN [1979]. It is called the *space of bounded deformation* and is denoted $\mathrm{BD}(\Omega)$. The properties of BD were analysed in SUQUET [1978, 1979], TEMAM and STRANG [1980a] and TEMAM [1981b]. These results were used to analyse the mathematical problem of plasticity in TEMAM [1981a, 1983], TEMAM and STRANG [1980b] and SUQUET [1981] among others.

In the attempt to prove the existence of collapse flow CHRISTIANSEN [1980a] and SUQUET [1981] independently concluded that even $\mathrm{BD}(\Omega)$ is not quite large enough. The collapse flow may have discontinuities along the boundary of Ω, and the associated energy dissipation is not measured if the strain rate tensor is defined only by distributional derivatives, as in $\mathrm{BD}(\Omega)$. A proof of duality and existence of collapse fields incorporating this is given in CHRISTIANSEN [1986].

CHAPTER II

The Duality Problem

4. Why so different from elasticity?

In this section we try to point out some of the mathematical characteristics of limit analysis without referring to Sobolev space theory. The purpose is to make it possible to understand the nature of the collapse solution without knowing the technical mathematical details.

Figure 1.2 in the introduction is an attempt to illustrate the difference between elastic and plastic behaviour of a solid. This section is an attempt to point out the profound theoretical differences between the underlying mathematical models.

In elasticity, both linear and nonlinear, the solution is uniquely determined as being the displacement field minimizing an energy functional. The *collapse solution* in limit analysis is a triple $(\lambda^*, \boldsymbol{\sigma}^*, \boldsymbol{u}^*)$, where $\boldsymbol{\sigma}^*$ provides the maximum in the static principle Eq. (2.11), $\boldsymbol{u}^*$ gives the minimum in the kinematic principle Eq. (2.13), both values being equal to λ^*. This is the saddle point problem Eq. (2.14) for the bilinear form $a(\boldsymbol{\sigma}, \boldsymbol{u})$ in Eq. (2.4). We shall see that existence holds in a generalized sense only. Uniqueness of the collapse mechanism simply does not hold in general, either for stresses or for the flow. And at the moment of collapse there may be large rigid regions of the material with zero strain rates, a phenomenon not seen in elasticity due to the nature of solutions to elliptic boundary value problems.

Concerning the *existence* problem it turns out that $\boldsymbol{u}^*$ does not necessarily exist in the ordinary Sobolev space sense, i.e., as a locally integrable function in Ω, which satisfies the boundary conditions (see, e.g., SHOEMAKER [1979]). This is illustrated by the following example, which is similar to Example 2.3 in CHRISTIANSEN [1982].

EXAMPLE 4.1 (*An example in antiplane shear*). Consider an infinite layer of constant thickness in the y–z-plane between $x = -1$ and $x = 1$ as indicated in Fig. 4.1. Only forces in the y-direction are allowed, and they may only depend on x. The problem is then completely described by the following scalar functions on the interval $-1 \leqslant x \leqslant 1$:

$$\sigma = \sigma_{xy} = \sigma_{xy}(x),$$
$$u = u_y = u_y(x),$$

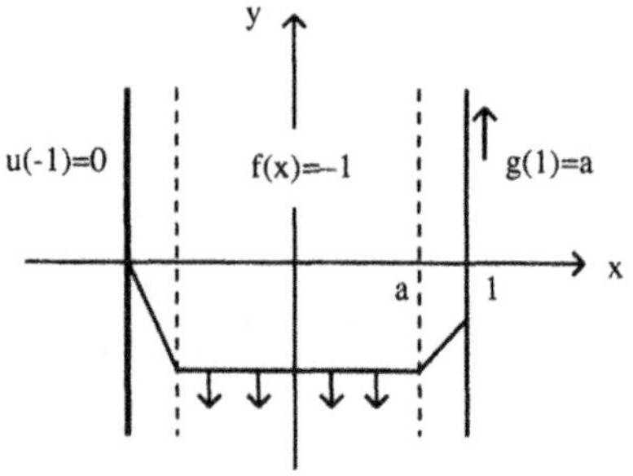

FIG. 4.1. A problem in antiplane shear.

$$f = f_y = f_y(x),$$
$$g = g_y = g_y(x).$$

The layer is kept fixed at $x = -1$ by imposing the boundary condition $u(-1) = 0$. We now solve the limit analysis problem for this example. The rate of internal work, Eq. (2.4), is

$$a(\sigma, u) = \int_{-1}^{1} \sigma u' \,\mathrm{d}x = -\int_{-1}^{1} \sigma' u \,\mathrm{d}x + \sigma(1)u(1). \tag{4.1}$$

As yield condition we use $|\sigma| \leqslant 1$. We consider the following external forces:

$$f(x) = \begin{cases} -1 & \text{for } |x| < a, \\ 0 & \text{for } |x| > a, \end{cases}$$
$$g(1) = a,$$

for some a, $0 < a \leqslant 1$. Thus one side of the layer at $x = -1$ is fixed. There is a uniform gravity-like force pulling down a central section of the layer, $-a < x < a$, and we are pulling up the other side of the layer at $x = 1$. The equilibrium equation, Eq. (2.6), for the pair $(\lambda f, \lambda g)$ is

$$a(\sigma, u) = \lambda F(u) = \lambda \left(-\int_{-a}^{a} u \,\mathrm{d}x + a u(1) \right) \tag{4.2}$$

for all u with $u(-1) = 0$. Combining Eq. (4.1) and Eq. (4.2) we get the classical form

$$\sigma'(x) = \begin{cases} \lambda & \text{for } |x| < a, \\ 0 & \text{for } |x| > a, \end{cases} \tag{4.3}$$
$$\sigma(1) = \lambda a.$$

The static principle takes the form

maximize λ
subject to Eq. (4.3) and $|\sigma| \leqslant 1$

The solution is easily seen to be

$$\lambda^* = \frac{1}{a}$$

with the uniquely determined collapse stresses

$$\sigma^*(x) = \begin{cases} -1 & \text{for } -1 \leqslant x \leqslant -a, \\ x/a & \text{for } -a \leqslant x \leqslant a, \\ +1 & \text{for } a \leqslant x \leqslant 1. \end{cases}$$

The dual solution u^* must satisfy Eq. (2.13) and Eq. (2.14):

$$\lambda^* = a(\sigma^*, u^*) = D(u^*)$$

or

$$\int_{-1}^{1} \sigma^* u^{*\prime} \, \mathrm{d}x = \sup_{|\sigma| \leqslant 1} \int_{-1}^{1} \sigma u^{*\prime} \, \mathrm{d}x = \|u^{*\prime}\|_M. \tag{4.4}$$

$\|u^{*\prime}\|_M$ denotes the norm of $u^{*\prime}$ as a measure on the *closed* interval $[-1, 1]$, i.e., the dual norm to the continuous functions on $[-1, 1]$. It is equal to the total variation of u^* on the interval $[-1, 1]$. Note that $u^{*\prime}$ is equal to the distributional derivative of u^* in the open interval $]-1, 1[$ plus a contribution from the end points of the interval. (In Eq. (4.1) and Eq. (4.4) we use the integral notation also to denote the formal duality between a measure and a continuous function on a closed interval.) Hence σ^* must be equal to the sign of $u^{*\prime}$, whenever $u^{*\prime}$ is nonzero (considered as a measure on $[-1, 1]$). This implies that u^* must satisfy the following:

$$\begin{aligned} &u^* \text{ is decreasing} && \text{for } -1 \leqslant x \leqslant 0, \\ &u^* \text{ is constant} && \text{for } -a < x < a, \\ &u^* \text{ is increasing} && \text{for } 0 \leqslant x \leqslant 1. \end{aligned} \tag{4.5}$$

(The intervals in the condition overlap, because $u(-a)$ must be greater than or equal to the constant value in the open interval $-a < x < a$ and similar for $u(a)$.) We see that the loaded region of the layer $[-a, a]$ moves as a solid block. In addition u^* is normalized to satisfy

$$1 = F(u^*) = -\int_{-a}^{a} u^* \, \mathrm{d}x + au^*(1) = a\,(-2u_0 + u^*(1))$$

or

$$u^*(1) = 2u_0 + \frac{1}{a}, \tag{4.6}$$

where u_0 denotes the constant value of u^* on $[-a, a]$. This leaves room for any collapse flow, which satisfies Eq. (4.5) and Eq. (4.6) in addition to the boundary condition $u(-1) = 0$. Thus there are solutions for each $u_0 \leqslant 0$.

For $a < 1$ there are both smooth and non-smooth collapse flows. For $a = 1$ there is only a one-parameter family of dual solutions, and every collapse flow has a discontinuity at $x = -1$, if $u_0 \neq 0$, or at $x = 1$, if $u_0 \neq -a^{-1}$, or both. The discontinuity at $x = -1$ appears to violate the boundary condition $u(-1) = 0$, but the corresponding work rate is included in the internal work rate $a(\sigma, u)$. On the other hand, the discontinuity at $x = 1$ causes a technical problem: The corresponding work rate is ignored in $a(\sigma, u)$, if u' is the derivative of u in the distributional sense. We must define u in such a way that a discontinuity along a loaded part of the boundary is significant in the energy dissipation rate. Hence the collapse flow must be defined as a pair (u^{Ω}, u^{T}), where u^{T} is a function defined on the loaded part of the boundary, and such that Eq. (4.1) holds for all stresses. More generally u^{Ω} and u^{T} are related through the equation of virtual work (rate) Eq. (2.4) and Green's formula Eq. (2.5).

An analogous example in plane stress is given in CHRISTIANSEN [1982, Section 3]. These examples show several important and unpleasant aspects of limit analysis, in addition to being some of the rare cases, which can be solved by hand. The strains in the collapse state are measures in the closed interval $[-1, 1]$, as emphasized by MOREAU [1976]. This means that the flow field is a function of bounded variation. In several dimensions this generalizes to the space BD of *bounded deformation* introduced in MATTHIES, STRANG and CHRISTIANSEN [1979]. For a vector field $\boldsymbol{u}$ to be of bounded deformation only the strain rates $\varepsilon_{ij}(\boldsymbol{u})$ are required to be measures, not all partial derivatives of $\boldsymbol{u}$. We have also established by simple calculations that the collapse flow does not always exist as an integrable function, but may have values on the loaded part of the boundary, which are neither given by boundary conditions, nor are they the limit of values from the inner of the domain occupied by the solid. If $\boldsymbol{u}$ has a discontinuity along the part of the boundary, where the boundary condition $\boldsymbol{u} = \boldsymbol{u}_S$ is given, then the associated work is included in the internal work rate Eq. (2.5) by adding a term of the form

$$\int_S (\boldsymbol{u} - \boldsymbol{u}_S)\cdot(\boldsymbol{\sigma}\cdot\boldsymbol{\nu}). \tag{4.7}$$

This is the relaxation method used in TEMAM and STRANG [1980b]. If, however, the discontinuity occurs at the loaded part of the boundary, the flow at the boundary, denoted by u^{T} in the example, must be treated as a part of the unknown solution. It is very satisfactory that the proof of existence of a collapse solution (in the next section) produces the pair $(\boldsymbol{u}^{\Omega}, \boldsymbol{u}^{T})$ as a purely mathematical object. The existence of the collapse flow as such a pair is suggested in CHRISTIANSEN [1976, 1980a] and with a slightly different notation in SUQUET [1980, 1981]. The existence problem is pinpointed and dealt with in a different way in TEMAM and STRANG [1980b] and TEMAM [1981a].

Another difficulty, not illustrated by the example, is that the stress tensor need not be continuous in Ω. Since the strain rates are measures, this means that the expression in Eq. (2.4) for the internal work rate is not necessarily well defined. As we shall see in the next section, the expression Eq. (2.5) is always well defined.

The example also illustrates non-uniqueness of the collapse flow. An example showing non-uniqueness of the collapse stresses is given in CHRISTIANSEN [1982, Example 3.1]. We see that the solution contains at least one rigid block of material without deformation, and there may be discontinuities. In some cases arbitrarily small perturbations (from $a = 1$ to $a < 1$) may change a problem with only discontinuous solutions to a problem which also has infinitely many smooth solutions.

From a mathematical point of view the above example is one-dimensional. True two-dimensional antiplane shear problems are studied in STRANG [1979a,b], and in STRANG and TEMAM [1982]. The solutions are similar to those above. Of particular interest is the following example suggested by Robert V. Kohn.

EXAMPLE 4.2 (*Two dimensional antiplane shear*). Consider an infinite pipe of constant square cross section in the x–y-plane: $\Omega = \{(x, y) \mid |x| \leqslant 1, |y| \leqslant 1\}$. The yield condition is $\|\boldsymbol{\sigma}\| \leqslant 1$, where $\boldsymbol{\sigma} = (\sigma_{xz}, \sigma_{yz})$. Apply zero volume forces $f = f_z(x, y) = 0$ in Ω and the following surface forces:

$$g = g_z = \begin{cases} 1 - y^2, & x = 1, \\ -1 + y^2, & x = -1, \\ 0, & y = \pm 1. \end{cases}$$

The limit multiplier is $\lambda^* = 1$. The collapse flow is given by $u = u_z = 0$ inside Ω, while the flow on the boundary u^T consists of a δ-function at $(1, 0)$ and a δ-function of opposite sign at $(-1, 0)$. Hence we must allow δ-distributions in the surface flow.

The two-dimensional problem in antiplane shear has been analysed in more general terms in DEMENGEL [1989, Section 2.1]. In particular, conditions are given, under which the collapse flow will exist in the space $\mathrm{BD}(\Omega)$.

Mathematically it is a complication that solutions do not exist in standard Sobolev spaces. From a numerical point of view it is bound to cause convergence problems that neither uniqueness nor regularity of the collapse fields can be expected. If the collapse flow involves discontinuities along the boundary, finite element spaces with the same property will give more accurate solutions. The fact that the solution may contain rigid blocks suggests that adaptive finite elements should be used. On the other hand, the nature of the solution may explain the early success of bound methods: If the rigid regions can be estimated or guessed, then good bounds may be obtained, even with hand calculations.

5. Existence of the collapse solution

In this section we give a precise mathematical frame for the duality problem of limit analysis. The existence theorem is the one given in CHRISTIANSEN [1986], but the presentation is somewhat different. The formulation is also closely related to the formulation in TEMAM and STRANG [1980b]. The static formulations are equivalent. The two formulations give the same limit multiplier and the existence of a collapse stress tensor. The main difference is our generalization of the velocity field to a pair $(\boldsymbol{u}^{\Omega}, \boldsymbol{u}^{T})$, which is significant for the existence of a collapse velocity field. In SUQUET [1980, 1981] $\boldsymbol{u}^{T}$ is introduced as the limit of a flow defined *outside* Ω.

Ω is assumed to be a bounded domain in $\mathbb{R}^N$, $N = 3$ or $N = 2$, with Lipschitz continuous boundary $\partial\Omega$ as defined in NEČAS [1967]. Then the outward normal $\boldsymbol{\nu}$ to Ω exists almost everywhere in $\partial\Omega$. Let S and T be open disjoint subsets of $\partial\Omega$, such that $\overline{S} \cup \overline{T} = \partial\Omega$. Assume that their common boundary is Lipschitz continuous in local coordinates on $\partial\Omega$. (For $N = 3$, S and T meet in a Lipschitz continuous curve. For $N = 2$ they meet in a finite number of points.) On S the flow $\boldsymbol{u}$ is prescribed, while T is free and subject to loading (see below).

The first step is to specify the spaces for stresses and plastic flow. The minimum problem of elasticity results in Hilbert space analysis within the frame of Sobolev spaces with $p = 2$. In limit analysis the duality between $\boldsymbol{\sigma}$ and $\boldsymbol{u}$ is non-symmetric, and there is nothing canonical about the choice $p = 2$. The starting point is the following facts:

- The yield condition on the stresses is given in terms of L^∞.
- The strain rates are bounded measures in $\overline{\Omega}$.
- The internal work rate is given by Eq. (2.4) and Eq. (2.5).
- The equilibrium equation, Eq. (2.3) or Eq. (2.6), must be satisfied with $\boldsymbol{f}$ and $\boldsymbol{g}$ in appropriate spaces.

As we shall see subsequently, these facts, combined with the results in TEMAM and STRANG [1980a] and TEMAM [1981b], settle the question of which Sobolev spaces to work with in limit analysis. Within these spaces the duality between the static and kinematic principles holds, and the saddle point in Eq. (2.14) exists.

In the previous section it was demonstrated that the distributional derivatives of the flow $\boldsymbol{u}$ are insufficient to describe the total energy dissipation. However, it is still preferable to reserve the notation $\boldsymbol{\varepsilon}(\boldsymbol{u})$ for the strain rates in the usual distributional sense. Hence formula (2.1) is specified as follows: For a locally integrable vector field $\boldsymbol{u}$ in Ω the associated strain rate tensor $\boldsymbol{\varepsilon}(\boldsymbol{u})$ is defined to have the following distributional derivatives as its components

$$\varepsilon_{ij}(\boldsymbol{u}) = \frac{1}{2}\left(\frac{\partial u_i}{\partial x_j} + \frac{\partial u_j}{\partial x_i}\right). \tag{5.1}$$

Let $M(\Omega)$ denote the space of bounded measures in Ω, i.e., the distributions, which are continuous with respect to the maximum norm on the test space $C_0^\infty(\Omega)$. The space of vector fields of bounded deformation $\mathrm{BD}(\Omega)$ introduced in MATTHIES, STRANG and CHRISTIANSEN [1979] is defined as

$$\mathrm{BD}(\Omega) = \left\{\boldsymbol{u} \in [L^1(\Omega)]^N \mid \varepsilon_{ij}(\boldsymbol{u}) \in M(\Omega),\ 1 \leqslant i, j \leqslant N\right\}. \tag{5.2}$$

$\mathrm{BD}(\Omega)$ is independent of the choice of coordinate system. For a coordinate-free definition, see TEMAM and STRANG [1980a].

BD(Ω) is a Banach space with the norm

$$\|\boldsymbol{u}\|_{\mathrm{BD}(\Omega)} = \sum_{i=1}^{N} \|u_i\|_{L^1(\Omega)} + \sum_{i,j=1}^{N} \|\varepsilon_{ij}(\boldsymbol{u})\|_{M(\Omega)} . \tag{5.3}$$

The following essential properties about BD(Ω) are proved in TEMAM and STRANG [1980a]. There the results are formulated with the assumption that $\partial\Omega$ is of class C^1, but as pointed out in CHRISTIANSEN [1986] the proofs given are also valid when $\partial\Omega$ is assumed to be Lipschitz continuous only. The distinction is of importance, since objects occupying geometries like, e.g., a cube are obvious applications.

THEOREM 5.1 (The trace theorem for BD). *Assume that $\Omega \subseteq \mathbb{R}^N$ is bounded with Lipschitz continuous boundary. There exists a unique continuous linear operator $\boldsymbol{\gamma}$ from* BD(Ω) *onto* $[L^1(\partial\Omega)]^N$*, such that*

$$\boldsymbol{\gamma}(\boldsymbol{u}) = \boldsymbol{u}|_{\partial\Omega} \quad \forall \boldsymbol{u} \in \mathrm{BD}(\Omega) \cap [C^0(\overline{\Omega})]^N. \tag{5.4}$$

For each i, j, and for every $\varphi \in C^1(\overline{\Omega})$ the following generalized Green's formula holds

$$\int_{\Omega} \left(u_j \frac{\partial\varphi}{\partial x_i} + u_i \frac{\partial\varphi}{\partial x_j} \right) + 2 \int_{\Omega} \varphi \varepsilon_{ij}(\boldsymbol{u}) = \int_{\partial\Omega} \varphi \left(\gamma_i(\boldsymbol{u})\nu_j + \gamma_j(\boldsymbol{u})\nu_i \right), \tag{5.5}$$

where $\boldsymbol{\nu}$ is the unit outward normal on $\partial\Omega$, and $\gamma_i(\boldsymbol{u})$ is the ith component of $\boldsymbol{\gamma}(\boldsymbol{u})$.

THEOREM 5.2 (Imbedding theorem for BD). *Assume that $\Omega \subseteq \mathbb{R}^N$ is bounded with Lipschitz continuous boundary. Let $q = N/(N-1)$. Then* BD(Ω) *is continuously imbedded in $[L^q(\Omega)]^N$. In other words:*

$$\mathrm{BD}(\Omega) \subseteq [L^q(\Omega)]^N \tag{5.6}$$

and

$$\sum_{i=1}^{N} \|u_i\|_{L^q(\Omega)} \leqslant C\|\boldsymbol{u}\|_{\mathrm{BD}(\Omega)} \quad \forall \boldsymbol{u} \in \mathrm{BD}(\Omega). \tag{5.7}$$

THEOREM 5.3. *Assume that $\Omega \subseteq \mathbb{R}^N$ is bounded with Lipschitz continuous boundary. If the N components of $\boldsymbol{u}$ are distributions in Ω, and if $\varepsilon_{ij}(\boldsymbol{u}) \in M(\Omega)$ for $1 \leqslant i, j \leqslant N$, then $\boldsymbol{u}$ belongs to* BD(Ω).

For the proofs of the above three theorems see TEMAM and STRANG [1980a].

We now define the stress space for the main case with unbounded yield set. Modifications for the (simpler) case with bounded yield set will be described later.

5.1. The stress space Σ

Let p be fixed in the interval $N < p < \infty$. The space Σ of stress tensors consists of all symmetric $N \times N$ tensors $\boldsymbol{\sigma}$ with the following properties

$$\boldsymbol{\sigma} \in [L^1(\Omega)]^{N^2}, \tag{5.8}$$

$$\boldsymbol{\sigma}^{\mathrm{D}} \in [L^\infty(\Omega)]^{N^2}, \tag{5.9}$$

$$\boldsymbol{\nabla}\cdot\boldsymbol{\sigma} \in [L^p(\Omega)]^N, \tag{5.10}$$

$$\boldsymbol{\nu}\cdot\boldsymbol{\sigma} \in [W^{1-1/p,p}(T)]^N. \tag{5.11}$$

Recall that $\boldsymbol{\sigma}^{\mathrm{D}}$ is the deviator of $\boldsymbol{\sigma}$ defined by Eq. (2.9). Σ is equipped with the norm

$$\|\boldsymbol{\sigma}\|_\Sigma = \|\boldsymbol{\sigma}^{\mathrm{D}}\|_{L^\infty(\Omega)} + \|\boldsymbol{\nabla}\cdot\boldsymbol{\sigma}\|_{L^p(\Omega)} + \|\boldsymbol{\nu}\cdot\boldsymbol{\sigma}\|_{W^{1-1/p,p}(T)}. \tag{5.12}$$

The yield condition is assumed to be of the form

$$\begin{aligned} &\boldsymbol{\sigma}^{\mathrm{D}}(\boldsymbol{x}) \in B(\boldsymbol{x}) \quad \forall \boldsymbol{x} \in \Omega, \\ &B(\boldsymbol{x}) \subseteq \mathbb{R}^{N^2} \text{ closed, bounded uniformly in } \boldsymbol{x} \text{ and convex for all } \boldsymbol{x} \in \Omega. \end{aligned} \tag{5.13}$$

$B(\boldsymbol{x})$ must have non-empty interior in the uniform sense that there exists a $\delta > 0$ independent of $\boldsymbol{x}$, such that the following implication holds:

$$|\sigma_{ij}| < \delta \ \forall i,j \ \Rightarrow \ \boldsymbol{\sigma} \in B(\boldsymbol{x}). \tag{5.14}$$

We also assume that $B(\boldsymbol{x})$ can be defined by inequalities which are piecewise continuous in $\boldsymbol{x}$.

Yield conditions of this form cover standard cases, such as the Von Mises and Tresca conditions in three-dimensional problems and in plane strain, but not plane stress problems or conditions such as the Coulomb yield condition for granular materials. The modifications for these cases are discussed in the next section.

REMARK 5.1. The choice of L^1 in Eq. (5.8) is arbitrary. Any L^q with q in the interval $1 \leqslant q < \infty$ would be equivalent. This follows from Theorem 5.4 below.

REMARK 5.2. The condition in Eq. (5.9) is consistent with the yield condition in Eq. (2.10).

REMARK 5.3. Condition (5.10) is motivated by the first term in Eq. (2.5) and Theorem 5.2. $p = N$ is conjugate to $q = N/(N-1)$, but since Ω is bounded we are free to choose $p > N$.

REMARK 5.4. It is proved in TEMAM [1977] that Eq. (5.8) and Eq. (5.10) imply the existence of $\boldsymbol{\nu}\cdot\boldsymbol{\sigma}$ on $\partial\Omega$ in a weaker sense than Eq. (5.11). Condition (5.11) is motivated by the second term on $\partial\Omega$ in Eq. (2.5) and the observation that the flow $\boldsymbol{u}$ on the loaded part of the boundary T may be a measure on the boundary (Example 4.2). The formal integral of the second term in Eq. (2.5) is well defined, if $\boldsymbol{\nu} \cdot \boldsymbol{\sigma}$ is continuous on T, which follows from Eq. (5.11) for $p > N$. (By Sobolev space theory functions in $W^{1-1/p,p}(\partial\Omega)$ are restrictions to $\partial\Omega$ of functions in $W^{1,p}(\Omega)$, and these are continuous for $p > N$. See, e.g., NEČAS [1967].)

The requirements in the definition of Σ restrict $\boldsymbol{\sigma}$ more than meets the eye:

THEOREM 5.4. *Let $\Omega \subseteq \mathbb{R}^N$ be bounded with Lipschitz continuous boundary. Let $p \geqslant N$ and assume that*

$$\boldsymbol{\sigma} \in [L^1(\Omega)]^{N^2}, \qquad \boldsymbol{\sigma}^{\mathrm{D}} \in [L^\infty(\Omega)]^{N^2}, \qquad \boldsymbol{\nabla}\cdot\boldsymbol{\sigma} \in [L^p(\Omega)]^N.$$

Then

$$\boldsymbol{\sigma} \in [L^q(\Omega)]^{N^2} \quad \forall q < \infty$$

and

$$\| \operatorname{tr}(\boldsymbol{\sigma}) \|_{L^q/\mathbb{R}} \leqslant C_q(\|\boldsymbol{\sigma}^{\mathrm{D}}\|_{L^\infty} + \|\boldsymbol{\nabla}\cdot\boldsymbol{\sigma}\|_{L^p}).$$

The norm on the left-hand side is the quotient norm modulo constants in $L^q(\Omega)$.

PROOF. See Lemma 2.1 in CHRISTIANSEN [1986]. □

5.2. *The flow space U*

We formulate the problem for the homogeneous boundary condition $\boldsymbol{u}=\mathbf{0}$ on S. The case of more general boundary conditions will be dealt with in the next section. The space U of displacement rates consists of all pairs $(\boldsymbol{u}^\Omega, \boldsymbol{u}^T)$ of the form

$$(\boldsymbol{u}^\Omega, \boldsymbol{u}^T) \in \mathrm{BD}(\Omega) \times [(W^{1-1/p,p}(T))^*]^N \tag{5.15}$$

which satisfy the following condition: The linear form

$$\boldsymbol{\sigma} \to -\int_\Omega (\boldsymbol{\nabla}\cdot\boldsymbol{\sigma})\cdot\boldsymbol{u}^\Omega + \int_T (\boldsymbol{\nu}\cdot\boldsymbol{\sigma})\cdot\boldsymbol{u}^T \tag{5.16}$$

defined for symmetric tensors $\boldsymbol{\sigma} \in [W^{1,p}(\Omega)]^{N^2}$ must be continuous with respect to the maximum norm on $\boldsymbol{\sigma}$. Thus $(\boldsymbol{u}^\Omega, \boldsymbol{u}^T)$ in Eq. (5.15) belongs to U if and only if there is a symmetric tensor of measures on $\overline{\Omega}$, $\boldsymbol{\mu} \in [(C(\overline{\Omega}))^*]^{N^2}$, such that for all symmetric $\boldsymbol{\sigma} \in [C^1(\overline{\Omega})]^{N^2}$ the following equality holds:

$$\sum_{i,j=1}^N \langle \mu_{ij}, \sigma_{ij} \rangle_{C(\overline{\Omega})^* \times C(\overline{\Omega})} = -\int_\Omega (\boldsymbol{\nabla}\cdot\boldsymbol{\sigma})\cdot\boldsymbol{u}^\Omega + \int_T (\boldsymbol{\nu}\cdot\boldsymbol{\sigma})\cdot\boldsymbol{u}^T. \tag{5.17}$$

U is an incomplete space equipped with the norm

$$\|\boldsymbol{u}\|_U = \|\boldsymbol{u}^\Omega\|_{\mathrm{BD}(\Omega)} + \|\boldsymbol{u}^T\|_{(W^{1-1/p,p}(T))^*}. \tag{5.18}$$

In the second integral of Eq. (5.16) we have kept integral notation for the duality $W^{1-1/p,p}(T) \times (W^{1-1/p,p}(T))^*$. In Eq. (5.17) we have tried to emphasize the generalized nature of the strain rate tensor $\boldsymbol{\mu}(\boldsymbol{u})$ by using the more correct duality notation.

In view of Theorem 5.2 and the trace theorem for the Sobolev space $W^{1,p}(\Omega)$, Eq. (5.16) defines a continuous linear form om $[W^{1,p}(\Omega)]^{N^2}$ with its "own" norm. The tensor $\boldsymbol{\mu} = \boldsymbol{\mu}(\boldsymbol{u})$ in Eq. (5.17) is the generalized strain rate tensor associated with the pair $\boldsymbol{u} = (\boldsymbol{u}^\Omega, \boldsymbol{u}^T)$. Clearly, $\boldsymbol{\mu}(\boldsymbol{u})$ and the distributional strain rate tensor $\boldsymbol{\varepsilon}(\boldsymbol{u}^\Omega)$ defined by Eq. (5.1) are equal as distributions in Ω. The difference between $\boldsymbol{\mu}(\boldsymbol{u})$ and $\boldsymbol{\varepsilon}(\boldsymbol{u})$ is characterized by the following theorem.

THEOREM 5.5. *Assume that $\Omega \subseteq \mathbb{R}^N$ is bounded with Lipschitz continuous boundary. Let $\boldsymbol{u} \in U$ with $\boldsymbol{\mu} = \boldsymbol{\mu}(\boldsymbol{u})$ and $\boldsymbol{\varepsilon} = \boldsymbol{\varepsilon}(\boldsymbol{u})$ defined by Eq. (5.17) and Eq. (5.1) respectively. Then the following equality holds for all symmetric tensors $\boldsymbol{\sigma} \in [C^1(\overline{\Omega})]^{N^2}$:*

$$\sum_{i,j=1}^{N} \langle \mu_{ij} - \varepsilon_{ij}, \sigma_{ij} \rangle_{C^*(\overline{\Omega}) \times C(\overline{\Omega})} = \int_T (\boldsymbol{\nu}\cdot\boldsymbol{\sigma})\cdot(\boldsymbol{u}^T - \gamma(\boldsymbol{u}^\Omega)) - \int_S (\boldsymbol{\nu}\cdot\boldsymbol{\sigma})\cdot\gamma(\boldsymbol{u}^\Omega), \tag{5.19}$$

where γ denotes the trace operator in Theorem 5.1. In particular, $\boldsymbol{\mu}(\boldsymbol{u})$ and $\boldsymbol{\varepsilon}(\boldsymbol{u})$ are equal as measures in $\overline{\Omega}$ if and only if

$$\gamma(\boldsymbol{u}^\Omega) = \boldsymbol{u}^T \text{ on } T \quad \text{and} \quad \gamma(\boldsymbol{u}^\Omega) = \boldsymbol{0} \text{ on } S. \tag{5.20}$$

PROOF. Equation (5.19) follows immediately from Eqs. (5.5) and (5.17). □

REMARK 5.5. Example 4.1 and Example 4.2 and further examples in CHRISTIANSEN [1982] show that it is necessary to consider the flow on the loaded part of the boundary separately in order to prove existence of a collapse flow field. In many applications it will hold that $\gamma(\boldsymbol{u}^\Omega) = \boldsymbol{u}^T$, in which case the collapse flow falls within the frame of TEMAM and STRANG [1980b] and KOHN and TEMAM [1983]. Intuitively, this will happen, when the surface shear forces are not too strong (according to Eq. (5.49) the normal components of $\gamma(\boldsymbol{u}^\Omega)$ and $\boldsymbol{u}^T$ are always identical for the collapse flow). For the special case of antiplane shear, this is proved in DEMENGEL [1989, Corollary 2.2].

REMARK 5.6. If we define the internal energy dissipation rate by Eq. (5.17), then the boundary condition $\boldsymbol{u} = \boldsymbol{0}$ on S is included in the definition of U in the same "relaxed" way as in TEMAM and STRANG [1980b]. The second term on the right-hand side of Eq. (5.19) is identical to Eq. (4.7), when $\boldsymbol{u}_S = \boldsymbol{0}$.

$\mathrm{BD}(\Omega)$ is a subset of U through the imbedding $\boldsymbol{u} \to (\boldsymbol{u}, \boldsymbol{\gamma}(\boldsymbol{u})|_T)$ for all $\boldsymbol{u} \in \mathrm{BD}(\Omega)$. In this case the generalized strain rate tensor is given by

$$\sum_{i,j=1}^{N} \langle \mu_{ij}, \sigma_{ij} \rangle_{(C(\overline{\Omega}))^* \times C(\overline{\Omega})} = \sum_{i,j=1}^{N} \langle \varepsilon_{ij}, \sigma_{ij} \rangle_{M(\Omega) \times C(\overline{\Omega})} - \int_S (\boldsymbol{\nu} \cdot \boldsymbol{\sigma}) \cdot \boldsymbol{\gamma}(\boldsymbol{u})$$

for all symmetric $\boldsymbol{\sigma} \in [C^1(\overline{\Omega})]^{N^2}$. If $\boldsymbol{u}$ satisfies the boundary condition in the sense that $\boldsymbol{\gamma}(\boldsymbol{u}) = \boldsymbol{0}$ on S, then the internal energy dissipation rate is given by the classical expression.

It is also instructive to identify a vector of bounded measures $\boldsymbol{m}$ on T with an element of U through the imbedding $\boldsymbol{m} \to (\boldsymbol{0}, \boldsymbol{m}) \in U$. The generalized strain rate tensor $\boldsymbol{\mu}$ is now given by the expression

$$\sum_{i,j=1}^{N} \langle \mu_{ij}, \sigma_{ij} \rangle_{C^*(\overline{\Omega}) \times C(\overline{\Omega})} = \langle \boldsymbol{m}, \boldsymbol{\nu} \cdot \boldsymbol{\sigma} \rangle_{M(T) \times C(\overline{T})}$$

for all symmetric $\boldsymbol{\sigma} \in [C(\overline{\Omega})]^{N^2}$.

5.3. *The duality of limit analysis*

On the product space $\Sigma \times U$ the internal work rate is defined by the right-hand side of Eq. (5.17):

$$a(\boldsymbol{\sigma}, \boldsymbol{u}) = -\int_\Omega (\boldsymbol{\nabla} \cdot \boldsymbol{\sigma}) \cdot \boldsymbol{u}^\Omega + \int_T (\boldsymbol{\nu} \cdot \boldsymbol{\sigma}) \cdot \boldsymbol{u}^T \quad \forall (\boldsymbol{\sigma}, \boldsymbol{u}) \in \Sigma \times U. \tag{5.21}$$

By Theorem 5.2 this is a continuous bilinear form on the product space $\Sigma \times U$.

The volume and surface forces are defined as a pair $(\boldsymbol{f}, \boldsymbol{g})$, where

$$\boldsymbol{f} \in [L^p(\Omega)]^N \quad \text{and} \quad \boldsymbol{g} \in [W^{1-1/p,p}(T)]^N \tag{5.22}$$

for any $p > N$. By Theorem 5.2 the work rate for the external forces is well defined as the following continuous linear form on U:

$$F(\boldsymbol{u}) = \int_\Omega \boldsymbol{f} \cdot \boldsymbol{u} + \int_T \boldsymbol{g} \cdot \boldsymbol{u}. \tag{5.23}$$

In the special case, where $T = \partial\Omega$, i.e., there is no boundary condition on $\boldsymbol{u}$, $(\boldsymbol{f}, \boldsymbol{g})$ must in addition satisfy the usual compatibility condition (the total load must be zero):

$$\int_\Omega \boldsymbol{f} + \int_{\partial\Omega} \boldsymbol{g} = 0. \tag{5.24}$$

REMARK 5.7. Since $p > N$ it follows from Eq. (5.22) that on smooth parts of T only continuous surface forces are allowed in this formulation (e.g., Remark 5.4). This restriction appears to be of the same nature as the exclusion of point loads in the corresponding plate model. If, for example, a piecewise constant load $\boldsymbol{g}$ is the limit of a sequence of continuous loads by use of an approximating unit, then the corresponding limit multipliers converge, and the limit may be interpreted as the limit multiplier for $\boldsymbol{g}$. This argument also justifies discretization and numerical solution of such problems. Physically, point loads and step loads are idealizations. In the elastic model they are admitted, but in the plastic model they are not.

We shall need the following theorem.

THEOREM 5.6. *Let $\Omega \subseteq \mathbb{R}^N$ be bounded with Lipschitz continuous boundary. Let S and T be open disjoint subsets of $\partial\Omega$ with $\overline{S} \cup \overline{T} = \partial\Omega$ and $\overline{S} \cap \overline{T}$ Lipschitz continuous in local coordinates on $\partial\Omega$. Let $\boldsymbol{f}$ and $\boldsymbol{g}$ be given as in Eq. (5.22). Then there exists at least one symmetric $\boldsymbol{\sigma} \in [W^{1,p}(\Omega)]^{N^2}$, such that the equilibrium equation on classical form is satisfied:*

$$-\boldsymbol{\nabla}\cdot\boldsymbol{\sigma} = \boldsymbol{f} \text{ in } \Omega \quad \text{and} \quad \boldsymbol{\nu}\cdot\boldsymbol{\sigma} = \boldsymbol{g} \text{ on } T. \tag{5.25}$$

PROOF. Using the regularity of T and Theorem 2.3.9 in NEČAS [1967] it can be seen that $\boldsymbol{g}$ may be extended to an element of $W^{1-1/p,p}(\partial\Omega)$, such that Eq. (5.24) is satisfied. In NEČAS [1966] it is proved that the system of elasticity is $W^{1,p}(\Omega)$-coercive. Hence the Neumann problem corresponding to the equations of equilibrium with $\boldsymbol{f}$ and $\boldsymbol{g}$ has a unique elastic solution $\boldsymbol{u} \in [W^{2,p}(\Omega)]^N$. The corresponding elastic stress tensor can be used as the solution. □

Finally it is convenient to introduce a short notation for the set of admissible stress tensors:

$$K = \{\boldsymbol{\sigma} \in \Sigma \mid \boldsymbol{\sigma}^{\mathrm{D}}(\boldsymbol{x}) \in B(\boldsymbol{x}) \text{ a.e. in } \Omega\} \tag{5.26}$$

$\boldsymbol{\sigma} \in \Sigma$ is admissible, if it satisfies the yield condition everywhere in Ω, except possibly for a subset of measure zero.

The *static principle* of limit analysis, Eq. (2.11), now takes the form

$$\lambda^* = \sup\{\lambda \mid \exists\boldsymbol{\sigma} \in K: a(\boldsymbol{\sigma},\boldsymbol{u}) = \lambda F(\boldsymbol{u})\ \forall\boldsymbol{u} \in U\} \tag{5.27}$$

$$= \sup_{\sigma\in K} \inf_{F(u)=1} a(\boldsymbol{\sigma},\boldsymbol{u}). \tag{5.28}$$

The equality between Eq. (5.27) and Eq. (5.28) follows from a standard algebraic argument: The inner infimum in Eq. (5.28) is, for fixed $\boldsymbol{\sigma}$, the infimum of a linear functional $a(\boldsymbol{\sigma},\cdot)$ over an affine hyperplane. This value is different from $-\infty$ only if the linear functional is constant on the hyperplane. In this case, $a(\boldsymbol{\sigma},\cdot)$ is proportional to the linear functional F, which defines the hyperplane, i.e., $a(\boldsymbol{\sigma},\boldsymbol{u}) = \lambda F(\boldsymbol{u})\ \forall\boldsymbol{u} \in U$ for some real λ.

The *kinematic principle* is

$$\lambda^* = \inf_{F(u)=1} D(\boldsymbol{u}) = \inf_{F(u)=1} \sup_{\sigma\in K} a(\boldsymbol{\sigma},\boldsymbol{u}), \tag{5.29}$$

where

$$D(\boldsymbol{u}) = \sup_{\sigma \in K} a(\boldsymbol{\sigma}, \boldsymbol{u}) \; \forall \boldsymbol{u} \in U. \tag{5.30}$$

$D(\boldsymbol{u})$ is called the total energy dissipation rate associated with $\boldsymbol{u}$.

The duality theorem of limit analysis states that the values of Eq. (5.28) and Eq. (5.29) are equal, and that the extremal values are attained as maxima, respectively minima.

THEOREM 5.7 (Duality theorem of limit analysis). *Assume that $\Omega \subseteq \mathbb{R}^N$ is bounded with Lipschitz continuous boundary. Let Σ be defined by Eqs.* (5.8–5.12), *U by Eqs.* (5.15–5.16), *and let K satisfy Eq.* (5.26) *and Eqs.* (5.13–5.14). *The bilinear form $a(\boldsymbol{\sigma}, \boldsymbol{u})$ on $\Sigma \times U$ is given by Eq.* (5.21), *and F by Eqs.* (5.22–5.23). *Then*

$$\max_{\sigma \in K} \min_{F(u)=1} a(\boldsymbol{\sigma}, \boldsymbol{u}) = \min_{F(u)=1} \max_{\sigma \in K} a(\boldsymbol{\sigma}, \boldsymbol{u}). \tag{5.31}$$

If $\boldsymbol{\sigma}^*$ maximizes the left-hand side, and $\boldsymbol{u}^*$ minimizes the right-hand side, then $(\boldsymbol{\sigma}^*, \boldsymbol{u}^*)$ satisfies the inequality

$$a(\boldsymbol{\sigma}, \boldsymbol{u}^*) \leqslant a(\boldsymbol{\sigma}^*, \boldsymbol{u}^*) = a(\boldsymbol{\sigma}^*, \boldsymbol{u}) = \lambda^* \tag{5.32}$$

for all $\boldsymbol{\sigma} \in K$ and all $\boldsymbol{u} \in U$ with $F(\boldsymbol{u}) = 1$. Here λ^* is the value of Eq. (5.31). *In particular, $(\boldsymbol{\sigma}^*, \boldsymbol{u}^*)$ is a saddle point for $a(\boldsymbol{\sigma}, \boldsymbol{u})$.*

PROOF. We first prove the duality part of Eq. (5.31) by proving the equality

$$\sup_{\sigma \in K} \inf_{F(u)=1} a(\boldsymbol{\sigma}, \boldsymbol{u}) = \min_{F(u)=1} \sup_{\sigma \in K} a(\boldsymbol{\sigma}, \boldsymbol{u}). \tag{5.33}$$

We apply Theorem 2.1 in CHRISTIANSEN [1980a] to the bilinear form $-a(\boldsymbol{\sigma}, \boldsymbol{u})$ on $\Sigma \times U$ and the convex sets $K \subseteq \Sigma$ and $C \subseteq U$ given by

$$C = \{\boldsymbol{u} \in U \mid F(\boldsymbol{u}) = 1\}.$$

C is closed in U, and due to Eq. (5.14) $\boldsymbol{\sigma} = \boldsymbol{0}$ is an interior point of K. Equation (5.33) will follow by establishing the following two implications (a) and (b):

(a) If Ψ is a continuous linear functional on U, which satisfies $\inf_{u \in C} \Psi(\boldsymbol{u}) > -\infty$, then there exists $\boldsymbol{\sigma} \in \Sigma$, such that

$$\Psi(\boldsymbol{u}) = a(\boldsymbol{\sigma}, \boldsymbol{u}) \quad \forall \boldsymbol{u} \in U. \tag{5.34}$$

(b) If Φ is a continuous linear functional on Σ, which satisfies $\sup_{\sigma \in K} \Phi(\boldsymbol{\sigma}) < \infty$ and the condition

$$a(\boldsymbol{\sigma}, \boldsymbol{u}) = 0 \; \forall \boldsymbol{u} \in U \;\; \Rightarrow \;\; \Phi(\boldsymbol{\sigma}) = 0,$$

then there exists $\boldsymbol{u} \in U$ such that

$$\Phi(\boldsymbol{\sigma}) = a(\boldsymbol{\sigma}, \boldsymbol{u}) \quad \forall \boldsymbol{\sigma} \in \Sigma. \tag{5.35}$$

If Ψ in (a) is bounded from below on the affine hyperplane given by $F(\boldsymbol{u}) = 1$, then by the argument following Eq. (5.28), we have

$$\Psi(\boldsymbol{u}) = tF(\boldsymbol{u}) \quad \forall \boldsymbol{u} \in U$$

for some real number t. Let $\boldsymbol{\sigma} \in [W^{1,p}(\Omega)]^{N^2}$ be a symmetric stress field satisfying Eq. (5.25) as stated in Theorem 5.6. Then

$$\Psi(\boldsymbol{u}) = tF(\boldsymbol{u}) = a(t\boldsymbol{\sigma}, \boldsymbol{u}) \quad \forall \boldsymbol{u} \in U.$$

This proves that the implication (a) holds.

Now let Φ satisfy the conditions in (b). By definition of $a(\boldsymbol{\sigma}, \boldsymbol{u})$ the following implication holds for Φ:

$$(\boldsymbol{\nabla}\cdot\boldsymbol{\sigma} = \boldsymbol{0} \text{ in } \Omega \text{ and } \boldsymbol{\nu}\cdot\boldsymbol{\sigma} = \boldsymbol{0} \text{ on } T) \Rightarrow \Phi(\boldsymbol{\sigma}) = 0. \tag{5.36}$$

Consider the linear map

$$D: [W^{1,p}(\Omega)]^{N^2,\text{sym}} \to [L^p(\Omega) \times W^{1-1/p,p}(T)]^N$$

defined on the symmetric tensors in $[W^{1,p}(\Omega)]^{N^2}$ by

$$D(\boldsymbol{\sigma}) = (-\boldsymbol{\nabla}\cdot\boldsymbol{\sigma}, \boldsymbol{\nu}\cdot\boldsymbol{\sigma}).$$

By Theorem 5.6, D is onto and hence an open map. Condition (5.36) implies that the null space of D is contained in the null space of Φ. Hence, there exists a continuous linear functional η on $[L^p(\Omega) \times W^{1-1/p,p}(T)]^N$, such that

$$\Phi(\boldsymbol{\sigma}) = \eta(D(\boldsymbol{\sigma})) \; \forall \boldsymbol{\sigma} \in [W^{1,p}(\Omega)]^{N^2,\text{sym}} \tag{5.37}$$

η may be identified with a pair $(\boldsymbol{u}^\Omega, \boldsymbol{u}^T)$ of the form

$$(\boldsymbol{u}^\Omega, \boldsymbol{u}^T) \in [L^q(\Omega) \times (W^{1-1/p,p}(T))^*]^N, \tag{5.38}$$

where $1/p + 1/q = 1$. Equation (5.37) now reads

$$\Phi(\boldsymbol{\sigma}) = -\int_\Omega (\boldsymbol{\nabla}\cdot\boldsymbol{\sigma})\cdot\boldsymbol{u}^\Omega + \int_T (\boldsymbol{\nu}\cdot\boldsymbol{\sigma})\cdot\boldsymbol{u}^T \tag{5.39}$$

for all symmetric $\boldsymbol{\sigma} \in [W^{1,p}(\Omega)]^{N^2}$. We claim that Eq. (5.39) holds for all $\boldsymbol{\sigma} \in \Sigma$: Let $\boldsymbol{\sigma} \in \Sigma$. From Theorem 5.6 applied to $\boldsymbol{f} = \boldsymbol{\nabla}\cdot\boldsymbol{\sigma} \in [L^p(\Omega)]^N$ and $\boldsymbol{\nu}\cdot\boldsymbol{\sigma} \in [W^{1-1/p,p}(T)]^N$ we conclude that there exists a symmetric $\boldsymbol{\sigma}_0 \in [W^{1,p}(\Omega)]^{N^2}$, such that $\boldsymbol{\nabla}\cdot(\boldsymbol{\sigma} - \boldsymbol{\sigma}_0) = \boldsymbol{0}$ in Ω and $\boldsymbol{\nu}\cdot(\boldsymbol{\sigma} - \boldsymbol{\sigma}_0) = \boldsymbol{0}$ on T. By Eq. (5.36) $\Phi(\boldsymbol{\sigma}) = \Phi(\boldsymbol{\sigma}_0)$, and thus Eq. (5.39) holds for $\boldsymbol{\sigma}$, as well as for $\boldsymbol{\sigma}_0$.

It remains to prove that $\boldsymbol{u}^\Omega$ belongs to $\text{BD}(\Omega)$. Part of the assumption (b) on Φ is $\sup_{\sigma\in K} \Phi(\boldsymbol{\sigma}) < \infty$. By Eq. (5.14) K contains a ball with respect to the L^∞-norm on $\boldsymbol{\sigma}$, so Φ is continuous on $[W^{1,p}(\Omega)]^{N^2,\text{sym}}$ equipped with the maximum norm. Hence, there exists bounded measures $\mu_{ij} = \mu_{ji} \in C^*(\overline{\Omega})$, such that

$$\Phi(\boldsymbol{\sigma}) = \sum_{i,j=1}^{N} \langle \mu_{ij}, \sigma_{ij} \rangle_{C^*(\overline{\Omega})\times C(\overline{\Omega})} \tag{5.40}$$

for all symmetric $\boldsymbol{\sigma} \in [W^{1,p}(\Omega)]^{N^2}$. Considering first $\boldsymbol{\sigma}$ with compact support in Ω we conclude from Eqs. (5.40) and (5.39) that

$$\mu_{ij} = \varepsilon_{ij}(\boldsymbol{u}^\Omega) \text{ as distributions in } \Omega \tag{5.41}$$

and hence that $\boldsymbol{u}^{\Omega} \in \mathrm{BD}(\Omega)$. Using again Eqs. (5.40) and (5.39) we see that $(\boldsymbol{u}^{\Omega}, \boldsymbol{u}^{T}) \in U$, and that

$$\Phi(\boldsymbol{\sigma}) = a(\boldsymbol{\sigma}, \boldsymbol{u}) \quad \forall \boldsymbol{\sigma} \in \Sigma.$$

This proves condition (b) and hence also Eq. (5.33).

It remains to verify that the maximum on the left-hand side of Eq. (5.31) is attained. Let λ^* be the value of Eq. (5.31), or equivalently of Eq. (5.27), and let $\boldsymbol{\sigma}_k \in K$ be a sequence, which satisfies

$$a(\boldsymbol{\sigma}_k, \boldsymbol{u}) = \lambda_k F(\boldsymbol{u}) \quad \forall \boldsymbol{u} \in U, \qquad \lambda_k \to \lambda^*. \tag{5.42}$$

The sequence of deviators $\boldsymbol{\sigma}_k^{\mathrm{D}}$ is bounded in L^∞ and hence there is a subsequence, also denoted $\boldsymbol{\sigma}_k$, such that $\boldsymbol{\sigma}_k^{\mathrm{D}}$ is weak* convergent in L^∞ as the dual of L^1. Let $\boldsymbol{\sigma} \in [L^\infty(\Omega)]^{N^2}$ be the limit. In particular, $\boldsymbol{\sigma}_k^{\mathrm{D}}$ converges to $\boldsymbol{\sigma}$ in the distributional sense. If $\boldsymbol{u}$ has components in $C_0^\infty(\Omega)$ and $\boldsymbol{\nabla}\cdot\boldsymbol{u} = 0$, then

$$a(\boldsymbol{\sigma}_k^{\mathrm{D}}, \boldsymbol{u}) = a(\boldsymbol{\sigma}_k, \boldsymbol{u}) = \lambda_k F(\boldsymbol{u}).$$

From Eq. (5.42) it follows that $\boldsymbol{\sigma}$ satisfies the following condition:

$$\begin{aligned} &\text{If } \boldsymbol{u} \in [C_0^\infty(\Omega)]^N \text{ with } \boldsymbol{\nabla}\cdot\boldsymbol{u} = 0,\\ &\text{then } \langle \boldsymbol{\nabla}\cdot\boldsymbol{\sigma} + \lambda^*\boldsymbol{f}, \boldsymbol{u}\rangle_{\mathcal{D}'\times\mathcal{D}} = 0. \end{aligned} \tag{5.43}$$

$\mathcal{D}'\times\mathcal{D}$ denotes the duality between distributions and testfunctions. Equation (5.43) is a familiar property in the analysis of the Navier–Stokes equation. From TEMAM [1977, Proposition 1.1, p. 14] we conclude that there exists a distribution $\varphi \in \mathcal{D}'(\Omega)$, such that (in the theory of Navier–Stokes equation φ is the pressure)

$$\boldsymbol{\nabla}\cdot\boldsymbol{\sigma} + \lambda^*\boldsymbol{f} = \boldsymbol{\nabla}\varphi = \boldsymbol{\nabla}\cdot(\varphi\boldsymbol{I}) \text{ in } \mathcal{D}'(\Omega). \tag{5.44}$$

Using NEČAS [1966, Theorem 1, p. 108] we see that $\varphi \in L^2(\Omega)$ (in fact by Theorem 5.4 $\varphi \in L^q(\Omega)$ for all $q < \infty$). Let

$$\boldsymbol{\sigma}^* = \boldsymbol{\sigma} - \varphi\boldsymbol{I}.$$

The components of $\boldsymbol{\sigma}^*$ are in $L^2(\Omega)$. $\boldsymbol{\sigma}^*$ satisfies the yield condition, since its deviator is $\boldsymbol{\sigma}$. Also $\boldsymbol{\nabla}\cdot\boldsymbol{\sigma} = -\lambda^*\boldsymbol{f}$ has components in $L^p(\Omega)$. Hence, $\boldsymbol{\sigma}^*$ is a maximizing stress tensor for Eq. (5.27), if we can prove that $\boldsymbol{\nu}\cdot\boldsymbol{\sigma}^* = \lambda^*\boldsymbol{g}$ on T. By standard Sobolev space theory, $\boldsymbol{\nu}\cdot\boldsymbol{\sigma}^*$ exists in $W^{-1/2,2}(\partial\Omega)$, and the following Green's formula holds for any $\boldsymbol{u} \in [C^1(\overline{\Omega})]^N$ with $\boldsymbol{u} = \boldsymbol{0}$ on S:

$$\begin{aligned} \int_\Omega \sum_{i,j} \sigma_{ij}^* \varepsilon_{ij}(\boldsymbol{u}) &= -\int_\Omega (\boldsymbol{\nabla}\cdot\boldsymbol{\sigma}^*)\cdot\boldsymbol{u} + \int_T (\boldsymbol{\nu}\cdot\boldsymbol{\sigma}^*)\cdot\boldsymbol{u} \\ &= \lambda^* \int_\Omega \boldsymbol{f}\cdot\boldsymbol{u} + \int_T (\boldsymbol{\nu}\cdot\boldsymbol{\sigma}^*)\cdot\boldsymbol{u}. \end{aligned} \tag{5.45}$$

If in addition $\boldsymbol{u}$ satisfies $\boldsymbol{\nabla}\cdot\boldsymbol{u} = 0$, the following holds:

$$\int_\Omega \sum_{i,j} \sigma_{ij}^* \varepsilon_{ij}(\boldsymbol{u}) = \int_\Omega \sum_{i,j} \sigma_{ij}\varepsilon_{ij}(\boldsymbol{u}) = \lim_{k\to\infty} \lambda_k F(\boldsymbol{u})$$
$$= \lambda^* \int_\Omega \boldsymbol{f}\cdot\boldsymbol{u} + \lambda^* \int_T \boldsymbol{g}\cdot\boldsymbol{u}. \tag{5.46}$$

From Eqs. (5.45) and (5.46) we conclude that the following implication holds for $\boldsymbol{\sigma}^*$:

$$\left(\boldsymbol{u} \in [W^{1,2}(\Omega)]^N, \boldsymbol{u} = \boldsymbol{0} \text{ on } S,\ \boldsymbol{\nabla}\cdot\boldsymbol{u} = 0\right) \Rightarrow \int_T (\boldsymbol{\nu}\cdot\boldsymbol{\sigma}^* - \lambda^*\boldsymbol{g})\cdot\boldsymbol{u} = 0.$$

In FOIAS and TEMAM [1978] it is proved that

$$\{\boldsymbol{\gamma}(\boldsymbol{u}) \mid \boldsymbol{u} \in [W^{1,2}(\Omega)]^N, \boldsymbol{\nabla}\cdot\boldsymbol{u} = 0\} = \{\boldsymbol{v} \in [W^{1/2,2}(\partial\Omega)]^N \mid \int_{\partial\Omega} \boldsymbol{\nu}\cdot\boldsymbol{v} = 0\},$$

where $\boldsymbol{\gamma}$ denotes the trace operator from $[W^{1,2}(\Omega)]^N$ onto $]W^{1/2,2}(\partial\Omega)]^N$. Thus the null space of the functional

$$\boldsymbol{v} \to \int_T (\boldsymbol{\nu}\cdot\boldsymbol{\sigma}^* - \lambda^*\boldsymbol{g})\cdot\boldsymbol{v}$$

defined on the space

$$\{\boldsymbol{v} \in [W^{1/2,2}(\partial\Omega)]^N \mid \boldsymbol{v} = \boldsymbol{0} \text{ on } S\}$$

contains the null space of the functional

$$\boldsymbol{v} \to \int_T \boldsymbol{\nu}\cdot\boldsymbol{v}.$$

We conclude that

$$\boldsymbol{\nu}\cdot\boldsymbol{\sigma}^* - \lambda^*\boldsymbol{g} = t\boldsymbol{\nu} = t\boldsymbol{\nu}\cdot\boldsymbol{I} \quad \text{on } T$$

for some real constant t. But since φ in Eq. (5.44) is only determined up to an additive constant, we may replace φ by $\varphi + t$. The resulting $\boldsymbol{\sigma}^*$ maximizes Eq. (5.27), concluding the proof. □

From Eq. (5.32) it follows that

$$a(\boldsymbol{\sigma}^*, \boldsymbol{u}^*) = D(\boldsymbol{u}^*) = \max_{\sigma\in K} a(\boldsymbol{\sigma}, \boldsymbol{u}^*). \tag{5.47}$$

Now suppose for the moment that $a(\boldsymbol{\sigma}, \boldsymbol{u})$ is given by the expression (2.4). We recall that Eq. (2.4) is not well defined on $\Sigma \times U$, but according to the results in Section 7 Eq. (2.4) may be interpreted as $a(\boldsymbol{\sigma}, \boldsymbol{u})$, if $\boldsymbol{u}$ is divergence free and thus, in particular, for $\boldsymbol{u} = \boldsymbol{u}^*$. Then Eq. (5.47) implies that at each point $\boldsymbol{x}$ in

Ω, where the collapse strain tensor $\boldsymbol{\varepsilon}^* = \boldsymbol{\varepsilon}(\boldsymbol{u}^*)$ is nonzero, the collapse stress tensor $\boldsymbol{\sigma}^*$ must be at the yield surface at a point, such that $\boldsymbol{\varepsilon}^*(\boldsymbol{x})$ is normal to the yield surface at $\boldsymbol{\sigma}^*(\boldsymbol{x})$. Conversely, if at some point $\boldsymbol{x} \in \Omega$ the collapse stress tensor $\boldsymbol{\sigma}^*$ is not at the yield surface, then the collapse strain rates are zero, and there is no local deformation in the collapse state. This is the principle of *complementary slackness* in limit analysis. The subset of Ω, where $\boldsymbol{\sigma}^*$ is at the yield surface and where local deformations may occur, is usually denoted the *plastified region* or with some ambiguity the *plastic region*. This region need not be uniquely determined for a given problem.

5.4. The reduced problem

We have earlier noted that $D(\boldsymbol{u})$ defined by Eq. (5.30) equals $+\infty$, if $\boldsymbol{u}$ is not divergence free. If, on the other hand, $\boldsymbol{u}$ is divergence free, then $a(\boldsymbol{\sigma}, \boldsymbol{u}) = a(\boldsymbol{\sigma}^{\mathrm{D}}, \boldsymbol{u})$ for all stress fields. This suggests that the saddle point problem of limit analysis could be formulated equivalently within a product of smaller spaces than Σ and U. This observation turns out to be useful both mathematically and numerically. Two complications must be handled, though: From a mathematical viewpoint, $a(\boldsymbol{\sigma}^{\mathrm{D}}, \boldsymbol{u})$ is not necessarily well defined, and from a numerical viewpoint, the condition $\nabla \cdot \boldsymbol{u} = 0$ increases the number of constraints, rather than reducing the size of the problem. The latter problem can be handled by using the finite element technique developed for the Navier–Stokes equation. The former is dealt with below.

We must specify what it means that $\boldsymbol{u} \in U$ is incompressible. From Eq. (5.17) we see that $D(\boldsymbol{u}) = +\infty$, unless $\boldsymbol{u}$ satisfies the following condition

$$\langle \mathrm{tr}(\boldsymbol{\mu}), \varphi \rangle_{C(\overline{\Omega})^* \times C(\overline{\Omega})} = 0 \quad \forall \varphi \in C^1(\overline{\Omega})$$

or equivalently

$$\mathrm{tr}(\boldsymbol{\mu}) = \sum_{i=1}^{N} \mu_{ii} = 0 \quad \text{in } C(\overline{\Omega})^*, \tag{5.48}$$

where $\boldsymbol{\mu} = \boldsymbol{\mu}(\boldsymbol{u})$ is the generalized strain rate tensor associated with $\boldsymbol{u} = (\boldsymbol{u}^{\Omega}, \boldsymbol{u}^{T})$. From Theorem 5.5 we see that Eq. (5.48) is equivalent to the following conditions

$$\begin{aligned} &\nabla \cdot \boldsymbol{u}^{\Omega} = 0 \quad \text{in } \Omega \text{ (as distribution)}, \\ &\boldsymbol{\nu} \cdot (\boldsymbol{u}^{T} - \gamma(\boldsymbol{u}^{\Omega})) = 0 \quad \text{on } T, \\ &\boldsymbol{\nu} \cdot \gamma(\boldsymbol{u}^{\Omega}) = 0 \quad \text{on } S. \end{aligned} \tag{5.49}$$

This is the generalization of incompressible flow to the pair $(\boldsymbol{u}^{\Omega}, \boldsymbol{u}^{T})$. In addition to $\boldsymbol{u}^{\Omega}$ being divergence free in the classical sense the *normal component* of the flow must be continuous along the boundary.

We conclude that the value of the kinematic principle, Eq. (5.29), remains unchanged, if U is replaced by the subspace

$$U_0 = \{\boldsymbol{u} \in U \mid \mathrm{tr}(\boldsymbol{\mu}) = 0 \text{ in } C(\overline{\Omega})^*\}. \tag{5.50}$$

For $\boldsymbol{u} \in U_0$ and $\boldsymbol{\sigma} \in [C^1(\overline{\Omega})]^{N^2}$ we get from Eq. (5.17)

$$a(\boldsymbol{\sigma}, \boldsymbol{u}) = a(\boldsymbol{\sigma}^{\mathrm{D}}, \boldsymbol{u}).$$

Hence U_0 is in duality with the quotient space

$$\Sigma_0 = \Sigma/\{\varphi \boldsymbol{I} \mid \varphi \in L^1(\Omega)\}. \tag{5.51}$$

The corresponding yield set in Σ_0 is

$$K_0 = K/\{\varphi \boldsymbol{I} \mid \varphi \in L^1(\Omega)\}. \tag{5.52}$$

The components of Σ_0 will be denoted $\tilde{\boldsymbol{\sigma}}$. Note that $a(\tilde{\boldsymbol{\sigma}}, \boldsymbol{u})$ is only well defined on $\Sigma_0 \times U_0$ through an appropriate choice of representative of $\tilde{\boldsymbol{\sigma}}$.

We may now formulate the duality problem of limit analysis on the product space $\Sigma_0 \times U_0$. The static formulation of the "reduced" problem is

$$\sup_{\tilde{\sigma} \in K_0} \inf_{\substack{u \in U_0 \\ F(u)=1}} a(\tilde{\boldsymbol{\sigma}}, \boldsymbol{u}) \tag{5.53}$$

or equivalently

$$\sup\{\lambda \mid \exists \tilde{\sigma} \in K_0 \colon a(\tilde{\boldsymbol{\sigma}}, \boldsymbol{u}) = \lambda F(\boldsymbol{u}) \ \forall \boldsymbol{u} \in U_0\}. \tag{5.54}$$

The kinematic form is

$$\inf_{\substack{u \in U_0 \\ F(u)=1}} D_0(\boldsymbol{u}) \tag{5.55}$$

with

$$D_0(\boldsymbol{u}) = \sup_{\tilde{\sigma} \in K_0} a(\tilde{\boldsymbol{\sigma}}, \boldsymbol{u}) = D(\boldsymbol{u}) \quad \forall \boldsymbol{u} \in U_0. \tag{5.56}$$

The relation between the static formulations Eq. (5.27) and Eq. (5.54) is expressed by the following theorem, which is of independent interest. It is essentially proved, although not explicitly stated, in TEMAM and STRANG [1980b].

THEOREM 5.8. *Assumptions as in Theorem 5.7. If $\tilde{\boldsymbol{\sigma}} \in \Sigma_0$ satisfies*

$$a(\tilde{\boldsymbol{\sigma}}, \boldsymbol{u}) = F(\boldsymbol{u}) \quad \forall \boldsymbol{u} \in U_0$$

then there is a representative $\boldsymbol{\sigma}$ of $\tilde{\boldsymbol{\sigma}}$, such that $\boldsymbol{\sigma}$ is in equilibrium with $(\boldsymbol{f}, \boldsymbol{g})$:

$$a(\boldsymbol{\sigma}, \boldsymbol{u}) = F(\boldsymbol{u}) \quad \forall \boldsymbol{u} \in U$$

or equivalently

$$-\boldsymbol{\nabla} \cdot \boldsymbol{\sigma} = \boldsymbol{f} \text{ in } \Omega \quad \text{and} \quad \boldsymbol{\nabla} \cdot \boldsymbol{\sigma} = \boldsymbol{g} \text{ on } T.$$

PROOF. This was essentially proved in the existence part of the previous proof. Equation (5.43) with $\lambda^* = 1$ holds for any representative of $\tilde{\boldsymbol{\sigma}}$. We now construct $\boldsymbol{\sigma}$ exactly as we constructed $\boldsymbol{\sigma}^*$. □

THEOREM 5.9 (Reduced duality of limit analysis). *We make the same assumptions as in Theorem 5.7. Let U_0 be given by Eq.* (5.50) *and Σ_0 by* (5.51). *Then $\boldsymbol{u} \in U$ minimizes Eq.* (5.29) *if and only if it minimizes Eq.* (5.55), *and $\boldsymbol{\sigma} \in K$ maximizes Eq.* (5.27) *if and only if it belongs to a class $\tilde{\boldsymbol{\sigma}} \in K_0$, which is optimal for Eq.* (5.54). *In particular, the duality between Eqs.* (5.54) *and* (5.55) *holds and the value is identical to λ^*, the value of Eq.* (5.31).

PROOF. By definition $D_0(\boldsymbol{u}) = D(\boldsymbol{u})$ for $\boldsymbol{u} \in U_0$. If $\boldsymbol{u}^* \in U$ minimizes Eq. (5.30), then $\lambda^* = D(\boldsymbol{u}^*) < \infty$, which by Eq. (5.17) implies that $\mu(\boldsymbol{u}^*)$ satisfies Eq. (5.48). In that case $\boldsymbol{u}^*$ belongs to U_0 and hence minimizes Eq. (5.55). It follows that Eqs. (5.29) and (5.55) have the same optimal solutions.

If $\tilde{\boldsymbol{\sigma}}$ is feasible for Eq. (5.54) with some value of λ, then it follows immediately from Theorem 5.8 that $\tilde{\boldsymbol{\sigma}}$ has a representative $\boldsymbol{\sigma}$, which is feasible for Eq. (5.27) with the same value of λ. Since the opposite direction is obvious the proof is complete. □

REMARK 5.8. In computations with divergence-free flow we may choose the discrete stresses in the obvious space of representatives

$$\Sigma_0' = \{\boldsymbol{\sigma} \in \Sigma \mid \operatorname{tr}(\boldsymbol{\sigma}) = 0\}.$$

With a suitable choice of finite element spaces $a(\boldsymbol{\sigma}, \boldsymbol{u})$ will be well defined and computable on the discrete subspace of $\Sigma_0' \times U_0$.

6. Generalizations and variations

In the previous section we made some simplifying assumptions for the sake of clarity. We shall now point out the changes for the more general case. Also the duality theorem was proved for yield conditions formulated in the deviator $\boldsymbol{\sigma}^{\mathrm{D}}$ alone. We shall discuss other types of yield conditions.

6.1. *Boundary flow constrained to a subspace*

Until now, we have assumed the boundary to be either completely fixed, or free and subject to loading. In applications we may face more general boundary conditions, such as $\boldsymbol{\nu} \cdot \boldsymbol{u} = 0$ for a part of the boundary, which can move only in directions tangent to the surface. $\boldsymbol{\nu}$ denotes the outward normal to Ω. Other examples are $\boldsymbol{\nu} \cdot \boldsymbol{u} = c$, where c is an unknown part of the solution. Also it may be required that the tangential component $\boldsymbol{u} - (\boldsymbol{u} \cdot \boldsymbol{\nu})\boldsymbol{\nu}$ vanishes. The special case, where the conditions are of the form $u_i = 0$ on part of the boundary with one of the coordinate axes as normal is treated in CHRISTIANSEN [1980a]. This case occurs, when a problem is reduced by symmetry.

We shall not try to formulate the duality theorem in such generality that the above mentioned examples of boundary conditions are covered. In the applications, which have come up, it has been obvious how to formulate and prove the theorem. It is no additional difficulty to handle such boundary conditions numerically.

6.2. *Presence of a pre-load*

There may be other loads present than the load for which the limit multiplier is to be determined. A typical example is the force of gravity. In the literature this is called a pre-load or a "dead load". The duality theorem of limit analysis is easily modified to cover also this case: Let $(\boldsymbol{f}_0, \boldsymbol{g}_0)$ be the pre-load. The equation of virtual work is now

$$a(\boldsymbol{\sigma}, \boldsymbol{u}) = \lambda F(\boldsymbol{u}) + F_0(\boldsymbol{u}) \quad \forall \boldsymbol{u} \in U, \tag{6.1}$$

where $F_0(\boldsymbol{u})$ is defined by

$$F_0(\boldsymbol{u}) = \int_\Omega \boldsymbol{f}_0 \cdot \boldsymbol{u} + \int_T \boldsymbol{g}_0 \cdot \boldsymbol{u}. \tag{6.2}$$

We define the following affine–linear form

$$a_1(\boldsymbol{\sigma}, \boldsymbol{u}) = a(\boldsymbol{\sigma}, \boldsymbol{u}) - F_0(\boldsymbol{u}) \quad \forall (\boldsymbol{\sigma}, \boldsymbol{u}) \in \Sigma \times U. \tag{6.3}$$

The duality between the static and kinematic principles is now of the same form as in the previous section with $a_1(\boldsymbol{\sigma}, \boldsymbol{u})$ instead of $a(\boldsymbol{\sigma}, \boldsymbol{u})$. It is again a matter of simple verification to see that the duality theorem still holds.

6.3. *Bounded yield set*

In the previous section the duality theorem was proved for yield conditions of the form Eq. (5.13), which are appropriate for example for isotropic materials. In particular, the Von Mises condition, Eq. (2.8), which applies to metal-like materials, is of this form. Also in the plane strain model with such materials the yield condition is of the form Eq. (5.13).

We now consider the case of a bounded yield set. We assume that the yield condition is of the following form:

$$\begin{aligned} &\boldsymbol{\sigma}(\boldsymbol{x}) \in B(\boldsymbol{x}) \quad \forall \boldsymbol{x} \in \Omega, \\ &B(\boldsymbol{x}) \subseteq \mathbb{R}^{N^2} \text{ closed, bounded uniformly in } \boldsymbol{x} \text{ and convex for all } \boldsymbol{x} \in \Omega. \end{aligned} \tag{6.4}$$

Also Eq. (5.14) is still assumed to hold. In the two-dimensional plane stress model the yield condition is of this form for all materials. For example the Von Mises condition in plane stress is

$$\sigma_{xx}^2 + \sigma_{yy}^2 - \sigma_{xx}\sigma_{yy} + 3\sigma_{xy}^2 \leqslant \sigma_0^2. \tag{6.5}$$

This condition clearly implies that each component of $\boldsymbol{\sigma}$ belongs to $L^\infty(\Omega)$. Hence in the definition of the stress space Σ conditions (5.8)–(5.9) may be replaced by the condition

$$\boldsymbol{\sigma} \in [L^\infty(\Omega)]^{N^2}. \tag{6.6}$$

The norm Eq. (5.12) on Σ is replaced by the norm

$$\|\boldsymbol{\sigma}\|_\Sigma = \|\boldsymbol{\sigma}\|_{L^\infty(\Omega)} + \|\boldsymbol{\nabla}\cdot\boldsymbol{\sigma}\|_{L^p(\Omega)} + \|\boldsymbol{\nu}\cdot\boldsymbol{\sigma}\|_{W^{1-1/p,p}(T)}. \tag{6.7}$$

The duality theorem of limit analysis now holds without further due.

THEOREM 6.1. *Assume that* $\Omega \subseteq \mathbb{R}^N$ *is bounded with Lipschitz continuous boundary. Let* Σ *be defined by Eq.* (6.6) *and Eqs.* (5.10)–(5.11), *and* U *by Eqs.* (5.15)–(5.16) *(as before). Let*

$$K = \{\boldsymbol{\sigma} \in \Sigma \mid \boldsymbol{\sigma}(\boldsymbol{x}) \in B(\boldsymbol{x}) \text{ a.e. in } \Omega\},$$

where $B(\boldsymbol{x})$ *satisfies Eqs.* (6.4) *and* (5.14). *Then the conclusions of Theorem* 5.7 *hold.*

PROOF. All steps in the proof of Theorem 5.7 can be repeated without modification, except for the existence of a maximizing stress tensor $\boldsymbol{\sigma}^*$, which is now much simpler: There is a sequence $\{\boldsymbol{\sigma}_k\}$ in K satisfying Eq. (5.42). This sequence (and not only its deviators) has a weak* convergent subsequence, and the limit is easily seen to be a maximizing stress tensor. □

In limit analysis with bounded yield set the collapse flow field is not divergence free. The total energy dissipation is finite for any flow field. Mathematically this is a much simpler analysis, and according to experience it is also easier to handle numerically. This could be the reason why the case of bounded yield set has been over-emphasized in the literature on limit analysis.

6.4. Yield conditions for granular materials

We now turn to a type of yield condition, which is more difficult to handle than both of the above mentioned, but which is too important for applications to be ignored. It applies to granular materials like soil and concrete (not reinforced). Characteristic for these materials is that they can resist compression, but not tension, or at least only to a lesser extend. This means that stress tensors of the form $\boldsymbol{\sigma} = \phi \boldsymbol{I}$ are admissible only if $\phi(\boldsymbol{x}) \leqslant \sigma_0$ for all $\boldsymbol{x}$ in Ω, where $\sigma_0 \geqslant 0$.

The classical yield condition for granular materials goes back to COULOMB [1773]. A more recent reference is KALISZKY [1975]. On any section with normal $\boldsymbol{n}$ the shear stress τ_n and the normal stress σ_n must satisfy the inequality

$$\mu\sigma_n + |\tau_n| \leqslant c, \tag{6.8}$$

where μ and c are non-negative constants. μ is the *coefficient of internal friction*, and c is a measure of *cohesion* in the material. If $c = 0$ the material is without cohesion. In a frictionless material $\mu = 0$, and in this case the Coulomb condition becomes equivalent to the Tresca yield condition, which bounds the maximum shear stress at every point of the material. The Tresca yield condition can be written in the form Eq. (5.13), (KALISZKY [1975, p. 66]), so the case of a frictionless material has been covered in Section 5.

The Coulomb condition may also be formulated in terms of the principal stresses σ_{I}, σ_{II} and σ_{III} (the eigenvalues of the stress tensor). Let φ be given by

$\mu = \tan\varphi$. Then the Coulomb yield condition can be written as three inequalities:

$$\begin{aligned} &|\sigma_{\mathrm{I}} - \sigma_{\mathrm{II}}| + (\sigma_{\mathrm{I}} + \sigma_{\mathrm{II}})\sin\varphi - 2c\cos\varphi \leqslant 0, \\ &|\sigma_{\mathrm{II}} - \sigma_{\mathrm{III}}| + (\sigma_{\mathrm{II}} + \sigma_{\mathrm{III}})\sin\varphi - 2c\cos\varphi \leqslant 0, \\ &|\sigma_{\mathrm{I}} - \sigma_{\mathrm{III}}| + (\sigma_{\mathrm{I}} + \sigma_{\mathrm{III}})\sin\varphi - 2c\cos\varphi \leqslant 0. \end{aligned} \tag{6.9}$$

In Eq. (6.9) an inequality can only be active, if it involves the maximum and minimum principal stresses.

From a computational viewpoint it is a complication that the principal stresses must be determined at each node of the discretization, before the yield condition can be applied. In the case of plane strain the extremal principal stresses are known, and the Coulomb condition may be expressed in rectangular coordinates of our choice:

$$\sqrt{(\sigma_{xx} - \sigma_{yy})^2 + 4\sigma_{xy}^2} + (\sigma_{xx} + \sigma_{yy})\sin\varphi - 2c\cos\varphi \leqslant 0. \tag{6.10}$$

For other yield conditions of this type we refer to KALISZKY [1975, Section 8.3]. Common for them is the property that the cone

$$\{\varphi \boldsymbol{I} \mid \varphi(\boldsymbol{x}) \leqslant \sigma_0 \ \forall \boldsymbol{x} \in \Omega\}$$

is contained in the set of admissible stress tensors. Hence a necessary condition for the total energy dissipation rate to satisfy $D(\boldsymbol{u}) < \infty$ is

$$\boldsymbol{\nabla}\cdot\boldsymbol{u} \geqslant 0 \quad \text{in } \Omega. \tag{6.11}$$

In particular, the collapse flow $\boldsymbol{u}^*$ must satisfy Eq. (6.11). In analogy with Eq. (5.48) the precise form of this semi-incompressibility condition for $\boldsymbol{u} \in U$ is

$$\operatorname{tr}(\boldsymbol{\mu}) = \sum_{i=1}^{N} \mu_{ii} \geqslant 0 \quad \text{in } C(\overline{\Omega})^*, \tag{6.12}$$

where $\boldsymbol{\mu}$ is the generalized strain rate tensor for $\boldsymbol{u} = (\boldsymbol{u}^{\Omega}, \boldsymbol{u}^{T})$.

Theorem 5.7 does not hold as stated for such yield conditions. If the condition (5.14) holds, the proof of Eq. (5.33) is valid without change. Hence the duality between the static and the kinematic principles holds, and a minimizing flow field exists for the kinematic principle. However, the existence proof of a collapse stress field does not hold. To our knowledge this analysis has not been made.

If the condition (5.14) does not hold, arbitrarily small forces may cause yield. This seems to be an extreme case without much practical interest. For the Coulomb yield condition it occurs, if $c = 0$ in Eq. (6.8), and in this case the limit multiplier is either 0 or $+\infty$ for any load distribution.

7. Green's formula in the collapse state

In this section we shall try to make sense of the expression in Eq. (2.4) for the internal energy dissipation rate, or equivalently, analyse the validity of

Green's formula for the collapse fields $(\boldsymbol{\sigma}^*, \boldsymbol{u}^*)$. On $\Sigma \times U$, $a(\boldsymbol{\sigma}, \boldsymbol{u})$ is defined by Eq. (5.21), which is identical to Eq. (2.5). By Eq. (5.17) a generalized Green's formula holds for $\boldsymbol{u} \in U$ and $\boldsymbol{\sigma}$ with components in $C^1(\overline{\Omega})$. However, the left-hand side of Eq. (5.17), which is the generalized form of Eq. (2.4), is not defined for $(\boldsymbol{\sigma}, \boldsymbol{u}) \in \Sigma \times U$; μ_{ij} is a measure, but σ_{ij} need not be continuous.

Recall that $\boldsymbol{\varepsilon}(\boldsymbol{u})$ denotes the classical (distributional) strain rate tensor defined by Eq. (5.1). We first analyse $\boldsymbol{\varepsilon}(\boldsymbol{u})$ under the extra assumption that the divergence $\boldsymbol{\nabla}\cdot\boldsymbol{u}$ belongs to $L^q(\Omega)$ for some q in the interval $1 < q < \infty$. Let $\boldsymbol{u}$ satisfy

$$\boldsymbol{u} \in \mathrm{BD}(\Omega), \qquad \boldsymbol{\nabla}\cdot\boldsymbol{u} \in L^q(\Omega) \tag{7.1}$$

and let $\boldsymbol{\sigma}$ be symmetric and satisfy

$$\boldsymbol{\sigma} \in [L^2(\Omega)]^{N^2}, \qquad \boldsymbol{\sigma}^{\mathrm{D}} \in [L^\infty(\Omega)]^{N^2}, \qquad \boldsymbol{\nabla}\cdot\boldsymbol{\sigma} \in [L^N(\Omega)]^N. \tag{7.2}$$

Define the following distribution, which we shall denote by $(\boldsymbol{\sigma}^{\mathrm{D}}, \boldsymbol{\varepsilon}^{\mathrm{D}}(\boldsymbol{u}))$:

$$\int_\Omega (\boldsymbol{\sigma}^{\mathrm{D}}, \boldsymbol{\varepsilon}^{\mathrm{D}}(\boldsymbol{u}))\, \varphi = -\frac{1}{N}\int_\Omega \operatorname{tr}(\boldsymbol{\sigma})(\boldsymbol{\nabla}\cdot\boldsymbol{u})\varphi - \int_\Omega (\boldsymbol{\nabla}\cdot\boldsymbol{\sigma})\cdot\boldsymbol{u}\,\varphi$$
$$- \int_\Omega \langle \boldsymbol{\sigma}, \boldsymbol{u} \otimes \boldsymbol{\nabla}\varphi \rangle \quad \forall \varphi \in C_0^\infty(\Omega). \tag{7.3}$$

We have used the notation

$$\langle \boldsymbol{\sigma}, \boldsymbol{u} \otimes \boldsymbol{\nabla}\varphi \rangle = \sum_{ij} \sigma_{ij} u_i \frac{\partial \varphi}{\partial x_j}.$$

In Eq. (7.3) each term on the right-hand side is well defined by Theorem 5.4 and Theorem 5.2. It must be emphasized that in general "$(\boldsymbol{\sigma}^{\mathrm{D}}, \boldsymbol{\varepsilon}^{\mathrm{D}}(\boldsymbol{u}))$" is not an inner product as suggested by the notation. It is a convenient, but inaccurate notation, which is justified only when $\boldsymbol{\sigma}$ and $\boldsymbol{u}$ are sufficiently smooth. In that case Eq. (7.3) is a consequence of Green's formula. The following theorem is proved in KOHN and TEMAM [1983, Theorem 3.2]. It is the closest to a regularity theorem for the collapse state we get.

THEOREM 7.1. *Assume that $\Omega \subseteq \mathbb{R}^N$ is bounded with Lipschitz continuous boundary. Let $1 < q < \infty$ and let $\boldsymbol{u}$ and $\boldsymbol{\sigma}$ satisfy Eqs.* (7.1) *and* (7.2) *respectively. Then the following statements hold:*

(a) *The distribution $(\boldsymbol{\sigma}^{\mathrm{D}}, \boldsymbol{\varepsilon}^{\mathrm{D}}(\boldsymbol{u}))$ defined by Eq.* (7.3) *is a bounded measure on Ω and it satisfies*

$$\left| \int_\Omega (\boldsymbol{\sigma}^{\mathrm{D}}, \boldsymbol{\varepsilon}^{\mathrm{D}}(\boldsymbol{u}))\varphi \right| \leqslant \|\boldsymbol{\sigma}^{\mathrm{D}}\|_{L^\infty} \int_\Omega |\varphi| \|\boldsymbol{\varepsilon}^{\mathrm{D}}(\boldsymbol{u})\| \quad \forall \varphi \in C^0(\overline{\Omega}). \tag{7.4}$$

(b) *If the boundary of Ω can be defined locally by C^2-functions, then there is* *a uniquely defined distribution "$(\boldsymbol{\sigma}\cdot\boldsymbol{\nu},\boldsymbol{u})$" on $\partial\Omega$, which satisfies*

$$\int_{\partial\Omega}(\boldsymbol{\sigma}\cdot\boldsymbol{\nu},\boldsymbol{u})\varphi=\int_{\Omega}(\boldsymbol{\sigma}^{\mathrm{D}},\boldsymbol{\varepsilon}^{\mathrm{D}}(\boldsymbol{u}))\varphi+\frac{1}{N}\int_{\Omega}(\boldsymbol{\nabla}\cdot\boldsymbol{u})\,\mathrm{tr}(\boldsymbol{\sigma})\varphi$$
$$+\int_{\Omega}(\boldsymbol{\nabla}\cdot\boldsymbol{\sigma})\cdot\boldsymbol{u}\,\varphi+\int_{\Omega}\langle\boldsymbol{\sigma},\boldsymbol{u}\otimes\boldsymbol{\nabla}\varphi\rangle\quad\forall\varphi\in C^{\infty}(\overline{\Omega}).\tag{7.5}$$

Furthermore $(\boldsymbol{\sigma}\cdot\boldsymbol{\nu},\boldsymbol{u})$ belongs to the dual of $C^1(\partial\Omega)$, and Eq. (7.5) holds for all $\varphi\in C^1(\overline{\Omega})$.

(c) *If $\{\boldsymbol{u}_k\}$ is a sequence in* $\mathrm{BD}(\Omega)$ *with $\boldsymbol{\nabla}\cdot\boldsymbol{u}_k\in L^q(\Omega)$, which converges to $\boldsymbol{u}$ in the following sense*

$$\begin{aligned}&\boldsymbol{u}_k\to\boldsymbol{u} &&\text{in}\quad [L^{N/N-1}(\Omega)]^N,\\ &\boldsymbol{\nabla}\cdot\boldsymbol{u}_k\to\boldsymbol{\nabla}\cdot\boldsymbol{u} &&\text{in}\quad L^q(\Omega),\\ &\textstyle\int_\Omega\|\boldsymbol{\varepsilon}^{\mathrm{D}}(\boldsymbol{u}_k)\|\to\int_\Omega\|\boldsymbol{\varepsilon}^{\mathrm{D}}(\boldsymbol{u})\|\end{aligned}\tag{7.6}$$

then

$$\int_{\Omega}(\boldsymbol{\sigma}^{\mathrm{D}},\boldsymbol{\varepsilon}^{\mathrm{D}}(\boldsymbol{u}_k))\varphi\to\int_{\Omega}(\boldsymbol{\sigma}^{\mathrm{D}},\boldsymbol{\varepsilon}^{\mathrm{D}}(\boldsymbol{u}))\varphi\quad\forall\varphi\in C^{0}(\overline{\Omega}).\tag{7.7}$$

(d) *If $\{\boldsymbol{\sigma}_k\}$ is a sequence satisfying Eq.* (7.2), *and which converges to $\boldsymbol{\sigma}$ in the following sense*

$$\begin{aligned}&\boldsymbol{\sigma}_k\to\boldsymbol{\sigma} &&\text{in}\quad [L^2(\Omega)]^{N^2},\\ &\boldsymbol{\nabla}\cdot\boldsymbol{\sigma}_k\to\boldsymbol{\nabla}\cdot\boldsymbol{\sigma} &&\text{in}\quad [L^N(\Omega)]^N,\\ &\|\boldsymbol{\sigma}_k^{\mathrm{D}}\|_{L^\infty}\leqslant C, &&C\text{ independent of }k\end{aligned}\tag{7.8}$$

then

$$\int_{\Omega}(\boldsymbol{\sigma}_k^{\mathrm{D}},\boldsymbol{\varepsilon}^{\mathrm{D}}(\boldsymbol{u}))\varphi\to\int_{\Omega}(\boldsymbol{\sigma}^{\mathrm{D}},\boldsymbol{\varepsilon}^{\mathrm{D}}(\boldsymbol{u}))\varphi\quad\forall\varphi\in C^{0}(\overline{\Omega}).\tag{7.9}$$

PROOF. See KOHN and TEMAM [1983, Theorem 3.2]. □

In Eq. (7.5) $(\boldsymbol{\sigma}\cdot\boldsymbol{\nu},\boldsymbol{u})$ is not in general an inner product as suggested by the notation. However, if $\boldsymbol{\sigma}\in\Sigma$, and $\boldsymbol{u}\in U$ with $\gamma(\boldsymbol{u})=\boldsymbol{0}$ on S and $\boldsymbol{\nabla}\cdot\boldsymbol{u}\in L^q(\Omega)$, $q>1$, then by Eq. (5.11) and Theorem 5.1 we may conclude that $(\boldsymbol{\sigma}^{\mathrm{D}},\boldsymbol{\varepsilon}^{\mathrm{D}}(\boldsymbol{u}))$ satisfies the identity

$$\int_\Omega (\boldsymbol{\sigma}^{\mathrm{D}}, \boldsymbol{\varepsilon}^{\mathrm{D}}(\boldsymbol{u}))\varphi = -\int_\Omega \langle \boldsymbol{\sigma}, \boldsymbol{u} \otimes \boldsymbol{\nabla}\varphi \rangle - \frac{1}{N}\int_\Omega \operatorname{tr}(\boldsymbol{\sigma})(\boldsymbol{\nabla}\cdot\boldsymbol{u})\varphi$$

$$-\int_\Omega (\boldsymbol{\nabla}\cdot\boldsymbol{\sigma})\cdot\boldsymbol{u}\,\varphi + \int_T (\boldsymbol{\sigma}\cdot\boldsymbol{\nu})\cdot\boldsymbol{\gamma}(\boldsymbol{u})\varphi \quad \forall \varphi \in C^1(\overline{\Omega}). \tag{7.10}$$

If $(\boldsymbol{\sigma}, \boldsymbol{u}) = (\boldsymbol{\sigma}^*, \boldsymbol{u}^*)$ is the saddle point in Theorem 5.7, then $\boldsymbol{\nabla}\cdot\boldsymbol{u} = 0$ and $\boldsymbol{\varepsilon}(\boldsymbol{u}) = \boldsymbol{\varepsilon}^{\mathrm{D}}(\boldsymbol{u})$. If furthermore $\boldsymbol{\gamma}(\boldsymbol{u}) = \boldsymbol{0}$ on S (the boundary condition without relaxation), then we may apply Eq. (7.10) with $\varphi = 1$ and get

$$\int_\Omega (\boldsymbol{\sigma}, \boldsymbol{\varepsilon}(\boldsymbol{u})) = \int_\Omega (\boldsymbol{\sigma}^{\mathrm{D}}, \boldsymbol{\varepsilon}^{\mathrm{D}}(\boldsymbol{u}))$$

$$= -\int_\Omega (\boldsymbol{\nabla}\cdot\boldsymbol{\sigma})\cdot\boldsymbol{u} + \int_T (\boldsymbol{\sigma}\cdot\boldsymbol{\nu})\cdot\boldsymbol{\gamma}(\boldsymbol{u}). \tag{7.11}$$

This is Green's formula in the case $\boldsymbol{\mu}(\boldsymbol{u}) = \boldsymbol{\varepsilon}(\boldsymbol{u})$, or equivalently $\boldsymbol{\gamma}(\boldsymbol{u}^\Omega) = \boldsymbol{0}$ on S and $\boldsymbol{\gamma}(\boldsymbol{u}^\Omega) = \boldsymbol{u}^T$ on T, where $\boldsymbol{\mu}(\boldsymbol{u})$ is the generalized strain rate tensor. Anticipating the more general case, we now state the above result with different assumptions, moving the regularity conditions from the domain Ω to the pair $(\boldsymbol{\sigma}, \boldsymbol{u})$. The technical details are not essential, and our assumptions may be stronger than necessary. We shall assume that one of the following two conditions holds:

$$\boldsymbol{\sigma} \in W^{1,p}(T_\delta) \quad \text{for some } \delta > 0, \tag{7.12}$$

where $T_\delta = \{\boldsymbol{x} \in \Omega \mid \operatorname{dist}(\boldsymbol{x}, T) < \delta\}$ is a neighbourhood of T in Ω.

$$\boldsymbol{\gamma}(\boldsymbol{u}^\Omega) \in W_0^{1-(1/r),r}(T) \quad \text{for some } r > 1. \tag{7.13}$$

THEOREM 7.2. *We make the same assumption as in Theorem 5.7. Let* $(\boldsymbol{\sigma}, \boldsymbol{u}) = (\boldsymbol{\sigma}^*, \boldsymbol{u}^*) \in \Sigma \times U$ *be the saddle point in Eq.* (5.32). *Assume that* $\boldsymbol{\gamma}(\boldsymbol{u}^\Omega) = \boldsymbol{0}$ *on* S *and that Eq.* (7.12) *or Eq.* (7.13) *holds. Then* $(\boldsymbol{\sigma}, \boldsymbol{\varepsilon}(\boldsymbol{u})) = (\boldsymbol{\sigma}^{\mathrm{D}}, \boldsymbol{\varepsilon}^{\mathrm{D}}(\boldsymbol{u}))$ *is a bounded measure on* Ω, *which satisfies Eq.* (7.4).

PROOF. See CHRISTIANSEN [1986, Corollary 3.4]. □

In the general case we need a similar result for the generalized strain rate tensor $\boldsymbol{\mu}(\boldsymbol{u})$ for $\boldsymbol{u} = (\boldsymbol{u}^\Omega, \boldsymbol{u}^T)$. We replace condition (7.13) by the analogous condition on $\boldsymbol{u}^T$:

$$\boldsymbol{u}^T \in W_0^{1-(1/r),r}(T) \quad \text{for some } r > 1. \tag{7.14}$$

THEOREM 7.3. *Make the same assumptions as in Theorem 5.7 and let* $(\boldsymbol{\sigma}, \boldsymbol{u}) = (\boldsymbol{\sigma}^*, \boldsymbol{u}^*)$ *be the saddle point in Eq.* (5.32). *We define* $(\boldsymbol{\sigma}, \boldsymbol{\mu}) = (\boldsymbol{\sigma}^{\mathrm{D}}, \boldsymbol{\mu}^{\mathrm{D}})$ *weakly by the equation*

$$\int_{\overline{\Omega}} (\boldsymbol{\sigma}^{\mathrm{D}}, \boldsymbol{\mu}^{\mathrm{D}})\varphi = -\int_{\Omega} (\boldsymbol{\nabla}\cdot\boldsymbol{\sigma})\cdot\boldsymbol{u}^{\Omega}\varphi - \int_{\Omega} \langle \boldsymbol{\sigma}, \boldsymbol{u}^{\Omega} \otimes \boldsymbol{\nabla}\varphi \rangle + \int_{T} (\boldsymbol{\sigma}\cdot\boldsymbol{\nu})\cdot\boldsymbol{u}^{T}\varphi \tag{7.15}$$

for all $\varphi \in C^1(\overline{\Omega})$. *If condition* (7.12) *or condition* (7.14) *holds, then* $(\boldsymbol{\sigma}, \boldsymbol{\mu})$ *is a bounded measure on* $\overline{\Omega}$, *and it satisfies*

$$\left| \int_{\overline{\Omega}} (\boldsymbol{\sigma}, \boldsymbol{\mu})\varphi \right| \leqslant \|\boldsymbol{\sigma}^{\mathrm{D}}\|_{L^\infty} \int_{\overline{\Omega}} |\varphi|\,|\boldsymbol{\mu}| \quad \forall \varphi \in C^0(\overline{\Omega}). \tag{7.16}$$

PROOF. See CHRISTIANSEN [1986, Theorem 3.1]. □

Applying Eq. (7.15) for $\boldsymbol{\sigma}$ smooth and $\varphi = 1$ shows that $(\boldsymbol{\sigma}, \boldsymbol{\mu})$ is an extension of the left-hand side of Eq. (5.17) and that Green's formula holds in the collapse state.

In Theorem 7.3 $(\boldsymbol{\sigma}^*, \boldsymbol{u}^*)$ can be replaced by any other pair $(\boldsymbol{\sigma}, \boldsymbol{u})$ as long as $\boldsymbol{u}$ is divergence free in the sense that Eq. (5.48) holds.

If the conclusion of Theorem 7.3 holds, then it follows from Eq. (5.47) that

$$\int_{\overline{\Omega}} (\boldsymbol{\sigma}^*, \boldsymbol{\mu}(\boldsymbol{u}^*)) = \sup_{\sigma \in C(\overline{\Omega}) \cap K} \int_{\overline{\Omega}} \sum_{ij} \sigma_{ij} \mu_{ij}(\boldsymbol{u}^*).$$

This is the generalized form of the principle of complementary slackness. If $\boldsymbol{\sigma}^*$ and $\boldsymbol{\mu}^*$ can be interpreted pointwise, then $\boldsymbol{\sigma}^*(\boldsymbol{x})$ must at each point $\boldsymbol{x}$ maximize the expression

$$\sum_{i,j} \sigma_{ij} \mu_{ij}^*$$

subject to the yield condition at the point $\boldsymbol{x}$.

CHAPTER III

Discretization by Finite Elements

8. Formulation of the discrete problem

The duality problem of limit analysis is well suited for discretization by the finite element method. In its simplest form $\Sigma_h \subseteq \Sigma$ and $U_h \subseteq U$ are finite-dimensional subspaces, and $K \subseteq \Sigma$ is replaced by $K_h \subseteq \Sigma_h$. In general K_h need not be contained in K. Then the discrete version of the problem is:

$$\begin{aligned} \lambda_h^* &= \max\{\lambda \mid \exists \boldsymbol{\sigma}_h \in K_h\colon a(\boldsymbol{\sigma}_h, \boldsymbol{u}_h) = \lambda F(\boldsymbol{u}_h)\ \forall \boldsymbol{u}_h \in U_h\} \\ &= \max_{\sigma_h \in K_h} \min_{F(u_h)=1} a(\boldsymbol{\sigma}_h, \boldsymbol{u}_h) \qquad (8.1) \\ &= \min_{F(u_h)=1} \max_{\sigma_h \in K_h} a(\boldsymbol{\sigma}_h, \boldsymbol{u}_h) \\ &= \min_{F(u_h)=1} D_h(\boldsymbol{u}_h). \end{aligned}$$

The above duality is proved in Theorem 8.1.

It is natural to choose mixed finite elements to discretize stresses and flow simultaneously. If only $\boldsymbol{\sigma}$ is discretized in Eq. (2.11) or Eq. (5.27) the equilibrium equation must be satisfied exactly for some $\boldsymbol{\sigma}_h \in \Sigma_h$, which can be the case only for special loads. If only $\boldsymbol{u}$ is discretized in Eq. (2.13) or Eq. (5.29), we will get $D(\boldsymbol{u}) = +\infty$, unless $\boldsymbol{u}$ is divergence free.

Let $\mathcal{T}_h$ be a triangulation of Ω (CIARLET [1978, 1991]) and let Σ_h and U_h be corresponding finite element spaces representing Σ and U. The discrete fields for $\boldsymbol{\sigma}$ and $\boldsymbol{u}$ will be denoted $\boldsymbol{\sigma}_h$ and $\boldsymbol{u}_h$ respectively. Frequently we will use different finite element functions to define the single coordinates of $\boldsymbol{\sigma}_h$ and $\boldsymbol{u}_h$, but for practical reasons they should be based on the same triangulation. We ignore the complications which arise when Ω is a domain with curved boundary. These are not specific to limit analysis, so we may refer to CIARLET [1978, Section 4.4] or CIARLET [1991, Chapter VI].

It is convenient to think in terms of conforming elements, i.e., $\Sigma_h \subseteq \Sigma$ and $U_h \subseteq U$. However, non-conforming elements (see, e.g., STRANG and FIX [1973, Chapter 4] or CIARLET [1991, Chapter 5]) may be used, if only the energy functions $a(\boldsymbol{\sigma}_h, \boldsymbol{u}_h)$ and $F(\boldsymbol{u}_h)$ are well defined. Computational experience indicates that piecewise constant elements for $\boldsymbol{\sigma}_h$ paired with piecewise bilinear elements for $\boldsymbol{u}_h$ is a useful combination.

Due to the essential boundary condition on $\boldsymbol{u}$, and in order to compute $a(\boldsymbol{\sigma}_h, \boldsymbol{u}_h)$ by Eq. (2.4), it is convenient to choose globally continuous element functions for $\boldsymbol{u}_h$. Then the partial derivatives of $\boldsymbol{u}_h$ are integrable functions (CIARLET [1991, Theorem 5.1]). Since the collapse fields $\boldsymbol{\sigma}^*$ and $\boldsymbol{u}^*$ are typically discontinuous there may not be obtained significantly better accuracy by using more smooth element functions.

For simplicity we will assume that the energy functions $a(\boldsymbol{\sigma}_h, \boldsymbol{u}_h)$ and $F(\boldsymbol{u}_h)$ can be computed exactly on Σ_h and U_h by the expressions (2.4) and (2.2). The effect of using numerical integration is not specific to limit analysis, so we may refer to CIARLET [1991, Chapter IV].

The yield condition $\boldsymbol{\sigma}(\boldsymbol{x}) \in B(\boldsymbol{x})$ will be imposed only through the nodal values of $\boldsymbol{\sigma}_h(\boldsymbol{x})$: If $\{\boldsymbol{x}_\nu\}$ are the nodal points for the value of $\boldsymbol{\sigma}_h(\boldsymbol{x})$, then the discrete set of admissible stresses K_h is defined as

$$K_h = \{\boldsymbol{\sigma}_h \in \Sigma_h \mid \boldsymbol{\sigma}_h(\boldsymbol{x}_\nu) \in B(\boldsymbol{x}_\nu) \ \forall \boldsymbol{x}_\nu\}. \tag{8.2}$$

Assume for the moment that $B(\boldsymbol{x}) = B$ is independent of $\boldsymbol{x}$. Then Eq. (8.2) implies that $\boldsymbol{\sigma}_h(\boldsymbol{x}) \in B \ \forall \boldsymbol{x} \in \Omega$ for piecewise linear and multilinear elements (linear n-simplices and n-rectangles of degree 2 in the notation of CIARLET [1991, Chapter II]). However, for other element types, such as quadratic or bi-quadratic elements, the yield condition may be violated between nodes, even if it is satisfied at all nodes. We shall return to this problem in Section 9. Examples of element combinations in plane strain are:

- Linear elements for both stresses and velocity. The nodes are the vertices of the triangles.
- Bilinear elements (rectangles of degree 2) for both stresses and velocity. The nodes are the vertices of the rectangles.
- Bilinear elements (rectangles of degree 2) for velocity and piecewise constant elements for stresses. The nodal values for the stresses are the values in each rectangle.
- Linear elements for velocity and piecewise constant elements for stresses. The nodal values for the stresses are the values in each triangle. This combination has too many degrees of freedom for the stresses compared to the velocity. In our experience such lack of balance causes ill-conditioning. This element combination cannot be recommended.
- C^1-bi-cubic rectangles (Bogner–Fox–Schmit elements, see CIARLET [1991, p. 92]) for the *stream function* and bi-quadratic rectangles for the stresses. This combination is used to implement divergence-free flow. The nodal values for the flow are the values of the stream function, its gradient (i.e., the flow) and the mixed second-order derivative at the vertices. The nodal values for the stresses are values of the stress components at the vertices, at the midpoints of the edges and at the center of the rectangles.

Since the notation for the discrete problem is somewhat complicated, we start by giving a specific example.

EXAMPLE 8.1 (*Linear elements for both $\boldsymbol{\sigma}_h$ and $\boldsymbol{u}_h$ in plane strain*). Let $\mathcal{T}_h$ be a triangulation of $\Omega \subseteq \mathbb{R}^2$ into triangles with node set $\mathcal{N}$. Associated with each node $\nu \in \mathcal{N}$ is a piecewise linear scalar function ϕ_ν with value equal to 1 at node ν and equal to 0 at all other nodes. Each of the three components in the stress tensor, σ_{11}, σ_{22} and σ_{12}, is given by a scalar function, which is a linear combination of node functions of type ϕ_ν. Hence associated with each node $\nu \in \mathcal{N}$ there are the following three basis functions for the discrete stress space Σ_h:

$$\boldsymbol{\phi}_\nu^{11} = \begin{bmatrix} \phi_\nu & 0 \\ 0 & 0 \end{bmatrix}, \qquad \boldsymbol{\phi}_\nu^{22} = \begin{bmatrix} 0 & 0 \\ 0 & \phi_\nu \end{bmatrix}, \qquad \boldsymbol{\phi}_\nu^{12} = \begin{bmatrix} 0 & \phi_\nu \\ \phi_\nu & 0 \end{bmatrix}.$$

The discrete stress tensors $\boldsymbol{\sigma}_h \in \Sigma_h$ may now be written

$$\boldsymbol{\sigma}_h = \sum_{\nu \in \mathcal{N}} \left(\xi_\nu^{11} \boldsymbol{\phi}_\nu^{11} + \xi_\nu^{22} \boldsymbol{\phi}_\nu^{22} + \xi_\nu^{12} \boldsymbol{\phi}_\nu^{12} \right).$$

Similarly there are two basis functions for the discrete velocity space U_h associated with each node $\mu \in \mathcal{N}$:

$$\boldsymbol{\psi}_\mu^1 = \begin{bmatrix} \phi_\mu \\ 0 \end{bmatrix}, \qquad \boldsymbol{\psi}_\mu^2 = \begin{bmatrix} 0 \\ \phi_\mu \end{bmatrix}.$$

Each $\boldsymbol{u}_h \in U_h$ may be written as a linear combination

$$\boldsymbol{u}_h = \sum_{\mu \in \mathcal{N}} \left(\eta_\mu^1 \boldsymbol{\psi}_\mu^1 + \eta_\mu^2 \boldsymbol{\psi}_\mu^2 \right).$$

We shall now describe the construction in the example in more general terms. Let Σ_h and U_h be the finite element representations of Σ and U respectively. Let $\mathcal{N}_\sigma$ and $\mathcal{N}_u$ denote the corresponding set of nodes for the single components of $\boldsymbol{\sigma}_h \in \Sigma_h$ and $\boldsymbol{u}_h \in U_h$ respectively. If several degrees of freedom are associated with the same node (as is the case for example for cubic Hermite elements defined in CIARLET [1991, p. 84]), then the node is counted once for each degree of freedom. Let ϕ_ν be the scalar basis function associated with the node $\nu \in \mathcal{N}_\sigma$. Based on ϕ_ν we define one basis function for Σ_h corresponding to each independent component of the stress tensor. In full three-dimensional problems there are six basis functions based on ϕ_ν:

$$\boldsymbol{\phi}_\nu^{11} = \begin{bmatrix} \phi_\nu & 0 & 0 \\ 0 & 0 & 0 \\ 0 & 0 & 0 \end{bmatrix}, \qquad \boldsymbol{\phi}_\nu^{12} = \begin{bmatrix} 0 & \phi_\nu & 0 \\ \phi_\nu & 0 & 0 \\ 0 & 0 & 0 \end{bmatrix}, \qquad \text{etc.} \tag{8.3}$$

In plane problems there are only three basis stress functions for each node, corresponding to the three independent components of a symmetric 2×2 tensor. The complete basis for Σ_h is the set of all basis tensor element functions:

$$\{\boldsymbol{\phi}_\nu^{ij} \mid i \leqslant j, \nu \in \mathcal{N}_\sigma\}.$$

Every discrete stress tensor $\boldsymbol{\sigma}_h \in \Sigma_h$ may be written as a linear combination of these basis functions:

$$\boldsymbol{\sigma}_h = \sum_{\nu \in \mathcal{N}_\sigma} \sum_{i \leqslant j} \xi_\nu^{ij} \boldsymbol{\phi}_\nu^{ij}. \tag{8.4}$$

The coordinate ξ_ν^{ij} is the nodal value of the component $\sigma_{h,ij}$ corresponding to the node ν.

Similarly, for every node $\mu \in \mathcal{N}_u$ we define a basis function for U_h corresponding to each velocity component at that node. If ψ_μ denotes the scalar basis function associated with the node μ we get the following functions for U_h in three-dimensional problems:

$$\boldsymbol{\psi}_\mu^1 = \begin{bmatrix} \psi_\mu \\ 0 \\ 0 \end{bmatrix}, \quad \boldsymbol{\psi}_\mu^2 = \begin{bmatrix} 0 \\ \psi_\mu \\ 0 \end{bmatrix}, \quad \boldsymbol{\psi}_\mu^3 = \begin{bmatrix} 0 \\ 0 \\ \psi_\mu \end{bmatrix}. \tag{8.5}$$

Disregarding for the moment boundary conditions on $\boldsymbol{u}$ the complete basis for U_h is

$$\{\boldsymbol{\psi}_\mu^k \mid k = 1, 2, 3,\ \mu \in \mathcal{N}_u\}.$$

Any discrete velocity $\boldsymbol{u}_h \in U_h$ may now be written

$$\boldsymbol{u}_h = \sum_{\mu \in \mathcal{N}_u} \sum_{k} \eta_\mu^k \boldsymbol{\psi}_\mu^k, \tag{8.6}$$

where η_μ^k is the nodal value of the velocity component $u_{h,k}$ corresponding to the node μ.

After insertion of the expressions (8.4) and (8.6) the energy functions may be expressed as

$$F(\boldsymbol{u}_h) = \sum_{\mu \in \mathcal{N}_u} \sum_{k} \eta_\mu^k F(\boldsymbol{\psi}_\mu^k) \tag{8.7}$$

and

$$a(\boldsymbol{\sigma}_h, \boldsymbol{u}_h) = \sum_{\nu \in \mathcal{N}_\sigma} \sum_{i \leqslant j} \sum_{\mu \in \mathcal{N}_u} \sum_{k} \xi_\nu^{ij} \eta_\mu^k\, a(\boldsymbol{\phi}_\nu^{ij}, \boldsymbol{\psi}_\mu^k). \tag{8.8}$$

Thus $F(\boldsymbol{u}_h)$ is a linear form and $a(\boldsymbol{\sigma}_h, \boldsymbol{u}_h)$ a bilinear form in the "long vectors" (ξ_ν^{ij}) and (η_μ^k), which means that F corresponds to a vector and a to a matrix. It is a computational necessity (at least for the classical algorithms in numerical linear algebra) that the coordinates of these vectors are given by one-dimensional numberings, i.e., by one-to-one mappings

$$n = n(\nu, i, j) \in \{1, 2, \ldots, N\} \tag{8.9}$$

and

$$m = m(\mu, k) \in \{1, 2, \ldots, M\}. \tag{8.10}$$

With such numberings we may write the nodal values as

$$x_n = \xi_\nu^{ij}, \quad \text{where } n = n(\nu, i, j) \tag{8.11}$$

and

$$y_m = \eta_\mu^k, \quad \text{where } m = m(\mu, k). \tag{8.12}$$

The vectors $\boldsymbol{x} = (x_n)_{n=1}^N \in \mathbb{R}^N$ and $\boldsymbol{y} = (y_m)_{m=1}^M \in \mathbb{R}^M$ are in unique correspondence with the discrete fields $\boldsymbol{\sigma}_h \in \Sigma_h$ and $\boldsymbol{u}_h \in U_h$ respectively. By insertion in Eqs. (8.7) and (8.8) we get

$$F(\boldsymbol{u}_h) = \sum_{m=1}^M y_m b_m = \boldsymbol{b}^{\mathrm{T}} \boldsymbol{y}, \tag{8.13}$$

where

$$b_m = F(\boldsymbol{\psi}_\mu^k), \quad m = m(\mu, k) \tag{8.14}$$

and

$$a(\boldsymbol{\sigma}_h, \boldsymbol{u}_h) = \sum_{m=1}^M \sum_{n=1}^N y_m x_n a_{mn} = \boldsymbol{y}^{\mathrm{T}} \boldsymbol{A} \boldsymbol{x} = \boldsymbol{x}^{\mathrm{T}} (\boldsymbol{A}^{\mathrm{T}} \boldsymbol{y}). \tag{8.15}$$

$\boldsymbol{A}$ is the $M \times N$ matrix with entries

$$a_{mn} = a(\boldsymbol{\phi}_\nu^{ij}, \boldsymbol{\psi}_\mu^k) \tag{8.16}$$

with $m = m(\mu, k)$ and $n = n(\nu, i, j)$ given by Eqs. (8.9) and (8.10). These numberings will typically be chosen in order to obtain a certain matrix structure. As usual, when working with the finite element method, the components of Eqs. (8.14) and (8.16) are computed through integration element by element and then assembled.

The matrix product notation $\boldsymbol{b}^{\mathrm{T}} \boldsymbol{y}$ is used for the ordinary inner product in $\mathbb{R}^N$. Also the components of "long" vectors and matrices are denoted by indices m, n instead of i, j to avoid confusion with the use of i and j as indices to space coordinates, such as in σ_{ij} and $\boldsymbol{\phi}_\nu^{ij}$.

With the notation Eqs. (8.13) and (8.15) the discrete version of the duality problem can be formulated as a finite-dimensional mathematical programming problem in the variables $\boldsymbol{x} \in \mathbb{R}^N$ and $\boldsymbol{y} \in \mathbb{R}^M$. Inserting Eqs. (8.13) and (8.15) in Eq. (8.1) gives

$$\lambda_h^* = \max\{\lambda \mid \exists \boldsymbol{x} \in K_{\mathrm{d}} : \boldsymbol{A}\boldsymbol{x} = \lambda \boldsymbol{b}\} \tag{8.17}$$

$$= \max_{x \in K_{\mathrm{d}}} \min_{b^{\mathrm{T}} y = 1} \boldsymbol{y}^{\mathrm{T}} \boldsymbol{A} \boldsymbol{x}$$

$$= \min_{b^{\mathrm{T}} y = 1} \max_{x \in K_{\mathrm{d}}} \boldsymbol{x}^{\mathrm{T}} (\boldsymbol{A}^{\mathrm{T}} \boldsymbol{y})$$

$$= \min_{b^{\mathrm{T}} y = 1} D_{\mathrm{d}}(\boldsymbol{y}). \tag{8.18}$$

We have introduced the notation

$$D_{\mathrm{d}}(\boldsymbol{y}) = \max_{x \in K_{\mathrm{d}}} \boldsymbol{x}^{\mathrm{T}} (\boldsymbol{A}^{\mathrm{T}} \boldsymbol{y}) \tag{8.19}$$

and

$$K_{\mathrm{d}} = \{\boldsymbol{x} \in \mathbb{R}^N \mid \boldsymbol{\sigma}_h \in K_h\}, \tag{8.20}$$

where $\boldsymbol{x}$ and $\boldsymbol{\sigma}_h$ are related through Eqs. (8.4) and (8.11), and K_h is defined by Eq. (8.2). (The subscript "d" is mnemonic for "discrete".) We shall always refer to the discrete static formulation (8.17) as the *primal problem* and to the discrete kinematic formulation (8.18) as the *dual problem*.

The nonlinear constraints $\boldsymbol{x} \in K_{\mathrm{d}}$ in the primal problem is of a particularly simple structure: Recall that the vector (x_n) is just a numbering of the variables ξ_ν^{ij} through Eq. (8.11). There is one constraint for each node ν, and this constraint only involves the "short" vector (ξ_ν^{ij}) with six components in three-dimensional problems. This sparse and perfectly banded structure of the nonlinear constraints is well suited for nonlinear optimization algorithms, such as the projected Lagrangian algorithm described in MURTAGH and SAUNDERS [1982].

EXAMPLE 8.2 (*Linear elements in plane strain*). Consider a problem in plane strain with the variables

$$\boldsymbol{\sigma} = \begin{bmatrix} \sigma_{11} & \sigma_{12} \\ \sigma_{12} & \sigma_{22} \end{bmatrix} \quad \text{and} \quad \boldsymbol{u} = \begin{bmatrix} u_1 \\ u_2 \end{bmatrix}.$$

The Von Mises yield condition in plane strain is

$$(\sigma_{11} - \sigma_{22})^2 + 4\sigma_{12}^2 \leqslant \tfrac{4}{3}\sigma_0^2. \tag{8.21}$$

Let ν be any numbering of the nodes, i.e., the vertices in the triangulation of Ω, and let the numbering in Eq. (8.9) be defined by

$$\begin{aligned} n(\nu,1,1) &= 3(\nu-1)+1, \\ n(\nu,2,2) &= 3(\nu-1)+2, \\ n(\nu,1,2) &= 3(\nu-1)+3. \end{aligned}$$

This corresponds to ordering the nodal values of $\boldsymbol{\sigma}_h$ by taking the triplets $(\sigma_{11}, \sigma_{22}, \sigma_{12})$ sequentially vertex by vertex. With this ordering the yield condition on $\boldsymbol{\sigma}_h$ is implemented as follows:

$$(x_{3(\nu-1)+1} - x_{3(\nu-1)+2})^2 + 4(x_{3(\nu-1)+3})^2 \leqslant \tfrac{4}{3}\sigma_0^2 \quad \forall \nu.$$

K_{d} in Eq. (8.20) is the set of $\boldsymbol{x} \in \mathbb{R}^N$, which satisfy this condition. In particular, N equals three times the number of vertices in the triangulation, and the yield condition is implemented as $N/3$ nonlinear inequalities, each involving three variables.

We now prove the duality between the discrete formulations Eqs. (8.17) and (8.18).

THEOREM 8.1 (Discrete duality theorem). *Let $\mathbf{A}$ be an $M \times N$ matrix and $\boldsymbol{b} \in \mathbb{R}^M$. Let $K_0 \subseteq \mathbb{R}^N$ be convex and compact, and assume that $\boldsymbol{x} = \mathbf{0}$ is an interior* *point of K_0. Let $V \subseteq \mathbb{R}^N$ be a subspace and define $K_d = K_0 + V$. Then the problems in Eqs.* (8.17) *and* (8.18) *have the same value, and there exists a saddle point $(\boldsymbol{x}^*, \boldsymbol{y}^*)$ for the bilinear form $\boldsymbol{y}^T\boldsymbol{A}\boldsymbol{x}$:*

$$(\boldsymbol{y}^*)^T\boldsymbol{A}\boldsymbol{x} \leqslant (\boldsymbol{y}^*)^T\boldsymbol{A}\boldsymbol{x}^* = \lambda_h^* = \boldsymbol{y}^T\boldsymbol{A}\boldsymbol{x}^* \tag{8.22}$$

for all $\boldsymbol{x} \in K_d$ and $\boldsymbol{y} \in \mathbb{R}^M$ with $\boldsymbol{b}^T\boldsymbol{y} = 1$.

Furthermore, if $\boldsymbol{b} \notin R_A = \{\boldsymbol{A}\boldsymbol{x} \mid \boldsymbol{x} \in \mathbb{R}^N\}$, then the common value of Eqs. (8.17) *and* (8.18) *is zero.*

PROOF. In the case $\boldsymbol{b} \in R_A$ there exists $\boldsymbol{x} \in \mathbb{R}^N$ such that $\boldsymbol{b} = \boldsymbol{A}\boldsymbol{x}$. Then the proof goes as the proof of Theorem 5.7, although much simpler in this finite-dimensional case.

If $\boldsymbol{b} \notin R_A$, the value of Eq. (8.17) is clearly equal to 0. Now $N_{A^T} = (R_A)^\perp$ where N_{A^T} and $(R_A)^\perp$ denote the null-space of the transpose of $\boldsymbol{A}$, and the orthogonal to R_A, respectively. Hence there is $\boldsymbol{y} \in N_{A^T}$, such that $\boldsymbol{b}^T\boldsymbol{y} = 1$, proving that also the value of Eq. (8.18) equals 0 in this case. □

The duality in Eqs. (8.17)–(8.18) is just a reformulation of Eq. (8.1). If we insert Eqs. (8.13) and (8.15), Eq. (8.22) may be expressed in the discrete stresses and flow corresponding to $\boldsymbol{x}$ and $\boldsymbol{y}$:

$$a(\boldsymbol{\sigma}_h, \boldsymbol{u}_h^*) \leqslant (\boldsymbol{\sigma}_h^*, \boldsymbol{u}_h^*) = \lambda_h^* = a(\boldsymbol{\sigma}_h^*, \boldsymbol{u}_h) \tag{8.23}$$

for all $\boldsymbol{\sigma}_h \in K_h$ and $\boldsymbol{u}_h \in U_h$ with $F(\boldsymbol{u}_h) = 1$.

The exceptional case $\boldsymbol{b} \notin R_A$ in Theorem 8.1 has an important implication: We must choose the finite element spaces Σ_h and U_h in such a way that $\boldsymbol{b} \in R_A$. Otherwise the discrete limit multiplier λ_h^* will equal zero, independent of the element size.

With one exception, to be pointed out below, the condition $\boldsymbol{b} \in R_A$ should be ensured by satisfying the following *consistency condition*:

The matrix $\boldsymbol{A}$ defined by Eq. (8.16) must have full row rank M.

This consistency condition is independent of the orderings of Eqs. (8.9) and (8.10). It may equivalently be formulated as a condition on the discrete spaces Σ_h and U_h:

$$a(\boldsymbol{\sigma}_h, \boldsymbol{u}_h) = 0 \ \forall \boldsymbol{\sigma}_h \in \Sigma_h \text{ must imply that } \boldsymbol{u}_h = \boldsymbol{0} \tag{8.24}$$

Loosely speaking there must be sufficiently many degrees of freedom for the stresses relative to the velocity.

In the very special case with no boundary conditions at all on $\boldsymbol{u}$, rigid body motions are admissible, and Eq. (8.24) cannot be satisfied. In this case the condition $\boldsymbol{b} \in R_A = (N_{A^T})^\perp$ is simply the discrete analogue of the condition (5.24).

Unfortunately the consistency condition (8.24) is not sufficient to guarantee good solutions for the collapse fields for stress and flow. If there are too many degrees of freedom for the stresses $\boldsymbol{\sigma}_h$ relative to the velocity $\boldsymbol{u}_h$, the collapse stresses will be poorly determined, and vice versa. In our experience the balance between primal variables $\boldsymbol{x}$ and dual variables $\boldsymbol{y}$ is quite delicate. In order to illustrate this point we return to the example in antiplane shear.

EXAMPLE 8.3 (*Antiplane shear*). This is a modification of Example 4.1. We use the same geometry and notation. Now both sides of the layer are kept fixed by imposing the boundary condition $u(-1) = u(1) = 0$. Hence there are only volume forces, and they are given by

$$f(x) = \begin{cases} -1 & \text{for } -1 < x < a, \\ 1 & \text{for } a < x < 1 \end{cases}$$

for some constant a in the interval $0 < a < 1$. The static principle is

$$\lambda^* = \max\{\lambda \mid \sigma'(x) = -\lambda f(x),\ |\sigma(x)| \leqslant 1\}$$

with the solution

$$\lambda^* = \frac{2}{1+a}$$

and

$$\sigma^*(x) = 1 - \lambda^*|x-a| \quad \forall x \in [-1, 1].$$

The collapse flow is

$$u^*(x) = \begin{cases} -1/(1+a) & \text{for } -1 < x < a, \\ 0 & \text{for } a < x < 1. \end{cases}$$

Both σ^* and u^* are uniquely determined. The region $-1 < x < a$ is moving "down" as a solid block. The rest does not move. The entire deformation takes place at $x = -1$ and $x = a$.

Now discretize the problem as described earlier in this section. Use linear elements for both σ_h and u_h. To fix ideas let $a = \frac{1}{2}$ and $h = \frac{1}{2}$. In this case the exact collapse stress σ^* actually belongs to the discrete space Σ_h, so we are tempted to expect that it coincides with the discrete collapse stresses σ_h^*. Not so! The discrete equilibrium equation in Eq. (8.1) is

$$a(\sigma_h, u_h) = \lambda F(u_h) \quad \forall u_h \in U_h$$

or equivalently (recall that $u(-1) = u(1) = 0$)

$$-\int_{-1}^{1} \sigma_h' \phi = \lambda \int_{-1}^{1} f\phi \quad \forall \phi,$$

where ϕ is any of the "hat-functions" that constitute the basis of the linear element space U_h. If σ_h alternates from node to node between two values, then σ_h' will simply change sign from interval to interval. For such a σ_h we will have

$$\int_{-1}^{1} \sigma_h' \phi = 0 \quad \forall \phi$$

for all basis functions ϕ and thus for all $u_h \in U_h$. Due to the boundary condition $u(-1) = u(1) = 0$ there are no basis functions for U_h associated with the nodes $x = -1$ and $x = 1$. This means that any such alternating σ_h is in discrete equilibrium with the zero load, in addition to the constant stresses.

Now let $(\lambda^*, \sigma^*, u^*)$ be the exact solution found previously. σ^* belongs to Σ_h for $a = \frac{1}{2}$ and $h = \frac{1}{2}$ and is in equilibrium with the load $\lambda^* f$ (both exact and discrete). Let $\sigma_h \in \Sigma_h$ be given by $\sigma_h(-1) = \sigma_h(0) = \sigma_h(1) = 1$ and $\sigma_h(-\frac{1}{2}) = \sigma_h(\frac{1}{2}) = -1$, so that it is in discrete equilibrium with zero. Then for small positive values of t the stress function $\sigma^* + t\sigma_h$ will still be in discrete equilibrium with $\lambda^* f$, and it will satisfy $|\sigma^* + t\sigma_h| \leqslant 1 - t < 1$. It follows that $\lambda^*/(1-t)$ is an admissible load multiplier for the discrete problem and that $\lambda_h^* > \lambda^*$. The discrete collapse stress function σ_h^* will satisfy $|\sigma_h^*(x)| = 1$ in the intervals adjacent to the points $x = -1$ and $x = \frac{1}{2}$, and in these intervals deformation may take place. The discrete collapse flow is no longer uniquely determined, and the plastified region may expand from two points for the continuous problem to intervals of positive length in the discrete case.

The point of the above example is that the discrete problem may have more unwanted features than the corresponding continuous problem. It may display a higher degree of non-uniqueness, both for stresses and flow, and it may spread out an otherwise narrow and well defined plastified region. In our experience these features play a significant role in two-dimensional problems.

The negative features in the example do not occur, if the piecewise linear elements for σ_h are replaced by piecewise constant elements. The set of stresses, which are in discrete equilibrium with the zero load, is a one-dimensional subspace consisting only of the constant functions as in the continuous case. We shall return to this element combination in Example 10.1 in a different context.

It is an important property of the exact collapse flow field $\boldsymbol{u}^*$ that it is divergence free: $\nabla \cdot \boldsymbol{u}^* = 0$. We shall now examine in which sense this condition is satisfied by the discrete collapse flow. $\boldsymbol{u}_h^*$ satisfies the condition

$$D_h(\boldsymbol{u}_h^*) = \max_{\sigma_h \in K_h} a(\boldsymbol{\sigma}_h, \boldsymbol{u}_h^*) = \lambda_h^* < \infty, \tag{8.25}$$

where K_h is given by Eq. (8.2). Inserting $\boldsymbol{\sigma}_h = \phi_h \boldsymbol{I}$ in Eq. (8.25) yields

$$\int_\Omega \phi_h \nabla \cdot \boldsymbol{u}_h^* = 0 \quad \forall \phi_h, \tag{8.26}$$

where ϕ_h denotes any finite element function of the type used to represent the single components of the stress tensors $\boldsymbol{\sigma}_h$.

The exact interpretation of Eq. (8.26) depends on the finite element spaces used. With piecewise linear elements for $\boldsymbol{u}_h$ and piecewise constant elements for $\boldsymbol{\sigma}_h$, and hence for ϕ_h, Eq. (8.26) means that $\nabla \cdot \boldsymbol{u}_h^* = 0$ exactly, since it is constant on each triangle. (As mentioned earlier this element combination has too many degrees of freedom for stresses relative to velocity.) If other elements for $\boldsymbol{u}_h$ are combined with piecewise constant stresses, we can conclude only that $\nabla \cdot \boldsymbol{u}_h^*$ has zero average over each element. When piecewise linear elements are used for both $\boldsymbol{\sigma}_h$ and $\boldsymbol{u}_h$, Eq. (8.26) implies that $\nabla \cdot \boldsymbol{u}_h^*$ has zero average over the union of the elements adjacent to each vertex. In general we may conclude that $\nabla \cdot \boldsymbol{u}_h^*$ is zero in some average sense only. In some of the applications we shall see that this zero mean may result from undesirable and non-physical fluctuations in the numerical solution.

In the next section we shall discuss the possibility of imposing discrete incompressibility as constraints.

9. Elements for divergence-free flow

In this section we shall discretize the reduced duality problem in Eqs. (5.54)–(5.55). It only applies to yield conditions of the form in Eq. (5.13). The expected gain is a smaller discrete problem considering only divergence-free flow and stresses with trace zero, $\Sigma\sigma_{ii} = 0$, cf. Remark 5.8. We also expect an easier discrete problem, since the set of admissible stress tensors is bounded in the reduced problem.

Finite element spaces for divergence-free flow are known from the numerical solution of the Navier–Stokes equation and are described in TEMAM [1977, Section 4.4–4.5]. They can be obtained in two ways: For certain elements discrete (but not exact) incompressibility can be obtained by removing degrees of freedom. Unfortunately, the resulting space does not have a standard finite element basis. Application of these elements corresponds to imposing Eq. (8.26) as a set of linear constraints on $\boldsymbol{u}_h$. TEMAM [1977, Section 4] gives specific examples of velocity spaces in the plane: Piecewise quadratic elements (over triangles) satisfying Eq. (8.26) for ϕ_h piecewise constant and a slightly larger space satisfying Eq. (8.26) for ϕ_h piecewise linear. However, the obvious combinations with piecewise constant and piecewise linear elements respectively, do not satisfy condition (8.24), since Σ_h has smaller dimension than U_h (see Table 15.1). To our knowledge computations in limit analysis with this approach have not been reported, but it certainly deserves further investigation.

The other approach is based on the observation that the divergence of the curl of any vector field is zero. This approach is briefly described in CHRISTIANSEN [1991]. Instead of choosing a finite element space for the flow $\boldsymbol{u}$ itself, we discretize the vector field $\boldsymbol{\Psi}$ to $\boldsymbol{\Psi}_h$, i.e., the components of $\boldsymbol{\Psi}_h$ belong to some finite element space. The discrete flow $\boldsymbol{u}_h$ is then given by the curl of $\boldsymbol{\Psi}_h$:

$$\boldsymbol{u}_h = \nabla \times \boldsymbol{\Psi}_h. \tag{9.1}$$

One advantage of this method is that $\nabla \cdot \boldsymbol{u}_h = 0$ is satisfied exactly. Care must be taken that $\boldsymbol{u}_h$ satisfies the boundary conditions. An example is given below. Now we pay the price for the expected gain: In order to get $\boldsymbol{u}_h$ continuous, the components of $\boldsymbol{\Psi}_h$ must be of class C^1, i.e., have continuous derivatives. In the plane this implies the use of elements like Argyris triangle or Bell's triangle (CIARLET [1991, Section 9]). If the geometry permits a triangulation into rectangles (in two or three dimensions) the tensor product of the standard cubic C^1-elements (Bogner–Fox–Schmit rectangle, CIARLET [1991, Section 9]) is a very convenient choice.

With C^1-elements for $\boldsymbol{\Psi}_h$ the dimension of the discrete flow space U_h is so large that piecewise constant or linear elements cannot be used for the stresses. Condition (8.24) cannot possibly be satisfied. More degrees of freedom are required for the stresses. This can be obtained by using quadratic or bi-quadratic elements (CIARLET [1991, Section 6–7]). With these elements the yield condition may be violated between nodes, even though it is imposed at the nodes. As earlier mentioned, this may be considered part of the discretization error, and we have no reservations against such elements. Having specified the finite element spaces the rest is a matter of numbering variables and computing integrals as described in Section 8. A specific example will be given below.

In plane problems this approach is particularly attractive: $\boldsymbol{\Psi}_h$ has only one non-zero component, which we shall denote Ψ_h, and Eq. (9.1) reduces to

$$\boldsymbol{u}_h = \left(\frac{\partial \Psi_h}{\partial x_2}, -\frac{\partial \Psi_h}{\partial x_1} \right). \tag{9.2}$$

The stress tensor of trace zero satisfies $\sigma_{22} = -\sigma_{11}$ and may be identified with the vector (σ_1, σ_2):

$$\boldsymbol{\sigma}_h = \begin{bmatrix} \sigma_1 & \sigma_2 \\ \sigma_2 & -\sigma_1 \end{bmatrix}. \tag{9.3}$$

With this notation the Von Mises yield condition in plane strain, Eq. (8.21), becomes

$$\sigma_1^2 + \sigma_2^2 \leqslant \tfrac{1}{3}\sigma_0^2. \tag{9.4}$$

The left-hand side is the Euclidean norm of (σ_1, σ_2). This is the simplest possible yield condition.

EXAMPLE 9.1 (*Example* 1.1 *revisited*). Consider the plane strain problem in Example 1.1. Ψ_h is given by bi-cubic C^1-elements over rectangles (actually squares). There are four basis functions associated with each vertex μ, corresponding to the following nodal values of Ψ_h at the node μ (see CIARLET [1991, p. 92]):

$$\Psi, \qquad \frac{\partial \Psi}{\partial x_1}, \qquad \frac{\partial \Psi}{\partial x_2}, \qquad \frac{\partial^2 \Psi}{\partial x_1 \partial x_2}.$$

We shall denote the corresponding four basis functions ψ_μ^{00}, ψ_μ^{10}, ψ_μ^{01} and ψ_μ^{11}. If V_ν denotes an arbitrary vertex, then the bi-cubic C^1-basis function $\psi_\mu^{\alpha\beta}$ is completely characterized by the following identities:

$$\psi_\mu^{00}(V_\nu) = \delta_{\mu\nu}, \quad \frac{\partial}{\partial x_1}\psi_\mu^{00}(V_\nu) = 0, \quad \frac{\partial}{\partial x_2}\psi_\mu^{00}(V_\nu) = 0, \quad \frac{\partial^2}{\partial x_1 \partial x_2}\psi_\mu^{00}(V_\nu) = 0,$$

$$\psi_\mu^{10}(V_\nu) = 0, \quad \frac{\partial}{\partial x_1}\psi_\mu^{10}(V_\nu) = \delta_{\mu\nu}, \quad \frac{\partial}{\partial x_2}\psi_\mu^{10}(V_\nu) = 0, \quad \frac{\partial^2}{\partial x_1 \partial x_2}\psi_\mu^{10}(V_\nu) = 0,$$

$$\psi_\mu^{01}(V_\nu) = 0, \qquad \frac{\partial}{\partial x_1}\psi_\mu^{01}(V_\nu) = 0, \quad \frac{\partial}{\partial x_2}\psi_\mu^{01}(V_\nu) = \delta_{\mu\nu}, \quad \frac{\partial^2}{\partial x_1 \partial x_2}\psi_\mu^{01}(V_\nu) = 0,$$

$$\psi_\mu^{11}(V_\nu) = 0, \qquad \frac{\partial}{\partial x_1}\psi_\mu^{11}(V_\nu) = 0, \quad \frac{\partial}{\partial x_2}\psi_\mu^{11}(V_\nu) = 0, \quad \frac{\partial^2}{\partial x_1 \partial x_2}\psi_\mu^{11}(V_\nu) = \delta_{\mu\nu}.$$

Recall that $\psi_\mu = \psi_\mu(x_1, x_2)$ is just the product of two cubic C^1-element functions in x_1 and x_2, respectively.

With a notation analogous to Eq. (8.6) Ψ_h may be written

$$\Psi_h = \sum_{\mu \in \mathcal{N}_u} \sum_{\alpha=0}^{1} \sum_{\beta=0}^{1} \eta_\mu^{\alpha\beta} \; \psi_\mu^{\alpha\beta}. \tag{9.5}$$

The boundary condition on $\boldsymbol{u}$ comes from the symmetry:

$$u_2 = 0 \quad \text{for } x_2 = 0,$$
$$u_1 = 0 \quad \text{for } x_1 = 0 \text{ and } 0 \leqslant x_2 \leqslant a,$$

where $x_2 = a$ marks the tip of the cut. By Eq. (9.2) this boundary condition may be implemented by the requirement

$$\Psi_h(x_1, x_2) = 0 \quad \text{for} \quad (x_1, x_2) \in \{x_2 = 0\} \cup \{x_1 = 0, \ 0 \leqslant x_2 \leqslant a\}$$

or equivalently

$$\eta_\mu^{00} = 0 \quad \text{for all nodes} \quad \mu \in \{x_2 = 0\} \cup \{x_1 = 0, \ 0 \leqslant x_2 \leqslant a\}.$$

The values of $\boldsymbol{u}_h$ are easily extracted from Eq. (9.5) using Eq. (9.2):

$$\boldsymbol{u}_h(V_\mu) = (\eta_\mu^{01}, -\eta_\mu^{10}).$$

It turns out that the space of piecewise bi-quadratic elements for the stress components $\sigma_1 = \sigma_{11} = -\sigma_{22}$ and $\sigma_2 = \sigma_{12}$ has a dimension which is compatible with the space for $\boldsymbol{u}_h$. The nodes are the vertices, the midpoints of the sides of the rectangles and the midpoint of the rectangles (see CIARLET [1991, p. 77]). Associated with each node is a basis function characterized completely by being equal to 1 at "its own" node and 0 at all other nodes. Let $\mathcal{N}_\sigma$ denote this set of nodes, and let ϕ_ν be the associated bi-quadratic basis functions. Corresponding to each node we have the following two basis functions for the reduced space of stresses in Eq. (9.3):

$$\boldsymbol{\phi}_\nu^1 = \begin{bmatrix} \phi_\nu & 0 \\ 0 & -\phi_\nu \end{bmatrix}, \qquad \boldsymbol{\phi}_\nu^2 = \begin{bmatrix} 0 & \phi_\nu \\ \phi_\nu & 0 \end{bmatrix}.$$

The discrete stress tensor may now be written

$$\boldsymbol{\sigma}_h = \sum_{\nu \in \mathcal{N}_\sigma} (\xi_\nu^1 \boldsymbol{\phi}_\nu^1 + \xi_\nu^2 \boldsymbol{\phi}_\nu^2). \tag{9.6}$$

The yield condition (9.4) is imposed as follows:

$$\left(\xi_\nu^1\right)^2 + \left(\xi_\nu^2\right)^2 \leqslant \tfrac{1}{3}\sigma_0^2 \quad \forall \nu \in \mathcal{N}_\sigma. \tag{9.7}$$

We now introduce numberings of the variables and "long" vectors $\boldsymbol{x}$ and $\boldsymbol{y}$ as in Eqs. (8.9)–(8.12) and compute the components of $\boldsymbol{A}$ and $\boldsymbol{b}$ defined as in Eqs. (8.13)–(8.16). Note that the integrals in Eq. (8.16) involve second-order derivatives of Ψ_h, but since Ψ_h is made of C^1-elements, the second-order derivatives are perfectly integrable. Also note that all integrals can be computed as products of one-dimensional integrals. Each integral is simple, but there is quite a manifold of them. They may with advantage be computed using a program for symbolic calculation.

The solution of the discrete problem and the results for this example will be described in Section 14.

10. Bounds and convergence

In this section we examine the relationship between the solutions of the exact continuous problem in Eq. (5.31) and the approximate discrete problem in Eq. (8.1). Let it be said immediately that the situation is far from as satisfactory as for elliptic problems. There is no regularity theorem for the solution; in fact, the collapse fields may not even be continuous. There is no energy norm, which is minimized by the solution, and there is no equivalent of Céa's lemma (CIARLET [1991, Theorem 13.1]). Still it must be possible to obtain better convergence results than those reported here, but that remains to be done.

According to our experience the convergence of λ_h^* will be linear:

$$|\lambda_h^* - \lambda^*| \leqslant ch \tag{10.1}$$

for some constant c. We shall briefly say that $\lambda_h^* - \lambda^*$ is $\mathrm{O}(h)$. This convergence is not improved by choosing elements of higher order, since the collapse fields are not smooth. Fortunately the linear convergence also seems to be independent of our ability to prove it.

In numerical analysis error estimates like Eq. (10.1) are frequently combined with the possibility of using Richardson extrapolation to improve the results or to estimate the actual size of the difference $\lambda_h^* - \lambda^*$. This is based on the following condition, which is stronger than Eq. (10.1):

$$\lim_{h\to 0} \frac{\lambda_h^* - \lambda^*}{h} = c. \tag{10.2}$$

It is well known that conditions like Eq. (10.2) can be checked from computed values of λ_h^*, i.e., λ^* need not be known. In limit analysis Eq. (10.2) typically does not not hold. Consider again the example in antiplane shear:

EXAMPLE 10.1 (*Example* 8.3 *revisited*). We now solve the problem in Example 8.3 using piecewise constant elements for σ_h and piecewise linear elements for u_h. The equation for discrete equilibrium with the load λf is

$$\int_{-1}^{1} \sigma_h u_h' = \lambda \int_{-1}^{1} f u_h \quad \forall u_h \in U_h$$

or equivalently

$$\int_{-1}^{1} \sigma_h \phi_i' = \lambda \int_{-1}^{1} f\, \phi_i\ , \quad i = 1, \ldots, N-1, \tag{10.3}$$

where ϕ_i is the linear "hat-function" at node $x_i = -1 + ih,\ h = 2/N$, and N is the number of intervals. Let s_i denote the (constant) value of σ_h in the interval $x_{i-1} < x < x_i$ for $i = 1, \ldots, N$. Then Eq. (10.3) can be written

$$s_i - s_{i+1} = \lambda \int_{x_{i-1}}^{x_{i+1}} f\phi_i, \quad i = 1, \ldots, N-1.$$

Note that this determines σ_h uniquely up to an additive constant in contrast to Example 8.3. Let the index k be determined by the location of a, such that $x_k \leqslant a < x_{k+1}$, and let $d = a - x_k$. (Recall that $f(x) = -1$ for $-1 < x < a$ and $f(x) = +1$ for $a < x < 1$, and that $a > 0$.) Then the discrete equilibrium equation takes the form

$$\begin{aligned} s_{i+1} - s_i &= \lambda h \quad \text{for } i < k, \\ s_{k+1} - s_k &= \lambda h^{-1}(2hd - d^2), \\ s_{k+2} - s_{k+1} &= -\lambda h^{-1}\left(2h(h-d) - (h-d)^2\right), \\ s_{i+1} - s_i &= -\lambda h \quad \text{for } i > k+1. \end{aligned}$$

The maximum value for λ is achieved with $s_1 = -1$ and $s_{k+1} = 1$:

$$\lambda_h^* = \frac{2}{1 + a + d - h - h^{-1}d^2}.$$

The continuous problem was solved in Example 8.3:

$$\lambda^* = \frac{2}{1+a}.$$

The error on λ_h^* is

$$\lambda_h^* - \lambda^* = 2h \frac{1 - (d/h) + (d/h)^2}{(1+a)(1 + a + d - h - d^2/h)}.$$

We see that

$$\liminf_{h \to 0} \frac{\lambda_h^* - \lambda^*}{h} = \tfrac{3}{4} \limsup_{h \to 0} \frac{\lambda_h^* - \lambda^*}{h}.$$

Hence Eq. (10.1) is satisfied, but Eq. (10.2) is not. The error primarily depends on the location of singularities relative to the finite element grid.

It is our experience that the behaviour illustrated in the example is typical in limit analysis. Equation (10.1) usually holds, but Eq. (10.2) does not.

Error estimates on the discrete limit multiplier λ_h^* are based on the saddle point properties of the exact and the discrete solutions. In the following

$(\lambda^*, \boldsymbol{\sigma}^*, \boldsymbol{u}^*)$ denotes a solution to the continuous problem, i.e., a triple satisfying Eq. (5.32), and $(\lambda_h^*, \boldsymbol{\sigma}_h^*, \boldsymbol{u}_h^*)$ denotes a solution to the discrete problem in Eq. (8.1), i.e., a triple satisfying Eq. (8.23).

THEOREM 10.1. *Let $(\lambda^*, \boldsymbol{\sigma}^*, \boldsymbol{u}^*)$ and $(\lambda_h^*, \boldsymbol{\sigma}_h^*, \boldsymbol{u}_h^*)$ denote an exact and a discrete solution, respectively. If $\boldsymbol{\sigma}_h^*$ satisfies the exact yield condition, i.e., $\boldsymbol{\sigma}_h^* \in K$, then*

$$a(\boldsymbol{\sigma}_h - \boldsymbol{\sigma}^*, \boldsymbol{u}_h^*) \leqslant \lambda_h^* - \lambda^* \leqslant a(\boldsymbol{\sigma}_h^*, \boldsymbol{u}_h - \boldsymbol{u}^*) \tag{10.4}$$

for all $\boldsymbol{\sigma}_h \in K_h$ and all $\boldsymbol{u}_h \in U_h$ with $F(\boldsymbol{u}_h) = 1$.

PROOF. In Eq. (5.32) insert $\boldsymbol{\sigma} = \boldsymbol{\sigma}_h^*$ and $\boldsymbol{u} = \boldsymbol{u}_h^*$ and subtract from Eq. (8.23). □

The condition $\boldsymbol{\sigma}_h^* \in K$ in the above theorem will automatically be satisfied, if $K_h \subseteq K$. Since the yield condition is only imposed at the nodes, this condition may fail to hold for two reasons: The yield condition may vary between nodes, and certain element functions may achieve their maximal value between nodes. Problems caused by changes in the yield condition with $\boldsymbol{x} \in \Omega$ should be treated on an ad hoc basis, so consider the case of a homogeneous material. Then $K_h \subseteq K$ will hold for piecewise constant, linear and multilinear elements: For these elements the convex yield condition will be satisfied everywhere, if it is satisfied at the nodes. For quadratic and cubic elements this is not the case. The effect of this on the discrete limit multiplier λ_h^* depends on the specific elements for both $\boldsymbol{\sigma}_h$ and $\boldsymbol{u}_h$ and on the collapse fields. Under quite weak conditions this effect can be seen to be of order $O(h)$ (bounded by a multiple of h), where h is a linear measure of the element size. Hence the effect may be considered equivalent to a discretization error.

In order to prove convergence of λ_h^* from Eq. (10.4) we need *stability conditions*, which bound $\boldsymbol{\sigma}_h^*$ and $\boldsymbol{u}_h^*$ in some norms and *regularity conditions*, which guarantee that $\boldsymbol{\sigma}^*$ and $\boldsymbol{u}^*$ may be approximated by $\boldsymbol{\sigma}_h \in \Sigma_h$ and $\boldsymbol{u}_h \in U_h$, respectively. Examples of such conditions are

$$\|\boldsymbol{u}_h^*\|_{\mathrm{BD}(\Omega)} \leqslant C \quad \forall h \tag{10.5}$$

paired with the regularity condition

$$\boldsymbol{\sigma}^* \in W^{1,\infty}(\Omega) \tag{10.6}$$

or

$$\|\boldsymbol{\varepsilon}(\boldsymbol{u}_h^*)\|_{C(\Omega)} \leqslant C \quad \forall h \tag{10.7}$$

paired with a weaker regularity condition on $\boldsymbol{\sigma}^*$. We may then write the left-hand side of Eq. (10.4) as

$$\lambda_h^* - \lambda^* \geqslant \int_\Omega \sum_{i,j} (\sigma_{h,ij} - \sigma_{ij}^*)\varepsilon_{ij}(\boldsymbol{u}_h^*). \tag{10.8}$$

We shall always assume that $\mathcal{T}_h$ is a *regular* finite element triangulation as defined in CIARLET [1991, Chapter III]. Also, we restrict ourselves to the case $\boldsymbol{u} \in \mathrm{BD}(\Omega)$, i.e., $\boldsymbol{\mu}(\boldsymbol{u}) = \boldsymbol{\varepsilon}(\boldsymbol{u})$.

THEOREM 10.2. *Assume continuous elements for both* $\boldsymbol{\sigma}_h$ *and* $\boldsymbol{u}_h$. *If Eqs.* (10.5) *and* (10.6) *hold, then*

$$\lambda_h^* - \lambda^* \geqslant -ch \tag{10.9}$$

for some positive constant c independent of h.

PROOF. We use Eq. (10.6) and CIARLET [1991, Theorem 16.1] combined with Sobolev's imbedding theorem to conclude that

$$\min_{\sigma_h \in \Sigma_h} \|\boldsymbol{\sigma}_h - \boldsymbol{\sigma}^*\|_{C(\overline{\Omega})} \leqslant ch$$

for some constant c. Since $\boldsymbol{\sigma}_h$ in Eq. (10.8) is arbitrary, Eq. (10.9) follows from Eq. (10.5) by definition of the norm on $\mathrm{BD}(\Omega)$ (see Eq. (5.3)). □

THEOREM 10.3. *Assume continuous elements for* $\boldsymbol{u}_h$, *but allow also piecewise constant elements for* $\boldsymbol{\sigma}_h$. *Assume that Eq.* (10.7) *holds, and that* $\boldsymbol{\sigma}^*$ *is "piecewise* $W^{1,\infty}$*" in the following sense:* Ω *can be divided into a finite number of regions* Ω_k, *such that*

(a) *the components of* $\boldsymbol{\sigma}^*$ *belong to* $L^\infty(\Omega)$ *and to* $W^{1,\infty}(\Omega_k)$ *for each* Ω_k;

(b) *the total volume of elements overlapping with more than one* Ω_k *tends to zero as* $O(h)$, *as* $h \to 0$.

Then there is a positive constant c such that

$$\lambda_h^* - \lambda^* \geqslant -ch$$

independent of h.

PROOF. We estimate the right-hand side of Eq. (10.8). The integral over elements contained in one Ω_k is $O(h)$ by the argument in the previous proof. The integral over the remaining elements is bounded by a constant times their total volume, which by assumption is $O(h)$. □

In order to obtain the opposite bound we need a stability condition on $\boldsymbol{\sigma}_h^*$:

$$\|\boldsymbol{\sigma}_h^*\|_{L^\infty} \leqslant C_1 \quad \text{and} \quad \|\boldsymbol{\nabla}\cdot\boldsymbol{\sigma}_h^*\|_{L^\infty} \leqslant C_2 \qquad \forall h. \tag{10.10}$$

THEOREM 10.4. *Assume continuous elements for* $\boldsymbol{\sigma}_h$ *and* $\boldsymbol{u}_h$. *If* $\boldsymbol{u}^*$ *is "piecewise* $W^{1,\infty}$*" as defined in the previous theorem and Eq.* (10.10) *holds, then*

$$\lambda_h^* - \lambda^* \leqslant ch. \tag{10.11}$$

PROOF. We write the right-hand side inequality in Eq. (10.4) as

$$\lambda_h^* - \lambda^* \leqslant -\int_\Omega (\boldsymbol{\nabla}\cdot\boldsymbol{\sigma}_h^*)\cdot(\boldsymbol{u}_h - \boldsymbol{u}^*) + \int_T (\boldsymbol{\nu}\cdot\boldsymbol{\sigma}_h^*)\cdot(\boldsymbol{u}_h - \boldsymbol{u}^*).$$

The proof now goes much as the previous proof. □

Condition (10.5) is rather weak, because $\|\boldsymbol{u}\|_{\mathrm{BD}}$ is dominated by the total energy dissipation rate $D(\boldsymbol{u})$ defined by Eq. (5.30). ($\|\boldsymbol{u}_h\|_{\mathrm{BD}}$ need not be dominated by the discrete form $D_h(\boldsymbol{u}_h)$ defined in Eq. (8.1), so Eq. (10.5) is not guaranteed to hold.) On the other hand condition (10.6) implies that $\boldsymbol{\sigma}^*$ is continuous, clearly a restrictive condition, which can only be checked a posteriori, and even that may not be trivial. Condition (10.7) is very restrictive. The problem with these conditions is that they are formulated in terms of either $\boldsymbol{\sigma}$ or $\boldsymbol{u}$. Good conditions should reflect the fact that singularity in one is paired with regularity in the other.

These error estimates are special cases of the following theorem, which follows directly from Eq. (10.4).

THEOREM 10.5. *Assume that* $\|\cdot\|_1$ *and* $\|\cdot\|_2$ *are norms on* Σ *and* U, *respectively, such that*

$$|a(\boldsymbol{\sigma},\boldsymbol{u})| \leqslant \|\boldsymbol{\sigma}\|_1 \, \|\boldsymbol{u}\|_2 \quad \forall(\boldsymbol{\sigma},\boldsymbol{u}) \in \Sigma \times U$$

and let $(\lambda_h^*, \boldsymbol{\sigma}_h^*, \boldsymbol{u}_h^*)$ *be a sequence of discrete solutions with* $\boldsymbol{\sigma}_h^* \in K$.

(a) *If* $\|\boldsymbol{\sigma}_h^*\|_1 \leqslant C_1$, *then*

$$\lambda_h^* - \lambda^* \leqslant C_1 \min_{F(u_h)=1} \|\boldsymbol{u}_h - \boldsymbol{u}^*\|_2 .$$

(b) *If* $\|\boldsymbol{u}_h^*\|_2 \leqslant C_2$, *then*

$$\lambda_h^* - \lambda^* \geqslant -C_2 \min_{\sigma_h \in K_h} \|\boldsymbol{\sigma}_h - \boldsymbol{\sigma}^*\|_1 .$$

The above error estimates are one-sided. Clearly the best one-sided estimates are of the form $\lambda_h^* \geqslant \lambda^*$ or $\lambda_h^* \leqslant \lambda^*$, which brings us back to the so-called bound methods mentioned in Section 3. If only $\boldsymbol{\sigma}$ (or $\boldsymbol{u}$) is discretized, there results a one-sided approximation of λ^*. In some cases this may also be obtained with the mixed finite element methods advocated here:

A discretization $\Sigma_h \times U_h$ is a *lower bound method* if $K_h \subseteq K$ (K_h is defined by Eq. (8.2)), and if the following implication holds for $\boldsymbol{\sigma}_h \in \Sigma_h$:

$$\text{If } a(\boldsymbol{\sigma}_h,\boldsymbol{u}_h) = F(\boldsymbol{u}_h) \ \forall \boldsymbol{u}_h \in U_h, \ \text{ then } a(\boldsymbol{\sigma}_h,\boldsymbol{u}) = F(\boldsymbol{u}) \ \forall \boldsymbol{u} \in U.$$

In fact it suffices that the implication holds for $\boldsymbol{\sigma}_h^*$, which will then be an admissible stress tensor in exact equilibrium with the load multiplier λ_h^*, implying $\lambda_h^* \leqslant \lambda^*$. This property depends not only on the discretization, but also on the actual forces $(\boldsymbol{f},\boldsymbol{g})$.

A discretization $\Sigma_h \times U_h$ is an *upper bound method,* if the discrete energy dissipation rate $D_h(\boldsymbol{u}_h)$ is exact on U_h, i.e., if the following equality holds:

$$\max_{\sigma_h \in K_h} a(\boldsymbol{\sigma}_h,\boldsymbol{u}_h) = \max_{\sigma \in K} a(\boldsymbol{\sigma},\boldsymbol{u}_h) \quad \forall \boldsymbol{u}_h \in U_h.$$

In this case it immediately follows from Eqs. (5.29) and (8.1) that $\lambda_h^* \geqslant \lambda^*$. It is easy to see that if the yield condition is independent of $\boldsymbol{x} \in \Omega$, then a discretization with piecewise linear elements for $\boldsymbol{u}_h$ and piecewise constant elements for

$\boldsymbol{\sigma}_h$ is an upper bound method (see CHRISTIANSEN [1981, Theorem 6]). This was explicitly seen in Example 10.1.

We do not know conditions, which guarantee convergence of the collapse fields $(\boldsymbol{\sigma}_h^*, \boldsymbol{u}_h^*)$ to $(\boldsymbol{\sigma}^*, \boldsymbol{u}^*)$. The fields will in general not be uniquely determined, and the degree of uniqueness need not be the same for the continuous and the discrete problems. The second best is to guarantee that if a sequence of discrete solutions $(\boldsymbol{\sigma}_h^*, \boldsymbol{u}_h^*)$ converges in some sense, then the limit is a solution to the continuous problem.

THEOREM 10.6. *Consider a sequence* $\mathcal{H}$ *of h-values satisfying* $\Sigma_{h_1} \subseteq \Sigma_{h_2} \subseteq \Sigma$ *and* $U_{h_1} \subseteq U_{h_2} \subseteq U$ *for* $h_2 < h_1$. *Assume that* $K_h = K \cap \Sigma_h$ $\forall h \in \mathcal{H}$ *and let* $C = \{\boldsymbol{u} \in U \mid F(\boldsymbol{u}) = 1\}$. *Let* $(\lambda_h^*, \boldsymbol{\sigma}_h^*, \boldsymbol{u}_h^*)$ *be a sequence of solutions to the discrete problem* (8.23) *for* $h \in \mathcal{H}$. *Assume that* $\lambda_h^* \to \lambda^0$ *and that* $(\boldsymbol{\sigma}_h^*, \boldsymbol{u}_h^*)$ *converges to* $(\boldsymbol{\sigma}^0, \boldsymbol{u}^0) \in K \times C$ *in the following weak sense:*

$$\lim_{h \to 0} a(\boldsymbol{\sigma}_h^*, \boldsymbol{u}) = a(\boldsymbol{\sigma}^0, \boldsymbol{u}) \quad \forall \boldsymbol{u} \in U, \tag{10.12}$$

$$\lim_{h \to 0} a(\boldsymbol{\sigma}, \boldsymbol{u}_h^*) = a(\boldsymbol{\sigma}, \boldsymbol{u}^0) \quad \forall \boldsymbol{\sigma} \in \Sigma. \tag{10.13}$$

Assume furthermore that $(\boldsymbol{\sigma}^0, \boldsymbol{u}^0)$ *satisfies the following weak approximation property:*

$$\forall \boldsymbol{u} \in C \; \forall \varepsilon > 0 \; \exists h \in \mathcal{H} \; \exists \boldsymbol{u}_h \in C \cap U_h \colon |a(\boldsymbol{\sigma}^0, \boldsymbol{u}_h - \boldsymbol{u})| < \varepsilon, \tag{10.14}$$

$$\forall \boldsymbol{\sigma} \in B \; \forall \varepsilon > 0 \; \exists h \in \mathcal{H} \; \exists \boldsymbol{\sigma}_h \in K \cap \Sigma_h \colon |a(\boldsymbol{\sigma}_h - \boldsymbol{\sigma}, \boldsymbol{u}^0)| < \varepsilon. \tag{10.15}$$

Then $(\lambda^0, \boldsymbol{\sigma}^0, \boldsymbol{u}^0)$ *is a solution to the continuous problem* (5.32).

PROOF. For fixed $h_1 \in \mathcal{H}$ and any $\boldsymbol{\sigma}_{h_1} \in K \cap \Sigma_{h_1}$ it follows from Eqs. (10.13) and (8.23) that

$$a(\boldsymbol{\sigma}_{h_1}, \boldsymbol{u}^0) = \lim_{h \to 0} a(\boldsymbol{\sigma}_{h_1}, \boldsymbol{u}_h^*) \leqslant \lim_{h \to 0} \lambda_h^* = \lambda^0$$

From Eq. (10.15) we conclude

$$a(\boldsymbol{\sigma}, \boldsymbol{u}^0) \leqslant \lambda^0 \quad \forall \boldsymbol{\sigma} \in K.$$

Similarly it follows from Eqs. (10.12) and (8.23) that for $h_1 \in \mathcal{H}$ and any $\boldsymbol{u}_{h_1} \in$

$C \cap U_{h_1}$

$$a(\boldsymbol{\sigma}^0, \boldsymbol{u}_{h_1}) = \lim_{h \to 0} a(\boldsymbol{\sigma}_h^*, \boldsymbol{u}_{h_1}) = \lim_{h \to 0} \lambda_h^* = \lambda^0,$$

which combined with Eq. (10.14) implies

$$a(\boldsymbol{\sigma}^0, \boldsymbol{u}) = \lambda^0 \quad \forall \boldsymbol{u} \in C.$$

Hence the triple $(\lambda^0, \boldsymbol{\sigma}^0, \boldsymbol{u}^0)$ satisfies Eq. (5.32). □

CHAPTER IV

Solution of the Discrete Problem

11. General remarks

Using a mixed finite element method the problem of limit analysis was discretized to the dual forms Eqs. (8.17)–(8.18):

$$\lambda_h^* = \max\{\lambda \mid \exists \boldsymbol{x} \in K_\mathrm{d} : \boldsymbol{A}\boldsymbol{x} = \lambda \boldsymbol{b}\} \tag{11.1}$$

$$= \min\{D_\mathrm{d}(\boldsymbol{y}) \mid \boldsymbol{b}^\mathrm{T}\boldsymbol{y} = 1\}, \tag{11.2}$$

where $\boldsymbol{A}$, $\boldsymbol{b}$, K_d and D_d are defined by the discretization Eqs. (8.2)–(8.20). Recall that in general K_d is convex in $\mathbb{R}^N$ and contains the sum of a ball and a subspace as illustrated in Example 8.2. Then $D_\mathrm{d}(\boldsymbol{y})$ is finite, only if $\boldsymbol{A}^\mathrm{T}\boldsymbol{y}$ is orthogonal to this subspace.

We shall rewrite the discrete problem in standard mathematical programming notation. The discrete static formulation Eq. (11.1) will be referred to as the *primal problem*:

$$(\mathrm{P}): \quad \begin{array}{l} \max \lambda, \\ \boldsymbol{A}\boldsymbol{x} = \boldsymbol{b}\lambda, \\ \boldsymbol{x} \in K_\mathrm{d}. \end{array} \tag{11.3}$$

The discrete kinematic formulation Eq. (11.2) is the corresponding *dual problem*

$$(\mathrm{D}): \quad \begin{array}{l} \min D_\mathrm{d}(\boldsymbol{y}), \\ \boldsymbol{b}^\mathrm{T}\boldsymbol{y} = 1. \end{array} \tag{11.4}$$

The problem in Eqs. (11.3)–(11.4) is an interesting mathematical optimization problem in its own right. Even for moderate size grids it is non-trivial, and for realistic problems and grids it is a real challenge. Not only is it large and sparse; in our experience it is also ill conditioned. We do not claim to have a "perfect" method yet. In fact it is not obvious which of our options, if any, will survive as "the best method". However, it does appear that we are able to solve realistic problems in limit analysis with reasonable accuracy. At least the limit multiplier and one of the collapse fields can be determined, but we cannot always determine collapse fields for both stresses and plastic flow simultaneously.

We shall illustrate the methods on a classical test problem in plane strain. Three-dimensional problems can also be solved, but the plane strain problem

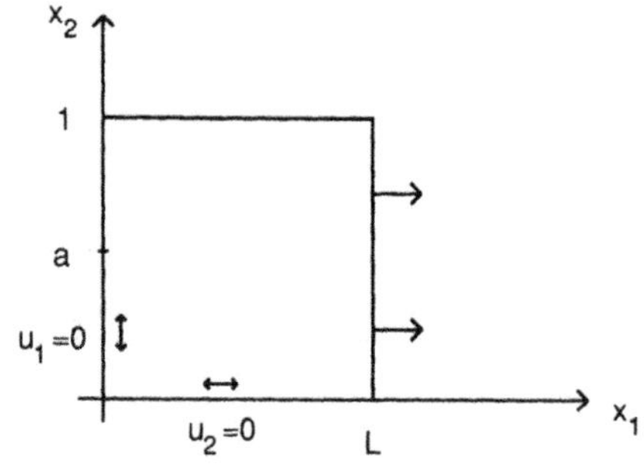

FIG. 11.1. The test problem reduced by symmetry.

with its two-dimensional convenience offers the typical difficulties of the general case: Unbounded yield set, ill conditioning, non-uniqueness, etc. Our test problem has been used in laboratory tests and reported in text books on material science. It was presented to this author by professor McLintock, MIT, in 1975 as a challenge to the numerical techniques in limit analysis. In this sense the problem serves two purposes: Numerical limit analysis has come of age, and realistic problems can be solved. Furthermore, the problem is well suited to reveal the virtues and shortcomings of potential solution methods.

EXAMPLE 11.1 (*The reference test problem*). The problem is described in Example 1.1. In the plane strain model a rectangular block of material is given external symmetric thin cuts of variable depth. The volume forces are zero, and the surface forces consist of a uniform pull at the ends. For symmetry reasons the problem is reduced to the first quadrant (see Fig. 11.1). Actually, due to non-uniqueness non-symmetric collapse fields exist, but by reflection and superposition it is easy to see that there will always be a symmetric collapse solution. The geometry of the reduced problem is seen in Fig. 11.1. The boundary conditions are

$$\begin{aligned} &u_2 = 0 \quad \text{for } x_2 = 0, \\ &u_1 = 0 \quad \text{for } x_1 = 0 \text{ and } 0 \leqslant x_2 \leqslant a. \end{aligned} \tag{11.5}$$

The Von Mises yield condition for a homogeneous material in plane strain is normalized to

$$\tfrac{1}{4}(\sigma_{11} - \sigma_{22})^2 + \sigma_{12}^2 \leqslant 1. \tag{11.6}$$

The problem is analysed in MARTIN [1975, Section 12.3] and KACHANOV [1971, p. 195]. The following bounds are given for the case $a = \frac{1}{3}$, i.e., the cuts extend two thirds to the center of the bar:

$$\tfrac{2}{3} \leqslant \lambda^* \leqslant \tfrac{4}{3}.$$

These bounds are based on exact upper and lower bound solutions. We find the true value for this case to be near $\lambda^* = 0.92$ (see Table 12.1 and Table 14.1).

Only computations with uniform finite element meshes are performed. It will be obvious from the results that the use of adaptive mesh refinement is even more useful in limit analysis than for elastic problems, due to the rigid regions. However, since there is ill-conditioning and non-uniqueness in limit analysis, caution must be taken not to influence the collapse solution by choosing a mesh with built-in preferences.

It should be emphasized that the results reported here, with one exception which will be pointed out, have been obtained with general purpose linear programming or nonlinear optimization software.

The following example shows a complete discretization of the test problem using piecewise constant element functions for the stresses and bilinear elements over rectangles for the velocity. This element combination has provided the best results so far. Unfortunately it can only be applied in simple geometries. The more generally applicable linear–linear element combination over triangles also works, but in our experience it does not determine the stress field in the collapse state quite as well.

EXAMPLE 11.2 (*The test problem with constant–bilinear elements*). Consider the test problem of Example 11.1. We describe the discretization with piecewise constant element function for the stresses and bilinear elements over rectangles for the flow.

Let $\mathcal{T}_h$ be a uniform division of the domain

$$\Omega = \left\{(x_1, x_2) \in \mathbb{R}^2 \mid 0 \leqslant x_1 \leqslant L, 0 \leqslant x_2 \leqslant 1\right\}$$

into squares with side $h = 1/N$. Each component of the discrete stress tensor is a linear combination of the functions ϕ_ν, where ϕ_ν is equal to 1 in the element (square) with index ν and equal to 0 elsewhere. In the notation of Section 8 the "nodes" for the discrete stress space Σ_h are the squares of $\mathcal{T}_h$. The basis for Σ_h consists of the tensors

$$\boldsymbol{\phi}_\nu^{11} = \begin{bmatrix} \phi_\nu & 0 \\ 0 & 0 \end{bmatrix}, \qquad \boldsymbol{\phi}_\nu^{22} = \begin{bmatrix} 0 & 0 \\ 0 & \phi_\nu \end{bmatrix}, \qquad \boldsymbol{\phi}_\nu^{12} = \begin{bmatrix} 0 & \phi_\nu \\ \phi_\nu & 0 \end{bmatrix}.$$

The discrete tensors $\boldsymbol{\sigma}_h \in \Sigma_h$ are of the form

$$\boldsymbol{\sigma}_h = \sum_{\nu \in \mathcal{N}_\sigma} \left(\xi_\nu^{11} \boldsymbol{\phi}_\nu^{11} + \xi_\nu^{22} \boldsymbol{\phi}_\nu^{22} + \xi_\nu^{12} \boldsymbol{\phi}_\nu^{12}\right), \tag{11.7}$$

where ξ_ν^{ij} is the constant value of $\sigma_{h,ij}$ in the square with index ν. The yield condition (11.6) takes the form of the following explicit set of nonlinear inequalities:

$$\tfrac{1}{4}\left(\xi_\nu^{11} - \xi_\nu^{22}\right)^2 + \left(\xi_\nu^{12}\right)^2 \leqslant 1 \quad \forall \nu \in \mathcal{N}_\sigma. \tag{11.8}$$

The node set $\mathcal{N}_u$ for the flow is the set of vertices of $\mathcal{T}_h$. Let ψ_μ denote the bilinear element function with value 1 at vertex $\mu \in \mathcal{N}_u$ and 0 at all other nodes. The basis for the discrete flow space consists of the vector functions

$$\boldsymbol{\psi}_\mu^1 = \begin{bmatrix} \psi_\mu \\ 0 \end{bmatrix}, \qquad \boldsymbol{\psi}_\mu^2 = \begin{bmatrix} 0 \\ \psi_\mu \end{bmatrix}.$$

The discrete flow $\boldsymbol{u}_h \in U_h$ can be written

$$\boldsymbol{u}_h = \sum_{\mu \in \mathcal{N}_u} \left(\eta_\mu^1 \boldsymbol{\psi}_\mu^1 + \eta_\mu^2 \boldsymbol{\psi}_\mu^2 \right), \tag{11.9}$$

where η_μ^k is the value of $u_{h,k}$ at the vertex with index μ. Due to the boundary condition in Eq. (11.5) the following components η_μ^k are fixed to zero, or equivalently the corresponding basis functions $\boldsymbol{\psi}_\mu^k$ are absent in Eq. (11.9):

$$\begin{aligned} \eta_\mu^2 &= 0 \text{ for vertices } \mu \text{ with } x_2 = 0, \\ \eta_\mu^1 &= 0 \text{ for vertices } \mu \text{ with } x_1 = 0 \text{ and } 0 \leqslant x_2 \leqslant a. \end{aligned} \tag{11.10}$$

In this example there are no volume forces, $\boldsymbol{f} = \boldsymbol{0}$, while the surface forces are given by $\boldsymbol{g} = (1, 0)$ at the right-hand boundary with $x_1 = L$ and $\boldsymbol{g} = \boldsymbol{0}$ elsewhere. Hence the terms in Eq. (8.7) are:

$$\begin{aligned} F(\boldsymbol{\psi}_\mu^1) &= h && \text{for vertices with } x_1 = L \text{ and } 0 < x_2 < 1, \\ F(\boldsymbol{\psi}_\mu^1) &= \tfrac{1}{2}h && \text{for the vertex with } x_1 = L \text{ and } x_2 = 0, \\ F(\boldsymbol{\psi}_\mu^1) &= \tfrac{1}{2}h && \text{for the vertex with } x_1 = L \text{ and } x_2 = 1, \\ F(\boldsymbol{\psi}_\mu^1) &= 0 && \text{for all other vertices,} \\ F(\boldsymbol{\psi}_\mu^2) &= 0 && \text{for all vertices.} \end{aligned}$$

These are the components of $\boldsymbol{b}$ in Eqs. (11.3)–(11.4).

The integrals in Eq. (8.8) are easily computed using that the bilinear basis functions $\psi_\mu(x, y)$ are of the form $\psi_\mu(x, y) = \psi_1(x)\psi_2(y)$, where ψ_j is a standard linear element function in one variable. Recall that

$$a(\boldsymbol{\sigma}, \boldsymbol{u}) = \sum_{i,j=1}^{2} \int_\Omega \sigma_{ij} \frac{\partial u_i}{\partial x_j}.$$

Since each basis function $\boldsymbol{\phi}_\nu^{ij}$ is only different from zero in element ν, there is nothing to assemble in the stiffness matrix Eq. (8.8), so we may write down the matrix elements directly. (Special for piecewise constant elements.) Let ν be any of the squares in $\mathcal{T}_h$ and let μ be the vertex of the lower left-hand corner. Then we find:

$$\begin{aligned} a(\boldsymbol{\phi}_\nu^{11}, \boldsymbol{\psi}_\mu^1) &= -\tfrac{1}{2}h, & a(\boldsymbol{\phi}_\nu^{11}, \boldsymbol{\psi}_\mu^2) &= 0, \\ a(\boldsymbol{\phi}_\nu^{22}, \boldsymbol{\psi}_\mu^1) &= 0, & a(\boldsymbol{\phi}_\nu^{22}, \boldsymbol{\psi}_\mu^2) &= -\tfrac{1}{2}h, \\ a(\boldsymbol{\phi}_\nu^{12}, \boldsymbol{\psi}_\mu^1) &= -\tfrac{1}{2}h, & a(\boldsymbol{\phi}_\nu^{12}, \boldsymbol{\psi}_\mu^2) &= -\tfrac{1}{2}h. \end{aligned}$$

The values for the other three vertices are the same except for permutations and sign variation. As usual this part of the discretization is easier to do than to describe.

It remains to number the variables (ξ_ν^{ij}) and (η_ν^k), or equivalently to choose the order of columns and rows in the matrix $\boldsymbol{A}$ in Eq. (8.15) or Eq. (11.3). Since the length satisfies $L \geqslant 1$ the elements and vertices are both ordered "bottom-up, left-to-right" to obtain smallest band width of $\boldsymbol{A}$. For the same reason the stress components $\boldsymbol{\phi}_\nu^{ij}$ are numbered element by element, i.e., the mapping (8.9) is given by

$$n(\nu,1,1) = 3(\nu-1)+1,$$
$$n(\nu,2,2) = 3(\nu-1)+2,$$
$$n(\nu,1,2) = 3(\nu-1)+3,$$

where ν is the element number. The column corresponding to λ, i.e., the $\boldsymbol{b}$ vector found above, is added as the last column, i.e., the variable λ gets the number $n = 3N^2L+1$. Similarly the velocity components are numbered by the mapping (8.12), which we define as follows:

$$m(\mu,1) = 2(\mu-1)+1,$$
$$m(\mu,2) = 2(\mu-1)+2.$$

Modern mathematical optimization software permits incorporating the boundary conditions on $\boldsymbol{u}$ by fixing the variables in Eq. (11.10). Otherwise the numbering $m(\mu,k)$ must be modified to leave out the corresponding rows (see CHRISTIANSEN and KORTANEK [1990]).

This completes the setup of the convex programming problem Eqs. (11.3)–(11.4). The following sections will describe several ways to solve it.

Other element combinations, such as bilinear–bilinear or linear–linear, for which we shall report results, are implemented with small modifications to the above example. Preferably one should try to program the logic and leave the details to the program, as is the standard now for elasticity problems.

12. Linear programming methods

Traditionally the discrete problem has been solved using linear programming. The main reasons are:

- Many structural problems can adequately be formulated with linear yield conditions.
- Good software, based on the simplex method, was available for linear programming, long before similar software for convex programming was competitive.
- The duality between the static and kinematic formulations was first established within the linear programming frame.

The price to pay for these virtues is, at first glance, modest. A new error is introduced, when the convex yield condition $\boldsymbol{x} \in K_{\mathrm{d}}$ is replaced by a piecewise linear condition $\boldsymbol{Bx} \leqslant \boldsymbol{c}$. This error is easily controlled and within the typical tolerance of applications.

After linearization of the yield condition the problems (11.3)–(11.4) turn into a pair of dual LP-problems:

$$(\mathrm{P}_{\mathrm{lin}}): \quad \begin{aligned} &\max \lambda, \\ &\boldsymbol{Ax} = \boldsymbol{b}\lambda, \\ &\boldsymbol{Bx} \leqslant \boldsymbol{c}. \end{aligned} \tag{12.1}$$

The equality constraints correspond to the equilibrium equation, while the inequalities are the linearization of the yield condition. The dual of the problem in Eq. (12.1) is

$$(\mathrm{D}_{\mathrm{lin}}): \quad \begin{aligned} &\min \boldsymbol{c}^{\mathrm{T}}\boldsymbol{y}_2, \\ &\boldsymbol{A}^{\mathrm{T}}\boldsymbol{y}_1 = \boldsymbol{B}^{\mathrm{T}}\boldsymbol{y}_2, \\ &\boldsymbol{b}^{\mathrm{T}}\boldsymbol{y}_1 = 1, \\ &\boldsymbol{y}_2 \geqslant 0. \end{aligned} \tag{12.2}$$

Of course Eq. (12.2) can also be obtained directly from Eq. (11.4) by inserting the linear yield condition $\boldsymbol{Bx} \leqslant \boldsymbol{c}$ in the definition of $D_{\mathrm{d}}(\boldsymbol{y})$.

Either of these forms may be fed to your favourite LP-solver. With various problems and LP-codes both forms have turned out slightly better conditioned than the other, so we have no recommendation as to which form should be used. Adding linear constraints in the yield condition corresponds to adding rows in Eq. (12.1) and columns in Eq. (12.2). This should favour the dual form Eq. (12.2), but our experience is not unanimous.

Before reporting on computational results we shall try to describe in detail how the linearization of the Von Mises yield condition in plane strain may be performed.

EXAMPLE 12.1 (*Linearized Von Mises condition in plane strain*). The normalized Von Mises yield condition in plane strain is

$$E_1: \quad \tfrac{1}{4}(\sigma_{11} - \sigma_{22})^2 + \sigma_{12}^2 \leqslant 1. \tag{12.3}$$

E_1 is an ellipse in the $(\sigma_{11} - \sigma_{22}, \sigma_{12})$ plane. It is shown in Fig. 12.1, which also shows the approximation of the ellipse by the intersection of 16 half-planes. Let this approximating convex polygon be denoted E_{lin}. Clearly $E_1 \subseteq E_{\mathrm{lin}}$. Let E_β denote the inflated ellipse (for $\beta > 1$):

$$E_\beta: \quad \tfrac{1}{4}(\sigma_{11} - \sigma_{22})^2 + \sigma_{12}^2 \leqslant \beta^2. \tag{12.4}$$

For a given number of lines we want to choose these lines in such a way that $E_{\mathrm{lin}} \subseteq E_\beta$ for β as small as possible. Then the following relation between the

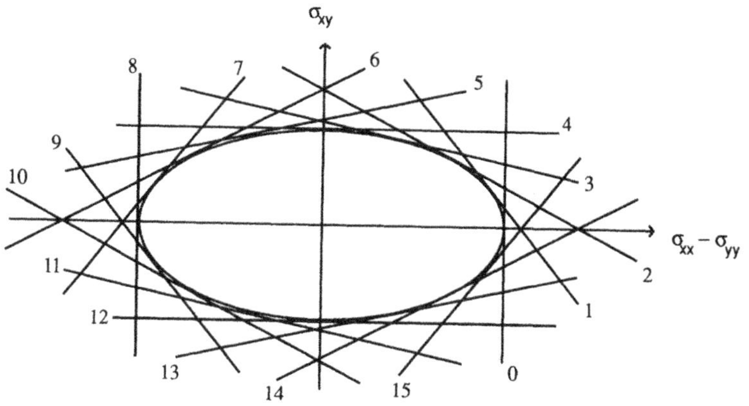

FIG. 12.1. Linearization of the Von Mises yield condition in plane strain.

exact limit multiplier λ^*, corresponding to E_1, and the approximate value λ^*_{lin}, computed from E_{lin}, will hold:

$$\frac{1}{\beta}\lambda^*_{\text{lin}} \leqslant \lambda^* \leqslant \lambda^*_{\text{lin}}. \tag{12.5}$$

The optimal choice of K lines can be constructed with a simple trick (CHRISTIANSEN and KORTANEK [1991, p. 52]): Inscribe the unit circle in a regular K-polygon. Then rescale by a factor 2 in the direction of the $\sigma_{11} - \sigma_{22}$ axis. The circle maps into the ellipse E_1, and the regular polygon maps into the optimal circumscribed polygon. It is convenient to let the number of lines K be divisible by 4 and include the four lines parallel to the axes. The error is estimated by Eq. (12.5) with the following values of β:

$$\begin{aligned} \beta &= \sqrt{2} && \text{for } K = 4,\\ \beta &= 1.0824 && \text{for } K = 8,\\ \beta &= 1.0196 && \text{for } K = 16. \end{aligned}$$

The linear constraints are explicitly given by

$$(\sigma_{11} - \sigma_{22})\cos\frac{2\pi k}{K} + 2\sigma_{12}\sin\frac{2\pi k}{K} \leqslant 2, \quad k = 0, \ldots, K-1. \tag{12.6}$$

For $K = 16$, Eq. (12.5) gives an error interval of about 2%. This exceeds the needs for most applications, but is adequate for testing. In three space dimensions it will clearly be necessary to settle for less.

At any given point of the material at most two of the constraints in Eq. (12.6) can be active. Based on physical considerations we expect that most of the constraints can be omitted in large regions of the material. Unfortunately this seems to hold only for certain solution methods.

EXAMPLE 12.2 (*LP-formulation for the test problem*). Consider again the test problem in Example 11.1 discretized with constant–bilinear elements as in Example 11.2. In the LP-formulations (12.1)–(12.2) $\boldsymbol{A}$ and $\boldsymbol{b}$ are the same as found in Example 11.2. The yield condition (11.8) is replaced by the linear inequalities (12.6):

$$\left(\xi_\nu^{11} - \xi_\nu^{22}\right) \cos \frac{2\pi k}{K} + 2\xi_\nu^{12} \sin \frac{2\pi k}{K} \leqslant 2 \tag{12.7}$$

for $k = 0, \ldots, K-1$ and all squares ν in $\mathcal{T}_h$. These are the constraints $\boldsymbol{Bx} \leqslant \boldsymbol{c}$ in Eq. (12.1). The relation between $\boldsymbol{x}$ and (ξ_ν^{ij}), i.e., the numbering (8.9) of variables is specified in Example 11.2. This completes the setup of the LP-problem (12.1). We shall add some efficiency considerations.

Traditional LP-codes may handle the problem (12.1) more efficiently if we help to minimize fill-in during the solution process. Each row in Eq. (12.7) only involves the variables associated with one element. Hence fill-in will be reduced, if this row is numbered adjacent to the velocity components associated with the vertices of that element. Similarly, if the problem must be brought to LP-standard form, the same consideration should be given to the slack variables. However, modern LP-codes for sparse problems may do this automatically.

In the remainder of this section we shall present results for the test problem obtained with three linear programming algorithms: The simplex method, the primal affine scaling algorithm and the dual affine scaling algorithm. The order of presentation is the chronological order of appearance and of application in limit analysis. It also happens to be the reverse order of performance in our evaluation.

12.1. Solution with the simplex method

Solutions for the test problem in Example 11.1 using the simplex method are reported in CHRISTIANSEN [1981]. Figure 12.2 visualizes the collapse flow field by the method described in Example 1.1: The velocity is multiplied by a suitable small time, and the resulting deformation is plotted. The result is shown for $a = \frac{1}{3}$ and $a = \frac{2}{3}$, both with $h = \frac{1}{15}$.

The solution in Fig. 12.2 is obtained with piecewise bilinear elements for both stresses and flow and with 8 lines in the linearized Von Mises yield condition ($K = 8$ in Eq. (12.6)). The flow field and the resulting deformations are physically acceptable, although there are some non-physical fluctuations in the x_1-coordinate in the upper left corner, which must be attributes of the solution method. In our experience such fluctuations in those parts of the material, which are expected to be rigid, tend to appear in the numerical solutions. Nevertheless, the solution clearly indicates rigid non-deformed regions in the upper part and at the lower left corner separated by a relatively narrow zone, where all the deformation occurs. This is the plastified (or "plastic") region.

The collapse stresses are not indicated on Fig. 12.2. They are *not physically acceptable* for this method. The value of $\boldsymbol{\sigma}_h^*$ shows strong variation from node to node. The nodes where $\boldsymbol{\sigma}_h^*$ is at the yield surface are distributed over the whole

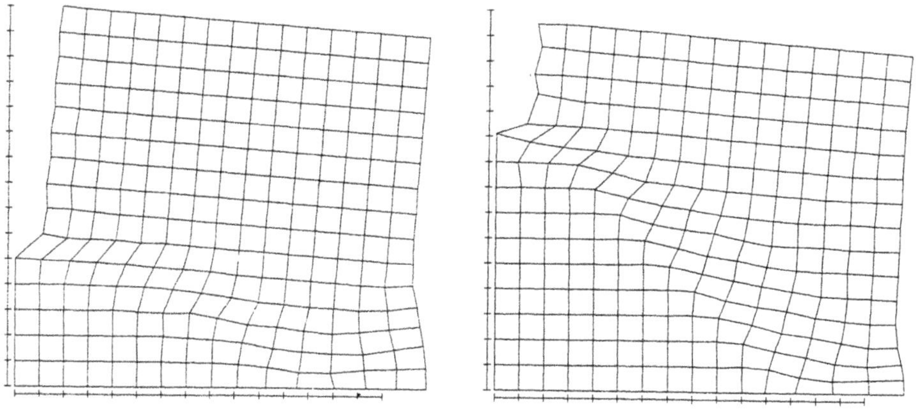

FIG. 12.2. Collapse deformation obtained with bilinear elements for stresses and flow. Linearized yield condition, simplex method.

domain in unsystematic fashion. We know from the principle of complementary slackness (see discussion following Eq. (5.47)) that $\boldsymbol{\sigma}_h^*$ must be at the yield surface in the plastified region, and that is indeed the case for this discrete solution. However, unsystematically scattered "plastified nodes" in the rest of the domain can only be attributed to the numerical solution method. There are two reasons for this deficiency:

(a) With this discretization, i.e., the choice of finite element spaces for stresses and flow, the discrete stresses are undetermined by the equilibrium equation $\boldsymbol{Ax} = \lambda \boldsymbol{b}$ and the variational principle. Loosely speaking, there are too many primal variables (degrees of freedom for $\boldsymbol{\sigma}_h$) relative to the number of dual variables (degrees of freedom for $\boldsymbol{u}_h$). This can be changed by using alternative finite element spaces, but then of course the opposite situation may (and to some extend will) occur.

(b) The simplex method, being an extreme point method, has a strong preference for solutions with variables at their bounds (non-basic variables) and inequalities at the limit of their range. Even if a whole face of the feasible set is optimal, the simplex method will choose an extreme point. This "arbitrary" selection has no mechanism to prevent non-physical fluctuations from node to node. In a manner of speaking, nature handles non-uniqueness far better than the simplex method does. Based on our experience with interior-point LP-methods and with convex programming methods (see later) we maintain that this phenomenon is mainly caused by the simplex method, and only to a lesser extent by the fact that we have replaced convex constraints by piecewise linear constraints. The extreme point nature of the simplex method, responsible for several of its virtues, is in this context a serious drawback.

Figure 12.3 shows the collapse velocity fields for $a = \frac{2}{3}$ and $h = \frac{1}{12}$, obtained with bilinear–bilinear elements (left) and constant–bilinear elements (right).

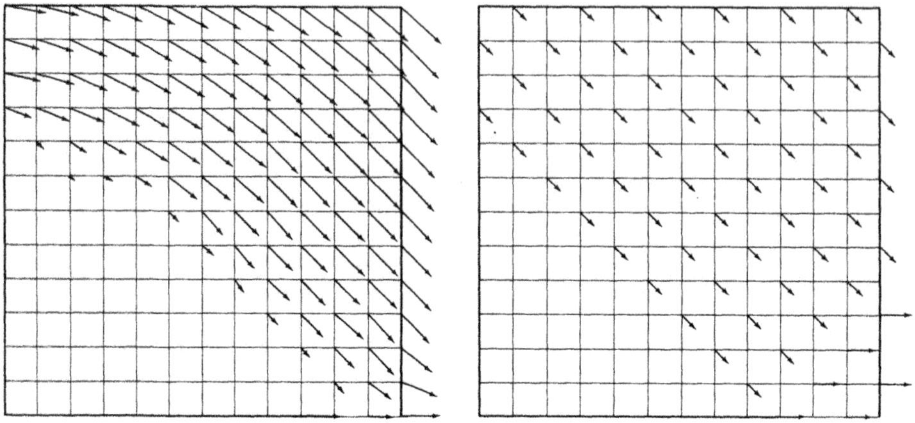

FIG. 12.3. Collapse velocity obtained with bilinear–bilinear elements (left) and constant–bilinear elements (right). Linearized yield condition, simplex method.

(Hence Fig. 12.3, (left) corresponds to Fig. 12.2, right, only with a different value of h.) The constant–bilinear element combination has fewer primal variables ($3N^2$ compared to $3(N+1)^2$ before) with the same number of dual variables ($2(N+1)^2$ minus $\left((N+1)+\frac{2}{3}N+1\right)$ lost to the boundary condition). However, the collapse stresses are still not sufficiently well determined to reveal physical information. On the other hand, the velocity field now also shows fluctuations: The velocity alternates from node to node between zero and values, which seem to reveal physical information. The human eye will easily "recognize" three rigid regions, but that is an interpretation. This is even more obvious in Fig. 12.4 with the solution for $L=2$, $a=\frac{1}{2}$ and $h=\frac{1}{12}$.

With bilinear–bilinear elements the discrete collapse flow $\boldsymbol{u}_h^*$ appears to be unique (different codes give identical solutions), while $\boldsymbol{\sigma}_h^*$ is not. With constant–bilinear elements neither $\boldsymbol{u}_h^*$ nor $\boldsymbol{\sigma}_h^*$ is uniquely determined.

In spite of the above, the constant–bilinear element combination cannot be rejected. With other optimization algorithms it will provide us with the best collapse fields, and for all algorithms (also the simplex method) it results in an optimization problem, which is more sparse and better conditioned than any other element combination. This means that computations with finer grids are possible.

Due to the poorly determined stresses it was not possible to eliminate constraints from the linearized yield condition Eq. (12.6) and thereby reduce the size of the LP-problem. The largest problems solved with the simplex method in CHRISTIANSEN [1981] are with a 24×24 grid and length $L=1$. The limiting factor with sparse LP-codes is neither computing time nor memory size. For fine grids the problem becomes so ill-conditioned that the LP-program cannot improve the objective value or verify optimum.

Based on the limited experience reported here we reluctantly conclude that

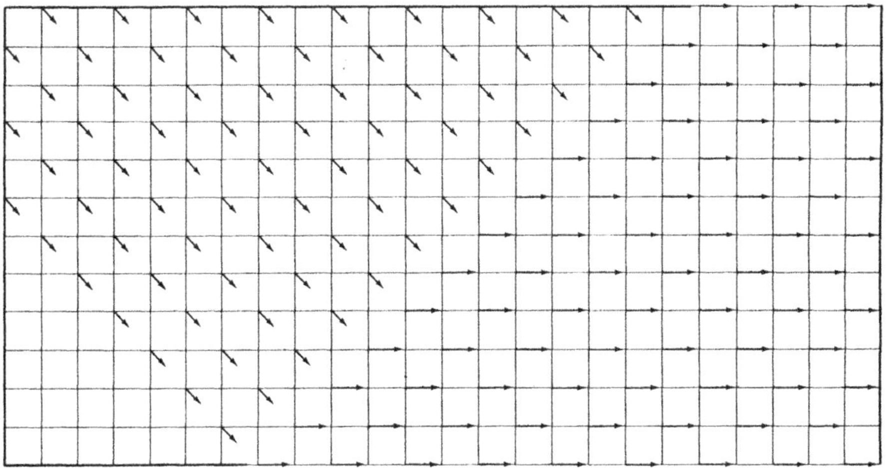

FIG. 12.4. Collapse velocity for $L = 2$ obtained with constant–bilinear elements. Linearized yield condition, simplex method.

after playing a key role in the development of limit analysis the simplex method is not well suited for problems in limit analysis for continuous media.

12.2. Solution with the primal affine scaling algorithm

We claimed above that the extreme point nature of the simplex method is responsible for its inability to determine the collapse fields for the test problem. In the wake of KARMARKAR [1984] several classes of interior point methods have surfaced. The simplest of these, the primal affine scaling algorithm, can be thought of as an affine version of Karmarkar's polynomial time algorithm, but was suggested as early as DIKIN [1967]. The algorithm takes the LP-problem on standard from, maintaining the equality constraints and keeping all variables strictly positive. In the process of convergence the feasible point is forced towards an optimal point, which may be an extreme point or, if a face is optimal, a point on an optimal face. This is the difference to the simplex method; the solution will only be an extreme point, if it is necessary in order to be optimal.

In KARMARKAR [1984] focus was on the efficiency of the interior point methods compared with the simplex method. In the context of limit analysis we are more interested in the qualitative properties of the method as consequence of the interior point versus extreme point principle.

For a description and analysis of the primal affine scaling algorithm we refer to BARNES [1986], VANDERBEI, MEKETON and FREEDMAN [1986], KORTANEK and MIAOGEN [1987], TSUCHIYA [1991]. The application to our test problem is reported in CHRISTIANSEN and KORTANEK [1990, 1991].

This algorithm makes it possible to determine collapse fields for both stresses and flow. With the piecewise constant–bilinear element combination and 16 lines

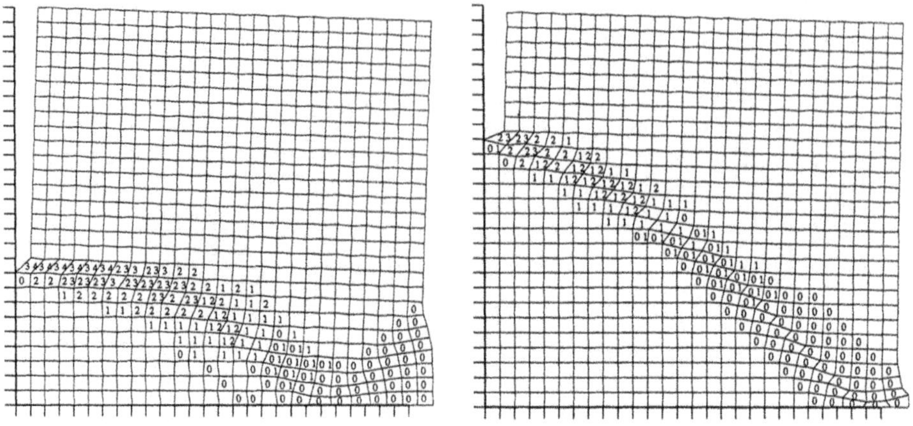

FIG. 12.5. Collapse fields obtained for $a = \frac{1}{3}$ and $a = \frac{2}{3}$, $h = \frac{1}{27}$ and constant–bilinear elements. Linearized yield condition and the primal affine scaling algorithm.

($K = 16$ in Eq. (12.6)) only the linear constraints with numbers 0–6 and 15 in Fig. 12.1 are active. The other half of the constraints may be removed (they are checked a posteriori), a considerable reduction in problem size. Figure 12.5 shows the collapse fields for $K = 16$, $L = 1$, $h = \frac{1}{27}$, $a = \frac{1}{3}$ and $a = \frac{2}{3}$. The stress field is visualized as follows: If an element is considered plastified, the number (referring to Fig. 12.1) of the active line or lines (at most 2) is written inside the element. For example $\sigma_{11} - \sigma_{22} = 2$ in squares with the number 0, and $\sigma_{12} = 1$ in squares with the number 4. In elements without a number the stresses are not at the linearized yield surface. Since the algorithm generates a sequence of interior points, the yield surface is never reached exactly. It is therefore necessary to interpret a constraint to be active if the computed stress tensor is close enough. More precisely, we say that the linear constraint Eq. (12.7) is δ-active, if

$$\left(\xi_\nu^{11} - \xi_\nu^{22}\right) \cos \frac{2\pi k}{K} + 2\xi_\nu^{12} \sin \frac{2\pi k}{K} > 2 - \delta. \tag{12.8}$$

Figure 12.5 shows the δ-active lines for $\delta = 10^{-3}$. The picture is almost the same for $\delta = 10^{-2}$ and $\delta = 10^{-4}$.

The collapse fields indicated in Fig. 12.5 are physically acceptable. There are no plastified elements scattered outside a well-defined region, and there are only very small fluctuations in the velocity. The values for the collapse multiplier λ_h^* were computed for $N \leqslant 30$, $h = 1/N$. They coincide with the results for the method in the next subsection and will be given later.

For the linear–linear element combination over triangles the stresses are not so well determined. Plastified nodes are scattered outside the plastified region, and no linear constraints can be removed. On the other hand the collapse velocity is without any fluctuations.

As is the case for the simplex method the limiting factor for the problem size, which can be solved, is the deterioration of accuracy near the solution. The primal affine scaling algorithm is particularly sensitive to loss of accuracy. In each iteration the scaled gradient must be projected onto the null space of the linear constraints in order to satisfy the equality constraints in the LP-standard form. Loss of accuracy in this projection accumulates and is not recovered by the algorithm. This is a weak point of the primal affine scaling algorithm.

It is to be expected that a linearization of convex constraints will cause ill conditioning near the optimal solution, and a natural conclusion is to abandon the linear approach altogether. However, we shall see that the ill-conditioning is also present with convex constraints. And in the next subsection we present another linear method, which seems to handle the problem extremely well.

12.3. Solution with the dual affine scaling algorithm

We find the following results surprising both from a mathematical programming and from a limit analysis point of view. The reason is that an algorithm, which should not work at all on this problem, provides the best results so far.

The dual affine scaling algorithm is essentially the same as the primal algorithm just described, applied to the dual of the standard LP-form. Consider the general LP-problem on standard form

$$\begin{aligned} &\min \boldsymbol{c}^{\mathrm{T}}\boldsymbol{x}, \\ &\boldsymbol{A}\boldsymbol{x} = \boldsymbol{b}, \\ &\boldsymbol{x} \geqslant \boldsymbol{0}. \end{aligned} \tag{12.9}$$

The dual problem is

$$\begin{aligned} &\max \boldsymbol{b}^{\mathrm{T}}\boldsymbol{y}, \\ &\boldsymbol{A}^{\mathrm{T}}\boldsymbol{y} \leqslant \boldsymbol{c} \end{aligned} \tag{12.10}$$

or equivalently

$$\begin{aligned} &\max \boldsymbol{b}^{\mathrm{T}}\boldsymbol{y}, \\ &\boldsymbol{A}^{\mathrm{T}}\boldsymbol{y} + \boldsymbol{s} = \boldsymbol{c}, \\ &\boldsymbol{s} \geqslant \boldsymbol{0}, \end{aligned} \tag{12.11}$$

where $\boldsymbol{s}$ is the vector of slack variables. Equation (12.11) is of the same form as Eq. (12.9), except that only some of the variables, namely $\boldsymbol{s}$, are restricted to be nonnegative. The components of $\boldsymbol{y}$ are free. The primal affine scaling algorithm can be modified to handle this form: Only the slack variables are subject to scaling, and the algorithm generates a sequence of points $(\boldsymbol{y}, \boldsymbol{s})$ with the $\boldsymbol{s}$-components strictly positive.

An obvious advantage of the form (12.11) over (12.9) is that it suffices to compute a new $\boldsymbol{y}$ at each iteration, which satisfies $\boldsymbol{A}^{\mathrm{T}}\boldsymbol{y} < \boldsymbol{c}$. Then we may define $\boldsymbol{s} = \boldsymbol{c} - \boldsymbol{A}^{\mathrm{T}}\boldsymbol{y} > 0$, and the linear constraints will be satisfied with no accumulated loss of accuracy. Hence the modified primal algorithm applied to Eq. (12.11) is less sensitive to round-off error than the primal algorithm applied to Eq. (12.9).

Now, if some variables in the original problem are free (i.e., unrestricted in sign; by duplication of variables such problems may still be written in the form (12.9)), then a corresponding number of inequalities in Eq. (12.10) will be inequality pairs of the form

$$\boldsymbol{a}_j^{\mathrm{T}}\boldsymbol{y} \leqslant c_j, \qquad \boldsymbol{a}_j^{\mathrm{T}}\boldsymbol{y} \geqslant c_j.$$

The corresponding slack variables will necessarily be zero. There are no feasible points for Eq. (12.11) with $\boldsymbol{s}$ strictly positive. We say that the feasible set for Eq. (12.11) has empty interior, and in this case the primal algorithm simply does not apply.

In limit analysis most of the variables are free, so how can the dual affine scaling algorithm be applied? Like the simplex method, the interior point methods need a "phase 1", where an interior feasible starting point is found. This is done by solving an associated problem with an objective function, which will force the variables towards feasibility. In ADLER, KARMARKAR, RESENDE and VEIGA [1989] the objective function for the original problem is included with some weight in phase 1 in order to produce a *good* starting point, such that fewer iterations will be necessary in phase 2. This idea has been improved in ANDERSEN [1993] to the extent that the optimal solution (to the original problem) will be found in phase 1, if the feasible set has empty interior. So phase 2 never starts!

The dual algorithm just described is very efficient in general (ANDERSEN [1993]). It is still amazing that, for the linearized problem of limit analysis, phase 1 alone works so well (ANDERSEN and CHRISTIANSEN [1995]). It is more accurate, more efficient in use of cpu-time and memory than other methods, including smooth convex optimization algorithms.

Results for the test problem obtained with (phase 1 of) the dual affine scaling algorithm are seen in Figs. 12.6–12.8. The piecewise constant–bilinear element combination is used, and all 16 lines of Eq. (12.7) are included. With this algorithm it turns out that there is no significant saving by removing lines. Figure 12.6 shows the case $a = \frac{1}{3}$ with $h = \frac{1}{198}$, i.e., a 198×198 grid. (All other algorithms, which we have tried, including convex methods, have had serious problems with ill-conditioning no later than $h = \frac{1}{30}$.) The top part shows the collapse deformation, and on the bottom part the stress field is visualized. The plastified elements are indicated by a "dot" instead of line numbers, for obvious reasons. The active lines are as one would expect from Fig. 12.5. The field for both stresses and flow are informative, although the rigid regions show the same fluctuations in the flow as observed earlier. With the linear–linear element combination these fluctuations are not seen, but the stresses are not so well determined, and the problem is more ill-conditioned allowing only more coarse grids.

On Figs. 12.7 and 12.8 corresponding solutions for the cases $a = \frac{2}{3}$, $h = \frac{1}{150}$, and $L = 2$, $a = \frac{1}{2}$, $h = \frac{1}{100}$ are shown. The "extra" band of plastified elements and the "extra" deformation layers for $L = 2$ are unexpected, but it must be recalled that the exact collapse field is probably not unique.

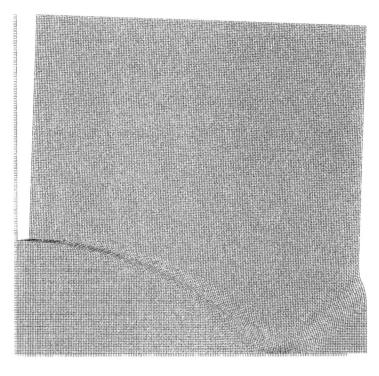

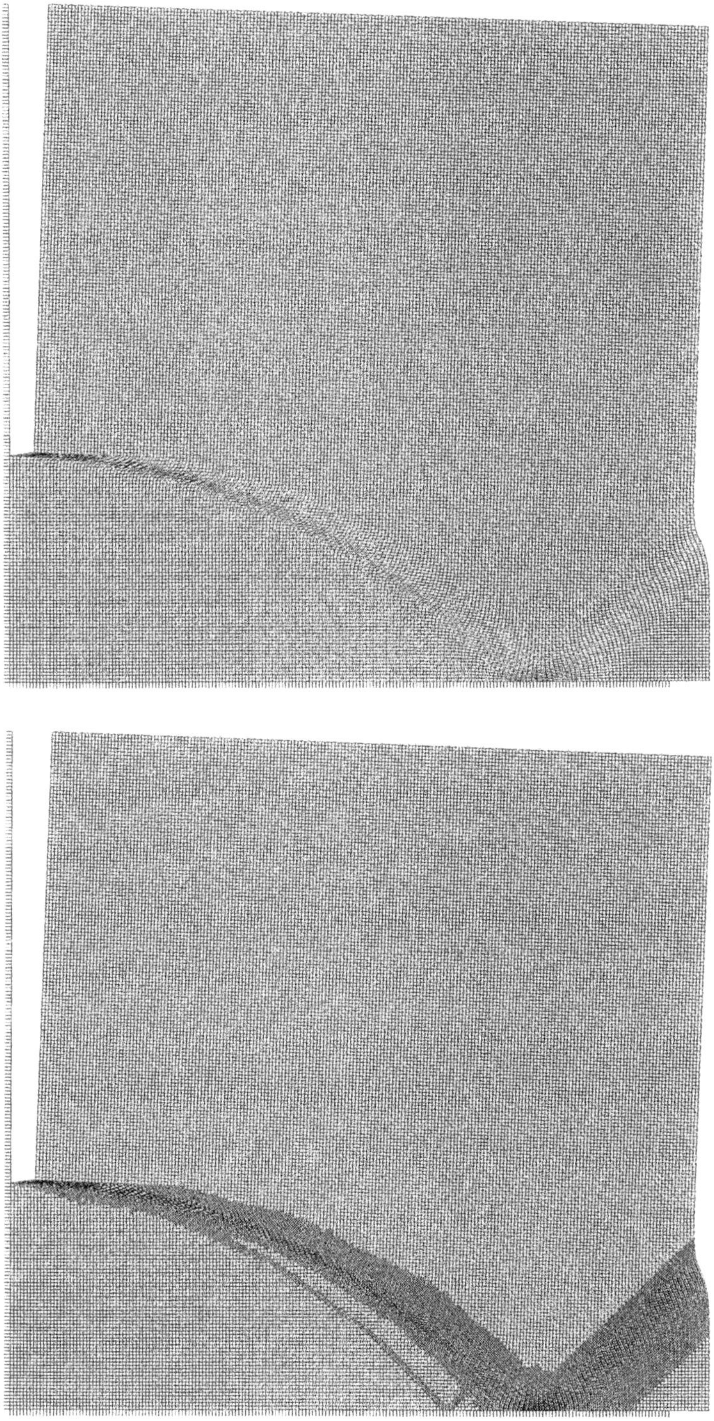

FIG. 12.6. Collapse fields for $L = 1$, $a = \frac{1}{3}$ and $h = \frac{1}{198}$ obtained with the linearized yield condition and the dual affine scaling method.

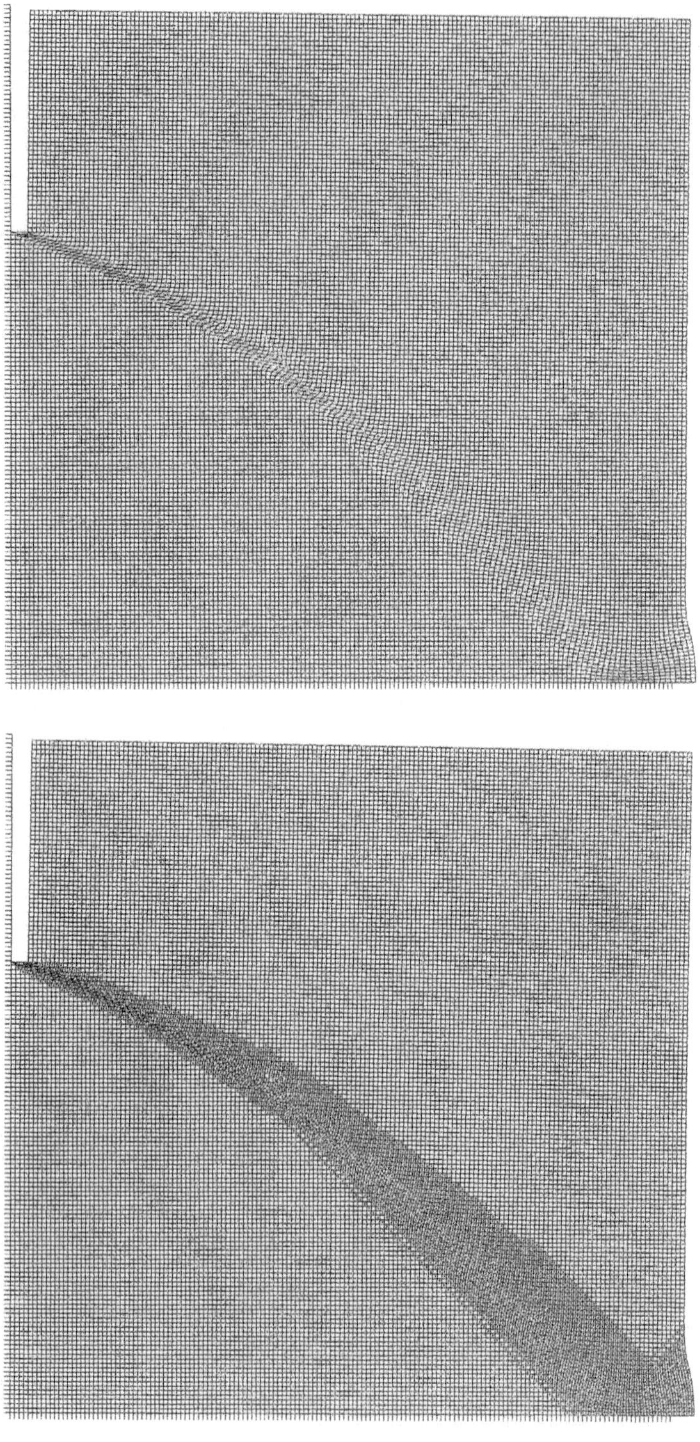

FIG. 12.7. Collapse fields for $L = 1$, $a = \frac{2}{3}$ and $h = \frac{1}{150}$ obtained with the linearized yield condition and the dual affine scaling method.

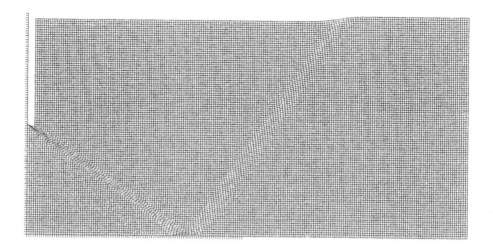

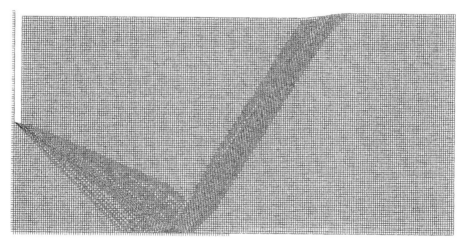

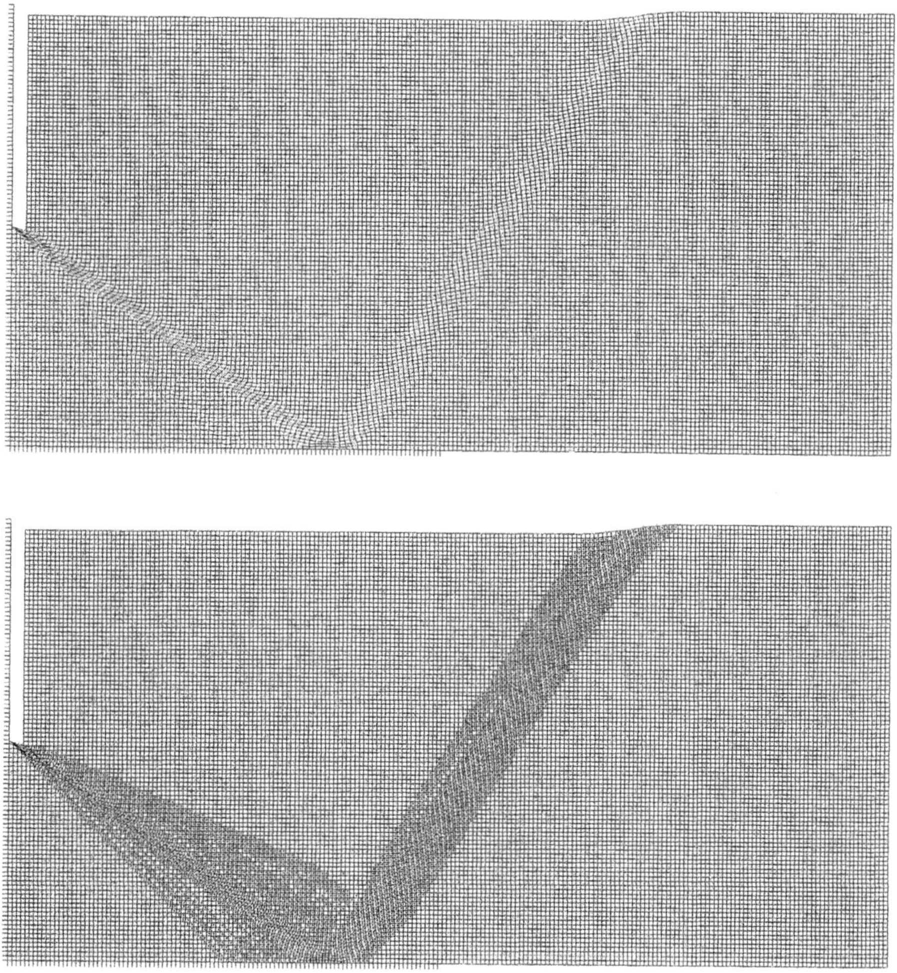

FIG. 12.8. Collapse fields for $L = 2$, $a = \frac{1}{2}$ and $h = \frac{1}{100}$ obtained with the linearized yield condition and the dual affine scaling method.

Clearly much effort used to compute the rigid regions could be saved by using adaptive elements. That would almost certainly influence the computed collapse fields, and numerical experiments are needed.

The computed values for the limit multiplier are shown in Fig. 13.2 (open triangles) together with values from other methods. The values with the same yield condition, but the linear–linear element combination, are also shown (black triangles), but only for coarse grids. For finer grids they coincide with the values from the constant–linear element combination. They should converge to the same limit, which also appears to be the case. On the other hand, the values

TABLE 12.1. Results for $L = 1$, $a = \frac{1}{3}$, $\frac{1}{2}$ and $\frac{2}{3}$ obtained with the dual affine scaling algorithm. The computed convergence order and the values from extrapolation to order 1 are also shown.

h^{-1}	$a = \frac{1}{3}$			$a = \frac{1}{2}$			$a = \frac{2}{3}$		
	λ_h^*	order	extr.	λ_h^*	order	extr.	λ_h^*	order	extr.
12	0.9826			1.2117			1.4826		
24	0.9584		0.9342	1.1776		1.1434	1.4400		1.3973
36	0.9493	0.79	0.9311	1.1656	0.90	1.1415	1.4250	0.90	1.3949
48	0.9448	1.01	0.9312	1.1596	1.03	1.1417	1.4177	1.11	1.3960
60	0.9421	0.97	0.9311	1.1561	1.07	1.1421	1.4134	1.03	1.3962
72	0.9403	1.19	0.9315	1.1537	0.82	1.1416	1.4104	0.73	1.3952
84	0.9391	1.08	0.9316	1.1519	0.97	1.1415	1.4082	1.08	1.3954
96	0.9381	1.08	0.9317	1.1506	0.96	1.1414	1.4066	1.00	1.3954
150	0.9359	1.10	0.9319	1.1473	1.02	1.1415	1.4025	0.90	1.3951
198	0.9349	1.06	0.9320						

computed with the linearized yield condition with 16 lines are the upper bound of an error interval of length 2%, independent of h, relative to the values for the exact convex Von Mises condition, also shown in the figure (open circles).

Tables 12.1 and 12.2 list some computed values for λ_h^*, experimental convergence orders based on these values and the extrapolated values to order 1. (For definition and method see CHRISTIANSEN and PETERSEN [1989].) Both the computed orders and the extrapolated values indicate that Eq. (10.1) holds, but that Eq. (10.2) does not. The values for order and extrapolation in the table are computed from the λ_h^*-values in the table (although we have computed many more values). If we had used all λ_h^*-values, it would have been more apparent that Eq. (10.2) does not hold. Larger intervals between the h-values has the effect of emphasizing the dominating linear term in the error.

13. Convex programming methods

We now return to the convex programming problem in Eqs. (11.3)–(11.4). In this section we concentrate on the form (11.3), which is a smooth convex programming problem: The objective function is linear, and the constraints consist of sparse linear equalities (equilibrium) and very sparse smooth nonlinear inequalities (the yield condition). The algorithm of ROBINSON [1972] and, in particular, the implementation of MURTAGH and SAUNDERS [1982], commercially available in the program package MINOS, is designed to solve such problems efficiently.

The discretization of the test problem with the piecewise constant–bilinear

TABLE 12.2. Same as in Table 12.1 for $L = 2$.

h^{-1}	$a = \frac{1}{3}$			$a = \frac{2}{3}$		
	λ_h^*	order	extr.	λ_h^*	order	extr.
12	1.2012			1.6667		
24	1.1713		1.1414	1.6662		
36	1.1607	0.89	1.1394	1.6614		1.6517
48	1.1552	0.93	1.1389	1.6588	0.80	1.6510
60	1.1519	0.93	1.1386	1.6570	0.42	1.6498
69	1.1502	0.99	1.1386	1.6557	0.56	1.6491
99	1.1465	0.87	1.1382	1.6537	0.61	1.6484

h^{-1}	$a = \frac{1}{2}$		
	λ_h^*	order	extr.
10	1.4722		
20	1.4481		1.4240
30	1.4374	0.50	1.4161
40	1.4316	0.73	1.4140
50	1.4280	0.92	1.4136
60	1.4256	1.06	1.4138
70	1.4239	1.01	1.4138
100	1.4209	0.93	1.4137

element combination is described in Example 11.2. MINOS solves the test problem more efficiently and with at least the same accuracy as any simplex code, which we have tried on the linearized problem. (One of these was the LP-mode of MINOS.) Were it not for the results obtained with the dual affine scaling algorithm described in the previous section the LP-approach would hardly be justified. Now the situation is not clear.

MINOS was applied both to the constant–bilinear and the linear–linear element combinations. In contrast to our experience with the linearized yield condition, the discretization with linear elements does not give rise to a more ill conditioned problem for MINOS than the constant–bilingear combination. We could compute with the same element size (30×30 grid). The values for the limit multiplier are equally good, the collapse stress field is almost as good, and the collapse deformation field has no fluctuations in the rigid regions.

The collapse fields for the test problem with piecewise linear elements for both

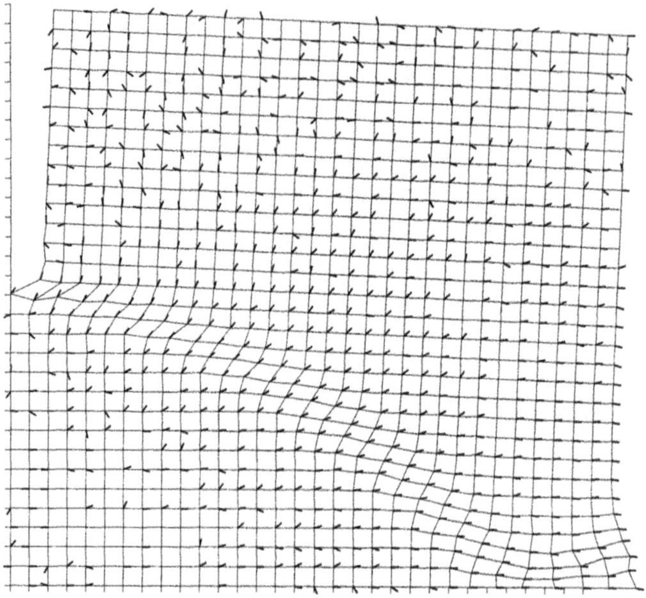

FIG. 13.1. Collapse fields for the test problem with $a = \frac{1}{2}$, $h = \frac{1}{30}$ obtained with MINOS for the exact Von Mises condition and linear–linear elements.

stresses and flow, obtained with MINOS, are shown in Fig. 13.1. The stresses are indicated by drawing the vector $\left(\frac{1}{2}(\sigma_{xx} - \sigma_{yy}), \sigma_{xy}\right)$ at each node. The deformation field is displayed as described earlier. The stress field is not as well determined as with the interior point methods for the linearized problem. The stress tensor fluctuates in a non-physical manner from node to node in the (expected) rigid regions. Again these fluctuations are caused by the extreme point nature of the algorithm. MINOS solves a sequence of linearly constrained problems operating with a basis as in the simplex method. The nonlinear objective function (in our case the nonlinear terms come from the linearization of the constraints) eventually may force variables to be *superbasic*, i.e., neither basic nor equal to zero. Since the optimization problem in limit analysis is ill conditioned and has a very "flat" optimum, the algorithm cannot force enough variables to be superbasic. It would be very interesting to try a projected Lagrangian algorithm as in MINOS, with the principle of basic variables replaced by the idea from interior (or exterior) point methods.

The values for the limit multiplier obtained with the exact Von Mises condition are seen in Fig. 13.2 (open circles).

EXAMPLE 13.1. The linear–linear element combination was used to discretize the plane strain problem indicated in Fig. 13.3. The triangulation is obtained by drawing a diagonal in each quadrangle. The structure is subject to a uniform

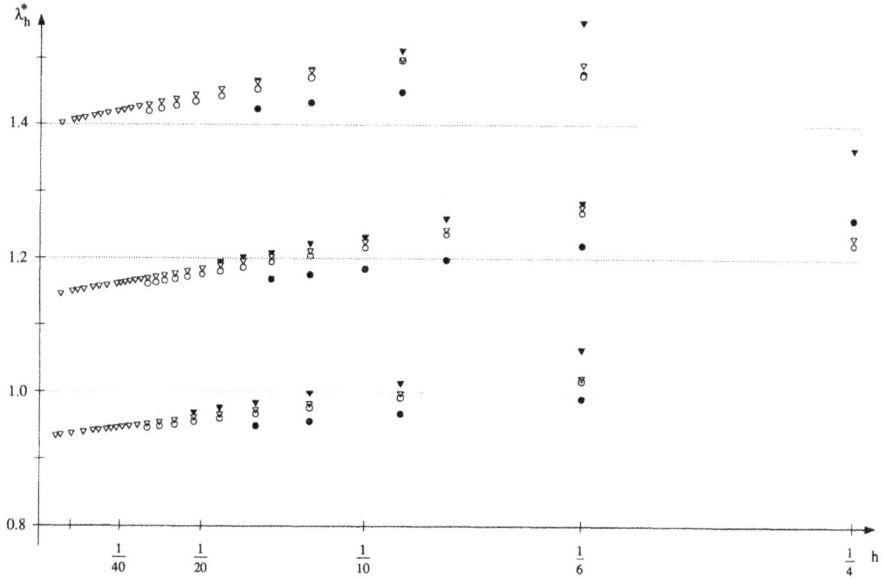

FIG. 13.2. Computed values for the limit multiplier for the test problem with $L = 1$, $a = \frac{1}{3}$, $a = \frac{1}{2}$ and $a = \frac{2}{3}$: ∇ linearized Von Mises condition (12.7) with $K = 16$, and constant–bilinear elements. ▼ linearized Von Mises condition (12.7) with $K = 16$, and linear–linear elements. ○ exact Von Mises condition and constant–bilinear elements. ● exact Von Mises condition and the elements described in Section 14.

vertical load on the top surface. The weight of the material itself is ignored, but might just as easily have been included as described in Section 6. The boundary conditions are $\boldsymbol{u} = \mathbf{0}$ at the lower edge of the frame, and $u_x = 0$ at the left and right sides of the frame. Thus the structure is assumed to be part of a larger structure, which is obtained by reflection in the left and the right side of the frame. The resulting problem was solved by MINOS for two different yield conditions: The Von Mises condition (11.6) and the Coulomb condition (6.10).

The discrete collapse stress field for the Von Mises condition is indicated in Fig. 13.3 by the same method as in Fig. 13.1. We can identify regions with compression, tension and shear, but there appears to be non-physical fluctuations in the stress field, causing the stress tensor to be at the yield surface at too many nodes. The corresponding collapse flow field is visualized in the usual way in Fig. 13.4.

The discrete problem with the Coulomb yield condition turns out to be more ill conditioned than with the Von Mises condition, and more computational experience is clearly necessary. The collapse solution for the Coulomb condition is seen in Fig. 13.5. In order to avoid the non-physical fluctuations, the stresses have been computed at the center of each diagonal. The stress tensor at this point will be at the yield surface, only if it is at the yield surface, and has almost

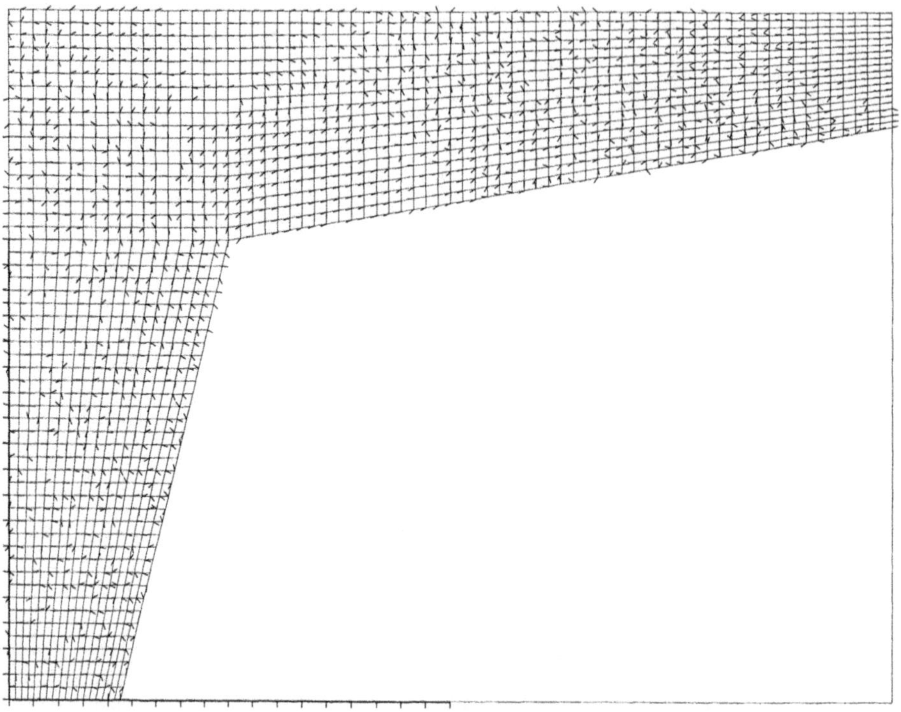

FIG. 13.3. Original geometry and collapse stresses obtained with MINOS for the exact Von Mises condition and linear–linear elements.

identical values, at the adjacent nodes. This has the effect of "purifying" the plastified region and must be thought of as an interpretation, not as a computational result.

14. Divergence-free elements

In Section 9 we described how the reduced form with divergence-free flow and trace-zero stress tensor could be discretized for the test problem in Example 11.1. The discrete problem is still of the form Eqs. (11.3)–(11.4), but now K_d is bounded. The dual property is that the objective function $D_d(\boldsymbol{y})$ in Eq. (11.4) is now finite for all $\boldsymbol{y} \in \mathbb{R}^M$. However, $D_d(\boldsymbol{y})$ is not differentiable (we shall give the precise form shortly), so standard algorithms do not apply efficiently to the problem in Eq. (11.4).

The finite element spaces for $\boldsymbol{\sigma}_h$ and the stream function Ψ_h and their bases are described in Example 9.1. The matrix elements in Eq. (8.16) are computed element by element as usual. In order to illustrate the simple, but rather te-

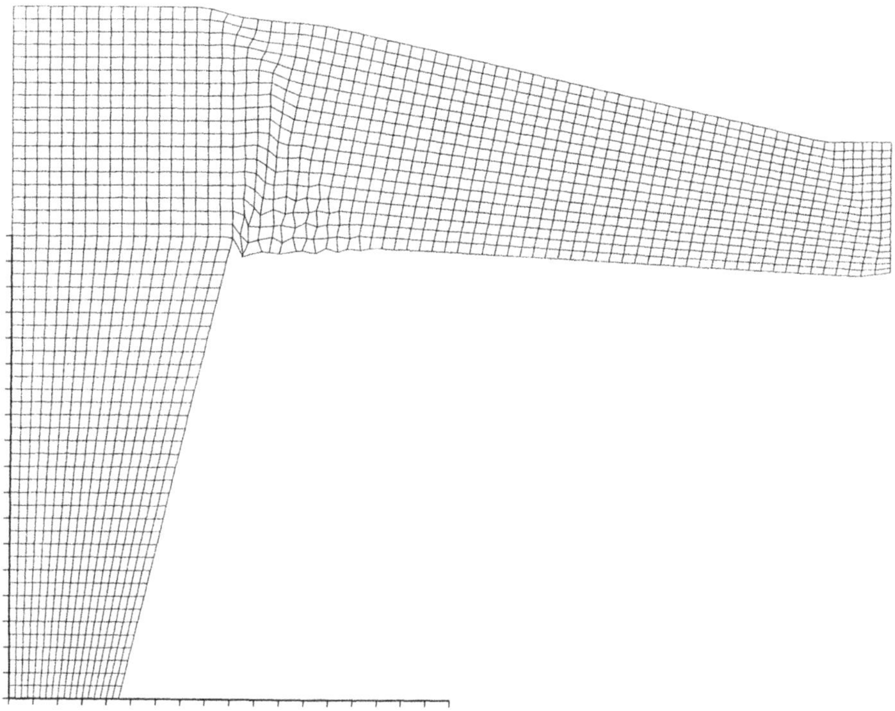

FIG. 13.4. Collapse flow for the problem in Fig. 13.3.

dious procedure, let us indicate how the integral over one element (square) is computed in the innermost of ten nested loops:

Loop 1–2: Lower left vertex of the square is specified.

Loop 3–6: Node for $\boldsymbol{\sigma}_h$ (i.e., a pair of columns) is specified.

Loop 7–8: Node for Ψ_h is specified.

Loop 9–10: Type of basis function (i.e., a row) is specified. Now the contribution to two matrix elements is computed.

The loops 3–10 only consist of two steps each. In the following we assume that $\boldsymbol{A}$ and $\boldsymbol{b}$ have been computed. The condition $\boldsymbol{x} \in K_{\mathrm{d}}$ is identical to Eq. (9.7).

14.1. Solution of the primal problem

The primal problem Eq. (11.3) can be solved by MINOS (MURTAGH and SAUNDERS [1982]) as in the previous section. The problem is formally the same. The constraint matrix $\boldsymbol{A}$ and $\boldsymbol{b}$ are different, and K_{d} defined by Eq. (9.7) is bounded.

The collapse fields for the test problem with $a = \frac{1}{2}$ obtained with this approach are visualized in Fig. 14.1. Since the velocity is not linear on the grid lines for this discretization the deformed grid appears curved between nodes. The stresses are indicated by drawing the vector (σ_1, σ_2) (see Eq. (9.3)) at each node, where

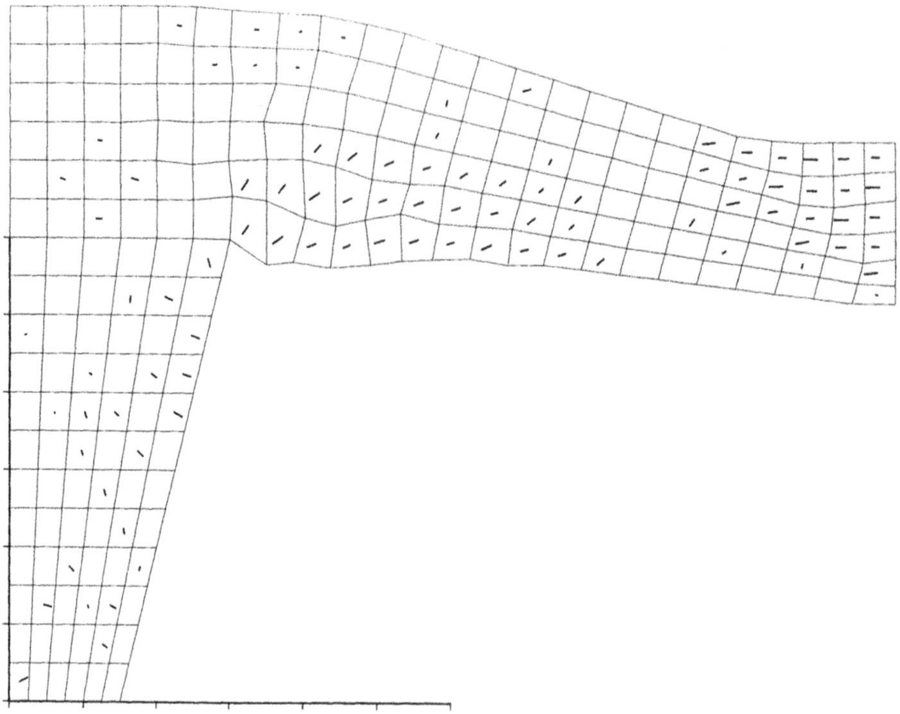

FIG. 13.5. Collapse fields for the Coulomb yield condition.

it is at the yield surface: $\sigma_1^2 + \sigma_2^2 = 1$. We see that the plastified region is not well localized. When comparing the results with other element combinations, it must be taken into consideration that this discretization has four times as many degrees of freedom per element for both stresses and velocity. Hence these results for an $N \times N$ grid should be compared with earlier results for a $2N \times 2N$ grid.

The values for the collapse multiplier λ_h^* are seen in Table 14.1. With these more smooth finite element functions the error is still $O(h)$, but in addition Eq. (10.2) appears to hold. This may justify the use of smooth elements, even if it does not change the order of the error. For general geometry this will require use of Argyris' triangle or Bell's triangle.

14.2. Solution of the dual problem

Our main motivation for trying finite elements for divergence-free flow was to solve the reduced dual problem, formally identical to Eq. (11.4):

$$\begin{aligned} &\min D_{\mathrm{d}}(\boldsymbol{y}), \\ &\boldsymbol{b}^{\mathrm{T}}\boldsymbol{y} = 1. \end{aligned} \tag{14.1}$$

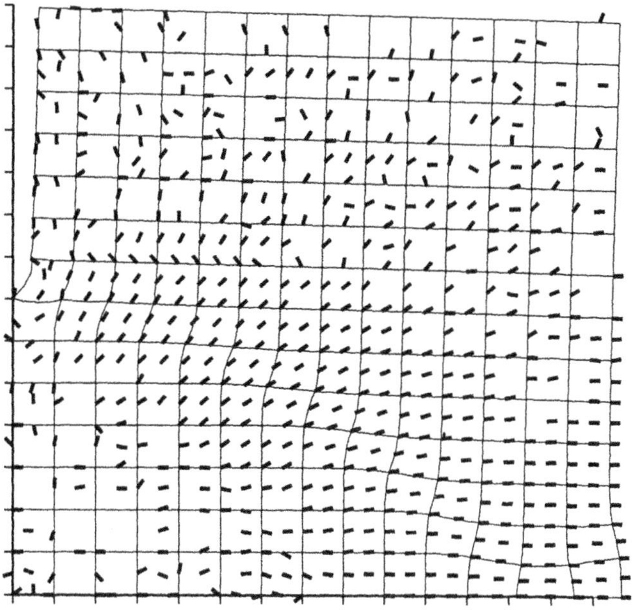

FIG. 14.1. Collapse fields for divergence-free elements, $a = \frac{1}{2}$ and $h = \frac{1}{14}$, obtained with MINOS.

In Eq. (14.1) $D_{\mathrm{d}}(\boldsymbol{y})$ is finite for all $\boldsymbol{y} \in \mathbb{R}^M$ after the discretization in Example 9.1. As mentioned earlier $D_{\mathrm{d}}(\boldsymbol{y})$ is not differentiable, but a closer look reveals a structure for which there may still be an efficient algorithm: According to Eq. (8.19) we have

$$D_{\mathrm{d}}(\boldsymbol{y}) = \max_{x \in K_{\mathrm{d}}} \boldsymbol{x}^{\mathrm{T}}(\boldsymbol{A}^{\mathrm{T}}\boldsymbol{y}),$$

where

$$\boldsymbol{x}^{\mathrm{T}}(\boldsymbol{A}^{\mathrm{T}}\boldsymbol{y}) = \sum_{n=1}^{N} x_n(\boldsymbol{A}^{\mathrm{T}}\boldsymbol{y})_n.$$

Recall that the vector (x_n) is just a linear ordering of the variables (ξ_ν^i), which define the discrete stress tensor through Eq. (9.6). If we keep the indices (ν, i) for the components of $\boldsymbol{x}$ and $\boldsymbol{A}^{\mathrm{T}}\boldsymbol{y}$ we have

$$\boldsymbol{x}^{\mathrm{T}}(\boldsymbol{A}^{\mathrm{T}}\boldsymbol{y}) = \sum_{\nu \in \mathcal{N}_\sigma} \left(\sum_{i=1}^{2} \xi_\nu^i (\boldsymbol{A}^{\mathrm{T}}\boldsymbol{y})_\nu^i \right).$$

By insertion of the normalized yield condition

$$\left(\xi_\nu^1\right)^2 + \left(\xi_\nu^2\right)^2 \leqslant 1 \quad \text{for all nodes } \nu$$

TABLE 14.1. Results for $L = 1$, $a = \frac{1}{3}, \frac{2}{3}$ and $a = \frac{1}{2}$ obtained with MINOS for the element combination described in Section 14.

h^{-1}	$a = \frac{1}{3}$			$a = \frac{2}{3}$		
	λ_h^*	order	extr.	λ_h^*	order	extr.
3	1.0555			1.5309		
6	0.9898		0.9242	1.4775		1.4242
9	0.9670	0.93	1.9214	1.4493	0.18	1.3929
12	0.9559	1.06	0.9224	1.4336	0.68	1.3862
15	0.9493	1.05	0.9229	1.4238	0.86	1.3846

h^{-1}	$a = \frac{1}{2}$		
	λ_h^*	order	extr.
4	1.2598		
6	1.2193		1.1383
8	1.1973	0.77	1.1314
10	1.1839	0.94	1.1303
12	1.1750	1.02	1.1305
14	1.1687	1.03	1.1307

we get

$$D_{\mathrm{d}}(\boldsymbol{y}) = \sum_{\nu \in \mathcal{N}_\sigma} \sqrt{\left(\left(\boldsymbol{A}^{\mathrm{T}}\boldsymbol{y}\right)_\nu^1\right)^2 + \left(\left(\boldsymbol{A}^{\mathrm{T}}\boldsymbol{y}\right)_\nu^2\right)^2}.$$

$D_{\mathrm{d}}(\boldsymbol{y})$ is a sum of norms of two-dimensional vectors. Each term only involves two rows of $\boldsymbol{A}^{\mathrm{T}}$, i.e., two columns of $\boldsymbol{A}$. These are precisely the two columns associated with the node ν and corresponding to the value of the two components of $\boldsymbol{\sigma}_h$ at that node. Now let $\boldsymbol{A}_\nu$ denote the $M \times 2$ matrix, which consists of these two columns, i.e., the two columns corresponding to the primal variables ξ_ν^1 and ξ_ν^2. Then we get the following expression for the objective function in the dual problem (14.1):

$$D_{\mathrm{d}}(\boldsymbol{y}) = \sum_{\nu \in \mathcal{N}_\sigma} \sqrt{\left(\boldsymbol{A}_\nu^{\mathrm{T}}\boldsymbol{y}\right)_1^2 + \left(\boldsymbol{A}_\nu^{\mathrm{T}}\boldsymbol{y}\right)_2^2} = \sum_{\nu \in \mathcal{N}_\sigma} \|\boldsymbol{A}_\nu^{\mathrm{T}}\boldsymbol{y}\|. \tag{14.2}$$

The norm is the ordinary Euclidian norm in $\mathbb{R}^2$.

FIG. 14.2. Collapse fields for $a = \frac{1}{3}$ and $h = \frac{1}{90}$ obtained by solving the minimization problem (14.3).

The problem (14.1) may now be written

$$\min_{b^{\mathrm{T}}y=1} \sum_{\nu \in \mathcal{N}_\sigma} \|\boldsymbol{A}_\nu^{\mathrm{T}} \boldsymbol{y}\|. \tag{14.3}$$

This belongs to a class of problems for which a second-order method was developed in CALAMAI and CONN [1980, 1982] and OVERTON [1983]. The sum of norms in Eq. (14.3) is not differentiable, when one of the terms $\|\boldsymbol{A}_\nu^{\mathrm{T}} \boldsymbol{y}\|$ vanishes. Typically, in particular in limit analysis, many of these terms are zero at the optimal vector $\boldsymbol{y}$, a severe obstruction for the efficiency of standard algorithms.

The problem is solved as follows: When a term $\|\boldsymbol{A}_\nu^{\mathrm{T}} \boldsymbol{y}\|$ becomes small, it is tentatively set to zero. If this does not increase the value of the objective function, the term is kept equal to zero by adding the linear constraint $\boldsymbol{A}_\nu^{\mathrm{T}} \boldsymbol{y} = 0$, and the term is deleted from the objective function. In the resulting subspace for $\boldsymbol{y}$ the objective function is now smooth, until either another term gets small and the above step is repeated, or a minimum on the present subspace for $\boldsymbol{y}$ is found. Thus a finite sequence of smooth convex linearly constrained problems is solved by a second-order method. When a minimum subject to a number of constraints of the form $\boldsymbol{A}_\nu^{\mathrm{T}}\boldsymbol{y} = 0$ is found, there are two possibilities: Either we have found the minimum for Eq. (14.3), or one (or more) of the constraints $\boldsymbol{A}_\nu^{\mathrm{T}}\boldsymbol{y} = 0$ must be removed again. (Unfortunately a term put equal to zero during the algorithm need not be zero in the optimal solution.) An optimality check determines which of the two situations has occurred. The principle of the method has a neat interpretation in limit analysis: The nodes ν for which $\boldsymbol{A}_\nu^{\mathrm{T}}\boldsymbol{y} = 0$ constitute the rigid region for the discrete problem. So the algorithm is trying to decrease $D_h(\boldsymbol{u}_h) = D_{\mathrm{d}}(\boldsymbol{y})$ by expanding the rigid region with the option to reduce it later, if need be.

There is a complication in limit analysis, which is not seen in the problems that originally motivated this method. The set of columns from the matrices $\boldsymbol{A}_\nu$ in the constraints is linearly dependent. This causes problems for the optimality check and the elimination of linear constraints.

Computing the limit load by minimizing a sum of norms has been applied to two problems in limit analysis with bounded yield condition in OVERTON [1984] and GAUDRAT [1991]. The above mentioned rank deficiency problem was handled on an ad hoc basis. The general method described in this Section was applied to our reference test problem (to be published) using the algorithm in OVERTON [1983], but results were obtained only for coarse grids (12×12), due to the rank deficiency problem. Very recently ANDERSEN [1995] has designed an efficient large scale algorithm for minimizing a sum of norms. With this algorithm we obtained results for much finer grids, confirming the first-order convergence observed in Table 14.1.

The collapse fields for a 90×90 grid are shown in Fig. 14.2. Recall that this should be compared with earlier results for a 180×180 grid. With this method it was also possible to solve the classical punch problem from HILL [1950, p. 254] of punching a die into a medium occupying a half space. To our knowledge this problem has not previously been solved using numerical methods.

15. Concluding remarks

If the test problem is indicative of the behaviour of solution methods, as we think it is, limit analysis is ready to be used along with the standard elastic analysis. Plane problems can certainly be solved, and three-dimensional problems can be approached with confidence.

Surprisingly fine grids can be handled by the dual affine scaling algorithm with the linearized yield condition, both with the constant–bilinear element combination over rectangles and with linear–linear elements over triangles. Also the convex problem can be solved with MINOS, using linear–linear elements, for more coarse, but still adequate, grids. These methods are ready to be applied. The approach with elements for divergence-free flow looks good, but must be extended to more general element shapes.

For the tested algorithms, except the dual affine scaling algorithm and the

TABLE 15.1. Dimensions of discrete spaces for some natural element combinations.

	Elements $\sigma_h - u_h$		$\dim(\Sigma_h)$	$\dim(U_h)$	ratio $\approx$	Expe-rience
1	constant–bilinear	□	$3N^2$	$2(N+1)^2$	$\frac{3}{2}$	good
2	bilinear–bilinear	□	$3(N+1)^2$	$2(N+1)^2$	$\frac{3}{2}$	good
3	linear–linear	△	$3(N+1)^2$	$2(N+1)^2$	$\frac{3}{2}$	good
4	constant–linear	△	$6N^2$	$2(N+1)^2$	3	bad
5	constant–quadratic	△	$6N^2$	$2(2N+1)^2$	$\frac{3}{4}$	
6	biquadratic–bicubic C^1	□	$2(2N+1)^2$	$4(N+1)^2$	(2)	good
7	bilinear–bicubic C^1	□	$2(N+1)^2$	$4(N+1)^2$	$(\frac{1}{2})$	bad
8	const.–quadr. "$\boldsymbol{\nabla} = 0$"	△	$2(2N+1)^2$	$2(2N+1)^2$	$\frac{2}{3}$	
9	linear–quadr. "$\boldsymbol{\nabla} = 0$"	△	$2(N+1)^2$	$2(2N+1)^2 - 4N^2$	$\frac{1}{2}$	
10	cubic–Argyris	△	$18N^2 + 12N + 2$	$9N^2 + 14N + 6$	(2)	
11	type 3′–Argyris	△	$14N^2 + 12N + 2$	$9N^2 + 14N + 6$	$(\frac{14}{9})$	
12	quadratic–Bell's	△	$2(2N+1)^2$	$6N^2 + 12N + 6$	$(\frac{4}{3})$	

algorithms for Eq. (14.3), the factor that limits the problem size, and hence the grid size, is the deterioration of accuracy near the optimal solution. That is our excuse for not reporting use of time and memory for the algorithms. We refer to the papers cited. It appears that the development of speed and memory size of computers has made it possible to solve problems so large that floating point accuracy becomes a bottleneck, at least for some applications.

Table 15.1 lists some of the more natural element combinations and the dimensions of the discrete spaces for the test problem for $L = 1$ and $h = 1/N$. The boundary conditions on $\boldsymbol{u}$ have been ignored, so the true dimension of U_h is $\mathrm{O}(N)$ smaller than indicated.

For the divergence-free C^1-elements in rows 6, 7, 10, 11, 12 the column $\dim(U_h)$ gives the dimension for the discrete stream function Ψ_h (see Example 9.1). Row 8 indicates constant elements for the stresses combined with quadratic elements for the flow satisfying (see TEMAM [1977, Section 4.4])

$$\int_T \boldsymbol{\nabla}\cdot\boldsymbol{u}_h = 0 \quad \forall T \in \mathcal{T}_h.$$

Then Eq. (8.26) is satisfied for any $\boldsymbol{u}_h \in U_h$. Row 9 indicates linear elements for stresses and quadratic elements for velocity satisfying

$$\int_T \phi\ \boldsymbol{\nabla}\cdot\boldsymbol{u}_h = 0 \quad \forall T \in \mathcal{T}_h,$$

where ϕ denotes any linear function on the triangle T. In particular Eq. (8.26) is satisfied for any $\boldsymbol{u}_h \in U_h$.

The element combinations 5, 7, 8 and 9 in the table cannot satisfy condition (8.24), since $\dim(\Sigma_h) < \dim(U_h)$. Hence they should not work. This has been confirmed for row 7. To our knowledge experience with the combinations in rows 5, 8–11 has not been reported. Based on the dimensions the element combinations 10 and 11 should work.

15.1. On the solution to the test problem

After spending so much effort finding approximate solutions to the test problem it seems reasonable to offer some comments on it.

We expected the collapse flow to consist of three rigid regions sliding along narrow zones, possibly slip-planes. Only near the boundary $x_2 = 0$, where the three regions meet, was a more complicated flow expected. The fine-grid solutions with the dual affine scaling method suggest a more interesting flow with deformation zones of finite width. The collapse flow may even be continuous.

In order to test the various solution methods with respect to uniqueness, localization of plastified region etc., they were also applied to the case with no cut at all. A simple solution is easily found by hand: Constant stresses and deformation (not flow) and $\lambda^* = 2$, but the fields are highly non-unique. All methods gave $\lambda_h^* = 2$ with various collapse fields, most of them similar or equal to the simple exact solution.

The problem with these solutions, both exact and discrete, is that they are not observed in real life. There we do not see uniform collapse fields, but a phenomenon called *necking*: The deformation is localized to a region where the material shrinks in the x_2-direction and stretches in the x_1-direction. In order to find such a solution numerically the following boundary condition was added to the test problem with $L = 1$:

$$u_2 = 0 \quad \text{for } x_1 = 1. \tag{15.1}$$

Restricting the flow $\boldsymbol{u}$ cannot possibly decrease the limit multiplier λ^*. In fact it was unchanged, $\lambda^* = 2$, as expected. The new collapse flow field displayed necking (it was forced to), but since λ^* was not changed the new flow field was also a solution to the original problem without the extra boundary condition. We conclude that the numerical method can find the "right" solution, but in the absence of uniqueness the numerical solution method will usually not have the same preferences as nature. There are more surprises in limit analysis than in elasticity.

CHAPTER V

Limit Analysis for Plates

16. The continuous problem

Also in the plate model the problem of limit analysis can be formulated as an infinite-dimensional mathematical programming problem. Formally the main difference is that there is one more derivative in the bilinear form for the internal work rate.

Consider a plate occupying the area Ω in the $x_1 - x_2$ plane. We only consider transversal loads $(0, 0, f)$, $f = f(x_1, x_2)$. Let $u = u_3$ be the transversal displacement rate. At a distance x_3, measured with sign, from the mid-plane of the plate we get (HODGE [1959]):

$$u_i = -x_3 \frac{\partial u}{\partial x_i}, \quad i = 1, 2,$$

and hence

$$\varepsilon_{ij} = -x_3 \frac{\partial^2 u}{\partial x_i \partial x_j}, \quad i, j = 1, 2.$$

In the plate model the other components of the strain rate tensor are neglected. This implies, in particular, that transversal shear is a violation of the model.

The internal work rate may be written

$$\begin{aligned} a(\boldsymbol{\sigma}, u) &= -\int_{\Omega} \left(\int_{x_3=-H/2}^{H/2} \sum_{i,j=1}^{2} \sigma_{ij} x_3 \frac{\partial^2 u}{\partial x_i \partial x_j} \, \mathrm{d}x_3 \right) \mathrm{d}x_1 \, \mathrm{d}x_2 \\ &= -\int_{\Omega} \sum_{i,j=1}^{2} m_{ij} \frac{\partial^2 u}{\partial x_i \partial x_j} \, \mathrm{d}x_1 \, \mathrm{d}x_2, \end{aligned}$$

where

$$m_{ij} = \int_{-H/2}^{H/2} x_3 \sigma_{ij} \, \mathrm{d}x_3, \quad i, j = 1, 2$$

are the *bending moments*. H is the thickness of the plate and may depend on

(x_1, x_2). Hence the internal work rate is expressed in the bending moments $\boldsymbol{m} = (m_{ij})_{i,j=1}^2$ and the transversal displacement rate u by the bilinear form

$$b(\boldsymbol{m}, u) = -\int_\Omega \sum_{i,j=1}^2 m_{ij} \frac{\partial^2 u}{\partial x_i \partial x_j} \, dx_1 \, dx_2. \tag{16.1}$$

The yield condition on the stress tensor can immediately be translated to the tensor $\boldsymbol{m}$ of bending moments. For example, the Von Mises condition for a homogeneous plate is

$$m_{11}^2 - m_{11}m_{22} + m_{22}^2 + 3m_{12}^2 \leqslant M_0^2. \tag{16.2}$$

The Tresca yield condition is

$$\max\left(|m_{\rm I}|, |m_{\rm II}|, |m_{\rm I} - m_{\rm II}|\right) \leqslant M_0, \tag{16.3}$$

where $m_{\rm I}$ and $m_{\rm II}$ are the *principal moments*.

The work rate of the transversal load $f(x_1, x_2)$ may be written as a linear form in u:

$$F(u) = \int_\Omega fu \, dx_1 \, dx_2. \tag{16.4}$$

The equilibrium equation for the bending moments is the equation of virtual work rate. The expressions in Eqs. (16.1) and (16.4) must be equal for all u satisfying the boundary conditions:

$$b(\boldsymbol{m}, u) = F(u) \quad \text{for all } u. \tag{16.5}$$

We shall consider only homogeneous boundary conditions, since the non-homogeneous case is not particular to limit analysis. The boundary $\partial\Omega$ consists of finitely many sections, each of which is either *supported* ($u = 0$) or non-supported, and either *clamped* ($\partial u/\partial \nu = 0$) or non-clamped. $\boldsymbol{\nu}$ denotes the outward normal to Ω, and $\partial u/\partial \nu$ the derivative of u in that direction. This gives four possible combinations for each section of the boundary.

Let $S_0 \subseteq \partial\Omega$ denote the supported part of the boundary:

$$u = 0 \quad \text{on } S_0. \tag{16.6}$$

Condition (16.6) is an *essential boundary condition*, to be imposed explicitly. The corresponding *natural boundary condition* on $T_0 = \partial\Omega \backslash S_0$ is, in the homogeneous case with no load or twisting moments applied at the boundary:

$$\boldsymbol{\nu} \cdot (\boldsymbol{\nabla} \cdot \boldsymbol{m}) = 0, \qquad \frac{\partial}{\partial \tau}(\boldsymbol{\nu} \cdot \boldsymbol{m} \cdot \boldsymbol{\tau}) = 0 \qquad \text{on } T_0 = \partial\Omega \backslash S_0. \tag{16.7}$$

$\boldsymbol{\tau}$ is the unit vector with angle $+\pi/2$ to the normal $\boldsymbol{\nu}$. Being a natural boundary condition, Eq. (16.7) is not imposed explicitly. Handling of inhomogeneous natural boundary conditions is mentioned below.

Similarly, let $S_1 \subseteq \partial\Omega$ denote the clamped part of the boundary (supported or not):

$$\frac{\partial u}{\partial \nu} = 0 \quad \text{on } S_1. \tag{16.8}$$

Equation (16.8) is a natural boundary condition and is not imposed explicitly. The corresponding essential boundary condition is in the homogeneous case

$$m_\nu = \boldsymbol{\nu}\cdot(\boldsymbol{m}\cdot\boldsymbol{\nu}) = 0 \quad \text{on } T_1 = \partial\Omega\backslash S_1. \tag{16.9}$$

If $\boldsymbol{m}$ and u are sufficiently smooth, Green's formula can be used repeatedly to rewrite the internal energy dissipation rate Eq. (16.1) as follows:

$$\begin{aligned}
b(\boldsymbol{m}, u) &= -\int_\Omega \sum_{i,j} m_{ij} \frac{\partial^2 u}{\partial x_i \partial x_j}\, dx_1\, dx_2 \\
&= \int_\Omega (\boldsymbol{\nabla}\cdot\boldsymbol{m})\cdot\boldsymbol{\nabla} u\, dx_1\, dx_2 - \int_{\partial\Omega} (\boldsymbol{\nu}\cdot\boldsymbol{m})\cdot\boldsymbol{\nabla} u\, ds \\
&= \int_\Omega (\boldsymbol{\nabla}\cdot\boldsymbol{m})\cdot\boldsymbol{\nabla} u\, dx_1\, dx_2 - \int_{\partial\Omega} (\boldsymbol{\nu}\cdot\boldsymbol{m})\cdot\left(\boldsymbol{\nu}\frac{\partial u}{\partial \nu} + \boldsymbol{\tau}\frac{\partial u}{\partial \tau}\right) ds \\
&= \int_\Omega (\boldsymbol{\nabla}\cdot\boldsymbol{m})\cdot\boldsymbol{\nabla} u\, dx_1\, dx_2 + \int_{\partial\Omega} \frac{\partial}{\partial \tau}(\boldsymbol{\nu}\cdot\boldsymbol{m}\cdot\boldsymbol{\tau})\, u\, ds \qquad (16.10) \\
&= \int_\Omega (\boldsymbol{\nabla}\cdot\boldsymbol{m})\cdot\boldsymbol{\nabla} u\, dx_1\, dx_2 \qquad (16.11) \\
&= -\int_\Omega \boldsymbol{\nabla}\cdot(\boldsymbol{\nabla}\cdot\boldsymbol{m})\, u\, dx_1\, dx_2 + \int_{\partial\Omega} \boldsymbol{\nu}\cdot(\boldsymbol{\nabla}\cdot\boldsymbol{m})\, u\, ds \\
&= -\int_\Omega \boldsymbol{\nabla}\cdot(\boldsymbol{\nabla}\cdot\boldsymbol{m})\, u\, dx_1\, dx_2. \qquad (16.12)
\end{aligned}$$

In addition to Green's formula we have applied Eqs. (16.8)–(16.9) in Eq. (16.10) and Eqs. (16.6)–(16.7) in Eq. (16.11) and in Eq. (16.12).

Inserting Eq. (16.12) into Eq. (16.5) gives the classical equilibrium equation for plates

$$\boldsymbol{\nabla}\cdot(\boldsymbol{\nabla}\cdot\boldsymbol{m}) = \frac{\partial^2 m_{11}}{\partial x_1^2} + 2\frac{\partial^2 m_{12}}{\partial x_1 \partial x_2} + \frac{\partial^2 m_{22}}{\partial x_2^2} = -f. \tag{16.13}$$

Conversely, if we impose the essential boundary condition Eq. (16.9) on $\boldsymbol{m}$, and require that Eq. (16.5) holds for all u satisfying Eq. (16.6), then for sufficiently smooth $\boldsymbol{m}$ and u we conclude that Eqs. (16.7) and (16.8) hold in addition to Eq. (16.13). We also see that inhomogeneous natural boundary conditions on the moments, i.e., nonzero boundary loads, are handled by including these loads in the expression (16.4). Inhomogeneous boundary conditions on u are irrelevant for a rigid plastic plate.

From a mathematical and numerical point of view it is preferable to use the most symmetric form Eq. (16.11) with one derivative on both $\boldsymbol{m}$ and u. Then the essential boundary conditions are meaningful, and standard finite elements can be used.

The problem of limit analysis for plastic plates may now be formulated in complete analogy with Section 5. Let K denote the set of bending moment tensors $\boldsymbol{m}$, which satisfy the yield condition and the boundary condition (16.9), and let U be the space of transversal displacement rates u, which satisfy the boundary condition (16.6). Then the static principle of limit analysis takes the form

$$\begin{aligned}\lambda^* &= \max\{\lambda \mid \exists \boldsymbol{m} \in K\colon b(\boldsymbol{m}, u) = \lambda F(u)\ \forall u \in U\} \\ &= \max_{\boldsymbol{m}\in K} \min_{F(u)=1} b(\boldsymbol{m}, u),\end{aligned} \tag{16.14}$$

where $F(u)$ is defined by Eq. (16.4), and $b(\boldsymbol{m}, u)$ by Eq. (16.11).

The kinematic principle can be written

$$\begin{aligned}\lambda^* &= \min_{F(u)=1} \max_{\boldsymbol{m}\in K} b(\boldsymbol{m}, u) \\ &= \min_{F(u)=1} D(u),\end{aligned} \tag{16.15}$$

where

$$D(u) = \sup_{\boldsymbol{m}\in K} b(\boldsymbol{m}, u). \tag{16.16}$$

In the next section we prove that, under certain conditions, the duality between the static and kinematic formulations holds, and that collapse solutions for $\boldsymbol{m}$ and u exist.

17. Duality of limit analysis

We shall formulate a mathematical frame for the duality problem for plastic plates. The result is not as general as one could wish, since it requires the entire boundary of the plate to be supported. We believe this restriction to be mainly of proof technical nature.

Assume that the plate is supported at the entire boundary (and only there)

$$u = 0 \quad \text{on } \partial\Omega, \tag{17.1}$$

i.e., $S_0 = \partial\Omega$ in Eq. (16.6). It may be clamped along part of or all of its boundary, as indicated by Eqs. (16.8) and (16.9). Where it is not clamped, i.e., it is simply supported, we impose the essential boundary condition (16.9).

Let $p > 2$, such that $W^{1,p}(\Omega)$ is continuously imbedded in $L^\infty(\Omega)$ (in fact in $C(\overline{\Omega})$), and let $1/p + 1/q = 1$. The space X of bending moments consists of all symmetric 2×2 tensors $\boldsymbol{m} = (m_{ij})$ with the following properties

$$m_{ij} \in W^{1,p}(\Omega), \tag{17.2}$$

$$m_\nu = \boldsymbol{\nu}\cdot(\boldsymbol{m}\cdot\boldsymbol{\nu}) = 0 \quad \text{on } \partial\Omega \backslash S_1, \tag{17.3}$$

where S_1 is the clamped part of the boundary. X is equipped with the $W^{1,p}$-norm.

The space Y of transversal displacement rates consists of those $u \in W_0^{1,q}(\Omega)$ for which there exists a symmetric 2×2 tensor of measures $\boldsymbol{\mu} = (\mu_{ij})$, $\mu_{ij} \in C^*(\overline{\Omega})$, such that

$$\sum_{i,j=1}^{2} \langle \mu_{ij}, m_{ij} \rangle_{C^* \times C} = \langle \boldsymbol{\nabla} \cdot (\boldsymbol{\nabla} \cdot \boldsymbol{m}), u \rangle_{W^{-1,p} \times W_0^{1,q}} \quad \forall \boldsymbol{m} \in X. \tag{17.4}$$

Y is equipped with the $W_0^{1,q}$-norm. The measures μ_{ij} are the generalized second-order derivatives of u. This generalization is necessary in order to incorporate the energy dissipation rate from a hinge (i.e., a discontinuity in the derivatives of u) along the clamped part of the boundary. The natural boundary condition (16.8) is incorporated through Eq. (17.4).

On $X \times Y$ the bilinear form $b(\boldsymbol{m}, u)$ is defined by the expression in Eq. (17.4).

The transversal load f must satisfy

$$f \in W^{-1,p}(\Omega) = (W_0^{1,q}(\Omega))^* \tag{17.5}$$

and the external work rate is given by the linear form

$$F(u) = \langle f, u \rangle_{W^{-1,p} \times W_0^{1,q}} \quad \forall u \in Y. \tag{17.6}$$

THEOREM 17.1. *Assume that Ω is bounded with Lipschitz continuous boundary. For any $f \in W^{-1,p}(\Omega)$ there exists $\boldsymbol{m} \in X$, such that*

$$\boldsymbol{\nabla} \cdot (\boldsymbol{\nabla} \cdot \boldsymbol{m}) = f \quad \textit{in } \Omega.$$

PROOF. We may find a solution $\phi \in W^{1,p}(\Omega)$ to the Dirichlet problem

$$\Delta \phi = f \quad \text{in } \Omega,$$

$$\phi = 0 \quad \text{on } \partial\Omega.$$

Let $m_{11} = m_{22} = \phi$ $m_{12} = 0$. Then

$$\boldsymbol{\nabla} \cdot (\boldsymbol{\nabla} \cdot \boldsymbol{m}) = \Delta \phi = f \quad \text{in } \Omega$$

and

$$\boldsymbol{\nu} \cdot (\boldsymbol{m} \cdot \boldsymbol{\nu}) = \phi \boldsymbol{\nu} \cdot \boldsymbol{\nu} = 0 \quad \text{on } \partial\Omega.$$ □

The yield condition is assumed to be of the form

$$\begin{aligned} &\boldsymbol{m}(\boldsymbol{x}) \in B(\boldsymbol{x}) \quad \forall x \in \Omega, \\ &B(\boldsymbol{x}) \text{ closed, convex and bounded uniformly in } x \in \Omega, \\ &\exists \delta > 0\colon |m_{ij}| < \delta \ \forall i, j \ \Rightarrow \ \boldsymbol{m} \in B(\boldsymbol{x}) \ \forall \boldsymbol{x} \in \Omega. \end{aligned} \tag{17.7}$$

Let K be the set of admissible moment tensors:

$$K = \{ \boldsymbol{m} \in X \mid \boldsymbol{m}(\boldsymbol{x}) \in B(\boldsymbol{x}) \ \forall \boldsymbol{x} \in \Omega \}. \tag{17.8}$$

THEOREM 17.2 (Duality theorem for plates). *Assume that Ω is bounded with*

Lipschitz continuous boundary. Let X, Y, $b(\boldsymbol{m},u)$, $F(u)$ and K be defined as above. Then the duality between Eq. (16.14) *and Eq.* (16.15) *holds. More precisely:*

$$\sup_{\boldsymbol{m}\in K}\min_{F(u)=1} b(\boldsymbol{m},u) = \min_{F(u)=1}\sup_{\boldsymbol{m}\in K} b(\boldsymbol{m},u). \tag{17.9}$$

The supremum on the left-hand side of Eq. (17.9) *is attained in the following weak sense: There exist $\boldsymbol{m}^* \in [L^\infty(\Omega)]^3$, which satisfies $\boldsymbol{m}^*(\boldsymbol{x}) \in B(\boldsymbol{x})$ almost everywhere in Ω and*

$$\boldsymbol{\nabla}\cdot(\boldsymbol{\nabla}\cdot\boldsymbol{m}^*) \in W^{-1,p}(\Omega),$$
$$b(\boldsymbol{m}^*,u) = \lambda^* F(u) \quad \forall u \in Y,$$

where λ^* is the value of Eq. (17.9). *If u^* minimizes the right-hand side o Eq.* (17.9), *we have the following saddle point inequality for $(\boldsymbol{m}^*,u^*)$:*

$$b(\boldsymbol{m},u^*) \leqslant \lambda^* = b(\boldsymbol{m}^*,u^*) = b(\boldsymbol{m}^*,u) \tag{17.10}$$

for all $\boldsymbol{m} \in K$ and all $u \in Y$ with $F(u) = 1$.

PROOF. The proof is similar to the proof of Theorem 5.7, only considerably simpler because of the boundary condition $u = 0$ on $\partial\Omega$ and the bounded yield set. We apply Theorem 2.1 in CHRISTIANSEN [1980a] to the bilinear form $-b(\boldsymbol{m},u)$ on $X \times Y$ and the convex sets $K \subseteq X$ and $C \subseteq Y$ given by

$$C = \{u \in Y \mid F(u) = 1\}.$$

C is closed in Y, and due to the third condition in Eq. (17.7), and the fact that $p > 2$, $\boldsymbol{m} = \boldsymbol{0}$ is an interior point of K. Equation (17.9) will follow by establishing the following two implications (a) and (b):

(a) If Ψ is a continuous linear functional on Y, which satisfies $\inf_{u\in C}\Psi(u) > -\infty$, then there exists $\boldsymbol{m} \in X$, such that

$$\Psi(u) = b(\boldsymbol{m},u)\ \forall u \in Y.$$

(b) If Φ is a continuous linear functional on X, which satisfies $\sup_{\boldsymbol{m}\in K}\Phi(\boldsymbol{m}) < \infty$ and the condition

$$b(\boldsymbol{m},u) = 0 \quad \forall u \in Y \Rightarrow \Phi(\boldsymbol{m}) = 0,$$

then there exists $u \in Y$, such that

$$\Phi(\boldsymbol{m}) = b(\boldsymbol{m},u) \quad \forall \boldsymbol{m} \in X.$$

If Ψ in (a) is bounded from below on the affine hyperplane C, then Ψ is proportional to F, i.e., there exists a real number t, such that

$$\Psi(u) = tF(u) \quad \forall u \in Y.$$

From Theorem 17.1 we conclude that there is $\boldsymbol{m} \in X$, such that

$$F(u) = b(\boldsymbol{m},u) \quad \forall u \in Y.$$

Hence (a) holds.

Now let Φ satisfy the conditions in (b). Due to Eq. (17.7) K contains a ball for the maximum-norm, and since Φ by assumption is bounded on this ball there exist measures $\mu_{ij} \in C^*(\overline{\Omega})$, such that $\mu_{12} = \mu_{21}$ and

$$\Phi(\boldsymbol{m}) = \sum_{i,j} \langle \mu_{ij}, m_{ij} \rangle_{C^* \times C} \quad \forall \boldsymbol{m} \in X. \tag{17.11}$$

The second condition on Φ implies that the null space of the map

$$T : \boldsymbol{m} \to \boldsymbol{\nabla} \cdot (\boldsymbol{\nabla} \cdot \boldsymbol{m})$$

is contained in the null space of Φ. Since by Theorem 17.1 T maps X *onto* $W^{-1,p}(\Omega)$ and thus is an open map, we conclude that Φ may be factorized over the null space of T:

$$\Phi(\boldsymbol{m}) = \langle u, T(\boldsymbol{m}) \rangle_{W_0^{1,q} \times W^{-1,p}} \quad \forall \boldsymbol{m} \in X$$

for some $u \in \left(W^{-1,p}(\Omega)\right)^* = W_0^{1,q}(\Omega)$. Comparing with Eq. (17.11) we see that $u \in Y$. Thus (b) holds, and Eq. (17.9) is proved.

Let $(\boldsymbol{m}^k)$ be a maximizing sequence for the left-hand side of Eq. (17.9), i.e., a sequence in K, which satisfies

$$b(\boldsymbol{m}^k, u) = \lambda_k F(u) \quad \forall u \in Y, \quad \lambda_k \to \lambda^*,$$

where λ^* is the value of Eq. (17.9). By weak* compactness of K in L^∞ there exists $\boldsymbol{m}^*$, which due to the convexity of $B(\boldsymbol{x})$ satisfies $\boldsymbol{m}^*(\boldsymbol{x}) \in B(\boldsymbol{x})$ almost everywhere in Ω. Since $\boldsymbol{m}^*$ is also a distributional limit we have

$$\boldsymbol{\nabla} \cdot (\boldsymbol{\nabla} \cdot \boldsymbol{m}^*) = \lambda^* f \in W^{-1,p}(\Omega).$$

This completes the proof. □

18. Discretization of the plate problem

The discretization of the plate problem is formally the same as in Section 8. However, the bilinear form $b(\boldsymbol{m}, u)$ and the essential boundary conditions on both $\boldsymbol{m}$ and u strongly suggest that continuous finite element functions are used. An obvious choice is piecewise linear elements for both $\boldsymbol{m}$ and u, or in simple geometries piecewise bilinear elements. The bilinear form $b(\boldsymbol{m}_h, u_h)$ is computed by the expression (16.11) with first order derivatives on both $\boldsymbol{m}$ and u.

$$\begin{aligned} b(\boldsymbol{m}, u) &= \int_\Omega (\boldsymbol{\nabla} \cdot \boldsymbol{m}) \cdot \boldsymbol{\nabla} u \, \mathrm{d}x_1 \, \mathrm{d}x_2 \\ &= \int_\Omega \left(\frac{\partial m_{11}}{\partial x_1} \frac{\partial u}{\partial x_1} + \frac{\partial m_{12}}{\partial x_2} \frac{\partial u}{\partial x_1} + \frac{\partial m_{12}}{\partial x_1} \frac{\partial u}{\partial x_2} + \frac{\partial m_{22}}{\partial x_2} \frac{\partial u}{\partial x_2} \right) \mathrm{d}x_1 \, \mathrm{d}x_2. \end{aligned} \tag{18.1}$$

Let $\mathcal{T}_h$ be a triangulation of Ω in the sense of CIARLET [1991, Chapter II]. Based on this triangulation we choose finite element spaces (e.g., piecewise linear element functions) for u and each component of $\boldsymbol{m}$. Let X_h and Y_h denote the resulting spaces for the discrete bending moments $\boldsymbol{m}_h$ and transversal displacement rate u_h, respectively. With continuous finite element functions $X_h \subseteq X$ and $Y_h \subseteq Y$ will always be satisfied (CIARLET [1991, Theorem 5.1]), where X and Y are the spaces defined in Section 17. Problems concerning curved boundary will not be discussed, since they are not specific to limit analysis. We refer to CIARLET [1991, Chapter VI].

The yield condition (17.7) on $\boldsymbol{m}_h$ will be imposed only at the nodes: If $\{\boldsymbol{x}_\nu\}$ are the nodal points for the value $\boldsymbol{m}_h(\boldsymbol{x})$, then the discrete set of admissible moment tensors K_h is defined as

$$K_h = \{\boldsymbol{m}_h \in X_h \mid \boldsymbol{m}_h(\boldsymbol{x}_\nu) \in B(\boldsymbol{x}_\nu)\ \forall \boldsymbol{x}_\nu\}. \tag{18.2}$$

As pointed out in Section 8, $K_h \subseteq K$ may not be satisfied for higher-order elements.

As a matter of convenience we assume that the work rate for the transversal load f given by Eq. (16.4), or formally by Eq. (17.6), can be computed exactly for $u_h \in Y_h$. The effect of using numerical integration is not specific to limit analysis, and we refer to CIARLET [1991, Chapter IV].

The discrete problem for plates may now be written

$$\begin{aligned} \lambda_h^* &= \max\{\lambda \mid \exists \boldsymbol{m}_h \in K_h : b(\boldsymbol{m}_h, u_h) = \lambda F(u_h)\ \forall u_h \in Y_h\} && (18.3)\\ &= \max_{\boldsymbol{m}_h \in K_h} \min_{F(u_h)=1} b(\boldsymbol{m}_h, u_h) \\ &= \min_{F(u_h)=1} \max_{\boldsymbol{m}_h \in K_h} b(\boldsymbol{m}_h, u_h) \\ &= \min_{F(u_h)=1} D_h(u_h), && (18.4) \end{aligned}$$

where

$$D_h(u_h) = \max_{\boldsymbol{m}_h \in K_h} b(\boldsymbol{m}_h, u_h). \tag{18.5}$$

As in Section 8 we shall indicate how the discrete problem may be written in a form suitable for mathematical programming. Let $\mathcal{N}_m$ and $\mathcal{N}_u$ be the node sets for the finite element representation of $\boldsymbol{m}_h$ and u_h, respectively. With first-order derivatives on both m and u in the bilinear form $b(\boldsymbol{m}, u)$ these node sets will typically be the same, except for the boundary conditions Eqs. (16.6) and (16.9). Let ϕ_ν be the scalar basis function associated with the node $\nu \in \mathcal{N}_m$. Then there are the following three basis functions for X_h associated with the node ν:

$$\boldsymbol{\phi}_\nu^{11} = \begin{bmatrix} \phi_\nu & 0 \\ 0 & 0 \end{bmatrix}, \quad \boldsymbol{\phi}_\nu^{12} = \begin{bmatrix} 0 & \phi_\nu \\ \phi_\nu & 0 \end{bmatrix}, \quad \boldsymbol{\phi}_\nu^{22} = \begin{bmatrix} 0 & 0 \\ 0 & \phi_\nu \end{bmatrix}. \tag{18.6}$$

Every discrete moment tensor $\boldsymbol{m}_h \in X_h$ may be written as a linear combination of the basis functions:

$$\boldsymbol{m}_h = \sum_{\nu \in \mathcal{N}_m} \left(\xi_\nu^{11} \boldsymbol{\phi}_\nu^{11} + \xi_\nu^{12} \boldsymbol{\phi}_\nu^{12} + \xi_\nu^{22} \boldsymbol{\phi}_\nu^{22}\right). \tag{18.7}$$

The coefficient ξ_ν^{ij} is the value of $m_{h,ij}$ corresponding to the nodal value ν.

For each node $\mu \in \mathcal{N}_u$ there is just one basis function ψ_μ for Y_h, since u_h is a scalar function. Hence each $u_h \in Y_h$ is of the form

$$u_h = \sum_{\mu \in \mathcal{N}_u} \eta_\mu \psi_\mu. \tag{18.8}$$

Equations (18.7) and (18.8) must be modified according to the boundary conditions.

The yield condition (18.2) can now be written

$$\begin{bmatrix} \xi_\nu^{11} & \xi_\nu^{12} \\ \xi_\nu^{12} & \xi_\nu^{22} \end{bmatrix} \in B(\boldsymbol{x}_\nu) \quad \forall \nu \in \mathcal{N}_m, \tag{18.9}$$

while the energy functions become

$$F(u_h) = \sum_{\mu \in \mathcal{N}_u} \eta_\mu F(\psi_\mu), \tag{18.10}$$

$$\begin{aligned} b(\boldsymbol{m}_h, u_h) = \sum_{\nu \in \mathcal{N}_m} \sum_{\mu \in \mathcal{N}_u} \big(&\xi_\nu^{11} \eta_\mu b(\boldsymbol{\phi}_\nu^{11}, \psi_\mu) + \xi_\nu^{12} \eta_\mu b(\boldsymbol{\phi}_\nu^{12}, \psi_\mu) \\ &+ \xi_\nu^{22} \eta_\mu b(\boldsymbol{\phi}_\nu^{22}, \psi_\mu) \big). \end{aligned} \tag{18.11}$$

After insertion of Eq. (18.1) the single terms in Eq. (18.11) become

$$b(\boldsymbol{\phi}_\nu^{11}, \psi_\mu) = \int_\Omega \frac{\partial \phi_\nu}{\partial x_1} \frac{\partial \psi_\mu}{\partial x_1} \, \mathrm{d}x_1 \, \mathrm{d}x_2, \tag{18.12}$$

$$b(\boldsymbol{\phi}_\nu^{22}, \psi_\mu) = \int_\Omega \frac{\partial \phi_\nu}{\partial x_2} \frac{\partial \psi_\mu}{\partial x_2} \, \mathrm{d}x_1 \, \mathrm{d}x_2, \tag{18.13}$$

$$b(\boldsymbol{\phi}_\nu^{12}, \psi_\mu) = \int_\Omega \left(\frac{\partial \phi_\nu}{\partial x_2} \frac{\partial \psi_\mu}{\partial x_1} + \frac{\partial \phi_\nu}{\partial x_1} \frac{\partial \psi_\mu}{\partial x_2} \right) \mathrm{d}x_1 \, \mathrm{d}x_2. \tag{18.14}$$

These are the elements of the matrix corresponding to $b(\boldsymbol{m}_h, u_h)$. As usual they should be assembled from the contributions from each element separately.

As in Section 8 we shall assume that the coefficients ξ_ν^{ij} and η_μ have been ordered in a linear numbering:

$$n = n(\nu, i, j) \in \{1, 2, \ldots, N\}, \tag{18.15}$$

$$m = m(\mu) \in \{1, 2, \ldots, M\}. \tag{18.16}$$

With such an ordering the coefficients (ξ_ν^{ij}) and (η_μ) may be identified with vectors $\boldsymbol{x} = (x_n) \in \mathbb{R}^N$ and $\boldsymbol{y} = (y_m) \in \mathbb{R}^M$:

$$x_n = \xi_\nu^{ij}, \quad \text{where } n = n(\nu, i, j), \tag{18.17}$$

$$y_m = \eta_\mu, \quad \text{where } m = m(\mu). \tag{18.18}$$

The work rate functions Eqs. (18.10) and (18.11) may now be written

$$F(u_h) = \sum_{m=1}^{M} y_m b_m = \boldsymbol{b}^{\mathrm{T}}\boldsymbol{y}, \tag{18.19}$$

where

$$b_m = F(\psi_\mu), \quad m = m(\mu)$$

and

$$b(\boldsymbol{m}_h, u_h) = \sum_{m=1}^{M}\sum_{n=1}^{N} y_m\, x_n\, a_{mn} = \boldsymbol{y}^{\mathrm{T}}\boldsymbol{A}\boldsymbol{x} = \boldsymbol{x}^{\mathrm{T}}\left(\boldsymbol{A}^{\mathrm{T}}\boldsymbol{y}\right). \tag{18.20}$$

$\boldsymbol{A}$ is the $M \times N$ matrix

$$a_{mn} = b(\boldsymbol{\phi}_\nu^{ij}, \psi_\mu), \quad m = m(\mu), \;\; n = n(\nu, i, j).$$

In the variables $\boldsymbol{x}$ and $\boldsymbol{y}$ the discrete problem (18.3)–(18.4) for plate bending is formally identical to the general discrete problem (8.17)–(8.18). For later reference we shall rewrite it here. Let

$$K_{\mathrm{d}} = \{\boldsymbol{x} \in \mathbb{R}^N \mid \boldsymbol{m}_h \in K_h\}, \tag{18.21}$$

where $\boldsymbol{x}$ and $\boldsymbol{m}_h$ are related through Eqs. (18.7) and (18.17), and K_h is defined in Eq. (18.2). Then

$$\lambda_h^* = \max\{\lambda \mid \exists \boldsymbol{x} \in K_{\mathrm{d}} : \boldsymbol{A}\boldsymbol{x} = \lambda \boldsymbol{b}\} \tag{18.22}$$

$$= \max_{x \in K_{\mathrm{d}}} \min_{b^{\mathrm{T}}y=1} \boldsymbol{y}^{\mathrm{T}}\boldsymbol{A}\boldsymbol{x}$$

$$= \min_{b^{\mathrm{T}}y=1} \min_{x \in K_{\mathrm{d}}} \boldsymbol{x}^{\mathrm{T}}(\boldsymbol{A}^{\mathrm{T}}\boldsymbol{y})$$

$$= \min_{b^{\mathrm{T}}y=1} D_{\mathrm{d}}(\boldsymbol{y}), \tag{18.23}$$

where

$$D_{\mathrm{d}}(\boldsymbol{y}) = \max_{x \in K_{\mathrm{d}}} \boldsymbol{x}^{\mathrm{T}}(\boldsymbol{A}^{\mathrm{T}}\boldsymbol{y}). \tag{18.24}$$

The discrete duality theorem, Theorem 8.1, applies, so we may conclude that the duality between Eqs. (18.22) and (18.23), or equivalently between Eqs. (18.3) and (18.4) holds, and that there exists a saddle point $(\boldsymbol{m}_h^*, u_h^*)$:

$$b(\boldsymbol{m}_h, u_h^*) \leqslant \lambda_h^* = b(\boldsymbol{m}_h^*, u_h^*) = b(\boldsymbol{m}_h^*, u_h) \tag{18.25}$$

for all $\boldsymbol{m}_h \in K_h$ and $u_h \in Y_h$ with $F(u_h) = 1$.

In complete analogy with Theorem 10.1 we have:

THEOREM 18.1. *Let $(\lambda^*, \boldsymbol{m}^*, u^*)$ denote an exact solution, i.e., a triple satisfying Eq. (17.10) and let $(\lambda_h^*, \boldsymbol{m}_h^*, u_h^*)$ be a discrete solution satisfying Eq. (18.25). If $\boldsymbol{m}_h^*$ satisfies the exact yield condition $\boldsymbol{m}_h^* \in K$, then*

$$b(\boldsymbol{m}_h - \boldsymbol{m}^*, u_h^*) \leqslant \lambda_h^* - \lambda^* \leqslant b(\boldsymbol{m}_h^*, u_h - u^*) \tag{18.26}$$

for all $\boldsymbol{m}_h \in K_h$ *and all* $u_h \in Y_h$ *with* $F(u_h) = 1$.

PROOF. In Eq. (17.10) insert $\boldsymbol{m} = \boldsymbol{m}_h^*$ and $u = u_h^*$ and subtract from Eq. (18.25). □

Equation (18.26) is formally identical to Eq. (10.4). However, the bilinear form $b(\boldsymbol{m}, u)$ is of higher order than $a(\boldsymbol{\sigma}, \boldsymbol{u})$ in Eq. (10.4), so a greater variety of norms may be used, and a convergence order greater than 1 is quite possible. Following CHRISTIANSEN [1980b, Section 4] we shall give the following general estimates, which are direct consequences of Eq. (18.26).

THEOREM 18.2. *Assume that* $\|\cdot\|_1$ *and* $\|\cdot\|_2$ *are norms on* X *and* Y *respectively, such that*

$$|b(\boldsymbol{m}, u)| \leqslant \|\boldsymbol{m}\|_1 \, \|u\|_2 \quad \forall (\boldsymbol{m}, u) \in X \times Y,$$

and let $(\lambda_h^*, \boldsymbol{m}_h^*, u_h^*)$ *be a sequence of discrete solutions with* $\boldsymbol{m}_h^* \in K$.

(a) *If* $\|\boldsymbol{m}_h^*\|_1 \leqslant C_1$, *then*

$$\lambda_h^* - \lambda^* \leqslant C_1 \min_{F(u_h)=1} \|u_h - u^*\|_2 .$$

(b) *If* $\|u_h^*\|_2 \leqslant C_2$, *then*

$$\lambda_h^* - \lambda^* \geqslant -C_2 \min_{m_h \in K_h} \|\boldsymbol{m}_h - \boldsymbol{m}^*\|_1 .$$

19. Solution of the discrete problem

We shall refer to the discrete static formulation (18.22) as the primal problem

$$(\mathrm{P}): \quad \begin{aligned} &\max \lambda, \\ &\boldsymbol{A}\boldsymbol{x} = \lambda \boldsymbol{b}, \\ &\boldsymbol{x} \in K_{\mathrm{d}} \end{aligned} \tag{19.1}$$

and to the discrete kinematic formulation (18.23) as the dual problem

$$(\mathrm{D}): \quad \begin{aligned} &\min D_{\mathrm{d}}(\boldsymbol{y}), \\ &\boldsymbol{b}^{\mathrm{T}} \boldsymbol{y} = 1. \end{aligned} \tag{19.2}$$

These problems are formally identical to Eqs. (11.3) and (11.4). However, the differences in the linear constraint matrix $\boldsymbol{A}$ and the yield set K_{d}, which is now bounded, are significant. In our experience the plate bending problem is considerably easier to handle numerically. Already HODGE and BELYTSCHKO [1968] and RANAWEERA and LECKIE [1970] used convex programming methods to compute limit loads for plates (bound methods), and in CHRISTIANSEN and LARSEN [1983] the conclusion is that the linear programming approach is not competitive with nonlinear methods. Since then the power of the software available for nonlinear optimization has increased significantly. For example MINOS (MURTAGH and SAUNDERS [1982]) can be applied immediately to the form Eq. (19.1) for general plate bending problems. This approach is ready for application.

Alternatively, consider the dual problem (19.2). The objective function $D_{\mathrm{d}}(\boldsymbol{y})$ is given by Eq. (18.24):

$$D_{\mathrm{d}}(\boldsymbol{y}) = \max_{x \in K_{\mathrm{d}}} \boldsymbol{x}^{\mathrm{T}}(\boldsymbol{A}^{\mathrm{T}}\boldsymbol{y}). \tag{19.3}$$

Recall that $\boldsymbol{x} = (x_n)$ is a numbering of the physically more meaningful variables (ξ_ν^{ij}) defined by Eqs. (18.7) and (18.17). In these variables we may write

$$\boldsymbol{x}^{\mathrm{T}}(\boldsymbol{A}^{\mathrm{T}}\boldsymbol{y}) = \sum_\nu \left(\xi_\nu^{11}(\boldsymbol{A}^{\mathrm{T}}\boldsymbol{y})_\nu^{11} + \xi_\nu^{22}(\boldsymbol{A}^{\mathrm{T}}\boldsymbol{y})_\nu^{22} + \xi_\nu^{12}(\boldsymbol{A}^{\mathrm{T}}\boldsymbol{y})_\nu^{12}\right). \tag{19.4}$$

In this expression

$$(\boldsymbol{A}^{\mathrm{T}}\boldsymbol{y})_\nu^{ij} = \sum_\mu b(\boldsymbol{\phi}_\nu^{ij}, \psi_\mu)\eta_\mu, \tag{19.5}$$

where the relation between $\boldsymbol{y} = (y_m)$ and (η_μ) is given by Eq. (18.8).

In order to be explicit, consider the normalized Von Mises yield condition:

$$m_{11}^2 - m_{11}m_{22} + m_{22}^2 + 3m_{12}^2 \leqslant 1. \tag{19.6}$$

Expressed in the variables (ξ_ν^{ij}) the condition is

$$(\xi_\nu^{11})^2 - \xi_\nu^{11}\xi_\nu^{22} + (\xi_\nu^{22})^2 + 3(\xi_\nu^{12})^1 \leqslant 1 \quad \text{for all nodes } \nu. \tag{19.7}$$

Thus $D_{\mathrm{d}}(\boldsymbol{y})$ in Eq. (19.3) may be computed by maximizing the sum in Eq. (19.4) subject to the constraints (19.7). Since the constraints do not couple different nodes ν, each term in the sum (19.4) is maximized. For convenience let $\boldsymbol{z} = \boldsymbol{A}^{\mathrm{T}}\boldsymbol{y} \in \mathbb{R}^N$, or in the notation of Eq. (19.5) $z_\nu^{ij} = (\boldsymbol{A}^{\mathrm{T}}\boldsymbol{y})_\nu^{ij}$. Then each term in Eq. (19.4) has the maximal value

$$\max_{\xi^{\mathrm{T}}Q\xi \leqslant 1} \left(\xi^{11}z_\nu^{11} + \xi^{22}z_\nu^{22} + \xi^{12}z_\nu^{12}\right),$$

where $\boldsymbol{\xi} = (\xi^{11}, \xi^{22}, \xi^{12})$, and Q is the symmetric positive definite matrix

$$\boldsymbol{Q} = \begin{bmatrix} 1 & -\frac{1}{2} & 0 \\ -\frac{1}{2} & 1 & 0 \\ 0 & 0 & 3 \end{bmatrix}, \qquad \boldsymbol{Q}^{-1} = \frac{1}{3}\begin{bmatrix} 4 & 2 & 0 \\ 2 & 4 & 0 \\ 0 & 0 & 1 \end{bmatrix}.$$

If we ignore boundary conditions on the moments, $D_{\mathrm{d}}(\boldsymbol{y})$ is explicitly given by the expression

$$D_{\mathrm{d}}(\boldsymbol{y}) = \sum_\nu \sqrt{\boldsymbol{z}_\nu^{\mathrm{T}}\boldsymbol{Q}^{-1}\boldsymbol{z}_\nu}$$

$$= \frac{1}{\sqrt{3}} \sum_{\nu} \sqrt{4\left((z_\nu^{11})^2 + z_\nu^{11} z_\nu^{22} + (z_\nu^{22})^2\right) + (z_\nu^{12})^2} \tag{19.8}$$

$$= \frac{1}{\sqrt{3}} \sum_{\nu} \|\boldsymbol{C} \boldsymbol{z}_\nu\|, \tag{19.9}$$

where $\boldsymbol{C}$ is the Cholesky factor of $3\boldsymbol{Q}^{-1}$:

$$\boldsymbol{C} = \begin{bmatrix} 2 & 1 & 0 \\ 0 & \sqrt{3} & 0 \\ 0 & 0 & 1 \end{bmatrix}. \tag{19.10}$$

Now let $\boldsymbol{A}_\nu$ be the $M \times 3$-matrix consisting of the three columns from $\boldsymbol{A}$ associated with the node ν, i.e., $(a_\nu)_{ij} = a_{\nu,i,j}$ for all nodes ν and $(i,j) = (1,1), (2,2), (1,2)$. Then $\boldsymbol{z}_\nu = \boldsymbol{A}_\nu^{\mathrm{T}} \boldsymbol{y}$ for all nodes ν, and Eq. (19.9) may be written

$$D_{\mathrm{d}}(\boldsymbol{y}) = \frac{1}{\sqrt{3}} \sum_{\nu} \left\| \boldsymbol{C} \boldsymbol{A}_\nu^{\mathrm{T}} \boldsymbol{y} \right\|. \tag{19.11}$$

Hence $D_{\mathrm{d}}(\boldsymbol{y})$ is the sum of Euclidian norms of the vectors $(\boldsymbol{C}\boldsymbol{A}_\nu^{\mathrm{T}})\boldsymbol{y}$. The dual problem (19.2) is to minimize this sum subject to the single linear constraint $\boldsymbol{b}^{\mathrm{T}} \boldsymbol{y} = 1$.

With the dual problem written in the form

$$\min_{b^{\mathrm{T}} y = 1} \frac{1}{\sqrt{3}} \sum_{\nu} \left\| (\boldsymbol{C}\boldsymbol{A}_\nu^{\mathrm{T}})\, \boldsymbol{y} \right\| \tag{19.12}$$

it can be solved with the algorithm in OVERTON [1983].

CHRISTIANSEN and LARSEN [1983] solved the dual problem (19.2) with the objective function given by the expression (19.8) using the algorithm GOLDFARB [1969]. This is an adaptation for linear constraints of the variable metric method and assumes differentiability of the object function, a requirement not met by the problem (19.2). For the test problem reported here, the differentiability problem was taken care of on an ad hoc basis, and the method can not be expected to work in general. The two more recent methods discussed above should be used.

In the following we report some of the results from CHRISTIANSEN and LARSEN [1983]. For a full discussion we refer to that paper. The results have also been obtained using the two more general methods described above: MINOS applied to the form (19.1) and Overton's algorithm applied to the form (19.12). For the case of a uniform load it is easy to obtain good accuracy, and the convergence order of λ_h^* as h tends to zero appears to be greater than 1. Figure 19.1 shows the collapse velocity for a simply supported square plate with a uniform load. The transversal displacement rate u^* has been multiplied by a suitable time scale, in order to visualize the deformation. The tensor of bending moments is at the yield surface at every node, i.e., the whole plate is plastified.

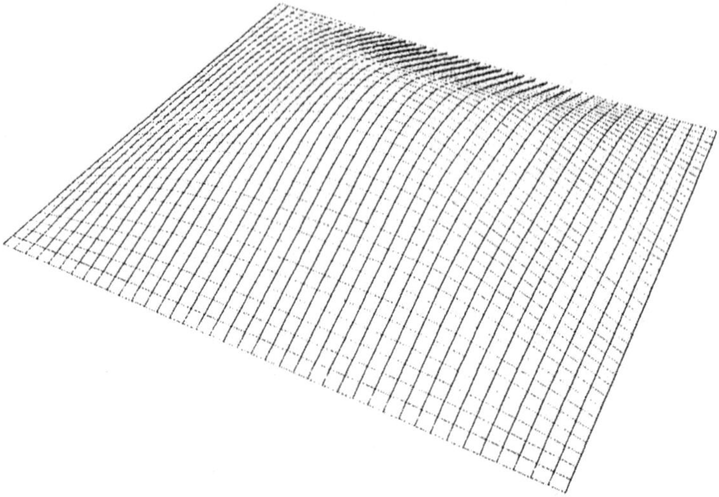

FIG. 19.1. Collapse deformation for the simply supported square plate with uniform load.

TABLE 19.1. λ_h^* for square ($L = 1$) and rectangular ($L = 2$) plates with uniform load.

h^{-1}	$L = 1$		$L = 2$	
	Simply supported	Clamped	Simply supported	Clamped
8	24.6574	42.0719	14.8363	26.1332
12	24.8634	43.0202	14.8924	26.4403
16	24.9280	43.4108	14.9116	26.5689
20	24.9590	43.6139	14.9204	26.6367
24	24.9765	43.7358	14.9253	26.6775
28	24.9873	43.8157	14.9283	26.7042
32	24.9945	43.8716		
36	24.9995	43.9124		
40	25.0031	43.9433		

The concept of a point load is in violation of the plate model, since shear in the z-direction cannot be neglected. Technically speaking, a point load does not belong to the space $W^{-1,p}(\Omega)$ of admissible loads for plastic plates. From a physical point of view a point load is the limit of loads, which are concentrated on a small finite area, which shrinks into a point, while keeping the total load fixed. The limit is independent of the actual sequence of loads. The sequence of

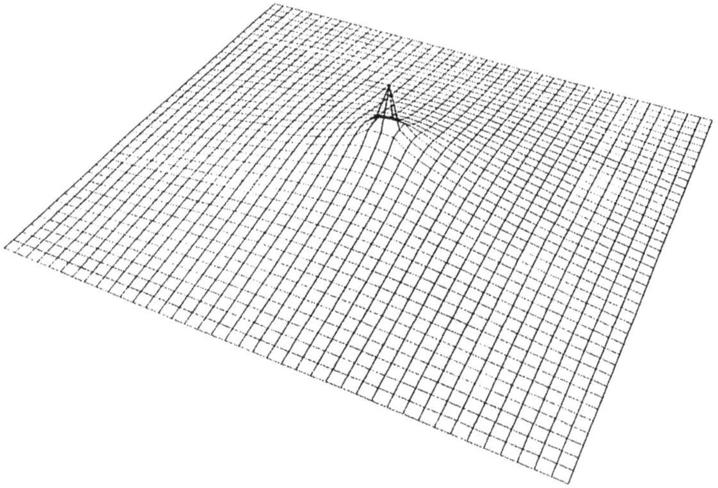

FIG. 19.2. Collapse deformation for the simply supported square plate with a point load at the center.

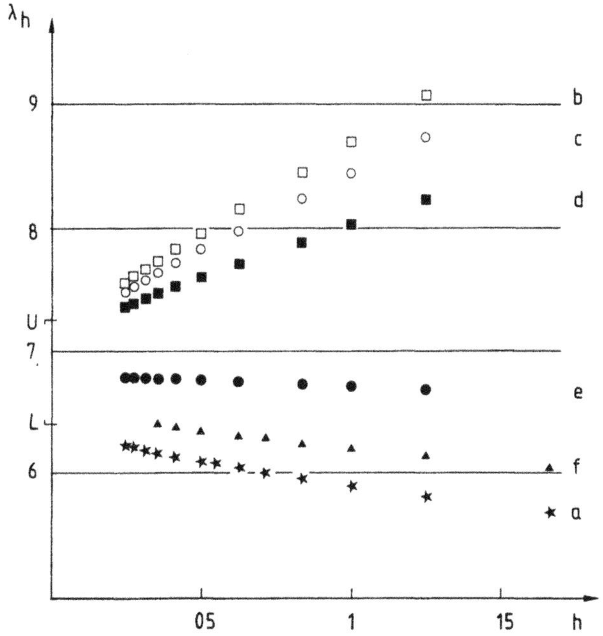

FIG. 19.3. λ_h^* for the simply supported plate with various discretizations of a point load: (a) Discrete point load. (b) Uniform load concentrated at a central square of side $2h$, i.e., over the four elements adjacent to the point. (c) Uniform load concentrated at a circle of radius h. (d) Uniform load concentrated at a square of side $h\sqrt{2}$ tilted 45 degrees. (e) Uniform load concentrated at a square of side $\frac{1}{2}h$. (f) Same as (a), but for the rectangular plate of length $L = 2$.

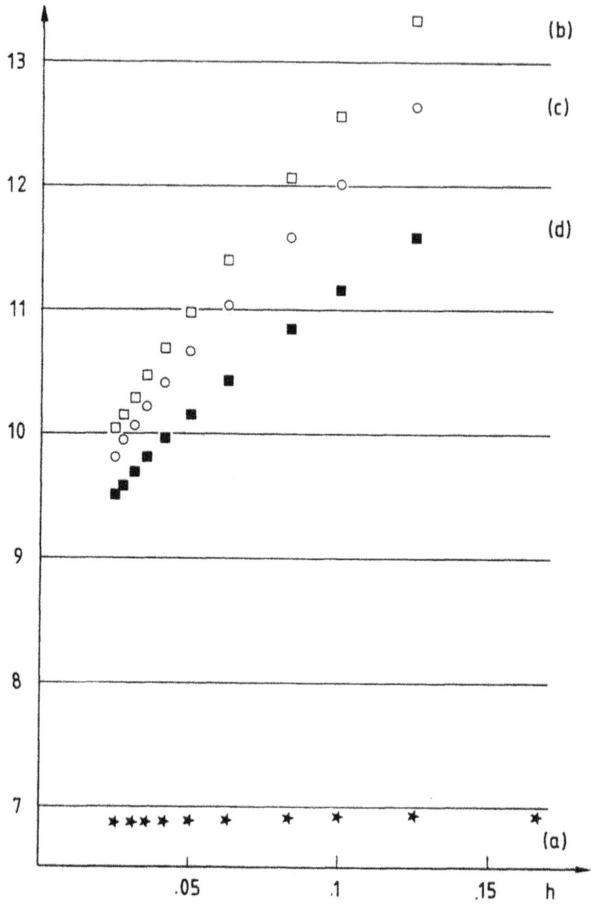

Fig. 19.4. Same as Fig. 19.3, (a)–(d), for the clamped plate. Values for the rectangular case coincides with (a).

collapse fields need not converge, and Theorem 17.2 does not apply. Even with these limitations the concept of a point load is considered useful for applications.

There are several ways to discretize a point load. One of them is a discrete point load at one node, but keeping in mind that a point load is a limit case, any sequence of discrete loads, which shrink with the grid size, will give the same limit of collapse multipliers. This can be used to obtain better approximations than with a discrete point load.

Figure 19.3 shows the discrete collapse multipliers for five discretizations of a central point load applied to the square and rectangular simply supported plate. Clearly the choice of discretization is important. The limit as $h \to 0$ appears to be the same, slightly less than seven, for the square and the rectangular plate.

The marks L and U on the λ-axis indicate the lower and upper bound for λ_h^* given in CASCIARO and DICARLO [1974].

The corresponding values for the clamped cases are shown in Fig. 19.4. The values for the rectangular plate ($L = 2$) with a discrete point load are identical to those for the square plate ((a) in Fig. 19.4). The results strongly indicate that the exact limit multiplier λ^* is the same in the four cases considered: Clamped and simply supported, square and rectangular plate.

Acknowledgement

I wish to give credit to the following students at Odense University, who have contributed to the computational experience reported here: Knud Dalgaard Andersen, Michael Bjørnskov, Allan Porse Kristiansen, Christian Nissen, Erling Kristian Purkær.

References

ADLER, I., N. KARMARKAR, M.G.C. RESENDE and G. VEIGA (1989), An implementation of Karmarkar's algorithm for linear programming, *Math. Programming* **44**, 297–335.

ANDERHEGGEN, E. and H. KNÖPFEL (1972), Finite element limit analysis using linear programming, *Internat. J. Solids and Structures* **8**, 1413–1431.

ANDERSEN, K.D. (1993), An infeasible dual affine scaling method for linear programming, *COAL Bulletin* **22**, 20–28.

ANDERSEN, K.D. (1995), An efficient Newton barrier method for minimizing a sum of Euclidean norms, *SIAM J. Optim.,* to appear.

ANDERSEN, K.D. and E. CHRISTIANSEN (1995), Limit analysis with the dual affine scaling algorithm, *J. Comput. Appl. Math.,* to appear.

ARGYRIS, J.H. (1967), Continua and discontinua, in: *Proceedings Conference on Matrix Methods in Structural Mechanics*, Wright-Patterson Air Force Base, OH 1965.

BARNES, E.R. (1986), A variation on Karmarkar's algorithm for solving linear programming problems, *Math. Programming* **36**, 174–182.

CALAMAI, P.H. and A.R. CONN (1980), A stable algorithm for solving the multifacility location problem involving Euclidean distances, *SIAM J. Sci. and Statist. Comput.* **1**, 512–526.

CALAMAI, P.H. and A.R. CONN (1982), A second order method for solving the continuous multifacility location problem, in: G.A. WATSON, ed., *Numerical Analysis: Proceedings Ninth Biennial Conference, Dundee, Scotland*, Lecture Notes in Mathematics 912 (Springer, Berlin), 1–25.

CAPURSO, M. (1971), Limit analysis of continuous media with piecewise linear yield condition, *Meccanica* **6**, 53–58.

CASCIARO, R. and A. DICARLO (1974), Mixed F.E. models in limit analysis, in: J.T. ODEN, ed., *Computational Methods in Nonlinear Mechanics* (Texas Inst. for Computational Mechanics, Austin, TX), 171–181.

CHARNES, A. and H.J. GREENBERG (1951), Plastic collapse and linear programming, *Bull. Amer. Math. Soc.* **57**, 480.

CHARNES, A., C.E. LEMKE and O.C. ZIENKIEWICZ (1959), Virtual work, linear programming and plastic limit analysis, *Proc. Roy. Soc. London Ser. A* **251**, 110–116.

CHRISTIANSEN, E. (1976), Limit analysis in plasticity; theory and approximation by finite elements, PhD. Thesis, MIT, Cambridge, MA.

CHRISTIANSEN, E. (1980a), Limit analysis in plasticity as a mathematical programming problem, *Calcolo* **17**, 41–65.

CHRISTIANSEN, E. (1980b), Limit analysis for plastic plates, *SIAM J. Math. Anal.* **11**, 514–522.

CHRISTIANSEN, E. (1981), Computation of limit loads, *Internat. J. Numer. Meth. Engrg.* **17**, 1547–1570.

CHRISTIANSEN, E. (1982), Examples of collapse solutions in limit analysis, *Utilitas Math.* **22**, 77–102.

CHRISTIANSEN, E. (1986), On the collapse solution in limit analysis, *Arch. Rational Mech. Anal.* **91**, 119–135.

CHRISTIANSEN, E. (1991), Limit analysis with unbounded convex yield condition, in: R. VICHNEVETSKY and J.J.H. MILLER, eds. *Proceedings IMACS 13th World Congress on Computational and Applied Mathematics* (Criterion Press, Dublin), 129–130.

CHRISTIANSEN, E. and K.O. KORTANEK (1990), Computing material collapse displacement fields on a CRAY X-MP/48 by the Primal Affine Scaling Algorithm, *Ann. Oper. Res.* **22**, 355–376.

CHRISTIANSEN, E. and K.O. KORTANEK (1991), Computation of the collapse state in limit analysis using the LP Primal Affine Scaling Algorithm, *J. Comput. Appl. Math.* **34**, 47–63.

CHRISTIANSEN, E. and S. LARSEN (1983), Computations in limit analysis for plastic plates, *Internat. J. Numer. Meth. Engrg.* **19**, 169–184.

CHRISTIANSEN, E. and H.G. PETERSEN (1989), Estimation of convergence orders in repeated Richardson extrapolation, *BIT* **29**, 48–59.

CIARLET, P.G. (1978), *The Finite Element Method for Elliptic Problems* (North-Holland, Amsterdam).

CIARLET, P.G. (1991), Basic error estimates for elliptic problems, in: P.G. CIARLET and J.L. LIONS, eds., *Handbook of Numerical Analysis, Vol. II* (North-Holland, Amsterdam), 17–351.

COHN, M.Z. (1979), Introduction to engineering plasticity by mathematical programming, in: M.Z. COHN and G. MAIER, eds., *Proceedings 1977 NATO Advanced Study Institute, University of Waterloo* (Pergamon Press, New York).

COULOMB, C.A. (1773), Essai sur une application des règles de maximis et minimis à quelques problèmes de statique relatifs a l'architecture, *Mém. Math. Phys. Acad. Roy. Sci.* **7**, 343–382.

DEMENGEL, F. (1989), Compactness theorems for spaces of functions with bounded derivatives and applications to limit analysis problems in plasticity, *Arch. Rational Mech. Anal.* **105**, 123–161.

DIKIN, I.I. (1967), Iterative solution of problems of linear and quadratic programming, *Soviet Math. Dokl.* **8**, 674–675.

DRUCKER, D.C, W. PRAGER and H.J. GREENBERG (1952), Extended limit design theorems for continuous media, *Quart. Appl. Math.* **9**, 381–389.

DUVAUT, G. and J.L. LIONS (1972), Les Inéquations en Mécanique et en Physique (Dunod, Paris). [English translation: Inequalities in Mechanics and Physics (Springer, Berlin, 1976)].

FOIAS, C. and R. TEMAM (1978), Remarques sur les équations de Navier-Stokes stationnaires et les phénomènes successifs de bifurcations, *Ann. Scuola Norm. Sup. Pisa Cl. Sci. (4)* **5**, 29–63.

GAUDRAT, V.F. (1991), A Newton type algorithm for plastic limit analysis, *Comput. Methods. Appl. Mech. Engrng.* **88**, 207–224.

GOLDFARB, D. (1969), Extension of Davidon's variable metric method to maximization under linear inequality and equality constraints, *SIAM J. Appl. Math.* **17**, 739–764.

GREENBERG, H.J. and W. PRAGER (1952), On limit design of beams and frames, *Trans. Amer. Soc. Civil Engrg.* **117**, 447.

GVOZDEV, A.A. (1938), The determination of the value of the collapse load for statically indeterminate systems undergoing plastic deformation, in: *Proceedings of the Conference on Plastic Deformations* Dec. 1936, Akademiia Nauk SSSR, Moscow-Leningrad. [*Internat. J. Mech. Sci.* **1**, 322 (1960)].

HAYES, D.G. and P.V. MARCAL (1964), Determination of upper bounds for problems in plane stress using finite element techniques, *Internat. J. Mech. Sci.* **9**, 245–251.

HILL, R. (1950), *The Mathematical Theory of Plasticity* (Oxford University Press, Oxford).

HODGE, P.G. (1959), *Plastic Analysis of Structures* (McGraw-Hill, New York).

HODGE, P.G. and T. BELYTSCHKO (1968), Numerical methods for the limit analysis of plates, *J. Appl. Mech.* **35**, 796–802.

KACHANOV, L.M. (1971), *Foundations of the Theory of Plasticity* (American Elsevier, New York).

KALISZKY, S. (1975), *Képlékenységtan. Elmélet és mérnöki alkalmazások* (Akadémiai Kiadó, Budapest). [English translation: *Plasticity. Theory and Engineering Applications* (Elsevier, Amsterdam, 1989)].

KARMARKAR, N.K. (1984), A new polynomial-time algorithm for linear programming, *Combinatorica* **4**, 373–395.

KAZINCZY, G. (1914), Bemessung von statisch unbestimmten Konstruktionen unter Berücksichtigung der bleibenden Formänderungen, *Betonszemle* 4, 5, 6.

KIST, N.C. (1917), Leidt een sterkteberekening, die uitgaat van de evenredigheid van kracht en vormverandering, tot een goede constructie van ijzeren bruggen en gebouwen? Inaugural Dissertation, Polytechnic Institute, Delft, Holland.

KOHN, R. and R. TEMAM (1983), Dual spaces of stresses and strains, with applications to Hencky plasticity, *Appl. Math. Optim.* **10**, 1–35.

KORTANEK, K.O. and S. MIAOGEN (1987), Convergence results and numerical experiments on a linear programming hybrid algorithm, *European J. Oper. Res.* **32**, 47–61.

MARTIN, J.B. (1975), *Plasticity* (MIT Press, Cambridge, MA).

MATTHIES, H., G. STRANG and E. CHRISTIANSEN (1979), The saddle point of a differential program, in: R. GLOWINSKI, E. RODIN and O.C. ZIENKIEWICZ, eds., *Energy Methods in Finite Element Analysis* (Wiley, New York).

MOREAU, J.J. (1976), Application of convex analysis to the treatment of elasto-plastic systems, in: P. GERMAIN and B. NAYROLES, eds., *Applications of Methods of Functional Analysis to Problems of Mechanics*, Lecture Notes in Mathematics 503 (Springer, Berlin), 56–89.

MURTAGH, B.A. and M.A. SAUNDERS (1982), A projected Lagrangian algorithm and its implementation for sparse nonlinear constraints, in: A.G. BUCKLEY and J.-L. GOFFIN, eds., *Algorithms for Constrained Minimization of Smooth Nonlinear Functions, Math. Programming Study* **16** (North-Holland, Amsterdam), 84–117.

NAYROLES, B. (1970), Essai de théorie fonctionnelle des structures rigides plastiques parfaites, *J. Mécan.* **9**, 491–506.

NEAL, B.G. (1956), *The Plastic Methods of Structural Analysis* (Chapman and Hall, London).

NEČAS, J. (1966), Sur les normes équivalentes dans $W_p^k(\Omega)$ et sur la coercivité des formes formellement positives, in: *Équations aux Dérivées Partielles* (Presses de l'Université de Montréal, Montréal).

NEČAS, J. (1967), *Les Méthodes Directes en Théorie des Équations Elliptiques* (Masson, Paris).

NEČAS, J. and I. HLAVÁČEK (1981), *Mathematical Theory of Elastic and Elasto-Plastic Bodies: An Introduction* (Elsevier, Amsterdam).

OVERTON, M.L. (1983), A quadratically convergent method for minimizing a sum of Euclidean norms, *Math. Programming* **27**, 34–63.

OVERTON, M.L. (1984), Numerical solution of a model problem from collapse load analysis, in: R. GLOWINSKI and J.L. LIONS eds., *INRIA Conf. Comput. Methods Engrg. Appl. Sci.* (North-Holland, Amsterdam), 421–437.

RANAWEERA, M.P. and F.A. LECKIE (1970), Bound methods in limit analysis, in: H. TOTTENHAM and C. BREBBIA, eds., *Finite Element Techniques in Structural Mechanics* (Stress Analysis Publishers, Southampton).

ROBINSON, S.M. (1972), A quadratically convergent algorithm for general nonlinear programming problems, *Math. Programming* **3**, 145–156.

SAVE, M. (1979), Fundamentals of rigid-plastic analysis and design, in: M.Z. COHN and G. MAIER, eds., *Proceedings 1977 NATO Advanced Study Institute, University of Waterloo* (Pergamon Press, New York).

SHOEMAKER, E.M. (1979), On nonexistence of collapse solutions in rigid-perfect plasticity, *Utilitas Math.* **16**, 3–13.

STRANG, G. (1979a), A family of model problems in plasticity, in: R. GLOWINSKI and J.L. LIONS, eds., *Computing Methods in Applied Sciences and Engineering*, Lecture Notes in Computer Sciences 704 (Springer, Berlin), 292–305.

STRANG, G. (1979b), A minimax problem in plasticity theory, in: M.Z. NASHED ed., *Functional Analysis Methods in Numerical Analysis*, Lecture Notes in Mathematics 701 (Springer, Berlin), 319–333.

STRANG, G. and G.J. FIX (1973), *An Analysis of the Finite Element Method*, (Prentice Hall, Englewood Cliffs, NJ).

STRANG, G. and R. TEMAM (1982), A problem in capillarity and plasticity, *Math. Programming Stud.* **17**, 91–102.

SUQUET, P.M. (1978), Sur un nouveau cadre fonctionnel pour les équations de la plasticité, *C.R. Acad. Sci. Paris Sér. A* **286**, 1129–1132.

SUQUET, P.M. (1979), Un espace fonctionnel pour les équations de la plasticité, *Ann. Fac. Sci. Toulouse* **1**, 77–87.

SUQUET, P.M. (1980), Existence and regularity of solutions for plasticity problems, in: S. NEMAT NASSER, ed., *Variational Methods in the Mechanics of Solids* (Bergman Press, New York).

Suquet, P.M. (1981), Sur les équations de la plasticité: Existence et régularité des solutions, *J. Méc.* **20**, 3–39.

Temam, R. (1977), *Navier-Stokes Equations* (North-Holland, Amsterdam).

Temam, R. (1981a), Existence theorems for the variational problems of plasticity, in: M. Atteia, D. Bancel and I. Gumowski, eds., *Nonlinear Problems of Analysis in Geometry and Mechanics* (Pitman, London).

Temam, R. (1981b), On the continuity of the trace of vector functions with bounded deformation, *Appl. Anal.* **11**, 291–302.

Temam, R. (1983), *Problèmes Mathématiques en Plasticité* (Gauthier-Villars, Paris).

Temam, R. and G. Strang (1980a), Functions of bounded deformation, *Arch. Rational Mech. Anal.* **75**, 7–21.

Temam, R. and G. Strang (1980b), Duality and relaxation in the variational problems of plasticity, *J. Méc.* **19**, 493–527.

Tsuchiya, T. (1991), Global convergence property of the affine scaling methods for degenerate linear programming problems, *Math. Programming* **52**, 377–404.

Vanderbei, R.J., M.S. Meketon and B.A. Freedman (1986), A modification of Karmarkar's algorithm, *Algorithmica* **1**, 395–407.

Zavelani-Rossi, A. (1979), Collapse load analysis of discretized continua, in: M.Z. Cohn and G. Maier, eds., *Proceedings 1977 NATO Advanced Study Institute, University of Waterloo* (Pergamon Press, New York).

List of Symbols

Continuum mechanics notation

Ω	domain occupied by material.
$\overline{\Omega}$	closure of Ω.
$\partial\Omega$	the boundary of Ω.
N	spatial dimension of problem (and of Ω).
S	fixed part of the boundary $\partial\Omega$.
T	loaded part of the boundary $\partial\Omega$.
$\boldsymbol{\nu}$	outward normal to Ω.
$\boldsymbol{f} = \boldsymbol{f}(\boldsymbol{x})$	field of volume forces in Ω.
$\boldsymbol{g} = \boldsymbol{g}(\boldsymbol{x})$	field of surface forces on T.
$\boldsymbol{u} = \boldsymbol{u}(\boldsymbol{x})$	field of plastic flow (displacement rates).
U	space of plastic flow fields $\boldsymbol{u}$.
U_0	subspace of U with divergence zero.
$\boldsymbol{u}^{\Omega}$	flow in the interior of Ω.
$\boldsymbol{u}^{T}$	flow on the loaded boundary $T \subseteq \Omega$.
$\boldsymbol{\varepsilon} = \boldsymbol{\varepsilon}(\boldsymbol{u})$	strain rate tensor with components ε_{ij}.
$\boldsymbol{\sigma} = \boldsymbol{\sigma}(\boldsymbol{x})$	field of stresses with components σ_{ij}.
$\sigma_{\mathrm{I}}, \sigma_{\mathrm{II}}, \sigma_{\mathrm{III}}$	the principal stresses.
$\mathrm{tr}(\boldsymbol{\sigma})$	trace of $\boldsymbol{\sigma}$, i.e. the sum of diagonal elements.
$\boldsymbol{\sigma}^{\mathrm{D}}$	deviator of the tensor $\boldsymbol{\sigma}$.
$\boldsymbol{I}$	the identity tensor with components (δ_{ij}).
Σ	space of stress tensor fields $\boldsymbol{\sigma}$.
Σ_0	quotient space $\Sigma/\{\varphi\boldsymbol{I} \mid \varphi \in L^1(\Omega)\}$.
$\tilde{\boldsymbol{\sigma}}$	elements of the quotient space Σ_0.
$B(\boldsymbol{x})$	set of admissible stress tensors at point $\boldsymbol{x} \in \Omega$.
$\boldsymbol{\sigma}(\boldsymbol{x}) \in B(\boldsymbol{x})$	yield condition at point $\boldsymbol{x} \in \Omega$.
K	set of stress fields satisfying the yield condition at every point.
K_0	quotient set $K/\{\varphi\boldsymbol{I} \mid \varphi \in L^1(\Omega)\} \subseteq \Sigma_0$.
$F(\boldsymbol{u})$	work rate for external forces.
$a(\boldsymbol{\sigma}, \boldsymbol{u})$	internal work rate in the material.
$D(\boldsymbol{u})$	total energy dissipation rate associated with $\boldsymbol{u}$.
λ	load multiplier.
$\lambda^*, \boldsymbol{\sigma}^*, \boldsymbol{u}^*$	$\lambda, \boldsymbol{\sigma}, \boldsymbol{u}$ as above in the collapse state.
$\boldsymbol{\nabla} \cdot \boldsymbol{u}$	divergence of vector field ($\boldsymbol{\nabla}$ is also applied to tensors).
$\boldsymbol{u} \otimes \boldsymbol{v}$	tensor with components $u_i v_j$.

$\boldsymbol{m}$	bending moments in plate problems.
u	transversal displacement rate for plates.
$b(\boldsymbol{m}, u)$	internal work rate for plates.
$m_{\mathrm{I}}, m_{\mathrm{II}}$	principal bending moments.
X	the space of plate bending moments.
Y	the space of transversal displacement rates for plates.
f	the transversal load distribution for plates.

Discrete variables

In general subscript 'h' indicates discrete variables.

U_h	space of discrete displacement rate fields $\boldsymbol{u}_h$.
Σ_h	space of discrete stress tensors $\boldsymbol{\sigma}_h$.
K_h	discrete stress tensors $\boldsymbol{\sigma}_h$, satisfying the yield condition at all nodes.
$D_h(\boldsymbol{u}_h)$	discrete analog of $D(\boldsymbol{u})$, usually different from $D(\boldsymbol{u}_h)$.
λ_h^*	the discrete limit multiplier.
$\mathcal{T}_h$	finite element triangulation of Ω.
$\mathcal{N}_\sigma$	set of nodes for the single components of $\boldsymbol{\sigma}_h$.
$\mathcal{N}_u$	set of nodes for the single components of $\boldsymbol{u}_h$.
$\boldsymbol{\phi}_\nu^{ij}$	basis functions for Σ_h corresponding to the node ν.
ξ_ν^{ij}	coefficients of $\boldsymbol{\sigma}_h$ w.r.t. the basis $\{\boldsymbol{\phi}_\nu^{ij}\}$.
$\boldsymbol{\psi}_\mu^k$	basis functions for U_h corresponding to the node μ.
η_μ^k	coefficients of $\boldsymbol{u}_h$ w.r.t. the basis $\boldsymbol{\psi}_\mu^k$.
$\boldsymbol{x} = (x_n)_{n=1}^N$	vector, whose coordinates are $\{\boldsymbol{\xi}_\nu^{ij}\}$ in some linear ordering.
$\boldsymbol{y} = (y_m)_{m=1}^M$	vector, whose coordinates are $\{\boldsymbol{\eta}_\nu^k\}$ in some linear ordering.
$\boldsymbol{b} \in \mathbb{R}^M$	vector with coordinates $b_m = F(\boldsymbol{\psi}_\mu^k)$.
$\boldsymbol{b}^{\mathrm{T}}\boldsymbol{y}$	the discrete expression for $F(\boldsymbol{u}_h)$.
$\boldsymbol{A}$	$M \times N$ 'stiffness' matrix with entries $a_{mn} = a(\boldsymbol{\phi}_\nu^{ij}, \boldsymbol{\psi}_\mu^k)$.
$\boldsymbol{y}^{\mathrm{T}}\boldsymbol{A}\boldsymbol{x}$	the discrete expression for $a(\boldsymbol{\sigma}_h, \boldsymbol{u}_h)$.
$D_{\mathrm{d}}(\boldsymbol{y})$	the discrete expression for $D_h(\boldsymbol{u}_h)$.
K_{d}	$\{\boldsymbol{x} \in \mathbb{R}^N \mid \boldsymbol{\sigma}_h \in K_h\}$.
$\boldsymbol{\Psi}_h$	the discrete stream function.
$\boldsymbol{A}_\nu$	the matrix consisting of the columns of $\boldsymbol{A}$, which are associated with the node ν.

Function spaces and norms

$C^m(\Omega)$	m times continuously differentiable functions in Ω.
$C^m(\overline{\Omega})$	subspace of $C^m(\Omega)$ with continuous derivatives in $\overline{\Omega}$.
$C_0^\infty(\Omega)$	subset of $C^\infty(\Omega)$ with compact support in Ω.

$L^p(\Omega)$	the measurable functions f in Ω with $\int_\Omega |f|^p < \infty$.
$\|f\|_p$	$\left(\int_\Omega |f|^p\right)^{1/p}$ (the norm in $L^p(\Omega)$).
$W^{m,p}(\Omega)$	functions in $L^p(\Omega)$ with all derivatives of order $\leqslant m$ in $L^p(\Omega)$.
$\|\cdot\|_{W^{m,p}}$	norm on $W^{m,p}(\Omega)$.
$W_0^{m,p}(\Omega)$	closure of $C_0^\infty(\Omega)$ in $W^{m,p}(\Omega)$.
$M(\Omega)$	the space of bounded measures in Ω.
$C(\overline{\Omega})^*$	$= \left(C^0(\overline{\Omega})\right)^*$: the space of bounded measures in $\overline{\Omega}$.
$\mathrm{BD}(\Omega)$	the space of bounded deformation in Ω.
$\boldsymbol{\gamma}(\boldsymbol{u})$	trace ("restriction") of $\boldsymbol{u} \in \mathrm{BD}(\Omega)$ to $[L^1(\partial\Omega)]^N$.
V^N	vector functions with N components from a space V.
$\|\cdot\|_V$	norm on the vector space V.
V^*	the dual of the normed space $(V, \|\cdot\|_V)$.
$\langle\cdot,\cdot\rangle_{V\times V^*}$	duality between V and V^*.

Subject Index

Numerical Methods for Unilateral Problems in Solid Mechanics

J. Haslinger

Faculty of Mathematics and Physics
Charles University
Ke Karlovu 5
120 00 Praha 2
Czech Republic

I. Hlaváček

Mathematical Institute
Academy of Sciences of the Czech Republic
Žitná 25
111 67 Praha 1
Czech Republic

J. Nečas

Faculty of Mathematics and Physics
Charles University
Sokolovská 83
186 00 Praha 8
Czech Republic

HANDBOOK OF NUMERICAL ANALYSIS, VOL. IV
Finite Element Methods (Part 2)—Numerical Methods for Solids (Part 2)
Edited by P.G. Ciarlet and J.L. Lions

Contents

Preface

In both engineering and physics literature we find many problems solved by intuitive, ad hoc methods, regardless of the fact that the possibility exists of formulating and solving these problems in the framework of the theory of variational inequalities, and thus penetrating more profoundly to the very core of the problems.

The theory of variational inequalities is a relatively young mathematical discipline. Apparently, the original impulse came from unilateral problems in solid mechanics, specifically, from the paper by FICHERA [1964] on the solution of the *Signorini* problem in the theory of elasticity. Later, LIONS and STAMPACCHIA [1967] laid the foundations of the theory itself.

Time-dependent variational inequalities have primarily been treated in the work of J.L. Lions and H. Brezis. The diverse applications of variational inequalities theory are the topics of a well-known monograph by DUVAUT and LIONS [1976]. Numerical analysis of various problems formulated in terms of variational inequalities was presented in the book by GLOWINSKI, LIONS and TRÉMOLIÈRES [1981].

The purpose of this article is to survey the numerical analysis of variational inequalities, which stem from solid mechanics. The article comprises six chapters.

In Chapter I we consider elliptic second order problems, where the unknown is a single real scalar function in a given domain of an n-dimensional space $\mathbb{R}^n$, $n = 2, 3$. Using simple models we illustrate the primal, mixed and dual variational formulations. Approximations of the above-mentioned problems are studied in Chapter II.

Chapter III contains an analysis of the one-sided contact of elastic bodies, both without friction and with friction of Coulomb's type. An extension of the contact problems to some elasto-plastic bodies is presented in Chapter IV. From various mathematical models of elasto-plastic bodies we choose the following: perfect plasticity and plasticity with strain hardening. Both existence problems and approximations via finite-element methods are studied in Chapter V.

Chapter VI is devoted to the bending of elastic plates with unilateral obstacles either on the boundary or in the interior of the domain occupied by the plate.

This article is a revised and updated edition of some parts of our book, Numerical Solution of Variational Inequalities (Springer-Verlag, New York, 1988).

We are aware that the contents of the article are not as exhaustive as its title may promise. We have tried to include the most significant mathematical results in the field up to the year 1992, hoping that our choice is complete enough to cover at least the main developmental trends.

CHAPTER I

Variational Inequalities of the Elliptic Type. Dual and Mixed Variational Approach

We start with several physical problems where mathematical modelling is based on a simple variational inequality for a scalar quantity. A finite element approximation of the model is described. Then, we summarize known results on the existence and uniqueness of the solution based on variational inequalities. Introducing the so-called inequality of the second kind, we study the continuous dependence of its solution on variations of convex, lower continuous functionals. This enables us to unify the penalty and regularization approaches, frequently used for the numerical treatment of these problems. Great attention is devoted to the duality approach, by means of which we can derive alternative variational formulations. We end the chapter by presenting existence results for the so-called semicoercive inequalities.

1. Examples of unilateral boundary value problems

EXAMPLE 1.1. Let us assume a homogeneous membrane represented by a domain $\Omega \subset \mathbb{R}^2$, loaded by a force f in the vertical direction. Let $v(x,y)$ denote the deflection of the membrane at a point $(x,y) \in \Omega$. We shall assume that the potential energy of the deformed membrane is proportional to the change of the area of its surface, i.e. the potential energy $P(v)$ is given by ($v_x = \partial v/\partial x, v_y = \partial v/\partial y$):

$$P(v) = \int_\Omega \sqrt{1 + v_x^2 + v_y^2}\, \mathrm{d}x\, \mathrm{d}y - \operatorname{meas}\Omega \approx \tfrac{1}{2} \int_\Omega |\nabla v|^2\, \mathrm{d}x\, \mathrm{d}y,$$

when higher order terms are neglected.

The work of external forces, corresponding to v is done by

$$E(v) = \int_\Omega f v\, \mathrm{d}x\, \mathrm{d}y$$

and the total potential energy $T(v) = P(v) - E(v)$. From the Lagrange principle of minimizing the total potential energy, the equilibrium state of the membrane is realized by a function u minimizing T over a class of functions v with finite energy, i.e. $T(v) \in \mathbb{R}^1$ and with a prescribed value g on the boundary $\partial\Omega$. More precisely

$$u \in V \text{ such that } T(u) \leqslant T(v) \quad \forall v \in V, \tag{1.1}$$

where

$$V = \{v \in H^1(\Omega) \,|\, v = g \text{ on } \partial\Omega\}.$$

Writing down formally Euler's equation for (1.1), we arrive at the following Dirichlet boundary value problem for the deflection u:

$$\begin{aligned} \Delta u \equiv \frac{\partial^2 u}{\partial x^2} + \frac{\partial^2 u}{\partial y^2} &= -f \quad \text{in } \Omega, \\ u = g \quad &\text{on } \partial\Omega. \end{aligned} \tag{1.2}$$

Let us assume now that the deflection of the membrane is restricted from below by a rigid obstacle represented by a body, occupying the set $Q = \{[x, y, z] \in \mathbb{R}^3 \,|\, z \leqslant \psi(x, y)\}$. The function ψ which represents the surface of the obstacle is such that $\psi \leqslant g$ on $\partial\Omega$. The set of admissible deflections is now given as follows:

$$K = \{v \in V \,|\, v \geqslant \psi \text{ a.e. in } \Omega\}.$$

Let ψ be such that K is non-empty. It is easy to verify that K is convex, closed subset of V, which is no longer a linear set (if v_1, $v_2 \in K$ and α, $\beta \in \mathbb{R}^1$ then $\alpha v_1 + \beta v_2 \notin K$, in general). According to the Lagrange principle of minimizing the potential energy, the deflection u corresponding to the equilibrium state is characterized by

$$u \in K\colon\ T(u) \leqslant T(v) \text{ for any } v \in K, \tag{1.3}$$

where T is the same as before. Then (1.3) is equivalent to (see CIARLET [1991, Volume II, p. 25]):

$$\text{find } u \in K\colon\ (\nabla u, \nabla v - \nabla u)_{0,\Omega} \geqslant (f, v - u)_{0,\Omega} \quad \forall v \in K \tag{1.4}$$

and (1.4) has a unique solution for any $f \in L^2(\Omega)$ (see CIARLET [1991, Volume II, Theorems 1.1 and 1.2, pp. 24–25]).

Let us suppose that the solution $u \in H^2(\Omega) \cap K$. Then applying Green's formula to the left hand side of (1.4), we obtain

$$\int_\Omega -\Delta u(v - u)\,\mathrm{d}x\,\mathrm{d}y + \int_{\partial\Omega} \frac{\partial u}{\partial n}(v - u)\,\mathrm{d}s \geqslant \int_\Omega f(v - u)\,\mathrm{d}x\,\mathrm{d}y \quad \forall v \in K. \tag{1.5}$$

As $v - u = 0$ on $\partial\Omega$, the integral along the boundary $\partial\Omega$ vanishes. Let $\varphi \in C_0^\infty(\Omega)$ be a nonnegative function in Ω. Then $v = u + t\varphi \in K$ for any $t \geqslant 0$. Substituting this element into (1.5) we get

$$\int_\Omega -\Delta u \varphi \, dx \, dy \geqslant \int_\Omega f \varphi \, dx \, dy$$

for any $\varphi \in C_0^\infty(\Omega)$, $\varphi \geqslant 0$ in Ω. Hence

$$-\Delta u \geqslant f \quad \text{a.e. in } \Omega.$$

The domain Ω can now be divided as follows:

$$\Omega = \Omega_0 \cup \Omega_+,$$

where

$$\Omega_0 = \{A \in \Omega \mid u(A) = \psi(A)\},$$

$$\Omega_+ = \{A \in \Omega \mid u(A) > \psi(A)\}.$$

Let us assume that $\psi \in C(\overline{\Omega})$. As $H^2(\Omega) \hookrightarrow C(\overline{\Omega})$, the set Ω_+ is open.

Let $\tilde{A} \in \Omega_+$ be given. Then there exists a neighbourhood $U_\delta(\tilde{A}) \subset \Omega_+$. If $\varphi \in C_0^\infty(U_\delta(\tilde{A}))$, then the function $v = u \pm t\varphi$ belongs to K provided $t > 0$ is sufficiently small. Substituting v into (1.5), we obtain

$$(\pm t) \int_{U_\delta(\tilde{A})} -\Delta u \varphi \, dx \, dy \geqslant \pm t \int_{U_\delta(\tilde{A})} f \varphi \, dx \, dy$$

which holds for any $\varphi \in C_0^\infty(U_\delta(\tilde{A}))$. Therefore

$$-\Delta u = f \quad \text{a.e. in } U_\delta(\tilde{A})$$

and also a.e. in Ω_+.

Summing up, we have proved that $u \in H^2(\Omega) \cap K$, being the solution of (1.4), satisfies the following set of relations

$$\begin{aligned} &-\Delta u \geqslant f \quad \text{a.e. in } \Omega, \\ &u \geqslant \psi \qquad \text{a.e. in } \Omega, \\ &\text{if } u(A) > \psi(A) \text{ then } -\Delta u(A) = f(A). \end{aligned} \tag{1.6}$$

Instead of (1.6), we can write also

$$\begin{aligned} &-\Delta u \geqslant f, \; u \geqslant \psi \; \in \Omega, \\ &(u - \psi)(-\Delta u - f) = 0 \quad \text{a.e. in } \Omega. \end{aligned}$$

REMARK 1.1. The sets Ω_0 and Ω_+ are called the *coincidence* and the *noncoincidence* sets, respectively. The boundary of the noncoincidence set in Ω,

$$\Gamma = \partial\Omega_+ \cap \Omega$$

is called the *free boundary* of the problem. Let us mention that the partition of Ω into Ω_0 and Ω_+ is not known a priori. Hence Γ is one of the unknowns of the problem (a detailed study of these problems can be found in FRIEDMAN [1982], KINDERLEHLER and STAMPACHIA [1980]).

EXAMPLE 1.2. In the present example, the constraint will be prescribed only on a part of the boundary instead of on the whole domain.

Let us assume a semi-permeable wall of negligible thickness which allows the fluid to escape freely out of $\Omega \subset \mathbb{R}^2$ but prevents inflow of the fluid. Let u and h be the inside and outside pressure, respectively.

Assume $x = (x_1, x_2) \in \Gamma$ on the semipermeable part of the boundary. Then

(i) if $u(x) < h(x)$ then $\partial u(x)/\partial n = 0$, i.e. there is no outflow;

(ii) if $u(x) \geqslant h(x)$ then $\partial u(x)/\partial n \leqslant 0$, i.e. there may be outflow at $x \in \Gamma$. The assumption on the thickness of the wall results in the fact that u cannot be greater than h at any point $x \in \Gamma$ (see DUVAUT and LIONS [1976]). If we assume the stationary case, only, the inside pressure u satisfies the following conditions:

$$\begin{aligned} &-\Delta u = f && \text{in } \Omega \\ &u = 0 && \text{on } \Gamma_1 \\ &u \leqslant h, \ \frac{\partial u}{\partial n} \leqslant 0 && \text{on } \Gamma_2 \\ &u(x) < h(x) \Rightarrow \frac{\partial u}{\partial n}(x) = 0 \quad (x \in \Gamma_2). \end{aligned} \tag{1.7}$$

Here Γ_1 and Γ_2 are nonempty, disjoint and open parts of $\partial\Omega = \overline{\Gamma}_1 \cup \overline{\Gamma}_2$ with Γ_2 representing the semi-permeable wall.

Next we shall give the variational formulation of (1.7). To this end we introduce

$$V = \{v \in H^1(\Omega) \,|\, v = 0 \text{ a.e. on } \Gamma_1\},$$
$$K = \{v \in V \,|\, v \leqslant h \text{ a.e. on } \Gamma_2\}.$$

Let us suppose that h defined a.e. on Γ_2 is such that K is nonempty. As in the previous example one can show that K is the closed and convex subset of V.

Let us consider the problem:

$$\text{find } u \in K \text{ such that } (\nabla u, \nabla v - \nabla u)_{0,\Omega} \geqslant (f, v - u)_{0,\Omega} \quad \forall v \in K. \tag{1.8}$$

Arguing in the same way as before, we can prove the existence and the uniqueness of the u, satisfying (1.8) and the equivalence of (1.8) with the problem:

$$\text{find } u \in K \text{ such that } J(u) \leqslant J(v) \quad \forall v \in K, \tag{1.8'}$$

where $J(v) = \frac{1}{2}(\nabla v, \nabla v)_{0,\Omega} - (f, v)_{0,\Omega}$.

Let the solution $u \in H^2(\Omega) \cap K$. Then applying Green's formula to the left hand side of (1.8) and assuming the boundary conditions on Γ_1, we obtain ($\mathrm{d}x = \mathrm{d}x_1\,\mathrm{d}x_2$):

$$\int_{\Omega} (-\Delta u)(v-u)\,\mathrm{d}x + \int_{\Gamma_2} \frac{\partial u}{\partial n}(v-u)\,\mathrm{d}s \geqslant \int_{\Omega} f(v-u)\,\mathrm{d}x \quad \forall v \in K. \tag{1.9}$$

Substituting $v = u \pm t\varphi$ into (1.9) with $\varphi \in C_0^\infty(\Omega)$ we see that

$$\pm t \int_{\Omega} (-\Delta u)\varphi\,\mathrm{d}x = \pm t \int_{\Omega} f\varphi\,\mathrm{d}x \; \forall \varphi \in C_0^\infty(\Omega). \tag{1.10}$$

Hence $-\Delta u = f$ a.e. in Ω.

This and (1.9) yield

$$\int_{\Gamma_2} \frac{\partial u}{\partial n}(v-u)\,\mathrm{d}s \geqslant 0 \quad \forall v \in K. \tag{1.11}$$

Let $\varphi \in C^\infty(\overline{\Omega})$ be such that $\varphi \leqslant 0$ on Γ_2 and $\varphi = 0$ on Γ_1. Then the function $v = u + t\varphi \in K$ for any $t > 0$. Its substitution into (1.11) leads to

$$\int_{\Gamma_2} \frac{\partial u}{\partial n}\varphi\,\mathrm{d}s \geqslant 0 \Longleftrightarrow \frac{\partial u}{\partial n} \leqslant 0 \text{ a.e. on } \Gamma_2. \tag{1.12}$$

The part Γ_2 can be divided into two parts: $\Gamma_2 = \Gamma_{20} \cup \Gamma_{2+}$, where

$$\Gamma_{20} = \{x \in \Gamma_2 \mid u(x) = h(x)\}, \qquad \Gamma_{2+} = \{x \in \Gamma_2 \mid u(x) < h(x)\}.$$

Since $u \in H^2(\Omega) \hookrightarrow C(\overline{\Omega})$ because of our assumption, we see that Γ_{2+} is open in $\partial\Omega$ assuming that h is continuous. Let $\tilde{x} \in \Gamma_{2+}$. Then there exists $\delta > 0$ such that a neighbourhood $U_\delta(\tilde{x}) \cap \partial\Omega \subset \Gamma_{2+}$. Let $\varphi \in C^\infty(\overline{\Omega})$ be such that $\operatorname{supp}\varphi|_{\partial\Omega} \subset U_\delta(\tilde{x}) \cap \partial\Omega$. Then $v = u \pm t\varphi$ belongs to K for any $t > 0$ sufficiently small. With such a choice of v, (1.11) leads to

$$\int_{U_\delta(\tilde{x})\cap\partial\Omega} \frac{\partial u}{\partial n}\varphi\,\mathrm{d}s = 0 \quad \forall \varphi \in C^\infty(\overline{\Omega}),\ \operatorname{supp}\varphi|_{\partial\Omega} \subset U_\delta(\tilde{x}) \cap \partial\Omega.$$

Therefore $\partial u/\partial n = 0$ a.e. in $U_\delta(\tilde{x}) \cap \partial\Omega$ and hence on the whole Γ_{2+}.

Let us mention that the partition of Γ_2 into Γ_{20} and Γ_{2+} is one of the unknowns of our problem.

More interesting applications of our problem involving an obstacle on the boundary can arise in solid mechanics, such as contact problems involving deformable bodies (see later).

EXAMPLE 1.3. Many physical problems are described by an elliptic boundary value problem formulated on an unknown domain Ω. The term "unknown" means that the boundary conditions on some part of $\partial\Omega$ are overdetermined and for an a priori given shape of Ω, the problem is not well posed. By using a suitable transformation, some of these problems can be converted into an obstacle type problem, defined on a fixed domain $\hat{\Omega}$. Then, the searched boundary $\partial\Omega$ is in a certain relation with the free boundary of the new problem formulated on $\hat{\Omega}$. Let us mention here one of the most typical problems of this type, namely the filtration problem through porous media.

Let us assume a dam made of a porous material with vertical walls. The dam separates two reservoirs of water at levels $y = H$ and $y = h$ (here we assume the two dimensional model, only). The unknowns are the pressure of the water and a curve, separating the wet from the dry part of the dam. The problem can be mathematically formulated as follows (RODRIGUES [1987], CHIPOT [1984]):

Let a, h and H satisfying $a > 0, 0 < h < H$ be given. We look for a decreasing function φ defined in $[0, a]$, such that $\varphi(0) = H$, $\varphi(a) > h$ and the function u defined in Ω, where

$$\overline{\Omega} = \{[x, y] \in \mathbb{R}^2, 0 \leqslant y \leqslant \varphi(x),\ x \in [0, a]\}$$

is such that

$$\begin{aligned}
&\Delta u = 0 \quad \text{in } \Omega,\\
&u(0, y) = H \qquad \text{for all } y \in [0, H],\\
&u(a, y) = \begin{cases} h & \text{for } 0 \leqslant y \leqslant h,\\ y & \text{for } h \leqslant y \leqslant \varphi(a),\end{cases}\\
&u_y(x, 0) = 0 \quad \forall x \in (0, a),\\
&\left.\begin{aligned} &u(x, y) = y,\\ &\frac{\partial u}{\partial n}(x, y) = 0,\end{aligned}\right\} \quad \text{for } y = \varphi(x).
\end{aligned} \tag{1.13}$$

On the unknown part of the boundary, given by the graph $[\varphi]$ of φ, two conditions have to be satisfied. This makes the problem ill-posed. In a pioneering paper, BAIOCCHI [1971] proposed the introduction of a new variable w related to u by means of

$$w(x, y) = \begin{cases} \displaystyle\int_y^{\varphi(x)} (u(x, t) - t)\,\mathrm{d}t & \text{for } y \in (0, \varphi(x)),\\ 0 & \text{otherwise.}\end{cases}$$

It can be proved that the function w is the unique solution of the elliptic variational inequality:

$$\text{find } w \in K\text{: } (\nabla w, \nabla v - \nabla w)_{0,\hat{\Omega}} \geqslant (-1, v - w)_{0,\hat{\Omega}} \quad \forall v \in K, \tag{1.14}$$

where $\hat{\Omega} = (0, a) \times (0, H)$ and

$$K = \{v \in H^1(\hat{\Omega}) \mid v = g \text{ on } \partial\hat{\Omega},\ v \geqslant 0 \text{ a.e. on } \hat{\Omega}\}$$

and g is a continuous function on the boundary $\partial\hat{\Omega}$ given by

$$\begin{aligned}
&g(0, y) = \tfrac{1}{2}(H - y)^2 \quad \text{if } y \in (0, H),\\
&g(a, y) = \tfrac{1}{2}(h - y)^2 \quad \text{if } y \in (0, h),\\
&g(x, 0) = \tfrac{1}{2}H^2\left(1 - \frac{x}{a}\right) + \tfrac{1}{2}h^2\frac{x}{a},\\
&g = 0 \quad \text{elsewhere on } \partial\Omega.
\end{aligned}$$

We see that (1.14) is an example of the unilateral boundary value problem, presented in Example 1.1. The relationship between the solutions of (1.13) and (1.14) is given by

$$\begin{aligned}
&u(x, y) = y - w_y(x, y),\\
&\Omega = \{[x, y] \in \hat{\Omega} \mid w(x, y) > 0\}.
\end{aligned}$$

2. Preliminaries concerning the approximation of variational inequalities

As an illustration let us recall Example 1.1, which is now formulated on a *polygonal* domain in $\mathbb{R}^2$. Let $\{\mathcal{T}_h\}$, $h \to 0+$ be a *regular* family of triangulations of $\bar{\Omega}$, satisfying (H1)–(H3) (CIARLET [1991, Volume II, p. 131]). Let $\{N_i\}$ and $\{M_i\}$ denote the boundary and interior nodes of $\mathcal{T}_h$, respectively. With any $\mathcal{T}_h \in \{\mathcal{T}_h\}$ the following sets will be associated:

$$\begin{aligned}
&V_h = \{v_h \in C(\bar{\Omega}) \mid v_h|_{T_i} \in P_1(T_i)\ \forall T_i \in \mathcal{T}_h\},\\
&V_{h,g} = \{v_h \in V_h \mid v_h(N_i) = g(N_i)\ \forall N_i \in \{N_i\}\},\\
&K_h = \{v_h \in V_{h,g} \mid v_h(M_i) \geqslant \psi(M_i)\ \forall M_i \in \{M_i\}\}.
\end{aligned}$$

Here $P_1(T)$ denotes the space of linear polynomials defined on the triangle T.

Instead of (1.4) we assume the problem

$$\text{find } u_h \in K_h\text{: } (\nabla u_h, \nabla v_h - \nabla u_h)_{0,\Omega} \geqslant (f, v_h - u_h)_{0,\Omega} \quad \forall v_h \in K_h, \tag{2.1}$$

or equivalently

$$\text{find } u_h \in K_h \text{ such that } T(u_h) \leqslant T(v_h) \quad \forall v_h \in K_h, \tag{2.1'}$$

where

$$T(v) = \tfrac{1}{2}(\nabla v, \nabla v)_{0,\Omega} - (f, v)_{0,\Omega}.$$

We show how problem (2.1′) can be solved in practice. Let $h > 0$ be fixed and $\tilde{g} \in V_{h,g}$ be a function satisfying

$$\tilde{g}(M_i) = 0 \quad \forall M_i \in \{M_i\}.$$

Then the set $V_{h,g}$ can be split as follows:

$$V_{h,g} = \tilde{g} + V_{h,0},$$

where $V_{h,0}$ is a subspace of V_h of functions, vanishing on $\partial\Omega$. The solution u_h of (2.1) (or (2.1′)) can also be split and written in the form $u_h = \tilde{g} + \tilde{u}_h$, where

$$\tilde{u}_h \in K_{h,0} = \{v_h \in V_{h,0} \mid v_h(M_i) \geqslant \psi(M_i)\ \forall M_i \in \{M_i\}\}$$

and $\tilde{u}_h$ solves

$$(\nabla\tilde{u}_h, \nabla v_h - \nabla\tilde{u}_h)_{0,\Omega} \geqslant (f, v_h - \tilde{u}_h)_{0,\Omega} - (\nabla\tilde{g}, \nabla v_h - \nabla\tilde{u}_h)_{0,\Omega} \tag{2.2}$$

or equivalently

$$\tilde{u}_h \in K_{h,0}\colon \mathscr{J}(\tilde{u}_h) \leqslant \mathscr{J}(v_h) \quad \forall v_h \in K_{h,0}, \tag{2.2′}$$

where

$$\mathscr{J}(v_h) = \tfrac{1}{2}(\nabla v_h, \nabla v_h)_{0,\Omega} - (f, v_h)_{0,\Omega} + (\nabla\tilde{g}, \nabla v_h)_{0,\Omega}.$$

Let $\varphi_1, \ldots, \varphi_n$ be a Courant basis of $V_{h,0}$, $n = \operatorname{card}\{M_i\}$, i.e. $\varphi_i \in V_{h,0}$, $\varphi_i(M_j) = \delta_{ij}$. The space $V_{h,0}$ can be identified with $\mathbb{R}^n$ by means of the isomorphism $\mathscr{T} : V_{h,0} \to \mathbb{R}^n$ defined by

$$\mathscr{T} v_h = \vec{\alpha} = (\alpha_1, \ldots, \alpha_n) \in \mathbb{R}^n, \quad v_h \in V_{h,0},$$

with $\alpha_1, \ldots, \alpha_n$ being the coordinates of v_h with respect to $\varphi_1, \ldots, \varphi_n$. Denote

$$\begin{aligned}\mathscr{L}(\vec{\alpha}) &= \mathscr{J}(\mathscr{T}^{-1}\vec{\alpha}) = \tfrac{1}{2}(A\vec{\alpha}, \vec{\alpha}) - (\vec{\mathscr{F}}, \vec{\alpha})\\ &= \tfrac{1}{2}\sum_{i,j=1}^{n} a_{ij}\alpha_i\alpha_j - \sum_{i=1}^{n} \mathscr{F}_i\alpha_i,\ A = (a_{ij})_{i,j=1}^{n},\ \vec{\mathscr{F}} = (\mathscr{F}_i)_{i=1}^{n}\end{aligned}$$

and

$$\mathscr{K} = \{\vec{\alpha} \in \mathbb{R}^n \mid \mathscr{T}^{-1}\vec{\alpha} \in K_{h,0}\},$$

where $\mathscr{T}^{-1}$ denotes the inverse of $\mathscr{T}$. It is easy to see that

$$a_{ij} = (\nabla\varphi_i, \nabla\varphi_j)_{0,\Omega}, \qquad \mathscr{F}_i = (f, \varphi_i)_{0,\Omega} - (\nabla\tilde{g}, \nabla\varphi_i)_{0,\Omega}$$

and $\mathscr{K} = \{\vec{\alpha} \in \mathbb{R}^n \mid \alpha_i \geqslant \psi(M_i),\ i = 1, \ldots, n\}$.

The equivalent algebraic expression of (2.2′) is

$$\text{find } \vec{\alpha}^* = (\alpha_1^*, \ldots, \alpha_n^*) \in \mathscr{K} \text{ such that } \mathscr{L}(\vec{\alpha}^*) \leqslant \mathscr{L}(\vec{\alpha}) \quad \forall \vec{\alpha} \in \mathscr{K}. \tag{2.3}$$

This is a quadratic programming problem, i.e. the minimization problem for the quadratic function $\mathscr{L}$ over the convex set $\mathscr{K}$ given by linear constraints. Some of the methods, which enable the numerical realization of (2.3) will be presented in the forthcoming chapters. When $\vec{\alpha}^*$, solving (2.3), is found, the corresponding solution is $u_h = \tilde{g} + \sum_{j=1}^n \alpha_j^* \varphi_j$.

Let $\{u_h\}$, $u_h \in K_h$ be a sequence of the solutions of (2.1) (or (2.1′)). A natural question arises, namely if $u_h \to u$, $h \to 0+$ in a suitable norm, eventually, what is the rate of the convergence, expressed in terms of h? Such questions will be discussed in the next section.

3. Elliptic variational inequalities. Different variational formulations. Existence results

First we define an abstract elliptic inequality of the first and second kind and present some results concerning the existence and the uniqueness of its solution. Then we shall discuss alternative variational formulations of elliptic inequalities and their mutual relations. Results are standard and the majority of them will be presented without proofs.

In the following, V will denote a real Hilbert space, V' its dual with the duality pairing $\langle\,,\,\rangle$. The norm in V will be denoted by $\|\ \|$ and the dual norm by $\|\ \|_*$. Let $a: V \times V \to \mathbb{R}^1$ be a bilinear form. Let K be a *nonempty, closed and convex subset of* V.

DEFINITION 3.1. A triplet $\{K, a, f\}$, $f \in V'$, is called an *abstract elliptic variational inequality of the first kind.* A function $u \in V$ is called a *solution* of $\{K, a, f\}$ iff

$$u \in K\colon\ a(u, v - u) \geqslant \langle f, v - u\rangle \quad \forall v \in K. \tag{3.1}$$

In order to prove the existence and the uniqueness of the solution of $\{K, a, f\}$, assumptions concerning a have to be added. Next we shall suppose that the bilinear form a is *bounded* on V:

$$\exists M > 0\colon\ |a(y, z)| \leqslant M\|y\|\|z\| \quad \forall y, z \in V \tag{3.2a}$$

and V-*elliptic* on V:

$$\exists \alpha > 0\colon\ a(y, y) \geqslant \alpha\|y\|^2 \quad \forall y \in V. \tag{3.2b}$$

REMARK 3.1. In applications it may happen that instead of (3.2b), a weaker assumption, namely

$$\exists \alpha > 0\colon\ a(y, y) \geqslant \alpha|y|^2 \quad \forall y \in V \tag{3.2b′}$$

holds, where $|\cdot|$ denotes a seminorm in V. This case requires a special treatment and will be mentioned at the end of this section. If $K = V$, then (3.1) reduces to a classical linear elliptic equation $\{V, a, f\}$, the solution u of which satisfies

$$a(u,v) = \langle f, v\rangle \quad \forall v \in V.$$

If K is a convex cone containing the zero element of V, then the problem (3.1) is equivalent to the following one:

$$u \in K\colon\ a(u,v) \geqslant \langle f, v\rangle \quad \forall v \in K,\ a(u,u) = \langle f, u\rangle.$$

THEOREM 3.1. *Let $a : V \times V \to \mathbb{R}^1$ satisfy* (3.2a) *and* (3.2b). *Then there exists a unique solution u of $\{K, a, f\}$ for any $f \in V'$. If u_i are solutions of $\{K, a, f_i\}$, $i = 1, 2$, then*

$$\|u_1 - u_2\| \leqslant \frac{M}{\alpha}\|f_1 - f_2\|_*.$$

REMARK 3.2. It is possible to replace (3.2b) by a weaker assumption, namely

$$\exists \alpha > 0\colon\ a(y_1 - y_2, y_1 - y_2) \geqslant \alpha\|y_1 - y_2\|^2 \quad \forall y_1, y_2 \in K.$$

The statement of Theorem 3.1 still holds.

If the bilinear form a is also *symmetric* on V, i.e. $a(y,z) = a(z,y)$ holds for any $y, z \in V$, the problem $\{K, a, f\}$ is equivalent to the following one:

$$\text{find } u \in K\colon\ \mathcal{J}(u) \leqslant \mathcal{J}(v) \quad \forall v \in K, \tag{3.3}$$

where $\mathcal{J} : V \to \mathbb{R}^1$ is the quadratic functional

$$\mathcal{J}(v) = \tfrac{1}{2}a(v,v) - \langle f, v\rangle, \tag{3.4}$$

i.e. $u \in K$ solves $\{K, a, f\}$ if and only if it minimizes $\mathcal{J}$ over K.

REMARK 3.3. Denote by $I_K : V \to \{0, +\infty\}$ the indicator function of K, i.e.:

$$I_K(v) = \begin{cases} 0 & \text{if } v \in K,\\ +\infty & \text{elsewhere.}\end{cases}$$

Then (3.1) is formally equivalent to

$$\text{find } u \in V\colon\ a(u, v-u) + I_K(v) - I_K(u) \geqslant \langle f, v-u\rangle \quad \forall v \in V. \tag{3.5}$$

This motivates the following extension of $\{K, a, f\}$. Let $j : V \to \mathbb{R}^1 \cup \{-\infty, \infty\}$ be a convex, *lower semicontinuous* and proper (i.e. $j(v) \not\equiv -\infty$) functional on

V. The quadruplet $\{V,a,j,f\}$ is said to be *an abstract elliptic inequality of the second kind*. The function $u \in V$ is said to be a solution of $\{V,a,j,f\}$ iff

$$a(u,v-u)+j(v)-j(u) \geqslant \langle f,v-u\rangle \quad \forall v \in V. \tag{3.6}$$

If $j = I_K$ is the indicator function of K, $\{V,a,j,f\}$ reduces to $\{K,a,f\}$. If moreover $a: V\times V \to \mathbb{R}^1$ is symmetric on V, the inequality $\{V,a,j,f\}$ is equivalent to the minimization of $\mathcal{J}: V \to \mathbb{R}^1 \cup \{-\infty,+\infty\}$ over V, where

$$\mathcal{J}(v) = \tfrac{1}{2}a(v,v)+j(v)-\langle f,v\rangle. \tag{3.4$'$}$$

Let us mention that if a and j satisfy the above mentioned assumptions, the problem $\{V,a,j,f\}$ has exactly one solution u.

Denote by $A \in \mathcal{L}(V,V')$ the mapping, defined by means of

$$\langle Ay,z\rangle = a(y,z) \quad \forall y,z \in V.$$

An equivalent expression of (3.6) is

$$\text{find } u \in V\colon f - Au \in \partial j(u), \tag{3.6$'$}$$

where $\partial j(u)$ stands for the subgradient of the convex function j at the point u (EKELAND and TEMAM [1976]).

To solve problems governed by nonlinear differential operators, yet another extension is necessary. Let $A: W \to W'$ be a mapping from a *reflexive Banach space* W into its dual W', K a *nonempty, closed* and *convex* subset of W. Consider the problem

$$\text{find } u \in K\colon \langle A(u),v-u\rangle_{W'\times W} \geqslant \langle f,v-u\rangle_{W'\times W} \quad \forall v \in K, \tag{3.7}$$

where $f \in W'$ and $\langle\ ,\ \rangle_{W'\times W}$ denotes a duality pairing between W' and W. As far as the existence and the uniqueness of the solution of (3.7) concerns we have the following.

THEOREM 3.2. *Let $A: W \to W'$ be a locally Lipschitz continuous and coercive mapping on a reflexive Banach space W:*

$$\|A(y)-A(z)\|_* \leqslant M(r)\|y-z\|_W \quad \forall y,z \in B_r, \tag{3.8}$$

$$\langle A(y)-A(z),y-z\rangle_{W'\times W} \geqslant \alpha(r)\|y-z\|_W^2 \quad \forall y,z \in B_r, \tag{3.9}$$

where

$$B_r = \{y \in W \mid \|y\|_W \leqslant r\},\ 0<\alpha(r)\leqslant M(r) \quad \forall r>0,$$

$\alpha(r)$ is non-increasing and such that $r\alpha(r) \to \infty$ as $r \to \infty$. Then (3.7) *has a unique solution u.*

We have seen that in the case of a symmetric and V-elliptic bilinear form, the inequality $\{K, a, f\}$ is closely related to a minimization problem for $\mathcal{J}$, given by (3.4). This result can be extended to more general situations as follows.

THEOREM 3.3. *Let $\Psi : W \to \mathbb{R}^1$ be a functional defined on a reflexive Banach space W, K a nonempty closed and convex subset of W. Let us assume that Ψ is Gâteaux differentiable at any point $y \in W$ and its differential $D\Psi(y, z)$ is continuous on any line segment (in the variable y) in W and moreover let*

$$D\Psi(y+z, z) - D\Psi(y, z) \geqslant \alpha(\|z\|_W)\|z\|_W^2 \quad \forall y, z \in W, \tag{3.10}$$

where the function $\alpha(r)$ has the same properties as formulated in Theorem 3.2. Then there exists a unique minimizer u of Ψ on K:

$$u \in K: \Psi(u) \leqslant \Psi(v) \quad \forall v \in K \tag{3.11}$$

which can be also characterized by the inequality

$$u \in K: D\Psi(u, v-u) \geqslant 0 \quad \forall v \in K. \tag{3.12}$$

PROOF. Any functional, satisfying the assumption of the theorem is strictly convex, weakly lower semicontinuous and coercive on W. The existence and the uniqueness of a minimizer as well as the equivalence of (3.11) and (3.12) is a classic result of the calculus of variations (CÉA [1971]). □

EXAMPLE 3.1. Let $W = W_0^{1,p}(\Omega)$, $p \geqslant 2$ and let $K \subset W$ be a non-empty closed and convex subset of W. Let

$$\Psi(v) = \frac{1}{p}\int_\Omega |\nabla v(x)|^p \,\mathrm{d}x - \langle f, v\rangle_{W' \times W}, \quad f \in W'.$$

It is easy to verify that Ψ is strictly convex, continuous and coercive on W (see CIARLET [1978]) and

$$D\Psi(y, z) = \int_\Omega |\nabla y|^{p-2} \nabla y \nabla z \,\mathrm{d}x - \langle f, z\rangle_{W' \times W}.$$

Hence there is a unique minimizer u of Ψ on W or equivalently, a unique $u \in K$, satisfying

$$\int_\Omega |\nabla u|^{p-2} \nabla u \nabla (v-u) \,\mathrm{d}x \geqslant \langle f, v-u\rangle_{W' \times W} \quad \forall v \in K.$$

An important question arises, namely how the solution of a variational inequality depends on the given data. This problem will be studied in detail in the next chapter. So far we know that the solution of $\{K, a, f\}$ is a Lipschitz continuous function of f (see Theorem 3.1). Consider, now, an elliptic inequality of the second kind $\{V, a, j, f\}$, whose (unique) solution will be denoted by u. Also, let $\{V, a, j_\varepsilon, f\}$, $\varepsilon > 0$, be a family of elliptic inequalities, having the unique solution u_ε. Below we formulate sufficient conditions on j_ε which guarantee $u_\varepsilon \to u$ as $\varepsilon \to 0+$.

Let us assume that

$$j_\varepsilon : V \to \mathbb{R}^1 \text{ is convex, lower continuous and finite on } V \text{ for any } \varepsilon > 0, \tag{3.13}$$

$$\exists \chi \in V' \text{ and } \mu \in \mathbb{R}^1\colon\ j_\varepsilon(v) \geqslant \chi(v) + \mu \quad \forall v \in V,\ \forall \varepsilon > 0, \tag{3.14}$$

$$\lim_{\varepsilon \to 0+} j_\varepsilon(v) = j(v) \quad \forall v \in V, \tag{3.15}$$

$$u_\varepsilon \rightharpoonup v \text{ (weakly) in } V \Rightarrow \liminf_{\varepsilon \to 0+} j_\varepsilon(u_\varepsilon) \geqslant j(v), \tag{3.16}$$

where u_ε are the solutions of $\{V, a, j_\varepsilon, f\}$.

THEOREM 3.4. *Let* (3.13)–(3.16) *be satisfied. Then if* $\varepsilon \to 0+$

$$\|u_\varepsilon - u\| \to 0, \qquad j_\varepsilon(u_\varepsilon) \to j(u), \tag{3.17}$$

where u and u_ε are the solutions of $\{V, a, j, f\}$ and $\{V, a, j_\varepsilon, f\}$, respectively.

PROOF. Let $\bar{v} \in V$ be such that $j(\bar{v})$ is finite. Then

$$a(u_\varepsilon, \bar{v} - u_\varepsilon) + j_\varepsilon(\bar{v}) - j_\varepsilon(u_\varepsilon) \geqslant \langle f, \bar{v} - u_\varepsilon \rangle.$$

From (3.2a), (3.2b), (3.14) and (3.15) the boundedness of $\{u_\varepsilon\}$ follows:

$$\exists c = \text{const.} > 0\colon\ \|u_\varepsilon\| \leqslant c \quad \forall \varepsilon > 0.$$

Hence there exists a subsequence of $\{u_\varepsilon\}$ (denoted by the same symbol) and an element $u \in V$ such that

$$u_\varepsilon \rightharpoonup u \text{ (weakly) in } V \text{ as } \varepsilon \to 0+. \tag{3.18}$$

Letting $\varepsilon \to 0+$ in

$$a(u_\varepsilon, v - u_\varepsilon) + j_\varepsilon(v) - j_\varepsilon(u_\varepsilon) \geqslant \langle f, v - u_\varepsilon \rangle \quad \forall v \in V \tag{3.19}$$

and taking into account (3.15), (3.16) and (3.18) as well as the weak lower continuity of the mapping $v \mapsto a(v, v)$ we arrive at

$$a(u, v - u) + j(v) - j(u) \geqslant \langle f, v - u \rangle \quad \forall v \in V.$$

Hence u is a solution of $\{V, a, j, f\}$. As u is unique, the whole sequence $\{u_\varepsilon\}$ tends weakly to u.

We now prove $j_\varepsilon(u_\varepsilon) \to j(u)$. First of all (3.16) and (3.18) yield

$$\liminf_{\varepsilon \to 0+} j_\varepsilon(u_\varepsilon) \geqslant j(u). \tag{3.20}$$

Substituting $v = u$ into (3.19) and using (3.15) we arrive at

$$\limsup_{\varepsilon \to 0+} j_\varepsilon(u_\varepsilon) \leqslant j(u).$$

From this and (3.20), the result follows.

Let us show that $u_\varepsilon \to u$ strongly. From (3.2b) it follows:

$$\begin{aligned} \alpha\|u - u_\varepsilon\|^2 &\leqslant a(u - u_\varepsilon, u - u_\varepsilon) = a(u, u - u_\varepsilon) - a(u_\varepsilon, u - u_\varepsilon) \\ &\leqslant a(u, u - u_\varepsilon) + j_\varepsilon(u) - j_\varepsilon(u_\varepsilon) + \langle f, u_\varepsilon - u\rangle \to 0 \end{aligned}$$

by virtue of (3.15), (3.18) and the second formula of (3.17). □

REMARK 3.4. If j_ε is directionally differentiable on V, then $\{V, a, j_\varepsilon, f\}$ is equivalent to the nonlinear equation:

$$\text{find } u_\varepsilon \in V \text{ such that } a(u_\varepsilon, v) + j_\varepsilon'(u_\varepsilon, v) = \langle f, v\rangle \quad \forall v \in V. \tag{3.21}$$

The symbol $j_\varepsilon'(u, v)$ stands for the derivative of j_ε at the point u and the direction v.

Now we present two important applications of the previous result.

EXAMPLE 3.2. Let us consider the elliptic inequality of the second kind $\{V, a, j, f\}$, where

$$V = H^1(\Omega), \qquad a(u, v) = (u, v)_{1,\Omega}, \qquad j(v) = \int_{\partial\Omega} |v|\, ds,$$

$$\langle f, v\rangle = (f, v)_{0,\Omega}, \qquad f \in L^2(\Omega), \quad \Omega \subset \mathbb{R}^2.$$

Let $u \in V$ be a solution of $\{V, a, j, f\}$:

$$(u, v - u)_{1,\Omega} + j(v) - j(u) \geqslant (f, v - u)_{0,\Omega} \quad \forall v \in V. \tag{3.22}$$

Applying Green's formula to (3.22) one can deduce that u solves the following boundary value problem:

$$-\Delta u + u = f \quad \text{in } \Omega,$$

$$\left|\frac{\partial u}{\partial n}\right| \leqslant 1 \quad \text{a.e. on } \partial\Omega,$$

$$\left|\frac{\partial u}{\partial n}(x)\right| < 1 \Rightarrow u(x) = 0, \qquad \operatorname{sign}\frac{\partial u}{\partial n}(x)\operatorname{sign} u(x) \neq 1.$$

As $\underline{a}$ is symmetric on V, the problem (3.22) is equivalent to the minimization of

$$\mathscr{J}(v) = \tfrac{1}{2}\|v\|^2_{1,\Omega} + j(v) - (f, v)_{0,\Omega}$$

on V. The functional $\mathscr{J}$ is convex but nondifferentiable because of the presence of the sublinear term j. To overcome this difficulty we shall use the following approach: let $j_\varepsilon : V \to \mathbb{R}^1$, $\varepsilon > 0$ be a functional defined by

$$j_\varepsilon(v) = \frac{1}{1+\varepsilon} \int_{\partial\Omega} |v|^{1+\varepsilon}\, \mathrm{d}s$$

and consider the problem $\{V, a, j_\varepsilon, f\}$. It is easy to see that (3.13)–(3.16) are satisfied. The condition (3.16) is valid for any weakly convergent sequence $\{v_\varepsilon\}$ and not only for $\{u_\varepsilon\}$. Hence

$$u_\varepsilon \to u, \quad \varepsilon \to 0+ .$$

As j_ε is Gâteaux differentiable in V, u_ε is the unique solution of the following nonlinear elliptic equation:

$$(u_\varepsilon, v)_{1,\Omega} + \int_{\partial\Omega} |u_\varepsilon|^{\varepsilon-1} u_\varepsilon v\, \mathrm{d}s = \langle f, v\rangle \quad \forall v \in V$$

or

$$-\Delta u_\varepsilon + u_\varepsilon = f \quad \text{in } \Omega,$$
$$\frac{\partial u_\varepsilon}{\partial n} + |u_\varepsilon|^{\varepsilon-1} u_\varepsilon = 0 \quad \text{on } \partial\Omega.$$

Using this approach we have converted the nonsmooth problem $\{V, a, j, f\}$ into a smooth one. This approach is called a *regularization* technique of the problem.

Another very important application of Theorem 3.4 is the following.

EXAMPLE 3.3 (Penalty approach). Let us suppose that the bilinear form $a : V \times V \to \mathbb{R}^1$ is symmetric on V. Hence $\{K, a, f\}$ is equivalent to the minimization of $\mathscr{J}$, given by (3.4) over K. Let $p : V \to \mathbb{R}^1$ be a functional with the following properties:

$$p(v) \geqslant 0 \text{ on } V \text{ and } p(v) = 0 \text{ iff } v \in K, \tag{3.23}$$

$p(v)$ is convex, weakly lower semicontinuous and directionally differentiable on V. (3.24)

Let us introduce

$$\mathcal{J}_\varepsilon(v) = \tfrac{1}{2}a(v,v) + \frac{1}{\varepsilon}p(v) - \langle f, v\rangle = \mathcal{J}(v) + \frac{1}{\varepsilon}p(v), \quad \varepsilon > 0.$$

The functional p, satisfying (3.23) and (3.24) will be called a *penalty functional* of K.

Instead of the constrained minimization problem for $\mathcal{J}$ we shall assume the *unconstrained* minimization problem for $\mathcal{J}_\varepsilon$:

$$\text{find } u_\varepsilon \in V\colon\ \mathcal{J}_\varepsilon(u_\varepsilon) \leqslant \mathcal{J}_\varepsilon(v) \quad \forall v \in V \tag{3.25}$$

or equivalently

$$\text{find } u_\varepsilon \in V\colon\ a(u_\varepsilon, v) + \frac{1}{\varepsilon}p'(u_\varepsilon, v) = \langle f, v\rangle \quad \forall v \in V. \tag{3.25'}$$

The relation between $\{K, a, f\}$ and (3.25) is shown in the following.

THEOREM 3.5. *For any $\varepsilon > 0$ there exists a unique solution of* (3.25) *and $u_\varepsilon \to u$ in V, $\varepsilon \to 0+$, where $u \in K$ is a unique solution of $\{K, a, f\}$.*

PROOF. As $\mathcal{J}_\varepsilon$ is strictly convex, coercive and weakly lower semicontinuous on V, the solution u_ε exists and is unique. Using the notations of Theorem 3.4, set

$$j(v) \equiv I_K(v) \text{ (the indicator function of } K),$$
$$j_\varepsilon(v) \equiv \frac{1}{\varepsilon}p(v) \text{ (the penalty functional)}\,.$$

It is readily seen that (3.13)–(3.15) hold. Hence $\{u_\varepsilon\}$ is bounded. We shall verify (3.16). Let $\{u_\varepsilon\}$ be a sequence of solutions tending weakly to v and $\bar{v} \in K$ be a fixed element. Then

$$0 \leqslant \frac{1}{\varepsilon}p(u_\varepsilon) \leqslant \mathcal{J}_\varepsilon(\bar{v}) - \mathcal{J}(u_\varepsilon) = \mathcal{J}(\bar{v}) - \mathcal{J}(u_\varepsilon),$$

follows from (3.23) and (3.25). Hence

$$0 \leqslant p(u_\varepsilon) \leqslant \varepsilon(\mathcal{J}(\bar{v}) - \mathcal{J}(u_\varepsilon)) \to 0 \text{ as } \varepsilon \to 0+\,. \tag{3.26}$$

On the other hand $\liminf_{\varepsilon\to 0+} p(u_\varepsilon) \geqslant p(v)$. From this, (3.26) and (3.23) it follows that $v \in K$. Hence $\liminf_{\varepsilon\to 0+} \varepsilon^{-1}p(u_\varepsilon) \geqslant 0 = p(v)$. □

EXAMPLE 3.4. Let $\{K, a, f\}$ be the same as in Example 1.2. Then

$$p(v) = \tfrac{1}{2}\int_{\Gamma_2} ([v - h]^+)^2 \,\mathrm{d}s$$

is the corresponding penalty functional satisfying (3.23) and (3.24). The symbol $[\beta]^+$ denotes the positive part of a real number β. The solution of (3.25) (or (3.25′)) satisfy the following mixed boundary value problem:

$$-\Delta u_\varepsilon = f \quad \text{in } \Omega,$$
$$u_\varepsilon = 0 \quad \text{on } \Gamma_1,$$
$$\frac{\partial u_\varepsilon}{\partial n} + \frac{1}{\varepsilon}[u_\varepsilon - h]^+ = 0 \quad \text{on } \Gamma_2.$$

REMARK 3.5. If the bilinear form a is not symmetric on V, one can still define (3.25′) which will be a penalized version of $\{K, a, f\}$. Assuming that (3.23) and (3.24) hold, one has $u_\varepsilon \to u$, where u_ε and u are solutions of (3.25′) and $\{K, a, f\}$, respectively (the proof follows immediately from Theorem 3.5, where no symmetry of a is used). In our applications however, the bilinear form a will always be symmetric.

Using a penalty approach, the constrained minimization problem can be replaced by a sequence of unconstrained ones. There is another way of achieving the same result, namely use of the Lagrange multiplier technique (see EKELAND and TEMAM [1976], CÉA [1971], GLOWINSKI, LIONS and TRÉMOLIÈRES [1981]). We shall see that this approach also enables us to transform a nonsmooth minimization problem into a smooth one.

Let V and Y be two real Hilbert spaces, norms of which will be denoted by $\| \; \|$ and $\| \; \|_Y$, respectively. Let $K \subset V$ be a nonempty, closed, convex subset, $L : v \mapsto \mathbb{R}^1$, $v \in V$ be a given functional (not necessarily quadratic). The main purpose of this approach is to rewrite the original minimization problem of L on K into a new form, namely,

$$\inf_{v \in K} L(v) = \inf_{v \in A} \sup_{\mu \in B} \mathscr{L}(v, \mu) \tag{3.27}$$

for a suitable choice of sets $A \subset V$, $B \subset Y$ and a function $\mathscr{L} : A \times B \to \mathbb{R}^1$. The function $\mathscr{L}$ related to L by means of (3.27) is called the *Lagrange function* or the Lagrangian of the problem. Below we present two typical examples, which will be frequently used in what follows.

EXAMPLE 3.5. Let $\Lambda \subset Y$ be a closed, convex cone containing the zero element of Y and $\Phi : V \times \Lambda \to \mathbb{R}^1$ a positively 1-homogeneous functional with respect to the second variable:

$$\Phi(v, \varrho\mu) = \varrho\Phi(v, \mu) \quad \forall (v, \mu) \in V \times \Lambda, \; \forall \varrho \geqslant 0. \tag{3.28}$$

Let us suppose that the following characterization of K holds:

$$v \in K \text{ iff } \Phi(v, \mu) \leqslant 0 \; \forall \mu \in \Lambda. \tag{3.29}$$

On the basis of (3.28) and (3.29) one can easily verify that

$$\sup_{\mu\in\Lambda} \Phi(v,\mu) = \begin{cases} 0 & \text{iff } v \in K \\ +\infty & \text{otherwise,} \end{cases}$$

i.e. $\sup_{\mu\in\Lambda} \Phi(v,\mu)$ is the indicator function of K.

Consequently, we may write

$$\inf_{v\in K} L(v) = \inf_{v\in V} \sup_{\mu\in\Lambda} \{L(v) + \Phi(v,\mu)\},$$

i.e. $A = V$, $B = \Lambda$ and $\mathcal{L}(v,\mu) = L(v) + \Phi(v,\mu)$.

EXAMPLE 3.6. Let $L(v) = \mathcal{J}(v) + j(v)$ be a functional defined on a subspace V of $H^1(\Omega)$, where $\mathcal{J}$ is the quadratic functional (3.4) and $j(v) = \int_\Omega |v|\,dx$. Then L is not differentiable on V. On the other hand

$$j(v) = \sup_{|\mu|\leqslant 1} \int_\Omega \mu v\,dx$$

so that

$$\inf_{v\in V} L(v) = \inf_{v\in V} \sup_{\mu\in B} \{\mathcal{J}(v) + \int_\Omega \mu v\,dx\}.$$

Here $A = V$, $B = \{\mu \in L^\infty(\Omega) \mid |\mu| \leqslant 1 \text{ a.e. in } \Omega\}$, $\mathcal{L}(v,\mu) = \mathcal{J}(v) + \int_\Omega \mu v\,dx$. The functional $\mathcal{L}$ is now smooth as a function of two variables v and μ.

Let the functionals L and $\mathcal{L}$ be related by means of (3.27). Problem (3.27) will be called the *primal variational* formulation , while the problem

$$\sup_{\mu\in B} \inf_{v\in A} \mathcal{L}(v,\mu)$$

will be called *the dual variational formulation.*

REMARK 3.6. The primal formulation is carried out by making the minimization of L a function of the (*primal*) variable v, while the dual formulation involves the maximization problem for the functional $\tilde{\mathcal{S}}(\mu) = \inf_{v\in A} \mathcal{L}(v,\mu)$ that is a function of the second (*dual*) variable μ. Setting $\mathcal{S} = -\tilde{\mathcal{S}}$ we replace the problem $\sup_{\mu\in B} \tilde{\mathcal{S}}(\mu)$ by $\inf_{\mu\in B} \mathcal{S}(\mu)$. The functional $\mathcal{S}$ is said to be *dual* to L.

EXAMPLE 3.7. Consider the elliptic inequality $\{K, a, f\}$, where $K \subseteq V$,

$$V = \{v \in H^1(\Omega) \mid v = 0 \text{ on } \Gamma_1\},$$
$$K = \{v \in V \mid v \geqslant 0 \text{ on } \Gamma_2\},$$
$$a(u,v) = (\nabla u, \nabla v)_{0,\Omega}, \qquad \langle f, v\rangle = (f,v)_{0,\Omega},\ f \in L^2(\Omega).$$

Γ_1 and Γ_2 are nonempty, disjoint and open parts of $\partial\Omega$, $\partial\Omega = \overline{\Gamma}_1 \cup \overline{\Gamma}_2$. As the bilinear form a is symmetric, the problem $\{K, a, f\}$ is equivalent to the minimization of $\mathcal{J}(v) = \frac{1}{2}\|\nabla v\|^2_{0,\Omega} - (f, v)_{0,\Omega}$ over K:

$$\text{find } u \in K \text{ such that } \mathcal{J}(u) = \inf_{v \in K} \mathcal{J}(v).$$

Introduce a new variable $q \in [L^2(\Omega)]^2$ by means of

$$q = \nabla v \quad \forall v \in V.$$

Then

$$\inf_{v \in K} \mathcal{J}(v) = \inf_{\substack{v \in K \\ q \in [L^2(\Omega)]^2}} \sup_{\mu \in [L^2(\Omega)]^2} \{(q, q)_{0,\Omega} + (\mu, q - \nabla v)_{0,\Omega} - (f, v)_{0,\Omega}\}.$$

Here the role of the primal variable is played by $(v, q) \in A$ and $A = K \times [L^2(\Omega)]^2$, $B = [L^2(\Omega)]^2$. The dual formulation reads as follows:

$$\sup_{\mu \in [L^2(\Omega)]^2} \inf_{\substack{v \in K \\ q \in [L^2(\Omega)]^2}} \{(q, q)_{0,\Omega} + (\mu, q - \nabla v)_{0,\Omega} - (f, v)_{0,\Omega}\}.$$

A direct calculation yields the following expression for the dual functional:

$$\mathcal{S}(\mu) = \tfrac{1}{2}\|\mu\|^2_{0,\Omega} + I_{K_f}(\mu), \quad \mu \in [L^2(\Omega)]^2,$$

where I_{K_f} is the indicator function of a closed convex subset K_f of $[L^2(\Omega)]^2$, which is defined by

$$K_f = \{\mu \in [L^2(\Omega)]^2 \mid (\mu, \nabla v)_{0,\Omega} \geqslant (f, v)_{0,\Omega} \ \forall v \in K\}.$$

It is easy to find that $\mu \in K_f$ if and only if $\operatorname{div} \mu + f = 0$ in Ω and the flux $\mu n \geqslant 0$ on Γ_2 in the sense of distributions. The explicit form of the dual formulation is

$$\inf_{\mu \in K_f} \mathcal{S}(\mu).$$

$\mathcal{S}$ restricted on K_f is the quadratic functional. From its form it is readily seen that there is a unique λ minimizing $\mathcal{S}$ on K_f. Moreover, $\lambda = \nabla u$, where u solves $\{K, a, f\}$. For more details see HLAVÁČEK, HASLINGER, NEČAS and LOVÍŠEK [1988].

We see from the previous example that in some cases there exists a relationship between solutions of the primal and dual formulations. This phenomenon is not random. Below we present some results, explaining the mutual relationship between solutions of both variational formulations. A detailed proof of all results presented below can be found in EKELAND and TEMAM [1976]. We start with a definition.

Let A and B be two non-empty sets.

DEFINITION 3.2. A pair $(u, \lambda) \in A \times B$ is said to be a *saddle point* of $\mathcal{H} : A \times B \to \mathbb{R}^1$ on $A \times B$ iff

$$\mathcal{H}(u, \mu) \leqslant \mathcal{H}(u, \lambda) \leqslant \mathcal{H}(v, \lambda) \quad \forall (v, \mu) \in A \times B.$$

The following characterization of (u, λ) holds:

THEOREM 3.6. *A pair* $(u, \lambda) \in A \times B$ *is a saddle-point of* $\mathcal{H}$ *on* $A \times B$ *iff*

$$\min_{v \in A} \sup_{\mu \in B} \mathcal{H}(v, \mu) = \max_{\mu \in B} \inf_{v \in A} \mathcal{H}(v, \mu). \tag{3.30}$$

The minimum and the maximum in (3.30) *is attained at* u *and* λ*, respectively and the common value equals* $\mathcal{H}(u, \lambda)$.

REMARK 3.7. Let $\mathcal{H} = \mathcal{L}$, where $\mathcal{L}$ is related to L by means of (3.27). Then the first component of the saddle point of $\mathcal{L}$ is the solution of the primal formulation and the second component solves the dual formulation.

In order to guarantee the existence, and eventually the uniqueness, of a saddle-point, we need supplementary conditions, concerning A, B and $\mathcal{H}$.

Next we shall suppose that

(i) A and B are *nonempty, closed, convex* subsets of V and Y, respectively.

(ii) $\forall \mu \in B$, $v \mapsto \mathcal{H}(v, \mu)$ is *convex* and *weakly lower semicontinuous.*

(iii) $\forall v \in A$, $\mu \mapsto \mathcal{H}(v, \mu)$ is *concave* and *weakly upper semicontinuous.*

THEOREM 3.7. *Let the assumptions* (i)–(iii) *be satisfied and* A *and* B *be bounded. Then there exists a saddle point* (u, λ) *of* $\mathcal{H}$ *on* $A \times B$.

If A and B are not bounded, some coerciveness assumptions have to be added as follows:

THEOREM 3.8. *Let* (i)–(iii) *be satisfied and assume moreover,*

$$\exists \mu_0 \in B \textit{ such that } \lim_{\substack{\|v\| \to \infty \\ v \in A}} \mathcal{H}(v, \mu_0) = +\infty, \tag{3.31}$$

$$\exists v_0 \in A \textit{ such that } \lim_{\substack{\|\mu\|_Y \to \infty \\ \mu \in B}} \mathcal{H}(v_0, \mu) = -\infty. \tag{3.32}$$

Then there exists a saddle point of $\mathcal{H}$ *on* $A \times B$.

REMARK 3.8. If only one of the sets A and B is unbounded, then the coerciveness of $\mathcal{H}$ with respect to the unbounded set has to be required.

REMARK 3.9. Let $(u, \lambda) \in A \times B$ be a saddle point of $\mathcal{H}$ on $A \times B$. If in (ii) "convex" is replaced by "strictly convex", the first component u is uniquely determined. Analogously, if in (iii) "concave" is replaced by "strictly concave", then λ is unique.

In practical applications, one usually has the coerciveness property (3.31) but the absence of (3.32). Instead of (3.32), a weaker assumption may hold, as follows.

THEOREM 3.9. *Let* (i)–(iii) *be satisfied. Moreover, let A be bounded or there exists $\mu_0 \in B$ such that*

$$\lim_{\substack{\|v\|\to\infty \\ v\in A}} \mathcal{H}(v, \mu_0) = +\infty, \tag{3.33}$$

B be bounded or

$$\lim_{\substack{\|\mu\|_Y\to\infty \\ \mu\in B}} \inf_{v\in A} \mathcal{H}(v, \mu) = -\infty. \tag{3.34}$$

Then there is a saddle point of $\mathcal{H}$ on $A \times B$.

Assume that the functional $\mathcal{H}$ can be written as a sum of $\mathcal{H}_1$ and $\mathcal{H}_2$: $\mathcal{H} = \mathcal{H}_1 + \mathcal{H}_2$, where $\mathcal{H}_1$ is a smooth part of $\mathcal{H}$ that is Gâteaux differentiable on $A \times B$ and both $\mathcal{H}_1$ and $\mathcal{H}_2$ satisfy (ii) and (iii). Then one has another characterization of the saddle point (u, λ).

THEOREM 3.10. *Let $\mathcal{H} = \mathcal{H}_1 + \mathcal{H}_2$, where $\mathcal{H}_1$ and $\mathcal{H}_2$ have properties, indicated above. Then $(u, \lambda) \in A \times B$ is a saddle point of $\mathcal{H}$ on $A \times B$ if and only if*

$$\partial_v \mathcal{H}_1(u, \lambda)(v - u) + \mathcal{H}_2(v, \lambda) - \mathcal{H}_2(u, \lambda) \geqslant 0 \quad \forall v \in A, \tag{3.35}$$

$$\partial_\mu \mathcal{H}_1(u, \lambda)(\mu - \lambda) + \mathcal{H}_2(u, \mu) - \mathcal{H}_2(u, \lambda) \leqslant 0 \quad \forall \mu \in B, \tag{3.36}$$

where $\partial_v \mathcal{H}_1$, $\partial_\mu \mathcal{H}_1$ denote the partial Gâteaux derivative of $\mathcal{H}_1$ with respect to v and μ, respectively.

An important application of Theorem 3.9 will be presented in the following example.

EXAMPLE 3.8. Let V and Y be two real Hilbert spaces, the norms of which are denoted by $\| \ \|$ and $\| \ \|_Y$, respectively. Let $b : V \times Y \to \mathbb{R}^1$ be a bilinear form, satisfying

$$\exists \overline{M} > 0 \text{ such that } |b(v, \mu)| \leqslant \overline{M}\|v\| \, \|\mu\|_Y \quad \forall (v, \mu) \in V \times Y, \tag{3.37}$$

$$\exists \beta > 0 \text{ such that } \sup_{\substack{v\in V \\ v\neq 0}} \frac{b(v, \mu)}{\|v\|} \geqslant \beta\|\mu\|_Y \quad \forall \mu \in Y. \tag{3.38}$$

Let $\Lambda \subset Y$ be a closed, convex cone, containing the zero element of Y, $g \in Y'$ is an element of the dual space to Y with the duality pairing denoted by $[\,,\,]$. Then it is easy to see that

$$K = \{v \in V \mid b(v, \mu) \leqslant [g, \mu] \ \forall \mu \in \Lambda\} \tag{3.39}$$

is a nonempty, closed and convex subset of V. Let $\mathcal{J}$ be the quadratic functional (3.4).

Consider the problem

$$\text{find } u \in K\colon \ \mathcal{J}(u) \leqslant \mathcal{J}(v) \quad \forall v \in K, \tag{3.40}$$

with K defined by (3.39). Setting $\Phi(v, \mu) = b(v, \mu) - [g, \mu]$, we see that (3.28) and (3.29) are satisfied. Hence

$$\inf_{v \in K} \mathcal{J} = \inf_{v \in V} \sup_{\mu \in \Lambda} \{\mathcal{J}(v) + b(v, \mu) - [g, \mu]\} = \inf_{v \in V} \sup_{\mu \in \Lambda} \mathcal{L}(v, \mu).$$

We show that the Lagrangian $\mathcal{L}$ has a unique saddle point on $V \times \Lambda$, by verifying the assumptions of Theorem 3.9.

First of all

$$\lim_{\substack{\|v\| \to \infty \\ v \in V}} \mathcal{L}(v, \mu) = +\infty$$

holds for any $\mu \in \Lambda$. Let $\overline{\mu} \in \Lambda$ be fixed and assume the relationship

$$\inf_{v \in V} \mathcal{L}(v, \overline{\mu}) = \mathcal{J}(u_{\overline{\mu}}) + b(u_{\overline{\mu}}, \overline{\mu}) - [g, \overline{\mu}], \tag{3.41}$$

where $u_{\overline{\mu}} \in V$ is a unique solution of

$$a(u_{\overline{\mu}}, v) + b(v, \overline{\mu}) = \langle f, v \rangle \quad \forall v \in V, \tag{3.42}$$

Substituting $v = u_{\overline{\mu}}$ into (3.42) and then eliminating terms $-\langle f, u_{\overline{\mu}} \rangle + b(u_{\overline{\mu}}, \overline{\mu})$ in (3.41), we obtain

$$\inf_{v \in V} \mathcal{L}(v, \overline{\mu}) \leqslant -\frac{\alpha}{2} \|u_{\overline{\mu}}\|^2 + \|g\|_{Y'} \|\overline{\mu}\|_Y, \tag{3.43}$$

where $\| \ \|_{Y'}$ denotes the dual norm in Y'. From (3.42), (3.1) and (3.38) it follows that

$$\beta \|\overline{\mu}\|_Y \leqslant M \|u_{\overline{\mu}}\| + \|f\|_*.$$

This together with (3.43) implies that (3.34) holds. The existence of a saddle point follows from Theorem 3.9.

The saddle point is unique. Indeed, as $\mathcal{L}$ is strictly convex as a function of the first variable, the uniqueness of the first component is a consequence of Remark 3.9.

Let (u, λ_i), $i = 1, 2$ be a saddle-point of $\mathcal{L}$ on $V \times \Lambda$. Then

$$a(u, v) + b(v, \lambda_1) = \langle f, v \rangle,$$
$$a(u, v) + b(v, \lambda_2) = \langle f, v \rangle$$

holds for any $v \in V$. Subtracting the second equation from the first one we obtain

$$b(v, \lambda_1 - \lambda_2) = 0 \quad \forall v \in V$$

which implies $\lambda_1 = \lambda_2$ as a consequence of (3.38).

We can summarize the previous result as follows.

THEOREM 3.11. *Let V and Y be two Hilbert spaces, $b: V \times Y \to \mathbb{R}^1$ a bilinear form satisfying* (3.37) *and* (3.38). *Let $\Lambda \subseteq Y$ be a closed, convex cone, containing the zero element of Y and $K \subseteq V$ a closed, convex subset, defined by* (3.39). *Then there exists a unique saddle point (u, λ) of $\mathscr{L}(v, \lambda) = \mathscr{J}(v) + b(v, \mu) - [g, \mu]$ on $V \times \Lambda$, where $g \in Y'$ and $\mathscr{J}$ is the quadratic functional* (3.4).

REMARK 3.10. Let all assumptions of the previous example be satisfied. Then the unique saddle point (u, λ) of $\mathscr{L}$ on $V \times \Lambda$ can be characterized as follows:

$$\begin{aligned}
&(u, \lambda) \in V \times \Lambda \text{ is such that}\\
&a(u, v) + b(v, \lambda) = \langle f, v \rangle \quad \forall v \in V,\\
&b(u, \mu - \lambda) \leqslant [g, \mu - \lambda] \quad \forall \mu \in \Lambda.
\end{aligned} \tag{3.44}$$

The system (3.44) is also meaningful in the case when the bilinear form a is not symmetric. In this case (3.44) is not related to a saddle point formulation. One can show however that there is a unique pair $(u, \lambda) \in V \times \Lambda$ satisfying (3.44), provided that (3.37) and (3.38) hold (see BREZZI, HAGER and RAVIART [1977]).

DEFINITION 3.3. Let the functionals L and $\mathscr{L}$ be related by means of (3.27). The problem of finding a saddle point of $\mathscr{L}$ on $A \times B$ will be called the *mixed variational formulation* of (3.27).

REMARK 3.11. The same physical problem can be studied using different variational formulations. Which of them will really be preferred for the numerical realization of the problem will depend on the characteristics of the problem itself (one type of the variational formulation may lead to a mathematical model of a simpler structure than others). The choice of the variational formulation should also be influenced by the nature of the physical quantity, which is of central importance us (usually knowledge of the stress distribution is more important than knowledge of the displacement field in many problems concerning the mechanics of solids). One can therefore expect that the direct use of the variational formulation in terms of stresses will lead to better approximation of stresses compared with their approximation through a displacement formulation. Application of the primal and dual formulations makes possible a direct approximation of the primal and dual quantities, respectively. The mixed variational formulation enables us to make simultaneous approximations of both the primal and dual quantities.

EXAMPLE 3.9. Let us consider the inequality from Example 3.7. Denote by $H^{1/2}(\Gamma_2)$ the space of restrictions of traces of all functions belonging to V:

$$H^{1/2}(\Gamma_2) = \{\varphi : \partial\Omega \to \mathbb{R}^1 \mid \exists v \in V \text{ such that } \varphi = v \text{ on } \Gamma_2\}$$

and by

$$H_+^{1/2}(\Gamma_2) = \{\varphi \in H^{1/2}(\Gamma_2)\colon \varphi \geqslant 0 \text{ on } \Gamma_2\}$$

a cone of nonnegative functions from $H^{1/2}(\Gamma_2)$. Let $H^{-1/2}(\Gamma_2)$ be the dual of $H^{1/2}(\Gamma_2)$, $\langle\, ,\, \rangle_{\Gamma_2}$ the corresponding duality pairing and let Λ be a cone of nonnegative functionals on $H^{1/2}(\Gamma_2)$:

$$\Lambda = \{\mu \in H^{-1/2}(\Gamma_2) | \langle \mu, \varphi\rangle_{\Gamma_2} \geqslant 0 \ \forall \varphi \in H_+^{1/2}(\Gamma_2)\}.$$

Using the separation theorem (see CÉA [1971]) one can prove that

$$v \in K \text{ if and only if } v \in V \text{ and } \langle \mu, v\rangle_{\Gamma_2} \geqslant 0 \quad \forall \mu \in \Lambda. \tag{3.44'}$$

Hence the function $\Phi(v,\mu) \equiv -\langle \mu, v\rangle_{\Gamma_2}$ possesses properties (3.28) and (3.29). Consequently we may write

$$\begin{aligned}\inf_{v\in K} \mathcal{J}(v) &= \inf_{v\in V} \sup_{\mu\in\Lambda} \{\mathcal{J}(v) - \langle \mu, v\rangle_{\Gamma_2}\} \\ &= \inf_{v\in V} \sup_{\mu\in\Lambda} \mathcal{L}(v,\mu).\end{aligned}$$

According to our classification, the mixed variational formulation of $\{K, a, f\}$ is defined as the problem of finding a saddle point of $\mathcal{L}$ on $V \times \Lambda$. Let us study the mixed formulation in detail. Let $u \in K$ be a solution of $\{K, a, f\}$. Then using Green's formula, one can show that u solves the following problem:

$$\begin{aligned}&-\Delta u = f \quad \text{in } \Omega,\\ &u = 0 \quad \text{on } \Gamma_1,\\ &\frac{\partial u}{\partial n} \in \Lambda, \qquad u \in H_+^{1/2}(\Gamma_2), \qquad \langle \frac{\partial u}{\partial n}, u\rangle_{\Gamma_2} = 0.\end{aligned}$$

In what follows we shall prove that $(u, \partial u/\partial n) \in V \times \Lambda$ is a unique saddle point of $\mathcal{L}$ on $V \times \Lambda$.

Let $(w, \lambda) \in V \times \Lambda$ be a saddle point. Then from Theorem 3.10 it follows that

$$(\nabla w, \nabla v)_{0,\Omega} = (f, v)_{0,\Omega} + \langle \lambda, v\rangle_{\Gamma_2} \quad \forall v \in V, \tag{3.45}$$

$$\langle \mu - \lambda, w\rangle_{\Gamma_2} \geqslant 0 \quad \forall \mu \in \Lambda. \tag{3.46}$$

Applying Green's formula,

$$(\nabla w, \nabla v)_{0,\Omega} = (-\Delta w, v)_{0,\Omega} + \langle \frac{\partial w}{\partial n}, v\rangle_{\Gamma_2}$$

which holds for any $w \in H^1(\Omega)$ such that $\Delta w \in L^2(\Omega)$ and any $v \in V$. We see from (3.45) that w and λ are related by $\partial w/\partial n = \lambda$ on Γ_2.

As Λ is a convex cone containing zero, (3.46) is equivalent to

$$\langle \lambda, w\rangle_{\Gamma_2} = 0, \tag{3.47}$$

$$\langle \mu, w\rangle_{\Gamma_2} \geqslant 0 \quad \forall \mu \in \Lambda, \tag{3.48}$$

i.e. $w \in K$. Next we show that w solves $\{K, a, f\}$. Substituting $v - w$ with $v \in K$ instead of v into (3.45) we have

$$(\nabla w, \nabla v - \nabla w)_{0,\Omega} = (f, v - w)_{0,\Omega} + \langle \lambda, v - w\rangle_{\Gamma_2} \geqslant (f, v - w)_{0,\Omega},$$

because of (3.47) and (3.44′). Hence $w = u$ and $\lambda = \partial u/\partial n$ with $u \in K$ being the solution of $\{K, a, f\}$. Consequently there is no more than one saddle-point of $\mathscr{L}$ on $V \times \Lambda$.

Now, let us show that the pair $(u, \partial u/\partial n) \in V \times \Lambda$ is a saddle-point of $\mathscr{L}$ on $V \times \Lambda$ by verifying the inequalities characterizing a saddle-point. First of all

$$\mathscr{L}(u, \frac{\partial u}{\partial n}) = \mathscr{J}(u) - \langle \frac{\partial u}{\partial n}, u\rangle_{\Gamma_2} \geqslant \mathscr{J}(u) - \langle \mu, u\rangle_{\Gamma_2} = \mathscr{L}(u, \mu) \tag{3.49}$$

holds for any $\mu \in \Lambda$ as follows from (3.44′). Let us show that

$$\mathscr{L}(u, \frac{\partial u}{\partial n}) \leqslant \mathscr{L}(v, \frac{\partial u}{\partial n}) \quad \forall v \in V. \tag{3.50}$$

This will complete the proof. Using the fact that

$$\|\nabla u\|_{0,\Omega}^2 = (f, u)_{0,\Omega},$$

the left hand side of (3.50) is equal to $-\frac{1}{2}\|\nabla u\|_{0,\Omega}^2$ and consequently (3.50) is equivalent to

$$\begin{aligned}
&\tfrac{1}{2}\|\nabla v\|_{0,\Omega}^2 + \tfrac{1}{2}\|\nabla u\|_{0,\Omega}^2 - (f, v)_{0,\Omega} - \langle \frac{\partial u}{\partial n}, v\rangle_{\Gamma_2} \geqslant 0 \\
&\quad \Leftrightarrow \tfrac{1}{2}\|\nabla u - \nabla v\|_{0,\Omega}^2 + (\nabla u, \nabla v)_{0,\Omega} - (f, v)_{0,\Omega} - \langle \frac{\partial u}{\partial n}, v\rangle_{\Gamma_2} \geqslant 0 \\
&\quad \Leftrightarrow \tfrac{1}{2}\|\nabla u - \nabla v\|_{0,\Omega}^2 \geqslant 0 \quad \forall v \in V.
\end{aligned}$$

The last equivalence follows from (3.44) and Green's formula. Consequently $(u, \partial u/\partial n) \in V \times \Lambda$ is a saddle-point.

Next we derive an explicit form of the dual variational formulation, corresponding to the choice of Lagrange function. Denote

$$\tilde{\mathscr{S}}(\mu) = \inf_{v \in V} \mathscr{L}(v, \mu), \quad \mu \in \Lambda$$

the dual functional, i.e.

$$\tilde{\mathscr{S}}(\mu) = \mathscr{L}(z(\mu), \mu),$$

where $z(\mu) \in V$ is a unique solution of

$$(\nabla z(\mu), \nabla v)_{0,\Omega} = (f, v)_{0,\Omega} + \langle \mu, v \rangle_{\Gamma_2} \quad \forall v \in V. \tag{3.51}$$

One can split $z(\mu)$ into $\tilde{z}$ and $\hat{z}$ and write $z(\mu) = \tilde{z} + \hat{z}$, where $\tilde{z}$, $\hat{z} \in V$ are unique solutions of

$$(\nabla \hat{z}, \nabla v)_{0,\Omega} = (f, v)_{0,\Omega} \quad \forall v \in V, \tag{3.52}$$

$$(\nabla \tilde{z}, \nabla v)_{0,\Omega} = \langle \mu, v \rangle_{\Gamma_2} \quad \forall v \in V. \tag{3.53}$$

Denote by $G : V' \to V$ the Green's operator corresponding to the bilinear form a. Instead of (3.52) and (3.53) we can write

$$\hat{z} = G(f) \quad \text{and} \quad \tilde{z} = G(\mu).$$

Since $z = z(\mu) \in V$ is the minimizer of $\mathscr{L}(v, \mu)$ and $\mu \in \Lambda$, we obtain

$$\begin{aligned} \mathscr{L}(z, \mu) &= -\tfrac{1}{2}\langle \mu, z \rangle_{\Gamma_2} - \tfrac{1}{2}(f, z)_{0,\Omega} \\ &= -\tfrac{1}{2}\langle \mu, \hat{z} \rangle_{\Gamma_2} - \tfrac{1}{2}\langle \mu, \tilde{z} \rangle_{\Gamma_2} - \tfrac{1}{2}(f, \hat{z})_{0,\Omega} - \tfrac{1}{2}(f, \tilde{z})_{0,\Omega}. \end{aligned} \tag{3.54}$$

Inserting $v = \tilde{z}$ into (3.52) and by using the symmetry of the scalar product and (3.53), we arrive at

$$\langle \mu, \hat{z} \rangle_{\Gamma_2} = (\nabla \hat{z}, \nabla \tilde{z})_{0,\Omega} = (f, \tilde{z})_{0,\Omega}.$$

Thus, (3.54) can be written as

$$\mathscr{L}(z, \mu) = -\tfrac{1}{2}\langle \mu, G(\mu) \rangle_{\Gamma_2} - \langle \mu, G(f) \rangle_{\Gamma_2} - \tfrac{1}{2}(f, G(f))_{0,\Omega}.$$

Let $b : H^{-1/2}(\Gamma_2) \times H^{-1/2}(\Gamma_2) \to \mathbb{R}^1$ be a bilinear form

$$b(\mu, \nu) \equiv \langle \mu, G(\nu) \rangle_{\Gamma_2}. \tag{3.55}$$

and $p : H^{-1/2}(\Gamma_2) \to \mathbb{R}^1$ a linear functional

$$p(\mu) = -\langle \mu, G(f) \rangle_{\Gamma_2}. \tag{3.56}$$

Then $\tilde{\mathscr{S}}(\mu) = -\tfrac{1}{2}b(\mu, \mu) + p(\mu) - \tfrac{1}{2}(f, G(f))_{0,\Omega}$.

Finally denoting $\mathscr{S}(\mu) \equiv -\tilde{\mathscr{S}}(\mu) - \tfrac{1}{2}(f, G(f))_{0,\Omega}$, the dual variational formulation of $\{K, a, f\}$ reads as follows:

$$\text{find } \lambda \in \Lambda \text{ such that } \mathscr{S}(\lambda) \leqslant \mathscr{S}(\mu) \quad \forall \mu \in \Lambda. \tag{3.57}$$

In accordance with the theory presented before, there exists a unique solution λ of (3.57) and $\lambda = \partial u / \partial n$ on Γ_2, where $u \in K$ solves $\{K, a, f\}$. Note that (3.57) is a variational formulation in terms of quantities defined on the boundary only.

REMARK 3.12. From the previous example we see that the proof of the existence and uniqueness of the saddle-point can be carried out without the use of Theorem 3.11. First of all we derive the "physical meaning" of the components of a saddle-point. As the second step we show that a point discovered in this way is really a saddle-point of the corresponding Lagrangian.

At the end of this chapter we briefly discuss the case of the so-called semicoercive variational inequalities. We deal with inequalities of the first kind only.

Let X be the set of all solutions of the variational inequality $\{K,a,f\}$. If the bilinear form $a : V \times V \to \mathbb{R}^1$ is non-negative on V, i.e. $a(v,v) \geqslant 0\ \forall v \in V$, then X is closed and convex but possibly empty.

REMARK 3.13. If $a : V \times V \to \mathbb{R}^1$ is a bounded, symmetric and non-negative bilinear form, the corresponding functional $\mathcal{J}$ given by (3.4) is lower weakly semicontinuous and convex on V but not necessarily coercive on V.

Next we shall suppose that $a : V \times V \to \mathbb{R}^1$ is *semicoercive* on V, i.e.

$$\exists \alpha = \text{const} > 0:\ a(v,v) \geqslant \alpha |v|^2 \quad \forall v \in V, \tag{3.58}$$

where $|\ |$ denotes a seminorm on V. If no additional assumptions are made on $f \in V'$, it may happen that $X = \emptyset$. To see that we shall consider the following model example.

EXAMPLE 3.10. Let $\{K,a,f\}$ be the variational inequality, where $V = H^1(\Omega)$,

$$K = \{v \in V \mid v \geqslant 0 \text{ on } \partial\Omega\}, \qquad a(y,v) = (\nabla y, \nabla v)_{0,\Omega},$$
$$\langle f,v\rangle = (f,v)_{0,\Omega}, \quad f \in L^2(\Omega).$$

Then $a(v,v) = |v|^2_{1,\Omega}$ satisfies (3.58).

Let $u \in K$ be a solution of $\{K,a,f\}$. Substituting $v = u+1 \in K$ into the definition of $\{K,a,f\}$, we are led to $(f,1)_{0,\Omega} = (\nabla u, \nabla 1)_{0,\Omega} \leqslant 0$ and consequently

$$(f,1)_{0,\Omega} \leqslant 0 \tag{3.59}$$

is the *necessary* condition for the existence of u.

Next we prove that the strict inequality

$$(f,1)_{0,\Omega} < 1 \tag{3.60}$$

guarantees the existence and the uniqueness of the solution. Let $u_1, u_2 \in K$ be solutions of $\{K,a,f\}$:

$$(\nabla u_i, \nabla v - \nabla u_i)_{0,\Omega} \geqslant (f, v-u_i)_{0,\Omega} \quad \forall v \in K.$$

Substituting first the function u_2, then u_1 for v and then subtracting resulting inequalities, we obtain

$$|u_1 - u_2|_{1,\Omega}^2 \leqslant 0.$$

Hence $u_2 = u_1 + c$, where $c \in \mathbb{R}^1$. If $c \neq 0$ then

$$\mathcal{J}(u_1 + c) = \mathcal{J}(u_1) = (f, u_1 + c)_{0,\Omega} = (f, u_1)_{0,\Omega} \Leftrightarrow (f, 1)_{0,\Omega} = 0$$

which contradicts our assumption on f. Hence $c = 0$ and $u_1 = u_2$. The next step is to show that the quadratic functional $\mathcal{J}$ is coercive on K, provided (3.60) holds. Let $\Gamma_0 \subset \partial\Omega$ be a non-empty open part of $\partial\Omega$ and define

$$\overline{v} = \frac{1}{\operatorname{meas} \Gamma_0} \int_{\Gamma_0} v \, ds \quad \forall v \in H^1(\Omega).$$

Let $\tilde{v} = v - \overline{v}$. Then $\int_{\Gamma_0} \tilde{v} \, ds = 0$ and moreover there exists a positive constant c such that (see NEČAS [1967])

$$|\tilde{v}|_{1,\Omega} \geqslant c\|\tilde{v}\|_{1,\Omega}.$$

Now

$$\begin{aligned} \mathcal{J}(v) &= \tfrac{1}{2}|\tilde{v}|_{1,\Omega}^2 - (f, \tilde{v})_{0,\Omega} - \overline{v}(f,1)_{0,\Omega} \\ &\geqslant \tfrac{1}{2}c^2\|\tilde{v}\|_{1,\Omega}^2 - c_1\|\tilde{v}\|_{1,\Omega} - \overline{v}(f,1)_{0,\Omega}. \end{aligned} \tag{3.61}$$

Let $v \in K$, $\|v\|_{1,\Omega} \to \infty$. Then $\overline{v} \geqslant 0$ and at least one of the norms $\|\tilde{v}\|_{1,\Omega}$, $\|\overline{v}\|_{1,\Omega} = \overline{v}(\operatorname{meas}\Omega)^{1/2}$ increase to ∞. Then $\mathcal{J}(v) \to \infty$, for $\|v\|_{1,\Omega} \to \infty$, $v \in K$ follows from (3.60) and (3.61).

The existence of a solution is a consequence of a more general theoretical result (FICHERA [1972]): let $K \subset V$ be a *non-empty, closed, convex cone* containing the zero element of V, $a : V \times V \to \mathbb{R}^1$ a bounded and semicoercive bilinear form, $\mathcal{R} = \{v \in V \mid a(v,v) = 0\}$. Denote by

$$\mathcal{R}^* = \{z \in K \cap \mathcal{R} \mid -z \in K\}.$$

Then the following theorem holds.

THEOREM 3.12. *Let*

$$\langle f, y\rangle \leqslant 0 \quad \forall y \in K \cap \mathcal{R}, \tag{3.62}$$

$$\langle f, y\rangle < 0 \quad \forall y \in K \cap \mathcal{R}\backslash\mathcal{R}^*. \tag{3.63}$$

Then there exists $u \in K$, minimizing $\mathcal{J}(v) = \frac{1}{2}a(v,v) - \langle f, v\rangle$ on K. Any other solution can be written in the form $\tilde{u} = u + y$, where $y \in V \cap \mathcal{R}$ is such that $u + y \in K$ and $\langle f, y\rangle = 0$.

Applying this theorem in the previous example we see that

$$\mathcal{R} = P_0(\text{polynomials of degree 0 in } \Omega)$$

and

$$\mathcal{R}^* = \{0\}.$$

The condition (3.63) reads as (3.60) in our illustrative example. Also, the uniqueness of the solution immediately follows.

Other conditions guaranteeing the existence (eventually the uniqueness) of variational inequalities are presented in Chapter 3.

Bibliography and comments to Chapter I

The mathematical theory of elliptic inequalities was developed in the sixties and seventies especially by the French and Italian mathematical schools: LIONS and STAMPACCHIA [1967], LIONS [1969], FICHERA [1972]. See also KINDERLEHRER and STAMPACCHIA [1980], FRIEDMAN [1982]. Applications of variational inequalities in solid mechanics are broadly discussed in DUVAUT and LIONS [1976], PANAGIOTOPOULOS [1985] and other problems of mathematical physics in ELLIOT and OCKENDON [1980], CHIPOT [1984], RODRIGUES [1987]. As far as semicoercive inequalities are concerned, we refer to FICHERA [1972], BAIOCHI, GASTALDI and TOMARELLI [1986]. Dual and mixed variational formulations of variational problems are systematically studied in EKELAND and TEMAM [1976], see also CÉA [1971].

CHAPTER II

Approximate Solution of Variational Inequalities

4. Approximation of the elliptic inequalities

The aim of the present chapter is to study the approximation of variational inequalities, using the Galerkin procedure. We use the approach introduced by FALK [1974]. The convergence results follow from the corresponding error estimates. The same approach is adopted for the approximation of mixed formulations.

4.1. Approximation of the primal formulation

Let V be a real Hilbert space, V' its dual and $K \subset V$ a non-empty, closed and convex subset. Let $a : V \times V \to \mathbb{R}^1$ be a bounded and V-elliptic bilinear form and $f \in V'$. By $u \in K$ we denote the unique solution of the elliptic inequality $\{K, a, f\}$:

$$u \in K\colon\ a(u, v-u) \geqslant \langle f, v-u\rangle \quad \forall v \in K. \tag{4.1}$$

Let $\{V_h\}, h \in (0,1)$ be a family of finite dimensional subspaces of V, $\dim V_h = n(h)$, $n(h) \to +\infty$ if $h \to 0+$. Let $K_h \subset V_h$ be a non-empty, closed and convex subset of V_h. By u_h we denote the solution of the elliptic inequality $\{K_h, a, f\}$:

$$u_h \in K_h\colon\ a(u_h, v_h - u_h) \geqslant \langle f, v_h - u_h\rangle \quad \forall v_h \in K_h. \tag{4.2}$$

If the bilinear form $a : V \times V \to \mathbb{R}^1$ is symmetric on V, the solution u_h of (4.2) can be characterized as the minimizer of $\mathcal{J}$ over K_h:

$$u_h \in K_h \text{ is such that } \mathcal{J}(u_h) \leqslant \mathcal{J}(v_h) \quad \forall v_h \in K_h, \tag{4.2'}$$

where $\mathcal{J}(v) = \frac{1}{2}a(v,v) - \langle f, v\rangle$.

Next we shall study the mutual relation between u and u_h and we shall formulate conditions under which $u_h \to u$ in V when $h \to 0+$.

LEMMA 4.1. *It holds that*

$$\begin{aligned} a(u-u_h, u-u_h) \leqslant{}& \langle f, u - v_h\rangle + \langle f, u_h - v\rangle + a(u_h - u, v_h - u) \\ &+ a(u, v - u_h) + a(u, v_h - u) \quad \forall v \in K,\ \forall v_h \in K_h. \end{aligned} \tag{4.3a}$$

PROOF. From (4.1) and (4.2) we immediately obtain

$$\begin{aligned} &a(u,u) \leqslant a(u,v) + \langle f, u-v\rangle \quad \forall v \in K,\\ &a(u_h,u_h) \leqslant a(u_h,v_h) + \langle f, u_h-v_h\rangle \quad \forall v_h \in K_h. \end{aligned} \tag{4.4}$$

Then

$$\begin{aligned} a(u-u_h,u-u_h) &\leqslant a(u,u) - a(u,u_h) - a(u_h,u) + a(u_h,u_h)\\ &\leqslant a(u,v) + \langle f,u-v\rangle - a(u,u_h) - a(u_h,u)\\ &\quad + a(u_h,v_h) + \langle f,u_h - v_h\rangle \quad \forall v \in K,\ \forall v \in K_h \end{aligned} \tag{4.5}$$

making use of (4.4). A simple calculation shows that the right hand sides of the inequalities (4.3a) and (4.5) are equal. □

REMARK 4.1. If $K_h \subset K$, then by choosing $v = u_h$ in (4.3a) we obtain the following simpler estimation of $a(u-u_h,u-u_h)$:

$$\begin{aligned} &a(u-u_h,u-u_h) \leqslant \langle f,u-v_h\rangle + a(u_h-u,v_h-u) + a(u,v_h-u)\\ &\quad \forall v_h \in K_h. \end{aligned} \tag{4.3b}$$

REMARK 4.2. Let $j: V \to \mathbb{R}^1 \cup \{-\infty,+\infty\}$ be a convex, lower semicontinuous and proper functional. Let $u \in V$ and $u_h \in V_h$ be the unique solutions of the elliptic inequalities of the second kind, $\{V,a,j,f\}$ and $\{V_h,a,j,f\}$, respectively:

$$u \in V\colon\ a(u,v-u) + j(v) - j(u) \geqslant \langle f,v-u\rangle \quad \forall v \in V,$$

$$u_h \in V_h\colon\ a(u_h,v_h-u_h) + j(v_h) - j(u_h) \geqslant \langle f,v_h-u_h\rangle \quad \forall v_h \in V_h.$$

Using exactly the same approach as before one can easily check that

$$\begin{aligned} a(u-u_h,u-u_h) &\leqslant \langle f,u-v_h\rangle + j(v_h) - j(u)\\ &\quad + a(u_h-u,v_h-u) + a(u,v_h-u) \quad \forall v_h \in V_h. \end{aligned} \tag{4.3c}$$

Relations (4.3a) and (4.3c) can be used to obtain

- the error estimates of $\|u-u_h\|$;
- the convergence statement $u_h \to u$, $h \to 0+$.

Next we shall analyse the convergence statement in more detail. To this end we introduce the following definition.

DEFINITION 4.1. A family $\{K_h\}$, $h \in (0,1)$ where $K_h \subset V_h$ are non-empty, closed and convex subsets of V_h is said to be an *approximation of* K iff

(i) $\forall v \in K\ \exists v_h \in K_h$: $v_h \to v$, $h \to 0+$ in V;

(ii) if $\{v_h\}$, $v_h \in K_h$ is such that $v_h \rightharpoonup v$, $h \to 0+$ in $V \Rightarrow v \in K$.

THEOREM 4.1. *Let $u \in K$ and $u_h \in K_h$ be solutions of $\{K, a, f\}$ and $\{K_h, a, f\}$, respectively. Let the family $\{K_h\}$ satisfy* (i) *and* (ii). *Then*

$$\varepsilon(h) \equiv \|u - u_h\| \to 0, \quad h \to 0+ .$$

PROOF. It is easy to see that the sequence $\{u_h\}$ is bounded. Hence, there is a subsequence $\{u_{h'}\} \subset \{u_h\}$ and an element $u^* \in V$ such that

$$u_{h'} \rightharpoonup u^*, \quad h' \to 0+ . \tag{4.6}$$

From (ii) it follows that $u^* \in K$. From (i) it follows that there exists a sequence $\{\overline{v}_h\}$, $\overline{v}_h \in K_h$ such that

$$\overline{v}_h \to u, \quad h \to 0+ . \tag{4.7}$$

Inserting $v := u^*$ and $v_h := \overline{v}_h$ into the right hand side of (4.3a) we arrive at

$$a(u_{h'} - u, u_{h'} - u) \to 0, \quad h' \to 0+$$

by making use of (4.6) and (4.7). The rest of the proof now follows from the V-ellipticity of the bilinear form a and the fact that the solution u of (4.1) is unique. □

REMARK 4.3. If $K_h \subset K$ for any $h \in (0, 1)$ then the condition (ii) is automatically satisfied.

REMARK 4.4. If the bilinear form a is semicoercive on V (see (3.58)), the situation concerning the convergence of u_h to u is more involved. Let V and H be two Hilbert spaces where $V \subset H$ with compact embedding. Let

$$\|v\|^2 = |v|^2 + \|v\|_H^2 \quad \forall v \in V,$$

where $\| \ \|_H$ is a norm in H. If the solutions u and u_h of (4.1) and (4.2), respectively, are unique and the sequence $\{\|u_h\|\}$ is bounded, the assertion of Theorem 4.1 remains valid under the same assumptions concerning the family $\{K_h\}$.

In order to show how (4.3a) can be used to establish the rate of convergence of u_h to u we shall assume the following elliptic inequality $\{K, a, f\}$:

$$K = \{v \in H_0^1(\Omega) \,|\, v \geqslant \psi \text{ a.e. in } \Omega\},$$

$$a(u, v) = (\nabla u, \nabla v)_{0,\Omega} \ \langle f, v\rangle = (f, v)_{0,\Omega},$$

$f \in L^2(\Omega)$, $\psi \in C(\overline{\Omega})$, $\psi \leqslant 0$ on $\partial\Omega$. Let us suppose that $\Omega \subset \mathbb{R}^2$ is a polygonal domain and $\{\mathcal{T}_h\}$ is a regular family of triangulations of $\overline{\Omega}$. With any $\mathcal{T}_h$ the following convex set K_h will be associated:

$$K_h = \{v_h \in C(\overline{\Omega}) \,|\, v_h|_{T_i} \in P_1(T_i) \ \forall T_i \in \mathcal{T}_h, \ v_h = 0 \text{ on } \partial\Omega,$$
$$v_h(M_i) \geqslant \psi(M_i) \ \forall i\},$$

where M_i are interior nodes of $\mathcal{T}_h$.

THEOREM 4.2. *Let us suppose that the solution* $u \in H^2(\Omega) \cap K$ *and* $\psi \in H^2(\Omega)$. *Then*

$$\|u - u_h\|_{1,\Omega} = \mathrm{O}(h), \quad h \to 0+ .$$

PROOF. It is easy to see that $K_h \not\subset K$, in general. From Green's formula and (4.3a) it follows that

$$\begin{aligned} |u - u_h|_{1,\Omega}^2 &= a(u - u_h, u - u_h) \\ &\leqslant (-\Delta u - f, v - u_h)_{0,\Omega} + (-\Delta u - f, v_h - u)_{0,\Omega} \\ &\quad + a(u_h - u, v_h - u) \\ &\equiv I_1 + I_2 + I_3 \end{aligned} \tag{4.8}$$

holds for any $v_h \in K_h$ and $v \in K$. Let $v_h = r_h u$, where $r_h u$ denotes the piecewise linear approximation of u. Using standard approximation results we see that

$$\begin{aligned} |I_2 + I_3| &\leqslant c\|v_h - u\|_{0,\Omega} + \tfrac{1}{2}|u_h - u|_{1,\Omega}^2 + \tfrac{1}{2}|v_h - u|_{1,\Omega}^2 \\ &\leqslant ch^2\|u\|_{2,\Omega}^2 + \tfrac{1}{2}|u_h - u|_{1,\Omega}^2. \end{aligned} \tag{4.9}$$

As far as the choice of v is concerned, we take

$$v \equiv \sup\{\psi, u_h\}.$$

It is easy to show that $v \in K$ and moreover

$$(-\Delta u - f, v - u_h)_{0,\Omega} = (-\Delta u - f, \psi - u_h)_{0,\Omega^-},$$

where

$$\Omega^- = \{x \in \Omega |\, u_h(x) \leqslant \psi(x)\}.$$

As $u_h(M_i) \geqslant \psi(M_i)$ at all interior nodes of $\mathcal{T}_h$ and u_h is piecewise linear, then also $u_h \geqslant r_h\psi$ in Ω. Therefore,

$$\begin{aligned} (-\Delta u - f, \psi - u_h)_{0,\Omega^-} &\leqslant (-\Delta u - f, \psi - r_h\psi)_{0,\Omega^-} \\ &\leqslant c(\|u\|_{2,\Omega} + \|f\|_{0,\Omega})\|\psi - r_h\psi\|_{0,\Omega} \\ &\leqslant ch^2(\|u\|_{2,\Omega} + \|f\|_{0,\Omega})\|\psi\|_{2,\Omega}, \end{aligned} \tag{4.10}$$

using the fact that $-\Delta u \geqslant f$ a.e. in Ω. The proof of Theorem 4.2 follows from (4.8)–(4.10). □

REMARK 4.5. Let $\psi \in C(\overline{\Omega})$, $\psi \leqslant 0$ on $\partial\Omega$. Then system $\{K_h\}$, constructed above, satisfies (i) and (ii) from Definition 4.1. Hence $\varepsilon(h) \equiv \|u - u_h\|_{1,\Omega} \to 0$, $h \to 0+$ without any regularity assumptions on u.

REMARK 4.6. Another approach, which allows error analysis, has been introduced by MOSCO and STRANG [1974]. Let $\{K, a, f\}$ be an elliptic inequality, $\{K_h, a, f\}$ its approximation and $K_h \subset K$ $\forall h \in (0,1)$, where $a : V \times V \to \mathbb{R}^1$ is a bounded and V-elliptic bilinear form. Then

$$\|u - u_h\| \leqslant c \inf_{w_h \in \mathcal{U}_h} \|u - w_h\|,$$

where $\mathcal{U}_h = \{w_h \in K_h \mid 2u - w_h \in K\}$. For the application of this approach see also HLAVÁČEK, HASLINGER, NEČAS and LOVÍŠEK [1988].

4.2. Approximation of mixed formulations

Let V and Y be two real Hilbert spaces, V' and Y' their duals, $\langle\, ,\rangle$ and $[\, ,]$ the duality pairings between V' and V and Y' and Y, respectively. The norms in V and Y will be denoted by $\|\ \|$ and $\|\ \|_Y$, respectively. Let $b : V \times Y \to \mathbb{R}^1$ be a bilinear form, satisfying (3.37) and (3.38), $a : V \times V \to \mathbb{R}^1$ be V-elliptic, bounded and symmetric on V. Set

$$\mathcal{L}(v, \mu) = \tfrac{1}{2}a(v, v) - \langle f, v\rangle + b(v, \mu) - [g, \mu], \quad (v, \mu) \in V \times Y,$$

where $f \in V'$, $g \in Y'$ are given.

Let $K \subseteq V$, $\Lambda \subseteq Y$ be non-empty, closed convex sets. In the following, we shall assume that Λ is either

a *convex cone,* containing the zero element of Y and $K = V$ (4.11a)

or

a *bounded convex subset* of Y. (4.11b)

Let $(u, \lambda) \in K \times \Lambda$ be a saddle point of $\mathcal{L}$ on $K \times \Lambda$:

$$\mathcal{L}(u, \mu) \leqslant \mathcal{L}(u, \lambda) \leqslant \mathcal{L}(v, \lambda) \quad \forall (v, \mu) \in K \times \Lambda$$

or equivalently (see Theorem 3.10):

$$\begin{aligned} &a(u, v - u) + b(v - u, \lambda) \geqslant \langle f, v - u\rangle \quad \forall v \in K, \\ &b(u, \mu - \lambda) \leqslant [g, \mu - \lambda] \quad \forall \mu \in \Lambda. \end{aligned} \tag{4.12}$$

THEOREM 4.3. *Let the bilinear form b satisfy* (3.37) *and* (3.38). *Then there exists a unique saddle-point* $(u, \lambda) \in K \times \Lambda$ *of* $\mathcal{L}$ *on* $K \times \Lambda$.

PROOF. See Example 3.8. □

REMARK 4.7. If (4.11b) is satisfied and b satisfy (3.37) only, there exists a saddle-point of $\mathcal{L}$ on $K \times \Lambda$, the first component of which is uniquely determined. This is a direct consequence of Theorem 3.9.

REMARK 4.8. Let (4.11a) be satisfied. Then (4.12) is the mixed formulation of the problem, the primal formulation of which reads as follows:

$$u \in K\colon \mathcal{J}(u) = \min_K \mathcal{J}(v),$$

where $K = \{v \in V \mid b(v,\mu) \leqslant [g,\mu] \quad \forall \mu \in \Lambda\}$ and

$$\mathcal{J}(v) = \tfrac{1}{2}a(v,v) - \langle f,v\rangle.$$

Indeed, $j(v) \equiv \sup_\Lambda\{b(v,\mu) - [g,\mu]\}$ is the indicator function of K.

Let $\{V_h\}$ and $\{Y_H\}$, h, $H \in (0,1)$ be two systems of the finite-dimensional subspaces of V and Y, respectively. Let K_h and Λ_H be non-empty, closed, convex subsets of V_h and Y_H, respectively. With any pair $(h,H) \in (0,1)$ we associate the set $K_h \times \Lambda_H$. Analogously to the continuous case we assume that Λ_H is either

a *convex cone*, containing the zero element of Y (4.13a)

or

a *uniformly bounded convex subset* of Y, i.e. there exists a positive number $c > 0$ such that $\|\mu_H\|_Y \leqslant c \quad \forall \mu_H \in Y_H, \ \forall H \in (0,1)$. (4.13b)

By the *approximation* of (4.12) we mean the problem of finding a saddle-point $(u_h, \lambda_H) \in K_h \times \Lambda_H$ of $\mathcal{L}$ on $K_h \times \Lambda_H$:

$$\mathcal{L}(u_h,\mu_H) \leqslant \mathcal{L}(u_h,\lambda_H) \leqslant \mathcal{L}(v_h,\lambda_H) \quad \forall (v_h,\mu_H) \in K_h \times \Lambda_H$$

or equivalently

$$\begin{aligned} &a(u_h, v_h - u_h) + b(v_h - u_h, \lambda_H) \geqslant \langle f, v_h - u_h\rangle \quad \forall v_h \in K_h,\\ &b(u_h, \mu_H - \lambda_H) \leqslant [g, \mu_H - \lambda_H] \quad \forall \mu_H \in \Lambda_H. \end{aligned} \tag{4.14}$$

Interpretation of (4.14). Let

$$j_H(v_h) = \sup_{\Lambda_H}\{b(v_h,\mu_H) - [g,\mu_H]\}.$$

Then the first component u_h of the saddle point (u_h, λ_H) is the solution of

$$\mathcal{J}(u_h) + j_H(u_h) \leqslant \mathcal{J}(v_h) + j_H(v_h) \quad \forall v_h \in K_h.$$

If (4.13a) is satisfied, then $j_H(v_h)$ is the indicator function of the closed convex set K_{hH} given by

$$K_{hH} = \{v_h \in V_h | \, b(v_h, \mu_H) \leqslant [g, \mu_H] \; \forall \mu_H \in \Lambda_H\}.$$

Concerning the existence of a solution of (4.14), we have the following.

THEOREM 4.4. *Let us assume that either* (4.13b) *or* (4.13a) *is satisfied and in the later case K_{hH} is a set with a non-empty interior. Then there exists a solution (u_h, λ_H) of* (4.14), *the first component of which is uniquely determined.*

For the proof see HLAVÁČEK, HASLINGER, NEČAS and LOVÍŠEK [1988, Lemmas 5.5 and 5.6, Section 1.1.52].

Now we shall study the mutual relationship between (4.14) and (4.12) when $h, H \to 0+$. To this end we present three lemmas, the proof of which can be found in HASLINGER [1981].

LEMMA 4.2. *Let (u, λ) and (u_h, λ_H) be solutions of* (4.12) *and* (4.14), *respectively. Then*

$$\begin{aligned}\|u - u_h\|^2 \leqslant c[&\|u - v_h\|^2 + \|\lambda - \mu_H\|_Y^2 + A_1(v_h) + A_2(v) \\ &+ \{b(u, \lambda_H - \mu) - [g, \lambda_H - \mu]\} \\ &+ \{b(u, \lambda - \mu_H) - [g, \lambda - \mu_H]\} + \|\lambda - \lambda_H\|_Y^2]\end{aligned}$$

holds for any $v_h \in K_h$, $v \in K$, $\mu_H \in \Lambda_H$, $\mu \in \Lambda$, where

$$\begin{aligned}A_1(v_h) &= a(u, v_h - u) + b(v_h - u, \lambda) + \langle f, u - v_h \rangle, \\ A_2(v) &= a(u, v - u_h) + b(v - u_h, \lambda) + \langle f, u_h - v \rangle\end{aligned}$$

and c is a positive constant independent of h, $H \in (0, 1)$.

LEMMA 4.3. *Let* (4.11a) *and* (4.13a) *be fulfilled and moreover, let there exists a positive constant $\tilde{\beta} > 0$ independent of h, $H \in (0, 1)$ such that*

$$\sup_{V_h} \frac{b(v_h, \mu_H)}{\|v_h\|} \geqslant \tilde{\beta} \|\mu_H\|_Y \quad \forall \mu_H \in \Lambda_H, \; \forall H \in (0, 1). \tag{4.15}$$

Then there exists a positive constant c independent of h, H such that

$$\begin{aligned}\|u - u_h\|^2 \leqslant c[&\|u - v_h\|^2 + \|\lambda - \mu_H\|_Y^2 \\ &+ \{b(u, \lambda_H - \mu) - [g, \lambda_H - \mu]\} \\ &+ \{b(u, \lambda - \mu_H) - [g, \lambda - \mu_H]\}], \\ \|\lambda - \lambda_H\|_Y \leqslant c\{&\|u - u_h\| + \|\lambda - \mu_H\|_Y\}\end{aligned} \tag{4.16}$$

hold for any $v_h \in V_h$, $\mu \in \Lambda$, $\mu_H \in \Lambda_H$.

REMARK 4.9. If $\Lambda_H \subset \Lambda$ for any $H \in (0,1)$, then by setting $\mu = \lambda_H$, the first equation of (4.16) reduces to

$$\|u - u_h\|^2 \leqslant c[\|u - v_h\|^2 + \|\lambda - \mu_H\|_Y^2 + \{b(u, \lambda - \mu_H) - [g, \lambda - \mu_H]\}] \quad \forall \mu_H \in \Lambda_H, \ \forall v_h \in V_h.$$

LEMMA 4.4. *Let* (4.11b) and (4.13b) *be fulfilled. Then if* $K = V$, $K_h = V_h$ *and condition* (4.15) *is satisfied,* (4.16) *holds, or*

$$\begin{aligned} \|u - u_h\|^2 \leqslant c[&\|u - v_h\|^2 + \|\lambda - \mu_H\|_Y^2 + A_1(v_h) + A_2(v) \\ &+ \|u - v_h\| + \{b(u, \lambda_H - \mu) - [g, \lambda_H - \mu]\} \\ &+ \{b(u, \lambda - \mu_H) - [g, \lambda - \mu_H]\}] \\ \forall v_h \in K_h, \ v \in K, \ &\mu \in \Lambda, \ \mu_H \in \Lambda_H, \end{aligned} \tag{4.17}$$

with $A_1(v_h)$ *and* $A_2(v)$ *having the same meaning as in Lemma* 4.2.

REMARK 4.10. If $K_h \subset K$ and $\Lambda_H \subset \Lambda$ $\forall h, H \in (0,1)$ then we can insert $v = u_h$ and $\mu = \lambda_H$ into (4.17). This choice of v and μ yields $A_2(v) = 0$ and $b(u, \lambda_H - \mu) - [g, \lambda_H - \mu] = 0$.

The error estimates presented above can be used either for the analysis of the convergence rate of u_h to u and λ_H to λ, respectively, provided both u and λ are sufficiently smooth, or for the proof of the following convergence results. Next we shall suppose that $h \to 0+$ iff $H \to 0+$.

THEOREM 4.5. *Let* (4.11b) *and* (4.13b) *be fulfilled and let*

$$\forall v \in K \ \exists v_h \in K_h\colon v_h \to v, \ h \to 0+ \ \text{in } V, \tag{4.18}$$

$$\forall \mu \in \Lambda \ \exists \mu_H \in \Lambda_H\colon \mu_H \to \mu, \ H \to 0+ \ \text{in } Y, \tag{4.19}$$

$$v_h \in K_h, \ v_h \rightharpoonup v, \ h \to 0+ \ \text{in } V \Rightarrow v \in K, \tag{4.20}$$

$$\mu_H \in \Lambda_H, \ \mu_H \rightharpoonup \mu, \ H \to 0+ \ \text{in } Y \Rightarrow \mu \in \Lambda. \tag{4.21}$$

Let the solution (u, λ) *of* (4.12) *be unique. Then*

$$u_h \to u, \ h \to 0+ \ \text{in } V,$$
$$\lambda_H \rightharpoonup \lambda, \ H \to 0+ \ \text{in } Y.$$

PROOF. First of all it is easy to see that $\{u_h\}$ and $\{\lambda_H\}$ are bounded in corresponding spaces. Consequently there exist subsequences $\{u_{h'}\} \subset \{u_h\}$ and $\{\lambda_{H'}\} \subset \{\lambda_H\}$ and a pair $(u^*, \lambda^*) \in V \times Y$ such that

$$u_{h'} \rightharpoonup u^*, \ h' \to 0+, \qquad \lambda_{H'} \rightharpoonup \lambda^*, \ H' \to 0+.$$

From (4.20) and (4.21) it follows that $u^* \in K$, $\lambda^* \in \Lambda$. The next step is to show that (u^*, λ^*) solves (4.12).

Let $(v, \mu) \in K \times \Lambda$ be an arbitrary fixed element. Then (4.18) and (4.19) imply the existence of sequences $\{v_h\}$, $\{\mu_H\}$, $v_h \in K_h$, $\mu_H \in \Lambda_H$ such that

$$v_h \to v, \ h \to 0+, \qquad \mu_H \to \mu, \ H \to 0+ \ .$$

The pair $(u_{h'}, \lambda_{H'})$, being the solution of (4.14), satisfies

$$a(u_{h'}, u_{h'} - v_{h'}) + b(u_{h'} - v_{h'}, \lambda_{H'}) \leqslant \langle f, u_{h'} - v_{h'} \rangle,$$

$$b(u_{h'}, \mu_{H'} - \lambda_{H'}) \leqslant [g, \mu_{H'} - \lambda_{H'}].$$

Passing to the limit with h', $H' \to 0+$ in the previous inequalities, we arrive at

$$a(u^*, u^* - v) + b(u^* - v, \lambda^*) \leqslant \langle f, u^* - v \rangle,$$

$$b(u^*, \mu - \lambda^*) \leqslant [g, \mu - \lambda^*].$$

As $(v, \mu) \in K \times \Lambda$ is arbitrarily chosen, the pair (u^*, λ^*) is a solution of (4.12). As a consequence of the assumption we have $(u^*, \lambda^*) = (u, \lambda)$. Moreover, not only the subsequences, but the entire sequences $\{u_h\}$ and $\{\lambda_H\}$ tend weakly to u and λ, respectively. It is also easy to prove that u_h tends to u strongly. This follows from the first equation of (4.16), (4.18) and (4.19). □

REMARK 4.11. If $K_h \subset K \ \forall h \in (0,1)$ or $\Lambda_H \subset \Lambda \ \forall H \in (0,1)$ then the condition (4.20) or (4.21), respectively, is automatically fulfilled.

THEOREM 4.6. *Let* (4.11a), (4.13a) *and* (4.15) *be satisfied and moreover, let*

$$\forall v \in V \ \exists v_h \in V_h\colon v_h \to v, \ h \to 0+ \ \text{in } V, \tag{4.22}$$

$$\forall \mu \in \Lambda \ \exists \mu_H \in \Lambda_H\colon \mu_H \to \mu, \ H \to 0+ \ \text{in } Y, \tag{4.23}$$

$$\mu_H \in \Lambda_H, \ \mu_H \rightharpoonup \mu, \ H \to 0+ \ \text{in } Y \Rightarrow \mu \in \Lambda, \tag{4.24}$$

$$\begin{aligned} &\text{there exists a bounded sequence } \{\overline{v}_h\}, \\ &\quad \overline{v}_h \in K_{hH}, \ h \to 0+, \ H \to 0+ \ . \end{aligned} \tag{4.25}$$

Let the solution (u, λ) *of* (4.12) *be unique. Then*

$$\begin{aligned} &u_h \to u, \ h \to 0+ \ \text{in } V, \\ &\lambda_H \to \lambda, H \to 0+ \ \text{in } Y. \end{aligned}$$

PROOF. We shall show that the sequences $\{u_h\}$ and $\{\lambda_H\}$ are bounded. The rest of the proof is exactly the same as in the preceding theorem. From the interpretation of (4.14) it follows that $u_h \in K_{hH}$ is a solution of the problem:

$$a(u_h, v_h - u_h) \geqslant \langle f, v_h - u_h \rangle \quad \forall v_h \in K_{hH}.$$

Inserting $v_h := \bar{v}_h$, where $\bar{v}_h$ satisfies (4.25), we obtain the boundedness of $\{u_h\}$. From this and the second equation of (4.16), the boundedness of $\{\lambda_H\}$ follows. □

REMARK 4.12. If $\Lambda_H \subset \Lambda\ \forall H \in (0,1)$, then (4.24) is automatically fulfilled.

In order to illustrate the use of the previous results, let us consider the elliptic inequality $\{K, a, f\}$, with

$$K = \{v \in H^1(\Omega) \mid v \geqslant g \text{ a.e. on } \partial\Omega\},$$
$$a(u,v) = (u,v)_{1,\Omega}\ \langle f, v\rangle = (f,v)_{0,\Omega},$$

where $f \in L^2(\Omega)$, $g \in C(\partial\Omega) \cap H^{1/2}(\partial\Omega)$ are given functions. The mixed formulation of $\{K, a, f\}$ reads as follows (see Example 3.9)

$$\begin{aligned} &\text{find } (u,\lambda) \in H^1(\Omega) \times \Lambda(\partial\Omega) \text{ such that} \\ &(u,v)_{1,\Omega} - \langle\lambda, v\rangle = (f,v)_{0,\Omega} \quad \forall v \in H^1(\Omega), \\ &\langle \mu - \lambda, u\rangle \geqslant \langle \mu - \lambda, g\rangle \quad \forall \mu \in \Lambda(\partial\Omega), \end{aligned} \tag{4.26}$$

where

$$\Lambda(\partial\Omega) = \{\mu \in H^{-1/2}(\partial\Omega) \mid \langle\mu, v\rangle \geqslant 0\ \forall v \in H^{1/2}(\partial\Omega),\ v \geqslant 0 \text{ on } \partial\Omega\}$$

and $\langle\ ,\ \rangle$ stands for the duality pairing between $H^{-1/2}(\partial\Omega)$ and $H^{1/2}(\partial\Omega)$. As in Example 3.9 it is easy to prove that the first component u solves $\{K, a, f\}$ and the second component $\lambda = \partial u/\partial n$ on $\partial\Omega$.

Approximation of (4.26). Let $\Omega \subset \mathbb{R}^2$ be a polygonal domain, $\{\mathcal{T}_h\}$ be a regular family of triangulations of $\bar{\Omega}$ and $h = \max \operatorname{diam} T_i$, $T_i \in \mathcal{T}_h$. Let $\{\mathcal{T}_H\}$ be a family of partitions of $\partial\Omega$, the nodes of which will be denoted by $a_i, i = 1, \ldots, M(H)$:

$$\partial\Omega = \bigcup_{i=1}^{M} \overline{a_i a_{i+1}}, \qquad a_1 = a_{M+1}$$

and $H_i = \operatorname{length} \overline{a_i a_{i+1}}$, $H = \max H_i$. Next we shall consider a regular family $\{\mathcal{T}_H\}$, $H \to 0+$ in the following sense: there exists a constant $\alpha_0 > 0$ such that

$$H/H_i \leqslant \alpha_0 \quad \forall i = 1, \ldots, M(H).$$

Let us introduce the finite dimensional spaces:

$$V_h = \{v_h \in C(\bar{\Omega}) \mid v_h|_{T_i} \in P_1(T_i)\ \forall T_i \in \mathcal{T}_h\},$$
$$\tilde{\Lambda}_H = \{\mu_H \in L^2(\partial\Omega) \mid \mu_H|_{\overline{a_i a_{i+1}}} \in P_0(\overline{a_i a_{i+1}})\ i = 1, \ldots, M\}.$$

The approximation of (4.26) is defined as follows:

$$\begin{aligned}&\text{find } (u_h, \lambda_H) \in V_h \times \Lambda_H \text{ such that}\\&(u_h, v_h)_{1,\Omega} - \langle \lambda_H, v_h\rangle = (f, v_h)_{0,\Omega} \quad \forall v_h \in V_h,\\&\langle \mu_H - \lambda_H, u_h\rangle \geqslant \langle \mu_H - \lambda_H, g\rangle \quad \forall \mu_H \in \Lambda_H\end{aligned} \tag{4.27}$$

with

$$\langle \mu_H, v_h\rangle = \int_{\partial\Omega} \mu_H v_h \,\mathrm{d}s \quad \forall v_h \in V_h, \quad \forall \mu_H \in \tilde{\Lambda}_H,$$

$$\Lambda_H = \{\mu_H \in \tilde{\Lambda}_H \mid \mu_H \geqslant 0 \text{ on } \partial\Omega\}.$$

From Theorem 4.4, the existence of a solution (u_h, λ_H) of (4.27) follows. Moreover, the first component u_h is unique and can be characterized by means of

$$u_h \in K_{hH}\colon \mathcal{J}(u_h) \leqslant \mathcal{J}(v_h) \quad \forall v_h \in K_{hH},$$

where

$$\mathcal{J}(v) = \tfrac{1}{2}\|v\|_{1,\Omega}^2 - (f, v)_{0,\Omega},$$

$$K_{hH} = \{v_h \in V_h \mid \pi_i(v_h) \geqslant \pi_i(g) \ \forall i = 1, \ldots, M\}.$$

Here $\pi_i(v)$ denotes the mean value of v on $\overline{a_i a_{i+1}}$.

The crucial point is the verification of condition (4.15) from Lemma 4.3. The following holds.

LEMMA 4.5. *Let h/H be sufficiently small. Then there exists a positive constant β such that*

$$\sup_{V_h} \frac{\langle \mu_H, v_h\rangle}{\|v_h\|_{1,\Omega}} \geqslant \beta \|\mu_H\|_{-1/2,\partial\Omega}$$

holds for any $\mu_H \in \tilde{\Lambda}_H$.

PROOF. Let $\mu \in H^{-1/2}(\partial\Omega)$. Then, following GIRAULT and RAVIART [1979],

$$\|\mu\|_{-1/2,\partial\Omega} = \sup_{H^1(\Omega)} \frac{\langle \mu, v\rangle}{\|v\|_{1,\Omega}} = \|\bar{v}\|_{1,\Omega},$$

where $\bar{v} \in H^1(\Omega)$ is the unique solution of the Neumann problem:

$$-\Delta v + v = 0 \quad \text{in } \Omega,$$

$$\frac{\partial v}{\partial n} = \mu \quad \text{on } \partial\Omega.$$

Let $\mu = \mu_H \in \tilde{\Lambda}_H$. Then $\mu_H \in H^{-1/2+\varepsilon}(\partial\Omega)$ $\forall \varepsilon \in\langle 0, 1)$ and the following regularity result holds: $\bar{v} \in H^{1+\varepsilon}(\Omega)$ and

$$\|\bar{v}\|_{1+\varepsilon,\Omega} \leqslant c(\varepsilon)\|\mu_H\|_{-1/2+\varepsilon,\partial\Omega}. \tag{4.28}$$

Furthermore, we write

$$\sup_{V_h} \frac{\langle \mu_H, v_h \rangle}{\|v_h\|_{1,\Omega}} \geqslant \frac{\langle \mu_H, \bar{v}_h \rangle}{\|\bar{v}_h\|_{1,\Omega}} = \|\bar{v}_h\|_{1,\Omega}, \tag{4.29}$$

where $\bar{v}_h \in V_h$ is the finite-element approximation of $\bar{v}$. The triangle inequality $\|\bar{v}_h\|_{1,\Omega} \geqslant \|\bar{v}\|_{1,\Omega} - \|\bar{v}_h - v\|_{1,\Omega}$ and (4.29) imply

$$\|\mu_H\|_{-1/2,\partial\Omega} = \|\bar{v}\|_{1,\Omega} \leqslant \sup_{V_h} \frac{\langle \mu_H, v_h \rangle}{\|v_h\|_{1,\Omega}} + \|\bar{v}_h - \bar{v}\|_{1,\Omega}. \tag{4.30}$$

Using well-known error estimates as well as (4.28) and the inverse inequality between $H^{-1/2}(\partial\Omega)$ and $H^{-1/2+\varepsilon}(\partial\Omega)$ for $\mu_H \in \tilde{\Lambda}_H$, we have

$$\begin{aligned} \|\bar{v} - \bar{v}_h\|_{1,\Omega} &\leqslant c(\varepsilon)h^{\varepsilon}\|\bar{v}\|_{1+\varepsilon,\Omega} \leqslant c(\varepsilon)h^{\varepsilon}\|\mu_H\|_{-1/2+\varepsilon,\partial\Omega} \\ &\leqslant c(\varepsilon)\left(\frac{h}{H}\right)^{\varepsilon}\|\mu_H\|_{-1/2,\partial\Omega}. \end{aligned}$$

From this and (4.30) the assertion of the lemma follows. □

Consequently if the ratio h/H is sufficiently small, then the second component λ_H of the solution of (4.27) is unique. Using Lemma 4.3 the following convergence rates can be established (see HASLINGER and LOVÍŠEK [1980]).

THEOREM 4.7. *Let the solution u of $\{K, a, f\}$ be such that $u \in H^2(\Omega) \cap K$, moreover let traces u, $g \in H^{1,\infty}(\overline{a_i, a_{i+1}})$ $\forall i = 1, \ldots, M$, and the set of points from $\partial\Omega$, where u changes from $u > g$ to $u = g$ is finite. If the ratio h/H is sufficiently small, then*

$$\|u - u_h\|_{1,\Omega} \leqslant c(h + H),$$
$$\|\lambda - \lambda_H\|_{-1/2,\partial\Omega} \leqslant c(h + H),$$

where c is a positive constant, which does not depend on h, H.

REMARK 4.13. If there is no regularity information on u and the condition (4.15) holds then

$$\|u - u_h\|_{1,\Omega} \to 0, \quad h \to 0+,$$
$$\|\lambda - \lambda_H\|_{-1/2,\partial\Omega} \to 0, \quad H \to 0+$$

as follows from Theorem 4.6 and the fact that the system $\{\Lambda_H\}$ is dense in Λ.

REMARK 4.14. The condition, requiring the ratio h/H to be sufficiently small means that the triangulation $\mathcal{T}_h$ of $\overline{\Omega}$ is finer then the partition $\mathcal{T}_H$ of $\partial\Omega$.

To illustrate that some relation between h and H has to be satisfied in order to guarantee the validity of condition (4.15), let us assume that the partition $\mathcal{T}_H$ of $\partial\Omega$ is generated by the boundary nodes of $\mathcal{T}_H$. Moreover, let us assume that the nodes $a_1, \ldots, a_M$ form an equidistant partition of $\partial\Omega$. Then the relation

$$\langle \mu_H, v_h \rangle = 0 \quad \forall v_h \in V_h$$

is equivalent to the system of linear algebraic equations:

$$\begin{aligned} \mu_1 + \mu_2 &= 0, \\ \mu_2 + \mu_3 &= 0, \\ \mu_1 + \mu_M &= 0, \end{aligned}$$

where $\mu_i = \mu_H|_{\overline{a_i a_{i+1}}}$. If M is an even number, then the system also has a non-trivial solution. Thus, condition (4.15) cannot be fulfilled.

Bibliography and comments to Chapter II

A detailed analysis of the approximation of variational inequalities is contained in GLOWINSKI, LIONS and TRÉMOLIÈRES [1981], HLAVÁČEK, HASLINGER, NEČAS and LOVÍŠEK [1988]. Here we discussed the approximation in the norm of the basic space of functions with finite energy. The convergence of approximate solutions in L^∞-norm has been studied in BAIOCCHI [1977], CORTEY-DUMONT [1985b], FINZI VITA [1982]. On the basis of these results, the behaviour of approximate free boundaries can be established, see NOCHETO [1986], BREZZI and CAFARELLI [1983]. The finite element approximation of mixed variational formulations of elliptic inequalities with strongly monotonic operators is analysed in GROSSMAN and ROSS [1992]. The approximation of semicoercive inequalities is studied in GWINNER [1991]. The approximation of variational inequalities with non-coercive operators is presented in CORTEY-DUMONT [1985a].

CHAPTER III

Contact Problems in Elasticity

A unilateral problem of historical importance is the famous problem of Signorini, who formulated, as early as 1933, the case of one-sided contact of one body with a perfectly rigid and smooth foundation (SIGNORINI [1933]). The problem was studied by FICHERA [1964, 1972], who gave the proof of the existence and regularity of the weak solution, and discussed the problem of the nonuniqueness of the solution.

It is worth mentioning that Fichera's first paper [1964] alone made a decisive impact on the development of a mathematical theory of variational inequalities —see, e.g., LIONS and STAMPACCHIA [1967].

If we suppose that friction occurs at the surface of the contact, and that this friction is governed, for example, by Coulomb's law, we obtain the so-called Signorini problem with friction. For a long time it was not known if this problem had a solution (see DUVAUT and LIONS [1976]). The proof of existence was given by NEČAS, JARUŠEK and HASLINGER [1980], who assumed that the coefficient of friction was small.

In this chapter, we first generalize the Signorini problem by introducing a formulation and an approximate solution of contact problems for two elastic bodies without friction. Further, we will study problems with friction of Coulomb's type. Throughout the chapter, approximate solutions are defined for displacements by means of the finite element method with piecewise linear functions on the triangulation of the given domains.

5. Formulation of contact problems

The one-sided contact of elastic bodies has been analysed in a numbers of papers of technical rather than mathematical nature. The authors of these papers have given no formulation of the continuous problem, starting instead directly from the finite-dimensional formulation by the finite-element method.

In order to be able to analyse the approximate solutions, we first introduce definitions of the continuous contact problems. To this end we shall use variational inequalities, proceeding similarly to FICHERA [1964], PANAGIOTOPOULOS [1985], DUVAUT and LIONS [1976], KIKUCHI and ODEN [1988], and others (see HLAVÁČEK, HASLINGER, NEČAS and LOVÍŠEK [1988]).

For the sake of simplicity, let us consider contact *without friction*. First we give the classic formulation, that is, a system of differential equations and boundary conditions. Then we introduce a variational formulation and prove the equivalence of both the formulations in a certain sense.

Throughout the chapter, we will consider:

- the plane problem,
- two bounded bodies,
- the theory of small strain,
- the linear stress-strain relation for generally inhomogenous, anisotropic elastic materials,
- the constant temperature field.

Let the bodies occupy bounded domains Ω', $\Omega'' \subset \mathbb{R}^2$ with Lipschitz boundaries. The superscript of one or two dashes will indicate in the sequel the correspondence to the body Ω' or Ω'', respectively. Let $x = (x_1, x_2)$ be Cartesian coordinates. We look for the vector field of displacement $u = (u_1, u_2)$ on the set $\Omega' \cup \Omega''$, that is, $u' = (u'_1, u'_2)$ on Ω' and $u'' = (u''_1, u''_2)$ on Ω'', and for the corresponding tensor field of strain

$$e_{ij}(u) = \tfrac{1}{2}\left(\frac{\partial u_i}{\partial x_j} + \frac{\partial u_j}{\partial x_i}\right), \quad i,j = 1,2. \tag{5.1}$$

The stress tensor is determined by the generalized Hooke's law

$$\tau_{ij} = c_{ijkm} e_{km}, \quad i,j = 1,2, \tag{5.2}$$

where the repeated index always means summation over 1,2.

Let the coefficients c_{ijkm} be bounded and measurable functions of $x \in \Omega' \cup \Omega''$,

$$c_{ijkm} = c_{kmij} = c_{jikm}, \tag{5.3}$$

and let there exist a positive constant c_0 such that

$$c_{ijkm}(x) e_{ij} e_{km} \geqslant c_0 e_{ij} e_{ij} \tag{5.4}$$

holds for all symmetric matrices e_{ij} and almost all $x \in \Omega' \cup \Omega''$.

The stress tensor satisfies the equations of equilibrium

$$\frac{\partial \tau_{ij}}{\partial x_j} + F_i = 0, \quad i,j = 1,2\,, \tag{5.5}$$

where F_i are components of the body forces vector.

We assume that the body Ω' is fixed by its part Γ_u,

$$u = 0 \quad \text{on } \Gamma_u \subset \partial\Omega'. \tag{5.6}$$

On some parts of boundaries the surface load is given, that is,

$$\tau_{ij}^M n_j^M = P_i^M \quad \text{on } \Gamma_\tau^M \subset \partial\Omega^M, \quad M = \text{'}, \text{''}, \; i,j = 1,2\,, \tag{5.7}$$

where n^M denotes the outer unit normal to $\partial\Omega^M$, while P_i^M are the components of the surface load.

Let the conditions of the "classic" (two-sided) contact be prescribed on a part $\Gamma_0 \subset \partial\Omega''$:

$$u_n = 0, \qquad T_t = 0 \qquad \text{on } \Gamma_0 \subset \partial\Omega'', \tag{5.8}$$

where

$$u_n = u_i n_i, \quad T_t = \tau_{ij} n_j t_i, \quad t = (t_1, t_2) = (-n_2, n_1)$$

are the normal component of the displacement and the tangential component of the stress vector, respectively. (For simplicity, we have not used the superscripts of two dashes.)

The conditions (5.8) occur, for example, on the axis of symmetry of the given problem and they enable us to derive a solution of the problem which satisfies only half of the given system.

Unilateral contact can occur on other parts of the boundary $\partial\Omega' \cup \partial\Omega''$. In the following, we will distinguish two types of contact problems: (1) with a bounded zone of contact, and (2) with an increasing zone of contact (see HLAVÁČEK, HASLINGER, NEČAS and LOVÍŠEK [1988]).

5.1. *Problems with a bounded zone of contact*

First, let us consider the case when the zone of contact during the process of deformation cannot expand beyond a certain domain, which is determined by the geometric shape of the bodies in the vicinity of the set $\partial\Omega' \cap \partial\Omega''$—see Fig. 5.1.

We define the zone of contact

$$\Gamma_K = \partial\Omega' \cap \partial\Omega'',$$

so that we have decompositions

$$\partial\Omega' = \overline{\Gamma}_u \cup \overline{\Gamma}'_\tau \cup \Gamma_K, \qquad \partial\Omega'' = \overline{\Gamma}_0 \cup \overline{\Gamma}''_\tau \cup \Gamma_K, \tag{5.9}$$

where Γ_u, Γ'_τ, Γ''_τ, Γ_0 are pairwise disjoint open parts of the boundaries. Let Γ_u and Γ_K have positive lengths. The other parts either have positive length or are empty.

We say that *one-sided contact* occurs on Γ_K, if

$$u'_n + u''_n \leqslant 0 \quad \text{on } \Gamma_K, \tag{5.10}$$

where

$$u_n^M = u_i^M n_i^M, \quad M = {}', {}'', \; n' = -n''.$$

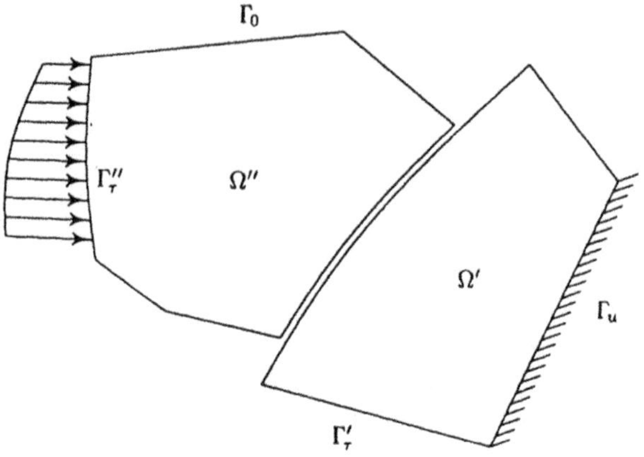

FIG. 5.1. A bounded zone of contact.

Condition (5.10) essentially represents the condition of nonpenetration of the bodies into each other. (For the derivation see HLAVÁČEK, HASLINGER, NEČAS and LOVÍŠEK [1988, pp. 112–113].)

Further, let us consider the contact forces. By the Action and Reaction law we have

$$T'_n = T''_n, \qquad T'_t = T''_t, \qquad \text{on } \Gamma_K.$$

The assumption of vanishing friction implies that the tangential components vanish. The normal components evidently cannot be tensions. Hence,

$$T'_n = T''_n \leqslant 0, \qquad T'_t = T''_t = 0.$$

Altogether, we introduce the following boundary conditions on Γ_K:

$$u'_n + u''_n \leqslant 0, \qquad T'_n = T''_n \leqslant 0, \tag{5.11}$$

$$(u'_n + u''_n)T'_n = 0, \tag{5.12}$$

$$T'_t = T''_t = 0. \tag{5.13}$$

Condition (5.12) results from the following argument: at the points where no contact occurs, that is, where $u'_n + u''_n < 0$, no contact force can arise either, that is, $T'_n = T''_n = 0$.

REMARK 5.1. Provided one of the bodies becomes perfectly rigid, the system (5.11)–(5.13) reduces to the system of boundary conditions of the Signorini problem (see SIGNORINI [1933], FICHERA [1964, 1972], DUVAUT and LIONS [1976]).

DEFINITION 5.1. A sufficiently regular function u is called a *classical solution of problem $\mathcal{P}_1$ with a bounded zone of contact* if it satisfies Eqs. (5.1), (5.2), (5.5)

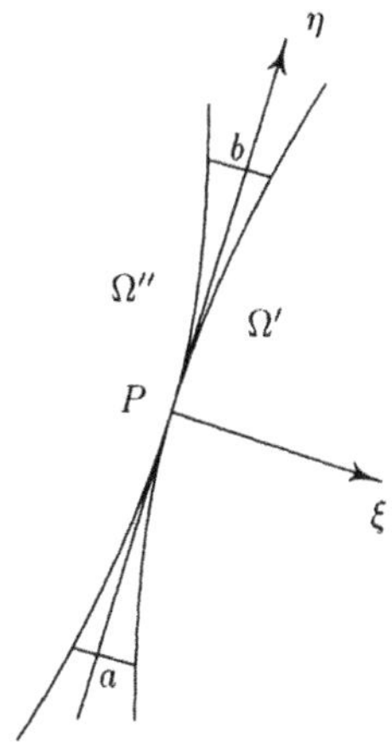

FIG. 5.2. An increasing zone of contact.

in $\Omega' \cup \Omega''$, boundary conditions (5.6) on Γ_u, (5.7) on Γ_τ' and Γ_τ'', (5.8) on Γ_0 and (5.11), (5.12), (5.13) on Γ_K.

5.2. *Problems with increasing zone of contact*

If the bodies Ω' and Ω'' have smooth boundaries in the vicinity of their intersection $\partial\Omega' \cap \partial\Omega''$, the contact zone may expand during the deformation process.

In these cases, our definition of one-sided contact would not be suitable, and requires some modifications.

As a typical case, let us consider the situation in Figure 5.2. Let us fix the local coordinate system (ξ, η) in such a way that the ξ-axis has the direction n'' and the η-axis has the direction of the common tangent at a certain point $P \in \partial\Omega' \cap \partial\Omega''$ at the center of the contact zone. The figure shows the state before deformation.

Let us now estimate those parts of $\partial\Omega'$ and $\partial\Omega''$ that could come in contact during the deformation process:

$$\Gamma_K^M = \{(\xi, \eta) \mid a \leqslant \eta \leqslant b,\ \xi = f^M(\eta)\}, \quad M = {}', {}'',$$

where the f^M are smooth in the interval $[a, b]$. (This interval must be a priori estimated so as to guarantee that it contains the projection of the possible contact zone.)

Analogous to the derivation of condition (5.10), we obtain the condition of *nonpenetration* in the form

$$u_\xi'' - u_\xi' \leqslant \varepsilon(\eta) \quad \forall \eta \in [a, b], \tag{5.14}$$

where $\varepsilon(\eta) = f'(\eta) - f''(\eta)$ is the distance of both boundaries before deformation and u_ξ', u_ξ'' are the ξ-components of the displacement vector.

Similarly to the preceding section, we further derive the following set of boundary conditions:

$$-T'_\xi(\cos\alpha')^{-1} = T''_\xi(\cos\alpha'')^{-1} \leqslant 0, \tag{5.15}$$

$$T'_\eta = T''_\eta = 0, \tag{5.16}$$

$$(u''_\xi - u'_\xi - \varepsilon)T''_\xi = 0, \tag{5.17}$$

which are valid for all points of $\Gamma'_K \cup \Gamma''_K$ with the same η-coordinate, $\eta \in [a, b]$. Here we use the notation

$$(\cos\alpha^M)^{-1} = \left[1 + \left(\frac{\partial f^M}{\partial\eta}\right)^2\right]^{1/2}, \quad M = {}', {}''.$$

α^M being the angle between the η-axis and the tangent to Γ_K^M.

Condition (5.17) results by the following argument: if no contact occurs, that is, if $u''_\xi - u'_\xi - \varepsilon < 0$, then the contact forces vanish: $T''_\xi = T'_\xi = 0$. Condition (5.16) represents an approximation of the condition of zero friction—it neglects the terms $T_\xi^M \sin\alpha^M$.

Let the decompositions

$$\partial\Omega' = \overline{\Gamma}_u \cup \overline{\Gamma}'_\tau \cup \Gamma'_K, \qquad \partial\Omega'' = \overline{\Gamma}_0 \cup \overline{\Gamma}''_\tau \cup \Gamma''_K$$

be valid with $\Gamma_K^M \cap \overline{\Gamma}_\tau^M = \emptyset$, $M = {}', {}''$.

DEFINITION 5.2. A sufficiently regular function u is called a *classical solution of problem $\mathcal{P}_2$ with an increasing zone of contact* if u fulfills Eqs. (5.1), (5.2) and (5.5) in $\Omega' \cup \Omega''$, the boundary conditions (5.6) on Γ_u, (5.7) on $\Gamma'_\tau \cup \Gamma''_\tau$, (5.8) on Γ_0 and (5.14)–(5.17) on $\Gamma'_K \cup \Gamma''_K$.

5.3. *Variational formulations*

Problems $\mathcal{P}_1$ and $\mathcal{P}_2$ can be associated with variational formulations on the basis of the principle of the minimization of potentional energy. Let us first introduce the space of the displacement function with finite energy

$$\mathcal{H}^1 = \{u \mid u = (u', u'') \in [H^1(\Omega')]^2 \times [H^1(\Omega'')]^2\}$$

with a standard norm $\|\!|\cdot|\!\|_1$ and the space of virtual displacements

$$V = \{u \in \mathcal{H}^1 \mid u' = 0 \text{ on } \Gamma_u,\ u''_n = 0 \text{ on } \Gamma_0\}.$$

Let us define the functional of potential energy

$$\mathcal{J}(v) = \tfrac{1}{2}A(v, v) - L(v), \tag{5.18}$$

where

$$A(u,v) = \int_{\Omega' \cup \Omega''} c_{ijkm} e_{km}(u) e_{ij}(v)\, dx, \tag{5.19}$$

$$L(v) = \int_{\Omega' \cup \Omega''} F_i v_i\, dx + \int_{\Gamma_\tau' \cup \Gamma_\tau''} P_i v_i\, ds. \tag{5.20}$$

The set of admissible displacements K for problem $\mathcal{P}_1$ with a bounded zone of contact is introduced by

$$K = \{v \in V \mid v_n' + v_n'' \leqslant 0 \text{ on } \Gamma_K\}. \tag{5.21}$$

DEFINITION 5.3. A function $u \in K$ is called a *weak (variational) solution of problem $\mathcal{P}_1$ with a bounded zone of contact*, if

$$\mathcal{J}(u) \leqslant \mathcal{J}(v) \quad \forall v \in K. \tag{5.22}$$

THEOREM 5.1. *Every classical solution of problem $\mathcal{P}_1$ is its weak solution. If a weak solution is sufficiently smooth, then it is a classical solution as well.*

For the proof, see HLAVÁČEK, HASLINGER, NEČAS and LOVÍŠEK [1988, Section 2.1.3].

Let us now consider problem $\mathcal{P}_2$ with an increasing zone of contact. Define the set K_ε of admissible displacements by

$$K_\varepsilon = \{v \in V \mid v_\xi'' - v_\xi' \leqslant \varepsilon \text{ for a.e. } \eta \in [a,b]\}.$$

DEFINITION 5.4. A function $u \in K_\varepsilon$ is called a *weak (variational) solution of problem $\mathcal{P}_2$ with an increasing zone of contact*, if

$$\mathcal{J}(u) \leqslant \mathcal{J}(v) \quad \forall v \in K_\varepsilon. \tag{5.23}$$

THEOREM 5.2. *Every classical solution of problem $\mathcal{P}_2$ is a weak solution. If a weak solution is sufficiently smooth, then it is a classical solution as well.*

The proof can be found again in HLAVÁČEK, HASLINGER, NEČAS and LOVÍŠEK [1988, Section 2.1.3].

6. Existence and uniqueness of solution

In this section we discuss the conditions guaranteeing the existence and uniqueness of weak solution of problems $\mathcal{P}_1$ and $\mathcal{P}_2$.

6.1. *Problem with a bounded zone of contact*

First we introduce the subspace of displacements of rigid bodies

$$\mathbb{R} = \{z \in \mathcal{H}^1 \,|\, z = (z', z''),\, z_1^M = a_1^M - b^M x_2,\, z_2^M = a_2^M + b^M x_1,\, M = ', ''\},$$

where a_i^M and b^M are arbitrary real constants.

Evidently $e_{ij}(z) = 0$ for every $z \in \mathbb{R}$ and hence

$$A(v, z) = 0 \quad \forall z \in \mathbb{R}.$$

Conversely, if $p \in \mathcal{H}^1$, $e_{ij}(p) = 0$, $\forall i, j = 1, 2$, then $p \in \mathbb{R}$. (Proof is found in NEČAS and HLAVÁČEK [1981].)

LEMMA 6.1. *Let there exist a weak solution of problem* $\mathcal{P}^1$. *Then*

$$L(y) \leqslant 0 \quad \forall y \in K \cap \mathbb{R}. \tag{6.1}$$

PROOF. Substituting $v = u + y$ with $y \in K \cap \mathbb{R}$ in condition

$$D\mathcal{J}(u, v - u) = A(u, v - u) - L(v - u) \geqslant 0 \quad \forall v \in K, \tag{6.2}$$

then $v \in K$, and

$$0 = A(u, y) \geqslant L(y).$$

□

THEOREM 6.1. *Let* $V \cap \mathbb{R} = \{0\}$ *or*

$$L(z) \neq 0 \quad \forall z \in V \cap \mathbb{R} \dot{-} \{0\}. \tag{6.3}$$

Then there is at most one weak solution of problem $\mathcal{P}_1$.

PROOF. Let u^1, u^2 be two weak solutions. Using (6.2) we can write

$$A(u^1, u^2 - u^1) \geqslant L(u^2 - u^1),$$
$$A(u^2, u^1 - u^2) \geqslant L(u^1 - u^2).$$

The sum of these inequalities yields

$$A(u^1 - u^2, u^2 - u^1) \geqslant 0.$$

Denoting $u^1 - u^2 = z$, we have $A(z, z) \leqslant 0$. Now Eq. (5.4) implies that $e_{ij}(z) = 0\ \forall i, j$, hence $z \in V \cap \mathbb{R}$. If $V \cap \mathbb{R} = \{0\}$, then $z = 0$ and the solution is unique.

If $z \neq 0$, denote $u^2 = u$, $u^1 = u + z$. Then

$$A(u, z) = A(z, z) = 0,$$
$$\mathcal{J}(u) = \mathcal{J}(u + z) \Rightarrow L(u) = L(u + z) \Rightarrow L(z) = 0,$$

which contradicts assumption (6.3). Hence, again $z = 0$.

□

Let us now present a general result on the existence of a weak solution of problem $\mathcal{P}_1$.

Define the set of "two-sided" admissible displacements of rigid bodies as

$$\mathbb{R}^* = \{z \in K \cap \mathbb{R} \mid z \in \mathbb{R}^* \Rightarrow -z \in \mathbb{R}^*\}. \tag{6.4}$$

We immediately see that

$$\mathbb{R}^* = \{z \in V \cap \mathbb{R} \mid z'_n + z''_n = 0 \text{ on } \Gamma_K\}.$$

THEOREM 6.2. *Let*

$$L(y) \leqslant 0 \quad \forall y \in K \cap \mathbb{R}, \tag{6.5}$$

$$L(y) < 0 \quad \forall y \in K \cap \mathbb{R} \dot{-} \mathbb{R}^*. \tag{6.6}$$

Then there exists a weak solution u of problem $\mathcal{P}_1$. Any other solution $\tilde{u}$ can be written in the form $\tilde{u} = u + y$, where $y \in V \cap \mathbb{R}$ is a function such that $u + y \in K$, $L(y) = 0$.

PROOF. Can be obtained on the basis of a general abstract theorem following FICHERA [1972] (see Theorem 1.II, ibid.). □

We also present another less general result together with its proof.

THEOREM 6.3. *Assume that*

$$\mathbb{R}^* = \{0\}, \qquad \mathbb{R} \cap V \neq \{0\}, \tag{6.7}$$

$$L(y) \neq 0 \quad \forall y \in \mathbb{R} \cap V \dot{-} \{0\}. \tag{6.8}$$

and let either $\mathbb{R} \cap K = \{0\}$ or

$$\mathbb{R} \cap K \neq \{0\}, \tag{6.9}$$

$$L(y) < 0 \quad \forall y \in \mathbb{R} \cap V \dot{-} \{0\}. \tag{6.10}$$

Then the functional $\mathcal{J}$ is coercive on K and there exists a unique weak solution of problem $\mathcal{P}_1$.

REMARK 6.1. Assumption (6.8) can be fulfilled only if $\dim \mathbb{R}_V \leqslant 1$, where $\mathbb{R}_V = \mathbb{R} \cap V$. Indeed, for $\Gamma_0 = \emptyset$, $\dim \mathbb{R}_V = 3$, the identity

$$L(y) = a_1 V''_1 + a_2 V''_2 + bM'' = 0$$

holds for every vector (a_1, a_2, b) orthogonal to (V''_1, V''_2, M'') in the space $\mathbb{R}^3$. (Here V''_i and M'' stand for the components of the force and the moment resultants, respectively, of the load acting on the body Ω''.)

An example satisfying $\dim \mathbb{R}_V = 1$, (6.7) and (6.9) is in Fig. 6.1. Another example with $\dim \mathbb{R}_V = 1$, which satisfies (6.7) and $\mathbb{R} \cap K = \{0\}$, is shown in Fig. 6.2.

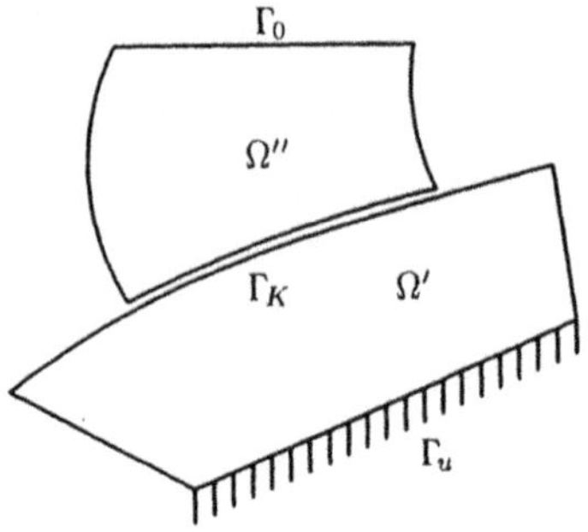

FIG. 6.1. An example for $\dim \mathbb{R}_V = 1$, (6.7) and (6.9).

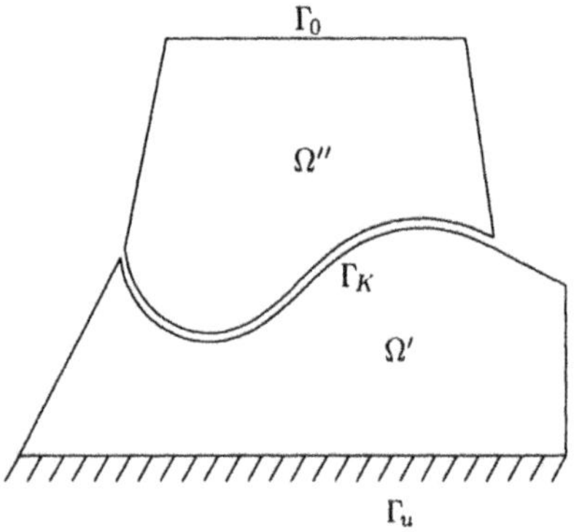

FIG. 6.2. An example for $\dim \mathbb{R}_V = 1$, (6.7) and $\mathbb{R} \cap K = \{0\}$.

PROOF (of Theorem 6.3). (1) Let us first consider the case $\mathbb{R} \cap K = \{0\}$. We shall use the following *Abstract Theorem* 1 (see NEČAS [1975, Theorem 2.2]):

ABSTRACT THEOREM 1. *Let $|u|$ be a seminorm in a Hilbert space H with a norm $\|u\|$. Let us define a subspace*

$$\mathcal{R} = \{u \in H \mid |u| = 0\}.$$

Assume $\dim \mathcal{R} < \infty$ *and*

$$C_1\|u\| \leqslant |u| + \|P_{\mathcal{R}}u\| \leqslant C_2\|u\| \quad \forall u \in H, \tag{6.11}$$

where $P_{\mathcal{R}}$ is the orthogonal projection to $\mathcal{R}$.

Let K be a closed convex subset of H containing the origin, $K \cap \mathcal{R} = \{0\}$. Let $\beta : H \to \mathbb{R}^1$ be a penalty functional whose Gâteaux derivative satisfies

$$D\beta(tu, v) = tD\beta(u, v) \quad \forall t > 0, \ u, v \in H,$$

and let

$$\beta(u) = 0 \Leftrightarrow u \in K.$$

Then

$$|u|^2 + \beta(u) \geqslant c\|u\|^2 \quad \forall u \in H \tag{6.12}$$

holds with some positive constant c.

We apply this theorem to our case with $H := V$, $\mathcal{R} := \mathbb{R}_V \equiv \mathbb{R} \cap V$, defining $|v|$ as follows

$$|v|^2 = \int_\Omega e_{ij}(v)e_{ij}(v)\,\mathrm{d}x,$$

and

$$\beta(u) = \tfrac{1}{2}\int_{\Gamma_K} [(u_n' + u_n'')^+]^2\,\mathrm{d}s.$$

In order to verify (6.11), we use an inequality of Korn's type (see HLAVÁČEK and NEČAS [1970]) and the decomposition

$$V = Q \oplus \mathbb{R}_V.$$

Thus, we obtain for all $u \in V$ the inequality

$$\begin{aligned}\|\!|u|\!\|_1^2 &= \|\!|P_Q u|\!\|_1^2 + \|\!|P_{R_V} u|\!\|_1^2 \leqslant C|P_Q u|^2 + \|\!|P_{R_V} u|\!\|_1^2 \\ &= C|u|^2 + \|\!|P_{R_V} u|\!\|_1^2,\end{aligned}$$

which implies the left-hand part of (6.11). The right-hand part is obvious.

Now (6.12) implies

$$|u|^2 \geqslant C\|\!|u|\!\|_1^2 \quad \forall u \in K,$$

which easily yields that $\mathcal{J}$ is coercive on K. Hence, there exists a weak solution of the problem $\mathcal{P}_1$, since $\mathcal{J}$ is weakly lower semicontinuous, being convex and continuous in V, and K is weakly closed.

If u^1 and u^2 are two weak solutions, then proceeding in the same way as in the proof of Theorem 6.1, we obtain

$$\begin{aligned}&y = u^1 - u^2 \in \mathbb{R}_V,\\ &\mathcal{J}(u^1) = \mathcal{J}(u^2) \Rightarrow L(u^1) = L(u^2) \Rightarrow L(y) = 0.\end{aligned}$$

By assumption (6.8), we conclude $y = 0$.

(2) Let us consider the case (6.9), (6.10). We shall use *Abstract Theorem* 2 (see NEČAS [1975, Theorem 2.3]).

ABSTRACT THEOREM 2. *Let the assumptions of Abstract Theorem* 1 *be fulfilled except for* $K \cap \mathcal{R} = \{0\}$ *(that is, we assume that* $K \cap \mathcal{R} \neq \{0\}$*). Further, let* f *be a continuous linear functional on* H*, such that*

$$f(y) < 0 \quad \forall y \in K \cap \mathcal{R} \dot{-} \{0\}.$$

Then

$$|u|^2 + \beta(u) - f(u) \geqslant C_1\|u\| - C_2 \quad \forall u \in H \tag{6.13}$$

holds with some positive constants C_1 *and* C_2.

We can apply this theorem with the same H, $\mathcal{R}$, $|\cdot|$ and β as in (1), with, in addition,

$$f(v) := L(v).$$

Then (6.13) implies the coerciveness of $\mathcal{J}$ on K. The existence and uniqueness of the weak solution is then proved in the same way as above. □

REMARK 6.2. The simplest case is the so-called coercive case with $\mathbb{R} \cap V = \{0\}$. Then an inequality of Korn's type holds, namely

$$|\!|\!| v |\!|\!|_1 \leqslant C|v| \quad \forall v \in V,$$

so that $\mathcal{J}$ is coercive on the whole space V. The rest of the proof of existence and uniqueness is straightforward.

6.2. *Problem with increasing zone of contact*

We introduce a result for the case when $\dim \mathbb{R}_V = 1$.

THEOREM 6.4. *Let us assume that* Γ_0 *consists of line segments parallel to the* x_1*-axis,* $\cos(\xi, x_1) > 0$ *(see Fig.* 6.3*) and*

$$V_1'' = \int_\Omega F_1'' \,\mathrm{d}x + \int_{\Gamma_\tau''} P_1'' \,\mathrm{d}s > 0. \tag{6.14}$$

Then $\mathcal{J}$ *is coercive on* K_ε *and there exists a unique solution of problem* $\mathcal{P}_2$.

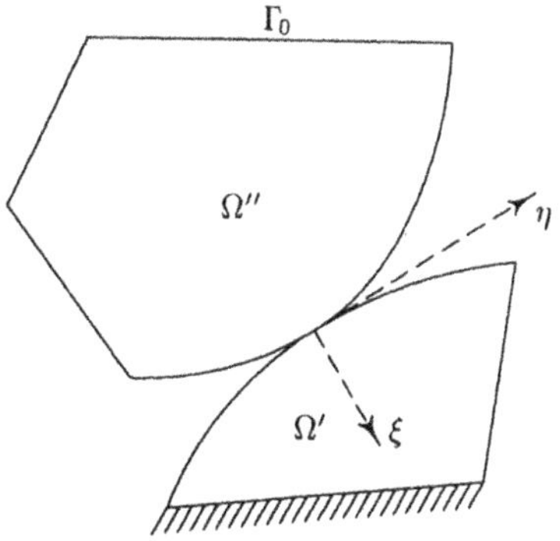

FIG. 6.3. An example for Theorem 6.4.

PROOF. Let us set

$$p_0(v) = \int_a^b (v''_\xi - v'_\xi)\,\mathrm{d}\eta, \quad V_p = \{v \in V \mid p_0(v) = 0\}.$$

Then

$$\mathbb{R} \cap V_p = 0.$$

Indeed,

$$\mathbb{R} \cap V_p \subset \mathbb{R}_V = \{z = (z', z'') \mid z' = 0,\ z'' = (c, 0),\ c \in \mathbb{R}^1\}.$$

The identity $p_0(v) = 0$ yields

$$0 = \int_a^b z''_\xi\,\mathrm{d}\eta = c \int_a^b \cos(\xi, x_1)\,\mathrm{d}\eta \Rightarrow c = 0.$$

Let $v \in V$. Define $y \in \mathbb{R}_V$ by the relations

$$y' = 0, \qquad y''_1 = p_0(v)d^{-1}, \qquad y''_2 = 0,$$

where

$$d = \int_a^b \cos(\xi, x_1)\,\mathrm{d}\eta.$$

It is easily verified that the difference $P_v = v - y$ satisfies

$$p_0(P_v) = p_0(v) - p_0(y) = p_0(v) - \int_a^b p_0(v)d^{-1}\cos(\xi, x_1)\,\mathrm{d}\eta = 0,$$

hence, $P_v \in V_p$.

With the help of an inequality of Korn's type (see HLAVÁČEK and NEČAS [1970])

$$|v| \geqslant c\|v\|_1 \quad \forall v \in V_p,$$

we can write

$$\mathcal{J}(v) = \tfrac{1}{2}A(P_v, P_v) - L(P_v) - L(y) \geqslant C_1\|P_v\|_1^2 - C_2\|P_v\|_1 - y_1''V_1''. \tag{6.15}$$

When $\|v\|_1 \to \infty$ then at least one of the norms $\|P_v\|_1$, $\|y\|_1$ tends to infinity. Moreover,

$$v \in K_\varepsilon \Rightarrow p_0(v) \leqslant \int_a^b \varepsilon \, \mathrm{d}\eta < +\infty, \tag{6.16}$$

$$\|y\|_1 = |y_1''|\Big(\int_{\Omega''} \mathrm{d}x\Big)^{1/2} = |p_0(v)|d^{-1}(\operatorname{meas}\Omega'')^{1/2}. \tag{6.17}$$

(1) Let $\|y\|_1 \to \infty$. Then (6.16) together with (6.17) imply $-p_0(v) \to +\infty$, and hence $-y_1'' \to +\infty$. As

$$C_1\|P_v\|_1^2 - C_2\|P_v\|_1 \geqslant C_3 > -\infty,$$

we conclude from (6.15) and (6.14) that $\mathcal{J}(v) \to +\infty$.

(2) Let $\|P_v\|_1 \to +\infty$. Then (6.16), (6.14) yield

$$\mathcal{J}_1(P_v) = C_1\|P_v\|_1^2 - C_2\|P_v\|_1 \to +\infty,$$

$$\mathcal{J}_2(y) = -y_1''V_1'' = -p_0(v)d^{-1}V_1'' \geqslant -d^{-1}V_1'' \int_a^b \varepsilon \, \mathrm{d}\eta > -\infty.$$

By virtue of (6.15) we have

$$\mathcal{J}(v) \geqslant \mathcal{J}_1(P_v) + \mathcal{J}_2(y) \to +\infty.$$

Thus, we have proved that $\mathcal{J}$ is coercive on K_ε. Since K_ε is a closed subset of V and the functional $\mathcal{J}$ is convex and continuous on V, a solution of problem $\mathcal{P}_2$ exists.

Uniqueness is a consequence of (6.14). Indeed, we first prove—as in the proof of Theorem 6.1—that the two solutions u^1 and u^2 differ from each other by an element $z \in \mathbb{R}_V$ with $L(z) = 0$. On the other hand, however, $L(z) = cV_1''$, $c \in \mathbb{R}^1$. Condition (6.14) implies $c = 0$, that is, $z = 0$. □

7. Contact problems with friction

7.1. Definitions and preliminary results

In Sections 5 and 6 we were concerned with contact problems of two elastic bodies.

Roughly speaking, those problems enter into the framework of the classical calculus of variations: one looks for the minimum of a functional on a closed convex set.

The contact problem without friction is expressed by the condition, that the tangential component of the stress vector on the contact zone is $T_t = 0$. It is clear, that this condition does not fully comport with the real situation. We shall consider for the sake of simplicity the contact between an elastic body Ω and a perfectly rigid foundation. Extension to the contact of two elastic bodies is possible, see JARUŠEK [1983].

Nonzero friction is expressed by *Coulomb's law*: on the contact surface Γ_K, we assume

$$u_n \leqslant 0, \qquad T_n \leqslant 0, \qquad u_n T_n = 0 \tag{7.1}$$

and

$$|T_t| \leqslant \mathscr{F}|T_n|, \qquad (\mathscr{F}|T_n| - |T_t|)u_t = 0, \qquad u_t T_t \leqslant 0, \tag{7.2}$$

where $\mathscr{F}$ is the friction coefficient, $\mathscr{F} \geqslant 0$ on Γ_K, $\mathscr{F}$ is smooth enough on $\overline{\Gamma}_K$ and $u_t = u - nu_n$ is the tangential component of the displacement vector.

Condition (7.2) expresses the obvious law, that the modulus of the tangential component does not exceed a multiple of the modulus of the normal component and if it is attained, then the body can slip off in the direction of the tangent force from the body to the foundation. The condition (7.2) is very simple and natural but the theory is much more difficult than for the case without friction. To the knowledge of the present authors, there is no theory established for finite elasticity, i.e. for the big geometry. The case without friction for finite elasticity is treated in the papers CIARLET and NEČAS [1984, 1987].

Let us first consider the problem with *given friction*, when the unknown normal component T_n is replaced by a given *slip stress* $g_n \geqslant 0$. In this case (7.2) is replaced by

$$|T_t| \leqslant \mathscr{F}g_n, \qquad (\mathscr{F}g_n - |T_t|)u_t = 0, \qquad u_t T_t \leqslant 0 \quad \text{on } \Gamma_K. \tag{7.3}$$

Let us pass to the variational formulation of this problem. Thus, let $\Omega \subset \mathbb{R}^2$ be a domain with a Lipschitzian boundary $\partial\Omega$ and let $\partial\Omega = \overline{\Gamma}_u \cup \overline{\Gamma}_P \cup \overline{\Gamma}_K$, where Γ_u, Γ_P, Γ_K are open disjoint subsets of $\partial\Omega$; moreover let us assume that Γ_u and Γ_K are nonempty. Let $u^0 \in [H^1(\Omega)]^2$, $F \in [L^2(\Omega)]^2$, $g_n \in L^2(\Gamma_K)$, $g_n \geqslant 0$, $P \in [L^2(\Gamma_P)]^2$. Let a closed convex set of *admissible displacements* K be given by

$$K = \{v \in [H^1(\Omega)]^2 \mid v = u^0 \text{ on } \Gamma_u,\ v_n \leqslant 0 \text{ on } \Gamma_K\}.$$

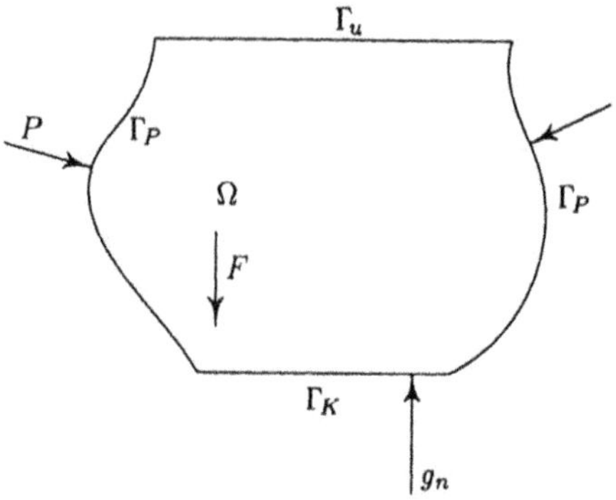

FIG. 7.1. A friction problem with a given slip stress g_n.

A function $u \in K$ is called a *weak solution of the Signorini problem with friction for a given* g_n, if

$$\forall v \in K\colon\ a(u, v-u) + \int_{\Gamma_K} \mathcal{F} g_n(|v_t| - |u_t|)\,\mathrm{d}s \geqslant \int_{\Omega} F_i(v_i - u_i)\,\mathrm{d}x + \int_{\Gamma_P} P_i(v_i - u_i)\,\mathrm{d}s. \tag{7.4}$$

Here

$$a(u, v) = \int_{\Omega} c_{ijkm} e_{ij}(u) e_{km}(v)\,\mathrm{d}x,$$

where $c_{ijkm} \in L^\infty(\Omega)$ and define the generalized linear Hooke's law, see (5.2).

We will prove formal equivalence of the classical and weak formulation of our problem.

We have, of course,

$$\tau_{ij} = c_{ijkm} e_{km}. \tag{7.5}$$

Taking for $v = u \pm \varphi$, $\varphi \in [C_0^\infty(\Omega)]^2$, we have from (7.4)

$$a(u, \varphi) = \int_{\Omega} F_i \varphi_i\,\mathrm{d}x,$$

hence

$$\frac{\partial \tau_{ij}}{\partial x_j} + F_i = 0 \quad \text{in } \Omega. \tag{7.6}$$

If we take $v = \pm\varphi$, $\varphi \in [C^\infty(\overline{\Omega})]^2$, $\varphi = 0$ on $\Gamma_K \cup \Gamma_u$, we get

$$a(u,\varphi) - \int_\Omega F_i\varphi_i \,\mathrm{d}x - \int_{\Gamma_P} P_i\varphi_i \,\mathrm{d}s = 0,$$

which implies

$$\tau_{ij}n_j = P_i \quad \text{on } \Gamma_P. \tag{7.7}$$

If we integrate by parts in (7.4) and use (7.6), (7.7), we get from (7.4)

$$\int_{\Gamma_K} T_t(v_t - u_t)\,\mathrm{d}s + \int_{\Gamma_K} T_n(v_n - u_n)\,\mathrm{d}s + \int_{\Gamma_K} \mathscr{F}g_n(|v_t| - |u_t|)\,\mathrm{d}s \geqslant 0 \quad \forall v \in K. \tag{7.8}$$

Let us take $v = u \pm \psi$, where $\psi_n = 0$ on Γ_K and $\psi = 0$ on Γ_u. Then

$$\pm \int_{\Gamma_K} T_t\psi_t \,\mathrm{d}s + \int_{\Gamma_K} \mathscr{F}g_n(|u_t \pm \psi_t| - |u_t|)\,\mathrm{d}s \geqslant 0,$$

hence

$$\left|\int_{\Gamma_K} T_t\psi_t \,\mathrm{d}s\right| \leqslant \int_{\Gamma_K} \mathscr{F}g_n|\psi_t|\,\mathrm{d}s,$$

which implies $|T_t| \leqslant \mathscr{F}g_n$. It follows that

$$T_tu_t + \mathscr{F}g_n|u_t| \geqslant 0 \quad \text{on } \Gamma_K.$$

Taking ψ such that $\psi = 0$ on Γ_u, $\psi_n = 0$ on Γ_K, $\psi_t = -u_t$ on Γ_K with $v = u + \psi$, we get from (7.8)

$$-\int_{\Gamma_K} T_tu_t \,\mathrm{d}s - \int_{\Gamma_K} \mathscr{F}g_n|u_t|\,\mathrm{d}s \geqslant 0$$

which gives finally $T_tu_t + Fg_n|u_t| = 0$ and so (7.3) completely.

Problem (7.4) can be solved once more by a standard variational method.

THEOREM 7.1. *There exists a unique $u \in K$ minimizing the functional*

$$\mathscr{J}(v) = \tfrac{1}{2}a(v,v) + \int_{\Gamma_K} \mathscr{F}g_n|v_t|\,\mathrm{d}s - \int_\Omega F_iv_i\,\mathrm{d}x - \int_{\Gamma_P} P_iv_i\,\mathrm{d}s$$

and u is the unique solution to the problem (7.4).

PROOF. The functional $\mathcal{J}(v)$ is continuous and strictly convex on K, hence it is *lower weakly continuous*: if $v_n \rightharpoonup v$, then $\liminf_{n\to\infty} \mathcal{J}(v_n) \geqslant \mathcal{J}(v)$. The functional is also *coercive* on K. Hence, there exists a unique minimum $u \in K$. For $\forall v \in K$, $0 < \varepsilon < 1$, we have

$$\mathcal{J}(u + \varepsilon(v - u)) - \mathcal{J}(u) \geqslant 0. \tag{7.9}$$

But

$$\int_{\Gamma_K} \mathcal{F} g_n(|u_t + \varepsilon(v_t - u_t)| - |u_t|)\,\mathrm{d}s \leqslant \varepsilon \int_{\Gamma_K} \mathcal{F} g_n(|v_t| - |u_t|)\,\mathrm{d}s$$

and dividing (7.9) by ε and letting $\varepsilon \to 0$, we get (7.4). □

7.2. *Contact problem with Coulomb's law*

Let us suppose that having solved problem (7.4), it is clear what $T_n(u)$ for our solution means. Consequently, if we define the mapping $g_n \mapsto T_n(u)$ then the fixed point is a solution to our problem.

Let $v \in V = \{v \in [H^1(\Omega)]^2 \mid v = 0$ in the neighbourhood of $\overline{\Gamma}_u \cup \overline{\Gamma}_P$ and $v_t = 0$ on $\Gamma_K\}$. Let u be a unique solution to the problem (7.4). Then by definition, which is formally clear from the integration by parts,

$$(T_n(u), v_n) = a(u, v) - \int_\Omega F_i v_i \,\mathrm{d}x \quad \forall v \in V. \tag{7.10}$$

We shall suppose in the following that Γ_K is smooth enough. Because $v_n \in H^{1/2}(\partial\Omega)$, where $H^{1/2}(\partial\Omega)$ is the space of traces of functions from $H^1(\Omega)$, $T_n(u) \in (H^{1/2}(\partial\Omega))' = H^{-1/2}(\partial\Omega)$.

THEOREM 7.2. *It holds that*

$$(T_n(u), v_n) \geqslant 0 \quad \forall v \in V, \ v_n \leqslant 0 \ on \ \Gamma_K,$$

i.e.,

$$T_n(u) \leqslant 0 \quad on \ \Gamma_K.$$

PROOF. Let $v \in V$, $v_n \leqslant 0$ on Γ_K, i.e. $v \in K$. We have $(T_n(u), v_n) = a(u, v) - \int_\Omega F_i v_i \,\mathrm{d}x$. Put $\overline{v} = u + v \in K$. It follows from (7.4) with $\overline{v}$:

$$a(u, v) - \int_\Omega F_i v_i \,\mathrm{d}x \geqslant 0.$$ □

Let us suppose some restrictions on the friction coefficient: in general, we suppose that $\mathcal{F}$ has compact support in Γ_K. If Γ_K is one component of $\partial\Omega$, it may be that $\mathcal{F}$ is bounded on Γ_K only, i.e. $\mathcal{F} = \text{const} > 0$ is a possible choice.

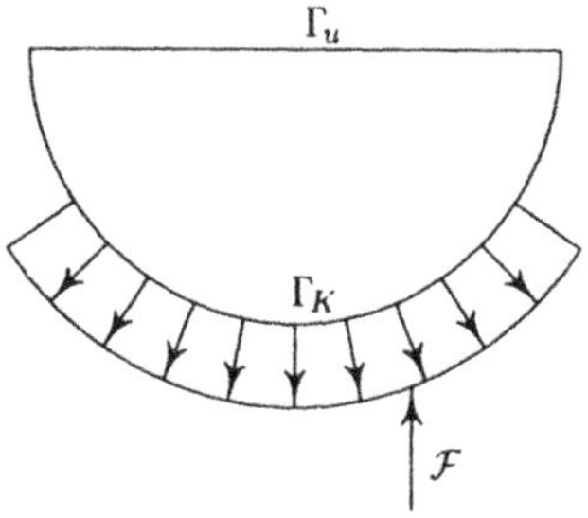

FIG. 7.2. An example of a restricted friction coefficient.

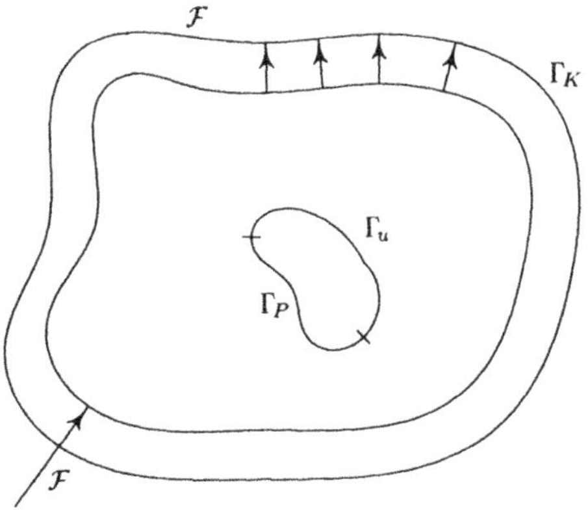

FIG. 7.3. Another example of the distribution of the friction coefficient.

Under the conditions mentioned, the following theorem on the regularity of the solution to the problem (7.4) is proven in HLAVÁČEK, HASLINGER, NEČAS and LOVÍŠEK [1988] for the case in Fig. 7.3 and in JARUŠEK [1983] for the general case.

THEOREM 7.3. *There exist constants* $c_1 > 0$ *and* $c_2 > 0$ *such that*

$$\int_{\Gamma_K'} |T_n(u)|^2 \, ds \leqslant c_1 t (\max_{\Gamma_K} \mathcal{F}) \int_{\Gamma_K} |g_n|^2 \, ds + c_2 (\|u^0\|_{1,\Omega}^2 + \|F\|_{1,\Omega}^2 + \|P\|_{0,\Gamma_P}^2).$$

(*The friction coefficient is* $t\mathcal{F}$, $\Gamma_K' \subset \overline{\Gamma}_K' \subset \Gamma_K$ *in the general case and* $\Gamma_K' = \Gamma_K$ *if* Γ_K' *is one of the components of* $\partial\Omega$.)

Looking at problem (7.4), under our conditions, we can easily extend the existence of a unique solution to the case $g_n \in (H_0^{1/2}(\Gamma_K))'$, where $H_0^{1/2}(\Gamma_K)$ is the closure in the space of the traces to the space $H^1(\Omega)$, vanishing in the neighbourhood of $\overline{\Gamma}_P \cup \overline{\Gamma}_u$. We leave it to the reader. We have the following.

THEOREM 7.4. *The mapping* $g_n \mapsto u$ *from* $(H_0^{1/2}(\Gamma_K))'$ *to* $[H^1(\Omega)]^2$ *is* $\frac{1}{2}$*-Hölderian.*

PROOF. Let $g_n^1, g_n^2 \in (H_0^{1/2}(\Gamma_K))' = H^{-1/2}(\Gamma_K)$ and let u^1, u^2 be the corresponding solutions. Then,

$$a(u^2 - u^1, u^2 - u^1) \leqslant (\mathscr{F}g_n^2 - \mathscr{F}g_n^1, |u_t^2| - |u_t^1|)$$
$$\leqslant c\|g_n^2 - g_n^1\|_{-1/2,\Gamma_K}(\|u^2\|_{1,\Omega} + (\|u^1\|_{1,\Omega}). \qquad \square$$

COROLLARY 7.1. *The mapping* $g_n \mapsto -T_n(u)$ *from* $L^2(\Gamma_K')$ *to* $L^2(\Gamma_K')$ ($\Gamma_K' \supset$ *support* $\mathscr{F}$) *is weakly continuous, i.e. if* $g_n^m \rightharpoonup g_n$ *in* $L^2(\Gamma_K')$ *then* $T_n(u^m) \rightharpoonup T_n(u)$ *in* $L^2(\Gamma_K')$.

PROOF. If $g_n^m \rightharpoonup g_n$ in $L^2(\Gamma_K')$, then $\mathscr{F}g_n^m \to \mathscr{F}g_n$ in $H^{-1/2}(\Gamma_K)$, which follows from the compact embedding $L^2(\Gamma_K') \hookrightarrow\hookrightarrow H^{-1/2}(\Gamma_K')$. Hence $T_n(u^m) \to T_n(u)$ in $H^{-1/2}(\Gamma_K)$. Because $H^{1/2}(\Gamma_K)$ is dense in $L^2(\Gamma_K)$ and we have Theorem 7.3 we get the result. $\square$

THEOREM 7.5. *There exists a fixed point* $g_n = -T_n(u(g_n))$*, i.e. the solution to the problem satisfying Coulomb's law, provided the coefficient of friction is small enough.*

PROOF. One uses the weak version of the Schauder's theorem; for details, see HLAVÁČEK, HASLINGER, NEČAS and LOVÍŠEK [1988]. $\square$

REMARK 7.1. It is possible to "mollify" Coulomb's law as it is done in KIKUCHI and ODEN [1988]. In this case the theory is standard.

REMARK 7.2. If $\Gamma_u = \emptyset$, there must appear conditions on the direction of the total force, see JARUŠEK [1984].

REMARK 7.3. The question how small the coefficient of the friction has to be is very important. In the paper NEČAS, JARUŠEK and HASLINGER [1980] for $\Omega \subset \mathbb{R}^2$, it is proven, considering $\tau_{ij} = \lambda\delta_{ij}e_{kk} + 2\mu e_{ij}$, that

$$\max_{\bar{\Gamma}_K} \mathscr{F}(x) < \sqrt{\frac{2\mu}{\lambda} + 3\mu}$$

is enough. This is largely satisfactory in applications.

8. Approximation of contact problems by finite elements. Frictionless case

The aim of this section is to study the approximation of contact problems of two elastic bodies when the influence of friction is neglected. We shall derive the rate of convergence of approximate solutions, provided the exact solution is smooth enough. If there is no additional information on the smoothness of the exact solution, the convergence itself will be established.

The continuous model is presented in Sections 5 and 6. We shall study both the case with a bounded zone of contact and the case with an increasing one.

8.1. *Approximation of the problem with a bounded zone of contact*

The meaning of the symbols are the same as in Sections 5 and 6.

First, let us assume that Ω', Ω'' are domains with *polygonal boundaries* $\partial\Omega'$, $\partial\Omega''$. Then the contact part Γ_K can be written as

$$\overline{\Gamma}_K = \bigcup_{i=1}^{n} \overline{\Gamma}_{K,i},$$

where $\Gamma_{K,i}$ is a straight line segment with an initial point A_i and the end point A_{i+1}. Let $\mathcal{T}_h'$ and $\mathcal{T}_h''$ be triangulations of $\overline{\Omega}'$ and $\overline{\Omega}''$, satisfying the usual requirements on the mutual position of triangles and consistent with the decompositions of $\partial\Omega'$ and $\partial\Omega''$. Moreover, the nodes, lying on $\overline{\Gamma}_K$ belong to both triangulations. The pair $\mathcal{T}_h \equiv \{\mathcal{T}_h', \mathcal{T}_h''\}$ defines a decomposition of $\overline{\Omega} = \overline{\Omega}' \cup \overline{\Omega}''$. $\mathcal{T}_h$ is said to be *regular* if both $\mathcal{T}_h'$ and $\mathcal{T}_h''$ are regular. With any $\mathcal{T}_h$ we associate a finite dimensional space V_h, given by

$$V_h = \{v_h \in [C(\overline{\Omega}')]^2 \times [C(\overline{\Omega}'')]^2 \cap V \mid v_h|_T \in [P_1(T)]^2 \ \forall T \in \mathcal{T}_h\}.$$

Let a_j^i, $j = 1, \ldots, m_i$ be vertices of $\mathcal{T}_h$ lying on $\overline{\Gamma}_{K,i}$ ($a_1^i \equiv A_i$, $a_{m_i}^i \equiv A_{i+1}$), $i = 1, \ldots, m$ and let n^i be the unit vector of the outer normal of the side $\Gamma_{K,i}$ with respect to Ω'.

We define

$$K_h = \{v_h \in V_h \mid n^i(v_h' - v_h'')(a_j^i) \leqslant 0, \ i = 1, \ldots, m, \ j = 1, \ldots, m_i\}. \tag{8.1}$$

Now, let us consider the domains Ω' and Ω'' with *more general boundaries* rather than simply polygonal. For the sake of the simplicity of our analysis, we restrict ourselves to the case where only Γ_K is curved (see Fig. 8.1). Let ψ be a continuous concave (or convex) function, defined on $[a, b]$, the graph of which coincides with $\overline{\Gamma}_K$. On $\overline{\Gamma}_K$ let us choose $m+1$ points $A_1, \ldots, A_{m+1}$ such that $A_1 = (a, \psi(a))$ and $A_{m+1} = (b, \psi(b))$. Let A_i, $A_{i+1} \in \overline{\Gamma}_K$ and $S \in \Omega' \cup \Omega''$. By a curved element T we mean the closed set bounded by line segments SA_i, SA_{i+1} and by the arc A_iA_{i+1}. The minimal inner angle of the straight triangle $A_iA_{i+1}S$ will be called the minimal interior angle of the curved element T. The triangulation $\mathcal{T}_h = (\mathcal{T}_h', \mathcal{T}_h'')$ of $\overline{\Omega} = \overline{\Omega}' \cup \overline{\Omega}''$ consists on the one hand of the curved elements along Γ_K and on the other hand of straight triangular elements inside Ω', Ω''. $\mathcal{T}_h$ is said to be regular for $h \to 0+$, if the minimal inner angle

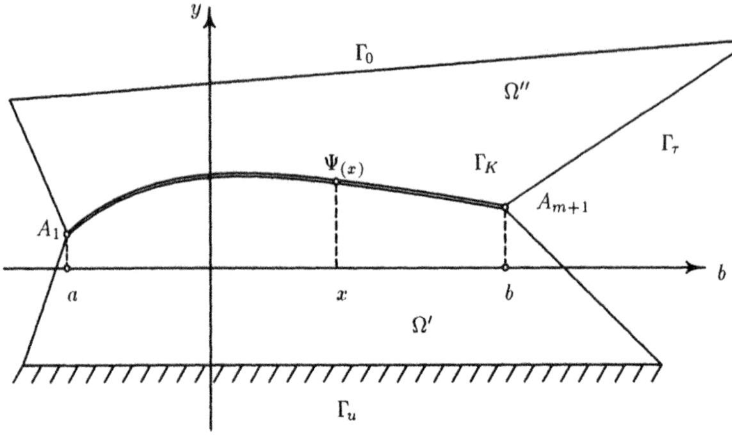

FIG. 8.1. A typical configuration of a bounded zone of contact.

θ of all $T \in \mathcal{T}_h$ is bounded from below by a constant $\theta_0 > 0$. With any $\mathcal{T}_h$ we associate the finite-dimensional space

$$V_h = \{v_h \in [C(\overline{\Omega}')]^2 \times [C(\overline{\Omega}'')]^2 \cap V \mid v_h|_T \in [P_1(T)]^2\ T \in \mathcal{T}_h\} \tag{8.2}$$

and

$$K_h = \{v_h \in V_h \mid n(v_h' - v_h'')(A_i) \leqslant 0,\ \forall i = 1, \ldots, m+1\}, \tag{8.3}$$

with V_h, defined by (8.2). Here n denotes the vector of the outer normal, oriented with respect to Ω'.

DEFINITION 8.1. An element $u_h \in K_h$ such that

$$\mathcal{J}(u_h) \leqslant \mathcal{J}(v_h) \quad \forall v_h \in K_h \tag{$(\mathcal{P}_1)_h$}$$

is called an *approximation of the contact problem with a bounded zone of contact.* The set K_h is defined either by (8.1) or by (8.3) and $\mathcal{J}$ by (5.18)–(5.20).

Next we shall derive the error estimates between u and u_h, the solutions of $(\mathcal{P}_1)$ and $(\mathcal{P}_1)_h$, respectively, provided that the solution u is *smooth enough.* We use the general results of Section 4. Using the notations of Section 5.3, the relation (4.3a) can be written in the following form:

$$\begin{aligned} c_0|u - u_h|^2 \leqslant A(u - u_h, u - u_h) &\leqslant L(u - v_h) + L(u_h - v) \\ &+ A(u_h - u, v_h - u) + A(u, v - u_h) + A(u, v_h - u), \end{aligned} \tag{8.4}$$

holds for any $v_h \in K_h$ and any $v \in K$, where $|v|^2 = (e_{ij}(v), e_{ij}(v))_{0,\Omega}$.

First we start with the case of *polygonal domains*, which is simpler from the standpoint of error analysis. Indeed, in this case the set K_h, defined by means of (8.1) represents the inner approximation of K, i.e. $K_h \subset K \ \forall h \in (0,1)$. Consequently, the simplified version of (8.4) will be used (see also (4.3b)):

$$\begin{aligned} c_0|u-u_h|^2 &\leqslant A(u-u_h, u-u_h) \\ &\leqslant L(u-v_h) + A(u_h-u, v_h-u) + A(u, v_h-u). \end{aligned} \tag{8.4'}$$

We have the following.

THEOREM 8.1. *Let there exist solutions u and u_h of problems $(\mathcal{P}_1)$ and $(\mathcal{P}_1)_h$, respectively, and let $u \in \mathcal{H}^2(\Omega) \cap K$, u', $u'' \in [W^{1,\infty}(\Gamma_{K,i})]^2$, $i = 1, \ldots, m$, $T_n'(u) \equiv T_n''(u) \in L^\infty(\Gamma_K)$. Further, let us assume that the set of points at which the change of $u_n' - u_n'' < 0$ to $u_n' - u_n'' = 0$ occurs, is finite. Then*

$$|u-u_h| = ch\{|u|_2^2 + \sum_{i=1}^m \|T_n(u)\|_{\infty,\Gamma_{K_i}} \cdot (|u'|_{1,\infty,\Gamma_{K_i}} + |u''|_{1,\infty,\Gamma_{K_i}})\}^{1/2}, \tag{8.5}$$

provided the system $\{\mathcal{T}_h\}$, $h \to 0+$ is regular.

PROOF. Integrating by parts, we obtain

$$L(u-v_h) + A(u, v_h-u) = (T_n(u), (v_{hn}' - u_n') - (v_{hn}'' - u_n''))_{0,\Gamma_K}$$

taking into account (5.5)–(5.8), (5.13) and the definitions of A and L. Hence (8.4′) can be written in the form:

$$c_0|u-u_h|^2 \leqslant A(u_h-u, v_h-u) + (T_n(u), (v_{hn}' - u_n') - (v_{hn}'' - u_n''))_{0,\Gamma_K} \tag{8.6}$$

holds for any $v_h \in K_h$. For the function v_h we choose the piecewise linear Lagrangian interpolation of the solution u, which we denote by $r_h u = (r_h u', r_h u'')$. It is readily seen that $r_h u \in K_h$. From the regularity assumption $u \in \mathcal{H}^2 \cap K$ it follows that

$$\begin{aligned} |A(u_h-u, r_h u - u)| &\leqslant \tfrac{1}{2}|u-u_h|^2 + c|u - r_h u|_{1,\Omega}^2 \\ &\leqslant \tfrac{1}{2}|u-u_h|^2 + ch^2|u|_{2,\Omega}^2. \end{aligned} \tag{8.7}$$

The remaining assumptions on the behavior of traces u', u'' and T_n on the contact zone Γ_K imply

$$\begin{aligned} &|(T_n(u), (r_{hn}u' - u_n') - (r_{hn}u'' - u_n''))_{0,\Gamma_K}| \\ &\quad\leqslant ch^2 \sum_{i=1}^m \|T_n(u)\|_{\infty,\Gamma_{K_i}} (|u'|_{1,\infty,\Gamma_{K_i}} + |u''|_{1,\infty,\Gamma_{K_i}}) \end{aligned} \tag{8.8}$$

(for details see HLAVÁČEK, HASLINGER, NEČAS and LOVÍŠEK [1988, p. 139]). From (8.6)–(8.8) the assertion of the theorem follows. □

REMARK 8.1. The existence of solutions of $(\mathscr{P}_1)$ and $(\mathscr{P}_1)_h$ was presumed. Sufficient conditions on the existence of solutions of $(\mathscr{P}_1)$ were formulated in Section 6. The uniqueness of solutions for any one of the problems is not required. As $K_h \subset K$ for any $h > 0$ and $\mathbb{R}_V \subset V_h$, the sufficient conditions for the existence and uniqueness of a solution of problem $(\mathscr{P}_1)_h$ are those which guarantee the existence and uniqueness of a solution of $(\mathscr{P}_1)$.

REMARK 8.2. Assuming that $u \in \mathscr{H}^2 \cap K$, $T_n(u) \in L^2(\Gamma_K)$ one can prove

$$|u - u_h| = \mathrm{O}(h^{3/4}), \quad h \to 0+. \tag{8.9}$$

REMARK 8.3. In the coercive case, when Korn's inequality holds on the space V, we can write the norm in $\mathscr{H}^1$ on the left hand side of (8.5) and (8.9). At that time, $(\mathscr{P}_1)$ and $(\mathscr{P}_1)_h$ have unique solutions u and u_h, respectively. In what follows, we shall study the convergence of u_h to the solution u without any additional regularity assumptions. To this end we need the following density result.

LEMMA 8.1. *Assume that $\overline{\Gamma}_K \cap \overline{\Gamma}_u = \emptyset$, $\overline{\Gamma}_K \cap \overline{\Gamma}_0 = \emptyset$ and let there exist only a finite number of boundary points of $\overline{\Gamma}_\tau \cap \overline{\Gamma}_K$, $\overline{\Gamma}_u \cap \overline{\Gamma}_\tau$, $\overline{\Gamma}_\tau \cap \overline{\Gamma}_0$. Then the set*

$$\mathscr{M} \equiv K \cap [C^\infty(\overline{\Omega}')]^2 \times [C^\infty(\overline{\Omega}'')]^2$$

is dense in K in the $\mathscr{H}^1$-norm.

PROOF. See HLAVÁČEK, HASLINGER, NEČAS and LOVÍŠEK [1988, p. 141]. □

From this and Remark 4.4, we immediately obtain the following.

THEOREM 8.2. *Let $\mathscr{J}$ be coercive on K and let $(\mathscr{P}_1)$ have exactly one solution u. Further, let all the assumptions of Lemma* 8.1 *be fulfilled. Then*

$$\|u - u_h\|_1 \to 0, \quad h \to 0+$$

for any regular system of triangulations $\{\mathscr{T}_h\}$, $h \to 0+$.

PROOF. On the basis of Lemma 8.1 it is readily seen that the system $\{K_h\}$ satisfies condition (i) from Definition 4.1. The rest of the proof is a consequence of Remark 4.4. □

Now let us pass to the case of domains with *curved boundaries*. The situation is more involved as the sets K_h defined by (8.3) *are not* subsets of K and therefore the relation (8.4) has to be used in order to get the a priori error estimate. To this end we shall assume a family $\{\mathscr{T}_h\}$, leading to the following approximation property of the piecewise linear Lagrangian interpolation $r_h u$ on V_h, given by (8.2):

$$\|v - r_h v\|_{0,\partial\Omega} \leqslant ch^{3/2}\|v\|_{2,\Omega} \tag{8.10}$$

holds for any $v \in H^2(\Omega)$ with a constant c independent of u, h. The family $\{\mathcal{T}_h\}$, satisfying (8.10) is discussed in NITSCHE [1971].

Then the following holds.

THEOREM 8.3. *Let problems* $(\mathcal{P}_1)$ *and* $(\mathcal{P}_1)_h$ *have solutions* u *and* u_h, *respectively. Let* $u \in \mathcal{H}^2 \cap K$, $T_n(u) \in L^2(\Gamma_K)$ *and let the norms* $|\!|\!|u_h|\!|\!|_1$ *remain bounded. Assume that a system of triangulations* $\{\mathcal{T}_h\}$ *possesses property* (8.10). *Finally, let the function* ψ *describing* Γ_K *be three times continuously differentiable on* $[a,b]$. *Then*

$$|u - u_h| = \mathrm{O}(h^{3/4}), \quad h \to 0+. \tag{8.11}$$

PROOF (Sketch of the proof). As already mentioned, we use (8.4). Similarly to the proof of Theorem 8.1, using integration by parts, we can write (8.4) in the following form:

$$c_0|u - u_h|^2 \leqslant A(u_h - u, v_h - u) + (T_n(u), (v_n' - u_{hn}') - (v_n'' - u_{hn}''))_{0,\Gamma_K}$$
$$+ (T_n(u), (v_{hn}' - u_n') - (v_{hn}'' - u_n''))_{0,\Gamma_K} \quad \forall v \in K, \ \forall v_h \in K_h.$$

The first term on the right side of the above inequality can be estimated in the same way as in the proof of Theorem 8.1 setting $v_h = r_h u$. The third term can be estimated as follows:

$$|(T_n(u), (r_{hn}u' - u_n') - (r_{hn}u'' - u_n''))_{0,\Gamma_K}| \leqslant ch^{3/2}|\!|\!|u|\!|\!|_2 \tag{8.12}$$

taking into account (8.10). The most complicated estimate is that of the term $(T_n(u), (v_n' - u_{hn}') - (v_n'' - u_{hn}''))_{0,\Gamma_K}$. It can be shown that there exists a function $v \in K$ such that

$$|(T_n(u), (v_n' - u_{hn}') - (v_n'' - u_{hn}''))_{0,\Gamma_K}| \leqslant ch^{3/2}|\!|\!|u_h|\!|\!|_1. \tag{8.13}$$

For the detailed proof see HLAVÁČEK, HASLINGER, NEČAS and LOVÍŠEK [1988]. The error estimate (8.11) now follows from (8.12) and (8.13). □

REMARK 8.4. If Korn's inequality holds on the whole space V, all the assumptions of Theorem 8.3 are fulfilled. $(\mathcal{P}_1)$ and $(\mathcal{P}_1)_h$ have exactly one solution each, u and u_h, respectively, and moreover, $|\!|\!|u_h|\!|\!|_1$ is bounded. The situation is considerably more complicated when

$$A(v,v) \geqslant c_0|v|^2 \quad \forall v \in V,$$

i.e. in the semicoercive case. In such a case the coerciveness of $\mathcal{J}$ on K does not imply the same property of $\mathcal{J}$ on K_h in view of the fact that $K_h \not\subset K$, in general.

But if $\mathcal{J}$ is coercive on $\bigcup_{h>0} K_h$ then $|||u_h|||_1$ is bounded. This can be guaranteed by a special geometry of Γ_K; for example, let Γ_K contain a segment I and let us define

$$K_I = \{v \in V \mid v_n' - v_n'' \leqslant 0 \text{ on } I\}.$$

In such a case $K_h \subset K_I$ for any $h > 0$ and if $\mathcal{J}$ is coercive on K_I, the same property holds on $\bigcup_{h>0} K_h$.

REMARK 8.5. If we have no information on the regularity of u, a result similar to the one from Theorem 8.2 can be established.

8.2. *Approximation of the problem with an increasing zone of contact*

Here, we shall discuss the approximation of the problem $(\mathcal{P}_2)$, introduced in Section 5.2. First we describe the construction of the finite-element space. For the sake of simplicity, we shall suppose in the following that only Γ_K' and Γ_K'' are curved and that the functions f' and f'' describing these arcs are twice continuously differentiable in $[a, b]$. Curved elements are defined in the same way as in the previous section. For the construction of the finite dimensional space on the curved element we use the isoparametric technique, as follows:

Let $\hat{T}$ be the reference triangle with vertices $(0,0)$, $(1,0)$, $(0,1)$. Let A_i, $A_{i+1} \in \Gamma_K'$, $S \in \Omega'$ (say) and let $x = \varphi(s)$, $y = \psi(s)$, $s \in [0,1]$, φ, $\psi \in C^2([0,1])$ be parametric equations of the arc A_iA_{i+1}. Let T denote the curved element determinated by the points A_i, A_{i+1} and S. If the diameter of T is not large, there exists a one-to-one mapping $F_T : \mathbb{R}^2 \to \mathbb{R}^2$ of $\hat{T}$ onto T and, moreover, this mapping is continuously differentiable in each of its variables.

Let $\hat{P} = P_1(\hat{T})$ be the set of linear polynomials defined on $\hat{T}$. Then the set of functions defined on the curved element T is given by

$$P(T) = \{p \mid \exists \hat{p} \in \hat{P} \colon p = \hat{p} \circ F_T^{-1}\}, \tag{8.14}$$

where F_T^{-1} is the inverse mapping of F_T.

The triangulation $\mathcal{T}_h = (\mathcal{T}_h', \mathcal{T}_h'')$ of the set $\overline{\Omega} = \overline{\Omega}' \cup \overline{\Omega}''$ consists on the one hand of curved elements along Γ_K' and Γ_K'' and on the other hand, of inner triangular elements. The elements along Γ_K' and Γ_K'' are constructed as follows: let $\{C_j\}_{j=1}^m$ be a partition of $[a, b]$, $C_1 \equiv a$, $C_m = b$. The points of intersection of perpendiculars at C_j with the arcs Γ_K' and Γ_K'' will be denoted by A_j and B_j, respectively (see Fig. 8.2). With any $\mathcal{T}_h$ we associate the space

$$V_h = \{v_h \in V \mid v_h|_T \in [P(T)]^2 \; \forall T \in \mathcal{T}_h\},$$

where $P(T) = P_1(T)$ provided T is a triangle and $P(T)$ is defined by (8.14) when T is a curved element. Finally, let

$$K_{\varepsilon h} = \{v_h \in V_h \mid v_{h\xi}''(B_j) - v_{h\xi}'(A_j) \leqslant \varepsilon(C_j),\; j = 1, \ldots, m\}, \tag{8.15}$$

where $\varepsilon = f'' - f'$.

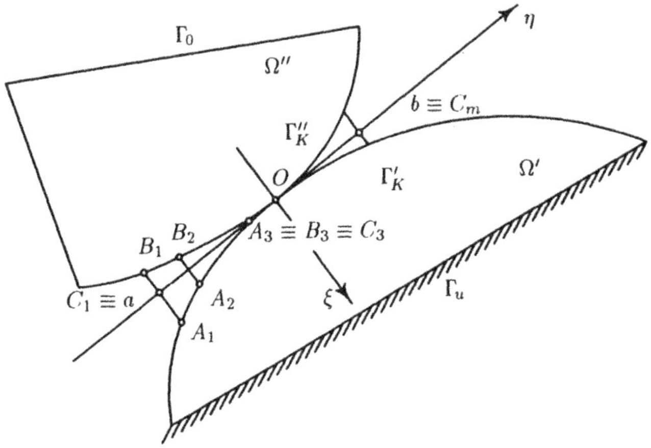

FIG. 8.2. A typical configuration with an increasing zone of contact.

DEFINITION 8.2. A function $u_h \in K_{\varepsilon h}$ is called an *approximation of the contact problem with an increasing zone of contact*, if it satisfies

$$\mathscr{J}(u_h) \leqslant \mathscr{J}(v_h) \quad \forall v_h \in K_{\varepsilon h}. \tag{$(\mathscr{P}_2)_h$}$$

As far as the rate of convergence of u_h to u is concerned, we have the following.

THEOREM 8.4. *Let* $(\mathscr{P}_2)$ *and* $(\mathscr{P}_2)_h$ *have solutions* u *and* u_h, *respectively. Let* $u \in \mathscr{H}^2 \cap K_\varepsilon$, $u'_\xi \in W^{1,\infty}(\Gamma'_K), u''_\xi \in W^{1,\infty}(\Gamma''_K), f', f'' \in C^2([a,b])$. *Further, let us assume that the number of points in* $[a,b]$ *at which the inequality* $u''_\xi - u'_\xi < \varepsilon$ *changes into the equality* $u''_\xi - u'_\xi = \varepsilon$ *is finite. Then*

$$|u - u_h| \leqslant c(u, f', f'')h$$

holds for an arbitrary regular system $\{\mathscr{T}_h\}$, $h \to 0+$.

PROOF (Sketch of the proof). As $K_{\varepsilon h}$ is an external approximation of K_ε, we use (4.3a), again. Analogously to the proof of Theorem 8.3, using integration by parts, we may write

$$\begin{aligned} c_0|u - u_h|^2 \leqslant{} & A(u_h - u, v_h - u) + (T'_\xi(u), v'_{h\xi} - u'_\xi)_{0,\Gamma'_K} \\ & + (T''_\xi(u), v''_{h\xi} - u''_\xi)_{0,\Gamma''_K} + (T'_\xi(u), v'_\xi - u'_{h\xi})_{0,\Gamma'_K} \\ & + (T''_\xi(u), v''_\xi - u''_{h\xi})_{0,\Gamma''_K} \quad \forall v_h \in K_{\varepsilon h},\ \forall v \in K_\varepsilon. \end{aligned} \tag{8.16}$$

For the function v_h we choose the corresponding P-interpolation of the exact solution u, i.e. $v_h = u_I$, where

$$u_I|_T = \hat{\pi}(u|_T \circ F_T) \circ F_T^{-1},$$

where $T = F_T(\hat{T})$ and $\hat{\pi}$ is the linear Lagrangian interpolation on $\hat{T}$. Using classical approximation results, we obtain

$$|A(u_h - u, u_I - u)| \leqslant \tfrac{1}{2}|u_h - u|^2 + ch^2 \|u\|_2^2. \tag{8.17}$$

The sum of the second and the third terms on the right hand side of (8.16) can be estimated as follows:

$$|(T'_\xi(u), u'_{I\xi} - u'_\xi)_{0,\Gamma'_K} + (T''_\xi(u), u''_{I\xi} - u''_\xi)_{0,\Gamma''_K}| \leqslant c(u, f', f'')h^2 \tag{8.18}$$

taking into account the assumptions concerning the regularity and the behaviour of u'_ξ and u''_ξ on Γ'_K and Γ''_K, respectively. It remains to estimate the last two terms on the right hand side of (8.16), considering a suitable choice of $v \in K_\varepsilon$. To this end let us denote

$$\mathcal{U}_h(\eta) = u''_{h\xi}(f''(\eta), \eta) - u'_{h\xi}(f'(\eta), \eta), \quad \eta \in (a, b)$$

and define a function W^h on $[a, b]$ by

$$W^h(\eta) = \inf_{\eta \in [a,b]} [\mathcal{U}_h(\eta), \varepsilon(\eta)].$$

Then it is easy to check that

$$\left| \int_a^b T_\xi(u)(\mathcal{U}_h - W^h)\, \mathrm{d}\eta \right| \leqslant ch^2 |\varepsilon|_{2,[a,b]}, \tag{8.19}$$

where $T_\xi(u) \equiv T''_\xi(u)(\cos\alpha'')^{-1} = -T'_\xi(u)(\cos\alpha')^{-1}$.

The meanings of symbols α' and α'' have been introduced in Section 5.2. Let $v = (v', v'') \in V$ fulfil $v'' \equiv 0$ on Ω'' and $-v'_\xi = W^h$. Then $v \in K_\varepsilon$ and the sum of the last terms can be written and estimated as follows:

$$\left| \int_a^b T_\xi(u)[(v''_\xi - v'_\xi) - (u''_{h\xi} - u'_{h\xi})]\, \mathrm{d}\eta \right| = \left| \int_a^b T_\xi(u)(W^h - \mathcal{U}_h)\, \mathrm{d}\eta \right| \leqslant ch^2 |\varepsilon|_{2,[a,b]}$$

making use of (8.19). The assertion of the theorem follows from this estimate with (8.17) and (8.18) (for a more detailed proof see HLAVÁČEK, HASLINGER, NEČAS and LOVÍŠEK [1988]). □

For the same reasons as in the case of the contact problem with a bounded zone of contact, we will formulate the convergence result for the approximate solutions u_h to the nonregular solution u. A fundamental role is played by the following density result, the proof of which can be found in HLAVÁČEK, HASLINGER, NEČAS and LOVÍŠEK [1988].

LEMMA 8.2. *Let* $f^M \in C^m((a-\delta, b+\delta))$, $m \geqslant 1$, $\delta > 0$, $\overline{\Gamma}_K^M \cap \overline{\Gamma}_u = \emptyset, \overline{\Gamma}_K^M \cap \overline{\Gamma}_0 = \emptyset$, $M = {}'$, $''$ *and let the intersections* $\overline{\Gamma}_u \cap \overline{\Gamma}_\tau$, $\overline{\Gamma}_0 \cap \overline{\Gamma}_\tau$ *consist of a finite number of points. Let* $v \in K_\varepsilon$ *fulfil the condition* $v''_\xi - v'_\xi \leqslant f' - f''$ *in* $(a-\delta, b+\delta)$. *Then* v *belongs to the closure (in the norm of* $\mathcal{H}^1$*) of the set*

$$K_\varepsilon \cap [C^m(\overline{\Omega}')]^2 \times [C^m(\overline{\Omega}'')]^2.$$

Having in mind this lemma it is not difficult to prove the following convergence result.

THEOREM 8.5. *Let* $(\mathcal{P}_2)$ *have exactly one solution* u *and let* $\mathcal{J}$ *be coercive on* $\bigcup_{h>0} K_{\varepsilon h}$. *Let all the assumptions of Lemma 8.2 be fulfilled with* $m = 2$. *Then*

$$|\!|\!| u - u_h |\!|\!|_1 \to 0, \quad h \to 0+$$

for any regular system $\{\mathcal{T}_h\}$.

9. Approximation of the contact problem with given friction

The aim of the present part is to extend the results of Section 8 to contact problems involving friction. We start with a simple model with given slip stress making use of different variational formulations of the mixed type. At the end we present results concerning approximations of the contact problems with friction obeying Coulomb's law.

Analogously to Section 7 we shall deal with the approximation of the so called Signorini problem, i.e. the contact between an elastic body Ω and a perfectly rigid foundation. We start with the problem with given friction by recalling notations.

Let the elastic body be represented by a bounded domain $\Omega \subset \mathbb{R}^2$, the boundary $\partial\Omega$ of which consists of three parts Γ_u, Γ_P and Γ_K which are disjoint and open in $\partial\Omega$:

$$\partial\Omega = \overline{\Gamma}_u \cup \overline{\Gamma}_P \cup \overline{\Gamma}_K.$$

Let Γ_u and Γ_K be nonempty. On Γ_u the displacement will be prescribed, while surface tractions $P = (P_1, P_2)$ are applied on Γ_P. On Γ_K unilateral conditions (7.1) and the friction law (7.3) will be assumed. For the sake of simplifying the notation, the coefficient of friction $\mathcal{F} \equiv 1$ and $g \equiv 1$ on Γ_K.

Let us put

$$V = \{v \in [H^1(\Omega)]^2 \mid v = 0 \text{ on } \Gamma_u\},$$
$$K = \{v \in V \mid v_n \leqslant 0 \text{ on } \Gamma_K\}.$$

According to Section 7, a function $u \in K$ will be called a *variational solution* of the Signorini problem with given friction $g \equiv 1$ if it satisfies

$$a(u, v-u) + j(v) - j(u) \geqslant L(v-u) \quad \forall v \in K, \tag{9.1}$$

where

$$a(u,v) = \int_\Omega c_{ijkl} e_{ij}(u) e_{kl}(v)\,\mathrm{d}x,$$

$$j(v) = \int_{\Gamma_K} |v_t|\,\mathrm{d}s,$$

$$L(v) = \int_\Omega F_i v_i\,\mathrm{d}x + \int_{\Gamma_P} P_i v_i\,\mathrm{d}s, \quad F \in [L^2(\Omega)]^2, \; P \in [L^2(\Gamma_P)]^2.$$

An equivalent formulation of (9.1) is the problem

$$\text{find } u \in K\colon\ \mathcal{J}(u) \leqslant \mathcal{J}(v) \quad \forall v \in K, \tag{9.2}$$

where $\mathcal{J}(v) = \frac{1}{2}a(v,v) + j(v) - L(v)$ is the total potential energy functional.

The main difficulty from the point of view of the numerical realization is the presence of the nondifferentiable term $j(v)$. To overcome this difficulty, we use the duality approach, illustrated in Example 3.6 for the scalar case. Below we present two variants of the mixed formulation of (9.2).

Variant 1. (*Dualization of the nondifferentiable term* $j(v)$). It is easy to see that

$$j(v) = \sup_{\mu \in \Lambda} (\mu, v_t)_{0,\Gamma_K},$$

where

$$\Lambda = \{\mu \in L^\infty(\Gamma_K) \mid |\mu| \leqslant 1 \text{ on } \Gamma_K\},$$

and $(\cdot,\cdot)_{0,\Gamma_K}$ denotes the scalar product in $L^2(\Gamma_K)$.

Hence

$$\inf_{v \in K} \mathcal{J}(v) = \inf_{v \in K} \sup_{\mu \in \Lambda} \mathcal{L}(v, \mu), \tag{9.3}$$

where $\mathscr{L}:[H^1(\Omega)]^2 \times \Lambda \to \mathbb{R}^1$ is the Lagrangian function given by

$$\mathscr{L}(v,\mu) = \tfrac{1}{2}a(v,v) + (\mu, v_t)_{0,\Gamma_K} - L(v).$$

In the spirit of (9.3) we shall consider the problem of finding a saddle-point (w,λ) of $\mathscr{L}$ on $K \times \Lambda$:

$$\mathscr{L}(w,\mu) \leqslant \mathscr{L}(w,\lambda) \leqslant \mathscr{L}(v,\lambda) \quad \forall v \in K,\ \forall \mu \in \Lambda \tag{9.4}$$

or equivalently,

$$\begin{aligned}
&\text{find } (w,\lambda) \in K \times \Lambda \text{ such that}\\
&a(w, v-w) + (\lambda, v_t - w_t)_{0,\Gamma_K} \geqslant L(v-w) \quad \forall v \in K,\\
&(\mu - \lambda, w_t)_{0,\Gamma_K} \leqslant 0 \quad \forall \mu \in \Lambda.
\end{aligned} \tag{9.5}$$

The relation between (9.2) and (9.5) is expressed as follows.

THEOREM 9.1. *There exists a unique solution* (w,λ) *of* (9.5) *and it satisfies*

$$w = u, \qquad \lambda = -T_t(u),$$

where $u \in K$ *is the unique solution of* (9.2).

PROOF. The existence of a saddle-point (w,λ) of $\mathscr{L}$ follows from Remark 3.8. Moreover, $w = u$, where u solves (9.2). By applying Green's formula to (9.5) we get $-\lambda = T_t(w) = T_t(u)$. □

Problem (9.5) is called the *mixed formulation* (I) *of the Signorini problem* with given friction. The advantage of this formulation is that the nondifferentiable functional $\mathscr{J}$ is replaced by a functional $\mathscr{L}$ which is smooth in both variables.

Below we describe an approximation of the Signorini problem with given friction, based on the mixed formulation (9.5). To this end let us suppose that $\Omega \subset \mathbb{R}^2$ is a *polygonal domain*, let $\{\mathscr{T}_h\}$, $h \to 0+$ be a regular system of triangulations of $\overline{\Omega}$, satisfying the usual requirements on the mutual position of two triangles belonging to $\mathscr{T}_h$. Furthermore, we shall assume that Γ_K consists of a *single segment*. We associate each $\mathscr{T}_h$ with a finite dimensional space V_h of piecewise linear vector functions

$$V_h = \{v_h \in [C(\overline{\Omega})]^2 \mid v_h|_T \in [P_1(T)]^2\ \forall T \in \mathscr{T}_h,\ v_h = 0 \text{ on } \Gamma_u\}$$

and with a convex, closed subset $K_h \subset V_h$

$$K_h = \{v_h \in V_h \mid (v_h \cdot n)(a_i) \leqslant 0\ \forall i = 1,\dots,m\},$$

where $a_1,\dots,a_m$ are the nodes of $\mathscr{T}_h$, lying on $\overline{\Gamma}_K \backslash \overline{\Gamma}_u$. It is easy to see that $K_h \subset K\ \forall h \in (0,1)$.

Let $\{\mathcal{T}_H\}$, $H \in (0,1)$ be a partition of $\overline{\Gamma}_K$, whose nodes we denote by $b_1, \ldots, b_M$, $H = \max |b_{i+1} - b_i|$. These nodes need not coincide with the nodes $a_1, \ldots, a_m$. Next, we shall write $h = H$ if and only if $m = M$ and $a_i = b_i$, $\forall i = 1, \ldots, m$. Let

$$L_H = \{\mu_H \in L^2(\Gamma_K) \mid \mu_H|_{\overline{b_i b_{i+1}}} \in P_0(\overline{b_i b_{i+1}})\}$$

and

$$\Lambda_H = \{\mu_H \in L_H \,|\, |\mu_H| \leqslant 1 \text{ on } \Gamma_K\}.$$

As an approximation to (9.4) we solve the problem of finding a saddle-point (w_h, λ_H) of $\mathcal{L}$ on $K_h \times \Lambda_H$:

$$\mathcal{L}(w_h, \mu_H) \leqslant \mathcal{L}(w_h, \lambda_H) \leqslant \mathcal{L}(v_h, \lambda_H) \quad \forall (v_h, \mu_H) \in K_h \times \Lambda_H \tag{9.6}$$

or equivalently

$$\begin{aligned} &a(w_h, v_h - w_h) + (\lambda_H, v_{ht} - w_{ht})_{0,\Gamma_K} \geqslant L(v_h - w_h) \quad \forall v_h \in K_h, \\ &(\mu_H - \lambda_H, w_{ht})_{0,\Gamma_K} \leqslant 0 \quad \forall \mu_H \in \Lambda_H. \end{aligned} \tag{9.7}$$

Using the results of Section 3, especially Remarks 3.8 and 3.9 we immediately have the following.

THEOREM 9.2. *Problem* (9.6) *has a solution* (w_h, λ_H) *for all h,* $H > 0$. *Moreover, its component* w_h *is uniquely determined.*

REMARK 9.1. Taking into account Theorem 9.1 we see that w_h may be taken for an approximation of the displacement field u and $-\lambda_H$ for an approximation of $T_t(u)$ on Γ_K.

As far as the relation between (w, λ) and (w_h, λ_H) is concerned, one gets the following.

THEOREM 9.3. *Let* $h \to 0+$ *if and only if* $H \to 0+$ *and let the intersection* $\overline{\Gamma}_K \cap \overline{\Gamma}_u$ *consist of a finite number of points. Then*

$$\begin{aligned} &w_h \to w \quad \text{in } [H^1(\Omega)]^2, \\ &\lambda_H \rightharpoonup \lambda \text{ (weakly)} \quad \text{in } L^2(\Gamma_K). \end{aligned}$$

PROOF. We shall apply Theorem 4.5. As K_h and Λ_H are internal approximations of K and Λ, respectively, conditions (4.20) and (4.21) are satisfied. Also condition (4.19) is clearly satisfied. If the intersection of $\overline{\Gamma}_u \cap \overline{\Gamma}_K$ contains only a finite number of points, the set $[C^\infty(\overline{\Omega})]^2 \cap K$ is dense in K (see HLAVÁČEK and LOVÍŠEK [1977]). All the assumptions of Theorem 4.5 are satisfied. □

Below we present another mixed formulation, which is based on the simultaneous dualization of the kinematic constraint $v_n \leqslant 0$ on Γ_K and of the nondifferentiable term j. To this end we introduce new notations and establish auxiliary results, which will be needed later. Henceforward we shall suppose that $\Gamma_P = \emptyset$. This assumption is solely for technical reasons. It simplifies the mathematical analysis.

Let $S(\Omega)$ denote the space of symmetric matrices

$$S(\Omega) = \{\sigma = (\sigma_{ij})_{i,j=1}^2 \in [L^2(\Omega)]^4,\ \sigma_{ij} = \sigma_{ji} \text{ a.e. in } \Omega\}$$

and

$$H(\operatorname{div}, \Omega) = \{\sigma \in S(\Omega)\colon \operatorname{div} \sigma = \left(\frac{\partial \sigma_{1j}}{\partial x_j}, \frac{\partial \sigma_{2j}}{\partial x_j}\right) \in [L^2(\Omega)]^2\}.$$

Both, $S(\Omega)$ and $H(\operatorname{div}, \Omega)$ are Hilbert spaces, equipped with the norm

$$(\sigma, \tau)_S = (\sigma_{ij}, \tau_{ij})_{0,\Omega}$$

and

$$(\sigma, \tau)_{H(\operatorname{div},\Omega)} \equiv (\sigma_{ij}, \tau_{ij})_{0,\Omega} + \left(\frac{\partial \sigma_{ij}}{\partial x_j}, \frac{\partial \tau_{ik}}{\partial x_k}\right)_{0,\Omega},$$

respectively. Let V be the space of virtual displacements

$$V = \{v \in [H^1(\Omega)]^2 \mid v_i = 0 \text{ on } \Gamma_u,\ i = 1, 2\},$$

equipped with the norm

$$|\!|\!| v |\!|\!| = \{(v_i, v_i)_{0,\Omega} + (e_{ij}(v), e_{ij}(v))_{0,\Omega}\}^{1/2}. \tag{9.8}$$

By $[H^{1/2}(\Gamma_K)]^2$ we denote the space of all traces of functions belonging to V, i.e.

$$\phi \in [H^{1/2}(\Gamma_K)]^2 \text{ iff } \exists v \in V\colon \phi = v \text{ on } \Gamma_K.$$

The norm in $[H^{1/2}(\Gamma_K)]^2$ is introduced as follows:

$$\|\phi\|_{1/2,\Gamma_K} = \inf_{\substack{v \in V \\ v=\phi \text{ on } \Gamma_K}} |\!|\!| v |\!|\!|. \tag{9.9}$$

Let $\delta : V \to [L^2(\Gamma_K)]^2$ be the mapping, defined by

$$\delta v = (v_n, v_t),$$

where $v_n = v \cdot n$ and $v_t = v \cdot t$ are the normal and tangential components, respectively, of the displacement field v.

By W we denote the image $\delta(V)$, i.e.

$$\mu \in W \quad \text{iff} \quad \exists v \in V\colon \mu = (\mu_1, \mu_2) = (v_n, v_t).$$

If Ω is a domain with a sufficiently smooth boundary $\partial\Omega$, then $W = [H^{1/2}(\Gamma_K)]^2$. Although this is not our case, W and $[H^{1/2}(\Gamma_K)]^2$ are isomorphic. Indeed, let $\phi \in [H^{1/2}(\Gamma_K)]^2$. Then the element $\mu = (\mu_1, \mu_2) = (\phi \cdot n, \phi \cdot t)$ belongs to W. Next, we shall write $\mu = \beta\phi$, $\beta : [H^{1/2}(\Gamma_K)]^2 \to W$ and the norm in W can be introduced as follows:

$$\|\mu\|_W = \|\beta^{-1}\mu\|_{1/2,\Gamma_K}. \tag{9.10}$$

Denote by $[H^{-1/2}(\Gamma_K)]^2$ and W', the dual space to $[H^{1/2}(\Gamma_K)]^2$ and W, respectively. The corresponding duality pairings will be denoted by $\langle\,,\rangle$ and $\langle\!\langle\,,\rangle\!\rangle$, respectively. Then the following Green's formulae hold.

LEMMA 9.1. (i) *There exists a unique mapping* $T = (T_1, T_2) \in \mathcal{L}(H(\mathrm{div}, \Omega), [H^{-1/2}(\Gamma_K)]^2)$ *such that*

$$(\tau_{ij}, e_{ij}(v))_{0,\Omega} + (\tau_{ij,j}, v_i)_{0,\Omega} = \langle T(\tau), v\rangle \quad \forall \tau \in H(\mathrm{div}, \Omega), \ \forall v \in V. \tag{9.11}$$

(ii) *There exists a unique mapping* $\tilde{T} = (T_n, T_t) \in \mathcal{L}(H(\mathrm{div}, \Omega), W')$ *such that*

$$(\tau_{ij}, e_{ij}(v))_{0,\Omega} + (\tau_{ij,j}, v_i)_{0,\Omega} = \langle\!\langle \tilde{T}(\tau), \delta v\rangle\!\rangle \quad \forall \tau \in H(\mathrm{div}, \Omega), \ \forall v \in V. \tag{9.12}$$

PROOF. For the proof see AUBIN [1972]. □

In what follows, we introduce suitable norms in the dual spaces $[H^{-1/2}(\Gamma_K)]^2$ and W'. Let the symbol $\|\phi^*\|_{-1/2,\Gamma_K}$, $\phi^* \in [H^{-1/2}(\Gamma_K)]^2$ stand for the usual dual norm of ϕ^*:

$$\|\phi^*\|_{-1/2,\Gamma_K} = \sup_{\phi} \frac{\langle \phi^*, \phi\rangle}{\|\phi\|_{1/2,\Gamma_K}},$$

where $\|\cdot\|_{1/2,\Gamma_K}$ is defined by (9.9). This norm, however, can be evaluated only with difficulty. To this end we present an equivalent expression for $\|\cdot\|_{-1/2,\Gamma_K}$, which will appear to be useful later on.

LEMMA 9.2. *It holds that*

$$\|\phi^*\|_{-1/2,\Gamma_K} = \|\sigma\|_{H(\mathrm{div},\Omega)} = |\!|\!| u(\phi^*) |\!|\!|, \tag{9.13}$$

where $u(\phi^*) \in V$ is the unique solution of the following problem

$$\text{find } u \in V\colon (e_{ij}(u), e_{ij}(v))_{0,\Omega} + (u_i, v_i)_{0,\Omega} = \langle \phi^*, v\rangle \quad \forall v \in V,$$

$\sigma = e(u(\phi^*))$, $|\!|\!|\cdot|\!|\!|$ *is defined by* (9.8) *and* $\|\sigma\|_{H(\mathrm{div},\Omega)} = (\sigma,\sigma)^{1/2}_{H(\mathrm{div},\Omega)}$.

As $[H^{1/2}(\Gamma_K)]^2$ and W are mutually isomorphic, the same holds for $[H^{-1/2}(\Gamma_K)]^2$ and W'. Consequently, the dual norm in W' can be introduced as follows:

$$\|\mu\|_{W'} = \|\beta^*\mu\|_{-1/2,\Gamma_K} \quad \forall \mu \in W',$$

where $\beta^* : W' \mapsto [H^{-1/2}(\Gamma_K)]^2$ is the adjoint of β.

Now, we are ready to present the mixed variational formulation of the Signorini problem with given friction, based on the simultaneous dualization of the unilateral condition $v_n \leqslant 0$ on Γ_K and the non-differentiable term j.

Variant 2. Define

$$\Lambda = \Lambda_1 \times \Lambda_2 \subseteq W',$$

where

$$\Lambda_1 = \{\mu_1 \mid \mu = (\mu_1,\mu_2) \in W'\colon \langle\!\langle \mu, v\rangle\!\rangle \geqslant 0 \ \forall v \in V,\ v_n \leqslant 0,\ v_t = 0 \text{ on } \Gamma_K\},$$
$$\Lambda_2 = \{\mu_2 \in L^2(\Gamma_K) \mid |\mu_2| \leqslant 1 \text{ on } \Gamma_K\}.$$

Let $\mathscr{L} : V \times \Lambda \to \mathbb{R}^1$ be the Lagrangian, defined through the relation

$$\mathscr{L}(v,\mu) = \tfrac{1}{2}(\tau_{ij}(v), e_{ij}(v))_{0,\Omega} - \langle\!\langle \mu, \delta v\rangle\!\rangle - (F_i, v_i)_{0,\Omega}, \quad (v,\mu) \in V \times \Lambda,$$

where

$$\tau_{ij}(v) = c_{ijkl}e_{kl}(v).$$

By the *mixed variational formulation* (II) of the Signorini problem with given friction we mean the problem of finding a saddle-point $(w, \lambda_1, \lambda_2) \equiv (w, \lambda) \in V \times \Lambda$, $\lambda = (\lambda_1, \lambda_2)$ of $\mathscr{L}$ on $V \times \Lambda$:

$$\mathscr{L}(w,\mu) \leqslant \mathscr{L}(w,\lambda) \leqslant \mathscr{L}(v,\lambda) \quad \forall v \in V,\ \forall \mu \in \Lambda. \tag{9.14}$$

The relation between (9.1) and (9.14) is expressed as follows.

THEOREM 9.4. *There exists a unique saddle-point $(w, \lambda_1, \lambda_2)$ of $\mathscr{L}$ on $V \times \Lambda_1 \times \Lambda_2$ and it holds that*

$$w = u, \qquad \lambda_1 = T_n(u), \qquad \lambda_2 = T_t(u), \tag{9.15}$$

where $u \in K$ solves (9.1).

PROOF. An equivalent formulation of (9.14) is

$$\begin{aligned}&\text{find } (w,\lambda) \in V \times \Lambda \text{ such that}\\&(\tau_{ij}(w), e_{ij}(v))_{0,\Omega} - \langle\langle \lambda, \delta v\rangle\rangle = (F_i, v_i)_{0,\Omega} \quad \forall v \in V,\\&\langle\langle \mu - \lambda, \delta w\rangle\rangle \geqslant 0 \quad \forall \mu \in \Lambda.\end{aligned} \tag{9.16}$$

From this and Green's formula (9.12), we immediately obtain the statement of the theorem. □

Next we shall analyze an approximation of the Signorini problem with given friction based on the mixed formulation (9.14). To this end let us suppose that Ω is a *polygonal domain*. As before, we shall assume a regular family $\{\mathcal{T}_h\}$ and $\{\mathcal{T}_H\}$ of triangulations of $\overline{\Omega}$ and partitions of $\overline{\Gamma}_K$, respectively. The following sets of functions defined on $\overline{\Omega}$ and on $\overline{\Gamma}_K$ will be associated with any $\mathcal{T}_h$ and $\mathcal{T}_H$ respectively:

$$V_h = \{v_h \in [C(\overline{\Omega})]^2 \mid v_h|_T \in [(P_1(T)]^2 \ \forall T \in \mathcal{T}_h,\ v_h = 0 \text{ on } \Gamma_u\} \tag{9.17}$$

$$W'_H = \{\mu_H \in [L^2(\Gamma_K)]^2 \mid \mu_H|_{\overline{b_{i-1}b_i}} \in [P_0(\overline{b_{i-1}b_i})]^2 \ \forall i = 1,\ldots,M\},$$

$$\begin{aligned}\Lambda_{1H} = \{&\mu_{1H} \in L^2(\Gamma_K) \mid \mu_{1H}|_{\overline{b_{i-1}b_i}} \in P_0(\overline{b_{i-1}b_i})\\&\forall i = 1,\ldots,M,\ \mu_{1H} \leqslant 0 \text{ on } \Gamma_K\},\end{aligned} \tag{9.18}$$

$$\begin{aligned}\Lambda_{2H} = \{&\mu_{2H} \in L^2(\Gamma_K) \mid \mu_{2H}|_{\overline{b_{i-1}b_i}} \in P_0(\overline{b_{i-1}b_i})\\&\forall i = 1,\ldots,M,\ |\mu_{2H}| \leqslant 1 \text{ on } \Gamma_K\}.\end{aligned} \tag{9.19}$$

Finally set $\Lambda_H \equiv \Lambda_{1H} \times \Lambda_{2H} \subseteq W'_H$.

As an approximation to problem (9.14) we solve the problem of finding a saddle-point $(w_h, \lambda_{1H}, \lambda_{2H}) \equiv (w_h, \lambda_H) \in V_h \times \Lambda_H$ of $\mathcal{L}$ on $V_h \times \Lambda_H$:

$$\mathcal{L}(w_h, \mu_H) \leqslant \mathcal{L}(w_h, \lambda_H) \leqslant \mathcal{L}(v_h, \lambda_H) \quad \forall v_h \in V_h,\ \forall \mu_H \in \Lambda_H \tag{9.20}$$

or equivalently

$$\begin{aligned}&\text{find } (w_h,\lambda_H) \in V_h \times \Lambda_H \text{ such that}\\&(\tau_{ij}(w_h), e_{ij}(v_h))_{0,\Omega} - \langle\langle \lambda_H, \delta v_h\rangle\rangle = (F_i, v_{ih})_{0,\Omega} \quad \forall v_h \in V_h,\\&\langle\langle \mu_H - \lambda_H, \delta w_h\rangle\rangle \geqslant 0 \quad \forall \mu_H \in \Lambda_H.\end{aligned} \tag{9.21}$$

Here the duality $\langle\langle\, ,\, \rangle\rangle$ is represented by the $[L^2(\Gamma_K)]^2$-scalar product, i.e. $\langle\langle \mu_H, \delta v_h\rangle\rangle \equiv (\mu_{1H}, v_{hn})_{0,\Gamma_K} + (\mu_{2H}, v_{ht})_{0,\Gamma_K}$. It is easy to prove the following.

THEOREM 9.5. *To every h, $H \in (0,1)$ there exists a solution (w_h, λ_H) of (9.20), the first component of which is uniquely determined.*

REMARK 9.2. The first component w_h will be taken for an approximation of the displacement field u, and λ_{1H} and λ_{2H} for approximations of $T_n(u)$ and $T_t(u)$ on Γ_K, respectively.

REMARK 9.3. Let us introduce the convex subset K_{hH} of V_h as follows:

$$K_{hH} = \{v_h \in V_h \mid (\mu_{1H}, v_{hn})_{0,\Gamma_K} \geqslant 0 \ \forall \mu_{1H} \in \Lambda_{1H}\}.$$

It is easy to see that K_{hH} contains all functions from V_h such that the mean values of their normal components on any $\overline{b_{i-1}b_i}$ are non-positive. Thus, K_{hH} can be viewed as an external approximation of K. From the results of Section 3 it follows that the first component w_h is a solution of the variational inequality

$$\begin{aligned}&(\tau_{ij}(w_h), e_{ij}(v_h) - e_{ij}(w_h))_{0,\Omega} - (\lambda_{2H}, v_{ht} - w_{ht})_{0,\Gamma_K}\\&\quad \geqslant (F_i, v_{ih} - w_{ih})_{0,\Omega} \quad \forall v_h \in K_{hH}.\end{aligned}$$

Let us suppose that $h \to 0+$ iff $H \to 0+$. As far as the convergence of the approximate solutions (w_h, λ_H) is concerned, one can prove the following.

THEOREM 9.6. *For any $v \in K$ let there exist a sequence $\{v_h\}$, $v_h \in K_{hH}$ such that $v_h \to v$ as $h \to 0$ in V. Then*

$$\begin{aligned}&w_h \to u, \quad h \to 0+ \quad \text{in } [H^1(\Omega)]^2,\\&\lambda_{2H} \rightharpoonup \lambda_2, \quad H \to 0+ \quad \text{in } L^2(\Gamma_K) \text{ (weakly)},\end{aligned}$$

where $u \in K$ solves (9.1) *and* $\lambda_2 = T_t(u)$.

PROOF. For the proof see HASLINGER and HLAVÁČEK [1982]. □

Next we shall study in more detail estimates of the differences between solutions of (9.14) and (9.20). In order to guarantee the uniqueness of the components λ_{1H}, λ_{2H}, we shall suppose that the following version of the stability condition (4.15) is satisfied: there exists a positive number β independent of h, $H > 0$ and such that

$$\sup_{V_h} \frac{\langle\langle \mu_H, \delta v_h \rangle\rangle}{|||v_h|||} \geqslant \beta \|\mu_H\|_{W'} \quad \forall \mu_H \in \Lambda_H. \tag{9.22}$$

Below we formulate a sufficient assumption under which the stability condition (9.22) holds. To this end we shall consider the elliptic boundary value problem:

$$\begin{aligned}&\text{find } w \in V \text{ such that}\\&(e_{ij}(w), e_{ij}(v))_{0,\Omega} + (w_i, v_i)_{0,\Omega} = \langle \phi^*, v \rangle \quad \forall v \in V.\end{aligned} \tag{9.23}$$

We say that problem (9.23) is *regular* if there exists a positive number $\varepsilon > 0$ such that for every $\phi^* \in [H^{-1/2+\varepsilon}(\Gamma_K)]^2$, the solution $w \in V \cap [H^{1+\varepsilon}(\Omega)]^2$ and

$$\|w\|_{1+\varepsilon,\Omega} \leqslant c(\varepsilon)\|\phi^*\|_{-1/2+\varepsilon,\Gamma_K}$$

holds with a positive constant c, depending on ε only. Then the following holds.

LEMMA 9.3. *Let the problem* (9.23) *be regular and the ratio h/H be sufficiently small. Then* (9.22) *holds.*

PROOF. The proof is parallel to that of Lemma 4.1 in the scalar case. □

Using the approach of Section 4, one can prove the following estimates of errors $\|w - w_h\|_{1,\Omega}$ and $\|\lambda - \lambda_H\|_{W'}$, where (w, λ) and (w_h, λ_H) are solutions of (9.14) and (9.20), respectively.

LEMMA 9.4. *Let* (9.22) *be satisfied. Then*

$$\|w - w_h\|_{1,\Omega}^2 \leqslant c \inf_{V_h \times \Lambda_H} \{\|w - v_h\|_{1,\Omega}^2 + \|\lambda - \mu_H\|_{W'}^2 + \langle\langle \mu_H - \lambda, \delta w \rangle\rangle\}, \quad (9.24)$$

$$\|\lambda - \lambda_H\|_{W'} \leqslant c\{\|w - w_h\|_{1,\Omega} + \inf_{\Lambda_H} \|\lambda - \mu_H\|_{W'}\}. \quad (9.25)$$

PROOF. For the detailed proof see HASLINGER and HLAVÁČEK [1982]. □

On the basis of (9.24) and (9.25) one can prove the following.

THEOREM 9.7. *Let all the assumptions of Lemma* 9.3 *be satisfied. If $w \in [H^{1+q}(\Omega)]^2$ for some $q > 0$ and $\lambda_1 \in L^2(\Gamma_K)$, then*

$$\|w - w_h\|_{1,\Omega} = O(H^{\tilde{q}}), \quad H \to 0+, \quad (9.26)$$

$$\|\lambda - \lambda_H\|_{W'} = O(H^{\tilde{q}}), \quad H \to 0+, \quad (9.27)$$

where $\tilde{q} = \min(\frac{1}{4}, q)$.

PROOF. In (9.24) let us insert $v_h = r_h w$, where $r_h w \in V_h$ is the piecewise linear Lagrange interpolate of w. We have

$$\|w - r_h w\|_{1,\Omega}^2 \leqslant ch^{2q} \|w\|_{1+q,\Omega}^2. \quad (9.28)$$

Since $\lambda_1 \in L^2(\Gamma_K)$ and $\lambda_2 \in L^\infty(\Gamma_K)$, we have $\beta^* \lambda = \overline{\mu} = (\overline{\mu}_1, \overline{\mu}_2) \in [L^2(\Gamma_K)]^2$. Let $\overline{\mu}_H \in W'_H$ be such that $\overline{\mu}_H|_{\overline{b_{i-1}b_i}}$ is the $[L^2(\overline{b_{i-1}b_i})]^2$ projection of $\overline{\mu}|_{\overline{b_{i-1}b_i}}$ onto $[P_0(\overline{b_{i-1}b_i})]^2$, $i = 1, \ldots, M$. Then $\beta^{*-1}\overline{\mu}_H|_{\overline{b_{i-1}b_i}}$ is the $[L^2(\overline{b_{i-1}b_i})]^2$ projection of $\lambda|_{\overline{b_{i-1}b_i}}$ onto $[P_0(\overline{b_{i-1}b_i})]^2$, as well. Moreover, $\beta^* \overline{\mu}_H \in \Lambda_H$. Inserting $\mu_H = \beta^{*-1}\overline{\mu}_H$ into the right hand side of (9.24) we obtain

$$\|\lambda - \beta^{*-1}\overline{\mu}_H\|_{W'}^2 = \|\beta^* \lambda - \overline{\mu}_H\|_{-1/2,\Gamma_K}^2 = O(H). \quad (9.29)$$

Using (9.29), we may write

$$\begin{aligned} |\langle\langle \beta^{*-1}\overline{\mu}_H - \lambda, \delta w \rangle\rangle| &\leqslant \|\beta^{*-1}\overline{\mu}_H - \lambda\|_{W'} \|\delta w\|_W \\ &= \|\overline{\mu}_H - \beta^* \lambda\|_{-1/2,\Gamma_K} \|\delta w\|_W = O(H^{1/2}). \end{aligned} \quad (9.30)$$

Combining (9.28), (9.29) and (9.30) and the fact that $h \leqslant cH$, $c > 0$, we arrive at (9.26). The estimate (9.27) now follows from (9.25), (9.26) and (9.29). □

More precise estimates can be obtained under stronger regularity assumptions on w, from the following.

THEOREM 9.8. *Let the stability condition* (9.22) *hold. Let the solution u of* (9.1) *belong to $[H^2(\Omega)]^2 \cap K$. Moreover, let the set of points of Γ_K, where u_n and u_t change from $u_n < 0$ to $u_n = 0$ and from $u_t = 0$ to $u_t \neq 0$, respectively, be finite. Then*

$$\|u - w_h\|_{1,\Omega} = \mathrm{O}(H^{1/2}), \quad H \to 0+,$$
$$\|\lambda - \lambda_H\|_{W'} = \mathrm{O}(H^{1/2}), \quad H \to 0+ .$$

PROOF. Let us mention that $u = w$, where w is the first component of the saddle-point (9.14). We proceed in the same way as in the proof of Theorem 9.7. On the basis of assumptions concerning the behaviour of $w = u$ on the contact part Γ_K one can prove that the error estimate (9.30) can be improved, namely

$$|\langle\langle \beta^{*-1}\overline{\mu}_H - \lambda, \delta w\rangle\rangle| \leqslant \mathrm{O}(H). \tag{9.31}$$

From this, (9.28) and (9.29), the assertion of the theorem follows. □

Let $(w, \lambda) \in V \times \Lambda$ be the solution of the mixed variational formulation (II). Then

$$\mathscr{L}(w, \lambda) = \min_{v \in V} \sup_{\mu \in \Lambda} \mathscr{L}(v, \mu) = \max_{\mu \in \Lambda} \inf_{v \in V} \mathscr{L}(v, \mu) \tag{9.32}$$

making use of Theorem 3.6. The elimination of the displacement field v in the second equality in (9.32) leads to a variational formulation (the so-called *reciprocal*) in term of quantities, defined on Γ_K only. The approach presented below will be parallel to the one presented in Example 3.9 for a scalar case.

Let us set

$$\tilde{\mathscr{S}}(\mu_1, \mu_2) = \inf_{v \in V} \mathscr{L}(v, \mu). \tag{9.33}$$

We derive the explicit form of $\tilde{S}$. Let $\mu = (\mu_1, \mu_2) \in \Lambda = \Lambda_1 \times \Lambda_2$ be fixed. Then the minimum of (9.33) is obtained from the function $z \in V$, which solves the following problem

$$(\tau_{ij}(z), e_{ij}(v))_{0,\Omega} = \langle\langle \mu, \delta v\rangle\rangle + (F_i, v_i)_{0,\Omega} \quad \forall v \in V. \tag{9.34}$$

In view of the linearity of (9.34), one can split z into $\hat{z}$ and $\tilde{z}$, $z = \hat{z} + \tilde{z}$, where $\hat{z}, \tilde{z} \in V$ are the unique solutions of

$$(\tau_{ij}(\hat{z}), e_{ij}(v))_{0,\Omega} = (F_i, v_i)_{0,\Omega} \quad \forall v \in V, \tag{9.35}$$
$$(\tau_{ij}(\tilde{z}), e_{ij}(v))_{0,\Omega} = \langle\langle \mu, \delta v\rangle\rangle \quad \forall v \in V. \tag{9.36}$$

Let $G: V' \to V$ be the Green's operator, corresponding to the bilinear form a defined at the beginning of this section. For brevity, instead of (9.35) and (9.36) we shall write $\hat{z} = G(F)$ and $\tilde{z} = G(\mu)$, respectively. Since $z \in V$ is a minimizer of $\mathscr{L}(v, \mu)$, we have

$$\begin{aligned}\mathscr{L}(z, \mu) &= -\tfrac{1}{2}\langle\langle \mu, \delta z\rangle\rangle - \tfrac{1}{2}(F_i, z_i)_{0,\Omega} \\ &= -\tfrac{1}{2}\langle\langle \mu, \delta \hat{z}\rangle\rangle - \tfrac{1}{2}\langle\langle \mu, \delta \tilde{z}\rangle\rangle - \tfrac{1}{2}(F_i, \hat{z}_i)_{0,\Omega} - \tfrac{1}{2}(F_i, \tilde{z}_i)_{0,\Omega}. \end{aligned} \tag{9.37}$$

Inserting $v = \tilde{z}$ in (9.35), using the symmetry of the bilinear form a and (9.36), we obtain

$$\langle\langle \mu, \delta \hat{z}\rangle\rangle = (\tau_{ij}(\tilde{z}), e_{ij}(\hat{z}))_{0,\Omega} = (\tau_{ij}(\hat{z}), e_{ij}(\tilde{z}))_{0,\Omega} = (F_i, \tilde{z}_i)_{0,\Omega}.$$

Thus (9.37) can be written as follows:

$$\mathscr{L}(z, \mu) = -\tfrac{1}{2}\langle\langle \mu, \delta(G(\mu))\rangle\rangle - \langle\langle \mu, \delta(G(F))\rangle\rangle - \tfrac{1}{2}(F_i, \hat{z}_i)_{0,\Omega}.$$

Denote the bilinear form by $\beta: W' \times W' \to \mathbb{R}^1$, defined as follows:

$$\beta(\mu, \nu) = \langle\langle \mu, \delta(G(\nu))\rangle\rangle, \quad (\mu, \nu) \in W' \times W'$$

and let $f: W' \to \mathbb{R}^1$ be the linear form, defined by

$$f(\mu) = -\langle\langle \mu, \delta(G(F))\rangle\rangle.$$

Then the functional $\tilde{\mathscr{S}}$ can be expressed as follows:

$$\tilde{\mathscr{S}}(\mu) = -\tfrac{1}{2}\beta(\mu, \mu) + f(\mu) - \tfrac{1}{2}(F_i, \hat{z}_i)_{0,\Omega}.$$

As the last term does not depend on μ, it can be omitted. Finally, let us set

$$\mathscr{S}(\mu) = -\tilde{\mathscr{S}}(\mu) - \tfrac{1}{2}(F_i, \hat{z}_i)_{0,\Omega} = \tfrac{1}{2}\beta(\mu, \mu) - f(\mu).$$

By the *reciprocal variational formulation* of the Signorini problem with given friction we mean the problem

$$\begin{aligned}&\text{find } \lambda \in \Lambda \text{ such that} \\ &\mathscr{S}(\lambda) \leqslant \mathscr{S}(\mu) \quad \forall \mu \in \Lambda.\end{aligned} \tag{9.38}$$

The relation between (9.2) and (9.38) is given by the following.

THEOREM 9.9. *There exists a unique solution* $\lambda = (\lambda_1, \lambda_2)$ *of* (9.38). *Moreover,*

$$\lambda_1 = T_n(u), \qquad \lambda_2 = T_t(u),$$

where $u \in K$ *solves* (9.2).

PROOF. The assertion immediately follows from Theorem 9.4 and (9.32). □

REMARK 9.4. Let us consider the problem with given friction for a given $g \in H_+^{-1/2}(\Gamma_K)$. The weak formulation is expressed by (7.4). Moreover, let us assume that Γ_K is a straight line segment. In such a case $W' = [H^{-1/2}(\Gamma_K)]^2$ and the reciprocal variational formulation is given by (9.38). The convex set $\Lambda \subseteq [H^{-1/2}(\Gamma_K)]^2$ is now defined as follows:

$$\Lambda = \Lambda(g) = \{(\mu_1, \mu_2) \in [H^{-1/2}(\Gamma_K)]^2 | \langle \mu_1, v_n \rangle \geqslant 0, \\ \langle \mu_2, v_t \rangle + \langle \mathscr{F} g, |v_t| \rangle \geqslant 0 \ \forall v \in K\}.$$

The second inequality, appearing in the definition of $\Lambda(g)$ is the weak form of

$$|\mu_2| \leqslant \mathscr{F} g \quad \text{on } \Gamma_K.$$

The convex set Λ depends on g. From this point of view, the Signorini problem obeying Coulomb's law of friction can be viewed as a *quasivariational inequality* for the functional $\mathscr{S}$.

Let V_h, Λ_{1H} and Λ_{2H} be defined by (9.17)–(9.19). The usual Ritz–Galerkin procedure, which consists in replacing (9.38) by

$$\begin{aligned} &\text{find } \lambda_H = (\lambda_{1H}, \lambda_{2H}) \in \Lambda_H = \Lambda_{1H} \times \Lambda_{2H} \text{ such that} \\ &\mathscr{S}(\lambda_H) \leqslant \mathscr{S}(\mu_H) \quad \forall \mu_H \in \Lambda_H, \end{aligned} \tag{9.39}$$

is seldom applicable, as the explicit form of the Green's operator is not known, in general. Instead of G, a suitable approximation G_h has to be used. We can use the inverse of the stiffness matrix A_h, corresponding to the bilinear form a on the space V_h.

As an approximation to (9.38) we solve the problem of finding $\lambda_H \in \Lambda_H$ such that

$$\mathscr{S}_h(\lambda_H) \leqslant \mathscr{S}_h(\mu_H) \quad \forall \mu_H \in \Lambda_H, \tag{9.40}$$

where

$$\mathscr{S}_h(\mu_H) = \tfrac{1}{2}(\mu_{1H}, G_h(\mu_{1H}, \mu_{2H}) \cdot n)_{0,\Gamma_K} + \tfrac{1}{2}(\mu_{2H}, G_h(\mu_{1H}, \mu_{2H}) \cdot t)_{0,\Gamma_K} \\ + (\mu_{1H}, G_h(F_h) \cdot n)_{0,\Gamma_K} + (\mu_{2H}, G_h(F_h) \cdot t)_{0,\Gamma_K}.$$

Here $F_h \in V_h$ is defined by means of

$$(F_{ih}, v_{ih})_{0,\Omega} = (F_i, v_{ih})_{0,\Omega} \quad \forall v_h \in V_h.$$

REMARK 9.5. Using the same procedure as in the continuous case it is possible to show that (9.40) can be obtained from (9.20) by eliminating the approximate displacement field w_h. The mutual relation between (9.38) and its approximation (9.40) is studied in HASLINGER and PANAGIOTOPOULOS [1984].

10. Approximation of the Signorini problem obeying Coulomb's law of friction

This section deals with the finite-element approximation of the problem, the continuous form of which was presented in Section 7. Next, we shall assume that the body is represented by a polygonal domain Ω such that Γ_K is *represented by a straight line segment parallel to the x_2-axis*. Moreover, we shall assume that $\Gamma_P = \emptyset$, i.e. no surface tractions are applied. The weak solution of the Signorini problem obeying Coulomb's law of friction has been defined as a fixed point of the mapping $\Phi : \Lambda \to \Lambda$, with $\Lambda \equiv H_+^{-1/2}(\Gamma_K) = \{\mu \in H^{-1/2}(\Gamma_K),\ \mu \geqslant 0\}$, defined as follows: $\Phi(g) = -T_n(u)$, where $g \in \Lambda$ and $u \in K$ is the unique solution of the variational inequality

$$u = u(g) \in K\colon\ (\tau_{ij}(u), e_{ij}(v-u))_{0,\Omega} + \langle \mathcal{F}g, |v_t| - |u_t| \rangle \geqslant (F_i, v_i - u_i)_{0,\Omega} \quad \forall v \in K. \tag{10.1}$$

T_n is the normal component of the stress vector T. The meaning of all symbols is the same as in Sections 7 and 9.

The approximation of the Signorini problem obeying Coulomb's law will be defined by a suitable approximation of the mapping Φ, based on a mixed variational formulation.

Let $\{\mathcal{T}_h\}$ and $\{\mathcal{T}_H\}$ be families of triangulations of $\overline{\Omega}$ and of partitions of $\overline{\Gamma}_K$, respectively. With any $\mathcal{T}_h$ and $\mathcal{T}_H$ we associate sets V_h and Λ_H

$$V_h = \{v_h \in [C(\overline{\Omega})]^2 \mid v_h|_{T_i} \in [P_1(T_i)]^2,\ v_h = 0 \text{ on } \Gamma_u\}$$

and

$$\Lambda_H = \{\mu_H \in L^2(\Gamma_K) \mid \mu_H|_{\overline{b_{i-1}b_i}} \in P_0(\overline{b_{i-1}b_i})\ \forall i = 1, \ldots, M,\ \mu_H \geqslant 0 \text{ on } \Gamma_K\},$$

respectively.

For any $g_H \in \Lambda_H$ we shall assume the problem of finding $u_h = u_h(g_H) \in V_h$ and $\lambda_H \in \Lambda_H$ such that

$$\begin{aligned} &a(u_h, v_h - u_h) + \langle \lambda_H, v_{hn} - u_{hn} \rangle + \langle \mathcal{F}g_H, |v_{ht}| - |u_{ht}| \rangle \\ &\quad \geqslant (F_i, v_{ih} - u_{ih})_{0,\Omega} \quad \forall v_h \in V_h, \\ &\langle \mu_H - \lambda_H, u_{hn} \rangle \leqslant 0 \quad \forall \mu_H \in \Lambda_H. \end{aligned} \tag{10.2}$$

The symbol $\langle\ ,\ \rangle$ means the $L^2(\Gamma_K)$-scalar product. The problem (10.2) is a mixed variational formulation of (10.1). Indeed, let

$$K_{hH} = \{v_h \in V_h \mid \langle \mu_H, v_{hn} \rangle \leqslant 0\ \forall \mu_H \in \Lambda_H\}.$$

Then K_{hH} contains functions from V_h such that the mean values of their normal components on any $\overline{b_{i-1}b_i}$ are non-positive (see also Section 9). Then it is easy to see that $u_h \in V_h$, being the solution of (10.2) also solves the problem

$$u_h \in K_{hH}: a(u_h, v_h - u_h) + \langle \mathscr{F} g_H, |v_{ht}| - |u_{ht}| \rangle \geqslant (F_i, v_{ih} - u_{ih})_{0,\Omega} \quad \forall v_h \in K_{hH} \tag{10.3}$$

or equivalently

$$\text{find } u_h \in K_{hH} \text{ such that } \mathscr{J}(u_h) \leqslant \mathscr{J}(v_h) \quad \forall v_h \in K_{hH}, \tag{10.3'}$$

where

$$\mathscr{J}(v_h) = \tfrac{1}{2}a(v_h, v_h) + \langle \mathscr{F} g_H, |v_{ht}| \rangle - (F_i, v_{ih})_{0,\Omega}.$$

Thus, λ_H is the Lagrange multiplier associated with the constraint $u_h \in K_{hH}$. The element $-\lambda_H$ can be interpreted as the discrete normal stress along Γ_K.

As far as the existence of the solution of (10.2) is concerned, we have the following.

THEOREM 10.1. *For any h, $H > 0$ there exists a solution (u_h, λ_H) of* (10.3) *with u_h uniquely determined.*

In order to guarantee the uniqueness of the second component λ_H, we shall suppose throughout this section the validity of the following condition:

$$\langle \mu_H, v_{h1} \rangle = 0 \ \forall v_h \in \mathring{V}_h \Leftrightarrow \mu_H = 0, \tag{10.4}$$

where

$$\mathring{V}_h = \{ v_h = (v_{h1}, v_{h2}) \in V_h \mid v_{h2} \equiv 0 \text{ in } \Omega \}$$

(let us recall that Γ_K is parallel to the x_2-axis). Consequently $v_{h1} = v_{hn}$, $v_{h2} = v_{ht} \equiv 0$ for $v_h \in \mathring{V}_h$). Then one can easily prove the following.

THEOREM 10.2. *Let* (10.4) *be satisfied. Then λ_H is unique.*

As a consequence of (10.4), a mapping $\Phi_H : \Lambda_H \mapsto \Lambda_H$ can be defined by means of the relation

$$\Phi_H(g_H) = \lambda_H. \tag{10.5}$$

Analogously to Section 7 we introduce the following.

DEFINITION 10.1. As an approximation to the Signorini problem obeying Coulomb's law of friction we find any pair (u_h, λ_H) solving (10.2) such that $\Phi_H(\lambda_H) = \lambda_H$, i.e. λ_H is the *fixed point* of Φ_H in Λ_H.

Before proving the existence of such a fixed point, we present several observations, concerning the properties of Φ_H.

Let (u_h, λ_H) and $(\overline{u}_h, \overline{\lambda}_H)$ be the solutions of (10.2) with given friction g_H and $\overline{g}_H$, respectively. Then a direct calculation shows that

$$\|u_h - \overline{u}_h\|_{1,\Omega} \leqslant \frac{c_1}{\alpha}[\mathscr{F}]\|g_H - \overline{g}_H\|_{0,\Gamma_K}, \tag{10.6}$$

where $[\mathscr{F}] = \max_{\Gamma_K} \mathscr{F}(x)$, c_1 is the norm of the mapping $\gamma : V \to L^2(\Gamma_K)$ and α is the constant of ellipticity of the bilinear form a. Consequently, the mapping $\Psi_H : \Lambda_H \to V_h$, defined as

$$\Psi_H(g_H) = u_h$$

is continuous. The same property will now be established for the mapping Φ_H. To this end we introduce the following notation:

$$\|\mu_H\|_{-1/2,h} = \sup_{\mathring{V}_h} \frac{\langle \mu_H, v_{h1}\rangle}{\|v_{h1}\|_{1,\Omega}}. \tag{10.7}$$

If (10.4) is satisfied, (10.7) is a mesh-dependent norm in the finite dimensional space L_H, where

$$L_H = \{\mu_H \in L^2(\Gamma_K) \mid \mu_H|_{\overline{b_{i-1}b_i}} \in P_0(\overline{b_{i-1}b_i}),\ i = 1, \ldots, M\}.$$

THEOREM 10.3. *The mapping* $\Phi_H : \Lambda_H \to \Lambda_H$ *defined by* (10.5) *is continuous.*

PROOF. Let (u_h, λ_H) and $(\overline{u}_h, \overline{\lambda}_H)$ be solutions of (10.2) with given friction g_H and $\overline{g}_H$, respectively. Then

$$a(u_h, v_h) + \langle \lambda_H, v_{h1}\rangle = (F_i, v_{ih})_{0,\Omega} \quad \forall v_h \in \mathring{V}_h,$$
$$a(\overline{u}_h, v_h) + \langle \overline{\lambda}_H, v_{h1}\rangle = (F_i, v_{ih})_{0,\Omega} \quad \forall v_h \in \mathring{V}_h.$$

Subtracting the second equation from the first one, we obtain

$$\langle \lambda_H - \overline{\lambda}_H, v_{h1}\rangle \leqslant M\|u_h - \overline{u}_h\|_{1,\Omega}\|v_h\|_{1,\Omega} = M\|u_h - \overline{u}_h\|_{1,\Omega}\|v_{h1}\|_{1,\Omega},$$

where M is the norm of the bilinear form a. From (10.7) we immediately obtain

$$\|\lambda_H - \overline{\lambda}_H\|_{-1/2,h} \leqslant \frac{Mc_1}{\alpha}[\mathscr{F}]\|g_H - \overline{g}_H\|_{0,\Gamma_K}, \tag{10.8}$$

if (10.6) has been taken into account. From this the assertion of the theorem follows. □

As L_H is a finite dimensional space, there exists a positive constant β, depending on h, H in general such that

$$\beta\|\mu_H\|_{-1/2,\Gamma_K} \leqslant \|\mu_H\|_{-1/2,h} \quad \forall \mu_H \in L_H. \tag{10.9}$$

The dual norm on the left hand side of (10.9) is defined as follows:

$$\|\mu_H\|_{-1/2,\Gamma_K} = \sup_{v \in \mathring{V}} \frac{\langle \mu_H, v_1 \rangle}{\|v_1\|_{1,\Omega}},$$

where

$$\mathring{V} = \{v = (v_1, v_2) \in V \mid v_2 \equiv 0 \text{ in } \Omega\}.$$

Denote by $B_\varepsilon(r)$, $r > 0$, $\varepsilon \in \mathbb{R}^1$ the ball in the $H^\varepsilon(\Gamma_K)$-topology with its center at the origin and its radius equal to r.

THEOREM 10.4. *Let*

$$[\mathscr{F}] \leqslant \alpha\beta/(Mc_2), \tag{10.10}$$

where c_2 is the norm of the trace mapping $\gamma : V \to H^{1/2}(\Gamma_K)$.
Then for any r such that

$$r \geqslant r_0 \equiv (\alpha + M)\|F\|_{0,\Omega}/(\alpha\beta - Mc_2[\mathscr{F}]) > 0 \tag{10.11}$$

it holds that

$$\Phi_H(B_{-1/2}(r) \cap \Lambda_H) \subset B_{-1/2}(r) \cap \Lambda_H. \tag{10.12}$$

PROOF. Let $(u_h, \lambda_H) \in V_h \times \Lambda_H$ be a solution of (10.2). Substituting $v_h = 0, 2u_h$ into the first inequality in (10.2), we immediately get

$$a(u_h, u_h) + \langle \mathscr{F} g_H, |u_{ht}| \rangle = (F_i, u_{ih})_{0,\Omega}. \tag{10.13}$$

Korn's inequality, the trace theorem and the inequality

$$\|\, |u_{ht}|\, \|_{1/2,\Gamma_K} \leqslant \|u_{ht}\|_{1/2,\Gamma_K}$$

used in (10.13), yield

$$\|u_h\|_{1,\Omega} \leqslant \frac{1}{\alpha}(c_2[\mathscr{F}]\|g_H\|_{-1/2,\Gamma_K} + \|F\|_{0,\Omega}). \tag{10.14}$$

Restricting ourselves to $v_h \in \mathring{V}_h$ only, we have

$$a(u_h, v_h) + \langle \lambda_H, v_{h1} \rangle = (F_i, v_{ih})_{0,\Omega}$$

and consequently, making use of (10.9), we arrive at

$$\beta\|\lambda_H\|_{-1/2,\Gamma_K} \leqslant \|\lambda_H\|_{-1/2,h} \leqslant M\|u_h\|_{1,\Omega} + \|F\|_{0,\Omega}.$$

This, combined with (10.14) gives

$$\|\lambda_H\|_{-1/2,\Gamma_K} \leqslant Mc_2/(\alpha\beta)[\mathscr{F}]\|g_H\|_{-1/2,\Gamma_K} + \left(\frac{M}{\alpha}+1\right)\Big/\beta\,\|F\|_{0,\Omega}. \qquad (10.15)$$

If the coefficient of friction $\mathscr{F}$ is such that $Mc_2/(\alpha\beta)[\mathscr{F}] < 1$, i.e. (10.10) is satisfied, then a direct calculation leads to (10.11) and (10.12). □

CONSEQUENCE 10.1. *Neglecting the term* $\langle \mathscr{F}g_h, |u_{ht}|\rangle$ *in* (10.13) *we immediately get*

$$\|u_h\|_{1,\Omega} \leqslant \frac{1}{\alpha}\|F\|_{0,\Omega}.$$

This and (10.15) *yields*

$$\|\lambda_H\|_{-1/2,\Gamma_K} \leqslant \left(\frac{M}{\alpha}+1\right)\Big/\beta\,\|F\|_{0,\Omega}.$$

From this, Theorem 10.3 *and Brouwer's fixed point theorem we can immediately derive the existence of at least one solution of the discrete Signorini problem with Coulomb's law of friction for any value of the coefficient of friction* $\mathscr{F}$. *This is a great difference between the continuous model and the discrete one.*

Next, we shall suppose that the following stability condition is satisfied:

the constant β, appearing in (10.9) is *independent* of h, $H > 0$. (10.16)

REMARK 10.1. Let (10.16) be satisfied. Then the relations (10.10) and (10.11) do not depend on h and H.

REMARK 10.2. From (10.8), (10.16) and the inverse inequality

$$\|\mu_H\|_{0,\Gamma_K} \leqslant \bar{c}H^{-1/2}\|\mu_H\|_{-1/2,\Gamma_K} \quad \forall \mu_H \in L_H$$

we obtain

$$\|\lambda_H - \lambda_{\overline{H}}\|_{-1/2,\Gamma_K} \leqslant Mc_1\bar{c}/(\alpha\beta)H^{-1/2}[\mathscr{F}]\|g_H - \bar{g}_H\|_{-1/2,\Gamma_K}.$$

If $[\mathscr{F}]$ is sufficiently small, namely if $[\mathscr{F}] < \alpha\beta H^{1/2}/(Mc_1\bar{c})$, then the mapping $\Phi_H : \Lambda_H \to \Lambda_H$ is *contractive* and its unique fixed point can be computed by the method of successive approximations. Unfortunately, the bounds for $\mathscr{F}$ guaranteeing this depend on H, even if (10.16) is satisfied.

Next, we shall study the relation between fixed points of the mapping Φ : $\Lambda \to \Lambda$ introduced in Section 7 and fixed points of the mapping $\Phi_H : \Lambda_H \to \Lambda_H$ introduced at the beginning of this section. In other words, we shall analyze the mutual relationships between the discrete and continuous models of the Signorini problem obeying Coulomb's law of friction.

To this end, we shall suppose that the problem (10.1) is *regular* in the following sense: there exists $\varepsilon_0 \in (0, \frac{1}{2})$ such that for any $\varepsilon \in (0, \varepsilon_0\rangle$ and any $g \in H_+^{-1/2+\varepsilon}(\Gamma_K)$, the corresponding solution $u \equiv u(g) \in [H^{1+\varepsilon}(\Omega)]^2$ and the following estimates hold:

$$\|u\|_{1+\varepsilon,\Omega} \leqslant c_1\|g\|_{-1/2+\varepsilon,\Gamma_K} + c\|F\|_{0,\Omega}, \tag{10.17}$$

$$\|T_n(u)\|_{-1/2+\varepsilon,\Gamma_K} \leqslant c_1\|g\|_{-1/2+\varepsilon,\Gamma_K} + c\|F\|_{0,\Omega}, \tag{10.18}$$

where c is an absolute positive constant and $c_1 > 0$ depends on ε and $\mathcal{F}$ in such a way that $c_1 \to 0$ if $[\mathcal{F}] \to 0$ (ε fixed). Moreover, we shall assume that if $F \in [H^1(\Omega)]^2$ and $g \in L^2(\Gamma_K)$, $g \geqslant 0$ on Γ_K, then $T_n(u) \in L^2(\Gamma_K)$ and

$$\|T_n(u)\|_{0,\Gamma_K} \leqslant c_1\|g\|_{0,\Gamma_K} + c\|F\|_{1,\Omega} \tag{10.19}$$

with c_1, c having the same properties as before (for a discussion of (10.17)–(10.19) see HLAVÁČEK, HASLINGER, NEČAS and LOVÍŠEK [1988], NEČAS, JARUŠEK and HASLINGER [1980], JARUŠEK [1983]).

REMARK 10.3. If (10.18) is satisfied, then the stability condition (10.16) is satisfied, provided the ratio h/H is small enough.

On the basis of (10.17)–(10.19) the following basic result can be proven.

THEOREM 10.5. *Let* (10.17)–(10.19) *and the stability condition* (10.16) *be satisfied. Let $\{\mathcal{T}_h\}$ and $\{\mathcal{T}_H\}$ be such that there exist positive numbers τ_1 and τ_2 independent of h and H such that $\tau_1 \leqslant h/H \leqslant \tau_2$. Then for small values of $\mathcal{F}$ there exists a number r_0 independent of h and H such that the mapping $\Phi_H : \Lambda_H \to \Lambda_H$ defined by* (10.5) *maps the set $B_{-1/2+\varepsilon_0}(r_0) \cap \Lambda_H$ into itself. Moreover, $\{\|u_h\|_{1+\varepsilon_0,\Omega}\}$ is bounded.*

For the proof see HASLINGER [1983].

REMARK 10.4. As a consequence of Theorem 10.5, there exists a fixed point λ_H of Φ_H in the set $B_{-1/2+\varepsilon_0}(r_0) \cap \Lambda_H$.

Next, let $(u_h, \lambda_H) \in V_h \times \Lambda_H$ denote a solution of the discrete Signorini problem obeying Coulomb's law of friction such that $\lambda_H \in B_{-1/2+\varepsilon_0}(r_0) \cap \Lambda_H$.

Then the following result holds.

THEOREM 10.6. *Let all assumptions of Theorem 10.5 be satisfied. Then there exist subsequences* $\{u_{h'}\} \subset \{u_h\}$, $\{\lambda_{H'}\} \subset \{\lambda_H\}$ *such that*

$$u_{h'} \to u, \quad h' \to 0+ \quad \textit{in } [H^1(\Omega)]^2,$$
$$\lambda_{H'} \to \lambda, \quad H' \to 0+ \quad \textit{in } H^{-1/2}(\Gamma_K),$$

where $\lambda \in \Lambda$ *is a fixed point of* Φ *and* $u = u(\lambda) \in K$ *is a solution of the Signorini problem obeying Coulomb's law of friction.*

PROOF. On the basis of Theorem 10.5 we may assume that

$$u_{h'} \to u, \quad h' \to 0+ \ \text{ in } [H^{1+\varepsilon}(\Omega)]^2 \quad \forall \varepsilon < \varepsilon_0, \tag{10.20}$$
$$\lambda_{H'} \to \lambda, \quad H' \to 0+ \ \text{ in } H^{-1/2}(\Gamma_K), \tag{10.21}$$

taking into account the compactness of the imbedding of $[H^{1+\varepsilon_0}(\Omega)]^2$ into $[H^{1+\varepsilon}(\Omega)]^2$ $\forall \varepsilon < \varepsilon_0$ and of $H^{-1/2+\varepsilon_0}(\Gamma_K)$ into $H^{-1/2}(\Gamma_K)$. As

$$\| \, |u_{ht}| \, \|_{1/2+\varepsilon,\Gamma_K} \leqslant \|u_{ht}\|_{1/2+\varepsilon,\Gamma_K} \leqslant c\|u_h\|_{1+\varepsilon,\Omega} \leqslant c,$$

there exist a subsequence of $\{|u_{h't}|\}$ (still denoted by the same symbol) and an element $\chi \in H^{1/2+\varepsilon}(\Gamma_K)$ such that

$$|u_{h't}| \to \chi \quad \text{in } H^{1/2}(\Gamma_K).$$

It is easy to see that $\chi = |u_t|$ on Γ_K.

Let $v \in [H^{1+\varepsilon_0}(\Omega)]^2$ and $\mu \in \Lambda \cap L^2(\Gamma_K)$. Then there exist sequences $\{v_h\}$ and $\{\mu_H\}$, $v_h \in V_h$ and $\mu_H \in \Lambda_H$ such that

$$v_h \to v, \quad h \to 0+ \ \text{ in } [H^{1+\varepsilon}(\Omega)]^2 \ \forall \varepsilon < \varepsilon_0, \tag{10.22}$$
$$\mu_H \to \mu, \quad H \to 0+ \ \text{ in } L^2(\Gamma_K). \tag{10.23}$$

Passing to the limit in $(10.2)_1$ with h', $H' \to 0+$ and taking into account (10.20)–(10.22) and the fact that $g_H = \lambda_H$, we obtain that

$$a(u, v-u) + \langle \lambda, v_n - u_n \rangle + \langle \mathscr{F}\lambda, |v_t| - |u_t| \rangle \geqslant (F_i, v_i - u_i)_{0,\Omega} \tag{10.24}$$

holds for any $v \in [H^{1+\varepsilon_0}(\Omega)]^2$. Analogously, passing to the limit with h', $H' \to 0+$ in $(10.2)_3$ we arrive at

$$\langle \mu - \lambda, u_n \rangle \leqslant 0 \quad \forall \mu \in \Lambda \cap L^2(\Gamma_K).$$

Thus $u \in K$. Now, restricting ourselves to $v \in K \cap [H^{1+\varepsilon_0}(\Omega)]^2$ in (10.24) and using the fact that $\lambda \in \Lambda$, we obtain

$$a(u, v-u) + \langle \mathscr{F}\lambda, |v_t| - |u_t| \rangle \geqslant (F_i, v_i - u_i)_{0,\Omega} \ \forall v \in K \cap [H^{1+\varepsilon_0}(\Omega)]^2. \tag{10.25}$$

Applying Green's formula (9.12) we see that $\lambda = -T_n(u)$. From this and (10.25) we arrive at the assertion of the theorem. □

11. Numerical realization of contact problems

This section is devoted to the numerical realization of contact problems. For the sake of simplicity we shall restrict ourselves to the Signorini problem for a plane elastic body, unilaterally supported by a rigid foundation. We start with the frictionless case. Using finite elements, the problems $(\mathcal{P}_1)_h$ and $(\mathcal{P}_2)_h$, presented in Section 8 lead to quadratic programming problems: to find the minimum of quadratic function

$$\mathcal{J}(x) = \tfrac{1}{2}(x, Ax) - (f, x) \tag{11.1}$$

and the closed convex subset of $\mathbb{R}^n$

$$\mathcal{K} = \{x \in \mathbb{R}^n \mid Bx \leqslant d\}, \tag{11.2}$$

where A is an $n \times n$ stiffness matrix, $f \in \mathbb{R}^n$ the vector arising by the integration of the body forces and surface fraction, B is generally a rectangular matrix of type $m \times n$ and $d \in \mathbb{R}^n$ is a given vector. The matrix A is symmetric and positive definite in the coercive case or positive semidefinite in the semicoercive case. Let us have a more detailed look at the definition of the closed, convex set $\mathcal{K}$. When dealing with problem $(\mathcal{P}_1)$, the vector $d \equiv 0$. A number of constraints, i.e. the number m depends on the number of nodes of the triangulation, lie on Γ_K in the following manner: if Γ_K is a single line segment or if Γ_K is curved, then m is equal to the number of the nodes on $\overline{\Gamma}_K \backslash \overline{\Gamma}_u$. If Γ_K is a broken line consisting of p segments, $p > 1$, then the total number of constraints is greater by $p - 1$. At each vertex, q constraints are prescribed. In problem $(\mathcal{P}_2)$ the number of constraints equals the number of nodes on $\overline{\Gamma}_K \backslash \overline{\Gamma}_u$ and $d = (\varepsilon((c_1), \ldots, \varepsilon(c_m))^{\mathrm{T}}$, where $\varepsilon = f'' - f'$. In our case f'' is the function describing the contact zone of the deformable body, while f' describes the surface of the rigid foundation. Each row of the matrix B contains at most two nonzero entries (components of the vector n or ξ), while in each column of B there is at most one, or in the case of $(\mathcal{P}_1)$ with Γ_K a broken line, at most two such entries. A popular method for solving the previous quadratic programming problem is the conjugate gradient method, described, for example, in CIARLET [1989] or PSENICNYJ and DANILIN [1975] and used for the numerical realization of unilateral problems in MAY [1986], JEUZETTE and SONZOGNI [1989]. The advantage of this method lies in the fact that it can be used in the case where the stiffness matrix is also positive semidefinite.

In DOSTÁL [1992], DILINTAS, LAURENT-GENGOUX and TRYSTAM [1988] the modification of the conjugate gradient method with preconditioning by a projector is described.

Another popular method for solving our problem is the SOR method with an additional projection (see GLOWINSKI, LIONS and TREMOLIÈRES [1981]). This method, however requires a special structure of constraints, namely the so called box constraints and this does not apply in our case. But simple transformation of the problem makes possible the use of the method. Indeed, let us assume that each node, lying on $\overline{\Gamma}_K\backslash\overline{\Gamma}_u$ is assigned one inequality constraint. Let $\mathscr{I}_i = \{k_1^i, k_2^i\}$, $i = 1,\ldots,m$ be the pair of indices of nonzero entries of the ith row of B. Introducing the variables $y = (y_1,\ldots,y_n)$ by

$$y_j = x_j, \quad j \neq k_2^i, \ i = 1,\ldots,m,$$
$$y_{k_2^i} = \sum_{j=1}^{2} b_{ik_j^i} x_{k_j^i}$$

or in the matrix form

$$y = Cx \Leftrightarrow x = C^{-1}y,$$

the minimization problem for (11.1) and (11.2), can be equivalently expressed as follows: find the minimum of the quadratic function

$$\tilde{\mathscr{J}}(y) = \tfrac{1}{2}(y, \mathscr{A}y) - (\mathscr{F}, y) \tag{11.3}$$

on the set

$$\tilde{\mathscr{K}} = \{y \in \mathbb{R}^n \mid y_{k_2^i} \leqslant d_{k_2^i}, \ i = 1,\ldots,m\}, \tag{11.4}$$

where $\mathscr{A} = (C^{-1})^{\mathrm{T}}AC^{-1}$, $\mathscr{F} = (C^{-1})^{\mathrm{T}}f$. The matrix $\mathscr{A}$ has the same properties as A. The quadratic programming problem given by (11.3) and (11.4) can be now realized by the SOR method with an additional projection provided the matrix $\mathscr{A}$ is regular.

It is possible to use penalty type methods for solving (11.1) and (11.2) (see KIKUCHI and ODEN [1988]) or Uzawa type methods (see later). Another popular method in the engineering community is the use of mathematical programming methods, which can be used to solve complementarity problems such as

$$\begin{aligned}
&\text{find } x \in \mathbb{R}^n, \ \lambda \in \mathbb{R}^n \text{ such that}\\
&x - M\lambda = q,\\
&x_j \geqslant 0, \ \lambda_j \geqslant 0 \quad \forall j = 1,\ldots,n,\\
&x_j\lambda_j = 0,
\end{aligned} \tag{11.5}$$

where A and M are $(n \times n)$ matrices and $q \in \mathbb{R}^n$ (see LEMKE [1978], JÚDICE and PIRES [1992]).

Recently, a parallel solution of variational inequalities based on Schwarz type algorithms has been discussed in HOFFMANN and ZOU [1992], KUZNETSOV and

NEITTAANMÄKI [1991]. To our knowledge, however, such methods have not been applied yet to the numerical realization of contact problems.

Another useful approach leading to an increase in the efficiency of the numerical methods presented before is the elimination of all components of the nodal displacement field, which correspond to nodes not belonging to $\overline{\Gamma}_K \backslash \overline{\Gamma}_u$. We have utilized the fact that the number of components of the nodal displacement field, subject to the condition on $\overline{\Gamma}_K \backslash \overline{\Gamma}_u$, is small compared to the total number of components of this field. We proceed as follows.

Let us assume that the enumeration of nodes in the domain Ω is chosen in such a way that the constrained components are listed last, i.e. $x = (x_1, x_2)$, $x \in \mathbb{R}^n$, $x_1 \in \mathbb{R}^p$, $x_2 \in \mathbb{R}^k$, $n = p + k$ and we look for the minimum of (11.1) on the set

$$\mathcal{K} = \{x = (x_1, x_2) \in \mathbb{R}^n \mid Bx_2 \leqslant 0\}, \tag{11.6}$$

where B is an $(m \times k)$ matrix.

The problem of finding the minimum $x^* = (x_1^*, x_2^*) \in \mathbb{R}^n$ of the function $\mathcal{J}$ on $\mathcal{K}$, given by (11.6) is equivalent to the problem of finding $x^* \in \mathcal{K}$, satisfying

$$(Ax^*, y - x^*) \geqslant (f, y - x^*) \quad \forall y \in \mathcal{K}. \tag{11.7}$$

Let us write $y = (y_1, y_2)^{\mathrm{T}}, f = (f_1, f_2)^{\mathrm{T}}, y_1, f_1 \in \mathbb{R}^p, y_2, f_2 \in \mathbb{R}^k$ in what follows. Analogously, we divide the matrix A into blocks,

$$A = \begin{pmatrix} A_{11} & A_{12} \\ A_{21} & A_{22} \end{pmatrix},$$

where A_{11} and A_{22} are square matrices of orders p and k, respectively, while A_{12} and A_{21} are rectangular matrices of types $(p \times k)$ and $(k \times p)$, respectively. It is easy to see that $A_{12} = A_{21}^{\mathrm{T}}$.

Now, let us choose y in (11.7) so that $y_1 = x_1^* + z_1$, $y_2 = x_2^*$, $z_1 \in \mathbb{R}^p$ arbitrary. Then $y \in \mathcal{K}$ and from (11.7) it follows that

$$A_{11}x_1^* + A_{12}x_2^* = f_1. \tag{11.8}$$

Now, let us choose $y \in \mathcal{K}$ in (11.7) so that $y_1 = x_1^*$, $y_2 = z_2$, $Bz_2 \leqslant 0$. Then from (11.7) it follows that

$$(z_2 - x_2^*)(A_{21}x_1^* + A_{22}x_1^*) \geqslant (z_2 - x_2^*)^{\mathrm{T}} f_2. \tag{11.9}$$

Substituting from (11.8) $x_1^* = A_{11}^{-1}(f_1 - A_{12}x_2^*)$ into (11.9), we obtain the following relation for x_2^*;

$$(z_2 - x_2^*)^{\mathrm{T}} \tilde{A} x_2^* \geqslant (z_2 - x_2^*)^{\mathrm{T}} \tilde{f}, \tag{11.10}$$

where $\tilde{f} = f_2 - A_{21}A_{11}^{-1}f_1$, $\tilde{A} = A_{22} - A_{21}A_{11}^{-1}A_{12}$.

Now, (11.10) implies that x_2^* is the minimum of the quadratic function

$$\tilde{\mathcal{J}}(x_2) = (x_2, \tilde{A}x_2)_{\mathbb{R}^k} - (\tilde{f}, x_2)_{\mathbb{R}^k} \tag{11.11}$$

on the closed convex set

$$\tilde{\mathcal{K}} = \{x_2 \in \mathbb{R}^k \mid Bx_2 \leqslant 0\}. \tag{11.12}$$

Taking into account the fact that k is usually much smaller than n in contact problems, the time saved is considerable. If we know x_2^*, we can calculate $x_1^* = A_{11}^{-1}(f_1 - A_{12}x_2^*)$. Thus, the minimization methods can be applied to the quadratic programming problem, given by (11.11) and (11.12). This approach is systematically studied in ANTES and PANAGIOTOPOULOS [1992].

Now, let us pass to the numerical realization of the Signorini problem involving friction. First, we shall concentrate on the realization of the Signorini problem with given friction, by means of which Coulomb's law of friction is defined.

The method of regularizing the nondifferentiable term j with the preconditioned conjugate gradient method has been used in RAOUS and BARBARIN [1992].

Another method for the numerical realization of the Signorini problem with a given friction is based on the mixed variational formulations, Variant 1 or 2, presented in Section 9. Here we restrict ourselves to Variant 1, the finite element approximation of which is given by (9.6) or (9.7). Since we have here the problem of finding a saddle point in a finite dimension, we will use Uzawa's method. The reader can find the theoretical background in FORTIN and GLOWINSKI [1982], GLOWINSKI [1984]. Each iterative step of Uzawa's method consists of two parts:

knowing $\lambda_H^n \in \Lambda_H$,

we calculate $w_h^n \in K_h$ as a solution of the minimization problem

$$\mathcal{L}(w_h^n, \lambda_H^n) \leqslant \mathcal{L}(v_h, \lambda_H^n) \quad \forall v_h \in K_h, \tag{11.13}$$

or equivalently

$$a(w_h^n, v_h - w_h^n) + (g\lambda_H^n, v_{ht} - w_{ht}^n)_{0,\Gamma_K} \geqslant L(v_h - w_h^n) \quad \forall v_h \in K_h \tag{11.13'}$$

and then replace λ_H^n by λ_H^{n+1} according to the rule

$$\lambda_H^{n+1} = P_{\Lambda_H}(\lambda_H^n + \rho g u_{ht}^n) \tag{11.14}$$

and return to (11.13). The symbol P_{Λ_H} stands for the projection onto a convex set Λ_H and $\rho > 0$ is a given parameter (step size). From the theory presented in FORTIN and GLOWINSKI [1982] and EKELAND and TEMAM [1976] it follows that there exist positive numbers $\rho_2 > \rho_1$ such that for all $\rho \in (\rho_1, \rho_2)$ we have

$$w_h^n \to w_h, \quad n \to \infty$$

and, if the second component λ_H of the saddle point of $\mathscr{L}$ on $K_h \times \Lambda_H$ is unique, then also

$$\lambda_H^n \to \lambda_H, \quad n \to \infty$$

(for details see HLAVÁČEK, HASLINGER, NEČAS and LOVÍŠEK [1988]). This iteration process leads to a sequence of solutions of quadratic programming problems which have the following important properties: (i) the stiffness matrix remains the same throughout the whole process, (ii) the linear term of $\mathscr{L}$ changes, during the iteration process, only those components, which in the given enumeration correspond to the components of w_h on $\overline{\Gamma}_K \backslash \overline{\Gamma}_u$. Consequently, the process of eliminating the free components of w_h, described above, can be used. A similar approach can be used in the case of Variant 2.

The numerical realization of the Signorini problem obeying Coulomb's law of friction can be solved by the method of successive approximations to find a fixed point of the mapping Φ, introduced in Section 7. This approach has been used in LICHT, PRATT and RAOUS [1991] and HASLINGER and PANAGIOTOPOULOS [1984]. The optimal control approach combined with the least square method for finding a fixed point of Φ has been studied in HASLINGER [1984].

The Coulomb friction model can be written in the form of a complementarity problem and then solved numerically by a Lemke type method (see KLARBRING [1986], ANTES and PANAGIOTOPOULOS [1992], RAOUS, CHABRAND and LEBON [1988], KLARBRING and BJÖRKMANN [1988]). In CURNIER and ALART [1988] a generalized Newton method was used for the numerical treatment of frictional contact problems.

Shape optimization in contact problems is analyzed in HASLINGER and NEITTAANMÄKI [1988].

CHAPTER IV

Contact of Elasto-Plastic Bodies

12. Deformation theory

The analysis of elastic bodies with a unilateral contact, presented in the previous chapter, can be extended to some cases obeying nonlinear constitutive law. One of the simplest mathematical models of such materials is the so-called *deformation theory of plasticity* (see KAČANOV [1948], MIHLIN [1971] or NEČAS and HLAVÁČEK [1981]).

In Section 12 we study a Signorini contact problem in the deformation theory of plasticity. The weak solution is defined on the basis of a variational inequality, which is in turn equivalent to the minimization of the potential energy. Then the so-called secant modules (Kačanov) iterative method is introduced, each step of which corresponds to a classical Signorini problem in elastostatics. Thus a finite-element analysis of the latter is obtained. We present the convergence of the secant modules method on a general abstract level.

12.1. Formulation of Signorini's contact problem in the deformation theory of plasticity

Let us consider a bounded domain $\Omega \subset \mathbb{R}^3$ with a Lipschitz boundary $\partial\Omega$ and assume that

$$\partial\Omega = \Gamma_u \cup \Gamma_\tau \cup \Gamma_K \cup \Gamma_M,$$

where Γ_u, Γ_τ and Γ_K are disjoint open subsets of $\partial\Omega$, $\Gamma_K \neq \emptyset$ and the surface measure of Γ_M vanishes.

Let the elasto-plastic body, occupying the domain Ω, be governed by the following Hencky–Mises stress-strain relation

$$\tau_{ij} = [k - \tfrac{2}{3}\mu(\gamma)]\delta_{ij}e_{mm} + 2\mu(\gamma)e_{ij}, \tag{12.1}$$

where k is a (constant) bulk modulus,

$$\begin{aligned}\gamma = \gamma(u) &= -\tfrac{2}{3}(\theta(u))^2 + 2e_{ij}(u)e_{ij}(u), \quad \theta(u) = \operatorname{div} u,\\ e_{ij}(u) &= \tfrac{1}{2}\left(\frac{\partial u_i}{\partial x_j} + \frac{\partial u_j}{\partial x_i}\right),\end{aligned} \tag{12.2}$$

u is a displacement and a repeated index implies summation over the range 1, 2, 3.

We assume that the function μ is continuously differentiable in $[0,\infty)$ and fulfills the following conditions

$$0<\mu_0\leqslant\mu(\gamma)\leqslant\tfrac{3}{2}k, \tag{12.3}$$

$$0<\alpha\leqslant\mu(\gamma)+2\gamma\frac{d\mu}{d\gamma}, \tag{12.4}$$

$$\frac{d\mu}{d\gamma}\leqslant 0. \tag{12.5}$$

We shall consider the following boundary conditions

$$\begin{aligned}&u=0 \quad \text{on } \Gamma_u,\\ &\tau_{ij}n_j=g_i \quad \text{on } \Gamma_\tau,\\ &u_n\leqslant 0,\qquad T_n\leqslant 0,\qquad u_nT_n=0 \qquad \text{on } \Gamma_K,\end{aligned}$$

where the surface load $g\in[L^2(\Gamma_\tau)]^3$ is given. (We adopt the notation of Section 5.)

Moreover, let a body force $F\in[L^2(\Omega)]^3$ be given. The weak solution of the problem minimizes the potential energy

$$\mathscr{L}(u)=\tfrac{1}{2}\int_\Omega k\theta^2(u)\,dx+\tfrac{1}{2}\int_\Omega\left[\int_0^{\gamma(u)}\mu(t)\,dt\right]dx-\int_\Omega F_iu_i\,dx-\int_{\Gamma_\tau}g_iu_i\,ds$$

over the set K, where

$$K=\{u\in[H^1(\Omega)]^3 \mid u=0 \text{ on } \Gamma_u,\ u_n\leqslant 0 \text{ on } \Gamma_K\}.$$

The latter problem is equivalent to the solution of the following variational inequality: find $u\in K$ such that

$$\begin{aligned}&\int_\Omega[(k-\tfrac{2}{3}\mu(\gamma))\theta(u)\theta(v-u)+2\mu(\gamma)e_{ij}(u)e_{ij}(v-u)]\,dx\\ &\quad-\int_\Omega F_i(v_i-u_i)\,dx-\int_{\Gamma_\tau}g_i(v_i-u_i)\,ds\geqslant 0 \quad \forall v\in K.\end{aligned} \tag{12.6}$$

12.2. Abstract convergence theorem for the secant modules method

Let a functional Φ be given on a Hilbert space H. Assume that Φ has the second Gâteaux differential $D^2\Phi(u;h,k)$ and the mapping $u\mapsto D^2\Phi(u;h,k)$ is continuous on every line segment. Assume further that

$$D^2\Phi(u;h,h)\geqslant m\|h\|^2 \tag{12.7}$$

holds for some positive constant m. Let a form $B(u;x,y)$ be given, which is bilinear and symmetric in x and y such that

$$B(u;x,x) \geqslant c_1\|x\|^2, \quad c_1 = \text{const} > 0, \tag{12.8}$$

$$|B(u;x,y)| \leqslant c_2\|x\|\|y\|, \tag{12.9}$$

$$B(u;u,v) = D\Phi(u,v), \tag{12.10}$$

$$\tfrac{1}{2}B(x;y,y) - \tfrac{1}{2}B(x;x,x) - \Phi(y) + \Phi(x) \geqslant 0 \tag{12.11}$$

hold for all x, y, u, $v \in H$.

Moreover, let K be a closed convex subset of H.

THEOREM 12.1. *Let the assumptions* (12.7)–(12.11) *be satisfied and let an element* $\varphi \in H$ *be given.*

Then the problem: find $u \in K$ *such that*

$$D\Phi(u, v-u) \geqslant (\varphi, v-u) \quad \forall v \in K. \tag{12.12}$$

has a unique solution.

Let $u_n \in K$, $n = 1, 2, \ldots,$ *be such that*

$$B(u_n; u_{n+1}, v - u_{n+1}) \geqslant (\varphi, v - u_{n+1}) \quad \forall v \in K \tag{12.13}$$

Then

$$\lim \|u - u_n\| = 0 \text{ as } n \to \infty.$$

PROOF. For the proof see the paper by NEČAS and HLAVÁČEK [1983] or the book by NEČAS and HLAVÁČEK [1981, Section 11.5]. □

12.3. Application to Signorini's contact problems

Let us assume that $\Gamma_u \neq \emptyset$. If we put

$$H = \{u \in [H^1(\Omega)]^3 \mid u = 0 \text{ on } \Gamma_u\},$$

$$\Phi = \mathscr{L},$$

$$B(u;v,w) = \int_\Omega [(k - \tfrac{2}{3}\mu(\gamma(u)))\theta(v)\theta(w) + 2\mu(\gamma(u))e_{ij}(v)e_{ij}(w)]\,\mathrm{d}x,$$

$$(\varphi, v) = \int_\Omega F_i v_i\,\mathrm{d}x + \int_{\Gamma_\tau} g_i v_i\,\mathrm{d}s,$$

then the conditions (12.7)–(12.11) are fulfilled (see NEČAS and HLAVÁČEK [1981, Sections 8.2 and 11.5]).

Indeed, conditions (12.9) and (12.10) are obvious. Conditions (12.7) and (12.8) can be verified, using (12.3), (12.4) and Korn's inequality. Condition (12.11) is a consequence of (12.5).

Let us define the secant modules approximation as the solution $u_{n+1} \in K$ of the following variational inequality

$$\int_\Omega [(k - \tfrac{2}{3}\mu(\gamma(u_n)))\theta(u_{n+1})\theta(v - u_{n+1}) + 2\mu(\gamma(u_n))e_{ij}(u_{n+1})e_{ij}(v - u_{n+1})]\,\mathrm{d}x$$

$$\geqslant \int_\Omega F_i(v - u_{n+1})_i\,\mathrm{d}x + \int_{\Gamma_\tau} g_i(v - u_{n+1})_i\,\mathrm{d}s \quad \forall v \in K.$$

Then Theorem 12.1 yields that the sequence u_n tends to the solution u of (12.6) in $[H^1(\Omega)]^3$.

The same result is also valid for the two-dimensional Signorini problem. Then the coefficient $(-\frac{2}{3})$ has to be replaced by (-1) in the definition of the form B and $3k/2$ in (12.3) should be replaced by k.

13. Contact of perfectly elasto-plastic bodies

We study the unilateral contact between two bodies, the material of which is perfectly elasto-plastic, obeying the *Hencky's law*. In this case, formulation in terms of stresses is more suitable than that in terms of displacements. Thus, we first extend the well-known Haar–Kármán principle to the boundary conditions of unilateral contact, for both a bounded zone of contact and an increasing zone of contact.

Approximations to the problems will be proposed on the basis of piecewise constant stress fields. Convergence of the method will be proven for any regular family of triangulations. For the proofs in detail, see HASLINGER and HLAVÁČEK [1982a].

Extended Haar–Kármán principle. We consider the same class of problems, as in Section 5, where only the stress–strain relations are replaced by Hencky's law. Thus, we shall restrict ourselves to plane problems, bounded bodies, small deformations, zero friction, zero initial strain and stress fields and a constant temperature field.

Let $\mathbb{R}_\sigma$ be the space of symmetric 2×2 matrices (strain or stress tensors). Let a *yield function* $f : \mathbb{R}_\sigma \to \mathbb{R}$ be given, which is convex and continuous in $\mathbb{R}_\sigma$.

Let the two bodies under consideration occupy bounded domains Ω' and Ω'' with Lipschitz boundaries. Henceforth, one or two primes denote that the quantity refers to the body Ω' or Ω'', respectively.

We introduce the following notation:

$$S = \{\tau : \Omega \to \mathbb{R}_\sigma \mid \tau_{ij} \in L^2(\Omega),\ i,j = 1,2\}, \quad \Omega = \Omega' \cup \Omega'',$$

$$\langle \sigma, e \rangle = \int_\Omega \sigma_{ij} e_{ij}\,\mathrm{d}x, \qquad \|\sigma\|_0 = \langle \sigma, \sigma \rangle^{1/2}.$$

In the space S we also introduce the energy inner product

$$(\sigma, \tau) = \langle c^{-1}\sigma, \tau\rangle, \qquad \|\sigma\| = (\sigma, \sigma)^{1/2},$$

where $c : S \to S$ is the isomorphism defined by the generalized Hooke's law, i.e.,

$$\sigma = ce \Leftrightarrow \sigma_{ij} = c_{ijkm}e_{km}.$$

Here $c_{ijkm} \in L^{\infty}(\Omega)$ and σ and e are stress and strain tensors, respectively. Moreover, we assume that a positive constant c_0 exists such that

$$\langle ce, e\rangle \geqslant c_0\|e\|_0^2 \quad \forall e \in S.$$

Let us define the set of plastically admissible stress fields

$$P = \{\tau \in S \mid f(\tau) \leqslant 1 \text{ a.e. in } \Omega' \cup \Omega''\}.$$

The set P is closed and convex in S.

Hencky's law can be stated in the following way (cf. DUVAUT and LIONS [1976] or MERCIER [1977]). Let $\Pi : S \to P$ be the projection onto the set P with respect to the energy inner product (σ, τ). Then the actual stress field σ is the projection of the "elastic" one, i.e.,

$$\sigma = \Pi ce(u), \tag{13.1}$$

where $e(u)$ is the strain tensor,

$$e_{ij}(u) = \frac{1}{2}\left(\frac{\partial u_i}{\partial x_j} + \frac{\partial u_j}{\partial x_i}\right) \tag{13.2}$$

and $u = \{u', u''\} \in [H^1(\Omega')]^2 \times [H^1(\Omega'')]^2$ is the actual displacement field.

13.1. Bounded zone of contact

First let us consider the configuration, which corresponds to the bounded zone of contact—see Section 5.1.

Thus we have the contact zone $\Gamma_K = \partial\Omega' \cap \partial\Omega''$ and the following decompositions

$$\partial\Omega' = \overline{\Gamma}_u \cup \overline{\Gamma}_\tau' \cup \Gamma_K, \qquad \partial\Omega'' = \overline{\Gamma}_0 \cup \overline{\Gamma}_\tau'' \cup \Gamma_K,$$

where Γ_u, Γ_0, Γ_τ' and Γ_τ'' are mutually disjoint open parts of the boundaries; let Γ_u and Γ_K have positive values. The remaining parts may be either positive or empty.

On Γ_u we consider the Dirichlet condition

$$u' = u_0',$$

where $u_0' \in [H^1(\Omega')]^2$ is given such that $u_{0n}' = 0$ on Γ_K.

The bilateral contact conditions (5.8) on Γ_0 and the tractions $P_i \in L^2(\Gamma_\tau)$ on $\Gamma_\tau' \cup \Gamma_\tau''$ are prescribed (see (5.7)). On Γ_K we consider the unilateral contact conditions, i.e.,

$$T_n'(\sigma) = T_n''(\sigma) \leqslant 0, \qquad u_n' + u_n'' \leqslant 0,$$
$$T_n'(\sigma)(u_n' + u_n'') = 0, \qquad T_t' = T_t'' = 0.$$

The stress field satisfies the equilibrium equations (5.5) with the body forces $F_i \in L^2(\Omega)$, $i = 1, 2$.

We shall use the space of virtual displacements

$$V = \{v \in [H^1(\Omega')]^2 \times [H^1(\Omega'')]^2 \mid v' = 0 \text{ on } \Gamma_u,\ v_n'' = 0 \text{ on } \Gamma_0\},$$

the cone of admissible virtual displacements

$$K = \{v \in V \mid v_n' + v_n'' \leqslant 0 \text{ on } \Gamma_K\}$$

and the set of statically admissible stress fields

$$K^+ = \{\tau \in S \mid \langle \tau, e(v)\rangle \geqslant L(v)\ \forall v \in K\},$$

where

$$L(v) = \int_{\Omega' \cup \Omega''} F_i v_i \, dx + \int_{\Gamma_\tau' \cup \Gamma_\tau''} P_i v_i \, ds.$$

Now we are able to prove the appropriate extension of the Haar–Kármán principle.

THEOREM 13.1. *Assume that fields of displacements u and of stresses σ (sufficiently smooth) exist such that* (13.1), (13.2), *the equilibrium equations* (5.5) *and all boundary conditions mentioned above are satisfied. Then the stress field σ is a solution of the following problem*:

$$\sigma = \arg \min_{\tau \in K^+ \cap P} \mathcal{S}(\tau), \tag{13.3}$$

where

$$\mathcal{S}(\tau) = \tfrac{1}{2}\|\tau\|^2 - \langle \tau, e(u_0)\rangle$$

and $u_0'' = 0$.

THEOREM 13.2. *If $K^+ \cap P$ is non-empty, then problem* (13.3) *has a unique solution.*

PROOF. The proof follows from the fact, that K^+ and P are convex and closed in S and $\mathcal{S}$ is quadratic and strictly convex. □

REMARK 13.1. The formulation in terms of *displacements* is more complicated, as it requires a departure from the range of Sobolev spaces, when defining functions with finite energy. Concerning this, see DUVAUT and LIONS [1976], MERCIER [1977], SUQUET [1978] for the classical boundary value problem and TOMARELLI [1988] for a problem of Signorini's type. The latter paper, however, does not involve any numerical analysis of the problem.

Approximation by piecewise constant stress fields. Assume that Γ_K, Γ_u and Γ_0 consist of straight elements only. Let us consider triangulations $\mathcal{T}_h^M$ of Ω^M, $(M = {}', {}'')$ such that their nodes on Γ_K coincide and the triangles $T \in \mathcal{T}_h^M$ adjacent to Γ_τ may have curved sides along the boundary. We introduce the space of piecewise linear virtual displacements

$$V_h = \{v \in V \mid v|_T \in [P_1(T)]^2 \ \forall T \in \mathcal{T}_h\},$$

where $\mathcal{T}_h = \mathcal{T}_h' \cup \mathcal{T}_h''$. Let h denote the maximal diameter of all triangles in $\mathcal{T}_h$. The minimal interior angle of all triangles in $\mathcal{T}_h$ will be denoted by θ_h. (If the triangle has a curved side, then the interior angles are defined by angles of the "straight" triangle with the same vertices.) We say that a family $\{\mathcal{T}_h\}$, $h \to 0+$, of triangulations is *regular*, if θ_h is bounded away from zero by a number θ, independent of h.

Introducing the space of piecewise constant tensor fields

$$S_h = \{\tau \in S \mid \tau_{ij}|_T \in P_0(T),\ i, j = 1, 2,\ \forall T \in \mathcal{T}_h\},$$

we define the following external approximation of the set K^+:

$$K_h^+ = \{\tau \in S_h \mid \langle \tau, e(v_h)\rangle \geqslant L(v_h)\ \forall v_h \in K_h\},$$

where

$$K_h = K \cap V_h = \{v \in V_h \mid v_n' + v_n'' \leqslant 0 \text{ on } \Gamma_K\}.$$

Instead of the problem (13.3) we shall solve the approximate problem

$$\sigma_h = \arg \min_{\tau \in K_h^+ \cap P} \mathcal{S}(\tau). \tag{13.4}$$

THEOREM 13.3. *Assume that* $K^+ \cap P \neq \emptyset$ *and there exists only a finite number of points* $\Gamma_K \cap \overline{\Gamma}_\tau$, $\overline{\Gamma}_\tau \cap \overline{\Gamma}_0$, $\overline{\Gamma}_\tau \cap \overline{\Gamma}_u$. *Let* σ *denote the solution of the problem* (13.3). *Then the approximate problem* (13.4) *is uniquely solvable for any h and*

$$\|\sigma_h - \sigma\|_0 \to 0 \quad \text{as } h \to 0+, \tag{13.5}$$

provided that the family of triangulations is regular.

PROOF. The set $K_h^+ \cap P$ is closed and convex. To show that it is non-empty, we consider the projection $r_h : S \to S_h$, which is defined by the following condition:

$$\langle \tau - r_h\tau, \chi_h \rangle = 0 \; \forall \chi_h \in S_h.$$

It is not difficult to show that

$$\tau \in K^+ \cap P \Rightarrow r_h\tau \in K_h^+ \cap P.$$

Hence the existence and uniqueness of the solution σ_h of (13.4) follows.

To prove the convergence (13.5), we employ the abstract Theorem 4.1 on the convergence of Ritz approximations. Thus it suffices to verify that

(i) $\exists\{\tau_h\}$, $\tau_h \in K_h^+ \cap P$, s.t. $\tau_h \to \sigma$ in S, as $h \to 0+$;

(ii) if $\tau_h \in K_h^+ \cap P$, $\tau_h \rightharpoonup \tau$ (weakly) in S, then $\tau \in K_h^+ \cap P$.

The first condition is satisfied with $\tau_h = r_h\sigma$. To prove condition (ii), we consider an arbitrary $v \in K$ and construct a sequence $\{v_h\}$, such that $v_h \in K_h$ and

$$v_h \to v \quad \text{in } V \text{ as } h \to 0+. \tag{13.6}$$

To this end we use the following result: the set

$$K^\infty = K \cap [C^\infty(\overline{\Omega}')]^2 \times [C^\infty(\overline{\Omega}'')]^2$$

is dense in K. (For the proof see Lemma 8.1 and HASLINGER and HLAVÁČEK [1982a, proof of Theorem 2.1]). Then (13.6) will be satisfied, if we set

$$v_h = I_h v_n,$$

where $\{v_n\}$ is the sequence of $v_n \in K^\infty$ such that

$$v_n \to v \quad \text{in } V \text{ as } n \to \infty$$

and $I_h : K^\infty \to K_h$ is the Lagrange linear interpolation over the triangulation $\mathcal{T}_h$.

Then it is readily seen that $e(v_h) \to e(v)$ in S. Since

$$\langle \tau_h, e(v_h) \rangle \geqslant L(v_h)$$

follows from the definition of K_h^+, passing to the limit with $h \to 0+$, we obtain $\tau \in K^+$. Since P is weakly closed, $\tau \in P$. Consequently, (ii) is verified. □

13.2. Increasing zone of contact

If the bodies Ω' and Ω'' have smooth boundaries in a neighbourhood of $\partial\Omega' \cap \partial\Omega''$, the contact zone can enlarge during the deformation process. We introduce (see Section 5.2 and Fig. 6.3) a new Cartesian system (ξ, η) at a point of $\partial\Omega' \cap$

$\partial\Omega''$, such that the ξ-axis coincides with the direction of n'' and define the function $\varepsilon(\eta) = f'(\eta) - f''(\eta)$, $\eta \in [a,b]$, which represents the gap between the bodies before deformation.

Recall that (see Section 5.2) Γ_K^M, $M = {}', {}''$ are defined as the graphs of $f^M(\eta)$, $\eta \in [a,b]$ and that we have the following decompositions

$$\partial\Omega' = \overline{\Gamma}_u \cup \overline{\Gamma}'_\tau \cup \Gamma'_K, \qquad \partial\Omega'' = \overline{\Gamma}_0 \cup \overline{\Gamma}''_\tau \cup \Gamma''_K,$$
$$\Gamma_\tau^M \cap \Gamma_K^M = \emptyset, \qquad M = {}', {}''.$$

The boundary conditions on Γ_u, Γ_0 and $\Gamma'_\tau \cup \Gamma''_\tau$ are the same as in the previous section. We assume that

$$u''_{on} = 0 \text{ on } \Gamma_0 \quad \text{and} \quad u''_{o\xi} - u'_{o\xi} = \varepsilon \text{ for a.a. } \eta \in [a,b].$$

The boundary conditions on Γ_K are now replaced by the following system of conditions

$$\begin{aligned} &u''_\xi - u'_\xi \leqslant \varepsilon, \\ &-T'_\xi(\sigma)(\cos\alpha')^{-1} = T''_\xi(\sigma)(\cos\alpha'')^{-1} \leqslant 0, \\ &T''_\xi(\sigma)(u''_\xi - u'_\xi - \varepsilon) = 0, \quad T''_\eta(\sigma) = T'_\eta(\sigma) = 0, \end{aligned} \tag{13.7}$$

which hold at almost all points of $\Gamma'_K \cup \Gamma''_K$ with the same coordinates $\eta \in [a,b]$. Here α^M, $M = {}', {}''$, denotes the angle between the η-axis and the tangent to Γ_K^M.

We introduce the set of admissible virtual displacements

$$K_\varepsilon = \{v \in V \mid v''_\xi - v'_\xi \leqslant \varepsilon \text{ for a.a. } \eta \in [a,b]\}$$

and the set of statically admissible stress fields

$$K_0^+ = \{\tau \in S \mid \langle \tau, e(v)\rangle \geqslant L(v)\ \forall v \in K_0\},$$

where K_0 is defined as K_ε with $\varepsilon = 0$.

Then an *analogue of Theorem* 13.1 (the extended Haar–Kármán principle) is valid, where the set $K^+ \cap P$ is replaced by $K_0^+ \cap P$ in (13.3) and the boundary conditions on Γ_K are replaced by the system (13.7).

The latter minimization problem is uniquely solvable, if $K_0^+ \cap P \neq \emptyset$, (i.e., an analogue of Theorem 13.2).

Approximation by piecewise constant stress fields. Let us consider triangulations $\mathcal{T}_h^M$ of Ω^M, $M = {}', {}''$, such that the triangles adjacent to the boundaries may have curved sides along the boundary and the vertices on Γ_K^M lie on straight lines parallel with the ξ-axis. If Γ_K^M contains a point of inflexion, then there is a vertex.

If a curved triangle T, adjacent to Γ_K^M is convex, it will be divided by the chord into a "straight" triangle T_0 and a segment T_S, so that $T = T_0 \cup T_S$. If $T_c \in \mathcal{T}_h$, adjacent to Γ_K^M is non-convex, then one of its sides is parallel with the ξ-axis. We define

$$\begin{aligned} V_h = \{v \in V \mid v|_{T_0} \in [P_1(T_0)]^2\ \forall T_0 \subset T \in \mathcal{T}_h \text{ adjacent to } \Gamma_K^M, \\ \left(\frac{\partial v}{\partial \xi}\right)_{T_S} = 0\ \forall T_S \subset T \in \mathcal{T}_h \text{ adjacent to } \Gamma_K^M, \\ \left(\frac{\partial v}{\partial \xi}\right)_{T_c} = 0 \text{ for all non-convex triangles } T_c \text{ adjacent to } \Gamma_K^M, \\ v|_T \in [P_1(T)]^2 \text{ for all remaining triangles } \}. \end{aligned}$$

In other words, V_h consists of piecewise linear functions, which are extended continuously onto the segments and onto the non-convex triangles adjacent to Γ_K^M by constants in the ξ-direction.

Moreover, let us introduce the following sets

$$\begin{aligned} S_h &= \{\tau \in S \mid \tau|_T \in [P_0(T^*)]^4\ \forall T^* = T, T_0, T_S, T_c \in \mathcal{T}_h\}, \\ K_{0h} &= \{v \in V_h \mid v''_\xi - v'_\xi \leqslant 0\ \forall \eta \in [a,b] \text{ on } \Gamma'_K \cup \Gamma''_K\}, \\ K^+_{0h} &= \{\tau \in S_h \mid \langle \tau, e(v_h)\rangle \geqslant L(v_h) \forall v_h \in K_{0h}\}. \end{aligned}$$

It is easy to see that if the condition $v''_\xi - v'_\xi \leqslant 0$ holds at the nodes of $\Gamma'_K \cup \Gamma''_K$, it holds on the whole interval $[a,b]$, by virtue of the definition of V_h.

Now we can define the approximate problem as follows:

$$\sigma_h = \arg \min_{\tau \in K^+_{0h} \cap P} \mathcal{S}(\tau). \tag{13.8}$$

THEOREM 13.4. *Assume that $K_0^+ \cap P \neq \emptyset$; there exists only a finite number of points $\overline{\Gamma}_\tau \cap \overline{\Gamma}_u$, $\overline{\Gamma}_\tau \cap \overline{\Gamma}_0$; $f^M \in C^2$ for $M = ', ''$ in a neighbourhood of the interval $[a,b]$ and $\Gamma'_K \cap \overline{\Gamma}_u = \emptyset$, $\Gamma''_K \cap \overline{\Gamma}_0 = \emptyset$.*

Then the approximate problem (13.8) *has a unique solution σ_h. If σ denotes the solution of the continuous problem (i.e. σ is the minimizer of $\mathcal{S}(\tau)$ over $K_0^+ \cap P$), then*

$$\|\sigma_h - \sigma\|_0 \to 0 \quad \textit{as } h \to 0+$$

holds for any regular family of triangulations.

PROOF. The proof is similar to that of Theorem 13.3. We employ the following theorem on the density:

Let the assumptions of Theorem 13.4 be satisfied. Then the set

$$K_0 \cap [C^2(\overline{\Omega}')]^2 \times [C^2(\overline{\Omega}'')]^2$$

is dense in K_0.

Furthermore, a Lagrange linear interpolation

$$I_h : [H^2(\Omega')]^2 \times [H^2(\Omega'')]^2 \to V_h$$

has to be analysed. We obtain that

$$\|I_h v - v\|_{1,\Omega' \cup \Omega''} \to 0 \quad \text{as } h \to 0+,$$

provided $f^M \in C^1([a,b])$ for $M = ', ''$ and the family of triangulations is regular. □

13.3. Solution of approximate problems

In approximate problems, the sets K_h^+ and K_{0h}^+ seem to cause difficulties, at a first glance. We can simplify the situation, however, by eliminating the test function v_h, as follows.

Let us consider problem (13.4) only, since a similar approach is applicable to problem (13.8). Let us denote

$$v_h(x) = \sum_{i=1}^{N} q_i \varphi_i(x),$$

where q_i are the nodal values at the nodes of the triangulation $\mathcal{T}_h$.

If we write down the conditions at the nodes of Γ_K, then precisely four components $\{q_{k_1}, q_{k_2}, q_{k_3}, q_{k_4}\}$ occur at each (double) node $A_k \in \Gamma_K$. In fact, assume for brevity that Γ_K is a single straight segment with $n_2'' \neq 0$. Then, the contact condition gives

$$\sum_{j=1}^{4} b_i q_{k_j} \leqslant 0,$$

where $b_j = n_j'$ for $j = 1, 2$ and $b_j = n_{j-2}''$ for $j = 3, 4$. Introducing a linear transformation $q = F_k y$, $F_k : \mathbb{R}^4 \to \mathbb{R}^4$, by means of

$$y_{k_j} = q_{k_j}, \quad j = 1, 2, 3 \quad \text{and} \quad y_{k_4} = \sum_{j=1}^{4} b_j q_{k_j},$$

we find out that F_k is regular. Let us consider the same transformation in each quadruplet $M_k = \{q_{k_1}, q_{k_2}, q_{k_3}, q_{k_4}\}$, $k = 1, \ldots, Q$, associated with each node $A_k \in \Gamma_K$. Setting also $y_p = q_p$ for $q_p \notin \bigcup_{k=1}^{Q} M_k$, $1 \leqslant p \leqslant N$, altogether we have $q = Fy$, $F : \mathbb{R}^N \to \mathbb{R}^N$ and

$$v_h \in K_h \Leftrightarrow q \in \mathcal{K}_q \Leftrightarrow y \in \mathcal{K}_y = \{y \in \mathbb{R}^N \mid y_{k_4} \leqslant 0, \ k = 1, \ldots, Q\}.$$

Next let ψ_T denote the characteristic function of the triangle $T \in \mathscr{T}_h$. Then we have

$$\tau_h \in S_h \Leftrightarrow \tau_h(x) = \sum_{T \in \mathscr{T}_h} \tau(T)\psi_T(x) \tag{13.9}$$

and denoting

$$t = (\tau_{11}(T_1), \tau_{22}(T_1), \tau_{12}(T_1), \tau_{11}(T_2), \tau_{22}(T_2), \ldots)^{\mathrm{T}} \tag{13.10}$$

the corresponding vector of coefficients, ($t \in \mathbb{R}^M$), we obtain

$$\langle \tau, e(v_h) \rangle = \sum_{T \in \mathscr{T}_h} \int_T \tau_{jk}(T) \sum_{i=1}^{N} q_i e_{jk}(\varphi_i)\,\mathrm{d}x = q^{\mathrm{T}} Et,$$

where E is an $N \times M$ matrix (note that $N < M$).

Since

$$L(v_h) = \sum_{i=1}^{N} q_i L(\varphi_i) = q^{\mathrm{T}} l,$$

where $l \in \mathbb{R}^N$ is a fixed vector, the condition $\tau \in K_h^+$ can finally be rewritten in the form

$$q^{\mathrm{T}}(l - Et) \leqslant 0 \quad \forall q \in \mathscr{K}_q.$$

In other words, the vector $l - Et$ belongs to the polar cone $\mathscr{K}_q^0$ of the cone $\mathscr{K}_q$. Employing the mapping F, we obtain an equivalent condition

$$(Fy)^{\mathrm{T}}(l - Et) \leqslant 0 \quad \forall y \in \mathscr{K}_y. \tag{13.11}$$

Let I^- be the set of all indices k_4, $k = 1, 2, \ldots, Q$ and $I_0 = \{1, \ldots, N\} - I^-$ the set of the remaining indices. Since the cone $\mathscr{K}_y$ is generated by the vectors

$$\{\pm e_j,\ j \in I_0,\ -e_m,\ m \in I^-\},$$

where e_j and e_m form an orthogonal basis in $\mathbb{R}^N$, (13.11) is equivalent to the following system:

$$g_j(t) \equiv (Fe_j)^{\mathrm{T}}(l - Et) = 0, \quad j \in I_0, \tag{13.12}$$

$$g_m(t) \equiv (Fe_m)^{\mathrm{T}}(l - Et) \geqslant 0, \quad m \in I^-. \tag{13.13}$$

Moreover, $\tau_h \in P$ if and only if

$$f(\tau_h(T)) \leqslant 1 \quad \forall T \in \mathscr{T}_h,$$

which may be written in the form

$$f_T(t) - 1 \leqslant 0 \quad \forall T \in \mathscr{T}_h. \tag{13.14}$$

Inserting (13.9) and (13.10) into the functional $\mathscr{S}(\tau_h)$, we are led to the following problem of nonlinear programming: $\mathscr{S}_0(t) = \min$ over the set of $t \in \mathbb{R}^M$, satisfying the constraints (13.12), (13.13) and (13.14).

REMARK 13.2. If Γ_K contains a vertex, we define

$$y_{k_j} = q_{k_j}, \quad j = 1,2, \qquad y_{k_3} = \sum_{j=1}^{4} b_j^{(1)} q_{k_j}, \qquad y_{k_4} = \sum_{j=1}^{4} b_j^{(2)} q_{k_j},$$

where $b_j^{(1)}$ and $b_j^{(2)}$ correspond to the normal vectors n', n'' on both sides of the vertex.

CHAPTER V

Problems of the Theory of Plasticity

In this chapter we deal with variational inequalities of evolution, that result from some problem of plasticity. We consider processes that depend on the history of loading. We also consider the so-called flow theory of plasticity. For more details see HLAVÁČEK, HASLINGER, NEČAS and LOVÍŠEK [1988]. We use the method of penalization which by itself has some features of an approximate method. It approximates the given problem by problems for an elasto-inelastic material with internal state variables (for details see NEČAS and HLAVÁČEK [1981]), which are of independent physical significance, being sensitive to creep and fading memory; our model is rate-independent and does not have this property. For references as well as for fundamental examples of stress–strain relations see HLAVÁČEK, HASLINGER, NEČAS and LOVÍŠEK [1988].

14. Prandtl–Reuss model of plastic flow

The model considered does not take account of hardening and is from some points of view more singular than when hardening occurs.

Let Ω be the considered domain with Lipschitz boundary $\partial\Omega$ and let a time interval $[0, T]$ be given. Let $F(t)$ be the vector function of the body forces and $g(t)$ the vector function of the given stress vector on the boundary. Here, the traction boundary problem is considered. Other problems can be solved in the same manner. We suppose $F \in C_0^1([0, T]; L^2(\Omega; \mathbb{R}^3))$ (i.e. $F(0) = 0$) $g \in C_0^1([0, T]; L^2(\partial\Omega; \mathbb{R}^3))$; for every time the conditions of total equilibrium have to be satisfied:

$$\int_{\Omega} F(t)\,\mathrm{d}x + \int_{\partial\Omega} g(t)\,\mathrm{d}S = 0, \tag{14.1}$$

$$\int_{\Omega} (x \times F(t))\,\mathrm{d}x + \int_{\partial\Omega} (x \times g(t))\,\mathrm{d}S = 0. \tag{14.2}$$

Let S be the subspace of $L^2(\Omega; \mathbb{R}^q)$ with elements $\sigma_{ij} = \sigma_{ji}$ and $H_0^1((0, T); S)$

the closure of $C_0^1([0,T];S)$ in the norm induced by the scalar product

$$\int_0^T \left(\int_\Omega \dot{\tau}_{ij}(t,x)\dot{\sigma}_{ij}(t,x)\,\mathrm{d}x \right) \mathrm{d}t := \int_0^T (\dot{\tau},\dot{\sigma})\,\mathrm{d}t. \tag{14.3}$$

We shall suppose, that the stress tensor fulfills the condition of equilibrium for every $t \in [0,T]$:

$$\frac{\partial \tau_{ij}}{\partial x_j} + F_i(t) = 0 \tag{14.4}$$

(more precisely see (14.13)) and that the strain tensor $\varepsilon_{ij} = \frac{1}{2}(\partial u_i/\partial x_j + \partial u_j/\partial x_i)$ can be written as the sum of two symmetric tensors

$$\varepsilon = e + p, \tag{14.5}$$

where e is the elastic and p the plastic part.

Further, let the function of plasticity $f(\sigma)$ be given (the yield function) such that $f(0)=0$, $f(\sigma)>0$ for $\sigma \neq 0$ and f is convex; we suppose $f \in C^2(\mathbb{R}^q)$ with bounded first and second derivatives in $\mathbb{R}^q$, i.e.

$$\left|\frac{\partial f}{\partial \sigma}\right| + \left|\frac{\partial^2 f}{\partial \tau \partial \sigma}\right| \leqslant c < \infty. \tag{14.6}$$

Between e and σ a linear Hooke's law is assumed

$$e_{ij} = A_{ijkl}(x)\sigma_{kl} \quad (e = A\sigma), \tag{14.7}$$

with $A_{ijkl} \in L^\infty(\Omega)$, $A_{ijkl} = A_{jikl} = A_{klij}$, $A_{ijkl}\sigma_{ij}\sigma_{kl} \geqslant c|\sigma|^2$, $c>0$.

We suppose that, for the solution in question,

$$f(\sigma) \leqslant \alpha_0, \quad \alpha_0 > 0 \tag{14.8}$$

and that

$$\dot{p}_{ij} = \lambda \frac{\partial f}{\partial \sigma_{ij}}, \quad \lambda \geqslant 0 \ \left(\dot{p} = \lambda \frac{\partial f}{\partial \sigma}\right), \tag{14.9}$$

(the condition of normality) with $\lambda > 0$ only for $f(\sigma) = \alpha_0$.

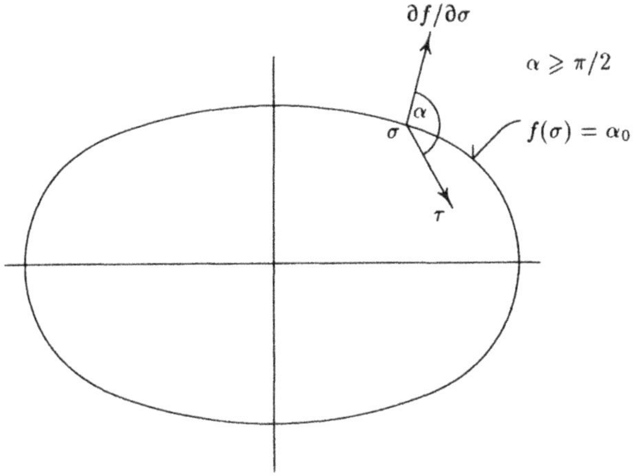

FIG. 14.1. A yield surface in the stress space.

Let us try to formulate first the problem in a formal way. First, on the boundary $\partial\Omega$, we suppose

$$\sigma_{ij}x_j = g_i. \tag{14.10}$$

If σ is a solution of our problem and τ another tensor satisfying (14.4), (14.8) and (14.10) then it follows from the compatibility condition that

$$\varepsilon_{ij} = \tfrac{1}{2}\left(\frac{\partial u_i}{\partial x_j} + \frac{\partial u_j}{\partial x_i}\right)\!: 0 = (\dot{\varepsilon}, \tau - \sigma) \text{ so } 0 = (\dot{e} + \dot{p}, \tau - \sigma) \tag{14.11}$$

and because of (14.9) $\dot{p}_{ij}(\tau_{ij} - \sigma_{ij}) \leqslant 0$ (see Fig. 14.1), it follows that

$$0 \leqslant \int_\Omega A_{ijkl}\dot{\sigma}_{kl}(\tau_{ij} - \sigma_{ij})\,\mathrm{d}x_i := [\dot{\sigma}, \tau - \sigma]. \tag{14.12}$$

So we define the weak *solution of the Prandtl–Reuss plastic flow*: for all $t \in [0, T)$:

$$(\sigma, e) = \int_{\partial\Omega} g_i v_i\,\mathrm{d}S + \int_\Omega F_i v_i\,\mathrm{d}x \quad \forall v \in W^{1,2}(\Omega, \mathbb{R}^3), \tag{14.13}$$

$$v \in W^{1,2}(\Omega, \mathbb{R}^3), \qquad f(\sigma(t)) \leqslant \alpha_0 \quad \text{a.e. in } \Omega, \tag{14.14}$$

$\sigma \in H_0^1((0, T); S)$ and for all $\tau \in H_0^1((0, T); S)$, satisfying (14.13) and (14.14) we have

$$\int_0^t [\dot{\sigma}, \tau - \sigma]\,\mathrm{d}t \geqslant 0. \tag{14.15}$$

THEOREM 14.1 (Uniqueness). *There exists at most one weak solution.*

PROOF. Let σ^1, σ^2 be two solutions. Then

$$\int_0^t [\dot{\sigma}^2, \sigma^1 - \sigma^2]\,dt \geqslant 0, \qquad \int_0^t [\dot{\sigma}^1, \sigma^2 - \sigma^1]\,dt \geqslant 0,$$

so

$$0 \geqslant \int_0^t [\dot{\sigma}^1 - \dot{\sigma}^2, \sigma^1 - \sigma^2]\,dt = \tfrac{1}{2}[\sigma^1(t) - \sigma^2(t), \sigma^1(t) - \sigma^2(t)] \geqslant 0. \qquad \square$$

THEOREM 14.2 (Existence). *Let there exist $\sigma_0 \in C_0^1([0,T]; S)$ satisfying* (14.13), *such that with some $\gamma > 0$: $f(\sigma^0(t) + \gamma\dot{\sigma}^0(t)) \leqslant \alpha_0$ for $t \in [0,T]$. (Security condition). Then there exists precisely one solution to the problem.*

PROOF. Let us replace (14.9) by

$$\dot{p} = \frac{1}{\varepsilon}(f - \alpha_0)^+ \frac{\partial f}{\partial \sigma}(1 + [(f - \alpha_0)^+]^2)^{-1/2}, \quad \varepsilon > 0. \tag{14.16}$$

Let us look for σ^ε, p^ε such that (14.11), (14.13) and (14.16) are satisfied. This problem can be reduced to an initial value problem for an abstract differential equation: Let $S_0 \subset S$ be the orthogonal complement to the space

$$E = \{\varepsilon \in S \mid \varepsilon_{ij} = \tfrac{1}{2}\left(\frac{\partial u_i}{\partial x_j} + \frac{\partial u_j}{\partial x_i}\right), \ u \in [H^1(\Omega)]^3\}$$

(closed because of Korn's inequality). Condition (14.1) is equivalent to

$$(\dot{\varepsilon}, \omega) = 0 \quad \forall \omega \in S_0. \tag{14.17}$$

Let P be the orthogonal projector of S to S_0. So $P(A\dot{\sigma}^\varepsilon + \dot{p}^\varepsilon) = 0$, which implies

$$PA\dot{\sigma}^\varepsilon = -P\left[\lambda^\varepsilon(\sigma)\frac{\partial f}{\partial \sigma}\right] \tag{14.18}$$

with $\lambda^\varepsilon(\sigma) = \frac{1}{\varepsilon}(f - \alpha_0)^+(1 + [(f - \alpha_0)^+]^2)^{-1/2}$.

Putting $\sigma^\varepsilon = \sigma^0 + \tilde{\sigma}^\varepsilon$, we have $\tilde{\sigma}^\varepsilon \in S_0$ and $B = PA$ is invertible in S_0, so

$$\dot{\tilde{\sigma}}^\varepsilon = -B^{-1}B\dot{\sigma}^0 - B^{-1}P\left[\lambda^\varepsilon(\sigma^0 + \tilde{\sigma}^\varepsilon)\frac{\partial f}{\partial \sigma}(\sigma^0 + \tilde{\sigma}^\varepsilon)\right] \tag{14.19}$$

and because of (14.6), there exists a unique solution to (14.19), $\tilde{\sigma} \in C_0^1([0,T]; S_0)$.

For $\varepsilon \to 0$ we get the solution of our problem: let us mention the point where the security condition comes in. It follows from (14.17) with $\omega = \gamma\dot{\tilde{\sigma}}^{\varepsilon} + \tilde{\sigma}^{\varepsilon}$,

$$\int_0^t [\dot{\sigma}^{\varepsilon}, \omega]\, dt + \int_0^t (\dot{p}^{\varepsilon}, \omega)\, dt = 0. \tag{14.20}$$

But $\dot{p} = \partial g/\partial\sigma$, where $\partial g/\partial\sigma$ is the derivative of the convex functional

$$g(\sigma) = \frac{1}{\varepsilon}\int (\{[(f(\sigma) - \alpha_0)^+]^2 + 1\}^{1/2} - 1)\, dx. \tag{14.21}$$

So

$$\begin{aligned}
&\int_0^t \left(\frac{\partial g}{\partial\sigma}(\sigma^{\varepsilon}), \gamma\dot{\tilde{\sigma}}^{\varepsilon} + \tilde{\sigma}^{\varepsilon}\right) dt \\
&\quad = \int_0^t \left(\frac{\partial g}{\partial\sigma}(\sigma^{\varepsilon}), \gamma\dot{\sigma}^{\varepsilon} + \sigma^{\varepsilon} - \gamma\dot{\sigma}^0 - \sigma^0\right) dt \\
&\quad = \gamma g(\sigma^{\varepsilon}(t)) + \int_0^t \left(\frac{\partial g}{\partial\sigma}(\sigma^{\varepsilon}) - \frac{\partial g}{\partial\sigma}(\gamma\dot{\sigma}^0 + \sigma^0), \sigma^{\varepsilon} - \gamma\dot{\sigma}^0 - \sigma^0\right) dt \\
&\quad \geqslant \gamma g(\sigma^{\varepsilon}(t)),
\end{aligned}$$

here we have used $\partial g(\gamma\dot{\sigma}^0 + \sigma^0)/\partial\sigma = 0$ and the fact that $\partial g/\partial\sigma$ is monotonic. So (14.20) implies

$$\gamma\int [\dot{\sigma}^{\varepsilon}, \dot{\sigma}^{\varepsilon}]\, dt + \tfrac{1}{2}[\sigma^{\varepsilon}(t), \sigma^{\varepsilon}(t)] - \int [\dot{\sigma}^{\varepsilon}, \gamma\dot{\sigma}^0 + \sigma^0]\, dt + \gamma g(\sigma^{\varepsilon}(t)) \leqslant 0 \tag{14.22}$$

and with $\varepsilon_n \to 0$, we can choose $\sigma^{\varepsilon_n} \rightharpoonup \sigma$ in $H_0^1((0, T); S)$. Further, because $\int_0^t [\dot{\sigma}^{\varepsilon_n}, \tau - \sigma^{\varepsilon_n}]\, dt \geqslant 0$, σ is the solution. □

REMARK 14.1. A typical yield function is

$$f(\sigma) = (\varepsilon^2 + \sigma'_{ij}\sigma'_{ij})^{1/2} - \varepsilon,$$

where $\sigma'_{ij} = \sigma_{ij} - \frac{1}{3}\delta_{ij}\sigma_{kk}$, $\varepsilon > 0$.

15. Plastic flow with hardening

We start with formal considerations. Let the yield function $f(\sigma)$ be given. We suppose $\dot{p} = 0$ for $f(\sigma) < \alpha_0$. Hardening means, that the level of yield surface in the course of the process increases to $f(\sigma) = \alpha > \alpha_0$; $\dot{p}$ again fulfils the normality condition, but after the fall of stress to the level $f(\sigma) < \alpha$, we have $\dot{p} = 0$. Put

$$\alpha(t) = \max\{\alpha_0, \max_{0 \leqslant \tau \leqslant t} f(\tau)\} \tag{15.1}$$

which is an *internal state variable*, describing the level of the yield surface. If we put

$$\mathcal{F}(\sigma, \alpha) = f(\sigma) - \alpha, \tag{15.2}$$

we have by assumption

$$\dot{p} = \frac{\partial \mathcal{F}}{\partial \sigma}\dot{\alpha}, \tag{15.3}$$

$$\dot{\alpha} = -\frac{\partial \mathcal{F}}{\partial \alpha}\dot{\alpha}, \tag{15.4}$$

$$\mathcal{F}(\sigma, \alpha) \leqslant 0. \tag{15.5}$$

This is *isotropic hardening* and expresses the fact that the elastic domain $f(\sigma) < \alpha$ increases during the process.

Let us consider now a tensor β, connected with the tensor p by the relation

$$\beta_{ij} = B_{ijkl} p_{kl}, \tag{15.6}$$

with $B_{ijkl} \in L^\infty(\Omega)$ and with standard symmetry and coercivity conditions.

Let us look for $\sigma(t)$, satisfying

$$f(\sigma - \beta) \leqslant \alpha_0 \tag{15.7}$$

and

$$\begin{aligned} &\dot{p} = \lambda \frac{\partial f}{\partial \sigma} \quad \text{with } \lambda \geqslant 0, \\ &\lambda = 0 \quad \text{for } f(\sigma - \beta) < \alpha_0. \end{aligned} \tag{15.8}$$

Putting $\gamma = B^{-1/2}\beta$ and $\mathcal{F}(\sigma, \gamma) = f(\sigma - B^{-1/2}\gamma) - \alpha_0$, we have

$$\mathcal{F}(\sigma, \gamma) \leqslant 0, \tag{15.9}$$

$$\dot{p} = \lambda \frac{\partial \mathcal{F}}{\partial \sigma}, \qquad \dot{\gamma} = -\lambda \frac{\partial \mathcal{F}}{\partial \gamma}. \tag{15.10}$$

This example describes *kinematic hardening* and expresses the fact that the center of the yield surface moves.

We can also consider the *combination* of both the isotropic and the kinematic hardening: we set

$$\alpha = \max\{\alpha_0, \max_{0 \leqslant \tau \leqslant t} f(\sigma(\tau) - B^{1/2}\gamma(\tau))\} \tag{15.11}$$

and putting

$$\mathscr{F}(\sigma, \gamma, \alpha) = f(\sigma - B^{1/2}\gamma) - \alpha, \tag{15.12}$$

we get by assumption

$$\mathscr{F}(\sigma, \gamma, \alpha) \leqslant 0, \qquad \dot{p} = \dot{\alpha}\frac{\partial \mathscr{F}}{\partial \sigma}, \qquad \dot{\gamma} = -\dot{\alpha}\frac{\partial \mathscr{F}}{\partial \gamma}, \qquad \dot{\alpha} = -\dot{\alpha}\frac{\partial \mathscr{F}}{\partial \alpha}. \tag{15.13}$$

Hence in all cases we get

$$\dot{p} = \lambda\frac{\partial \mathscr{F}}{\partial \sigma}, \qquad \dot{\gamma} = -\lambda\frac{\partial \mathscr{F}}{\partial \gamma}, \qquad \dot{\alpha} = -\lambda\frac{\partial \mathscr{F}}{\partial \alpha}, \tag{15.14}$$

where $\lambda \geqslant 0$ and $\lambda = 0$ provided $\mathscr{F}(\sigma, \gamma, \alpha) < 0$.

Further, we suppose the *hardening condition*:

$$\frac{\partial \mathscr{F}}{\partial \gamma_{ij}}\frac{\partial \mathscr{F}}{\partial \gamma_{ij}} + \frac{\partial \mathscr{F}}{\partial \alpha}\frac{\partial \mathscr{F}}{\partial \alpha} \geqslant c > 0. \tag{15.15}$$

We shall suppose that $\mathscr{F} \in C^2(\mathbb{R}^9 \times \mathbb{R}^9 \times \mathbb{R}^1)$, is convex and all its first and second derivatives are bounded.

We get from (15.14) that if $\mathscr{F}(\sigma, \gamma, \alpha) = 0$ and $\mathscr{F}(\tilde{\sigma}, \tilde{\gamma}, \tilde{\alpha}) \leqslant 0$, then

$$\dot{p}_{ij}(\tilde{\sigma}_{ij} - \sigma_{ij}) - \dot{\gamma}_{ij}(\tilde{\gamma}_{ij} - \gamma_{ij}) - \dot{\alpha}(\tilde{\alpha} - \alpha) \leqslant 0.$$

Of course, both σ and $\tilde{\sigma}$ have to satisfy (14.13). From this follows, as in the preceding section, the *fundamental inequality of plasticity*:

$$[\dot{\sigma}, \tilde{\sigma} - \sigma] + (\dot{\gamma}, \tilde{\gamma} - \gamma) + (\dot{\alpha}, \tilde{\alpha} - \alpha) \geqslant 0. \tag{15.16}$$

So we suppose that the generalized yield function $\mathscr{F}(\sigma, \gamma, \alpha)$ is given with the properties mentioned above, matrices A from (14.7) and B from (15.6), body forces $F \in C^1([0, T]; (L^2(\Omega))^3)$ and boundary forces $g \in C^1([0, T]; (L^2(\partial\Omega))^3)$, satisfying (14.1) and (14.2). We do not assume in general, that $F(0) = 0$, $g(0) = 0$ or that $\sigma(0) = 0$. Then $\sigma \in H^1((0, T); S)$ (the closure of $C^1([0, T]; S)$), $\gamma \in H^1((0, T); S)$, $\alpha \in H^1((0, T); L^2(\Omega))$ is a *weak solution to the plastic flow*, if σ satisfies (14.13) for all t, $\sigma(0) = \sigma_0$, $\gamma(0) = \gamma_0$, $\alpha(0) = \alpha_0$, $\mathscr{F}(\sigma_0, \gamma_0, \alpha_0) \leqslant 0$, and if for all t, $\mathscr{F}(\sigma(t), \gamma(t), \alpha(t)) \leqslant 0$ and for $\tilde{\alpha} \in H^1((0, T); S)$, $\tilde{\gamma} \in H^1((0, T); S)$, $\tilde{\alpha} \in H^1((0, T); L^2(\Omega))$, such that $\mathscr{F}(\tilde{\sigma}(t), \tilde{\gamma}(t), \tilde{\alpha}(t)) \leqslant 0$ and such that $\tilde{\sigma}$ satisfies (14.13), the inequality

$$\int_0^t [\dot{\sigma}, \tilde{\sigma} - \sigma]\,\mathrm{d}t + \int_0^t (\dot{\gamma}, \tilde{\gamma} - \gamma)\,\mathrm{d}t + \int_0^t (\dot{\alpha}, \tilde{\alpha} - \alpha)\,\mathrm{d}t \geqslant 0 \tag{15.17}$$

is satisfied.

We shall suppose also the following condition (which is satisfied in all examples):

$$\begin{aligned}&\text{if } \tau \in C([0,T];S), \text{ then there exist}\\ &\delta \in C([0,T];S) \text{ and } \beta \in C([0,T];L^2(\Omega)), \text{ such that}\\ &\mathcal{F}(\tau(t),\delta(t),\beta(t)) \leqslant 0 \quad \text{for all } t \in [0,T].\end{aligned} \tag{15.18}$$

THEOREM 15.1. *There exists at most one solution to the plastic flow problem.*

PROOF. As in Theorem 14.1. □

THEOREM 15.2 (Existence of a solution to plastic flow with hardening). *Let all assumptions be satisfied for the yield function $\mathcal{F}$ and the data. Suppose that* (15.15) *as well as* (15.18) *are satisfied. Then there exists precisely one solution to plastic flow with hardening. If we put $\dot{p} = \lambda(\partial\mathcal{F}/\partial\sigma)$, $p(0) = p_0$, $\varepsilon = A\sigma + p$, then the strain tensor is obtained from σ, γ, α; here*

$$\lambda = \frac{\dfrac{\partial F}{\partial\sigma}\dot{\sigma}}{\dfrac{\partial F}{\partial\gamma}\dfrac{\partial F}{\partial\gamma} + \dfrac{\partial F}{\partial\alpha}\dfrac{\partial F}{\partial\alpha}} = -\frac{\dot{\gamma}\dfrac{\partial F}{\partial\gamma} + \dot{\alpha}\dfrac{\partial F}{\partial\alpha}}{\dfrac{\partial F}{\partial\gamma}\dfrac{\partial F}{\partial\gamma} + \dfrac{\partial F}{\partial\alpha}\dfrac{\partial F}{\partial\alpha}},$$

if $\mathcal{F}(\sigma,\gamma,\alpha) = 0$ and $\lambda = 0$ otherwise. ε^0 is supposed compatible. We have

$$\dot{p} = \lambda\frac{\partial F}{\partial\sigma}, \qquad \dot{\gamma} = -\lambda\frac{\partial F}{\partial\gamma}, \qquad \dot{\alpha} = -\lambda\frac{\partial F}{\partial\alpha}. \tag{15.19}$$

PROOF. As in Theorem 14.2, we shall use a penalization functional. Put

$$g(\alpha,\gamma,\alpha) = \frac{1}{\varepsilon}\int_\Omega (\{(\mathcal{F}(\sigma,\gamma,\alpha)^+)^2 + 1\}^{1/2} - 1)\,dx. \tag{15.20}$$

Let us look for $p^\varepsilon \in C^1([0,T];S)$, $\gamma^\varepsilon \in C^1([0,T];S)$, $\alpha^\varepsilon \in C^1([0,T];L^2(\Omega))$, $p^\varepsilon(0) = p_0$, $\gamma^\varepsilon(0) = \gamma_0$, $\alpha^\varepsilon(0) = \alpha_0$, $\sigma^\varepsilon(0) = \sigma_0$.

Putting $\varepsilon = A\sigma + p$, we suppose (14.7). Further, we solve the system

$$\dot{p}^\varepsilon = D_\sigma g, \qquad \dot{\gamma}^\varepsilon = -D_\gamma g, \qquad \dot{\alpha}^\varepsilon = -D_\alpha g,$$

where $D_\sigma g$, $D_\gamma g$ and $D_\alpha g$ denote partial derivatives.

Proceeding as in the proof of Theorem 14.2, we get a unique solution. Let $\sigma_0 \in C^1([0,T];S)$ satisfy (14.13) and let δ and β be from (15.18) corresponding to $\dot{\sigma}_0 + \sigma_0$. We get with $\tau = \dot{\sigma}^\varepsilon + \sigma^\varepsilon - \dot{\sigma}^0 - \sigma^0$, $\gamma = \dot{\gamma}^\varepsilon + \gamma^\varepsilon - \delta$, $\alpha = \dot{\alpha}^\varepsilon + \alpha^\varepsilon - \beta$,

$$\begin{aligned}&\int_0^t [\dot{\sigma}^\varepsilon, \tau]\,\mathrm{d}t + \int_0^t (\dot{\gamma}^\varepsilon, \gamma)\,\mathrm{d}t + \int_0^t (\dot{\alpha}^\varepsilon, \alpha)\,\mathrm{d}t \\ &\quad + \int_0^t (D_\sigma g, \dot{\tau})\,\mathrm{d}t + \int_0^t (D_\gamma g, \gamma)\,\mathrm{d}t + \int_0^t (D_\alpha g, \alpha)\,\mathrm{d}t = 0.\end{aligned} \tag{15.21}$$

It follows from (15.21), because of $\mathcal{F}(\dot{\sigma}_0 + \sigma_0, \alpha, \beta) \leqslant 0$ and the convexity of $\mathcal{F}$ (for details see HLAVÁČEK, HASLINGER, NEČAS and LOVÍŠEK [1988]), that

$$\begin{aligned}&\int_0^t [\dot{\sigma}^\varepsilon, \dot{\sigma}^\varepsilon]\,\mathrm{d}t + \tfrac{1}{2}[\sigma^\varepsilon(t), \sigma^\varepsilon(t)] + \int_0^t (\dot{\gamma}^\varepsilon, \dot{\gamma}^\varepsilon)\,\mathrm{d}t + \tfrac{1}{2}(\gamma^\varepsilon(t), \gamma^\varepsilon(t)) \\ &\quad + \int_0^t (\dot{\alpha}^\varepsilon, \dot{\alpha}^\varepsilon)\,\mathrm{d}t + \tfrac{1}{2}(\alpha^\varepsilon(t), \alpha^\varepsilon(t)) + g(\sigma^\varepsilon(t), \gamma^\varepsilon(t), \alpha^\varepsilon(t)) \leqslant c\end{aligned} \tag{15.22}$$

with c independent of ε.

So we can suppose for $\varepsilon_n \to 0$, that $\sigma^{\varepsilon_n} \rightharpoonup \sigma$ in $H^1((0,T);S)$, $\gamma^{\varepsilon_n} \rightharpoonup \gamma$ in $H^1((0,T);S)$ and $\alpha^{\varepsilon_n} \rightharpoonup \alpha$ in $H^1((0,T);L^2(\Omega))$. It is easy to see that $\mathcal{F}(\sigma(t), \gamma(t), \alpha(t)) \leqslant 0$ and that (15.17) is satisfied. Indeed for any triplet of elements τ, δ, β satisfying the definition, we have

$$0 \leqslant \int_0^t [\dot{\sigma}^{\varepsilon_n}, \tau - \sigma^{\varepsilon_n}]\,\mathrm{d}t + \int_0^t (\dot{\gamma}^{\varepsilon_n}, \delta - \gamma^{\varepsilon_n})\,\mathrm{d}t + \int_0^t (\dot{\alpha}^{\varepsilon_n}, \beta - \alpha^{\varepsilon_n})\,\mathrm{d}t \tag{15.23}$$

and the assertion follows for $n \to \infty$.

If we put σ, γ, α to (15.23), we get $\sigma^{\varepsilon_n} \to \sigma$ in $C((0,T);S)$, $\gamma^{\varepsilon_n} \to \gamma$ in $C((0,T);S)$ and $\alpha^{\varepsilon_n} \to \alpha$ in $C((0,T);L^2(\Omega))$. We have from the construction

$$\dot{p}^{\varepsilon_n} = \lambda_n \frac{\partial F}{\partial \sigma}, \qquad \dot{\gamma}^{\varepsilon_n} = -\lambda_n \frac{\partial F}{\partial \gamma}, \qquad \dot{\alpha}^{\varepsilon_n} = -\lambda_n \frac{\partial F}{\partial \alpha},$$

so

$$\lambda_n = \frac{\dot{\gamma}^{\varepsilon_n} \dfrac{\partial F}{\partial \gamma} + \dot{\alpha}^{\varepsilon_n} \dfrac{\partial F}{\partial \alpha}}{\dfrac{\partial F}{\partial \gamma}\dfrac{\partial F}{\partial \gamma} + \dfrac{\partial F}{\partial \alpha}\dfrac{\partial F}{\partial \alpha}}.$$

Hence $\lambda_n \rightharpoonup \lambda$ and (15.19) is valid.

In a formal way "it is clear" that $\lambda = 0$ for $F(\sigma, \gamma, \alpha) < 0$. If $F(\sigma(t), \gamma(t), \alpha(t)) = 0$ then formally $\partial F(\dot{\sigma})/\partial\sigma + \partial F(\dot{\gamma})/\partial\gamma + \partial F(\dot{\alpha})/\partial\alpha = 0$ and we get the last relation. □

REMARK 15.1. If $F(\sigma, \gamma, \alpha) = f(\sigma) - \alpha$ and $\alpha(0) = \alpha_0 > 0$ then according to Theorem 15.2, $\dot{\alpha} = \lambda \geqslant 0$, $\dot{\alpha} = 0$ for $f(\sigma) < \alpha$, $\dot{\alpha} = \partial f(\dot{\sigma})/\partial\sigma$ for $f(\sigma) = \alpha$. So first, because $f(\sigma) \leqslant \alpha$, and α is not decreasing,

$$\alpha(t) \geqslant \max(\alpha_0, \max_{0 \leqslant \tau \leqslant t} f(\tau)). \tag{15.24}$$

On the other hand, $\alpha(t)$ increases only if $f(\sigma) = \alpha$; in this case (up to a set of measure zero in $(0, T)$) $\dot{\alpha} = \mathrm{d}(f(\sigma))/\mathrm{d}t$. So

$$\alpha(t) \leqslant \max(\alpha_0, \max_{0 \leqslant \tau \leqslant t} f(\tau));$$

hence

$$\alpha(t) = \max(\alpha_0, \max_{0 \leqslant \tau \leqslant t} f(\tau)).$$

The internal parameter α measures the maximum level of the yield surface and expresses a feature of the stress history in the point $x \in \Omega$.

16. Solution of the Prandtl–Reuss model by the equilibrium finite-element method

We shall show how to solve variational inequalities of the type (14.15). However, we restrict ourselves to the case where the body occupies a domain $\Omega \subset \mathbb{R}^n$, $n = 2, 3$, with a polyhedral (polygonal) boundary $\partial\Omega$ and to the displacement boundary value problem ($\partial\Omega = \Gamma_u$). We follow JOHNSON [1976b], who also proved the existence and uniqueness of the solution of this problem.

Let $\mathbb{R}_\sigma$ stand for the space of symmetric matrices of the type $(n \times n)$. Let us assume that we are given the yield function $f : \mathbb{R}_\sigma \to \mathbb{R}$, which is convex and continuous, and a constant $\alpha_0 > 0$. Let $f(0) = 0$. We introduce the set of admissible plastical stress fields

$$\mathbb{P} = \{\tau \in S \mid f(\tau(x)) \leqslant \alpha_0 \text{ a.e. in } \Omega\}.$$

Let the body forces be given in the form

$$F(t, x) = \gamma(t) F^0(x),$$

where $F^0 \in [C(\overline{\Omega})]^n$ and $\gamma \in C^2(I)$ is a nonnegative function on the interval $I = [0, T]$, $\gamma(0) = 0$.

We define the set of admissible static stress fields

$$E(t) = \{\tau \in S| \int_\Omega \tau_{ij}\varepsilon_{ij}(v)\,dx = \int_\Omega F_i(t)v_i\,dx \ \forall v \in [H_0^1(\Omega)]^n\}$$

and put

$$K(t) = E(t) \cap \mathbb{P}.$$

Analogously to the definition in Section 14, we say that $\sigma \in H_0^1(I, S)$ is a weak solution of the problem under consideration, if for almost all $t \in I$ we have

$$\sigma(t) \in K(t), \qquad [\dot{\sigma}(t), \tau - \sigma(t)] \geqslant 0 \quad \forall \tau \in K(t). \tag{16.1}$$

Let us recall that this definition differs from that in Section 14 in that (14.15) results only by integrating (16.1) with respect to time. Inequality (16.1) immediately corresponds to (14.12). Zero displacement is given on the boundary $\partial\Omega$.

Johnson proved the unique solvability of problem (16.1) (see JOHNSON [1976a]), under the following assumption:

$$\begin{aligned}&\text{there are } \overline{\chi} \in K(\bar{t}), \text{ where } \bar{t} = \arg\max_{t\in I} \gamma(t),\\ &\text{and positive constants } C \text{ and } d \text{ such that } |\overline{\chi}(x)| \leqslant C\\ &\text{for almost all } x \in \Omega \text{ and } \pm(1+d)\overline{\chi} \in \mathbb{P}.\end{aligned} \tag{16.2}$$

With the aim of defining approximate solutions of problem (16.1), we introduce finite-dimensional internal approximations of the set $E(t)$,

$$E_h(t) \subset E(t) \quad \forall t \in I, \ \ 0 < h < h_0.$$

The sets $E_h(t)$ can be constructed as sums of a particular solution of the equations of equilibrium and the subsets E_h^0 of selfequilibrated finite elements —see Remark 16.1 below. The set $\mathbb{P}$ can be approximated as well, for instance, by the sets

$$\mathbb{P}_h = \{\tau \in S \mid f_{i,h} \leqslant \alpha_0, \ i = 1, \ldots, M_h\},$$

where $f_{i,h}$ are certain linear functions such that $\mathbb{P}_h \subset \mathbb{P}$.

If we now define

$$K_h(t) = E_h(t) \cap \mathbb{P}_h,$$

then $K_h(t) \subset K(t)$ for all $t \in I$.

We introduce the discretization of the time interval I. Let N be a positive integer, $k = T/N$, $t_m = mk$, $m = 0, 1 \ldots, N$, $I_m = [t_{m-1}, t_m]$, $\tau^m = \tau(t_m)$, $\partial\tau^m = (\tau^m - \tau^{m-1})/k$.

Instead of the variational inequality (16.1) we shall solve the discrete problem: find $\sigma_{hk}^m \in K_h(t_m)$, $m = 1, \ldots, N$, such that

$$[\partial \sigma_{hk}^m, \tau - \sigma_{hk}^m] \geqslant 0 \quad \forall \tau \in K_h(t_m), \; m = 1, \ldots, N, \; \sigma_{hk}^0 = 0. \tag{16.3}$$

Let us assume that the following analog of assumption (16.2) holds:

$$\begin{aligned} &\text{there exist } \chi_h \in K_h(\bar{t}) \text{ and positive constants } C, \; d \\ &\text{independent of } h \text{ and such that } |\chi_h(x)| \leqslant C \text{ for almost all } x \in \Omega \\ &\text{and } \pm (1+d)\chi_h \in \mathbb{P}_h. \end{aligned} \tag{16.4}$$

Problem (16.3) is uniquely solvable. This follows from the fact that for all m, σ_{hk}^m is the minimizer of the strictly convex quadratic functional

$$\tfrac{1}{2}[\tau, \tau] - [\sigma_{hk}^{m-1}, \tau]$$

on the closed convex set $K_h(t_m)$.

Thus, at each time level t_m we have to solve the quadratic programming problem.

16.1. A priori error estimates

Let us estimate the error $\sigma^m - \sigma_{hk}^m$, where $\sigma^m = \sigma(t_m)$ is the exact solution of the original problem (16.1). To this end it is useful to assume that the partition of the interval I is ordered (possibly nonuniformly) in such a way that $\gamma(t)$ is a monotonic function in each subinterval I_m. However, since the proof of the a priori error estimate changes only inessentially, we will, in the following, consider an equidistant partition for the sake of simplicity.

First, for N-tuples $q = (q^1, \ldots, q^N)$, $q^m \in S$, let us define

$$\|q\|_{l^2(S)} = \left(\sum_{m=1}^{N} k \|q^m\|_S^2 \right)^{1/2}$$

LEMMA 16.1. *If* (16.4) *is valid, then there exist positive constants* C *and* k_0 *such that*

$$\|\partial \sigma_{hk}\|_{l^2(S)} \leqslant C$$

for $h < h_0$, $k < k_0$.

PROOF. See JOHNSON [1976a, Lemma 2] or HLAVÁČEK [1980]. □

Let us define

$$\varepsilon(h, k) = \inf_{\tau \in \mathscr{K}_{hk}} \|\sigma - \tau\|_{l^2(S)},$$

where

$$\mathscr{K}_{hk} = \{\tau = (\tau^1, \ldots, \tau^N) \mid \tau^m \in K_h(t_m),\ m = 1, \ldots, N\}.$$

The quantity $\varepsilon(h,k)$ is actually determined by the approximation properties of the sets $E_h(t_m)$, $\mathbb{P}_h$ and by the regularity of the solution σ, provided we have any information at all about the latter.

THEOREM 16.1. *Let assumptions* (16.2) *and* (16.4) *be fulfilled and let σ and σ_{hk} be solutions of problems* (16.1) *and* (16.3), *respectively. Then for k sufficiently small and $h < h_0$ we have*

$$\max_{1 \leqslant m \leqslant N} \|\sigma^m - \sigma^m_{hk}\|_S \leqslant C(\varepsilon^{1/2}(h,k) + k^{1/2}). \tag{16.5}$$

PROOF. We extend σ_{hk} to the whole interval I as follows:

$$\sigma_{hk}(t) = \lambda(t)\sigma^{m-1}_{hk} + (1 - \lambda(t))\sigma^m_{hk},$$

where

$$\lambda(t) = \frac{\gamma(t) - \gamma(t_m)}{\gamma(t_{m-1}) - \gamma(t_m)}, \quad t \in I_m \text{ provided } \gamma(t_m) \neq \gamma(t_{m-1}),$$
$$\lambda(t) = (t_m - t)/k, \quad t \in I_m \text{ provided } \gamma(t_m) = \gamma(t_{m-1}).$$

Then we easily check that $\sigma_{hk}(t) \in E(t)$ for all $t \in I$, taking into acount the monotonicity of the function γ in each subinterval I_m.

The last property also implies that $0 \leqslant \lambda(t) \leqslant 1$. Since $\mathbb{P}$ is convex, we also have $\sigma_{hk}(t) \in \mathbb{P}$ for all $t \in I$. Summarizing, we conclude that $\sigma_{hk}(t) \in K(t)$ for all $t \in I$.

Now, putting $\tau := \sigma_{hk}$ in (16.1), we obtain

$$[\dot{\sigma}, \sigma_{hk} - \sigma] \geqslant 0 \quad \text{a.e. in } I.$$

Integrating over I_m we arrive at the inequality

$$[\partial\sigma^m, \sigma^m_{hk} - \sigma^m] \geqslant k^{-1} \int_{I_m} [\dot{\sigma}, \sigma^m_{hk} - \sigma^m + \sigma(t) - \sigma_{hk}(t)]\, \mathrm{d}t. \tag{16.6}$$

Let us consider $\tau_h \in \mathscr{K}_{hk}$ such that

$$\|\sigma - \tau_h\|_{l^2(S)} \leqslant 2\varepsilon(h,k)$$

and substitute $\tau := \tau^m_h$ in (16.3). Thus, we obtain

$$[\partial\sigma^m_{hk}, \tau^m_h - \sigma^m_{hk}] \geqslant 0. \tag{16.7}$$

By virtue of (16.6), (16.7) we come to the following inequality for the error $e = \sigma - \sigma_{hk}$:

$$[\partial e^m, e^m] \leqslant [\partial \sigma_{hk}^m, \tau_h^m - \sigma^m] + |r_m|,$$

where r_m is the right-hand side of (16.6). Multiplying this inequality by k and summing over m yields

$$\max_m \|e_m\|_S^2 \leqslant C\|\partial\sigma_{hk}\|_{l^2(S)}\|\sigma - \tau_h\|_{l^2(S)} + 2k\sum_{m=1}^{N}|r_m|. \tag{16.8}$$

For r_m we may write

$$\begin{aligned}|r_m| &\leqslant Ck^{-1}\int_{I_m}\|\dot\sigma(t)\|_S\left[k\|\partial\sigma_{hk}^m\|_S + k^{1/2}\left(\int_{I_m}\|\dot\sigma\|_S^2\,\mathrm{d}s\right)^{1/2}\right]\mathrm{d}t\\ &\leqslant Ck^{1/2}\|\partial\sigma_{hk}^m\|_S\left[\int_{I_m}\|\dot\sigma\|_S^2\,\mathrm{d}t\right]^{1/2} + C\int_{I_m}\|\dot\sigma(t)\|_S^2\,\mathrm{d}t\\ &\leqslant C\int_{I_m}\|\dot\sigma\|_S^2\,\mathrm{d}t + Ck\|\partial\sigma_{hk}^m\|_S^2.\end{aligned}$$

Let us substitute into (16.8). Thus, we obtain the estimate

$$\begin{aligned}\max_m \|e^m\|_S^2 \leqslant{}& 2C\|\partial\sigma_{hk}\|_{l^2(S)}\varepsilon(h,k)\\ &+ Ck\left(\sum_{m=1}^{N}k\|\partial\sigma_{hk}^m\|_S^2 + \int_I\|\dot\sigma(t)\|_S^2\,\mathrm{d}t\right) \leqslant C_1(\varepsilon(h,k)+k),\end{aligned}$$

by simply taking into account the inequality

$$\int_I \|\dot\sigma(t)\|_S^2\,\mathrm{d}t < \infty$$

and applying Lemma 16.1. This completes the proof of Theorem 16.1. □

REMARK 16.1. If $n = 2$ and the solution σ is sufficiently smooth, we can prove that $\varepsilon(h,k) \leqslant Ch^2$ with C independent of h and k, using the equilibrium finite-element model, proposed by WATWOOD and HARTZ [1968].

This model consists of triangular block elements, which are formed by joining the vertices of a general triangle with its center of gravity. In each subtriangle we define three linear functions—the components of the selfequilibrated stress field. The stress vector is continuous when crossing any boundary between two subtriangles.

We choose the approximations

$$E_h(t) = \chi(t) + N_h(\Omega),$$

where $\chi(t)$ is a particular solution of the equations of equilibrium (14.4) and $N_h(\Omega)$ is the space of triangular block elements, mentioned above.

The estimate $O(h^2)$ for $\varepsilon(h,k)$ then follows on the basis of some results of HLAVÁČEK [1979] or JOHNSON and MERCIER [1978]. The proof is a special case of that of Theorem 17.2 in the following Section 17 and hence is omitted.

17. Solution of isotropic hardening using finite elements

In this section we will solve boundary value problems of the theory of plastic flow with isotropic hardening. These problems are described by relations (15.1)–(15.5) and lead to the time-dependent variational inequality (15.16), where of course the second term is dropped. We will follow the approach of HLAVÁČEK [1980], which is based on some results of JOHNSON [1976b, 1977, 1978].

We will consider a body in $\mathbb{R}^n$, $n = 2, 3$, occupying a domain Ω with a polyhedral (polygonal) boundary $\partial\Omega$. Let us denote $I = [0, T]$ and let $\mathbb{R}_\sigma$ be the space of symmetric $(n \times n)$ matrices.

We assume that a yield function $f : \mathbb{R}_\sigma \to \mathbb{R}$ is given, which is convex and continuous in $\mathbb{R}_\sigma$, continuously differentiable in $\mathbb{R}_\sigma \dot{-} Q$, where Q is either $\{0\}$ (for $n = 2$) or a one-dimensional subspace of $\mathbb{R}_\sigma$ (for $n = 3$), and satisfies

$$F(\lambda\sigma) = |\lambda| f(\sigma) \quad \forall \lambda \in \mathbb{R},\ \sigma \in \mathbb{R}_\sigma, \tag{17.1}$$

$$\left|\frac{\partial f}{\partial \sigma_{ij}}\right| < C \quad \forall i, j,\ \forall \sigma \in \mathbb{R}_\sigma \dot{-} Q \tag{17.2}$$

for some positive constant C.

As an example of a function satisfying the above assumptions let us mention the Von Mises yield function (for $n = 3$)

$$f(\sigma) = (\sigma_{ij}^D \sigma_{ij}^D)^{1/2},$$

where $\sigma_{ij}^D = \sigma_{ij} - \frac{1}{3}\delta_{ij}\sigma_{kk}$ is the stress deviator.

In case of plane stress ($n = 2$) the Von Mises function is reduced to

$$f(\sigma) = (\sigma_{11}^2 + \sigma_{22}^2 - \sigma_{11}\sigma_{22} + 3\sigma_{12}^2)^{1/2} \cdot (\tfrac{2}{3})^{1/2}.$$

Let us assume that

$$\partial\Omega = \bar{\Gamma}_u \cup \bar{\Gamma}_\sigma, \qquad \Gamma_u \cap \Gamma_\sigma = \emptyset,$$

where each of the sets Γ_u and Γ_σ is either empty or an open subset of $\partial\Omega$.

We assume that we are given the (reference) body force vector field $F^0 \in [C(\overline{\Omega})]^n$ and the (reference) surface loads $g^0 \in [L_2(\Gamma_\sigma)]^n$.

If $\Gamma_u = \emptyset$, let the conditions of the total equilibrium be fulfilled, that is, (14.1) and (14.2) for $n = 3$ or

$$\int_\Omega (x_1F_2^0 - x_2F_1^0)\,\mathrm{d}x + \int_{\partial\Omega} (x_1g_2^0 - x_2g_1^0)\,\mathrm{d}s = 0$$

(instead of (14.2)) for $n = 2$.

Let the actual body forces and surface loads be

$$\begin{aligned} F(t,x) &= \gamma(t)F^0(x) \quad \text{in } I \times \Omega, \\ g(t,x) &= \gamma(t)g^0(x) \quad \text{on } I \times \Gamma_\sigma, \end{aligned}$$

with $\gamma : I \to \mathbb{R}$ a nonnegative function from $C^2(I)$, such that

$$\exists t_1 > 0, \;\; \gamma(t) = 0 \quad \text{in } [0, t_1]. \tag{17.3}$$

We introduce the set of admissible statical stress fields:

$$\begin{aligned} E(t) &= E(F(t), g(t)) \\ &= \{\sigma \in S \mid \int_\Omega \sigma_{ij}\varepsilon_{ij}(v)\,\mathrm{d}x = \int_\Omega F_i(t)v_i\,\mathrm{d}x + \int_{\Gamma_\sigma} g_i(t)v_i\,\mathrm{d}s \;\; \forall v \in V\}, \end{aligned}$$

where

$$V = \{v \in [H^1(\Omega)]^n \mid v = 0 \text{ on } \Gamma_u\}.$$

Finally, let us recall the definition (see (15.2))

$$\mathcal{F}(\tau, \alpha) = f(\tau) - \alpha.$$

We denote

$$\begin{aligned} H &= S \times L^2(\Omega), \\ B &= \{(\tau, \alpha) \in \mathbb{R}_\sigma \times \mathbb{R} \mid \mathcal{F}(\tau, \alpha) \leqslant 0\}, \\ \mathbb{P} &= \{(\tau, \alpha) \in H \mid (\tau(x), \alpha(x)) \in B \text{ a.e. in } \Omega\}, \\ K(t) &= (E(t) \times L^2(\Omega)) \cap \mathbb{P}. \end{aligned} \tag{17.4}$$

Let the coefficients of the generalized Hooke's Law satisfy the second and third conditions of (14.7). Further, let a positive constant α_0 be given.

For couples (σ, α) we introduce a new symbol $\hat{\sigma}$, for example,

$$\hat{\sigma} = (\sigma, \alpha), \qquad \hat{\tau} = (\tau, \beta),$$

and we define the inner products with the corresponding norms,

$$\langle\hat\sigma,\hat\tau\rangle = \sigma_{ij}\tau_{ij} + \alpha\beta, \qquad |\hat\tau| = \langle\hat\tau,\hat\tau\rangle^{1/2},$$

$$(\hat\sigma,\hat\tau)_0 = \int_\Omega \langle\hat\sigma,\hat\tau\rangle\,\mathrm{d}x, \qquad \|\hat\sigma\| = (\hat\sigma,\hat\sigma)_0^{1/2},$$

$$\{\hat\sigma,\hat\tau\} = [\sigma,\tau] + (\alpha,\beta)_{0,\Omega}, \qquad |\!|\!|\hat\sigma|\!|\!| = \{\hat\sigma,\hat\sigma\}^{1/2}.$$

Similarly to the definition in Section 15, a weak solution will be a couple $\hat\sigma \equiv (\sigma,\alpha) \in H_0^1(I,S) \times H^1(I,L^2(\Omega))$ such that $\alpha(0) = \alpha_0$, $\hat\sigma(t) \in K(t)$ and

$$\{\mathrm{d}\hat\sigma(t)/\mathrm{d}t, \hat\tau - \hat\sigma(t)\} \geqslant 0 \ \forall\hat\tau \in K(t) \tag{17.5}$$

hold almost everywhere in I.

Let us point out that our definition corresponds to $\sigma(0) = 0$ and to zero displacements on Γ_u. The time-dependent variational inequality (15.17) from the definition is equivalent to (17.5) provided $\gamma \equiv 0$ (that is, without kinematic hardening). The existence and uniqueness of the solution of (17.5) was studied, for instance, by JOHNSON [1978] for $\partial\Omega = \Gamma_u$ and the case $\partial\Omega = \Gamma_\sigma$ is dealt with in NEČAS and HLAVÁČEK [1981].

With the aim of solving problem (17.5) approximately, we first introduce finite-dimensional (internal) approximations of the set $E(t)$:

$$E_h(t) = \chi(t) + E_h^0, \quad 0 < h < h_0,$$

where $\chi \in H_0^1(I,S)$ is a fixed stress field such that $\chi(t) \in E(t)$ a.e. in I and $E_h^0 \subset E(0,0)$ is the finite-dimensional subspace of the selfequilibrated stress fields. Then, evidently $E_h(t) \subset E(t)$.

The existence of the function χ will be established later in Lemma 17.2 and a choice of the subspace E_h^0 will be shown below in 17.1 (see also Remark 16.1).

Let $V_h \subset L^2(\Omega)$ be a finite-dimensional subspace—an approximation of $L^2(\Omega)$. Assume that V_h includes constant functions.

We define

$$K_h(t) = (E_h(t) \times V_h) \cap \mathbb{P},$$

so that $K_h(t) \subset K(t)$.

We again introduce a discretization of the time interval I, that is, $k = T/N$, N a positive integer, $t_m = mk$, $m = 0,1,\ldots,N$, $I_m = [t_{m-1},t_m]$, $\hat\tau^m = \hat\tau(t_m)$, $\partial\tau^m = (\hat\tau^m - \hat\tau^{m-1})/k$.

Instead of (17.5) we will solve the following *approximate problem*: find $\hat\sigma_{hk}^m \in K_h(t_m)$, such that for $m = 1,\ldots,N$ we have

$$\{\partial\hat\sigma_{hk}^m, \hat\tau - \hat\sigma_{hk}^m\} \geqslant 0 \ \forall\hat\tau \in K_h(t_m). \tag{17.6}$$

(We set $\hat\sigma_{hk}^0 \equiv (0,\alpha_0)$.)

Since $\hat{\sigma}^m_{hk}$ minimizes the strictly convex functional

$$\tfrac{1}{2}|||\hat{\sigma}|||^2 - \{\hat{\sigma}^{m-1}_{hk}, \hat{\sigma}\} \tag{17.6'}$$

on the closed convex set $K_h(t_m)$, there exists a unique solution provided $K_h(t_m) \neq \emptyset$. (Lemma 17.2. below yields a sufficient condition for $K_h(t_m) \neq \emptyset$, since $\xi(t) \in K_h(t)$.)

17.1. A priori error estimates

First, we prove an important auxiliary result. Define

$$\|q\|_{l^2(H)} = \left(\sum_{m=1}^{N} k\|q^m\|^2\right)^{1/2}$$

for N-tuples $q = (q^1, q^2, \ldots, q^N)$, $q^m \in H$.

PROPOSITION 17.1. *If $\Gamma_\sigma \neq \emptyset$, let there exist a function*

$$\chi^0 \in [L^\infty(\Omega)]^{n^2} \cap E(F^0, g^0). \tag{17.7}$$

Then there exist positive constants C and k_0 such that

$$\|\partial\hat{\sigma}^m_{hk}\|_{l^2(H)} \leqslant C \tag{17.8}$$

for $k \leqslant k_0$ and $0 < h \leqslant h_0$.

REMARK 17.1. Let F^0 be continuous on $\overline{\Omega}$. Then there is a $\chi^{(1)} \in S \cap [L^\infty(\Omega)]^{n^2}$ such that

$$\operatorname{div} \chi^{(1)} = -F^0 \quad \text{in } \Omega$$

($\chi^{(1)}$ can be obtained by a simple integration). Let the function $g^0 - \chi^{(1)} \cdot \nu$, where ν denotes the unit outward normal, be piecewise linear on Γ_σ with respect to a simplified partition of Γ_σ. Condition (17.7) is fulfilled. Indeed, there is a simplified partition of Ω and $\chi^{(2)} \in E^0_h$ (due to the structure of $N_h(\Omega)$—see the text below Theorem 17.1) such that

$$\chi^{(2)} \cdot \nu = g^0 - \chi^{(1)} \cdot \nu \quad \text{on } \Gamma_\sigma.$$

Putting $\chi^0 = \chi^{(1)} + \chi^{(2)}$ we obtain

$$\chi^0 \in [L^\infty(\Omega)]^{n^2}, \quad \operatorname{div} \chi^0 = -F^0 \quad \text{in } \Omega, \qquad \chi^0 \cdot \nu = g^0 \quad \text{on } \Gamma_\sigma,$$

which implies $\chi^0 \in E(F^0, g^0)$.

To prove Proposition 17.1, we need some auxiliary results:

LEMMA 17.1. *Let* (17.7) *be fulfilled. Then there is*

$$\xi(t) = (\chi(t), \zeta(t)) \in K(t) \quad \forall t \in I, \quad \xi(0) = (0, \alpha_0)$$

and there are positive constants C and δ_1 such that

$$\sup_{t \in I} \|d^j \xi / dt^j\| \leqslant C, \quad j = 0, 1, 2, \tag{17.9}$$

$$\operatorname{dist}(\xi(x,t), \partial B) \geqslant \delta_1 \quad \forall t \in I, \quad \textit{a.e. in } \Omega. \tag{17.10}$$

PROOF. If $\Gamma_\sigma = \emptyset$, we define

$$\chi^0 \in [L^\infty(\Omega)]^{n^2} \cap E(F^0)$$

which can be obtained by integrating the equations of equilibrium $\operatorname{div} \chi^0 = -F^0$ in Ω.

Put

$$\chi(t) = \gamma(t)\chi^0.$$

Then evidently $\chi(t) \in E(t)$ for any $t \in I$ and

$$f(\chi^0(x)) \leqslant C_1 \quad \text{a.e. in } \Omega$$

for a certain constant C_1, due to the boundedness of χ^0. Thus, if we put

$$\zeta(t) = \gamma(t)C_1 + \alpha_0,$$

then

$$\mathcal{F}(\chi, \zeta) = f(\chi) - \zeta = \gamma(t)[f(\chi^0) - C_1] - \alpha_0 \leqslant -\alpha_0 < 0 \tag{17.11}$$

for all $t \in I$ and almost all $x \in \Omega$. Hence, $\xi = (\chi, \zeta) \in \mathbb{P}$ $\forall t \in I$, $\xi(t) \in K(t)$, $\xi(0) = (0, \alpha_0)$. Since χ_0 is bounded and $\gamma \in C^2(I)$, we easily find that (17.9) holds.

In order to prove (17.10), let us observe that (17.11) and (17.2) imply that there is $\delta_1 > 0$ such that

$$\mathcal{F}(\xi(t) + \hat{\varrho}) \leqslant 0 \quad \forall \hat{\varrho} \in \mathbb{R}_\sigma \times \mathbb{R}, \ |\hat{\varrho}| \leqslant \delta_1.$$

Hence, $\xi(t) + \hat{\varrho} \in B$, which yields (17.10). □

Proposition 17.1 can be proved by the penalty method. We present here the main points of the proof. For the details, see HLAVÁČEK [1980] or HLAVÁČEK, HASLINGER, NEČAS and LOVÍŠEK [1988, Section 3.2.21].

We introduce the penalty functional

$$J_\mu(\hat\tau) = \frac{1}{2\mu}\|\hat\tau - \pi\hat\tau\|^2, \quad \mu > 0, \ \tau \in H,$$

where π is the projection to the closed convex set B in the space $\mathbb{R}_\sigma \times \mathbb{R}$. Notice that the Gâteaux derivative of J_μ is

$$J'_\mu(\hat\tau) = \frac{1}{\mu}(\hat\tau - \pi\hat\tau).$$

Define new approximations

$$\hat\sigma^m_{hk\mu} \in E_h(t_m) \times V_h, \quad m = 0, 1, \ldots, N$$

by the conditions (omitting for brevity the indices $hk\mu$ in the following)

$$\begin{aligned} &\{\partial\hat\sigma^m, \hat\tau\} + (J'_\mu(\hat\sigma^m), \hat\tau)_0 = 0 \quad \forall \hat\tau \in E^0_h \times V_h, \\ &\hat\sigma^0 = (0, \alpha_0), \quad m = 1, \ldots, N. \end{aligned} \tag{17.12}$$

Problem (17.12) is uniquely solvable for every m, since $\hat\sigma^m$ minimizes the coercive, strictly convex and continuous functional

$$\tfrac{1}{2}|||\hat\sigma|||^2 + kJ_\mu(\hat\sigma) - \{\hat\sigma^{m-1}, \hat\sigma\}$$

on the set $E_h(t_m) \times V_h$, which is closed and convex in H.

LEMMA 17.2. *Let the conditions* (17.7) *be fulfilled. Then there are positive constants C and k_0 such that for $k \leqslant k_0$, $0 < h \leqslant h_0$ and $\mu > 0$ the following estimates hold*

$$\max_{1\leqslant m\leqslant N} |||\hat\sigma^m_{hk\mu}||| \leqslant C,$$

$$\sum_{m=1}^{N} kJ_\mu(\hat\sigma^m_{hk\mu}) \leqslant C,$$

$$\sum_{m=1}^{N} k\|J'_\mu(\hat\sigma^m_{hk\mu})\|_{L^1} \leqslant C,$$

where $\|z\|_{L^1} = \int_\Omega |z|\,\mathrm{d}x$ *for* $z \in H$,

$$\|\partial\hat\sigma^m_{hk\mu}\|_{l^2(H)} \leqslant C.$$

PROOF. The proof, based on Lemma 17.1, uses the decomposition $\hat\sigma^m = \xi^m + \varrho^m$ and the discrete analogue of the Gronwall lemma. □

PROOF (of Proposition 17.1). Let $\mu \to 0$ for a sequence of positive numbers μ. Lemma 17.3 implies the existence of $\hat{\sigma}_{hk}$, $\hat{s}_{hk}$ and of a subsequence such that $\hat{\sigma}_{hk\mu} \rightharpoonup \hat{\sigma}_{hk}$ and $\partial\hat{\sigma}_{hk\mu} \rightharpoonup \hat{s}_{hk}$ (weakly) in the space $l^2(H)$.

It is not difficult to verify that $\hat{s}_{hk} = \partial\hat{\sigma}_{hk}$. We can show that $\hat{\sigma}_{hk}$ is a solution of problem (17.6), again using Lemma 17.3. Finally, we may write

$$\sum_{m=1}^{N} k\|\partial\hat{\sigma}_{hk}^m\|^2 \leqslant \sum_{m=1}^{N} k \liminf_{\mu\to 0} \|\partial\hat{\sigma}_{hk\mu}^m\|^2$$
$$\leqslant \liminf_{\mu\to 0} \|\partial\hat{\sigma}_{hk\mu}\|^2_{l^2(H)} \leqslant C.$$

THEOREM 17.1. *Denote*

$$\varepsilon(h,k) = \inf_{\hat{\tau}\in\mathcal{K}_{hk}} \|\hat{\sigma} - \hat{\tau}\|_{l^2(H)},$$

where

$$\mathcal{K}_{hk} = \{\hat{\tau} = (\hat{\tau}^1, \ldots, \hat{\tau}^N) \mid \hat{\tau}^m \in K_h(t_m),\ m = 1, \ldots, N\}.$$

Assume that (17.7) *is fulfilled. Then there exist positive constants C and k_0 such that for $k \leqslant k_0$ and $h < h_0$*

$$\max_{1\leqslant m\leqslant N} \|\hat{\sigma}^m - \hat{\sigma}_{hk}^m\| \leqslant C(\varepsilon^{1/2}(h,k) + k^{1/2}).$$

PROOF. The proof is analogous to that of Theorem 17.2. □

Let us now consider $n = 2$, that is, the *plane problem*, and evaluate the quantity $\varepsilon(h,k)$, provided piecewise linear triangular elements are used. However, to this end we must assume a certain regularity of the weak solution $\hat{\sigma}$ of problem (17.5).

Let the reference body forces F^0 be *constant* and the reference surface load g_0 be *piecewise linear* on Γ_σ. We use a regular system of triangulations $\{\mathcal{T}_h\}$, $0 < h \leqslant h_0$, of the domain Ω and the equilibrium finite-element model, proposed by WATWOOD and HARTZ [1968]. The stress approximation is defined by linear polynomials on subtriangles $K_i \subset K$, $i = 1, 2, 3$ (see Remark 17.1), which generate subspaces $N_h(\Omega) \subset S$ of selfequilibrated stress fields. Approximation properties of the subspaces $N_h(\Omega)$ were studied in JOHNSON and MERCIER [1978] and HLAVÁČEK [1979]. We will use the following result (see HLAVÁČEK [1979, Theorem 2.3]):

Let $\tau \in S \cap [C^2(\overline{\Omega})]^4$. There is a linear mapping

$$r_h : E(0,0) \cap [C^2(\overline{\Omega})]^4 \to N_h(\Omega)$$

satisfying, on each triangle $K \in \mathcal{T}_h$, the estimate

$$\max_{i=1,2,3} \|\tau - (r_h\tau)^i\|_{[C(K_i)]^4} \leqslant Ch_K^2 \|\tau\|_{[C^2(K)]^4}, \tag{17.13}$$

where $(r_h\tau)^i$ is the restriction of $r_h\tau$ on K_i, h_K is the maximal side of K and C is independent of τ, h_K and K. Let us define spaces of finite elements

$$E_h^0 = N_h(\Omega) \cap E(0,0) = \{\tau \in N_h(\Omega) \mid \tau \cdot \nu = 0 \text{ on } \Gamma_\sigma\},$$
$$V_h = \{\beta \in L^2(\Omega) \mid \beta|_{K_i} \in P_1(K_i)\ \forall K_i \subset K \in \mathcal{T}_h\}.$$

Under the assumptions imposed on F^0 and g^0 we can find a triangulation $\mathcal{T}_{h_0}$ and an auxiliary function χ^0 that is piecewise linear with respect to $\mathcal{T}_{h_0}$ (see Remark 17.1). Then $\chi(t_m) = \gamma(t_m)\chi^0$ is also piecewise linear. In the following, we will assume that the system of triangulations $\{\mathcal{T}_h\}$ results from refining the original triangulation $\mathcal{T}_{h_0}$, subdivided into subtriangles $K_i^0 \subset K^0 \in \mathcal{T}_{h_0}$.

THEOREM 17.2. *Let the solution $\hat{\sigma} \equiv (\sigma, \alpha)$ be such that for all $t \in I$, $\sigma_0(t) \equiv \sigma(t) - \chi(t) \in [C^2(K^0)]^4$ and $\alpha(t) \in H^2(K_i^0)$, $K_i^0 \subset K^0$ holds in each triangle $K^0 \in \mathcal{T}_{h_0}$ and*

$$\sup_{t\in I} \|\sigma_0(t)\|_{[C^2(K^0)]^4} < \infty,$$
$$\sup_{t\in I} \|\alpha(t)\|_{2,K_i^0} < \infty, \quad i = 1,2,3.$$

Then

$$\varepsilon(h,k) \leqslant Ch^2,$$

where $C = C(\sigma_0, \alpha)$ is independent of h, k.

PROOF. Recall that $\hat{\tau}^m = (\tau^m, \beta^m)$, $\tau^m = \chi^m + \tau_0^m$ and $\sigma^m = \chi^m + \sigma_0^m$ where $\tau_0^m \in E_h^0$ and $\sigma_0^m \subseteq E(0,0)$. Hence, we can write for all $m = 1, \ldots, N$ (omitting the index m):

$$\|\hat{\sigma} - \hat{\tau}\| = \|\sigma_0 - \tau_0\|_S^2 + \|\alpha - \beta\|_{0,\Omega}^2. \tag{17.14}$$

Put $\tau_0 = r_h\sigma_0$. Then the definition of the mapping r_h (see HLAVÁČEK [1979]) implies that $r_h\sigma_0 \cdot \nu = 0$ on Γ_σ, hence $r_h\sigma_0 \in E_h^0$.

Consider an arbitrary subtriangle $K_i \subset K$, $K \in \mathcal{T}_h$ and denote its vertices by a_j. Then, provided $\beta \in P_1(K_i)$ and

$$\beta(a_j) \geqslant f(\chi + r_h\sigma_0)(a_j) \equiv d_j, \quad j = 1,2,3,$$

we have

$$\beta \geqslant f(\chi + r_h\sigma_0) \quad \text{in } K_i.$$

This is a consequence of the linearity of $\chi + r_h\sigma_0$, β and of the convexity of the function f.

Let $\Pi\alpha \in P_1(K_i)$ denote the Lagrangian interpolation of the function α on K_i. Define $r \in P_1(K_i)$, $\beta_h \in P_1(K_i)$ by the relations

$$r(a_j) = [d_j - \alpha(a_j)]^+, \qquad \beta_h = \Pi\alpha + r.$$

Then, evidently $\beta_h(a_j) \geqslant d_j$, $j = 1, 2, 3$. The following estimate is well known (see CIARLET [1978])

$$\|\alpha - \Pi\alpha\|_{0,K_i} \leqslant Ch^2|\alpha|_{2,K_i}. \tag{17.15}$$

By using assumptions (17.1) and (17.2) and the estimate (17.13), we obtain

$$\begin{aligned}|d_j - f(\chi + \sigma_0)(a_j)| &\leqslant \|f(\chi + r_h\sigma_0) - f(\chi + \sigma_0)\|_{C(K_i)} \\ &\leqslant C\|\sigma_0 - r_h\sigma_0\|_{[C(K_i)]^4} \leqslant Ch^4\|\sigma_0\|_{[C^2(K^0)]^4} \equiv \varepsilon_1(h).\end{aligned}$$

This immediately yields

$$\|r\|_{0,K_i} \leqslant Ch\varepsilon_1(h). \tag{17.16}$$

Now (17.15) and (17.16) successively imply

$$\|\alpha - \beta_h\|^2_{0,\Omega} \leqslant Ch^4 \sum_{K^0 \in \mathcal{T}_{h_0}} \left(\sum_{i=1}^{3} |\alpha|^2_{2,K_i^0} + \|\sigma_0\|^2_{[C^2(K^0)]^4} \right). \tag{17.17}$$

It follows from (17.13) that

$$\|\sigma_0 - r_h\sigma_0\|^2_S \leqslant Ch^4 \sum_{K^0 \in \mathcal{T}_{h_0}} \|\sigma_0\|^2_{[C^2(K^0)]^4}). \tag{17.18}$$

By substituting (17.17) and (17.18) into (17.14), we derive

$$\begin{aligned}\|\hat{\sigma}^m - \hat{\tau}^m\|^2 &\leqslant Ch^4 \sum_{K^0 \in \mathcal{T}_{h_0}} \left[\|\sigma_0^m\|^2_{[C^2(K^0)]^4} + \sum_{i=1}^{3} |\alpha^m|^2_{2,K_i^0} \right] \\ &\leqslant Ch^4 \sum_{K^0 \in \mathcal{T}_{h_0}} \left[\sup_{t \in I} \|\sigma_0(t)\|^2_{[C^2(K^0)]^4} + \sum_{i=1}^{3} \sup_{t \in I} |\alpha(t)|^2_{2,K_i^0} \right] \\ &= h^4 C_1(\sigma_0, \alpha).\end{aligned}$$

Now it is easy to establish the estimate

$$\|\hat{\sigma} - \hat{\tau}\|^2_{l^2(H)} = \sum_{m=1}^{N} k\|\hat{\sigma}^m - \hat{\tau}^m\|^2 \leqslant C_1(\sigma_0, \alpha)Th^4,$$

which completes the proof of the theorem. □

REMARK 17.2. For a three-dimensional problem, we can use a four-faced element consisting of four tetrahedra, which is analogous to the triangular block-element. Then, estimates of the type (17.13) and (17.15) are valid (see KŘÍŽEK [1982] and CIARLET [1978, 1991]), and proceeding in the same way as in the proof of Theorem 17.2 we arrive at an analogous assertion.

An immediate consequence of Theorems 17.1 and 17.2 is the following

COROLLARY 17.1. *Let* (17.7) *and the assumptions of Theorem* 17.2 *be fulfilled. Then there are positive constants* C *and* k_0 *such that for all* $k \leqslant k_0$ *and* $h \leqslant h_0$

$$\max_{1\leqslant m\leqslant N} \|\hat{\sigma}^m - \hat{\sigma}^m_{hk}\| \leqslant C(h + k^{1/2}).$$

REMARK 17.3 (*Algorithm for the solution of the approximate problem* (17.6)) Defining E_h^0 and V_h as above, we obtain for $\tau_0^m \in E_h^0$:

$$(\chi^m + \tau_0^m, \beta^m) \in \mathbb{P} \Leftrightarrow \beta^m(a_j) \geqslant f(\chi^m + \tau_0^m)(a_j) \tag{17.19}$$

at all vertices $a_j \in K_i \subset K$ of all triangles $K \in \mathscr{T}_h$.

Hence, we have nonlinear constraints for the parameters β^m and τ_0^m. (In the case of the Von Mises yield function these constraints are quadratic.) At each time level it is thus necessary to minimize the quadratic functional (17.6′) with nonlinear constraints (17.19) and with linear constraints—equations—which guarantee the continuity of the stress vector on the boundaries between triangles. For this purpose, a suitable algorithm of nonlinear programming has to be chosen. Some numerical experiments have been carried out by KESTŘÁNEK [1986].

17.2. Convergence in the case of a nonregular solution

A natural question arises as to whether the approximations $\hat{\sigma}_{hk}$ converge, provided the solution $\hat{\sigma}$ fails to be sufficiently regular. We can prove a positive answer to this question for a broad class of domains and boundary conditions.

Let us again assume $n = 2$, the plane problem. Let us keep the assumptions on F^0 and g^0 from Section 17.1, so that we can find piecewise linear functions χ^0 and $\chi(t_m)$ with respect to the reference triangulation $\mathscr{T}_{h_0}$.

THEOREM 17.3. *Let us assume that if* $\Gamma_\sigma \neq \emptyset$, *there is a point* $A \in \mathbb{R}^2$ *such that, provided the origin of the coordinate system is at* A, *we obtain for* $\lambda = 1 + \varepsilon$ *and for sufficiently small positive* ε *that either*

$$[\lambda\overline{\Gamma}_\sigma \subset \mathbb{R}^2 - \overline{\Omega}]$$

or

$$[\lambda\overline{\Gamma}_\sigma \subset \Omega].$$

Here, $\lambda\overline{\Gamma}_\sigma$ *stands for the image of* $\overline{\Gamma}_\sigma$ *under the dilatation mapping* $x \mapsto \lambda x$.

Then

$$\lim_{h\to 0} \varepsilon(h,k) = 0 \tag{17.20}$$

holds for every fixed $k = T/N$.

PROOF. Consider a time moment t_m and omit the index m for simplicity. We use the following result on density: under the assumptions of Theorem 17.3, the set

$$E(0,0) \cap [C^\infty(\overline{\Omega})]^4$$

is dense in $E(0,0)$ (with respect to the norm in the space S). (See HLAVÁČEK [1979]). Consequently, there exists

$$\sigma_{0\varepsilon} \in E(0,0) \cap [C^\infty(\overline{\Omega})]^4, \quad \|\sigma_0 - \sigma_{0\varepsilon}\| \leqslant \varepsilon. \tag{17.21}$$

Regularizing α we obtain

$$\alpha_\varepsilon \in C^\infty(\overline{\Omega}), \quad \|\alpha - \alpha_\varepsilon\|_{0,\Omega} \leqslant \varepsilon. \tag{17.22}$$

Let us define functions

$$\mathscr{F} = f(\chi + \sigma_0) - \alpha, \qquad \mathscr{F}_\varepsilon = f(\chi + \sigma_{0\varepsilon}) - \alpha_\varepsilon.$$

Evidently $\mathscr{F} \leqslant 0$ a.e. in Ω, but $\mathscr{F}_\varepsilon$ generally does not satisfy such an inequality. Recall that χ is linear in each subtriangle $K_i^0 \subset K^0$. Choose $\hat{\tau} = (\chi + r_h\sigma_{0\varepsilon}, \beta_h)$, where $\beta_h = \Pi\alpha_\varepsilon + \varrho$. Here $\Pi_h\alpha_\varepsilon$ and ϱ are defined locally in each $K_i \subset K$ in the following way: $\Pi_h\alpha_\varepsilon$ is the linear Lagrange interpolation $\Pi_{K_i}\alpha_\varepsilon$ on K_i,

$$\varrho \in P_1(K_i), \qquad \varrho(a_j) = [d_j - \alpha_\varepsilon(a_j)]^+, \quad j = 1,2,3,$$

where a_j are the vertices of K_i and $d_j = f(\chi + r_h\sigma_{0\varepsilon})(a_j)$. It is easily seen that $\chi + r_h\sigma_{0\varepsilon} \in E(t_m)$ and

$$f(\chi + r_h\sigma_{0\varepsilon}) - \beta_h \leqslant 0 \quad \text{a.e. on } \Omega.$$

Let us derive an estimate for ϱ in $L^2(\Omega)$. For $j = 1,2,3$ we may write

$$0 \leqslant \varrho(a_j) \leqslant |d_j - f(\chi + \sigma_{0\varepsilon})(a_j)| + \mathscr{F}_\varepsilon^+(a_j),$$

since

$$-\alpha_\varepsilon(a_j) \leqslant \mathscr{F}_\varepsilon^+(a_j) - f(\chi + \sigma_{0\varepsilon})(a_j).$$

Further,

$$|d_j - f(\chi + \sigma_{0\varepsilon})(a_j)| \leqslant Ch^2\|\sigma_{0\varepsilon}\|_{[C^2(\overline{\Omega})]^4} \equiv \varepsilon_1(h, \sigma_{0\varepsilon}),$$

hence,

$$\begin{aligned}&\varrho \leqslant \varepsilon_1(h,\sigma_{0\varepsilon}) + \Pi_{K_i}\mathscr{F}_\varepsilon^+ \quad \text{on } K_i,\\ &\|\varrho\|_{0,K_i}^2 \leqslant 2\varepsilon_1^2 \operatorname{meas} K_i + 2\|\Pi_{K_i}\mathscr{F}_\varepsilon^+\|_{0,K_i}^2.\end{aligned} \tag{17.23}$$

It is not difficult to show that

$$\|\mathscr{F}_\varepsilon^+\|_{0,K^0} \leqslant C\varepsilon \tag{17.24}$$

holds for each triangle $K^0 \in \mathscr{T}_{h_0}$.

Indeed, by virtue of (17.1), (17.2) and (17.21) we obtain

$$\|f(\chi+\sigma_0) - f(\chi+\sigma_{0\varepsilon})\|_{0,K^0}^2 \leqslant C\|\sigma_0 - \sigma_{0\varepsilon}\|_S^2 \leqslant C\varepsilon^2.$$

Hence, we have

$$\|\mathscr{F} - \mathscr{F}_\varepsilon\|_{0,K^0} \leqslant \|f(\chi+\sigma_0) - f(\chi+\sigma_{0\varepsilon})\|_{0,K^0} + \|\alpha - \alpha_\varepsilon\|_{0,K^0} \leqslant C\varepsilon.$$

Finally, denoting $\Omega_1 = \operatorname{supp} \mathscr{F}_\varepsilon^+ \cap K^0$, we have

$$\|\mathscr{F}_\varepsilon^+\|_{0,K^0}^2 \leqslant \int_{\Omega_1} (\mathscr{F}_\varepsilon^+ - \mathscr{F})^2\,\mathrm{d}x = \|\mathscr{F}_\varepsilon - \mathscr{F}\|_{0,\Omega_1}^2 \leqslant C\varepsilon^2,$$

which yields (17.24).

Let us define $\Pi_h\mathscr{F}_\varepsilon^+$ locally on each $K_i \subset K \subset K^0$ as the Lagrange interpolation $\Pi_{K_i}\mathscr{F}_\varepsilon^+$. Then we can prove that

$$\|\mathscr{F}_\varepsilon^+ - \Pi_h\mathscr{F}_\varepsilon^+\|_{0,K^0} \leqslant \varepsilon_2(h), \qquad \lim_{h\to 0} \varepsilon_2(h) = 0. \tag{17.25}$$

Indeed, since $\mathscr{F}_\varepsilon \in C(K^0)$, we have $\mathscr{F}_\varepsilon^+ \in C(K^0)$ as well. For every $\eta > 0$ there is a polynomial p such that

$$\|\mathscr{F}_\varepsilon^+ - p\|_{C(K^0)} \leqslant \eta.$$

Further,

$$\begin{aligned}&\|\Pi_h p - \Pi_h\mathscr{F}_\varepsilon^+\|_{C(K_i)} \leqslant \|p - \mathscr{F}_\varepsilon^+\|_{C(K_i)} \leqslant \eta,\\ &\|p - \Pi_h p\|_{C(K_i)} \leqslant C_0 h^2 \|p\|_{C^2(K^0)}\end{aligned}$$

holds for every $K_i \subset K \subset K^0$. Hence, we can write

$$\begin{aligned}&\|\mathscr{F}_\varepsilon^+ - \Pi_h\mathscr{F}_\varepsilon^+\|_{C(K^0)}\\ &\quad\leqslant \|\mathscr{F}_\varepsilon^+ - p\|_{C(K^0)} + \|p - \Pi_h p\|_{C(K^0)} + \|\Pi_h p - \Pi_h\mathscr{F}_\varepsilon^+\|_{C(K^0)}\\ &\quad\leqslant 2\eta + C_0 h^2 \|p\|_{C^2(K^0)} \equiv \delta(h, K^0)\end{aligned}$$

and this implies (17.25) if we put

$$\varepsilon_2(h) = \max_{K^0 \in \mathcal{T}_{h_0}} [\delta(h, K^0)(\operatorname{meas} K^0)^{1/2}].$$

Now (17.24), (17.25) and the inequality (17.23) yield

$$\begin{aligned} \|\varrho\|_{0,K^0}^2 &\leqslant 2\varepsilon_1^2 \operatorname{meas} K^0 + 4(\|\mathcal{F}_\varepsilon^+\|_{0,K^0}^2 + \|\Pi_h \mathcal{F}_\varepsilon^+ - \mathcal{F}_\varepsilon^+\|_{0,K^0}^2) \\ &\leqslant 2\varepsilon_1^2 \operatorname{meas} K^0 + C(\varepsilon^2 + \varepsilon_2^2). \end{aligned}$$

Finally, we obtain

$$\begin{aligned} \|\hat{\sigma} - \hat{\tau}\|^2 &= \|\sigma_0 - r_h \sigma_{0\varepsilon}\|_S^2 + \|\alpha - \beta_h\|_{0,\Omega}^2 \\ &\leqslant 2(\|\sigma_0 - \sigma_{0\varepsilon}\|_S^2 + \|\sigma_{0\varepsilon} - r_h \sigma_{0\varepsilon}\|_S^2) \\ &\quad + 3(\|\alpha - \alpha_\varepsilon\|_{0,\Omega}^2 + \|\alpha_\varepsilon - \Pi_h \alpha_\varepsilon\|_{0,\Omega}^2 + \|\varrho\|_{0,\Omega}^2 \\ &\leqslant 2(\varepsilon^2 + Ch^4 \|\sigma_{0\varepsilon}\|_{[C^2(\overline{\Omega})]^4}^2) \\ &\quad + C(\varepsilon^2 + h^4 |\alpha_\varepsilon|_{2,\Omega}^2 + + h^4 \|\sigma_{0\varepsilon}\|_{[C^2(\overline{\Omega})]^4}^2 + \varepsilon^2 + \varepsilon_2^2(h)). \end{aligned}$$

Hence,

$$\lim_{h \to 0} \|\hat{\sigma}^m - \hat{\tau}^m\| = 0, \quad m = 1, \ldots, N.$$

Thus, for a fixed $k = T/N$ we obtain assertion (17.20). □

REMARK 17.4. The assumptions of Theorem 17.3 on Γ_σ can be weakened, if we use the Airy stress function to prove the density of $E(0,0) \cap [C^\infty(\overline{\Omega})]^4$ in $E(0,0)$.

REMARK 17.5. In three-dimensional problems, the analog of Theorem 17.3 can be established, if we apply four-faced block elements (see KŘÍŽEK [1982]).

18. Bibliography and comments to Chapter V

Another mathematical model of the elasto-plastic behaviour of solids with a hardening effect has been established using a variational inequality concept in the strain space instead of stress space. Such a model is preferable, for instance, in a neighbourhood of the maximum of the loading curve—see CASEY and NAGHDI [1983]. An algorithm is given for implementing a finite element approximation of this model in HLAVÁČEK, ROSENBERG, BEAGLES and WHITEMAN [1992].

Models and algorithms for the flow theory of plasticity have also been proposed by KORNEEV and LANGER [1984] and MIYOSHI [1985].

Some finite element analysis of the Hencky's model, using the Haar–Kármán principle, has been given in HLAVÁČEK [1981].

CHAPTER VI

Unilateral Problems for Elastic Plates

19. Introduction. Statement of the basic problems

The theory of bending of elastic homogeneous thin plates was developed in the nineteenth century on the basis of several geometric and static hypotheses. We have no intention of presenting a full derivation of the theory, instead referring the reader to the literature (see, e.g., the books of DUVAUT and LIONS [1976, Chapter IV], CIARLET [1990] or that of NEČAS and HLAVÁČEK [1981, (10.4.3)].

The classical Kirchhoff model consists of the following linear equation of the fourth order in a bounded domain $\Omega \subset \mathbb{R}^2$,

$$\Delta^2 u = p/D. \tag{19.1}$$

Here u denotes the *deflection* of the middle plane, p is the transverse *loading*,

$$D = \frac{Et^3}{12(1-\sigma^2)}$$

is the *bending rigidity*, E the Young's modulus, t the (constant) thickness of the plate and σ the Poisson's ratio, $\sigma \in (0, \frac{1}{2})$. Equation (19.1) with constant E, t and σ represents the equilibrium of a thin plate made of elastic, homogeneous and isotropic material.

Henceforth, we shall use the notation

$$u_{ij} = \partial^2 u/\partial x_i \partial x_j,$$

where the Cartesian coordinate system (x_1, x_2) in Ω is used.

Assume that the domain Ω has a Lipschitz boundary $\partial\Omega$ (see, e.g., the book of NEČAS and HLAVÁČEK [1981] for the definition). We denote the unit outward normal to $\partial\Omega$ by n and the unit tangential vector by $s = (-n_2, n_1)$.

On the boundary $\partial\Omega$ we define the *bending moment* per unit length of the arc of $\partial\Omega$ by the operator

$$Mu = -D[\sigma\Delta u + (1-\sigma)\partial^2 u/\partial n^2]$$

(where $\partial^2 u/\partial n^2 = u_{ij}n_i n_j$ and repeated subscripts imply summation within the range 1, 2) and the *effective shearing force* per unit length

$$Tu = D\{-\partial\Delta u/\partial n + (1-\sigma)\frac{\partial}{\partial s}[(u_{11} - u_{22})n_1 n_2 + (n_2^2 - n_1^2)u_{12}]\}.$$

We shall briefly outline the most important *classical* boundary conditions:

$$\begin{array}{llll} \textit{clamped edge} & u = 0, & \partial u/\partial n = 0, & (19.2) \\ \textit{simply supported edge} & u = 0, & Mu = 0, & (19.3) \\ \textit{free edge} & Mu = 0, & Tu = 0. & (19.4) \end{array}$$

Unilateral boundary conditions can be formulated in terms of the same quantities. Here we present the most frequent conditions.

Unilateral displacement:

$$u \geqslant 0,$$

$$\left\{\begin{matrix} u > 0 \Rightarrow Tu = 0 \\ u = 0 \Rightarrow Tu \geqslant 0 \end{matrix}\right\} \quad \text{or} \quad \{u \geqslant 0,\ Tu \geqslant 0,\ uTu = 0\} \tag{19.5}$$

while conditions for Mu or $\partial u/\partial n$ are of the classical type.

Unilateral rotation:

$$\partial u/\partial n \geqslant 0,$$

$$\left\{\begin{matrix} \partial u/\partial n > 0 \Rightarrow Mu = 0 \\ \partial u/\partial n = 0 \Rightarrow Mu \leqslant 0 \end{matrix}\right\} \quad \text{or} \quad \{\frac{\partial u}{\partial n} \geqslant 0,\ Mu \leqslant 0,\ \frac{\partial u}{\partial n}Mu = 0\} \tag{19.6}$$

while conditions for Tu or u are of the classical type.

Displacement with friction:

$$|Tu| < \mathscr{F} \Rightarrow u = 0,$$
$$|Tu| = \mathscr{F} \Rightarrow \text{there exists a real number } \lambda \geqslant 0 \text{ such that } u = -\lambda Tu \tag{19.7}$$

(where $\mathscr{F}$ is a given positive function), while conditions for Mu or $\partial u/\partial n$ are of the classical type.

Rotation with friction:

$$|Mu| < \mathscr{M} \Rightarrow \partial u/\partial n = 0,$$
$$|Mu| = \mathscr{M} \Rightarrow \text{there exists a real number } \lambda \geqslant 0 \text{ such that } \partial u/\partial n = \lambda Mu \tag{19.8}$$

(where $\mathscr{M}$ is a given positive function) while conditions for Tu or u are of the classical type.

There can be conditions for one of the preceding types (19.2)–(19.8) on one part of $\partial\Omega$ and conditions for other types on other parts of $\partial\Omega$.

In the conditions for types (19.5) and (19.6) one can consider also a "nonhomogeneous obstacle" for u and $\partial u/\partial n$, respectively, i.e.

$$u \geqslant g,$$

$$\begin{Bmatrix} u > g \Rightarrow Tu = 0 \\ u = g \Rightarrow Tu \geqslant 0 \end{Bmatrix} \quad \text{or} \quad \{u \geqslant g,\ Tu \geqslant 0,\ (u-g)Tu = 0\} \tag{19.5'}$$

(where g is a given function);

$$\partial u/\partial n \geqslant g,$$

$$\begin{Bmatrix} \partial u/\partial n > g \Rightarrow Mu = 0 \\ \partial u/\partial n = g \Rightarrow Mu \leqslant 0 \end{Bmatrix} \quad \text{or} \quad \{\frac{\partial u}{\partial n} \geqslant g,\ Mu \leqslant 0,\ \left(\frac{\partial u}{\partial n} - g\right) Mu = 0\}. \tag{19.6'}$$

Another important unilateral problem is that for unilateral displacement of the points of the middle plane (*inner obstacle problem*)

$$u \geqslant \varphi \quad \text{in } \Omega,$$

$$\begin{Bmatrix} u > \varphi \Rightarrow \Delta^2 u = p/D \\ u = \varphi \Rightarrow \Delta^2 u - p/D \geqslant 0 \end{Bmatrix}$$

$$\text{or} \quad \{u \geqslant \varphi,\ \Delta^2 u \geqslant \frac{p}{D},\ (u-\varphi)\left(\Delta^2 u - \frac{p}{D}\right) = 0\} \tag{19.9}$$

(where φ is a given function), while the conditions on the boundary $\partial\Omega$ are of the classical type.

20. Unilateral displacement problem

Let us consider the problem with boundary conditions of type (19.5).

20.1. Primal approach

Let the domain Ω be a union of parallelograms whose sides are parallel with two fixed directions. Moreover, let

$$\partial\Omega = \Gamma_u \cup \Gamma_K, \quad \operatorname{meas}\Gamma_u > 0, \quad \operatorname{meas}\Gamma_K > 0, \quad \Gamma_u \cap \Gamma_K = \emptyset.$$

We shall consider the boundary conditions of a clamped edge (19.2) on Γ_u and of unilateral displacement (19.5) on Γ_K.

We introduce the sets

$$\mathscr{V} = \{v \in C^\infty(\overline{\Omega}) \mid v = 0, \ \frac{\partial v}{\partial n} = 0 \text{ on } \Gamma_u\},$$
$$\mathscr{K} = \{v \in \mathscr{V} \mid v \geqslant 0 \text{ on } \Gamma_K\}$$

and let V be the closure of $\mathscr{V}$ and K the closure of $\mathscr{K}$ in the space $H^2(\Omega)$, respectively.

Let P_j, $j = 1, 2, \ldots, j_0$ be given constants, $X_j \in \overline{\Omega} \dot{-} \Gamma_u$ given points, $f_0 \in L^1(\Omega)$, γ a prescribed rectifiable curve, $\gamma \subset \Omega$, $f_1 \in L^1(\gamma)$.

We define the virtual work of the external loads by the following formula:

$$\langle f, v\rangle = \sum_{j=1}^{j_0} P_j v(X_j) + \int_\Omega f_0 v \, \mathrm{d}x + \int_\gamma f_1 v \, \mathrm{d}s. \tag{20.1}$$

It represents a linear continuous functional $f \in V'$. (Note that f corresponds to the *reduced* actual loading p/D in what follows.)

Let us define the following bilinear form on $H^2(\Omega) \times H^2(\Omega)$:

$$a(u, v) = \int_\Omega [u_{11}v_{11} + u_{22}v_{22} + \sigma(u_{11}v_{22} + u_{22}v_{11}) + 2(1-\sigma)u_{12}v_{12}] \, \mathrm{d}x. \tag{20.2}$$

Then the equilibrium of the plate is represented by the following variational inequality

$$\begin{aligned} &u \in K, \\ &a(u, v-u) \geqslant \langle f, v-u\rangle \quad \forall v \in K. \end{aligned} \tag{20.3}$$

The problem (20.3) has a unique solution. Indeed,

$$a(v, v) \geqslant (1-\sigma)|v|^2_{2,\Omega} \tag{20.4}$$

holds for all $v \in H^2(\Omega)$ and due to the boundary conditions on Γ_u, a positive constant c exists such that

$$|v|^2_{2,\Omega} \geqslant c\|v\|^2_{2,\Omega} \quad \forall v \in V \tag{20.5}$$

(see, e.g., NEČAS and HLAVÁČEK [1981, Lemma 11.3.2]). Consequently, the quadratic functional (total potential energy)

$$\mathscr{J}(v) = \tfrac{1}{2}a(v, v) - \langle f, v\rangle \tag{20.6}$$

is coercive and strictly convex on V. There exists a unique minimizer u of $\mathscr{J}$ on the convex closed set K, i.e., a unique solution of (20.3).

Approximate solution. Let the domain Ω be divided by a uniform partition $\mathcal{T}_{h_0}$ into a finite number of (open) parallelograms by means of two systems of equidistant straight lines parallel with the sides of $\partial\Omega$. Assume that $\mathcal{T}_{h_0}$ is consistent with the partition of the boundary conditions, i.e., the number of points from $\overline{\Gamma}_u \cap \overline{\Gamma}_K$ is finite and every point of this kind coincides with a node of $\mathcal{T}_{h_0}$.

In what follows, we shall consider only families $\{\mathcal{T}_h\}$, $h \to 0+$, of uniform partitions, which refine the reference partition $\mathcal{T}_{h_0}$. Thus any partition $\mathcal{T}_h$ consists of parallelograms G_i, $i = 1, 2, \ldots, I(h)$, where

$$h = \operatorname{diam} G_i \quad \forall i.$$

We may write

$$\Gamma_K = \bigcup_{j=1}^{n(h)} \overline{A_{j-1}^h A_j^h},$$

where A_j^h are nodes of $\mathcal{T}_h$.

We shall use the spaces $Q_3(G_i)$ of bicubic polynomials defined on a parallelogram G_i (see, e.g., CIARLET [1978, 1991, Theorem 9.4]).

If G_i is not rectangular, the space $Q_3(G_i)$ is defined via the affine mapping

$$x = F(y)\colon \begin{cases} x_1 = y_1 + y_2 \cos\alpha, \\ x_2 = y_2 \sin\alpha, \end{cases}$$

which maps a rectangle G_{0i} onto G_i. We set

$$q \in Q_3(G_i) \Leftrightarrow q \circ F = \hat{q} \in Q_3(G_{0i}).$$

Next let

$$V_h = \{v \in C^{(1)}(\overline{\Omega}) \mid v|_{G_i} \in Q_3(G_i),\ i = 1, 2 \ldots, I(h),$$
$$v = 0 \ \text{and}\ \frac{\partial v}{\partial n} = 0 \text{ on } \Gamma_u\},$$

$$K_h = \{v \in V_h \mid v(A_j^h) \geqslant 0,\ j = 0, 1, 2 \ldots, n(h)\}.$$

(Note that $V_h \subset V$.)

We shall employ some numerical integration formulae. Instead of $\langle f, v_h\rangle$ we introduce

$$\langle f, v_h\rangle_h = \sum_{j=1}^{j_0} P_j v_h(X_j) + \sum_{i=1}^{I(h)} \sum_{m=1}^{m_0} \hat{\omega}_m J_{F_i} f_0(F_i(\hat{b}_m)) v_h(F_i(\hat{b}_m)) + \int_\gamma f_1 v_h \,\mathrm{d}s, \tag{20.7}$$

where F_i is the affine mapping of a unit reference square $\hat{K}$ onto G_i. J_{F_i} is the (constant) Jacobian matrix of F_i, $\hat{\omega}_m$ are weights and $\hat{b}_m \in \hat{K}$ are nodes of the quadrature formula under consideration. We assume that we are able to evaluate the integral over γ. Otherwise, we can use an analogous numerical integration formula.

We define the following *approximate solution*:

$$\begin{aligned} &u_h \in K_h, \\ &a(u_h, v_h - u_h) \geqslant \langle f, v_h - u_h \rangle_h \quad \forall v_h \in K_h. \end{aligned} \tag{20.8}$$

Some sufficient conditions for the convergence of approximate solutions to the exact solution will be presented.

LEMMA 20.1. *Assume that open subdomains $D_k \subset \Omega$ exist such that:*

(i) $\overline{\Omega} = \bigcap_{k=1}^{k_0} \overline{D}_k$, $D_k \cap D_n = \emptyset$ *for* $k \neq n$;

(ii) *every boundary* ∂D_k *consists of segments lying on the grid of* $\mathcal{T}_{h_0}$;

(iii) $f_0 \in W^{m,\infty}(D_k)$ $\forall k$, *where* $m \geqslant 3$.

Let the quadratic formula in (20.7) *be exact for all polynomials* $p \in Q_{2m-3}(\hat{K})$. *Then*

$$|\langle f, v_h \rangle - \langle f, v_h \rangle_h| \leqslant Ch^m \|v_h\|_{2,\Omega}$$

holds for all partitions $\mathcal{T}_h$, *which refine* $\mathcal{T}_{h_0}$, *the constant* C *is independent of* h *and* $v_h \in V_h$.

PROOF. The proof is based on an analogue of Theorem 4.1.5 in CIARLET [1978] or Theorem 28.3 in CIARLET [1991]. □

We say that a family $\{\mathcal{T}_h\}$, $h \to 0+$, refining $\mathcal{T}_{h_0}$, is *regular* if there exists a positive constant C such that

$$h/\varrho \leqslant C \quad \forall G_i \in \mathcal{T}_h, \ \forall \mathcal{T}_h \in \{\mathcal{T}_h\}, \ h \to 0+,$$

where ϱ denotes the diameter of the maximal circle contained in G_i.

THEOREM 20.1. *The approximate problem* (20.8) *has a unique solution* u_h *for any partition* $\mathcal{T}_h$ *under consideration.*

If the assumptions of Lemma 20.1 *are fulfilled and the family* $\{\mathcal{T}_h\}$, $h \to 0+$ *of partitions is regular, then*

$$\lim_{h \to 0+} \|u_h - u\|_{2,\Omega} = 0,$$

where u *is the solution of the problem* (20.3).

PROOF (Sketch of the proof). Following the method of Falk, we obtain

(i)

$$(1-\sigma)|u-u_h|^2_{2,\Omega} \leqslant \langle f, u-v\rangle + \langle f, u_h - v_h\rangle_h + a(u_h-u, v_h-u)$$
$$+ a(u, v-u_h) + a(u, v_h-u) \quad \forall v \in K,\ \forall v_h \in K_h.$$

(ii) For any $v \in K$ there exists a sequence $\{v_h\}$ such that

$$v_h \in K_h \quad \text{and} \quad \lim_{h\to 0+} \|v-v_h\|_{2,\Omega} = 0.$$

(iii) If $v_h \in K_h$ and $v_h \rightharpoonup v$ (weakly) in V for $h \to 0+$, then $v \in K$.

On the basis of (ii), a sequence $\{v_h\}$ exists, converging to u in V.

Since $\mathcal{J}$ is coercive on V, the boundedness of the sequence $\{u_h\}$ follows. Then the subsequence converges weakly and the weak limit (denoted by v) belongs to K, by virtue of (iii). We employ the estimate (i), Lemma 20.1 and the inequality (20.5) to obtain strong convergence. □

REMARK 20.1. The proof is a direct application of a general convergence theorem for the Ritz–Galerkin method—see Chapter II, Theorem 4.1—completed with Lemma 20.1.

20.2. A mixed variational formulation for a semicoercive problem

Let us consider the case, when unilateral conditions of the type (19.5) are prescribed on the *whole boundary* $\partial\Omega$. This configuration leads to a "semicoercive" functional of the potential energy and the problem is solvable only if the resultants of the external loads have a proper sign—see (20.9), i.e. if the plate is "forced towards the obstacle".

Following COMODI [1985], we assume that: Ω is a convex polygon, Poisson's ratio $\sigma = 0$,

$$\begin{aligned} &\langle f, v\rangle = (f_0, v)_0 \quad \text{with given } f_0 \in L^2(\Omega), \\ &(f_0, p)_0 < 0 \quad \forall p \in K \cap P_1(\Omega) \dot{-} \{0\}. \end{aligned} \tag{20.9}$$

Then problem (20.3) is uniquely solvable, since

$$\mathcal{J}(v) \to +\infty \quad \text{as } \|v\|_{2,\Omega} \to \infty \tag{20.10}$$

if $v \in K$, but not if $v \in V$, in general.

Note that $V = H^2(\Omega)$, therefore (20.5) is not true and we are left with (20.4). Condition (20.10) then follows from (20.4) and (20.9). (The proof can be based on NEČAS [1975, Theorem 2.3].)

Instead of the variational inequality (20.3), COMODI [1985] proposed a mixed variational formulation, based on the so called Herrmann–Hellan model, extending the results of BREZZI and RAVIART [1977] to the unilateral boundary condition (19.5). The tensor-function of bending moments is computed simultaneously in addition to the deflection function. Using triangulations of Ω with piecewise linear approximations for the deflections and piecewise constant approximations for the moments, then the $O(h\,|\ln h|^{1/2})$ error estimate for the deflections in $H^1(\Omega)$ and $O(h)$, and for the moments in $[L^2(\Omega)]^4$ can be proved.

21. Unilateral rotation problem

Let us consider the boundary conditions (19.6) together with

$$u = 0 \quad \text{on } \partial\Omega, \tag{21.1}$$

where Ω is a bounded open domain of $\mathbb{R}^2$ with a regular boundary. We shall follow GLOWINSKI, LIONS and TRÉMOLIÈRES [1981, Section 4.2.2, p. 323].

Assume that the right-hand side is given by

$$\langle f, v\rangle = \sum_{j=1}^{j_0} P_j v(X_j) + \int_\Omega f_0 v \,\mathrm{d}x,$$

where $X_j \in \Omega$ and $f_0 \in L^2(\Omega)$. Let K be the closed convex subset of

$$V = H^2(\Omega) \cap H_0^1(\Omega)$$

defined by

$$K = \{v \in V \mid \frac{\partial v}{\partial n} \geqslant 0 \text{ a.e. on } \partial\Omega \equiv \Gamma\}.$$

Since the functional $\mathcal{J}$ (20.6) is again coercive and strictly convex on V by virtue of (20.4) and (20.5), there exists a unique solution u of the variational inequality (20.3).

Using Green's theorem and boundary condition (21.1), we derive that

$$\int_\Omega (v_{11}v_{22} - v_{12}^2)\,\mathrm{d}x = 0 \quad \forall v \in C^{(2)}(\overline{\Omega}), \text{ such that } v = 0 \text{ on } \partial\Omega,$$

so that

$$a(v, v) = \int_\Omega (v_{11}^2 + v_{22}^2 + 2v_{12}^2)\,\mathrm{d}x = \int_\Omega (\Delta v)^2\,\mathrm{d}x.$$

Therefore we may replace the bilinear form in (20.2) by the formula

$$a(u, v) = \int_\Omega \Delta u \Delta v \,\mathrm{d}x. \tag{21.2}$$

Following the argument of LIONS [1969, Chapter 2, Section 8.7.2] and using some results of LIONS and MAGENES [1968], it may be proved that the solution of (20.3) $u \in H^s(\Omega)$ for any $s < 3$. Then the trace $\gamma_0 \Delta u$ of Δu on Γ exists and belongs to $L^2(\Gamma)$.

Let us define $\mathscr{L} : V \times L^2(\Gamma) \to \mathbb{R}$ by

$$\mathscr{L}(v, \mu) = \mathscr{J}(v) - \int_\Gamma \mu v \, \mathrm{d}\Gamma \tag{21.3}$$

and

$$\Lambda = L^2_+(\Gamma) = \{\mu \in L^2(\Gamma) \mid \mu \geqslant 0 \text{ a.e.}\}.$$

THEOREM 21.1. *The functional $\mathscr{L}$ admits $(u, \gamma_0 \Delta u)$ as a unique saddle point on $V \times \Lambda$.*

REMARK 21.1. Since $u = 0$ on Γ, we can derive that

$$Mu = -D\frac{\partial^2 u}{\partial n^2} = -D\Delta u \quad \text{on } \Gamma.$$

Thus the second component of the saddle point has an important physical meaning—a multiple of the bending moment.

Having Theorem 21.1 in mind, we may apply to $\mathscr{L}$ a saddle-point search *algorithm of the Uzawa type*. Let λ^0 be chosen arbitrarily in Λ (e.g., zero).

Step 1. Having λ^m, solve the "classical" boundary value problem

$$\begin{aligned} &\Delta^2 u^m = f \quad \text{in } \Omega, \\ &u^m = 0, \ \Delta u^m = \lambda^m \quad \text{on } \Gamma. \end{aligned} \tag{21.4}$$

Step 2.

$$\lambda^{m+1} = P_\Lambda(\lambda^m - \varrho_m \frac{\partial u^m}{\partial n}) = (\lambda^m - \varrho_m \frac{\partial u^m}{\partial n})^+,$$

where ϱ_m is a positive parameter.

Then

$$u^m \to u \text{ in } V, \quad \text{as } m \to \infty,$$

provided

$$0 < r_0 \leqslant \varrho_m \leqslant r_1 < 2\sigma_0^2,$$

where

$$\sigma_0 = \inf_{v \in H^2 \cap H_0^1} \frac{\|\Delta v\|_{0,\Omega}}{\|\partial v/\partial n\|_{0,\Gamma}}.$$

This solution of the "classical" plate problem (21.4) factorizes into two Dirichlet problems for the Laplace operator, i.e.,

$$\begin{aligned} -\Delta p^m &= f \quad \text{in } \Omega, \\ p^m &= -\lambda^m \quad \text{on } \Gamma \end{aligned} \quad \text{and} \quad \begin{aligned} -\Delta u^m &= p^m \quad \text{in } \Omega, \\ u^m &= 0 \quad \text{on } \Gamma. \end{aligned} \tag{21.5}$$

REMARK 21.2. Reduction to second order problems (21.5) enables us to avoid the use of finite-element models, requiring the $C^{(1)}(\Omega)$-continuity (see CIARLET [1978, Chapter 2, § 2.2] or CIARLET [1991, Section 9]). For a related method, see BÉGIS and GLOWINSKI [1972] and BOURGAT [1976].

REMARK 21.3. If the boundary conditions (19.6) alone are considered on the whole boundary $\partial\Omega$, we are led to a "semicoercive" problem (cf. 20.2). A thorough analysis of such a problem has been given by COMODI [1986], who employed Morley's nonconforming finite elements CIARLET [1991, p. 309]).

22. Inner obstacle problem

Let us consider the problem corresponding to the conditions (19.9), i.e., unilateral displacements of the points of Ω.

22.1. A primal approach

Assume that:
- Ω is a parallelogram;
- the plate is *clamped* along the whole boundary, i.e., conditions (19.2) hold on $\partial\Omega$;
- the obstacle is given by a function $\varphi \in C(\overline{\Omega})$ such that φ is negative on $\partial\Omega$;
- external loads are given as in Section 20.1;
- a uniform partition $\mathcal{T}_{h_0}$ is given such that the assumptions of Lemma 20.1 are fulfilled.

If we introduce the set

$$K = \{v \in H_0^2(\Omega) \mid v \geqslant \varphi \text{ in } \Omega\}, \tag{22.1}$$

then the variational inequality (20.3) governs the equilibrium of the plate. There exists a unique solution of the latter inequality, since K is a convex closed subset of $H_0^2(\Omega)$ and (20.4) and (20.5) hold for $v \in V \equiv H_0^2(\Omega)$.

To define approximate solutions, we consider the same parallelogram partitions $\mathcal{T}_h$ of Ω as in Section 20.1 and the finite-element spaces $V_h \subset C^{(1)}(\overline{\Omega}) \cap V$

of piecewise bicubic polynomials. We introduce approximations K_h of K as follows

$$K_h = \{v \in V_h \mid v(A^h) \geqslant \varphi(A^h) \text{ for all interior nodes } A^h \subset \Omega\}. \tag{22.2}$$

Let approximate solutions $u_h \in K_h$ satisfy inequality (20.8). Theorem 20.1 can be verified, using the general theorem of convergence (see Chapter II, Theorem 4.1) and Lemma 20.1 (following the argument of HASLINGER [1978], where only $f \in L^2(\Omega)$ was considered without the use of numerical integration).

22.2. *A dual approach*

Another method of approximate solution was proposed in GLOWINSKI, LIONS and TRÉMOLIÈRES [1981, Section 4.2.2, Example 3, p. 326].

Assume that

- Ω is a bounded domain with a regular boundary Γ,
- $\varphi \in L^\infty(\Omega)$, φ is negative in a neighbourhood of Γ, $f \in [H_0^2(\Omega)]'$.

Denoting $[L^\infty(\Omega)]'$ as the topological dual of $L^\infty(\Omega)$ and $\langle\cdot,\cdot\rangle_\infty$ the bilinear form of the corresponding duality, we define

$$\Lambda = \{\mu \in [L^\infty(\Omega)]' \mid \langle\mu, v\rangle_\infty \geqslant 0 \ \forall v \in L^\infty(\Omega),\ v \geqslant 0 \text{ a.e.}\}$$

(i.e., Λ is the positive cone of $[L^\infty(\Omega)]'$).

We introduce

$$\mathscr{L} : H_0^2(\Omega) \times [L^\infty(\Omega)]' \to \mathbb{R}$$

by (cf. (20.6) for the definition of $\mathscr{J}$)

$$\mathscr{L}(v, \mu) = \mathscr{J}(v) + \langle\mu, \varphi - v\rangle_\infty. \tag{22.3}$$

Since there exists a $v_0 \in K$ such that

$$v_0 - \varphi \geqslant \delta \quad \text{a.e.},$$

where δ is some positive constant, it can be proved that $\mathscr{L}$ admits a saddle point (u, λ) on $H_0^2(\Omega) \times \Lambda$, where u is the solution of (20.3). For the proof, see EKELAND and TEMAM [1976, Chapter 3, Section 5].

To solve the saddle point problem, we discretize it and then use an *approximate form* of the following algorithm: $\lambda^0 \geqslant 0$ chosen arbitrarily. Knowing λ^n, we solve the "classical" boundary value problem

$$\begin{aligned} &\Delta^2 u^n = f + \lambda^n \quad \text{in } \Omega, \\ &u^n = 0, \qquad \partial u^n/\partial n = 0 \quad \text{on } \Gamma; \end{aligned} \tag{22.4}$$

$$\lambda^{n+1} = [\lambda^n + \varrho_n(\varphi - u^n)]^+, \quad \varrho_n > 0. \tag{22.5}$$

Problem (22.4) may itself be solved by the duality method, enabling factorization: $\mu^0 \in L^2(\Gamma)$ arbitrarily. Knowing μ^m, we calculate p^m, μ^{m+1} as follows:

$$-\Delta p^m = f + \lambda^n \quad \text{in } \Omega, \qquad -\Delta u^m = p^m \quad \text{in } \Omega,$$
$$p^m = \mu^m \quad \text{on } \Gamma, \qquad u^m = 0 \quad \text{on } \Gamma,$$

$$\mu^{m+1} = \mu^m + \kappa_m \partial u^m/\partial n, \quad \kappa_m > 0.$$

For the details and examples refer to GLOWINSKI, LIONS and TRÉMOLIÈRES [1981, Chapter 4, Section 4.7].

22.3. *A dual approach combined with the penalty method*

The same inner obstacle problem can be solved by a combination of a dual approach and the penalty method, as was proposed by SCHOLZ [1987] (He even considered two obstacles—one from below and one from above.)

Introducing the penalty functional

$$\beta(v) = (v - \varphi)^-$$

(i.e., the negative part of $(v - \varphi)$), we shall consider the following problem using the penalty method: find $u^\varepsilon \in H_0^2(\Omega)$ such that

$$(\Delta u^\varepsilon, \Delta v)_0 + \frac{1}{\varepsilon}(\beta(u^\varepsilon), v)_0 = (f, v)_0 \quad \forall v \in H_0^2(\Omega), \tag{22.6}$$

where ε is a positive parameter, $f \in L^2(\Omega)$.

If u is the solution of the original problem (20.3) and $u^\varepsilon \in H_0^2(\Omega) \cap H^3(\Omega)$, then

$$\|u - u^\varepsilon\|_{2,\Omega} \leqslant C\varepsilon^{1/4} \tag{22.7}$$

holds with C independent of ε.

To approximate problem (22.6), we can use a mixed finite-element method based on standard factorization (see CIARLET [1978, Chapter 7.1, pp. 383]), as follows.

Let $\mathcal{V}$ be the subspace of pairs $(v, p) \in H_0^1(\Omega) \times L^2(\Omega)$ such that

$$\int_\Omega \nabla v \cdot \nabla z \, \mathrm{d}x = (p, z)_0 \quad \forall z \in H^1(\Omega) \tag{22.8}$$

(which implies that $-\Delta v = p$ in Ω and $\partial v/\partial n = 0$ on $\partial\Omega$ in a weak sense).

Now the problem (22.6) can be reformulated: find $(u^\varepsilon, q^\varepsilon) \in \mathcal{V}$ such that

$$(q^\varepsilon, p)_0 + \frac{1}{\varepsilon}(\beta(u^\varepsilon), v)_0 = (f, v)_0 \quad \forall (v, p) \in \mathcal{V}. \tag{22.9}$$

Using the spaces $S_h \subset H^1(\Omega)$ of piecewise quadratic functions on triangulations of Ω, we define

$$S_h^0 = S_h \cap H_0^1(\Omega),$$

and the subspace $\mathcal{V}_h \subset S_h^0 \times S_h$ of pairs (v_h, p_h) such that

$$\int_\Omega \nabla v_h \cdot \nabla z_h \,\mathrm{d}x = (p_h, z_h)_0 \quad \forall z_h \in S_h.$$

The *discrete problem* using the penalty method is then: find $(u_h^\varepsilon, q_h^\varepsilon) \in \mathcal{V}_h$ such that

$$(q_h^\varepsilon, p_h)_0 + \frac{1}{\varepsilon}(\beta(u_h^\varepsilon), v_h)_0 = (f, v_h)_0 \quad \forall (v_h, p_h) \in \mathcal{V}_h. \tag{22.10}$$

The latter problem is uniquely solvable. One way to solve it is the use of an iteration method proposed by BREZIS and SIBONY [1968], see also SCHOLZ [1984].

If the solution of (22.6) $u^\varepsilon \in H^3(\Omega)$, then

$$\|q^\varepsilon - q_h^\varepsilon\|_{0,\Omega} + \varepsilon^{-1/2}\|\beta(u^\varepsilon) - \beta(u_h^\varepsilon)\|_{0,\Omega} \leqslant C(\varepsilon^{-1/2}h + \varepsilon^{-1}h^2) \tag{22.11}$$

holds with C independent of ε and h.

As a consequence of (22.7) and (22.11), for $\varepsilon = h^{4/3}$ we obtain the following error estimate:

$$\|\Delta u + q_h^\varepsilon\|_{0,\Omega} \leqslant Ch^{1/3}.$$

22.4. *A mixed formulation by factorization*

The inner obstacle problem with

(i) a simply supported boundary (19.3) and

(ii) a clamped boundary (19.2)

has been solved and analyzed also by FUSCIARDI and SCARPINI [1980]. The same factorization technique as in Section 22.3 has been used.

In case (i) the setting of the problem (for convex polygonal domain Ω and $f \in L^2(\Omega)$) is: find $(u, q) \in X$ such that

$$\int_\Omega \nabla q \cdot \nabla(v - u)\,\mathrm{d}x \geqslant (f, v - u)_0 \quad \forall v \in \overline{K}, \tag{22.12}$$

where

$$X = \{(v, p) \in [H_0^1(\Omega)]^2 \mid -\Delta v = p,\ v \geqslant \varphi \text{ a.e. in } \Omega\},$$
$$\overline{K} = \{v \in H_0^1(\Omega) \mid v \geqslant \varphi \text{ a.e. in } \Omega\}.$$

If problem (22.12) is described by means of piecewise quadratic functions on triangulations of Ω, for the approximations (u_h, q_h) we obtain a complementarity system of the following type:

$$U \geqslant \mathcal{A}, \qquad TU - b \geqslant 0, \qquad (TU - b)(U - \mathcal{A}) = 0$$

with T a positive definite matrix.

Assuming that the family of triangulations is strongly regular, the convergence

$$\left.\begin{array}{l} u_h \to u \text{ in } H_0^1(\Omega), \\ q_h \rightharpoonup q \text{ (weakly) in } L^2(\Omega) \end{array}\right. \quad \text{as } h \to 0,$$

can be proved.

In case (ii) of a clamped boundary, we define

$$\tilde{X} = \{(v,p) \in H_0^1(\Omega) \times H^1(\Omega) \mid (22.8) \text{ holds and } v \geqslant \varphi \text{ a.e. in } \Omega\},$$

$\tilde{K}$ is the set of the first components of pairs $(v,p) \in \tilde{X}$.

The problem is to find $(u,q) \in \tilde{X}$ such that (22.12) holds for all $v \in \tilde{K}$. To solve this problem, the authors proposed an iteration procedure introducing a sequence of complementarity systems in finite-dimensional spaces.

Bibliography and comments to Chapter VI

The inner obstacle plate problem has also been investigated by WESTBROOK [1990], who applied the so called discrete Kirchhoff triangular elements to the approximations K_h (22.2) of the set K. The method of pointwise successive overrelaxation with projection has been used together with some refined approximations of the set K. Such a refinement seems to be advisable when it is important to know the contact region.

The duality approach of Section 22.2 was applied by OUTRATA, JARUŠEK and BERAN [1981] in the solution of an optimal control problem. BOCK, HLAVÁČEK and LOVÍŠEK [1985] considered an optimal design problem (thickness optimization) of a plate with unilateral boundary displacement, using the approximations of Section 20.1.

The so-called Herrmann–Hellan mixed finite-element model (for deflections and bending moments) has been applied to inner obstacle problems by HLAVÁČEK [1993]. He extended some results of BREZZI and RAVIART [1977] and combined them with a dual approach of Section 22.2. In the convergence analysis, he employed some ideas of COMODI [1985] (cf. Section 20.2).

The Herrmann–Hellan model of mixed approximations has also been applied and analysed by BREZZI, JOHNSON and MERCIER [1977] in the bending problem of a clamped *elasto-plastic* plate. The formulation for an isotropic, perfectly elasto-plastic plate leads to a system of a variational inequality combined with an equation. A unilateral contact problem for a *cylindrical shell* has been solved in BOCK and LOVÍŠEK [1983] by triangular finite elements with quintic polynomials.

When the *large deflection* theory of elastic plates is considered, we are led to the nonlinear system of two Von Kármán equations, involving the deflection function and the Airy stress function. The unique solution exists only if the in-plane loading is small enough. For such a case and for the inner obstacle problem, SCARPINI [1985] presented a finite-element analysis, using a mixed model with affine elements and the generalized patch test from STUMMEL [1979]. Another method has been applied in OHTAKE, ODEN and KIKUCHI [1980].

References

ANTES, H. and P.D. PANAGIOTOPOULOS (1992), *The Boundary Integral Approach to Static and Dynamic Contact Problems, Equality and Inequality Methods,* Internat. Ser. Numer. Math. **108** (Birkhäuser, Basel).

AUBIN, J.P. (1972), *Approximation of Elliptic Boundary value Problems* (Wiley-Interscience, New York).

BAIOCCHI, C. (1971), Sur un problème à frontière libre traduisant le filtrage de liquides à travers des milieux poreux, *C. R. Acad. Sci. Paris Sér. I Math.* **273**, 1215–1217.

BAIOCCHI, C. (1977), Estimation d'erreur dans L^∞ pour les inéquations à obstacle, in: *Lecture Notes in Math.* **606** (Springer, Berlin), 27–34.

BAIOCCHI, C., F. GASTALDI and F. TOMARELLI (1986), Some existence results on noncoercive variational inequalities, *Ann. Scuola Norm. Sup. Pisa, Cl. Sci, Ser. IV* **13**, 617–659.

BÉGIS, D. and R. GLOWINSKI (1972), Dual numerical techniques for some variational problems involving biharmonic operators, in: A.V. Balakrishnan, ed., *Techniques of Optimisation* (Academic Press, New York) 159–174.

BOCK, I., I. HLAVÁČEK and J. LOVÍŠEK (1985), Optimal control of a variational inequality with applications to structural analysis, II and III, *Appl. Math. Optim.* **13**, 117–136.

BOCK, I. and J. LOVÍŠEK (1983), An analysis of a contact problem for a cylindrical shell: A primary and dual formulation, *Apl. Mat.* **28**, 408–429.

BOURGAT, J.F. (1976), Numerical study of a dual iterative method for solving a finite element approximation of the biharmonic equation, *Comput. Methods Appl. Mech. Engrg.* **9**, 203–218.

BREZIS, H. and M. SIBONY (1968), Méthodes d'approximation et d'itération pour les opérateurs monotones, *Arch. Rational Mech. Anal.* **28**, 59–82.

BREZZI, F. and C.A. CAFFARELLI (1983), Convergence of the discrete free boundaries for finite element approximations, *RAIRO Anal. Numér.* **17**, 385–395.

BREZZI, F., W.W. HAGER and P.A. RAVIART (1977), Error estimates for the finite element solution of variational inequalities, *Numer. Math.* **28**, 431–443.

BREZZI, F., C. JOHNSON and B. MERCIER (1977), Analysis of a mixed finite-element method for elasto-plastic plates, *Math. Comp.* **31**, 809–817.

BREZZI, F. and P.A. RAVIART (1977), Mixed finite element methods for fourth order elliptic equations, in: J.J.H. Miller, ed., *Topics in Numerical Analysis,* **III** (Academic Press, London) 33–56.

CASEY, G. and P.M. NAGHDI (1983), On the nonequivalence of the stress space and strain space formulation of plasticity theory, *ASME Trans. J. Appl. Mech.* **50**, 350–358.

CÉA, J. (1971), *Optimization, Théorie et Algorithmes* (Dunod, Paris).

CHIPOT, M. (1984), *Variational Inequalities and Flow in Porous Media,* Appl. Math. Sci. **52** (Springer, New York).

CIARLET, P.G. (1978), *The Finite Element Method for Elliptic Problems,* Stud. Math. Appl. **4** (North-Holland, Amsterdam).

CIARLET, P.G. (1989), *Introduction to the Numerical Linear Algebra and Optimization* (Cambridge University Press, Cambridge).

CIARLET, P.G. (1990), *Plates and Junctions in Elastic Multi-Structures: An Asymptotic Analysis* (Masson, Paris and Springer, Heidelberg).

CIARLET, P.G. (1991), Basic error estimates for elliptic problems, in: P.G. Ciarlet and J.L. Lions, ed.,

Handbook of Numerical Analysis, **II**. *Finite Element Methods* (Part 1) (North-Holland, Amsterdam) 17–352.

CIARLET, P.G. and J. NEČAS (1984), Unilateral problems in nonlinear three-dimensional elasticity, *Arch. Rational Mech. Anal.* **87**, 319–338.

CIARLET, P.G. and J. NEČAS (1987), Injectivity and self contact in non-linear elasticity, *Arch. Rational Mech. Anal.* **19**, 171–188.

COMODI, M.I. (1985), Approximation of a bending plate problem with a boundary unilateral constraint, *Numer. Math.* **47**, 435–458.

COMODI, M.I. (1986), Approximation of a fourth order variational inequality, *RAIRO Math. Model. Numer. Anal.* **20**, 5–24.

CORTEY-DUMONT, P. (1985a), Sur les inéquations variationnelles à opérateur non-coercif, *RAIRO Math. Model. Numer. Anal.* **19**, 195–212.

CORTEY-DUMONT, P. (1985b), On finite element approximation in the L^∞-norm of variational inequalities, *Numer. Math.* **47**, 45–57.

CURNIER, A. and P. ALART (1988), A generalized Newton method for contact problems with friction, *J. Theory Appl. Mech.* **7**, Special issue supplement no. 1, 67–82.

DILINTAS, G., LAURENT-GENGOUX and D. TRYSTAM (1988), A conjugate projected gradient method with preconditioning for unilateral problems, *Comput. Struct.* **29**, 675–680.

DOSTÁL, Z. (1992), Conjugate projector preconditioning for the solution of contact problems, *Internat. J. Numer. Meth. Engrg.* **33**, 1–7.

DUVAUT, G. and J.L. LIONS (1976), *Inequalities in Mechanics and Physics*, Ser. Compr. Stud. Math. **219** (Springer, Berlin).

EKELAND, I. and R. TEMAM (1976), *Convex Analysis and Variational Problems* (North-Holland, Amsterdam).

ELLIOTT, C. and J. OCKENDON (1980), *Weak and Variational Methods for Moving Boundary Value Problems* (Pitman, Boston).

FALK, R.S. (1974), Error estimates for the approximation of a class of variational inequalities, *Math. Comp.* **28**, 963–971.

FICHERA, G. (1964), Problemi elastostatici con vincoli unilaterali, il problema di Signorini con ambigue condizioni al contorno, *Mem. Accad. Naz. Lincei* **8** (7), 91–140.

FICHERA, G. (1972), Boundary value problems in elasticity with unilateral constraints, in: C. Truesdell, *Encyclopedia of Physics VI a/2, Mechanics of Solids* **II** (Springer, Berlin) 391–424.

FINZI VITA, S. (1982), L^∞-error estimates for variational inequalities with Hölder continuous obstacles, *RAIRO Anal. Numér.* **16**, 27–37.

FORTIN, M. and R. GLOWINSKI (1982), *Méthodes de Lagrangian Augmenté, Applications à la Résolution Numériques de Problèmes aux Limite.* Collection Methodes Mathématiques de l'informatique (Dunod, Paris).

FRIEDMAN, A. (1982), *Variational Principles and Free Boundary Problems* (Wiley, New York).

FUSCIARDI, A. and F. SCARPINI (1980), A mixed finite element solution of some biharmonic unilateral problem, *Numer. Funct. Anal. Optim.* **2**, 397–420.

GIRAULT, V. and P.A. RAVIART (1979), *Finite Element Approximation of the Navier-Stokes Equations*, Lecture Notes in Mathematics **749** (Springer, Berlin).

GLOWINSKI, R. (1984), *Numerical Methods for Nonlinear Variational Problems*, Springer Ser. Comput. Phys. (Springer, New York).

GLOWINSKI, R., J.L. LIONS and R. TRÉMOLIÈRES (1981), *Numerical Analysis of Variational Inequalities*, Stud. Math. Appl. (North-Holland, Amsterdam).

GROSSMAN, CH. and H.G. ROSS (1992), *Numerik Partieller Differential-Gleichungen*, (Teubner, Stuttgart).

GWINNER, J. (1991), Discretization of semicoercive variational inequalities, *Equationes Math.* **42**, 72–78.

HASLINGER, J. (1978), On numerical solution of a variational inequality of the 4th order by finite element method, *Apl. Mat.* **23**, 334–345.

HASLINGER, J. (1981), Mixed formulation of variational inequalities and its approximation, *Apl. Mat.* **26**, 462–475.

HASLINGER, J. (1983), Approximation of the Signorini problem with friction, obeying Coulomb's law, *Math. Methods Appl. Sci* **5**, 422–437.

HASLINGER, J. (1984), Least square method for solving contact problems with friction obeying Coulomb's law, *Apl. Mat.* **29**, 212–224.

HASLINGER, J. and I. HLAVÁČEK (1982a), Contact between elastic perfectly plastic bodies, *Apl. Math.* **27**, 27–45.

HASLINGER, J. and I. HLAVÁČEK (1982b), Approximation of the Signorini problem with friction by a mixed finite element method, *J. Math. Anal. Appl.* **86**, 99–122.

HASLINGER, J. and J. LOVÍŠEK (1980), Mixed variational formulation of unilateral problems, *Comment. Math. Univ. Carolin.* **21**, 231–246.

HASLINGER, J. and P. NEITTAANMÄKI (1988), *Finite Element Approximation for Optimal Shape Design, Theory and Applications* (Wiley, Chichester).

HASLINGER, J. and P.D. PANAGIOTOPOULOS (1984), Approximation of contact problems with friction by reciprocal variational formulations, *Proc. Roy. Soc. Edinburgh* **98**A, 365–383.

HLAVÁČEK, I. (1979), Convergence of an equilibrium finite element model for plane elastostatics, *Apl. Mat.* **24**, 427–457.

HLAVÁČEK, I. (1980), A finite element solution for plasticity with strain-hardening, *RAIRO Anal. Numér.* **14**, 347–368.

HLAVÁČEK, I. (1981), A finite element analysis for elasto-plastic bodies obeying Hencky's law, *Apl. Mat.* **26**, 449–461.

HLAVÁČEK, I. (1993), A mixed finite element method for plate bending with a unilateral inner obstacle, *Appl. Math.* **39**, 25–44.

HLAVÁČEK, I., J. HASLINGER, J. NEČAS and J. LOVÍŠEK (1988), *Numerical Solution of Variational Inequalities,* Springer Ser. Appl. Math. Sci. **66** (Springer, New York).

HLAVÁČEK, I. and J. LOVÍŠEK (1977), A finite element analysis for the Signorini problem in plane elasticity, *Apl. Mat.* **22**, 215–228.

HLAVÁČEK, I. and J. LOVÍŠEK (1980), Finite element analysis of the Signorini problem in semi-coercive cases, *Apl. Mat.* **25**, 273–285.

HLAVÁČEK, I. and J. NEČAS (1970), On inequalities of Korn's type, *Arch. Rational Mech. Anal.* **36**, 305–334.

HLAVÁČEK, I., J. ROSENBERG, A.E. BEAGLES and J.R. WHITEMAN (1992), Variational inequality formulation in strain space and finite element solution of an elasto-plastic problem with hardening, *Comput. Methods Appl. Mech. Engrg.* **94**, 93–112.

HOFFMANN, K.H. and J. ZOU (1992), Parallel algorithms of Schwarz variant for variational inequalities, *Numer. Funct. Anal. Optimiz.* **13** (15), 5–6, 449–462.

JARUŠEK, J. (1983), Contact problem with a bounded friction. Coercive case, *Czechoslovak Mat. J.* **33**, 254–278.

JARUŠEK, J. (1984), Contact problems with bounded friction. Semicoercive case, *Czechoslovak Mat. J.* **34**, 619–629.

JEUZETTE, J.P. and V. SONZOGNI (1989), A projected conjugate gradient method for structural stability analysis with linear constraints, *Comput. Struct.* **33**, 31–39.

JOHNSON, C. (1976a), Existence theorems for plasticity problems, *J. Math. Pure Appl.* **55**, 431–444.

JOHNSON, C. (1976b), On finite element method for plasticity problems, *Numer. Math.* **26**, 79–84.

JOHNSON, C. (1977), A mixed finite element method for plasticity problems with hardening, *SIAM J. Numer. Anal.* **14**, 575–583.

JOHNSON, C. (1978), On plasticity with hardening, *J. Math. Anal. Appl.* **62**, 325–336.

JOHNSON, C. and B. MERCIER (1978), Some equilibrium finite element methods for two-dimensional elasticity problems, *Numer. Math.* **30**, 103–116.

JÚDICE, J.J. and F.M. PIRES (1992), Basic set algorithm for a generalized linear complementarity problem, *J. Optim. Theory Appl.* **74**, 391–412.

KAČANOV, L.M. (1948), *Mechanics of Plastic Media* (in Russian) (Nauka, Moscow).

KESTŘÁNEK, Z. (1986), Variational inequalities in plasticity with strain-hardening-equilibrium finite element approach, *Apl. Mat.* **31**, 270–281.

KIKUCHI, N. and J.T. ODEN (1988), *Contact Problems in Elasticity: A Study of Variational Inequalities and Finite Element Methods,* SIAM Stud. Appl. Math. (SIAM, Philadelphia, PA).

KINDERLEHRER, D. and G. STAMPACCHIA (1980), *An Introduction to Variational Inequalities and Their Applications* (Academic Press, New York).

KLARBRING, A. (1986), General contact boundary conditions and the analysis of frictional systems, *Internat. J. Solids and Structures* **22**, 1377–1398.

KLARBRING, A. and G. BJÖRKMAN (1988), The treatment of problems in contact mechanics by mathematical programming, *J. Theory Appl. Mech.* **7**, Special issue supplement 83–96.

KORNEEV, V.G. and U. LANGER (1984), *Approximate Solution of Plastic Flow Problems* (Teubner, Leipzig).

KŘÍŽEK, M. (1982), An equilibrium finite element method in three-dimensional elasticity, *Apl. Mat.* **28**, 46–75.

KUZNETSOV, Y. and P. NEITTAANMÄKI (1991), Overlapping domain decomposition method for a unilateral boundary value problem, *Proceedings 13th IMACS World Conf. on Comput. and Applied Math.* (Criterion Press, Dublin) 1671–1673.

LEMKE, C.A. (1978), Some pivot schemes for linear complementarity problem, *Math. Programming Stud.* **7**, 15–35.

LICHT, C., E. PRATT and M. RAOUS (1991), *Remarks on a numerical method for unilateral contact including friction,* ISNM **101** (Birkhauser, Basel).

LIONS, J.L. (1969), *Quelques Méthodes de Résolution des Problèmes aux Limites non Linéaires* (Dunod, Paris).

LIONS, J.L. and E. MAGENES (1968), *Problèmes aux Limites non Homogénes* **1**, (Dunod, Paris).

LIONS, J.L. and G. STAMPACCHIA (1967), Variational inequalities, *Comm. Pure Appl. Math.* **20**, 493–519.

MAY, H.O. (1986), The conjugate gradient method for unilateral problems, *Comput. Struct.* **12**, 595–598.

MERCIER, B. (1977), Sur la théorie et l'analyse numérique de problèmes de plasticité, Thésis, Université Paris VI.

MIHLIN, S.G. (1971), *The Numerical Performance Implementation of Variational Methods* (Wolters-Noordhoff, Groningen).

MIYOSHI, T. (1985), *Foundation of the Numerical Analysis of Plasticity* (North-Holland, Amsterdam).

MOSCO, U. and G. STRANG (1974), One-sided approximations and variational inequalities, *Bull. Amer. Math. Soc.* **80**, 308–312.

NEČAS, J. (1967), *Les Méthodes Directes en Théorie des Equations Elliptiques* (Masson, Paris).

NEČAS, J. (1975), On regularity of solutions to nonlinear variational inequalities for second order elliptic systems, *Rend. Mat. 2 (6)* **8**, 481–498.

NEČAS, J. and I. HLAVÁČEK (1981), *The Mathematical Theory of Elastic and Elasto-plastic Bodies: An Introduction* (Elsevier, Amsterdam).

NEČAS, J. and I. HLAVÁČEK (1983), Solution of Signorini's contact problem in the deformation theory of plasticity by secant modules method, *Apl. Mat.* **28**, 199–214.

NEČAS, J., J. JARUŠEK and J. HASLINGER (1980), On the solution of the variational inequality to the Signorini problem with small friction, *Boll. Un. Mat. Ital. B* **17**, 796–811.

NITSCHE, J. (1971), Über ein Variationsprinzip zur Lösung von Dirichlet Problem bei Verwendung von Teilräumen, die keinen randbedingungen unterworfen sind, *Abh. Mat. Sem. Univ. Hamburg* **36**, 9–15.

NOCHETO, R.N. (1986), A note on the approximation of the free boundary by finite element method, *RAIRO Math. Model. Numer. Anal.* **20**, 355–368.

OHTAKE, K., J.T. ODEN and N. KIKUCHI (1980), Analysis of certain unilateral problems in von Kármán plate theory by a penalty method. I. A variational principle with penalty, *Comput. Methods Appl. Mech. Engrg.* **24**, 187–213. II. Approximation and numerical analysis, *Comput. Methods Appl. Mech. Engrg.* **24**, 317–337.

OUTRATA, J.V., J. JARUŠEK and Z. BERAN (1981), An application of the augmented Lagrangian approach to the optimal control of a biharmonic system with state-space constraints, *Problems Control Inform. Theory* **10**, 363–373.

PANAGIOTOPOULOS, P.D. (1985), *Inequality Problems in Mechanics and Applications* (Birkhauser, Basel).

PŠENIČNYJ, B.N. and J.M. DANILIN (1975), Čislennyje metody v elementarnych zadačach (in Russian) (Nauka, Moscow).

RAOUS, M. and S. BARBARIN (1992), Preconditioned conjugate gradient method for a unilateral problem with friction, in: A. Curnier, ed., *Proceedings Internat. Sympos. on Contact Mechanics, Lausanne, October* 7–9, ISBN 2-88074-253-6.

RAOUS, M., P. CHABRAND and F. LEBON (1988), Numerical methods for frictional contact problems and applications, *J. Theory Appl. Mech.* **7**, Special issue supplement no 1, 111–128.

RODRIGUES, J.F. (1987), *Obstacle Problems in Mathematical Physics*, Math. Stud. **134** (North-Holland, Amsterdam).

SCARPINI, F. (1985), A mixed finite element approximation of a von Kármán plate bending unilateral problem, *Numer. Funct. Anal. Optim.* **8**, 623–636.

SCHOLZ, R. (1984), Numerical solution of the obstacle problem by the penalty method, *Computing* **32**, 297–306.

SCHOLZ, R. (1987), Mixed finite element approximation of a fourth order variational inequality by the penalty method, *Numer. Funct. Anal. Optim.* **9**, 233–247.

SIGNORINI, A. (1933), Sopra alcune questioni di elastostatica, *Atti Soc. della Ital. per il Progresso della Scienze.*

STUMMEL, F. (1979), The generalized patch test, *SIAM J. Numer. Anal.* **16**, 449–476.

SUQUET, P.-M. (1978), Existence and regularity of solutions for plasticity problems, in: *Proc. IUTAM Congress* Evanston.

TOMARELLI, F. (1988), Signorini problem in Hencky plasticity, Inst. Anal. Numer. CNR, Publ. N. 640, Pavia.

WATWOOD, V.B., JR. and B.J. HARTZ (1968), An equilibrium stress field model for finite element solutions of two-dimensional elastostatic problems, *Internat. J. Solids and Structures* **4**, 857–873.

WESTBROOK, D.R. (1990), The obstacle problem for beams and plates, *J. Comput. Appl. Math.* **30**, 295–311.

List of Symbols

General notation

$\mathbb{R}^d$, $d \geqslant 1$: d-dimensional vector space over the field of real numbers.
$\{x \in M \mid A(x)\}$: general notation for a set.
$u(\cdot, b)$ or $u(a, \cdot)$: partial mappings $x \mapsto u(x, b)$ or $x \mapsto u(a, x)$.
$\operatorname{supp} u$: support of a function u.
$u|_A$: restriction of a function u to the set A.
$C(a)$, $C(a, b)$, $\ldots$: any constant, depending only on a, a and b, $\ldots$.
$\operatorname{cl} A$ or $\bar{A}$: closure of a set A.
∂A: boundary of a set A.
$\operatorname{int} A$ or $\mathring{A}$: interior of a set A.
$\operatorname{card}(A)$: number of elements of a set A.
$\operatorname{diam}(A)$: diameter of a set A.
$\Rightarrow$: implies.
$\exists$: there exists.
$\forall$: for all.
a.e.: almost everywhere.

Differential calculus

$$D^\alpha v = \frac{\partial^{|\alpha|} v}{\partial x_1^{|\alpha_1|} \partial x_2^{|\alpha_2|} \cdots \partial x_d^{|\alpha_d|}}, \quad |\alpha| = \sum_{i=1}^{d} \alpha_i.$$

$\operatorname{div} q = \sum_{i=1}^{d} \frac{\partial q_i}{\partial x_i}$: divergence of a vector-function q in $\mathbb{R}^d$.
$e(v) = (e_{ij}(v))_{i,j=1}^2$, $e_{ij}(v) = \frac{1}{2}(\partial v_i/\partial x_j + \partial v_j/\partial x_i)$: linearized strain tensor.
$\boldsymbol{\nabla} u = (\partial u/\partial x_i)_{i=1}^d$ also denoted $\operatorname{grad} u$: gradient of a function u in $\mathbb{R}^d$.
$\Delta u = \sum_{i=1}^{d} \partial^2 u/\partial x_i^2$.
$n = (n_i)_{i=1}^d$: unit (outward) normal vector.
$t = (t_i)_{i=1}^d$: unit tangential vector.
$\frac{\partial}{\partial n} = \sum n_i \partial/\partial x_i$: (outward) normal derivative operator.

Notation for vector spaces

$\|\cdot\|_X$: norm (in the space X).
$|\cdot|_X$: seminorm (in the space X).
$\mathscr{L}(X;Y)$: space of continuous linear mappings from X into Y.
X': dual of a space X.
$\langle\cdot,\cdot\rangle$: duality pairing between a space and its dual.
$x+Y$: $\{x+y \mid y\in Y\}$.
X/Y: quotient space of X by Y.
I: identity mapping.
$\hookrightarrow$: inclusion with continuous injection.
$\hookrightarrow\hookrightarrow$: inclusion with compact injection.
$\dim X$: dimension of a space X.
$X\times Y$: Cartesian product of the set X and Y.
$v_n\to v$: strong convergence of a sequence $\{v_n\}$.
$v_n\rightharpoonup v$: weak convergence of a sequence $\{v_n\}$.

Notation for specific vector spaces

$C^m(\overline{\Omega})$: space of functions whose derivatives up to order m are continuous in $\overline{\Omega}$.
$C(\overline{\Omega})=C^0(\overline{\Omega})$.
$C^\infty(\overline{\Omega})=\bigcap_{m=0}^{\infty} C^m(\overline{\Omega})$.
$C_0^\infty(\Omega)=\{C^\infty(\overline{\Omega}) \mid \operatorname{supp} v$ is a compact subset of $\Omega\}$.
$W^{m,p}(\Omega)=\{v\in L^p(\Omega) \mid D^\alpha v\in L^p(\Omega)\ \forall\alpha,\ |\alpha|\leqslant m\}$: Sobolev space.
$W_0^{m,p}(\Omega)$: closure of $C_0^\infty(\Omega)$ in $W^{m,p}(\Omega)$.
$\|v\|_{m,p,\Omega}=\{\sum_{|\alpha|\leqslant m}\int_\Omega |D^\alpha v|^p\,\mathrm{d}x\}^{1/p},\ 1\leqslant p<\infty.$
$|v|_{m,p,\Omega}=\{\sum_{|\alpha|=m}\int_\Omega |D^\alpha v|^p\,\mathrm{d}x\}^{1/p},\ 1\leqslant p<\infty.$
$H^m(\Omega)=W^{m,2}(\Omega)$.
$H_0^m(\Omega)=W_0^{m,2}(\Omega)$.
$\|v\|_{m,\Omega}\equiv\|v\|_{m,2,\Omega}$.
$|v|_{m,\Omega}\equiv|v|_{m,2,\Omega}$.
$\mathscr{H}^k=\{u \mid u=(u',u'')\in[H^k(\Omega')]^2\times[H^k(\Omega'')]^2\}$.
$|\!|\!|v|\!|\!|_k=(\|v'\|_{k,\Omega'}^2+\|v''\|_{k,\Omega''}^2)^{1/2},\ v\in\mathscr{H}^k.$
If $v\in\mathscr{H}^1$ then $|v|\equiv((e_{ij}(v'),e_{ij}(v'))_{0,\Omega'}^2+(e_{ij}(v''),e_{ij}(v''))_{0,\Omega''}^2)^{1/2}$.
$\|v\|_{m,\infty,\Omega}=\max_{|\alpha|\leqslant m}\{\operatorname{ess\,sup}_{x\in\Omega}|D^\alpha v(x)|\}$.
$|v|_{m,\infty,\Omega}=\max_{|\alpha|=m}\{\operatorname{ess\,sup}_{x\in\Omega}|D^\alpha v(x)|\}$.
$(u,v)_0=(u,v)_{0,\Omega}=\int_\Omega uv\,\mathrm{d}x$: scalar product in $L^2(\Omega)$.

Spaces of polynomials

P_k: space of all polynomials in $x_1, \ldots, x_d$ of degree $\leqslant k$.
Q_k: space of all polynomials in $x_1, \ldots, x_d$ of degree $\leqslant k$ with respect to each variable x_i, $1 \leqslant i \leqslant d$.

Notation in $\mathbb{R}^d$

$\|a\| = \left(\sum_{i=1}^{d} |a_i|^2\right)^{1/2}$: Euclidean norm of the vector a.

$a \cdot b = \sum_{i=1}^{d} a_i b_i$, also denoted $a_i b_i$: scalar inner product of the vectors a and b.

$a \times b = \begin{vmatrix} e_1 & e_2 & e_3 \\ a_1 & a_2 & a_3 \\ b_1 & b_2 & b_3 \end{vmatrix}$: vector product of the vectors a and b in $\mathbb{R}^3$, where $e_1 = (1,0,0)^{\mathrm{T}}$, $e_2 = (0,1,0)^{\mathrm{T}}$, $e_3 = (0,0,1)^{\mathrm{T}}$.

$\operatorname{meas}(A) = \int_A \mathrm{d}x$: Lebesgue measure of a set $A \subset \mathbb{R}^d$.

$a_{ij} b_{ij} = \sum_{i,j=1}^{d} a_{ij} b_{ij}$: summation convention.

Subject Index

Mathematical Modelling of Rods

L. Trabucho

Centro de Matemática e Aplicações Fundamentais
Universidade de Lisboa
Av. Prof. Gama Pinto 2
1699 Lisboa Codex, Portugal

J.M. Viaño

Departamento de Matemática Aplicada
Universidad de Santiago de Compostela
15706 Santiago de Compostela, Spain

HANDBOOK OF NUMERICAL ANALYSIS, VOL. IV
Finite Element Methods (Part 2)—Numerical Methods for Solids (Part 2)
Edited by P.G. Ciarlet and J.L. Lions

Contents

Preface

This work constitutes an up-to-date and unified account of the contributions of the authors and their collaborators to the field of asymptotic modelling of elastic rods and their relationship with other methods. It represents what we know about the mathematical derivation and justification of models describing the elastic behaviour of rod type structures obtained via asymptotic expansion methods, combined with techniques from functional analysis and numerical analysis.

The necessity of possessing powerful models governing the equilibrium and the quasi-static or dynamic behaviour of both thick- and thin-walled structures is undoubtedly a major technical and industrial problem. Specific beam models have by now been known for long. As examples we mention the classical theories of Bernoulli–Navier, Euler, Saint Venant, Timoshenko, Vlassov, and so on.

In spite of its long historical background, rod theory is not an old-fashioned topic. It is not even a "closed" subject! In fact, the present technological needs for the conception of simultaneously light, resistent and flexible structures (oil recovery platforms, space structures, towers, bridges, cables, antennae, robots, surgical instruments, and so on) together with the use of new materials and new design methods represent a wide field of applied mathematics.

On the other hand the fast development of both numerical software and numerical methods of analysis during the last decades has made the exploration of more sophisticated mathematical models possible. This also justifies the mathematical search for even more accurate models. It is in this direction that we would like our contribution to be understood.

This article is the result of the work of many people with whom we had the opportunity to collaborate. We thank them for all the improvements they introduced and that we learned from them. We would especially like to thank them for allowing us to put everything together in a book form. These are their names:

J.A. Álvarez-Dios—University of Santiago de Compostela and University of La Coruña, Spain.

L.J. Álvarez-Vázquez—University of Santiago de Compostela and University of Vigo, Spain.

J.C. Barros—Instituto Superior Técnico, Lisbon, Portugal.

A. Bermúdez—University of Santiago de Compostela, Spain.

D. Camotim—Instituto Superior Técnico, Lisbon, Portugal.

I.N. Figueiredo—University of Coimbra, Portugal.

M.L. Mascarenhas—University of Lisbon and Centro de Matemática e Aplicações Fundamentais, Lisbon, Portugal.

B. Miara—Ecole Supérieure d'Ingénieurs en Électrotechnique et Électronique, Noisy-le-Grand, France.

J.M. Rodríguez—University of Santiago de Compostela, Spain.

M.F. Veiga—Centro de Matemática e Aplicações Fundamentais, Lisbon, Portugal.

In some parts of the work we were also assisted by students A. Bravo, T. Fernández and M. Vilares (University of Santiago de Compostela, Spain). We express our sincere gratitude to them.

The work is divided into ten chapters which can be read almost independently. However, in order to avoid unnecessary repetitions some notations and techniques of the asymptotic method are only detailed in the first chapter. Consequently, its reading may be necessary for those not familiar with the application of the asymptotic method in elasticity theory. Due to the technical complexity of some expressions, we are sure that in spite of all our efforts to produce an error-free text, some misprints are likely to have been introduced in the typesetting of the manuscript. We apologize for them as well as for any other errors the text might contain, and we thank in advance those readers who would be so kind as to point them out to us.

We are specially grateful to our colleagues A. Bermúdez, coauthor of an original paper introducing the asymptotic method for elastic rods, and to M.L. Mascarenhas, for the great scientific influence that their collaboration exerted upon us and for all the interest they have always shown towards our work.

The collaboration between the authors on this research topic was initiated in 1986 when they were invited by Professor P.G. Ciarlet to spend one semester at the Laboratoire d'Analyse Numérique of the Université Pierre et Marie Curie, Paris, France. We thank him for this kind invitation and his continued interest in our research.

Since then and at several stages of the work we have had the opportunity of discussing it with different colleagues who greatly contributed to our knowledge on the subject. In particular we must thank C. Conca (University of Santiago de Chile, Chile) for some extremely fruitful discussions about "boundary layers", H. Le Dret (University Pierre et Marie Curie, Paris, France) for having taught us, among other things, how to evaluate the first order stress components using higher order terms, and A. Raoult (University Joseph Fourier, Grenoble, France) who not only read some of our original papers carefully and corrected some mistakes, but also showed us the relationship between the displacement and the mixed displacement-stress approach, which we have adopted ever since.

We also wish to express our gratitude to J.A. Álvarez-Dios for his suggestions towards improving the rendering into English of the early versions of the manuscript and, in general, to all our colleagues at both the Centro de Matemática e Aplicações Fundamentais (C.M.A.F., Lisbon), and the Departamento de Matemática Aplicada (Santiago de Compostela), for all their support during the elaboration of this manuscript.

Last, but not least, we express our sincere gratitude to Profs. P.G. Ciarlet and J.L. Lions for having invited us to contribute to the *Handbook of Numerical Analysis* series.

Luis Trabacho
Juan M. Viaño
December 1992

Several parts of this work and the preparation of the manuscript itself have received financial support from different institutions through the following research projects:

Métodos Assimptóticos e de Homogeneização para uma Classe de Problemas em Mecânica dos Sólidos. Junta Nacional de Investigação Científica e Tecnológica (J.N.I.C.T.), Project 87 593, 1987, Portugal.

Análise duma Classe de Problemas em Mecânica dos Sólidos – Analisis Asintótico y Numérico de una Clase de Problemas en Mecánica de Sólidos. Acção Integrada Luso–Espanhola, 1990, Portugal and Spain.

Analisis Teórico y Numérico de Nuevos Modelos en Elastoplasticidad de Vigas, Placas e Láminas. Dirección General de Investigación Científica y Técnica (D.G.I.C.Y.T.) and Xunta de Galicia, Project PB 87–0481, 1988–1990, Spain.

Simulación Numérica en Vigas Elásticas. Study-Contract, Universidad de Santiago de Compostela (Departamento de Matemática Aplicada) and I.B.M., 1991–1992, Spain.

Junctions in Elastic Multistructures. Project SCIENCE of the Commission of the European Communities, SC1* 0473–C(EDB), 1991–1993, University Pierre et Marie Curie (France), C.M.A.F. (Portugal), University of Santiago de Compostela (Spain).

Introduction

From the mechanical point of view three main categories of structures may be distinguished according to their spatial "dimensions": *three-dimensional solids, plates and shells, and rods.*

Three-dimensional solids are "massive" structures whose dimensions in all three spatial directions are of the same order of magnitude, such as in a sphere, a cube, an ellipsoid, a parallelepiped bounded by rectangles not very oblong, and so on. The computation of the displacement and stress fields in these solid structures as well as some of the methods of resolution constitute the objective of the *mathematical theory of elasticity*, the foundations of which may be seen in the classical books by LOVE [1929], TIMOSHENKO and GOODIER [1951], SOKOLNIKOFF [1956], LEKHNITSKII [1977], or in the more recent books of TRUESDELL and NOLL [1965], DUVAUT and LIONS [1972a], FICHERA [1972], GERMAIN [1972], GURTIN [1972], WANG and TRUESDELL [1973], VALID [1981], MARSDEN and HUGHES [1983], CIARLET [1988].

Mathematical modelling of elastic plates and shells: Classical and asymptotic methods

Plates and shells are three-dimensional solids that may be in a certain sense approximated by a two-dimensional surface when one of its spatial dimensions, (the thickness, or the ratio between the thickness and the minimum radius of curvature) is much "smaller" than the others. The most common examples of such structures are thin and thick plates, as well as shells of known mean surface (cylindrical, conical, spherical, elliptic, hyperbolic, and so on). These structures of very small thickness are widely used in all the industrial applications such as building, aeronautics, astronautics, in the navy and in many other technological applications.

In spite of the validity of the three-dimensional elasticity models for these cases, the geometric characteristics of this type of solids make them ill conditioned from a numerical point of view. Together with the difficulty of computing three-dimensional solutions, this is one of the reasons why the *two-dimensional models for plates and shells* are so important, that is to say, models in which the unknowns are expressed as a function of the middle surface (of the plate or of the shell). There is an extensive and specific literature about plates and shells and the list of references we give is not by any means exhaustive.

Within classical theories, the two-dimensional models are obtained from the three-dimensional elasticity equations by making a priori assumptions on the form of some of the unknowns (displacement, stress components) which try to account "in the best possible way" for the fact that the structure possesses a "small thickness". Several developments along this line may be found in NOVOZHILOV [1959], TIMOSHENKO and WOINOWSKY–KRIEGER [1959], BUDIANSKY and SANDERS [1967], LANDAU and LIFCHITZ [1967], NIORDSON [1969], KOITER [1970], DUVAUT and LIONS [1972a], NAGHDI [1972], FRAEJIS DE VEUBEKE [1979], DIKMEN [1982], and references therein.

A second type of methods that have a great influence in plate and shell modellings are the *asymptotic methods*. In a general sense these methods are based on a formal expansion of the solution in power series in terms of a small parameter that characterizes the problem. Methods to handle problems depending on a small parameter have been developed from the earlier works of VAN DYKE [1964], KAPLUN [1967] and COLE [1968]. More general and systematic expositions of asymptotic methods are to be found in LAGESTROM and CASTEN [1972], LIONS [1973], NAYFEH [1973], O'MALLEY [1974] and ECKHAUS [1979].

The *asymptotic methods* applied to plates and shells, in the framework of elasticity theory, are based on the the construction of the successive terms of an asymptotic expansion of the three-dimensional solution, in terms of the thickness ("small parameter" for the problem). The first contributions in this direction were those of FRIEDRICHS and DRESSLER [1961], GOLDENVEIZER [1962] and JOHN [1971].

In these works, the asymptotic expansion method is applied directly to the equilibrium equations of three-dimensional elasticity, that is, *not to their variational formulation*. Nevertheless, some a priori hypotheses are necessary and some difficulties associated to the boundary conditions also appear. Moreover, it has not been possible to justify these methods by an appropriate convergence analysis.

By the end of the seventies, *another application of the asymptotic method* appeared, that not only avoids the difficulties mentioned above, but also constitutes a fundamental tool to derive and mathematically justify, in a systematic way, the two-dimensional models associated to plate and shell theories. The success of the method seems to rely both on the fact that it is applied directly to the three-dimensional equations but in their *variational formulation* and on an appropriate *transformation to an equivalent problem* posed on a *fixed reference domain* which *does not depend* on the small parameter. One is then in a position to apply the general theory of asymptotic expansions of LIONS [1973], for variational problems depending on a small parameter.

The first results in this direction were obtained by CIARLET and DESTUYNDER [1979a,b] in order to justify the bi-harmonic model of Kirchhoff–Love for the bending of plates in the framework of linearized and nonlinear elasticity. The applications of the method soon considered the most well-known theories of plates and shells for a great variety of situations: linear and nonlinear cases, anisotropic and composite materials, dynamic and thermoelastic cases, homogenization, and so on. Without attempting to be exhaustive, we mention for the plate case the works of CIARLET and DESTUYNDER [1979b], CAILLERIE [1980], CIARLET [1980], DESTUYNDER [1980], RAOULT [1980], BLANCHARD [1981], DESTUYNDER [1981], CIARLET and KESAVAN [1981], DESTUYNDER [1982], BLANCHARD and CIARLET [1983], VIAÑO [1983], CAILLERIE [1984], KOHN and VOGELIUS [1984, 1985], RAOULT [1985], CIORANESCU and SAINT JEAN PAULIN [1986], DAVET [1986], DESTUYNDER [1986], KOHN and VOGELIUS [1986], BLANCHARD and FRANCFORT [1987], CAILLERIE [1987], CIARLET [1987], RAOULT [1988], CIARLET [1989], CIARLET and LE DRET [1989], QUINTELA-ESTEVEZ [1989], XIANG [1989]. CIARLET [1990], RAOULT [1990a], MIARA [1992a,b]. A complete analysis of the plate case, together with exhaustive bibliographic references, may be found in CIARLET [1990]. For the case of shells, the asymptotic method has been considered by DESTUYNDER [1980, 1985], CIARLET and PAUMIER [1986], FIGUEIREDO [1989, 1990], CIARLET and MIARA [1992], among others. We also mention the new perspectives pointed out in the work of SANCHEZ-PALENCIA [1989a,b, 1990].

Mathematical modelling of elastic rods

In a broad sense, the class of bodies that are called *rods* is formed by all solids for which *two of their characteristic dimensions are much smaller than the third one, its length*. In this group, we include a great variety of structural elements commonly used in a lot of practical applications, because of their lightness and their ease of manufactoring and transport, to name but a few. These comprise: "usual" rods, beams, bars, cables, axes, arches, pipelines, rails, antennae, and so on.

In a strict sense, we call a *rod* any three-dimensional solid occupying the volume generated by a plane connected domain when its centroid varies perpendicularly to a spatial curve, its axis. The plane surface is called the *cross section* of the rod. We also consider the case where the cross section may vary both in shape and size along the axis of the rod, in which case one should assume that this change takes place very slowly. We shall say the beam is *straight* whenever its axis is a straight line segment. Moreover, if in this case the cross section is constant, the beam is called *prismatic*; otherwise we simply call it a *curved rod* or a *curved beam*. If the cross section possesses one or more holes (i.e., it is not simply connected), we call it a *tube*.

Although we consider in this work the case of curved rods with a variable cross section, the greatest emphasis will be laid upon the *prismatic rods* not only because they are the most studied cases in the classical literature, but also because their study will nearly always contain all the ingredients of the most complex cases. Consequently, *unless otherwise mentioned*, the word *rod* or *beam* will be used in the classical sense of a right prismatic rod with a connected cross section (possibly with holes).

The main geometric characteristic of a rod is the fact that *the diameter of the cross section is much smaller than its length*. As for plates and shells, the *classical methods* take advantage of this geometric characteristic in order to introduce some a priori assumptions which greatly simplify the three-dimensional elasticity system in that it reduces to an easier-to-calculate, two- or one-dimensional model. These a priori assumptions depend on the type of applied loads, and in the literature it is customary to distinguish between the *stretching–bending* and the *torsion* cases.

The simplest case for the stretching and bending of rods is given in the *Bernoulli–Navier theory*, also called the *"engineering beam theory"*, which was formulated between 1691 and 1744. It is based on the *law of plane sections* or *Bernoulli's assumptions* stating that *plane sections perpendicular to the axis of the beam before deformation remain plane and perpendicular to the same axis after deformation*, and on Navier's assumption, proposed in 1823, stating that *there is no deformation of the cross section on its own plane.*

As a consequence, only the axial component is considered to be different from zero, among the six independent components of the stress tensor. In particular the shear stresses are assumed to vanish in contradiction with the existence of shear forces. In particular, the variation of the shear stress components along the cross section of relatively thick beams cannot be neglected. This is the reason why in 1921 S.P. Timoshenko introduced an amendment to the previous theory, stating that *plane sections perpendicular to the axis of the beam before deformation remain plane but not necessarily perpendicular to that same axis after deformation* (cf. TIMOSHENKO [1921]). Timoshenko's theory of bending, together with the introduction of the constant which bears his name, has been extremely useful in engineering in spite of somewhat obscure consequences of this assumption.

Regarding the study of the deformation of bars subjected to torsional loads, almost all the classical theories are based on the *pure torsion theory of Barré de Saint Venant* formulated around 1855, and based on the a priori assumption that *one may neglect all the deformations on the plane of the cross section as well as the stretching along the longitudinal axis*, that is, all the cross sections have the same warping. The oldest reference is SAINT VENANT [1855].

One of the most complete and accurate classical theories formulated in this way is due to B.Z. Vlassov and was formulated between 1940 and 1949. It is an elegant and powerful combination of mathematical analysis and mechanics which allowed the author to obtain some extremely simple and clear results. It includes *combined stretching and bending and torsion deformations* based also on *Navier's assumption* according to which *one may neglect the deformation of the cross section on its own plane although each cross section may possess its own rotation and warping*. It was only after a translation of VLASSOV [1961, 1962] that Vlassov's work became known in the West.

In none of the rod theories mentioned so far are the deformations of the cross section on its own plane taken into account. This deformation, which is due mainly to the *Poisson effect* and which is calculated by engineers from an a priori assumption on the planar displacement components is, somehow "added" to the displacement field given by the classical theories.

Moreover, the classical stretching (or buckling), bending and torsion theories assume in practice the validity of the *Saint Venant principle*. In one of the many ways it has been formulated it can be stated as: "two different distributions of force acting on the same portion of a body have essentially the same effects on parts of the body which are sufficiently far from the region of application, provided that these force distributions have the same resultant".

The consideration of the different a priori assumptions and the computation of the solutions associated with the classical rod theories mentioned above are the main object of most of the classical courses on strength of materials and are always part of the various monographs on the subject. We mention some of them, which should be consulted for completeness: TIMOSHENKO and GOODIER [1951], SOKOLNIKOFF [1956], VLASSOV [1961, 1962], GERMAIN [1962], FUNG [1965], TRUESDELL and NOLL [1965], LANDAU and LIFCHITZ [1967], GREEN and ZERNA [1968], GERMAIN [1972], DYM and SHAMES [1973], LEKHNITSKII [1977], ANTMAN and KENNEY [1981], NEČAS and HLAVÁČEK [1981], LAROZE [1988]

In the general case, the classical models are used for the computation of the displacement field in rods independently of the relative dimensions of the cross section. Due mainly to the progress in Aeronautics and Astronautics, in the last half of the century a great emphasis has been made in the use of rods having a cross section characterized by two dimensions with a different order of magnitude. As three-dimensional solids these rods have as a geometric characteristic the fact that *each of their dimensions in the direction of the coordinate axes is of a different order of magnitude*. VLASSOV [1962] called this type of solids *thin-walled beams* reserving the term *solid beams* for those rods whose cross section possesses both dimensions of the same order of magnitude.

Structures made of thin-walled beams are extensively used in industry, thus permitting a maximum of rigidity with a minimum weight, not to mention the aestetics of the construction. They can be found in a variety of applications such as the roof of industrial buildings, the main structure of a bridge, a hydraulic pipeline, the body of an aircraft or of a rocket, the lateral surface of a ship or a submarine, i.e., in almost all structural applications, and even in other fields such as biomechanics, for instance.

The main reason why these thin-walled structures need an independent treatment stems from the fact that both the shear deformations and stresses cannot be neglected as in the case of a solid cross section. This fact challenges both the validity of Bernoulli's assumption on the deformation of plane sections and Saint Venant's pure torsion theory.

As mentioned earlier, thin-walled beam theory was developed by VLASSOV [1961, 1962]. It is based on a more general and more natural a priori hypothesis than Bernoulli–Navier's assumption on plane sections: invariance of the section contour and absence of shear stresses in the middle surface, which constitute the basis for a new law of distribution of longitudinal stresses in the cross section called *law of sectorial areas* by the author. Besides the fundamental reference mentioned above in this paragraph, one may also consult ODEN [1967], ODEN and RIPPERGER [1981], MURRAY [1986] and the references therein.

In the preface of his book, VLASSOV [1962] makes the following remarks concerning the distinction between rods with solid and thin-walled cross sections:

> "This classification has a qualitative character, above all, for it is based on hypotheses that allow us to classify structures in each class according to the way they are calculated. For this reason it is not possible to establish a clear boundary between both classes. The same structure, subjected to different loading and working conditions and for different requirements on the precision of the calculations, may be considered as belonging to either class. This is the case of a closed thin-walled tubular rod subjected to both torsional and bending loads. If the profile is considered to be rigid, one may consider it as made of a solid cross section, but on the other hand, if the profile is supposed to be flexible (deformable) it should be considered a thin-walled beam. Moreover, in some very special, but very important problems, such as the deformation of rails, beams of rectangular cross section, among others, it is necessary to introduce the same type of corrections as in the thin-walled case in order to take into account the axial stresses due to the warping of the different sections."

Mathematical modelling of elastic rods: Asymptotic methods

In the mid-seventies and through the influence of several works in the theory of plates and shells, the application of *asymptotic expansion methods* for *obtaining and justifying different models of elastic beams* started. In these methods, *the main idea is to approximate the three-dimensional solution of the elasticity equations through the successive terms of a power series where the diameter of the cross section is taken as a small parameter.*

The first works in this direction are owed to RIGOLOT [1976, 1977a,b], who adapted to rods the ideas of GOLDENVEIZER [1962] for plates. The same idea is also used in WIDERA, FAN and AFSHARI [1989]. When problems are formulated as partial differential equations, the same difficulties as in the case of plates arise, namely: necessity of some a priori hypotheses, restrictions on the applied loads, difficulties associated to the boundary conditions and to the study of the convergence.

The combination of *variational theory and asymptotic expansion methods* applied to the three-dimensional elasticity problem for rods was initiated by BERMÚDEZ and VIAÑO [1984], who adapted the ideas set forth by CIARLET and DESTUYNDER [1979a,b] for plates. Since then this method has met great success in the generalization and mathematical justification of models for elastic rods.

The full application of the method consists of the following steps:

(i) As a starting point one regards the rod as a three-dimensional elastic body modelled by the equilibrium equations written in the variational form (principle of virtual work) together with a constitutive law (linear, nonlinear, thermoelastic, and so on). This model is assumed to be "exact" for the three-dimensional solid.

(ii) Next we "imbed" the beam under study into a family of rods whose cross sections are homothetic, with ratio ε, to a given reference section of constant area A, e.g., $A = 1$. Since *ε is of the same order as the diameter of the cross section, it is chosen as a small parameter in the problem.* For each rod in the family, one explicitly defines the system of applied forces as well as material properties (elasticity and conductivity coefficients, specific mass, and so on).

(iii) Through an appropriate *change of variable, one transforms the original problem into an equivalent one now posed in a fixed reference domain* which does not depend on the diameter of the cross section and to which it is possible to apply the general theory of asymptotic expansions for variational problems with a small parameter developed by LIONS [1973].

(iv) Whenever possible, the convergence of the problem is studied in order to justify the formal asymptotic expansion, or as many as possible of the successive terms. Although it is possible to evaluate any terms in the asymptotic development no matter how high their order, we have performed such computations up to the fourth term only, because it is the minimum required to recover all the classical theories and in order to keep cumbersome notation down to a reasonable limit.

(v) Finally, the inverse change of variable is considered and the results are written with respect to the original domain. In this way it is possible to obtain successive approximations of the three-dimensional elasticity problem for a given ε.

With this procedure, one then hopes to find not only the best known models but also to refine, or even to complement, them.

In this fashion, BERMÚDEZ and VIAÑO [1984] have mathematically justified Bernoulli–Navier's beam theory simply by identifying the first term in the asymptotic expansion for the linearized three-dimensional elasticity problem. With the same technique, starting from a general nonlinear constitutive law, CIMETIÈRE, GEYMONAT, LE DRET, RAOULT and TUTEK [1986] found nonlinear rod models and suggested the way to evaluate higher-order terms. Using these ideas in the linear case, TRABUCHO and VIAÑO [1987, 1989] calculated the first three terms of the displacement field together with the first two terms of the stress field. This allowed them to justify and to generalize the classical theories of Bernoulli–Navier, Saint Venant, Timoshenko and Vlassov (cf. TRABUCHO and VIAÑO [1987, 1988, 1989, 1990a,b]).

Using and adapting these ideas, the method has been successfully applied to a wide variety of rod problems in the framework of elasticity theory. It is these different applications together with their consequences that form this work. The topics under study are the following: isotropic, homogeneous, linearly elastic rods (BERMÚDEZ and VIAÑO [1984], TRABUCHO and VIAÑO [1987, 1988, 1989, 1990a,b]), rods of variable cross section (BERMÚDEZ and VIAÑO [1984], VEIGA [1993b]), static and dynamic linearly thermoelastic rods (BERMÚDEZ and VIAÑO [1984], ÁLVAREZ-VÁZQUEZ and VIAÑO [1989, 1991a,b,c, 1992a], ÁLVAREZ-

VÁZQUEZ [1991]), transversely anisotropic and nonhomogeneous rods (ÁLVAREZ-DIOS and VIAÑO [1988, 1991a,b, 1992a,b], ÁLVAREZ-DIOS [1992], SANCHEZ-HUBERT and SANCHEZ-PALENCIA [1991]), rods in unilateral contact with a rigid or elastic foundation (VIAÑO [1985a,b], ÁLVAREZ-DIOS [1986]), nonlinearly elastic rods (CIMETIÈRE, GEYMONAT, LE DRET, RAOULT and TUTEK [1986, 1988], CAMOTIM and TRABUCHO [1989]), dynamic nonlinearly elastic rods (ÁLVAREZ-VÁZQUEZ [1991], ÁLVAREZ-VÁZQUEZ and VIAÑO [1991d, 1992b]), rods made of composite materials (GEYMONAT, KRASUCKI and MARIGO [1987a,b], TUTEK [1987]).

Some of the results obtained were the starting point for new developments. Some of them have been included in this work. These are: homogenization of rods with a periodic multicellular cross section (TUTEK [1987], MASCARENHAS and TRABUCHO [1990, 1991]), asymptotic analysis of asymptotic models for thin-walled structures (BARROS [1989], BARROS, TRABUCHO and VIAÑO [1989], RODRIGUEZ [1990, 1993], RODRIGUEZ and VIAÑO [1991, 1992a,b,c]), numerical calculation of the coefficients depending on and characterizing the cross section and comparison between the asymptotic and the three-dimensional models (VILARES [1988], RODRIGUEZ [1990], TRABUCHO and VIAÑO [1990a], VIAÑO [1990], FERNANDEZ and VIAÑO [1992]), optimal Galerkin type methods applied to the approximation of the three-dimensional elasticity solution for rods (MIARA and TRABUCHO [1990, 1992], MASCARENHAS and TRABUCHO [1991]), optimal Galerkin type methods applied to the approximation of the three-dimensional elasticity solution for curved rods (FIGUEIREDO and TRABUCHO [1992, 1993]).

This broad scope of applications can be justified by the following properties inherent to the asymptotic expansion method:

(a) *General applicability*. In fact, because of its very general formulation it is easily adapted to the different constitutive laws (linear, nonlinear, thermoelastic, anisotropic, nonhomogeneous, and so on) *without imposing additional restrictions either on the system of applied loads or on the shape of the cross section*. Consequently, the three dimensional models obtained by the asymptotic expansion method should be considered as successive approximations of the three-dimensional elasticity equations valid both for "open" or "closed", solid or thin-walled cross sections. In this respect, it is worthwile to recall the previously mentioned observation by VLASSOV [1962], and to remark that, although the natural starting small parameter is the diameter of the cross section ε, *the true parameter intervening in the asymptotic expansion is the area of the cross section. This is the genuine "small" parameter.*

The validity of the asymptotic model for any cross section being thus assumed, a new way to obtain the classical models for thin-walled beams is to perform an *asymptotic analysis on the asymptotic model*, as one of the characteristic dimensions of the cross section becomes much smaller then the other (see references earlier mentioned).

(b) *Physical meaning*. The asymptotic expansion method gives, in a successive, explicit, and separate fashion, the various effects associated with stretching, bending, torsion, Poisson's effects and the deformation of the cross section within its own plane, and it associates with each one of these the respective order of magnitude as function of the applied loads. In this way Bernoulli–Navier's beam theory appears associated with the first order terms in the asymptotic expansion, while Saint Venant's and Timoshenko's theories are associated with second order terms. Also associated with second order terms are Poisson's effects and some additional warping effects which are not considered in the classical theories. The deformation of the cross section within its own plane, not considered in any classical theory, is associated here with fourth-order terms. In the same way, the stress tensor, bending moment and shear force components are associated with different terms of the asymptotic expansion. *A hierarchy is thus established among the classical theories*, and since the results obtained are for very general loading conditions and geometries, it is easy to assess the intrinsic importance of each term in its relation to the ability to "construct" a specific model adapted to a specific loading condition or to a specific geometry.

(c) *Mathematical rigour.* The asymptotic model does not use the classical a priori assumptions inherent to the classical rod theories, which may be unsuitable when high accuracy is needed. On the contrary, *the asymptotic method justifies these assumptions* for they are obtained as necessary and sufficient conditions for the *convergence results* which validate the asymptotic method (BERMÚDEZ and VIAÑO [1984], TUTEK and AGANOVIČ [1987], MASCARENHAS and TRABUCHO [1990], ÁLVAREZ-DIOS [1992], ÁLVAREZ-DIOS and VIAÑO [1991a,b]).

Accepting at the onset the validity of the three-dimensional model, the successive terms of the asymptotic expansion may be described as sums of products of functions of one variable by functions of two variables. This is also the case in plate theory and was one of the reasons that led to the application of the Galerkin type method to rods by MIARA and TRABUCHO [1990], MASCARENHAS and TRABUCHO [1991], MIARA and TRABUCHO [1992] and FIGUEIREDO and TRABUCHO [1992b], where the basis functions found in the method are precisely linear combinations of the warping and torsion functions, among others.

The coefficients of this linear combination have an important physical meaning for they may be the coordinates of the shear centre, area moments or area bimoments, for example. *The functions of two variables are either polynomials, or solutions of linear elliptic boundary value problems defined on the cross section of the beam (torsion, warping and Timoshenko's functions). The one-dimensional functions are given as solutions of second- and fourth-order boundary value problems, and they constitute the governing equations associated to each model.* Depending on the case under study, be it stationary, dynamic, linear, nonlinear, . . . , these one-dimensional problems may be elliptic, parabolic, hyperbolic, linear, nonlinear, . . . , respectively.

It is thus understandable that in the numerical approximation of these models, all the standard numerical techniques for approximating the solution of boundary value problems for partial differential equations are of interest, in spite of the fact that *the equilibrium equations are second- and fourth-order elliptic or-*

dinary differential equations, which are extensively considered in the classical literature. We do not specifically address the numerical aspects in this work, but we nevertheless give some references among the rather long list available. As references on the mathematical foundations of numerical methods for partial differential equations, we mention: ODEN and REDDY [1976], CIARLET [1978], FAIRWEATHER [1978], MITCHELL and GRIFFITHS [1980], ODEN and CAREY [1981-1984], BATHE [1982], CIARLET [1982], RAVIART and THOMAS [1983], AXELSSON and BAKER [1984], DAUTRAY and LIONS [1984], DHAT and TOUZOT [1984], GLOWINSKI [1984], THOMÉE [1984], HUGHES [1987], JOHNSON [1987], JOHNSON, NÄVERT and PITKÄRANTA [1984], THOMÉE [1990], FUJITA and SUZUKI [1991].

For a more specific numerical treatment in structural mechanics, we refer to: ZIENKIEWICZ [1971]. BREBBIA and CONNOR [1974], CIARLET [1974], BATHE and WILSON [1976], KAMAL and WOLF [1977], BERNADOU and DUCATEL [1978], HINTON and OWEN [1979], MEIROVITCH [1980], BATHE [1982], BERNADOU and BOISSERIE [1982], BERNADOU and DUCATEL [1982], ODEN and CAREY [1984, vol. IV], REDDY [1984], CRISFIELD [1986], KIKUCHI [1986].

We finally mention some references related to the more specific numerical analysis for specific applications in beam theory: BOURGAT, DUMAY and GLOWINSKI [1980], BERCOVIER [1982], BERMÚDEZ and VIAÑO [1983], BERMÚDEZ and FERNANDEZ [1986], COMPE, GEORGE, ROUSSELET and VIDRASCU [1986], SIMO and VU QUOC [1986], BERNADOU [1987], BOURGAT, LE TALLEC and MANI [1988], LE TALLEC, MANI and ROCHINHA [1992].

Nowadays, the asymptotic expansion method is also applied to other fields of great practical interest such as *mathematical modelling of junctions in elastic multi-structures*. As a typical example of a multi-structure we mention those formed by a "three-dimensional" part and a "two-dimensional" one, for example a plate clamped on a three-dimensional solid. The mathematical notion of junction is in fact very general for it may include the cases of folded plates and shells, I-, L- and T-shaped beams, plates and shells with stiffeners, plates and shells supported by rods, etc.

The asymptotic expansion methods for rods, plates and shells have their "meeting point" in the mathematical modelling of junctions from which a "mixed asymptotic method" is generated for the mathematical modelling of multi-structures. This approach is now well under way and will not be considered here, but we refer to the following works instead: CIARLET, LE DRET and NZENGWA [1987, 1989], BOURQUIN and CIARLET [1989], AUFRANC [1990], CIARLET [1990], BOURQUIN [1991], for the junctions between a three-dimensional solid and a plate; LE DRET [1989, 1990, 1991], BERNADOU, FAYOLLE and LÉNÉ [1989] for the junctions between two plates; FAYOLLE [1987] and GRUAIS [1990] for the junctions between plates and rods; LE DRET [1989, 1991], RODRIGUEZ and VIAÑO [1991, 1992a,b,c], LODS [1992], for junctions between two rods.

Finally, we must also mention an interesting and general approach to the mathematical modelling of junctions in elastic multi-structures given by LEGUILLON and SANCHEZ-PALENCIA [1990], MAMPASSI [1992], and MAMPASSI and SANCHEZ-PALENCIA [1992].

Overview of contents

Next, let us give a brief description of the topics treated in this work. More detailed information is to be found in the introductions of each individual chapter or section. The work is divided into ten chapters and fifty sections. Any reader unfamiliar with the application of the asymptotic method in elasticity theory should refer to Chapter I for notation and techniques. But for that the chapters are independent of one another and so it should be safe to read them separately.

Chapter I gives a complete description of the mathematical aspects associated with a clamped linear elastic beam. In Section 1, the variational formulation of the corresponding three-dimensional displacement problem is described. In Section 2, we introduce the change of variable to the reference domain together with the fundamental scalings on the unknowns and assumptions on the data, where the diameter of the cross section is the small parameter. The formal description of the asymptotic expansion of the displacement field together with the characterization of the first terms of the expansion are found in Section 3 and 4. Also to be found in Section 4 are the mathematical difficulties related to the boundary layer phenomenon associated with Saint Venant's principle and the clamping boundary condition. In Section 5 we characterize the asymptotic expansion of the displacement field for a weak clamping boundary condition. Section 6 is dedicated to the presentation of the (original) version of the asymptotic expansion method for the mixed displacement–stress formulation of the three-dimensional elasticity problem, and its relationship with the displacement approach. The characterization of the zeroth-, second- and fourth-order terms of the asymptotic expansion both for the displacement and the stress fields is the object of Sections 7 through 9. Finally, in Section 10, some convergence results for the displacement and stress fields, as well as bending and shear force components, are studied.

The main objective of Chapter II is to give an interpretation of the results obtained in Chapter I and to relate them to the classical linear beam theories. In order to do so we start by de-scaling the successive terms in the asymptotic expansion and thus finding, in this fashion, an asymptotic model of order zero (Section 11) and one of order two (Section 13). These models are compared to the classical theories of Bernoulli–Navier (Section 11), Vlassov (Section 12), Saint Venant (Section 14) and Timoshenko (Section 15). A correct definition of Timoshenko's constant is also introduced in Section 15. Several numerical examples are presented for the most common types of cross sections. All the results obtained are compared to the classical ones.

Chapter III is an introduction to the expanding field of thin-walled beam modelling. We restrict our attention to the rectangular case only, and another asymptotic analysis, where the thickness of the rectangle is now a smaller parameter is performed, but on the former asymptotic model itself. After a formal discussion, in Section 16, of several aspects of this second limit problem one observes that this analysis reduces to the study of the convergence and of the limit process for the Poisson equation in a thin rectangle, with either pure Dirichlet, or pure Neumann, boundary conditions. This is the object of Section 17. These results are then applied, in Section 18, to the warping and torsion functions for a thin rectangle and, in Section 19, to the functions and constants that appear in the generalization of Timoshenko's constant. A comparison of these results with those given by the classical theories is also made. In Section 20, some results related to Timoshenko's theory for thin-walled rectangular beams are obtained and compared to the numerical results of Section 15. We conclude this chapter by a discussion on the possibilities of applying the asymptotic expansion method in order to obtain general thin-walled models (Section 21).

Chapter IV is devoted to a generalization of the results presented in Chapters I and II to anisotropic, nonhomogeneous rods. In Section 22, the asymptotic method is applied to rods that are weakly clamped at one end and subjected to a loading on the other free end. The calculation of the first term in the asymptotic expansion, the convergence analysis, the interpretation of the coupling between stretching and bending in the model obtained together with its analysis for the case of fibre-reinforced rods are the objects of Section 23. The calculation of the second order terms leading to general theories with torsion, Timoshenko and Poisson's effects, is done separately for nonhomogeneous, isotropic rods and for homogeneous anisotropic rods, in Sections 24 and 25, respectively.

Chapter V consists of the application of the asymptotic analysis methods for obtaining rod models from three-dimensional linear thermoelastic models. In Section 26, the stationary thermoelastic case is considered, that is, where thermoelastic stresses are generated by a known temperature increment. The study of the three-dimensional problem, its asymptotic analysis, the calculation of the zeroth-order term, and the respective convergence results are presented. In Section 27, the asymptotic study of the fully coupled dynamic, linear, thermoelastic model is considered. To this end, the already mentioned change of scaling technique to transform the original problem into an equivalent one is used. Afterwards, the limit behaviour of this new problem is studied through the technique of a priori estimates and passage to the limit in the weak topology. Moreover, some sufficient conditions for strong convergence are also given.

Chapter VI is concerned with the application of the asymptotic expansion method to two particular cases. In Section 28 some unilateral contact models on a rigid or elastic foundation, without friction, are obtained. To this end, the same asymptotic expansion technique is applied to the study of the three-dimensional elastic problem, now posed as elliptic variational inequalities (Signorini's problem for the case of a rigid foundation) or as nonlinear elliptic variational equation for the case of an elastic foundation of the Winkler–Westergaard type. The models obtained in this fashion are compared to the classical ones. In Section 29, a variant of the asymptotic expansion method is applied to the case of a beam whose cross section may vary along its axis. The scalings on the unknowns and the assumptions on the data are defined, the zeroth and second order terms are computed and the bending–stretching–torsion models obtained are also interpreted.

The objective of Chapter VII is to obtain general models for the homogenized behaviour of multicellular rods with a periodic structure. Two approaches are given and it is proved that for axially perforated rods they both give the same limit models: the first consists in homogenizing the models obtained in the previous chapters after the asymptotic analysis has been performed (Sections 30 and 31); the second consists in homogenizing the three-dimensional linear elasticity model in the first place and then applying the asymptotic expansion method to this three-dimensional homogenized model (Section 32).

In Chapter VIII a Galerkin-type approximation of the three-dimensional linear elasticity solution is considered for straight rods. The scaled displacement field is approximated by linear combinations of products of functions depending on the variables associated with the cross section, and of functions depending on the axial coordinate only. Some error estimates are proved and the relationship between this approach and the asymptotic expansion method is established through some examples (Section 33). In Section 34, this technique is also applied to the multicellular case. A multicellular beam with a transversal periodic structure is again considered in Section 35, and the homogenization method is applied to the basis functions obtained after the application of the Galerkin method. In Section 36, this procedure is compared to the application of the Galerkin method to the three-dimensional homogenized elasticity problem. In the same section, it is also shown that the two techniques, homogenization and Galerkin, do not commute. Some error estimates are established, and it is concluded that although the approximations are not the same, they obey the same type of estimates, and consequently they have the same limit. In Sections 37, 38 and 39, the Galerkin method is also applied to the case of shallow curved rods. The three-dimensional elasticity problem is considered and an equivalent formulation is given for a straight rod with the same cross section, and afterwards for a straight reference rod with a unitary cross section. The techniques of Sections 33 to 36 are then applied and error estimates for the difference between the three-dimensional elasticity solution and the Galerkin approximation are established, now as a function also of the curvature and torsion tensors characterizing the curved axis of the rod.

Chapter IX is devoted to the application of the asymptotic expansion method to a nonlinearly elastic rod in the static case, the constitutive equation being that of a Saint Venant–Kirchhoff material. We follow the same lines as in Chapter I, that is, the three-dimensional nonlinear elasticity problem, for a clamped rod, is first considered in Section 40. Then we use the same fundamental scalings on the unknowns and assumptions on the data (Section 41). The asymptotic expansion technique is then applied to the displacement formulation in Section 42. In Sections 43 and 44, we derive the equilibrium equations associated with the zeroth order terms in the asymptotic expansion, and we establish their relationship with known nonlinear rod models. The chapter ends up with the equivalent study now in the mixed displacement–stress formulation.

The main objective of Chapter X is the mathematical justification of the well-known semilinear evolution equation modelling the buckling of a prismatic rod submitted to an axial loading. In Section 46, we begin by formulating the three-dimensional nonlinear dynamic elasticity problem, again for a rod made of a Saint Venant–Kirchhoff material and subjected to axial forces together with appropriate boundary conditions. In this fashion, we are able to pass to the limit as the diameter of the cross section goes to zero. The usual asymptotic analysis is then performed in Section 47, and the first term in the asymptotic expansion is computed in Section 48. The physical interpretation of the results obtained is given in Section 49, and in the final Section 50, these results are extended to more general loading conditions.

CHAPTER I

Asymptotic Expansion Method for a Linearly Elastic Clamped Rod

The purpose of this chapter is to generalize and justify from the mathematical point of view the classical equations used to model the stretching, bending and torsion deformations of a linearly elastic rod made of an isotropic material. To this end, we show that the solutions of the classical models are the $H^1(\Omega)$ limit, when the area of the cross section becomes small, of the corresponding three-dimensional equations of linearized elasticity, after a suitable rescaling.

The first convergence result of this type was obtained by BERMÚDEZ and VIAÑO [1984] for the case of a clamped beam, using a mixed displacement–stress formulation of the three-dimensional linearized elasticity problem. They adapted to this case the ideas of CIARLET and DESTUYNDER [1979a], DESTUYNDER [1980] and CIARLET and KESAVAN [1979, 1981] and they used the general asymptotic analysis techniques of LIONS [1973] for variational problems depending on a small parameter.

Using the same method, CIMETIÈRE, GEYMONAT, LE DRET, RAOULT and TUTEK [1988] studied the geometrically nonlinear case and suggested a new technique for the calculation of higher order terms in the asymptotic expansion. Using these ideas TRABUCHO and VIAÑO [1987, 1989] evaluated the first three terms of the displacement field together with the first two terms of the stress field associated with the asymptotic expansion proposed in BERMÚDEZ and VIAÑO [1984] for the linearized case. From these works the most well known stretching, bending and torsion theories for rods have been derived and justified, including those of Bernoulli–Navier (see BERMÚDEZ and VIAÑO [1984]), Saint Venant, Timoshenko and Vlassov (see TRABUCHO and VIAÑO [1987, 1988, 1989, 1990a, b]). These results are valid for both thick- and thin-walled beams, open and closed cross sections, varying cross sections, and they have been extended to the cases of nonhomogeneous and anisotropic materials, among others.

Although the first result of BERMÚDEZ and VIAÑO [1984] is based on a mixed displacement–stress approach (both the displacement and the stress fields are explicitly rescaled) we shall start the chapter using a displacement formulation suggested by RAOULT [1990b, 1991] which, besides being more natural, allows for a better understanding of the asymptotic expansion proposed for both the displacement and the stress fields.

This chapter is mainly concerned with the mathematical aspects of the asymptotic expansion method. Consequently, it may be considered as a previous requirement for the following applications. In the first place we expose the asymptotic expansion technique employed in order to study the limit of the three-dimensional elasticity problem when the cross sectional area goes to zero. In the second place, we characterize the successive terms of the asymptotic expansion and establish convergence results which validate the proposed asymptotic expansion.

In Section 1 the corresponding three-dimensional problem is studied. In Section 2 we introduce the change of variable to the reference domain together with the fundamental scalings on the data and the unknowns, for a displacement variational formulation taking the diameter of the cross section as the small parameter. The formal description of the asymptotic expansion of the displacement field together with the characterization of the first terms in the expansion are to be found in Sections 3 and 4. Also to be found in Section 4 are the mathematical difficulties related to the boundary layer phenomenon associated with Saint Venant's principle and the clamping boundary condition. In Section 5 we characterize the asymptotic expansion in the displacement field but now for a weak clamping boundary condition. Section 6 is dedicated to the presentation of the (original) version of the asymptotic expansion method for the mixed displacement-stress formulation of the three-dimensional elasticity problem, and its relationship with the displacement approach. The characterization of the zeroth-, second- and fourth-order terms of the asymptotic expansion both for the displacement and the stress fields is the object of Sections 7 through 9. Finally, in Section 10, some convergence results for the displacement and stress fields, as well as bending and shear force components are studied.

1. The three-dimensional equations of a linearly elastic clamped rod

Let ω be an open, bounded, connected subset of $\mathbb{R}^2$ having area A. Without any loss of generality we shall assume that $A = 1$. However, in some circumstances, and in order to keep the physical meaning of the equations, we shall write A instead of 1. Given $\varepsilon \leqslant 1$ let $\omega^\varepsilon = \varepsilon\omega$ in such a way that the area of ω^ε is $A^\varepsilon = \varepsilon^2 A$.

The main objective in this work is to study the elastic behaviour of a cylindrical rod which we identify with a three-dimensional solid occupying the *reference configuration*

$$\overline{\Omega}^\varepsilon = \omega^\varepsilon \times [0, L]. \tag{1.1}$$

The length of the rod is L, its cross sectional area is A^ε, and the main geometric feature is that ε is very small with respect to L.

For $\varepsilon \leqslant 1$ we introduce the following notations:

$$\gamma^\varepsilon = \partial\omega^\varepsilon, \qquad \Gamma^\varepsilon = \gamma^\varepsilon \times (0, L), \tag{1.2}$$

$$\Gamma_0^\varepsilon = \overline{\omega}^\varepsilon \times \{0\}, \qquad \Gamma_L^\varepsilon = \overline{\omega}^\varepsilon \times \{L\}. \tag{1.3}$$

We assumed that the boundary of ω^ε is sufficiently smooth and in order to consider a possibly multi-connected cross section we write:

$$\gamma^\varepsilon = \gamma_0^\varepsilon \cup \gamma_1^\varepsilon \cup \cdots \cup \gamma_p^\varepsilon, \tag{1.4}$$

where γ_0^ε denotes the exterior part of the boundary and $\gamma_1^\varepsilon, \ldots, \gamma_p^\varepsilon$ the boundaries of the bounded components of $\mathbb{R}^2 \setminus \overline{\omega}^\varepsilon$ which we denote by $\omega_1^\varepsilon, \ldots, \omega_p^\varepsilon$.

As it is customary in elasticity theory, Latin indices take their values in the set $\{1,2,3\}$ while Greek indices take their values in the set $\{1,2\}$. Moreover, the summation convention on the repeated indices shall also be employed.

An arbitrary point of $\overline{\Omega}^\varepsilon$ will be denoted by $\boldsymbol{x}^\varepsilon = (x_1^\varepsilon, x_2^\varepsilon, x_3^\varepsilon)$, the unit outer normal vector to the boundary Γ^ε by $\boldsymbol{n}^\varepsilon = (n_i^\varepsilon)$ and the differential operators $\partial/\partial x_i^\varepsilon$ and $n_i^\varepsilon(\partial/\partial x_i^\varepsilon)$ by ∂_i^ε and ∂_n^ε, respectively.

We remark that $\boldsymbol{n}^\varepsilon = (n_1^\varepsilon, n_2^\varepsilon, 0)$ in Γ^ε where $(n_1^\varepsilon, n_2^\varepsilon)$ is the unit outer normal vector to the set γ^ε. Furthermore, $\boldsymbol{n}^\varepsilon = (0,0,1)$ on Γ_L^ε and $\boldsymbol{n}^\varepsilon = (0,0,-1)$ on Γ_0^ε. Consequently, ∂_n^ε is the differential operator $n_\alpha^\varepsilon \partial_\alpha^\varepsilon$ and it will also stand for the normal derivative on $\gamma^\varepsilon = \partial\omega^\varepsilon$.

In order to simplify notation we shall also represent the first, second, etc., derivatives of a function $z(x_3)$ defined in $[0,L]$ by the symbols $\partial_3 z$, $\partial_{33} z$, and so on.

Whenever the meaning is clear, we identify ω^ε with the cross section $\omega^\varepsilon \times \{x_3^\varepsilon\}$, $x_3^\varepsilon \in [0,L]$. Moreover, we shall also omit the superscript $\varepsilon = 1$ and write $\omega = \omega^1$, $\Omega = \Omega^1 = \omega \times (0,L)$, $\gamma = \gamma^1 = \partial\omega$, and so on.

With no loss of generality, in all this work we assume that the coordinate system $Ox_1^\varepsilon x_2^\varepsilon x_3^\varepsilon$ is a *principal system of inertia* associated with solid Ω^ε. Consequently, axis Ox_3^ε passes through the centroid of each cross section $\omega^\varepsilon \times \{x_3^\varepsilon\}$ and one has

$$\int_{\omega^\varepsilon} x_\alpha^\varepsilon \, \mathrm{d}\omega^\varepsilon = \int_{\omega^\varepsilon} x_1^\varepsilon \, x_2^\varepsilon \, \mathrm{d}\omega^\varepsilon = 0. \tag{1.5}$$

In order to fix ideas, we suppose that the rod is clamped at both extremities $(\Gamma_0^\varepsilon \cup \Gamma_L^\varepsilon)$, although later on different boundary conditions will be considered. Moreover, we consider that volume forces act in the interior of Ω^ε and that surface forces act on its lateral surface Γ^ε, which we denote by $\boldsymbol{f}^\varepsilon = (f_i^\varepsilon) : \Omega^\varepsilon \to \mathbb{R}^3$ and by $\boldsymbol{g}^\varepsilon = (g_i^\varepsilon) : \Gamma^\varepsilon \to \mathbb{R}^3$, respectively.

The rod Ω^ε is made of a linearly elastic, homogeneous and isotropic material, whose Lamé constants are denoted by λ^ε and μ^ε and are related with Young's modulus E^ε and Poisson's ratio ν^ε through

$$\lambda^\varepsilon = \frac{\nu^\varepsilon E^\varepsilon}{(1+\nu^\varepsilon)(1-2\nu^\varepsilon)}, \qquad \mu^\varepsilon = \frac{E^\varepsilon}{2(1+\nu^\varepsilon)}. \tag{1.6}$$

The boundary value problem of linearized elasticity associated with rod Ω^ε is the following (see DUVAUT and LIONS [1972a,b], CIARLET [1988]):

Find $\boldsymbol{u}^\varepsilon = (u_i^\varepsilon) : \overline{\Omega}^\varepsilon \to \mathbb{R}^3$, such that

$$\begin{aligned}
&-\partial_j^\varepsilon[\lambda^\varepsilon e_{pp}^\varepsilon(\boldsymbol{u}^\varepsilon)\delta_{ij} + 2\mu^\varepsilon e_{ij}^\varepsilon(\boldsymbol{u}^\varepsilon)] = f_i^\varepsilon, \quad \text{in } \Omega^\varepsilon,\\
&\boldsymbol{u}^\varepsilon = \boldsymbol{0}, \quad \text{on } \Gamma_0^\varepsilon \cup \Gamma_L^\varepsilon,\\
&[\lambda^\varepsilon e_{pp}^\varepsilon(\boldsymbol{u}^\varepsilon)\delta_{ij} + 2\mu^\varepsilon e_{ij}^\varepsilon(\boldsymbol{u}^\varepsilon)]n_j^\varepsilon = g_i^\varepsilon, \quad \text{on } \Gamma^\varepsilon,
\end{aligned} \tag{1.7}$$

where

$$e_{ij}^\varepsilon(\boldsymbol{u}^\varepsilon) = \tfrac{1}{2}(\partial_i^\varepsilon u_j^\varepsilon + \partial_j^\varepsilon u_i^\varepsilon), \tag{1.8}$$

denotes the components of the linearized elasticity *strain tensor* $\boldsymbol{e}^\varepsilon(\boldsymbol{u}^\varepsilon) = (e_{ij}^\varepsilon(\boldsymbol{u}^\varepsilon))$.

Associated with the strain tensor is the linearized *stress tensor* $\boldsymbol{\sigma}^\varepsilon(\boldsymbol{u}^\varepsilon)$ given by

$$\boldsymbol{\sigma}^\varepsilon(\boldsymbol{u}^\varepsilon) = (\sigma_{ij}^\varepsilon(\boldsymbol{u}^\varepsilon)) = (\lambda^\varepsilon e_{pp}^\varepsilon(\boldsymbol{u}^\varepsilon)\delta_{ij} + 2\mu^\varepsilon e_{ij}^\varepsilon(\boldsymbol{u}^\varepsilon)). \tag{1.9}$$

Equation (1.9) is the constitutive law of linearized elasticity, also known as *Hooke's law* for a homogeneous, isotropic material.

The variational formulation of problem (1.7) is well known. Specifically (cf. CIARLET [1988, Theorem 6.3-1]) problem (1.7) is formally equivalent to finding solution $\boldsymbol{u}^\varepsilon = (u_i^\varepsilon)$ of the following variational equation:

$$\boldsymbol{u}^\varepsilon \in V(\Omega^\varepsilon) = \{\boldsymbol{v}^\varepsilon = (v_i^\varepsilon) \in [H^1(\Omega^\varepsilon)]^3 : \boldsymbol{v}^\varepsilon = \boldsymbol{0} \text{ on } \Gamma_0^\varepsilon \cup \Gamma_L^\varepsilon\}, \tag{1.10}$$

$$\int_{\Omega^\varepsilon} \sigma_{ij}^\varepsilon(\boldsymbol{u}^\varepsilon) e_{ij}^\varepsilon(\boldsymbol{v}^\varepsilon)\,\mathrm{d}\boldsymbol{x}^\varepsilon = \int_{\Omega^\varepsilon} f_i^\varepsilon v_i^\varepsilon\,\mathrm{d}\boldsymbol{x}^\varepsilon + \int_{\Gamma^\varepsilon} g_i^\varepsilon v_i^\varepsilon\,\mathrm{d}a^\varepsilon, \tag{1.11}$$

for all $\boldsymbol{v}^\varepsilon \in V(\Omega^\varepsilon)$.

In order for this problem to be well posed we assume that

$$f_i^\varepsilon \in L^2(\Omega^\varepsilon), \qquad g_i^\varepsilon \in L^2(\Gamma^\varepsilon).$$

Since the bilinear form $B^\varepsilon : V(\Omega^\varepsilon) \times V(\Omega^\varepsilon) \to \mathbb{R}$ defined by

$$B^\varepsilon(\boldsymbol{u}^\varepsilon, \boldsymbol{v}^\varepsilon) = \int_{\Omega^\varepsilon} \sigma_{ij}^\varepsilon(\boldsymbol{u}^\varepsilon) e_{ij}^\varepsilon(\boldsymbol{v}^\varepsilon)\,\mathrm{d}\boldsymbol{x}^\varepsilon, \tag{1.12}$$

is symmetric, continuous and $V(\Omega^\varepsilon)$–elliptic (due to Korn's inequality for the clamped case; cf. CIARLET [1988, Theorem 6.3-3]) and since the linear form $F^\varepsilon : V(\Omega^\varepsilon) \to \mathbb{R}$ defined by

$$F^\varepsilon(\boldsymbol{v}^\varepsilon) = \int_{\Omega^\varepsilon} f_i^\varepsilon v_i^\varepsilon\,\mathrm{d}\boldsymbol{x}^\varepsilon + \int_{\Gamma^\varepsilon} g_i^\varepsilon v_i^\varepsilon\,\mathrm{d}a^\varepsilon, \tag{1.13}$$

is continuous, problem (1.10), (1.11) has one and only one solution (cf. CIARLET [1988, Theorem 6.3-5]). This solution can also be characterized as the unique solution of the following *minimization problem*:

$$\boldsymbol{u}^\varepsilon \in V(\Omega^\varepsilon), \qquad J^\varepsilon(\boldsymbol{u}^\varepsilon) = \inf_{\boldsymbol{v}^\varepsilon \in V(\Omega^\varepsilon)} J^\varepsilon(\boldsymbol{v}^\varepsilon), \tag{1.14}$$

where

$$\begin{aligned} J^\varepsilon(\boldsymbol{v}^\varepsilon) &= \tfrac{1}{2}\int_{\Omega^\varepsilon} \sigma^\varepsilon_{ij}(\boldsymbol{v}^\varepsilon) e^\varepsilon_{ij}(\boldsymbol{v}^\varepsilon)\,\mathrm{d}\boldsymbol{x}^\varepsilon - \int_{\Omega^\varepsilon} f^\varepsilon_i v^\varepsilon_i \,\mathrm{d}\boldsymbol{x}^\varepsilon - \int_{\Gamma^\varepsilon} g^\varepsilon_i v^\varepsilon_i \,\mathrm{d}a^\varepsilon \\ &= \tfrac{1}{2}B^\varepsilon(\boldsymbol{v}^\varepsilon, \boldsymbol{v}^\varepsilon) - F^\varepsilon(\boldsymbol{v}^\varepsilon). \end{aligned} \tag{1.15}$$

We end this first section by listing some important references related to mathematical aspects of three-dimensional elasticity: GERMAIN [1972], GURTIN [1972], DUVAUT and LIONS [1972a], FICHERA [1972], NEČAS and HLAVÁČEK [1981], SALENÇON [1988], CIARLET [1988], DUVAUT [1990].

2. The fundamental scalings on the unknowns and assumptions on the data: The displacement approach

The major geometric feature of a three-dimensional rod is the fact that the measure of the largest cross sectional dimension is very small when compared to its length ($\varepsilon \ll L$). This fact has driven mechanists in search of simpler models.

Until recently these lower dimension models were derived after introducing a priori hypotheses, based on experience, in the three-dimensional elasticity equations. An example of this is the classical Bernoulli–Navier hypothesis stating that *plane sections before deformation and perpendicular to the neutral axis of the rod remain plane and perpendicular to the same axis after deformation*. Translated in terms of displacement and stress relationships these hypotheses lead to considerable simplifications in the three-dimensional elasticity equations and thus they allowed engineers to find simpler models, although eventually some contradictions may show up: incompatibility between the shear stresses and applied shear forces, failure to obey the boundary conditions or Hooke's law, and so on.

One of the purposes of this work is to study the behaviour of the solution $(\boldsymbol{\sigma}^\varepsilon, \boldsymbol{u}^\varepsilon)$ when ε goes to zero. In order to do so we use an asymptotic expansion technique with respect to ε. Although an *asymptotic expansion method with respect to small dimensions* has also been proposed by RIGOLOT [1972, 1976, 1977a,b], it is in the framework of BERMÚDEZ and VIAÑO [1984], inspired by the previous work of CIARLET and DESTUYNDER [1979a], which uses a change of variable in a variational formulation, that more general conclusions can be drawn.

The dependence of the solution $(\boldsymbol{\sigma}^\varepsilon, \boldsymbol{u}^\varepsilon)$ with respect to ε is rather complex. The technique of *change of variable to a fixed domain* (*reference rod*) and the subsequent *rescaling of the displacement and stress fields* allows us to define an equivalent problem to (1.10), (1.11) where parameter ε shows up in an explicit way in the rescaled equations. The general methods of LIONS [1973] for elliptic problems depending on a small parameter are then applied.

The *reference domain* is obtained by considering a *zoom* over the cross section along directions Ox_1^ε and Ox_2^ε, obtaining in this way an homothetic rod with respect to the original one:

$$\Omega = \omega \times (0, L), \quad \Omega = \Omega^1. \tag{2.1}$$

We denote

$$\gamma = \partial\omega = \gamma_0 \cup \gamma_1 \cup \cdots \cup \gamma_p, \qquad \gamma_i = \partial\omega_i,\ i = 1, \ldots, p\ , \tag{2.2}$$

$$\Gamma = \gamma \times (0, L), \qquad \Gamma_0 = \overline{\omega} \times \{0\}, \qquad \Gamma_L = \overline{\omega} \times \{L\}. \tag{2.3}$$

With every point $\boldsymbol{x} \in \overline{\Omega}$ we associate the point $\boldsymbol{x}^\varepsilon \in \overline{\Omega}^\varepsilon$ through the following transformation proposed by BERMÚDEZ and VIAÑO [1984]:

$$\begin{aligned} \Pi^\varepsilon : \boldsymbol{x} &= (x_1, x_2, x_3) \in \overline{\Omega} \\ &\to \boldsymbol{x}^\varepsilon = \Pi^\varepsilon(\boldsymbol{x}) = (x_1^\varepsilon, x_2^\varepsilon, x_3^\varepsilon) = (\varepsilon x_1, \varepsilon x_2, x_3) \in \overline{\Omega}^\varepsilon. \end{aligned} \tag{2.4}$$

Consequently we have

$$\Gamma^\varepsilon = \Pi^\varepsilon(\Gamma), \qquad \Gamma_0^\varepsilon = \Pi^\varepsilon(\Gamma_0), \qquad \Gamma_L^\varepsilon = \Pi^\varepsilon(\Gamma_L), \qquad \boldsymbol{n} = \boldsymbol{n}^\varepsilon, \tag{2.5}$$

where $\boldsymbol{n}$ is the unit outer normal vector to the set $\partial\Omega = \Gamma \cup \Gamma_0 \cup \Gamma_L$.

In order to obtain an equivalent problem to (1.10), (1.11) in set Ω, we associate with the displacement field $\boldsymbol{u}^\varepsilon$ and with functions $\boldsymbol{v}^\varepsilon$ in $V(\Omega^\varepsilon)$ the *scaled displacements* $\boldsymbol{u}(\varepsilon) = (u_i(\varepsilon))$ and the *scaled functions* $v = (v_i)$ through the following scalings, valid for all $\boldsymbol{x}^\varepsilon = \Pi^\varepsilon(\boldsymbol{x})$, $\boldsymbol{x} \in \overline{\Omega}$:

$$\begin{aligned} u_\alpha(\varepsilon)(\boldsymbol{x}) &= \varepsilon u_\alpha^\varepsilon(\boldsymbol{x}^\varepsilon), \qquad & v_\alpha(\boldsymbol{x}) &= \varepsilon v_\alpha^\varepsilon(\boldsymbol{x}^\varepsilon), \\ u_3(\varepsilon)(\boldsymbol{x}) &= u_3^\varepsilon(\boldsymbol{x}^\varepsilon), & v_3(\boldsymbol{x}) &= v_3^\varepsilon(\boldsymbol{x}^\varepsilon). \end{aligned} \tag{2.6}$$

Moreover, we also consider the following *hypotheses on the data*:

(i) Constants $\lambda^\varepsilon > 0$ and $\mu^\varepsilon > 0$ ($E^\varepsilon > 0, \nu^\varepsilon > 0$), are *independent of* ε:

$$\lambda^\varepsilon = \lambda, \quad \mu^\varepsilon = \mu \qquad (E^\varepsilon = E,\ \nu^\varepsilon = \nu). \tag{2.7}$$

(ii) There exist functions $f_i \in L^2(\Omega)$ and $g_i \in L^2(\Gamma)$, *independent of* ε, such that, for all $\boldsymbol{x}^\varepsilon = \Pi^\varepsilon(\boldsymbol{x}) \in \overline{\Omega}^\varepsilon$

$$f_\alpha^\varepsilon(\boldsymbol{x}^\varepsilon) = \varepsilon f_\alpha(\boldsymbol{x}), \qquad f_3^\varepsilon(\boldsymbol{x}^\varepsilon) = f_3(\boldsymbol{x}), \tag{2.8}$$

$$g_\alpha^\varepsilon(\boldsymbol{x}^\varepsilon) = \varepsilon^2 g_\alpha(\boldsymbol{x}), \qquad g_3^\varepsilon(\boldsymbol{x}^\varepsilon) = \varepsilon g_3(\boldsymbol{x}). \tag{2.9}$$

REMARK 2.1. Hypothesis (2.7) means that the material properties of the rod are kept through the rescaling. The scalings (2.6) and (2.8), (2.9) are the covariant and the contravariant transformations associated with the coordinate change (2.4). Consequently, the *scaled virtual work* of the system of applied forces is homogeneous in ε, that is,

$$\int_{\Omega^\varepsilon} f_i^\varepsilon v_i^\varepsilon \, \mathrm{d}\boldsymbol{x}^\varepsilon + \int_{\Gamma^\varepsilon} g_i^\varepsilon v_i^\varepsilon \, \mathrm{d}a^\varepsilon = \varepsilon^2 \left[\int_\Omega f_i v_i \, \mathrm{d}\boldsymbol{x} + \int_\Gamma g_i v_i \, \mathrm{d}a \right]. \tag{2.10}$$

One also concludes that if (2.7) holds then (2.10) also holds for other equivalent scalings. For instance, in BERMÚDEZ and VIAÑO [1984] and TRABUCHO and VIAÑO [1987, 1988, 1990a,b] the following scaling is used:

$$u_\alpha(\varepsilon)(\boldsymbol{x}) = \varepsilon^{r+1} u_\alpha^\varepsilon(\boldsymbol{x}^\varepsilon), \qquad u_3(\varepsilon)(\boldsymbol{x}) = \varepsilon^r u_3^\varepsilon(\boldsymbol{x}^\varepsilon), \tag{2.11}$$

$$f_\alpha^\varepsilon(\boldsymbol{x}^\varepsilon) = \varepsilon^{-r+1} f_\alpha(\boldsymbol{x}), \qquad f_3^\varepsilon(\boldsymbol{x}^\varepsilon) = \varepsilon^{-r} f_3(\boldsymbol{x}), \tag{2.12}$$

$$g_\alpha^\varepsilon(\boldsymbol{x}^\varepsilon) = \varepsilon^{-r+2} g_\alpha(\boldsymbol{x}), \qquad g_3^\varepsilon(\boldsymbol{x}^\varepsilon) = \varepsilon^{-r+1} g_3(\boldsymbol{x}), \tag{2.13}$$

where r is a real number in such a way that in (2.12) and (2.13) functions $f_i \in L^2(\Omega)$ and $g_i \in L^2(\Gamma)$ are independent of ε.

Using these scalings on the displacement field together with hypotheses (2.7)–(2.9), we reformulate problem (1.10), (1.11) in the following theorem which will be the basis for the *displacement approach*.

THEOREM 2.1. *The scaled displacement* $\boldsymbol{u}(\varepsilon)$ *is the unique solution of the following problem*:

$$\boldsymbol{u}(\varepsilon) \in V(\Omega) = \{\boldsymbol{v} = (v_i) \in [H^1(\Omega)]^3 \colon v_i = 0 \quad \text{on } \Gamma_0 \cup \Gamma_L\}, \tag{2.14}$$

$$\begin{aligned} &\int_\Omega (\lambda + 2\mu) e_{33}(\boldsymbol{u}(\varepsilon)) e_{33}(\boldsymbol{v}) \, \mathrm{d}\boldsymbol{x} \\ &+ \varepsilon^{-2} \int_\Omega [\lambda e_{pp}(\boldsymbol{u}(\varepsilon)) e_{33}(\boldsymbol{v}) + \lambda e_{33}(\boldsymbol{u}(\varepsilon)) e_{pp}(\boldsymbol{v}) + 4\mu e_{3\alpha}(\boldsymbol{u}(\varepsilon)) e_{3\alpha}(\boldsymbol{v})] \, \mathrm{d}\boldsymbol{x} \\ &+ \varepsilon^{-4} \int_\Omega [\lambda e_{pp}(\boldsymbol{u}(\varepsilon)) e_{\sigma\sigma}(\boldsymbol{v}) + 2\mu e_{\alpha\beta}(\boldsymbol{u}(\varepsilon)) e_{\alpha\beta}(\boldsymbol{v})] \, \mathrm{d}\boldsymbol{x} \\ &= \int_\Omega f_i v_i \, \mathrm{d}\boldsymbol{x} + \int_\Gamma g_i v_i \, \mathrm{d}a, \quad \textit{for all } \boldsymbol{v} \in V(\Omega), \end{aligned} \tag{2.15}$$

where

$$e_{ij}(\boldsymbol{v}) = \tfrac{1}{2}(\partial_i v_j + \partial_j v_i).$$

The proof of this result is straightforward if one takes into account the following relationship between the linearized strain tensor $\boldsymbol{e}^\varepsilon(\boldsymbol{u}^\varepsilon) = (e^\varepsilon_{ij}(\boldsymbol{u}(\varepsilon))$ and the *scaled linearized strain tensor* $\boldsymbol{e}(\boldsymbol{u}(\varepsilon)) = (e_{ij}(\boldsymbol{u}(\varepsilon)))$:

$$\begin{aligned} e_{\alpha\beta}(\boldsymbol{u}(\varepsilon))(\boldsymbol{x}) &= \varepsilon^2 e^\varepsilon_{\alpha\beta}(\boldsymbol{u}^\varepsilon)(\boldsymbol{x}^\varepsilon), \\ e_{3\beta}(\boldsymbol{u}(\varepsilon))(\boldsymbol{x}) &= \varepsilon e^\varepsilon_{3\beta}(\boldsymbol{u}^\varepsilon)(\boldsymbol{x}^\varepsilon), \\ e_{33}(\boldsymbol{u}(\varepsilon))(\boldsymbol{x}) &= e^\varepsilon_{33}(\boldsymbol{u}^\varepsilon)(\boldsymbol{x}^\varepsilon). \end{aligned} \tag{2.16}$$

The scaled displacement field $\boldsymbol{u}(\varepsilon)$ is also the solution of the following *minimization problem*:

$$\boldsymbol{u}(\varepsilon) \in V(\Omega), \qquad J(\varepsilon)(\boldsymbol{u}(\varepsilon)) = \inf_{\boldsymbol{v}\in V(\Omega)} J(\varepsilon)(\boldsymbol{v}), \tag{2.17}$$

where

$$\begin{aligned} J(\varepsilon)(\boldsymbol{v}) = {} & \tfrac{1}{2}\int_\Omega (\lambda + 2\mu) e_{33}(\boldsymbol{v}) e_{33}(\boldsymbol{v})\,\mathrm{d}\boldsymbol{x} - \int_\Omega f_i v_i\,\mathrm{d}\boldsymbol{x} - \int_\Gamma g_i v_i\,\mathrm{d}a \\ & + \varepsilon^{-2}\tfrac{1}{2}\int_\Omega [2\lambda e_{\rho\rho}(\boldsymbol{v}) e_{33}(\boldsymbol{v}) + 4\mu e_{3\alpha}(\boldsymbol{v}) e_{3\alpha}(\boldsymbol{v})]\,\mathrm{d}\boldsymbol{x} \\ & + \varepsilon^{-4}\tfrac{1}{2}\int_\Omega [2\lambda e_{\rho\rho}(\boldsymbol{v}) e_{\zeta\zeta}(\boldsymbol{v}) + 4\mu e_{\alpha\beta}(\boldsymbol{v}) e_{\alpha\beta}(\boldsymbol{v})]\,\mathrm{d}\boldsymbol{x}. \end{aligned} \tag{2.18}$$

We finish this section with an equivalent formulation of problem (2.14), (2.15) which together with the results of the next section constitutes the basis of another application of the asymptotic expansion method, designated as the *displacement–stress approach*. Specifically, from (1.9), (2.11) and (2.15) the following *scaled principle of virtual work*, and *scaled constitutive law* may be deduced:

$$\int_\Omega \sigma_{ij}(\varepsilon) e_{ij}(\boldsymbol{v})\,\mathrm{d}\boldsymbol{x} = \int_\Omega f_i v_i\,\mathrm{d}\boldsymbol{x} + \int_\Gamma g_i v_i\,\mathrm{d}a, \quad \text{for all } \boldsymbol{v} \in V(\Omega), \tag{2.19}$$

$$\begin{aligned} \sigma_{\alpha\beta}(\varepsilon) &= \varepsilon^{-4}[\lambda e_{\rho\rho}(\boldsymbol{u}(\varepsilon))\delta_{\alpha\beta} + 2\mu e_{\alpha\beta}(\boldsymbol{u}(\varepsilon))] + \varepsilon^{-2}\lambda e_{33}(\boldsymbol{u}(\varepsilon))\delta_{\alpha\beta}, \\ \sigma_{3\beta}(\varepsilon) &= \varepsilon^{-2} 2\mu e_{3\beta}(\boldsymbol{u}(\varepsilon)), \\ \sigma_{33}(\varepsilon) &= \varepsilon^{-2}\lambda e_{\rho\rho}(\boldsymbol{u}(\varepsilon)) + (\lambda + 2\mu) e_{33}(\boldsymbol{u}(\varepsilon)). \end{aligned} \tag{2.20}$$

From (2.16) we conclude that the following relationship between the stress components $\sigma^\varepsilon_{ij}(\boldsymbol{x}^\varepsilon) \in L^2(\Omega^\varepsilon)$ and the scaled stress components $\sigma_{ij}(\varepsilon)(\boldsymbol{x}) \in L^2(\Omega)$ holds:

$$\begin{aligned} \sigma_{\alpha\beta}(\varepsilon)(\boldsymbol{x}) &= \varepsilon^{-2}\sigma^\varepsilon_{\alpha\beta}(\boldsymbol{x}^\varepsilon), \\ \sigma_{3\beta}(\varepsilon)(\boldsymbol{x}) &= \varepsilon^{-1}\sigma^\varepsilon_{3\beta}(\boldsymbol{x}^\varepsilon), \\ \sigma_{33}(\varepsilon)(\boldsymbol{x}) &= \sigma^\varepsilon_{33}(\boldsymbol{x}^\varepsilon). \end{aligned} \tag{2.21}$$

REMARK 2.2. From (2.21) we conclude that the transformation used keeps both the left- and right-hand sides of the principle of virtual work invariant, that is,

$$\int_{\Omega^\varepsilon} \sigma_{ij}^\varepsilon(\boldsymbol{u}^\varepsilon)e_{ij}^\varepsilon(\boldsymbol{v}^\varepsilon)\,d\boldsymbol{x}^\varepsilon = \varepsilon^2 \int_{\Omega} \sigma_{ij}(\varepsilon)(\boldsymbol{u}(\varepsilon))e_{ij}(\boldsymbol{v}(\varepsilon))\,\mathrm{d}\boldsymbol{x} \tag{2.22}$$

which should be compared with (2.10).

If we instead were to use (2.11)–(2.13) then (2.21) would be multiplied by ε^r. Moreover, if hypothesis (2.7) is substituted by the more general one:

$$\lambda^\varepsilon = \varepsilon^t\lambda, \qquad \mu^\varepsilon = \varepsilon^t\mu, \qquad t \in \mathbb{R}, \tag{2.23}$$

then one should require, in scalings (2.6), that the system of applied forces verifies the following conditions (cf. CIARLET [1990, Section 1.9]):

$$f_\alpha^\varepsilon(\boldsymbol{x}^\varepsilon) = \varepsilon^{t+1}f_\alpha(\boldsymbol{x}), \qquad f_3^\varepsilon(\boldsymbol{x}^\varepsilon) = \varepsilon^t f_3(\boldsymbol{x}),$$
$$g_\alpha^\varepsilon(\boldsymbol{x}^\varepsilon) = \varepsilon^{t+2}g_\alpha(\boldsymbol{x}), \qquad g_3^\varepsilon(\boldsymbol{x}^\varepsilon) = \varepsilon^{t+1}g_3(\boldsymbol{x}).$$

In this case, an exponent t will also show up in the powers of ε in expression (2.21). Admitting (2.23) one would still have the additional possibility of keeping (2.8), (2.9) and multiplying (2.6) by ε^t.

REMARK 2.3. Functions $\sigma_{ij}(\varepsilon)$ show up in (2.20) as a simple convenience in notation. However, they may also be seen as the components of a *scaled stress tensor* $\boldsymbol{\sigma}(\varepsilon) = (\sigma_{ij}(\varepsilon)) : \overline{\Omega} \to \mathbb{R}_S^9 = \{\boldsymbol{\tau} = (\tau_{ij}) \in \mathbb{R}^9\colon \tau_{ij} = \tau_{ji}\}$. It is this way of looking at $\sigma_{ij}(\varepsilon)$ that is the basis of the *displacement–stress approach* (see Section 6).

3. The asymptotic expansion method

A first important observation related to the transformation of problem (1.10), (1.11) into problem (2.14), (2.15) is that the scaling induces a partition of the set $\{1,2,3\}$ into $\{1,2\} \cup \{3\}$, for the various expressions showing up in (2.15). This may be considered as a first step towards the transformation of the three-dimensional linearized elasticity model of the rod into a one-dimensional one, as the size of the cross section, on plane Ox_1x_2, becomes small with respect to the beam's length. Another important observation is that as a natural consequence of the transformation employed, parameter $\varepsilon = [\text{meas}(\omega^\varepsilon)]^{1/2}$ shows up in Eq. (2.15) in an explicit way and in the *polynomial* form.

This polynomial form with respect to the "small parameter" ε leads naturally to the use of *asymptotic expansion* techniques whose general ideas can be seen in LIONS [1973], ECKHAUS [1979].

The adaptation of this asymptotic expansion method to the case under study consists in the following steps:

(i) Write $\boldsymbol{u}(\varepsilon)$ a priori as a *formal expansion*:

$$\boldsymbol{u}(\varepsilon) = \boldsymbol{u}^0 + \varepsilon\boldsymbol{u}^1 + \varepsilon^2\boldsymbol{u}^2 + \varepsilon^3\boldsymbol{u}^3 + \varepsilon^4\boldsymbol{u}^4 + \text{h.o.t.} \tag{3.1}$$

where h.o.t. stands for "higher order terms", and this expansion takes into account the fact that the number of terms required is left unspecified for the moment.

(ii) Substitute expansion (3.1) for $\boldsymbol{u}(\varepsilon)$ into the scaled three-dimensional problem (2.14), (2.15) and set the successive powers of ε^q, $q \geqslant -4$ equal to zero.

(iii) Identify the successive terms in the asymptotic expansion (3.1), assuming additional hypotheses, on $\boldsymbol{u}^0, \boldsymbol{u}^1, \boldsymbol{u}^2, \ldots,$ if necessary. In the present case we require that

$$\boldsymbol{u}(\varepsilon) = \sum_{p=0}^{4} \varepsilon^p \boldsymbol{u}^p + \text{ h.o.t.}, \quad \boldsymbol{u}^p \in V(\Omega),\ 0 \leqslant p \leqslant 4.$$

Substituting (3.1) into (2.20) we obtain the following formal expansions on $\sigma_{ij}(\varepsilon)$:

$$\begin{aligned}
\sigma_{\alpha\beta}(\varepsilon) &= \varepsilon^{-4}\sigma_{\alpha\beta}^{-4} + \varepsilon^{-3}\sigma_{\alpha\beta}^{-3} + \varepsilon^{-2}\sigma_{\alpha\beta}^{-2} + \cdots, \\
\sigma_{3\alpha}(\varepsilon) &= \varepsilon^{-2}\sigma_{3\alpha}^{-2} + \varepsilon^{-1}\sigma_{3\alpha}^{-1} + \sigma_{3\alpha}^{0} + \cdots, \\
\sigma_{33}(\varepsilon) &= \varepsilon^{-2}\sigma_{33}^{-2} + \varepsilon^{-1}\sigma_{33}^{-1} + \sigma_{33}^{0} + \cdots,
\end{aligned} \tag{3.2}$$

where

$$\begin{aligned}
\sigma_{\alpha\beta}^{-4} &= \lambda e_{\rho\rho}(\boldsymbol{u}^0)\delta_{\alpha\beta} + 2\mu e_{\alpha\beta}(\boldsymbol{u}^0), \\
\sigma_{\alpha\beta}^{-3} &= \lambda e_{\rho\rho}(\boldsymbol{u}^1)\delta_{\alpha\beta} + 2\mu e_{\alpha\beta}(\boldsymbol{u}^1), \\
\sigma_{\alpha\beta}^{-2} &= \lambda e_{\rho\rho}(\boldsymbol{u}^2)\delta_{\alpha\beta} + 2\mu e_{\alpha\beta}(\boldsymbol{u}^2) + \lambda e_{33}(\boldsymbol{u}^0)\delta_{\alpha\beta}, \\
\sigma_{\alpha\beta}^{p} &= \lambda e_{\rho\rho}(\boldsymbol{u}^{p+4})\delta_{\alpha\beta} + 2\mu e_{\alpha\beta}(\boldsymbol{u}^{p+4}) + \lambda e_{33}(\boldsymbol{u}^{p+2})\delta_{\alpha\beta}, \quad p \geqslant -2,
\end{aligned} \tag{3.3}$$

$$\sigma_{3\beta}^{p} = 2\mu e_{3\beta}(\boldsymbol{u}^{p+2}), \quad p \geqslant -2, \tag{3.4}$$

$$\begin{aligned}
\sigma_{33}^{-2} &= \lambda e_{\rho\rho}(\boldsymbol{u}^0), \\
\sigma_{33}^{-1} &= \lambda e_{\rho\rho}(\boldsymbol{u}^1), \\
\sigma_{33}^{p} &= \lambda e_{\rho\rho}(\boldsymbol{u}^{p+2}) + (\lambda + 2\mu)e_{33}(\boldsymbol{u}^p), \quad p \geqslant 0.
\end{aligned} \tag{3.5}$$

Substituting (3.2)–(3.5) into (2.19) we deduce that

$$\varepsilon^{-4}\int_\Omega \sigma_{\alpha\beta}^{-4} e_{\alpha\beta}(\boldsymbol{v})\,\mathrm{d}\boldsymbol{x} + \varepsilon^{-3}\int_\Omega \sigma_{\alpha\beta}^{-3} e_{\alpha\beta}(\boldsymbol{v})\,\mathrm{d}\boldsymbol{x}$$

$$+ \varepsilon^{-2} \int_\Omega \sigma_{ij}^{-2} e_{ij}(\boldsymbol{v})\,\mathrm{d}\boldsymbol{x} + \varepsilon^{-1} \int_\Omega \sigma_{ij}^{-1} e_{ij}(\boldsymbol{v})\,\mathrm{d}\boldsymbol{x}$$

$$+ \varepsilon^{0} \int_\Omega \sigma_{ij}^{0} e_{ij}(\boldsymbol{v})\,\mathrm{d}\boldsymbol{x} + \sum_{p\geqslant 1} \varepsilon^{p} \int_\Omega \sigma_{ij}^{p} e_{ij}(\boldsymbol{v})\,\mathrm{d}\boldsymbol{x}$$

$$= \int_\Omega f_i v_i\,\mathrm{d}\boldsymbol{x} + \int_\Gamma g_i v_i\,\mathrm{d}a, \quad \text{for all } \boldsymbol{v} \in V(\Omega). \tag{3.6}$$

Setting the factors of the different powers of ε equal to zero, we obtain for all $\boldsymbol{v} \in V(\Omega)$,

$$(\varepsilon^{-4}) \quad \int_\Omega \sigma_{\alpha\beta}^{-4} e_{\alpha\beta}(\boldsymbol{v})\,\mathrm{d}\boldsymbol{x} = 0, \tag{3.7}$$

$$(\varepsilon^{-3}) \quad \int_\Omega \sigma_{\alpha\beta}^{-3} e_{\alpha\beta}(\boldsymbol{v})\,\mathrm{d}\boldsymbol{x} = 0, \tag{3.8}$$

$$(\varepsilon^{-2}) \quad \int_\Omega \sigma_{ij}^{-2} e_{ij}(\boldsymbol{v})\,\mathrm{d}\boldsymbol{x} = 0, \tag{3.9}$$

$$(\varepsilon^{-1}) \quad \int_\Omega \sigma_{ij}^{-1} e_{ij}(\boldsymbol{v})\,\mathrm{d}\boldsymbol{x} = 0, \tag{3.10}$$

$$(\varepsilon^{0}) \quad \int_\Omega \sigma_{ij}^{0} e_{ij}(\boldsymbol{v})\,\mathrm{d}\boldsymbol{x} = \int_\Omega f_i v_i\,\mathrm{d}\boldsymbol{x} + \int_\Gamma g_i v_i\,\mathrm{d}a, \tag{3.11}$$

$$(\varepsilon^{p}) \quad \int_\Omega \sigma_{ij}^{p} e_{ij}(\boldsymbol{v})\,\mathrm{d}\boldsymbol{x} = 0, \quad p \geqslant 1. \tag{3.12}$$

Equations (3.7)–(3.12) with components σ_{ij}^q defined by (3.3)–(3.5) summarize all the available information in order to evaluate and characterize terms $\boldsymbol{u}^p \in V(\Omega)$, $p \geqslant 0$. This will be the object of the next sections.

4. Cancellation of the factors of ε^q, $-4 \leqslant q \leqslant -1$. Some properties of u^0 and u^2. Boundary layer phenomenon

In this section we characterize the scaled stress components $\boldsymbol{\sigma}^q$, $-4 \leqslant q \leqslant -1$, and also $\boldsymbol{u}^p$, $0 \leqslant p \leqslant 3$, as a consequence. The main idea is to try to show that (3.7)–(3.12) are equivalent to saying that $\boldsymbol{\sigma}^q = 0$, $-4 \leqslant q \leqslant -1$. This in turn would imply that the asymptotic expansion of $\boldsymbol{\sigma}(\varepsilon)$ is in reality of the form $\boldsymbol{\sigma}(\varepsilon) = \boldsymbol{\sigma}^0 + \varepsilon\boldsymbol{\sigma}^1 + \varepsilon^2\boldsymbol{\sigma}^2 + \cdots$ as it is a priori assumed in the works of BERMÚDEZ and VIAÑO [1984], CIMETIÈRE, GEYMONAT, LE DRET, RAOULT and TUTEK [1986, 1988] and TRABUCHO and VIAÑO [1987, 1989]. However, it is not possible to reach this conclusion with Dirichlet boundary conditions (both ends strongly clamped), except for very special geometries and loading. This is due to the fact that (3.7)–(3.9) impose restrictions on $\boldsymbol{u}^2$ which are not always compatible with the requirement $\boldsymbol{u}^2 \in V(\Omega)$, and it is the reason why in order to evaluate higher order terms one has to consider a *weakly clamping boundary condition* which enlarges the space of admissible displacements. Nevertheless, the zeroth order term verifies the Dirichlet condition in both cases. This is the main idea followed by CIMETIÈRE, GEYMONAT, LE DRET, RAOULT and TUTEK [1988] and TRABUCHO and VIAÑO [1987, 1989], and subsequent works.

We now study Eqs. (3.7)–(3.10) in a systematic and elegant way suggested by RAOULT [1990b, 1991].

THEOREM 4.1. *For functions* $\sigma_{\alpha\beta}^{-4}$ *and* σ_{ij}^{-2} *defined by* (3.3)–(3.5), *Eqs.* (3.7) *and* (3.9) *are equivalent to*

$$\sigma_{\alpha\beta}^{-4} = 0, \qquad \sigma_{3\beta}^{-2} = 0, \qquad \sigma_{33}^{-2} = 0, \tag{4.1}$$

$$\int_\Omega \sigma_{\alpha\beta}^{-2} e_{\alpha\beta}(\boldsymbol{v})\, \mathrm{d}\boldsymbol{x} = 0, \quad \textit{for all } \boldsymbol{v} \in V(\Omega). \tag{4.2}$$

PROOF. From (3.3) and setting $\boldsymbol{v} = \boldsymbol{u}^0$ in (3.7) we have

$$\int_\Omega [\lambda e_{\rho\rho}(\boldsymbol{u}^0) e_{\zeta\zeta}(\boldsymbol{u}^0) + 2\mu e_{\alpha\beta}(\boldsymbol{u}^0) e_{\alpha\beta}(\boldsymbol{u}^0)]\, \mathrm{d}\boldsymbol{x} = 0,$$

from which we conclude that $e_{\alpha\beta}(\boldsymbol{u}^0) = 0$ in $L^2(\Omega)$ (λ and μ are positive). Consequently $\sigma_{\alpha\beta}^{-4} = \sigma_{33}^{-2} = 0$.

On the other hand, taking $\boldsymbol{v} = \boldsymbol{u}^0$ in (3.9) one has

$$\int_\Omega 2\sigma_{3\beta}^{-2} e_{3\beta}(\boldsymbol{u}^0)\, \mathrm{d}\boldsymbol{x} = \int_\Omega 4\mu e_{3\beta}(\boldsymbol{u}^0) e_{3\beta}(\boldsymbol{u}^0)\, \mathrm{d}\boldsymbol{x} = 0,$$

thus, $e_{3\beta}(\boldsymbol{u}^0) = 0$ in $L^2(\Omega)$ and therefore, $\sigma_{3\beta}^{-2} = 0$.

Finally (4.2) can be deduced from (3.9) taking into account (4.1).

The converse is immediate. □

In a completely analogous way one may prove the following result:

THEOREM 4.2. *Functions* $\sigma_{\alpha\beta}^{-3}$ *and* σ_{ij}^{-1} *defined in* (3.3)–(3.5) *satisfy Eqs.* (3.8)–(3.10) *if and only if*

$$\sigma_{\alpha\beta}^{-3} = 0, \qquad \sigma_{3\beta}^{-1} = 0, \qquad \sigma_{33}^{-1} = 0, \tag{4.3}$$

$$\int_\Omega \sigma_{\alpha\beta}^{-1} e_{\alpha\beta}(\boldsymbol{v})\, \mathrm{d}\boldsymbol{x} = 0, \quad \textit{for all } \boldsymbol{v} \in V(\Omega). \tag{4.4}$$

The two previous theorems may be summarized in the following two corollaries which partially characterize $\boldsymbol{u}^0$ and $\boldsymbol{u}^1$.

COROLLARY 4.1. *For* $\boldsymbol{\sigma}^q$, $-4 \leqslant q \leqslant -1$ *defined in* (3.3)–(3.5) *one has*
(i) *Equations* (3.7) *and* (3.9) *are equivalent to*

$$e_{\alpha\beta}(\boldsymbol{u}^0) = e_{3\beta}(\boldsymbol{u}^0) = 0, \tag{4.5}$$

$$\int_{\Omega} \sigma_{\alpha\beta}^{-2} e_{\alpha\beta}(\boldsymbol{v})\,\mathrm{d}\boldsymbol{x} = 0. \tag{4.6}$$

(ii) *Equations* (3.8) *and* (3.10) *are equivalent to*

$$e_{\alpha\beta}(\boldsymbol{u}^1) = e_{3\beta}(\boldsymbol{u}^1) = 0, \tag{4.7}$$

$$\int_{\Omega} \sigma_{\alpha\beta}^{-1} e_{\alpha\beta}(\boldsymbol{v})\,\mathrm{d}\boldsymbol{x} = 0. \tag{4.8}$$

The method employed in Theorem 4.1 to conclude that $\sigma_{\alpha\beta}^{-4} = 0$ *cannot be applied in general* to Eq. (4.6) in order to conclude that $\sigma_{\alpha\beta}^{-2} = 0$ due to the presence of the term $\lambda e_{33}(\boldsymbol{u}^0)\delta_{\alpha\beta}$ showing up in the definition of $\sigma_{\alpha\beta}^{-2}$. However, this will still be possible if one is able to prove the existence of a function $\boldsymbol{\xi}$ such that

$$\boldsymbol{\xi} \in V(\Omega), \qquad \lambda e_{\rho\rho}(\boldsymbol{\xi})\delta_{\alpha\beta} + 2\mu e_{\alpha\beta}(\boldsymbol{\xi}) = -\lambda e_{33}(\boldsymbol{u}^0)\delta_{\alpha\beta}. \tag{4.9}$$

It is clear that the existence of such a function depends both on the form of $\boldsymbol{u}^0$ and on the space $V(\Omega)$. We begin by studying the characterization of $\boldsymbol{u}^0$. The following result due to BERMÚDEZ and VIAÑO [1984] characterizes the elements $\boldsymbol{u}^0 \in V(\Omega)$ satisfying (4.5).

THEOREM 4.3. *A displacement field* $\boldsymbol{u}^0 \in V(\Omega)$ *satisfies*

$$e_{\alpha\beta}(\boldsymbol{u}^0) = e_{3\beta}(\boldsymbol{u}^0) = 0, \tag{4.10}$$

if and only if $\boldsymbol{u}^0$ *is of the following form*:

$$\begin{aligned} u_\alpha^0(x_1,x_2,x_3) &= z_\alpha^0(x_3), \quad z_\alpha^0 \in H_0^2(0,L), \\ u_3^0(x_1,x_2,x_3) &= \underline{u}_3^0(x_3) - x_\alpha \partial_3 z_\alpha^0(x_3), \quad \underline{u}_3^0 \in H_0^1(0,L). \end{aligned} \tag{4.11}$$

PROOF. From Eq. (4.10) one has

$$\partial_1 u_1^0 = 0, \qquad \partial_2 u_2^0 = 0, \qquad \partial_1 u_2^0 + \partial_2 u_1^0 = 0.$$

Consequently,

$$u_1^0(x_1,x_2,x_3) = \varphi_1^0(x_2,x_3), \qquad u_2^0(x_1,x_2,x_3) = \varphi_2^0(x_1,x_3),$$

where functions φ_α^0 satisfy

$$\partial_2\varphi_1^0(x_2,x_3) + \partial_1\varphi_2^0(x_1,x_3) = 0.$$

From this we deduce the existence of a function $z^0(x_3)$ such that

$$\partial_2\varphi_1^0(x_2,x_3) = -\partial_1\varphi_2^0(x_1,x_3) = z^0(x_3).$$

Integrating we conclude that there exist functions $z_\alpha^0(x_3)$ such that

$$u_1^0(x_1,x_2,x_3) = \varphi_1^0(x_2,x_3) = z_1^0(x_3) + x_2 z^0(x_3),$$
$$u_2^0(x_1,x_2,x_3) = \varphi_2^0(x_1,x_3) = z_2^0(x_3) - x_1 z^0(x_3).$$

Since $\boldsymbol{u}^0 \in V(\Omega)$, from condition $u_\alpha^0 = 0$ on $\Gamma_0 \cup \Gamma_L$ we deduce that z^0, $z_\alpha^0 \in H_0^1(0,L)$.

Condition $e_{3\beta}(\boldsymbol{u}^0) = 0$ now reads

$$\partial_1 u_3^0 = -\partial_3 u_1^0 = -\partial_3 z_1^0 - x_2\partial_3 z^0,$$
$$\partial_2 u_3^0 = -\partial_3 u_2^0 = -\partial_3 z_2^0 + x_1\partial_3 z^0.$$

From the above equations we have $\partial_{\alpha\beta}u_3^0 \in L^2(\Omega)$ and differentiating these expressions one obtains

$$\partial_{12}u_3^0 = -\partial_3 z^0 = \partial_{21}u_3^0 = \partial_3 z^0, \qquad \partial_{11}u_3^0 = \partial_{22}u_3^0 = 0.$$

As a consequence $\partial_3 z^0 = 0$, and since $z^0 \in H_0^1(0,L)$ one has $z^0 = 0$. We are then left with $\partial_\alpha u_3^0 = -\partial_3 z_\alpha^0$ which upon integration leads to the existence of a function $\underline{u}_3^0$, of the variable x_3 only, and such that

$$u_3^0(x_1,x_2,x_3) = \underline{u}_3^0(x_3) - x_\alpha\partial_3 z_\alpha^0(x_3).$$

From condition $u_3^0 = 0$ in $\Gamma_0 \cup \Gamma_L$ one concludes that $\underline{u}_3^0 \in H_0^1(0,L)$ and that $\partial_3 z_\alpha^0 \in H_0^1(0,L)$.

Since the converse may be verified by substituting (4.11) into (4.10), the proof of this theorem is now complete. □

REMARK 4.1. From now on, and whenever (4.11) holds true, we shall not distinguish between u_α^0 and z_α^0. We shall simply say that u_α^0 depends only on x_3 and that

$$u_\alpha^0 \in H_0^2(0,L), \qquad u_3^0 = \underline{u}_3^0 - x_\alpha\partial_3 u_\alpha^0, \qquad \underline{u}_3^0 \in H_0^1(0,L). \tag{4.12}$$

In the engineering literature a displacement field of the form (4.12) is known as a Bernoulli–Navier displacement field. We denote by $V_{\rm BN}(\Omega)$ the subspace of $V(\Omega)$ formed by the displacement fields of the Bernoulli–Navier type, that is,

$$V_{\rm BN}(\Omega) = \{\boldsymbol{v} = (v_i) \in [H^1(\Omega)]^3 : v_\alpha \in H_0^2(0,L),\ v_3 = \underline{v}_3 - x_\alpha \partial_3 v_\alpha,\ \underline{v}_3 \in H_0^1(0,L)\}. \tag{4.13}$$

The previous theorem shows that equivalently we may write

$$V_{\rm BN}(\Omega) = \{\boldsymbol{v} = (v_i) \in V(\Omega) : e_{\alpha\beta}(\boldsymbol{v}) = e_{3\beta}(\boldsymbol{v}) = 0\}. \tag{4.14}$$

We consider that $V_{\rm BN}(\Omega)$ is equipped with the following norm:

$$\|\boldsymbol{v}\|_{\rm BN} = \{\|v_1\|^2_{H_0^2(0,L)} + \|v_2\|^2_{H_0^2(0,L)} + \|\underline{v}_3\|^2_{H_0^1(0,L)}\}^{1/2}. \tag{4.15}$$

Consequently, the following canonic application is an isometric isomorphism:

$$j : \boldsymbol{v} = (v_1, v_2, v_3) \in [H_0^2(0,L)]^2 \times H_0^1(0,L) \rightarrow (v_1, v_2, v_3 - x_\alpha \partial_3 v_\alpha) \in V_{\rm BN}(\Omega).$$

On the other hand, it is possible to prove that j is also continuous if $V_{\rm BN}(\Omega)$ is equipped with the induced $\|\cdot\|_{1,\Omega}$ norm of V. Thus, the open mapping theorem (cf. YOSIDA [1980, p. 75]) guarantees the continuity of the inverse application of j and finally that $\|\cdot\|_{\rm BN}$ is equivalent to the $\|\cdot\|_{1,\Omega}$ norm in $V_{\rm BN}(\Omega)$.

From the zeroth order Eq. (3.11) we shall obtain variational equations for the displacement components u_α^0 and $\underline{u}_3^0$. However, its immediate use is not possible since in the definition of $\boldsymbol{\sigma}^0$ the yet unknown terms $e_{ij}(\boldsymbol{u}^2)$ show up. In order to obtain some information about $\boldsymbol{u}^2$ we go back to Eq. (4.6) but now we take into account the fact that $\boldsymbol{u}^0 \in V_{\rm BN}(\Omega)$. The following results are due to CIMETIÈRE, GEYMONAT, LE DRET, RAOULT and TUTEK [1986, 1988] and TRABUCHO and VIAÑO [1987, 1989 p.236], although the way they are presented here was suggested by RAOULT [1990b, 1991].

THEOREM 4.4. *Condition $\sigma_{\alpha\beta}^{-2} = 0$ is equivalent to the existence of $\boldsymbol{u}^2 \in V(\Omega)$ such that*

$$e_{\alpha\beta}(\boldsymbol{u}^2) = -\frac{\lambda}{2(\lambda+\mu)} e_{33}(\boldsymbol{u}^0)\delta_{\alpha\beta}. \tag{4.16}$$

PROOF. It is enough to remark that (4.16) is equivalent to

$$S_{\alpha\beta}(\boldsymbol{u}^2) = \lambda e_{\rho\rho}(\boldsymbol{u}^2)\delta_{\alpha\beta} + 2\mu e_{\alpha\beta}(\boldsymbol{u}^2) = -\lambda e_{33}(\boldsymbol{u}^0)\delta_{\alpha\beta}.$$ □

In general the solution of (4.16) may not exist and even if it exists it may not be simple. However, in the present case, since $\boldsymbol{u}^0 \in V_{\rm BN}(\Omega)$ it is possible to characterize element $\boldsymbol{u}^2$ satisfying (4.16). In fact we have the following.

THEOREM 4.5. *Let* $\boldsymbol{u}^0 \in V_{\mathrm{BN}}(\Omega)$, $u_3^0 = \underline{u}_3^0 - x_\alpha \partial_3 u_\alpha^0$, *then every element* $(u_\alpha^2) \in [L^2(\Omega)]^2$ *satisfying* (4.16) *is of the form*:

$$u_1^2 = z_1^2 + x_2 z^2 - \frac{\lambda}{2(\lambda+\mu)}[x_1 \partial_3 \underline{u}_3^0 - \tfrac{1}{2}(x_1^2 - x_2^2)\partial_{33} u_1^0 - x_1 x_2 \partial_{33} u_2^0],$$
$$u_2^2 = z_2^2 - x_1 z^2 - \frac{\lambda}{2(\lambda+\mu)}[x_2 \partial_3 \underline{u}_3^0 - x_1 x_2 \partial_{33} u_1^0 - \tfrac{1}{2}(x_2^2 - x_1^2)\partial_{33} u_2^0], \tag{4.17}$$

where z_α^2, $z^2 \in H^1(0,L)$, *are arbitrary functions depending only on the variable* x_3.

PROOF. Equations (4.16) are equivalent to

$$\partial_1 u_1^2 = \partial_2 u_2^2 = -\frac{\lambda}{2(\lambda+\mu)}(\partial_3 \underline{u}_3^0 - x_\alpha \partial_{33} u_\alpha^0), \tag{4.18}$$
$$\partial_2 u_1^2 + \partial_1 u_2^2 = 0. \tag{4.19}$$

A direct integration gives

$$u_1^2 = -\frac{\lambda}{2(\lambda+\mu)}[x_1 \partial_3 \underline{u}_3^0 - \tfrac{1}{2}x_1^2 \partial_{33} u_1^0 - x_1 x_2 \partial_{33} u_2^0] + \phi_1(x_2, x_3), \tag{4.20}$$
$$u_2^2 = -\frac{\lambda}{2(\lambda+\mu)}[x_2 \partial_3 \underline{u}_3^0 - x_1 x_2 \partial_{33} u_1^0 - \tfrac{1}{2}x_2^2 \partial_{33} u_2^0] + \phi_2(x_1, x_3). \tag{4.21}$$

Substituting these expressions in (4.19) we obtain

$$-\frac{\lambda}{2(\lambda+\mu)} x_1 \partial_{33} u_2^0(x_3) - \partial_1 \phi_2(x_1, x_3)$$
$$= \frac{\lambda}{2(\lambda+\mu)} x_2 \partial_{33} u_1^0(x_3) + \partial_2 \phi_1(x_2, x_3).$$

Consequently, there exists a function z^2, depending only on x_3, such that

$$\partial_2 \phi_1 = z^2 - \frac{\lambda}{2(\lambda+\mu)} x_2 \partial_{33} u_1^0, \qquad \partial_1 \phi_2 = -z^2 - \frac{\lambda}{2(\lambda+\mu)} x_1 \partial_{33} u_2^0, \tag{4.22}$$

which has a general solution of the form

$$\phi_1 = x_2 z^2 - \frac{\lambda}{4(\lambda+\mu)} x_2^2 \partial_{33} u_1^0 + z_1^2,$$
$$\phi_2 = -x_1 z^2 - \frac{\lambda}{4(\lambda+\mu)} x_1^2 \partial_{33} u_2^0 + z_2^2, \tag{4.23}$$

where functions z_α^2 depend only on variable x_3. From (4.20), (4.21) and (4.23) the result now follows. □

COROLLARY 4.2. *A necessary and sufficient condition in order for problem* (4.16) *to possess (infinitely many) solutions* $(u_\alpha^2) \in [H^1(\Omega)]^2$ *is that* $\boldsymbol{u}^0 \in V_{\mathrm{BN}}(\Omega)$ *be such that* $\underline{u}_3^0 \in H^2(0,L)$, $u_\alpha^0 \in H^3(0,L)$. *In this case* u_α^2 *is of the form* (4.17) *and* $u_\alpha^2 = 0$ *in* $\Gamma_0 \cup \Gamma_L$ *if and only if* $\underline{u}_3^0 \in H_0^2(0,L)$, $u_\alpha^0 \in H_0^3(0,L)$, z_α^2, $z^2 \in H_0^1(0,L)$.

PROOF. We remark that in the general case $u_\alpha^2 \in L^2[0,L;H^1(\Omega)]$. The regularity on $\underline{u}_3^0$ and u_α^0 is then necessary and sufficient in order to increase the regularity on x_3. Since now $(u_\alpha^2) \in [H^1(\Omega)]^2$, from (4.17) we obtain that $\int_\omega u_\alpha^2\, d\omega \in H^1(0,L)$ and $\int_\omega (x_2 u_1^2 - x_1 u_2^2)\, d\omega \in H^1(0,L)$. Taking into account (1.5) we deduce that $\partial_3 \underline{u}_3^0$ and $\partial_{33} u_\alpha^0 \in H^1(0,L)$.

On the other hand, given a solution $(u_\alpha^2) \in [H^1(\Omega)]^2$ we have $u_\alpha^2 = 0$ on $\Gamma_0 \cup \Gamma_L$ if and only if for $a = 0$ and $a = L$ and for all $(x_1,x_2) \in \omega$ one has

$$\begin{aligned} &z_1^2(a) + x_2 z^2(a) \\ &\quad - \frac{\lambda}{2(\lambda+\mu)}\big[x_1 \partial_3 \underline{u}_3^0(a) - \tfrac{1}{2}(x_1^2 - x_2^2)\partial_{33} u_1^0(a) - x_1 x_2 \partial_{33} u_2^0(a)\big] = 0, \\ &z_2^2(a) - x_1 z^2(a) \\ &\quad - \frac{\lambda}{2(\lambda+\mu)}\big[x_2 \partial_3 \underline{u}_3^0(a) - x_1 x_2 \partial_{33} u_1^0(a) - \tfrac{1}{2}(x_2^2 - x_1^2)\partial_{33} u_2^0(a)\big] = 0, \end{aligned} \tag{4.24}$$

which admits only the trivial solution. □

From Theorem 4.4 and Corollary 4.2 we may deduce the following.

COROLLARY 4.3. *Element* $\sigma_{\alpha\beta}^{-2} = 0$ *if and only if* $\boldsymbol{u}^0$ *is such that* $\underline{u}_3^0 \in H_0^2(0,L)$ *and* $u_\alpha^0 \in H_0^3(0,L)$. *In this case, components* u_α^2 *of* $\boldsymbol{u}^2 \in V(\Omega)$ *are of the form* (4.17) *with* z_α^2, $z^2 \in H_0^1(0,L)$.

In the general case when $\boldsymbol{u}^0 \notin V_{\mathrm{BN}}(\Omega)$ the information one is able to obtain for u_α^2 just reduces to the following (see (3.3) and (4.6)):

$$S_{\alpha\beta}(\boldsymbol{u}^2) = \lambda e_{\rho\rho}(\boldsymbol{u}^2)\delta_{\alpha\beta} + 2\mu e_{\alpha\beta}(\boldsymbol{u}^2) = -\lambda e_{33}(\boldsymbol{u}^0)\delta_{\alpha\beta} + S_{\alpha\beta}^{-2}, \tag{4.25}$$

where $S_{\alpha\beta}^{-2}$ is a (nontrivial) solution of (4.6).

We observe that from (3.5) one has

$$\sigma_{33}^0 = \lambda e_{\rho\rho}(\boldsymbol{u}^2) + (\lambda + 2\mu) e_{33}(\boldsymbol{u}^0),$$

and using (4.25)

$$\sigma_{33}^0 = \frac{\lambda}{2(\lambda+\mu)} S_{\rho\rho}^{-2} + \frac{\mu(3\lambda+2\mu)}{\lambda+\mu} e_{33}(\boldsymbol{u}^0). \tag{4.26}$$

Setting $\boldsymbol{v} \in V_{\text{BN}}(\Omega)$ in (3.11), we deduce that

$$\begin{aligned} &\boldsymbol{u}^0 \in V_{\text{BN}}(\Omega), \\ &\int_\Omega \frac{\mu(3\lambda + 2\mu)}{\lambda + \mu} e_{33}(\boldsymbol{u}^0) e_{33}(\boldsymbol{v})\, \mathrm{d}\boldsymbol{x} \\ &\quad = -\frac{\lambda}{2(\lambda+\mu)} \int_\Omega S_{\rho\rho}^{-2} e_{33}(\boldsymbol{v})\, \mathrm{d}\boldsymbol{x} \\ &\quad + \int_\Omega f_i v_i \,\mathrm{d}\boldsymbol{x} + \int_\Gamma g_i v_i \,\mathrm{d}a, \quad \text{for all } \boldsymbol{v} \in V_{\text{BN}}(\Omega). \end{aligned} \tag{4.27}$$

Since it is not possible to evaluate $S_{\rho\rho}^{-2}$, in the general case, we are left with no way to proceed any further. We are then faced with an important *difficulty* in order to find the asymptotic expansion (3.1) *in the space* $V(\Omega)$ defined by (2.14). This discussion shows that the difficulty comes from the fact that *in general the solutions of* (4.16) *are not zero at the extremities of the rod* $\Gamma_0 \cup \Gamma_L$. This phenomenon inherent to the asymptotic expansion techniques is designated by "boundary layer problem" (cf. LIONS [1973]).

The form of $\boldsymbol{u}^2$ given in (4.17) and the difficulty in satisfying the boundary conditions were remarked for the first time in the engineering literature by Saint Venant (1856), (cf. FRAEIJS DE VEUBEKE [1979], NEČAS and HLAVÁČEK [1981]). This is related to the fact that in general, a strong clamping of the rod is physically impossible. In order to overcome this difficulty, we shall assume a weakly clamping condition, also known as an "average clamping". In the present circumstances this would give the following conditions for the displacement components u_α^2:

$$\int_{\omega\times\{a\}} u_\alpha^2 \,\mathrm{d}\omega = 0, \qquad \int_{\omega\times\{a\}} (x_2 u_1^2 - x_1 u_2^2)\,\mathrm{d}\omega = 0, \qquad \text{at } a = 0, L. \tag{4.28}$$

We remark that now, in order for (4.28) to hold, it would be enough to consider in (4.17) z_α^2 and $z^2 \in H^1(0,L)$ and verifying for $a = 0$ and L:

$$z_\alpha^2(a) = -\frac{\lambda}{4(\lambda+\mu)}(I_\alpha - I_\beta)\partial_{33}u_\alpha^0, \quad \alpha \neq \beta \quad \text{(no sum on } \alpha\text{)}, \tag{4.29}$$

$$z^2(a) = \frac{\lambda}{2(\lambda+\mu)(I_1+I_2)}[H_2\partial_{33}u_1^0(a) - H_1\partial_{33}u_2^0(a)], \tag{4.30}$$

where the following constants depending on the geometry of the cross section are used, (see (7.21)):

$$I_\alpha = \int_\omega x_\alpha^2 \,\mathrm{d}\omega, \qquad H_\alpha = \tfrac{1}{2}\int_\omega x_\alpha (x_1^2 + x_2^2)\,\mathrm{d}\omega. \tag{4.31}$$

These ideas suggest that the difficulties in the calculation of $\boldsymbol{u}^2$ will disappear when using a weakly clamped condition. This is equivalent to substituting the space of admissible displacements $V(\Omega)$ by a larger one where the first terms of the asymptotic expansion may be evaluated in an easier way. This is the idea followed in CIMETIÈRE, GEYMONAT, LE DRET, RAOULT and TUTEK [1988], and in TRABUCHO and VIAÑO [1987, 1988, 1989, 1990] and which will be presented in the next section still in the context of the "displacement approach".

Before ending this section we remark that *if we consider that* $S_{\alpha\beta}^{-2} = 0$ in (4.25), that is, $\sigma_{\alpha\beta}^{-2} = 0$, then from (4.27) we obtain a variational formulation for the displacement component $\boldsymbol{u}^0$. In fact in that case the following result follows.

THEOREM 4.6. *If* $\sigma_{\alpha\beta}^{-2} = 0$, *then* $\boldsymbol{u}^0 \in V_{\mathrm{BN}}(\Omega)$ *is determined in a unique way by the following variational problems (no sum on* α*):*

$$\int_0^L \frac{\mu(3\lambda+2\mu)}{\lambda+\mu} I_\alpha \partial_{33} u_\alpha^0 \partial_{33} v^0 \, \mathrm{d}x_3 = \int_0^L \left[\int_\omega f_\alpha \, \mathrm{d}\omega + \int_\gamma g_\alpha \, \mathrm{d}\gamma \right] v^0 \, \mathrm{d}x_3 - \int_0^L \left[\int_\omega x_\alpha f_3 \, \mathrm{d}\omega + \int_\gamma x_\alpha g_3 \, \mathrm{d}\gamma \right] \partial_3 v^0 \, \mathrm{d}x_3, \quad \textit{for all } v^0 \in H_0^2(0,L), \tag{4.32}$$

$$\int_0^L \frac{\mu(3\lambda+2\mu)}{\lambda+\mu} \partial_3 \underline{u}_3^0 \partial_3 \underline{v}_3^0 \, \mathrm{d}x_3 = \int_0^L \left[\int_\omega f_3 \, \mathrm{d}\omega + \int_\gamma g_3 \, \mathrm{d}\gamma \right] \underline{v}_3^0 \, \mathrm{d}x_3, \quad \textit{for all } \underline{v}_3^0 \in H_0^1(0,L). \tag{4.33}$$

PROOF. Choose in (4.27), $\boldsymbol{v} \in V_{\mathrm{BN}}(\Omega)$ in such a way that $v_3 = -x_\alpha \partial_3 v^0$, $v^0 \in H_0^2(0,L)$, and afterwards such that $v_3 = \underline{v}_3^0$, $\underline{v}_3^0 \in H_0^1(0,L)$. □

REMARK 4.2. This result was previously obtained in BERMÚDEZ and VIAÑO [1984] using a mixed formulation of the initial boundary value problem (1.7)–(1.9). In this work an asymptotic expansion of the type

$$(\boldsymbol{\sigma}(\varepsilon), \boldsymbol{u}(\varepsilon)) = (\boldsymbol{\sigma}^0, \boldsymbol{u}^0) + \varepsilon^2(\boldsymbol{\sigma}^2, \boldsymbol{u}^2) + \text{h.o.t.},$$

was assumed, *which implicitly contains the hypothesis* $\sigma_{\alpha\beta}^{-2} = 0$.

5. The weakly clamping conditions. The asymptotic expansion method in the displacement approach

In this section we briefly describe the asymptotic expansion method for the three-dimensional linearized elasticity equation of a *weakly clamped rod*. This amounts to considering the following space of admissible displacements:

$$V(\Omega^\varepsilon) = \{\boldsymbol{v}^\varepsilon = (v_i^\varepsilon) \in [H^1(\Omega^\varepsilon)]^3 \colon \int_{\omega^\varepsilon \times \{a\}} v_i^\varepsilon \, \mathrm{d}\omega^\varepsilon = 0,$$

$$\int_{\omega^\varepsilon \times \{a\}} (x_j^\varepsilon v_i^\varepsilon - x_i^\varepsilon v_j^\varepsilon) \, \mathrm{d}\omega^\varepsilon = 0, \quad \text{at } a = 0, L\} \tag{5.1}$$

and the corresponding variational formulation for the displacement components is

$$\begin{aligned} &\boldsymbol{u}^\varepsilon \in V(\Omega^\varepsilon), \\ &\int_{\Omega^\varepsilon} [\lambda^\varepsilon e_{pp}^\varepsilon(\boldsymbol{u}^\varepsilon)\delta_{ij} + 2\mu^\varepsilon e_{ij}^\varepsilon(\boldsymbol{u}^\varepsilon)] e_{ij}^\varepsilon(\boldsymbol{v}^\varepsilon) \, \mathrm{d}\boldsymbol{x}^\varepsilon \\ &\quad = \int_{\Omega^\varepsilon} f_i^\varepsilon v_i^\varepsilon \, \mathrm{d}\boldsymbol{x}^\varepsilon + \int_{\Gamma^\varepsilon} g_i^\varepsilon v_i^\varepsilon \, \mathrm{d}a^\varepsilon, \quad \text{for all } \boldsymbol{v}^\varepsilon \in V(\Omega^\varepsilon). \end{aligned} \tag{5.2}$$

REMARK 5.1. The boundary value problem associated with (5.2) is (1.7) but with different boundary conditions in $\Gamma_0 \cup \Gamma_L$.

REMARK 5.2. From (1.5) we see that the weakly clamping conditions for $\boldsymbol{v}^\varepsilon \in V(\Omega^\varepsilon)$ may be equivalently written in the form

$$\int_{\omega^\varepsilon \times \{a\}} v_i^\varepsilon \, \mathrm{d}\omega^\varepsilon = 0, \qquad \int_{\omega^\varepsilon \times \{a\}} (x_2^\varepsilon v_1^\varepsilon - x_1^\varepsilon v_2^\varepsilon) \, \mathrm{d}\omega^\varepsilon = 0,$$

$$\int_{\omega^\varepsilon \times \{a\}} x_\alpha^\varepsilon v_3^\varepsilon \, d\omega^\varepsilon = 0.$$

As can be seen, everything that was written in Sections 2 and 3 is still valid for problem (5.2) with now $V(\Omega^\varepsilon)$ defined by (5.1) and consequently

$$V(\Omega) = \{\boldsymbol{v} = (v_i) \in [H^1(\Omega)]^3 \colon \int_{\omega \times \{a\}} v_i \, \mathrm{d}\omega = 0,$$

$$\int_{\omega \times \{a\}} (x_j v_i - x_i v_j) \, \mathrm{d}\omega = 0, \quad \text{at } a = 0, L\}. \tag{5.3}$$

In what concerns Section 4, Theorems 4.1–4.4 and Corollary 4.1 still hold (in the same form) for $V(\Omega)$ defined by (5.3). The proofs are exactly the same except for Theorem 4.3, which will be adapted in Theorem 5.2. We summarize these results in the following.

THEOREM 5.1. *Let* $V(\Omega)$ *be defined by* (5.3) *and let* $\boldsymbol{u}(\varepsilon)$ *be the unique solution of problem* (2.14), (2.15). *Then, for the asymptotic expansion* (3.1) *of* $\boldsymbol{u}(\varepsilon)$ *the following properties hold*:

$$e_{\alpha\beta}(\boldsymbol{u}^0) = e_{3\beta}(\boldsymbol{u}^0) = 0, \tag{5.4}$$

$$\int_\Omega \sigma^{-2}_{\alpha\beta} e_{\alpha\beta}(\boldsymbol{v})\,\mathrm{d}\boldsymbol{x} = 0, \quad \textit{for all } \boldsymbol{v} \in V(\Omega), \tag{5.5}$$

$$e_{\alpha\beta}(\boldsymbol{u}^1) = e_{3\beta}(\boldsymbol{u}^1) = 0, \tag{5.6}$$

$$\int_\Omega \sigma^{-1}_{\alpha\beta} e_{\alpha\beta}(\boldsymbol{v})\,\mathrm{d}\boldsymbol{x} = 0, \quad \textit{for all } \boldsymbol{v} \in V(\Omega). \tag{5.7}$$

We shall now show that Theorem 4.3 is also valid for $V(\Omega)$ defined by (5.3). This is a *remarkable property* for it shows that the Bernoulli–Navier displacement space $V_{\mathrm{BN}}(\Omega)$ *does not depend on the boundary conditions in the extremities of the beam.*

THEOREM 5.2. *Let* $V(\Omega)$ *be defined by* (5.3) *and let* $\boldsymbol{u}^0 \in V(\Omega)$. *Then* $\boldsymbol{u}^0$ *satisfies*

$$e_{\alpha\beta}(\boldsymbol{u}^0) = e_{3\beta}(\boldsymbol{u}^0) = 0, \tag{5.8}$$

if and only if $\boldsymbol{u}^0$ *is of the following form*:

$$\begin{aligned} u^0_\alpha(x_1, x_2, x_3) &= z^0_\alpha(x_3), \quad z^0_\alpha \in H^2_0(0, L), \\ u^3_o(x_1, x_2, x_3) &= \underline{u}^0_3(x_3) - x_\alpha \partial_3 z^0_\alpha(x_3), \quad \underline{u}^0_3 \in H^1_0(0, L). \end{aligned} \tag{5.9}$$

PROOF. Using the same arguments as in Theorem 4.3 we conclude that u^0_α is of the following form:

$$\begin{aligned} u^0_1(x_1, x_2, x_3) &= z^0_1(x_3) + x_2 z^0(x_3), \\ u^0_2(x_1, x_2, x_3) &= z^0_2(x_3) - x_1 z^0(x_3), \end{aligned}$$

where functions z^0_α and z^0 depend only on variable x_3. Since $\boldsymbol{u}^0 \in V(\Omega)$, we have

$$\int_\omega u^0_\alpha\,\mathrm{d}\omega \in H^1_0(0, L), \qquad \int_\omega (x_2 u^0_1 - x_1 u^0_2)\,\mathrm{d}\omega \in H^1_0(0, L).$$

Using property (1.5) we conclude that z^0_α and z^0 belong to $H^1_0(0, L)$.

In a similar way we obtain

$$u_3^0(x_1, x_2, x_3) = \underline{u}_3^0(x_3) - x_\alpha \partial_3 z_\alpha^0(x_3),$$

and condition $\boldsymbol{u}^0 \in V(\Omega)$ leads to

$$\underline{u}_3^0 = \int_\omega u_3^0 \, d\omega \in H_0^1(0, L). \qquad \square$$

The weak boundary conditions considered allow us to improve the results of Theorems 4.4 and 4.5 as well as Corollary 4.2. In fact one has the following.

THEOREM 5.3. *Let $\boldsymbol{u}^0 \in V_{\mathrm{BN}}(\Omega)$ be such that $\underline{u}_3^0 \in H^2(0, L)$ and $u_\alpha^0 \in H^3(0, L)$. Then, $\boldsymbol{u}^2 \in V(\Omega)$ is such that u_α^2 is of the form (4.17) with z_α^2, $z^2 \in H^1(0, L)$ verifying (4.29) and (4.30). Moreover, $\boldsymbol{u}^2$ verifies (4.16) and consequently $\sigma_{\alpha\beta}^{-2} =$ 0. For any $\boldsymbol{u}^0 \in V_{\mathrm{BN}}(\Omega)$ the converse is also true.*

PROOF. From Theorem 4.5 we obtain that the following problem:

$$\boldsymbol{\xi} \in V(\Omega), \qquad e_{\alpha\beta}(\boldsymbol{\xi}) = -\frac{\lambda}{2(\lambda + \mu)} e_{33}(\boldsymbol{u}^0)\delta_{\alpha\beta}, \tag{5.10}$$

possesses infinitely many solutions. In fact, Theorem 4.5 assures us that it is necessary and sufficient to take ξ_α of the form:

$$\begin{aligned} \xi_1 &= z_1^2 + x_2 z^2 - \frac{\lambda}{2(\lambda + \mu)}[x_1 \partial_3 \underline{u}_3^0 - \tfrac{1}{2}(x_1^2 - x_2^2)\partial_{33} u_1^0 - x_1 x_2 \partial_{33} u_2^0], \\ \xi_2 &= z_2^2 - x_1 z^2 - \frac{\lambda}{2(\lambda + \mu)}[x_2 \partial_3 \underline{u}_3^0 - x_1 x_2 \partial_{33} u_1^0 - \tfrac{1}{2}(x_2^2 - x_1^2)\partial_{33} u_2^0]. \end{aligned} \tag{5.11}$$

In order to assure that $\boldsymbol{\xi} \in V(\Omega)$ it is sufficient to take z_α^2, $z^2 \in H^1(0, L)$ satisfying (4.29) and (4.30) for $a = 0$, and $a = L$.

From (5.10) we are now able to conclude that

$$S_{\alpha\beta}(\boldsymbol{\xi}) = \lambda e_{\rho\rho}(\boldsymbol{\xi})\delta_{\alpha\beta} + 2\mu e_{\alpha\beta}(\boldsymbol{\xi}) = -\lambda e_{33}(\boldsymbol{u}^0)\delta_{\alpha\beta},$$

and consequently, from (3.3) we obtain

$$\sigma_{\alpha\beta}^{-2} = S_{\alpha\beta}(\boldsymbol{u}^2 - \boldsymbol{\xi}).$$

Now choosing $\boldsymbol{v} = \boldsymbol{u}^2 - \boldsymbol{\xi}$, in (5.5), we have

$$\int_\Omega [\lambda e_{\rho\rho}(\boldsymbol{u}^2 - \boldsymbol{\xi}) e_{\zeta\zeta}(\boldsymbol{u}^2 - \boldsymbol{\xi}) + 2\mu e_{\alpha\beta}(\boldsymbol{u}^2 - \boldsymbol{\xi}) e_{\alpha\beta}(\boldsymbol{u}^2 - \boldsymbol{\xi})] \, d\boldsymbol{x} = 0,$$

thus, $e_{\alpha\beta}(\boldsymbol{u}^2 - \boldsymbol{\xi}) = 0$. We then conclude that $\sigma^{-2}_{\alpha\beta} = 0$ and moreover that $\boldsymbol{u}^2$ is also a solution of (5.10) and consequently it must also be of the form (5.11) (namely (4.17)). The converse is immediate. □

From (5.10) and (3.5) we have

COROLLARY 5.1.

$$\sigma^0_{33} = \frac{\mu(3\lambda + 2\mu)}{\lambda + \mu} e_{33}(\boldsymbol{u}^0) = \frac{\mu(3\lambda + 2\mu)}{\lambda + \mu}(\partial_3 \underline{u}^0_3 - x_\alpha \partial_{33} u^0_\alpha). \tag{5.12}$$

Using the same reasoning as in Theorem 4.6 we obtain

COROLLARY 5.2. *The components* $\underline{u}^0_3 \in H^1_0(0, L)$ *and* $u^0_\alpha \in H^2_0(0, L)$ *are the unique solutions of the following variational problems (no sum on* α*):*

$$\begin{aligned} &\int_0^L \frac{\mu(3\lambda + 2\mu)}{\lambda + \mu} I_\alpha \partial_{33} u^0_\alpha \partial_{33} v^0 \,\mathrm{d}x_3 \\ &\quad = \int_0^L \left[\int_\omega f_\alpha \,\mathrm{d}\omega + \int_\gamma g_\alpha \,\mathrm{d}\gamma \right] v^0 \,\mathrm{d}x_3 \\ &\quad - \int_0^L \left[\int_\omega x_\alpha f_3 \,\mathrm{d}\omega + \int_\gamma x_\alpha g_3 \, g\gamma \right] \partial_3 v^0 \,\mathrm{d}x_3, \quad \textit{for all } v^0 \in H^2_0(0, L), \end{aligned} \tag{5.13}$$

$$\begin{aligned} &\int_0^L \frac{\mu(3\lambda + 2\mu)}{\lambda + \mu} \partial_3 \underline{u}^0_3 \partial_3 \underline{v}^0_3 \,\mathrm{d}x_3 \\ &\quad = \int_0^L \left[\int_\omega f_3 \,\mathrm{d}\omega + \int_\gamma g_3 \,\mathrm{d}\gamma \right] \underline{v}^0_3 \,\mathrm{d}x_3, \quad \textit{for all } \underline{v}^0_3 \in H^1_0(0, L). \end{aligned} \tag{5.14}$$

With respect to the regularity assumed in Theorem 5.3 for $\underline{u}^0_3$ and u^0_α we recall the following classical result of existence and regularity (cf. BREZIS [1983], for example).

THEOREM 5.4. *Let*

$$f_i \in L^2(\Omega), \qquad g_i \in L^2(\Gamma),$$

then problems (5.13) *and* (5.14) *possess a unique solution in the space* $H^2_0(0, L)$ *and* $H^1_0(0, L)$*, respectively. Moreover,*

$$u^0_\alpha \in H^3(0, L), \qquad \underline{u}^0_3 \in H^2(0, L).$$

We observe that the variational problems for the zeroth order components u_α^0 and $\underline{u}_3^0$ are independent of the fact of having a weak or a strong clamping condition in the three-dimensional linearized elasticity model.

In the initial work of BERMÚDEZ and VIAÑO [1984] the variational formulation (5.13), (5.14) is also obtained, but in the framework of a mixed displacement–stress approach, which in a certain sense is equivalent to the displacement approach presented here, although less rigorous in its initial setting (cf. Section 6).

We shall also see that Eq. (5.12) gives precisely the zeroth order axial stress component of the asymptotic expansion of the scaled stress tensor in the mixed displacement–stress approach.

Evaluating now the next term in the asymptotic expansion, and using the same technique as for the zeroth order term, from (5.6) and (5.7) we obtain the following result.

THEOREM 5.5. *The displacement and stress components associated to odd powers of ε in the asymptotic expansions* (3.1)–(3.5) *are such that* $\boldsymbol{u}^1 \in V_{\mathrm{BN}}(\Omega)$, $u_3^1 = \underline{u}_3^1 - x_\alpha \partial_3 u_\alpha^1$, $\underline{u}_3^1 \in H_0^1(0,L)$, $u_\alpha^1 \in H_0^2(0,L)$, *and if* $\underline{u}_3^1 \in H^2(0,L)$ *and* $u_\alpha^1 \in H^3(0,L)$, *then* $\boldsymbol{u}^3 \in V(\Omega)$ *is of the form*:

$$\begin{aligned} u_1^3 &= z_1^3 + x_2 z^3 - \frac{\lambda}{2(\lambda+\mu)}[x_1\partial_3\underline{u}_3^1 - \tfrac{1}{2}(x_1^2 - x_2^2)\partial_{33}u_1^1 - x_1x_2\partial_{33}u_2^1],\\ u_2^3 &= z_2^3 - x_1 z^3 - \frac{\lambda}{2(\lambda+\mu)}[x_2\partial_3\underline{u}_3^1 - x_1x_2\partial_{33}u_1^1 - \tfrac{1}{2}(x_2^2 - x_1^2)\partial_{33}u_2^1], \end{aligned} \tag{5.15}$$

where z_α^3 *and* $z^3 \in H^1(0,L)$ *are such that for* $a=0$ *and* $a=L$

$$\begin{aligned} z_\alpha^3(a) &= -\frac{\lambda}{4(\lambda+\mu)}(I_\alpha - I_\beta)\partial_{33}u_\alpha^1(a), \quad \alpha \neq \beta \textit{(no sum on } \alpha),\\ z^3(a) &= \frac{\lambda}{2(\lambda+\mu)(I_1+I_2)}[H_2\partial_{33}u_1^1(a) - H_1\partial_{33}u_2^1(a)]. \end{aligned} \tag{5.16}$$

Moreover,

$$e_{\alpha\beta}(\boldsymbol{u}^3) = -\frac{\lambda}{2(\lambda+\mu)}e_{33}(\boldsymbol{u}^1)\delta_{\alpha\beta}, \quad \sigma_{\alpha\beta}^{-1} = 0, \tag{5.17}$$

$$\sigma_{33}^1 = \frac{\mu(3\lambda+2\mu)}{\lambda+\mu}e_{33}(\boldsymbol{u}^1) = \frac{\mu(3\lambda+2\mu)}{\lambda+\mu}(\partial_3\underline{u}_3^1 - x_\alpha\partial_{33}u_\alpha^1). \tag{5.18}$$

COROLLARY 5.3.

$$\begin{aligned} \boldsymbol{u}^1 &= \boldsymbol{u}^3 = \cdots = \boldsymbol{u}^{2m+1} = \mathbf{0}, \quad m \geqslant 0.\\ \boldsymbol{\sigma}^1 &= \boldsymbol{\sigma}^3 = \cdots = \boldsymbol{\sigma}^{2m+1} = \mathbf{0}, \quad m \geqslant 0. \end{aligned} \tag{5.19}$$

PROOF. If we proceed as in Theorem 4.6, substituting (5.17) into the first order Eq. (3.12) we obtain $\underline{u}_3^1 = 0$ and $u_\alpha^1 = 0$ which implies $\boldsymbol{u}^1 = \mathbf{0}$. From (3.3) with $p=-1$ we get

$$0 = \sigma_{\alpha\beta}^{-1} = \lambda e_{pp}(\boldsymbol{u}^3)\delta_{\alpha\beta} + 2\mu e_{\alpha\beta}(\boldsymbol{u}^3),$$

therefore, $e_{\alpha\beta}(\boldsymbol{u}^3) = 0$ and $\sigma_{33}^1 = 0$.

Taking $\boldsymbol{v} = \boldsymbol{u}^3$ in (3.12) with $p = 1$, we obtain

$$0 = \int_\Omega 2\sigma^1_{3\beta} e_{3\beta}(\boldsymbol{u}^3)\,\mathrm{d}\boldsymbol{x} = \int_\Omega 4\mu e_{3\beta}(\boldsymbol{u}^3) e_{3\beta}(\boldsymbol{u}^3)\,\mathrm{d}\boldsymbol{x} = 0.$$

Consequently, $e_{3\beta}(\boldsymbol{u}^3) = 0$ and $\sigma^1_{3\beta} = 0$ also. We are then left with

$$e_{\alpha\beta}(\boldsymbol{u}^3) = e_{3\beta}(\boldsymbol{u}^3) = 0, \qquad \int_\Omega \sigma^1_{\alpha\beta} e_{\alpha\beta}(\boldsymbol{v})\,\mathrm{d}\boldsymbol{x} = 0, \quad \text{for all } \boldsymbol{v} \in V(\Omega),$$

from which we conclude that $\boldsymbol{u}^3 = \boldsymbol{0}$ and $\sigma^1_{\alpha\beta} = 0$, by applying the same reasoning as for $\boldsymbol{u}^1$ and $\sigma^{-1}_{\alpha\beta}$. The result is completed by induction on m. □

6. The asymptotic expansion method in the displacement–stress approach

In the previous section we showed that the *formal asymptotic expansion* of $\boldsymbol{u}(\varepsilon)$ in $V(\Omega)$ does not contain odd powers of ε, that is,

$$\boldsymbol{u}(\varepsilon) = \boldsymbol{u}^0 + \varepsilon^2 \boldsymbol{u}^2 + \varepsilon^4 \boldsymbol{u}^4 + \text{h.o.t.} \tag{6.1}$$

Simultaneously, we have shown (cf. Theorems 5.1, 5.3, 5.5 and Corollary 5.3) that the formal asymptotic expansion in the stress tensor, induced by that of the displacement field via the constitutive law, contains neither negative nor odd powers of ε, that is,

$$\boldsymbol{\sigma}(\varepsilon) = \boldsymbol{\sigma}^0 + \varepsilon^2 \boldsymbol{\sigma}^2 + \varepsilon^4 \boldsymbol{\sigma}^4 + \text{h.o.t.} \tag{6.2}$$

From (3.11) and (3.12) we conclude that $\boldsymbol{\sigma}^p$, $p = 0, 2, 4, \ldots$ satisfies the following equations:

$$\begin{aligned} \int_\Omega \sigma^0_{ij} e_{ij}(\boldsymbol{v})\,\mathrm{d}\boldsymbol{x} &= \int_\Omega \sigma^0_{ij} \partial_j v_i\,\mathrm{d}\boldsymbol{x} \\ &= \int_\Omega f_i v_i\,\mathrm{d}\boldsymbol{x} + \int_\Gamma g_i v_i\,\mathrm{d}a, \quad \text{for all } \boldsymbol{v} \in V(\Omega), \end{aligned} \tag{6.3}$$

$$\int_\Omega \sigma^p_{ij} e_{ij}(\boldsymbol{v})\,\mathrm{d}\boldsymbol{x} = \int_\Omega \sigma^p_{ij} \partial_j v_i\,\mathrm{d}\boldsymbol{x} = 0, \text{for all } \boldsymbol{v} \in V(\Omega),\ p = 2, 4, \ldots . \tag{6.4}$$

It is now possible to conclude that we would have obtained exactly the same Eqs. (6.3) and (6.4), had we required at the onset an asymptotic expansion, for both the scaled displacement and stress fields, of the form (6.1) and (6.2), respectively.

This important observation is the key to another application of the asymptotic expansion method to the three-dimensional linearized elasticity problem. Following the terminology of CIARLET [1990] we designate it by *the displacement–stress approach*. Historically, this was the first to be considered and was introduced in BERMÚDEZ and VIAÑO [1984] as an adaptation to the rod case of the same technique used by CIARLET and DESTUYNDER [1979a] for the plate case.

In order to describe it we start by reformulating the three-dimensional linearized elasticity problem for a linearly elastic rod. We follow the basic ideas of a mixed variational formulation (Hellinger–Reissner): in the first place the *stress tensor* is considered as one of the *unknowns* of the problem; secondly *the constitutive equation* (1.9) (Hooke's law) is written in the inverse form, that is, the linearized strain tensor is written as a function of the stress tensor.

Consequently, from the inversion of (1.9), from (1.10) and (1.11), we deduce that the displacement field $\boldsymbol{u}^\varepsilon = (u_i^\varepsilon)$ and the stress field $\boldsymbol{\sigma}^\varepsilon = (\sigma_{ij}^\varepsilon)$ are a solution of the following variational problem:

$$\boldsymbol{u}^\varepsilon \in V(\Omega^\varepsilon), \qquad \boldsymbol{\sigma}^\varepsilon \in \Sigma(\Omega^\varepsilon) = \{\boldsymbol{\tau}^\varepsilon = (\tau_{ij}^\varepsilon) \in [L^2(\Omega^\varepsilon)]^9 \colon \tau_{ij}^\varepsilon = \tau_{ji}^\varepsilon\}, \tag{6.5}$$

$$e_{ij}^\varepsilon(\boldsymbol{u}^\varepsilon) = \tfrac{1}{2}(\partial_i^\varepsilon u_j^\varepsilon + \partial_j^\varepsilon u_i^\varepsilon) = -\frac{\lambda^\varepsilon}{2\mu^\varepsilon(3\lambda^\varepsilon + 2\mu^\varepsilon)}\sigma_{pp}^\varepsilon \delta_{ij} + \frac{1}{2\mu^\varepsilon}\sigma_{ij}^\varepsilon, \tag{6.6}$$

$$\int_{\Omega^\varepsilon} \sigma_{ij}^\varepsilon e_{ij}^\varepsilon(\boldsymbol{v}^\varepsilon)\,\mathrm{d}\boldsymbol{x}^\varepsilon = \int_{\Omega^\varepsilon} f_i^\varepsilon v_i^\varepsilon\,\mathrm{d}\boldsymbol{x}^\varepsilon + \int_{\Gamma^\varepsilon} g_i^\varepsilon v_i^\varepsilon\,\mathrm{d}a^\varepsilon, \quad \text{for all } \boldsymbol{v}^\varepsilon \in V(\Omega^\varepsilon). \tag{6.7}$$

REMARK 6.1. This description is valid independently of the boundary conditions. Therefore, if not explicitly stated, $V(\Omega^\varepsilon)$ is an admissible displacement space of the type (1.10) or (5.1), or even of any other type usually encountered in the mechanics literature.

REMARK 6.2. In order to be consistent with the original works of CIARLET and DESTUYNDER [1979a], BERMÚDEZ and VIAÑO [1984] and with the usual notation in rod theory, we use constants E^ε and ν^ε which are related to Lamé's constants λ^ε and μ^ε through

$$E^\varepsilon = \frac{\mu^\varepsilon(3\lambda^\varepsilon + 2\mu^\varepsilon)}{\lambda^\varepsilon + \mu^\varepsilon}, \qquad \nu^\varepsilon = \frac{\lambda^\varepsilon}{2(\lambda^\varepsilon + \mu^\varepsilon)}, \quad E^\varepsilon > 0, \quad 0 < \nu^\varepsilon < \tfrac{1}{2}. \tag{6.8}$$

REMARK 6.3. In the engineering literature it is also customary to consider the following elements of $L^2(0, L)$:

$$\begin{aligned} &\text{bending moments:} \quad && m_\beta^\varepsilon = \int_{\omega^\varepsilon} x_\beta^\varepsilon \sigma_{33}^\varepsilon(\boldsymbol{x}^\varepsilon)\,\mathrm{d}\omega^\varepsilon, \\ &\text{shear forces:} && q_\beta^\varepsilon = \int_{\omega^\varepsilon} \sigma_{3\beta}^\varepsilon(\boldsymbol{x}^\varepsilon)\,\mathrm{d}\omega^\varepsilon, \end{aligned} \tag{6.9}$$

which, besides having an important physical meaning, will also be used in what follows.

Considering the variational formulation of (6.6) we obtain a mixed formulation of the Hellinger–Reissner type, that is, $\boldsymbol{u}^\varepsilon$ and $\boldsymbol{\sigma}^\varepsilon$ are the unique solution of the following variational problem (equivalent to (1.10), (1.11)):

$$(\boldsymbol{u}^\varepsilon, \boldsymbol{\sigma}^\varepsilon) \in V(\Omega^\varepsilon) \times \Sigma(\Omega^\varepsilon), \tag{6.10}$$

$$\int_{\Omega^\varepsilon} \left\{ \frac{1+\nu^\varepsilon}{E^\varepsilon} \sigma_{ij}^\varepsilon - \frac{\nu^\varepsilon}{E^\varepsilon} \sigma_{pp}^\varepsilon \delta_{ij} \right\} \tau_{ij}^\varepsilon \, \mathrm{d}\boldsymbol{x}^\varepsilon = \int_{\Omega^\varepsilon} e_{ij}^\varepsilon(\boldsymbol{u}^\varepsilon) \tau_{ij}^\varepsilon \, \mathrm{d}\boldsymbol{x}^\varepsilon,$$

$$\text{for all } \boldsymbol{\tau}^\varepsilon \in \Sigma(\Omega^\varepsilon), \tag{6.11}$$

$$\int_{\Omega^\varepsilon} \sigma_{ij}^\varepsilon e_{ij}^\varepsilon(\boldsymbol{v}^\varepsilon) \, \mathrm{d}\boldsymbol{x}^\varepsilon = \int_{\Omega^\varepsilon} f_i^\varepsilon v_i^\varepsilon \, \mathrm{d}\boldsymbol{x}^\varepsilon + \int_{\Gamma^\varepsilon} g_i^\varepsilon v_i^\varepsilon \, \mathrm{d}a^\varepsilon,$$

$$\text{for all } \boldsymbol{v}^\varepsilon \in V(\Omega^\varepsilon). \tag{6.12}$$

The existence and unicity of a solution for (6.10)–(6.12) is a consequence of an abstract result of BREZZI [1974] and of Korn's inequality in $V(\Omega^\varepsilon)$ (cf. CIARLET [1988], DUVAUT and LIONS [1972a]).

As in Section 2 to the displacement field $\boldsymbol{u}^\varepsilon = (u_i^\varepsilon) : \overline{\Omega}^\varepsilon \to \mathbb{R}^3$ we associate a scaled displacement vector field $\boldsymbol{u}(\varepsilon) = (u_i(\varepsilon)) : \overline{\Omega} \to \mathbb{R}^3$ defined by the scalings

$$\begin{aligned} u_\alpha(\varepsilon)(\boldsymbol{x}) &= \varepsilon u_\alpha^\varepsilon(\boldsymbol{x}^\varepsilon), \\ u_3(\varepsilon)(\boldsymbol{x}) &= u_3^\varepsilon(\boldsymbol{x}^\varepsilon), \end{aligned} \quad \text{for all } \boldsymbol{x}^\varepsilon = \Pi^\varepsilon(\boldsymbol{x}) \in \overline{\Omega}^\varepsilon. \tag{6.13}$$

In a similar way for any function $\boldsymbol{v}^\varepsilon = (v_i^\varepsilon) \in V(\Omega^\varepsilon)$ we associate the scaled function $\boldsymbol{v} = (v_i) : \overline{\Omega} \to \mathbb{R}^3$ defined by

$$\begin{aligned} v_\alpha(\boldsymbol{x}) &= \varepsilon v_\alpha^\varepsilon(\boldsymbol{x}^\varepsilon), \\ v_3(\boldsymbol{x}) &= v_3^\varepsilon(\boldsymbol{x}^\varepsilon), \end{aligned} \quad \text{for all } \boldsymbol{x}^\varepsilon = \Pi^\varepsilon(\boldsymbol{x}) \in \overline{\Omega}^\varepsilon. \tag{6.14}$$

Moreover, to the stress tensor $\boldsymbol{\sigma}^\varepsilon = (\sigma_{ij}^\varepsilon) : \overline{\Omega}^\varepsilon \to \mathbb{R}_s^9$ we associate the *scaled stress tensor field* $\boldsymbol{\sigma}(\varepsilon) = (\sigma_{ij}(\varepsilon)) : \overline{\Omega} \to \mathbb{R}_s^9$ defined by the scalings

$$\begin{aligned} \sigma_{\alpha\beta}(\varepsilon)(\boldsymbol{x}) &= \varepsilon^{-2} \sigma_{\alpha\beta}^\varepsilon(\boldsymbol{x}^\varepsilon), \\ \sigma_{3\beta}(\varepsilon)(\boldsymbol{x}) &= \varepsilon^{-1} \sigma_{3\beta}^\varepsilon(\boldsymbol{x}^\varepsilon), \\ \sigma_{33}(\varepsilon)(\boldsymbol{x}) &= \sigma_{33}^\varepsilon(\boldsymbol{x}^\varepsilon), \end{aligned} \quad \text{for all } \boldsymbol{x}^\varepsilon = \Pi^\varepsilon(\boldsymbol{x}) \in \overline{\Omega}^\varepsilon, \tag{6.15}$$

and to any second order tensor $\boldsymbol{\tau}^\varepsilon = (\tau_{ij}^\varepsilon) : \overline{\Omega}^\varepsilon \to \mathbb{R}_s^9$ we associate the *scaled tensor field* $\boldsymbol{\tau} = (\tau_{ij}) : \overline{\Omega} \to \mathbb{R}_s^9$ defined by the scalings

$$\begin{aligned} \tau_{\alpha\beta}(\boldsymbol{x}) &= \varepsilon^{-2} \tau_{\alpha\beta}^\varepsilon(\boldsymbol{x}^\varepsilon), \\ \tau_{3\beta}(\boldsymbol{x}) &= \varepsilon^{-1} \tau_{3\beta}^\varepsilon(\boldsymbol{x}^\varepsilon), \\ \tau_{33}(\boldsymbol{x}) &= \tau_{33}^\varepsilon(\boldsymbol{x}^\varepsilon), \end{aligned} \quad \text{for all } \boldsymbol{x}^\varepsilon = \Pi^\varepsilon(\boldsymbol{x}) \in \overline{\Omega}^\varepsilon. \tag{6.16}$$

Transformations (6.13)–(6.16) were introduced in BERMÚDEZ and VIAÑO [1984] as a natural consequence of the change of variable (2.4). They may be seen as the covariant and contravariant transformation for the tensor fields associated to the change of variable used, their physical meaning being that the equilibrium between the work done by the system of applied forces on the displacement field and the internal virtual work in the rod does not depend on the coordinate system employed, that is,

$$\int_{\Omega^\varepsilon} f_i^\varepsilon v_i^\varepsilon \, \mathrm{d}\boldsymbol{x}^\varepsilon + \int_{\Gamma^\varepsilon} g_i^\varepsilon v_i^\varepsilon \, \mathrm{d}a^\varepsilon = \varepsilon^2 \left[\int_\Omega f_i v_i \, \mathrm{d}\boldsymbol{x} + \int_\Gamma g_i v_i \, \mathrm{d}a \right],$$

$$\int_{\Omega^\varepsilon} \sigma_{ij}^\varepsilon e_{ij}^\varepsilon(\boldsymbol{v}^\varepsilon) \, \mathrm{d}\boldsymbol{x}^\varepsilon = \varepsilon^2 \int_\Omega \sigma_{ij}(\varepsilon) e_{ij}(\boldsymbol{v}) \, \mathrm{d}\boldsymbol{x}.$$

REMARK 6.4. (i) Scalings (6.13), (6.14) are the same as in the "displacement approach" (cf. (2.6)) and moreover, (6.15) coincides with the induced transformation in the displacement approach (cf. (2.21) and Remark 2.3).

(ii) Scalings (6.15) induce the following transformations for the bending moments and for the shear forces:

$$\begin{aligned} m_\beta(\varepsilon) &= \int_\omega x_\beta \sigma_{33}(\varepsilon) \, \mathrm{d}\omega = \varepsilon^{-3} m_\beta^\varepsilon, \\ q_\beta(\varepsilon) &= \int_\omega \sigma_{3\beta}(\varepsilon) \, \mathrm{d}\omega = \varepsilon^{-3} q_\beta^\varepsilon. \end{aligned} \tag{6.17}$$

Finally, we consider the *same hypotheses on the data* (system of applied forces and material properties) as in Section 2 (cf. (2.7)–(2.9)).

Combining the scalings both on the displacement and the stress fields together with a change of variable technique, we reformulate the original problem (6.10)–(6.12) in an equivalent way, posed now on a fixed set $\overline{\Omega}$ and which in the terminology of CIARLET [1990] we call *the scaled three-dimensional equations of a clamped rod in a mixed-displacement approach.* It consists in a *scaled inverse constitutive equation* written in the variational form and in a *scaled principle of virtual work*, and it is but yet another way of looking at problem (2.19), (2.20).

THEOREM 6.1. *The scaled displacement field $\boldsymbol{u}(\varepsilon) = (u_i(\varepsilon))$ and the scaled stress field $\boldsymbol{\sigma}(\varepsilon) = (\sigma_{ij}(\varepsilon))$ are the unique solution of the following problem:*

$$(\boldsymbol{u}(\varepsilon), \boldsymbol{\sigma}(\varepsilon)) \in V(\Omega) \times \Sigma(\Omega),$$

$$\int_\Omega \frac{1}{E} \sigma_{33}(\varepsilon) \tau_{33} \, \mathrm{d}\boldsymbol{x}$$

$$+\varepsilon^2 \int_\Omega \left\{ \frac{2(1+\nu)}{E} \sigma_{3\beta}(\varepsilon)\tau_{3\beta} - \frac{\nu}{E}[\sigma_{33}(\varepsilon)\tau_{\mu\mu} + \sigma_{\mu\mu}(\varepsilon)\tau_{33}] \right\} \mathrm{d}\boldsymbol{x}$$

$$+\varepsilon^4 \int_\Omega \left[\frac{1+\nu}{E} \sigma_{\alpha\beta}(\varepsilon) - \frac{\nu}{E}\sigma_{\mu\mu}(\varepsilon)\delta_{\alpha\beta} \right] \tau_{\alpha\beta}\, \mathrm{d}\boldsymbol{x}$$

$$= \int_\Omega e_{ij}(\boldsymbol{u}(\varepsilon))\tau_{ij}\, \mathrm{d}\boldsymbol{x}, \quad \textit{for all } \boldsymbol{\tau} \in \Sigma(\Omega), \tag{6.18}$$

$$\int_\Omega \sigma_{ij}(\varepsilon) e_{ij}(\boldsymbol{v})\, \mathrm{d}\boldsymbol{x} = \int_\Omega f_i v_i\, \mathrm{d}\boldsymbol{x} + \int_\Gamma g_i v_i\, \mathrm{d}a, \quad \textit{for all } \boldsymbol{v} \in V(\Omega), \tag{6.19}$$

In order to simplify the notation we introduce the bilinear forms $a_p(\cdot,\cdot) : \Sigma(\Omega) \times \Sigma(\Omega) \to \mathbb{R}$, $p = 0, 2, 4$; $b(\cdot,\cdot) : \Sigma(\Omega) \times V(\Omega) \to \mathbb{R}$ and the linear form $F(\cdot) : V(\Omega) \to \mathbb{R}$, defined for all elements $\boldsymbol{\sigma}, \boldsymbol{\tau} \in \Sigma(\Omega)$ and $\boldsymbol{v} \in V(\Omega)$ as follows:

$$a_0(\boldsymbol{\sigma}, \boldsymbol{\tau}) = \int_\Omega \frac{1}{E} \sigma_{33}\tau_{33}\, \mathrm{d}\boldsymbol{x}, \tag{6.20}$$

$$a_2(\boldsymbol{\sigma}, \boldsymbol{\tau}) = \int_\Omega \left[\frac{2(1+\nu)}{E} \sigma_{3\beta}\tau_{3\beta} - \frac{\nu}{E}(\sigma_{33}\tau_{\mu\mu} + \sigma_{\mu\mu}\tau_{33}) \right] \mathrm{d}\boldsymbol{x}, \tag{6.21}$$

$$a_4(\boldsymbol{\sigma}, \boldsymbol{\tau}) = \int_\Omega \left[\frac{1+\nu}{E} \sigma_{\alpha\beta} - \frac{\nu}{E}\sigma_{\mu\mu}\delta_{\alpha\beta} \right] \tau_{\alpha\beta}\, \mathrm{d}\boldsymbol{x}, \tag{6.22}$$

$$b(\boldsymbol{\tau}, \boldsymbol{v}) = -\int_\Omega \tau_{ij} e_{ij}(\boldsymbol{v})\, \mathrm{d}\boldsymbol{x}, \tag{6.23}$$

$$F(\boldsymbol{v}) = -\int_\Omega f_i v_i\, \mathrm{d}\boldsymbol{x} - \int_\Gamma g_i v_i\, \mathrm{d}a. \tag{6.24}$$

Then, problem (6.18), (6.19) can be stated as

$$(\boldsymbol{u}(\varepsilon), \boldsymbol{\sigma}(\varepsilon)) \in V(\Omega) \times \Sigma(\Omega),$$

$$a_0(\boldsymbol{\sigma}(\varepsilon), \boldsymbol{\tau}) + \varepsilon^2 a_2(\boldsymbol{\sigma}(\varepsilon), \boldsymbol{\tau}) + \varepsilon^4 a_4(\boldsymbol{\sigma}(\varepsilon), \boldsymbol{\tau}) + b(\boldsymbol{\tau}, \boldsymbol{u}(\varepsilon)) = 0,$$
$$\text{for all } \boldsymbol{\tau} \in \Sigma(\Omega), \tag{6.25}$$

$$b(\boldsymbol{\sigma}(\varepsilon), \boldsymbol{v}) = F(\boldsymbol{v}), \quad \text{for all } \boldsymbol{v} \in V(\Omega). \tag{6.26}$$

As in the displacement approach we find a polynomial dependence with respect to the small parameter ε, which once again leads us to apply the asymptotic expansion method in the following way:

(i) Write $\boldsymbol{u}(\varepsilon)$ and $\boldsymbol{\sigma}(\varepsilon)$ as an a priori asymptotic expansion:

$$\begin{aligned} \boldsymbol{u}(\varepsilon) &= \boldsymbol{u}^0 + \varepsilon^2\boldsymbol{u}^2 + \varepsilon^4\boldsymbol{u}^4 + \text{h.o.t.} \\ \boldsymbol{\sigma}(\varepsilon) &= \boldsymbol{\sigma}^0 + \varepsilon^2\boldsymbol{\sigma}^2 + \varepsilon^4\boldsymbol{\sigma}^4 + \text{h.o.t.} \end{aligned} \tag{6.27}$$

REMARK 6.5. This hypothesis induces the following formal asymptotic expansions:

$$m_\beta(\varepsilon) = m_\beta^0 + \varepsilon^2 m_\beta^2 + \varepsilon^4 m_\beta^4 + \text{h.o.t.}$$
$$q_\beta(\varepsilon) = q_\beta^0 + \varepsilon^2 q_\beta^2 + \varepsilon^4 q_\beta^4 + \text{h.o.t.},$$

where

$$m_\beta^{2p} = \int_\omega x_\beta \sigma_{33}^{2p}\,\mathrm{d}\omega, \qquad q_\beta^{2p} = \int_\omega \sigma_{3\beta}^{2p}\,\mathrm{d}\omega. \tag{6.28}$$

(ii) Substitute this formal expansion in expressions (6.25) and (6.26) and set the factors of the successive powers of ε to zero, both for the scaled constitutive law (6.25) and for the scaled principle of virtual work (6.26). The result can be summarized in the following theorem.

THEOREM 6.2. *If the successive terms of the asymptotic expansion* (6.27) *satisfy* $(\boldsymbol{u}^{2p}, \boldsymbol{\sigma}^{2p}) \in V(\Omega) \times \Sigma(\Omega)$, *then for all* $\boldsymbol{\tau} \in \Sigma(\Omega)$ *and all* $\boldsymbol{v} \in V(\Omega)$ *we have*

$$\begin{aligned} &a_0(\boldsymbol{\sigma}^0, \boldsymbol{\tau}) + b(\boldsymbol{\tau}, \boldsymbol{u}^0) = 0, \\ &b(\boldsymbol{\sigma}^0, \boldsymbol{v}) = F(\boldsymbol{v}), \end{aligned} \tag{6.29}$$

$$\begin{aligned} &a_0(\boldsymbol{\sigma}^2, \boldsymbol{\tau}) + b(\boldsymbol{\tau}, \boldsymbol{u}^2) = -a_2(\boldsymbol{\sigma}^0, \boldsymbol{\tau}), \\ &b(\boldsymbol{\sigma}^2, \boldsymbol{v}) = 0, \end{aligned} \tag{6.30}$$

$$\begin{aligned} &a_0(\boldsymbol{\sigma}^4, \boldsymbol{\tau}) + b(\boldsymbol{\tau}, \boldsymbol{u}^4) = -a_2(\boldsymbol{\sigma}^2, \boldsymbol{\tau}) - a_4(\boldsymbol{\sigma}^0, \boldsymbol{\tau}), \\ &b(\boldsymbol{\sigma}^4, \boldsymbol{v}) = 0, \end{aligned} \tag{6.31}$$

$$\begin{aligned} &a_0(\boldsymbol{\sigma}^{2p}, \boldsymbol{\tau}) + b(\boldsymbol{\tau}, \boldsymbol{u}^{2p}) = -a_2(\boldsymbol{\sigma}^{2p-2}, \boldsymbol{\tau}) - a_4(\boldsymbol{\sigma}^{2p-4}, \boldsymbol{\tau}), \\ &b(\boldsymbol{\sigma}^{2p}, \boldsymbol{v}) = 0, \quad (p \geqslant 2). \end{aligned} \tag{6.32}$$

(iii) The third step in the asymptotic expansion method is to use equations (6.29)–(6.32) in order to identify as many terms as possible, assuming, if necessary, additional properties in the previous terms.

In the displacement approach we saw that in order to identify $\boldsymbol{u}^0$ and σ_{33}^0 it was necessary to know $\boldsymbol{u}^1$ and $\boldsymbol{u}^2$. In the displacement–stress approach the identification of $\boldsymbol{u}^0$ and σ_{33}^0 are a direct consequence of Eq. (6.29) and does not require the knowledge of any other terms. In fact the following result owed to BERMÚDEZ and VIAÑO [1984] holds.

THEOREM 6.3. *The following property holds*:

$$\boldsymbol{u}^0 \in V_{\mathrm{BN}}(\Omega)\colon u_3^0 = \underline{u}_3^0 - x_\alpha \partial_3 u_\alpha^0, \qquad \underline{u}_3^0 \in H_0^1(0, L), \qquad u_\alpha^0 \in H_0^2(0, L), \tag{6.33}$$

$$\sigma_{33}^0 = E(\partial_3 \underline{u}_3^0 - x_\alpha \partial_{33} u_\alpha^0), \tag{6.34}$$

where u_α^0 is the unique solution of the following problem (no sum on α):

$$u_\alpha^0 \in H_0^2(0,L),$$

$$\int_0^L EI_\alpha \partial_{33} u_\alpha^0 \partial_{33} v_\alpha^0 \, dx_3$$

$$= \int_0^L \left[\int_\omega f_\alpha \, d\omega + \int_\gamma g_\alpha \, d\gamma \right] v^0 \, dx_3$$

$$- \int_0^L \left[\int_\omega x_\alpha f_3 \, d\omega + \int_\gamma x_\alpha g_3 \, d\gamma \right] \partial_3 v^0 \, dx_3, \quad \textit{for all } v^0 \in H_0^2(0,L), \tag{6.35}$$

and $\underline{u}_3^0$ is the unique solution of the following problem:

$$\underline{u}_3^0 \in H_0^1(0,L),$$

$$\int_0^L E \partial_3 \underline{u}_3^0 \partial_3 \underline{v}^0 \, dx_3 = \int_0^L \left[\int_\omega f_3 \, d\omega + \int_\gamma g_3 \, d\gamma \right] \underline{v}^0 \, dx_3, \tag{6.36}$$

for all $\underline{v}^0 \in H_0^1(0,L)$.

PROOF. In the first place we observe that the first equation of (6.29) is equivalent to the following conditions:

$$e_{\alpha\beta}(\boldsymbol{u}^0) = e_{3\beta}(\boldsymbol{u}^0) = 0, \tag{6.37}$$

$$e_{33}(\boldsymbol{u}^0) = \frac{1}{E}\sigma_{33}^0, \tag{6.38}$$

which are equivalent to (6.33) and (6.34), respectively (cf. Theorem 4.3). Equations (6.35) and (6.36) are obtained using the second expression in (6.29) restricted to $V_{\mathrm{BN}}(\Omega)$ (cf. Theorem 4.6 and Corollary 5.2). □

COROLLARY 6.1. *The bending moment component m_β^0 is given by*

$$m_\beta^0 = -EI_\beta \partial_{33} u_\beta^0 \quad (\textit{no sum on } \beta). \tag{6.39}$$

As expected, the same equations as in the displacement approach are obtained.

We remark once again that Theorem 6.3 is also valid for the strong clamping conditions at the ends, that is, for $V(\Omega)$ defined by (2.14) and for $V(\Omega)$ defined by (5.3). This is a consequence of the fact that Theorem 4.3 holds in both cases.

We may then conclude that the "displacement–stress approach" is conceptually simpler than the "displacement approach" since Eqs. (6.33)–(6.35) are obtained at a lower computational cost: it is enough to set to zero the coefficient of ε^0 both in the constitutive equation and in the principle of virtual work, while in the displacement approach it was necessary to prove that all the factors of ε^q, $-4 \leqslant q \leqslant 0$ were zero.

However, one must remember that the asymptotic expansion hypothesis on the displacement field $\boldsymbol{u}(\varepsilon) = \boldsymbol{u}^0 + \varepsilon^2\boldsymbol{u}^2 + \varepsilon^4\boldsymbol{u}^4 + \text{h.o.t.}$ induces an asymptotic expansion in the stress field $\boldsymbol{\sigma}(\varepsilon)$ which possesses factors of negative powers of ε (cf. (3.2)) and one *must show* that they vanish. This fundamental information is given to us from the displacement approach, and can only be proved to hold in the weakly clamping case (cf. Sections 4 and 5).

Since we already know that the asymptotic expansion of $\boldsymbol{\sigma}(\varepsilon)$ starts with a zeroth order term, and possesses only terms with even powers of ε, it turns out that it is easier to use the inverse constitutive equation, which only has positive powers of ε.

7. Characterization of the zeroth order stress term and second order displacement fields: Statement of the theorem

Equations (6.29) determine components $\boldsymbol{u}^0$ and σ^0_{33} in a unique way, no need to have any information about $\boldsymbol{u}^1$ and $\boldsymbol{u}^2$. However, Eqs. (6.29) are not enough to determine $\sigma^0_{3\beta}$ and $\sigma^0_{\alpha\beta}$. This lack of unicity already mentioned in BERMÚDEZ and VIAÑO [1984] represents a fundamental difference with respect to the plate case (cf. CIARLET and DESTUYNDER [1979a] and DESTUYNDER [1980]). Imposing the compatibility conditions with $\boldsymbol{u}^2$ and $\boldsymbol{u}^4$ (Eqs. (6.30) and (6.31)) TRABUCHO and VIAÑO [1987, 1989] showed that $\sigma^0_{3\beta}$ and $\sigma^0_{\alpha\beta}$ are uniquely determined whenever $V(\Omega)$ is defined as in (5.3). This idea had also been used in CIMETIÈRE, GEYMONAT, LE DRET, RAOULT and TUTEK [1986, 1988] for the nonlinear case.

The complete identification of $(\boldsymbol{u}^0, \boldsymbol{\sigma}^0)$ gives in addition the complete characterization of $\boldsymbol{u}^2$ and of σ^2_{33}. Moreover, the same technique allows us to characterize $\boldsymbol{u}^4$, σ^4_{33} and $\sigma^2_{3\beta}$. In the remaining part of this section, all the calculations and mathematical aspects are owed to TRABUCHO and VIAÑO [1989].

From now on we assume a weakly clamping condition, that is,

$$V(\Omega) = \{\boldsymbol{v} = (v_i) \in [H^1(\Omega)]^3 \colon \int_{\omega\times\{a\}} v_i \,\mathrm{d}\omega = 0,$$
$$\int_{\omega\times\{a\}} (x_j v_i - x_i v_j)\,\mathrm{d}\omega = 0, \text{ at } a = 0, L\}. \tag{7.1}$$

Taking into account Remark 5.2 we conclude that

$$V(\Omega) = W_2(\Omega) \times W_1(\Omega), \qquad \Sigma(\Omega) = L^2(\Omega) \times [L^2(\Omega)]^2 \times [L^2(\Omega)]^4_s, \tag{7.2}$$

where

$$W_1(\Omega) = \{\Phi \in H^1(\Omega)\colon \int_{\omega\times\{a\}} \Phi \,\mathrm{d}\omega = \int_{\omega\times\{a\}} x_\alpha \Phi \,\mathrm{d}\omega = 0,\ a = 0, L\}, \tag{7.3}$$

$$W_2(\Omega) = \{\boldsymbol{\Psi} = (\Psi_\alpha) \in [H^1(\Omega)]^2\colon \int_{\omega\times\{a\}} \Psi_\alpha \,\mathrm{d}\omega = 0,$$
$$\int_{\omega\times\{a\}} (x_2\Psi_1 - x_1\Psi_2)\,\mathrm{d}\omega = 0,\ a = 0, L\}, \tag{7.4}$$

$$[L^2(\Omega)]^4_s = \{\boldsymbol{\tau} = (\tau_{\alpha\beta}) \in [L^2(\Omega)]^4\colon\ \tau_{\alpha\beta} = \tau_{\beta\alpha}\}. \tag{7.5}$$

As a consequence, Eqs. (6.29)–(6.31) are equivalent to the following set of equations:

$$\int_\Omega \frac{1}{E}\sigma^0_{33}\tau_{33}\,\mathrm{d}\boldsymbol{x} - \int_\Omega \partial_3 u^0_3 \tau_{33}\,\mathrm{d}\boldsymbol{x} = 0, \quad \text{for all } \tau_{33} \in L^2(\Omega), \tag{7.6}$$

$$-\int_\Omega (\partial_3 u^0_\beta + \partial_\beta u^0_3)\tau_{3\beta}\,\mathrm{d}\boldsymbol{x} = 0, \quad \text{for all } (\tau_{3\beta}) \in [L^2(\Omega)]^2, \tag{7.7}$$

$$-\int_\Omega \tfrac{1}{2}(\partial_\alpha u^0_\beta + \partial_\beta u^0_\alpha)\tau_{\alpha\beta}\,\mathrm{d}\boldsymbol{x} = 0, \quad \text{for all } (\tau_{\alpha\beta}) \in [L^2(\Omega)]^4_s, \tag{7.8}$$

$$\int_\Omega (\sigma^0_{33}\partial_3 v_3 + \sigma^0_{3\beta}\partial_\beta v_3)\,\mathrm{d}\boldsymbol{x} = \int_\Omega f_3 v_3\,\mathrm{d}\boldsymbol{x} + \int_\Gamma g_3 v_3\,\mathrm{d}a,$$
$$\text{for all } v_3 \in W_1(\Omega), \tag{7.9}$$

$$\int_\Omega (\sigma^0_{\alpha\beta}\partial_\alpha v_\beta + \sigma^0_{3\beta}\partial_3 v_\beta)\,\mathrm{d}\boldsymbol{x} = \int_\Omega f_\beta v_\beta\,\mathrm{d}\boldsymbol{x} + \int_\Gamma g_\beta v_\beta\,\mathrm{d}a,$$
$$\text{for all } (v_\beta) \in W_2(\Omega), \tag{7.10}$$

$$\int_\Omega \frac{1}{E}\sigma^2_{33}\tau_{33}\,\mathrm{d}\boldsymbol{x} - \int_\Omega \partial_3 u^2_3 \tau_{33}\,\mathrm{d}\boldsymbol{x} = \int_\Omega \frac{\nu}{E}\sigma^0_{\mu\mu}\tau_{33}\,\mathrm{d}\boldsymbol{x},$$
$$\text{for all } \tau_{33} \in L^2(\Omega), \tag{7.11}$$

$$-\int_\Omega (\partial_3 u^2_\beta + \partial_\beta u^2_3)\tau_{3\beta}\,\mathrm{d}\boldsymbol{x} = -\int_\Omega \frac{2(1+\nu)}{E}\sigma^0_{3\beta}\tau_{3\beta}\,\mathrm{d}\boldsymbol{x},$$
$$\text{for all } (\tau_{3\beta}) \in [L^2(\Omega)]^2, \tag{7.12}$$

$$-\int_\Omega \tfrac{1}{2}(\partial_\alpha u^2_\beta + \partial_\beta u^2_\alpha)\tau_{\alpha\beta}\,\mathrm{d}\boldsymbol{x} = \int_\Omega \frac{\nu}{E}\sigma^0_{33}\tau_{\alpha\beta}\delta_{\alpha\beta}\,\mathrm{d}\boldsymbol{x},$$
$$\text{for all } (\tau_{\alpha\beta}) \in [L^2(\Omega)]^4_s, \tag{7.13}$$

$$\int_{\Omega} (\sigma_{33}^2 \partial_3 v_3 + \sigma_{3\beta}^2 \partial_\beta v_3)\,\mathrm{d}\boldsymbol{x} = 0, \quad \text{for all } v_3 \in W_1(\Omega), \tag{7.14}$$

$$\int_{\Omega} (\sigma_{\alpha\beta}^2 \partial_\alpha v_\beta + \sigma_{3\beta}^2 \partial_3 v_\beta)\,\mathrm{d}\boldsymbol{x} = 0, \quad \text{for all } (v_\beta) \in W_2(\Omega), \tag{7.15}$$

$$\int_{\Omega} \frac{1}{E}\sigma_{33}^4 \tau_{33}\,\mathrm{d}\boldsymbol{x} - \int_{\Omega} \partial_3 u_3^4 \tau_{33}\,\mathrm{d}\boldsymbol{x} = \int_{\Omega} \frac{\nu}{E}\sigma_{\mu\mu}^2 \tau_{33}\,\mathrm{d}\boldsymbol{x},$$
$$\text{for all } \tau_{33} \in L^2(\Omega), \tag{7.16}$$

$$-\int_{\Omega} (\partial_3 u_\beta^4 + \partial_\beta u_3^4)\tau_{3\beta}\,\mathrm{d}\boldsymbol{x} = -\int_{\Omega} \frac{2(1+\nu)}{E}\sigma_{3\beta}^2 \tau_{3\beta}\,\mathrm{d}\boldsymbol{x},$$
$$\text{for all } (\tau_{3\beta}) \in [L^2(\Omega)]^2, \tag{7.17}$$

$$-\int_{\Omega} \tfrac{1}{2}(\partial_\alpha u_\beta^4 + \partial_\beta u_\alpha^4)\tau_{\alpha\beta}\,\mathrm{d}\boldsymbol{x}$$
$$= -\int_{\Omega} \frac{1+\nu}{E}\sigma_{\alpha\beta}^0 \tau_{\alpha\beta}\,\mathrm{d}\boldsymbol{x} + \int_{\Omega} \frac{\nu}{E}(\sigma_{33}^2 + \sigma_{\mu\mu}^0)\tau_{\alpha\beta}\delta_{\alpha\beta}\,\mathrm{d}\boldsymbol{x},$$
$$\text{for all } (\tau_{\alpha\beta}) \in [L^2(\Omega)]_s^4, \tag{7.18}$$

$$\int_{\Omega} (\sigma_{33}^4 \partial_3 v_3 + \sigma_{3\beta}^4 \partial_\beta v_3)\,\mathrm{d}\boldsymbol{x} = 0, \quad \text{for all } v_3 \in W_1(\Omega), \tag{7.19}$$

$$\int_{\Omega} (\sigma_{\alpha\beta}^4 \partial_\alpha v_\beta + \sigma_{3\beta}^4 \partial_3 v_\beta)\,\mathrm{d}\boldsymbol{x} = 0, \quad \text{for all } (v_\beta) \in W_2(\Omega). \tag{7.20}$$

We now introduce some notations and study some auxiliary problems that show up in a natural way in what follows. Some of these functions and constants are known in classical beam theory and consequently shall be referred to by their usual names. Some others show up in a natural way in the application of the asymptotic expansion method. Both these constants and functions depend only on the geometry of the cross section and, in a certain sense, characterize it.

We denote the *second moment of area* of the cross section with respect to axis Ox_β by I_α, $(\alpha \neq \beta)$ and by H_i the *bimoments of area*, defined as follows:

$$I_\alpha = \int_\omega x_\alpha^2\,\mathrm{d}\omega, \qquad H_\alpha = \int_\omega \tfrac{1}{2}x_\alpha(x_1^2 + x_2^2)\,\mathrm{d}\omega, \qquad H_3 = \int_\omega \tfrac{1}{4}(x_1^2 + x_2^2)^2\,\mathrm{d}\omega. \tag{7.21}$$

Functions $\Phi_{\alpha\beta}$ and δ_α are defined in the following way:

$$(\Phi_{\alpha\beta}(x_1, x_2)) = \begin{pmatrix} \tfrac{1}{2}(x_1^2 - x_2^2) & x_1 x_2 \\ x_1 x_2 & \tfrac{1}{2}(x_2^2 - x_1^2) \end{pmatrix},$$
$$\delta_1(x_1, x_2) = -x_2, \qquad \delta_2(x_1, x_2) = x_1. \tag{7.22}$$

These functions were already used in the displacement approach (see (4.17)) and are related with Poisson's effects in Saint Venant's torsion theory (cf. FRAEJIS DE VEUBEKE [1979]).

Moreover, associated with the torsion theory are the *warping and torsion functions*. The warping function w of ω is the unique solution of the following elliptic boundary value problem:

$$
\begin{aligned}
&-\partial_{\alpha\alpha} w = 0 \quad \text{in } \omega, \\
&\partial_n w = x_2 n_1 - x_1 n_2 \quad \text{on } \gamma, \\
&\int_\omega w \, \mathrm{d}\omega = 0.
\end{aligned}
\tag{7.23}
$$

From its variational formulation together with the divergence theorem we obtain that w may be given as the unique solution of

$$
\begin{aligned}
&w \in H^1(\omega), \\
&\int_\omega \partial_\alpha w \partial_\alpha \varphi \, \mathrm{d}\omega = \int_\omega (x_2 \partial_1 \varphi - x_1 \partial_2 \varphi) \, \mathrm{d}\omega, \quad \text{for all } \varphi \in H^1(\omega), \\
&\int_\omega w \, \mathrm{d}\omega = 0.
\end{aligned}
\tag{7.24}
$$

Setting successively $\varphi = x_\alpha$, $\varphi = x_1 x_2$ and using (1.5) we have

$$
\int_\omega \partial_\alpha w \, \mathrm{d}\omega = 0, \qquad \int_\omega (x_2 \partial_1 w + x_1 \partial_2 w) \, \mathrm{d}\omega = I_2 - I_1. \tag{7.25}
$$

We introduce the *sectorial moments of area* I_β^w and the *warping constant* J_w, associated with the warping function w:

$$
I_\beta^w = 2 \int_\omega x_\beta w \, \mathrm{d}\omega, \tag{7.26}
$$

$$
J_w = \int_\omega w^2 \, \mathrm{d}\omega. \tag{7.27}
$$

Let

$$
\begin{aligned}
p_1 &= \partial_1 w - x_2, \\
p_2 &= \partial_2 w + x_1.
\end{aligned}
\tag{7.28}
$$

Then from (7.23) we have

$$
\begin{aligned}
&\partial_\alpha p_\alpha = 0 \quad \text{in } \omega, \\
&p_\alpha n_\alpha = 0 \quad \text{on } \gamma,
\end{aligned}
\tag{7.29}
$$

which implies (cf. GIRAULT and RAVIART [1981, Theorem 3.1]) that there exists a "stream" function Ψ such that $\boldsymbol{p} = \mathbf{rot}\ \Psi$, that is,

$$\begin{aligned} p_1 &= \partial_1 w - x_2 = \partial_2 \Psi, \\ p_2 &= \partial_2 w + x_1 = -\partial_1 \Psi. \end{aligned} \tag{7.30}$$

Moreover, Ψ is constant on each component γ_k of the boundary of ω and if we impose it to be zero on γ_0, then it is given as the unique solution of the following elliptic boundary value problem:

$$\begin{aligned} &-\partial_{\alpha\alpha}\Psi = 2 \quad \text{in } \omega, \\ &\Psi = 0 \quad \text{on } \gamma_0, \\ &\partial_\tau \Psi = 0 \quad \text{on } \gamma, \\ &\int_{\gamma_k} \partial_n \Psi \,\mathrm{d}\gamma = 2A(\omega_k), \quad k = 1, 2, \ldots, p, \end{aligned} \tag{7.31}$$

where $\partial_\tau \Psi$ denotes the tangential derivative of Ψ and $A(\omega_k)$ the area of hole ω_k, $k = 1, 2, \ldots, p$. We remark that problem (7.31) is equivalent to

$$\begin{aligned} &-\partial_{\alpha\alpha}\Psi = 2 \quad \text{in } \omega, \\ &\Psi = 0, \\ &\partial_n \Psi = -p_2 n_1 + p_1 n_2 \quad \text{on } \gamma_k, \ k = 1, 2, \ldots, p. \end{aligned} \tag{7.32}$$

Function Ψ is also known as *Prandtl's potential function*. We define the following *torsional constants* associated with this function:

$$I_1^\Psi = -\int_\omega x_2^2 \partial_2 \Psi \,\mathrm{d}\omega, \qquad I_2^\Psi = \int_\omega x_1^2 \partial_1 \Psi \,\mathrm{d}\omega, \tag{7.33}$$

$$J = -\int_\omega x_\alpha \partial_\alpha \Psi \,\mathrm{d}\omega. \tag{7.34}$$

Denoting by c_k the (constant) trace of Ψ on boundary γ_k the following alternative expressions for the torsion constants are obtained:

$$\begin{aligned} I_1^\Psi &= 2\int_\omega x_2 \Psi \,\mathrm{d}\omega + 2c_k \int_{\omega_k} x_2 \,\mathrm{d}\omega, \\ I_2^\Psi &= -2\int_\omega x_1 \Psi \,\mathrm{d}\omega - 2c_k \int_{\omega_k} x_1 \,\mathrm{d}\omega, \\ J &= 2\int_\omega \Psi \,\mathrm{d}\omega + 2c_k A(\omega_k). \end{aligned} \tag{7.35}$$

From the definition of J, (7.30) and (7.24) we deduce the relation

$$J = I_1 + I_2 - \int_\omega [(\partial_1 w)^2 + (\partial_2 w)^2]\,\mathrm{d}\omega. \tag{7.36}$$

The following functions do not appear explicitly in classical literature. Function η_β is the unique solution of the following boundary value problem:

$$\begin{aligned} &-\partial_{\alpha\alpha}\eta_\beta = -2x_\beta \quad \text{in } \omega,\\ &\partial_n \eta_\beta = 0 \quad \text{on } \gamma,\\ &\int_\omega \eta_\beta\,\mathrm{d}\omega = 0. \end{aligned} \tag{7.37}$$

which has the variational formulation

$$\begin{aligned} &\eta_\beta \in H^1(\omega),\\ &\int_\omega \partial_\alpha \eta_\beta \partial_\alpha \varphi\,\mathrm{d}\omega = -2\int_\omega x_\beta \varphi\,\mathrm{d}\omega, \quad \text{for all } \varphi \in H^1(\omega),\\ &\int_\omega \eta_\beta\,\mathrm{d}\omega = 0. \end{aligned} \tag{7.38}$$

Setting $\varphi = x_\gamma$ and $\varphi = x_\gamma x_\gamma$, we deduce that

$$\begin{aligned} &\int_\omega \partial_\beta \eta_\beta\,\mathrm{d}\omega = -2I_\beta \quad (\text{no sum on } \beta),\\ &\int_\omega \partial_\alpha \eta_\beta\,\mathrm{d}\omega = 0 \quad (\alpha \neq \beta),\\ &\int_\omega x_\alpha \partial_\alpha \eta_\beta\,\mathrm{d}\omega = -2H_\beta, \end{aligned} \tag{7.39}$$

and setting $\varphi = w$ and $\varphi = \eta_\beta$, it follows that

$$I_\beta^w = \int_\omega (x_1 \partial_2 \eta_\beta - x_2 \partial_1 \eta_\beta)\,\mathrm{d}\omega. \tag{7.40}$$

Function θ_β is the unique solution of the following boundary value problem:

$$\begin{aligned} &-\partial_{\alpha\alpha}\theta_\beta = 2x_\beta \quad \text{in } \omega,\\ &\partial_n \theta_\beta = -\Phi_{\alpha\beta} n_\alpha \quad \text{on } \gamma,\\ &\int_\omega \theta_\beta\,\mathrm{d}\omega = 0, \end{aligned} \tag{7.41}$$

the variational formulation of which is

$$\begin{aligned}&\theta_\beta \in H^1(\omega),\\ &\int_\omega (\partial_\alpha\theta_\beta + \Phi_{\beta\alpha})\partial_\alpha\varphi \,\mathrm{d}\omega = 0, \quad \text{for all } \varphi \in H^1(\omega),\\ &\int_\omega \theta_\beta = 0.\end{aligned} \tag{7.42}$$

Setting $\varphi = x_\gamma$ in (7.42), we get $\int_\omega \partial_\alpha\theta_\beta \,\mathrm{d}\omega = -\int_\omega \Phi_{\alpha\beta}\,\mathrm{d}\omega$ or equivalently,

$$\begin{aligned}&\int_\omega \partial_1\theta_1 \,\mathrm{d}\omega = \tfrac{1}{2}(I_2 - I_1), \qquad \int_\omega \partial_2\theta_2 \,\mathrm{d}\omega = \tfrac{1}{2}(I_1 - I_2),\\ &\int_\omega \partial_1\theta_2 \,\mathrm{d}\omega = \int_\omega \partial_2\theta_1 \,\mathrm{d}\omega = 0.\end{aligned} \tag{7.43}$$

Choosing $\varphi = w$ in (7.42), $\varphi = \theta_\beta$ in (7.24) and using (7.30) we get

$$I_1^\Psi - H_2 = \int_\omega (x_1\partial_2\theta_1 - x_2\partial_1\theta_1)\,\mathrm{d}\omega, \tag{7.44}$$

$$I_2^\Psi + H_1 = \int_\omega (x_1\partial_2\theta_2 - x_2\partial_1\theta_2)\,\mathrm{d}\omega. \tag{7.45}$$

Associated with functions η_β and θ_β the following constants are introduced which we designate by *Timoshenko's constants* because they will show up in a generalization of this classical constant appearing in Timoshenko's beam theory.

$$\begin{aligned}&L^\eta_{\alpha\beta} = \int_\omega x_\alpha\eta_\beta\,\mathrm{d}\omega, \qquad L^\theta_{\alpha\beta} = \int_\omega x_\alpha\theta_\beta\,\mathrm{d}\omega,\\ &K^\eta_{\alpha\beta} = \int_\omega \Phi_{\alpha\mu}\partial_\mu\eta_\beta\,\mathrm{d}\omega, \qquad K^\theta_{\alpha\beta} = \int_\omega \Phi_{\alpha\mu}\partial_\mu\theta_\beta\,\mathrm{d}\omega.\end{aligned} \tag{7.46}$$

REMARK 7.1. Since the data in the corresponding problems are infinitely regular, the regularity of functions w, Ψ, η_β, θ_β depends only on the regularity of the boundary γ of ω. The following elementary properties will be also useful in the sequel:

$$K^\eta_{\alpha\beta} = 2L^\theta_{\alpha\beta}, \qquad K^\theta_{\alpha\beta} = -\int_\omega \partial_\mu\theta_\alpha\partial_\mu\theta_\beta\,\mathrm{d}\omega.$$

If ω is symmetric with respect to axis Ox_1 then

$$I_1^w = I_1^\Psi = L^\eta_{12} = L^\theta_{12} = L^\eta_{21} = L^\theta_{21} = H_2 = 0.$$

If ω is symmetric with respect to axis Ox_2 then

$$I_2^w = I_2^\Psi = L_{12}^\eta = L_{12}^\theta = L_{21}^\eta = L_{21}^\theta = H_1 = 0.$$

If ω is symmetric with respect to both axes then

$$I_\beta^w = I_\beta^\Psi = H_\beta = 0, \qquad L_{\alpha\beta}^\eta = L_{\alpha\beta}^\theta = K_{\alpha\beta}^\eta = K_{\alpha\beta}^\theta = 0, \qquad \alpha \neq \beta.$$

We finally introduce functions $\overline{\rho}$, $\overline{w}$, $\overline{\eta}_\beta$ and $\overline{\theta}_\beta$, which will be used in the characterization of $\boldsymbol{u}^4$ (cf. Section 9) and are defined in a unique way by the following variational problems:

$$\begin{aligned}
&\overline{\rho} \in H^1(\omega),\\
&\int_\omega \partial_\alpha \overline{\rho} \partial_\alpha \varphi \, \mathrm{d}\omega = \int_\omega [\tfrac{1}{2}(x_1^2 + x_2^2) - \tfrac{1}{2}(I_1 + I_2)]\varphi \, \mathrm{d}\omega, \quad \text{for all } \varphi \in H^1(\omega),\\
&\int_\omega \overline{\rho} \, \mathrm{d}\omega = 0,
\end{aligned} \tag{7.47}$$

$$\begin{aligned}
&\overline{w} \in H^1(\omega),\\
&\int_\omega \partial_\alpha \overline{w} \partial_\alpha \varphi \, \mathrm{d}\omega = \int_\omega w\varphi \, \mathrm{d}\omega, \quad \text{for all } \varphi \in H^1(\omega),\\
&\int_\omega \overline{w} \, \mathrm{d}\omega = 0,
\end{aligned} \tag{7.48}$$

$$\begin{aligned}
&\overline{\eta}_\beta \in H^1(\omega),\\
&\int_\omega \partial_\alpha \overline{\eta}_\beta \partial_\alpha \varphi \, \mathrm{d}\omega = \int_\omega \eta_\beta \varphi \, \mathrm{d}\omega, \quad \text{for all } \varphi \in H^1(\omega),\\
&\int_\omega \overline{\eta}_\beta \, \mathrm{d}\omega = 0,
\end{aligned} \tag{7.49}$$

$$\begin{aligned}
&\overline{\theta}_\beta \in H^1(\omega),\\
&\int_\omega \partial_\alpha \overline{\theta}_\beta \partial_\alpha \varphi \, \mathrm{d}\omega = \int_\omega \theta_\beta \varphi \, \mathrm{d}\omega, \quad \text{for all } \varphi \in H^1(\omega),\\
&\int_\omega \overline{\theta}_\beta \, \mathrm{d}\omega = 0.
\end{aligned} \tag{7.50}$$

It is interesting to remark that functions $\overline{w}$, $\overline{\eta}_\beta$ and $\overline{\theta}_\beta$ satisfy

$$\partial_{\alpha\alpha\beta\beta} \overline{w} = 0, \qquad \partial_{\alpha\alpha\beta\beta} \overline{\eta}_\mu = -2x_\mu, \qquad \partial_{\alpha\alpha\beta\beta} \overline{\theta}_\mu = 2x_\mu. \tag{7.51}$$

Setting $\varphi = w$ in (7.47)–(7.50), and using (7.24) together with an integration by parts, we obtain

$$\int_\omega (x_2\partial_1\overline{\rho} - x_1\partial_2\overline{\rho})\,\mathrm{d}\omega = \tfrac{1}{2}\int_\omega (x_1^2 + x_2^2)w\,\mathrm{d}\omega, \tag{7.52}$$

$$J_w = \int_\omega w^2\,\mathrm{d}\omega = \int_\omega (x_2\partial_1\overline{w} - x_1\partial_2\overline{w})\,\mathrm{d}\omega, \tag{7.53}$$

$$\int_\omega (x_2\partial_1\overline{\eta}_\beta - x_1\partial_2\overline{\eta}_\beta)\,\mathrm{d}\omega = \int_\omega \eta_\beta w\,\mathrm{d}\omega, \tag{7.54}$$

$$\int_\omega (x_2\partial_1\overline{\theta}_\beta - x_1\partial_2\overline{\theta}_\beta)\,\mathrm{d}\omega = \int_\omega \theta_\beta w\,\mathrm{d}\omega. \tag{7.55}$$

Finally, we shall also consider the following constants:

$$\begin{aligned} &\overline{H}_\beta = \int_\omega x_\beta\overline{\rho}\,\mathrm{d}\omega, \qquad \overline{I}_\beta^w = 2\int_\omega x_\beta\overline{w}\,\mathrm{d}\omega,\\ &\overline{L}_{\alpha\beta}^\eta = \int_\omega x_\alpha\overline{\eta}_\beta\,\mathrm{d}\omega, \qquad \overline{L}_{\alpha\beta}^\theta = \int_\omega x_\alpha\overline{\theta}_\beta\,\mathrm{d}\omega. \end{aligned} \tag{7.56}$$

Our main objective now is to determine $\sigma_{3\beta}^0$ and $\sigma_{\alpha\beta}^0$ from Eqs. (7.6)–(7.20). We shall then have calculated all the components of the displacement and stress fields $\boldsymbol{u}^0$ and $\boldsymbol{\sigma}^0$. In spite of the fact that the proof is done in several distinct steps, which have a certain interest by themselves, we would rather summarize all the information concerning $(\boldsymbol{u}^0, \boldsymbol{\sigma}^0)$ in a single theorem (cf. TRABUCHO and VIAÑO [1989]).

THEOREM 7.1. *Let the system of applied forces be such that*

$$\begin{aligned} &f_\alpha \in L^2(\Omega), \qquad g_\alpha \in L^2(\Gamma),\\ &f_3 \in H^1[0,L;L^2(\omega)], \qquad g_3 \in H^1[0,L;L^2(\gamma)]. \end{aligned} \tag{7.57}$$

Then, Eqs. (7.6)–(7.18) *determine in a unique way element* $(\boldsymbol{u}^0, \boldsymbol{\sigma}^0) \in V(\Omega) \times \Sigma(\Omega)$, $\boldsymbol{u}^2 \in V(\Omega)$ *and* $\sigma_{33}^2 \in L^2(\Omega)$, *through the following characterization*:

(i)

$$\boldsymbol{u}^0 \in V_{\mathrm{BN}}(\Omega), \qquad u_3^0 = \underline{u}_3^0 - x_\alpha\partial_3 u_\alpha^0, \qquad \underline{u}_3^0 \in H_0^1(0,L), \qquad u_\alpha^0 \in H_0^2(0,L), \tag{7.58}$$

where functions u_α^0 *and* $\underline{u}_3^0$ *are defined in the following way*:

(a) *Displacements* u_β^0 *are the unique solution of the following variational problem*:

$$u_\beta^0 \in H_0^2(0,L),$$
$$\int_0^L EI_\beta \partial_{33}u_\beta^0 \partial_{33} v \, \mathrm{d}x_3 = \int_0^L \tilde{F}_\beta v \, \mathrm{d}x_3, \tag{7.59}$$
for all $v \in H_0^2(0,L)$ (*no sum on* β),

where

$$\tilde{F}_\beta = \int_\omega f_\beta \, \mathrm{d}\omega + \int_\gamma g_\beta \, \mathrm{d}\gamma + \int_\omega x_\beta \partial_3 f_3 \, \mathrm{d}\omega + \int_\gamma x_\beta \partial_3 g_3 \, \mathrm{d}\gamma. \tag{7.60}$$

(b) *The stretching component* $\underline{u}_3^0$ *is the unique solution of the following problem*:

$$\underline{u}_3^0 \in H_0^1(0,L),$$
$$\int_0^L E\partial_3 \underline{u}_3^0 \partial_3 v \, \mathrm{d}x_3 = \int_0^L F_3 v \, \mathrm{d}x_3, \quad \text{for all } v \in H_0^1(0,L), \tag{7.61}$$

with F_3 *given by*

$$F_3 = \int_\omega f_3 \, \mathrm{d}\omega + \int_\gamma g_3 \, \mathrm{d}\gamma. \tag{7.62}$$

(ii) *The stress component* σ_{33}^0 *is a function of* $\boldsymbol{u}^0$, *given by*

$$\sigma_{33}^0 = E\partial_3 u_3^0 = E(\partial_3 \underline{u}_3^0 - x_\alpha \partial_{33} u_\alpha^0). \tag{7.63}$$

and the respective bending moment component is

$$m_\beta^0 = \int_\omega x_\beta \sigma_{33}^0 = -EI_\beta \partial_{33} u_\beta^0 \quad (\textit{no sum on } \beta). \tag{7.64}$$

(iii) *Displacements* u_i^2 *are of the form*

$$\begin{aligned} u_1^2 &= z_1^2 + x_2 z^2 - \nu(x_1 \partial_3 \underline{u}_3^0 - \Phi_{1\beta}\partial_{33}u_\beta^0), \\ u_2^2 &= z_2^2 - x_1 z^2 - \nu(x_2 \partial_3 \underline{u}_3^0 - \Phi_{2\beta}\partial_{33}u_\beta^0), \end{aligned} \tag{7.65}$$

$$\begin{aligned} u_3^2 = {} & \underline{u}_3^2 - x_\alpha \partial_3 z_\alpha^2 - w\partial_3 z^2 + \nu[\tfrac{1}{2}(x_1^2 + x_2^2) - \tfrac{1}{2}(I_1 + I_2)]\partial_{33}\underline{u}_3^0 \\ & + [(1+\nu)\eta_\alpha + \nu\theta_\alpha]\partial_{333}u_\alpha^0 + \frac{2(1+\nu)}{E} w^0, \end{aligned} \tag{7.66}$$

where functions z^2, w^0, $\underline{u}_3^2$ *and* z_α^2 *are characterized in the following problems* (a)–(d):

(a) *The angle of twist* z^2 *depends only on* x_3 *and is the unique solution of*

$$\begin{aligned}
&z^2 \in H^1(0,L),\\
&\int_0^L \frac{EJ}{2(1+\nu)} \partial_3 z^2 \partial_3 v \,\mathrm{d}x_3 = \int_0^L M_3^2 v \,\mathrm{d}x_3, \quad \textit{for all } v \in H_0^1(0,L),\\
&z^2(a) = \frac{\nu}{(I_1+I_2)}[H_2\partial_{33}u_1^0(a) - H_1\partial_{33}u_2^0(a)] \quad \textit{at } a = 0, L,
\end{aligned} \tag{7.67}$$

where

$$\begin{aligned}
M_3^2 &= \int_\omega (x_2 f_1 - x_1 f_2)\,\mathrm{d}\omega + \int_\gamma (x_2 g_1 - x_1 g_2)\,\mathrm{d}\gamma + \int_\omega w\partial_3 f_3 \,\mathrm{d}\omega\\
&\quad + \int_\gamma w\partial_3 g_3 \,\mathrm{d}\gamma - \frac{E}{2(1+\nu)}[(1+\nu)I_\alpha^w + \nu I_\alpha^\Psi]\partial_{3333} u_\alpha^0.
\end{aligned} \tag{7.68}$$

(b) *The additional warping* w^0 *is the unique solution of the following problem:*

$$\begin{aligned}
&w^0 \in H^1[0,L;H^1(\omega)] \quad \textit{and for all } x_3 \in [0,L],\\
&\int_\omega \partial_\beta w^0(x_3)\partial_\beta \varphi \,\mathrm{d}\omega\\
&\quad = \int_\omega f_3(x_3)\varphi \,\mathrm{d}\omega + \int_\gamma g_3(x_3)\varphi \,\mathrm{d}\gamma\\
&\quad\quad - \left[\int_\omega f_3(x_3)\,\mathrm{d}\omega + \int_\gamma g_3(x_3)\,\mathrm{d}\gamma\right] \int_\omega \varphi \,\mathrm{d}\omega, \quad \textit{for all } \varphi \in H^1(\omega),\\
&\int_\omega w^0(x_3)\,\mathrm{d}\omega = 0.
\end{aligned} \tag{7.69}$$

(c) *The stretching component* $\underline{u}_3^2$ *depends only on* x_3 *and uniquely solves the following problem:*

$$\begin{aligned}
&\underline{u}_3^2 \in H_0^1(0,L),\\
&\int_0^L E\partial_3 \underline{u}_3^2 \partial_3 v \,\mathrm{d}x_3 = \int_0^L \nu G_3^2 \partial_3 v \,\mathrm{d}x_3, \quad \textit{for all } v \in H_0^1(0,L),
\end{aligned} \tag{7.70}$$

where

$$G_3^2 = -\tfrac{1}{2}E(I_1 + I_2)\partial_{333}\underline{u}_3^0 + EH_\alpha\partial_{3333}u_\alpha^0 - \int_\omega x_\alpha f_\alpha \,\mathrm{d}\omega - \int_\gamma x_\alpha g_\alpha \,\mathrm{d}\gamma$$
$$- \int_\omega \tfrac{1}{2}(x_1^2 + x_2^2)\partial_3 f_3 \,\mathrm{d}\omega - \int_\gamma \tfrac{1}{2}(x_1^2 + x_2^2)\partial_3 g_3 \,\mathrm{d}\gamma. \qquad (7.71)$$

(d) *The bending component z_α^2 depends only on the variable x_3 and is the unique solution of the following variational problem (no sum on α, $\beta \neq \alpha$):*

$$z_\alpha^2 \in H^2(0,L) \quad \textit{and for all } v \in H_0^2(0,L):$$
$$\int_0^L EI_\alpha\partial_{33}z_\alpha^2\partial_{33}v \,\mathrm{d}x_3 = \int_0^L M_\alpha^2\partial_{33}v \,\mathrm{d}x_3,$$
$$z_\alpha^2(a) = -\tfrac{1}{2}\nu(I_\alpha - I_\beta)\partial_{33}u_\alpha^0(a) \quad \textit{at } a = 0, L\ ,$$
$$\partial_3 z_\alpha^2(a) = \frac{1}{I_\alpha}\{\frac{2(1+\nu)}{E}\int_\omega x_\alpha w^0(a)\,\mathrm{d}\omega - \tfrac{1}{2}I_\alpha^w\partial_3 z^2(a) + \nu H_\alpha\partial_{33}\underline{u}_3^0(a)$$
$$+ [(1+\nu)L_{\alpha\beta}^\eta + \nu L_{\alpha\beta}^\theta]\partial_{333}u_\beta^0(a)\} \quad \textit{at } a = 0, L\ , \qquad (7.72)$$

where

$$M_\alpha^2 = E\{(1+\nu)L_{\alpha\beta}^\eta + \nu L_{\alpha\beta}^\theta + \tfrac{1}{2}\nu K_{\alpha\beta}^\eta + \frac{\nu^2}{2(1+\nu)}(K_{\alpha\beta}^\theta + H_3\delta_{\alpha\beta})\}\partial_{3333}u_\beta^0$$
$$+ \nu EH_\alpha\partial_{333}\underline{u}_3^0 - \frac{E}{2(1+\nu)}[(1+\nu)I_\alpha^w + \nu I_\alpha^\Psi]\partial_{33}z^2$$
$$- \int_\omega [(1+\nu)\eta_\alpha + \nu\theta_\alpha]\partial_3 f_3 \,\mathrm{d}\omega - \int_\gamma [(1+\nu)\eta_\alpha + \nu\theta_\alpha]\partial_3 g_3 \,\mathrm{d}\gamma$$
$$+ \int_\omega \nu\Phi_{\alpha\beta}f_\beta \,\mathrm{d}\omega + \int_\gamma \nu\Phi_{\alpha\beta}g_\beta \,\mathrm{d}\gamma. \qquad (7.73)$$

(iv) *The shear stress components $\sigma_{3\beta}^0$ are uniquely determined by:*

$$\sigma_{31}^0 = \frac{E}{2(1+\nu)}\{-\partial_2\Psi\partial_3 z^2 + [(1+\nu)\partial_1\eta_\beta + \nu(\partial_1\theta_\beta + \Phi_{1\beta})]\partial_{333}u_\beta^0\} \qquad (7.74)$$

$$\sigma_{32}^0 = \frac{E}{2(1+\nu)}\{\partial_1\Psi\partial_3 z^2 + [(1+\nu)\partial_2\eta_\beta + \nu(\partial_2\theta_\beta + \Phi_{2\beta})]\partial_{333}u_\beta^0\} + \partial_2 w^0, \qquad (7.75)$$

and the shear force components $q^0_\beta = \int_\omega \sigma^0_{3\beta}$ *are given by*

$$q^0_\beta = -EI_\beta \partial_{333} u^0_\beta + \int_\omega x_\beta f_3 \, d\omega + \int_\gamma x_\beta g_3 \, d\gamma \quad (\textit{no sum on } \beta). \tag{7.76}$$

(v) *The plane stress components* $\sigma^0_{\alpha\beta}$ *are given by*

$$\sigma^0_{\alpha\beta} = S_{\alpha\beta}(\underline{\boldsymbol{u}}^4) + S^0_{\alpha\beta}, \tag{7.77}$$

where

$$S^0_{\alpha\beta} = \frac{\nu E}{(1+\nu)(1-2\nu)} \partial_3 (u^2_3 - \underline{u}^2_3 + x_\mu \partial_3 z^2_\mu) \delta_{\alpha\beta}, \tag{7.78}$$

$$S_{\alpha\beta}(\underline{\boldsymbol{u}}^4) = \frac{E}{(1+\nu)} e_{\alpha\beta}(\underline{\boldsymbol{u}}^4) + \frac{\nu E}{(1+\nu)(1-2\nu)} e_{\mu\mu}(\underline{\boldsymbol{u}}^4) \delta_{\alpha\beta}, \tag{7.79}$$

and $\underline{\boldsymbol{u}}^4 = (\underline{u}^4_\alpha)$ *is the unique solution of the following plane elasticity problem*:

$$\begin{aligned}
&\underline{\boldsymbol{u}}^4 \in L^2[0,L;(H^1(\omega))^2], \quad \textit{and a.e. in } (0,L) \textit{ for all } \boldsymbol{\varphi} \in [H^1(\omega)]^2: \\
&\int_\omega S_{\alpha\beta}(\underline{\boldsymbol{u}}^4(x_3)) e_{\alpha\beta}(\boldsymbol{\varphi}) \, d\omega = \int_\omega f_\beta(x_3)\varphi_\beta \, d\omega + \int_\gamma g_\beta(x_3)\varphi_\beta \, d\gamma \\
&\qquad\qquad + \int_\omega \partial_3 \sigma^0_{3\beta}(x_3)\varphi_\beta \, d\omega - \int_\omega S^0_{\alpha\beta} e_{\alpha\beta}(\boldsymbol{\varphi}) \, d\omega, \\
&\int_\omega \underline{u}^4_\alpha(x_3) \, d\omega = \int_\omega [x_2 \underline{u}^4_1(x_3) - x_1 \underline{u}^4_2(x_3)] \, d\omega = 0.
\end{aligned} \tag{7.80}$$

(vi) *The axial stress* σ^2_{33} *and the second order bending moment components are given by*

$$\begin{aligned}
\sigma^2_{33} &= E\partial_3 u^2_3 + \nu\sigma^0_{\mu\mu} \\
&= E\{\partial_3 \underline{u}^2_3 - x_\alpha \partial_{33} z^2_\alpha - w\partial_{33} z^2 + \nu[\tfrac{1}{2}(x_1^2 + x_2^2) - \tfrac{1}{2}(I_1 + I_2)]\partial_{333}\underline{u}^0_3 \\
&\quad + [(1+\nu)\eta_\alpha + \nu\theta_\alpha]\partial_{3333} u^0_\alpha\} + 2(1+\nu)\partial_3 w^0 + \nu\sigma^0_{\mu\mu},
\end{aligned} \tag{7.81}$$

$$m^2_\beta = -EI_\beta \partial_{33} z^2_\beta + M^2_\beta \quad (\textit{no sum on } \beta). \tag{7.82}$$

The proof of Theorem 7.1 will be presented in several steps in the next section. It is an adaptation of the results in TRABUCHO and VIAÑO [1989].

8. Characterization of the zeroth order stress and second order displacement fields: Proof of the theorem

Step 1. *Characterization of* $\boldsymbol{u}^0$, σ^0_{33} *and* m^0_α. Equations (7.58)–(7.64) were already established in Theorem 6.3 and included in Theorem 7.1 for completeness. □

Step 2. Displacements u_β^2 *are of the form* (7.65) *with* z^2, $z_\alpha^2 \in H^1(0,L)$ *and such that* $z^2(a)$, $z_\alpha^2(a)$, $a = 0$, *and* $a = L$, *are given by the boundary conditions in* (7.67) *and in* (7.72), *respectively.* This result was already established. It is enough to take into account that from (7.13) and (7.63) we deduce (4.16). Consequently, the result is a transcription of Theorems 5.2 and 5.3. □

Step 3. Displacement u_3^2 *is of the form* (7.66) *with* $\underline{u}_3^2 \in H^1(0,L)$ *and* $I_\beta \partial_3 z_\beta^2 + \frac{1}{2} I_\beta^w \partial_3 z^2 \in H^1(0,L)$ (*no sum on* β). From (7.12) we obtain the following relation between $\sigma_{3\beta}^0$ and u_i^2:

$$\sigma_{3\beta}^0 = \frac{E}{2(1+\nu)}(\partial_3 u_\beta^2 + \partial_\beta u_3^2). \tag{8.1}$$

Substituting this expression into (7.9) and using Green's formula, we conclude that u_3^2 is a solution of the following variational problem:

$$u_3^2 \in W_1(\Omega) \quad \text{and for all } v_3 \in W_1(\Omega):$$

$$\int_\Omega \partial_\beta u_3^2 \partial_\beta v_3 \,\mathrm{d}\boldsymbol{x} = \frac{2(1+\nu)}{E}\left[\int_\Omega f_3 v_3 \,\mathrm{d}\boldsymbol{x} + \int_\Gamma g_3 v_3 \,\mathrm{d}a + \int_\Omega \partial_3 \sigma_{33}^0 v_3 \,\mathrm{d}\boldsymbol{x}\right] - \int_\Omega \partial_3 u_\beta^2 \partial_\beta v_3 \,\mathrm{d}\boldsymbol{x}. \tag{8.2}$$

Choosing in (8.2) test functions of the form $v_3(x_1,x_2,x_3) = \phi(x_1,x_2)\psi(x_3)$ with $\phi \in H^1(\omega)$ and $\psi \in H_0^1(0,L)$, we find that u_3^2 is a solution of the following problem, which is a weaker version of (8.2):

$$u_3^2 \in L^2[0,L;H^1(\omega)] \quad \text{and a.e. in } (0,L), \text{ for all } \phi \in H^1(\omega):$$

$$\int_\omega \partial_\beta u_3^2 \partial_\beta \phi \,\mathrm{d}\omega = \frac{2(1+\nu)}{E}\left[\int_\omega f_3 \phi \,\mathrm{d}\omega + \int_\gamma g_3 \phi \,\mathrm{d}\gamma + \int_\omega \partial_3 \sigma_{33}^0 \phi \,\mathrm{d}\omega\right] - \int_\omega \partial_3 u_\beta^2 \partial_\beta \phi \,\mathrm{d}\omega. \tag{8.3}$$

A necessary and sufficient condition in order for u_3^2 to exist is given by

$$\int_\Omega f_3 \,\mathrm{d}\omega + \int_\Gamma g_3 \,\mathrm{d}\gamma + \int_\omega \partial_3 \sigma_{33}^0 \,\mathrm{d}\omega = 0 \quad \text{a.e. in } (0,L), \tag{8.4}$$

which may be obtained directly from (7.61)–(7.63).

Consequently, the solution u_3^2 of (8.3) is unique up to an additive function $\underline{u}_3^2 \in L^2(0,L)$. In order to characterize u_3^2 we start by developing the second member of (8.3). From (7.61)–(7.63) and (7.65), we deduce that it may be written in the following way:

$$\frac{2(1+\nu)}{E}\left[\int_{\Omega} f_3\phi \,\mathrm{d}\omega + \int_{\gamma} g_3\phi \,\mathrm{d}\gamma - \left(\int_{\Omega} f_3 \,\mathrm{d}\omega + \int_{\gamma} g_3\right)\int_{\omega}\phi \,\mathrm{d}\omega\right]$$
$$- \left[\int_{\omega} \partial_\beta\phi \,\mathrm{d}\omega\right]\partial_3 z_\beta^2 - \left[\int_{\omega}(x_2\partial_1\phi - x_1\partial_2\phi)\,\mathrm{d}\omega\right]\partial_3 z^2$$
$$+ \nu\left[\int_{\omega} x_\beta\partial_\beta\phi \,\mathrm{d}\omega\right]\partial_{33}\underline{u}_3^0$$
$$- \left[\nu\left(\int_{\omega}\Phi_{\alpha\beta}\partial_\beta\phi \,\mathrm{d}\omega\right) + 2(1+\nu)\left(\int_{\omega} x_\alpha\phi \,\mathrm{d}\omega\right)\right]\partial_{333}u_\alpha^0. \tag{8.5}$$

Now using properties (7.24) for function w, (7.38) for functions η_β, (7.42) for functions θ_β, and definition (7.69) of function w^0 we conclude that function

$$U = u_3^2 - [(1+\nu)\eta_\alpha + \nu\theta_\alpha]\partial_{333}u_\alpha^0 + w\partial_3 z^2 - \frac{2(1+\nu)}{E}w^0, \tag{8.6}$$

is a solution of the problem

$$U \in L^2[0,L;H^1(\omega)] \quad \text{and a.e. in } (0,L), \text{ for all } \phi \in H^1(\omega):$$
$$\int_{\omega}\partial_\beta U\partial_\beta\phi \,\mathrm{d}\omega = -\left[\int_{\omega}\partial_\beta\phi \,\mathrm{d}\omega\right]\partial_3 z_\beta^2 + \nu\left[\int_{\omega} x_\beta\partial_\beta\phi \,\mathrm{d}\omega\right]\partial_{33}\underline{u}_3^0. \tag{8.7}$$

Using the identities

$$\int_{\omega}\partial_\beta\phi \,\mathrm{d}\omega = \int_{\omega}\partial_\alpha(x_\beta)\partial_\alpha\phi \,\mathrm{d}\omega, \tag{8.8}$$

$$\int_{\omega} x_\beta\partial_\beta\phi \,\mathrm{d}\omega = \int_{\omega}\partial_\beta\{\tfrac{1}{2}(x_1^2+x_2^2) - \tfrac{1}{2}(I_1+I_2)\}\partial_\beta\phi \,\mathrm{d}\omega, \tag{8.9}$$

we obtain that

$$U + x_\beta\partial_3 z_\beta^2 - \nu[\tfrac{1}{2}(x_1^2+x_2^2) - \tfrac{1}{2}(I_1+I_2)]\partial_{33}\underline{u}_3^0 = \underline{u}_3^2, \tag{8.10}$$

where $\underline{u}_3^2 \in L^2(0,L)$ depends only on variable x_3. Expression (7.66) now follows from (8.6) and (8.10).

REMARK 8.1. Functions x_β and $\frac{1}{2}(x_1^2+x_2^2)-\frac{1}{2}(I_1+I_2)$ in (8.8) and (8.9) are unique up to additive constants and the present choice is justified because we have

$$\int_\omega x_\alpha \, d\omega = \int_\omega [\tfrac{1}{2}(x_1^2+x_2^2)-\tfrac{1}{2}(I_1+I_2)] \, d\omega = 0$$

which eases up the calculations concerning the boundary conditions for u_3^2. In any case, these constants would have been evaluated in the calculation of $\underline{u}_3^2$ and the final result would have been the same.

We shall now establish some necessary and sufficient conditions in order that $u_3^2 \in W_1(\Omega)$. This fact implies that both $\int_\omega u_3^2 \, d\omega$ and $\int_\omega x_\beta u_3^2 \, d\omega$ are in the space $H_0^1(0,L)$; specifically we have

$$\underline{u}_3^2 \in H_0^1(0,L), \tag{8.11}$$

$$\begin{aligned}&\{-I_\beta \partial_3 z_\beta^2 - \tfrac{1}{2} I_\beta^w \partial_3 z^2 + \frac{2(1+\nu)}{E}\int_\omega x_\beta w^0 \, d\omega + \nu H_\beta \partial_{33}\underline{u}_3^0 \\ &\quad + [(1+\nu)L_{\beta\alpha}^\eta + \nu L_{\beta\alpha}^\theta]\partial_{333}u_\alpha^0\} \in H_0^1(0,L) \quad (\text{no sum on } \beta).\end{aligned} \tag{8.12}$$

On the other hand, using regularity results for variational problems (7.59) and (7.61), (assumption (7.57) is considered here), together with (7.69), we have

$$\partial_{333}u_\alpha^0, \qquad \partial_{33}\underline{u}_3^0, \qquad \int_\omega x_\alpha w^0 \, d\omega \in H^1(0,L). \tag{8.13}$$

Consequently, (8.11) and the following conditions are necessary in order that $u_3^2 \in W_1(\Omega)$:

$$[I_\beta \partial_3 z_\beta^2 + \tfrac{1}{2} I_\beta^w \partial_3 z^2] \in H^1(0,L) \quad (\text{no sum on } \beta), \tag{8.14}$$

$$\begin{aligned}&[I_\beta \partial_3 z_\beta^2 + \tfrac{1}{2} I_\beta^w \partial_3 z^2](a) \\ &\quad = \frac{2(1+\nu)}{E}\int_\omega x_\beta w^0(a) \, d\omega + \nu H_\beta \partial_{33}\underline{u}_3^0(a) \\ &\qquad + [(1+\nu)L_{\beta\alpha}^\eta + \nu L_{\beta\alpha}^\theta]\partial_{333}u_\alpha^0(a), \quad \text{at } a = 0, L \ (\text{no sum on } \beta).\end{aligned} \tag{8.15}$$

Actually, conditions (8.11), (8.14) and (8.15) are not sufficient in order to get $u_3^2 \in W_1(\Omega)$. As a matter of fact, from (8.13) we see that a necessary and sufficient condition is given by (8.11), (8.15) together with

$$x_\beta \partial_3 z_\beta^2 + w \partial_3 z^2 \in H^1(\Omega). \tag{8.16}$$

This will be established in Step 5 and then u_3^2 given by (7.66) will be the solution of (8.2). □

Step 4. Calculation of $\sigma_{3\beta}^0$ and q_β^0. After having proved (7.65), (7.66), expressions (7.74), (7.75) for $\sigma_{3\beta}^0$ are an immediate consequence of (8.1). In fact, upon substitution we get

$$\sigma_{31}^0 = \frac{E}{2(1+\nu)}(x_2 - \partial_1 w)\partial_3 z^2 + \underline{\sigma}_{31}^0, \tag{8.17}$$

$$\sigma_{32}^0 = \frac{E}{2(1+\nu)}(-x_1 - \partial_2 w)\partial_3 z^2 + \underline{\sigma}_{32}^0, \tag{8.18}$$

where,

$$\underline{\sigma}_{3\beta}^0 = \frac{E}{2(1+\nu)}[\nu \Phi_{\alpha\beta} + (1+\nu)\partial_\beta \eta_\alpha + \nu \partial_\beta \theta_\alpha]\partial_{333} u_\alpha^0 + \partial_\beta w^0. \tag{8.19}$$

Expressions (8.17)–(8.19) give (7.74), (7.75) if we just take into account Eq. (7.30). Moreover, using (7.22), together with properties (7.39) and (7.43) we obtain from (7.74), (7.75):

$$q_\beta^0 = \int_\omega \sigma_{3\beta}^0 \, \mathrm{d}\omega = -EI_\beta \partial_{333} u_\beta^0 + \int_\omega \partial_\beta w^0 \, \mathrm{d}\omega \quad (\text{no sum on } \beta). \tag{8.20}$$

Expression (7.76) for q_β^0 is obtained from (8.20) and from the following property of w^0 (see (7.69)):

$$\int_\omega \partial_\beta w^0 \, \mathrm{d}\omega = \int_\omega \partial_\alpha(x_\beta)\partial_\alpha w^0 \, \mathrm{d}\omega = \int_\omega x_\beta f_3 \, \mathrm{d}\omega + \int_\gamma x_\beta g_3 \, \mathrm{d}\gamma. \tag{8.21}$$

REMARK 8.2. It is possible to find q_β^0 from (7.9) and (7.63). In order to do so, we only need to take $v_3 = x_\beta v^0$ with $v^0 \in H_0^1(0, L)$ in (7.9) as shown in BERMÚDEZ and VIAÑO [1984]. This fact proves that the expression for $\sigma_{3\beta}^0$ calculated above is compatible with (7.63). □

Step 5. Function z^2 solves (7.67), (7.68), $z_\alpha^2 \in H^2(0, L)$ and $\partial_3 z_\alpha^2$ *satisfies the boundary conditions in* (7.72). In order to characterize z^2 as the solution of boundary value problem (7.67) we set $v_1 = x_2 v^2$, $v_2 = -x_1 v^2$ in (7.10) and obtain for all $v^2 \in H_0^1(0, L)$

$$\int_0^L \left[\int_\omega (x_2 \sigma_{31}^0 - x_1 \sigma_{32}^0) \, \mathrm{d}\omega \right] \partial_3 v^2 \, \mathrm{d}x_3$$
$$= \int_0^L \left[\int_\omega (x_2 f_1 - x_1 f_2) \, \mathrm{d}\omega \right] v^2 \, \mathrm{d}x_3 + \int_0^L \left[\int_\gamma (x_2 g_1 - x_1 g_2) \, \mathrm{d}\gamma \right] v^2 \, \mathrm{d}x_3. \tag{8.22}$$

Substituting (8.17), (8.18) and then (7.30) into this equation, leads to

$$\frac{E}{2(1+\nu)}\left[\int_\omega -x_\beta\partial_\beta\Psi\,\mathrm{d}\omega\right]\int_0^L \partial_3 z^2\partial_3 v^2\,\mathrm{d}x_3$$
$$=\int_0^L\left[\int_\omega (x_2 f_1 - x_1 f_2)\,\mathrm{d}\omega\right]v^2\,\mathrm{d}x_3+\int_0^L\left[\int_\gamma (x_2 g_1 - x_1 g_2)\,\mathrm{d}\gamma\right]v^2\,\mathrm{d}x_3$$
$$-\int_0^L\left[\int_\omega (x_2\underline{\sigma}^0_{31} - x_1\underline{\sigma}^0_{32})\,\mathrm{d}\omega\right]\partial_3 v^2\,\mathrm{d}x_3,\quad \text{for all } v^2\in H^1_0(0,L). \tag{8.23}$$

In order to check that (8.23) is the same as (7.67), (7.68) we only need to apply Green's formula to the last term ($\sigma^0_{3\beta}\in H^1[0,L;L^2(\omega)]$) and use properties (7.40), (7.44), (7.45) together with the following ones obtained from (7.69) with $\phi = w$ and from (7.24) with $\phi = w^0(x_3)$:

$$\int_\omega (x_2\partial_1 w^0 - x_1\partial_2 w^0)\,\mathrm{d}\omega = \int_\omega \partial_\alpha w\partial_\alpha w^0\,\mathrm{d}\omega = \int_\omega f_3 w\,\mathrm{d}\omega + \int_\gamma g_3 w\,\mathrm{d}\gamma. \tag{8.24}$$

Thus, with Step 2 in mind, we have shown that z^2 satisfies boundary value problem (7.67), (7.68) which completely determines z^2 from the data and from u^0_α. The second member M^2_3 of (7.67), (7.68) is in $L^2(0,L)$ and, consequently, the regularity $z^2\in H^2(0,L)$ is obtained. From (8.14), established in Step 3, we conclude that $z^2_\alpha\in H^2(0,L)$. It is now possible to obtain the boundary conditions for $\partial_3 z^2$ announced in (7.72) directly from (8.15). □

Step 6. Calculation of $\sigma^0_{\alpha\beta}$. From (7.10) with $v_\beta(x_1,x_2,x_3)=\phi_\beta(x_1,x_2)\psi(x_3)$, $\phi_\beta\in H^1(\omega)$, $\psi\in H^1_0(0,L)$ we deduce that $\sigma^0_{\alpha\beta}$ is a solution of the following problem:

$$(\sigma^0_{\alpha\beta})\in L^2[0,L;(L^2(\omega))^4_s]\quad\text{and a.e. in } (0,L),\ \text{for all } \boldsymbol{\phi}=(\phi_\beta)\in[H^1(\omega)]^2:$$
$$\int_\omega \sigma^0_{\alpha\beta}(x_3)e_{\alpha\beta}(\boldsymbol{\phi})\,\mathrm{d}\omega$$
$$=\int_\omega f_\beta(x_3)\phi_\beta\,\mathrm{d}\omega+\int_\gamma g_\beta(x_3)\phi_\beta\,\mathrm{d}\gamma+\int_\omega \partial_3\sigma^0_{3\beta}(x_3)\phi_\beta\,\mathrm{d}\omega. \tag{8.25}$$

In order to obtain (8.25) we apply Green's formula to the term $\int_\Omega \sigma^0_{3\beta}\partial_3 v_\beta\,\mathrm{d}\boldsymbol{x}$ which is allowed because after Step 4 and (7.74), (7.75), we have $\sigma^0_{3\beta}\in H^1[0,L;L^2(\omega)]$. Problem (8.25) admits infinitely many solutions if and only if the following compatibility condition is satisfied a.e. in $(0,L)$ (see DUVAUT and

LIONS [1972a]):

$$\int_\omega f_\beta \phi_\beta \, d\omega + \int_\gamma g_\beta \phi_\beta \, d\gamma + \int_\omega \partial_3 \sigma^0_{3\beta} \phi_\beta \, d\omega = 0,$$
$$\text{for all } \boldsymbol{\phi} = (\phi_\beta) \in [H^1(\omega)]^2, \text{ such that } e_{\alpha\beta}(\boldsymbol{\phi}) = 0, \tag{8.26}$$

or equivalently

$$\int_\omega f_\beta \, d\omega + \int_\gamma g_\beta \, d\gamma + \int_\omega \partial_3 \sigma^0_{3\beta} \, d\omega = 0, \tag{8.27}$$

$$\int_\omega (x_2 f_1 - x_1 f_2) \, d\omega + \int_\gamma (x_2 g_1 - x_1 g_2) \, d\gamma + \int_\omega (x_2 \partial_3 \sigma^0_{31} - x_1 \partial_3 \sigma^0_{32}) \, d\omega = 0. \tag{8.28}$$

On the other hand, from (7.76) we obtain

$$\int_\omega \partial_3 \sigma^0_{3\beta} \, d\omega = \partial_3 q^0_\beta = -EI_\beta \partial_{3333} u^0_\beta$$
$$+ \int_\omega x_\beta \partial_3 f_3 \, d\omega + \int_\gamma x_\beta \partial_3 g_3 \, d\gamma \quad \text{(no sum on } \beta). \tag{8.29}$$

Substituting the strong interpretation of (7.59) into the equation above, we get (8.27). In order to verify (8.28), we use (7.67) under its equivalent form (8.23), which upon integrating by parts on both sides gives

$$\int_\omega (x_2 \partial_3 \underline{\sigma}^0_{31} - x_1 \partial_3 \underline{\sigma}^0_{32}) \, d\omega$$
$$= \frac{E}{2(1+\nu)} \left[\int_\omega x_\beta \partial_\beta \Psi \, d\omega \right] \partial_{33} z^2 - \int_\omega (x_2 f_1 - x_1 f_2) \, d\omega$$
$$+ \int_\gamma (x_2 g_1 - x_1 g_2) \, d\gamma. \tag{8.30}$$

On the other hand, from (8.17), (8.18) and (7.30), we get

$$\int_\omega (x_2 \partial_3 \sigma^0_{31} - x_1 \partial_3 \sigma^0_{32}) \, d\omega$$
$$= \int_\omega (x_2 \partial_3 \underline{\sigma}^0_{31} - x_1 \partial_3 \underline{\sigma}^0_{32}) \, d\omega - \frac{E}{2(1+\nu)} \left[\int_\omega x_\beta \partial_\beta \Psi \, d\omega \right] \partial_{33} z^2, \tag{8.31}$$

from which we obtain (8.28) after substituting (8.30) into (8.31).

In order to evaluate $\sigma^0_{\alpha\beta}$ *we shall consider the compatibility with equations* (7.11) and (7.18), from which we get

$$\begin{aligned} &\frac{1}{E}(\sigma^2_{33} - \nu\sigma^0_{\mu\mu}) = \partial_3 u^2_3, \\ &\frac{1+\nu}{E}\sigma^0_{\alpha\beta} - \frac{\nu}{E}(\sigma^2_{33} + \sigma^0_{\mu\mu})\delta_{\alpha\beta} = e_{\alpha\beta}(\boldsymbol{u}^4). \end{aligned} \tag{8.32}$$

Eliminating σ^2_{33}, we find

$$\sigma^0_{\alpha\beta} - \nu\sigma^0_{\mu\mu}\delta_{\alpha\beta} = \frac{E}{(1+\nu)}e_{\alpha\beta}(\boldsymbol{u}^4) + \frac{\nu E}{(1+\nu)}\partial_3 u^2_3 \delta_{\alpha\beta}, \tag{8.33}$$

and, finally, after a contraction in the indices followed by a substitution back into (8.33), we obtain

$$\sigma^0_{\alpha\beta} = S_{\alpha\beta}(\boldsymbol{u}^4) + \frac{\nu E}{(1+\nu)(1-2\nu)}\partial_3 u^2_3 \delta_{\alpha\beta}, \tag{8.34}$$

where we denote, for all $\boldsymbol{\phi} = (\phi_\beta) \in [H^1(\omega)]^2$,

$$S_{\alpha\beta}(\boldsymbol{\phi}) = \frac{E}{(1+\nu)}e_{\alpha\beta}(\boldsymbol{\phi}) + \frac{\nu E}{(1+\nu)(1-2\nu)}e_{\mu\mu}(\boldsymbol{\phi})\delta_{\alpha\beta}. \tag{8.35}$$

Using notations (7.78) we obtain

$$\sigma^0_{\alpha\beta} = S_{\alpha\beta}(\boldsymbol{u}^4) + S^0_{\alpha\beta} + \frac{\nu E}{(1+\nu)(1-2\nu)}\partial_3(\underline{u}^0_3 - x_\mu \partial_3 z^2_\mu)\delta_{\alpha\beta}. \tag{8.36}$$

Let $\boldsymbol{\xi}^4 = (\xi^4_\alpha) \in L^2[0, L; [H^1(\omega)]^2]$ be such that

$$S_{\alpha\beta}(\boldsymbol{\xi}^4) = \frac{\nu E}{(1+\nu)(1-2\nu)}\partial_3(x_\mu \partial_3 z^2_\mu - \underline{u}^2_3)\delta_{\alpha\beta}. \tag{8.37}$$

The existence of $\boldsymbol{\xi}^4 = (\xi^4_\alpha)$ may be proved by direct integration using the technique of Theorem 4.5. In fact, (8.37) is equivalent to

$$\begin{aligned} &\partial_1 \xi^4_1 = \partial_2 \xi^4_2 = \nu(x_\mu \partial_3 z^2_\mu - \partial_3 \underline{u}^2_3), \\ &\partial_1 \xi^4_2 + \partial_2 \xi^4_1 = 0, \end{aligned} \tag{8.38}$$

which has a general solution of the form (cf. Theorem 4.5)

$$\begin{aligned} &\xi^4_1 = z^4_1 + x_2 z^4 - \nu(x_1 \partial_3 \underline{u}^2_3 - \Phi_{\beta 1}\partial_{33} z^2_\beta), \\ &\xi^4_2 = z^4_2 - x_1 z^4 - \nu(x_2 \partial_3 \underline{u}^2_3 - \Phi_{\beta 2}\partial_{33} z^2_\beta), \end{aligned} \tag{8.39}$$

with z^4_α, $z^4 \in L^2(0, L)$ depending only on variable x_3.

Setting

$$\underline{u}_\alpha^4 = u_\alpha^4 - \xi_\alpha^4, \tag{8.40}$$

from (8.36) we obtain

$$\sigma_{\alpha\beta}^0 = S_{\alpha\beta}(\underline{u}^4) + S_{\alpha\beta}^0. \tag{8.41}$$

It only remains to show that $(\underline{u}_\alpha^4)$ is the solution of (7.80). In order to do so, we substitute (8.41) in (8.25) and obtain that $(\underline{u}_\alpha^4)$ is the unique solution of (7.80) up to a planar rigid body motion $(\Xi_\alpha^4)(x_3)$ such that $S_{\alpha\beta}(\Xi^4) = 0$. The last condition in (7.80) determines $(\underline{u}_\alpha^4)$ in a unique way and does not affect the expression for $\sigma_{\alpha\beta}^0$.

REMARK 8.3. This presentation of Step 6 is a corrected version of TRABUCHO and VIAÑO [1989] and was suggested by RAOULT [1990b, 1991].

REMARK 8.4. It is interesting to notice that Eq. (7.10) is satisfied for all $(v_\beta) \in W_2(\Omega)$ if and only if components $\sigma_{3\beta}^0$ verify the following condition:

$$\int_{\omega\times\{L\}} \sigma_{3\beta}^0 v_\beta \, d\omega - \int_{\omega\times\{0\}} \sigma_{3\beta}^0 v_\beta \, d\omega = 0, \quad \text{for all } (v_\beta) \in W_2(\Omega), \tag{8.42}$$

which is equivalent to saying that $\sigma_{3\beta}(0)$ and $\sigma_{3\beta}(L)$ are of the following form:

$$\begin{aligned} &\sigma_{31}^0(0) = A_1^0 + x_2 A^0, \qquad \sigma_{32}^0(0) = A_2^0 - x_1 A^0, \\ &\sigma_{31}^0(L) = A_1^L + x_2 A^L, \qquad \sigma_{32}^0(L) = A_2^L - x_1 A^L. \end{aligned} \tag{8.43}$$

Consequently, constants A_α^0, A_α^L, A^0 and A^L satisfy

$$A_\beta^0 = \int_\omega \sigma_{3\beta}^0(0) \, d\omega, \qquad A_\beta^L = \int_\omega \sigma_{3\beta}^0(L) \, d\omega,$$

$$A^0 = \frac{1}{I_1 + I_2} \int_\omega [x_2 \sigma_{31}^0(0) - x_1 \sigma_{32}^0(0)] \, d\omega,$$

$$A^L = \frac{1}{I_1 + I_2} \int_\omega [x_2 \sigma_{31}^0(L) - x_1 \sigma_{32}^0(L)] \, d\omega.$$

The verification of these conditions imposes additional restrictions on the applied loads and/or on the geometry, which, in general, do not hold. Therefore the equilibrium Eq. (6.29) is not satisfied, in general, by $\boldsymbol{\sigma}^0 = (\sigma_{ij}^0)$ calculated above. However, this equation is satisfied for all $\boldsymbol{v} = (v_1, v_2, v_3) \in V(\Omega)$ such that $v_\alpha = 0$ on $\Gamma_0 \cup \Gamma_L$. This is a consequence of a boundary layer effect near the ends, which, from a mechanical point of view, is associated with Saint Venant's principle (cf. SOKOLNIKOFF [1956]). □

Step 7. Calculation of σ_{33}^2 and m_β^2. Expression (7.81) for σ_{33}^2 follows directly from (7.11) taking into account (7.77) and (7.66). From (7.81) we get

$$\begin{aligned} m_\beta^2 = & -EI_\beta \partial_{33} z_\beta^2 - \tfrac{1}{2} EI_\beta^w \partial_{33} z^2 + \nu E H_\beta \partial_{333} \underline{u}_3^0 \\ & + E\left[(1+\nu) L_{\beta\alpha}^\eta + \nu L_{\beta\alpha}^\theta\right] \partial_{3333} u_\alpha^0 \\ & + 2(1+\nu) \int_\omega x_\beta \partial_3 w^0 \, \mathrm{d}\omega + \nu \int_\omega x_\beta \sigma_{\mu\mu}^0 \, \mathrm{d}\omega \quad \text{(no sum on } \beta), \end{aligned} \tag{8.44}$$

and the final expression (7.82) for m_β^2 is obtained from the following calculation for the last term in (8.44). From (7.80) with $\varphi_1 = \Phi_{\beta 1}$, $\varphi_2 = \Phi_{\beta 2}$ and as a consequence of $\partial_1 \Phi_{\beta 1} = x_\beta$, $\partial_2 \Phi_{\beta 2} = x_\beta$, $\partial_1 \Phi_{12} = x_2$, $\partial_1 \Phi_{22} = -x_1$, $\partial_2 \Phi_{11} = -x_2$ and $\partial_2 \Phi_{21} = x_1$, we have

$$\int_\omega x_\beta \sigma_{\mu\mu}^0 \, \mathrm{d}\omega = \int_\omega \Phi_{\beta\alpha} f_\alpha \, \mathrm{d}\omega + \int_\gamma \Phi_{\beta\alpha} g_\alpha \, \mathrm{d}\gamma + \int_\omega \Phi_{\beta\alpha} \partial_3 \sigma_{3\alpha}^0 \, \mathrm{d}\omega. \tag{8.45}$$

Now using (7.74) and (7.75), we obtain

$$\begin{aligned} \int_\omega \Phi_{\beta\alpha} \partial_3 \sigma_{3\alpha}^0 \, \mathrm{d}\omega = & \frac{E}{2(1+\nu)} \{ -I_\beta^\Psi \partial_{33} z^2 + [(1+\nu) K_{\beta\alpha}^\eta + \nu K_{\beta\alpha}^\theta \\ & + \nu H_3 \delta_{\beta\alpha}] \partial_{3333} u_\alpha^0 \} + \int_\omega \partial_{3\alpha} w^0 \Phi_{\beta\alpha} \, \mathrm{d}\omega. \end{aligned} \tag{8.46}$$

Substituting (8.45) and (8.46) into (8.44) and using (7.73) we finally obtain (7.82), after using the following equalities obtained from (7.38), (7.42) and (7.69):

$$\begin{aligned} \int_\omega 2x_\beta \partial_3 w^0 \, \mathrm{d}\omega & = \partial_3 \left[\int_\omega 2x_\beta w^0 \, \mathrm{d}\omega\right] = -\partial_3 \left[\int_\omega \partial_\alpha \eta_\beta \partial_\alpha w^0 \, \mathrm{d}\omega\right] \\ & = -\int_\omega \eta_\beta \partial_3 f_3 \, \mathrm{d}\omega - \int_\gamma \eta_\beta \partial_3 g_3 \, \mathrm{d}\gamma, \end{aligned}$$

$$\begin{aligned} \int_\omega \partial_{3\alpha} w^0 \Phi_{\beta\alpha} \, \mathrm{d}\omega & = \partial_3 \left[\int_\omega \Phi_{\beta\alpha} \partial_\alpha w^0 \, \mathrm{d}\omega\right] = -\partial_3 \left[\int_\omega \partial_\alpha \theta_\beta \partial_\alpha w^0 \, \mathrm{d}\omega\right] \\ & = -\int_\omega \theta_\beta \partial_3 f_3 \, \mathrm{d}\omega - \int_\gamma \theta_\beta \partial_3 g_3 \, \mathrm{d}\gamma. \end{aligned}$$

It is interesting to remark that M_α^2 can be written as a function of the data f_i and g_i using the strong interpretations of problems (7.59), (7.61) and (7.67). □

Step 8. *Variational problem for* z_α^2. Setting $v_3 = x_\alpha \partial_3 v^2$, $v^2 \in H_0^2(0,L)$ in Eq. (7.14) leads to

$$\int_0^L m_\alpha^2 \partial_{33} v^2 \,\mathrm{d}x_3 + \int_0^L \left[\int_\omega \sigma_{3\alpha}^2 \,\mathrm{d}\omega\right] \partial_3 v^2 \,\mathrm{d}x_3 = 0, \quad \text{for all } v^2 \in H_0^2(0,L). \tag{8.47}$$

On the other hand, from (7.15) with $v_\beta = v^2$, $v^2 \in H_0^2(0,L)$, we have

$$\int_0^L \left[\int_\omega \sigma_{3\alpha}^2 \,\mathrm{d}\omega\right] \partial_3 v^2 \,\mathrm{d}x_3 = 0, \quad \text{for all } v^2 \in H_0^2(0,L). \tag{8.48}$$

Substituting (8.48) into (8.47) and taking into account expression (7.82) for m_α^2 we obtain Eq. (7.72). □

Step 9. *Variational problem for* $\underline{u}_3^2$. Introducing once again $v_3 = v^2 \in H_0^1(0,L)$ into (7.14), we deduce that

$$\int_0^L \left[\int_\omega \sigma_{33}^2 \,\mathrm{d}\omega\right] \partial_3 v^2 \,\mathrm{d}x_3 = 0, \quad \text{for all } v_2 \in H_0^1(0,L), \tag{8.49}$$

but from (7.81) we have

$$\int_\omega \sigma_{33}^2 \,\mathrm{d}\omega = E \partial_3 \underline{u}_3^2 + \nu \int_\omega \sigma_{\mu\mu}^0 \,\mathrm{d}\omega. \tag{8.50}$$

Therefore, from (8.49) we obtain

$$\int_0^L E \partial_3 \underline{u}_3^2 \partial_3 v^2 \,\mathrm{d}x_3 = -\nu \int_0^L \left[\int_\omega \sigma_{\mu\mu}^0 \,\mathrm{d}\omega\right] \partial_3 v^2 \,\mathrm{d}x_3, \quad \text{for all } v^2 \in H_0^1(0,L),$$

and consequently, Eqs. (7.70), (7.71) follow if

$$G_3^2 = -\int_\omega \sigma_{\mu\mu}^0 \,\mathrm{d}\omega, \tag{8.51}$$

which we shall now prove. In fact, from (7.10) with $v_\beta = x_\beta v^0$, $v^0 \in H_0^1(0,L)$, we have

$$\int_0^L \left[\int_\omega \sigma_{\mu\mu}^0 \,\mathrm{d}\omega\right] v^0 \,\mathrm{d}x_3 + \int_0^L \left[\int_\omega x_\beta \sigma_{3\beta}^0 \,\mathrm{d}\omega\right] \partial_3 v^0 \,\mathrm{d}x_3$$
$$= \int_0^L \left[\int_\omega x_\beta f_\beta \,\mathrm{d}\omega + \int_\gamma x_\beta g_\beta \,\mathrm{d}\gamma\right] v^0 \,\mathrm{d}x_3, \quad \text{for all } v^0 \in H_0^1(0,L), \tag{8.52}$$

which after an integration by parts gives,

$$\int_\omega \sigma^0_{\mu\mu}\,\mathrm{d}\omega = \int_\omega x_\beta f_\beta\,\mathrm{d}\omega + \int_\gamma x_\beta g_\beta\,\mathrm{d}\gamma + \int_\omega x_\beta \partial_3 \sigma^0_{3\beta}\,\mathrm{d}\omega \quad \text{a.e. in } (0, L). \tag{8.53}$$

In order to evaluate the last term in (8.53) without using (7.74), (7.75) we set $v_3 = \frac{1}{2}(x_1^2 + x_2^2)v^0$, $v^0 \in H_0^1(0, L)$ in (7.9) and obtain, for all $v^0 \in H_0^1(0, L)$,

$$\int_0^L \left[\int_\omega x_\beta \sigma^0_{3\beta}\,\mathrm{d}\omega\right] v^0\,\mathrm{d}x_3 = \int_0^L \left[\int_\omega \tfrac{1}{2}(x_1^2 + x_2^2) f_3\,\mathrm{d}\omega + \int_\gamma \tfrac{1}{2}(x_1^2 + x_2^2) g_3\,\mathrm{d}\gamma + \int_\omega \tfrac{1}{2}(x_1^2 + x_2^2)\partial_3 \sigma^0_{33}\,\mathrm{d}\omega\right] \mathrm{d}x_3. \tag{8.54}$$

Thus, we may write a.e. in $(0, L)$

$$\int_\omega x_\beta \partial_3 \sigma^0_{3\beta}\,\mathrm{d}\omega = \int_\omega \tfrac{1}{2}(x_1^2 + x_2^2)\partial_3 f_3\,\mathrm{d}\omega + \int_\gamma \tfrac{1}{2}(x_1^2 + x_2^2)\partial_3 g_3\,\mathrm{d}\gamma + \int_\omega \tfrac{1}{2}(x_1^2 + x_2^2)\partial_{33}\sigma^0_{33}\,\mathrm{d}\omega. \tag{8.55}$$

Finally, (8.51) follows from (8.53), (8.55) and (7.63). Using the strong interpretation of (7.59) and (7.61) it is possible to give an expression for G_3^2 that only uses the data f_i and g_i. □

9. Characterization of the second order stress and fourth order displacement fields

We shall now consider the calculation of the next terms in the asymptotic expansion. In order to do so, we shall need some additional regularity on the data. The complete result is stated in the following theorem.

THEOREM 9.1. *Let the system of applied forces be such that*

$$\begin{aligned} &f_3 \in H^3[0, L; L^2(\omega)], \qquad g_3 \in H^3[0, L; L^2(\gamma)], \\ &f_\alpha \in H^2[0, L; L^2(\omega)], \qquad g_\alpha \in H^2[0, L; L^2(\gamma)]. \end{aligned} \tag{9.1}$$

Then functions $(\boldsymbol{u}^0, \boldsymbol{\sigma}^0) \in V(\Omega) \times \Sigma(\Omega)$, $(\boldsymbol{u}^2, \sigma^2_{3i}) \in V(\Omega) \times [L^2(\Omega)]^3$, $\boldsymbol{u}^4 \in V(\Omega)$, $n^4 = \int_\omega \sigma^4_{33}\,\mathrm{d}\omega \in L^2(0, L)$ *and* $p^2 = \int_\omega \sigma^2_{\mu\mu}\,\mathrm{d}\omega \in L^2(0, L)$ *are determined in a unique way from Eqs.* (7.6)–(7.20). *This characterization is obtained as explained under* (i)–(iv) *below.*

(i) *Elements* $(\boldsymbol{u}^0, \boldsymbol{\sigma}^0) \in V(\Omega) \times \Sigma(\Omega)$, $\boldsymbol{u}^2 \in V(\Omega)$, *and* $\sigma_{33}^2 \in L^2(\Omega)$ *are given by Theorem* 7.1.

(ii) *Displacements* u_i^4 *are of the form*

$$\begin{aligned} u_1^4 &= z_1^4 + x_2 z^4 + \nu(\Phi_{1\beta}\partial_{33} z_\beta^2 - x_1 \partial_3 \underline{u}_3^2) + \underline{u}_1^4, \\ u_2^4 &= z_2^4 - x_1 z^4 + \nu(\Phi_{2\beta}\partial_{33} z_\beta^2 - x_2 \partial_3 \underline{u}_3^2) + \underline{u}_2^4, \end{aligned} \tag{9.2}$$

$$\begin{aligned} u_3^4 = {} & \underline{u}_3^4 - x_\alpha \partial_3 z_\alpha^4 - w \partial_3 z^4 + \tilde{\underline{u}}_3^4 + \frac{4(1+\nu)^2}{E} w^{00} \\ & + 2\nu(1+\nu)\overline{p}\partial_{3333}\underline{u}_3^0 + \nu\{\tfrac{1}{2}(x_1^2 + x_2^2) - \tfrac{1}{2}(I_1 + I_2)\}\partial_{33}\underline{u}_3^2 \\ & + [(1+\nu)\eta_\alpha + \nu\theta_\alpha]\partial_{333} z_\alpha^2 - 2(1+\nu)\overline{w}\partial_{333} z^2 \\ & + 2(1+\nu)[(1+\nu)\overline{\eta}_\alpha + \nu\overline{\theta}_\alpha]\partial_{33333} u_\alpha^0, \end{aligned} \tag{9.3}$$

where functions $\underline{u}_\alpha^4$ *were already defined in* (7.80) *and where functions* $\tilde{\underline{u}}_3^4$, w^{00}, z^4, z_α^4 *and* $\underline{u}_3^4$ *are given by* (a)–(e) *below.*

(a) *Function* $\tilde{\underline{u}}_3^4$ *is the unique solution of the following boundary value problem*:

$$\begin{aligned} & \tilde{\underline{u}}_3^4 \in L^2[0, L; H^1(\omega)] \quad \text{and a.e. in } (0, L): \\ & \int_\omega \partial_\beta \tilde{\underline{u}}_3^4(x_3) \partial_\beta \phi \, d\omega = \frac{2\nu(1+\nu)}{E} \partial_3 \Bigg[\int_\omega \sigma_{\mu\mu}^0(x_3) \phi \, d\omega - \int_\omega \sigma_{\mu\mu}^0(x_3) \, d\omega \int_\omega \phi \, d\omega \\ & \qquad - \int_\omega \underline{u}_\beta^4(x_3) \partial_\beta \phi \, d\omega \Bigg], \quad \text{for all } \phi \in H^1(\omega), \\ & \int_\omega \tilde{\underline{u}}_3^4(x_3) \, d\omega = 0. \end{aligned} \tag{9.4}$$

(b) *The secondary warping* w^{00} *is the unique solution of the following problem*

$$\begin{aligned} & w^{00} \in L^2[0, L; H^1(\omega)] \quad \text{and a.e. in } (0, L): \\ & \int_\omega \partial_\beta w^{00}(x_3) \partial_\beta \phi \, d\omega = \int_\omega \partial_{33} w^0(x_3) \phi \, d\omega, \quad \text{for all } \phi \in H^1(\omega), \\ & \int_\omega w^{00}(x_3) \, d\omega = 0. \end{aligned} \tag{9.5}$$

(c) *The angle of twist* z^4 *depends only on* x_3 *and it is defined by the following boundary value problem*:

$$
\begin{aligned}
& z^4 \in H^1(0,L), \\
& \frac{EJ}{2(1+\nu)} \int_0^L \partial_3 z^4 \partial_3 v^4 \, \mathrm{d}x_3 = \int_0^L [EJ_w \partial_{333} z^2 + M_3^4] \partial_3 v^4 \, \mathrm{d}x_3, \\
& \quad \textit{for all } v^4 \in H_0^1(0,L), \\
& z^4(a) = \frac{\nu}{(I_1+I_2)} [H_2 \partial_{33} z_1^2(a) - H_1 \partial_{33} z_2^2(a)] \quad \textit{at } a = 0, L,
\end{aligned} \tag{9.6}
$$

where

$$
\begin{aligned}
M_3^4 = & -\nu \left[\int_\omega \sigma_{\mu\mu}^0 w \, \mathrm{d}\omega \right] - \frac{E}{2(1+\nu)} \partial_3 \left[\int_\omega (\partial_2 \underline{u}_1^4 - \partial_1 \underline{u}_2^4) \Psi \, \mathrm{d}\omega \right] \\
& - 2(1+\nu) \partial_{33} \left[\int_\omega w^0 w \right] - \tfrac{1}{2} \nu E \left\{ \int_\omega (x_1^2 + x_2^2) w \, \mathrm{d}\omega \right\} \partial_{3333} \underline{u}_3^0 \\
& + \frac{E}{2(1+\nu)} [(1+\nu) I_\alpha^w + \nu I_\alpha^\Psi] \partial_{333} z_\alpha^2 \\
& - E \left[(1+\nu) \int_\omega \eta_\alpha w \, \mathrm{d}\omega + \nu \int_\omega \theta_\alpha w \, \mathrm{d}\omega \right] \partial_{33333} u_\alpha^0 .
\end{aligned} \tag{9.7}
$$

(d) *The stretching component* $\underline{u}_3^4$ *depends only on* x_3 *and it is the unique solution of the following problem*

$$
\begin{aligned}
& \underline{u}_3^4 \in H_0^1(0,L), \\
& \int_0^L E \partial_3 \underline{u}_3^4 \partial_3 v \, \mathrm{d}x_3 = \nu \int_0^L G_3^4 \partial_3 v \, \mathrm{d}x_3, \quad \textit{for all } v^4 \in H_0^1(0,L),
\end{aligned} \tag{9.8}
$$

where

$$
G_3^4 = - \int_\omega \tfrac{1}{2} (x_1^2 + x_2^2) \partial_{33} \sigma_{33}^2 \, \mathrm{d}\omega. \tag{9.9}
$$

(e) *Displacement* z_α^4 *depends only on* x_3 *and is the unique solution of the following variational problem (no sum on* α*)*:

$$
\begin{aligned}
& z_\alpha^4 \in H^2(0,L) \quad \textit{and for all } v^4 \in H_0^2(0,L): \\
& \int_0^L E I_\alpha \partial_{33} z_\alpha^4 \partial_{33} v^4 \, \mathrm{d}x_3 = \int_0^L M_\alpha^4 \partial_{33} v^4 \, \mathrm{d}x_3, \\
& z_\alpha^4(a) = -\tfrac{1}{2} \nu (I_\alpha - I_\beta) \partial_{33} z_\alpha^2(a) \quad \textit{at } a = 0, L, \\
& \partial_3 z_\alpha^4(a) = A_\alpha^4(a) \quad \textit{at } a = 0, L,
\end{aligned} \tag{9.10}
$$

where

$$A_\alpha^4 = \frac{1}{I_\alpha}\Big\{ -\tfrac{1}{2}I_\alpha^w \partial_3 z^4 + \frac{4(1+\nu)^2}{E}\int_\omega x_\alpha w^{00}\,\mathrm{d}\omega + [(1+\nu)L_{\alpha\beta}^\eta + \nu L_{\alpha\beta}^\theta]\partial_{333}z_\beta^2$$
$$+ 2\nu(1+\nu)\overline{H}_\alpha \partial_{333}\underline{u}_3^0 - (1+\nu)\overline{I}_\alpha^w \partial_{333}z^2$$
$$+ 2(1+\nu)[(1+\nu)\overline{L}_{\alpha\beta}^\eta + \nu\overline{L}_{\alpha\beta}^\theta]\partial_{33333}u_\beta^0\Big\}, \tag{9.11}$$

and where

$$M_\alpha^4 = -\tfrac{1}{2}EI_\alpha^w \partial_{33}z^4 + 2\nu E(1+\nu)\overline{H}_\alpha\partial_{33333}\underline{u}_3^0 + E[(1+\nu)L_{\alpha\beta}^\eta + \nu L_{\alpha\beta}^\theta]\partial_{3333}z_\beta^2$$
$$- E(1+\nu)\overline{I}_\alpha^w\partial_{3333}z^2 + 2E(1+\nu)[(1+\nu)\overline{L}_{\alpha\beta}^\eta + \nu\overline{L}_{\alpha\beta}^\theta]\partial_{333333}u_\beta^0$$
$$+ \partial_3\left[E\int_\omega x_\alpha \underline{\tilde{u}}_3^4\,\mathrm{d}\omega + 4(1+\nu)^2\int_\omega x_\alpha w^{00}\,\mathrm{d}\omega\right] + \nu\int_\omega \Phi_{\alpha\beta}\partial_3\sigma_{3\beta}^2. \tag{9.12}$$

(iii) *The stress components $\sigma_{3\beta}^2$ are uniquely determined by*

$$\sigma_{31}^2 = -\frac{E}{2(1+\nu)}\partial_2\Psi\partial_3 z^4 - E\partial_1\overline{w}\partial_{333}z^2 + \underline{\sigma}_{31}^2,$$
$$\sigma_{32}^2 = \frac{E}{2(1+\nu)}\partial_1\Psi\partial_3 z^4 - E\partial_2\overline{w}\partial_{333}z^2 + \underline{\sigma}_{32}^2, \tag{9.13}$$

where

$$\underline{\sigma}_{3\beta}^2 = \frac{E}{2(1+\nu)}\Big\{\partial_3\underline{u}_\beta^4 + \partial_\beta\underline{\tilde{u}}_3^4 + \frac{4(1+\nu)^2}{E}\partial_\beta w^{00}$$
$$+ 2\nu(1+\nu)\partial_\beta\overline{p}\partial_{3333}\underline{u}_3^0 + [(1+\nu)\partial_\beta\eta_\alpha + \nu(\partial_\beta\theta_\alpha + \Phi_{\beta\alpha})]\partial_{333}z_\alpha^2$$
$$+ 2(1+\nu)[(1+\nu)\partial_\beta\overline{\eta}_\alpha + \nu(\partial_\beta\overline{\theta}_\alpha)]\partial_{33333}u_\alpha^0\Big\}. \tag{9.14}$$

As a consequence we have the following expressions for the shear force components:

$$q_\beta^2 = \int_\omega \sigma_{3\beta}^2\,\mathrm{d}\omega = -EI_\beta\partial_{333}z_\beta^2 + \partial_3 M_\beta^2 = \partial_3 m_\beta^2. \tag{9.15}$$

(iv) *Finally, elements n^4, m_β^4 and p^2 are given by*

$$p^2 = \int_\omega \sigma_{\mu\mu}^2\,\mathrm{d}\omega = -G_3^4, \tag{9.16}$$

$$n^4 = \int_\omega \sigma_{33}^4\,\mathrm{d}\omega = E\partial_3\underline{u}_3^4 + \nu p^2, \tag{9.17}$$

$$m_\beta^4 = \int_\omega x_\beta\sigma_{33}^4\,\mathrm{d}\omega = -EI_\beta\partial_{33}z_\beta^4 + M_\beta^4 \quad (\textit{no sum on } \beta). \tag{9.18}$$

PROOF. Assumptions (9.1) imply (7.57) and, therefore, Theorem 7.1 is still valid from which conclusion (i) follows. The rest of the proof is done with the same notation as before and for simplicity we shall decompose it into several steps, as in Theorem 7.1.

Step 1. *Displacements* u_β^4 *are of the form* (9.2), *with* z^4, $z_\alpha^4 \in H^1(0,L)$ *and such that* $z^4(a)$, $z_\alpha^4(a)$ *satisfy the boundary conditions in* (9.6) *and in* (9.10), *respectively.* The form (9.2) for displacement components u_β^4 follows directly from (8.39), (8.40) already proved in Theorem 7.1.

The following conditions are necessary in order that $(u_\alpha) \in W_2(\Omega)$:

$$\int_\omega u_\alpha^4 \,\mathrm{d}\omega = \tfrac{1}{2}\nu(I_\alpha - I_\beta)\partial_{33}z_\alpha^2 + z_\alpha^4 \in H_0^1(0,L), \quad (\alpha \neq \beta, \text{ no sum on } \alpha), \tag{9.19}$$

$$\int_\omega (x_2u_1^4 - x_1u_2^4)\,\mathrm{d}\omega = -\nu[H_2\partial_{33}z_1^2 - H_1\partial_{33}z_2^2](I_1 + I_2)z^4 \in H_0^1(0,L). \tag{9.20}$$

However, if we assume (9.1), then the regularity on the data leads to $z_\alpha^2 \in H^4(0,L)$, $(\underline{u}_\alpha^4) \in W_2(\Omega)$ and therefore, conditions (9.19)–(9.20) are also sufficient. In this case, the fact that z_α^4, $z^4 \in H^1(0,L)$ verify the boundary conditions in (9.6) and (9.10), respectively, is a direct consequence of (9.19) and (9.20).

REMARK 9.1. Substituting (8.34) into (7.10) we obtain that (u_α^4) must be a solution of the following plane elasticity problem:

$$\begin{aligned}
&(u_\alpha^4) \in W_2(\Omega),\\
&\int_\Omega S_{\alpha\beta}(\boldsymbol{u}^4)e_{\alpha\beta}(\boldsymbol{v})\,\mathrm{d}\boldsymbol{x}\\
&\quad = \int_\Omega f_\beta v_\beta \,\mathrm{d}\boldsymbol{x} + \int_\Gamma g_\beta v_\beta \,\mathrm{d}a - \int_\Omega \sigma_{3\beta}^0 \partial_3 v_\beta \,\mathrm{d}\boldsymbol{x}\\
&\qquad - \frac{\nu E}{(1+\nu)(1-2\nu)}\int_\Omega \partial_3 u_3^2 e_{\mu\mu}(\boldsymbol{v})\,\mathrm{d}\boldsymbol{x}, \quad \text{for all } \boldsymbol{v} = (v_\alpha) \in W_2(\Omega).
\end{aligned} \tag{9.21}$$

A necessary and sufficient condition for the existence of a solution of (9.21) is given by (8.42). Consequently, in the general case, the (u_β^4) which we have just calculated cannot be a solution of (9.21). It is not difficult to show that (u_β^4) given by (9.2) satisfies Eq. (9.21) for all $(v_\beta) \in W_2(\Omega)$ such that $v_\beta = 0$ on $\Gamma_0 \cup \Gamma_L$. We just have a boundary layer problem similar to the one already mentioned in Remark 8.4. □

Step 2. *Displacement* u_3^4 *is of the form* (9.3) *with* $\underline{u}_3^4 \in H_0^1(0,L)$ *and with* $I_\beta \partial_3 z_\beta^4 + \frac{1}{2} I_\beta^w \partial_3 z^4 \in H^1(0,L)$ (*no sum on* β). From Eq. (7.14), setting $v_3(x_1,x_2,x_3) = \phi(x_1,x_2)\psi(x_3)$ with $\phi \in H^1(\omega)$ and $\psi \in H_0^1(0,L)$, we find that $\sigma_{3\beta}^2$ must be a solution of the problem:

$$\begin{aligned} &\sigma_{3\beta}^2 \in L^2[0,L;(L^2(\omega))^2] \quad \text{and a.e. in } (0,L), \text{ for all } \phi \in H^1(\omega): \\ &\int_\omega \sigma_{3\beta}^2(x_3)\partial_\beta \phi \,\mathrm{d}\omega = \int_\omega \partial_3 \sigma_{33}^2(x_3)\phi \,\mathrm{d}\omega. \end{aligned} \tag{9.22}$$

A necessary and sufficient condition in order for $(\sigma_{3\beta})$ to exist is

$$\int_\omega \partial_3 \sigma_{33}^2 \,\mathrm{d}\omega = 0 \quad \text{a.e. in } (0,L), \tag{9.23}$$

but this is a consequence of (7.81), (8.51) and (7.70). Futhermore, (9.23) may be obtained by an integration by parts in (8.49). On the other hand, from (7.17) we have

$$\sigma_{3\beta}^2 = \frac{E}{2(1+\nu)}(\partial_3 u_\beta^4 + \partial_\beta u_3^4), \tag{9.24}$$

and substituting into (9.22) we obtain that u_3^4 must be a solution of

$$\begin{aligned} &u_3^4 \in L^2[0,L;H^1(\omega)] \quad \text{and a.e. in } (0,L): \\ &\int_\omega \partial_\beta u_3^4 \partial_\beta \phi \,\mathrm{d}\omega = \frac{2(1+\nu)}{E}\int_\omega \partial_3 \sigma_{33}^2 \phi \,\mathrm{d}\omega - \int_\omega \partial_3 u_\beta^4 \partial_\beta \phi \,\mathrm{d}\omega, \\ &\quad \text{for all } \phi \in H^1(\omega). \end{aligned} \tag{9.25}$$

From (7.81), the strong interpretation of (7.70) and from (8.51) we have

$$\begin{aligned} \int_\omega \partial_3 \sigma_{33}^2 \phi \,\mathrm{d}\omega = &- E\partial_{333} z_\alpha^2 \int_\omega x_\alpha \phi \,\mathrm{d}\omega - E\partial_{333} z^2 \int_\omega w\phi \,\mathrm{d}\omega \\ &+ \nu E \partial_{3333} \underline{u}_3^0 \int_\omega \{\tfrac{1}{2}(x_1^2+x_2^2) - \tfrac{1}{2}(I_1+I_2)\}\phi \,\mathrm{d}\omega \\ &+ E\partial_{33333} u_\alpha^0 \int_\omega [(1+\nu)\eta_\alpha + \nu\theta_\alpha]\phi \,\mathrm{d}\omega \\ &+ 2(1+\nu)\int_\omega \partial_{33} w^0 \phi \,\mathrm{d}\omega + \nu \int_\omega \partial_3 \sigma_{\mu\mu}^0 \phi \,\mathrm{d}\omega \\ &- \nu\left(\int_\omega \partial_3 \sigma_{\mu\mu}\right)\int_\omega \phi \,\mathrm{d}\omega. \end{aligned} \tag{9.26}$$

In the same way, from (9.2) we obtain

$$\partial_3 u_\beta^4 \partial_\beta \phi = \left(\int_\omega \partial_\beta \phi \, \mathrm{d}\omega\right) \partial_3 z_\beta^4 + \nu \left(\int_\omega \Phi_{\beta\alpha} \partial_\beta \phi \, \mathrm{d}\omega\right) \partial_{333} z_\alpha^2$$
$$- \nu \left(\int_\omega x_\beta \partial_\beta \phi \, \mathrm{d}\omega\right) \partial_{33} \underline{u}_3^2$$
$$+ \left[\int_\omega (x_2 \partial_1 \phi - x_1 \partial_2 \phi) \, \mathrm{d}\omega\right] \partial_3 z^4 + \int_\omega \partial_3 \underline{u}_\beta^4 \partial_\beta \phi \, \mathrm{d}\omega. \tag{9.27}$$

Now using the definition of functions $\tilde{\underline{u}}_3^4$, w^{00}, w, η_β, θ_β, $\overline{\rho}$, $\overline{w}$, $\overline{\eta}_\beta$ and $\overline{\theta}_\beta$ it is possible to prove that the general solution u_3^4 of (9.25) is given by (9.3) where $\underline{u}_3^4$ is a function of variable x_3 only. Conditions $\int_\omega u_3^4 \, \mathrm{d}\omega \in H_0^1(0, L)$ and $\int_\omega x_\beta u_3^4 \, \mathrm{d}\omega \in H_0^1(0, L)$ are necessary in order to obtain $u_3^4 \in W_1(\Omega)$. Therefore, (with no sum on β) we obtain

$$\underline{u}_3^4 \in H_0^1(0, L), \tag{9.28}$$

$$\Big\{ - I_\beta \partial_3 z_\beta^4 - \tfrac{1}{2} I_\beta^w \partial_3 z^4 + \frac{4(1+\nu)^2}{E} \int_\omega x_\beta w^{00} \, \mathrm{d}\omega + [(1+\nu) L_{\alpha\beta}^\eta + \nu L_{\alpha\beta}^\theta] \partial_{333} z_\alpha^2$$
$$+ 2\nu(1+\nu) \overline{H}_\beta \partial_{333} \underline{u}_3^0 - (1+\nu) \overline{I}_\beta^w \partial_{333} z^2$$
$$+ 2(1+\nu)[(1+\nu) \overline{L}_{\alpha\beta}^\eta + \nu \overline{L}_{\alpha\beta}^\theta] \partial_{33333} u_\alpha^0 \Big\} \in H_0^1(0, L). \tag{9.29}$$

Using regularity results for variational problems (7.59), (7.61), (7.67), (7.72), (7.69) and (9.5) we have, from hypotheses (9.1),

$$\partial_{33333} u_\alpha^0, \ \partial_{333} \underline{u}_3^0, \ \partial_{333} z^2, \ \partial_{333} z_\alpha^2, \ \int_\omega x_\beta w^{00} \, \mathrm{d}\omega \in H^1(0, L). \tag{9.30}$$

Consequently, besides (9.28), the following necessary conditions in order that $u_3^4 \in W_1(\Omega)$ are obtained (no sum on β):

$$I_\beta \partial_3 z_\beta^4 + \tfrac{1}{2} I_\beta^w \partial_3 z^4 \in H^1(0, L), \tag{9.31}$$

and, at $a = 0$ and L,

$$[I_\beta \partial_3 z_\beta^4 + \tfrac{1}{2} I_\beta^w \partial_3 z^4](a)$$
$$= \frac{4(1+\nu)^2}{E} \int_\omega x_\beta w^{00}(a) \, \mathrm{d}\omega + [(1+\nu) L_{\alpha\beta}^\eta + \nu L_{\alpha\beta}^\theta] \partial_{333} z_\alpha^2(a)$$
$$+ 2\nu(1+\nu) \overline{H}_\beta \partial_{333} \underline{u}_3^0(a) - (1+\nu) \overline{I}_\beta^w \partial_{333} z^2(a)$$
$$+ 2(1+\nu)[(1+\nu) \overline{L}_{\alpha\beta}^\eta + \nu \overline{L}_{\alpha\beta}^\theta] \partial_{33333} u_\alpha^0(a). \tag{9.32}$$

Actually, conditions (9.28), (9.31) and (9.32) are not sufficient in order to have $u_3^4 \in W_1(\Omega)$. In fact, from (9.29) and (9.30) a necessary and sufficient condition is given by (9.28), (9.32) together with

$$x_\beta \partial_3 z_\beta^4 + w\partial_3 z^4 \in H^1(\Omega) \quad \text{(no sum on } \beta). \tag{9.33}$$

REMARK 9.2. Substituting (9.24) directly into Eq. (7.14) we obtain that u_3^4 must be a solution of the following problem:

$$\begin{aligned}
&u_3^4 \in W_1(\Omega) \quad \text{and for all } v_3 \in W_1(\Omega): \\
&\int_\Omega \partial_\beta u_3^4 \partial_\beta v_3 \,\mathrm{d}\boldsymbol{x} = -\frac{2(1+\nu)}{E} \int_\Omega \sigma_{33}^2 \partial_3 v_3 \,\mathrm{d}\boldsymbol{x} - \int_\Omega \partial_3 u_\beta^4 \partial_\beta v_3 \,\mathrm{d}\boldsymbol{x}.
\end{aligned} \tag{9.34}$$

A necessary and sufficient condition for the existence of u_3^4 is given by

$$\int_{\omega\times\{L\}} \sigma_{33}^2 v_3 \,\mathrm{d}\omega - \int_{\omega\times\{0\}} \sigma_{33}^2 v_3 \,\mathrm{d}\omega = 0, \quad \text{for all } v_3 \in W_1(\Omega), \tag{9.35}$$

which is equivalent to saying that σ_{33}^2 is of the following form:

$$\sigma_{33}^2(0) = A^0 + x_\alpha A_\alpha^0, \qquad \sigma_{33}^2(L) = A^L + x_\alpha A_\alpha^L, \tag{9.36}$$

where A^0, A^L, A_α^0 and A_α^L are constants.

In general, these conditions are not satisfied as can be seen from (7.81). Consequently, u_3^4 calculated above cannot be a solution of (9.34). However, u_3^4 given by (9.3) satisfies (9.34) for all $v_3 \in W_1(\Omega)$ such that $v_3 = 0$ on $\Gamma_0 \cup \Gamma_L$; a new boundary layer phenomenon showing up near the ends of the beam. □

Step 3. *Calculation of* $\sigma_{3\beta}^2$ *and of* q_β^2. Expressions (9.13) follow directly from (9.24) after using (7.30) in a similar way as in Step 4 of Section 8. It is possible to obtain expression (9.15) for the shear force components q_β^2 directly from (9.13) by just using the properties of the functions intervening in (9.15). However, this expression can also be obtained in an easier way from Eq. (7.14). In fact, assumptions (9.1) imply that m_β^2 obtained in (7.82) belongs to the space $H^2(0,L)$ and consequently (9.15) follows from (7.14) simply by setting $v_3 = x_\beta v^2$ for any $v^2 \in H_0^1(0,L)$. □

Step 4. *Function* z^4 *solves* (9.6), (9.7), $z_\alpha^4 \in H^2(0,L)$, *and* $\partial_3 z_\alpha^4$ *satisfies the boundary conditions in* (9.10), (9.11). In order to show that z^4 solves problem (9.6), (9.7), we set $v_1 = x_2 v^4$, $v_2 = -x_1 v^4$, with $v^4 \in H_0^1(0,L)$, in (7.15). We then get, for all $v^4 \in H_0^1(0,L)$,

$$\int_0^L \left[\int_\omega (x_2 \sigma_{31}^2 - x_1 \sigma_{32}^2) \,\mathrm{d}\omega \right] \partial_3 v^4 \,\mathrm{d}x_3 = 0. \tag{9.37}$$

Substituting (9.13) into (9.37) and using (7.53) we have

$$z^4 \in H^1(0,L) \quad \text{and for all } v^4 \in H_0^1(0,L):$$

$$\int_0^L \frac{EJ}{2(1+\nu)} \partial_3 z^4 \partial_3 v^4 \, dx_3$$
$$= \int_0^L EJ_w \partial_{333} z^2 \partial_3 v^4 \, dx_3 - \int_0^L \left[\int_\omega (x_2 \underline{\sigma}_{31}^2 - x_1 \underline{\sigma}_{32}^2) \, d\omega \right] \partial_3 v^4 \, dx_3. \tag{9.38}$$

Problem (9.6) is then obtained from the following equality:

$$M_3^4 = -\int_\omega (x_2 \underline{\sigma}_{31}^2 - x_1 \underline{\sigma}_{32}^2) \, d\omega, \tag{9.39}$$

which can be proved using properties (7.52)–(7.55) together with the following ones obtained by setting $\phi = w$ in (9.4) and in (9.5):

$$\int_\omega (x_2 \partial_1 \underline{\tilde{u}}_3^4 - x_1 \partial_2 \underline{\tilde{u}}_3^4) \, d\omega + \int_\omega (x_2 \partial_3 \underline{u}_1^4 - x_1 \partial_3 \underline{u}_2^4) \, d\omega$$
$$= \frac{2\nu(1+\nu)}{E} \partial_3 \left[\int_\omega \sigma_{\mu\mu}^0 w \, d\omega + \int_\omega (\partial_2 \underline{u}_1^4 - \partial_1 \underline{u}_2^4) \Psi \, d\omega \right], \tag{9.40}$$

$$\int_\omega (x_2 \partial_1 w^{00} - x_1 \partial_2 w^{00}) \, d\omega = \partial_{33} \int_\omega w^0 w \, d\omega. \tag{9.41}$$

The regularity results in (9.6) and in (9.7) give $z^4 \in H^2(0,L)$. Therefore, from (9.31) and (9.32) we conclude that $z_\alpha^4 \in H^2(0,L)$ and that $\partial_3 z_\alpha^4$ satisfies the boundary conditions in (9.10), (9.11). □

Step 5. Calculation of q_β^4, n^4 and p^2. From (7.15) with $v_\beta = x_\beta v^2$, $v^2 \in H_0^1(0,L)$, we deduce that

$$\int_0^L \left[\int_\omega \sigma_{\mu\mu}^2 \, d\omega \right] v^2 \, dx_3 + \int_0^L \left(\int_\omega x_\beta \sigma_{3\beta}^2 \, d\omega \right) \partial_3 v^2 \, dx_3 = 0,$$
$$\text{for all } v^2 \in H_0^1(0,L). \tag{9.42}$$

Thus,

$$p^2 = \int_\omega x_\beta \partial_3 \sigma_{3\beta}^2 \, d\omega \quad \text{a.e. in } (0,L). \tag{9.43}$$

Now setting $v_3 = \frac{1}{2}(x_1^2 + x_2^2)v^2$, with $v^2 \in H_0^1(0, L)$ in (7.14), we obtain

$$\int_0^L \left[\int_\omega x_\beta \sigma_{3\beta}^2 \, d\omega\right] v^2 \, dx_3 = \int_0^L \left[\int_\omega \tfrac{1}{2}(x_1^2 + x_2^2)\partial_3 \sigma_{33}^2 \, d\omega\right] v^2 \, dx_3, \quad \text{for all } v^2 \in H_0^1(0, L), \tag{9.44}$$

and (9.16) may be obtained from (9.43) and (9.44). On the other hand, from (7.16) we have

$$\sigma_{33}^4 = E\partial_3 u_3^4 + \nu \sigma_{\mu\mu}^2, \tag{9.45}$$

from which (9.17) follows since $\int_\omega \partial_3 u_3^4 \, d\omega = \partial_3 \underline{u}_3^4$.

In an analogous way we get (9.18) from (9.45) using the following equality which is a consequence of (7.15):

$$\int_\omega x_\beta \sigma_{\mu\mu}^2 \, d\omega = \int_\omega \partial_3 \sigma_{3\beta}^2 \Phi_{\alpha\beta} \, d\omega. \tag{9.46}$$

□

Step 6. *Equation for* z_α^4. Setting $v_3 = x_\alpha \partial_3 v^4$ in Eq. (7.19), the following equality holds for all $v^4 \in H_0^2(0, L)$:

$$\int_0^L m_\alpha^4 \partial_{33} v^4 \, dx_3 + \int_0^L q_\alpha^2 \partial_3 v^4 \, dx_3 = 0. \tag{9.47}$$

In the same way, from (7.20) with $v_\beta = v^4$, for any $v^4 \in H_0^2(0, L)$, we get

$$\int_0^L q_\beta^2 \partial_3 v^4 \, dx_3 = 0, \quad \text{for all } v^4 \in H_0^2(0, L). \tag{9.48}$$

Equation (9.10) is now a consequence of (9.47), (9.48) and of (9.18). □

Step 7. *Equation for* $\underline{u}_3^4$. From (7.19) with $v_3 = v^4$, $v^4 \in H_0^1(0, L)$ we deduce that

$$\int_0^L n^4 \partial_3 v^4 \, dx_3 = 0, \quad \text{for all } v^4 \in H_0^1(0, L). \tag{9.49}$$

Now substituting (9.16) and (9.17) into (9.49), Eqs. (9.8) and (9.9) are obtained.

This completes the proof of Theorem 9.1. □

REMARK 9.3. Although calculations will become more and more complex, the technique employed can be pursued in order to evaluate the next terms of the asymptotic expansion. It is clear that in order to do so, some additional regularity on the data is needed and that new boundary layer phenomena will show up.

REMARK 9.4. It is now possible to see that for the initial problem (rod strongly clamped at both ends with $V(\Omega)$ defined by (2.14)), all the calculations made so far in Sections 7, 8 and 9 can be done exactly in the same way for this case. As proved in Section 6 one gets $\boldsymbol{u}^0 \in V(\Omega)$ but the same is not true for $\boldsymbol{u}^2$. In fact, conditions given for z_α^2, z^2 and $\underline{u}_3^2$ are necessary in order to have $u_3^2 \in V(\Omega)$, but in general, they are not sufficient. Once again, we are in the presence of a boundary layer phenomenon near the extremities of the beam which justifies (or is justified by) Saint Venant's principle.

10. Convergence of the scaled displacements and stresses

In this section we establish that *the scaled displacement field* $\boldsymbol{u}(\varepsilon)$ *converges in the space* $V(\Omega)$ and that *the scaled axial stress component* $\sigma_{33}(\varepsilon)$ *converges in* $L^2(\Omega)$ *as* $\varepsilon \to 0$. In order to do so, we directly study problem (6.18), (6.19) using classical techniques.

Moreover, besides showing that the scaled displacement converges in $[H^1(\Omega)]^3$, and that the scaled axial stress component $\sigma_{33}(\varepsilon)$ converges in $L^2(\Omega)$ we shall show that *the limits are exactly the first terms* $\boldsymbol{u}^0$ *and* σ_{33}^0 *of the formal asymptotic expansion* (6.27). This is the most important result of the convergence study, for it allows us to justify the asymptotic expansions considered for $\boldsymbol{u}(\varepsilon)$ and $\sigma_{33}(\varepsilon)$ in the spaces $V(\Omega)$ and $L^2(\Omega)$, respectively.

The proof of the convergence of sequences $(\boldsymbol{u}(\varepsilon))_{\varepsilon>0}$ in $V(\Omega)$ and $(\sigma_{33}(\varepsilon))_{\varepsilon>0}$ in $L^2(\Omega)$ is done in several steps:

(i) a priori estimates of $(\boldsymbol{u}(\varepsilon))_{\varepsilon>0}$ in $V(\Omega)$ and of $(\sigma_{33}(\varepsilon))_{\varepsilon>0}$ in $L^2(\Omega)$,

(ii) weak convergence in $V(\Omega) \times L^2(\Omega)$ of at least a subsequence. Weak convergence of the whole sequence $(\boldsymbol{u}(\varepsilon))_{\varepsilon>0}$,

(iii) strong convergence.

As a by-product one obtains that the sequences

$$(\varepsilon\sigma_{3\beta}(\varepsilon))_{\varepsilon>0} \text{ and } (\varepsilon^2\sigma_{\alpha\beta}(\varepsilon))_{\varepsilon>0}$$

converge strongly to zero in $L^2(\Omega)$. Moreover, the sequence $(m_\beta(\varepsilon))_{\varepsilon>0}$ *converges strongly* in $L^2(0, L)$ to m_β^0, defined by (6.39), and $(q_\beta(\varepsilon))_{\varepsilon>0}$ *converges weakly in* $L^2(0, L)$ *to* q_β^0, as defined by (7.76).

The weak convergence results of $\boldsymbol{u}(\varepsilon)$, $\sigma_{33}(\varepsilon)$, $\varepsilon\sigma_{3\beta}(\varepsilon)$, $\varepsilon^2\sigma_{\alpha\beta}(\varepsilon)$, $m_\beta(\varepsilon)$ and $q_\beta(\varepsilon)$ were obtained by BERMÚDEZ and VIAÑO [1984]. The strong convergence was proved by TUTEK and AGANOVIČ [1986]. In both cases a technique similar to that of CIARLET and KESAVAN [1981], for an eigenvalue problem in plate theory, is employed (cf. also DESTUYNDER [1980] and VIAÑO [1983]). We shall use the displacement–stress approach. An adaptation for the displacement approach is also possible (cf. CIARLET [1990, Section 3.3] for the plate case).

We denote by $|\cdot|_{0,\Omega}$ the usual norm in $L^2(\Omega)$ and in $\Sigma(\Omega)$ and by $\|\cdot\|_{1,\Omega}$ the usual norm in $H^1(\Omega)$ and in $V(\Omega)$. The weak and strong convergences will be denoted by $\rightharpoonup$ and $\to$, respectively. In this section C stands for a constant which may have a different value, from situation to situation, and $V(\Omega)$ *represents any of the spaces defined by* (2.14) *or* (5.3).

Korn's inequality, assures that the mapping

$$\boldsymbol{v} \in V(\Omega) \to |\boldsymbol{e}(\boldsymbol{v})|_{0,\Omega} = \left[\sum_{1\leqslant i,j\leqslant 3} |e_{ij}(\boldsymbol{v})|^2_{0,\Omega}\right]^{1/2}, \tag{10.1}$$

is a norm in $V(\Omega)$, equivalent to the $\|\boldsymbol{v}\|_{1,\Omega}$ norm. Consequently there exists a constant $C_0(\Omega)$, such that

$$|\boldsymbol{e}(\boldsymbol{v})|_{0,\Omega} \geqslant C_0(\Omega)\|\boldsymbol{v}\|_{1,\Omega}, \quad \text{for all } \boldsymbol{v} \in V(\Omega). \tag{10.2}$$

From this equation we derive the following (BABUŠKA–BREZZI) property:

$$\sup_{\boldsymbol{\tau}\in\Sigma(\Omega)} \frac{\int_\Omega e_{ij}(\boldsymbol{v})\tau_{ij}\, d\boldsymbol{x}}{|\boldsymbol{\tau}|_{0,\Omega}} \geqslant C_0(\Omega)\|\boldsymbol{v}\|_{1,\Omega}, \quad \text{for all } \boldsymbol{v} \in V(\Omega). \tag{10.3}$$

In order to show this for any $\boldsymbol{v} \in V(\Omega)$ we take $\boldsymbol{\tau} = e(\boldsymbol{v}) \in \Sigma(\Omega)$. In this section, for any element $\boldsymbol{\tau} \in \Sigma(\Omega)$ we use the following notation:

$$|\tau_{3\beta}|_{0,\Omega} = |\tau_{3\beta}|_{[L^2(\Omega)]^2},$$
$$|\tau_{\alpha\beta}|_{0,\Omega} = |\tau_{\alpha\beta}|_{[L^2(\Omega)]^4_s}.$$

THEOREM 10.1. *There exists a constant C, independent of ε, such that for all* $0 < \varepsilon \leqslant 1$ *the solution* $(\boldsymbol{u}(\varepsilon), \boldsymbol{\sigma}(\varepsilon)) \in V(\Omega) \times \Sigma(\Omega)$ *of problem* (6.18), (6.19) *verifies*

$$|\sigma_{33}(\varepsilon)|_{0,\Omega} \leqslant C, \qquad \varepsilon|\sigma_{3\beta}(\varepsilon)|_{0,\Omega} \leqslant C, \qquad \varepsilon^2|\sigma_{\alpha\beta}(\varepsilon)|_{0,\Omega} \leqslant C, \tag{10.4}$$
$$\|\boldsymbol{u}(\varepsilon)\|_{1,\Omega} \leqslant C. \tag{10.5}$$

PROOF. Setting $\boldsymbol{\tau} = \boldsymbol{\sigma}(\varepsilon)$ in (6.25) and $\boldsymbol{v} = \boldsymbol{u}(\varepsilon)$ in (6.26) we obtain

$$a_0(\boldsymbol{\sigma}(\varepsilon), \boldsymbol{\sigma}(\varepsilon)) + \varepsilon^2 a_2(\boldsymbol{\sigma}(\varepsilon), \boldsymbol{\sigma}(\varepsilon)) + \varepsilon^4 a_4(\boldsymbol{\sigma}(\varepsilon), \boldsymbol{\sigma}(\varepsilon)) = -F(\boldsymbol{u}(\varepsilon)). \tag{10.6}$$

Let $\tilde{\boldsymbol{\sigma}}(\varepsilon) \in \Sigma(\Omega)$ be the following element:

$$\tilde{\sigma}_{33}(\varepsilon) = \sigma_{33}(\varepsilon), \qquad \tilde{\sigma}_{3\beta}(\varepsilon) = \varepsilon\sigma_{3\beta}(\varepsilon), \qquad \tilde{\sigma}_{\alpha\beta}(\varepsilon) = \varepsilon^2\sigma_{\alpha\beta}(\varepsilon).$$

Substituting in (10.6), we obtain

$$\int_{\Omega} \left[\frac{1+\nu}{E} \tilde{\sigma}_{ij}(\varepsilon) - \frac{\nu}{E} \tilde{\sigma}_{pp}(\varepsilon) \delta_{ij} \right] \tilde{\sigma}_{ij}(\varepsilon)\, \mathrm{d}\boldsymbol{x} = \int_{\Omega} f_i u_i(\varepsilon)\, \mathrm{d}\boldsymbol{x} + \int_{\Gamma} g_i u_i(\varepsilon)\, \mathrm{d}a. \tag{10.7}$$

Using properties (6.8) for the elasticity coefficients and the fact that $f_i \in L^2(\Omega)$ and $g_i \in L^2(\Gamma)$ one obtains the existence of two constants C_1 and C_2 such that

$$\int_{\Omega} \left[\frac{1+\nu}{E} \tilde{\sigma}_{ij}(\varepsilon) - \frac{\nu}{E} \tilde{\sigma}_{pp}(\varepsilon) \delta_{ij} \right] \tilde{\sigma}_{ij}(\varepsilon)\, \mathrm{d}\boldsymbol{x} \geqslant C_1 |\tilde{\boldsymbol{\sigma}}(\varepsilon)|_{0,\Omega}^2,$$

$$\int_{\Omega} f_i u_i(\varepsilon)\, \mathrm{d}\boldsymbol{x} + \int_{\Gamma} g_i u_i(\varepsilon) \leqslant C_2 [|\boldsymbol{f}|_{0,\Omega} + |\boldsymbol{g}|_{0,\Gamma}] \, \|\boldsymbol{u}(\varepsilon)\|_{1,\Omega}.$$

Therefore,

$$|\tilde{\boldsymbol{\sigma}}(\varepsilon)|_{0,\Omega}^2 \leqslant C \|\boldsymbol{u}(\varepsilon)\|_{1,\Omega}, \tag{10.8}$$

where constant C depends only on Ω, E, $\nu, \boldsymbol{f}$ and $\boldsymbol{g}$.

On the other hand, for any $\boldsymbol{\tau} \in \Sigma(\Omega)$, from Eq. (6.25) we obtain

$$\int_{\Omega} e_{ij}(\boldsymbol{u}(\varepsilon)) \tau_{ij}\, \mathrm{d}\boldsymbol{x} = a_0(\boldsymbol{\sigma}(\varepsilon), \boldsymbol{\tau}) + \varepsilon^2 a_2(\boldsymbol{\sigma}(\varepsilon), \boldsymbol{\tau}) + \varepsilon^4 a_4(\boldsymbol{\sigma}(\varepsilon), \boldsymbol{\tau}). \tag{10.9}$$

Taking into account definitions (6.20)–(6.23) and the fact that $0 < \varepsilon \leqslant 1$ from (10.9) we conclude that

$$\int_{\Omega} e_{ij}(\boldsymbol{u}(\varepsilon)) \tau_{ij}\, \mathrm{d}\boldsymbol{x} \leqslant C |\tilde{\boldsymbol{\sigma}}(\varepsilon)|_{0,\Omega} |\boldsymbol{\tau}|_{0,\Omega}, \quad \text{for all } \boldsymbol{\tau} \in \Sigma(\Omega). \tag{10.10}$$

Combining (10.10) with the BABUŠKA–BREZZI condition (10.3), we obtain

$$\|\boldsymbol{u}(\varepsilon)\|_{1,\Omega} \leqslant C |\tilde{\boldsymbol{\sigma}}(\varepsilon)|_{0,\Omega}. \tag{10.11}$$

The result now follows from (10.11) and (10.8). □

REMARK 10.1. Defining $\tilde{\boldsymbol{e}}(\boldsymbol{u}(\varepsilon))$ in the following way:

$$\begin{aligned}
\tilde{\boldsymbol{e}}_{33}(\varepsilon) &= e_{33}(\boldsymbol{u}(\varepsilon)), \\
\tilde{\boldsymbol{e}}_{3\beta}(\varepsilon) &= \varepsilon^{-1} e_{3\beta}(\boldsymbol{u}(\varepsilon)), \\
\tilde{\boldsymbol{e}}_{\alpha\beta}(\varepsilon) &= \varepsilon^{-2} e_{\alpha\beta}(\boldsymbol{u}(\varepsilon)),
\end{aligned} \tag{10.12}$$

from Eq. (2.20) we obtain the following identity:

$$\tilde{\sigma}_{ij}(\varepsilon) = \frac{E}{(1+\nu)}\tilde{e}_{ij}(\boldsymbol{u}(\varepsilon)) + \frac{\nu E}{(1+\nu)(1-2\nu)}\tilde{e}_{pp}(\boldsymbol{u}(\varepsilon))\delta_{ij}. \tag{10.13}$$

From Theorem 10.1 we conclude that for all $0 < \varepsilon \leqslant 1$

$$|e_{33}(\varepsilon)|_{0,\Omega} \leqslant C, \qquad \varepsilon^{-1}|e_{3\beta}(\varepsilon)|_{0,\Omega} \leqslant C, \qquad \varepsilon^{-2}|e_{\alpha\beta}(\varepsilon)|_{0,\Omega} \leqslant C. \tag{10.14}$$

The next two theorems summarize the weak convergence results of BERMÚDEZ and VIAÑO [1984].

THEOREM 10.2. *There exists a subsequence, denoted with the same index ε, and there exist elements $\boldsymbol{\Psi} = (\Psi_{ij}) \in \Sigma(\Omega)$, $\zeta_\beta \in L^2(0,L)$ and $\rho_\beta \in L^2(0,L)$ such that, when $\varepsilon \to 0$*

$$\sigma_{33}(\varepsilon) \rightharpoonup \Psi_{33}, \qquad \varepsilon\sigma_{3\beta}(\varepsilon) \rightharpoonup \Psi_{3\beta}, \qquad \varepsilon^2\sigma_{\alpha\beta}(\varepsilon) \rightharpoonup \Psi_{\alpha\beta}, \qquad \text{in } L^2(\Omega), \tag{10.15}$$

$$m_\beta(\varepsilon) \rightharpoonup \zeta_\beta, \qquad q_\beta(\varepsilon) \rightharpoonup \rho_\beta, \qquad \text{in } L^2(\Omega), \tag{10.16}$$

Moreover, the following properties hold:

$$\int_\Omega \Psi_{3\beta}\partial_\beta v \,\mathrm{d}\boldsymbol{x} = 0, \quad \text{for all } v \in W_1(\Omega), \tag{10.17}$$

$$\int_\Omega \Psi_{\alpha\beta}\partial_\beta v_\beta \,\mathrm{d}\boldsymbol{x} = 0, \quad \text{for all } (v_\beta) \in W_2(\Omega), \tag{10.18}$$

$$\int_\omega \Psi_{\alpha\beta}\,\mathrm{d}\omega = \int_\omega x_\gamma \Psi_{\alpha\beta}\,\mathrm{d}\omega = 0, \tag{10.19}$$

$$\zeta_\beta = \int_\omega x_\beta \Psi_{33}\,\mathrm{d}\omega, \tag{10.20}$$

$$\int_0^L \rho_\beta v^0 \,\mathrm{d}x_3 + \int_0^L \zeta_\beta \partial_3 v^0 \,\mathrm{d}x_3 = \int_0^L \left[\int_\omega x_\beta f_3 \,\mathrm{d}\omega + \int_\gamma x_\beta g_3 \,d\gamma\right] v^0 \,\mathrm{d}x_3, \quad \text{for all } v^0 \in H_0^1(0,L). \tag{10.21}$$

PROOF. Bounds (10.4) and (10.5) guarantee us the existence of a subsequence $(\boldsymbol{u}(\varepsilon), \boldsymbol{\sigma}(\varepsilon))_{\varepsilon>0}$, which we denote in the same way, such that $(\boldsymbol{u}(\varepsilon))_{\varepsilon>0}$ converges weakly in $[H^1(\Omega)]^3$ and that the weak convergences in (10.15) hold. It only remains to prove (10.17)–(10.19).

Taking $\boldsymbol{v} = (v_1, v_2, 0)$ in (6.19) and multiplying by ε^2 one has

$$\varepsilon^2 \int_\Omega \sigma_{\alpha\beta}(\varepsilon) e_{\alpha\beta}(\boldsymbol{v})\,\mathrm{d}\boldsymbol{x} + 2\varepsilon^2 \int_\Omega \sigma_{3\beta}(\varepsilon) e_{3\beta}(\boldsymbol{v})\,\mathrm{d}\boldsymbol{x}$$
$$= \varepsilon^2 \left[\int_\Omega f_\beta v_\beta\,\mathrm{d}\boldsymbol{x} + \int_\Gamma g_\beta v_\beta\,\mathrm{d}a \right],$$

from which one obtains (10.18) by taking the limit when $\varepsilon \to 0$. In a similar way, taking $\boldsymbol{v} = (0, 0, v_3)$ in Eq. (6.19), and then multiplying by ε, gives

$$\varepsilon \int_\Omega \sigma_{3\beta}(\varepsilon) \partial_\beta v_3\,\mathrm{d}\boldsymbol{x} + \varepsilon \int_\Omega \sigma_{33}(\varepsilon) \partial_3 v_3\,\mathrm{d}\boldsymbol{x} = \varepsilon \left[\int_\Omega f_3 v_3\,\mathrm{d}\boldsymbol{x} + \int_\Gamma g_3 v_3\,\mathrm{d}a \right].$$

Taking the limit as $\varepsilon \to 0$ gives (10.17). In order to show (10.19) it is enough to take in (10.18) functions of the form $v_\beta = x_\beta v^0$, $v_\beta = \frac{1}{2} x_\gamma^2 v^0$ and $v_\beta = x_1 x_2 v^0$, with $v^0 \in H_0^1(0, L)$.

By definition the weak convergence $\sigma_{33}(\varepsilon) \rightharpoonup \Psi_{33}$ means that

$$\text{as } \varepsilon \to 0, \quad \int_\Omega \sigma_{33}(\varepsilon) \tau_{33}\,\mathrm{d}\boldsymbol{x} \to \int_\Omega \Psi_{33} \tau_{33}\,\mathrm{d}\boldsymbol{x} \quad \text{for all } \tau_{33} \in L^2(\Omega).$$

Setting $\tau_{33} = x_\beta \tau_{33}^0$, $\tau_{33}^0 \in L^2(0, L)$ one gets, as $\varepsilon \to 0$,

$$\int_0^L m_\beta(\varepsilon) \tau_{33}^0\,\mathrm{d}x_3 \to \int_0^L \left[\int_\omega x_\beta \Psi_{33}\,\mathrm{d}\omega \right] \mathrm{d}x_3, \quad \text{for all } \tau_{33}^0 \in L^2(\Omega),$$

which shows that $m_\beta(\varepsilon) \rightharpoonup \zeta_\beta = \int_\omega x_\beta \Psi_{33}\,\mathrm{d}\omega$ in $L^2(0, L)$.

In order to obtain (10.21) we consider in Eq. (6.9) $\boldsymbol{v} = (0, 0, x_\beta v^0)$, $v^0 \in H_0^1(0, L)$. We have

$$\int_\Omega \sigma_{3\beta}(\varepsilon) v^0\,\mathrm{d}\boldsymbol{x} + \int_\Omega x_\beta \sigma_{33}(\varepsilon) \partial_3 v^0\,\mathrm{d}\boldsymbol{x} = \int_\Omega x_\beta f_3 v^0\,\mathrm{d}\boldsymbol{x} + \int_\Gamma x_\beta g_3 v^0\,\mathrm{d}a,$$

or, equivalently,

$$\int_0^L q_\beta(\varepsilon) v^0\,\mathrm{d}x_3 = -\int_0^L m_\beta(\varepsilon) \partial_3 v^0\,\mathrm{d}x_3$$
$$+ \int_0^L \left[\int_\omega x_\beta f_3\,\mathrm{d}\omega + \int_\gamma x_\beta g_3\,\mathrm{d}\gamma \right] v^0\,\mathrm{d}x_3.$$

Convergence $q_\beta \rightharpoonup \rho_\beta$ in $L^2(0,L)$ together with (10.21) are now a consequence of the weak convergence $m_\beta(\varepsilon) \rightharpoonup \zeta_\beta$. □

REMARK 10.2. (i) The strong interpretation of (10.17), (10.18) allows us to write

$$\begin{aligned} &\partial_\beta \Psi_{3\beta} = 0, \quad \text{in } \Omega, \qquad &\partial_\beta \Psi_{\alpha\beta} = 0, \quad \text{in } \Omega, \\ &\Psi_{3\beta} n_\beta = 0, \quad \text{on } \Gamma, \qquad &\Psi_{\alpha\beta} n_\beta = 0, \quad \text{on } \Gamma. \end{aligned}$$

This gives a formal characterization of $\Psi_{i\beta}$ in each cross section $\omega \times \{x_3\}$. The lack of unicity for these systems establishes a fundamental difference with respect to the plate case which can be studied in a similar way (cf. DESTUYNDER [1980], CIARLET [1990]).

(ii) Using bounds (10.14) one concludes the existence of a subsequence of $(\boldsymbol{u}(\varepsilon))_{\varepsilon>0}$ and $\boldsymbol{\varsigma} \in \Sigma(\Omega)$ such that

$$e_{33}(\boldsymbol{u}(\varepsilon)) \rightharpoonup \varsigma_{33}, \qquad \varepsilon^{-1} e_{3\beta}(\boldsymbol{u}(\varepsilon)) \rightharpoonup \varsigma_{3\beta}, \qquad \varepsilon^{-2} e_{\alpha\beta}(\boldsymbol{u}(\varepsilon)) \rightharpoonup \varsigma_{\alpha\beta}, \\ \text{in } L^2(\Omega). \tag{10.22}$$

Moreover, from (10.13) we obtain

$$\Psi_{ij} = \frac{E}{(1+\nu)} \varsigma_{ij} + \frac{\nu E}{(1+\nu)(1-2\nu)} \varsigma_{pp} \delta_{ij}. \tag{10.23}$$

We shall now establish the weak convergence of sequence $(\boldsymbol{u}(\varepsilon))_{\varepsilon>0}$ to the first term of the asymptotic expansion $\boldsymbol{u}^0 \in V_{\mathrm{BN}}(\Omega)$.

THEOREM 10.3. *Sequence* $(\boldsymbol{u}(\varepsilon))_{\varepsilon>0}$ *satisfies as* $\varepsilon \to 0$

$$\boldsymbol{u}(\varepsilon) \rightharpoonup \boldsymbol{u}^0 \quad \text{in } [H^1(\Omega)]^3, \tag{10.24}$$

where $\boldsymbol{u}^0 \in V_{\mathrm{BN}}(\Omega)$ *is the Bernoulli–Navier displacement field characterized by Eqs.* (6.33)–(6.36) *and which is the first term in the asymptotic expansion of the scaled displacement field* (6.27).

PROOF. Denote by $(\boldsymbol{u}(\varepsilon))_{\varepsilon>0}$ the subsequence, from the previous theorem, which converges weakly in $[H^1(\Omega)]^3$, and let $\boldsymbol{u} \in V(\Omega)$ be its limit. We show that $\boldsymbol{u} = \boldsymbol{u}^0$.

In fact, taking the limit when $\varepsilon \to 0$ in (6.18) gives

$$\frac{1}{E}\int_\Omega \Psi_{33}\tau_{33}\,\mathrm{d}\boldsymbol{x} - \frac{\nu}{E}\int_\Omega \Psi_{\mu\mu}\tau_{33}\,\mathrm{d}\boldsymbol{x} = \int_\Omega e_{ij}(\boldsymbol{u})\tau_{ij}\,\mathrm{d}\boldsymbol{x}, \quad \text{for all } \boldsymbol{\tau} \in \Sigma(\Omega),$$

which is equivalent to writing

$$e_{\alpha\beta}(\boldsymbol{u}) = e_{3\beta}(\boldsymbol{u}) = 0, \tag{10.25}$$

$$\Psi_{33} = E e_{33}(\boldsymbol{u}) + \nu \Psi_{\mu\mu}. \tag{10.26}$$

Applying Theorem 5.2, we conclude that (10.25) is equivalent to saying that $\boldsymbol{u}$ is a displacement field of the Bernoulli–Navier type, that is,

$$\boldsymbol{u} \in V_{\mathrm{BN}}(\Omega), \qquad u_\alpha \in H_0^2(0,L), \qquad u_3 = \underline{u}_3 - x_\alpha \partial_3 u_\alpha, \qquad \underline{u}_3 \in H_0^1(0,L).$$

Consequently, Ψ_{33} may now be written in the form

$$\Psi_{33} = E(\partial_3 \underline{u}_3 - x_\alpha \partial_{33} u_\alpha) + \nu \Psi_{\mu\mu}. \tag{10.27}$$

Let us now consider the restriction of Eq. (6.19) to $\boldsymbol{v} \in V_{\mathrm{BN}}(\Omega)$: $v_\alpha \in H_0^2(0,L), v_3 = \underline{v}_3 - x_\alpha \partial_3 v_\alpha, \underline{v}_3 \in H_0^1(0,L)$. Then since $e_{3\beta}(\boldsymbol{v}) = e_{\alpha\beta}(\boldsymbol{v}) = 0$, we obtain

$$\int_\Omega \sigma_{33}(\varepsilon) e_{33}(\boldsymbol{v})\, \mathrm{d}\boldsymbol{x} = \int_\Omega f_i v_i \,\mathrm{d}\boldsymbol{x} + \int_\Gamma g_i v_i \,\mathrm{d}a, \quad \text{for all } \boldsymbol{v} \in V_{\mathrm{BN}}(\Omega).$$

Passing to the limit when $\varepsilon \to 0$ gives

$$\int_\Omega \Psi_{33}(\varepsilon) e_{33}(\boldsymbol{v})\, \mathrm{d}\boldsymbol{x} = \int_\Omega f_i v_i \,\mathrm{d}\boldsymbol{x} + \int_\Gamma g_i v_i \,\mathrm{d}a, \quad \text{for all } \boldsymbol{v} \in V_{\mathrm{BN}}(\Omega).$$

Substituting (10.27) in the previous equation and using properties (10.19) we obtain that $\underline{u}_3$ and u_α are solutions of

$$\begin{aligned} &\underline{u}_3 \in H_0^1(0,L), \\ &\int_0^L E \partial_3 \underline{u}_3 \partial_3 \underline{v}_3 \,\mathrm{d}x_3 = \int_0^L \left[\int_\omega f_3 \,\mathrm{d}\omega + \int_\gamma g_3 \,\mathrm{d}\gamma \right] \underline{v}_3 \,\mathrm{d}x_3, \\ &\quad\text{for all } \underline{v}_3 \in H_0^1(0,L), \end{aligned} \tag{10.28}$$

$$\begin{aligned} &u_\alpha \in H_0^2(0,L), \\ &\int_0^L E I_\alpha \partial_{33} u_\alpha \partial_{33} v \\ &\quad = \int_0^L \left[\int_\omega f_\alpha \,\mathrm{d}\omega + \int_\gamma g_\alpha \,\mathrm{d}\gamma \right] v \,\mathrm{d}x_3 \\ &\quad - \int_0^L \left[\int_\omega x_\alpha f_3 \,\mathrm{d}\omega + \int_\gamma x_\alpha g_3 \,\mathrm{d}\gamma \right] \partial_3 v \,\mathrm{d}x_3, \quad \text{for all } v \in H_0^2(0,L). \end{aligned} \tag{10.29}$$

From the uniqueness of solution of (10.28) and (10.29) and comparing with (6.33)–(6.36) we conclude that $\underline{u}_3 = \underline{u}_3^0$ and $u_\alpha = u_\alpha^0$ and therefore $\boldsymbol{u} = \boldsymbol{u}^0$.

For any subsequence of $(\boldsymbol{u}(\varepsilon), \boldsymbol{\sigma}(\varepsilon))_{\varepsilon>0}$ such that $(\boldsymbol{u}(\varepsilon), \tilde{\boldsymbol{\sigma}}(\varepsilon))_{\varepsilon>0}$ is weakly convergent in $V(\Omega) \times \Sigma(\Omega)$ we may repeat the reasoning of Theorems 10.2 and 10.3. The unicity of the limit $\boldsymbol{u}^0$ implies that the whole sequence $(\boldsymbol{u}(\varepsilon))_{\varepsilon>0}$ converges to $\boldsymbol{u}^0$. □

Problems (10.17) and (10.18) possess infinitely many solutions. Consequently, the weak convergence of the whole sequence $(\tilde{\boldsymbol{\sigma}}(\varepsilon))_{\varepsilon>0}$ *cannot be obtained with the same arguments* as for $(\boldsymbol{u}(\varepsilon))_{\varepsilon>0}$. This circumstance establishes a substantial difference with respect to the plate case in which the limits of the equivalent of $\tilde{\sigma}(\varepsilon)$ are uniquely determined (cf. DESTUYNDER [1980], CIARLET [1990]). Nevertheless, TUTEK and AGANOVIC [1986] showed the *strong convergence* of the whole sequences $(\tilde{\boldsymbol{\sigma}}(\varepsilon))_{\varepsilon>0}$ and $(\boldsymbol{u}(\varepsilon))_{\varepsilon>0}$. Their results can be summarized in the following theorem.

THEOREM 10.4. *For* $0 < \varepsilon \leqslant 1$, *let* $(\boldsymbol{u}(\varepsilon), \boldsymbol{\sigma}(\varepsilon)) \in V(\Omega) \times \Sigma(\Omega)$ *be the solution of problem* (6.25), (6.26). *Let* $\boldsymbol{u}^0 \in V(\Omega)$ *and* $\sigma_{33}^0 \in L^2(\Omega)$ *be the elements defined by* (6.33)–(6.36). *Then, when* $\varepsilon \to 0$, *the following strong convergences hold*:

$$\|\boldsymbol{u}(\varepsilon) - \boldsymbol{u}^0\|_{1,\Omega} \to 0, \tag{10.30}$$

$$|\sigma_{33}(\varepsilon) - \sigma_{33}^0|_{0,\Omega} \to 0, \tag{10.31}$$

$$\varepsilon|\sigma_{3\beta}(\varepsilon)|_{0,\Omega} \to 0, \tag{10.32}$$

$$\varepsilon^2|\sigma_{\alpha\beta}(\varepsilon)|_{0,\Omega} \to 0. \tag{10.33}$$

PROOF. For any $\tilde{\boldsymbol{\tau}} \in \Sigma(\Omega)$ one has

$$a_0(\tilde{\boldsymbol{\tau}}, \tilde{\boldsymbol{\tau}}) + a_2(\tilde{\boldsymbol{\tau}}, \tilde{\boldsymbol{\tau}}) + a_4(\tilde{\boldsymbol{\tau}}, \tilde{\boldsymbol{\tau}}) \geqslant C|\tilde{\boldsymbol{\tau}}|_{0,\Omega}^2. \tag{10.34}$$

Consider the subsequence satisfying (10.15) and let

$$\boldsymbol{S}(\varepsilon) = \boldsymbol{\sigma}(\varepsilon) - \tilde{\boldsymbol{\sigma}}^0, \qquad \tilde{\sigma}_{33}^0 = \sigma_{33}^0, \qquad \tilde{\sigma}_{i\beta}^0 = 0.$$
$$\tilde{S}_{33}(\varepsilon) = S_{33}(\varepsilon), \qquad \tilde{S}_{3\beta}(\varepsilon) = \varepsilon S_{3\beta}(\varepsilon), \qquad \tilde{S}_{\alpha\beta}(\varepsilon) = \varepsilon^2 S_{\alpha\beta}(\varepsilon).$$

From problems (6.25), (6.26), (6.29) and applying inequality (10.34) one obtains

$$\begin{aligned}
C[|\sigma_{33}(\varepsilon) - \sigma_{33}^0|_{0,\Omega}^2 &+ \varepsilon^2|\sigma_{3\beta}(\varepsilon)|_{0,\Omega}^2 + \varepsilon^4|\sigma_{\alpha\beta}(\varepsilon)|_{0,\Omega}^2] = C|\tilde{\boldsymbol{S}}(\varepsilon)|_{0,\Omega}^2 \\
&\leqslant a_0(\tilde{\boldsymbol{S}}(\varepsilon), \tilde{\boldsymbol{S}}(\varepsilon)) + a_2(\tilde{\boldsymbol{S}}(\varepsilon), \tilde{\boldsymbol{S}}(\varepsilon)) + a_4(\tilde{\boldsymbol{S}}(\varepsilon), \tilde{\boldsymbol{S}}(\varepsilon)) \\
&= a_0(\boldsymbol{\sigma}(\varepsilon) - \tilde{\boldsymbol{\sigma}}^0, \boldsymbol{\sigma}(\varepsilon) - \tilde{\boldsymbol{\sigma}}^0) + \varepsilon^2 a_2(\boldsymbol{\sigma}(\varepsilon) - \tilde{\boldsymbol{\sigma}}^0, \boldsymbol{\sigma}(\varepsilon) - \tilde{\boldsymbol{\sigma}}^0) \\
&\quad + \varepsilon^4 a_4(\boldsymbol{\sigma}(\varepsilon) - \tilde{\boldsymbol{\sigma}}^0, \boldsymbol{\sigma}(\varepsilon) - \tilde{\boldsymbol{\sigma}}^0) \\
&= -b(\boldsymbol{\sigma}(\varepsilon) - \tilde{\boldsymbol{\sigma}}^0, \boldsymbol{u}(\varepsilon)) + b(\boldsymbol{\sigma}(\varepsilon) - \tilde{\boldsymbol{\sigma}}^0, \boldsymbol{u}^0) - \varepsilon^2 a_2(\tilde{\boldsymbol{\sigma}}^0, \boldsymbol{\sigma}(\varepsilon) - \tilde{\boldsymbol{\sigma}}^0) \\
&\quad - \varepsilon^4 a_4(\tilde{\boldsymbol{\sigma}}^0, \boldsymbol{\sigma}(\varepsilon) - \tilde{\boldsymbol{\sigma}}^0) \\
&= -b(\boldsymbol{\sigma}(\varepsilon), \boldsymbol{u}(\varepsilon) - \boldsymbol{u}^0) + b(\tilde{\boldsymbol{\sigma}}^0, \boldsymbol{u}(\varepsilon) - \boldsymbol{u}^0) - \varepsilon^2 a_2(\tilde{\boldsymbol{\sigma}}^0, \boldsymbol{\sigma}(\varepsilon) - \tilde{\boldsymbol{\sigma}}^0)
\end{aligned}$$

$$= \int_\Omega f_i(u_i(\varepsilon) - u_i^0)\,\mathrm{d}\boldsymbol{x} + \int_\Gamma g_i(u_i(\varepsilon) - u_i^0)\,\mathrm{d}a$$
$$- \int_\Omega \sigma_{33}^0 e_{33}(\boldsymbol{u}(\varepsilon) - \boldsymbol{u}^0)\,\mathrm{d}\boldsymbol{x} + \varepsilon^2 \frac{\nu}{E} \int_\Omega \sigma_{33}^0 \sigma_{\mu\mu}(\varepsilon)\,\mathrm{d}\boldsymbol{x}. \tag{10.35}$$

From the weak convergence results (10.15), together with Theorem 10.3, one concludes that expression (10.35) converges to

$$\mathscr{L} = \frac{\nu}{E} \int_\Omega \sigma_{33}^0 \Psi_{\mu\mu}\,\mathrm{d}\boldsymbol{x}. \tag{10.36}$$

The form of σ_{33}^0 (cf. (6.34)) and properties (10.19) imply that $\mathscr{L} = 0$ and, consequently, convergences (10.31)–(10.33) hold for this subsequence.

Conclusions (10.15)–(10.19) and (10.31)–(10.33) may be drawn for any weakly convergent subsequence of $(\tilde{\boldsymbol{\sigma}}(\varepsilon))_{\varepsilon>0} \in \Sigma(\Omega)$, $0 < \varepsilon \leqslant 1$, giving $\tilde{\boldsymbol{\sigma}}^0$ as the only limit (i.e., $\tilde{\sigma}_{33}^0 = \sigma_{33}^0$, and $\tilde{\sigma}_{i\beta} = 0$). Consequently, the whole sequence $(\tilde{\boldsymbol{\sigma}}(\varepsilon))_{\varepsilon>0}$ converges strongly and (10.31)–(10.33) hold.

In order to show (10.30) we consider the following identity which is obtained from (6.25) and (6.29):

$$-b(\boldsymbol{\tau}, \boldsymbol{u}(\varepsilon) - \boldsymbol{u}^0) = a_0(\boldsymbol{\sigma}(\varepsilon) - \boldsymbol{\sigma}^0, \boldsymbol{\tau}) + \varepsilon^2 a_2(\boldsymbol{\sigma}(\varepsilon), \boldsymbol{\tau}) + \varepsilon^4 a_4(\boldsymbol{\sigma}(\varepsilon), \boldsymbol{\tau}). \tag{10.37}$$

Still, from (6.25) we have for all $\boldsymbol{\tau} \in \Sigma(\Omega)$

$$a_0(\boldsymbol{\sigma}(\varepsilon) - \boldsymbol{\sigma}^0, \boldsymbol{\tau}) + \varepsilon^2 a_2(\boldsymbol{\sigma}(\varepsilon), \boldsymbol{\tau}) + \varepsilon^4 a_4(\boldsymbol{\sigma}(\varepsilon), \boldsymbol{\tau})$$
$$\leqslant C[|\sigma_{33}(\varepsilon) - \sigma_{33}^0|_{0,\Omega}^2 + \varepsilon^2 |\sigma_{3\beta}(\varepsilon)|_{0,\Omega}^2 + \varepsilon^4 |\sigma_{\alpha\beta}(\varepsilon)|_{0,\Omega}^2]^{1/2} |\boldsymbol{\tau}|_{0,\Omega}. \tag{10.38}$$

Combining (10.37) and (10.38) with (10.31)–(10.33) we deduce that

$$\frac{\int_\Omega e_{ij}(\boldsymbol{u}(\varepsilon) - \boldsymbol{u}^0)\tau_{ij}\,\mathrm{d}\boldsymbol{x}}{|\boldsymbol{\tau}|_{0,\Omega}} \to 0, \quad \text{as } \varepsilon \to 0.$$

Conclusion (10.30) is now a consequence of (10.3). □

REMARK 10.3. (i) From the previous proof one sees that the choice $\tilde{\sigma}_{i\beta}^0 = 0$ is not essential. In fact, if one takes instead $\tilde{\sigma}_{i\beta}^0 = \sigma_{i\beta}^0$ — the zeroth order terms of the asymptotic expansion evaluated in Section 7 — the same proof would give us

$$\varepsilon|\sigma_{3\beta}(\varepsilon) - \sigma_{3\beta}^0|_{0,\Omega} \to 0, \qquad \varepsilon^2|\sigma_{\alpha\beta}(\varepsilon) - \sigma_{\alpha\beta}^0|_{0,\Omega} \to 0. \tag{10.39}$$

However apparently elegant, this result is of little use. In the plate case DESTUYNDER [1980] shows that the scaled stress tensor converges in spaces larger than $L^2(\Omega)$. In the rod case the convergence in a "natural" space containing $L^2(\Omega)$ was analyzed only recently by LE DRET [1993].

(ii) It is interesting to remark that in the proof one only uses the first equation of (6.29) and *never the second one*:

$$b(\boldsymbol{\sigma}^0, \boldsymbol{v}) = F(\boldsymbol{v}), \quad \text{for all } \boldsymbol{v} \in V(\Omega).$$

Consequently, the "boundary layer" phenomena mentioned in Remarks 8.4 and 9.1 does not affect the convergence results just proved. Moreover, this is actually one of the reasons why the results are valid in any of the spaces defined either by (2.14) or (5.3).

From the previous theorem we see that it is possible to consider in Theorem 10.2:

$$\Psi_{\alpha\beta} = \Psi_{3\beta} = 0, \qquad \Psi_{33} = \sigma_{33}^0, \qquad \zeta_\beta = m_\beta^0 = -EI_\beta \partial_{33} u_\beta^0.$$

From the unicity of the limit m_β^0 we conclude that the entire sequence $(m_\beta(\varepsilon))_{\varepsilon>0}$ converges weakly to m_β^0 in $L^2(0,L)$. Actually, the convergence is also strong as can be seen from (10.31) and from the following inequality:

$$|m_\beta(\varepsilon) - m_\beta^0|_{0,L}^2 = \int_0^L \left[\int_\omega x_\beta(\sigma_{33}(\varepsilon) - \sigma_{33}^0)\, d\omega \right]^2 dx_3 \leqslant I_\beta |\sigma_{33}(\varepsilon) - \sigma_{33}^0|_{0,\Omega}^2.$$

Moreover, from (10.21) we conclude that $\rho_\beta \in L^2(0,L)$ verifies

$$\int_0^L \rho_\beta v^0 \, dx_3 = -\int_0^L m_\beta^0 \partial_3 v^0 \, dx_3 + \int_0^L \left[\int_\omega x_\beta f_3 \, d\omega + \int_\gamma x_\beta g_3 \, d\gamma \right] v^0 \, dx_3,$$
$$\text{for all } v^0 \in H_0^1(0,L). \tag{10.40}$$

From the fact that $u_\beta^0 \in H^3(0,L)$, (cf. (6.35)), one obtains $m_\beta^0 \in H^1(0,L)$ and consequently from (10.40) one concludes

$$\rho_\beta = q_\beta^0 = \partial_3 m_\beta^0 + \int_\omega x_\beta f_3 \, d\omega + \int_\gamma x_\beta g_3 \, d\gamma.$$

Once again, the uniqueness of the limit implies that the entire sequence $(q_\beta(\varepsilon))_{\varepsilon>0}$ converges weakly to q_β^0.

On the other hand, setting $\boldsymbol{v} = (0, 0, x_\beta v^0)$, $v^0 \in H_0^1(0,L)$ in Eq. (6.19), we obtain

$$\int_0^L q_\beta(\varepsilon) v^0 \, dx_3 + \int_0^L m_\beta(\varepsilon) \partial_3 v^0 \, dx_3$$
$$= \int_0^L \left[\int_\omega x_\beta f_3 \, d\omega + \int_\gamma x_\beta g_3 \, d\gamma \right] v^0 \, dx_3, \quad \text{for all } v^0 \in H_0^1(0,L).$$

Now assuming that $m_\beta(\varepsilon) \in H^1(0, L)$ we have

$$q_\beta(\varepsilon) = \partial_3 m_\beta(\varepsilon) + \int_\omega x_\beta f_3 \, d\omega + \int_\gamma x_\beta g_3 \, d\gamma,$$

and, consequently, the sequence $(\partial_3 m_\beta(\varepsilon))_{\varepsilon>0}$ converges weakly in $L^2(0, L)$ to $\partial_3 m^0_\beta$. We have then proved the following result, which constitutes an improvement of an analogous one in BERMÚDEZ and VIAÑO [1984, Corollary 6.1].

THEOREM 10.5. *The sequences* $(m_\beta(\varepsilon))_{\varepsilon>0}$ *and* $(q_\beta(\varepsilon))_{\varepsilon>0}$ *verify*

$$m_\beta(\varepsilon) \to m^0_\beta, \quad \text{in } L^2(0, L),$$
$$q_\beta(\varepsilon) \rightharpoonup q^0_\beta, \quad \text{in } L^2(0, L),$$

where m^0_β *and* q^0_β *are defined by* (7.64) *and* (7.76), *respectively. Moreover, if* $m_\beta(\varepsilon) \in H^1(0, L)$ *then*

$$m_\beta(\varepsilon) \rightharpoonup m^0_\beta \text{ in } H^1(0, L).$$

□

CHAPTER II

Linear Asymptotic Models. Comparison with Classical Theories

The main objective of this chapter is to give an interpretation of the results obtained in Chapter I and to relate them with the classical linear beam theories. In order to do so we start by descaling the successive terms in the asymptotic expansion; we then propose an asymptotic model of order one (Section 11) and another one of order two (Section 12). These models are compared to the classical theories of Bernoulli–Navier (Section 11), Vlassov (Section 13), Saint Venant (Section 14) and Timoshenko (Section 15). Also in Section 15 a correct definition of Timoshenko's constant is introduced. Several numerical examples are presented for the most common types of cross sections. All the results obtained are compared to the classical values.

11. Asymptotic first order one-dimensional equations of a linearly elastic clamped rod. Comparison with Bernoulli–Navier's theory

In the previous sections we showed that when the cross sectional area is small the scaled displacement and stress fields $(\boldsymbol{u}(\varepsilon), \boldsymbol{\sigma}(\varepsilon))$ are approximated in Ω by the first terms of the asymptotic expansion (6.27). Associated with the different powers of ε we may distinguish the following successive approximations:

$$\begin{aligned} &\text{order 0:} \quad (\boldsymbol{u}^0, \boldsymbol{\sigma}^0), \\ &\text{order 2:} \quad (\boldsymbol{u}^0, \boldsymbol{\sigma}^0) + \varepsilon^2(\boldsymbol{u}^2, \boldsymbol{\sigma}^2), \\ &\text{order 4:} \quad (\boldsymbol{u}^0, \boldsymbol{\sigma}^0) + \varepsilon^2(\boldsymbol{u}^2, \boldsymbol{\sigma}^2) + \varepsilon^4(\boldsymbol{u}^4, \boldsymbol{\sigma}^4). \end{aligned} \tag{11.1}$$

Consequently, the "descaling" of the unknowns through (2.6) and (2.21) leads to successive approximations of the true displacement and stress fields in Ω^ε.

From scalings (2.6) and (2.21), in a natural way, we define components $\boldsymbol{u}^{p\varepsilon} \in V(\Omega^\varepsilon)$ and $\boldsymbol{\sigma}^{p\varepsilon} \in \Sigma(\Omega^\varepsilon)$, $p = 0, 2, 4, \ldots$ through the following "descalings", for all $\boldsymbol{x}^\varepsilon = \Pi^\varepsilon(\boldsymbol{x}) \in \overline{\Omega}^\varepsilon$:

$$u_\alpha^{p\varepsilon}(\boldsymbol{x}^\varepsilon) = \varepsilon^{-1+p} u_\alpha^p(\boldsymbol{x}), \qquad u_3^{p\varepsilon}(\boldsymbol{x}^\varepsilon) = \varepsilon^p u_3^p(\boldsymbol{x}),$$
$$\sigma_{\alpha\beta}^{p\varepsilon}(\boldsymbol{x}^\varepsilon) = \varepsilon^{2+p}\sigma_{\alpha\beta}^p(\boldsymbol{x}), \qquad \sigma_{3\beta}^{p\varepsilon}(\boldsymbol{x}^\varepsilon) = \varepsilon^{1+p}\sigma_{3\beta}^p(\boldsymbol{x}), \qquad \sigma_{33}^{p\varepsilon}(\boldsymbol{x}^\varepsilon) = \varepsilon^p \sigma_{33}^p(\boldsymbol{x}). \tag{11.2}$$

The successive approximations of the three-dimensional solution $(\boldsymbol{u}^\varepsilon, \boldsymbol{\sigma}^\varepsilon)$ in Ω^ε are then defined by

$$\begin{aligned} &\text{order 0:} \quad (\boldsymbol{u}^{0\varepsilon}, \boldsymbol{\sigma}^{0\varepsilon}),\\ &\text{order 2:} \quad (\boldsymbol{u}^{0\varepsilon}, \boldsymbol{\sigma}^{0\varepsilon}) + (\boldsymbol{u}^{2\varepsilon}, \boldsymbol{\sigma}^{2\varepsilon}),\\ &\text{order 4:} \quad (\boldsymbol{u}^{0\varepsilon}, \boldsymbol{\sigma}^{0\varepsilon}) + (\boldsymbol{u}^{2\varepsilon}, \boldsymbol{\sigma}^{2\varepsilon}) + (\boldsymbol{u}^{4\varepsilon}, \boldsymbol{\sigma}^{4\varepsilon}). \end{aligned} \tag{11.3}$$

As in (6.17), we define in a similar way the following descalings:

$$\begin{aligned} m_\beta^{p\varepsilon}(\boldsymbol{x}^\varepsilon) &= \varepsilon^{3+p} m_\beta^p(\boldsymbol{x}),\\ q_\beta^{p\varepsilon}(\boldsymbol{x}^\varepsilon) &= \varepsilon^{3+p} q_\beta^p(\boldsymbol{x}), \end{aligned} \tag{11.4}$$

and one has the following approximations for the bending moment and shear force components:

$$\begin{aligned} &\text{order 0:} \quad m_\beta^{0\varepsilon}, \qquad q_\beta^{0\varepsilon},\\ &\text{order 2:} \quad m_\beta^{0\varepsilon} + m_\beta^{2\varepsilon}, \qquad q_\beta^{0\varepsilon} + q_\beta^{2\varepsilon},\\ &\text{order 4:} \quad m_\beta^{0\varepsilon} + m_\beta^{2\varepsilon} + m_\beta^{4\varepsilon}, \qquad q_\beta^{0\varepsilon} + q_\beta^{2\varepsilon} + q_\beta^{4\varepsilon}. \end{aligned} \tag{11.5}$$

As was to be expected, the "descaling" introduces only different powers of ε in different places for the terms characterizing $\boldsymbol{u}^{p\varepsilon}$ and $\boldsymbol{\sigma}^{p\varepsilon}$ but *the general form of the equations remains unchanged and it is directly obtained from the equations for* $\boldsymbol{u}^p$ *and* $\boldsymbol{\sigma}^p$.

In the characterization of $\boldsymbol{u}^{p\varepsilon}$ and $\boldsymbol{\sigma}^{p\varepsilon}$ as functions of $V(\Omega^\varepsilon)$ and $\Sigma(\Omega^\varepsilon)$ the functions and constants defined in Section 7 (cf. (7.21)–(7.26)) show up, *but now referred to the cross section* ω^ε. In order to define them, it is enough to add ε as an exponent in all the equations, operators and sets showing up in (7.21)–(7.26). For instance

$$\Phi_{11}^\varepsilon(x_1^\varepsilon, x_2^\varepsilon) = \tfrac{1}{2}[(x_1^\varepsilon)^2 - (x_2^\varepsilon)^2]$$

and the warping function for section ω^ε is the the unique solution of the following problem

$$w^\varepsilon \in H^1(\omega^\varepsilon),$$
$$\int_{\omega^\varepsilon} \partial_\alpha^\varepsilon w^\varepsilon \partial_\alpha^\varepsilon \varphi^\varepsilon \, \mathrm{d}\omega^\varepsilon = \int_{\omega^\varepsilon} (x_2^\varepsilon \partial_1^\varepsilon \varphi^\varepsilon - x_1^\varepsilon \partial_2^\varepsilon \varphi^\varepsilon)\, \mathrm{d}\omega^\varepsilon, \quad \text{for all } \varphi^\varepsilon \in H^1(\omega^\varepsilon),$$
$$\int_{\omega^\varepsilon} w^\varepsilon \, \mathrm{d}\omega^\varepsilon = 0.$$

In this section, we uniquely characterize the zeroth order terms $\boldsymbol{u}^{0\varepsilon}$, $\sigma_{33}^{0\varepsilon}$, $m_\beta^{0\varepsilon}$, $q_\beta^{0\varepsilon}$ and the results are compared with those obtained from the classical Bernoulli–Navier beam theory.

From now on, since for any function z defined in $[0, L]$ one has $\partial_3^\varepsilon z = \partial_3 z$, whenever convenient we identify both notations with the standard one $\mathrm{d}z/\mathrm{d}x_3$.

The study of terms $\boldsymbol{u}^{2\varepsilon}$, $\sigma_{i\beta}^{0\varepsilon}$, $\sigma_{33}^{2\varepsilon}$, $m_\beta^{2\varepsilon}$, $q_\beta^{2\varepsilon}$ together with its interpretation, in the framework of a general second order model, is done in the next section.

For the sake of brevity, we omit expressions for $\sigma_{3\beta}^{2\varepsilon}$ and the fourth order terms $\boldsymbol{u}^{4\varepsilon}$ and $m_\beta^{4\varepsilon}$, which can be obtained directly from Theorem 9.1 (which corresponds to the case $\varepsilon = 1$).

THEOREM 11.1 (Zeroth order terms $\boldsymbol{u}^{0\varepsilon}$, $\sigma_{33}^{0\varepsilon}$, $m_\beta^{0\varepsilon}$, $q_\beta^{0\varepsilon}$). *Let the system of applied forces be such that*

$$f_i^\varepsilon \in L^2(\Omega^\varepsilon), \qquad g_i^\varepsilon \in L^2(\Gamma^\varepsilon).$$

Then, the zeroth order terms $\boldsymbol{u}^{0\varepsilon}$, $\sigma_{33}^{0\varepsilon}$, $m_\beta^{0\varepsilon}$, $q_\beta^{0\varepsilon}$, *are uniquely characterized as follows:*

(i) *The displacement field* $\boldsymbol{u}^{0\varepsilon}$ *is of the Bernoulli–Navier type:*

$$\begin{aligned} &\boldsymbol{u}^{0\varepsilon} \in V_{\mathrm{BN}}(\Omega^\varepsilon): \; u_3^{0\varepsilon} = \underline{u}_3^{0\varepsilon} - x_\alpha^\varepsilon \partial_3^\varepsilon u_\alpha^{0\varepsilon}, \\ &\underline{u}_3^{0\varepsilon} \in H_0^1(0, L), \qquad u_\alpha^{0\varepsilon} \in H_0^2(0, L), \end{aligned} \tag{11.6}$$

where the axial stretching $\underline{u}_3^{0\varepsilon}$ *and the bending* $u_\alpha^{0\varepsilon}$ *components are the unique solutions of the following one-dimensional problems*:

$$\begin{aligned} &\underline{u}_3^{0\varepsilon} \in H_0^1(0, L), \\ &\int_0^L E^\varepsilon A^\varepsilon \partial_3 \underline{u}_3^{0\varepsilon} \partial_3^\varepsilon \underline{v}^0 \, \mathrm{d}x_3^\varepsilon \\ &\quad = \int_0^L [\int_{\omega^\varepsilon} f_3^\varepsilon \, d\omega^\varepsilon + \int_{\gamma^\varepsilon} g_3^\varepsilon \, \mathrm{d}\gamma^\varepsilon] \underline{v}^0 \, \mathrm{d}x_3^\varepsilon, \quad \textit{for all } \underline{v}^0 \in H_0^1(0, L), \end{aligned} \tag{11.7}$$

$$\begin{aligned} &u_\alpha^{0\varepsilon} \in H_0^2(0, L), \\ &\int_0^L E^\varepsilon I_\alpha^\varepsilon \partial_{33}^\varepsilon u_\alpha^{0\varepsilon} \partial_{33}^\varepsilon v^0 \, \mathrm{d}x_3^\varepsilon \\ &\quad = \int_0^L \left[\int_{\omega^\varepsilon} f_\alpha^\varepsilon \, \mathrm{d}\omega^\varepsilon + \int_{\gamma^\varepsilon} g_\alpha^\varepsilon \, \mathrm{d}\gamma^\varepsilon \right] v^0 \, \mathrm{d}x_3^\varepsilon \\ &\quad - \int_0^L [\int_{\omega^\varepsilon} x_\alpha^\varepsilon f_3^\varepsilon \, \mathrm{d}\omega^\varepsilon + \int_{\gamma^\varepsilon} x_\alpha^\varepsilon g_3^\varepsilon \, \mathrm{d}\gamma^\varepsilon] \partial_3^\varepsilon v^0 \, \mathrm{d}x_3^\varepsilon, \end{aligned} \tag{11.8}$$

for all $v^0 \in H_0^2(0, L)$.

(ii) *The axial stress component* $\sigma_{33}^{0\varepsilon}$ *is obtained from* $u_i^{0\varepsilon}$ *in the following way*:

$$\sigma_{33}^{0\varepsilon} = E^\varepsilon(\partial_3^\varepsilon \underline{u}_3^{0\varepsilon} - x_\alpha^\varepsilon \partial_{33}^\varepsilon u_\alpha^{0\varepsilon}). \tag{11.9}$$

(iii) *The bending moment* $m_\beta^{0\varepsilon}$ *and shear force* $q_\beta^{0\varepsilon}$ *components are given by* (*no sum on* β)

$$m_\beta^{0\varepsilon} = \int_{\omega^\varepsilon} x_\beta^\varepsilon \sigma_{33}^{0\varepsilon}\, d\omega^\varepsilon = -E^\varepsilon I_\beta^\varepsilon \partial_{33}^\varepsilon u_\beta^{0\varepsilon}, \tag{11.10}$$

$$q_\beta^{0\varepsilon} = \int_{\omega^\varepsilon} \sigma_{3\beta}^{0\varepsilon}\, d\omega^\varepsilon = -E^\varepsilon I_\beta^\varepsilon \partial_{333}^\varepsilon u_\beta^{0\varepsilon} + \int_{\omega^\varepsilon} x_\beta^\varepsilon f_3^\varepsilon\, d\omega^\varepsilon + \int_{\gamma^\varepsilon} x_\beta^\varepsilon g_3^\varepsilon\, d\gamma^\varepsilon. \tag{11.11}$$

One of the *most important conclusions* of the present asymptotic analysis is the following (cf. BERMÚDEZ and VIAÑO [1984]): *Without any a priori assumption, either of a mechanic or geometric nature, from the previous theorem one finds the classical equations of the linearized Bernoulli–Navier beam theory.*

In fact, any sufficiently smooth solution of the variational problem (11.7) is also a solution of the following boundary value problem:

$$-E^\varepsilon A^\varepsilon \frac{d^2 \underline{u}_3^{0\varepsilon}}{dx_3^2} = F_3^\varepsilon, \quad \text{in } (0, L), \tag{11.12}$$

$$\underline{u}_3^{0\varepsilon}(0) = \underline{u}_3^{0\varepsilon}(L) = 0, \tag{11.13}$$

where

$$F_3^\varepsilon = \int_{\omega^\varepsilon} f_3^\varepsilon\, d\omega^\varepsilon + \int_{\gamma^\varepsilon} g_3^\varepsilon\, d\gamma^\varepsilon. \tag{11.14}$$

Equation (11.12) is the classical equation of the axial deformation for a linearized isotropic beam (cf., for instance, DYM and SHAMES [1973, Section 4.3], SOKOLNIKOFF [1956, Section 30], GERMAIN [1962, Section VII-6]). Conditions (11.13) correspond to the clamping case.

In a similar way, in order to give the strong interpretation of the variational equation (11.8) we assume that

$$f_3^\varepsilon \in H^1[0, L; L^2(\omega^\varepsilon)], \qquad g_3^\varepsilon \in H^1[0, L; L^2(\gamma^\varepsilon)], \tag{11.15}$$

and we introduce the following notation for the resultant of the applied loads and moments in each cross section:

$$F_\beta^\varepsilon = \int_{\omega^\varepsilon} f_\beta^\varepsilon\, d\omega^\varepsilon + \int_{\gamma^\varepsilon} g_\beta^\varepsilon\, d\gamma^\varepsilon, \tag{11.16}$$

$$M_\beta^\varepsilon = \int_{\omega^\varepsilon} x_\beta^\varepsilon f_3^\varepsilon\, d\omega^\varepsilon + \int_{\gamma^\varepsilon} x_\beta^\varepsilon g_3^\varepsilon\, d\gamma^\varepsilon. \tag{11.17}$$

Then, the regular solution of (11.8) is also a solution of the following one-dimensional fourth order boundary value problem:

$$E^\varepsilon I_\beta^\varepsilon \frac{\mathrm{d}^4 u_\beta^{0\varepsilon}}{\mathrm{d}x_3^4} = F_\beta^\varepsilon + \frac{\mathrm{d}M_\beta^\varepsilon}{\mathrm{d}x_3}, \quad \text{in } (0, L), \tag{11.18}$$

$$u_\beta^{0\varepsilon}(0) = u_\beta^{0\varepsilon}(L) = 0, \qquad \frac{\mathrm{d}u_\beta^{0\varepsilon}}{\mathrm{d}x_3}(0) = \frac{\mathrm{d}u_\beta^{0\varepsilon}}{\mathrm{d}x_3}(L) = 0. \tag{11.19}$$

Problem (11.18), (11.19) constitutes the classical model for the transverse bending of a clamped beam (usually $M_\beta^\varepsilon = 0$) (cf. LANDAU and LIFCHITZ [1967, Eq. (20.4)], GERMAIN [1962, Section IX.3, Eq. (32)], DYM and SHAMES [1973, Eq. (4.10)]).

Expressions (11.10) and (11.11) for the bending moments and the shear forces, respectively, coincide with the standard equations in literature (cf. LANDAU and LIFCHITZ [1967, Eqs. (20.3) and (20.5)]. Finally, we remark that we have also obtained the Bernoulli–Navier displacement field (cf. (11.6)) and expression (11.9) for the axial stress *without any* a priori hypothesis.

REMARK 11.1. One immediately sees that functions $u_\beta^{0\varepsilon}$ and $m_\beta^{0\varepsilon}$ are the unique solution of the following mixed variational problem (no sum on β):

$$\begin{aligned}
&(u_\beta^{0\varepsilon}, m_\beta^{0\varepsilon}) \in H_0^2(0, L) \times L^2(0, L),\\
&\int_0^L m_\beta^{0\varepsilon}\, p \,\mathrm{d}x_3^\varepsilon + \int_0^L E^\varepsilon I_\beta^\varepsilon \partial_{33}^\varepsilon u_\beta^{0\varepsilon}\, p \,\mathrm{d}x_3^\varepsilon = 0,\\
&\quad \text{for all } p \in L^2(0, L),\\
&-\int_0^L m_\beta^{0\varepsilon} \partial_{33}^\varepsilon v \,\mathrm{d}x_3^\varepsilon = \int_0^L F_\beta^\varepsilon v \,\mathrm{d}x_3^\varepsilon - \int_0^L M_\beta^\varepsilon \partial_3^\varepsilon v \,\mathrm{d}x_3^\varepsilon, \quad \text{for all } v \in H_0^2(0, L),
\end{aligned} \tag{11.20}$$

which represents the *mixed Hellinger–Reissner variational principle* for a clamped beam.

In a similar way, if we denote the axial normal force resultant by:

$$n^{0\varepsilon} = \int_{\omega^\varepsilon} \sigma_{33}^{0\varepsilon} \,\mathrm{d}\omega^\varepsilon, \tag{11.21}$$

we obtain the following mixed variational problem for the axial deformations:

$$(\underline{u}_3^{0\varepsilon}, n^{0\varepsilon}) \in H_0^1(0,L) \times L^2(0,L),$$

$$\begin{aligned}
&\int_0^L n^{0\varepsilon}\, p \,\mathrm{d}x_3^\varepsilon - \int_0^L E^\varepsilon A^\varepsilon \partial_3^\varepsilon \underline{u}_3^{0\varepsilon}\, p\,\mathrm{d}x_3^\varepsilon = 0, \quad \text{for all } p \in L^2(0,L),\\
&\int_0^L n^{0\varepsilon} \partial_3^\varepsilon \underline{v}\,\mathrm{d}x_3^\varepsilon = \int_0^L F_3^\varepsilon \underline{v}\,\mathrm{d}x_3^\varepsilon, \quad \text{for all } \underline{v} \in H_0^1(0,L),
\end{aligned} \tag{11.22}$$

12. Asymptotic second order general model of a linearly elastic clamped rod

As Theorem 11.1, the following result is a simple corollary of Theorem 7.1 and of transformations (11.2) and (11.4).

THEOREM 12.1 (Second order terms $\boldsymbol{u}^{2\varepsilon}$, $\sigma_{33}^{2\varepsilon}$, $m_\beta^{2\varepsilon}$, $q_\beta^{2\varepsilon}$ and zeroth order stress components $\sigma_{i\beta}^{0\varepsilon}$). *Let the system of applied forces be such that*

$$\begin{aligned}
&f_\alpha^\varepsilon \in L^2(\Omega^\varepsilon), \qquad f_3^\varepsilon \in H^1[0,L;L^2(\omega^\varepsilon)],\\
&g_\alpha^\varepsilon \in L^2(\Gamma^\varepsilon), \qquad g_3^\varepsilon \in H^1[0,L;L^2(\gamma^\varepsilon)].
\end{aligned} \tag{12.1}$$

Then, besides the previous expressions characterizing $\boldsymbol{u}^{0\varepsilon}$, $\sigma_{33}^{0\varepsilon}$, $m_\beta^{0\varepsilon}$ and $q_\beta^{0\varepsilon}$, we have the following characterization of $\boldsymbol{u}^{2\varepsilon}$, $\sigma_{33}^{2\varepsilon}$, $m_\beta^{2\varepsilon}$, $q_\beta^{2\varepsilon}$ and $\sigma_{i\beta}^{0\varepsilon}$.

(i) *Displacements $u_i^{2\varepsilon}$ are of the form*

$$u_1^{2\varepsilon} = z_1^{2\varepsilon} + x_2^\varepsilon z^{2\varepsilon} - \nu^\varepsilon (x_1^\varepsilon \partial_3^\varepsilon \underline{u}_3^{0\varepsilon} - \Phi_{1\beta}^\varepsilon \partial_{33}^\varepsilon u_\beta^{0\varepsilon}), \tag{12.2}$$

$$u_2^{2\varepsilon} = z_2^{2\varepsilon} - x_1^\varepsilon z^{2\varepsilon} - \nu^\varepsilon (x_2^\varepsilon \partial_3^\varepsilon \underline{u}_3^{0\varepsilon} - \Phi_{2\beta}^\varepsilon \partial_{33}^\varepsilon u_\beta^{0\varepsilon}), \tag{12.3}$$

$$\begin{aligned}
u_3^{2\varepsilon} =\, &\underline{u}_3^{2\varepsilon} - x_\alpha^\varepsilon \partial_3^\varepsilon z_\alpha^{2\varepsilon} - w^\varepsilon \partial_3^\varepsilon z^{2\varepsilon} + \nu^\varepsilon \left[\tfrac{1}{2}(x_1^{2\varepsilon} + x_2^{2\varepsilon}) - \frac{1}{2A^\varepsilon}(I_1^\varepsilon + I_2^\varepsilon)\right] \partial_{33}^\varepsilon \underline{u}_3^{0\varepsilon}\\
&+ [(1+\nu^\varepsilon)\eta_\alpha^\varepsilon + \nu^\varepsilon \theta_\alpha^\varepsilon]\partial_{333}^\varepsilon u_\alpha^{0\varepsilon} + \frac{2(1+\nu^\varepsilon)}{E^\varepsilon} w^{0\varepsilon},
\end{aligned} \tag{12.4}$$

where functions $z^{2\varepsilon}$, $w^{0\varepsilon}$, $\underline{u}_3^{2\varepsilon}$ and $z_\alpha^{2\varepsilon}$ are characterized in the following way:

(a) *The angle of twist $z^{2\varepsilon}$ depends only on x_3^ε and is the unique solution of the following one dimensional problem:*

$$z^{2\varepsilon} \in H^1(0,L),$$

$$\int_0^L \frac{E^\varepsilon J^\varepsilon}{2(1+\nu^\varepsilon)} \partial_3^\varepsilon z^{2\varepsilon} \partial_3^\varepsilon v\,\mathrm{d}x_3^\varepsilon = \int_0^L M_3^{2\varepsilon} v\,\mathrm{d}x_3^\varepsilon, \quad \textit{for all } v \in H_0^1(0,L), \tag{12.5}$$

$$z^{2\varepsilon}(a) = \frac{\nu^\varepsilon}{(I_1^\varepsilon + I_2^\varepsilon)}[H_2^\varepsilon \partial_{33}^\varepsilon u_1^{0\varepsilon}(a) - H_1^\varepsilon \partial_{33}^\varepsilon u_2^{0\varepsilon}(a)] \quad \textit{at } a = 0, L,$$

where

$$M_3^{2\varepsilon} = \int_{\omega^\varepsilon} (x_2^\varepsilon f_1^\varepsilon - x_1^\varepsilon f_2^\varepsilon)\,\mathrm{d}\omega^\varepsilon + \int_{\gamma^\varepsilon} (x_2^\varepsilon g_1^\varepsilon - x_1^\varepsilon g_2^\varepsilon)\,\mathrm{d}\gamma^\varepsilon + \int_{\omega^\varepsilon} w^\varepsilon \partial_3^\varepsilon f_3^\varepsilon\,\mathrm{d}\omega^\varepsilon$$
$$+ \int_{\gamma^\varepsilon} w^\varepsilon \partial_3^\varepsilon g_3^\varepsilon\,\mathrm{d}\gamma^\varepsilon - \frac{E^\varepsilon}{2(1+\nu^\varepsilon)}[(1+\nu^\varepsilon)I_\alpha^{w^\varepsilon} + \nu^\varepsilon I_\alpha^{\Psi^\varepsilon}]\partial_{3333}^\varepsilon u_\alpha^{0\varepsilon}. \tag{12.6}$$

(b) *The additional warping $w^{0\varepsilon}$ is the unique solution of the following problem*:

$$w^{0\varepsilon} \in H^1[0,L;H^1(\omega^\varepsilon)] \quad \textit{and for all } x_3^\varepsilon \in [0,L],$$
$$\int_{\omega^\varepsilon} \partial_\beta^\varepsilon w^{0\varepsilon}(x_3^\varepsilon)\partial_\beta^\varepsilon \varphi^\varepsilon\,\mathrm{d}\omega^\varepsilon$$
$$= \int_{\omega^\varepsilon} f_3^\varepsilon(x_3^\varepsilon)\varphi^\varepsilon\,\mathrm{d}\omega^\varepsilon + \int_{\gamma^\varepsilon} g_3^\varepsilon(x_3^\varepsilon)\varphi^\varepsilon\,\mathrm{d}\gamma^\varepsilon$$
$$- \frac{1}{A^\varepsilon}\left[\int_{\omega^\varepsilon} f_3^\varepsilon(x_3^\varepsilon)\,\mathrm{d}\omega^\varepsilon + \int_{\gamma^\varepsilon} g_3^\varepsilon(x_3^\varepsilon)\,\mathrm{d}\gamma^\varepsilon\right]^\varepsilon \int_\omega \varphi^\varepsilon\,\mathrm{d}\omega^\varepsilon, \tag{12.7}$$
$$\textit{for all } \varphi^\varepsilon \in H^1(\omega^\varepsilon),$$
$$\int_{\omega^\varepsilon} w^{0\varepsilon}(x_3^\varepsilon)\,\mathrm{d}\omega^\varepsilon = 0.$$

(c) *The stretching component $\underline{u}_3^{2\varepsilon}$ depends only on x_3^ε and uniquely solves the following problem*:

$$\underline{u}_3^{2\varepsilon} \in H_0^1(0,L),$$
$$\int_0^L E^\varepsilon A^\varepsilon \partial_3^\varepsilon \underline{u}_3^{2\varepsilon}\partial_3^\varepsilon v\,\mathrm{d}x_3^\varepsilon = \int_0^L \nu^\varepsilon G_3^{2\varepsilon}\partial_3^\varepsilon v\,\mathrm{d}x_3^\varepsilon, \quad \textit{for all } v \in H_0^1(0,L), \tag{12.8}$$

where

$$G_3^{2\varepsilon} = -\tfrac{1}{2}E^\varepsilon(I_1^\varepsilon + I_2^\varepsilon)\partial_{333}^\varepsilon \underline{u}_3^{0\varepsilon} + E^\varepsilon H_\alpha^\varepsilon \partial_{3333}^\varepsilon u_\alpha^{0\varepsilon} - \int_{\omega^\varepsilon} x_\alpha^\varepsilon f_\alpha^\varepsilon\,\mathrm{d}\omega^\varepsilon - \int_{\gamma^\varepsilon} x_\alpha^\varepsilon g_\alpha^\varepsilon\,\mathrm{d}\gamma^\varepsilon$$
$$- \int_{\omega^\varepsilon} \tfrac{1}{2}[(x_1^\varepsilon)^2 + (x_2^\varepsilon)^2]\partial_3^\varepsilon f_3^\varepsilon\,\mathrm{d}\omega^\varepsilon - \int_{\gamma^\varepsilon} \tfrac{1}{2}[(x_1^\varepsilon)^2 + x_2^\varepsilon)^2]\partial_3^\varepsilon g_3^\varepsilon\,\mathrm{d}\gamma^\varepsilon. \tag{12.9}$$

(d) *The bending component $z_\alpha^{2\varepsilon}$ depends only on the variable x_3^ε and is the unique solution of the following variational problem (no sum on α, $\beta \neq \alpha$):*

$$z_\alpha^{2\varepsilon} \in H^2(0,L) \textit{ and for all } v \in H_0^2(0,L):$$

$$\int_0^L E^\varepsilon I_\alpha^\varepsilon \partial_{33}^\varepsilon z_\alpha^{2\varepsilon} \partial_{33}^\varepsilon v \,\mathrm{d}x_3^\varepsilon = \int_0^L M_\alpha^{2\varepsilon} \partial_{33}^\varepsilon v \,\mathrm{d}x_3^\varepsilon$$

$$z_\alpha^{2\varepsilon}(a) = -\frac{1}{2A^\varepsilon}\nu^\varepsilon (I_\alpha^\varepsilon - I_\beta^\varepsilon)\partial_{33}^\varepsilon u_\alpha^{0\varepsilon}(a) \text{ at } a = 0, L,$$

$$\partial_3^\varepsilon z_\alpha^{2\varepsilon}(a) = \frac{1}{I_\alpha^\varepsilon}\Big\{ \frac{2(1+\nu^\varepsilon)}{E^\varepsilon}\int_{\omega^\varepsilon} x_\alpha^\varepsilon w^{0\varepsilon}(a)\,\mathrm{d}\omega^\varepsilon - \tfrac{1}{2} I_\alpha^{w^\varepsilon}\partial_3^\varepsilon z^{2\varepsilon}(a) + \nu^\varepsilon H_\alpha^\varepsilon \partial_{33}^\varepsilon \underline{u}_3^{0\varepsilon}(a)$$
$$+[(1+\nu^\varepsilon)L_{\alpha\beta}^{\eta^\varepsilon} + \nu^\varepsilon L_{\alpha\beta}^{\theta^\varepsilon}]\partial_{333}^\varepsilon u_\beta^{0^\varepsilon}(a)\Big\} \quad \text{at } a = 0, L, \tag{12.10}$$

where

$$M_\alpha^{2\varepsilon} = E^\varepsilon\{(1+\nu^\varepsilon)L_{\alpha\beta}^{\eta^\varepsilon} + \nu^\varepsilon L_{\alpha\beta}^{\theta^\varepsilon} + \tfrac{1}{2}\nu^\varepsilon K_{\alpha\beta}^{\eta^\varepsilon}$$
$$+ \frac{(\nu^\varepsilon)^2}{2(1+\nu^\varepsilon)}(K_{\alpha\beta}^{\theta^\varepsilon} + H_3^\varepsilon \delta_{\alpha\beta})\}\partial_{3333}^\varepsilon u_\beta^{0\varepsilon}$$
$$+ \nu^\varepsilon E^\varepsilon H_\alpha^\varepsilon \partial_{333}^\varepsilon \underline{u}_3^{0\varepsilon} - \frac{E^\varepsilon}{2(1+\nu^\varepsilon)}[(1+\nu^\varepsilon)I_\alpha^{w^\varepsilon} + \nu^\varepsilon I_\alpha^{\Psi^\varepsilon}]\partial_{33}^\varepsilon z^{2\varepsilon}$$
$$- \int_\omega [(1+\nu^\varepsilon)\eta_\alpha^\varepsilon + \nu^\varepsilon\theta_\alpha^\varepsilon]\partial_3^\varepsilon f_3^\varepsilon \,\mathrm{d}\omega^\varepsilon - \int_{\gamma^\varepsilon}[(1+\nu^\varepsilon)\eta_\alpha^\varepsilon + \nu^\varepsilon\theta_\alpha^\varepsilon]\partial_3^\varepsilon g_3^\varepsilon \,\mathrm{d}\gamma^\varepsilon$$
$$+ \int_{\omega^\varepsilon} \nu^\varepsilon \Phi_{\alpha\beta}^\varepsilon f_\beta^\varepsilon \,\mathrm{d}\omega^\varepsilon + \int_{\gamma^\varepsilon} \nu^\varepsilon \Phi_{\alpha\beta}^\varepsilon g_\beta^\varepsilon \,\mathrm{d}\gamma^\varepsilon. \tag{12.11}$$

(ii) *The shear stress components* $\sigma_{3\beta}^{0\varepsilon}$ *are uniquely determined by*

$$\sigma_{31}^{0\varepsilon} = \frac{E^\varepsilon}{2(1+\nu^\varepsilon)}\{-\partial_2^\varepsilon \Psi^\varepsilon \partial_3^\varepsilon z^{2\varepsilon} + [(1+\nu^\varepsilon)\partial_1^\varepsilon \eta_\beta^\varepsilon + \nu^\varepsilon(\partial_1^\varepsilon\theta_\beta^\varepsilon + \Phi_{1\beta}^\varepsilon)]\partial_{333}^\varepsilon u_\beta^{0\varepsilon}\}$$
$$+ \partial_1^\varepsilon w^{0\varepsilon}, \tag{12.12}$$

$$\sigma_{32}^{0\varepsilon} = \frac{E^\varepsilon}{2(1+\nu^\varepsilon)}\{\partial_1^\varepsilon \Psi^\varepsilon \partial_3^\varepsilon z^{2\varepsilon} + [(1+\nu^\varepsilon)\partial_2^\varepsilon \eta_\beta^\varepsilon + \nu^\varepsilon(\partial_2^\varepsilon\theta_\beta^\varepsilon + \Phi_{2\beta}^\varepsilon)]\partial_{333}^\varepsilon u_\beta^{0\varepsilon}\}$$
$$+ \partial_2^\varepsilon w^{0\varepsilon}. \tag{12.13}$$

(iii) *The plane stress components* $\sigma_{\alpha\beta}^{0\varepsilon}$ *are given by*

$$\sigma_{\alpha\beta}^{0\varepsilon} = S_{\alpha\beta}^\varepsilon(\underline{\boldsymbol{u}}^{4\varepsilon}) + S_{\alpha\beta}^{0\varepsilon}, \tag{12.14}$$

where

$$S_{\alpha\beta}^{0\varepsilon} = \frac{\nu^\varepsilon E^\varepsilon}{(1+\nu^\varepsilon)(1-2\nu^\varepsilon)}\partial_3^\varepsilon(u_3^{2\varepsilon} - \underline{u}_3^{2\varepsilon} + x_\mu^\varepsilon \partial_3^\varepsilon z_\mu^{2\varepsilon})\delta_{\alpha\beta}, \tag{12.15}$$

$$S_{\alpha\beta}^\varepsilon(\underline{\boldsymbol{u}}^{4\varepsilon}) = \frac{E^\varepsilon}{(1+\nu^\varepsilon)}e_{\alpha\beta}^\varepsilon(\underline{\boldsymbol{u}}^{4\varepsilon}) + \frac{\nu^\varepsilon E^\varepsilon}{(1+\nu^\varepsilon)(1-2\nu^\varepsilon)}e_{\mu\mu}^\varepsilon(\underline{\boldsymbol{u}}^{4\varepsilon})\delta_{\alpha\beta}, \tag{12.16}$$

and $\underline{\boldsymbol{u}}^{4\varepsilon} = (\underline{u}_\alpha^{4\varepsilon})$ *is the unique solution of the following plane elasticity problem*:

$$\underline{\boldsymbol{u}}^{4\varepsilon} \in L^2[0, L; (H^1(\omega^\varepsilon))^2], \text{ and a.e. in } (0, L), \text{ for all } \boldsymbol{\varphi}^\varepsilon \in [H^1(\omega^\varepsilon)]^2:$$

$$\int_{\omega^\varepsilon} S_{\alpha\beta}^\varepsilon(\underline{\boldsymbol{u}}^{4\varepsilon}) e_{\alpha\beta}^\varepsilon(\boldsymbol{\varphi}^\varepsilon)\,\mathrm{d}\omega^\varepsilon = \int_{\omega^\varepsilon} f_\beta^\varepsilon \varphi_\beta^\varepsilon\,\mathrm{d}\omega^\varepsilon + \int_{\gamma^\varepsilon} g_\beta^\varepsilon \varphi_\beta^\varepsilon\,\mathrm{d}\gamma^\varepsilon + \int_{\omega^\varepsilon} \partial_3^\varepsilon \sigma_{3\beta}^{0\varepsilon} \varphi_\beta^\varepsilon\,\mathrm{d}\omega^\varepsilon - \int_{\omega^\varepsilon} S_{\alpha\beta}^{0\varepsilon} e_{\alpha\beta}^\varepsilon(\boldsymbol{\varphi}^\varepsilon)\, d\omega^\varepsilon,$$

$$\int_{\omega^\varepsilon} \underline{u}_\alpha^{4\varepsilon}\,\mathrm{d}\omega^\varepsilon = \int_{\omega^\varepsilon} [x_2^\varepsilon \underline{u}_1^{4\varepsilon} - x_1^\varepsilon \underline{u}_2^{4\varepsilon}]\,\mathrm{d}\omega^\varepsilon = 0. \tag{12.17}$$

(iv) *The axial stress component* $\sigma_{33}^{2\varepsilon}$ *is given by*

$$\begin{aligned}\sigma_{33}^{2\varepsilon} &= E^\varepsilon \partial_3^\varepsilon u_3^{2\varepsilon} + \nu^\varepsilon \sigma_{\mu\mu}^{0\varepsilon} \\ &= E^\varepsilon \{\partial_3^\varepsilon \underline{u}_3^{2\varepsilon} - x_\alpha^\varepsilon \partial_{33}^\varepsilon z_\alpha^{2\varepsilon} - w^\varepsilon \partial_{33}^\varepsilon z^{2\varepsilon} - \nu^\varepsilon [\tfrac{1}{2}[(x_1^\varepsilon)^2 + (x_2^\varepsilon)^2] \\ &\quad - \tfrac{1}{2}(I_1^\varepsilon + I_2^\varepsilon)] \partial_{333}^\varepsilon \underline{u}_3^{0\varepsilon} + [(1+\nu^\varepsilon)\eta_\alpha^\varepsilon + \nu^\varepsilon \theta_\alpha^\varepsilon] \partial_{3333}^\varepsilon u_\alpha^{0\varepsilon}\} \\ &\quad + 2(1+\nu^\varepsilon)\partial_3^\varepsilon w^{0\varepsilon} + \nu^\varepsilon \sigma_{\mu\mu}^{0\varepsilon}.\end{aligned} \tag{12.18}$$

(v) *The bending moment* $m_\beta^{2\varepsilon} = \int_{\omega^\varepsilon} x_\beta^\varepsilon \sigma_{33}^{2\varepsilon}\,\mathrm{d}\omega^\varepsilon$ *and the shear force* $q_\beta^{2\varepsilon} = \int_{\omega^\varepsilon} \sigma_{3\beta}^{2\varepsilon}\,\mathrm{d}\omega^\varepsilon$ *components are given by the following expressions:*

$$m_\beta^{2\varepsilon} = -E^\varepsilon I_\alpha^\varepsilon \partial_{33}^\varepsilon z_\beta^{2\varepsilon} + M_\beta^{2\varepsilon} \quad (\text{no sum on } \beta), \tag{12.19}$$

$$q_\beta^{2\varepsilon} = \partial_3^\varepsilon m_\beta^{2\varepsilon} = -E^\varepsilon I_\beta^\varepsilon \partial_{333}^\varepsilon z_\beta^{2\varepsilon} + \partial_3^\varepsilon M_\beta^{2\varepsilon} \quad (\text{no sum on } \beta). \tag{12.20}$$

The *other important conclusion* that one can draw from this characterization of the second order terms is (cf. TRABUCHO and VIAÑO [1987, 1988, 1989, 1990a,b]): *without any a priori hypotheses on the geometry or on the system of applied forces, the second order approximation given by the asymptotic expansion method allows us to formulate a general stretching-bending-torsion elastic beam theory*, justifying both the a priori hypotheses and the *basic equations* of the classical beam theories, namely: Saint Venant for simple torsion and with Poisson's effects; Timoshenko's bending theory and Vlassov's beam theory.

The *general second order asymptotic stretching–bending–torsion model consists in a second order approximation for the displacement field, axial stresses, bending moment and shear force components, and a first order approximation for shear and plane stress components*. Consequently, it is characterized by the fields $\hat{\boldsymbol{u}}^\varepsilon$, $\hat{\boldsymbol{\sigma}}^\varepsilon$, $\hat{\boldsymbol{m}}^\varepsilon$ and $\hat{\boldsymbol{q}}^\varepsilon$ defined in the following way:

$$\begin{aligned}&\hat{\boldsymbol{u}}^\varepsilon = \boldsymbol{u}^{0\varepsilon} + \boldsymbol{u}^{2\varepsilon}, &&\hat{\sigma}_{33}^\varepsilon = \sigma_{33}^{0\varepsilon} + \sigma_{33}^{2\varepsilon}, \\ &\hat{\sigma}_{3\beta}^\varepsilon = \sigma_{3\beta}^{0\varepsilon}, &&\hat{\sigma}_{\alpha\beta}^\varepsilon = \sigma_{\alpha\beta}^{0\varepsilon}, \\ &\hat{m}_\beta^\varepsilon = m_\beta^{0\varepsilon} + m_\beta^{2\varepsilon}, &&\hat{q}_\beta^\varepsilon = q_\beta^{0\varepsilon} + q_\beta^{2\varepsilon}.\end{aligned} \tag{12.21}$$

Defining the "main bending and stretching components" by $\underline{\hat{u}}_\alpha^\varepsilon$ and $\underline{\hat{u}}_3^\varepsilon$, respectively, with

$$\underline{\hat{u}}_\alpha^\varepsilon = u_\alpha^{0\varepsilon} + z_\alpha^{2\varepsilon}, \qquad \underline{\hat{u}}_3^\varepsilon = \underline{u}_3^{0\varepsilon} + \underline{u}_3^{2\varepsilon}, \tag{12.22}$$

the displacement components $\hat{u}_i^\varepsilon$ may be written in the following way:

$$\begin{aligned} \hat{u}_1^\varepsilon &= \underline{\hat{u}}_1^\varepsilon + x_2^\varepsilon z^{2\varepsilon} - \nu^\varepsilon(x_1^\varepsilon \partial_3^\varepsilon \underline{u}_3^{0\varepsilon} - \Phi_{1\beta}^\varepsilon \partial_{33}^\varepsilon u_\beta^{0\varepsilon}), \\ \hat{u}_2^\varepsilon &= \underline{\hat{u}}_2^\varepsilon - x_1^\varepsilon z^{2\varepsilon} - \nu^\varepsilon(x_2^\varepsilon \partial_3^\varepsilon \underline{u}_3^{0\varepsilon} - \Phi_{2\beta}^\varepsilon \partial_{33}^\varepsilon u_\beta^{0\varepsilon}), \end{aligned} \tag{12.23}$$

$$\begin{aligned} \hat{u}_3^\varepsilon &= \underline{\hat{u}}_3^\varepsilon - x_\alpha^\varepsilon \partial_3^\varepsilon \underline{\hat{u}}_\alpha^\varepsilon - w^\varepsilon \partial_3^\varepsilon z^{2\varepsilon} \\ &\quad + \nu^\varepsilon\{[\tfrac{1}{2}[(x_1^\varepsilon)^2 + (x_2^\varepsilon)^2] - \frac{1}{2A^\varepsilon}(I_1^\varepsilon + I_2^\varepsilon)]\}\partial_{33}^\varepsilon \underline{u}_3^{0\varepsilon} \\ &\quad + [(1+\nu^\varepsilon)\eta_\alpha^\varepsilon + \nu^\varepsilon \theta_\alpha^\varepsilon]\partial_{333}^\varepsilon u_\alpha^{0\varepsilon} + \frac{2(1+\nu^\varepsilon)}{E^\varepsilon} w^{0\varepsilon}. \end{aligned} \tag{12.24}$$

This form of the displacement field coincides with the a priori assumptions used in the most general classical beam theories which are thus justified and generalized. In fact, the only terms that are not present in the classical theories are the secondary warping $w^{0\varepsilon}$ and the stretching term $[(1+\nu^\varepsilon)\eta_\alpha^\varepsilon + \nu^\varepsilon\theta_\alpha^\varepsilon]\partial_{333}^\varepsilon u_\alpha^{0\varepsilon}$. If one omits these terms then the displacement field for the extension-bending-torsion theory of Saint Venant with Poisson's effects is found (cf. FRAEJIS DE VEUBEKE [1979, Eqs. (6.35) and (6.52)]). Poisson's effects are included in the terms multiplying ν^ε in (12.23), (12.24) and they are completely determined by the zeroth order displacement components $u_\alpha^{0\varepsilon}$ and $\underline{u}_3^{0\varepsilon}$. Consequently, they may be "added" in any theory that does not contain them per se (cf. GERMAIN [1972, Section VII-6, Eq. (22)]).

The remaining terms are more familiar. Those involving $\underline{\hat{u}}_i^\varepsilon$ are of the Bernoulli–Navier type and those involving $z^{2\varepsilon}$ represent displacements due to twist and warping of the cross section.

Later on we shall see that Vlassov's and Timoshenko's beam theories use *a priori* hypotheses leading to displacement fields slightly different (simpler) than those in (12.23), (12.24) (cf. VLASSOV [1962, Chapter X, Eq. (1.1)] and DYM and SHAMES [1973, Eq. (4.30)]). Nevertheless, a complete justification of these theories may be given (cf. Sections 13 and 15 and also TRABUCHO and VIAÑO [1990a,b]).

Another fact necessary to point out is that (12.21)–(12.24) *do not obey Hooke's law*. In fact, applying Hooke's law to the displacement field (12.23), (12.24) gives the following stress field:

$$\tilde{\sigma}_{\alpha\beta}^\varepsilon = \frac{\nu^\varepsilon E^\varepsilon}{(1+\nu^\varepsilon)(1-2\nu^\varepsilon)} \partial_3^\varepsilon u_3^{2\varepsilon} \delta_{\alpha\beta}, \tag{12.25}$$

$$\tilde{\sigma}_{3\beta}^\varepsilon = \sigma_{3\beta}^{0\varepsilon}, \tag{12.26}$$

$$\tilde{\sigma}_{33}^\varepsilon = \sigma_{33}^{0\varepsilon} + \frac{(1-\nu^\varepsilon)E^\varepsilon}{(1+\nu^\varepsilon)(1-2\nu^\varepsilon)} \partial_3^\varepsilon u_3^{2\varepsilon}. \tag{12.27}$$

A direct comparison of (12.25)–(12.27) with (8.34), (12.12), (12.13) and (12.18) gives the following relations:

$$\begin{aligned}
\hat{\sigma}^{\varepsilon}_{\alpha\beta} &= \sigma^{0\varepsilon}_{\alpha\beta} = \tilde{\sigma}^{\varepsilon}_{\alpha\beta} + S^{\varepsilon}_{\alpha\beta}(\boldsymbol{u}^{4\varepsilon}),\\
\hat{\sigma}^{\varepsilon}_{3\beta} &= \sigma^{0\varepsilon}_{3\beta} = \tilde{\sigma}^{\varepsilon}_{3\beta},\\
\hat{\sigma}^{\varepsilon}_{33} &= \sigma^{0\varepsilon}_{33} + \sigma^{2\varepsilon}_{33} = \tilde{\sigma}^{\varepsilon}_{33} + \nu^{\varepsilon} S^{\varepsilon}_{\mu\mu}(\boldsymbol{u}^{4\varepsilon}).
\end{aligned} \tag{12.28}$$

Taking into account (8.37) – (8.41) these expressions may be written in the following equivalent way:

$$\begin{aligned}
\hat{\sigma}^{\varepsilon}_{\alpha\beta} &= \sigma^{0\varepsilon}_{\alpha\beta} = \tilde{\sigma}^{\varepsilon}_{\alpha\beta} + \frac{\nu^{\varepsilon}E^{\varepsilon}}{(1+\nu^{\varepsilon})(1-2\nu^{\varepsilon})}(\partial^{\varepsilon}_{3}\underline{u}^{2\varepsilon}_{3} - x^{\varepsilon}_{\mu}\partial^{\varepsilon}_{33}z^{2\varepsilon}_{\mu})\delta_{\alpha\beta} + S^{\varepsilon}_{\alpha\beta}(\underline{\boldsymbol{u}}^{4\varepsilon}),\\
\hat{\sigma}^{\varepsilon}_{3\beta} &= \sigma^{0\varepsilon}_{3\beta} = \tilde{\sigma}^{\varepsilon}_{3\beta},\\
\hat{\sigma}^{\varepsilon}_{33} &= \sigma^{0\varepsilon}_{33} + \sigma^{2\varepsilon}_{33} = \tilde{\sigma}^{\varepsilon}_{33} + \frac{2E^{\varepsilon}(\nu^{\varepsilon})^2}{(1+\nu^{\varepsilon})(1-2\nu^{\varepsilon})}(\partial^{\varepsilon}_{3}\underline{u}^{2\varepsilon}_{3} - x^{\varepsilon}_{\mu}\partial^{\varepsilon}_{33}z^{2\varepsilon}_{\mu}) + \nu^{\varepsilon}S^{\varepsilon}_{\mu\mu}(\underline{\boldsymbol{u}}^{4\varepsilon}).
\end{aligned} \tag{12.29}$$

Equation (12.28) is extremely important, for it shows that *in the asymptotic model the shear stresses coincide with those obtained by applying Hooke's law to the displacement field* (12.23), (12.24) *and the plane stresses introduce an additional term based on the fourth order displacement components.* However, due to the difficulty in evaluating $\underline{\boldsymbol{u}}^{4\varepsilon}$ for the general case, in the next sections we shall consider just the term $S^{0\varepsilon}_{\alpha\beta}$ and $\tilde{\boldsymbol{\sigma}}^{\varepsilon}$ obtained applying Hooke's law to the displacement field.

We also remark that the term $[(1+\nu^{\varepsilon})\partial^{\varepsilon}_{\alpha}\eta^{\varepsilon}_{\beta} + \nu^{\varepsilon}\partial^{\varepsilon}_{\alpha}\theta^{\varepsilon}_{\beta}]\partial^{\varepsilon}_{333}u^{0\varepsilon}_{\beta}$ showing up in the expression for $\sigma^{\varepsilon}_{3\alpha}$ is a generalization of the corresponding term in the classical theory (cf. SOKOLNIKOFF [1956, Section 54]).

Our next objective is to show that although the equations obtained are of the form of those showing up in the classical engineering literature, they are in fact much more general. They include, for example, the thick- and thin-walled beam cases, the deformation of the cross section on its own plane, a generalization of Timoshenko's constants, a coupled Timoshenko's beam theory, and some other effects not present in the classical cases.

As in the engineering literature, we shall proceed in a formal way and consider just the equilibrium equations written in $(0, L)$. The corresponding boundary conditions can be obtained from the variational problems verified by the respective displacement components and are enumerated in Theorem 12.1.

In order to be able to interpret these variational problems in a strong way we now consider that the system of applied loads is such that

$$\begin{aligned}
&f^{\varepsilon}_{3} \in H^3[0,L;L^2(\omega^{\varepsilon})], \qquad g^{\varepsilon}_{3} \in H^3[0,L;L^2(\gamma^{\varepsilon})],\\
&f^{\varepsilon}_{\alpha} \in H^2[0,L;L^2(\omega^{\varepsilon})], \qquad g^{\varepsilon}_{\alpha} \in H^2[0,L;L^2(\gamma^{\varepsilon})].
\end{aligned}$$

Whenever convenient Eqs. (11.12) and (11.18) will be used in order to express $\partial^{\varepsilon}_{33}\underline{u}^{0\varepsilon}_{3}$ and $\partial^{\varepsilon}_{3333}u^{0\varepsilon}_{\alpha}$ as functions of the loading. Moreover, we introduce the following (engineering) notation for the loading dependent terms showing up in the previous theorem:

$$M_3^\varepsilon = \int_{\omega^\varepsilon} (x_2^\varepsilon f_1^\varepsilon - x_1^\varepsilon f_2^\varepsilon)\, \mathrm{d}\omega^\varepsilon + \int_{\gamma^\varepsilon} (x_2^\varepsilon f_1^\varepsilon - x_1^\varepsilon f_2^\varepsilon)\, \mathrm{d}\gamma^\varepsilon, \tag{12.30}$$

$$R^\varepsilon = \int_{\omega^\varepsilon} w^\varepsilon f_3^\varepsilon\, \mathrm{d}\omega^\varepsilon + \int_{\gamma^\varepsilon} w^\varepsilon g_3^\varepsilon\, \mathrm{d}\gamma^\varepsilon, \tag{12.31}$$

$$P^\varepsilon = \int_{\omega^\varepsilon} x_\beta^\varepsilon f_\beta^\varepsilon\, \mathrm{d}\omega^\varepsilon + \int_{\gamma^\varepsilon} x_\beta^\varepsilon g_\beta^\varepsilon\, \mathrm{d}\gamma^\varepsilon, \tag{12.32}$$

$$N_\alpha^\varepsilon = \int_{\omega^\varepsilon} \Phi_{\alpha\beta}^\varepsilon f_\beta^\varepsilon\, \mathrm{d}\omega^\varepsilon + \int_{\gamma^\varepsilon} \Phi_{\alpha\beta}^\varepsilon g_\beta^\varepsilon\, \mathrm{d}\gamma^\varepsilon, \tag{12.33}$$

$$Q^\varepsilon = \int_{\omega^\varepsilon} \tfrac{1}{2}[(x_1^\varepsilon)^2 + (x_2^\varepsilon)^2] f_3^\varepsilon\, \mathrm{d}\omega^\varepsilon + \int_{\gamma^\varepsilon} \tfrac{1}{2}[(x_1^\varepsilon)^2 + (x_2^\varepsilon)^2] g_3^\varepsilon\, \mathrm{d}\gamma^\varepsilon, \tag{12.34}$$

$$S_\alpha^{\eta^\varepsilon} = \int_{\omega^\varepsilon} \eta_\alpha^\varepsilon f_3^\varepsilon\, \mathrm{d}\omega^\varepsilon + \int_{\gamma^\varepsilon} \eta_\alpha^\varepsilon g_3^\varepsilon\, \mathrm{d}\gamma^\varepsilon,$$

$$S_\alpha^{\theta^\varepsilon} = \int_{\omega^\varepsilon} \theta_\alpha^\varepsilon f_3^\varepsilon\, \mathrm{d}\omega^\varepsilon + \int_{\gamma^\varepsilon} \theta_\alpha^\varepsilon g_3^\varepsilon\, \mathrm{d}\gamma^\varepsilon. \tag{12.35}$$

From Theorems 11.1 and 12.1 one concludes that the stretching $\underline{\hat{u}}_3^\varepsilon$, the bending components $\underline{\hat{u}}_\alpha^\varepsilon$ and the torsion angle $z^{2\varepsilon}$ of the second order displacement field (12.23), (12.24) satisfy the following general equations:

Stretching:

$$-E^\varepsilon A^\varepsilon \frac{\mathrm{d}^2\underline{\hat{u}}_3^\varepsilon}{\mathrm{d}x_3^2} = F_3^\varepsilon - \frac{\nu^\varepsilon(I_1^\varepsilon + I_2^\varepsilon)}{2A^\varepsilon}\frac{\mathrm{d}^2F_3^\varepsilon}{\mathrm{d}x_3^2} - \frac{\nu^\varepsilon H_\alpha^\varepsilon}{I_\alpha^\varepsilon}\Big(\frac{\mathrm{d}F_\alpha^\varepsilon}{\mathrm{d}x_3} + \frac{\mathrm{d}^2M_\alpha^\varepsilon}{\mathrm{d}x_3^2}\Big) + \nu^\varepsilon\Big(\frac{\mathrm{d}P^\varepsilon}{\mathrm{d}x_3} + \frac{\mathrm{d}^2Q^\varepsilon}{\mathrm{d}x_3^2}\Big). \tag{12.36}$$

Torsion angle:

$$-\frac{E^\varepsilon J^\varepsilon}{2(1+\nu^\varepsilon)}\frac{\mathrm{d}^2z^{2\varepsilon}}{\mathrm{d}x_3^2} = M_3^\varepsilon + \frac{\mathrm{d}R_3^\varepsilon}{\mathrm{d}x_3} - \frac{1}{2(1+\nu^\varepsilon)I_\alpha^\varepsilon}[(1+\nu^\varepsilon)I_\alpha^{w^\varepsilon} + \nu^\varepsilon I_\alpha^{\Psi^\varepsilon}]\left(F_\alpha^\varepsilon + \frac{\mathrm{d}M_\alpha^\varepsilon}{\mathrm{d}x_3}\right). \tag{12.37}$$

Bending:

$$E^\varepsilon I_\alpha^\varepsilon \frac{\mathrm{d}^4\underline{\hat{u}}_\alpha^\varepsilon}{\mathrm{d}x_3^4} = F_\alpha^\varepsilon + \frac{\mathrm{d}M_\alpha^\varepsilon}{\mathrm{d}x_3} - T_{\alpha\beta}^\varepsilon\Big(\frac{\mathrm{d}^2F_\beta^\varepsilon}{\mathrm{d}x_3^2} + \frac{\mathrm{d}^3M_\beta^\varepsilon}{\mathrm{d}x_3^3}\Big)$$

$$-\frac{1}{J^\varepsilon}[(1+\nu^\varepsilon)I_\alpha^{w^\varepsilon}+\nu^\varepsilon I_\alpha^{\Psi^\varepsilon}]\left(\frac{d^2M_3^\varepsilon}{dx_3^2}-\frac{d^3R^\varepsilon}{dx_3^3}\right)-\frac{\nu^\varepsilon H_\alpha^\varepsilon}{A^\varepsilon}\frac{d^3F_3^\varepsilon}{dx_3^3}$$
$$-\left[(1+\nu^\varepsilon)\frac{d^3S_\alpha^{\eta^\varepsilon}}{dx_3^3}+\nu^\varepsilon\frac{d^3S_\alpha^{\theta^\varepsilon}}{dx_3^3}\right]+\nu^\varepsilon\frac{d^2N_\alpha^\varepsilon}{dx_3^2},\quad \text{(no sum on } \alpha), \tag{12.38}$$

where

$$\begin{aligned}T_{\alpha\beta}^\varepsilon = &-\frac{1}{I_\beta^\varepsilon}\{(1+\nu^\varepsilon)L_{\alpha\beta}^{\eta^\varepsilon}+\nu^\varepsilon L_{\alpha\beta}^{\theta^\varepsilon}\\ &+\frac{\nu^\varepsilon}{2(1+\nu^\varepsilon)}[(1+\nu^\varepsilon)K_{\alpha\beta}^{\eta^\varepsilon}+\nu^\varepsilon K_{\alpha\beta}^{\theta^\varepsilon}+\nu^\varepsilon H_3^\varepsilon\delta_{\alpha\beta}]\\ &-\frac{1}{2(1+\nu^\varepsilon)J^\varepsilon}[(1+\nu^\varepsilon)I_\alpha^{w^\varepsilon}+\nu^\varepsilon I_\alpha^{\Psi^\varepsilon}][(1+\nu^\varepsilon)I_\beta^{w^\varepsilon}+\nu^\varepsilon I_\beta^{\Psi^\varepsilon}]\}.\end{aligned} \tag{12.39}$$

Constants $T_{\alpha\beta}^\varepsilon$, which depend only on the geometry of the cross section ω^ε and on Poisson's ratio ν^ε, constitute a generalization of the classical *Timoshenko constant* (cf. Section 15) and therefore matrix $\boldsymbol{T}^\varepsilon = (T_{\alpha\beta}^\varepsilon)$ is called *Timoshenko's matrix* (cf. TRABUCHO and VIAÑO [1990a]).

In Eqs. (12.36)–(12.38) a great number of mechanical effects are superimposed, and their relative importance depends on each specific situation. The consideration of the different effects is what makes some theories different from the others. For example, except for the first term on the right-hand side of Eq. (12.36), all the terms represent a contribution to the axial stretching of the beam from the loading and are due to Poisson's effects. If ν^ε is very small or if one adopts Navier's hypotheses then all these terms are neglected in the classical theories.

In a similar way, for the torsion angle equation (12.37) three different effects are considered: rotation due to the applied torsion moment (M_3^ε), rotation due to the (nonuniform) distribution of the axial loads (dR_ε/dx_3), along the cross section, and finally a rotation due to the fact that the resultant of the applied loads may not pass through the shear centre. In fact, it can be shown that the cooordinates of the shear centre $(\hat{x}_1^\varepsilon, \hat{x}_2^\varepsilon)$ for the planar case are given as

$$\begin{aligned}\hat{x}_1^\varepsilon &= -\frac{1}{2(1+\nu^\varepsilon)I_2^\varepsilon}[(1+\nu^\varepsilon)I_2^{w^\varepsilon}+\nu^\varepsilon I_2^{\Psi^\varepsilon}],\\ \hat{x}_2^\varepsilon &= \frac{1}{2(1+\nu^\varepsilon)I_1^\varepsilon}[(1+\nu^\varepsilon)I_1^{w^\varepsilon}+\nu^\varepsilon I_1^{\Psi^\varepsilon}],\end{aligned} \tag{12.40}$$

(cf. TRABUCHO and VIAÑO [1988], SOKOLNIKOFF [1956] or TIMOSHENKO AND GOODIER [1951]) and consequently, the third term in (12.37) represents the torsion moment originated by the eccentricity of the shear centre with respect to the centroid of the cross section.

From Remark 7.1 one concludes that if the cross section possesses two axes of symmetry or if it possesses only one axis of symmetry and if the resultant of the applied loads acts along that axis, then, this third term vanishes.

Finally, there are also several terms associated with Poisson's effect in Eq. (12.38). Moreover, the terms associated with $T^{\varepsilon}_{\alpha\beta}$ are related to a generalization of Timoshenko's beam theory and represent a contribution to bending due to the nonuniform distribution of the shear stresses on the cross section (cf. Section 15). The term $\mathrm{d}^2 M^{\varepsilon}_3/\mathrm{d}x_3^2 - \mathrm{d}R^{\varepsilon}/\mathrm{d}x_3$ is related to the additional bending caused by the applied moments, and springing from the fact that the shear centre may not coincide with the centroid of the cross section. This contribution disappears if, for instance, the axis under consideration is an axis of symmetry (cf. Remark 7.1).

REMARK 12.1. One should have in mind that it is always possible to consider a more precise asymptotic theory. It is enough to consider the fourth order terms evaluated in Theorem 9.1, for example.This model would be characterized by the following quantities:

$$\begin{aligned}
&\overline{\boldsymbol{u}}^{\varepsilon} = \boldsymbol{u}^{0\varepsilon} + \boldsymbol{u}^{2\varepsilon} + \boldsymbol{u}^{4\varepsilon}, \qquad \overline{\sigma}^{\varepsilon}_{33} = \sigma^{0\varepsilon}_{33} + \sigma^{2\varepsilon}_{33},\\
&\overline{\sigma}^{\varepsilon}_{3\beta} = \sigma^{0\varepsilon}_{3\beta}, \qquad \overline{\sigma}^{\varepsilon}_{\alpha\beta} = \sigma^{0\varepsilon}_{\alpha\beta},\\
&\overline{m}^{\varepsilon}_{\beta} = m^{0\varepsilon}_{\beta} + m^{2\varepsilon}_{\beta} + m^{4\varepsilon}_{\beta}, \qquad \overline{q}^{\varepsilon}_{\beta} = q^{0\varepsilon}_{\beta} + q^{2\varepsilon}_{\beta}.
\end{aligned} \tag{12.41}$$

However, the model that we actually propose here is the one given directly by the asymptotic expansion method in Theorem 12.1. The rearranging of the equations in the form (12.36)–(12.39) is done only in order to show that all the major classical theories are included in these second order terms, and this is also the reason why we shall not study (12.41) in detail.

In the next sections we analyze the most well-known classical theories and a comparative study with the general second order asymptotic model is done. This comparison will be established at two levels:

(i) the form of the displacement and of the stress fields with respect to (12.21)–(12.24),

(ii) the form of the respective stretching, bending and torsion equations with respect to the asymptotic second order approximation equations (12.36)–(12.39).

REMARK 12.2. Without any loss of generality and in order to simplify the notation this comparative analysis will be performed for the reference cross section, i.e., $\varepsilon = 1$.

REMARK 12.3. A numerical comparison between the second order asymptotic model (solved by two- and one-dimensional finite elements) and the three-dimensional model (solved by three-dimensional finite element methods) has just been initiated by FERNANDEZ and VIAÑO [1992].

13. Comparison with Vlassov's beam theory for thick rods

The classical beam theory formulated by B.Z. Vlassov around 1940 is one of the most important for the study of combined effects of bending and torsion. It is especially used in the engineering literature for the calculation of aircraft and aerospace structures. We shall point out the major features of this theory, for rods with relatively thick cross sections, and compare with the results obtained from the asymptotic expansion method. The major reference for Vlassov's theory is the monograph VLASSOV [1962, Chapter X].

For a system of applied forces f_i^ε and g_i^ε, the a priori hypotheses, on the displacement field in Vlassov's beam theory are (cf. (12.23)–(12.24))

$$\begin{aligned}
\tilde{u}_1(x_1,x_2,x_3) &= \underline{\tilde{u}}_1(x_3) + x_2 z(x_3),\\
\tilde{u}_2(x_1,x_2,x_3) &= \underline{\tilde{u}}_2(x_3) - x_1 z(x_3),\\
\tilde{u}_3(x_1,x_2,x_3) &= \underline{\tilde{u}}_3(x_3) - x_\alpha v_\alpha(x_3) - w(x_1,x_2)v_3(x_3),
\end{aligned} \tag{13.1}$$

where the unknown functions $\underline{\tilde{u}}_i$, v_i and z depend only on variable x_3.

In his work Vlassov uses, for any cross section, the following approximate expression for the warping function:

$$w(x_1,x_2) \approx x_1x_2 - \frac{J_1}{I_1}x_1 - \frac{J_2}{I_2}x_2, \tag{13.2}$$

where

$$J_\alpha = \int_\omega x_\alpha^2 x_\beta \,\mathrm{d}\omega \quad (\alpha \neq \beta). \tag{13.3}$$

This fact accounts for some simplifications which forbid a direct comparison with the results obtained from the asymptotic expansion method. Since this is not a restriction in deriving Vlassov's theory, we have rederived Vlassov's results but with the warping function w as defined in (7.23). This is the reason why the equations we are going to write for Vlassov's beam theory do not coincide with Vlassov's own results, nevertheless, they reduce to Vlassov's equations (cf. VLASSOV [1962, Chapter X, Eqs. (1.18)]) whenever assumption (13.2) is used.

These equations, which were derived by TRABUCHO and VIAÑO [1990b] (see also RODRÍGUEZ [1990]) are as follows:

Stretching:

$$-EA\frac{\mathrm{d}^2\underline{\tilde{u}}_3}{\mathrm{d}x_3^2} = F_3, \tag{13.4}$$

Torsion angle:

$$EC_w^0\frac{\mathrm{d}^4 z}{\mathrm{d}x_3^4} - \rho\mu J\frac{\mathrm{d}^2 z}{\mathrm{d}x_3^2} = \rho M_3 - \frac{EC_w^0}{\mu(I_1+I_2)}\frac{\mathrm{d}^2 M_3}{\mathrm{d}x_3^2} + \rho\left[\frac{\mathrm{d}R}{\mathrm{d}x_3} - \frac{I_1^w}{2I_1}\left(F_1 + \frac{\mathrm{d}M_1}{\mathrm{d}x_3}\right) - \frac{I_2^w}{2I_2}\left(F_2 + \frac{\mathrm{d}M_2}{\mathrm{d}x_3}\right)\right]. \tag{13.5}$$

Bending:

$$EI_\alpha \frac{\mathrm{d}^4\tilde{\underline{u}}_\alpha}{\mathrm{d}x_3^4} + \frac{EI_\alpha^w}{2\rho}\frac{\mathrm{d}^4 z}{\mathrm{d}x_3^4} = F_\alpha + \frac{\mathrm{d}M_\alpha}{\mathrm{d}x_3} - \frac{2(1+\nu)I_\alpha}{A}\frac{\mathrm{d}^2F_\alpha}{\mathrm{d}x_3^2} - \frac{EI_\alpha^w}{2\mu\rho(I_1+I_2)}\frac{\mathrm{d}^2M_3}{\mathrm{d}x_3^2} \quad \text{(no sum on } \alpha\text{)}, \tag{13.6}$$

where the following classical constants were used:

$$\rho = 1 - \frac{J}{(I_1+I_2)}, \qquad \mu = \frac{E}{2(1+\nu)}, \tag{13.7}$$

$$C_w^0 = J_w - \frac{(I_1^w)^2}{4I_1} - \frac{(I_2^w)^2}{4I_2}. \tag{13.8}$$

Constant C_w^0 is called the *warping constant.*

REMARK 13.1. The above equations are not valid for $\rho = 0$. This is the case if (and only if) ω is a circle (see (7.36)). In this case, $w = 0$, $I_\alpha^w = 0$, $R = 0$, and the equations reduce to the following:

$$\begin{aligned} &-EA\frac{\mathrm{d}^2\tilde{\underline{u}}_3}{\mathrm{d}x_3^2} = F_3,\\ &\mu J\frac{\mathrm{d}^2 z}{\mathrm{d}x_3^2} = M_3,\\ &EI_\alpha\frac{\mathrm{d}^4\tilde{\underline{u}}_\alpha}{\mathrm{d}x_3^4} = F_\alpha + \frac{\mathrm{d}M_\alpha}{\mathrm{d}x_3} - \frac{2(1+\nu)I_\alpha}{A}\frac{\mathrm{d}^2F_\alpha}{\mathrm{d}x_3^2}. \end{aligned} \tag{13.9}$$

From a direct comparison between Eqs. (13.4) and (13.6) with (12.36) and (12.38) we conclude that Vlassov's model does not include some stretching and bending effects that are present in the asymptotic model.

The fact that we have a fourth order equation for the angle of torsion is a consequence of the a priori hypotheses, which, although not including the deformation of the cross section on its own plane, are not of the Bernoulli–Navier type for the stretching and bending components. The elimination of functions v_α in the equations introduces this fourth order term (cf. VLASSOV [1962, p. 570]).

This is not the case for the displacement field of the asymptotic expansion model, which is of the Bernoulli–Navier type for the stretching component. Consequently, if we wish to compare both models we must proceed in Eq. (12.37) for the torsion angle as in Vlassov's beam theory. Moreover, since in the classical theory constant C_w^0, which includes J_w, plays an important role, inspired by Eq. (9.6) (which involves the fourth derivative of z^2) we introduce the following *generalized warping constant*, (cf. TRABUCHO and VIAÑO [1990b]), including both the primary and secondary warping:

$$C_w^\nu = J_w - \frac{[(1+\nu)I_1^w + \nu I_1^\Psi]^2}{4(1+\nu)^2 I_1} - \frac{[(1+\nu)I_2^w + \nu I_2^\Psi]^2}{4(1+\nu)^2 I_2}. \tag{13.10}$$

We remark that for $\nu = 0$ constant C_w^ν coincides with C_w^0 defined in (13.8), which justifies the notation. A simple formal derivation of Eq. (12.37) shows that the angle of torsion of the asymptotic theory satisfies

$$\begin{aligned} &EC_w^\nu \frac{\mathrm{d}^4 z^2}{\mathrm{d}x_3^4} - \rho\mu J \frac{\mathrm{d}^2 z^2}{\mathrm{d}x_3^2} \\ &= \rho M_3 - \frac{EC_w^\nu}{\mu J}\left(\frac{\mathrm{d}^2 M_3}{\mathrm{d}x_3^2} + \frac{\mathrm{d}^3 R}{\mathrm{d}x_3^3}\right) \\ &+ \rho\left[\frac{\mathrm{d}R}{\mathrm{d}x_3} - \frac{(1+\nu)I_\alpha^w + \nu I_\alpha^\Psi}{2(1+\nu)I_\alpha}\left(F_\alpha + \frac{\mathrm{d}M_\alpha}{\mathrm{d}x_3}\right)\right] \\ &+ \frac{C_w^\nu[(1+\nu)I_\alpha^w + \nu I_\alpha^\Psi]}{JI_\alpha}\left(\frac{\mathrm{d}^2 F_\alpha}{\mathrm{d}x_3^2} + \frac{\mathrm{d}^3 M_\alpha}{\mathrm{d}x_3^3}\right). \end{aligned} \tag{13.11}$$

The similarity between Eq. (13.11) with Vlassov's equation for the angle of torsion (13.5) is now clear, especially when $\nu \to 0$.

REMARK 13.2. Vlassov's theory uses Hooke's law for the calculation of the stress field. Consequently, the differences between the displacement fields of both models are those already exposed in (12.25) and (12.28) (cf. VLASSOV [1962, Chapter X, Eqs. (1.4)–(1.7)]).

REMARK 13.3. Besides Vlassov's beam theory, several other models including also the combined effects of stretching, bending and torsion are obtained, using, for example, Saint Venant's semi inverse method (cf. FRAEJIS DE VEUBEKE [1979, Chapter 6]). The comparison with the asymptotic model can be done with similar arguments.

In the following sections we shall compare the general asymptotic model with Saint Venant's and Timoshenko's classical theories which do not include these combined effects.

14. Comparison with Saint Venant's pure torsion theory

Saint Venant's pure torsion theory considers the case of a beam clamped at one extremity and subjected to a torsion moment at the other end (cf. SOKOLNIKOFF [1956, Section 33], NEČAS and HLAVÁČEK [1981, p. 225]). This simple case has been compared with the asymptotic model in TRABUCHO and VIAÑO [1988] and in ÁLVAREZ-DIOS and VIAÑO [1991a,b] for the anisotropic case, which is also considered in Sections 24 and 25.

In order to illustrate the torsion effects in the asymptotic model for a clamped beam, we consider the beam subjected only to a (constant) bending moment M_3 distributed in the volume and along the lateral surface. This situation is also considered in SOKOLNIKOFF [1956, Section 64] and from a classical point of view it corresponds to having

$$\begin{aligned} & f_3 = 0, \qquad g_3 = 0, \\ & f_\beta(x_1, x_2, x_3) = f_\beta(x_1, x_2), \qquad g_\beta(x_1, x_2, x_3) = g_\beta(x_1, x_2), \\ & F_\beta = \int_\omega f_\beta \,\mathrm{d}\omega + \int_\gamma g_\beta \,\mathrm{d}\gamma = 0, \\ & M_3 = \int_\omega (x_2 f_1 - x_1 f_2)\,\mathrm{d}\omega + \int_\gamma (x_2 g_1 - x_1 g_2)\,\mathrm{d}\gamma, \\ & P = \int_\omega x_\beta f_\beta \,\mathrm{d}\omega + \int_\gamma x_\beta g_\beta \,\mathrm{d}\gamma. \end{aligned} \tag{14.1}$$

From the asymptotic model (12.21)–(12.24) and taking into account Theorem 12.1, we obtain for this case $u_\alpha^0 = 0$, $\underline{u}_3^0 = 0$, $\sigma_{33}^0 = 0$, and consequently

$$\begin{aligned} & \hat{u}_1 = z_1^2 + x_2 z^2, \\ & \hat{u}_2 = z_2^2 - x_1 z^2, \\ & \hat{u}_3 = \underline{u}_3^2 - x_\alpha \partial_3 z_\alpha^2 - w \partial_3 z^2, \end{aligned} \tag{14.2}$$

where functions z_α^2, $\underline{u}_3^2$ and z^2 are the unique solutions of the following problems:

$$\begin{aligned} & -\frac{EJ}{2(1+\nu)} \frac{\mathrm{d}^2 z^2}{\mathrm{d}x_3^2} = M_3, \quad \text{in } (0, L), \\ & z^2(0) = z^2(L) = 0, \end{aligned} \tag{14.3}$$

$$\begin{aligned} & -EA \frac{\mathrm{d}^2 \underline{u}_3^2}{\mathrm{d}x_3^2} = \nu P, \quad \text{in } (0, L), \\ & \underline{u}_3^2(0) = \underline{u}_3^2(L) = 0, \end{aligned} \tag{14.4}$$

$$\begin{aligned} & EI_\alpha \frac{\mathrm{d}^4 z_\alpha^2}{\mathrm{d}x_3^4} = -\frac{E}{2(1+\nu)} [(1+\nu) I_\alpha^w + \nu I_\alpha^\Psi] \frac{\mathrm{d}^2 z^2}{\mathrm{d}x_3^2} + \nu \frac{\mathrm{d}^2 N_\alpha^2}{\mathrm{d}x_3^2}, \quad \text{in } (0, L), \\ & z_\alpha^2(0) = z_\alpha^2(L) = 0, \\ & \frac{\mathrm{d}z_\alpha^2}{\mathrm{d}x_3}(a) = \tfrac{1}{2} I_\alpha^w \frac{\mathrm{d}z^2}{\mathrm{d}x_3}(a), \quad (a = 0, L). \end{aligned} \tag{14.5}$$

Moreover, for the stress components we have (cf. (12.25)–(12.29)):

$$\hat{\sigma}_{31} = \sigma^0_{31} = \frac{E}{2(1+\nu)}(-\partial_1 w + x_2)\partial_3 z^2,$$
$$\hat{\sigma}_{32} = \sigma^0_{32} = \frac{E}{2(1+\nu)}(-\partial_2 w - x_1)\partial_3 z^2, \tag{14.6}$$
$$\hat{\sigma}_{\alpha\beta} = \sigma^0_{\alpha\beta} = \frac{\nu E}{(1+\nu)(1-2\nu)}\partial_3 \hat{u}_3 \delta_{\alpha\beta} + \frac{\nu E}{(1+\nu)(1-2\nu)}(\partial_3 \underline{u}^2_3 - x_\mu \partial_{33} z^2_\mu)\delta_{\alpha\beta} + S_{\alpha\beta}(\underline{\boldsymbol{u}}^4), \tag{14.7}$$
$$\hat{\sigma}_{33} = \sigma^0_{33} + \sigma^2_{33} = \frac{(1-\nu)E}{(1+\nu)(1-2\nu)}\partial_3 \hat{u}_3 + \frac{2E\nu^2}{(1+\nu)(1-2\nu)}(\partial_3 \underline{u}^2_3 - x_\mu \partial_{33} z^2_\mu) + \nu S_{\mu\mu}(\underline{\boldsymbol{u}}^4). \tag{14.8}$$

In Eqs. (14.4) and (14.5) some Poisson's effects are included both in stretching and bending through terms νP and $\nu(\mathrm{d}^2 N^2_\alpha/\mathrm{d}x^2_3)$ respectively, and also an additional bending effect from the fact that the twist centre may not coincide with the centroid of the cross section. In the present case, since M_3 is considered constant, all these terms vanish, which does not imply that $z^2_\alpha = 0$ due to the boundary condition in (14.5). Nevertheless, since Poisson's effects are neglected in the classical theory of Saint Venant we consider only one part of the displacement field (or equivalently set $\underline{u}^2_3 = 0$, $z^2_\alpha = 0$) and obtain

$$\bar{u}_1 = x_2 z^2, \qquad \bar{u}_2 = -x_1 z^2, \qquad \bar{u}_3 = -w\partial_3 z^2. \tag{14.9}$$

We remark that in the particular case where ω has two axes of symmetry and $P = 0$, $\mathrm{d}^2 N^2_\alpha/\mathrm{d}x^2_3 = 0$, one has $\bar{u}_\alpha = \hat{u}_\alpha$.

Solving (14.3), one has

$$z^2(x_3) = -\frac{(1+\nu)M_3}{EJ}x_3(x_3 - L),$$
$$\bar{u}_1(x_1, x_2, x_3) = -\frac{(1+\nu)M_3}{EJ}x_2 x_3(x_3 - L),$$
$$\bar{u}_2(x_1, x_2, x_3) = \frac{(1+\nu)M_3}{EJ}x_1 x_3(x_3 - L), \tag{14.10}$$
$$\bar{u}_3(x_1, x_2, x_3) = \frac{(1+\nu)M_3}{EJ}w(x_1, x_2)(2x_3 - L).$$

In a similar way, solving (14.6)–(14.8), with $\hat{\boldsymbol{u}}$ replaced by $\overline{\boldsymbol{u}}$, gives

$$
\begin{aligned}
\overline{\sigma}_{31}(x_1,x_2,x_3) &= \frac{M_3}{2J}[\partial_1 w(x_1,x_2) - x_2](2x_3 - L),\\
\overline{\sigma}_{32}(x_1,x_2,x_3) &= \frac{M_3}{2J}[\partial_2 w(x_1,x_2) + x_1](2x_3 - L),\\
\overline{\sigma}_{\alpha\beta}(x_1,x_2,x_3) &= \frac{2\nu M_3}{(1-2\nu)J} w(x_1,x_2)\delta_{\alpha\beta} + S_{\alpha\beta}(\underline{\boldsymbol{u}}^4),\\
\overline{\sigma}_{33}(x_1,x_2,x_3) &= \frac{2(1-\nu)M_3}{(1-2\nu)J} + \nu S_{\mu\mu}(\underline{\boldsymbol{u}}^4),
\end{aligned}
\tag{14.11}
$$

where

$$
\begin{aligned}
&\underline{\boldsymbol{u}}^4 = (\underline{u}^4_\alpha) \in [H^1(\omega)]^2,\\
&\int_\omega [\frac{E}{1+\nu} e_{\alpha\beta}(\underline{\boldsymbol{u}}^4) + \frac{\nu E}{(1+\nu)(1-2\nu)} e_{\mu\mu}(\underline{\boldsymbol{u}}^4)\delta_{\alpha\beta}] e_{\alpha\beta}(\boldsymbol{\varphi})\,\mathrm{d}\omega\\
&\quad = \int_\omega f_\beta \varphi_\beta \,\mathrm{d}\omega + \int_\gamma g_\beta \varphi_\beta \,\mathrm{d}\gamma - \frac{2\nu M_3}{(1-2\nu)J}\int_\omega w(\partial_1\varphi_1 + \partial_2\varphi_2)\,\mathrm{d}\omega\\
&\quad + \frac{M_3}{J}\int_\omega [(\partial_1 w - x_2)\varphi_1 + (\partial_2 w + x_1)\varphi_2]\,\mathrm{d}\omega, \quad \text{for all } \boldsymbol{\varphi} = (\varphi_\alpha) \in [H^1(\omega)]^2.\\
&\int_\omega \underline{u}^4_\alpha \,\mathrm{d}\omega = \int_\omega (x_2 \underline{u}^4_1 - x_1 \underline{u}^4_2)\,\mathrm{d}\omega = 0.
\end{aligned}
\tag{14.12}
$$

Expressions (14.10) and (14.11) are a generalization of the corresponding equations in SOKOLNIKOFF [1956, Eqs. (64.6)–(64.10)].

REMARK 14.1. The case where M_3 is a function of x_3 can be considered in a similar way. The only difficulty might be the solution of (14.3)–(14.5), which can be done numerically. The classical treatment of these cases is of a somewhat greater complexity (cf. FRAEJIS DE VEUBEKE [1979, Section 6.17]).

15. Comparison with Timoshenko's beam theory. New Timoshenko constants

The theory formulated by TIMOSHENKO [1921] provides a simple way to take into account the additional contribution to bending deformations due to the nonuniform shear stress distribution along the cross section. This effect, which is not included in the classical theory of Bernoulli–Navier, cannot be neglected for relatively short beams with relatively large transverse cross sections. Moreover, these stresses are also involved in the main mechanism associated with delamination in multilayered structures.

We shall now summarize Timoshenko's beam theory following DYM and SHAMES [1973] and FUNG [1965]. For the sake of simplicity and since Timoshenko's theory is only concerned with bending effects, we assume that the system of applied forces satisfies the following properties:

$$f_3 = 0, \qquad g_3 = 0, \qquad M_3 = 0, \tag{15.1}$$

and, consequently, the only loading effects are those due to the linear transverse force densities F_β. In this case, the kinematic a priori hypotheses associated with Timoshenko's beam theory are (cf. DYM and SHAMES [1973, Section 4.5]):

(i) The transverse displacements depend only on x_3, that is,

$$\tilde{u}_\alpha(x_1, x_2, x_3) = \hat{u}_\alpha(x_3). \tag{15.2}$$

(ii) The axial displacement u_3 is of the form

$$\tilde{u}_3(x_1, x_2, x_3) = -x_\alpha(\partial_3 \hat{u}_\alpha - \hat{v}_\alpha), \tag{15.3}$$

where $\hat{v}_\alpha$ is a function of x_3 only, which must be determined.

(iii) The shear stress components $\sigma_{3\alpha}$ are given by

$$\tilde{\sigma}_{3\alpha}(x_1, x_2, x_3) = \tilde{\sigma}_{3\alpha}(x_3) = \frac{Ek}{2(1+\nu)} \hat{v}_\alpha, \tag{15.4}$$

where k is Timoshenko's constant, depending on the material the beam is made of and on the shape of the cross section. From (i)–(iii) the Navier–Cauchy equilibrium equations for Timoshenko's beam theory become (cf. DYM and SHAMES [1973, p. 191])

$$\hat{\beta}_\alpha = \partial_3 \hat{u}_\alpha - \hat{v}_\alpha, \tag{15.5}$$

$$EI_\alpha \partial_{33} \hat{\beta}_\alpha + \frac{EkA}{2(1+\nu)}(\partial_3 \hat{u}_\alpha - \hat{\beta}_\alpha) = 0 \quad (\text{no sum on } \alpha) \tag{15.6}$$

$$\frac{EkA}{2(1+\nu)} \partial_3(\partial_3 \hat{u}_\alpha - \hat{\beta}_\alpha) = -F_\alpha. \tag{15.7}$$

For the case of a cantilevered beam, for example, we must add the boundary conditions:

$$\hat{\beta}_\alpha(a) = 0, \qquad \hat{u}_\alpha(a) = 0, \qquad \text{at } a = 0, L. \tag{15.8}$$

Differentiating with respect to x_3 in (15.6) and (15.7) we are able to decouple the system and obtain the classical equations of Timoshenko's beam theory:

$$EI_\alpha \partial_{333} \hat{\beta}_\alpha = F_\alpha \quad (\text{no sum on } \alpha) \tag{15.9}$$

$$EI_\alpha \partial_{3333} \hat{u}_\alpha = F_\alpha - \frac{2(1+\nu)I_\alpha}{kA} \partial_{33} F_\alpha, \quad (\text{no sum on } \alpha) \tag{15.10}$$

which must be completed with the corresponding boundary conditions.

Several shortcomings of this theory must be mentioned. For example, from (15.2), (15.3) and Hooke's law we obtain

$$\tilde{\sigma}_{3\alpha} = \frac{E}{2(1+\nu)}\hat{v}_\alpha,$$

which does not agree with (15.4) and besides is constant in each cross section. Consequently, although the displacement field associated with Timoshenko's beam theory already includes the additional bending deformation due to the shear stress distribution, the shear stress itself is not correctly determined. This is due to the introduction of the factor k in order to account for the nonuniform shear stress distribution along a cross section of the beam, while still retaining the one-dimensional approach. Moreover, it is not clear how this factor should be calculated, either. TIMOSHENKO [1921] stated that constant k should depend on the shape of the cross section and proposed $k = \frac{2}{3}$ for the rectangular case. MINDLIN [1951] suggests that this value can be selected in such a way that the solution of (15.6) agrees with certain exact solutions of the three-dimensional equations. *Most of the definitions make it a function of the shape of the cross section and of Poisson's ratio.*

In order to illustrate this dependence we reproduce in Table 15.1 a list of values of k, which were taken from DYM and SHAMES [1973].

We also remark that these constants are used independently of the loading direction and do not take possible coupled bending effects into account. Moreover, for most of the cases the indicated constants are used regardless of the absolute and relative dimensions of the cross section.

Another major drawback of Timoshenko's theory resides in the fact that even when no surface loads are applied on a part of Γ we always have a shear stress contribution given by $\sigma_{3\alpha}n_\alpha$, in contradiction with the equilibrium equations of elasticity.

From the general asymptotic model (12.21) TRABUCHO and VIAÑO [1990a] proposed a generalization of this theory which gives a correct interpretation of all these phenomena. In this section we present a summary of the main results.

Whenever the system of applied forces satisfies conditions (15.1), in the general asymptotic model one has $\underline{u}_3^0 = 0$ and the following equations for u_α^0, z^2, z_α^2 and $\underline{u}_3^2$ are obtained (cf. Theorem 12.1 and Eqs. (12.36)–(12.39)):

$$\begin{aligned} &EI_\alpha \frac{\mathrm{d}^4 u_\alpha^0}{\mathrm{d}x_3^4} = F_\alpha, \quad \text{in } (0,L), \\ &u_\alpha^0(0) = u_\alpha^0(L) = 0, \quad \frac{\mathrm{d}u_\alpha^0}{\mathrm{d}x_3}(0) = \frac{\mathrm{d}u_\alpha^0}{\mathrm{d}x_3}(L) = 0, \end{aligned} \tag{15.11}$$

$$\begin{aligned} &-\frac{EJ}{2(1+\nu)}\frac{\mathrm{d}^2 z^2}{\mathrm{d}x_3^2} = -\frac{1}{2(1+\nu)I_\alpha}[(1+\nu)I_\alpha^w + \nu I_\alpha^\Psi]F_\alpha, \quad \text{in } (0,L), \\ &z^2(a) = \frac{\nu}{I_1+I_2}[H_2\partial_{33}u_1^0(a) - H_1\partial_{33}u_2^0(a)], \quad a = 0, L, \end{aligned} \tag{15.12}$$

TABLE 15.1. Classical Timoshenko's constants for solid and hollow cross sections.

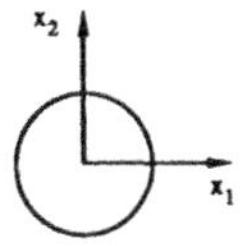

Circle $\frac{6(1+\nu)}{7+6\nu}$

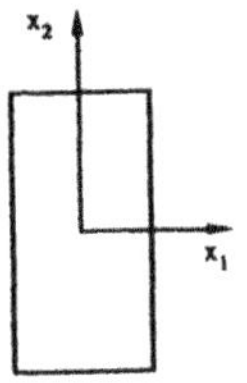

Rectangle $\frac{10(1+\nu)}{12+11\nu}$

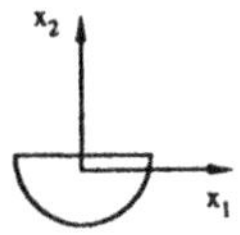

Half circle $\frac{(1+\nu)}{1.305+1.273\nu}$

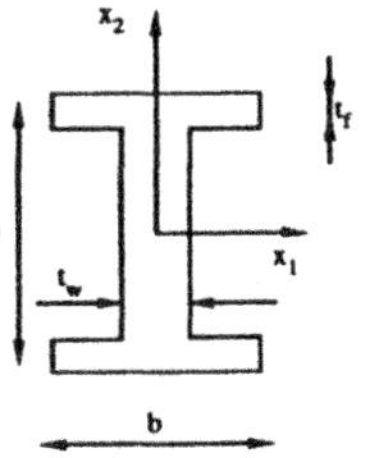

I shaped beam $\frac{10(1+\nu)c_1}{c_2\nu+c_3}$

$c_1=(1+3m)^2$

$c_2=11+66m+135m^2+90m^3+5n^2(8m+9m^2)$

$c_3=12+72m+150m^2+90m^3+30n^2(m+m^2)$

$m=2bt_f/ht_w, \quad n=b/h$

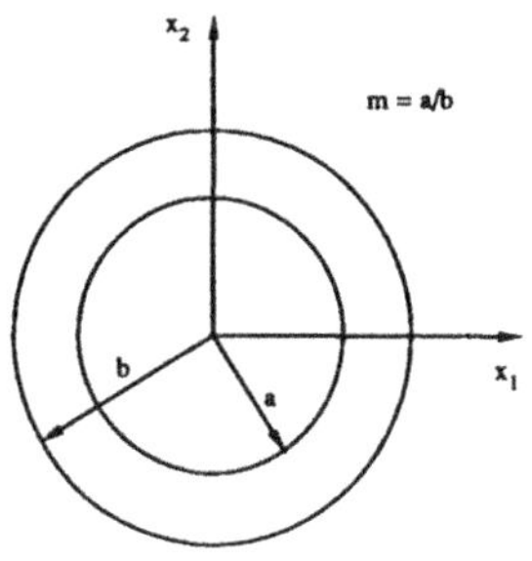

Hollow cylinder $\frac{6(1+\nu)(1+m)^2}{(7+6\nu)(1+m)^2+(20+12\nu)m^2}$

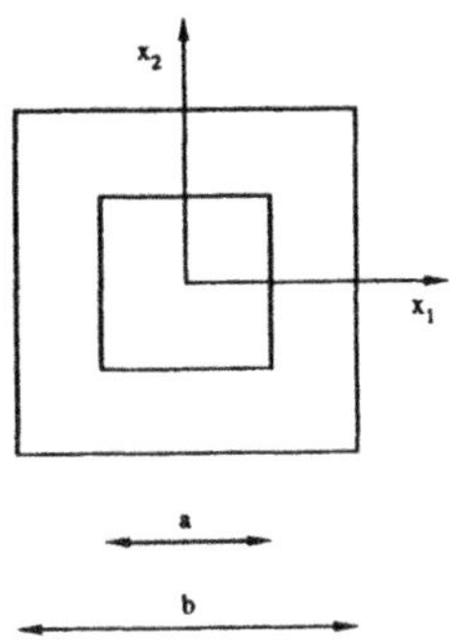

Thin-walled square tube $\frac{20(1+\nu)}{48+39\nu}$

Thin-walled tube $\frac{2(1+\nu)}{4+\nu}$

$$
\begin{aligned}
&-EA\frac{\mathrm{d}^2\underline{u}_3^2}{\mathrm{d}x_3^2} = -\frac{\nu H_\alpha}{I_\alpha}\frac{\mathrm{d}F_\alpha}{\mathrm{d}x_3} + \nu\frac{\mathrm{d}P}{\mathrm{d}x_3}, \quad \text{in } (0,L),\\
&\underline{u}_3^2(0) = \underline{u}_3^2(L) = 0,
\end{aligned} \tag{15.13}
$$

$$
\begin{aligned}
&EI_\alpha\frac{\mathrm{d}^4 z_\alpha^2}{\mathrm{d}x_3^4} = -T_{\alpha\beta}\frac{\mathrm{d}^2F_\beta}{\mathrm{d}x_3^2} + \nu\frac{\mathrm{d}^2N_\alpha}{\mathrm{d}x_3^2}, \quad \text{in } (0,L),\\
&z_\alpha^2(a) = -\frac{\nu}{2A}(I_\alpha - I_\beta)\partial_{33}u_\alpha^0(a), \quad a = 0, L, \quad (\alpha \neq \beta),\\
&\partial_3 z_\alpha^2(a) = \frac{1}{I_\alpha}\{-\tfrac{1}{2}I_\alpha^w\partial_3 z^2(a) + [(1+\nu)L_{\alpha\beta}^\eta + \nu L_{\alpha\beta}^\theta]\partial_{333}u_\beta^0(a)\}, \quad a = 0, L.
\end{aligned} \tag{15.14}
$$

Moreover, for the tangential stresses we obtain the following expressions:

$$
\begin{aligned}
\sigma_{31}^0 &= \frac{E}{2(1+\nu)}\{-\partial_2\Psi\partial_3 z^2 + [(1+\nu)\partial_1\eta_\beta + \nu\partial_1\theta_\beta + \nu\Phi_{1\beta}]\partial_{333}u_\beta^0\},\\
\sigma_{32}^0 &= \frac{E}{2(1+\nu)}\{\partial_1\Psi\partial_3 z^2 + [(1+\nu)\partial_2\eta_\beta + \nu\partial_2\theta_\beta + \nu\Phi_{2\beta}]\partial_{333}u_\beta^0\}.
\end{aligned} \tag{15.15}
$$

From (15.11)–(15.15) we observe that the asymptotic model contains Poisson's effects both on bending and stretching components and also geometric torsional effects related to the excentricity of the resultant of the applied loads with respect to the shear centre. As these effects are not present in the classical theory, we must recover the classical theory of Timoshenko by just considering a displacement field $\overline{\boldsymbol{u}}$ obtained from $\hat{\boldsymbol{u}}$ on neglecting these effects, that is (cf. TRABUCHO and VIAÑO [1990a]):

$$
\begin{aligned}
\overline{u}_\alpha &= u_\alpha^0 + \overline{z}_\alpha + \nu\Phi_{\alpha\beta}\partial_{33}u_\beta^0,\\
\overline{u}_3 &= -x_\alpha\partial_3(u_\alpha^0 + \overline{z}_\alpha) + [(1+\nu)\eta_\alpha + \nu\theta_\alpha]\partial_{333}u_\alpha^0,
\end{aligned} \tag{15.16}
$$

where u_α^0 is the solution of (15.11) and $\overline{z}_\alpha$ is the solution of the following problem:

$$
\begin{aligned}
&EI_\alpha\frac{\mathrm{d}^4\overline{z}_\alpha}{\mathrm{d}x_3^4} = -T_{\alpha\beta}\frac{\mathrm{d}^2F_\beta}{\mathrm{d}x_3^2}, \quad \text{in } (0,L),\\
&\overline{z}_\alpha(a) = -\frac{\nu}{2A}(I_\alpha - I_\beta)\partial_{33}u_\alpha^0(a), \quad a = 0, L, \quad (\alpha \neq \beta),\\
&\partial_3\overline{z}_\alpha(a) = \frac{1}{I_\alpha}[(1+\nu)L_{\alpha\beta}^\eta + \nu L_{\alpha\beta}^\theta]\partial_{333}u_\beta^0(a), \quad a = 0, L.
\end{aligned} \tag{15.17}
$$

REMARK 15.1. The displacement field $\overline{\boldsymbol{u}}$ coincides with $\hat{\boldsymbol{u}}$ if, for instance, the cross section possesses two axes of symmetry and if $\mathrm{d}P/\mathrm{d}x_3 = \mathrm{d}^2N_\alpha/\mathrm{d}x_3^2 = 0$.

Proceeding with the stress field in a similar way, we obtain (cf. (12.29))

$$\begin{aligned}\overline{\sigma}_{33} &= E\partial_3 u_3^0 + \frac{(1-\nu)E}{(1+\nu)(1-2\nu)}\partial_3\overline{u}_3^2 - \frac{2\nu^2 E}{(1+\nu)(1-2\nu)}x_\mu\partial_{33}\overline{z}_\mu + \nu S_{\mu\mu}(\underline{\boldsymbol{u}}^4)\\ &= -Ex_\alpha\partial_{33}(u_\alpha^0 + \overline{z}_\alpha) + \frac{(1-\nu)E}{(1+\nu)(1-2\nu)}[(1+\nu)\eta_\alpha + \nu\theta_\alpha]\partial_{3333}u_\alpha^0\\ &\quad + \nu S_{\mu\mu}(\underline{\boldsymbol{u}}^4),\end{aligned} \tag{15.18}$$

$$\begin{aligned}\overline{\sigma}_{\alpha\beta} &= \frac{\nu E}{(1+\nu)(1-2\nu)}\partial_3\overline{u}_3\delta_{\alpha\beta} - \frac{\nu E}{(1+\nu)(1-2\nu)}x_\mu\partial_{33}\overline{z}_\mu\delta_{\alpha\beta} + S_{\alpha\beta}(\underline{\boldsymbol{u}}^4)\\ &= -\frac{\nu E}{(1+\nu)(1-2\nu)}x_\mu\partial_{33}u_\mu^0 - \frac{2\nu E}{(1+\nu)(1-2\nu)}x_\mu\partial_{33}\overline{z}_\mu\delta_{\alpha\beta}\\ &\quad + \frac{\nu E}{(1+\nu)(1-2\nu)}[(1+\nu)\eta_\mu + \nu\theta_\mu]\partial_{3333}u_\mu^0\delta_{\alpha\beta} + S_{\alpha\beta}(\underline{\boldsymbol{u}}^4),\end{aligned} \tag{15.19}$$

$$\overline{\sigma}_{3\alpha} = \frac{E}{2(1+\nu)}[(1+\nu)\partial_\alpha\eta_\beta + \nu\partial_\alpha\theta_\beta + \nu\Phi_{\alpha\beta}]\partial_{333}u_\beta^0. \tag{15.20}$$

Besides Poisson's effects, from (15.16) we obtain the a priori hypothesis (15.2) and a correct hypothesis for the displacement component u_3 which *gives a meaning to the classical hypothesis* (15.3): it is enough to consider $\hat{u}_\alpha = u_\alpha^0 + \overline{z}_\alpha$ and substitute $x_\alpha\hat{v}_\alpha$ for $[(1+\nu)\eta_\alpha + \nu\theta_\alpha]\partial_{3333}u_\alpha^0$. Equation (15.20) gives also a sense to the classical hypothesis (15.4) and *shows how the shear stress $\sigma_{3\beta}$ varies on the cross section, which was one of the goals of Timoshenko's beam theory. Moreover, Eq.* (15.20) *is compatible with Hooke's law eliminating in this way one of the contradictions of Timoshenko's theory.*

Moreover, from Eqs. (15.11) and (15.17) we deduce that the total bending component $\hat{u}_\alpha = u_\alpha^0 + \overline{z}_\alpha$ is a solution of the following differential equation:

$$EI_\alpha\frac{\mathrm{d}^4\hat{u}_\alpha}{\mathrm{d}x_3^4} = F_\alpha - T_{\alpha\beta}\frac{\mathrm{d}^2F_\beta}{\mathrm{d}x_3^2}, \qquad \text{in } (0,L) \text{ (no sum on } \alpha). \tag{15.21}$$

Equation (15.21) constitutes a generalization of Eq. (15.10) where coupled bending effects, not present in the classical theory, are taken into account through the presence of matrix $(T_{\alpha\beta})$. Consequently, in order to be able to compare (15.21) with (15.10) we assume that simple bending takes place, that is, $\partial_{33}F_\beta = 0$, $(\beta \neq \alpha)$. In this case, the following expression provides *a precise definition for the constant that should be considered for calculating the additional bending deformations along direction Ox_α* when the coupling effect due to loads acting along direction Ox_β is to be neglected:

$$\hat{k}_\alpha = \frac{2(1+\nu)I_\alpha}{T_{\alpha\alpha}A}, \qquad \text{no sum on } \alpha. \tag{15.22}$$

Nevertheless, one has $T_{11} \neq T_{22}$ in the general case. Consequently, (15.22) represents an improvement with respect to the classical theory which assumes the same constant for either direction.

TABLE 15.2. Functions and constants for a circular cross section of radius R.

$w=0,$		$\eta_\alpha = \frac{1}{4}x_\alpha[x_1^2+x_2^2-3R^2]$
$\Psi = -\frac{1}{2}[x_1^2+x_2^2-R^2],$		$\theta_\alpha = -\frac{1}{4}x_\alpha[x_1^2+x_2^2-R^2]$
$I_\alpha = \frac{1}{4}\pi R^4$	$L_{11}^\eta = -\frac{7}{48}\pi R^6$	$K_{11}^\eta = \frac{1}{24}\pi R^6$
$I_\alpha^w = 0$	$L_{22}^\eta = -\frac{7}{48}\pi R^6$	$K_{22}^\eta = \frac{1}{24}\pi R^6$
$I_\alpha^\Psi = 0$	$L_{12}^\eta = L_{21}^\eta = 0$	$K_{12}^\eta = K_{21}^\eta = 0$
$J = \frac{1}{2}\pi R^4$	$L_{11}^\theta = \frac{1}{48}\pi R^6$	$K_{11}^\theta = -\frac{1}{24}\pi R^6$
$H_\alpha = 0$	$L_{22}^\theta = \frac{1}{48}\pi R^6$	$K_{22}^\theta = -\frac{1}{24}\pi R^6$
$H_3 = \frac{1}{12}\pi R^6$	$L_{12}^\theta = L_{21}^\theta = 0$	$K_{12}^\theta = K_{21}^\theta = 0$

$$T_{11} = T_{22} = \frac{R^2(7+12\nu+4\nu^2)}{12(1+\nu)}$$
$$T_{12} = T_{21} = 0.00000000$$
$$\hat{k}_1 = \hat{k}_2 = \frac{6+12\nu+6\nu^2}{7+12\nu+4\nu^2}$$

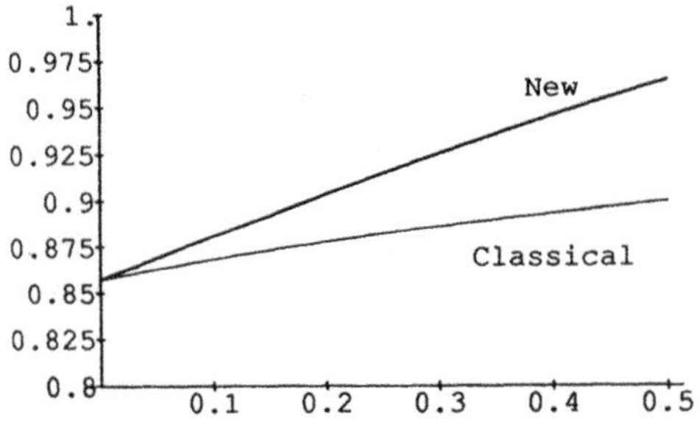

FIG. 15.1. Circular cross section.

From definition (12.39) we see that the evaluation of Timoshenko's matrix $\boldsymbol{T} = (T_{\alpha\beta})$ for a specific cross section made of a specific material requires the solution of six boundary value problems for Laplace's operator in the cross section. For the case of a circular cross section, there is an analytic solution available for the calculation of $\hat{k}_\alpha$. The results are summarized in Table 15.2. We remark the good agreement with the classical constant, obtained by completely different methods.

Both the new and classical constants are represented as a function of Poisson's ratio ν in Fig. 15.1. They both coincide for $\nu = 0$ and the maximum difference is obtained for $\nu = 0.5$.

For the case of a circular tube, with interior radius a and exterior radius b, there is also an analytical solution which is summarized in Table 15.3 where polar

TABLE 15.3. Functions and constants for a circular tube.

$$w = 0 \qquad \Psi = \tfrac{1}{2}(b^2 - r^2)$$
$$\eta_1 = \tfrac{1}{4}\cos\theta\{-3[(a^2+b^2)r + \frac{a^2b^2}{r}] + r^3\}$$
$$\theta_1 = \tfrac{1}{4}\cos\theta[(a^2+b^2)r + \frac{a^2b^2}{r} - r^3]$$
$$\eta_2 = \tfrac{1}{4}\sin\theta\{-3[(a^2+b^2)r + \frac{a^2b^2}{r}] + r^3\}$$
$$\theta_2 = \tfrac{1}{4}\sin\theta[(a^2+b^2)r + \frac{a^2b^2}{r} - r^3]$$

$$I_\alpha = \tfrac{1}{4}\pi(b^4 - a^4) \quad I_\alpha^w = 0 \quad I_\alpha^\Psi = 0$$
$$J = \tfrac{1}{2}\pi(b^4 - a^4) \quad H_\alpha = 0$$
$$H_3 = \tfrac{1}{12}\pi(b^2 - a^2)(b^4 + a^2b^2 + a^4)$$

$$L_{11}^\eta = L_{22}^\eta = -\tfrac{1}{48}\pi(b^2 - a^2)(7b^4 + 34a^2b^2 + 7a^4)$$
$$K_{11}^\eta = K_{22}^\eta = \tfrac{1}{48}\pi(b^2 - a^2)(b^4 + 10a^2b^2 + a^4)$$
$$L_{12}^\eta = L_{21}^\eta = 0 \quad K_{12}^\eta = K_{21}^\eta = 0$$

$$L_{11}^\theta = L_{22}^\theta = \tfrac{1}{48}\pi(b^2 - a^2)(b^4 + 10a^2b^2 + a^4)$$
$$K_{11}^\theta = K_{22}^\theta = -\tfrac{1}{24}\pi(b^2 - a^2)(b^4 + 4a^2b^2 + a^4)$$
$$L_{12}^\theta = L_{21}^\theta = 0 \quad K_{12}^\theta = K_{21}^\theta = 0$$

$$\hat{k}_1 = \hat{k}_2 = \frac{6(1+\nu)^2(1+m^2)^2}{(7+12\nu+4\nu^2)(1+m^2)^2 + m^2(20+24\nu+8\nu^2)}, \quad m = \frac{a}{b}$$

coordinates were used. These results must be compared with those from Table 15.1, with the value for a circular crown ($m = 0$) and with those for a thin-walled tube ($m \simeq 1$). Function $(1+\nu)\eta_1 + \nu\theta_1$ is also considered by WIDERA, FAN and AFSHARI [1989, p. 38, Eqs. (20)–(21)], for this type of cross section.

For a generic cross section the calculation of the constant can be done using a numerical technique in order to evaluate functions w, Ψ, η_α and θ_α and then a numerical quadrature formula in order to compute the integrals in the definition of $T_{\alpha\beta}$. The dimensions of the cross section are very important because *constant* $\hat{k}_\alpha$ *depends on Poisson's ratio and on the shape of the cross section* (relative dimensions) *but* it is homothetic invariant for a given geometry, that is, *it does not depend on the absolute dimensions of a given cross section.* We have the following.

THEOREM 15.1. *Given a certain geometry, the new Timoshenko's constant* $\hat{k}_\alpha$ *is homothetic invariant, relatively to the centroid of the cross section.*

PROOF. Let $h > 0$ and let $\overline{\omega}$ be the set obtained from ω through the homothety $y_\alpha = hx_\alpha$. The dependence on y and on $\overline{\omega}$ shall be denoted with a bar. One then

TABLE 15.4. Semicircular cross section ($R = 1$).

$I_1^w = 0.06635056$	$I_2^w = 0.00000000$
$I_1^\Psi = 0.00703070$	$I_2^\Psi = 0.00000000$
$I_1 = 0.39219778$	$I_2 = 0.10964360$
$L_{11}^\eta = -0.22880905$	$L_{22}^\eta = -0.02005027$
$L_{12}^\eta = 0.00000000$	$L_{21}^\eta = 0.00000000$
$L_{11}^\theta = 0.01650639$	$L_{22}^\theta = 0.01047391$
$L_{12}^\theta = 0.00000000$	$L_{21}^\theta = 0.00000000$
$K_{11}^\eta = 0.03376054$	$K_{22}^\eta = 0.02084043$
$K_{12}^\eta = 0.00000000$	$K_{21}^\eta = 0.00000000$
$K_{11}^\theta = -0.05813010$	$K_{22}^\theta = -0.03056944$
$K_{12}^\theta = 0.00000000$	$K_{21}^\theta = 0.00000000$
$H_1 = 0.00000000$	$H_2 = -0.01326206$
$H_3 = 0.06426122$	$J = 0.29631603$

$$T_{11} = \frac{14 + 30\nu + 15\nu^2}{23(1+\nu)}$$
$$T_{22} = \frac{1 + \nu - \nu^2}{6(1+\nu)}$$
$$T_{12} = T_{21} = 0$$
$$\hat{k}_1 = \frac{1 + 2\nu + \nu^2}{1.205 + 2.581\nu + 1.360\nu^2}$$
$$\hat{k}_2 = \frac{1 + 2\nu + \nu^2}{1.309 + 1.254\nu - 1.149\nu^2}$$

has

$$\begin{aligned}
&\overline{\Phi}_{\alpha\beta} = h^2 \Phi_{\alpha\beta}, \qquad \overline{I}_\alpha = h^4 I_\alpha, \qquad \overline{H}_3 = h^6 H_3,\\
&\overline{w}(hx_1, hx_2) = h^2 w(x_1, x_2), \qquad \overline{\Psi}(hx_1, hx_2) = h^2 \Psi(x_1, x_2),\\
&\overline{\eta}_\beta(hx_1, hx_2) = h^3 \eta_\beta(x_1, x_2), \qquad \overline{\theta}_\beta(hx_1, hx_2) = h^3 \theta_\beta(x_1, x_2),\\
&\overline{I}_\beta^w = h^5 I_\beta^w, \qquad \overline{I}_\beta^\Psi = h^5 I_\beta^\Psi, \qquad \overline{J} = h^4 J,\\
&\overline{L}_{\alpha\beta}^\eta = h^6 L_{\alpha\beta}^\eta, \qquad \overline{L}_{\alpha\beta}^\theta = h^6 L_{\alpha\beta}^\theta,\\
&\overline{K}_{\alpha\beta}^\eta = h^6 K_{\alpha\beta}^\eta, \qquad \overline{K}_{\alpha\beta}^\theta = h^6 K_{\alpha\beta}^\theta,
\end{aligned}$$

and finally,

$$\overline{T}_{\alpha\beta} = h^2 T_{\alpha\beta}, \qquad \overline{\hat{k}}_\alpha = \hat{k}_\alpha. \qquad \square$$

Using the finite element method with linear triangular elements and exact quadrature formulas for the different polynomial expressions showing up in the definition of matrix $T_{\alpha\beta}$, several cases are considered in TRABUCHO and VIAÑO

TABLE 15.5. I-shaped beam ($h = 5$, $b = 4$, $t_w = 2$, $t_f = 0.5$).

$I_1^w = 0.00000000$	$I_2^w = 0.00000000$
$I_1^\Psi = 0.00000000$	$I_2^\Psi = 0.00000000$
$I_1 = 8.00000000$	$I_2 = 31.00000000$
$L_{11}^\eta = -14.88317100$	$L_{22}^\eta = -227.90495000$
$L_{12}^\eta = 0.00000000$	$L_{21}^\eta = 0.00000000$
$L_{11}^\theta = 9.17661890$	$L_{22}^\theta = -25.50602800$
$L_{12}^\theta = 0.00000000$	$L_{21}^\theta = 0.00000000$
$K_{11}^\eta = 18.31755100$	$K_{22}^\eta = -52.47956700$
$K_{12}^\eta = 0.00000000$	$K_{21}^\eta = 0.00000000$
$K_{11}^\theta = -37.81388500$	$K_{22}^\theta = -49.81746700$
$K_{12}^\theta = 0.00000000$	$K_{21}^\theta = 0.00000000$
$H_1 = 0.00000000$	$H_2 = 0.00000000$
$H_3 = 51.59578400$	$J = 9.84146960$

$$T_{11} = \frac{1.86 + 1.4290\nu - 1.293\nu^2}{(1+\nu)}$$

$$T_{22} = \frac{7.35 + 16.373\nu + 8.992\nu^2}{(1+\nu)}$$

$$T_{12} = T_{21} = 0$$

$$\hat{k}_1 = \frac{1 + 2\nu + \nu^2}{1.395 + 1.071\nu - 0.970\nu^2}$$

$$\hat{k}_2 = \frac{1 + 2\nu + \nu^2}{1.423 + 3.169\nu + 1.740\nu^2}$$

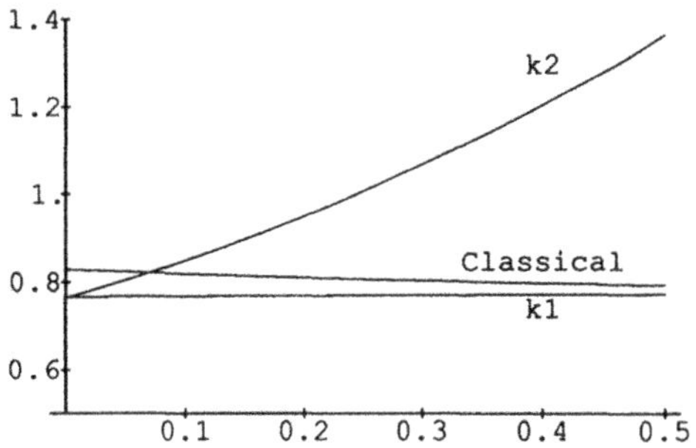

FIG. 15.2. Semicircular cross section.

[1990a] and in VILARES [1988]. Both constants are compared as a function of Poisson's ratio. The results for a semi-circular cross section of unitary radius and for an I-shaped beam with relative dimensions $h = 5$, $b = 4$, $t_w = 2$, $t_f = 0.5$ are shown in Tables 15.4 and 15.5 and in Figs. 15.2 and 15.3. The good agreement for small values of ν is to be noticed.

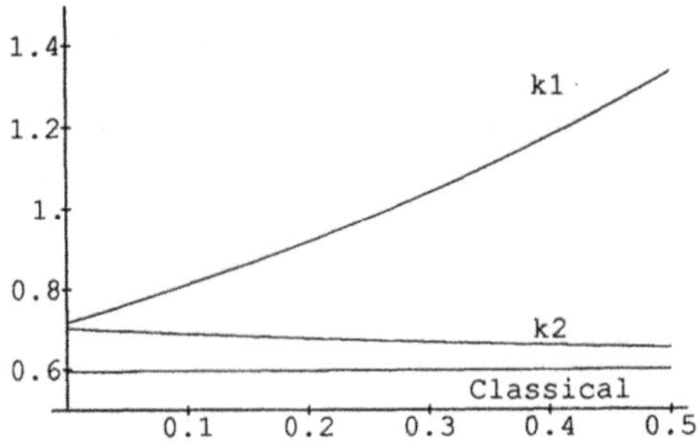

FIG. 15.3. I-shaped cross section.

TABLE 15.6. Square tube.

a	b	N
10	0	$12.04 + 22.01\,\nu + 8.585\,\nu^2$
10	2	$13.30 + 23.70\,\nu + 9.160\,\nu^2$
10	5	$18.30 + 30.20\,\nu + 11.40\,\nu^2$
10	8	$22.40 + 36.20\,\nu + 13.70\nu^2$

TABLE 15.7. Square tube with square holes.

Holes	N
4	$14.80 + 29.90\,\nu + 10.70\,\nu^2$
16	$14.90 + 27.66\,\nu + 11.49\,\nu^2$
64	$14.40 + 26.70\,\nu + 11.12\,\nu^2$

Tables 15.6, 15.7 and 15.8 represent the calculation for rectangular cross sections with holes. In Table 15.6 we consider a square cross section of dimension a, with a hole of dimension b (cf. Table 15.1). Writing the new constant in the form

$$\hat{k}_1 = \hat{k}_2 = \frac{10(1+\nu)^2}{N}$$

the different computed expressions of N, for the different cross sections, are shown. The values obtained should be compared with the classical expressions $10(1+\nu)/(12+11\nu)$ (solid square) and $20(1+\nu)/(48+39\nu)$ (thin square tube).

This study can now be generalized to the case of rectangular tubes of exterior dimensions $a \times b$ (a parallel to axis $0x_2$) with thickness e. Once again writing the

TABLE 15.8. Rectangular tubes.

a	e	N_1	N_2
10	4	$13.30 + 23.70\,\nu + 9.16\,\nu^2$	$13.30 + 23.70\,\nu + 9.16\,\nu^2$
20	4	$30.55 + 23.07\,\nu - 13.38\,\nu^2$	$12.83 + 26.90\,\nu + 13.99\,\nu^2$
40	4	$108.3 - 0.044\,\nu - 130.2\,\nu^2$	$12.32 + 27.17\,\nu + 14.84\,\nu^2$
100	4	$679.5 - 263.5\,\nu - 1064\,\nu^2$	$12.07 + 27.05\,\nu + 14.98\,\nu^2$
10	1	$22.40 + 36.20\,\nu + 13.70\,\nu^2$	$22.40 + 36.20\,\nu + 13.70\,\nu^2$
20	1	$44.42 + 47.10\,\nu + 2.50\,\nu^2$	$15.67 + 31.07\,\nu + 15.39\,\nu^2$
40	1	$112.3 + 56.20\,\nu - 57.11\,\nu^2$	$13.25 + 28.50\,\nu + 15.29\,\nu^2$
100	1	$515.3 - 12.02\,\nu - 529.8\,\nu^2$	$12.25 + 27.32\,\nu + 15.07\,\nu^2$

TABLE 15.9. Unitary square cross section.

$I_1^w = 0.00000000$	$I_2^w = 0.00000000$
$I_1^\Psi = 0.00000000$	$I_2^\Psi = 0.00000000$
$I_1 = 0.08333301$	$I_2 = 0.08333301$
$L_{11}^\eta = -0.16687270$	$L_{22}^\eta = -0.16687270$
$L_{12}^\eta = 0.00000000$	$L_{21}^\eta = 0.00000000$
$L_{11}^\theta = 0.00142491$	$L_{22}^\theta = 0.00142491$
$L_{12}^\theta = 0.00000000$	$L_{21}^\theta = 0.00000000$
$K_{11}^\eta = 0.00275738$	$K_{22}^\eta = 0.00275738$
$K_{12}^\eta = 0.00000000$	$K_{21}^\eta = 0.00000000$
$K_{11}^\theta = -0.00580318$	$K_{22}^\theta = -0.00580318$
$K_{12}^\theta = 0.00000000$	$K_{21}^\theta = 0.00000000$
$H_1 = 0.00000000$	$H_2 = 0.00000000$
$H_3 = 0.00972218$	$J = 0.13990354$

$$T_{11} = T_{22} = \frac{1 + 1.832\nu + 0.714\nu^2}{5(1+\nu)}$$

$$T_{12} = T_{21} = 0$$

$$\hat{k}_1 = \hat{k}_2 = \frac{10 + 20\nu + 10\nu^2}{12 + 22\nu + 8.585\nu^2}$$

generalized constants in the form $\hat{k}_\alpha = 10(1+\nu)^2/N_\alpha$ the results are represented in Table 15.8 (for the case $b = 10$).

In Table 15.7 we summarize the numerical results obtained for a multicellular square cross section. We considered the exterior dimension to be 8 and the square holes equally distributed. The cases of 4, 16 and 64 holes with sizes 1, 0.5, and 0.25, respectively, are analyzed. We remark that when the number of holes increases $\hat{k}_\alpha(\nu)$ tends to a limit value which is not easy to grasp. A rigorous mathematical analysis of this question will be made in Sections 30 and

TABLE 15.10. Rectangular cross section.

Dim. (b/t)	N_1	N_2
1	$12.01 + 22.01\,\nu + 8.59\,\nu^2$	$12.01 + 22.01\,\nu + 8.59\,\nu^2$
2	$12.04 + 6.97\,\nu - 19.17\,\nu^2$	$12.01 + 25.76\,\nu + 13.63\,\nu^2$
2.4	$12.04 - 1.84\,\nu - 38.15\,\nu^2$	$12.01 + 26.14\,\nu + 14.07\,\nu^2$
2.5	$12.04 - 4.32\,\nu - 43.62\,\nu^2$	$12.01 + 26.21\,\nu + 14.15\,\nu^2$
3	$12.04 - 18.13\,\nu - 75.51\,\nu^2$	$12.00 + 26.45\,\nu + 14.42\,\nu^2$
5	$12.10 - 98.79\,\nu - 279.23\,\nu^2$	$12.00 + 26.80\,\nu + 14.79\,\nu^2$
10	$12.10 - 476.98\,\nu - 1315.87\,\nu^2$	$12.00 + 26.95\,\nu + 14.95\,\nu^2$

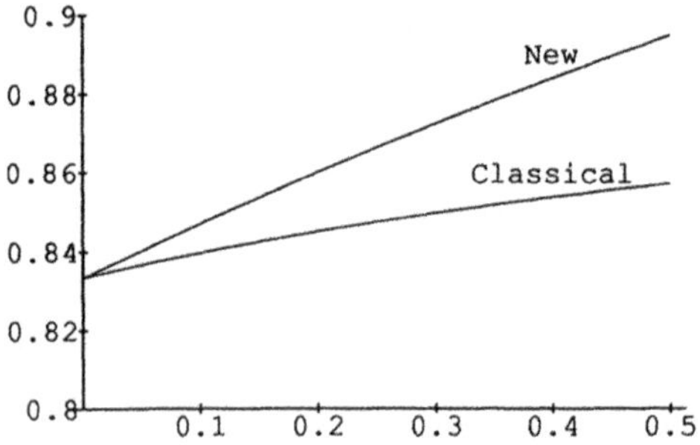

FIG. 15.4. Square cross section.

31 using homogenization techniques. Some results in this way may be seen in MASCARENHAS and TRABUCHO [1991].

As the last case under study we shall consider a deceptively simple one, because of the important consequences it brings about and which will be developed in the next section. More precisely, we study the dependence of Timoshenko's constant $\hat{k}_\alpha$ with respect to the relative dimensions of a rectangular cross section. In the first place a square cross section of unitary dimensions was considered. The respective results are shown in Table 15.9 and the comparison with the classical constant (cf. Table 15.1) is done in Fig. 15.4.

In Table 15.10 the results obtained for $\hat{k}_\alpha = 10(1+\nu)^2/N_\alpha$ are represented for different dimensions of the cross section. The dependence of $\hat{k}_1$ and $\hat{k}_2$ with respect to Poisson's ratio is represented in Figs. 15.5 and 15.6, respectively.

From these results spring two major facts which deserve a deeper study. The first one is related to constant $\hat{k}_2$. As the ratio between the sizes of the rectangle $r = t/b$ goes to zero, it looks as if constant $\hat{k}_2 = \hat{k}_2^r(\nu)$ converges to

$$\hat{k}_2^0(\nu) = \frac{10(1+\nu)^2}{12 + 27\nu + 15\nu^2}. \tag{15.23}$$

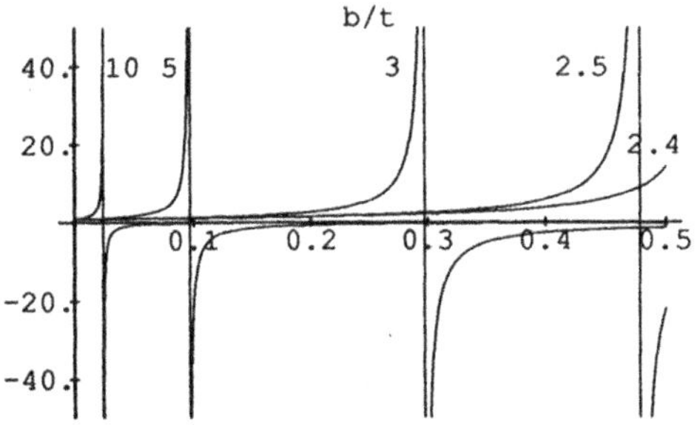

FIG. 15.5. Rectangular cross section. Constant $\hat{k}_1$.

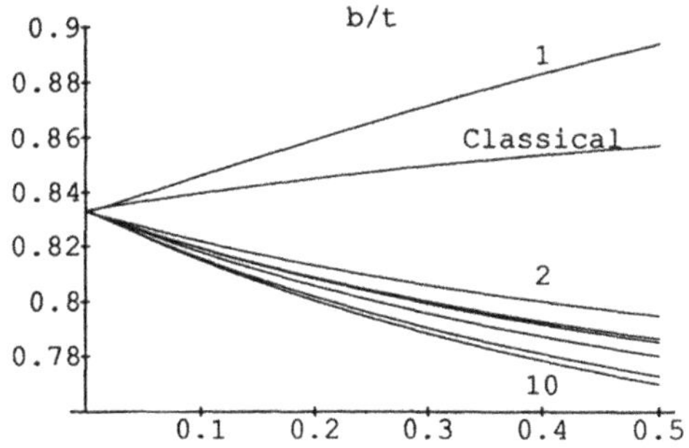

FIG. 15.6. Rectangular cross section. Constant $\hat{k}_2$.

The second fact concerns constant $\hat{k}_1$. The graph in Fig. 15.5 shows that when b is smaller or equal to 2.5, there is a critical value of Poisson's ratio ν_r for which $\hat{k}_1 = \hat{k}_1^r(\nu)$ presents a singularity, in the sense that

$$\lim_{\nu \to \nu_r^-} \hat{k}_1^r(\nu) = +\infty, \qquad \lim_{\nu \to \nu_r^+} \hat{k}_1^r(\nu) = -\infty. \tag{15.24}$$

We also observe that when $r \to 0$, ν_r also goes to zero. Consequently, for a given material with Poisson's ratio ν_r, there exists a relationship between the relative dimensions of the cross section ($r = t/b$) for which Timoshenko's and Bernoulli–Navier's beam models (asymptotically) coincide. We may then wonder what is the relationship between r and ν_r or in other words, what is the domain of validity of Bernoulli–Navier's theory?

These questions will be analyzed in the next sections. Once again using the asymptotic expansion method as $t \to 0$ (b fixed) it is possible to give a mathematical framework in order to interpret the previous numerical results (cf. TRABUCHO and VIAÑO [1989], BARROS [1989], RODRÍGUEZ [1990], RODRÍGUEZ and VIAÑO [1991, 1992a,b,c]).

CHAPTER III

Asymptotic Modelling in Thin-Walled Rods – An Introduction

This chapter is an introduction to the large field of modelling thin-walled rods by asymptotic techniques. We restrict our attention to the rectangular cross section only and a second asymptotic analysis, having now as a smaller parameter the thickness of the rectangle, is performed on the asymptotic model itself. This chapter is based on the formal results of BARROS [1989] and BARROS, TRABUCHO and VIAÑO [1989], and on the convergence results by RODRÍGUEZ [1990, 1993] and RODRÍGUEZ and VIAÑO [1991, 1992a,b,c] for the Poisson's equation in a multirectangular domain, which improve those obtained by LE DRET [1991].

After a formal discussion of several aspects of this second limit problem, which is done in Section 16, one observes that everything reduces to the study the convergence and the limiting process for Poisson's equation in a thin rectangle either with pure Dirichlet or Neumann type boundary conditions. This analysis is performed in Section 17. These results are then applied to the warping and torsion functions for a thin rectangle in Section 18, and to the functions and constants showing up in the generalization of Timoshenko's constant in Section 19. A comparison of these results with those given by the classical theories is also done. In Section 20 some results related to Timoshenko's theory for thin-walled rectangular beams are obtained and compared to the numerical results of Section 15. The chapter closes down with a discussion on the possibilities of applying the asymptotic expansion method in order to obtain general thin-walled models (Section 21).

16. Formal asymptotic approach of bending and torsion properties for rods with a thin rectangular cross section

The contents of the present section are strongly motivated by the peculiar properties demonstrated by the numerical analysis of Timoshenko's constants on a rectangular cross section (cf. Section 15, Eqs. (15.23) and (15.24)).

This section constitutes a summary of the works of BARROS [1989] and BARROS, TRABUCHO and VIAÑO [1989] who obtained the first results concerning the formal asymptotic approximation of functions and constants characterizing the geometry of thin rectangular cross sections. The mathematical justification of these results is presented in Sections 17, 18 and 19.

In order to fix ideas, let ω^t denote a rectangular cross section of dimensions $b \times t$, that is,

$$\omega^t = (-\tfrac{1}{2}t, \tfrac{1}{2}t) \times (-\tfrac{1}{2}b, \tfrac{1}{2}b). \tag{16.1}$$

We assume that b is fixed and that t is much smaller than b. A generic point in ω^t is denoted by $\boldsymbol{x}^t = (x_1^t, x_2^t)$ and all the functions and constants concerning ω^t will be written with exponent t: w^t, Ψ^t, J^t, $\hat{k}_\alpha^t$, and so on. Moreover, the boundary of ω^t is denoted by γ^t and $\partial_a lpha^t$ stands for differentiation with respect to x_α^t. The outward unit normal to γ^t is denoted by $\boldsymbol{n}^t = (n_1^t, n_2^t)$.

The *main purpose* of this section is to study the behaviour of constants $\hat{k}_\alpha^t$, when t becomes very small with respect to b, and to try to justify properties (15.23) and (15.24), from a mathematical point of view.

From Eq. (15.22) we conclude that *the dependence of constant* $\hat{k}_\alpha^t$ (see (12.39)) *with respect to t is very complex*. In fact, $\hat{k}_\alpha^t$ depends of the following ten constants characterizing the geometry of the cross section:

$$A^t, \quad I_\alpha^t, \quad L_{\alpha\alpha}^{\eta^t}, \quad L_{\alpha\alpha}^{\theta^t}, \quad K_{\alpha\alpha}^{\eta^t},$$
$$K_{\alpha\alpha}^{\theta^t}, \quad H_3^t, \quad J^t, \quad I_\alpha^{w^t}, \quad I_\alpha^{\Psi^t}.$$

From their definition we immediately obtain

$$A^t = bt, \tag{16.2}$$
$$I_1^t = \tfrac{1}{12}bt^3, \tag{16.3}$$
$$I_2^t = \tfrac{1}{12}b^3t, \tag{16.4}$$
$$I_\alpha^{w^t} = I_\alpha^{\Psi^t} = 0, \tag{16.5}$$
$$H_3^t = \frac{b^5t}{320} + \frac{b^3t^3}{288} + \frac{bt^5}{320}. \tag{16.6}$$

Consequently, *the problem reduces to the study of* the behaviour of constants $L_{\alpha\alpha}^{\eta^t}$, $L_{\alpha\alpha}^{\theta^t}$, $K_{\alpha\alpha}^{\eta^t}$, $K_{\alpha\alpha}^{\theta^t}$ and J^t as $t \to 0$, or equivalently, *the asymptotic behaviour of the six functions* w^t, Ψ^t, η_α^t and θ_α^t as $t \to 0$.

In this way one may know the asymptotic behaviour of not only $\hat{k}_\alpha^t$ but also of other constants which play an important role in structural mechanics, as for instance the torsion constant J^t or the area bimoments J_w^t (which in this particular case coincide with the warping constant $C_w^{0t} = C_w^{\nu t}$ (cf. (13.10) and (13.8)).

In the engineering literature the torsion constant J^t for a rectangular cross section is, approximately, given by (cf. ODEN [1967 p. 43–44], VLASSOV [1962, p. 59]):

$$J^t \simeq c(t)bt^3, \tag{16.7}$$

where $c(t)$ is a constant given in Table 16.1.

TABLE 16.1. Torsional constants for a rectangle.

b/t	∞	10.0	5.0	3.0	2.5	2.0	1.5	1.2	1.0
$c(t)$	0.33	0.31	0.29	0.26	0.25	0.23	0.20	0.17	0.14

For a very thin rectangular cross section, i.e., when $b/t > 10$, the torsion constant becomes

$$J^t \simeq \frac{bt^3}{3}. \tag{16.8}$$

In a similar way the bimoment of area $J_w^t = (C_w^{\nu t} = C_w^{0t})$ showing up in Vlassov's theory for the angle of torsion (cf. (13.5)) and for a thin rectangular cross section is approximately given in the engineering literature by (cf. ODEN [1967, p. 205]):

$$J_w^t \simeq \frac{b^3t^3}{144}. \tag{16.9}$$

Since functions w, Ψ, η_α and θ_α (together with their associated constants) were introduced regardless of the form of the cross section, they should also be appropriate for thin rectangular cross sections. Due to the fact that $t \ll b$ one may think of getting explicit approximate expressions for these functions and associated constants, of the type (16.8) and (16.9). The purpose of the present study is to justify such type of expressions. We show that they are an asymptotic approximation of exact formulae as $t \to 0$. As a by-product we shall also be able to obtain new asymptotic expressions for constants $L_{\alpha\beta}^{\eta^t}$, $L_{\alpha\beta}^{\theta^t}$, $K_{\alpha\beta}^{\eta^t}$, $K_{\alpha\beta}^{\theta^t}$ and consequently for $\hat{k}_\alpha^t$.

In order to study the asymptotic behaviour of w^t, Ψ^t, η_α^t and θ_α^t when $t \to 0$ we use the same asymptotic technique as for the elasticity problem (cf. Sections 2 and 3), that is:

(i) change of variable to a fixed domain ω and consequently transformation of the original problem into an equivalent scaled problem now posed in a domain independent of t,

(ii) use the classical techniques in order to study the scaled problem: convergence as $t \to 0$, asymptotic expansions, characterization of the limit,

(iii) descale the solution and identify the limit obtained in the original domain ω^t.

In order to do so we define the following spaces:

$$V_1(\omega^t) = H_0^1(\omega^t), \qquad V_2(\omega^t) = \{\varphi^t \in H^1(\omega^t): \int_{\omega^t} \varphi^t \,\mathrm{d}\omega^t = 0\}, \tag{16.10}$$

equipped with the usual norms.

Functions Ψ^t, w^t, η^t_α and θ^t_α are given as the unique solutions of the following boundary value problems (cf. (7.28), (7.31), (7.37) and (7.41)):

$$\begin{aligned} &\Psi^t \in V_1(\omega^t), \\ &\int_{\omega^t} \partial^t_\alpha \Psi^t \partial^t_\alpha \varphi^t \,\mathrm{d}\omega^t = \int_{\omega^t} 2\varphi^t \,\mathrm{d}\omega^t, \quad \text{for all } \varphi^t \in V_1(\omega^t), \end{aligned} \tag{16.11}$$

$$\begin{aligned} &w^t \in V_2(\omega^t), \\ &\int_{\omega^t} \partial^t_\alpha w^t \partial^t_\alpha \varphi^t \,\mathrm{d}\omega^t = \int_{\gamma^t} (x^t_2 n^t_1 - x^t_1 n^t_2)\varphi^t \,\mathrm{d}\gamma^t, \quad \text{for all } \varphi^t \in V_2(\omega^t), \end{aligned} \tag{16.12}$$

$$\begin{aligned} &\eta^t_\beta \in V_2(\omega^t), \\ &\int_{\omega^t} \partial^t_\alpha \eta^t_\beta \partial^t_\alpha \varphi^t \,\mathrm{d}\omega^t = -\int_{\omega^t} 2x^t_\beta \varphi^t \,\mathrm{d}\omega^t, \quad \text{for all } \varphi^t \in V_2(\omega^t), \end{aligned} \tag{16.13}$$

$$\begin{aligned} &\theta^t_\beta \in V_2(\omega^t), \\ &\int_{\omega^t} \partial^t_\alpha \theta^t_\beta \partial^t_\alpha \varphi^t \,\mathrm{d}\omega^t = \int_{\omega^t} 2x^t_\beta \varphi^t \,\mathrm{d}\omega^t - \int_{\gamma^t} \Phi^t_{\beta\alpha} n^t_\alpha \varphi^t \,\mathrm{d}\gamma^t, \quad \text{for all } \varphi^t \in V_2(\omega^t). \end{aligned} \tag{16.14}$$

In order to apply the asymptotic expansion method, we consider the *reference domain*

$$\omega = (-\tfrac{1}{2}, \tfrac{1}{2}) \times (-\tfrac{1}{2}b, \tfrac{1}{2}b),$$

and consequently the associated change of variable is defined by the transformation:

$$\Pi^t : (x_1, x_2) \in \overline{\omega} \to \Pi^t(x_1, x_2) = (x^t_1, x^t_2) = (tx_1, x_2) \in \overline{\omega}^t. \tag{16.15}$$

We also define the following spaces:

$$V_1(\omega) = H^1_0(\omega), \qquad V_2(\omega) = \{\varphi \in H^1(\omega) \colon \int_\omega \varphi \,\mathrm{d}\omega = 0\}. \tag{16.16}$$

Denoting by γ the boundary of ω and by $\boldsymbol{n} = (n_1, n_2)$ its outward unit normal, we define the following partition of boundary γ:

$$\begin{aligned} \gamma_v &= \{(x_1, x_2) \in \gamma \colon x_1 = \pm\tfrac{1}{2}\} = \{-\tfrac{1}{2}, \tfrac{1}{2}\} \times [-\tfrac{1}{2}b, \tfrac{1}{2}b], \\ \gamma_h &= \{(x_1, x_2) \in \gamma \colon x_2 = \pm\tfrac{1}{2}b\} = [-\tfrac{1}{2}, \tfrac{1}{2}] \times \{-\tfrac{1}{2}b, \tfrac{1}{2}b\}. \end{aligned} \tag{16.17}$$

To any function $\Psi^t \in V_1(\omega^t)$, $\Phi^t \in V_2(\omega^t)$ and $\varphi^t \in H^1(\omega^t)$ we associate the functions $\Psi(t) \in V_1(\omega)$, $\Phi(t) \in V_2(\omega)$ and $\varphi \in H^1(\omega)$, respectively defined by

$$\begin{aligned} &\Psi(t)(x_1, x_2) = \Psi^t(x_1^t, x_2^t), \\ &\Phi(t)(x_1, x_2) = \Phi^t(x_1^t, x_2^t), \\ &\varphi(x_1, x_2) = \varphi^t(x_1^t, x_2^t). \end{aligned} \tag{16.18}$$

One has

$$\begin{aligned} &\partial_1 \varphi(t)(x_1, x_2) = t\partial_1^t \varphi^t(x_1^t, x_2^t), \\ &\partial_2 \varphi(t)(x_1, x_2) = \partial_2^t \varphi^t(x_1^t, x_2^t). \end{aligned} \tag{16.19}$$

From these relations, we obtain that functions $\Psi(t)$, $w(t)$, $\eta_\beta(t)$ and $\theta_\beta(t)$, obtained from Ψ^t, w^t, η_β^t and θ_β^t through the change of variable (16.18), are the unique solutions of the following boundary value problems:

$$\begin{aligned} &\Psi(t) \in V_1(\omega), \\ &\int_\omega \partial_1 \Psi(t) \partial_1 \varphi \,\mathrm{d}\omega + t^2 \int_\omega \partial_2 \Psi(t) \partial_2 \varphi \,\mathrm{d}\omega \\ &\quad = 2t^2 \int_\omega \varphi \,\mathrm{d}\omega, \quad \text{for all } \varphi \in V_1(\omega), \end{aligned} \tag{16.20}$$

$$\begin{aligned} &w(t) \in V_2(\omega), \\ &\int_\omega \partial_1 w(t) \partial_1 \varphi \,\mathrm{d}\omega + t^2 \int_\omega \partial_2 w(t) \partial_2 \varphi \,\mathrm{d}\omega \\ &\quad = t \int_{\gamma_v} x_2 n_1 \varphi \,\mathrm{d}\gamma - t^3 \int_{\gamma_h} x_1 n_2 \varphi \,\mathrm{d}\gamma, \quad \text{for all } \varphi \in V_2(\omega), \end{aligned} \tag{16.21}$$

$$\begin{aligned} &\eta_1(t) \in V_2(\omega), \\ &\int_\omega \partial_1 \eta_1(t) \partial_1 \varphi \,\mathrm{d}\omega + t^2 \int_\omega \partial_2 \eta_1(t) \partial_2 \varphi \,\mathrm{d}\omega \\ &\quad = -2t^3 \int_\omega x_1 \varphi \,\mathrm{d}\omega, \quad \text{for all } \varphi \in V_2(\omega), \end{aligned} \tag{16.22}$$

$$\begin{aligned} &\eta_2(t) \in V_2(\omega), \\ &\int_\omega \partial_1 \eta_2(t) \partial_1 \varphi \,\mathrm{d}\omega + t^2 \int_\omega \partial_2 \eta_2(t) \partial_2 \varphi \,\mathrm{d}\omega \\ &\quad = -2t^2 \int_\omega x_2 \varphi \,\mathrm{d}\omega, \quad \text{for all } \varphi \in V_2(\omega), \end{aligned} \tag{16.23}$$

$$
\begin{aligned}
&\theta_1(t) \in V_2(\omega),\\
&\int_\omega \partial_1\theta_1(t)\partial_1\varphi\,\mathrm{d}\omega + t^2\int_\omega \partial_2\theta_1(t)\partial_2\varphi\,\mathrm{d}\omega =\\
&\quad = t^3\left[2\int_\omega x_1\varphi\,\mathrm{d}\omega - \tfrac{1}{2}\int_{\gamma_v} x_1^2 n_1\varphi\,\mathrm{d}\gamma - \int_{\gamma_h} x_1x_2n_2\varphi\,\mathrm{d}\gamma\right]\\
&\qquad + \frac{t}{2}\int_{\gamma_v} x_2^2 n_1\varphi\,\mathrm{d}\gamma, \quad \text{for all } \varphi\in V_2(\omega),
\end{aligned}
\tag{16.24}
$$

$$
\begin{aligned}
&\theta_2(t) \in V_2(\omega),\\
&\int_\omega \partial_1\theta_2(t)\partial_1\varphi\,\mathrm{d}\omega + t^2\int_\omega \partial_2\theta_2(t)\partial_2\varphi\,\mathrm{d}\omega\\
&\quad = \frac{t^4}{2}\int_{\gamma_h} x_1^2 n_2\varphi\,\mathrm{d}\gamma\\
&\qquad + t^2\left[2\int_\omega x_2\varphi\,\mathrm{d}\omega - \tfrac{1}{2}\int_{\gamma_h} x_2^2 n_2\varphi\,\mathrm{d}\gamma - \int_{\gamma_v} x_1x_2n_1\varphi\,\mathrm{d}\gamma\right],\\
&\qquad \text{for all } \varphi\in V_2(\omega).
\end{aligned}
\tag{16.25}
$$

An asymptotic expansion method for problems of the type (16.20)–(16.25) for *cylindrical domains* can be studied in LIONS [1973, Chapter IV, Example 2.1] and LE DRET [1991], for the case of a Dirichlet boundary condition on γ_h and a Neumann type boundary condition on γ_v. Although in the present case we have either pure Dirichlet or pure Neumann boundary conditions on the whole boundary γ, we are faced with the same boundary layer type problems which prevent the verification of the boundary conditions on γ_h.

The first step in the application of the asymptotic expansion method consists in the identification of the successive terms in a *formal* asymptotic expansion of functions $\Psi(t)$, $w(t)$, $\eta(t)$ and $\theta(t)$ in $V_\alpha(\omega)$ together with the identification of the respective boundary layer phenomena. This has been done in BARROS [1989] and in BARROS, TRABUCHO and VIAÑO [1989]. In this section we show the calculations for two typical cases concerning functions $\Psi(t)$-Dirichlet boundary condition- and $w(t)$-Neumann boundary condition.

The second part of the asymptotic expansion method concerning *the convergence results in appropriate spaces* (not in V_α due to the boundary layer) is treated in RODRÍGUEZ [1990] and RODRÍGUEZ and VIAÑO [1991, 1992a,b,c] as is summarized in the next section. In this way the results of LE DRET [1991] are generalized. For more general types of cross sections we refer to RODRÍGUEZ [1993].

Considering a formal asymptotic expansion of $\Psi(t)$, solution of problem

(16.20), in the form

$$\Psi(t) = \Psi_0 + t\Psi_1 + t^2\Psi_2 + \text{h.o.t.}, \quad \Psi_i \in V_1(\omega), \tag{16.26}$$

and substituting back into (16.20), one obtains the following set of equations verified by the successive terms $\Psi_i \in V_1(\omega)$ and for all $\varphi \in V_1(\omega)$:

$$\int_\omega \partial_1 \Psi_0 \partial_1 \varphi \, \mathrm{d}\omega = 0, \tag{16.27}$$

$$\int_\omega \partial_1 \Psi_1 \partial_1 \varphi \, \mathrm{d}\omega = 0, \tag{16.28}$$

$$\int_\omega \partial_1 \Psi_2 \partial_1 \varphi \, \mathrm{d}\omega = -\int_\omega \partial_2 \Psi_0 \partial_2 \varphi \, \mathrm{d}\omega + 2\int_\omega \varphi \, \mathrm{d}\omega, \tag{16.29}$$

$$\int_\omega \partial_1 \Psi_p \partial_1 \varphi \, \mathrm{d}\omega = -\int_\omega \partial_2 \Psi_{p-2} \partial_2 \varphi \, \mathrm{d}\omega \quad (p \geqslant 2). \tag{16.30}$$

Equation (16.27) implies that Ψ_0 depends only on variable x_2. From the boundary condition $\Psi_0 = 0$ on γ_v we conclude that $\Psi_0 = 0$. In a similar way, from (16.28), we obtain that $\Psi_1 = 0$. Consequently, Eq. (16.29) becomes

$$\int_\omega \partial_1 \Psi_2 \partial_1 \varphi \, \mathrm{d}\omega = 2\int_\omega \varphi \, \mathrm{d}\omega, \tag{16.31}$$

from which we have

$$-\partial_{11}\Psi_2 = 2.$$

Integrating, we conclude that Ψ_2 is of the following form:

$$\Psi_2(x_1, x_2) = -x_1^2 + c_1(x_2)x_1 + c_0(x_2),$$

where functions c_0 and c_1 depend only on variable x_2 and are such that

$$c_0,\ c_1 \in H^1(-\tfrac{1}{2}b, \tfrac{1}{2}b).$$

From condition $\Psi_2 = 0$ on γ_h, we have

$$c_0,\ c_1 \in H_0^1(-\tfrac{1}{2}b, \tfrac{1}{2}b)$$

and from condition $\Psi_2 = 0$ on γ_v, we deduce

$$c_1(x_2) = 0, \qquad c_0(x_2) = \tfrac{1}{4}, \qquad x_2 \in (-\tfrac{1}{2}b, \tfrac{1}{2}b).$$

Consequently,

$$\Psi_2(x_1, x_2) = -(x_1^2 - \tfrac{1}{4}).$$

However, *function Ψ_2 does not belong to $V_1(\omega)$ since it does not vanish on γ_h*. This is a boundary layer phenomenon (cf. LIONS [1973, Chapter III, Remarque 2.3]) showing that an asymptotic expansion of the type (16.26) cannot converge to Ψ in $V_1(\omega)$. The convergence result must hold in a larger space (see Theorem 17.1).

REMARK 16.1. Since $\partial_2 \Psi_2 = 0$, from (16.30), one has

$$\int_\omega \partial_1 \Psi_p \partial_1 \varphi \, d\omega = 0, \quad \text{for all } \varphi \in V_1(\omega), \ (p \geqslant 3),$$

consequently, $\Psi_p = 0$, for $p \geqslant 3$. Thus, *if we neglect the boundary condition $\Psi(t) = 0$ on γ_h* we obtain that $\Psi(t)$ is approximated by

$$\Psi(t)(x_1, x_2) \simeq -t^2(x_1^2 - \tfrac{1}{4}). \tag{16.32}$$

In a similar way for function $w(t)$ the unique solution of problem (16.21), considering a *formal expansion* in $V_2(\omega)$ of the form

$$w(t) = w_0 + t w_1 + t^2 w_2 + \text{h.o.t.}, \quad w_i \in V_2(\omega), \tag{16.33}$$

and substituting into (16.21) we obtain the following set of equations, for all $\varphi \in V_2(\omega)$:

$$\int_\omega \partial_1 w_0 \partial_1 \varphi \, d\omega = 0, \tag{16.34}$$

$$\int_\omega \partial_1 w_1 \partial_1 \varphi \, d\omega = \int_{\gamma_0} x_2 n_1 \varphi \, d\gamma, \tag{16.35}$$

$$\int_\omega \partial_1 w_2 \partial_1 \varphi \, d\omega = -\int_\omega \partial_2 w_0 \partial_2 \varphi \, d\omega, \tag{16.36}$$

$$\int_\omega \partial_1 w_3 \partial_1 \varphi \, d\omega = -\int_\omega \partial_2 w_1 \partial_2 \varphi \, d\omega - \int_{\gamma_h} x_1 n_2 \varphi \, d\gamma, \tag{16.37}$$

$$\int_\omega \partial_1 w_p \partial_1 \varphi \, d\omega = -\int_\omega \partial_2 w_{p-2} \partial_2 \varphi \, d\omega \quad (p \geqslant 4). \tag{16.38}$$

Equation (16.34) means that w_0 depends only on variable x_2 and consequently

$$w_0 \in V_2(-\tfrac{1}{2}b, \tfrac{1}{2}b) = \{z \in H^1(-\tfrac{1}{2}b, \tfrac{1}{2}b) \colon \int_{-b/2}^{b/2} z(x_2) \, dx_2 = 0\}. \tag{16.39}$$

Setting $\varphi = \varphi^0$, $\varphi^0 \in V_2(-\frac{1}{2}b, \frac{1}{2}b)$ into Eq. (16.36) we obtain

$$\int_{-b/2}^{b/2} \partial_2 w_0 \partial_2 \varphi^0 \, \mathrm{d}x_2 = 0, \quad \text{for all } \varphi^0 \in V_2(-\tfrac{1}{2}b, \tfrac{1}{2}b), \tag{16.40}$$

which together with (16.39) gives $w_0 = 0$.

Equation (16.35) is formally equivalent to

$$\partial_{11} w_1 = 0, \qquad \partial_1 w_1 n_1 = x_2 n_1, \qquad \text{on } \gamma_v,$$

from which we conclude that w_1 is of the following form:

$$w_1(x_1, x_2) = x_1 x_2 + c_0(x_2).$$

Condition $w_1 \in V_2(\omega)$ implies that $c_0 \in V_2(-\frac{1}{2}b, \frac{1}{2}b)$.

Setting $\varphi = \varphi^0$, $\varphi^0 \in V_2(-\frac{1}{2}b, \frac{1}{2}b)$ into Eq. (16.37) we obtain c_0 must be a solution of (16.40). Consequently, $c_0 = 0$ and

$$w_1(x_1, x_2) = x_1 x_2.$$

Equations (16.36) and (16.38) for $p = 4$ give now

$$\int_\omega \partial_1 w_2 \partial_1 \varphi \, \mathrm{d}\omega = 0,$$

$$\int_\omega \partial_2 w_2 \partial_2 \varphi \, \mathrm{d}\omega = -\int_\omega \partial_1 w_4 \partial_1 \varphi \, \mathrm{d}\omega,$$

from which as before we conclude that $w_2 = 0$.

Substituting these results into Eq. (16.37) we obtain

$$\begin{aligned}\int_\omega \partial_1 w_3 \partial_1 \varphi \, \mathrm{d}\omega &= -\int_\omega x_1 \partial_2 \varphi \, \mathrm{d}\omega - \int_{\gamma_h} x_1 n_2 \varphi \, \mathrm{d}\gamma \\ &= -2\int_{\gamma_h} x_1 n_2 \varphi \, \mathrm{d}\gamma, \quad \text{for all } \varphi \in V_2(\omega).\end{aligned} \tag{16.41}$$

However, *problem* (16.41) *does not have a solution* in $V_2(\omega)$. In fact, (16.41) is formally equivalent to

$$\begin{aligned}&\partial_{11} w_3 = 0, \quad \text{in } \omega, \\ &\int_{\gamma_v} \partial_1 w_3 n_1 \varphi \, \mathrm{d}\gamma = -2\int_{\gamma_h} x_1 n_2 \varphi \, \mathrm{d}\gamma, \quad \text{for all } \varphi \in V_2(\omega).\end{aligned} \tag{16.42}$$

Consequently, w_3 must be of the following form:

$$w_3(x_1, x_2) = c_1(x_2)x_1 + c_0(x_2),$$

where functions c_0 and c_1 depend only on variable x_2. From the second condition in (16.42) we conclude that $c_1(x_2)$ must verify the following condition:

$$\int_{\gamma_v} c_1(x_2)n_1\varphi \, d\gamma = -2\int_{\gamma_h} x_1 n_2 \varphi \, d\gamma, \quad \text{for all } \varphi \in V_2(\omega). \tag{16.43}$$

But *this is impossible*. In order to see this we take φ of the following form:

$$\varphi \in V_2(\omega), \qquad \varphi = 0 \quad \text{in } \gamma_v, \qquad \varphi(x_1, \tfrac{1}{2}b) = 0, \qquad -\tfrac{1}{2} \leqslant x_1 \leqslant \tfrac{1}{2},$$
$$\varphi(x_1, -\tfrac{1}{2}b) = \begin{cases} x_1 & \text{if } -\tfrac{1}{4} \leqslant x_1 \leqslant \tfrac{1}{4}, \\ 0 & \text{otherwise.} \end{cases}$$

The nonexistence of solution in space $V_2(\omega)$ is once again due to a boundary layer phenomenon. *The asymptotic expansion* (16.33) *is not convergent in* $V_2(\omega)$.

REMARK 16.2. If one weakens the second condition in (16.42) and replaces it by

$$\partial_1 w_3 n_1 = 0, \quad \text{on } \gamma_v,$$

then, one has $w_3 = 0$ and from (16.38) one concludes that $w_p = 0$, $p \geqslant 2$. Thus, *if one neglects the boundary condition* $\partial_n w(t) = x_2 n_1 - x_1 n_2$ *on* γ_h we *formally* obtain that $w(t)$ is approximated by

$$w(t) \simeq t x_1 x_2. \tag{16.44}$$

An analogous calculation for all the other functions $\eta_\beta(t)$ and $\theta_\beta(t)$ was done by BARROS, TRABUCHO and VIAÑO [1989] and by BARROS [1989]. The results are as follows:

$$\begin{aligned}
&\Psi(t) \simeq t^2(\tfrac{1}{4} - x_1^2),\\
&w(t) \simeq t x_1 x_2,\\
&\eta_1(t) \simeq t^3\Big(\frac{x_1^3}{3} - \frac{x_1}{4}\Big),\\
&\eta_2(t) \simeq \frac{x_2^3}{3} - b^2\frac{x_2}{4},\\
&\theta_1(t) \simeq t\frac{x_1 x_2^2}{2} + t^3\Big(-\frac{x_1^3}{2} + \frac{x_1}{4}\Big),\\
&\theta_2(t) \simeq -\frac{x_2^3}{6} - t^2\frac{x_1^2 x_2}{2}.
\end{aligned} \tag{16.45}$$

Once again we point out the *formal character of these calculations, of which the final obtainment depends on the way the boundary conditions on γ_h were weakened*. Its validity *will be established in a rigorous way through a convergence result* (cf. Section 18). Nevertheless, these results give a simple and clear way to justify the classical constants used in engineering literature.

Now using the inverse transformation of (16.18) we obtain that functions Ψ^t, w^t, η^t_β and θ^t_β are formally approximated in ω^t, as t becomes very small, by

$$\begin{aligned}
\Psi^t &\simeq \frac{t^2}{4} - (x_1^t)^2,\\
w^t &\simeq x_1^t x_2^t,\\
\eta_1^t &\simeq \frac{(x_1^t)^3}{3} - t^2\frac{x_1^t}{4},\\
\eta_2^t &\simeq \frac{(x_2^t)^2}{3} - b^2\frac{x_2^t}{4},\\
\theta_1^t &\simeq \frac{x_1^t}{2}[(x_2^t)^2 - (x_1^t)^2] + t^2\frac{x_1^t}{4},\\
\theta_2^t &\simeq -\frac{(x_2^t)^3}{6} - \frac{(x_1^t)^2 x_2^t}{2}.
\end{aligned} \tag{16.46}$$

REMARK 16.3. We remark that only η_1 and η_2 are exact solutions of the respective boundary value problems. This is due to the homogeneous (Neumann) boundary condition for these two functions.

Now using the approximations (16.46) we may calculate a still *formal approximation of the constants characterizing the geometry of the cross section*, valid for t very small. One then has (cf. BARROS [1989])

$$A^t = bt, \tag{16.47}$$

$$I_1^t = \frac{bt^3}{12}, \qquad I_2^t = \frac{b^3t}{12}, \qquad H_3^t = \frac{b^5t}{320} + \frac{b^3t^3}{288} + \frac{bt^5}{320}, \tag{16.48}$$

$$J^t \simeq \frac{bt^3}{3}, \qquad J_w^t \simeq \frac{b^3t^3}{144}, \tag{16.49}$$

$$L_{11}^{\eta^t} = -\frac{bt^5}{60}, \qquad L_{22}^{\eta^t} = -\frac{b^5t}{60}, \tag{16.50}$$

$$K_{11}^{\eta^t} = \frac{b^3t^3}{144} - \frac{bt^5}{240}, \qquad K_{22}^{\eta^t} = \frac{b^3t^3}{144} - \frac{b^5t}{240}, \tag{16.51}$$

$$L_{11}^{\theta^t} \simeq \frac{7bt^5}{480} + \frac{b^3t^3}{288}, \qquad L_{22}^{\theta^t} \simeq -\frac{b^5t}{480} + \frac{b^3t^3}{288}, \tag{16.52}$$

$$K_{11}^{\theta^t} \simeq -\frac{b^5t}{320} + \frac{b^3t^3}{288} + \frac{bt^5}{960}, \qquad K_{22}^{\theta^t} \simeq -\frac{b^5t}{320} - \frac{b^3t^3}{144} + \frac{bt^5}{320}. \tag{16.53}$$

In spite of the formal aspect of all these calculations, we remark that in (16.49) we recovered the classical torsion and warping constants for thin rectangular cross sections (cf. (16.8) and (16.9)). It thus seems that the asymptotic expansion method is an appropriate tool for the study of the properties of thin-walled cross sections. We remark, however, that all the information is already contained in the functions and constants derived via the asymptotic expansion method, which is valid for any type of cross section. The main purpose of these calculations is to show that they include and generalize the constants used in engineering literature. In the next section we give a rigorous mathematical proof of these results.

17. Convergence and limit problem for Poisson's equation in a thin rectangle with Dirichlet or Neumann boundary conditions

As was pointed out in the previous section, the mathematical justification of the asymptotic expansion method just used needs a convergence result that cannot be established in $V_\alpha(\omega)$ (that is in $\|\cdot\|_{1,\omega}$).

The previous analysis for the Dirichlet problem suggests that the space where convergence holds must be such that the trace boundary condition on γ_h does not make sense. This is also an indication that the derivatives with respect to variable x_2 cannot be controlled (in the L^2 sense). The natural space for the convergence study seems to be

$$H = \{\varphi \in L^2(\omega)\colon\ \partial_1\varphi \in L^2(\omega),\ \varphi = 0 \text{ on } \gamma_v\}, \tag{17.1}$$

equipped with the norm

$$\|\varphi\|_H = [\|\varphi\|^2_{0,\omega} + \|\partial_1\varphi\|^2_{0,\omega}]^{1/2}. \tag{17.2}$$

The space H is isometric to $H^1_0[-\frac{1}{2},\frac{1}{2}; L^2(-\frac{1}{2}b,\frac{1}{2}b)]$ with the natural identification $\varphi(x_1)(x_2) = \varphi(x_1,x_2)$ (cf. BREZIS [1983], LIONS and MAGENES [1968]). Consequently

$$H = \{\varphi \in H^1[-\tfrac{1}{2},\tfrac{1}{2}; L^2(-\tfrac{1}{2}b,\tfrac{1}{2}b)]\colon\ \varphi(-\tfrac{1}{2},x_2) = \varphi(\tfrac{1}{2},x_2) = 0, x_2 \in (-\tfrac{1}{2}b,\tfrac{1}{2}b)\}.$$

Moreover, the application

$$\varphi \in H \to |\varphi|_H = \|\partial_1\varphi\|_{0,\omega}, \tag{17.3}$$

is a norm equivalent to the $\|\cdot\|_H$ norm.

The convergence results for the approximations (16.45) can be obtained as particular cases of Theorems 17.1 and 17.2. They establish the convergence in the space H (as $t \to 0$) of the solution of a problem of the Dirichlet type (as (16.20)) and of the Neumann type (as (16.21)), respectively, together with the identification of the limit problems. These results are owed to RODRÍGUEZ [1990] and to RODRÍGUEZ and VIAÑO [1991, 1992a,b].

REMARK 17.1. In this and the next two sections all the convergences are referred to $t \to 0$.

THEOREM 17.1. *Consider a family of functions $(F(t))_{t>0}$ in $L^2(\omega)$ and let $\Psi(t)$ be the unique solution of the following problem:*

$$\begin{aligned} &\Psi(t) \in H_0^1(\omega),\\ &\int_\omega \partial_1 \Psi(t)\partial_1 \varphi \,\mathrm{d}\omega + t^2 \int_\omega \partial_2 \Psi(t)\partial_2 \varphi \,\mathrm{d}\omega \\ &\quad = t^2 \int_\omega F(t)\varphi \,\mathrm{d}\omega, \quad \text{for all } \varphi \in H_0^1(\omega). \end{aligned} \tag{17.4}$$

If there exists a real k such that

$$t^{-k}F(t) \to \tilde{F}, \quad \text{in } L^2(\omega), \tag{17.5}$$

then,

$$\begin{aligned} &t^{-k}\Psi(t) \to 0, && \text{in } H^1(\omega),\\ &t^{-k-2}\Psi(t) \to \tilde{\Psi}, && \text{in } H \end{aligned} \tag{17.6}$$

where $\tilde{\Psi}$ is the unique solution of the following variational problem:

$$\begin{aligned} &\tilde{\Psi} \in H,\\ &\int_\omega \partial_1 \tilde{\Psi}\partial_1 \varphi \,\mathrm{d}\omega = \int_\omega \tilde{F}\varphi \,\mathrm{d}\omega, \quad \text{for all } \varphi \in H. \end{aligned} \tag{17.7}$$

PROOF. We hereinafter summarize the proof owed to RODRÍGUEZ [1990] and to RODRÍGUEZ and VIAÑO [1991, 1992a,b].

Taking $\varphi = t^{-k}\Psi(t)$ in (17.4), one has

$$t^{-k}\|\partial_1 \Psi(t)\|_{0,\omega}^2 + t^{-k+2}\|\partial_2 \Psi(t)\|_{0,\omega}^2 = t^{-k}\int_\omega F(t)\Psi(t)\,\mathrm{d}\omega,$$

from which we conclude that the sequence $t^{-k}\Psi(t)$ is bounded in $H_0^1(\omega)$. Therefore, there exists a subsequence, denoted in the same way, converging weakly in $H_0^1(\omega)$ to a function that we denote by $\tilde{\Psi}_0$. From (17.4) one obtains

$$\begin{aligned} &t^{-k}\int_\omega \partial_1 \Psi(t)\partial_1 \varphi \,\mathrm{d}\omega + t^{-k+2}\int_\omega \partial_2 \Psi(t)\partial_2 \varphi \,\mathrm{d}\omega \\ &\quad = t^{-k+2}\int_\omega F(t)\varphi \,\mathrm{d}\omega, \quad \text{for all } \varphi \in H_0^1(\omega). \end{aligned}$$

Passing to the limit one gets

$$\begin{aligned} &\tilde{\Psi}_0 \in H_0^1(\omega),\\ &\int_\omega \partial_1 \tilde{\Psi}_0\partial_1 \varphi \,\mathrm{d}\omega = 0, \quad \text{for all } \varphi \in H_0^1(\omega). \end{aligned}$$

This implies that $\tilde{\Psi}_0$ depends only on variable x_2 and as it vanishes on the boundary of ω we must have $\tilde{\Psi}_0 = 0$.

Setting $\varphi = t^{-2k-2}\Psi(t)$ in (17.4) on obtains

$$t^{-2k-2}\|\partial_1 \Psi(t)\|_{0,\omega}^2 + t^{-2k}\|\partial_2 \Psi(t)\|_{0,\omega}^2 = t^{-2k} \int_\omega F(t)\Psi(t)\,\mathrm{d}\omega,$$

from which we conclude that for $t \leqslant 1$:

$$0 \leqslant t^{-2k}\|\Psi(t)\|_{1,\omega}^2 \leqslant t^{-2k} \int_\omega F(t)\Psi(t)\,\mathrm{d}\omega.$$

The right-hand side in this inequality converges to zero due to the weak convergence of $t^{-k}\Psi(t)$ to zero and the strong convergence of $t^{-k}F(t)$ in $L^2(\omega)$, (cf. BREZIS [1983, p. 35]). As this is valid for any convergent subsequence of $t^{-k}\Psi(t)$ one proves that the whole sequence strongly converges in $H^1(\omega)$ and that the limit is zero.

In order to establish the second convergence in (17.6) we consider the following equality obtained from (17.4) by setting $\varphi = t^{-k-2}\Psi(t)$ and taking into account the fact that $t^{-k}F(t)$ is bounded in $L^2(\omega)$:

$$t^{-k-2}\|\partial_1 \Psi(t)\|_{0,\omega}^2 + t^{-k}\|\partial_2 \Psi(t)\|_{0,\omega}^2 \leqslant C\|\Psi(t)\|_{0,\omega}.$$

From the Poincaré–Friedrichs inequality we have

$$\|\Psi(t)\|_{0,\omega} \leqslant C\|\partial_1 \Psi(t)\|_{0,\omega}.$$

Combining these two results one concludes that the sequence $t^{-k-2}\partial_1 \Psi(t)$ is bounded in $L^2(\omega)$. From the equivalence between the norms (17.2) and (17.3) one obtains that $t^{-k-2}\Psi(t)$ is a bounded sequence in H. As a consequence, there exists, at least, a subsequence of $t^{-k-2}\Psi(t)$, denoted in the same way, and which converges weakly in H. We denote by $\tilde{\Psi}$ its limit. From (17.4) one has

$$t^{-k-2} \int_\omega \partial_1 \Psi(t)\partial_1\varphi\,\mathrm{d}\omega + t^{-k} \int_\omega \partial_2\Psi(t)\partial_2\varphi\,\mathrm{d}\omega$$
$$= t^{-k} \int_\omega F(t)\varphi\,\mathrm{d}\omega, \quad \text{for all } \varphi \in H_0^1(\omega).$$

Passing to the limit and considering the $H^1(\omega)$-convergence of $t^{-k}\Psi(t)$, we deduce that

$$\tilde{\Psi} \in H,$$
$$\int_\omega \partial_1 \tilde{\Psi}\partial_1\varphi\,\mathrm{d}\omega = \int_\omega \tilde{F}\varphi\,\mathrm{d}\omega, \quad \text{for all } \varphi \in H_0^1(\omega).$$

Since $H_0^1(\omega)$ is dense in H we conclude, by continuity, that $\tilde{\Psi}$ solves problem (17.7).

This reasoning can be done for any convergent subsequence and thus, one concludes that the whole sequence $t^{-k-2}\Psi(t)$ converges weakly in H.

In order to show the strong convergence we use the equivalence between the norms (17.2) and (17.3). One has

$$C\|t^{-k-2}\Psi(t) - \tilde{\Psi}\|_H^2 \leqslant \|t^{-k-2}\partial_1\Psi(t) - \partial_1\tilde{\Psi}\|_{0,\omega}^2$$
$$= \|t^{-k-2}\partial_1\Psi(t)\|_{0,\omega}^2 - 2t^{-k-2}\int_\omega \partial_1\Psi(t)\partial_1\tilde{\Psi}\,\mathrm{d}\omega + \|\partial_1\tilde{\Psi}\|_{0,\omega}^2.$$

Setting $\varphi = t^{-2k-4}\Psi(t)$ in (17.4) one obtains

$$t^{-2k-4}\|\partial_1\Psi(t)\|_{0,\omega}^2 + t^{-2k-2}\|\partial_2\Psi(t)\|_{0,\omega}^2 = t^{-2k-2}\int_\omega F(t)\Psi(t)\,\mathrm{d}\omega.$$

Combining these results one gets

$$C\|t^{-k-2}\Psi(t) - \tilde{\Psi}\|_H^2$$
$$\leqslant t^{-2k-2}\int_\omega F(t)\Psi(t)\,\mathrm{d}\omega - 2t^{-k-2}\int_\omega \partial_1\Psi(t)\partial_1\tilde{\Psi}\,\mathrm{d}\omega + \|\partial_1\tilde{\Psi}\|_{0,\omega}^2.$$

From the weak convergence of $t^{-k-2}\Psi(t)$ to $\tilde{\Psi}$ in H and from hypothesis (17.5) we conclude that the right-hand side converges to

$$\mathscr{L} = \int_\omega \tilde{F}\tilde{\Psi}\,\mathrm{d}\omega - \|\partial_1\tilde{\Psi}\|_{0,\omega}^2.$$

Since we have already proved that $\tilde{\Psi}$ is a solution of (17.7) we conclude that $\mathscr{L} = 0$ and the strong convergence of $t^{-k-2}\Psi(t)$ to $\tilde{\Psi}$ follows. This concludes the proof of the theorem. □

REMARK 17.2. The $H^1(\omega)$ convergence for the mixed Dirichlet–Neumann problem was obtained in a similar way by LE DRET [1991, Theorem 2.1].

A similar result may be obtained for Neumann type problems with a minimal restriction on the data (which holds in our case) allowing to reduce this case to the previous one. In fact, the problem we study in ω is the scaled equivalent to the following Neumann problem posed in ω^t:

$$\Phi^t \in V_2(\omega^t),$$
$$\int_{\omega^t} \partial_\alpha\Phi^t\partial_\alpha\varphi^t\,\mathrm{d}\omega^t = \int_{\omega^t} h^t\varphi^t\,\mathrm{d}\omega^t + \int_{\gamma^t} g^t\varphi^t\,\mathrm{d}\gamma^t, \quad \text{for all } \varphi^t \in V_2(\omega^t), \tag{17.8}$$

with h^t and g^t of the following form:

$$h^t = \partial_a lpha^t f_\alpha^t, \qquad g^t = -f_\alpha^t n_\alpha^t, \qquad f_\alpha^t \in H^1(\omega^t). \tag{17.9}$$

We remark that every problem of the type (17.8) may be reduced to an equivalent which has data verifying (17.9). In order to do so it is enough to consider $\zeta^t \in H^1(\omega^t)$ such that $\partial_2^t \zeta^t n_1^t - \partial_1^t \zeta^t n_2^t = 0$ on γ^t and set

$$f_1^t = -\partial_1 \varphi^t + \partial_2 \zeta^t,$$
$$f_2^t = -\partial_2 \varphi^t - \partial_1 \zeta^t.$$

THEOREM 17.2. *Consider a family of functions $(f_\alpha(t))_{t>0} \in (H^1(\omega))^2$, and let $\Phi(t)$ be the unique solution of the following variational problem:*

$$\begin{aligned} &\Phi(t) \in V_2(\omega), \\ &\int_\omega \partial_1 \Phi(t) \partial_1 \varphi \, \mathrm{d}\omega + t^2 \int_\omega \partial_2 \Phi(t) \partial_2 \varphi \, \mathrm{d}\omega \\ &\quad = t \left[\int_\omega \partial_1 f_1(t) \varphi \, \mathrm{d}\omega - \int_{\gamma_v} f_1 n_1 \varphi \, \mathrm{d}\gamma \right] \\ &\quad + t^2 \left[\int_\omega \partial_2 f_2(t) \varphi \, \mathrm{d}\omega - \int_{\gamma_h} f_2 n_2 \varphi \, \mathrm{d}\gamma \right], \quad \text{for all } \varphi \in V_2(\omega). \end{aligned} \tag{17.10}$$

If there exists a real constant k such that

$$t^{-k} f_1(t) \to \tilde{f}_1, \qquad t^{-k-1} f_2(t) \to \tilde{f}_2, \qquad \text{in } H^1(\omega), \tag{17.11}$$

then,

$$t^{-k-1} \Phi(t) \to \tilde{\Phi}, \quad \text{in } H^1(\omega), \tag{17.12}$$

where $\tilde{\Phi}$ is the unique function verifying

$$\partial_1 \tilde{\Phi} = -\tilde{f}_1, \qquad \partial_2 \tilde{\Phi} = -\partial_1 \tilde{\Psi} - \tilde{f}_2, \qquad \int_\omega \tilde{\Phi} \, \mathrm{d}\omega = 0, \tag{17.13}$$

and $\tilde{\Psi}$ is the unique solution of the following problem:

$$\begin{aligned} &\tilde{\Psi} \in H, \\ &\int_\omega \partial_1 \tilde{\Psi} \partial_1 \varphi \, \mathrm{d}\omega = \int_\omega (\partial_1 \tilde{f}_2 - \partial_2 \tilde{f}_1) \varphi \, \mathrm{d}\omega, \quad \text{for all } \varphi \in H. \end{aligned} \tag{17.14}$$

PROOF (cf. RODRÍGUEZ [1990] and to RODRÍGUEZ and VIAÑO [1991, 1992a,b]). Let

$$p_1(t) = \partial_1 \Phi(t) + t f_1(t), \qquad p_2(t) = t^2 \partial_2 \Phi(t) + t^2 f_2(t).$$

Then, from (17.10) we have

$$\partial_\alpha p_\alpha(t) = 0, \qquad p_\alpha(t) n_\alpha = 0.$$

Thus, there exists a unique "stream" function $\Psi(t) \in H_0^1(\omega)$ such that (cf. GIRAULT and RAVIART [1981])

$$p_1(t) = \partial_2 \Psi(t), \qquad p_2(t) = -\partial_1 \Psi(t),$$

$$\int_\omega \partial_1 \Psi(t) \partial_1 \varphi \, d\omega + t^2 \int_\omega \partial_2 \Psi(t) \partial_2 \varphi \, d\omega$$

$$= t^2 \int_\omega [\partial_1 f_2(t) - t \partial_2 f_1(t)] \varphi \, d\omega, \quad \text{for all } \varphi \in H_0^1(\omega).$$

Hypothesis (17.11) implies that Theorem 17.1 holds for this case. Therefore, we obtain

$$\begin{aligned} &t^{-k-1} \Psi(t) \to 0, \quad \text{in } H_0^1(\omega), \\ &t^{-k-3} \Psi(t) \to \tilde{\Psi}, \quad \text{in } H, \end{aligned} \tag{17.15}$$

where $\tilde{\Psi}$ is the solution of (17.14). Taking into account the relationship between $\Psi(t)$ and $\Phi(t)$, together with (17.11) and (17.15), one gets

$$\begin{aligned} &t^{-k-1} \partial_1 \Phi(t) \to -\tilde{f}_1, \quad &&\text{in } L^2(\omega), \\ &t^{-k-1} \partial_2 \Phi(t) \to -\partial_1 \tilde{\Psi} - \tilde{f}_2, \quad &&\text{in } L^2(\omega). \end{aligned} \tag{17.16}$$

Since $|\cdot|_{1,\omega}$ is a norm in $V_2(\omega)$ equivalent to the $\|\cdot\|_{1,\omega}$ norm (cf. RABIER and THOMAS [1985, p. 66]) we deduce from (17.16) the convergence of $t^{-k-1}\Phi(t)$ in $H^1(\omega)$ to an element $\tilde{\Phi}$ which must necessarily verify (17.13). □

The following corollary, which is an immediate consequence of this theorem, will be used in the next section.

COROLLARY 17.1. *Let $\zeta(t)$, $g(t) \in H^1(\omega)$ be such that $\Phi(t) = \zeta(t) + g(t)$ is a solution of* (17.10) *with f_α verifying* (17.11). *Then,*

$$\zeta(t) = -g(t) + t^{k+1} \tilde{\Phi} + \varepsilon(t), \tag{17.17}$$

where $\tilde{\Phi}$ verifies (17.13) *and*

$$t^{-k-1} \varepsilon(t) \to 0, \quad \text{in } H^1(\omega).$$

18. Convergence of torsion and warping functions for thin rectangular cross sections

The application of the previous theorems to problems (16.20) and (16.21) allows us to justify the formal calculations summarized in (16.45), from the mathematical point of view. We remark that in Theorems 17.1 and 17.2 not only the convergence issue is treated in the appropiate spaces, but it is also shown how to compute the respective limit; consequently, previous knowledge of the limits (16.45) is not required.

The following results are owed to RODRÍGUEZ [1990] and RODRÍGUEZ and VIAÑO [1992a, b].

We recall that for a given function $\varphi^t : \overline{\omega}^t \to \mathbb{R}$, we denote by $\varphi(t) : \overline{\omega} \to \mathbb{R}$ the function $\varphi(t)(x_1, x_2) = \varphi^t(tx_1, x_2)$.

THEOREM 18.1. *In the rectangle ω^t, function Ψ^t verifies*

$$\Psi^t(x_1^t, x_2^t) = \tfrac{1}{4}t^2 - (x_1^t)^2 + \varepsilon_\Psi^t(x_1^t, x_2^t), \tag{18.1}$$

where $\varepsilon_\Psi^t(x_1^t, x_2^t)$ is such that

$$t^{-2}\varepsilon_\Psi(t) \to 0, \quad \text{in } H. \tag{18.2}$$

PROOF. It is enough to show that:

$$\Psi(t)(x_1, x_2) = t^2(\tfrac{1}{4} - x_1^2) + \varepsilon_\Psi(t)(x_1, x_2), \tag{18.3}$$

with $\varepsilon_\Psi(t)(x_1, x_2)$ verifying (18.2). But this is a consequence of Theorem 17.1 when applied to problem (16.20). In fact, hypothesis (17.5) holds with $k = 0$ and $\tilde{F} = 2$. One thus has

$$t^{-2}\Psi(t) \to \tilde{\Psi}, \quad \text{in } H, \tag{18.4}$$

where $\tilde{\Psi}$ is the unique solution of the following problem:

$$\begin{aligned} &\tilde{\Psi} \in H, \\ &\int_\omega \partial_1 \tilde{\Psi} \partial_1 \varphi \, \mathrm{d}\omega = \int_\omega 2\varphi \, \mathrm{d}\omega, \quad \text{for all } \varphi \in H. \end{aligned} \tag{18.5}$$

Problem (18.5) is equivalent to the following (in the distributional sense):

$$-\partial_{11}\tilde{\Psi} = 2, \quad \text{in } \omega \qquad \tilde{\Psi} = 0, \quad \text{on } \gamma_0,$$

whose unique solution is

$$\tilde{\Psi}(x_1, x_2) = \tfrac{1}{4} - x_1^2. \tag{18.6}$$

From (18.6) and (18.4) one obtains (18.3), which proves the result. □

THEOREM 18.2. *In rectangle ω^t, the warping function w^t is such that*

$$w^t(x_1^t, x_2^t) = x_1^t x_2^t + \varepsilon_w^t(x_1^t, x_2^t), \tag{18.7}$$

where ε_w^t verifies

$$t^{-1}\varepsilon_w(t) \to 0, \qquad \text{in } H^1(\omega). \tag{18.8}$$

PROOF. We apply Theorem (17.2) to problem (16.21) for function $w(t)$. In order to do so it is enough to consider

$$f_1(t)(x_1, x_2) = -x_2, \qquad f_2(t) = tx_1.$$

Hypothesis (17.11) holds with $k = 0$ and one has

$$t^{-1}w(t) \to \tilde{w}, \quad \text{in } H^1(\omega), \tag{18.9}$$

where $\tilde{w}$ is the unique solution of

$$\partial_1 \tilde{w} = x_2, \qquad \partial_2 \tilde{w} = -\partial_1 \tilde{\Psi} - x_1, \qquad \int_\omega \tilde{w}\, d\omega = 0, \tag{18.10}$$

with $\tilde{\Psi}$ the solution of

$$\begin{aligned} &\tilde{\Psi} \in H, \\ &\int_\omega \partial_1 \tilde{\Psi} \partial_1 \varphi \, d\omega = \int_\omega 2\varphi \, d\omega, \quad \text{for all } \varphi \in H. \end{aligned} \tag{18.11}$$

We remark that this is exactly (18.5) and consequently $\tilde{\Psi}$ is the function (18.6) from which we conclude that

$$\tilde{w}(x_1, x_2) = x_1 x_2. \tag{18.12}$$

Properties (18.7) and (18.8) are now a consequence of (18.12) and (18.9). □

REMARK 18.1. (i) These theorems confirm the formal asymptotic expansion calculations (cf. (16.32) and (16.44)).

(ii) We remark that the *approximated functions*

$$\tilde{\Psi}^t(x_1^t, x_2^t) = \tfrac{1}{4}t^2 - (x_1^t)^2, \qquad \tilde{w}^t(x_1^t, x_2^t) = x_1^t x_2^t,$$

satisfy relation (7.30), that is,

$$\partial_2^t \tilde{\Psi}^t = \partial_1^t \tilde{w}^t - x_2^t, \qquad -\partial_1^t \tilde{\Psi}^t = \partial_2^t \tilde{w}^t + x_1^t.$$

We are now in a position to justify, from a mathematical point of view, the expressions that show up in engineering literature for the torsion and warping constants for a thin rectangular cross section (cf. (16.6) and (16.7)). In fact, from Theorems 18.1 and 18.2 one has the following (cf. RODRÍGUEZ and VIAÑO [1992c]).

THEOREM 18.3. *The torsion constant J^t and the warping constant J^t_w for rectangle ω^t verify*

$$J^t = \frac{bt^3}{3} + o(t^3), \tag{18.13}$$

$$J^t_w = \frac{b^3t^3}{144} + o(t^3). \tag{18.14}$$

PROOF. From Theorem 18.1 we have

$$J^t = 2\int_{\omega} \Psi^t \, d\omega^t = \frac{bt^3}{3} + 2\int_{\omega^t} \varepsilon^t_{\Psi} \, d\omega^t,$$

$$t^{-3}\int_{\omega^t} \varepsilon^t_{\Psi} \, d\omega^t = t^{-2}\int_{\omega} \varepsilon_{\Psi}(t) \, d\omega \leqslant b^{1/2}t^{-2}\|\varepsilon_{\Psi}(t)\|_{0,\omega} \to 0.$$

In a similar way, for the warping constant one has

$$J^t_w = \int_{\omega^t} (w^t)^2 \, d\omega^t = t\int_{\omega} (w(t))^2 \, d\omega = t\|w(t)\|^2_{0,\omega}.$$

From Theorem 18.2 we obtain

$$t^{-1}w(t) \to x_1x_2, \quad \text{in } H^1(\omega).$$

Therefore,

$$t^{-3}J^t_w = \|t^{-1}w(t)\|^2_{0,\omega} \to \|x_1x_2\|^2_{0,\omega} = \frac{b^3}{144}. \qquad \square$$

In Table 18.1 we show the comparison between the values for the torsion constant J^t in rectangle ω^t using the asymptotic expression $bt^3/3$, the classical expression $c(t)bt^3$ (cf. Table 16.1), and the values obtained by the finite element method to evaluate function Ψ^t and by a numerical integration to calculate the torsion constant (J^*). The values of J^* depend on the mesh and their accuracy decreases as t becomes very small.

In Table 18.2 we show the comparison between the values for the warping constant using the asymptotic (also classical) expression $b^3t^3/144$ and those obtained via the finite element method (J^*_w).

TABLE 18.1. Numerical and asymptotic values of J^t for a rectangle ω^t

b/t	J^*	$c(t)bt^3$	$bt^3/3$
1.0	0.1399	0.1410	0.3333
1.5	0.2911	0.2940	0.5000
2.0	0.4545	0.4580	0.6666
2.5	0.6195	0.6225	0.8333
3.0	0.7854	0.7890	1.0000
4.0	1.1082	1.1080	1.3333
5.0	1.4349	1.4550	1.6666
10.0	3.0849	3.1200	3.3333
20.0	6.3739	6.6666	6.6666
40.0	12.2293	13.3333	13.3333

TABLE 18.2. Numerical and asymptotic values of J_w^t for a rectangle ω^t.

$\frac{b}{t}$	$\frac{b^3t^3}{144}$	J_w^*
1.0	0.006944	0.000134
2.0	0.055556	0.020257
3.0	0.187500	0.119123
4.0	0.444444	0.342234
5.0	0.868055	0.730487
10.0	6.944444	6.631916
20.0	55.555556	54.816360
40.0	444.444444	441.338083

19. Convergence of Timoshenko's functions and constants for thin rectangular cross sections

This section is dedicated to the convergence properties of the functions related to Timoshenko's beam theory: η_β^t and θ_β^t. These results are contained in RODRÍGUEZ and VIAÑO [1992c]. Functions η_β may be evaluated *exactly*.

THEOREM 19.1. *Functions η_β in rectangle ω^t may be evaluated exactly and are given by*

$$\eta_1^t(x_1^t, x_2^t) = \frac{(x_1^t)^3}{3} - \frac{t^2 x_1^t}{4}, \tag{19.1}$$

$$\eta_2^t(x_1^t, x_2^t) = \frac{(x_2^t)^3}{3} - \frac{b^2 x_2^t}{4}. \tag{19.2}$$

REMARK 19.1. (i) These functions are also obtained if one applies Theorem 17.2

to problems (16.22) and (16.23), respectively, with

$$f_1(t)(x_1,x_2) = t^2(\tfrac{1}{4} - x_1^2), \qquad f_2(t) = 0,$$
$$f_1(t) = 0, \qquad f_2(t) = \frac{b^2}{4} - x_2^2.$$

(ii) Comparing (19.1) and (19.2) with the corresponding formulas (16.46) we conclude that the formal asymptotic expansion is exact in this case.

THEOREM 19.2. *In rectangle* ω^t, *the function* θ_1^t *is such that*

$$\theta_1^t(x_1^t,x_2^t) = \tfrac{1}{2}x_1^t[(x_2^t)^2 - \tfrac{1}{3}(x_1^t)^2] + \varepsilon_{\theta_1}^t(x_1^t,x_2^t), \tag{19.3}$$

where $\varepsilon_{\theta_1}^t$ *satisfies*

$$t^{-1}\varepsilon_{\theta_1}(t) \to 0, \quad in\ H^1(\omega), \tag{19.4}$$

PROOF. The result follows by applying Corollary 17.1 to problem (16.24). In order to do so let us consider

$$\zeta_1^t(x_1^t,x_2^t) = \theta_1^t(x_1^t,x_2^t) + \tfrac{1}{6}(x_1^t)^3,$$

which is a solution of the following problem:

$$-\partial_{\alpha\alpha}\zeta_1^t = x_1^t, \quad \text{in } \omega^t,$$
$$\partial_n^t\zeta_1^t = \tfrac{1}{2}(x_2^t)^2 n_1^t - x_1^t x_2^t n_2^t, \quad \text{in } \gamma^t,$$
$$\int_{\omega^t} \zeta_1^t \, \mathrm{d}\omega^t = 0.$$

If we consider the following functions

$$f_1^t(x_1^t,x_2^t) = -\tfrac{1}{2}(x_2^t)^2, \qquad f_2^t(x_1^t,x_2^t) = x_1^t x_2^t,$$

it is possible to show that they verify

$$\partial_a lpha^t f_\alpha^t = x_1^t, \qquad -f_\alpha^t n_\alpha^t = \tfrac{1}{2}(x_2^t)^2 n_1^t - x_1^t x_2^t n_2^t.$$

Consequently, Theorem 17.2 and Corollary 17.1 may be applied with

$$f_1(t)(x_1,x_2) = -\tfrac{1}{2}x_2^2, \qquad f_2(t)(x_1,x_2) = tx_1x_2.$$

For $k = 0$ one has

$$\tilde{f}_1 = -\tfrac{1}{2}x_2^2, \qquad \tilde{f}_2 = x_1x_2,$$

thus,

$$\theta_1(t) = -\tfrac{1}{6}t^3x_1^3 + t\tilde{\Phi} + \varepsilon_{\theta_1}(t), \tag{19.5}$$

where

$$t^{-1}\varepsilon_{\theta_1}(t) \to 0, \quad \text{in } H^1(\omega),$$

and $\tilde{\Phi} \in H^1(\omega)$ is the unique function such that

$$\partial_1\tilde{\Phi} = \tfrac{1}{2}x_2^2, \qquad \partial_2\tilde{\Phi} = -\partial_1\tilde{\Psi} - x_1x_2, \qquad \int_\omega \tilde{\Phi}\,\mathrm{d}\omega = 0, \tag{19.6}$$

with $\tilde{\Psi} \in H$ verifying

$$\int_\omega \partial_1\tilde{\Psi}\partial_1\varphi\,\mathrm{d}\omega = \int_\omega 2x_2\varphi\,\mathrm{d}\omega, \quad \text{for all } \varphi \in H.$$

We then obtain

$$\tilde{\Psi}(x_1, x_2) = -x_2(x_1^2 - \tfrac{1}{4}), \qquad \tilde{\Phi}(x_1, x_2) = \tfrac{1}{2}x_2^2x_1.$$

The result now follows from (19.5) and (19.6). □

An equivalent result for θ_2^t may be obtained in a similar way.

THEOREM 19.3. *In rectangle ω^t function θ_2^t is such that*

$$\theta_2^t(x_1^t, x_2^t) = -\tfrac{1}{2}x_2^t[(x_1^t)^2 + \tfrac{1}{3}(x_2^t)^2] + \tfrac{1}{12}t^2x_2^t + \varepsilon_{\theta_2}^t(x_1^t, x_2^t), \tag{19.7}$$

where $\varepsilon_{\theta_2}^t$ satisfies

$$t^{-2}\varepsilon_{\theta_2}(t) \to 0, \quad \textit{in } H^1(\omega). \tag{19.8}$$

PROOF. The following function

$$\zeta_2^t(x_1^t, x_2^t) = \theta_2^t(x_1^t, x_2^t) + \tfrac{1}{6}(x_2^t)^3$$

satisfies the problem

$$\begin{aligned}
&-\partial_{\alpha\alpha}^t\zeta_2^t = x_2^t, \quad \text{in } \omega^t,\\
&\partial_n^t\zeta_2^t = -x_1^tx_2^tn_1^t + \tfrac{1}{2}(x_1^t)^2n_2^t, \quad \text{on } \gamma^t,\\
&\int_{\omega^t} \zeta_2^t\,\mathrm{d}\omega^t = 0.
\end{aligned}$$

Functions,

$$f_1^t(x_1^t, x_2^t) = x_1^t x_2^t, \qquad f_2^t(x_1^t, x_2^t) = -\tfrac{1}{2}(x_1^t)^2,$$

are such that

$$\partial_a lpha^t f_\alpha^t = x_2^t, \qquad -f_\alpha^t n_\alpha^t = -x_1^t x_2^t n_1^t + \tfrac{1}{2}(x_1^t)^2 n_2^t,$$

consequently, the behaviour of $\zeta_2(t)$ may be studied by applying Corollary 17.1 with

$$f_1(t)(x_1, x_2) = t x_1 x_2, \qquad f_2(t)(x_1, x_2) = -\tfrac{1}{2} t^2 x_1^2.$$

For $k = 1$ one has

$$\tilde{f}_1 = x_1 x_2, \qquad \tilde{f}_2 - \tfrac{1}{2} x_1^2,$$

thus,

$$\theta_2(t) = -\tfrac{1}{6} x_2^3 + t^2 \tilde{\Phi} + \varepsilon_{\theta_2}(t), \tag{19.9}$$

where

$$t^{-2} \varepsilon_{\theta_2}(t) \to 0, \quad \text{in } H^1(\omega), \tag{19.10}$$

and $\tilde{\Phi} \in H^1(\omega)$ is the unique function such that

$$\partial_1 \tilde{\Phi} = -x_1 x_2, \qquad \partial_2 \tilde{\Phi} = -\partial_1 \tilde{\Psi} + \tfrac{1}{2} x_1^2, \qquad \int_\omega \tilde{\Phi} \, \mathrm{d}\omega = 0,$$

with $\tilde{\Psi} \in H$ verifying

$$\int_\omega \partial_1 \tilde{\Psi} \partial_1 \varphi \, \mathrm{d}\omega = -\int_\omega 2 x_1 \varphi \, \mathrm{d}\omega, \quad \text{for all } \varphi \in H.$$

Solving the above problem, we obtain

$$\tilde{\Psi}(x_1, x_2) = \tfrac{1}{3} x_1 (x_1^2 - \tfrac{1}{4}), \qquad \tilde{\Phi}(x_1, x_2) = -\tfrac{1}{2} x_2 (x_1^2 - \tfrac{1}{6}).$$

The result now follows from (19.9) and (19.10). □

REMARK 19.2. The final expressions (16.46) for θ_1^t and for θ_2^t obtained from (16.45) are not exactly the same as those obtained from the convergence proof. In the case of θ_1^t there is a term missing, and for θ_2^t in the convergence analysis we find more terms than in the asymptotic expansion method. This confirms that the final result via the formal asymptotic expansion strongly depends on the way we relax the conditions on γ_h.

The last part of this section is devoted to the approximation of Timoshenko's constants for the thin rectangular region ω^t. In order to evaluate them we use (the exact) functions η^t_β and (the approximate) functions θ^t_β given in Theorems 19.2 and 19.3. We have the following.

THEOREM 19.4. *For the thin rectangular section ω^t we obtain*

$$\begin{aligned} L^{\eta^t}_{11} &= -\frac{bt^5}{60}, \\ L^{\eta^t}_{22} &= -\frac{b^5t}{60}, \\ K^{\eta^t}_{11} &= \frac{b^3t^3}{144} - \frac{bt^5}{240}, \\ K^{\eta^t}_{22} &= \frac{b^3t^3}{144} - \frac{b^5t}{240}. \end{aligned} \tag{19.11}$$

THEOREM 19.5. *For the thin rectangular section ω^t we obtain*

$$\begin{aligned} L^{\theta^t}_{11} &= \frac{b^3t^3}{288} - \frac{bt^5}{480}, \\ L^{\theta^t}_{22} &= \frac{b^3t^3}{288} - \frac{b^5t}{480}. \end{aligned} \tag{19.12}$$

PROOF. It is enough to consider the identity $K^{\eta^t}_{\alpha\beta} = 2L^{\theta^t}_{\alpha\beta}$. □

COROLLARY 19.1. *Functions $\varepsilon^t_{\theta_1}$ and $\varepsilon^t_{\theta_2}$ verify*

$$\begin{aligned} \int_{\omega^t} x^t_1 \varepsilon^t_{\theta_1}\, \mathrm{d}\omega^t &= 0, \\ \int_{\omega^t} x^t_2 \varepsilon^t_{\theta_2}\, \mathrm{d}\omega^t &= 0. \end{aligned} \tag{19.13}$$

PROOF. From Theorems 19.2 and 19.3 one has

$$L^{\theta^t}_{11} = \int_{\omega^t} x^t_1 \theta^t_1\, \mathrm{d}\omega^t = \frac{b^3t^3}{288} - \frac{bt^5}{480} + \int_{\omega^t} x^t_1 \varepsilon^t_{\theta_1}\, \mathrm{d}\omega^t,$$

$$L^{\theta^t}_{22} = \int_{\omega^t} x^t_2 \theta^t_2\, \mathrm{d}\omega^t = \frac{b^3t^3}{288} - \frac{b^5t}{480} + \int_{\omega^t} x^t_2 \varepsilon^t_{\theta_2}\, \mathrm{d}\omega^t.$$

Since expressions (19.12) are exact the result follows. □

REMARK 19.3. The formal asymptotic expressions (16.52) for $L_{11}^{\theta^t}$ do not coincide with the exact value.

The calculation of constants $K_{\alpha\beta}^{\theta^t}$ is slightly more complicated. The result is as follows.

THEOREM 19.6. *For the thin rectangular section* ω^t *we obtain*

$$K_{11}^{\theta^t} = -\frac{b^5 t}{320} + \frac{7b^3t^3}{288} - \frac{bt^5}{320} + \mathrm{o}(t^3), \tag{19.14}$$

$$K_{22}^{\theta^t} = -\frac{b^5 t}{320} + \frac{b^3t^3}{288} - \frac{7bt^5}{2880} + \mathrm{o}(t^5). \tag{19.15}$$

PROOF. From the definition of constant $K_{11}^{\theta^t}$, we have

$$\begin{aligned} K_{11}^{\theta^t} &= \int_{\omega^t} \Phi_{1\mu}^t \partial_\mu^t \theta_1^t \,\mathrm{d}\omega^t \\ &= \int_{\omega^t} \frac{1}{2}[(x_1^t)^2 - (x_2^t)^2]\partial_1^t \theta_1^t \,\mathrm{d}\omega^t + \int_{\omega^t} x_1^t x_2^t \partial_2^t \theta_1^t \,\mathrm{d}\omega^t. \end{aligned} \tag{19.16}$$

The term

$$-\tfrac{1}{2}\int_{\omega^t} (x_2^t)^2 \partial_1^t \theta_1^t \,\mathrm{d}\omega^t = -\tfrac{1}{2}\int_{\omega} x_2^2 \partial_1 \theta_1(t) \,\mathrm{d}\omega,$$

does not have the correct exponent in t in order to pass to the limit (cf. (19.4)). Therefore, it has to be evaluated explicitly. Using the variational formulation of θ_1^t one has

$$\int_{\omega^t} \partial_a lpha^t \theta_1^t \partial_a lpha^t \varphi^t \,\mathrm{d}\omega^t = \int_{\omega^t} 2x_1^t \varphi^t \,\mathrm{d}\omega^t - \int_{\gamma^t} \Phi_{1\alpha}^t n_\alpha^t \varphi^t \,\mathrm{d}\gamma^t.$$

Setting

$$\varphi^t = -\tfrac{1}{2} x_1^t (x_2^t)^2,$$

we obtain

$$\begin{aligned} &-\tfrac{1}{2}\int_{\omega^t} (x_2^t)^2 \partial_1^t \theta_1^t \,\mathrm{d}\omega^t \\ &\quad = \int_{\omega^t} x_1^t x_2^t \partial_2^t \theta_1^t \,\mathrm{d}\omega^t - \int_{\omega^t} (x_1^t x_2^t)^2 \,\mathrm{d}\omega^t + \tfrac{1}{2}\int_{\gamma^t} \Phi_{1\alpha}^t n_\alpha^t x_1^t (x_2^t)^2 \,\mathrm{d}\gamma^t. \end{aligned}$$

Substituting into (19.5) we get

$$K_{11}^{\theta^t} = \tfrac{1}{2}\int_{\omega^t} (x_1^t)^2 \partial_1^t \theta_1^t \,\mathrm{d}\omega^t + 2\int_{\omega^t} x_1^t x_2^t \partial_2^t \theta_1^t \,\mathrm{d}\omega^t$$

$$- \int_{\omega^t} (x_1^t x_2^t)^2 \,\mathrm{d}\omega^t + \tfrac{1}{2}\int_{\gamma^t} \Phi_{1\alpha}^t n_\alpha^t x_1^t (x_2^t)^2 \,\mathrm{d}\gamma^t$$

$$= \tfrac{1}{4}\int_{\omega^t} (x_1^t)^2[(x_2^t)^2 - (x_1^t)^2] \,\mathrm{d}\omega^t + \tfrac{1}{2}\int_{\omega^t} (x_1^t)^2 \partial_1^t \varepsilon_{\theta_1}^t \,\mathrm{d}\omega^t$$

$$+ 2\int_{\omega^t} (x_1^t x_2^t)^2 \,\mathrm{d}\omega^t + 2\int_{\omega^t} x_1^t x_2^t \partial_2^t \varepsilon_{\theta_1}^t \,\mathrm{d}\omega^t$$

$$- \int_{\omega^t} (x_1^t x_2^t)^2 \,\mathrm{d}\omega^t + \tfrac{1}{2}\int_{\gamma^t} \Phi_{1\alpha}^t n_\alpha^t x_1^t (x_2^t)^2 \,\mathrm{d}\gamma^t$$

$$= \tfrac{5}{4}\int_{\omega^t} (x_1^t x_2^t)^2 \,\mathrm{d}\omega^t - \tfrac{1}{4}\int_{\omega^t} (x_1^t)^4 \,\mathrm{d}\omega^t + \tfrac{1}{2}\int_{\gamma^t} \Phi_{1\alpha} n_\alpha^t x_1^t (x_2^t)^2 \,\mathrm{d}\gamma^t$$

$$+ \tfrac{1}{2}\int_{\omega^t} (x_1^t)^2 \partial_1^t \varepsilon_{\theta_1}^t \,\mathrm{d}\omega^t + 2\int_{\omega^t} x_1^t x_2^t \partial_2^t \varepsilon_{\theta_1}^t \,\mathrm{d}\omega^t .$$

Evaluating the integrals on the first three terms, we find

$$K_{11}^{\theta^t} = -\frac{b^5 t}{320} + \frac{7b^3 t^3}{288} - \frac{bt^5}{320} + \chi_{\theta_1}^t, \tag{19.17}$$

where,

$$\chi_{\theta_1}^t = \tfrac{1}{2} t^2 \int_{\omega} x_1^2 \partial_1 \varepsilon_{\theta_1}(t) \,\mathrm{d}\omega + 2t^2 \int_{\omega} x_1 x_2 \partial_2 \varepsilon_{\theta_1}(t) \,\mathrm{d}\omega. \tag{19.18}$$

As Theorem 19.2 assures that $t^{-3}\chi_{\theta_1}^t \to 0$, expression (19.14) is proved. In order to show that (19.15) also holds we use a similar argument.

$$K_{22}^{\theta^t} = \int_{\omega^t} \Phi_{2\mu}^t \partial_\mu^t \theta_2^t \,\mathrm{d}\omega^t$$

$$= \int_{\omega^t} x_1^t x_2^t \partial_1^t \theta_2^t \,\mathrm{d}\omega^t + \int_{\omega^t} \tfrac{1}{2}[(x_2^t)^2 - (x_1^t)^2] \partial_2^t \theta_2^t \,\mathrm{d}\omega^t . \tag{19.19}$$

Terms

$$\int_{\omega^t} x_1^t x_2^t \partial_1^t \theta_2^t \,\mathrm{d}\omega^t = t\int_{\omega} x_1 x_2 \partial_1 \theta_2(t) \,\mathrm{d}\omega,$$

$$\tfrac{1}{2}\int_{\omega^t} (x_2^t)^2 \partial_2^t \theta_2^t \,\mathrm{d}\omega^t = \tfrac{1}{2} t\int_{\omega} x_2^2 \partial_2 \theta_2(t) \,\mathrm{d}\omega,$$

do not have the correct exponent in t in order to pass to the limit (cf. (19.8)). We use the variational formulation of θ_2^t in order to evaluate them:

$$\int_{\omega^t} \partial_a lpha^t\, \theta_2^t \partial_\alpha \varphi^t \,\mathrm{d}\omega^t = \int_{\omega^t} 2x_2^t \varphi^t \,\mathrm{d}\omega^t - \int_{\gamma^t} \Phi_{2\alpha}^t n_\alpha^t \varphi^t \,\mathrm{d}\gamma^t.$$

Setting

$$\varphi^t = \tfrac{1}{2}(x_1^t)^2 x_2^t + \tfrac{1}{6}(x_2^t)^3,$$

we obtain

$$\int_{\omega^t} x_1^t x_2^t \partial_1^t \theta_2^t \,\mathrm{d}\omega^t + \int_{\omega^t} \tfrac{1}{2}[(x_1^t)^2 + (x_2^t)^2]\partial_2^t \theta_2^t \,\mathrm{d}\omega^t$$
$$= \int_{\omega^t} (x_1^t x_2^t)^2 \,\mathrm{d}\omega^t + \tfrac{1}{3}\int_{\omega^t} (x_2^t)^4 \,\mathrm{d}\omega^t - \int_{\gamma^t} \Phi_{2\alpha}^t n_\alpha^t [\tfrac{1}{2}(x_1^t)^2 x_2^t + \tfrac{1}{6}(x_2^t)^3] \,\mathrm{d}\gamma^t.$$

Substituting into (19.17) we get

$$K_{22}^{\theta^t} = -\int_{\omega^t} (x_1^t)^2 \partial_2^t \theta_2^t \,\mathrm{d}\omega^t + \int_{\omega^t} (x_1^t x_2^t)^2 \,\mathrm{d}\omega^t$$
$$+ \tfrac{1}{3}\int_{\omega^t} (x_2^t)^4 \,\mathrm{d}\omega^t - \int_{\gamma^t} \Phi_{2\alpha}^t n_\alpha^t [\tfrac{1}{2}(x_1^t)^2 x_2^t + \tfrac{1}{6}(x_2^t)^3] \,\mathrm{d}\gamma^t.$$

Now using Theorem 19.3, we obtain

$$K_{22}^{\theta^t} = \tfrac{1}{2}\int_{\omega^t} (x_1^t)^2[(x_1^t)^2 + (x_2^t)^2] \,\mathrm{d}\omega^t - \tfrac{1}{12}t^2 \int_{\omega^t} (x_1^t)^2 \,\mathrm{d}\omega^t + \int_{\omega^t} (x_1^t x_2^t)^2 \,\mathrm{d}\omega^t$$
$$+ \tfrac{1}{3}\int_{\omega^t} (x_2^t)^4 \,\mathrm{d}\omega^t - \int_{\gamma^t} \Phi_{2\alpha}^t n_\alpha^t [\tfrac{1}{2}(x_1^t)^2 x_2^t + \tfrac{1}{6}(x_2^t)^3] \,\mathrm{d}\gamma^t$$
$$- \int_{\omega^t} (x_1^t)^2 \partial_2^t \varepsilon_{\theta_2}^t \,\mathrm{d}\omega^t.$$

After some simplifications this expression becomes

$$K_{22}^{\theta^t} = \tfrac{1}{2}\int_{\omega^t} (x_1^t)^4 \,\mathrm{d}\omega^t + \frac{3}{2}\int_{\omega^t} (x_1^t x_2^t)^2 \,\mathrm{d}\omega^t - \tfrac{1}{12}t^2 \int_{\omega^t} (x_1^t)^2 \,\mathrm{d}\omega^t$$
$$+ \tfrac{1}{3}\int_{\omega^t} (x_2^t)^4 \,\mathrm{d}\omega^t - \int_{\gamma_v^t} [\tfrac{1}{6}x_1^t (x_2^t)^4 + \tfrac{1}{2}(x_1^t)^3 (x_2^t)^2] n_1^t \,\mathrm{d}\gamma^t$$
$$- \int_{\gamma_h^t} [\tfrac{1}{4}(x_2^t)^3 (x_1^t)^2 - \tfrac{1}{4}(x_1^t)^4 x_2^t + \tfrac{1}{12}(x_2^t)^5 - \tfrac{1}{12}(x_1^t)^2 (x_2^t)^3] n_2^t \,\mathrm{d}\gamma^t$$
$$- \int_{\omega^t} (x_1^t)^2 \partial_2^t \varepsilon_{\theta_2}^t \,\mathrm{d}\omega^t.$$

Evaluating these integrals we obtain

$$K_{22}^{\theta^t} = -\frac{b^5 t}{320} - \frac{b^3 t^3}{288} + \frac{7bt^5}{2880} + \chi_{\theta_2}^t, \tag{19.20}$$

where

$$\chi_{\theta_2}^t = -t^3 \int_\omega x_1^2 \partial_2 \varepsilon_{\theta_2}(t)\,\mathrm{d}\omega. \tag{19.21}$$

and the result now follows from (19.8) because we have $t^{-5}\chi_2^t \to 0$. □

Substituting the results of Theorems 19.4, 19.5 and 19.6 together with expressions (16.2)–(16.4) in the definition of Timoshenko's matrix (cf. (12.39)), we obtain the following result, where expressions (19.17)–(19.20) are used.

THEOREM 19.7. *For the thin rectangle ω^t elements $T_{\alpha\alpha}^t$ of Timoshenko's matrix verify*

$$T_{11}^t = -\frac{\nu(1+3\nu)b^2}{12(1+\nu)} + \frac{(4+5\nu)t^2}{20} - \frac{6\nu^2}{(1+\nu)bt^3}\chi_{\theta_1}^t, \tag{19.22}$$

$$T_{22}^t = \frac{(4+5\nu)b^2}{20} - \frac{\nu t^2}{12} - \frac{\nu^2 t^4}{30(1+\nu)b^2} - \frac{6\nu^2}{(1+\nu)b^3 t}\chi_{\theta_2}^t. \tag{19.23}$$

Substituting now (19.22) and (19.23) in the new definition of Timoshenko's constant (cf. Section 15),

$$\hat{k}_\alpha^t = \frac{2(1+\nu)I_\alpha^t}{T_{\alpha\alpha}^t A^t} \quad \text{(no sum on } \alpha),$$

the following approximation result for these constants is obtained.

THEOREM 19.8. *For the thin rectangular cross section ω^t, the new Timoshenko's constants, $\hat{k}_\alpha^t$, are given by*

$$\hat{k}_1^t = \frac{10(1+\nu)^2}{-5\nu(1+3\nu)(b/t)^2 + 3(1+\nu)(4+5\nu) - (360\nu^2/bt^5)\chi_{\theta_1}^t}, \tag{19.24}$$

$$\hat{k}_2^t = \frac{10(1+\nu)^2}{3(1+\nu)(4+5\nu) - 5\nu(1+\nu)(t/b)^2 - 2\nu^2(t/b)^4 - (360\nu^2/b^5 t)\chi_{\theta_2}^t}. \tag{19.25}$$

Now, considering the fact that $\chi_{\theta_2}^t = \mathrm{o}(t^5)$, we see that $\hat{k}_2^t$ takes the classical value $\frac{5}{6}$ when $\nu = 0$ (cf. Table 15.1). Moreover, its limit $\hat{k}_2^0$ taken when t goes to zero confirms the numerical results obtained in Section 15 (cf. (15.23) and Table 15.10). In fact one has the following.

THEOREM 19.9. *For the thin rectangular cross section ω^t, Timoshenko's constant verifies*

$$\lim_{\nu\to 0}\hat{k}_2^t = \tfrac{5}{6},$$

$$\lim_{t\to 0}\hat{k}_2^t = \frac{10(1+\nu)^2}{12+27\nu+15\nu^2}.$$

20. Critical rectangular cross sections in Timoshenko's beam theory

This section is devoted to the mathematical justification of the numerical results obtained in Section 15 concerning constant $\hat{k}_1^t$. More specifically, devoted to find the *critical dimensions in order for this constant to be infinity*, for a given Poisson's ratio (cf. (15.24) and Table 15.10). The results presented in this section are owed to RODRÍGUEZ [1990] and RODRÍGUEZ and VIAÑO [1992c].

Let us write constant $\hat{k}_1^t$ in the form

$$\hat{k}_1^t = \frac{10(1+\nu)^2}{X^t + Y^t\nu + Z^t\nu^2}. \tag{20.1}$$

From (19.24) one has

$$X^t = 12, \qquad Y^t = 27 - 5\left(\frac{b}{t}\right)^2, \qquad Z^t = 15 - 15\left(\frac{b}{t}\right)^2 - \frac{360}{bt^5}\chi_{\theta_1}^t. \tag{20.2}$$

This expression agrees with the numerical results in Table 15.10 and explains why the coefficients of ν and ν^2 are negative and decreasing as t goes to zero.

It is necessary to take into account that $\chi_{\theta_1}^t = o(t^3)$ and so the last two terms in the expression of Z^t are of the order $o(t^{-2})$. From the values in Table 15.10 it seems that $\chi_{\theta_1}^t$ is a monotonically decreasing function as $t \to 0$.

Since constant $\hat{k}_1^t$ depends on the ratio b/t, and not on variables b and t separately (cf. Theorem 15.1) and since $\chi_{\theta_1}^t = o(t^3)$ one may consider an asymptotic expansion of the form

$$\frac{360}{bt^5}\chi_{\theta_1}^t = \alpha(t/b)^{-1} + \beta + \gamma(t/b) + \delta(t/b)^2 + \text{h.o.t.}. \tag{20.3}$$

When t goes to zero the terms of the form $(t/b)^p$, $p \geqslant 1$ are not important when compared with $(t/b)^{-2}$ and $(t/b)^{-1}$. Let us analyze what happens if we just keep the first two terms in (20.3), that is, if

$$\frac{360}{bt^5}\chi_{\theta_1}^t \simeq \alpha(t/b)^{-1} + \beta. \tag{20.4}$$

Substituting into (20.2) and in (20.1) we obtain the following *approximated formula* for constant $\hat{k}_1^t$ when t is very small:

$$\hat{k}_1^t \simeq \frac{10(1+\nu)^2}{12+[27-5(b/t)^2]\nu+[15-\beta-15(b/t)^2-\alpha(b/t)]\nu^2}. \tag{20.5}$$

In order to evaluate the unknown parameters α and β we numerically interpolate the results in Table 15.10, thus obtaining an approximate value for quantity

$$15-\beta-15\left(\frac{b}{t}\right)^2-\alpha\left(\frac{b}{t}\right), \tag{20.6}$$

for different values of the ratio b/t. In order to increase the accuracy of the interpolation we choose $b/t = 2$ and $b/t = 3$, and obtain, respectively

$$2\alpha+\beta+45 = 19.167,$$
$$3\alpha+\beta+120 = 75.511,$$

therefore,

$$\alpha = -18.656, \qquad \beta = 11.479. \tag{20.7}$$

Consequently setting $r = t/b$ the following expression gives a valid approximation for Timoshenko's constant $\hat{k}_1^t$ for a material with Poisson's ratio ν,

$$\hat{k}_1^r(\nu) \simeq \frac{10(1+\nu)^2}{12+(27-5r^{-2})\nu+(3.521-15r^{-2}+18.656r^{-1})\nu^2}. \tag{20.8}$$

This approximation is more accurate for $r \leqslant 0.5$. If $0.5 < r \leqslant 1$ then one should interpolate in an interval containing the desired value of r.

If Poisson's ratio ν is such that the denominator of (20.8) is zero, then the constant becomes infinity and in Timoshenko's model (cf. (15.21) and (15.22)) the additional term giving the contribution to bending due to the nonuniform shear stress distribution on the cross section disappears. We are then left with Bernoulli–Navier's model. Consequently the *critical dimensions and the critical Poisson's ratio* giving the transition between the two theories for the thin rectangular cross section is given by the roots of the following equation:

$$F(r,\nu) = 12r^2+(27r^2-5)\nu+(3.521r^2+18.656r-15)\nu^2 = 0. \tag{20.9}$$

Given a certain material (ν fixed) and solving (20.9) as a function of r, we find the critical ratio r_ν for which $\hat{k}_1^r(\nu)$ becomes infinity. Since the only values with a physical meaning are $0 < \nu \leqslant \frac{1}{2}$ and $0 < r_\nu$, we obtain

$$r_\nu = \frac{-18.656\nu^2+[348.046\nu^4+20\nu(1+3\nu)(12+27\nu+3.521\nu^2)]^{1/2}}{2(12+27\nu+3.521\nu^2)}. \tag{20.10}$$

TABLE 20.1. Critical dimensions.

ν	$r_\nu = t/b$	b/t
0.01	0.06471	15.4538
0.05	0.14497	6.8981
0.10	0.20379	4.9069
0.20	0.28150	3.5524
0.25	0.31023	3.2234
0.30	0.33475	2.9873
0.33	0.34788	2.8747
0.40	0.37476	2.6683
0.45	0.39140	2.5549
0.50	0.40631	2.4612

In Table 20.1 we show some of the values of ν^r for different values of ν. We observe that the results obtained are in good agreement with the numerical results in the sense that there are no critical cross sections with $b/t < 2.4$ (cf. Fig. 15.5).

Conversely, solving (20.9) for variable ν, we find the critical Poisson's ratio ν_r for the given dimensions $r = t/b$. Taking into account that $0 < \nu \leqslant \frac{1}{2}$ we obtain

$$\nu_r = \frac{(5 - 27r^2) - [(27r^2 - 5)^2 - 48r^2(3.521r^2 + 18.656r - 15)]^{1/2}}{2(3.521r^2 + 18.656r - 15)}. \tag{20.11}$$

In Table 20.2 we represent some values of ν_r for given dimensions of the cross section. We remark that only when $r \leqslant 0.406$ (that is, $b/t \geqslant 2.463$) does one obtain admissible values for Poisson's ratio, which agree with the numerical results (cf. Fig. 15.5). Moreover, property (15.24) is now justified since for r very small the coefficient of ν^2 in (20.9) is negative.

REMARK 20.1. If in (20.4) we consider more terms in the approximation, the previous analysis can still be done. One just has to use a higher order interpolation for Table 15.10. In this case, the explicit obtainment of formula (20.10) would not be possible, in general, since Eq. (20.9) would be of a higher order on the variable r.

21. Asymptotic problems in thin-walled beams

Thin-walled beams, either with open or closed cross sections are of great importance in practice. One cross section may be considered thin on a purely geometric basis whenever it possesses two characteristic dimensions a and δ ($\varepsilon^2 = a\delta$) with one of them, say δ, much smaller than the other ($\delta \ll a$).

One of the most important classical theories for the behaviour of thin-walled beams is owed to VLASSOV [1962] (see also ODEN and RIPPERGER

TABLE 20.2. Critical Poisson's ratio.

$r = t/b$	b/t	ν_r
0.020	50.000	0.00096
0.040	25.000	0.00383
0.050	20.000	0.00598
0.100	10.000	0.02380
0.200	5.000	0.09619
0.250	4.000	0.15400
0.300	3.333	0.23122
0.333	3.000	0.29617
0.400	2.500	0.47820
0.406	2.463	0.49892

[1981] and MURRAY [1986]). Besides the classical hypotheses, a parametrization $(x_1(s), x_2(s))$, as a function of the arc length along the centre line of the profile of the cross section, is also introduced. Moreover, the functions and constants depending on the geometry of the cross section are approximated by functions of variable s or of the *sectorial area* $\omega(s)$, defined as the area enclosed between two radii having as origin a given point O and as extremity points $(x_1(0), x_2(0))$ and $(x_1(s), x_2(s))$ respectively. With this notation in mind the angle of torsion (θ), the stretching (ζ_3), and the bending (ζ_α) components of the displacement field verify

$$-EA\frac{\mathrm{d}^2\zeta_3}{\mathrm{d}x_3^2} = F_3, \tag{21.1}$$

$$EI_\alpha\frac{\mathrm{d}^4\zeta_\alpha}{\mathrm{d}x_3^4} = F_\alpha + \frac{\mathrm{d}M_\alpha}{\mathrm{d}x_3}, \quad \text{no sum on } \alpha, \tag{21.2}$$

$$EC_\omega\frac{\mathrm{d}^4\theta}{\mathrm{d}x_3^4} - GJ_d\frac{\mathrm{d}^2\theta}{\mathrm{d}x_3^2} = M_3 + \frac{\mathrm{d}R}{\mathrm{d}x_3}. \tag{21.3}$$

REMARK 21.1. We remark the analogy between Eqs. (21.1)–(21.3) and the equivalent equations for thick cross sections (cf. (13.4)–(13.8)).

Constant J_d in Eq. (21.3) denotes the torsion constant. In classical literature a general method to evaluate J_d is not proposed. For example, for open cross sections composed of n thin rectangular elements of dimensions $b_i \times t_i$ $(i = 1, \dots, n)$, the torsion constant is given as the sum of the corresponding constants for thin rectangular sections, that is

$$J_d = \tfrac{1}{3}\sum_{i=1}^{n} b_i t_i^3. \tag{21.4}$$

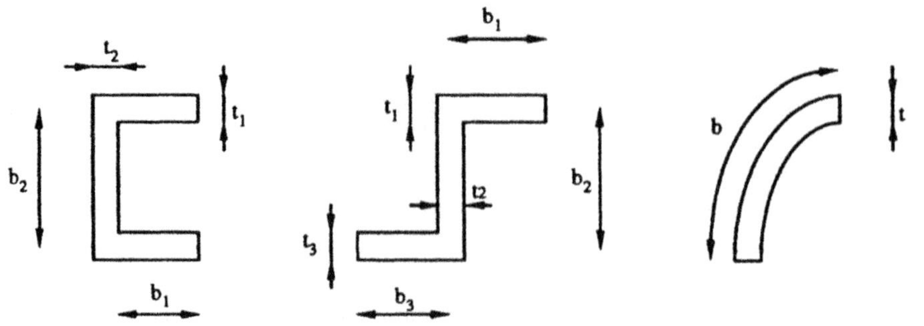

$$\frac{2b_1t_1^3+b_2t_2^3}{3} \qquad \frac{b_1t_1^3+b_2t_2^3+b_3t_3^3}{3} \qquad \frac{bt^3}{3}$$

FIG. 21.1. Typical open cross sections and their torsional constants J_d.

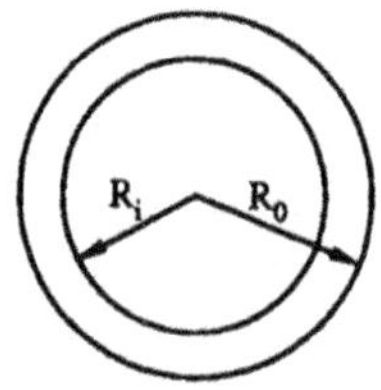

$$J_d = 2\Pi(R_0 - R_i)(\tfrac{R_0+R_i}{2})^3$$

FIG. 21.2. Practical torsional constant for a hollowed circular shaft.

In Fig. 21.1 several examples of thin-walled open cross sections are depicted together with their respective value of the torsion constant J_d.

For the case of thin-walled tubes constant J_d depends on the sectorial area $\omega(s)$ and it is customary to use (cf. ODEN [1967, p. 50])

$$J_d = \frac{4\Omega^2}{\oint \mathrm{d}s/t}, \tag{21.5}$$

where Ω is the total area enclosed by the centreline of the cross section profile, and t its thickness. Consequently, if *t is constant* one obtains the expression

$$J_d = \frac{4\Omega^2 t}{S}, \tag{21.6}$$

where S is the total perimeter of the profile. In Fig. 21.2 this formula is used to calculate the torsion constant for a hollow circular shaft. It is interesting to compare this value to the exact one given in Table 15.3.

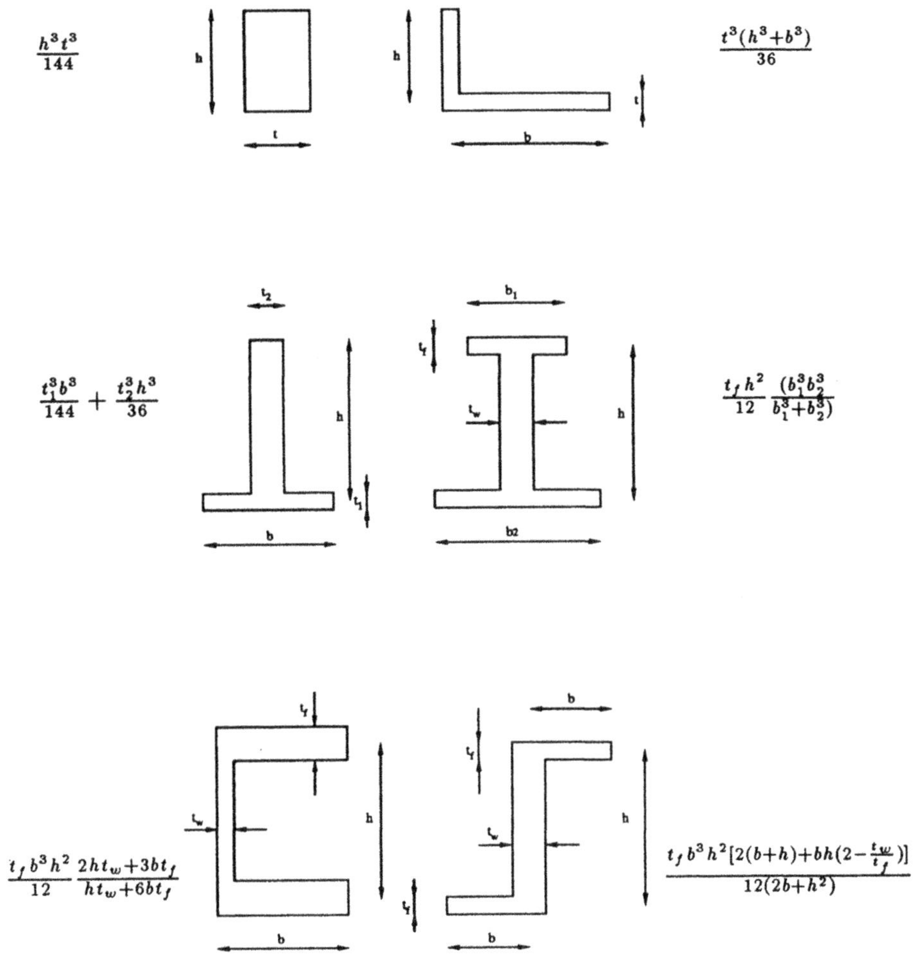

FIG. 21.3. Warping constants for thin-walled cross sections.

The case of thin multicellular cross section is more difficult (cf. ODEN [1967, p. 53], for example).

In engineering literature, constant C_ω in Eq. (21.3) has two different meanings. For some cases it stands for *the warping constant*, whose definition for a thin-walled cross section is (cf. VLASSOV [1962, p. 68], ODEN [1967, p. 193])

$$C_\omega = \oint (2\omega)^2 \, ds, \tag{21.7}$$

and for some other cases it represents the *secondary warping constant* (cf. ODEN [1967, p. 203]). In Fig. 21.3 some common values of this constant are shown for different geometries.

The asymptotic expansion method used to derive the general model (12.21)–(12.24) and (12.36)–(12.39) from the three-dimensional equation of linearized elasticity does not impose any restrictions on the relative dimensions or the shape of the cross section. Consequently, it is valid for both thin- or thick-walled beams, closed or opened cross sections, and so on. Thus, definition (7.34) for the torsion constant J and definitions (7.27), (13.8) and (13.10) for the warping constants J_w and C_w^0 have an universal character and they are also valid for these cases.

Consequently, the classical definitions of J_d and of C_ω for thin-walled cross sections must be *asymptotic limits of the exact constants taken when one of the cross sectional dimensions becomes very small with respect to the other.* A formal asymptotic expansion and a numerical computation using the finite element method for this kind of cross sections was done in BARROS [1989] and in BARROS, TRABUCHO and VIAÑO [1989] and RODRÍGUEZ [1990]. The results are also in excellent agreement with the classical values.

Therefore, a mathematical justification of the classical constants must be obtained from the asymptotic behaviour of functions Ψ, w, η_α, θ_α for this type of cross sections. *The nature of the problem is now the same as in the case of junctions in elastic multistructures* and the techniques of CIARLET [1987], CIARLET, LE DRET and NZENGWA [1989] and LE DRET [1989, 1991] may be adapted to this case. This was done by RODRÍGUEZ and VIAÑO [1992a,b,c] where a convergence result giving the mathematical justification of this fact, is obtained.

The generalization of these results to any type of open or closed thin-walled profile, gives a general and robust method to evaluate these constants, and justifies Vlassov's theory for thin-walled beams (cf. RODRÍGUEZ [1993]).

CHAPTER IV

Asymptotic Expansion Method for Nonhomogeneous Anisotropic Rods

As in classical beam theories for homogeneous isotropic materials, when one tries to review classical results for anisotropic beams, one finds very little systematization in the treatment of each problem, in the sense that, according to the nature of the applied forces considered, different a priori hypotheses are made, all of them based on either empirical experiences or similarity to the isotropic version of the problem. Besides, mainly due to the fact that Hooke's law is not easy to handle as in the isotropic case, results on anisotropic materials are much less known and far more difficult to come across than the corresponding results in isotropic materials. Actually, most of the classical texts on elasticity theory centre themselves on the isotropic case and only very sporadically do they get some insight on anisotropic rod theories. One of the few exceptions is the book of LEKHNITSKII [1977].

The asymptotic expansion method for homogeneous isotropic rods presented in Chapter I admits a natural extension to the case of linear elastic rods made of an anisotropic nonhomogeneous material. This extension was made by ÁLVAREZ-DIOS [1992], ÁLVAREZ-DIOS and VIAÑO [1988, 1991a,b, 1992a,b], following DESTUYNDER [1986] where the anisotropic plate case is considered.

We summarize their results in Sections 22 through 25 and compare the general asymptotic models obtained with the corresponding equations for the homogeneous isotropic case. In Section 22 we describe the asymptotic method for this case considering a rod weakly clamped at one extremity and subjected to a system of applied loads at the otherwise free end. In Section 23 the first term of the asymptotic expansion is calculated and convergence results are proved. Moreover, the *coupled* stretching–bending model obtained is also studied, particularly for fibre-reinforced rods.

The major differences with respect to the homogeneous isotropic case show up in the calculation of the second order term in the asymptotic expansion. In fact, it is impossible to find a solution for the second order displacement component, in the general case. This reflects the fact that for this case the three-dimensional elasticity solution is of a "fully" three-dimensional nature and cannot be decomposed in sums of products of functions of cross section variables times functions of axial variable.

It is nevertheless possible to find a solution for most of the interesting cases in practice, namely

(i) isotropic rods but transversely nonhomogeneous, with Poisson's ratio independent of cross section variables, including the case of axially reinforced rods (Section 24),

(ii) anisotropic, transversely homogeneous rods (Section 25),

(iii) anisotropic nonhomogeneous rods in pure torsion (Section 24).

Another difference, with respect to the isotropic case, both for the first order terms (stretching and bending), as well as for the second order terms (torsion, Poisson's effects and additional bending due to the nonuniform shear stress distribution on the cross section), resides in the fact that the unidimensional stretching and bending equations are now *coupled* and the elasticity coefficients cannot be separated from the definitions of the torsion, warping and Timoshenko's functions, except for particular cases of anisotropy such as transverse isotropy for a homogeneous material (cf. ÁLVAREZ-DIOS [1992] and ÁLVAREZ-DIOS and VIAÑO [1992b]).

The most important result is that without any a priori hypotheses on the data, the geometry or the material properties, one finds general models, valid for completely general loading and geometric situations. The classical models for anisotropic rods (cf. LEKHNITSKII [1977], SOKOLNIKOFF [1956]) are recovered as particular cases of the asymptotic models.

22. The asymptotic expansion method for anisotropic and nonhomogeneous elastic rods. Applied forces at one end

In this section we consider a nonhomogeneous anisotropic rod clamped at one end and submitted to a load on the otherwise free end (cantilever beam), together with applied body forces and surface tractions on the lateral surface, as before. This does not introduce any additional difficulty in the application of the asymptotic expansion method and is considered here only with the purpose of showing that the only difference lies in the appearance of additional terms and different boundary conditions at the displacement free end.

Considering the notations of Section 1, we assume that the end Γ_0^ε is weakly clamped (for strong clamping see Sections 1–4) and that extremity Γ_L^ε is submitted to the action of surface forces $\boldsymbol{h}^\varepsilon = (h_i^\varepsilon) : \Gamma_L^\varepsilon \to \mathbb{R}^3$ such that

$$h_i^\varepsilon \in L^2(\Gamma_L^\varepsilon). \tag{22.1}$$

We assume that the beam $\overline{\Omega}^\varepsilon$ is made of a linear elastic material having *elasticity coefficients* a_{ijkl}^ε, satisfying

$$a_{ijkl}^\varepsilon \in L^\infty(\Omega^\varepsilon), \tag{22.2}$$

and verifying the following symmetry and positiveness properties:

$$a^{\varepsilon}_{ijkl} = a^{\varepsilon}_{jikl} = a^{\varepsilon}_{ijlk} = a^{\varepsilon}_{klij}, \text{ a.e. in } \Omega^{\varepsilon}, \tag{22.3}$$

$$a^{\varepsilon}_{ijkl}\tau^{\varepsilon}_{ij}\tau^{\varepsilon}_{kl} > C^{\varepsilon}_{0}\tau^{\varepsilon}_{ij}\tau^{\varepsilon}_{ij}, \quad \text{for all } \boldsymbol{\tau}^{\varepsilon} \in \mathbb{R}^{9}_{s}, \text{ and a.e. in } \Omega^{\varepsilon}, \tag{22.4}$$

where C^{ε}_{0} is a positive constant.

Consequently the relation between the linearized strain tensor $e^{\varepsilon}_{ij}(\boldsymbol{u}^{\varepsilon})$ and the Cauchy stress tensor $\boldsymbol{\sigma}^{\varepsilon}$ (generalized Hooke's law) is written in the following form:

$$e^{\varepsilon}_{ij}(\boldsymbol{u}^{\varepsilon}) = a^{\varepsilon}_{ijkl}\sigma^{\varepsilon}_{kl}, \tag{22.5}$$

or equivalently,

$$\sigma^{\varepsilon}_{ij} = A^{\varepsilon}_{ijkl}e^{\varepsilon}_{kl}(\boldsymbol{u}^{\varepsilon}), \tag{22.6}$$

where the *elasticity constants* A^{ε}_{ijkl} satisfy also (22.2)–(22.4).

The space of admissible displacements is now,

$$V(\Omega^{\varepsilon}) = \{\boldsymbol{v}^{\varepsilon} = (v^{\varepsilon}_{i}) \in [H^{1}(\Omega^{\varepsilon})]^{3}: \int_{\omega^{\varepsilon}\times\{0\}} v^{\varepsilon}_{i}\,\mathrm{d}\omega^{\varepsilon} = \int_{\omega^{\varepsilon}\times\{0\}} (x^{\varepsilon}_{j}v^{\varepsilon}_{i} - x^{\varepsilon}_{i}v^{\varepsilon}_{j})\,\mathrm{d}\omega^{\varepsilon} = 0\}. \tag{22.7}$$

The displacement variational formulation of the three-dimensional linearized elasticity problem is (cf. DUVAUT and LIONS [1972a])

$$\begin{aligned} &\boldsymbol{u}^{\varepsilon} \in V(\Omega^{\varepsilon}),\\ &\int_{\Omega^{\varepsilon}} A^{\varepsilon}_{ijkl}e^{\varepsilon}_{kl}(\boldsymbol{u}^{\varepsilon})e^{\varepsilon}_{ij}(\boldsymbol{v}^{\varepsilon})\,\mathrm{d}\boldsymbol{x}^{\varepsilon}\\ &\quad = \int_{\Omega^{\varepsilon}} f^{\varepsilon}_{i}v^{\varepsilon}_{i}\,\mathrm{d}\boldsymbol{x}^{\varepsilon} + \int_{\Gamma^{\varepsilon}} g^{\varepsilon}_{i}v^{\varepsilon}_{i}\,\mathrm{d}a^{\varepsilon} + \int_{\Gamma^{\varepsilon}_{L}} h^{\varepsilon}_{i}v^{\varepsilon}_{i}\,\mathrm{d}a^{\varepsilon},\\ &\text{for all } \boldsymbol{v}^{\varepsilon} \in V(\Omega^{\varepsilon}). \end{aligned} \tag{22.8}$$

The existence and unicity of a solution to problem (22.8) is a consequence of Lax–Milgram's lemma, Korn's inequality and properties (22.2)–(22.4) for the elasticity coefficients A^{ε}_{ijkl} which guarantee the $V(\Omega^{\varepsilon})$-ellipticity of the bilinear form in (22.8).

Problem (22.8) is formally identical to problem (1.10), (1.11) for the isotropic homogeneous case. Consequently, the *change of variable* (2.4) to the reference domain Ω and *the same scalings* (2.6) *for the unknowns* will be used in order to study the behaviour of $\boldsymbol{u}^\varepsilon$ whenever ε goes to zero. With respect to *the hypotheses on the data*, we assume:

(i) the body forces $\boldsymbol{f}^\varepsilon$ and the surface tractions $\boldsymbol{g}^\varepsilon$ satisfy (2.8) and (2.9),

(ii) the forces applied at the free end Γ_L^ε are such that

$$h_\alpha^\varepsilon(\boldsymbol{x}^\varepsilon) = \varepsilon h_\alpha(\boldsymbol{x}), \qquad h_3^\varepsilon(\boldsymbol{x}^\varepsilon) = h_3(\boldsymbol{x}), \qquad \text{for all } \boldsymbol{x}^\varepsilon = \Pi^\varepsilon(\boldsymbol{x}) \in \Gamma_L^\varepsilon, \tag{22.9}$$

where functions $h_i \in L^2(\Gamma_L)$ are independent of ε,

(iii) the elasticity constants are "independent of the size of the cross section", that is,

$$A_{ijkl}(\boldsymbol{x}^\varepsilon) = A_{ijkl}(\boldsymbol{x}), \qquad a_{ijkl}(\boldsymbol{x}^\varepsilon) = a_{ijkl}(\boldsymbol{x}), \qquad \boldsymbol{x} \in \Omega, \tag{22.10}$$

where A_{ijkl} and a_{ijkl} are independent of ε,

(iv) longitudinal axis Ox_3^ε is a *principal direction of elasticity*, that is (cf. LEKHNITSKII [1977]),

$$a_{\alpha\beta\gamma 3}^\varepsilon = a_{\alpha 333}^\varepsilon = 0. \tag{22.11}$$

This hypothesis is also known as the *elastic symmetry of the plane* $Ox_1^\varepsilon x_2^\varepsilon$, and it is considered in the majority of classical theories. As a consequence, the shear stresses will be decoupled from the axial and plane components of the strain tensor. *This hypothesis is also necessary*, in the asymptotic expansion method, *in order to be able to show that the terms with odd powers of ε vanish*, thus allowing the calculation of the asymptotic terms with a methodology very similar to the isotropic case. The same hypothesis is considered in DESTUYNDER [1986] for the anisotropic plate case. If this hypothesis is not considered the nature of the asymptotic expansion (if it exists) and of the limit problems is essentially different, reflecting the deeper three-dimensional character of the solution.

Using all the notation defined in Section 2, and denoting by $V(\Omega)$ the space of admissible scaled displacements, defined by

$$V(\Omega) = \{\boldsymbol{v} = (v_i) \in [H^1(\Omega)]^3: \int_{\omega\times\{0\}} v_i \,\mathrm{d}\omega = \int_{\omega\times\{0\}} (x_j v_i - x_i v_j)\,\mathrm{d}\omega = 0\}, \tag{22.12}$$

it is possible to show the following result.

THEOREM 22.1. *The scaled displacement field* $\boldsymbol{u}(\varepsilon)$ *is the unique solution of the following variational problem*:

$$\boldsymbol{u}(\varepsilon) \in V(\Omega),$$

$$\int_\Omega A_{3333} e_{33}(\boldsymbol{u}(\varepsilon)) e_{33}(\boldsymbol{v})\,\mathrm{d}\boldsymbol{x}$$

$$+\varepsilon^{-2}\left[\int_\Omega A_{\alpha\beta 33}e_{\alpha\beta}(\boldsymbol{u}(\varepsilon))e_{33}(\boldsymbol{v})\,\mathrm{d}\boldsymbol{x}+\int_\Omega 4A_{3\alpha 3\beta}e_{3\alpha}(\boldsymbol{u}(\varepsilon))e_{3\beta}(\boldsymbol{v})\,\mathrm{d}\boldsymbol{x}\right.$$

$$\left.+\int_\Omega A_{\alpha\beta 33}e_{33}(\boldsymbol{u}(\varepsilon))e_{\alpha\beta}(\boldsymbol{v})\,\mathrm{d}\boldsymbol{x}\right]+\varepsilon^{-4}\int_\Omega A_{\alpha\beta\gamma\mu}e_{\alpha\beta}(\boldsymbol{u}(\varepsilon))e_{\gamma\mu}(\boldsymbol{v})\,\mathrm{d}\boldsymbol{x}$$

$$=\int_\Omega f_i v_i\,\mathrm{d}\boldsymbol{x}+\int_\Gamma g_i v_i\,\mathrm{d}a+\int_{\Gamma_L} h_i v_i\,\mathrm{d}a,\quad \textit{for all } \boldsymbol{v}\in V(\Omega). \tag{22.13}$$

REMARK 22.1. We have the following formulation, equivalent to (22.12), (22.13):

$$\begin{aligned}&\boldsymbol{u}(\varepsilon)\in V(\Omega),\\&\int_\Omega \sigma_{ij}(\varepsilon)e_{ij}(\boldsymbol{v})\,\mathrm{d}\boldsymbol{x}\\&\quad=\int_\Omega f_i v_i\,\mathrm{d}\boldsymbol{x}+\int_\Gamma g_i v_i\,\mathrm{d}a+\int_{\Gamma_L} h_i v_i\,\mathrm{d}a,\quad \text{for all } \boldsymbol{v}\in V(\Omega),\end{aligned} \tag{22.14}$$

where,

$$\begin{aligned}&\sigma_{\alpha\beta}(\varepsilon)=\varepsilon^{-4}A_{\alpha\beta\gamma\mu}e_{\gamma\mu}(\boldsymbol{u}(\varepsilon))+\varepsilon^{-2}A_{\alpha\beta 33}e_{33}(\boldsymbol{u}(\varepsilon)),\\&\sigma_{3\beta}(\varepsilon)=2\varepsilon^{-2}A_{3\alpha 3\beta}e_{3\alpha}(\boldsymbol{u}(\varepsilon)),\\&\sigma_{33}(\varepsilon)=A_{3333}e_{33}(\boldsymbol{u}(\varepsilon))+\varepsilon^{-2}A_{\alpha\beta 33}e_{\alpha\beta}(\boldsymbol{u}(\varepsilon)).\end{aligned} \tag{22.15}$$

We remark that (22.15) represents the *scaled constitutive law*.

Problem (22.12), (22.13) (resp. (22.14), (22.15)) represents the extension of problem (2.14), (2.15) (resp. (2.19), (2.20)) to the nonhomogeneous and anisotropic case. This generalized version keeps the same structure and consequently the asymptotic expansion method exposed in Section 3 for problem (2.14), (2.15) may also be applied to problem (22.12), (22.13).

Let us start by assuming an asymptotic expansion of the form

$$\boldsymbol{u}(\varepsilon)=\boldsymbol{u}^0+\varepsilon\boldsymbol{u}^1+\varepsilon^2\boldsymbol{u}^2+\varepsilon^3\boldsymbol{u}^3+\varepsilon^4\boldsymbol{u}^4+\text{h.o.t.}, \tag{22.16}$$

and substitute it into Eq. (22.13). Setting equal to zero the different factors of the sucessive powers of ε and using a similar analysis to the one developed in Sections 4 and 5, we are allowed to prove that the asymptotic expansion (22.16) does not contain odd powers of ε, that is,

$$\boldsymbol{u}(\varepsilon)=\boldsymbol{u}^0+\varepsilon^2\boldsymbol{u}^2+\varepsilon^4\boldsymbol{u}^4+\text{h.o.t.} \tag{22.17}$$

Through (22.15) we conclude that this asymptotic expansion induces the following expansion in the scaled stress field $\boldsymbol{\sigma}(\varepsilon)$:

$$\begin{aligned}\sigma_{\alpha\beta}(\varepsilon) &= \varepsilon^{-4}\sigma_{\alpha\beta}^{-4} + \varepsilon^{-2}\sigma_{\alpha\beta}^{-2} + \sigma_{\alpha\beta}^{0} + \text{h.o.t.},\\ \sigma_{3\beta} &= \varepsilon^{-2}\sigma_{3\beta}^{-2} + \sigma_{3\beta}^{0} + \varepsilon^{2}\sigma_{3\beta}^{2} + \text{h.o.t.},\\ \sigma_{33} &= \varepsilon^{-2}\sigma_{33}^{-2} + \sigma_{33}^{0} + \varepsilon^{2}\sigma_{33}^{2} + \text{h.o.t.},\end{aligned} \tag{22.18}$$

where

$$\begin{aligned}\sigma_{\alpha\beta}^{-4} &= A_{\alpha\beta\gamma\mu}e_{\gamma\mu}(\boldsymbol{u}^0),\\ \sigma_{\alpha\beta}^{p} &= A_{\alpha\beta\gamma\mu}e_{\gamma\mu}(\boldsymbol{u}^{p+4}) + A_{\alpha\beta33}e_{33}(\boldsymbol{u}^{p+2}), \quad p \geqslant -2,\\ \sigma_{3\beta}^{p} &= 2A_{3\alpha3\beta}3_{e\alpha}(\boldsymbol{u}^{p+2}), \quad p \geqslant -2,\\ \sigma_{33}^{-2} &= A_{\alpha\beta33}e_{\alpha\beta}(\boldsymbol{u}^0),\\ \sigma_{33}^{p} &= A_{3333}e_{33}(\boldsymbol{u}^p) + A_{\alpha\beta33}e_{\alpha\beta}(\boldsymbol{u}^{p+2}), \quad p \geqslant 0.\end{aligned} \tag{22.19}$$

Using the same technique as in Section 5 it is possible to show that the asymptotic expansion (22.18) does not contain terms with negative powers of ε, that is,

$$\boldsymbol{\sigma}(\varepsilon) = \boldsymbol{\sigma}^0 + \varepsilon^2\boldsymbol{\sigma}^2 + \varepsilon^4\boldsymbol{\sigma}^4 + \text{h.o.t.} \tag{22.20}$$

Summarizing, the major difference between the application of the asymptotic expansion method to the nonhomogeneous anisotropic and the homogeneous isotropic cases is solely related to the computational aspects. This computational task becomes simpler if one uses a mixed formulation of problem (22.8) and an equivalent asymptotic expansion method, which we denoted *the mixed displacement–stress approach* (cf. Section 6), which is valid after (22.20).

The mixed variational formulation of problem (22.8) is the following:

$$\boldsymbol{u}^\varepsilon \in V(\Omega^\varepsilon), \qquad \boldsymbol{\sigma}^\varepsilon \in \Sigma(\Omega^\varepsilon), \tag{22.21}$$

$$\int_{\Omega^\varepsilon} a_{ijkl}^\varepsilon \sigma_{kl}^\varepsilon \tau_{ij}^\varepsilon \, \mathrm{d}\boldsymbol{x}^\varepsilon - \int_{\Omega^\varepsilon} e_{ij}^\varepsilon(\boldsymbol{u}^\varepsilon)\tau_{ij}^\varepsilon \, \mathrm{d}\boldsymbol{x}^\varepsilon = 0, \quad \text{for all } \boldsymbol{\tau}^\varepsilon \in \Sigma(\Omega^\varepsilon), \tag{22.22}$$

$$\int_{\Omega^\varepsilon} \sigma_{ij}^\varepsilon e_{ij}^\varepsilon(\boldsymbol{v}^\varepsilon)\, \mathrm{d}\boldsymbol{x}^\varepsilon = \int_{\Omega^\varepsilon} f_i^\varepsilon v_i^\varepsilon \, \mathrm{d}\boldsymbol{x}^\varepsilon + \int_{\Gamma^\varepsilon} g_i^\varepsilon v_i^\varepsilon \, \mathrm{d}a^\varepsilon + \int_{\Gamma_L^\varepsilon} h_i^\varepsilon v_i^\varepsilon \, \mathrm{d}a^\varepsilon,$$
$$\text{for all } \boldsymbol{v}^\varepsilon \in V(\Omega^\varepsilon). \tag{22.23}$$

Introducing the same change of variable and scalings of Section 6 it is possible to prove the following result.

THEOREM 22.2. *The scaled displacement field* $\boldsymbol{u}(\varepsilon) = (u_i(\varepsilon))$ *and the scaled stress field* $\boldsymbol{\sigma}(\varepsilon) = (\sigma_{ij}(\varepsilon))$ *are the unique solution of the following scaled mixed variational problem*:

$$(\boldsymbol{u}(\varepsilon), \boldsymbol{\sigma}(\varepsilon)) \in V(\Omega) \times \Sigma(\Omega), \tag{22.24}$$

$$a_0(\boldsymbol{\sigma}(\varepsilon), \boldsymbol{\tau}) + \varepsilon^2 a_2(\boldsymbol{\sigma}(\varepsilon), \boldsymbol{\tau}) + \varepsilon^4 a_4(\boldsymbol{\sigma}(\varepsilon), \boldsymbol{\tau}) + b(\boldsymbol{\tau}, \boldsymbol{u}(\varepsilon)) = 0, \quad \text{for all } \boldsymbol{\tau} \in \Sigma(\Omega), \tag{22.25}$$

$$b(\boldsymbol{\sigma}(\varepsilon), \boldsymbol{v}) = F(\boldsymbol{v}), \quad \text{for all } \boldsymbol{v} \in V(\Omega), \tag{22.26}$$

where, for all $(\boldsymbol{\sigma}, \boldsymbol{\tau}, \boldsymbol{v}) \in \Sigma(\Omega) \times \Sigma(\Omega) \times V(\Omega)$ *the following symmetric, continuous, bilinear forms were defined:*

$$a_0(\boldsymbol{\sigma}, \boldsymbol{\tau}) = \int_\Omega a_{3333}\sigma_{33}\tau_{33}\,\mathrm{d}\boldsymbol{x}, \tag{22.27}$$

$$a_2(\boldsymbol{\sigma}, \boldsymbol{\tau}) = \int_\Omega [4a_{3\alpha3\beta}\sigma_{3\alpha}\tau_{3\beta}\,\mathrm{d}\boldsymbol{x} + a_{\alpha\beta33}(\sigma_{33}\tau_{\alpha\beta} + \sigma_{\alpha\beta}\tau_{33})]\,\mathrm{d}\boldsymbol{x}, \tag{22.28}$$

$$a_4(\boldsymbol{\sigma}, \boldsymbol{\tau}) = \int_\Omega a_{\alpha\beta\gamma\mu}\sigma_{\gamma\mu}\tau_{\alpha\beta}\,\mathrm{d}\boldsymbol{x}, \tag{22.29}$$

$$b(\boldsymbol{\tau}, \boldsymbol{v}) = -\int_\Omega \tau_{ij}e_{ij}(\boldsymbol{v})\,\mathrm{d}\boldsymbol{x}, \tag{22.30}$$

together with the following linear, continuous form:

$$F(\boldsymbol{v}) = -\int_\Omega f_i v_i\,\mathrm{d}\boldsymbol{x} - \int_\Gamma g_i v_i\,\mathrm{d}a - \int_{\Gamma_L} h_i v_i\,\mathrm{d}a. \tag{22.31}$$

Problem (22.24)–(22.26) is exactly of the same type as the corresponding one for the isotropic homogeneous case (cf. (6.25), (6.26)). Therefore, we shall apply the same asymptotic method in order to study the behaviour of $\boldsymbol{u}(\varepsilon)$ and $\boldsymbol{\sigma}(\varepsilon)$ as ε becomes very small. Theorem 6.2 remains valid for problem (22.24)–(22.26). We now write the equivalent of (6.29)–(6.32) for the present case. In order to do so we consider the following decomposition of the space $V(\Omega)$:

$$V(\Omega) = W_1(\Omega) \times W_2(\Omega), \tag{22.32}$$

where

$$W_1(\Omega) = \{\phi \in H^1(\Omega): \int_{\omega\times\{0\}} \phi\,\mathrm{d}\omega = \int_{\omega\times\{0\}} x_\alpha\phi\,\mathrm{d}\omega = 0\}, \tag{22.33}$$

$$W_2(\Omega) = \{\boldsymbol{\psi} = (\psi_\alpha) \in [H^1(\Omega)]^2: \int_{\omega\times\{0\}} \psi_\alpha\,\mathrm{d}\omega = \int_{\omega\times\{0\}} (x_2\psi_1 - x_1\psi_2)\,\mathrm{d}\omega = 0\}. \tag{22.34}$$

From (22.27)–(22.34) and Theorem 6.2 we obtain the following result.

THEOREM 22.3. *If the successive terms of the asymptotic expansions* (22.17) *and* (22.20) *verify* $(\boldsymbol{u}^p, \boldsymbol{\sigma}^p) \in V(\Omega) \times \Sigma(\Omega)$, $p = 0, 2, 4$, *then they satisfy the following set of equations*:

$$\int_\Omega a_{3333}\sigma_{33}^0 \tau_{33}\,\mathrm{d}\boldsymbol{x} - \int_\Omega \partial_3 u_3^0 \tau_{33}\,\mathrm{d}\boldsymbol{x} = 0, \quad \textit{for all } \tau_{33} \in L^2(\Omega), \tag{22.35}$$

$$\int_\Omega (\partial_3 u_\beta^0 + \partial_\beta u_3^0)\tau_{3\beta}\,\mathrm{d}\boldsymbol{x} = 0, \quad \textit{for all } (\tau_{3\beta}) \in [L^2(\Omega)]^2, \tag{22.36}$$

$$-\frac{1}{2}\int_\Omega (\partial_\alpha u_\beta^0 + \partial_\beta u_\alpha^0)\tau_{\alpha\beta}\,\mathrm{d}\boldsymbol{x} = 0, \quad \textit{for all } (\tau_{\alpha\beta}) \in [L^2(\Omega)]_s^4, \tag{22.37}$$

$$\int_\Omega (\sigma_{33}^0 \partial_3 v_3 + \sigma_{3\beta}^0 \partial_\beta v_3)\,\mathrm{d}\boldsymbol{x}$$
$$= \int_\Omega f_3 v_3\,\mathrm{d}\boldsymbol{x} + \int_\Gamma g_3 v_3\,\mathrm{d}a + \int_{\Gamma_L} h_3 v_3\,\mathrm{d}a, \tag{22.38}$$

for all $v_3 \in W_1(\Omega)$,

$$\int_\Omega (\sigma_{\alpha\beta}^0 \partial_\alpha v_\beta + \sigma_{3\beta}^0 \partial_3 v_\beta)\,\mathrm{d}\boldsymbol{x}$$
$$= \int_\Omega f_\beta v_\beta\,\mathrm{d}\boldsymbol{x} + \int_\Gamma g_\beta v_\beta\,\mathrm{d}a + \int_{\Gamma_L} h_\beta v_\beta\,\mathrm{d}a, \quad \textit{for all } (v_\beta) \in W_2(\Omega), \tag{22.39}$$

$$\int_\Omega a_{3333}\sigma_{33}^2 \tau_{33}\,\mathrm{d}\boldsymbol{x} - \int_\Omega \partial_3 u_3^2 \tau_{33}\,\mathrm{d}\boldsymbol{x}$$
$$= -\int_\Omega a_{\alpha\beta 33}\sigma_{\alpha\beta}^0 \tau_{33}\,\mathrm{d}\boldsymbol{x}, \quad \textit{for all } \tau_{33} \in L^2(\Omega), \tag{22.40}$$

$$\int_\Omega (\partial_3 u_\beta^2 + \partial_\beta u_3^2)\tau_{3\beta}\,\mathrm{d}\boldsymbol{x}$$
$$= \int_\Omega 4a_{3\alpha 3\beta}\sigma_{3\alpha}^0 \tau_{3\beta}\,\mathrm{d}\boldsymbol{x}, \quad \textit{for all } (\tau_{3\beta}) \in [L^2(\Omega)]^2, \tag{22.41}$$

$$-\frac{1}{2}\int_\Omega (\partial_\alpha u_\beta^2 + \partial_\beta u_\alpha^2)\tau_{\alpha\beta}\,\mathrm{d}\boldsymbol{x}$$
$$= -\int_\Omega a_{\alpha\beta 33}\sigma_{33}^0 \tau_{\alpha\beta}\,\mathrm{d}\boldsymbol{x}, \quad \textit{for all } (\tau_{\alpha\beta}) \in [L^2(\Omega)]_s^4, \tag{22.42}$$

$$\int_\Omega (\sigma^2_{33}\partial_3 v_3 + \sigma^2_{3\beta}\partial_\beta v_3)\,\mathrm{d}\boldsymbol{x} = 0, \quad \textit{for all } v_3 \in W_1(\Omega), \tag{22.43}$$

$$\int_\Omega (\sigma^2_{\alpha\beta}\partial_\alpha v_\beta + \sigma^2_{3\beta}\partial_3 v_\beta)\,\mathrm{d}\boldsymbol{x} = 0, \quad \textit{for all } (v_\beta) \in W_2(\Omega), \tag{22.44}$$

$$\int_\Omega a_{3333}\sigma^4_{33}\tau_{33}\,\mathrm{d}\boldsymbol{x} - \int_\Omega \partial_3 u^4_3 \tau_{33}\,\mathrm{d}\boldsymbol{x} = -\int_\Omega a_{\alpha\beta 33}\sigma^2_{\alpha\beta}\tau_{33}\,\mathrm{d}\boldsymbol{x}, \quad \textit{for all } \tau_{33} \in L^2(\Omega), \tag{22.45}$$

$$\int_\Omega (\partial_3 u^4_\beta + \partial_\beta u^4_3)\tau_{3\beta}\,\mathrm{d}\boldsymbol{x} = \int_\Omega 4a_{3\alpha 3\beta}\sigma^2_{3\alpha}\tau_{3\beta}\,\mathrm{d}\boldsymbol{x}, \quad \textit{for all } (\tau_{3\beta}) \in [L^2(\Omega)]^2, \tag{22.46}$$

$$-\frac{1}{2}\int_\Omega (\partial_\alpha u^4_\beta + \partial_\beta u^4_\alpha)\tau_{\alpha\beta}\,\mathrm{d}\boldsymbol{x} = -\int_\Omega (a_{\alpha\beta 33}\sigma^2_{33} + a_{\alpha\beta\gamma\mu}\sigma^0_{\gamma\mu})\tau_{\alpha\beta}\,\mathrm{d}\boldsymbol{x}, \quad \textit{for all } (\tau_{\alpha\beta}) \in [L^2(\Omega)]^4_s, \tag{22.47}$$

$$\int_\Omega (\sigma^4_{33}\partial_3 v_3 + \sigma^4_{3\beta}\partial_\beta v_3)\,\mathrm{d}\boldsymbol{x} = 0, \quad \textit{for all } v_3 \in W_1(\Omega), \tag{22.48}$$

$$\int_\Omega (\sigma^4_{\alpha\beta}\partial_\alpha v_\beta + \sigma^4_{3\beta}\partial_3 v_\beta)\,\mathrm{d}\boldsymbol{x} = 0, \quad \textit{for all } (v_\beta) \in W_2(\Omega). \tag{22.49}$$

As in the homogeneous and isotropic case, Eqs. (22.35)–(22.39) characterize in a unique way the scaled displacement component $\boldsymbol{u}^0$ and the scaled stress component σ^0_{33}, as well as the scaled bending moment and shear stress components m^0_β and q^0_β, respectively (cf. Section 23). The other components of $\boldsymbol{\sigma}^0$ and the second order displacement field $\boldsymbol{u}^2$ are evaluated using the compatibility with Eqs. (22.40)–(22.49). However, we are able to do this *only for some particular cases* (cf. Sections 24 and 25), which nevertheless include most of the interesting cases in practice.

23. First order coupled bending–stretching model for anisotropic and nonhomogeneous elastic rods. Convergence results. Fibre-reinforced beams

One of the objectives of this section is to determine the pair $(\boldsymbol{u}^0, \boldsymbol{\sigma}^0)$ from Eqs. (22.35)–(22.39). In order to do so the same technique as for the homogeneous isotropic case is used (cf. Section 6) and therefore details will be omitted here. We obtain a first order model which, in the general case, couples the bending u_α^0 and the stretching $\underline{u}_3^0$ components, and which generalizes classical models.

Following DESTUYNDER [1986] we introduce the following functions of $L^\infty(0,L)$:

$$l = \int_\omega \frac{1}{a_{3333}}\,\mathrm{d}\omega, \qquad e_\alpha = \int_\omega \frac{x_\alpha}{a_{3333}}\,\mathrm{d}\omega, \qquad h_{\alpha\beta} = \int_\omega \frac{x_\alpha x_\beta}{a_{3333}}\,\mathrm{d}\omega. \tag{23.1}$$

We denote by $H : [0,L] \to \mathbb{R}_s^9$ the function defined a.e. in $[0,L]$ by

$$H(x_3) = \begin{pmatrix} h_{11}(x_3) & h_{12}(x_3) & e_1(x_3) \\ h_{21}(x_3) & h_{22}(x_3) & e_2(x_3) \\ e_1(x_3) & e_2(x_3) & l(x_3) \end{pmatrix} \tag{23.2}$$

and by $H^{-1} : [0,L] \to \mathbb{R}_s^9$ the function such that $H^{-1}(x_3) = [H(x_3)]^{-1}$ a.e. in $(0,L)$.

We introduce the following spaces:

$$\begin{aligned} V_0^1(0,L) &= \{\phi \in H^1(0,L)\colon \phi(0) = 0\}, \\ V_0^2(0,L) &= \{\phi \in H^2(0,L)\colon \phi(0) = \partial_3\phi(0) = 0\}, \end{aligned} \tag{23.3}$$

and the linear forms $\tilde{F}_3 : V_0^1(0,L) \to \mathbb{R}$, $\tilde{F}_\beta : V_0^2(0,L) \to \mathbb{R}$, given by

$$\begin{aligned} \tilde{F}_3(\phi) &= \int_0^L \left[\int_\omega f_3\,\mathrm{d}\omega + \int_\gamma g_3\,\mathrm{d}\gamma\right] \phi\,\mathrm{d}x_3 + \left[\int_\omega h_3\,\mathrm{d}\omega\right]\phi(L), \\ \tilde{F}_\beta(\phi) &= \int_0^L \left[\int_\omega f_\beta\,\mathrm{d}\omega + \int_\gamma g_\beta\,\mathrm{d}\gamma\right] \phi\,\mathrm{d}x_3 \\ &\quad - \int_0^L \left[\int_\omega x_\beta f_3\,\mathrm{d}\omega + \int_\gamma x_\beta g_3\,\mathrm{d}\gamma\right] \partial_3\phi\,\mathrm{d}x_3 \\ &\quad + \left[\int_\omega h_\beta\,\mathrm{d}\omega\right]\phi(L) - \left[\int_\omega x_\beta h_3\,\mathrm{d}\omega\right]\partial_3\phi(L). \end{aligned} \tag{23.4}$$

In a natural way we shall also come across the following Bernoulli–Navier displacement space:

$$V_{BN}^0(\Omega) = \{\boldsymbol{v} = (v_i)\colon v_\alpha \in V_0^2(0,L), v_3 = \underline{v}_3 - x_\alpha\partial_3 v_\alpha, \underline{v}_3 \in V_0^1(0,L)\}, \tag{23.5}$$

endowed with the norm $\|\cdot\|_{BN}$ defined by (4.15).

The following canonical application is an isometric isomorphism:

$$\begin{aligned} j : v = (v_1, v_2, v_3) &\in [V_0^2(0, L)]^2 \times V_0^1(0, L) \\ &\to (v_1, v_2, v_3 - x_\alpha \partial_3 v_\alpha) \in V_{BN}^0(\Omega), \end{aligned}$$

and norm $\|\cdot\|_{BN}$ is equivalent to the norm $\|\cdot\|_{1,\Omega}$ in $V_{BN}^0(\Omega)$ (cf. Section 4).

Theorem 6.3 admits a direct generalization to the anisotropic nonhomogeneous case. The arguments used are those of BERMÚDEZ and VIAÑO [1984] and of DESTUYNDER [1986]. The result is owed to ÁLVAREZ-DIOS and VIAÑO [1991a,b] (cf. also ÁLVAREZ-DIOS [1992]):

THEOREM 23.1. *We suppose that* a_{3333} *and the applied forces are such that*

$$\begin{aligned} &\frac{1}{a_{3333}} \in H^2[0, L; L^2(\omega)], \\ &H, H^{-1} \in [H^2(0, L)]_s^9, \\ &f_\alpha \in L^2(\Omega), \qquad f_3 \in H^1[0, L; L^2(\omega)], \\ &g_\alpha \in L^2(\Gamma), \qquad g_3 \in H^1[0, L; L^2(\gamma)], \\ &h_i \in L^2(\Gamma_L) \equiv L^2(\omega), \end{aligned} \tag{23.6}$$

then every solution of problem (22.35)–(22.39) *must be of the following form*:

(i) $u^0 \in V_{BN}^0(\Omega)$, *i.e. displacements* u_α^0 *depend on* x_3 *only and displacement* u_3^0 *takes the form* $u_3^0 = \underline{u}_3^0 - x_\alpha \partial_3 u_\alpha^0$, *where* $\underline{u}_3^0$ *is a function of* x_3 *only.*

(ii) *Displacement components* u_α^0 *and* $\underline{u}_3^0$ *are determined as the unique solutions of the following variational problems (coupled in general)*:

$$\begin{aligned} &u_\alpha^0 \in V_0^2(0, L), \qquad \underline{u}_3^0 \in V_0^1(0, L), \\ &\int_0^L l \partial_3 \underline{u}_3^0 \partial_3 v \, \mathrm{d}x_3 - \int_0^L e_\alpha \partial_{33} u_\alpha^0 \partial_3 v \, \mathrm{d}x_3 = \tilde{F}_3(v), \quad \textit{for all } v \in V_0^1(0, L), \\ &- \int_0^L e_\beta \partial_3 \underline{u}_3^0 \partial_{33} v_\beta \, \mathrm{d}x_3 + \int_0^L h_{\alpha\beta} \partial_{33} u_\alpha^0 \partial_{33} v_\beta \, \mathrm{d}x_3 \\ &\quad = \tilde{F}_\beta(v_\beta), \quad \textit{for all } (v_\beta) \in [V_0^2(0, L)]^2. \end{aligned} \tag{23.7}$$

(iii) *The axial stress* σ_{33}^0 *is given by*

$$\sigma_{33}^0 = \frac{1}{a_{3333}} (\partial_3 \underline{u}_3^0 - x_\alpha \partial_{33} u_\alpha^0), \tag{23.8}$$

and consequently the bending moments are given by

$$
\begin{aligned}
m_\beta^0 &= \int_\omega x_\beta \sigma_{33}^0 \,\mathrm{d}\omega = e_\beta \partial_3 \underline{u}_3^0 - h_{\alpha\beta}\partial_{33} u_\alpha^0 \\
&= \int_{x_3}^{L}\int_0^{t}\left[\int_\omega f_\beta \,\mathrm{d}\omega + \int_\gamma g_\beta \,\mathrm{d}\gamma\right] \\
&\quad + \int_{x_3}^{L}\left[\int_\omega x_\beta f_3 \,\mathrm{d}\omega + \int_\gamma x_\beta g_3 \,\mathrm{d}\gamma\right] + \int_\omega x_\beta h_3 \,\mathrm{d}\omega \\
&\quad + (x_3 - L)\left[\int_0^{L}\left(\int_\omega f_\beta \,\mathrm{d}\omega + \int_\gamma g_\beta \,\mathrm{d}\gamma\right) + \int_\omega h_\beta \,\mathrm{d}\omega\right].
\end{aligned}
\tag{23.9}
$$

(iv) *The shear stress components* $\sigma_{3\beta}^0$ *are a solution of the following problem*:

$$
\begin{aligned}
&\sigma_{3\beta}^0 \in H^1[0,L;H^1(\omega)] \quad \textit{verifying a.e. in } (0,L): \\
&-\partial_\beta \sigma_{3\beta}^0 = f_3 + \partial_3 \sigma_{33}^0, \quad \textit{in } \omega, \\
&\sigma_{3\beta}^0 n_\beta = g_3, \quad \textit{on } \gamma,
\end{aligned}
\tag{23.10}
$$

therefore, the shear force and the torsion moments are given by

$$
\begin{aligned}
q_\beta^0 &= \int_\omega \sigma_{3\beta}^0 \,\mathrm{d}\omega = \partial_3 m_\beta^0 + \int_\omega x_\beta f_3 \,\mathrm{d}\omega + \int_\gamma x_\beta g_3 \,\mathrm{d}\gamma \\
&= -\int_0^{x_3}\left[\int_\omega f_\beta \,\mathrm{d}\omega + \int_\gamma g_\beta \,\mathrm{d}\gamma\right] + \int_0^{L}\left[\int_\omega f_\beta \,\mathrm{d}\omega + \int_\gamma g_\beta \,\mathrm{d}\gamma\right] + \int_\omega h_\beta \,\mathrm{d}\omega,
\end{aligned}
\tag{23.11}
$$

$$
\begin{aligned}
m_3^0 &= \int_\omega (x_2\sigma_{31}^0 - x_1\sigma_{32}^0)\,\mathrm{d}\omega \\
&= -\int_0^{x_3}\left[\int_\omega (x_2 f_1 - x_1 f_2)\,\mathrm{d}\omega + \int_\gamma (x_2 g_1 - x_1 g_2)\,\mathrm{d}\gamma\right] \\
&\quad + \int_0^{L}\left[\int_\omega (x_2 f_1 - x_1 f_2)\,\mathrm{d}\omega + \int_\gamma (x_2 g_1 - x_1 g_2)\,\mathrm{d}\gamma\right] \\
&\quad + \int_\omega (x_2 h_1 - x_1 h_2)\,\mathrm{d}\omega.
\end{aligned}
\tag{23.12}
$$

(v) *The plane components of the stress tensor* $\sigma^0_{\alpha\beta}$ *are a solution of the following problem*:

$$\begin{aligned}&\sigma^0_{\alpha\beta} \in L^2[0,L;H^1(\omega)] \quad \textit{verifying a.e. in } (0,L),\\ &-\partial_\alpha\sigma^0_{\alpha\beta} = f_\beta + \partial_3\sigma^0_{3\beta}, \quad \textit{in } \omega,\\ &\sigma^0_{\alpha\beta}n_\alpha = g_\beta, \quad \textit{on } \gamma.\end{aligned} \tag{23.13}$$

PROOF. Theorem 4.3 applied to $V(\Omega)$ defined by (22.12) guarantees that

$$V^0_{BN}(\Omega) = \{\boldsymbol{v} = (v_i) \in V(\Omega) : e_{\alpha\beta}(\boldsymbol{v}) = e_{3\beta}(\boldsymbol{v}) = 0\}.$$

Consequently Eqs. (22.36), (22.37) warrant $u^0 \in V^0_{BN}(\Omega)$ and so we obtain (i). Expression (23.8) is a direct consequence of Eq. (22.35).

Taking $v_3 = v^0_3 \in V^1_0(0,L)$ in (22.38) we obtain

$$\int_0^L \left[\int_\omega \sigma^0_{33}\,\mathrm{d}\omega\right] \partial_3 v^0_3 = \tilde{F}_3(v^0), \quad \text{for all } v^0 \in V^1_0(0,L), \tag{23.14}$$

which is the first equation of (23.7) since we have already proved (23.8). To obtain the second one, succesively taking $v_3 = x_\beta\partial_3 v^0_\beta$ and $v_\beta = v^0_\beta \in V^2_0(0,L)$, with $v^0_\beta \in V^2_0(0,L)$ in (22.38) and (22.39) we show, after a subtraction, that

$$-\int_0^L \left[\int_\omega x_\beta\sigma^0_{33}\,\mathrm{d}\omega\right] \partial_{33}v^0_\beta\,\mathrm{d}x_3 = \tilde{F}_\beta(v^0_\beta), \quad \text{for all } v^0_\beta \in V^2_0(0,L), \tag{23.15}$$

Finally using (23.8) we get the second equation of (23.7). In order to show existence and uniqueness of $(u^0_1, u^0_2, \underline{u}^0_3)$, let us consider the following equivalent formulation of (23.7):

$$\begin{aligned}&\boldsymbol{u}^0 = (u^0_i) = (u^0_1, u^0_2, \underline{u}^0_3 - x_\alpha\partial_3 u^0_\alpha) \in V^0_{BN}(\Omega), \quad \text{such that}\\ &\int_\omega \frac{1}{a_{3333}}\partial_3 u^0_3\partial_3 v_3\,\mathrm{d}\omega = \int_\omega f_i v_i\,\mathrm{d}\omega + \int_\gamma g_i v_i\,\mathrm{d}a + \int_{\Gamma_L} h_i v_i\,\mathrm{d}a,\\ &\text{for all } \boldsymbol{v} \in V^0_{BN}(\Omega).\end{aligned} \tag{23.16}$$

The following inequality, valid for all $\boldsymbol{v} \in V^0_{BN}(\Omega)$,

$$\int_\omega \frac{1}{a_{3333}}\partial_3 v_3\partial_3 v_3\,\mathrm{d}\omega \geqslant \frac{1}{\|a_{3333}\|_{\infty,\Omega}}\left\{\int_0^L (\partial_3\underline{v}_3)^2\,\mathrm{d}x_3 + I_\alpha\int_0^L (\partial_{33}v_\alpha)^2\,\mathrm{d}x_3\right\}, \tag{23.17}$$

proves the $V^0_{BN}(\Omega)$–ellipticity for the left-hand side of (23.16) and then uniqueness of the solution for problems (23.16) and (23.7). Using (23.8) and solving the one-dimensional boundary value problem (23.15), we get expressions (23.9) for the bending moment m^0_β. Besides, the strong interpretation of (23.14) gives

$$\int_\omega \partial_3 \sigma^0_{33}\,\mathrm{d}\omega + \int_\omega f_3\,\mathrm{d}\omega + \int_\gamma g_3\,\mathrm{d}\gamma = 0, \tag{23.18}$$

$$\int_\omega \sigma^0_{33}(L)\,\mathrm{d}\omega = \int_\omega h_3\,\mathrm{d}\omega. \tag{23.19}$$

Now we set out to get regularity results for the solution to (23.7). Obviously from hypotheses (23.6) on the applied forces we have

$$\begin{aligned} &- l\partial_3 \underline{u}^0_3 + e_\alpha \partial_{33} u^0_\alpha \in H^2(0,L),\\ &- e_\beta \partial_3 \underline{u}^0_3 + h_{\alpha\beta}\partial_{33} u^0_\alpha \in H^2(0,L). \end{aligned} \tag{23.20}$$

From (23.6) and the fact that if $u, v \in H^1(0,L)$ then $uv \in H^1(0,L)$ (BREZIS [1983, Corollary VII. 9]) we deduce

$$u^0_\alpha \in H^4(0,L), \qquad \underline{u}^0_3 \in H^3(0,L). \tag{23.21}$$

Consequently from (23.8) and (23.6) we obtain

$$\sigma^0_{33} \in H^2[0,L;L^2(\omega)]. \tag{23.22}$$

This regularity of σ^0_{33} and hypotheses (23.6) allows us to use Green's formula in Eq. (22.38) and so find that $(\sigma^0_{3\beta})$ must be a solution of problem (23.10). On the other hand, taking $(v_1, v_2) = (v^0_1, v^0_2) \in [V^1_0(0,L)]^2$ and $(v_1, v_2) = (x_2 v^0, -x_1 v^0)$, $v^0 \in V^1_0(0,L)$ successively in (22.39) we find that $(\sigma^0_{3\beta})$ must verify the following conditions:

$$\begin{aligned} &\int_0^L \left[\int_\omega \sigma^0_{3\beta}\,\mathrm{d}\omega\right] \partial_3 v^0_\beta\,\mathrm{d}x_3\\ &\quad = \int_0^L \left[\int_\omega f_\beta\,\mathrm{d}\omega + \int_\gamma g_\beta\,\mathrm{d}\gamma\right] v^0_\beta\,\mathrm{d}x_3 + \left[\int_\omega h_\beta\,\mathrm{d}\omega\right] v^0_\beta(L), \end{aligned} \tag{23.23}$$

for all $v^0_\beta \in V^1_0(0,L)$,

$$\int_0^L \left[\int_\omega (x_2\sigma^0_{31} - x_1\sigma^0_{32})\,\mathrm{d}\omega\right] \partial_3 v^0\,\mathrm{d}x_3$$

$$= \int_0^L \left[\int_\omega (x_2 f_1 - x_1 f_2)\,\mathrm{d}\omega + \int_\gamma (x_2 g_1 - x_1 g_2)\,\mathrm{d}\gamma \right] v^0 \,\mathrm{d}x_3$$

$$+ \left[\int_\omega (x_2 h_1 - x_1 h_2)\,\mathrm{d}\omega \right] v^0(L), \qquad \text{for all } v^0 \in V_0^1(0,L). \tag{23.24}$$

The strong interpretation of (23.23) and (23.24) leads us to

$$\int_\omega \partial_3 \sigma^0_{3\beta}\,\mathrm{d}\omega + \int_\omega f_\beta\,\mathrm{d}\omega + \int_\gamma g_\beta\,\mathrm{d}\gamma = 0, \tag{23.25}$$

$$\int_\omega \sigma^0_{3\beta}(L)\,\mathrm{d}\omega = \int_\omega h_\beta\,\mathrm{d}\omega, \tag{23.26}$$

$$\int_\omega \partial_3(x_2\sigma^0_{31} - x_1\sigma^0_{32})\,\mathrm{d}\omega + \int_\omega (x_2 f_1 - x_1 f_2)\,\mathrm{d}\omega + \int_\gamma (x_2 g_1 - x_1 g_2)\,\mathrm{d}\gamma = 0, \tag{23.27}$$

$$\int_\omega [x_2\sigma^0_{31}(L) - x_1\sigma^0_{32}(L)]\,\mathrm{d}\omega = \int_\omega (x_2 h_1 - x_1 h_2)\,\mathrm{d}\omega. \tag{23.28}$$

Solving the above one-dimensional differential equations, we obtain expressions (23.11) and (23.12) for the shear force and torsion moments, respectively.

A similar argument applied to Eq. (22.39) furnishes problem (23.13) for $\sigma^0_{\alpha\beta}$, which completes the proof. □

REMARK 23.1. The compatibility condition (23.18) is necessary and sufficient in order to guarantee the existence of infinitely many solutions to (23.10) that differ from one another in an element $\chi_\beta \in H^1(0,L;H^1(\omega))$ such that a.e. in $(0,L)$:

$$\partial_\beta \chi_\beta = 0, \quad \text{in } \omega,$$
$$\chi_\beta n_\beta = 0, \quad \text{on } \gamma.$$

To determine $\sigma^0_{3\beta}$ uniquely it is necessary to assume further conditions derived from compatibility with (22.41). In Sections 7 and 8 it is proved that for the homogeneous isotropic case these conditions are compatible with (23.10), (23.11) and (23.12). For the homogeneous anisotropic case one can see ÁLVAREZ-DIOS and VIAÑO [1992a,b].

In the compatibility of (23.10)–(23.12) with (22.41) the weak clamping boundary conditions play an essential role, and they cannot be fulfilled for strong-clamping conditions, a boundary layer phenomenon showing up. Nevertheless for weak-clamping conditions this boundary layer phenomenon manifests itself through the fact that the computed solution $\sigma_{3\beta}^0$ does not satisfy (22.38) in the most general case. In fact, in order for $(\sigma_{3\beta}^0)$ thus defined to satisfy (22.38) it is necessary for (σ_{33}^0) to verify the following conditions on both ends (which can be derived using Green's formula):

$$\int_{\Gamma_0} \sigma_{33}^0 v_3 \,\mathrm{d}a = 0, \qquad \int_{\Gamma_L} (h_3 - \sigma_{33}^0) v_3 \,\mathrm{d}a = 0, \qquad \text{for all } v_3 \in W_1(\Omega). \tag{23.29}$$

These conditions cannot be verified in general, unless the applied forces in $x_3 = 0$ and $x_3 = L$ meet certain compatibility requirements which we shall not go into. Nevertheless it is convenient to remark that any $(\sigma_{3\beta}^0)$ solution of (23.10)–(23.12) verifies (22.38) for any function $v_3 \in W_1^* = \{\phi \in H^1(\Omega)\colon \phi = 0 \text{ on } \Gamma_0, \phi \text{ constant on } \Gamma_L\}$, as can be deduced from (23.29) and (23.19).

We remark that this phenomenon does not occur in the homogeneous isotropic case because (23.29) is always true due to the particular form of σ_{33}^0 (cf. (6.34)).

In the same way, in order for problem (22.39) to admit a solution it is necessary that $\sigma_{3\beta}^0$ fulfils the following condition at both ends:

$$\int_{\Gamma_0} \sigma_{3\beta}^0 v_\beta \,\mathrm{d}a = 0, \qquad \int_{\Gamma_L} (h_\beta - \sigma_{3\beta}^0) v_\beta \,\mathrm{d}a = 0, \qquad \text{for all } (v_\beta) \in W_2(\Omega). \tag{23.30}$$

The fulfilment of these conditions necessarily implies restrictions on the form of $\sigma_{3\beta}^0$ at the ends Γ_0 and Γ_L, which consequently can be read as restrictions on the applied forces (see Remark 8.4 for a similar situation). Eq. (22.39) is verified for all $(v_\beta) \in W_2(\Omega)$ such that $v_\beta = 0$ on Γ_0 and v_β constant on Γ_L (see (23.26)).

The compatibility conditions (23.25) and (23.27) are necessary and sufficient to warrant the existence of infinitely many solutions to (23.13) (cf. DUVAUT and LIONS [1972a]). In order to determine $(\sigma_{\alpha\beta}^0)$, in a unique way, we must impose compatibility with (22.40) (see Sections 7 and 8 for the homogeneous isotropic case, Section 24 for two particular nonhomogeneous cases and ÁLVAREZ-DIOS and VIAÑO [1992a,b] for the homogeneous and anisotropic case).

The next step in the asymptotic analysis being carried on here is the establishment of *convergence results* justifying the formal asymptotic expansions. We shall use the technique described in Section 10 for the homogeneous isotropic case and we shall see that *some of the results do not hold for the nonhomogeneous case* (the anisotropy does not have any influence).

Properties (22.3) and (22.4) allow us to prove Theorem 10.1 for $(\boldsymbol{u}(\varepsilon), \boldsymbol{\sigma}(\varepsilon))$ solution of problem (22.24)–(22.31). That is, sequences $(\sigma_{33}(\varepsilon))_{\varepsilon>0}$, $(\varepsilon\sigma_{3\beta}(\varepsilon))_{\varepsilon>0}$,

$(\varepsilon^2\sigma_{\alpha\beta}(\varepsilon))_{\varepsilon>0}$, are uniformly bounded in $L^2(\Omega)$ and sequence $(\boldsymbol{u}(\varepsilon))_{\varepsilon>0}$ is uniformly bounded in $V(\Omega)$ $(0 < \varepsilon \leqslant 1)$. Therefore, we immediately obtain that Theorem 10.2 holds with a formally identical proof. *On the other hand the weak convergence in* $[H^1(\Omega)]^3$ of $\boldsymbol{u}(\varepsilon)$ to $\boldsymbol{u}^0$ (Theorem 10.3) *cannot be established* in the same way. In fact, from the uniform boundedness of $\boldsymbol{u}(\varepsilon)$ in $V(\Omega)$ we deduce the existence of at least a weakly convergent subsequence, still denoted by $\boldsymbol{u}(\varepsilon)$, such that

$$\boldsymbol{u}(\varepsilon) \rightharpoonup \boldsymbol{u}, \quad \text{in } V(\Omega). \tag{23.31}$$

For the general nonhomogeneous case, it is not possible to show that $\boldsymbol{u}$ coincides with $\boldsymbol{u}^0$ and therefore the techniques used in Theorems 10.3, 10.4 and 10.5 cannot be used. The following theorem is a modified version, for the present case, of Theorems 10.2 and 10.3. It allows us to compare the weak limit $(\boldsymbol{u}, \boldsymbol{\Psi})$, with the first term of the asymptotic expansion $(\boldsymbol{u}^0, \boldsymbol{\sigma}^0)$. It is owed to ÁLVAREZ-DIOS and VIAÑO [1991b].

THEOREM 23.2. *There exists at least a subsequence of* $(\boldsymbol{u}(\varepsilon), \boldsymbol{\sigma}(\varepsilon))_{\varepsilon>0}$, *denoted in the same way, and there exist elements* $\boldsymbol{u} \in V(\Omega)$ *and* $\boldsymbol{\Psi} \in \Sigma(\Omega)$ *such that, when* $\varepsilon \to 0$

$$\boldsymbol{u}(\varepsilon) \rightharpoonup \boldsymbol{u}, \quad \textit{weakly in } V(\Omega), \tag{23.32}$$

$$\sigma_{33} \rightharpoonup \Psi_{33}, \qquad \varepsilon\sigma_{3\beta} \rightharpoonup \Psi_{3\beta}, \qquad \varepsilon^2\sigma_{\alpha\beta} \rightharpoonup \Psi_{\alpha\beta}, \qquad \textit{weakly in } L^2(\Omega). \tag{23.33}$$

Functions Ψ_{ij} *and* u_i *satisfy the following properties:*

$$\int_\Omega \Psi_{3\beta}\partial_\beta v \,\mathrm{d}\boldsymbol{x} = 0, \quad \textit{for all } v \in W_1(\Omega), \tag{23.34}$$

$$\int_\Omega \Psi_{\alpha\beta}\partial_\alpha v_\beta \,\mathrm{d}\boldsymbol{x} = 0, \quad \textit{for all } (v_\beta) \in W_2(\Omega), \tag{23.35}$$

$$\boldsymbol{u} \in V^0_{BN}(\Omega), \qquad u_3 = \underline{u}_3 - x_\alpha\partial_3 u_\alpha, \qquad \underline{u}_3 \in V^1_0(0,L), \qquad u_\alpha \in V^2_0(0,L), \tag{23.36}$$

$$\boldsymbol{u} = \boldsymbol{u}^0 + \boldsymbol{w}, \tag{23.37}$$

$$\Psi_{33} = \sigma^0_{33} + \frac{1}{a_{3333}}[e_{33}(\boldsymbol{w}) - a_{\alpha\beta 33}\Psi_{\alpha\beta}], \tag{23.38}$$

where $\boldsymbol{w}$ *is the solution of the following problem:*

$$\boldsymbol{w} \in V^0_{BN}(\Omega),$$
$$\int_\Omega \frac{1}{a_{3333}} e_{33}(\boldsymbol{w}) e_{33}(\boldsymbol{v}) \,\mathrm{d}\boldsymbol{x} = \int_\Omega \frac{a_{\alpha\beta 33}}{a_{3333}} \Psi_{\alpha\beta} e_{33}(\boldsymbol{v}) \,\mathrm{d}\boldsymbol{x}, \tag{23.39}$$
$$\textit{for all } \boldsymbol{v} \in V^0_{BN}(\Omega).$$

Moreover, the corresponding subsequences $q_\beta(\varepsilon)$, $m_\beta(\varepsilon)$ and $q_3(\varepsilon) = \int_\omega \sigma_{33}(\varepsilon)\,\mathrm{d}\omega$ satisfy

$$q_\beta(\varepsilon) \rightharpoonup q^0_\beta, \qquad q_3(\varepsilon) \rightharpoonup q^0_3 = \int_\omega \sigma^0_{33}\,\mathrm{d}\omega, \qquad \textit{weakly in } L^2(0,L), \tag{23.40}$$

$$m_\beta(\varepsilon) \rightharpoonup m^0_\beta, \quad \textit{weakly in } L^2(0,L). \tag{23.41}$$

PROOF. From the previous uniform estimates we conclude the existence of a subsequence verifying (23.32) and (23.33). With the same arguments used in the proof of Theorem 10.2 one concludes (23.34) and (23.35). Moreover, the following weak convergences are also obtained:

$$q_i(\varepsilon) \rightharpoonup \rho_i, \qquad m_\beta(\varepsilon) \rightharpoonup r_\beta, \qquad \text{in } L^2(0,L), \tag{23.42}$$

where ρ_i and r_β satisfy the following properties:

$$\rho_3 = \int_\omega \Psi_{33}\,\mathrm{d}\omega, \qquad r_\beta = \int_\omega x_\beta \Psi_{33}\,\mathrm{d}\omega, \tag{23.43}$$

$$\begin{aligned} &\int_0^L \rho_\beta v^0\,\mathrm{d}x_3 + \int_0^L r_\beta \partial_3 v^0\,\mathrm{d}x_3 \\ &\quad = \int_0^L \left[\int_\omega x_\beta f_3\,\mathrm{d}\omega + \int_\gamma x_\beta g_3\,\mathrm{d}\gamma\right] v^0\,\mathrm{d}x_3 \\ &\quad + \left(\int_\omega x_\beta h_3\,\mathrm{d}\omega\right) v^0(L), \quad \text{for all } v^0 \in V^1_0(0,L). \end{aligned} \tag{23.44}$$

On the other hand, using the same ideas as in the proof of Theorem 10.3, and passing to the limit in Eq. (22.25), we obtain

$$\int_\Omega a_{3333}\Psi_{33}\tau_{33}\,\mathrm{d}\boldsymbol{x} + \int_\Omega a_{\alpha\beta 33}\Psi_{\alpha\beta}\tau_{33}\,\mathrm{d}\boldsymbol{x} = \int_\Omega e_{ij}(\boldsymbol{u})\tau_{ij}\,\mathrm{d}\boldsymbol{x}, \quad \text{for all } \boldsymbol{\tau} \in \Sigma(\Omega). \tag{23.45}$$

This equation is equivalent to

$$\begin{aligned} &e_{\alpha\beta}(\boldsymbol{u}) = e_{3\beta}(\boldsymbol{u}) = 0, \\ &e_{33}(\boldsymbol{u}) = a_{3333}\Psi_{33} + a_{\alpha\beta 33}\Psi_{\alpha\beta}, \end{aligned} \tag{23.46}$$

from which we conclude (23.36), (23.37) and (23.38) by using (23.8) and writing $\boldsymbol{w} = \boldsymbol{u} - \boldsymbol{u}^0$.

Restricting Eq. (22.26) to $V^0_{BN}(\Omega)$ and passing to the limit one obtains

$$\int_\Omega \Psi_{33} e_{33}(\boldsymbol{v})\,\mathrm{d}\boldsymbol{x} = -F(\boldsymbol{v}), \quad \text{for all } \boldsymbol{v} \in V^0_{BN}(\Omega). \tag{23.47}$$

Substituting (23.38) in (23.47) and taking into account that σ^0_{33} satisfies (23.16) we conclude that $\boldsymbol{w}$ is a solution of (23.39).

Let

$$\begin{aligned} \rho^w_3 &= \int_\omega \frac{1}{a_{3333}} [e_{33}(\boldsymbol{w}) - a_{\alpha\beta 33}\Psi_{\alpha\beta}]\,\mathrm{d}\omega, \\ r^w_\beta &= \int_\omega \frac{x_\beta}{a_{3333}} [e_{33}(\boldsymbol{w}) - a_{\gamma\mu 33}\Psi_{\gamma\mu}]\,\mathrm{d}\omega. \end{aligned} \tag{23.48}$$

Equation (23.39) is equivalent to the following two variational equalities:

$$\int_0^L \rho^w_3 \partial_3 \underline{v}_3\,\mathrm{d}x_3 = 0, \quad \text{for all } \underline{v}_3 \in V^1_0(0,L),$$

$$\int_0^L r^w_\beta \partial_{33} v_\beta\,\mathrm{d}x_3 = 0, \quad \text{for all } v_\beta \in V^2_0(0,L),$$

whose unique solution is

$$\rho^w_3 = 0, \quad r^w_\beta = 0. \tag{23.49}$$

From (23.48), (23.49), (23.38), (23.42) and (23.43) we obtain (23.41) together with the second part of (23.40). Finally, from (23.44) now we obtain

$$\begin{aligned} \int_0^L \rho_\beta v^0\,\mathrm{d}x_3 = {} & \int_0^L \left[\int_\omega x_\beta f_3\,\mathrm{d}\omega + \int_\gamma x_\beta g_3\,\mathrm{d}\gamma \right] v^0\,\mathrm{d}x_3 \\ & - \int_0^L m^0_\beta \partial_3 v^0\,\mathrm{d}x_3 + \left[\int_\omega x_\beta h_3\,\mathrm{d}\omega \right] v^0(L), \\ & \text{for all } v^0 \in V^1_0(0,L). \end{aligned} \tag{23.50}$$

Since $m^0_\beta \in H^1(0,L)$, applying Green's formula to (23.50) and taking (23.9) and (23.11) into account, we get $\rho_\beta = q^0_\beta$ from which (23.40) follows and this ends the proof of the theorem. □

REMARK 23.2. From the unicity of the limits q_i^0 and m_β^0 we conclude that the weak convergences (23.40) and (23.41) hold true for the whole sequences $q_i(\varepsilon)$ and $m_\beta(\varepsilon)$, respectively.

Since problems (23.34) and (23.35) admit infinitely many solutions and since functions $\Psi_{\alpha\beta}$ show up in the characterization of the limit $\boldsymbol{u}$ (see (23.39)), *in general* $\boldsymbol{u}$ *will be different from* $\boldsymbol{u}^0$. This *lack of unicity* does not allow us to use the same arguments as in the homogeneous isotropic case in order to obtain the weak (and strong) convergence of the whole sequence $\boldsymbol{u}(\varepsilon)$ (cf. Theorem 10.3 and 10.4). This seems to be still an open problem for the general case. The dependence of the coefficients $a_{\alpha\beta33}/a_{3333}$ with respect to x_1 and x_2 seems to prevent this convergence from being obtained. If this dependence is not present, then a similar result as in Theorem 10.4 may be obtained. It may be stated in the following way (cf. ÁLVAREZ-DIOS and VIAÑO [1991b] and ÁLVAREZ-DIOS [1992a,b]).

THEOREM 23.3. *If the material the beam is made of is such that coefficients* $a_{\alpha\beta33}/a_{3333}$ *do not depend on* x_1 *and* x_2, *then the sequence* $(\boldsymbol{u}(\varepsilon), \boldsymbol{\sigma}(\varepsilon)) \in V(\Omega) \times \Sigma(\Omega)$, *solution of* (22.24)–(22.26) *satisfies the following convergences, as* $\varepsilon \to 0$:

$$\|\boldsymbol{u}(\varepsilon) - \boldsymbol{u}^0\|_{1,\Omega} \to 0, \tag{23.51}$$

$$\|\sigma_{33}(\varepsilon) - \sigma_{33}^0\|_{0,\Omega} \to 0, \tag{23.52}$$

$$\varepsilon\|\sigma_{3\beta}(\varepsilon)\|_{0,\Omega} \to 0, \qquad \varepsilon^2\|\sigma_{\alpha\beta}(\varepsilon)\|_{0,\Omega} \to 0. \tag{23.53}$$

PROOF. For the subsequence verifying (23.32), (23.33) and in a similar way as in the proof of Theorem 10.4 we obtain (cf. (10.35)):

$$\begin{aligned} &C\{|\sigma_{33}(\varepsilon) - \sigma_{33}^0|_{0,\Omega}^2 + \varepsilon^2|\sigma_{3\beta}(\varepsilon)|_{0,\Omega}^2 + \varepsilon^4|\sigma_{\alpha\beta}(\varepsilon)|_{0,\Omega}^2\} \\ &\quad \leqslant \int_\Omega f_i(u_i(\varepsilon) - u_i^0)\,\mathrm{d}\boldsymbol{x} + \int_\Gamma g_i(u_i(\varepsilon) - u_i^0)\,\mathrm{d}a + \int_{\Gamma_L} h_i(u_i(\varepsilon) - u_i^0)\,\mathrm{d}a \\ &\quad \int_\Omega \sigma_{33}^0 e_{33}(\boldsymbol{u}(\varepsilon) - \boldsymbol{u}^0)\,\mathrm{d}\boldsymbol{x} - \varepsilon^2 \int_\Omega a_{\alpha\beta33}\sigma_{33}^0\sigma_{\alpha\beta}(\varepsilon)\,\mathrm{d}\boldsymbol{x}. \end{aligned} \tag{23.54}$$

From the weak convergences already established we conclude that when $\varepsilon \to 0$ (23.54) converges to

$$\mathscr{L} = -F(\boldsymbol{u} - \boldsymbol{u}^0) + \int_\Omega \sigma_{33}^0 e_{33}(\boldsymbol{u} - \boldsymbol{u}^0)\,\mathrm{d}\boldsymbol{x} - \int_\Omega a_{\alpha\beta33}\sigma_{33}^0\Psi_{\alpha\beta}\,\mathrm{d}\boldsymbol{x}. \tag{23.55}$$

Taking $\boldsymbol{v} = \boldsymbol{u}^0$ in (23.39) we prove that the last two terms on the right-hand side of (23.55) cancel and consequently:

$$\mathscr{L} = -F(\boldsymbol{w}). \tag{23.56}$$

As was already mentioned, in the general case one has $\mathscr{L} \neq 0$ and the proof cannot proceed as in Theorem 10.4. A sufficient condition is then that $\boldsymbol{w} = \mathbf{0}$, for any function $\Psi_{\alpha\beta}$ solution of (23.35).

The right-hand side of (23.39) may be written in an equivalent form as

$$\int_0^L \left[\int_\omega \frac{a_{\alpha\beta 33}}{a_{3333}} \Psi_{\alpha\beta}\, \mathrm{d}\omega \right] \partial_3 \underline{v}_3 \, \mathrm{d}x_3 - \int_0^L \left[\int_\omega x_\mu \frac{a_{\alpha\beta 33}}{a_{3333}} \Psi_{\alpha\beta}\, \mathrm{d}\omega \right] \partial_{33} v_\mu \, \mathrm{d}x_3,$$
$$\text{for all } \underline{v}_3 \in V_0^1(0,L), \text{ and all } v_\mu \in V_0^2(0,L). \tag{23.57}$$

If coefficients $a_{\alpha\beta 33}/a_{3333}$ do not depend either on x_1 or on x_2, then from property (10.19) satisfied by functions $\Psi_{\alpha\beta}$ we obtain that (23.57) is identically zero and that (23.39) implies $\boldsymbol{w} = \mathbf{0}$. Therefore, for this particular case $\mathscr{L} = 0$ and the result follows as in Theorem 10.4. □

REMARK 23.3. We observe that the previous theorem remains valid whenever coefficients $a_{\alpha\beta 33}/a_{3333}$ verify the following property:

$$\int_\omega \frac{a_{\alpha\beta 33}}{a_{3333}} \Psi_{\alpha\beta}\, \mathrm{d}\omega = \int_\omega x_\mu \frac{a_{\alpha\beta 33}}{a_{3333}} \Psi_{\alpha\beta}\, \mathrm{d}\omega = 0, \tag{23.58}$$

for any solution $\Psi_{\alpha\beta}$ of (23.35). A sufficient condition, in order for this to hold, is that the coefficients may be expressed in the form

$$\frac{a_{\alpha\beta 33}}{a_{3333}} = \partial_\alpha v_\beta^*, \qquad x_\mu \frac{a_{\alpha\beta 33}}{a_{3333}} = \partial_\alpha v_\beta^\mu, \qquad (v_\beta^*),\ (v_\beta^\mu) \in W_2(\Omega). \tag{23.59}$$

REMARK 23.4. From convergence (23.52) we conclude that under the conditions of the previous theorem the sequence $m_\beta(\varepsilon)$ strongly converges to m_β^0 in $L^2(0,L)$ (cf. Theorem 10.5).

As in the homogeneous isotropic case, the next step is the obtainment of a first order general model posed on the original beam Ω^ε by inverting the change of scales used. We shall no more stress this point because as in Section 11, the final result is "equivalent" to writing the exponent ε on all the forces, coefficients and functions showing up in Theorem 23.1, which may be seen as a *general first order model for the beam* Ω. In particular, expressions (23.8), (23.9) and (23.11) constitute a generalization of the classical expressions for the axial stress, bending moments, and shear forces (cf. (7.63), (7.64) and (7.76) for the homogeneous isotropic case). Moreover, the variational problem (23.7) constitutes a natural generalization of the stretching and bending Eqs. (7.59) and (7.61) for the homogeneous isotropic case. It is interesting to see the strong form of this model:

$$
\begin{aligned}
&-\partial_3(l\partial_3\underline{u}_3^0 - e_\alpha\partial_{33}u_\alpha^0) = \int_\omega f_3\,\mathrm{d}\omega + \int_\gamma g_3\,\mathrm{d}\gamma, \quad \text{in } (0,L),\\
&\partial_{33}(-e_\beta\partial_3\underline{u}_3^0 + h_{\alpha\beta}\partial_{33}u_\alpha^0)\\
&\quad = \int_\omega f_\beta\,\mathrm{d}\omega + \int_\gamma g_\beta\,\mathrm{d}\gamma + \int_\omega x_\beta\partial_3 f_3\,\mathrm{d}\omega + \int_\gamma x_\beta\partial_3 g_3\,\mathrm{d}\gamma, \quad \text{in } (0,L),\\
&(l\partial_3\underline{u}_3^0 - e_\alpha\partial_{33}u_\alpha^0)(L) = \int_\omega h_3\,\mathrm{d}\omega,\\
&(e_\beta\partial_3\underline{u}_3^0 - h_{\alpha\beta}\partial_{33}u_\alpha^0)(L) = \int_\omega x_\beta h_3\,\mathrm{d}\omega,\\
&\partial_3(e_\beta\partial_3\underline{u}_3^0 - h_{\alpha\beta}\partial_{33}u_\alpha^0)(L) = \int_\omega h_\beta\,\mathrm{d}\omega - \int_\omega x_\beta f_3(L)\,\mathrm{d}\omega - \int_\gamma x_\beta g_3(L)\,\mathrm{d}\gamma.
\end{aligned}
\tag{23.60}
$$

We notice that *in the general case the bending and stretching effects are coupled* in (23.7). Moreover, if $h_{12} \neq 0$ the transverse bending components are also coupled as the second equation in (23.7) shows.

We also remark that this model contains Bernoulli–Navier's model for the homogeneous isotropic case. In order to see this it is enough to remark that for this case the only nonzero elasticity coefficients are (no sum)

$$
a_{iiii} = \frac{1}{E}, \qquad a_{iijj} = -\frac{\nu}{E}, \qquad a_{ijij} = \frac{1+\nu}{2E}, \qquad (i \neq j). \tag{23.61}
$$

REMARK 23.5. For the isotropic case, one has

$$
\frac{a_{\alpha\beta 33}}{a_{3333}} = -\nu\delta_{\alpha\beta}, \tag{23.62}
$$

thus, a sufficient condition in order to have strong convergences (23.51)–(23.53) is that Poisson's ratio be independent of x_1 and x_2. We shall find this condition in the next section once again.

We end this section by analyzing model (23.7) for some particular but interesting cases (cf. ÁLVAREZ-DIOS and VIAÑO [1991b] and ÁLVAREZ-DIOS [1992a,b]). It is clear that in the general model bending and stretching effects may decouple when so to speak the material possesses enough elastic symmetry to have $e_\alpha = 0$. Furthermore, if we have $h_{12} = 0$, then bending effects along axis Ox_1 and Ox_2 also appear uncoupled. This is satisfied when we talk about isotropic materials, but it can also be true in more complex cases. In this section we shall give examples taking as a reference the case of fibre-reinforced beams made of homogeneous isotropic materials. By that we mean $\omega = \bigcup_{p=0}^{P} \omega_p^*$ where $\Omega_p = \omega_p^* \times (0, L)$ is made of an isotropic material having Young's modulus E_p and Poisson's ratio ν_p. We shall suppose ω to be simply connected as often occurs in practice, although this is not strictly necessary. Figures 23.1 (a)–(d) (from left to right and top to bottom), show several examples of this type of beams. Figures 23.1 (c) and 23.1 (d) show the very common case of the so-called multilayered beams, namely those in which the cross section ω can be divided in subsections ω_p^* whose shape is similar to that of the whole cross section.

In this way, for Fig. 23.1 (c) we can write

$$\omega_p^* = \left(-\tfrac{1}{2}, \tfrac{1}{2}\right) \times (a_p, a_{p+1}), \qquad -\tfrac{1}{2} = a_0 < a_1 < \cdots < a_P < a_{P+1} = \tfrac{1}{2}.$$

In a similar way for Fig. 23.1 (d), we have

$$\omega_p^* = r_p\omega \backslash r_{p-1}\omega \quad (p = 1, \ldots, P), \qquad 0 = r_{-1} < r_0 < \cdots < r_P = 1.$$

In fibre-reinforced beams made of homogeneous isotropic materials we have $E(\boldsymbol{x}) = E_p$, $\nu(\boldsymbol{x}) = \nu_p$, for all $\boldsymbol{x} \in \Omega_p$. Then for any reinforced beam we have

$$\begin{aligned} e_\alpha &= \sum_{p=0}^{P} E_p \int_{\omega_p^*} x_\alpha \, d\omega, \\ h_{\alpha\beta} &= \sum_{p=0}^{P} E_p \int_{\omega_p^*} x_\alpha x_\beta \, d\omega, \\ l &= \sum_{p=0}^{P} E_p |\omega_p^*|. \end{aligned} \tag{23.63}$$

From (23.7) we deduce that when $e_\alpha = 0$, bending component u_α^0 is independent of the stretch. This happens in some simple and interesting cases such as the following ones:

(i) For beams of the type of Fig. 23.1 (d), no matter the reinforcing materials and for both components since $e_1 = e_2 = 0$.

(ii) For beams of the type of Fig. 23.1 (c), we only have bending component u_2^0 coupled with the stretch since $e_1 = 0$ and $e_2 = \frac{1}{2} \sum_{p=0}^{P} E_p (a_{p+1}^2 - a_p^2)$.

(iii) For cases of Fig. 23.1 (a) and Fig. 23.1 (b) if $E_1 = E_3$ and/or $E_2 = E_4$ then $e_1 = 0$ (and/or $e_2 = 0$, respectively) and component u_1^0 (and/or u_2^0) is independent of the stretch. Should these not hold, these effects appear coupled. Besides, if $h_{12} = 0$ then the equations governing bending components u_1^0 and u_2^0 become independent. This happens for instance in multilayered beams (c), (d) and also in beam (b) if all reinforcements are made of the same material. The other coefficients intervening in the equations have the following values:

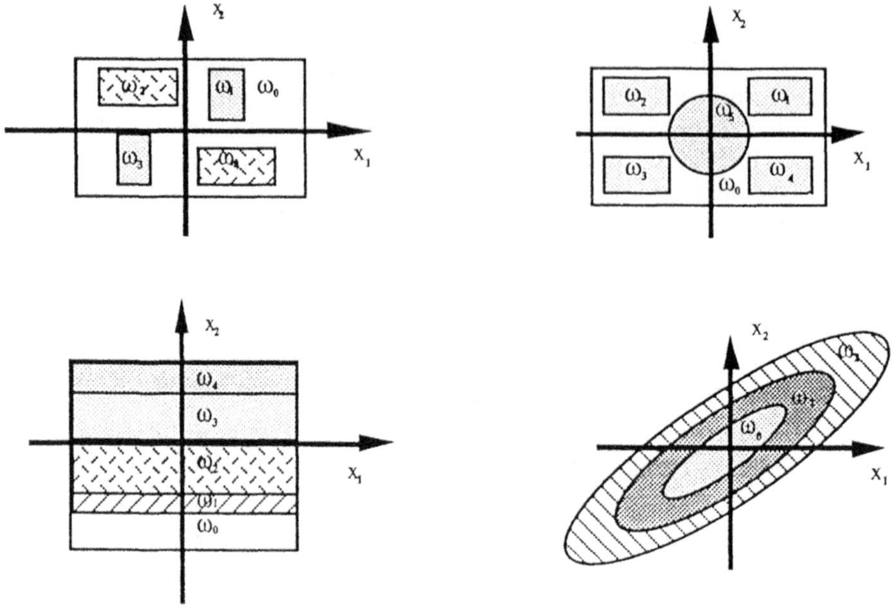

FIG. 23.1. Cross sections of fibre-reinforced beams; (a), (b), (c), (d) from left to right and from top to bottom.

(a) Beam in Fig. 23.1 (c).

$$
\begin{aligned}
h_{11} &= \frac{1}{12}\sum_{p=0}^{P} E_p(a_{p+1} - a_p), \\
h_{22} &= \frac{1}{3}\sum_{p=0}^{P} E_p(a_{p+1}^3 - a_p^3), \\
l &= \sum_{p=0}^{P} E_p(a_{p+1} - a_p).
\end{aligned}
\tag{23.64}
$$

(b) Beam in Fig. 23.1 (d) (no sum on α).

$$
\begin{aligned}
h_{\alpha\alpha} &= I_\alpha \sum_{p=0}^{P} E_p(r_p^2 - r_{p-1}^2), \\
l &= A \sum_{p0}^{P} E_p(r_p^2 - r_{p-1}^2).
\end{aligned}
\tag{23.65}
$$

In the case of multilayered beams of Fig. 23.1 (d) type, equations resulting from (23.7) appear uncoupled in bending and stretching effects and the model is in fact very similar to the standard Bernoulli–Navier bending model (see (6.35), (6.36)) except for the fact that Young's modulus has been replaced by the so called *composite modulus*:

$$\tilde{E} = \sum_{p=0}^{P} E_p(r_p^2 - r_{p-1}^2). \tag{23.66}$$

It is interesting to compare (23.66) to the formula arrived at in GREEN and NADGI [1990] for a two-layered, circular-sectioned cylindrical rod, that is

$$\tilde{E}_m = \sum_{p=0}^{P} E_p(r_p^2 - r_{p-1}^2) + \frac{4r_0^2(1-r_0^2)(\nu_1-\nu_0)^2}{\dfrac{(1-r_0^2)}{(K_0+(G_0/3))} + \dfrac{r_0^2}{(K_1+(G_1/3))}}, \tag{23.67}$$

where $G_p = E_p/[2(1+\nu_p)]$ is the shear modulus of the layer Ω_p and $K_p = E_p/[3(1-2\nu_p)]$ its modulus of compression. Other important aspects of asymptotic analysis of composite beams are considered in GEYMONAT, KRASUCKI and MARIGO [1987a,b].

In spite of the bending and stretching components being coupled in model (23.60), it does not take either torsion or Poisson's effects into account which are also considered in the simplest classical theories (cf. for example LEKHNITSKII [1977, p. 80]). As was asserted in TRABUCHO and VIAÑO [1989] and in Section 7, for the case of isotropic materials, those models can be obtained if one resorts to the second order terms in the asymptotic expansion. From (22.42) and (23.8) we deduce that, in particular, we must solve the following equations:

$$\partial_1 u_1^2 = \frac{a_{1133}}{a_{3333}}(\partial_3 \underline{u}_3^0 - x_\alpha \partial_{33} u_\alpha^0), \tag{23.68}$$

$$\partial_2 u_2^2 = \frac{a_{2233}}{a_{3333}}(\partial_3 \underline{u}_3^0 - x_\alpha \partial_{33} u_\alpha^0), \tag{23.69}$$

$$\partial_1 u_2^2 + \partial_2 u_1^2 = 2\frac{a_{1233}}{a_{3333}}(\partial_3 \underline{u}_3^0 - x_\alpha \partial_{33} u_\alpha^0). \tag{23.70}$$

In the most general case, with $(a_{\alpha\beta33}/a_{3333})$ being a function of x_1, x_2 and x_3, it is impossible to solve (23.68)–(23.70) explicitly. Nevertheless, there are three very interesting cases where it is possible to find out an explicit solution:

(i) *Anisotropic, transversely homogeneous beams.* In this case coefficients a_{ijkl} do not depend on variables (x_1, x_2) and (23.68)–(23.70) may be solved as in the isotropic case (Section 7). One has to introduce auxiliary functions depending on the geometry of the cross section and on the elasticity coefficients. In Section 25 we summarize the results obtained in this direction by ÁLVAREZ-DIOS [1992] and ÁLVAREZ-DIOS and VIAÑO [1992a,b]. These results generalize those of Sections 7 and 8 and represent generalizations of the most well-known classical models for anisotropic beams (cf. LEKHNITSKII [1977]).

(ii) *Isotropic transversely nonhomogeneous beams.* For this case coefficients a_{ijkl} satisfy (23.61) where now functions E and ν depend on the variables x_1, x_2, and x_3. In this case system (23.68)–(23.70) reduces to

$$\partial_\beta u^2_\alpha + \partial_\alpha u^2_\beta = -2\nu(\partial_3 \underline{u}^0_3 - x_\mu \partial_{33} u^0_\mu)\delta_{\alpha\beta}, \tag{23.71}$$

whose explicit solution presents the same difficulties as in (23.68)–(23.70). In the next section we summarize the results obtained by ÁLVAREZ-DIOS [1992] and ÁLVAREZ-DIOS and VIAÑO [1991a,b], for *the particular case where Poisson's ratio is independent of x_1 and x_2.*

(iii) *Anisotropic homogeneous beams in torsion.* No matter the dependence of the coefficients with respect to the space variable, system (23.68)–(23.69) becomes trivial whenever

$$e_{33}(\boldsymbol{u}^0) = \partial_3 \underline{u}^0_3 - x_\alpha \partial_{33} u^0_\alpha = 0. \tag{23.72}$$

A necessary and sufficient condition for this to hold is that the resultant of the system of applied forces be zero, that is, $\tilde{F}_i = 0$. Consequently the beam is subjected to applied torsion moments only. In the following section we consider this case for isotropic beams. The general case may be obtained, in a similar way, combining the results of the isotropic nonhomogeneous case (Section 24) with those of the anisotropic homogeneous case (Section 25).

24. Higher order models for transversely nonhomogeneous isotropic rods

In this section we consider that the elasticity coefficients a_{ijkl} satisfy (23.61) where Young's modulus $E(x_1, x_2, x_3)$ and Poisson's ratio $\nu(x_1, x_2, x_3)$ satisfy

$$E, \nu \in L^\infty(\Omega), \qquad E(\boldsymbol{x}) > 0, \qquad 0 < \nu(\boldsymbol{x}) < \tfrac{1}{2}, \quad \text{a.e. in } \Omega. \tag{24.1}$$

In this case the system of Eqs. (23.68)–(23.70) reduces to

$$\partial_\alpha u^2_\beta(x_1, x_2, x_3) = -2\nu(x_1, x_2, x_3)(\partial_3 \underline{u}^0_3(x_3) - x_\mu \partial_{33} u^0_\mu(x_3)) + \partial_\beta u^2_\alpha(x_1, x_2, x_3), \tag{24.2}$$

whose explicit integration cannot be performed without the introduction of restrictive hypotheses on Poisson's ratio or on the form of the term $(\partial_3 \underline{u}^0_3(x_3) - x_\mu \partial_{33} u^0_\mu(x_3))$.

The case of fibre reinforced beams in the previous section allows us to prove that even when the dependence of ν on x_1, x_2 is very simple indeed, solving (24.2) can be impossible in the considered spaces. Classical treatments of this problem are very few; nevertheless it is known that if $\nu_p \neq \nu_{p+1}$ for some $p = 0, \ldots, P$, then for problems of stretching and pure bending it is impossible to obtain continuous displacements, while no difficulty of this sort arises in problems of pure torsion of the Saint Venant type (cf. SOKOLNIKOFF [1956, p. 241]). A mathematical justification of this conclusion is embodied in the following result owed to ÁLVAREZ-DIOS and VIAÑO [1991b].

THEOREM 24.1. *For reinforced isotropic beams of the type studied in the previous section, if $\nu_p \neq \nu_q$ for some $p, q \in \{0, \dots, P\}$ such that $\partial\Omega_p \cap \partial\Omega_q \neq \emptyset$, and $\tilde{F}_i \neq 0$, then there exists no solution to problem* (24.2) *compatible with definition* (23.7) *of $\underline{u}_3^0$ and u_α^0.*

PROOF. In this case we are dealing with a problem posed in a domain Ω made up of subdomains Ω_p and (24.2) furnishes in this case

$$\partial_\alpha u_{\beta p}^2 + \partial_\beta u_{\alpha p}^2 = -2\nu_p(\underline{u}_3^0 - x_\mu \partial_3 u_\mu^0)\delta_{\alpha\beta}, \quad \text{in } \Omega_p, \ p = 0, \dots, P\,, \tag{24.3}$$

where index p denotes restriction to Ω_p. Since we are looking for $u_\beta^2 \in W_2(\Omega) \subset [H^1(\Omega)]^2$, this must be completed with the compatibility conditions,

$$u_{\alpha r}^2 = u_{\alpha s}^2, \quad \text{on } \partial\Omega_r \cap \partial\Omega_s, \ 0 \leqslant r, s \leqslant P, \ \partial\Omega_r \cap \partial\Omega_s \neq \emptyset. \tag{24.4}$$

We shall now prove the incompatibility of (24.4) with (24.3), which can be looked upon as a boundary-layer phenomenon due to the nonregularity of the elastic coefficients. A direct integration of (24.3) as in Theorem 4.5 gives us

$$\begin{aligned} u_{1p}^2 &= z_{1p} + x_2 z_p - \nu_p[x_1\partial_3\underline{u}_3^0 - \tfrac{1}{2}(x_1^2 - x_2^2)\partial_{33}u_1^0 - x_1x_2\partial_{33}u_2^0], \\ u_{2p}^2 &= z_{2p} - x_1 z_p - \nu_p[x_2\partial_3\underline{u}_3^0 - x_1x_2\partial_{33}u_1^0 - \tfrac{1}{2}(x_2^2 - x_1^2)\partial_{33}u_2^0)], \end{aligned} \tag{24.5}$$

where $z_{\alpha p}, z_p \in H^1(0, L)$. Then, verification of (24.4) would imply that

$$\begin{aligned} z_{1r} - z_{1s} + x_2(z_r - z_s) &= (\nu_r - \nu_s)[x_1\partial_3\underline{u}_3^0 - \tfrac{1}{2}(x_1^2 - x_2^2)\partial_{33}u_1^0 - x_1x_2\partial_{33}u_2^0], \\ z_{2r} - z_{2s} - x_1(z_r - z_s) &= (\nu_r - \nu_s)[x_2\partial_3\underline{u}_3^0 - x_1x_2\partial_{33}u_1^0 - \tfrac{1}{2}(x_2^2 - x_1^2)\partial_{33}u_2^0], \end{aligned} \tag{24.6}$$

If we suppose $\nu_r \neq \nu_s$ since the first members depend only on x_2, x_3, and x_1, x_3, respectively, we must have $\partial_3\underline{u}_3^0 = \partial_{33}u_\alpha^0 = 0$ which is contradictory with $\tilde{F}_i \neq 0$, (cf. (23.7)). □

The above result makes it clear that we shall only have the chance of solving (24.2) in $V(\Omega)$ in either one of the two following situations:

(a) Coefficient ν is constant on each cross section, that is, ν depends only on variable x_3.

(b) The applied forces are such that $\tilde{F}_i = 0$, that is, the beam is subjected only to torsion moments.

The first case will naturally lead us to a generalization of the theories of Saint Venant (with Poisson's effects), and of Timoshenko for isotropic rods with Young's modulus depending on variables x_1, x_2 and x_3.

The second case will enable us to obtain a generalization of Saint Venant's torsion theory for isotropic nonhomogeneous rods (see TRABUCHO and VIAÑO [1988] and Section 14 for the isotropic homogeneous case).

After having solved Eq. (24.2) the next steps are identical to those followed in Sections 7 and 8, for the homogeneous isotropic case. The difficulties are only of a computational nature. In what follows we summarize the results obtained by ÁLVAREZ-DIOS and VIAÑO [1991b].

As for the homogeneous case it is convenient to introduce some *auxiliary functions and constants depending* on the geometry of the cross section and *now, also on the elasticity coefficients*. Without loss of generality and in order to simplify the notation we give the definitions for $\varepsilon = 1$. Moreover, for simplicity, we also assume that $E = E(x_1, x_2)$ and $\nu = \nu(x_1, x_2)$ do not depend on variable x_3. The dependence on x_3 brings only some technical difficulties.

With previous notations in mind we introduce some new definitions. We denote by ζ_α the functions

$$\zeta_1(x_1, x_2) = -x_2, \tag{24.7}$$

$$\zeta_2(x_1, x_2) = x_1. \tag{24.8}$$

The *warping function* w is now the unique solution of

$$\begin{aligned}
&w \in H^1(\omega),\\
&\int_\omega \frac{E}{2(1+\nu)}(\partial_\beta w + \zeta_\beta)\partial_\beta \varphi \,\mathrm{d}\omega = 0, \quad \text{for all } \varphi \in H^1(\omega),\\
&\int_\omega w \,\mathrm{d}\omega = 0.
\end{aligned} \tag{24.9}$$

Let

$$p_\beta = \partial_\beta w + \zeta_\beta, \qquad r_\beta = \frac{E}{2(1+\nu)} p_\beta, \tag{24.10}$$

then, from (24.9) we deduce that $\boldsymbol{r} = (r_\beta)$ satisfies

$$\partial_\beta r_\beta, \quad \text{in } \omega, \qquad r_\beta n_\beta = 0, \quad \text{on } \gamma,$$

consequently, (see GIRAULT and RAVIART [1981, Theorem 3.1]) there exists a function $\Psi \in H^1(\omega)$ such that $\boldsymbol{r} = \operatorname{rot} \Psi$, that is

$$p_1 = \frac{2(1+\nu)}{E}\partial_2 \Psi, \qquad p_2 = -\frac{2(1+\nu)}{E}\partial_1 \Psi. \tag{24.11}$$

Moreover, from (24.9) we conclude that Ψ is the unique solution of the following problem (cf. (7.32)):

$$\begin{aligned}
&-\partial_{\alpha\alpha}\Psi = \partial_1 r_2 - \partial_2 r_1 = \operatorname{rot} \boldsymbol{r},\\
&\Psi = 0, \quad \text{on } \gamma_0,\\
&\partial_n \Psi = -r_2 n_1 + r_1 n_2, \quad \text{on } \gamma_k,\ k = 1, 2, \ldots, p.
\end{aligned} \tag{24.12}$$

Function Ψ denotes *Saint Venant's torsion function* for cross section ω, and for an isotropic nonhomogeneous material.

We now introduce function θ_α as the unique solution of the following variational problem:

$$\begin{aligned}
&\theta_\alpha \in H^1(\omega),\\
&\int_\omega \frac{E}{2(1+\nu)}(\partial_\beta \theta_\alpha + \Phi_{\beta\alpha})\partial_\beta \varphi \, \mathrm{d}\omega = 0, \quad \text{for all } \varphi \in H^1(\omega),\\
&\int_\omega \theta_\alpha \, d\omega = 0.
\end{aligned} \tag{24.13}$$

Associated with these functions we define the following constants (see Section 7 in order to compare with the homogeneous isotropic case):

$$\begin{aligned}
&\tilde{H}_\alpha = \frac{1}{2}\int_\omega E(x_1^2 + x_2^2)x_\alpha \, \mathrm{d}\omega,\\
&\tilde{H}_3 = \frac{1}{4}\int_\omega E(x_1^2 + x_2^2)^2 \, \mathrm{d}\omega,\\
&\tilde{B}_\alpha = \int_\omega E\{\frac{1}{2}(x_1^2 + x_2^2) - \frac{1}{2A}(I_1 + I_2)\}x_\alpha \, \mathrm{d}\omega = \tilde{H}_\alpha - \frac{1}{2A}(I_1 + I_2)e_\alpha,\\
&\tilde{B}_3 = \int_\omega E\{\frac{1}{2}(x_1^2 + x_2^2) - \frac{1}{2A}(I_1 + I_2)\} \, \mathrm{d}\omega = \frac{1}{2}(h_{11} + h_{22}) - \frac{1}{2A}(I_1 + I_2)l,\\
&I_\alpha^w = 2\int_\omega x_\alpha w \, \mathrm{d}\omega, \qquad \tilde{I}_\alpha^w = 2\int_\omega E x_\alpha w \, \mathrm{d}\omega, \qquad \tilde{C}^w = \int_\omega E w \, \mathrm{d}\omega,\\
&J = -\int_\omega x_\alpha \partial_\alpha \Psi \, \mathrm{d}\omega, \qquad I_1^\Psi = -\int_\omega x_2^2 \partial_2 \Psi \, \mathrm{d}\omega, \qquad I_2^\Psi = \int_\omega x_1^2 \partial_1 \Psi \, \mathrm{d}\omega,\\
&\tilde{C}_\alpha^\theta = \int_\omega E\theta_\alpha \, \mathrm{d}\omega,\\
&L_{\alpha\beta}^\theta = \int_\omega x_\alpha \theta_\beta \, \mathrm{d}\omega, \qquad \tilde{L}_{\alpha\beta}^\theta = \int_\omega E x_\alpha \theta_\beta \, \mathrm{d}\omega, \qquad \tilde{K}_{\alpha\beta}^\theta = \int_\omega E \Phi_{\alpha\mu} \partial_\mu \theta_\beta \, \mathrm{d}\omega.
\end{aligned} \tag{24.14}$$

REMARK 24.1. If one particularizes these definitions to the homogeneous case (E and ν constants) *these functions do not coincide exactly with those of Section 7 for the homogeneous isotropic case.* In fact functions w and θ_α are exactly the same as before but function Ψ from definition (24.12) would be equal to the corresponding function of the homogeneous case (see (7.32)) multiplied by coefficient $E/(2(1+\nu))$. Therefore, constants J and I_α^Ψ of (24.14) would also be equal to those of Section 7 multiplied also by $E/(2(1+\nu))$. Moreover, one has (E is constant)

$$\begin{aligned} &\tilde{H}_\alpha = EH_\alpha, \quad \tilde{H}_3 = EH_3, \quad \tilde{I}_\alpha^w = EI_\alpha^w, \quad \tilde{C}^w = \tilde{C}^\theta = 0, \\ &\tilde{L}_{\alpha\beta} = EL_{\alpha\beta}, \quad \tilde{K}_{\alpha\beta} = EK_{\alpha\beta}, \quad \tilde{B}_\alpha = EH_\alpha, \quad \tilde{B}_3 = 0. \end{aligned} \tag{24.15}$$

REMARK 24.2. For the general isotropic case one has

$$l = \int_\omega E \,\mathrm{d}\omega, \qquad e_\alpha = \int_\omega Ex_\alpha \,\mathrm{d}\omega, \qquad h_{\alpha\beta} = \int_\omega Ex_\alpha x_\beta \,\mathrm{d}\omega. \tag{24.16}$$

With these comments in mind it is immediately possible to enunciate the following theorem characterizing the second order displacement components and the zeroth order stress components of the asymptotic expansion for the isotropic nonhomogeneous case with a constant Poisson's ratio. This generalizes the corresponding result for the homogeneous case (Theorem 7.1) and its proof will be omited because it can be done as before (see ÁLVAREZ-DIOS [1992] and ÁLVAREZ-DIOS and VIAÑO [1991b]).

THEOREM 24.2. *Assume that the elasticity coefficients satisfy the isotropy hypotheses* (23.61) *and* (24.1), *with Young's modulus E depending only on variables x_1 and x_2 and with a constant Poisson's ratio in Ω. If the system of applied forces verifies hypothesis* (23.6), *then, Eqs.* (22.35)–(22.47) *determine in a unique way functions $(\boldsymbol{u}^0, \boldsymbol{\sigma}^0) \in V(\Omega) \times \Sigma(\Omega)$, $\boldsymbol{u}^2 \in V(\Omega)$ and $\sigma_{33}^2 \in L^2(\Omega)$ through the following characterizations*:

(i) *Displacement components $\boldsymbol{u}^0$ satisfy*

$$\boldsymbol{u}^0 \in V_{BN}^0(\Omega) \colon u_3^0 = \underline{u}_3^0 - x_\alpha \partial_3 u_\alpha^0, \quad \underline{u}_3^0 \in V_0^1(0,L), \quad u_\alpha^0 \in V_0^2(0,L), \tag{24.17}$$

where functions $\underline{u}_3^0$ and u_α^0 are the unique solution of the following coupled problem (*see* (23.7), (24.16)):

$$\begin{aligned} &\int_0^L \left[\int_\omega E \,\mathrm{d}\omega\right] \partial_3 \underline{u}_3^0 \partial_3 \underline{v}_3 \,\mathrm{d}x_3 - \int_0^L \left[\int_\omega Ex_\alpha \,\mathrm{d}\omega\right] \partial_{33} u_\alpha^0 \partial_3 \underline{v}_3 \,\mathrm{d}x_3 \\ &\quad = \tilde{F}_3(\underline{v}_3), \quad \text{for all } \underline{v}_3 \in V_0^1(0,L), \\ &-\int_0^L \left[\int_\omega Ex_\beta \,\mathrm{d}\omega\right] \partial_3 \underline{u}_3^0 \partial_{33} v_\beta \,\mathrm{d}x_3 + \int_0^L \left[\int_\omega Ex_\alpha x_\beta \,\mathrm{d}\omega\right] \partial_{33} u_\alpha^0 \partial_{33} v_\beta \,\mathrm{d}x_3 \\ &\quad = \tilde{F}_\beta(v_\beta), \quad \text{for all } v_\beta \in V_0^2(0,L). \end{aligned} \tag{24.18}$$

(ii) *The stress component* σ_{33}^0 *is obtained from* u_3^0 *by*

$$\sigma_{33}^0 = Ee_{33}(\boldsymbol{u}^0) = E(\partial_3 \underline{u}_3^0 - x_\alpha \partial_{33} u_\alpha^0). \tag{24.19}$$

(iii) *Displacement components* u_i^2 *take the form*

$$\begin{aligned} u_1^2 &= z_1^2 + x_2 z^2 - \nu[x_1 \partial_3 \underline{u}_3^0 - \Phi_{1\beta} \partial_{33} u_\beta^0], \\ u_2^2 &= z_2^2 - x_1 z^2 - \nu[x_2 \partial_3 \underline{u}_3^0 - \Phi_{2\beta} \partial_{33} u_\beta^0], \\ u_3^2 &= \underline{u}_3^2 - x_\alpha \partial_3 z_\alpha^2 - w \partial_3 z^2 + \tilde{w}^0 \\ &\quad + \nu[\frac{1}{2}(x_1^2 + x_2^2) - \frac{1}{2A}(I_1 + I_2)] \partial_{33} \underline{u}_3^0 + \nu \theta_\alpha \partial_{333} u_\alpha^0, \end{aligned} \tag{24.20}$$

where functions z_α^2, z^2, $\underline{u}_3^2$ *and* $\tilde{w}^0$ *are determined by*

(a) *The angle of torsion* z^2 *depends on* x_3 *only and is the unique solution to the problem*

$$\begin{aligned} & z^2 \in H^1(0, L), \\ & J \int_0^L \partial_3 z^2 \partial_3 v \, dx_3 = M_3^2(v), \quad \text{for all } v \in V_0^1(0, L), \\ & z^2(0) = \frac{\nu}{I_1 + I_2} [H_2 \partial_{33} u_1^0(0) - H_1 \partial_{33} u_2^0(0)], \end{aligned} \tag{24.21}$$

where

$$\begin{aligned} M_3^2(v) = \int_0^L & \Bigg\{ \int_\omega (x_2 f_1 - x_1 f_2) \, d\omega + \int_\gamma (x_2 g_1 - x_1 g_2) \, d\gamma + \int_\omega w \partial_3 f_3 \, d\omega \\ & + \int_\gamma w \partial_3 g_3 \, d\gamma + \tilde{C}^w \partial_{333} \underline{u}_3^0 - [\nu I_\alpha^\Psi + \tfrac{1}{2} \tilde{I}_\alpha^w] \partial_{3333} u_\alpha^0 \Bigg\} v \, dx_3 \\ & + \Bigg\{ \int_\omega (x_2 h_1 - x_1 h_2) \, d\omega - \tilde{C}^w \partial_{33} \underline{u}_3^0(L) + [\nu I_\alpha^\Psi + \frac{1}{2} \tilde{I}_\alpha^w] \partial_{333} u_\alpha^0(L) \\ & - \int_\omega w f_3(L) \, d\omega - \int_\gamma w g_3(L) \, d\gamma \Bigg\} v(L). \end{aligned} \tag{24.22}$$

(b) *The additional warping* $\tilde{w}^0$ *is the unique solution of the problem*

$\tilde{w}^0 \in H^1[0, L; H^1(\omega)]$ *such that, for all* $x_3 \in [0, L]$:

$$\int_\omega \frac{E}{2(1+\nu)} \partial_\beta \tilde{w}^0 \partial_\beta \varphi \, d\omega$$

$$= \int_\omega f_3 \varphi \, d\omega + \int_\gamma g_3 \varphi \, d\gamma$$

$$+ \left(\int_\omega E\varphi \, d\omega \right) \partial_{33} \underline{u}_3^0 - \left(\int_\omega E x_\alpha \varphi \, d\omega \right) \partial_{333} u_\alpha^0, \quad \textit{for all } \varphi \in H^1(\omega),$$

$$\int_\omega \tilde{w}^0 \, d\omega = 0. \tag{24.23}$$

(c) *Both the additional bending component* z_α^2 *and the additional stretching* $\underline{u}_3^2$ *depend only on* x_3 *and they are the unique solution to the following problem (coupled in general):*

$$z_\alpha^2 \in H^2(0,L), \quad \underline{u}_3^2 \in V_0^1(0,L) \quad \textit{such that:}$$

$$\int_0^L l \partial_3 \underline{u}_3^2 \partial_3 v \, dx_3 - \int_0^L e_\alpha \partial_{33} z_\alpha^2 \partial_3 v \, dx_3$$

$$= \int_0^L G_3^2 \partial_3 v \, dx_3, \quad \textit{for all } v \in V_0^1(0,L),$$

$$- \int_0^L e_\beta \partial_3 \underline{u}_3^2 \partial_{33} v_\beta \, dx_3 + \int_0^L h_{\alpha\beta} \partial_{33} z_\alpha^2 \partial_{33} v_\beta \, dx_3$$

$$= \int_0^L M_\beta^2 \partial_{33} v_\beta \, dx_3, \quad \textit{for all } (v_\beta) \in [V_0^2(0,L)]^2,$$

$$z_\alpha^2(0) = -\frac{\nu}{2A}(I_\alpha - I_\beta)\partial_{33} u_\alpha^0(0) \quad (\beta \neq \alpha, \textit{ no sum on } \alpha),$$

$$\partial_3 z_\alpha^2(0) = \frac{1}{I_\alpha} \left\{ \int_\omega x_\alpha \tilde{w}^0(0) - \tfrac{1}{2} I_\alpha^w \partial_3 z^2(0) + \nu H_\alpha \partial_{33} u_3^0(0) + \nu L_{\alpha\beta}^\theta \partial_{333} u_\beta^0(0) \right\}, \tag{24.24}$$

where

$$M_\alpha^2 = -\left(\tfrac{1}{2}\tilde{I}_\alpha^w + \nu I_\alpha^\Psi\right)\partial_{33} z^2 + \int_\omega E x_\alpha \partial_3 \tilde{w}^0 \, d\omega + \nu \tilde{B}_\alpha \partial_{333} \underline{u}_3^0$$

$$+ \left[\frac{\nu^2}{2(1+\nu)} \tilde{K}_{\alpha\beta}^\theta + \nu \tilde{L}_{\alpha\beta}^\theta + \frac{\nu^2}{2(1+\nu)} \tilde{H}_3 \delta_{\alpha\beta} \right] \partial_{3333} u_\beta^0$$

$$+ \nu \int_\omega \Phi_{\alpha\beta} f_\beta \, d\omega + \nu \int_\gamma \Phi_{\alpha\beta} g_\beta \, d\gamma + \nu \int_\omega \frac{E}{2(1+\nu)} \Phi_{\alpha\beta} \partial_{3\beta} \tilde{w}^0 \, d\omega, \tag{24.25}$$

$$G_3^2 = \tilde{C}^w \partial_{33} z^2 - \int_\omega E \partial_3 \tilde{w}^0 \, d\omega - \nu[\tilde{B}_3 + \tfrac{1}{2}(h_{11} + h_{22})] \partial_{333} \underline{u}_3^0$$
$$- \nu(\tilde{C}_\alpha^\theta - \tilde{H}_\alpha) \partial_{3333} u_\alpha^0 - \nu \int_\omega x_\alpha f_\alpha \, d\omega - \nu \int_\gamma x_\alpha g_\alpha \, d\gamma$$
$$- \nu \int_\omega \tfrac{1}{2}(x_1^2 + x_2^2) \partial_3 f_3 \, d\omega - \nu \int_\gamma \tfrac{1}{2}(x_1^2 + x_2^2) \partial_3 g_3 \, d\gamma. \qquad (24.26)$$

(d) *The tangential stress components* $\sigma_{3\beta}^0$ *are given by*

$$\sigma_{31}^0 = -\partial_2 \Psi \partial_3 z^2 + \frac{\nu E}{2(1+\nu)} [\Phi_{1\alpha} + \partial_1 \theta_\alpha] \partial_{333} u_\alpha^0 + \frac{E}{2(1+\nu)} \partial_1 \tilde{w}^0,$$
$$\sigma_{32}^0 = \partial_1 \Psi \partial_3 z^2 + \frac{\nu E}{2(1+\nu)} [\Phi_{2\alpha} + \partial_2 \theta_\alpha] \partial_{333} u_\alpha^0 + \frac{E}{2(1+\nu)} \partial_2 \tilde{w}^0. \qquad (24.27)$$

(iv) *The plane stress components* $\sigma_{\alpha\beta}^0$ *are obtained as follows:*

$$\sigma_{\alpha\beta}^0 = S_{\alpha\beta}(\underline{u}^4) + S_{\alpha\beta}^0, \qquad (24.28)$$

where

$$S_{\alpha\beta}(\underline{u}^4) = \frac{E}{1+\nu} e_{\alpha\beta}(\underline{u}^4) + \frac{\nu E}{(1-2\nu)(1+\nu)} e_{\mu\mu}(\underline{u}^4) \delta_{\alpha\beta}, \qquad (24.29)$$

$$S_{\alpha\beta}^0 = \frac{\nu E}{(1-2\nu)(1+\nu)} \partial_3 (u_3^2 - \underline{u}_3^2 + x_\mu \partial_3 z_\mu^2) \delta_{\alpha\beta}, \qquad (24.30)$$

and $\underline{u}^4 = (\underline{u}_\alpha^4)$ *is the unique solution of*

$$\underline{u}^4 = (\underline{u}_\alpha^4) \in L^2[0, L; (H^1(\omega))^2] \quad \textit{such that a.e. in } (0, L):$$
$$\int_\omega S_{\alpha\beta}(\underline{u}^4) e_{\alpha\beta}(\boldsymbol{\varphi}) \, d\omega = \int_\omega f_\beta \varphi_\beta \, d\omega + \int_\gamma g_\beta \varphi_\beta \, d\gamma$$
$$+ \int_\omega \partial_3 \sigma_{3\beta}^0 \varphi_\beta \, d\omega - \int_\omega S_{\alpha\beta}^0 e_{\alpha\beta}(\boldsymbol{\varphi}) \, d\omega, \quad \textit{for all } \boldsymbol{\varphi} = (\varphi_\beta) \in [H^1(\omega)]^2,$$
$$\int_\omega \underline{u}_\alpha^4 \, d\omega = \int_\omega (x_2 \underline{u}_1^4 - x_1 \underline{u}_2^4) \, d\omega = 0. \qquad (24.31)$$

(v) *The axial stress component* σ_{33}^2 *and the second order bending moment components are given by:*

$$\sigma_{33}^2 = E \partial_3 u_3^2 + \nu \sigma_{\mu\mu}^0, \qquad (24.32)$$
$$m_\beta^2 = e_\beta \partial_3 \underline{u}_3^2 - h_{\alpha\beta} \partial_{33} z_\alpha^2 + M_\beta^2. \qquad (24.33)$$

REMARK 24.3. *If* in the previous theorem *we assume E constant, we recover the results of Theorem* 7.1. The comparison is immediate except for the fact that when E is variable it is not possible to write explicitly functions η_α which are "hiding" as it were in the definition (24.23) of $\tilde{w}^0$. In fact if E is constant then from (24.23) one has

$$\int_\omega \frac{E}{(1+2\nu)} \partial_\beta \tilde{w}^0 \partial_\beta \varphi \,\mathrm{d}\omega = \int_\omega f_3 \varphi \,\mathrm{d}\omega + \int_\gamma g_3 \varphi \,\mathrm{d}\gamma - \left[\int_\omega f_3 \,\mathrm{d}\omega + \int_\gamma g_3 \,\mathrm{d}\gamma\right] \int_\omega \varphi \,\mathrm{d}\omega - E \partial_{333} u_\alpha^0 \int_\omega x_\alpha \varphi \,\mathrm{d}\omega. \tag{24.34}$$

Consequently, from (7.69) and (7.38) we have

$$\frac{E}{2(1+\nu)} \tilde{w}^0 = w^0 + \tfrac{1}{2} E \eta_\alpha \partial_{333} u_\alpha^0. \tag{24.35}$$

Substituting this equation into (24.20) and (24.24)–(24.27) we recover the expressions in Theorem 7.1.

REMARK 24.4 (*Generalization of Timoshenko's beam theory*). It is obvious that the bending and stretching obtained by adding up the corresponding first- and second-order terms will lead us to a more approximated model. Therefore let us define (cf. 15.21)

$$\hat{u}_\alpha = u_\alpha^0 + z_\alpha^2,$$
$$\underline{\hat{u}}_3 = \underline{u}_3^0 + \underline{u}_3^2.$$

Then, from (24.18) and (24.24) reasoning formally and doing away with boundary conditions we obtain that $(\hat{u}_1, \hat{u}_2, \underline{\hat{u}}_3)$ must verify the following equations (coupled in general):

$$\begin{aligned} &-\partial_3(l \partial_3 \underline{\hat{u}}_3) + \partial_3(e_\alpha \partial_{33} \hat{u}_\alpha) = \overline{F}_3 - \partial_3 G_3^2, \\ &-\partial_{33}(e_\beta \partial_3 \underline{\hat{u}}_3) + \partial_{33}(h_{\alpha\beta} \partial_{33} \hat{u}_\alpha) = \overline{F}_\beta + \partial_{33} M_\beta^2, \end{aligned} \tag{24.36}$$

with

$$\begin{aligned} \overline{F}_\beta &= \int_\omega f_\beta \,\mathrm{d}\omega + \int_\gamma g_\beta \,\mathrm{d}\gamma + \int_\omega x_\beta \partial_3 f_3 \,\mathrm{d}\omega + \int_\gamma x_\beta \partial_3 g_3 \,\mathrm{d}\gamma, \\ \overline{F}_3 &= \int_\omega f_3 \,\mathrm{d}\omega + \int_\gamma g_3 \,\mathrm{d}\gamma. \end{aligned} \tag{24.37}$$

Equations (24.36) constitute a generalization of Timoshenko's classical bending beam theory also including other secondary effects gathered in terms $\partial_3 G_3^2$ and $\partial_{33} M_\beta^2$. In order to convince ourselves of this fact, let us consider Timoshenko's classical bending case, namely, no axial loads and no applied moments:

$$\begin{aligned} &f_3 = 0, \qquad g_3 = 0, \qquad h_3 = 0, \\ &M_3 = \int_\omega (x_2 f_1 - x_1 f_2)\,\mathrm{d}\omega + \int_\gamma (x_2 g_1 - x_1 g_2)\,\mathrm{d}\gamma = 0, \\ &M_L = \int_\omega (x_2 h_1 - x_1 h_2)\,\mathrm{d}\omega = 0. \end{aligned} \tag{24.38}$$

Besides, let us also suppose that (cf. Section 15)

$$\begin{aligned} &\int_\omega \Phi_{\alpha\beta} f_\alpha\,\mathrm{d}\omega + \int_\gamma \Phi_{\alpha\beta} g_\alpha\,\mathrm{d}\gamma = 0, \\ &\int_\omega x_\alpha f_\alpha\,\mathrm{d}\omega + \int_\gamma x_\alpha g_\alpha\,\mathrm{d}\gamma = 0. \end{aligned} \tag{24.39}$$

If the cross section is such that $h_{12} = 0$ and $e_\alpha = 0$, then the bending equations turn into

$$\partial_{33}(h_{\alpha\alpha}\partial_{33}\hat{u}_\alpha) = \overline{F}_\alpha - T_{\alpha\beta}\partial_{33}\overline{F}_\beta \quad \text{(no sum on } \alpha), \tag{24.40}$$

where $T_{\alpha\beta}$ denotes the *Timoshenko's matrix*. One has

$$T_{\alpha\beta} = \tilde{T}_{\alpha\beta} + T^*_{\alpha\beta}, \tag{24.41}$$

where

$$\begin{aligned} \tilde{T}_{\alpha\beta} = -\frac{1}{h_{\beta\beta}} \Big\{ &\frac{\nu^2}{2(1+\nu)}\tilde{K}^\theta_{\alpha\beta} + \nu\tilde{L}^\theta_{\alpha\beta} + \frac{\nu^2}{2(1+\nu)}\tilde{H}_3\delta_{\alpha\beta} \\ &-\frac{1}{J}(\tfrac{1}{2}\tilde{I}^w_\alpha + \nu I^\Psi_\alpha)(\tfrac{1}{2}\tilde{I}^w_\beta + \nu I^\Psi_\beta)\Big\} \quad \text{(no sum on } \beta), \end{aligned} \tag{24.42}$$

and where $T^*_{\alpha\beta}$ stands for the coefficient of $-\partial_{33}\overline{F}_\beta$ arising from the following terms showing up in $\partial_{33}M_\beta^2$ and which cannot be written in an explicit way:

$$\partial_{333}\left[\int_\omega E x_\beta \tilde{w}^0\,\mathrm{d}\omega\right], \qquad \partial_{333}\left[\int_\omega \frac{\nu E}{2(1+\nu)}\Phi_{\beta\alpha}\partial_\alpha\tilde{w}^0\,\mathrm{d}\omega\right]. \tag{24.43}$$

In the homogeneous case and taking into account that (24.38) gives $w^0 = 0$, we obtain from (24.35)

$$\tilde{w}^0 = (1+\nu)\eta_\alpha\partial_{333}u^0_\alpha, \tag{24.44}$$

and consequently

$$T^*_{\alpha\beta} = -\frac{1}{I_\beta}[(1+\nu)L^\eta_{\alpha\beta} + \tfrac{1}{2}\nu K^\eta_{\alpha\beta}] \quad \text{(no sum on } \beta), \tag{24.45}$$

from which we conclude that $T_{\alpha\beta}$ coincides with definition (12.39), for E and ν constant. Thus, Eq. (24.40) constitutes a generalisation of Timoshenko's theory introduced in Section 15, for the isotropic nonhomogeneous case (ν constant).

We conclude this section analyzing the other case for which it is possible to solve Eq. (23.71) explicitly, that is, for the case where $e_{33}(\boldsymbol{u}^0) = \partial_3 \underline{u}^0_3 - x_\alpha \partial_{33} u^0_\alpha = 0$, and $\nu = \nu(x_1, x_2)$. This is equivalent to saying (see (24.18)) that the resultant of the applied loads $\tilde{F}_i$ is zero and therefore the beam is subjected only to torsion loads. Following a similar analysis as for Theorems 7.1 and 24.2 one obtains the following result (cf. ÁLVAREZ-DIOS and VIAÑO [1991b], ÁLVAREZ-DIOS [1992]).

THEOREM 24.3. *Let us consider that the beam is made of an isotropic nonhomogeneous material whose Young's modulus and Poisson's ratio are a function of x_1 and x_2. Moreover, let us assume that besides satisfying* (23.6), *the system of applied loads is such that*

$$\int_\omega f_i \,\mathrm{d}\omega + \int_\gamma g_i \,\mathrm{d}\gamma = 0, \qquad \int_\omega h_i \,\mathrm{d}\omega = 0, \tag{24.46}$$

$$\int_\omega x_\beta f_3 \,\mathrm{d}\omega + \int_\gamma x_\beta g_3 \,\mathrm{d}\gamma = 0, \qquad \int_\omega x_\beta h_3 \,\mathrm{d}\omega = 0. \tag{24.47}$$

Then, Eqs. (22.35)–(22.47) *determine in a unique way functions $(\boldsymbol{u}^0, \boldsymbol{\sigma}^0) \in V(\Omega) \times \Sigma(\Omega)$, $\boldsymbol{u}^2 \in V(\Omega)$ and $\sigma^2_{33} \in L^2(\Omega)$, through the following expressions:*

(i) *Components u^0_i and σ^0_{33} verify*

$$u^0_i = 0, \quad \sigma^0_{33} = 0, \tag{24.48}$$

(ii) *Displacement components u^2_i are of the form:*

$$\begin{aligned} u^2_1 &= z^2_1 + x_2 z^2, \\ u^2_2 &= z^2_2 - x_1 z^2, \\ u^2_3 &= \underline{u}^2_3 - x_\alpha \partial_3 z^2_\alpha - w \partial_3 z^2 + \tilde{w}^0, \end{aligned} \tag{24.49}$$

where functions z^2_α, z^2, $\underline{u}^2_3$ and $\tilde{w}^0$ are determined by:

(a) *The angle of torsion z^2 depends only on x_3 and is the unique solution to the following variational problem:*

$$\begin{aligned} &z^2 \in V^1_0(0, L), \\ &J \int_0^L \partial_3 z^2 \partial_3 v \,dx_3 = M^2_3(v), \quad \textit{for all } v \in V^1_0(0, L), \end{aligned} \tag{24.50}$$

where

$$M_3^2(v) = \int_0^L \left[\int_\omega (x_2 f_1 - x_1 f_2)\,\mathrm{d}\omega + \int_\gamma (x_2 g_1 - x_1 g_2)\,\mathrm{d}\gamma + \int_\omega w\partial_3 f_3\,\mathrm{d}\omega + \int_\gamma w\partial_3 g_3\,\mathrm{d}\gamma \right] v\,\mathrm{d}x_3 + \left[\int_\omega (x_2 h_1 - x_1 h_2)\,\mathrm{d}\omega - \int_\omega w f_3(L)\,\mathrm{d}\omega - \int_\gamma w g_3(L)\,\mathrm{d}\gamma \right] v(L). \tag{24.51}$$

(b) *The additional warping* $\tilde{w}^0$ *is the unique solution of the problem*

$$\begin{aligned} &\tilde{w}^0 \in H^1[0, L; H^1(\omega)] \quad \textit{such that for all } x_3 \in [0, L], \\ &\int_\omega \frac{E}{2(1+\nu)} \partial_\beta \tilde{w}^0 \partial_\beta \varphi\,\mathrm{d}\omega = \int_\omega f_3 \varphi\,\mathrm{d}\omega + \int_\gamma g_3 \varphi\,\mathrm{d}\gamma, \\ &\quad \textit{for all } \varphi \in H^1(\omega), \\ &\int_\omega \tilde{w}^0\,\mathrm{d}\omega = 0. \end{aligned} \tag{24.52}$$

(c) *Both the additional bending component* z_α^2 *and the additional stretching* $\underline{u}_3^2$ *depend on* x_3 *only and they are the unique solution to the following problem (coupled in general)*:

$$\begin{aligned} &z_\alpha^2 \in H^2(0, L) \cap V_0^1(0, L), \quad \underline{u}_3^2 \in V_0^1(0, L) \\ &\int_0^L l\partial_3 \underline{u}_3^2 \partial_3 v\,\mathrm{d}x_3 - \int_0^L e_\alpha \partial_{33} z_\alpha^2 \partial_3 v\,\mathrm{d}x_3 \\ &\quad = \int_0^L G_3^2 \partial_3 v\,\mathrm{d}x_3, \quad \textit{for all } v \in V_0^1(0, L), \\ &- \int_0^L e_\beta \partial_3 \underline{u}_3^2 \partial_{33} v_\beta\,\mathrm{d}x_3 + \int_0^L h_{\alpha\beta} \partial_{33} z_\alpha^2 \partial_{33} v_\beta\,\mathrm{d}x_3 \\ &\quad = \int_0^L M_\beta^2 \partial_{33} v_\beta\,\mathrm{d}x_3, \quad \textit{for all } (v_\beta) \in [V_0^2(0, L)]^2, \\ &\partial_3 z_\alpha^2(0) = \frac{1}{I_\alpha} \left[\int_\omega x_\alpha \tilde{w}^0(0)\,\mathrm{d}\omega - \tfrac{1}{2} I_\alpha^w \partial_3 z^2(0) \right] \quad (\textit{no sum on } \alpha), \end{aligned} \tag{24.53}$$

where

$$M_\alpha^2 = -\frac{1}{2}\tilde{I}_\alpha^w \partial_{33} z^2 + \int_\omega E x_\alpha \partial_3 \tilde{w}^0 \, d\omega + \int_\omega \nu x_\alpha \sigma_{\mu\mu}^0 \, d\omega, \tag{24.54}$$

$$G_3^2 = \tilde{C}^w \partial_{33} z^2 - \int_\omega E \partial_3 \tilde{w}^0 \, d\omega - \int_\omega \nu \sigma_{\mu\mu}^0 \, d\omega. \tag{24.55}$$

(iii) *The shear stress components* $\sigma_{3\beta}^0$ *are given by*

$$\begin{aligned} \sigma_{31}^0 &= -\partial_2 \Psi \partial_3 z^2 + \frac{E}{2(1+\nu)} \partial_1 \tilde{w}^0, \\ \sigma_{32}^0 &= \partial_1 \Psi \partial_3 z^2 + \frac{E}{2(1+\nu)} \partial_2 \tilde{w}^0. \end{aligned} \tag{24.56}$$

(iv) *Moreover, the plane stress components* $\sigma_{\alpha\beta}^0$, *the axial stress component* σ_{33}^2 *and the bending moments* $m_\beta^2 = \int_\omega x_\beta \sigma_{33}^2 \, d\omega$, *are given by the same type of expressions as in* (24.28)–(24.33).

REMARK 24.5. We have to pay attention to the fact that now ν is no longer necessarily constant, which makes it impossible to evaluate terms $\int_\omega \nu x_\alpha \sigma_{\mu\mu}^0 \, d\omega$, $\int_\omega \nu \sigma_{\mu\mu}^0 \, d\omega$ in (24.54), (24.55) in an explicit way. Everything else happens exactly as in Theorem 24.2 with $\underline{u}_3^0 = 0$ and $u_\alpha^0 = 0$.

REMARK 24.6. (*Generalization of Saint Venant's torsion theory*). Theorem 24.3 constitutes a generalisation of Saint Venant's isotropic torsion beam theory, also including transversal nonhomogeneity and other secondary effects such as:

(i) Influence on the angle of torsion z^2 of the applied moments on the free end, of the moments distributed along the lateral surface and of the distribution of the axial loads on the cross section (see (24.51)).

(ii) Secondary stretching and bending effects due to axial loads and the geometric distribution of the shear loads (see (24.52)–(24.55)).

Saint Venant's classical theory corresponds to the case in which only an applied moment on the free end is considered, so that

$$f_i = 0, \qquad g_i = 0, \qquad h_3 = 0, \qquad \int_\omega h_\beta \, d\omega = 0. \tag{24.57}$$

Let us denote by M the applied moment on the free end:

$$M = \int_\omega (x_2 h_1 - x_1 h_2) \, d\omega,$$

then, from (24.50), (24.51) we deduce that in this case the angle of torsion is proportional to the distance between the clamped end and it is given by

$$z^2(x_3) = -\alpha x_3, \qquad \alpha = \frac{M}{J}. \tag{24.58}$$

Besides from (24.56) we obtain that the shear stresses do not vary along the length of the rod, and they are given by

$$\sigma^0_{31} = \alpha \partial_2 \Psi, \qquad \sigma^0_{32} = -\alpha \partial_1 \Psi. \tag{24.59}$$

From (24.53)–(24.55), taking (24.41) into account, we conclude that $\sigma^0_{\alpha\beta} = 0$, $\underline{u}^2_3 = 0$, and, if we suppose that $I^w_\alpha = 0$, $z^2_\alpha = 0$.

To give it in a nutshell, for the case of Saint Venant's classical torsion theory, the asymptotic theory can be summed up in the displacement and stress fields defined by $\hat{u} = u^0 + u^2$, $\hat{\sigma}_{33} = \sigma^0_{33} + \sigma^2_{33}$, $\hat{\sigma}_{3\beta} = \sigma^0_{3\beta}$, $\hat{\sigma}_{\alpha\beta} = \sigma^0_{\alpha\beta}$. We then have

$$\begin{aligned} \hat{u}_1(x_1, x_2, x_3) &= -\alpha x_2 x_3, \\ \hat{u}_2(x_1, x_2, x_3) &= \alpha x_1 x_3, \\ \hat{u}_3(x_1, x_2, x_3) &= \alpha w(x_1, x_2), \end{aligned} \tag{24.60}$$

$$\hat{\sigma}_{31} = \alpha \partial_2 \Psi, \qquad \hat{\sigma}_{32} = -\alpha \partial_1 \Psi, \qquad \hat{\sigma}_{33} = \hat{\sigma}_{\alpha\beta} = 0. \tag{24.61}$$

Equations (24.60), (24.61) furnish a generalization of Saint Venant's classical torsion theory for the case of transversally nonhomogeneous isotropic beams (see NEČAS and HLAVÁČEK [1981, p. 225]).

We remark that the angle of torsion and the stresses depend only on the applied moment, the form of the cross section and on Lamé's coefficient $\mu = G = E/[2(1+\nu)]$. In literature we found a very similar theory for the case of multilayered cylindrical rods with cylindrical anisotropy in LEKHNITSKII [1977, p. 308]. Let us consider, in the first place, a rod of the type of the one depicted in Fig. 23.1 (d) where ω is a circle of radius $r_P = R$, ω_0 is a circle of radius $r = r_0 < R$ and ω^*_p, $1 \leqslant p \leqslant P$ is a circular crown of exterior radius r_p and interior radius r_{p-1}. We also assume that $\Omega_p = \omega^*_p \times (0, L)$ is made of an isotropic, homogeneous material with Young's modulus E_p and Poisson's ratio ν_p assumed to be constant.

From definition (24.12) we conclude that in this case Ψ is the solution of the following problem

$$\begin{aligned} & -\partial_{\alpha\alpha}\Psi = \frac{E}{(1+\nu)}, \quad \text{in } \omega, \\ & \Psi = 0, \quad \text{on } \gamma, \end{aligned} \tag{24.62}$$

where as in Section 23 we denote

$$E(x_1, x_2) = E_p, \qquad \nu(x_1, x_2) = \nu_p, \qquad (x_1, x_2) \in \omega.$$

Thus,

$$\int_\omega \partial_\alpha \Psi \partial_\alpha \varphi \, \mathrm{d}\omega = \int_\omega \frac{E}{(1+\nu)} \varphi \, \mathrm{d}\omega, \quad \text{for all } \varphi \in H_0^1(\omega).$$

Choosing,

$$\varphi(x_1, x_2) = \tfrac{1}{2}(R^2 - x_1^2 - x_2^2), \tag{24.63}$$

we obtain

$$J = -\int_\omega x_\alpha \partial_\alpha \Psi = \int_\omega \frac{E}{2(1+\nu)} (R^2 - x_1^2 - x_2^2) \, \mathrm{d}\omega,$$

that is (with $r_{-1} = 0$)

$$J = \pi \frac{E_0}{2(1+\nu_0)} r^2 (R^2 - \tfrac{1}{2} r^2) + \pi \sum_{p=1}^{P} \frac{E_p}{2(1+\nu_p)} (r_p^2 - r_{p-1}^2)[R^2 - \tfrac{1}{2}(r_p^2 + r_{p-1}^2)]. \tag{24.64}$$

If, in the previous situation, we consider the hollow case, that is, ω_0 is a hole and we have a tube of exterior and interior radii R and r, respectively, made of circular, concentric isotropic layers of thickness $r_p - r_{p-1}$, $p = 1, \ldots, P$, and if we denote by γ_0 and γ_1 the exterior and the interior boundaries, respectively, then, since ω_p^* is a circular crown we see that $w = 0$ satisfies (24.9) and consequently, from definition (24.12) we conclude that Ψ is the solution of the system

$$\begin{aligned} &-\partial_{\alpha\alpha} \Psi = \frac{E}{(1+\nu)}, \quad \text{in } \omega, \\ &\Psi = 0, \quad \text{on } \gamma_0, \\ &\partial_n \Psi = -\frac{E}{2(1+\nu)} x_\alpha n_\alpha = -\frac{E}{2(1+\nu)} r, \quad \text{on } \gamma_1. \end{aligned} \tag{24.65}$$

Therefore,

$$\int_\omega \partial_\alpha \Psi \partial_\alpha \varphi \, \mathrm{d}\omega = \int_\omega \frac{E}{(1+\nu)} \varphi \, \mathrm{d}\omega + \int_{\gamma_1} \frac{E}{2(1+\nu)} r \varphi \, \mathrm{d}\gamma,$$

for all $\varphi \in H^1(\omega)$, $\varphi = 0$ on γ_0.

Taking φ as in (24.63) one obtains

$$\begin{aligned} J &= -\int_\omega x_\alpha \partial_\alpha \Psi \\ &= \int_\omega \frac{E}{2(1+\nu)} (R^2 - x_1^2 - x_2^2) \, \mathrm{d}\omega + \frac{r}{2} \int_{\gamma_1} \frac{E}{2(1+\nu)} (R^2 - x_1^2 - x_2^2) \, \mathrm{d}\gamma, \end{aligned} \tag{24.66}$$

that is,

$$\begin{aligned} J &= \pi \frac{E_1}{2(1+\nu_1)} r^2 (R^2 - r^2) \\ &\quad + \pi \sum_{p=1}^{P} \frac{E_p}{2(1+\nu_p)} (r_p^2 - r_{p-1}^2)[R^2 - \tfrac{1}{2}(r_p^2 + r_{p-1}^2)], \end{aligned} \tag{24.67}$$

which may be compared with the corresponding expression in LEKHNITSKII [1977, Section 61], written in cylindrical coordinates.

25. A general asymptotic model for homogeneous anisotropic rods

In this section we summarize the results of ÁLVAREZ-DIOS and VIAÑO [1992a,b] and ÁLVAREZ-DIOS [1992] for the application of the asymptotic expansion method in order to obtain a general bending–torsion model for homogeneous, anisotropic rods.

The methodology to be followed is the same as in all the previous cases: assuming that all the elasticity coefficients a_{ijkl} are independent of x_1 and x_2, solve (23.68)–(23.70) and evaluate $\boldsymbol{u}^2$, $\sigma^0_{3\beta}$ and $\sigma^0_{\alpha\beta}$ from (22.35)–(22.47), using the usual technique. The determination of u_3^2 is now more complicated and it is necessary to generalize the definitions of the functions $(w, \Psi, \eta_\alpha, \theta_\alpha)$ because they now depend on the elasticity coefficients in a nontrivial way.

For simplicity, and without loss of generality, we also assume that the coefficients do not depend on variable x_3. It is also possible to consider that the only coefficients not depending on variables x_1 and x_2 are $a_{\alpha\beta 33}/a_{3333}$. We analyze here *the simplest case where coefficients a^ε_{ijkl} are constant in Ω^ε*. Consequently, property (22.10) may now be written as

$$a^\varepsilon_{ijkl} = a_{ijkl}, \qquad A^\varepsilon_{ijkl} = A_{ijkl}. \tag{25.1}$$

We remind that we still assume the hypothesis that Ox_3 is a principal elasticity direction, that is (see (22.11)),

$$a_{\alpha\beta\gamma 3} = a_{\alpha 333} = 0, \tag{25.2}$$

which implies the following equality valid for all $\boldsymbol{\tau} \in \mathbb{R}^9_s$:

$$a_{ijkl}\tau_{kl}\tau_{ij} = \begin{Bmatrix} \tau_{11} \\ \tau_{22} \\ 2\tau_{12} \\ 2\tau_{13} \\ 2\tau_{23} \\ \tau_{33} \end{Bmatrix}^{\mathrm{T}} \mathscr{A} \begin{Bmatrix} \tau_{11} \\ \tau_{22} \\ 2\tau_{12} \\ 2\tau_{13} \\ 2\tau_{23} \\ \tau_{33} \end{Bmatrix},$$

where $\mathscr{A}$ is the following 6×6 matrix:

$$\mathscr{A} = \begin{bmatrix} a_{1111} & a_{1122} & a_{1112} & 0 & 0 & a_{1133} \\ a_{2211} & a_{2222} & a_{2212} & 0 & 0 & a_{2233} \\ a_{1211} & a_{1222} & a_{1212} & 0 & 0 & a_{1233} \\ 0 & 0 & 0 & a_{1313} & a_{1323} & 0 \\ 0 & 0 & 0 & a_{2313} & a_{2323} & 0 \\ a_{3311} & a_{3322} & a_{3312} & 0 & 0 & a_{3333} \end{bmatrix}. \tag{25.3}$$

The positivity hypothesis (22.4) guarantees that matrix $\mathscr{A}$ is positive definite and thus

$$a_{3333} > 0, \qquad \begin{vmatrix} a_{1313} & a_{1323} \\ a_{2313} & a_{2323} \end{vmatrix} > 0, \qquad \begin{vmatrix} a_{1111} & a_{1122} & a_{1112} \\ a_{2211} & a_{2222} & a_{2212} \\ a_{1211} & a_{1222} & a_{1212} \end{vmatrix} > 0. \tag{25.4}$$

From (22.5) and hypothesis (25.2) one has

$$e^\varepsilon_{3\alpha} = e^\varepsilon_{3\alpha}(\boldsymbol{u}^\varepsilon) = 2a_{3\alpha 3\beta}\sigma^\varepsilon_{3\beta},$$

that is

$$\begin{Bmatrix} e^\varepsilon_{31} \\ e^\varepsilon_{32} \end{Bmatrix} = 2 \begin{bmatrix} a_{1313} & a_{1323} \\ a_{2313} & a_{2323} \end{bmatrix} \begin{Bmatrix} \sigma^\varepsilon_{31} \\ \sigma^\varepsilon_{32} \end{Bmatrix}.$$

Equivalently,

$$\begin{Bmatrix} \sigma^\varepsilon_{31} \\ \sigma^\varepsilon_{32} \end{Bmatrix} = \frac{1}{2} \begin{bmatrix} a_{1313} & a_{1323} \\ a_{2313} & a_{2323} \end{bmatrix}^{-1} \begin{Bmatrix} e^\varepsilon_{31} \\ e^\varepsilon_{32} \end{Bmatrix}.$$

On the other hand from (22.6) we have

$$\sigma^\varepsilon_{3\beta} = A_{3\beta ij} e^\varepsilon_{ij}.$$

Comparing these two equations one concludes that whenever (25.2) holds one has

$$A_{\alpha\beta\gamma 3} = A_{\alpha 333} = 0, \tag{25.5}$$

$$\begin{bmatrix} A_{1313} & A_{1323} \\ A_{2313} & A_{2323} \end{bmatrix} = \frac{1}{4} \begin{bmatrix} a_{1313} & a_{1323} \\ a_{2313} & a_{2323} \end{bmatrix}^{-1}. \tag{25.6}$$

We define

$$D = \begin{vmatrix} A_{1313} & A_{1323} \\ A_{2313} & A_{2323} \end{vmatrix} = A_{1313}A_{2323} - A_{1323}A_{2313}. \tag{25.7}$$

It will be useful to use the *reduced elastic constants* (cf. LEKHNITSKII [1977, Section 6]), defined by the following nonlinear relations:

$$b_{ijkl} = a_{ijkl} - \frac{a_{ij33}a_{kl33}}{a_{3333}}. \tag{25.8}$$

We observe that $b_{ij33} = 0$ and if hypothesis (25.2) holds then

$$b_{3\alpha3\beta} = a_{3\alpha3\beta}. \tag{25.9}$$

Adding to the first column of $\mathscr{A}$ the sixth column multiplied by $-a_{1133}/a_{3333}$, to the second the sixth multiplied by $-a_{2233}/a_{3333}$, and to the third the sixth multiplied by $-a_{1233}/a_{3333}$, we obtain the following positive definite matrix:

$$M = \begin{bmatrix} b_{1111} & b_{1122} & b_{1112} & 0 & 0 & a_{1133} \\ b_{2211} & b_{2222} & b_{2212} & 0 & 0 & a_{2233} \\ b_{1211} & b_{1222} & b_{1212} & 0 & 0 & a_{1233} \\ 0 & 0 & 0 & a_{1313} & a_{1323} & 0 \\ 0 & 0 & 0 & a_{2313} & a_{2323} & 0 \\ 0 & 0 & 0 & 0 & 0 & a_{3333} \end{bmatrix}.$$

Since $\det \mathscr{A} = \det M$ one has

$$\begin{vmatrix} b_{1111} & b_{1122} & b_{1112} \\ b_{2211} & b_{2222} & b_{2212} \\ b_{1211} & b_{1222} & b_{1212} \end{vmatrix} = \begin{vmatrix} a_{1111} & a_{1122} & a_{1112} \\ a_{2211} & a_{2222} & a_{2212} \\ a_{1211} & a_{1222} & a_{1212} \end{vmatrix} > 0. \tag{25.10}$$

We also define

$$\begin{bmatrix} B_{1111} & B_{1122} & 2B_{1112} \\ B_{2211} & B_{2222} & 2B_{2212} \\ 2B_{1211} & 2B_{1222} & 4B_{1212} \end{bmatrix} = \begin{bmatrix} b_{1111} & b_{1122} & b_{1112} \\ b_{2211} & b_{2222} & b_{2212} \\ b_{1211} & b_{1222} & b_{1212} \end{bmatrix}^{-1} \tag{25.11}$$

Whenever the coefficients are constant and with the notations introduced in (23.1) we have (no sum on α):

$$l = \frac{A}{a_{3333}}, \qquad e_\alpha = 0, \qquad h_{12} = 0, \qquad h_{\alpha\alpha} = \frac{I_\alpha}{a_{3333}}. \tag{25.12}$$

We now introduce functions $\Phi_{\alpha\beta}$, w, Ψ, η_α and θ_α which generalize to the anisotropic case the corresponding definitions for the isotropic case given in Section 24. We keep the same notation as for the homogeneous isotropic case although the new functions for the present case are the same of Section 7 affected by factors depending on E and ν (cf. Remark 24.1).

(a) Functions $\Phi_{\alpha\beta}$ are now defined as

$$\begin{aligned}
&\Phi_{11}(x_1, x_2) = \tfrac{1}{2}(-a_{1133}x_1^2 + a_{2233}x_2^2),\\
&\Phi_{12}(x_1, x_2) = -a_{1233}x_1^2 - a_{2233}x_1x_2,\\
&\Phi_{21}(x_1, x_2) = -a_{1133}x_1x_2 - a_{1233}x_2^2,\\
&\Phi_{22}(x_1, x_2) = \tfrac{1}{2}(a_{1133}x_1^2 - a_{2233}x_2^2) = -\Phi_{11}(x_1, x_2).
\end{aligned} \tag{25.13}$$

(b) The warping function w, and functions η_β, θ_β are the unique solutions to the following boundary value problems, respectively:

$$\begin{aligned}
&-\partial_\beta[A_{3\beta3\alpha}(\partial_\alpha w + \zeta_\alpha)] = 0, \quad \text{in } \omega,\\
&A_{3\beta3\alpha}(\partial_\alpha w + \zeta_\alpha)n_\beta = 0, \quad \text{on } \gamma,\\
&\int_\omega w \,\mathrm{d}\omega = 0,
\end{aligned} \tag{25.14}$$

$$\begin{aligned}
&-\partial_\beta[A_{3\beta3\alpha}\partial_\alpha \eta_\mu] = -x_\mu, \quad \text{in } \omega,\\
&A_{3\beta3\alpha}\partial_\alpha \eta_\mu n_\beta = 0, \quad \text{on } \gamma,\\
&\int_\omega \eta_\mu \,\mathrm{d}\omega = 0,
\end{aligned} \tag{25.15}$$

$$\begin{aligned}
&-\partial_\beta[A_{3\beta3\alpha}(\partial_\alpha \theta_\mu + \Phi_{\mu\alpha})] = 0, \quad \text{in } \omega,\\
&A_{3\beta3\alpha}(\partial_\alpha \theta_\mu + \Phi_{\mu\alpha})n_\beta = 0, \quad \text{on } \gamma,\\
&\int_\omega \theta_\mu \,\mathrm{d}\omega = 0.
\end{aligned} \tag{25.16}$$

(c) Saint Venant's torsion function is the unique function $\Psi \in H^1(\omega)$, $\Psi = 0$ on γ_0, such that

$$r_1 = \partial_2 \Psi, \qquad r_2 = -\partial_1 \Psi, \tag{25.17}$$

where

$$r_\beta = A_{3\beta3\alpha}p_\alpha, \tag{25.18}$$

$$p_\alpha = \partial_\alpha w + \zeta_\alpha. \tag{25.19}$$

This function Ψ is the unique solution of the following boundary value problem:

$$\begin{aligned}
&-\partial_{\alpha\alpha}\Psi = \partial_1 r_2 - \partial_2 r_1, \quad \text{in } \omega,\\
&\Psi = 0, \quad \text{on } \gamma_0,\\
&\partial_n \Psi = -r_2 n_1 + r_1 n_2, \quad \text{on } \gamma_k, \ k = 1, 2, \ldots, p.
\end{aligned} \tag{25.20}$$

It is possible to show that Ψ is constant on each of the border components γ_k, $k = 1, 2, \ldots, p$. We denote its value on γ_k by c_k.

We remark that all these definitions are also valid for the nonhomogeneous case. When coefficients $A_{3\beta 3\alpha}$ are constant the first equation of (25.14) is equivalent to

$$-\partial_\beta(A_{3\beta 3\alpha}\partial_\alpha w) = 0. \tag{25.21}$$

Moreover, for this case (25.20) is equivalent to

$$\begin{aligned}
&-\partial_\beta(A_{3\beta 3\alpha}\partial_\alpha \Psi) = 2D, \quad \text{in } \omega,\\
&\Psi = 0, \quad \text{on } \gamma_0,\\
&A_{3\beta 3\alpha}\partial_\alpha \Psi n_\beta = D(\partial_\tau w - x_\rho n_\rho), \quad \text{on } \gamma_k, \quad k = 1, 2, \ldots, p,
\end{aligned} \tag{25.22}$$

where $\partial_\tau w$ denotes the tangential derivative of w, ($\tau = (-n_2, n_1)$).

Associated with the above functions we define the following constants:

$$\begin{aligned}
H_\beta &= \frac{1}{2}\int_\omega a_{\alpha\gamma 33} x_\alpha x_\gamma x_\beta \, \mathrm{d}\omega,\\
X_{\alpha\beta} &= \int_\omega \Phi_{\alpha\beta} \, \mathrm{d}\omega,\\
Y_\beta &= \int_\omega (\Phi_{\beta 1} x_2 - \Phi_{\beta 2} x_1) \, \mathrm{d}\omega,\\
Z_{\alpha\beta\gamma\mu} &= \int_\omega \Phi_{\alpha\beta}\Phi_{\gamma\mu} \, \mathrm{d}\omega,
\end{aligned} \tag{25.23}$$

$$\begin{aligned}
I_\beta^w &= 2\int_\omega x_\beta w \, \mathrm{d}\omega,\\
J &= -\int_\omega x_\alpha \partial_\alpha \Psi \, \mathrm{d}\omega = 2\int_\omega \Psi \, \mathrm{d}\omega + 2c_k A(\omega_k),\\
I_1^\Psi &= -\int_\omega x_2^2 \partial_2 \Psi \, \mathrm{d}\omega = 2\int_\omega x_2 \Psi \, \mathrm{d}\omega + 2c_k \int_{\omega_k} x_2 \, \mathrm{d}\omega,\\
I_2^\Psi &= \int_\omega x_1^2 \partial_1 \Psi \, \mathrm{d}\omega = -2\int_\omega x_1 \Psi \, \mathrm{d}\omega - 2c_k \int_{\omega_k} x_1 \, \mathrm{d}\omega,\\
D_1 &= -\frac{1}{2} I_1^w - \int_\omega (\Phi_{11}\partial_2 \Psi - \Phi_{12}\partial_1 \Psi) \, \mathrm{d}\omega,\\
D_2 &= -\frac{1}{2} I_2^w - \int_\omega (\Phi_{21}\partial_2 \Psi - \Phi_{22}\partial_1 \Psi) \, \mathrm{d}\omega,
\end{aligned} \tag{25.24}$$

$$
\begin{aligned}
L^{\eta}_{\alpha\beta} &= \int_{\omega} x_{\alpha}\eta_{\beta}\,\mathrm{d}\omega,\\
L^{\theta}_{\alpha\beta} &= \int_{\omega} x_{\alpha}\theta_{\beta}\,\mathrm{d}\omega,\\
K^{\eta}_{\alpha\beta\lambda\mu} &= \int_{\omega} \Phi_{\lambda\mu}\partial_{\alpha}\eta_{\beta}\,\mathrm{d}\omega,\\
K^{\theta}_{\alpha\beta\lambda\mu} &= \int_{\omega} \Phi_{\lambda\mu}\partial_{\alpha}\theta_{\beta}\,\mathrm{d}\omega.
\end{aligned}
\tag{25.25}
$$

If ω is simply connected, constants D_{α} can be written in the following way:

$$D_1 = a_{2233}I_1^{\Psi} - a_{1233}I_2^{\Psi} - \tfrac{1}{2}I_1^{w}, \qquad D_2 = a_{1133}I_2^{\Psi} - a_{1233}I_1^{\Psi} - \tfrac{1}{2}I_2^{w}. \tag{25.26}$$

A combination of the techniques used for the homogeneous isotropic case (Theorem 7.1) together with those used for the nonhomogeneous anisotropic cases (Theorems 23.1 and 24.2) allows us to prove the next result. It includes the calculations of the zeroth and second order terms for a homogeneous anisotropic material. Terms $\boldsymbol{u}^0$ and σ^0_{33} are evaluated as in Theorem 23.1. The result is owed to ÁLVAREZ-DIOS and VIAÑO [1992a,b] (cf. also ÁLVAREZ-DIOS [1992]).

THEOREM 25.1. *Let the elasticity coefficients a_{ijkl} be constant in Ω (homogeneous material). If the system of applied forces verifies the regularity hypothesis (23.6), then Eqs. (22.35)–(22.47) determine functions $(\boldsymbol{u}^0, \boldsymbol{\sigma}^0) \in V(\Omega) \times \Sigma(\Omega)$, $\boldsymbol{u}^2 \in V(\Omega)$ and $\sigma^2_{33} \in L^2(\Omega)$ in a unique way, through the following characterizations*:

(i)

$$\boldsymbol{u}^0 \in V^0_{BN}(\Omega), \qquad u^0_3 = \underline{u}^0_3 - x_{\alpha}\partial_3 u^0_{\alpha}, \qquad \underline{u}^0_3 \in V^1_0(0,L), \qquad u^0_{\alpha} \in V^2_0(0,L), \tag{25.27}$$

where functions $\underline{u}^0_3$ and u^0_{α} are the unique solutions of the following variational problems:

$$
\begin{aligned}
&u^0_{\alpha} \in V^2_0(0,L),\\
&\int_0^L \frac{I_{\alpha}}{a_{3333}}\partial_{33}u^0_{\alpha}\partial_{33}v\,\mathrm{d}x_3 = \tilde{F}_{\alpha}(v), \quad \textit{for all } v \in V^2_0(0,L) \ (\textit{no sum on } \alpha),
\end{aligned}
\tag{25.28}
$$

$$
\begin{aligned}
&\underline{u}^0_3 \in V^1_0(0,L),\\
&\int_0^L \frac{A}{a_{3333}}\partial_3\underline{u}^0_3\partial_3 v\,\mathrm{d}x_3 = \tilde{F}_3(v), \quad \textit{for all } v \in V^1_0(0,L),
\end{aligned}
\tag{25.29}
$$

where the linear continuous forms $\tilde{F}_i(\cdot)$ are defined in (23.4).

(ii) *The axial stress component* σ_{33}^0 *and the bending moment components* m_β^0 *are given by (no sum on* β*):*

$$\sigma_{33}^0 = \frac{1}{a_{3333}}(\partial_3 \underline{u}_3^0 - x_\alpha \partial_{33} u_\alpha^0), \tag{25.30}$$

$$\begin{aligned} m_\beta^0 &= \int_\omega x_\beta \sigma_{33}^0 \, \mathrm{d}\omega = -\frac{I_\beta}{a_{3333}} \partial_{33} u_\beta^0 \\ &= \int_{x_3}^{L} \int_0^t \left[\int_\omega f_\beta \, \mathrm{d}\omega + \int_\gamma g_\beta \, \mathrm{d}\gamma \right] \mathrm{d}x_3 + \int_{x_3}^{L} \left[\int_\omega x_\beta f_3 \, \mathrm{d}\omega + \int_\gamma x_\beta g_3 \, \mathrm{d}\gamma \right] \mathrm{d}x_3 \\ &\quad + (x_3 - L) \left[\int_0^L \left(\int_\omega f_\beta \, \mathrm{d}\omega + \int_\gamma g_\beta \, \mathrm{d}\gamma \right) \mathrm{d}x_3 + \int_\omega h_\beta \, \mathrm{d}\omega \right] \mathrm{d}x_3. \end{aligned} \tag{25.31}$$

(iii) *Displacements* $u_i^2 \in V(\Omega)$ *have the following expressions:*

$$\begin{aligned} u_1^2 &= z_1^2 + x_2 z^2 + \frac{1}{a_{3333}}[a_{1\beta 33} x_\beta \partial_3 \underline{u}_3^0 + \Phi_{\beta 1} \partial_{33} u_\beta^0], \\ u_2^2 &= z_2^2 - x_1 z^2 + \frac{1}{a_{3333}}[a_{2\beta 33} x_\beta \partial_3 \underline{u}_3^0 + \Phi_{\beta 2} \partial_{33} u_\beta^0], \\ u_3^2 &= \underline{u}_3^2 - x_\alpha \partial_3 z_\alpha^2 - w \partial_3 z^2 + w^0 \\ &\quad + \frac{1}{a_{3333}} \left[\frac{1}{2A}(a_{1133} I_1 + a_{2233} I_2) - \frac{1}{2} a_{\alpha\beta 33} x_\alpha x_\beta \right] \partial_{33} \underline{u}_3^0 \\ &\quad + \frac{1}{a_{3333}} [\eta_\alpha + \theta_\alpha] \partial_{333} u_\alpha^0, \end{aligned} \tag{25.32}$$

where z_α^2, z^2, $\underline{u}_3^2$ *and* w^0 *are such that:*

(a) *The angle of torsion* z^2 *depends only on* x_3 *and is the unique solution of the following variational problem:*

$$\begin{aligned} & z^2 \in H^1(0, L), \\ & \int_0^L J \partial_3 z^2 \partial_3 v \, \mathrm{d}x_3 = M_3^2(v), \quad \textit{for all } v \in V_0^1(0, L), \\ & z^2(0) = \frac{1}{a_{3333}(I_1 + I_2)}[a_{1233}(I_1 - I_2) \partial_3 \underline{u}_3^0(0) - Y_\beta \partial_{33} u_\beta^0(0)], \end{aligned} \tag{25.33}$$

where

$$M_3^2(v) = \int_0^L \left\{ \int_\omega (x_2 f_1 - x_1 f_2) \, \mathrm{d}\omega + \left[\int_\gamma (x_2 g_1 - x_1 g_2) \, \mathrm{d}\gamma \right. \right.$$

$$+ \int_\omega w\partial_3 f_3 \,\mathrm{d}\omega + \int_\gamma w\partial_3 g_3 \,\mathrm{d}\gamma + \frac{D_\alpha}{a_{3333}} \partial_{3333} u_\alpha^0 \Bigg] v \Bigg\} \,\mathrm{d}x_3$$

$$+ \Bigg[\int_\omega (x_2 h_1 - x_1 h_2) \,\mathrm{d}\omega - \frac{D_\alpha}{a_{3333}} \partial_{333} u_\alpha^0 (L) - \int_\omega w f_3(L) \,\mathrm{d}\omega$$

$$- \int_\gamma w g_3(L) \,\mathrm{d}\gamma \Bigg] v(L), \quad \textit{for all } v \in V_0^1(0, L). \tag{25.34}$$

(b) *The additional warping w^0 is the unique solution of the following problem:*

$$w^0 \in H^1[0, L; H^1(\omega)] \quad \textit{such that for all } x_3 \in [0, L]:$$

$$\int_\omega A_{3\beta 3\alpha} \partial_\alpha w^0 \partial_\beta \varphi \,\mathrm{d}\omega$$

$$= \int_\omega f_3 \varphi \,\mathrm{d}\omega + \int_\gamma g_3 \varphi \,\mathrm{d}\gamma - \frac{1}{A} \Bigg[\int_\omega f_3 \,\mathrm{d}\omega + \int_\gamma g_3 \,\mathrm{d}\gamma \Bigg] \int_\omega \varphi \,\mathrm{d}\omega, \tag{25.35}$$

for all $\varphi \in H^1(\omega)$,

$$\int_\omega w^0 \,\mathrm{d}\omega = 0.$$

(c) *The additional stretch $\underline{u}_3^2$ depends only on x_3 and is the unique solution of the following problem:*

$$\underline{u}_3^2 \in V_0^1(0, L),$$

$$\int_0^L \frac{A}{a_{3333}} \partial_3 \underline{u}_3^2 \partial_3 v \,\mathrm{d}x_3 = \int_0^L G_3^2 \partial_3 v \,\mathrm{d}x_3, \quad \textit{for all } v \in V_0^1(0, L), \tag{25.36}$$

where

$$G_3^2 = \left(\frac{a_{1133}}{2a_{3333}^2} I_1 + \frac{a_{2233}}{2a_{3333}^2} I_2 \right) \partial_{333} \underline{u}_3^0 - \frac{H_\alpha}{a_{3333}^2} \partial_{3333} u_\alpha^0$$

$$+ \frac{a_{\alpha\beta 33}}{a_{3333}} \Bigg[\int_\omega x_\alpha f_\beta \,\mathrm{d}\omega + \int_\gamma x_\alpha g_\beta \,\mathrm{d}\gamma + \frac{1}{2} \int_\omega x_\alpha x_\beta \partial_3 f_3 \,\mathrm{d}\omega$$

$$+ \frac{1}{2} \int_\gamma x_\alpha x_\beta \partial_3 g_3 \,\mathrm{d}\gamma \Bigg]. \tag{25.37}$$

(d) *Displacements z^2_α depend only on variable x_3 and are given as the unique solution to the problem*

$$
\begin{aligned}
& z^2_\alpha \in H^2(0,L),\\
& \int_0^L \frac{I_\alpha}{a_{3333}} \partial_{33} z^2_\alpha \partial_{33} v \,\mathrm{d}x_3\\
& \quad = \int_0^L M^2_\alpha \partial_{33} v \,\mathrm{d}x_3, \quad \textit{for all } v \in V^2_0(0,L) \ (\textit{no sum on } \alpha),\\
& z^2_\alpha(0) = -\frac{X_{\beta\alpha}}{a_{3333}} \partial_{33} u^0_\beta(0),\\
& \partial_3 z^0_\alpha(0) = \frac{1}{I_\alpha}\Bigg\{ \int_\omega x_\alpha w^0(0)\,\mathrm{d}\omega - \frac{1}{2} I^w_\alpha \partial_3 z^2(0) - \frac{H_\alpha}{a_{3333}} \partial_{33} \underline{u}^0_3(0)\\
& \qquad + \frac{1}{a_{3333}} [L^\eta_{\alpha\beta} + L^\theta_{\alpha\beta}] \partial_{333} u^0_\beta(0) \Bigg\} \quad (\textit{no sum on } \alpha),
\end{aligned}
\tag{25.38}
$$

where

$$
\begin{aligned}
M^2_\alpha = {} & \frac{1}{a^2_{3333}} (L^\eta_{\alpha\beta} + L^\theta_{\alpha\beta}) \partial_{3333} u^0_\beta + \frac{A_{3\mu 3\rho}}{a^2_{3333}} (K^\eta_{\mu\beta\alpha\rho} + K^\theta_{\mu\beta\alpha\rho} + Z_{\beta\mu\alpha\rho}) \partial_{3333} u^0_\beta\\
& - \frac{H_\alpha}{a^2_{3333}} \partial_{333} \underline{u}^0_3 + \frac{D_\alpha}{a_{3333}} \partial_{33} z^2 + \frac{A_{3\mu 3\rho}}{a_{3333}} \int_\omega \Phi_{\alpha\mu} \partial_{3\rho} w^0 \,\mathrm{d}\omega\\
& + \frac{1}{a_{3333}} \left[\int_\omega \Phi_{\alpha\beta} f_\beta \,\mathrm{d}\omega + \int_\gamma \Phi_{\alpha\beta} g_\beta \,\mathrm{d}\gamma \right] + \frac{1}{a_{3333}} \int_\omega x_\alpha \partial_3 w^0 \,\mathrm{d}\omega.
\end{aligned}
\tag{25.39}
$$

(iv) *The shear stress $\sigma^0_{3\beta}$ and the shear force q^0_β components are uniquely determined by*

$$
\begin{aligned}
\sigma^0_{31} &= -\partial_2 \Psi \partial_3 z^2 + A_{3\beta 31} \underline{\sigma}^0_{3\beta},\\
\sigma^0_{32} &= \partial_1 \Psi \partial_3 z^2 + A_{3\beta 32} \underline{\sigma}^0_{3\beta},
\end{aligned}
\tag{25.40}
$$

where

$$
\underline{\sigma}^0_{3\beta} = \frac{1}{a_{3333}} (\partial_\beta \eta_\alpha + \partial_\beta \theta_\alpha + \Phi_{\alpha\beta}) \partial_{333} u^0_\alpha + \partial_\beta w^0. \tag{25.41}
$$

$$
q^0_\beta = \partial_3 m^0_\beta + \int_\omega x_\beta f_3 \,\mathrm{d}\omega + \int_\gamma x_\beta g_3 \,\mathrm{d}\gamma. \tag{25.42}
$$

(v) *The plane stress components $\sigma^0_{\alpha\beta}$ are obtained through the expression*

$$\sigma^0_{\alpha\beta} = S_{\alpha\beta}(\underline{\boldsymbol{u}}^4) + S^0_{\alpha\beta} = B_{\alpha\beta\gamma\delta} e_{\gamma\delta}(\underline{\boldsymbol{u}}^4) + S^0_{\alpha\beta}, \tag{25.43}$$

where $\underline{\boldsymbol{u}}^4 = (\underline{u}^4_\alpha)$ is the unique solution of the following problem:

$$\underline{\boldsymbol{u}}^4 = (\underline{u}^4_\alpha) \in (L^2[0, L; H^1(\omega)])^2, \quad \textit{such that a.e. in } (0, L)\text{:}$$

$$\int_\omega S_{\alpha\beta}(\underline{\boldsymbol{u}}^4) e_{\alpha\beta}(\boldsymbol{\varphi})\,\mathrm{d}\omega = \int_\omega f_\beta \varphi_\beta\,\mathrm{d}\omega + \int_\gamma g_\beta \varphi_\beta\,\mathrm{d}\gamma + \int_\omega \partial_3 \sigma^0_{3\beta} \varphi_\beta\,\mathrm{d}\omega$$
$$- \int_\omega S^0_{\alpha\beta} e_{\alpha\beta}(\boldsymbol{\varphi})\,\mathrm{d}\omega, \quad \textit{for all } (\varphi_\beta) \in [H^1(\omega)]^2,$$

$$\int_\omega \underline{u}^4_\alpha\,\mathrm{d}\omega = \int_\omega (x_2 \underline{u}^4_1 - x_1 \underline{u}^4_2)\,\mathrm{d}\omega = 0, \tag{25.44}$$

with

$$S^0_{\alpha\beta} = -B_{\alpha\beta\gamma\delta} \frac{a_{\gamma\delta 33}}{a_{3333}} \partial_3 (u^2_3 - \underline{u}^2_3 + x_\mu \partial_3 z^2_\mu). \tag{25.45}$$

(vi) *The axial stress component σ^2_{33} and the second order bending moment components $m^2_\beta = \int_\omega x_\beta \sigma^2_{33}\,\mathrm{d}\omega$ are given by*

$$\begin{aligned}
\sigma^2_{33} &= \frac{1}{a_{3333}} (\partial_3 u^2_3 - a_{\alpha\beta 33} \sigma^0_{\alpha\beta}) \\
&= \frac{1}{a_{3333}} (\partial_3 \underline{u}^2_3 - x_\alpha \partial_3 z^2_\alpha - w \partial_{33} z^2_\alpha + \partial_3 w^0) \\
&\quad + \frac{1}{a^2_{3333}} \left[\frac{1}{2A} (a_{1133} I_1 + a_{2233} I_2 - \frac{1}{2} a_{\alpha\beta 33} x_\alpha x_\beta \right] \partial_{333} \underline{u}^0_3 \\
&\quad + \frac{1}{a^2_{3333}} (\eta_\alpha + \theta_\alpha) \partial_{3333} u^0_\alpha - \frac{a_{\alpha\beta 33}}{a_{3333}} (S_{\alpha\beta}(\underline{\boldsymbol{u}}^4) + S^0_{\alpha\beta}),
\end{aligned} \tag{25.46}$$

$$m^2_\beta = -\frac{I_\beta}{a_{3333}} \partial_{33} z^2_\beta + M^2_\beta \quad (\textit{no sum on } \beta). \tag{25.47}$$

From this theorem one may obtain the expressions for the components $\boldsymbol{u}^{p\varepsilon}$ and $\boldsymbol{\sigma}^{p\varepsilon}$ of the displacement and stress fields referred to the real beam Ω^ε (see definitions (11.2) and (11.3)). These equations are analogous to those of Theorem 25.1 except for a superscript ε showing up in all terms. In a similar way as in Section 11 we may define the first and second order asymptotic models, respectively, by $(\boldsymbol{u}^{0\varepsilon}, \boldsymbol{\sigma}^{0\varepsilon})$ and $(\boldsymbol{u}^{0\varepsilon} + \boldsymbol{u}^{2\varepsilon}, \boldsymbol{\sigma}^{0\varepsilon} + \boldsymbol{\sigma}^{2\varepsilon})$. In order to simplify the notation we shall refer to the case $\varepsilon = 1$ and to $(\boldsymbol{u}^0, \boldsymbol{\sigma}^0)$ and $(\hat{\boldsymbol{u}}, \hat{\boldsymbol{\sigma}}) = (\boldsymbol{u}^0 + \boldsymbol{u}^2, \boldsymbol{\sigma}^0 + \boldsymbol{\sigma}^2)$ as the first and second order models, respectively. We shall also denote the second order model by the *general asymptotic model in the reference domain.*

As in the homogeneous, isotropic case (see Section 11) *the asymptotic model explains and generalizes the corresponding classical models and its validity does not depend on the shape of the cross section or on the system of applied loads.*

We observe, in the first place, that the anisotropy has no influence on the first order displacement and stress fields. In fact, Eqs. (25.28)–(25.30) are exactly the same as those for the isotropic case (see (7.58)–(7.63)) and constitute a Bernoulli–Navier type model. Coefficient $1/a_{3333}$ plays the role of Young's modulus E. On the other hand, as was pointed out in Sections 23 and 24, the nonhomogeneity of the material, couples components u_α^0 and $\underline{u}_3^0$.

It is on the second order terms that the asymptotic model takes torsion, Poisson, geometric bending, and other effects into account. We compare the asymptotic model $(\hat{\boldsymbol{u}}, \hat{\boldsymbol{\sigma}})$ with the most well-known classical models of engineering literature.

In what concerns the system of applied loads, we use the notation introduced in (11.16), (11.17), (12.30) together with the following ones:

$$F_i^L = \int_\omega h_i \, \mathrm{d}\omega, \qquad M_\beta^L = \int_\omega x_\beta h_3 \, \mathrm{d}\omega, \qquad M_3^L = \int_\omega (x_2 h_1 - x_1 h_2) \, \mathrm{d}\omega. \tag{25.48}$$

EXAMPLE 25.1 (*Stretching model for a rod acted by an axial force at one end*). This is one of the simplest problems in anisotropic elasticity. It was studied by W. Voight in 1928 (LEKHNITSKII [1977, p. 79]). We assume that at end Γ_L the rod is acted by a system of applied forces in such a way that its resultant P acts along the Ox_3 axis. Neglecting the weight of the rod, one has

$$f_i = 0, \qquad g_i = 0, \qquad h_\beta = 0, \qquad h_3 = \frac{P}{A}. \tag{25.49}$$

The following a priori hypotheses are assumed in the classical theory:

$$\sigma_{33} = \frac{P}{A}, \qquad \sigma_{3\beta} = 0, \qquad \sigma_{\alpha\beta} = 0, \tag{25.50}$$

which, through Hooke's law and considering that the centroid of the cross section $\omega \times \{0\}$ can neither translate nor rotate, allow us to determine the displacement components u_i as

$$\begin{aligned} u_1 &= \frac{P}{A}(a_{1133}x_1 + a_{1233}x_2), \\ u_2 &= \frac{P}{A}(a_{1233}x_1 + a_{2233}x_2), \\ u_3 &= \frac{P}{A}a_{3333}x_3. \end{aligned} \tag{25.51}$$

For a system of applied loads satisfying (25.41) Theorem 25.1 gives

$$
\begin{aligned}
&u_\alpha^0 = 0, \quad \underline{u}_3^0 = \frac{P}{A} a_{3333} x_3, \\
&z^2 = 0, \quad w^0 = 0, \quad \underline{u}_3^2 = 0, \quad z_\alpha^2 = 0, \\
&\sigma_{\alpha\beta}^0 = 0, \quad \sigma_{3\beta}^0 = 0, \quad \sigma_{33}^0 = \frac{P}{A},
\end{aligned}
\tag{25.52}
$$

therefore, $\hat{\boldsymbol{u}} = \boldsymbol{u}^0 + \boldsymbol{u}^2$ gives us (25.51). Moreover, the a priori hypotheses (25.50) are a consequence of the theorem.

These expressions show that not only does the beam stretch along the direction of the applied load but it also undergoes shear on transversal directions.

EXAMPLE 25.2 (*Stretching of a beam acted by its own weight only*). Let us now consider a beam in upright position and subjected to its own weight only. Denoting the specific weight of the material by p, one has

$$
f_\alpha = 0, \quad f_3 = p, \quad g_i = 0, \quad h_i = 0. \tag{25.53}
$$

In this case, the following a priori hypotheses are considered (cf. LEKHNITSKII [1977, p. 82]:

$$
\sigma_{33} = p(L - x_3), \quad \sigma_{3\beta} = 0, \quad \sigma_{\alpha\beta} = 0, \quad \text{in } \Omega, \tag{25.54}
$$

and in a similar way as before, one obtains

$$
\begin{aligned}
u_1 &= p(a_{1133}x_1 + a_{1233}x_2)(L - x_3), \\
u_2 &= p(a_{1233}x_1 + a_{2233}x_2)(L - x_3), \\
u_3 &= \tfrac{1}{2} p[a_{\alpha\beta 33} x_\alpha x_\beta + a_{3333} x_3 (2L - x_3)].
\end{aligned}
\tag{25.55}
$$

As a result the cross sections warp; nevertheless, the centroidal axis is still a straight line. From Theorem 25.1, one has

$$
u_\alpha^0 = 0, \quad \underline{u}_3^0 = \tfrac{1}{2} a_{3333} p x_3 (2L - x_3), \tag{25.56}
$$

$$
w^0 = 0, \quad \underline{u}_3^2 = 0, \tag{25.57}
$$

$$
z^2 = p a_{1233} \frac{(I_1 - I_2)L}{I_1 + I_2}, \quad z_1^2 = p \frac{H_1}{I_1} x_3, \quad z_2^2 = p \frac{H_2}{I_2} x_3. \tag{25.58}
$$

Finally, we obtain

$$
\begin{aligned}
\hat{u}_1 &= p(a_{1133}x_1 + a_{1233}x_2)(L - x_3) + p\left[\frac{H_1}{I_1} x_3 + \frac{(I_1 - I_2)L}{I_1 + I_2} a_{1233} x_2\right], \\
\hat{u}_2 &= p(a_{1233}x_1 + a_{2233}x_2)(L - x_3) + p\left[\frac{H_2}{I_2} x_3 - \frac{(I_1 - I_2)L}{I_1 + I_2} a_{1233} x_1\right],
\end{aligned}
\tag{25.59}
$$

$$
\begin{aligned}
\hat{u}_3 = \frac{1}{2} p \Bigg[&-2\frac{H_1}{I_1} x_1 - 2\frac{H_2}{I_2} x_2 + a_{\alpha\beta 33} x_\alpha x_\beta \\
&- \frac{1}{A}(a_{1133} I_1 + a_{2233} I_2) + a_{3333} x_3 (2L - x_3) \Bigg],
\end{aligned}
\tag{25.60}
$$

which include (25.55) together with some additional terms associated to the angle of torsion z^2 and to the bending components z^2_α defined in (25.58). We remark that these last terms vanish whenever ω possesses two axes of symmetry, because $H_\alpha = 0$ in this case. Moreover, the a priori hypotheses (25.55) are also obtained since one has

$$\sigma^0_{\alpha\beta} = 0, \quad \sigma^0_{3\beta} = 0, \quad \sigma^0_{33} = p(L - x_3). \tag{25.61}$$

EXAMPLE 25.3 (*Saint Venant's anisotropic torsion theory*). This is a particular case of a more general theory in LEKHNITSKII [1977, p. 263]. As for the corresponding isotropic case we consider the rod subjected to a system of applied forces at the end Γ_L, having a constant twisting moment M_3^L as resultant. We thus have

$$f_i = 0, \qquad g_i = 0, \qquad h_3 = 0, \qquad F^L_\beta = 0, \qquad M^L_3 \neq 0. \tag{25.62}$$

Experiments show that the bar undergoes a twisting deformation consisting on the rotation of each cross section with respect to axis Ox_3 combined with a warping for all the cross sections (except for the circular and ring type cases).

The state of stress is characterized by the fact that the only nonzero stresses are the shear stresses. If $\theta(x_3)$ stands for the angle of twist of the cross section $\omega \times \{x_3\}$, then we find that $\theta(x_3)$ is proportional to the distance to the clamped end Γ_0 and consequently, the following a priori hypotheses are considered in the classical theory:

$$\begin{aligned} &\theta(x_3) = -\alpha x_3, \\ &u_1(x_1, x_2, x_3) = x_2\theta(x_3) = -\alpha x_2 x_3, \\ &u_2(x_1, x_2, x_3) = -x_1\theta(x_3) = \alpha x_1 x_3. \end{aligned} \tag{25.63}$$

As mentioned before each cross section warps in such a way that if $w(x_1, x_2)$ denotes the warping function in ω one has

$$u_3(x_1, x_2, x_3) = -w(x_1, x_2)\partial_3\theta(x_3) = \alpha w(x_1, x_2). \tag{25.64}$$

Consequently, from Hooke's law we get

$$\sigma_{\alpha\beta} = 0, \qquad \sigma_{33} = 0, \qquad \sigma_{3\beta} = \alpha[A_{3\beta31}(\partial_1 - x_2) + A_{3\beta32}(\partial_2 w + x_1)]. \tag{25.65}$$

From the equilibrium equations we obtain that w is the unique solution of problem (25.14). Consequently, relations (25.17) and (25.18) hold, that is

$$\sigma_{31} = \alpha\partial_2\Psi, \qquad \sigma_{32} = -\alpha\partial_1\Psi. \tag{25.66}$$

Moreover, from the equilibrium condition at end Γ_L, α may be computed as, (cf. Remark 24.6 for the homogeneous case)

$$\alpha = \frac{M^L_3}{J}. \tag{25.67}$$

As in Remark 24.6, starting from the system of applied forces (25.62) we obtain from Theorem 25.1

$$
\begin{aligned}
&\boldsymbol{u}^0 = 0, \qquad \sigma^0_{\alpha\beta} = 0, \qquad \sigma^0_{33} = 0,\\
&z^2 = -\alpha x_3, \qquad \underline{u}^2_3 = 0, \qquad z^2_1 = \frac{1}{2}\alpha\frac{I^w_1}{I_1}x_3, \qquad z^2_2 = \frac{1}{2}\alpha\frac{I^w_2}{I_2}x_3,\\
&\sigma_{31} = \alpha\partial_2\Psi, \qquad \sigma_{32} = -\alpha\partial_1\Psi.
\end{aligned}
\tag{25.68}
$$

Hence,

$$
\begin{aligned}
\hat{u}_1 &= -\alpha x_2 x_3 + \frac{1}{2}\alpha\frac{I^w_1}{I_1}x_3,\\
\hat{u}_2 &= \alpha x_1 x_3 + \frac{1}{2}\alpha\frac{I^w_2}{I_2}x_3,\\
\hat{u}_3 &= \alpha w - \frac{1}{2}\alpha\frac{I^w_1}{I_2}x_2 - \frac{1}{2}\alpha\frac{I^w_2}{I_2}x_2.
\end{aligned}
\tag{25.69}
$$

As can be seen, from the classical conditions for Saint Venant's theory the asymptotic model also incorporates some additional terms associated with geometric bending and warping, besides including the classical theory.

EXAMPLE 25.4 (*Bending of an anisotropic bar by a moment applied at its end*). Let us consider that the beam is now subjected to a bending moment M^L_2 acting on the plane Ox_2x_3, so that the system of applied forces is now of the form

$$
f_i = 0, \qquad g_i = 0, \qquad h_\alpha = 0, \qquad F^L_3 = 0, \qquad M^L_1 = 0, \qquad M^L_2 \neq 0. \tag{25.70}
$$

In this case it is customary to consider the following a priori hypotheses for the stress tensor components (cf. LEKHNITSKII [1977, p. 90]):

$$
\sigma_{33} = \frac{M^L_2}{I_2}x_2, \qquad \sigma_{3\beta} = 0, \qquad \sigma_{\alpha\beta} = 0. \tag{25.71}
$$

As in the previous cases, a substitution into the equilibrium equations gives

$$
\begin{aligned}
u_1 &= \frac{M^L_2}{I_2}(a_{1133}x_1x_2 + a_{1233}x_2^2),\\
u_2 &= \frac{M^L_2}{2I_2}(-a_{1133}x_1^2 + a_{2233}x_2^2 - a_{3333}x_3^2),\\
u_3 &= \frac{M^L_2}{I_2}a_{3333}x_2x_3.
\end{aligned}
\tag{25.72}
$$

If the system of applied loads satisfies (25.70), using Theorem 25.1, we obtain from the asymptotic model

$$\underline{u}_3^0 = 0, \qquad u_1^0 = 0, \qquad u_2^0 = -\frac{M_2^L}{2I_2} a_{3333} x_3^2, \\ \underline{u}_3^2 = 0, \qquad w^0 = 0, \qquad z^2 = \frac{M_2^L Y_2}{I_2(I_1 + I_2)}, \qquad z_\alpha^2 = \frac{M_2^L X_{2\alpha}}{I_2}, \tag{25.73}$$

and thus

$$\hat{u}_1 = u_1^0 + u_1^2 = \frac{M_2^L}{I_2}(a_{1133} x_1 x_2 + a_{1233} x_2^2) + \frac{M_2^L}{I_2}\left[\frac{Y_2}{(I_1 + I_2)} x_2 + X_{21}\right],$$
$$\hat{u}_2 = u_2^0 + u_2^2 = \frac{M_2^L}{I_2}(-a_{1133} x_1^2 + a_{2233} x_2^2 - a_{3333} x_3^2) + \frac{M_2^L}{I_2}\left[-\frac{Y_2}{(I_1 + I_2)} x_1 + X_{22}\right],$$
$$\hat{u}_3 = u_3^0 + u_3^2 = \frac{M_2^L}{I_2} a_{3333} x_2 x_3.$$

These equations contain the classical displacement field together with some additional bending effects associated with terms Y_2 and $X_{2\beta}$. Once again the a priori hypotheses (25.70) are obtained as a by-product of Theorem 25.1.

EXAMPLE 25.5 (*A bending and torsion model for an anisotropic beam loaded on one of its ends*). Let us consider that now the resultant of the applied loads P is parallel to axis Ox_2, so that we have

$$f_i = 0, \qquad g_i = 0, \qquad h_1 = 0, \qquad h_3 = 0, \qquad F_2^L = P, \tag{25.74}$$

then, as for the isotropic case, the a priori hypotheses on the stress components σ_{33} are (cf. LEKHNITSKII [1977, p. 230]):

$$\sigma_{33} = \frac{P}{I_2} x_2 x_3, \tag{25.75}$$

and from the equilibrium equations one has

$$\partial_1 \sigma_{31} + \partial_2 \sigma_{32} = -\frac{P}{I_2} x_2. \tag{25.76}$$

Consequently, we assume $\sigma_{3\beta}$ to be of the form

$$\sigma_{31} = -\partial_2 \Psi + \tau_1, \qquad \sigma_{32} = \partial_1 \Psi + \tau_2, \tag{25.77}$$

where (τ_α) is a solution of

$$\partial_\alpha \tau_\alpha + \frac{P}{I_2} x_2 = 0, \tag{25.78}$$

subjected to different boundary conditions depending on the shape of the cross section. Considering now the remaining equilibrium equations together with

Hooke's law, the expressions for the displacement field are

$$u_1 = \frac{P}{I_2}(a_{1133}x_1x_2 + a_{1233}x_2^2)(x_3 - L) + Cx_2x_3,$$
$$u_2 = \frac{P}{I_2}[\tfrac{1}{6}a_{3333}x_3^2(3L - x_3) - \tfrac{1}{2}(a_{1133}x_1^2 - a_{2233}x_2^2)(x_3 - L)] - Cx_1x_3, \quad (25.79)$$
$$u_3 = \frac{P}{2I_2}a_{3333}x_3(x_3 - 2L)x_2 + W - W_0,$$

where C is a constant and W_0 is the value of function $W = W(x_1, x_2)$ at the origin. This function W must satisfy

$$\partial_1 W = 4a_{1313}\sigma_{31} + 4a_{1323}\sigma_{32} - \frac{P}{I_2}(a_{1133}x_1x_2 + a_{1233}x_2^2) - Cx_2,$$
$$\partial_2 W = 4a_{1323}\sigma_{31} + 4a_{2323}\sigma_{32} - \frac{P}{2I_2}(a_{2233}x_2^2 - a_{1133}x_1^2) + Cx_1. \quad (25.80)$$

From (25.80) and (25.75)–(25.77) the following equations for W and Ψ can be deduced:

$$-\partial_\beta(A_{3\beta3\alpha}\partial_\alpha W) = \frac{P}{I_2}(1 + A_{3\beta3\alpha}a_{\alpha\beta33})x_2, \quad (25.81)$$

$$-\partial_\beta(A_{3\beta3\alpha}\partial_\alpha \Psi) = 2D\,[\,\frac{P}{I_2}(a_{1133}x_1 + a_{1233}x_2) - C$$
$$- 2\partial_2(a_{13\alpha3}\tau_\alpha) + 2\partial_1(a_{23\alpha3}\tau_\alpha). \quad (25.82)$$

We refer to LEKHNITSKII [1977, p. 323] for further details concerning this classical model.

When the system of applied forces satisfies (25.74), we obtain for the asymptotic model, from Theorem 25.1

$$\underline{u}_3^0 = 0, \qquad u_1^0 = 0, \qquad u_2^0 = \frac{P}{6I_2}a_{3333}(3L - x_3)x_3^2, \quad (25.83)$$

$$\underline{u}_3^2 = 0, \qquad w^0 = 0, \qquad z^2 = cx_3 - d, \quad (25.84)$$

where

$$c = \frac{PD_2}{JI_2}, \qquad d = \frac{PY_2L}{I_2(I_1 + I_2)}. \quad (25.85)$$

Without making an explicit calculation for z_α^2 we can therefore write

$$\hat{u}_1 = u_1^0 + u_1^2 = \frac{P}{I_2}(a_{1133}x_1x_2 + a_{1233}x_2^2)(x_3 - L) + cx_2x_3 - dx_2 + z_1^2, \quad (25.86)$$

$$\hat{u}_2 = u_2^0 + u_2^2 = \frac{P}{I_2}[\tfrac{1}{6}a_{3333}(3L - x_3)x_3^2 + \tfrac{1}{2}(a_{1133}x_1^2 - a_{2233}x_2^2)(L - x_3)]$$
$$- cx_1x_3 + dx_1 + z_2^2, \quad (25.87)$$

$$\hat{u}_3 = u_3^0 + u_3^2 = \frac{P}{2I_2}a_{3333}x_2x_3(x_3 - 2L) - cw - \frac{P}{I_2}(\eta_2 + \theta_2) - x_\alpha\partial_3 z_\alpha^2. \quad (25.88)$$

These equations generalize the classical expressions (25.79) and include additional bending (z_α^2) and warping (w) effects. Constant W_0 does not show up in (25.88) due to the boundary conditions chosen. Comparing (25.79) with (25.86)–(25.88) we observe that constant c is precisely given in (25.85) and function W is in reality $-P/I_2(\eta_2 + \theta_2)$. This can be seen by comparing u_3 with $\hat{u}_3$ or by observing that (25.81) may be written as

$$-\partial_\beta(A_{3\beta 3\alpha}\partial_\alpha W) = \frac{P}{I_2}(x_2 - A_{3\beta 3\alpha}\partial_\beta \Phi_{2\alpha}), \tag{25.89}$$

which is satisfied by $-P/I_2(\eta_2 + \theta_2)$ (cf. (25.15)–(25.16)).

From all these examples we see that *the asymptotic model generalizes all the well-known classical cases* (see LEKHNITSKII [1977] for more interesting cases also included in the asymptotic model). This is done with no ambiguity and for a very general loading and geometry of the cross section. The systematic definition of the displacement and stress fields together with the different associated quantities allows us to clarify several of the issues that were not so well explained in classical literature. Moreover, it is also clear how an improvement of the asymptotic model itself could be done by including higher order terms.

These are the most important characteristics of the asymptotic model (given in Theorem 25.1) and one of the most important conclusions of the work presented so far.

CHAPTER V

Asymptotic Modelling of Rods in Linearized Thermoelasticity

This chapter contains the asymptotic analysis for rods in the framework of three-dimensional linearized thermoelasticity.

In Section 26 we consider an additional displacement field, in the deformation of rod, due to a *known* temperature distribution. We consider a linearized thermoelastic constitutive law (Duhamel–Neumann law), and following the same ideas and techniques put forward in the previous sections, for the isothermal case, we obtain *reduced models for thermoelastic bending, stretching and torsion of rods*. In this section we consider the stationary case where the temperature field is known a priori and is independent of the displacement field. The contents of this section are adapted from the works of BERMÚDEZ and VIAÑO [1984] and ÁLVAREZ-VÁZQUEZ and VIAÑO [1989].

In the previous sections we saw how the asymptotic expansion method could be applied to the obtainment and mathematical justification of a great number of models, governing the mechanical behaviour of beams, as the limit of a three-dimensional elasticity problem. In spite of the great number of applications presented and existing in the literature for other types of structures, such as plates and shells, very little attention has been given to the dynamic behaviour of rods. We mention the works of RAOULT [1980, 1985, 1988, 1992] and BLANCHARD and FRANCFORT [1987] for the plate case.

Section 27 is devoted to the justification of a linear evolution thermoelastic model, for rods, using asymptotic methods. The results presented there were initially obtained by ÁLVAREZ-VÁZQUEZ [1991] and ÁLVAREZ-VÁZQUEZ and VIAÑO [1991a,b,c, 1992a], following the early literature for plates.

The behaviour of the scaled problem when the area of the cross section becomes very small is studied using the classical techniques of a priori estimates and passage to the limit in the weak topology, thus justifying the limit problem. The strong convergence is proved under certain compatibility assumptions related to the initial data of the three-dimensional problem. Finally, we end up this section by applying the same technique to different boundary conditions (beam clamped in only one extremity, heat flux on the lateral surface).

26. The asymptotic expansion method for stationary thermoelastic rods

In order to simplify the notation we assume that the beam Ω^ε is made of an homogeneous and isotropic material with Young's modulus E^ε, Poisson's ratio ν^ε, linear thermal expansion coefficient α_T^ε and thermal conductivity coefficient k^ε. Using the same notations as in Sections 1–22 we consider that the extremity Γ_0^ε of the rod is weakly clamped and that the rod is subjected to a system of applied forces $\boldsymbol{f}^\varepsilon$, $\boldsymbol{g}^\varepsilon$ and $\boldsymbol{h}^\varepsilon$, as in Section 22.

In its undeformed configuration Ω^ε we consider the rod subjected to a uniform reference temperature T_0^ε and that when the loads are applied the temperature change in Ω^ε is known and given by $\Theta^\varepsilon = T^\varepsilon - T_0^\varepsilon$.

REMARK 26.1. One may assume that Θ^ε is the solution of the following heat equation with mixed boundary conditions:

$$\begin{aligned}
&-\partial_i^\varepsilon(k^\varepsilon \partial_i^\varepsilon \Theta^\varepsilon) = q^\varepsilon, \quad \text{in } \Omega^\varepsilon,\\
&\Theta^\varepsilon = 0, \quad \text{on } \Gamma_0^\varepsilon \cup \Gamma_D^\varepsilon,\\
&k^\varepsilon \partial_n^\varepsilon \Theta^\varepsilon = k^\varepsilon \partial_3^\varepsilon \Theta^\varepsilon = p_L^\varepsilon, \quad \text{in } \Gamma_L^\varepsilon,\\
&k^\varepsilon \partial_n^\varepsilon \Theta^\varepsilon = p^\varepsilon, \quad \text{in } \Gamma_N^\varepsilon,
\end{aligned} \tag{26.1}$$

where

$$\gamma^\varepsilon = \gamma_D^\varepsilon \cup \gamma_N^\varepsilon, \tag{26.2}$$

$$\gamma_D^\varepsilon \cap \gamma_N^\varepsilon = \emptyset, \tag{26.3}$$

$$\Gamma_D^\varepsilon = \gamma_D^\varepsilon \times (0, L), \qquad \Gamma_N^\varepsilon = \gamma_N^\varepsilon \times (0, L), \tag{26.4}$$

and where q^ε, p^ε and p_L^ε are given functions defined respectively in $L^2(\Omega^\varepsilon)$, $L^2(\Gamma_L^\varepsilon)$ and $L^2(\Gamma_N^\varepsilon)$.

The constitutive thermoelastic law is given by the Duhamel–Neumann relation (cf. DUVAUT and LIONS [1972b], CARLSON [1972]):

$$\sigma_{ij}^\varepsilon(\boldsymbol{u}^\varepsilon) = \frac{E^\varepsilon}{1+\nu^\varepsilon} e_{ij}^\varepsilon(\boldsymbol{u}^\varepsilon) + \frac{\nu^\varepsilon E^\varepsilon}{(1+\nu^\varepsilon)(1-2\nu^\varepsilon)} e_{kk}^\varepsilon(\boldsymbol{u}^\varepsilon)\delta_{ij} - \frac{\alpha_T^\varepsilon E^\varepsilon}{1-2\nu^\varepsilon} \Theta^\varepsilon \delta_{ij}. \tag{26.5}$$

The equilibrium equations are, as before

$$\begin{aligned}
&-\partial_j^\varepsilon \sigma_{ij}^\varepsilon(\boldsymbol{u}^\varepsilon) = f_i^\varepsilon, \quad \text{in } \Omega^\varepsilon,\\
&\sigma_{ij}^\varepsilon(\boldsymbol{u}^\varepsilon) n_j^\varepsilon = g_i^\varepsilon, \quad \text{in } \Gamma^\varepsilon,\\
&\sigma_{ij}^\varepsilon(\boldsymbol{u}^\varepsilon) n_j^\varepsilon = \sigma_{i3}^\varepsilon(\boldsymbol{u}^\varepsilon) = h_i^\varepsilon, \quad \text{in } \Gamma_L^\varepsilon.
\end{aligned} \tag{26.6}$$

The space of admissible functions is given by (cf. (22.7))

$$V(\Omega^\varepsilon) = \{\boldsymbol{v}^\varepsilon = (v_i^\varepsilon) \in [H^1(\Omega^\varepsilon)]^3: \int_{\omega^\varepsilon \times \{0\}} v_i^\varepsilon \, \mathrm{d}\omega^\varepsilon = \int_{\omega^\varepsilon \times \{0\}} (x_j^\varepsilon v_i^\varepsilon - x_i^\varepsilon v_j^\varepsilon) \, \mathrm{d}\omega^\varepsilon = 0\}. \tag{26.7}$$

The displacement variational formulation of the three-dimensional linearized elasticity problem is (cf. (1.10), (1.11) and (22.8))

$$\begin{aligned} &\boldsymbol{u}^\varepsilon \in V(\Omega^\varepsilon), \\ &\int_{\Omega^\varepsilon} \sigma_{ij}^\varepsilon(\boldsymbol{u}^\varepsilon) e_{ij}^\varepsilon(\boldsymbol{v}^\varepsilon) \, \mathrm{d}\boldsymbol{x}^\varepsilon \\ &\quad = \int_{\Omega^\varepsilon} f_i^\varepsilon v_i^\varepsilon \, \mathrm{d}\boldsymbol{x}^\varepsilon \\ &\quad + \int_{\Gamma^\varepsilon} g_i^\varepsilon v_i^\varepsilon \, \mathrm{d}a^\varepsilon + \int_{\Gamma_L^\varepsilon} h_i^\varepsilon v_i^\varepsilon \, \mathrm{d}a^\varepsilon, \quad \text{for all } \boldsymbol{v}^\varepsilon \in V(\Omega^\varepsilon). \end{aligned} \tag{26.8}$$

Substituting the constitutive equation into this equality, we have

$$\begin{aligned} &\int_{\Omega^\varepsilon} \left\{ \frac{E^\varepsilon}{1+\nu^\varepsilon} e_{ij}^\varepsilon(\boldsymbol{u}^\varepsilon) + \frac{\nu^\varepsilon E^\varepsilon}{(1+\nu^\varepsilon)(1-2\nu^\varepsilon)} e_{pp}^\varepsilon(\boldsymbol{u}^\varepsilon) \delta_{ij} \right\} e_{ij}^\varepsilon(\boldsymbol{u}^\varepsilon) \, \mathrm{d}\boldsymbol{x}^\varepsilon \\ &\quad = \int_{\Omega^\varepsilon} f_i^\varepsilon v_i^\varepsilon \, \mathrm{d}\boldsymbol{x}^\varepsilon + \int_{\Gamma^\varepsilon} g_i^\varepsilon v_i^\varepsilon \, \mathrm{d}a^\varepsilon + \int_{\Gamma_L^\varepsilon} h_i^\varepsilon v_i^\varepsilon \, \mathrm{d}a^\varepsilon \\ &\quad + \frac{\alpha_T^\varepsilon E^\varepsilon}{1-2\nu^\varepsilon} \int_{\Omega^\varepsilon} \Theta^\varepsilon e_{kk}^\varepsilon(\boldsymbol{v}^\varepsilon) \, \mathrm{d}\boldsymbol{x}^\varepsilon, \quad \text{for all } \boldsymbol{v}^\varepsilon \in V(\Omega^\varepsilon), \end{aligned} \tag{26.9}$$

from which we deduce existence and unicity of solution as a consequence of Lax–Milgram's Lemma (cf. CIARLET [1988, Theorem 6.3–3]). Moreover, $\boldsymbol{u}^\varepsilon$ is the unique solution of the following *minimization problem*:

$$\boldsymbol{u}^\varepsilon \in V(\Omega^\varepsilon), \qquad J(\boldsymbol{u}^\varepsilon) = \inf_{\boldsymbol{v}^\varepsilon \in V(\Omega^\varepsilon)} J(\boldsymbol{v}^\varepsilon), \tag{26.10}$$

where J^ε denotes the following energy functional:

$$\begin{aligned} J^\varepsilon(\boldsymbol{v}^\varepsilon) &= \frac{1}{2} \int_{\Omega^\varepsilon} \left\{ \frac{E^\varepsilon}{1+\nu^\varepsilon} e_{ij}^\varepsilon(\boldsymbol{v}^\varepsilon) e_{ij}^\varepsilon(\boldsymbol{v}^\varepsilon) + \frac{\nu^\varepsilon E^\varepsilon}{(1+\nu^\varepsilon)(1-2\nu^\varepsilon)} e_{pp}^\varepsilon(\boldsymbol{v}^\varepsilon) e_{qq}^\varepsilon(\boldsymbol{v}^\varepsilon) \right\} \mathrm{d}\boldsymbol{x}^\varepsilon \\ &\quad - \frac{\alpha_T^\varepsilon E^\varepsilon}{1-2\nu^\varepsilon} \int_{\Omega^\varepsilon} \Theta^\varepsilon e_{kk}^\varepsilon(\boldsymbol{v}^\varepsilon) \, \mathrm{d}\boldsymbol{x}^\varepsilon - \int_{\Omega^\varepsilon} f_i^\varepsilon v_i^\varepsilon \, \mathrm{d}\boldsymbol{x}^\varepsilon \\ &\quad - \int_{\Gamma^\varepsilon} g_i^\varepsilon v_i^\varepsilon \, \mathrm{d}a^\varepsilon - \int_{\Gamma_L^\varepsilon} h_i^\varepsilon v_i^\varepsilon \, \mathrm{d}a^\varepsilon. \end{aligned} \tag{26.11}$$

The corresponding mixed displacement–stress variational formulation associated to problem (26.5), (26.6) is (cf. (6.10)–(6.12))

$$(\boldsymbol{u}^\varepsilon, \boldsymbol{\sigma}^\varepsilon) \in V(\Omega^\varepsilon) \times \Sigma(\Omega^\varepsilon),$$

$$\int_{\Omega^\varepsilon} \left\{ \frac{1+\nu^\varepsilon}{E^\varepsilon} \sigma_{ij}^\varepsilon - \frac{\nu^\varepsilon}{E^\varepsilon} \sigma_{pp}^\varepsilon \delta_{ij} \right\} \tau_{ij}^\varepsilon \, d\boldsymbol{x}^\varepsilon - \int_{\Omega^\varepsilon} e_{ij}^\varepsilon(\boldsymbol{u}^\varepsilon) \tau_{ij}^\varepsilon \, d\boldsymbol{x}^\varepsilon$$

$$= \alpha_T^\varepsilon \int_{\Omega^\varepsilon} \Theta^\varepsilon \tau_{ij}^\varepsilon \delta_{ij} \, d\boldsymbol{x}^\varepsilon, \quad \text{for all } \boldsymbol{\tau} \in \Sigma(\Omega^\varepsilon), \tag{26.12}$$

$$\int_{\Omega^\varepsilon} \sigma_{ij}^\varepsilon e_{ij}^\varepsilon(\boldsymbol{v}^\varepsilon) \, d\boldsymbol{x}^\varepsilon = \int_{\Omega^\varepsilon} f_i^\varepsilon v_i^\varepsilon \, d\boldsymbol{x}^\varepsilon + \int_{\Gamma^\varepsilon} g_i^\varepsilon v_i^\varepsilon \, da^\varepsilon + \int_{\Gamma_L^\varepsilon} h_i^\varepsilon v_i^\varepsilon \, da^\varepsilon,$$

for all $\boldsymbol{v}^\varepsilon \in V(\Omega^\varepsilon)$.

In order to study the behaviour of $\boldsymbol{u}^\varepsilon$ and $\boldsymbol{\sigma}^\varepsilon$ as ε goes to zero as before, we use, a change of variable to a fixed domain Ω and the scalings (2.6). We begin by assuming that hypotheses (2.7)–(2.9) and (22.9) on the data hold. Moreover, we consider that

$$\alpha_T^\varepsilon = \alpha_T, \qquad k^\varepsilon = k, \tag{26.13}$$

$$\Theta^\varepsilon(\boldsymbol{x}^\varepsilon) = \Theta(\boldsymbol{x}), \quad \text{for all } \boldsymbol{x}^\varepsilon = \Pi^\varepsilon(\boldsymbol{x}) \in \Omega^\varepsilon, \tag{26.14}$$

where $\Theta \in L^2(\Omega)$ is a function independent of ε.

REMARK 26.2. As in Remark 1.2 the linearity of the problem allows us to consider classes of equivalent scalings. Thus, if one considers (2.11)–(2.13), then (26.14) should be replaced by $\Theta^\varepsilon(\boldsymbol{x}^\varepsilon) = \varepsilon^{-r}\Theta(\boldsymbol{x})$. Moreover, the same linearity allows us to obtain force and temperature fields of a more complex nature and of the following type:

$$\begin{aligned}
f_\alpha^\varepsilon(\boldsymbol{x}^\varepsilon) &= \varepsilon^{-r+1} f_\alpha(\varepsilon)(\boldsymbol{x}) = \varepsilon^{-r+1} f_\alpha^0(\boldsymbol{x}) + \varepsilon^{-r+2} f_\alpha^1(\boldsymbol{x}) + \varepsilon^{-r+3} f_\alpha^2(\boldsymbol{x}) + \cdots, \\
f_3^\varepsilon(\boldsymbol{x}^\varepsilon) &= \varepsilon^{-r} f_3(\varepsilon)(\boldsymbol{x}) = \varepsilon^{-r} f_3^0(\boldsymbol{x}) + \varepsilon^{-r+1} f_3^1(\boldsymbol{x}) + \varepsilon^{-r+2} f_3^2(\boldsymbol{x}) + \cdots, \\
g_\alpha^\varepsilon(\boldsymbol{x}^\varepsilon) &= \varepsilon^{-r+2} g_\alpha(\varepsilon)(\boldsymbol{x}) = \varepsilon^{-r+2} g_\alpha^0(\boldsymbol{x}) + \varepsilon^{-r+3} g_\alpha^1(\boldsymbol{x}) + \varepsilon^{-r+4} g_\alpha^2(\boldsymbol{x}) + \cdots, \\
g_3^\varepsilon(\boldsymbol{x}^\varepsilon) &= \varepsilon^{-r+1} g_3(\varepsilon)(\boldsymbol{x}) = \varepsilon^{-r+1} g_3^0(\boldsymbol{x}) + \varepsilon^{-r+2} g_3^1(\boldsymbol{x}) + \varepsilon^{-r+3} g_3^2(\boldsymbol{x}) + \cdots, \\
h_\alpha^\varepsilon(\boldsymbol{x}^\varepsilon) &= \varepsilon^{-r+1} h_\alpha(\varepsilon)(\boldsymbol{x}) = \varepsilon^{-r+1} h_\alpha^0(\boldsymbol{x}) + \varepsilon^{-r+2} h_\alpha^1(\boldsymbol{x}) + \varepsilon^{-r+3} h_\alpha^2(\boldsymbol{x}) + \cdots, \\
h_3^\varepsilon(\boldsymbol{x}^\varepsilon) &= \varepsilon^{-r} h_3(\varepsilon)(\boldsymbol{x}) = \varepsilon^{-r} h_3^0(\boldsymbol{x}) + \varepsilon^{-r+1} h_3^1(\boldsymbol{x}) + \varepsilon^{-r+2} h_3^2(\boldsymbol{x}) + \cdots, \\
\Theta^\varepsilon(\boldsymbol{x}^\varepsilon) &= \varepsilon^{-r} \Theta(\varepsilon)(\boldsymbol{x}) = \varepsilon^{-r} \Theta^0(\boldsymbol{x}) + \varepsilon^{-r+1} \Theta^1(\boldsymbol{x}) + \varepsilon^{-r+2} \Theta^2(\boldsymbol{x}) + \cdots.
\end{aligned} \tag{26.15}$$

REMARK 26.3. Let us assume that for the case mentioned in Remark 26.1 the data verify

$$q^\varepsilon(\boldsymbol{x}^\varepsilon) = \varepsilon^{-r} q(\boldsymbol{x}), \qquad p^\varepsilon(\boldsymbol{x}^\varepsilon) = \varepsilon^{-r+1} p(\boldsymbol{x}), \qquad p_L^\varepsilon(\boldsymbol{x}^\varepsilon) = \varepsilon^{-r} p_L(\boldsymbol{x}), \tag{26.16}$$

where q, p and p_L are functions independent of ε. Then, function $\Theta(\varepsilon)(\boldsymbol{x}) = \varepsilon^r \Theta^\varepsilon(\boldsymbol{x}^\varepsilon)$ is a solution of the following problem:

$$\begin{aligned} &k\partial_{33}\Theta(\varepsilon) + \varepsilon^{-2}k\partial_{\alpha\alpha}\Theta(\varepsilon) = q, \quad \text{in } \Omega,\\ &\Theta(\varepsilon) = 0, \quad \text{on } \Gamma_0 \cup \Gamma_D,\\ &k\partial_n\Theta(\varepsilon) = p_L, \quad \text{on } \Gamma_L,\\ &\varepsilon^{-2}k\partial_n\Theta(\varepsilon) = p, \quad \text{on } \Gamma_N, \end{aligned} \tag{26.17}$$

consequently, it is natural to consider an asymptotic expansion for $\Theta(\varepsilon)$ of the following form:

$$\Theta(\varepsilon) = \Theta^0 + \varepsilon\Theta^1 + \varepsilon^2\Theta^2 + \text{h.o.t.} \tag{26.18}$$

Substituting in (26.17) we finally obtain for each $x_3 \in (0, L)$

$$\begin{aligned} &k\partial_{\alpha\alpha}\Theta^0(x_3) = 0, \quad \text{in } \omega,\\ &\Theta^0(x_3) = 0, \quad \text{on } \gamma_D,\\ &k\partial_n\Theta^0(x_3) = p, \quad \text{on } \gamma_N, \end{aligned} \tag{26.19}$$

This example shows the necessity of considering temperature fields which can be expanded in the form (26.15).

The following theorem may be obtained in a completely analogous way to Theorems 2.1 and 22.1. In order to simplify the notation we use Lamé's coefficients λ and μ (cf. (1.6)).

THEOREM 26.1. *The scaled displacement field $\boldsymbol{u}(\varepsilon) = (u_i(\varepsilon))$ is the unique solution of the following problem*:

$$\begin{aligned} &\boldsymbol{u}(\varepsilon) \in V(\Omega),\\ &\int_\Omega (\lambda + 2\mu)e_{33}(\boldsymbol{u}(\varepsilon))e_{33}(\boldsymbol{v})\,\mathrm{d}\boldsymbol{x}\\ &\quad + \varepsilon^{-2}\int_\Omega [\lambda e_{\rho\rho}(\boldsymbol{u}(\varepsilon))e_{33}(\boldsymbol{v}) + \lambda e_{33}(\boldsymbol{u}(\varepsilon))e_{\rho\rho}(\boldsymbol{v}) + 4\mu e_{3\alpha}(\boldsymbol{u}(\varepsilon))e_{3\alpha}(\boldsymbol{v})]\,\mathrm{d}\boldsymbol{x}\\ &\quad + \varepsilon^{-4}\int_\Omega [\lambda e_{\rho\rho}(\boldsymbol{u}(\varepsilon))e_{\sigma\sigma}(\boldsymbol{v}) + 2\mu e_{\alpha\beta}(\boldsymbol{u}(\varepsilon))e_{\alpha\beta}(\boldsymbol{v})]\,\mathrm{d}\boldsymbol{x}\\ &= \int_\Omega f_i v_i\,\mathrm{d}\boldsymbol{x} + \int_\Gamma g_i v_i\,\mathrm{d}a + \int_{\Gamma_L} h_i v_i\,\mathrm{d}a + \int_\Omega (3\lambda + 2\mu)\alpha_T\Theta e_{33}(\boldsymbol{v})\,\mathrm{d}\boldsymbol{x}\\ &\quad + \varepsilon^{-2}\int_\Omega (3\lambda + 2\mu)\alpha_T\Theta e_{\zeta\zeta}(\boldsymbol{v})\,\mathrm{d}\boldsymbol{x}, \quad \textit{for all } \boldsymbol{v} \in V(\Omega). \end{aligned} \tag{26.20}$$

Problem (26.20) may be written in the following form, corresponding to the scaled principle of virtual work and to the scaled constitutive law, respectively (cf. (2.19), (2.20)):

$$\int_\Omega \sigma_{ij}(\varepsilon)e_{ij}(\boldsymbol{v})\,\mathrm{d}\boldsymbol{x} = \int_\Omega f_i v_i\,\mathrm{d}\boldsymbol{x} + \int_\Gamma g_i v_i\,\mathrm{d}a + \int_{\Gamma_L} h_i v_i\,\mathrm{d}a, \tag{26.21}$$

for all $\boldsymbol{v} \in V(\Omega)$,

$$\begin{aligned} \sigma_{\alpha\beta}(\varepsilon) &= \varepsilon^{-4}[\lambda e_{pp}(\boldsymbol{u}(\varepsilon))\delta_{\alpha\beta} + 2\mu e_{\alpha\beta}(\boldsymbol{u}(\varepsilon))] \\ &\quad + \varepsilon^{-2}[\lambda e_{33}(\boldsymbol{u}(\varepsilon)) - (3\lambda + 2\mu)\alpha_T\Theta]\delta_{\alpha\beta}, \\ \sigma_{3\beta}(\varepsilon) &= \varepsilon^{-2}2\mu e_{3\beta}(\boldsymbol{u}(\varepsilon)), \\ \sigma_{33}(\varepsilon) &= \varepsilon^{-2}\lambda e_{pp}(\boldsymbol{u}(\varepsilon)) + (\lambda + 2\mu)e_{33}(\boldsymbol{u}(\varepsilon)) - (3\lambda + 2\mu)\alpha_T\Theta. \end{aligned} \tag{26.22}$$

Proceeding in the usual way we assume an asymptotic expansion of the form

$$\boldsymbol{u}(\varepsilon) = \boldsymbol{u}^0 + \varepsilon\boldsymbol{u}^1 + \varepsilon^2\boldsymbol{u}^2 + \varepsilon^4\boldsymbol{u}^4 + \text{h.o.t.}, \tag{26.23}$$

which induces a formal expansion of the scaled stress field components $\sigma_{ij}(\varepsilon)$ in the form (cf. (3.2)–(3.5))

$$\begin{aligned} \sigma_{\alpha\beta}(\varepsilon) &= \varepsilon^{-4}\sigma_{\alpha\beta}^{-4} + \varepsilon^{-3}\sigma_{\alpha\beta}^{-3} + \varepsilon^{-2}\sigma_{\alpha\beta}^{-2} + \text{h.o.t.}, \\ \sigma_{3\beta}(\varepsilon) &= \varepsilon^{-2}\sigma_{3\beta}^{-2} + \varepsilon^{-1}\sigma_{3\beta}^{-1} + \varepsilon^{0}\sigma_{3\beta}^{0} + \text{h.o.t.}, \\ \sigma_{33}(\varepsilon) &= \varepsilon^{-2}\sigma_{33}^{-2} + \varepsilon^{-1}\sigma_{33}^{-1} + \varepsilon^{0}\sigma_{33}^{0} + \text{h.o.t.}, \end{aligned} \tag{26.24}$$

where

$$\begin{aligned} \sigma_{\alpha\beta}^{-4} &= \lambda e_{pp}(\boldsymbol{u}^0)\delta_{\alpha\beta} + 2\mu e_{\alpha\beta}(\boldsymbol{u}^0), \\ \sigma_{\alpha\beta}^{-3} &= \lambda e_{pp}(\boldsymbol{u}^1)\delta_{\alpha\beta} + 2\mu e_{\alpha\beta}(\boldsymbol{u}^1), \\ \sigma_{\alpha\beta}^{-2} &= \lambda e_{pp}(\boldsymbol{u}^2)\delta_{\alpha\beta} + 2\mu e_{\alpha\beta}(\boldsymbol{u}^2) + [\lambda e_{33}(\boldsymbol{u}^0) - (3\lambda + 2\mu)\alpha_T\Theta]\delta_{\alpha\beta}, \\ \sigma_{\alpha\beta}^{p} &= \lambda e_{pp}(\boldsymbol{u}^{p+4})\delta_{\alpha\beta} + 2\mu e_{\alpha\beta}(\boldsymbol{u}^{p+4}) + \lambda e_{33}(\boldsymbol{u}^{p+2})\delta_{\alpha\beta}, \quad p \geqslant -1, \end{aligned} \tag{26.25}$$

$$\sigma_{3\beta}^{p} = 2\mu e_{3\beta}(\boldsymbol{u}^{p+2}), \quad p \geqslant -2, \tag{26.26}$$

$$\begin{aligned} \sigma_{33}^{-2} &= \lambda e_{pp}(\boldsymbol{u}^0), \\ \sigma_{33}^{-1} &= \lambda e_{pp}(\boldsymbol{u}^1), \\ \sigma_{33}^{0} &= \lambda e_{pp}(\boldsymbol{u}^2) + (\lambda + 2\mu)e_{33}(\boldsymbol{u}^0) - (3\lambda + 2\mu)\alpha_T\Theta, \\ \sigma_{33}^{p} &= \lambda e_{pp}(\boldsymbol{u}^{p+2}) + (\lambda + 2\mu)e_{33}(\boldsymbol{u}^p), \quad p \geqslant 1. \end{aligned} \tag{26.27}$$

Except for the fact that in the scaled constitutive law (26.22) some additional terms, containing the temperature field Θ, show up, problem (26.21) is exactly of the same nature as the isothermal problem (2.19), (2.20). Consequently, the formal substitution of the asymptotic expansion (26.24) in (26.21) and the setting to zero of the different factors of the different powers of ε leads us to Eqs. (3.7)–(3.12) and, therefore, to the following result (cf. Theorems 4.1, 4.2, 5.1).

THEOREM 26.2. *For the asymptotic expansion* (26.23) *the following properties hold*:

$$\sigma_{\alpha\beta}^{-4} = \sigma_{\alpha\beta}^{-3} = 0 \qquad \sigma_{3\beta}^{-2} = \sigma_{3\beta}^{-1} = 0 \qquad \sigma_{33}^{-2} = \sigma_{33}^{-1} = 0. \tag{26.28}$$

$$e_{\alpha\beta}(\boldsymbol{u}^0) = e_{3\beta}(\boldsymbol{u}^0) = 0, \quad \textit{and consequently } \boldsymbol{u}^0 \in V_{BN}^0(\Omega), \textit{ i.e.},$$
$$u_3^0 = \underline{u}_3^0 - x_\alpha \partial_{33} u_\alpha^0, \qquad \underline{u}_3^0 \in V_0^1(0,L), \qquad u_\alpha \in V_0^2(0,L). \tag{26.29}$$

$$e_{\alpha\beta}(\boldsymbol{u}^1) = e_{3\beta}(\boldsymbol{u}^1) = 0. \tag{26.30}$$

$$\int_\Omega \sigma_{\alpha\beta}^{-2} e_{\alpha\beta}(\boldsymbol{v})\, \mathrm{d}\boldsymbol{x} = 0, \quad \textit{for all } \boldsymbol{v} \in V(\Omega). \tag{26.31}$$

$$\int_\Omega \sigma_{\alpha\beta}^{-1} e_{\alpha\beta}(\boldsymbol{v})\, \mathrm{d}\boldsymbol{x} = 0, \quad \textit{for all } \boldsymbol{v} \in V(\Omega), \tag{26.32}$$

Let us consider that $\boldsymbol{\zeta} \in V(\Omega)$ is a solution of the following problem (cf. (5.10)):

$$\boldsymbol{\zeta} \in V(\Omega),$$
$$S_{\alpha\beta}(\boldsymbol{\zeta}) = \lambda e_{\rho\rho}(\boldsymbol{\zeta})\delta_{\alpha\beta} + 2\mu e_{\alpha\beta}(\boldsymbol{\zeta}) = -[\lambda e_{33}(\boldsymbol{u}^0) - (3\lambda + 2\mu)\alpha_T \Theta]\delta_{\alpha\beta}. \tag{26.33}$$

From (26.25) one then has

$$\sigma_{\alpha\beta}^{-2} = S_{\alpha\beta}(\boldsymbol{u}^2 - \boldsymbol{\zeta}).$$

Taking $\boldsymbol{v} = \boldsymbol{u}^2 - \boldsymbol{\zeta}$ in (26.31) one obtains

$$\int_\Omega [\lambda e_{\rho\rho}(\boldsymbol{u}^2 - \boldsymbol{\zeta}) e_{\rho\rho}(\boldsymbol{u}^2 - \boldsymbol{\zeta}) + 2\mu e_{\alpha\beta}(\boldsymbol{u}^2 - \boldsymbol{\zeta}) e_{\alpha\beta}(\boldsymbol{u}^2 - \boldsymbol{\zeta})\, \mathrm{d}\boldsymbol{x} = 0,$$

and thus $e_{\alpha\beta}(\boldsymbol{u}^2 - \boldsymbol{\zeta}) = 0$. As a consequence, one has $\sigma_{\alpha\beta}^{-2} = 0$ and thus $\boldsymbol{u}^2$ is the solution of (26.33), which is equivalent to saying that

$$e_{\alpha\beta}(\boldsymbol{u}^2) = -\frac{1}{2(\lambda + \mu)}[\lambda e_{33}(\boldsymbol{u}^0) - (3\lambda + 2\mu)\alpha_T \Theta]\delta_{\alpha\beta}, \tag{26.34}$$

or also

$$\partial_1 u_1^2 = \partial_2 u_2^2 = -\frac{\lambda}{2(\lambda + \mu)}(\partial_3 \underline{u}_3^0 - x_\alpha \partial_{33} u_\alpha^0) + \frac{(3\lambda + 2\mu)}{2(\lambda + \mu)}\alpha_T \Theta,$$
$$\partial_1 u_2^2 + \partial_2 u_1^2 = 0. \tag{26.35}$$

As for the isothermal case, the characterization of the scaled displacement components u^2_β is obtained through a direct integration of (26.34). Considering Theorem 4.5 (and also Theorem 5.3) we conclude that the possibility of solving (26.35) directly depends exclusively on the form of the function Θ. In the following theorem a sufficient condition is given for this to be possible.

THEOREM 26.3. *Let us assume that*

$$\underline{u}^0_3 \in H^2(0,L), \quad u^0_\alpha \in H^3(0,L), \tag{26.36}$$

and that there exist functions Θ_α such that

$$\Theta_\alpha \in H^1(\Omega), \quad \partial_1\Theta_1 = \partial_2\Theta_2 = \Theta, \quad \partial_2\Theta_1 + \partial_1\Theta_2 = 0. \tag{26.37}$$

Then, problem (26.35) *admits infinitely many solutions $(u^2_\alpha) \in [H^1(\Omega)]^2$ of the following form*:

$$\begin{aligned} u^2_1 &= z^2_1 + x_2 z^2 - \frac{\lambda}{2(\lambda+\mu)}[x_1\partial_3\underline{u}^0_3 - \tfrac{1}{2}(x_1^2 - x_2^2)\partial_{33}u^0_1 - x_1x_2\partial_{33}u^0_2] \\ &\quad + \frac{3\lambda+2\mu)}{2(\lambda+\mu)}\alpha_T\Theta_1, \\ u^2_2 &= z^2_2 - x_1 z^2 - \frac{\lambda}{2(\lambda+\mu)}[x_2\partial_3\underline{u}^0_3 - x_1x_2\partial_{33}u^0_1 - \tfrac{1}{2}(x_2^2 - x_1^2)\partial_{33}u^0_2] \\ &\quad + \frac{3\lambda+2\mu)}{2(\lambda+\mu)}\alpha_T\Theta_2, \end{aligned} \tag{26.38}$$

with z^2_α, $z^2 \in H^1(0,L)$. Conversely, if $(u^2_\alpha) \in [H^1(\Omega)]^2$ is of the form (26.38) *and is a solution of* (26.35) *then, $\underline{u}^0_3$, u^0_α and Θ_α satisfy* (26.36) *and* (26.37).

PROOF. As in Theorem 4.5 and Corollary 4.2. □

COROLLARY 26.1. *If $\underline{u}^0_3$, u^0_α and Θ_α satisfy* (26.36), (26.37) *then displacement components u^2_β are of the form* (26.38) *with z^2_α, $z^2 \in H^1(0,L)$ verifying (no sum on α, $\beta \neq \alpha$)*:

$$z^2_\alpha(0) = -\frac{\lambda}{4(\lambda+\mu)}(I_\alpha - I_\beta)\partial_{33}u^0_\alpha(0) - \frac{3\lambda+2\mu}{2(\lambda+\mu)}\alpha_T\int_{\omega\times\{0\}}\Theta_\alpha\,\mathrm{d}\omega, \tag{26.39}$$

$$\begin{aligned} z^2(0) &= \frac{\lambda}{2(\lambda+\mu)(I_1+I_2)}[H_2\partial_{33}u^0_1(0) - H_1\partial_{33}u^0_2(0)] \\ &\quad - \frac{3\lambda+2\mu}{2(\lambda+\mu)(I_1+I_2)}\alpha_T\int_{\omega\times\{0\}}(x_2\Theta_1 - x_1\Theta_2)\,\mathrm{d}\omega, \end{aligned} \tag{26.40}$$

and

$$\sigma_{\alpha\beta}^{-2} = 0, \tag{26.41}$$

$$\sigma_{33}^{0} = \frac{\mu(3\lambda + 2\mu)}{\lambda + \mu}\left(e_{33}(\boldsymbol{u}^{0}) - \alpha_T \Theta\right). \tag{26.42}$$

PROOF. It is enough to verify that u_β^2 is necessarily of the form (26.38) in order to be a solution of (26.34). Moreover, $(u_\beta^2) \in W_2(\Omega)$ (cf. (22.34)) since $\boldsymbol{u}^2 \in V(\Omega)$. Properties (26.41) and (26.42) are a direct consequence of (26.25) and (26.27), respectively. □

REMARK 26.4. Condition (26.37) on the temperature field Θ is a restriction which allows the explicit calculation of the second order displacement components and therefore it is important in the models including the thermoelastic torsion effects. We also remark that in the example considered in Remarks 26.1 and 26.3 function Θ^0 solution of (26.19) satisfies this condition because it is harmonic (cf. RUDIN [1970]).

Proceeding as in Section 5 (Theorem 5.5 and Corollary 5.3) it is now possible to prove the following result.

THEOREM 26.4. *One has*

$$\begin{aligned} &\sigma_{\alpha\beta}^{-1} = 0, \\ &\boldsymbol{u}^{2m+1} = \mathbf{0}, \qquad \boldsymbol{\sigma}^{2m+1} = \mathbf{0}, \qquad m \geqslant 0. \end{aligned} \tag{26.43}$$

Consequently, if the temperature field satisfies hypothesis (26.37), everything behaves as if, at the onset, we had supposed an asymptotic expansion for the scaled displacement $\boldsymbol{u}(\varepsilon)$ and stress $\boldsymbol{\sigma}(\varepsilon)$ fields, of the form

$$(\boldsymbol{u}(\varepsilon), \boldsymbol{\sigma}(\varepsilon)) = (\boldsymbol{u}^0, \boldsymbol{\sigma}^0) + \varepsilon^2(\boldsymbol{u}^2, \boldsymbol{\sigma}^2) + \text{h.o.t.} \tag{26.44}$$

In order to obtain the equations characterizing the different components $\boldsymbol{u}^0$, $\boldsymbol{\sigma}^0$, $\boldsymbol{u}^2$, $\boldsymbol{\sigma}^2$, and so on, it is convenient to use the mixed displacement stress approach obtained by scaling (26.12).

Together with hypotheses (6.20)–(6.23) we also introduce the following notations:

$$F(\boldsymbol{v}) = -\int_\Omega f_i v_i \,\mathrm{d}\boldsymbol{x} - \int_\Gamma g_i v_i \,\mathrm{d}a - \int_{\Gamma_L} h_i v_i \,\mathrm{d}a, \quad \text{for all } \boldsymbol{v} \in V(\Omega), \tag{26.45}$$

$$G_0(\boldsymbol{\tau}) = -\alpha_T \int_\Omega \Theta \tau_{33} \,\mathrm{d}\boldsymbol{x}, \quad \text{for all } \boldsymbol{\tau} \in \Sigma(\Omega), \tag{26.46}$$

$$G_2(\boldsymbol{\tau}) = -\alpha_T \int_\Omega \Theta \tau_{\mu\mu} \,\mathrm{d}\boldsymbol{x}, \quad \text{for all } \boldsymbol{\tau} \in \Sigma(\Omega). \tag{26.47}$$

A change of variable in (26.12) gives the following result.

THEOREM 26.5. *The scaled displacement $\boldsymbol{u}(\varepsilon)$ and the scaled stress $\boldsymbol{\sigma}(\varepsilon)$ fields are the unique solution of the following problem*:

$$(\boldsymbol{u}(\varepsilon), \boldsymbol{\sigma}(\varepsilon)) \in V(\Omega) \times \Sigma(\Omega), \tag{26.48}$$

$$a_0(\boldsymbol{\sigma}(\varepsilon), \boldsymbol{\tau}) + \varepsilon^2 a_2(\boldsymbol{\sigma}(\varepsilon), \boldsymbol{\tau}) + \varepsilon^4 a_4(\boldsymbol{\sigma}(\varepsilon), \boldsymbol{\tau}) + b(\boldsymbol{\tau}, \boldsymbol{u}(\varepsilon)) = G_0(\boldsymbol{\tau}) + \varepsilon^2 G_2(\boldsymbol{\tau}), \quad \textit{for all } \boldsymbol{\tau} \in \Sigma(\Omega), \tag{26.49}$$

$$b(\boldsymbol{\sigma}(\varepsilon), \boldsymbol{v}) = F(\boldsymbol{v}), \quad \textit{for all } \boldsymbol{v} \in V(\Omega). \tag{26.50}$$

Substituting expansion (26.44) into (26.48)–(26.50) and considering the coefficients of the different powers in ε we obtain the following set of equations for terms $(\boldsymbol{u}^p, \boldsymbol{\sigma}^p)$, $p = 0, 2, 4$ where the spaces $W_\alpha(\Omega)$ are defined in (22.33) and (22.34):

$$\int_\Omega \frac{1}{E} \sigma_{33}^0 \tau_{33} - \int_\Omega \partial_3 u_3^0 \tau_{33} \, d\boldsymbol{x} = -\alpha_T \int_\Omega \Theta \tau_{33} \, d\boldsymbol{x}, \quad \text{for all } \tau_{33} \in L^2(\Omega), \tag{26.51}$$

$$-\int_\Omega (\partial_3 u_\beta^0 + \partial_\beta u_3^0) \tau_{3\beta} \, d\boldsymbol{x} = 0, \quad \text{for all } (\tau_{3\beta}) \in [L^2(\Omega)]^2, \tag{26.52}$$

$$-\int_\Omega \frac{1}{2} (\partial_\alpha u_\beta^0 + \partial_\beta u_\alpha^0) \tau_{\alpha\beta} \, d\boldsymbol{x} = 0, \quad \text{for all } (\tau_{\alpha\beta}) \in [L^2(\Omega)]_s^4, \tag{26.53}$$

$$\int_\Omega (\sigma_{33}^0 \partial_3 v_3 + \sigma_{3\beta}^0 \partial_\beta v_3) \, d\boldsymbol{x} = \int_\Omega f_3 v_3 \, d\boldsymbol{x} + \int_\Gamma g_3 v_3 \, da + \int_{\Gamma_L} h_3 v_3 \, da, \quad \text{for all } v_3 \in W_1(\Omega), \tag{26.54}$$

$$\int_\Omega (\sigma_{\alpha\beta}^0 \partial_\alpha v_\beta + \sigma_{3\beta}^0 \partial_3 v_\beta) \, d\boldsymbol{x} = \int_\Omega f_\beta v_\beta \, d\boldsymbol{x} + \int_\Gamma g_\beta v_\beta \, da + \int_{\Gamma_L} h_\beta v_\beta \, da, \quad \text{for all } (v_\beta) \in W_2(\Omega), \tag{26.55}$$

$$\int_\Omega \frac{1}{E} \sigma_{33}^2 \tau_{33} - \int_\Omega \partial_3 u_3^2 \tau_{33} \, d\boldsymbol{x} = \int_\Omega \frac{\nu}{E} \sigma_{\mu\mu}^0 \tau_{33} \, d\boldsymbol{x}, \quad \text{for all } \tau_{33} \in L^2(\Omega), \tag{26.56}$$

$$-\int_\Omega (\partial_3 u_\beta^2 + \partial_\beta u_3^2) \tau_{3\beta} \, d\boldsymbol{x} = -\int_\Omega \frac{2(1+\nu)}{E} \sigma_{3\beta}^0 \tau_{3\beta} \, d\boldsymbol{x}, \quad \text{for all } (\tau_{3\beta}) \in [L^2(\Omega)]^2, \tag{26.57}$$

$$-\int_\Omega \frac{1}{2}(\partial_\alpha u_\beta^2 + \partial_\beta u_\alpha^2)\tau_{\alpha\beta}\,\mathrm{d}\boldsymbol{x} = \int_\Omega \frac{\nu}{E}\sigma_{33}^0\tau_{\alpha\beta}\delta_{\alpha\beta}\,\mathrm{d}\boldsymbol{x} - \alpha_T\int_\Omega \Theta\tau_{\mu\mu}\,\mathrm{d}\boldsymbol{x},$$
$$\text{for all } (\tau_{\alpha\beta}) \in [L^2(\Omega)]_s^4, \tag{26.58}$$

$$\int_\Omega (\sigma_{33}^2\partial_3 v_3 + \sigma_{3\beta}^2\partial_\beta v_3)\,\mathrm{d}\boldsymbol{x} = 0, \quad \text{for all } v_3 \in W_1(\Omega), \tag{26.59}$$

$$\int_\Omega (\sigma_{\alpha\beta}^2\partial_\alpha v_\beta + \sigma_{3\beta}^2\partial_3 v_\beta)\,\mathrm{d}\boldsymbol{x} = 0, \quad \text{for all } (v_\beta) \in W_2(\Omega), \tag{26.60}$$

$$\int_\Omega \frac{1}{E}\sigma_{33}^4\tau_{33} - \int_\Omega \partial_3 u_3^4\tau_{33}\,\mathrm{d}\boldsymbol{x} = \int_\Omega \frac{\nu}{E}\sigma_{\mu\mu}^2\tau_{33}\,\mathrm{d}\boldsymbol{x}, \quad \text{for all } \tau_{33} \in L^2(\Omega), \tag{26.61}$$

$$-\int_\Omega (\partial_3 u_\beta^4 + \partial_\beta u_3^4)\tau_{3\beta}\,\mathrm{d}\boldsymbol{x}$$
$$= -\int_\Omega \frac{2(1+\nu)}{E}\int_\Omega \sigma_{3\beta}^2\tau_{3\beta}\,\mathrm{d}\boldsymbol{x}, \quad \text{for all } (\tau_{3\beta}) \in [L^2(\Omega)]^2, \tag{26.62}$$

$$-\int_\Omega \frac{1}{2}(\partial_\alpha u_\beta^4 + \partial_\beta u_\alpha^4)\tau_{\alpha\beta}\,\mathrm{d}\boldsymbol{x}$$
$$= -\int_\Omega \frac{1+\nu}{E}\sigma_{\alpha\beta}^0\tau_{\alpha\beta}\,\mathrm{d}\boldsymbol{x}$$
$$+ \int_\Omega \frac{\nu}{E}(\sigma_{33}^2 + \sigma_{\mu\mu}^0)\tau_{\alpha\beta}\delta_{\alpha\beta}\,\mathrm{d}\boldsymbol{x}, \quad \text{for all } (\tau_{\alpha\beta}) \in [L^2(\Omega)]_s^4, \tag{26.63}$$

$$\int_\Omega (\sigma_{ij}^4\partial_j v_i\,\mathrm{d}\boldsymbol{x} = 0, \quad \text{for all } \boldsymbol{v} \in V(\Omega). \tag{26.64}$$

Proceeding as in Sections 7–9, and imposing some additional hypotheses, it is possible to characterize the different terms in the asymptotic expansion. We remark that equation (26.58) is equivalent to (26.35) and consequently Theorem 26.3 provides a sufficient condition to solve it directly. The next result was obtained by BERMÚDEZ and VIAÑO [1984] — zeroth order terms — and by ÁLVAREZ-VÁZQUEZ and VIAÑO [1989] — higher order terms. The proof is essentially the same as in Theorem 7.1 if one takes forces h_i into account (see Sections 23, 24, 25) and also the temperature (see Theorem 26.3 and Corollary 26.1). For this reason we shall omit the proof. In what follows functions w, Ψ, η_α, θ_α and $\Phi_{\alpha\beta}$ and the associated constants are the same as in Section 7.

THEOREM 26.6. *Let the system of applied forces and the temperature field be such that*

$$\begin{aligned} &f_\alpha \in L^2(\Omega), \qquad f_3 \in H^1[0,L;L^2(\omega)],\\ &g_\alpha \in L^2(\Gamma), \qquad g_3 \in H^1[0,L;L^2(\gamma)].\\ &h_i \in L^2(\Gamma_L) \equiv L^2(\omega),\\ &\Theta \in H^2(0,L;L^2(\omega)], \end{aligned} \tag{26.65}$$

and let there exist functions Θ_α verifying

$$\Theta_\alpha \in H^2[0,L;H^1(\omega)], \qquad \partial_1\Theta_1 = \partial_2\Theta_2 = \Theta, \qquad \partial_1\Theta_2 + \partial_2\Theta_1 = 0. \tag{26.66}$$

Then, Eqs. (26.51)–(26.64) *determine elements* $(\boldsymbol{u}^0, \boldsymbol{\sigma}^0) \in V(\Omega) \times \Sigma(\Omega)$, $\boldsymbol{u}^2 \in V(\Omega)$ *and* $\sigma_{33}^2 \in L^2(\Omega)$, *in a unique way, through the following characterization:*

(i) *Displacements* $\boldsymbol{u}^0$ *verify*

$$\boldsymbol{u}^0 \in V_{BN}^0(\Omega), \qquad u_3^0 = \underline{u}_3^0 - x_\alpha\partial_3 u_\alpha^0, \qquad \underline{u}_3^0 \in V_0^1(0,L), \qquad u_\alpha^0 \in V_0^2(0,L), \tag{26.67}$$

where functions u_α^0 and $\underline{u}_3^0$ are characterized in the following way:

$$\begin{aligned} &u_\beta^0 \in V_0^2(0,L),\\ &\int_0^L EI_\beta\partial_{33}u_\beta^0\partial_{33}v^0\,\mathrm{d}x_3 = \tilde{F}_\beta(v^0) - E\alpha_T\int_0^L\left[\int_\omega x_\beta\Theta\,\mathrm{d}\omega\right]\partial_{33}v^0\,\mathrm{d}x_3, \end{aligned} \tag{26.68}$$

for all $v^0 \in V_0^2(0,L)$ (no sum on β),

$$\begin{aligned} &\underline{u}_3^0 \in V_0^1(0,L),\\ &\int_0^L EA\partial_3\underline{u}_3^0\partial_3 v\,\mathrm{d}x_3 = \tilde{F}_3(v)E\alpha_T\int_0^L\left[\int_\omega \Theta\,\mathrm{d}\omega\right]\partial_3 v\,\mathrm{d}x_3, \end{aligned} \tag{26.69}$$

for all $v \in V_0^1(0,L)$,

with

$$\begin{aligned} \tilde{F}_\beta(\varphi) = &\int_0^L\left[\int_\omega f_\beta\,\mathrm{d}\omega + \int_\gamma g_\beta\,\mathrm{d}\gamma\right]\varphi\,\mathrm{d}x_3\\ &-\int_0^L\left[\int_\omega x_\beta f_3\,\mathrm{d}\omega + \int_\gamma x_\beta g_3\,\mathrm{d}\gamma\right]\partial_3\varphi\,\mathrm{d}x_3 \end{aligned}$$

$$+\left[\int_\omega h_\beta \, d\omega\right]\varphi(L) - \left[\int_\omega x_\beta h_3 \, d\omega\right]\partial_3\varphi(L), \quad \textit{for all } \varphi \in V_0^2(0,L). \tag{26.70}$$

$$\tilde{F}_3(\varphi) = \int_0^L \left[\int_\omega f_3 \, d\omega + \int_\gamma g_3 \, d\gamma\right]\varphi \, dx_3 + \left[\int_\omega h_3 \, d\omega\right]\varphi(L),$$
$$\textit{for all } \varphi \in V_0^1(0,L). \tag{26.71}$$

(ii) *The axial stress and bending moment components are given by*

$$\sigma_{33}^0 = E(\partial_3 u_3^0 - \alpha_T \Theta) = E(\partial_3 \underline{u}_3^0 - x_\alpha \partial_{33} u_\alpha^0 - \alpha_T \Theta), \tag{26.72}$$

$$m_\beta^0 = \int_\omega x_\beta \sigma_{33}^0 \, d\omega = -EI_\beta \partial_{33} u_\beta^0 - E\alpha_T \int_\omega x_\beta \Theta \, d\omega, \quad \textit{no sum on } \beta. \tag{26.73}$$

(iii) *Displacements* $\boldsymbol{u}^2 \in V(\Omega)$ *are of the form*

$$u_1^2 = z_1^2 + x_2 z^2 - \nu(x_1 \partial_3 \underline{u}_3^0 - \Phi_{1\beta}\partial_{33}u_\beta^0) + (1+\nu)\alpha_T \Theta_1, \tag{26.74}$$

$$u_2^2 = z_2^2 - x_1 z^2 - \nu(x_2 \partial_3 \underline{u}_3^0 - \Phi_{2\beta}\partial_{33}u_\beta^0) + (1+\nu)\alpha_T \Theta_2, \tag{26.75}$$

$$u_3^2 = \underline{u}_3^2 - x_\alpha \partial_3 z_\alpha^2 - w\partial_3 z^2 - \nu\left[\tfrac{1}{2}(x_1^2 + x_2^2) - \frac{1}{2A}(I_1 + I_2)\right]\partial_{33}\underline{u}_3^0$$
$$+ [(1+\nu)\eta_\alpha + \nu\theta_\alpha]\partial_{333}u_\alpha^0 + \frac{2(1+\nu)}{E}w^0, \tag{26.76}$$

where functions z^2, w^0, $\underline{u}_3^2$ *and* z_α^2 *are characterized in the following problems* (a)–(d):

(a) *The angle of twist* z^2 *depends only on* x_3 *and is the unique solution of*

$$\begin{aligned}
&z^2 \in H^1(0,L),\\
&\int_0^L \frac{EJ}{2(1+\nu)}\partial_3 z^2 \partial_3 v \, dx_3 = M_3^2(v), \quad \textit{for all } v \in V_0^1(0,L),\\
&z^2(0) = \frac{\nu}{(I_1+I_2)}[H_2\partial_{33}u_1^0(0) - H_1\partial_{33}u_2^0(0)]\\
&\qquad - \frac{(1+\nu)}{I_1+I_2}\alpha_T \int_{\omega\times\{0\}} (x_2\Theta_1 - x_1\Theta_2) \, d\omega,
\end{aligned} \tag{26.77}$$

where

$$M_3^2(v) = \int_0^L \left\{\int_\omega (x_2 f_1 - x_1 f_2) \, d\omega + \int_\gamma (x_2 g_1 - x_1 g_2) \, d\gamma\right.$$

$$+\int_\omega w\partial_3 f_3\,\mathrm{d}\omega + \int_\gamma w\partial_3 g_3\,\mathrm{d}\gamma$$

$$\left.-\frac{E}{2(1+\nu)}[(1+\nu)I_\alpha^w + \nu I_\alpha^\Psi]\partial_{3333}u_\alpha^0\right\}v\,\mathrm{d}x_3$$

$$+\left\{\int_\omega (x_2h_1 - x_1h_2)\,\mathrm{d}\omega + \frac{E}{2(1+\nu)}[(1+\nu)I_\alpha^w + \nu I_\alpha^\Psi]\partial_{333}u_\alpha^0(L)\right.$$

$$\left.-\int_\omega wf_3(L)\,\mathrm{d}\omega + \int_\gamma wg_3(L)\,\mathrm{d}\gamma\right\}v(L)$$

$$+\int_0^L\left\{E\alpha_T\int_\omega w\partial_3\Theta\,\mathrm{d}\omega\right.$$

$$\left.+\tfrac{1}{2}E\alpha_T\int_\omega[\partial_2\Psi\partial_3\Theta_1 - \partial_1\Psi\partial_3\Theta_2]\,\mathrm{d}\omega\right\}\partial_3 v\,\mathrm{d}x_3. \qquad (26.78)$$

(b) *The additional warping* w^0 *is the unique solution of the following problem*: $w^0 \in H^1[0,L;H^1(\omega)]$ *and for all* $x_3 \in [0,L]$,

$$\int_\omega \partial_\beta w^0 \partial_\beta\varphi\,\mathrm{d}\omega$$

$$= \int_\omega f_3\varphi\,\mathrm{d}\omega + \int_\gamma g_3\varphi\,\mathrm{d}\gamma - E\alpha_T\int_\omega \partial_3\Theta\varphi\,\mathrm{d}\omega$$

$$-\frac{1}{A}\left[\int_\omega f_3\,\mathrm{d}\omega + \int_\gamma g_3\,\mathrm{d}\gamma - E\alpha_T\int_\omega\partial_3\Theta\,\mathrm{d}\omega\right]\int_\omega\varphi\,\mathrm{d}\omega \qquad (26.79)$$

$$-\tfrac{1}{2}E\alpha_T\int_\omega \partial_3\Theta_\beta\partial_\beta\varphi\,\mathrm{d}\omega, \quad \textit{for all } \varphi \in H^1(\omega),$$

$$\int_\omega w^0\,\mathrm{d}\omega = 0.$$

(c) *The stretching component* $\underline{u}_3^2$ *depends only on* x_3 *and uniquely solves the following problem*:

$$\underline{u}_3^2 \in V_0^1(0,L),$$

$$\int_0^L EA\partial_3\underline{u}_3^2\partial_3 v\,\mathrm{d}x_3 = \int_0^L \nu G_3^2\partial_3 v\,\mathrm{d}x_3, \quad \textit{for all } v \in V_0^1(0,L), \qquad (26.80)$$

where

$$G_3^2 = -\tfrac{1}{2}EA(I_1 + I_2)\partial_{333}\underline{u}_3^0 + \frac{1}{\nu}EH_\alpha\partial_{3333}u_\alpha^0$$
$$- \int_\omega x_\alpha f_\alpha \, d\omega - \int_\gamma x_\alpha g_\alpha \, d\gamma$$
$$- \int_\omega \frac{1}{2\nu}(x_1^2 + x_2^2)\partial_3 f_3 \, d\omega - \int_\gamma \tfrac{1}{2}(x_1^2 + x_2^2)\partial_3 g_3 \, d\gamma$$
$$- \frac{1}{2\nu}E\alpha_T \int_\omega \partial_{33}\Theta \, d\omega. \tag{26.81}$$

(d) *The bending component* z_α^2 *depends only on variable* x_3 *and is the unique solution of the following variational problem (no sum on* α, $\beta \neq \alpha$*)*:

$$z_\alpha^2 \in H^2(0, L),$$
$$\int_0^L EI_\alpha \partial_{33} z_\alpha^2 \partial_{33} v \, dx_3 = \int_0^L M_\alpha^2 \partial_{33} v \, dx_3, \quad \textit{for all } v \in V_0^2(0, L),$$
$$z_\alpha^2(0) = -\tfrac{1}{2}\nu(I_\alpha - I_\beta)\partial_{33}u_\alpha^0(0) - (1+\nu)\alpha_T \int_{\omega\times\{0\}} \Theta_\alpha \, d\omega, \tag{26.82}$$
$$\partial_3 z_\alpha^2(0) = \frac{1}{I_\alpha}\left\{ \frac{2(1+\nu)}{E} \int_\omega x_\alpha w^0(0) - \tfrac{1}{2}I_\alpha^w \partial_3 z^2(0) + \nu H_\alpha \partial_{33}\underline{u}_3^0(0) \right.$$
$$\left. + [(1+\nu)L_{\alpha\beta}^\eta + \nu L_{\alpha\beta}^\theta]\partial_{333}u_\beta^0(0) \right\},$$

where

$$M_\alpha^2 = E\{(1+\nu)L_{\alpha\beta}^\eta + \nu L_{\alpha\beta}^\theta + \tfrac{1}{2}\nu K_{\alpha\beta}^\eta + \frac{\nu^2}{2(1+\nu)}(K_{\alpha\beta}^\theta + H_3\delta_{\alpha\beta})\}\partial_{3333}u_\beta^0$$
$$+ \nu EH_\alpha \partial_{333}\underline{u}_3^0 - \frac{E}{2(1+\nu)}[(1+\nu)I_\alpha^w + \nu I_\alpha^\Psi]\partial_{33}z^2$$
$$- \int_\omega [(1+\nu)\eta_\alpha + \nu\theta_\alpha]\partial_3 f_3 \, d\omega - \int_\gamma [(1+\nu)\eta_\alpha + \nu\theta_\alpha]\partial_3 g_3 \, d\gamma$$
$$+ \int_\omega \nu\Phi_{\alpha\beta} f_\beta \, d\omega + \int_\gamma \nu\Phi_{\alpha\beta} g_\beta \, d\gamma + \tfrac{1}{2}E\alpha_T \int_\omega \partial_{33}\Theta_\beta \Phi_{\alpha\beta} \, d\omega. \tag{26.83}$$

(iv) *The shear stress components* $\sigma_{3\beta}^0$ *are uniquely determined by*:

$$\sigma_{31}^0 = \frac{E}{2(1+\nu)}\{-\partial_2\Psi\partial_3 z^2 + [(1+\nu)\partial_1\eta_\beta + \nu(\partial_1\theta_\beta + \Phi_{1\beta})]\partial_{333}u_\beta^0\}$$
$$+ \partial_1 w^0 + \tfrac{1}{2}E\alpha_T\partial_3\Theta_1, \tag{26.84}$$

$$\sigma_{32}^0 = \frac{E}{2(1+\nu)}\{\partial_1 \Psi \partial_3 z^2 + [(1+\nu)\partial_2 \eta_\beta + \nu(\partial_2\theta_\beta + \Phi_{2\beta})]\partial_{333}u_\beta^0\}$$
$$+ \partial_2 w^0 + \tfrac{1}{2} E\alpha_T \partial_3 \Theta_2, \tag{26.85}$$

and the shear force components $q_\beta^0 = \int_\omega \sigma_{3\beta}^0$ *are given by (no sum on* β*)*

$$q_\beta^0 = -EI_\beta \partial_{333} u_\beta^0 + \int_\omega x_\beta f_3 \, \mathrm{d}\omega + \int_\gamma x_\beta g_3 \, \mathrm{d}\gamma - E\alpha_T \int_\omega x_\beta \partial_3 \Theta \, \mathrm{d}\omega. \tag{26.86}$$

(v) *The plane stress components* $\sigma_{\alpha\beta}^0$ *are given by*

$$\sigma_{\alpha\beta}^0 = S_{\alpha\beta}(\underline{\boldsymbol{u}}^4) + S_{\alpha\beta}^0 \tag{26.87}$$

where

$$S_{\alpha\beta}^0 = \frac{\nu E}{(1+\nu)(1-2\nu)}\partial_3(u_3^2 - \underline{u}_3^2 + x_\mu \partial_3 z_\mu^2)\delta_{\alpha\beta}, \tag{26.88}$$

$$S_{\alpha\beta}(\underline{\boldsymbol{u}}^4) = \frac{E}{(1+\nu)} e_{\alpha\beta}(\underline{\boldsymbol{u}}^4) + \frac{\nu E}{(1+\nu)(1-2\nu)} e_{\mu\mu}(\underline{\boldsymbol{u}}^4)\delta_{\alpha\beta}, \tag{26.89}$$

and $\underline{\boldsymbol{u}}^4 = (\underline{u}_\alpha^4)$ *is the unique solution of the following plane elasticity problem:*

$$\begin{aligned}
&\underline{\boldsymbol{u}}^4 \in L^2[0,L;(H^1(\omega))^2], \quad \text{and a.e. in } (0,L), \text{ for all } \boldsymbol{\varphi} \in [H^1(\omega)]^2:\\
&\int_\omega S_{\alpha\beta}(\underline{\boldsymbol{u}}^4) e_{\alpha\beta}(\boldsymbol{\varphi})\, \mathrm{d}\omega = \int_\omega f_\beta \varphi_\beta \, \mathrm{d}\omega + \int_\gamma g_\beta \varphi_\beta \, \mathrm{d}\gamma\\
&\qquad + \int_\omega \partial_3 \sigma_{3\beta}^0 \varphi_\beta \, \mathrm{d}\omega - \int_\omega S_{\alpha\beta}^0 e_{\alpha\beta}(\varphi)\, \mathrm{d}\omega,\\
&\int_\omega \underline{u}_\alpha^4 \, \mathrm{d}\omega = \int_\omega [x_2 \underline{u}_1^4 - x_1 \underline{u}_2^4]\, \mathrm{d}\omega = 0.
\end{aligned} \tag{26.90}$$

(vi) *The axial stress* σ_{33}^2 *and the second order bending moment components are given by*

$$\begin{aligned}
\sigma_{33}^2 &= E\partial_3 u_3^2 + \nu \sigma_{\mu\mu}^0\\
&= E\Bigg\{\partial_3 \underline{u}_3^2 - x_\alpha \partial_{33} z_\alpha^2 - w\partial_{33} z^2 - \nu\left[\tfrac{1}{2}(x_1^2 + x_2^2) - \frac{1}{2A}(I_1 + I_2)\right]\partial_{333}\underline{u}_3^0\\
&\qquad + [(1+\nu)\eta_\alpha + \nu\theta_\alpha]\,\partial_{3333}u_\alpha^0\Bigg\} + 2(1+\nu)\partial_3 w^0 + \nu\sigma_{\mu\mu}^0,
\end{aligned} \tag{26.91}$$

$$m_\beta^2 = -EI_\beta \partial_{33} z_\beta^2 + M_\beta^2 \quad (\textit{no sum on } \beta). \tag{26.92}$$

Finally, we analyze the convergence of the solution $(\boldsymbol{u}(\varepsilon), \boldsymbol{\sigma}(\varepsilon))$ of the scaled problem (26.48)–(26.50) as $\varepsilon \to 0$. With a similar technique to the one used in Theorems 10.3 and 10.4 it is possible to prove the following weak convergence result initially owed to BERMÚDEZ and VIAÑO [1984, Theorem 6.2, Corollary 6.1].

THEOREM 26.7. *Let $\boldsymbol{f}$, $\boldsymbol{g}$, $\boldsymbol{h}$ and Θ verify hypothesis* (26.56). *For $0 < \varepsilon \leqslant 1$, let $(\boldsymbol{u}(\varepsilon), \boldsymbol{\sigma}(\varepsilon)) \in V(\Omega) \times \Sigma(\Omega)$ be the solution of the scaled problem* (26.48)–(26.50) *and let $\boldsymbol{u}^0 \in V(\Omega)$, $\sigma_{33}^0 \in L^2(\Omega)$, m_β^0, $q_\beta^0 \in L^2(0, L)$ be the elements defined in Theorem* 26.5. *Then, as $\varepsilon \to 0$ the following weak convergences hold*:

$$\begin{aligned} &\boldsymbol{u}(\varepsilon) \rightharpoonup \boldsymbol{u}^0, && \text{in } V(\Omega),\\ &m_\beta(\varepsilon) \rightharpoonup m_\beta^0, && \text{in } L^2(0, L),\\ &q_\beta(\varepsilon) \rightharpoonup q_\beta^0, && \text{in } L^2(0, L). \end{aligned} \tag{26.93}$$

Moreover, there exists at least a subsequence of $(\boldsymbol{\sigma}(\varepsilon))_{\varepsilon>0}$, denoted with the same index and there exists $\boldsymbol{\Psi} = (\Psi_{ij}) \in \Sigma(\Omega)$ such that

$$\sigma_{33}(\varepsilon) \rightharpoonup \Psi_{33}, \qquad \varepsilon\sigma_{3\beta}(\varepsilon) \rightharpoonup \Psi_{3\beta}, \qquad \varepsilon^2\sigma_{\alpha\beta}(\varepsilon) \rightharpoonup \Psi_{\alpha\beta}, \qquad \text{in } L^2(\Omega), \tag{26.94}$$

with the following properties:

$$\int_\Omega \Psi_{3\beta}\partial_\beta v \,\mathrm{d}\boldsymbol{x} = 0, \quad \text{for all } v \in W_1(\Omega), \tag{26.95}$$

$$\int_\Omega \Psi_{\alpha\beta}\partial_\alpha v_\beta \,\mathrm{d}\boldsymbol{x} = 0, \quad \text{for all } (v_\beta) \in W_2(\Omega), \tag{26.96}$$

$$\int_\omega \Psi_{\alpha\beta} \,\mathrm{d}\omega = \int_\omega x_\gamma \Psi_{\alpha\beta} \,\mathrm{d}\omega = 0, \tag{26.97}$$

$$\Psi_{33} = \sigma_{33}^0 + \nu\Psi_{\mu\mu}. \tag{26.98}$$

PROOF. The proof is done using the same arguments as in the related Theorems 10.2 and 10.3. Consequently we shall only give some details about the a priori estimates (10.4) and (10.5) since now they are obtained in a different way because the right-hand side of (26.49) is no longer zero. Following BERMÚDEZ and VIAÑO [1984, Theorem 6.1] we take $\boldsymbol{\tau} = \boldsymbol{\sigma}(\varepsilon)$ in (26.49) and $\boldsymbol{v} = \boldsymbol{u}(\varepsilon)$ in (26.50). One has

$$\begin{aligned} &a_0(\boldsymbol{\sigma}(\varepsilon), \boldsymbol{\sigma}(\varepsilon)) + \varepsilon^2 a_2(\boldsymbol{\sigma}(\varepsilon), \boldsymbol{\sigma}(\varepsilon)) + \varepsilon^4 a_4(\boldsymbol{\sigma}(\varepsilon), \boldsymbol{\sigma}(\varepsilon))\\ &\quad = G_0(\boldsymbol{\sigma}(\varepsilon)) + \varepsilon^2 G_2(\boldsymbol{\sigma}(\varepsilon)) - F(\boldsymbol{u}(\varepsilon)). \end{aligned} \tag{26.99}$$

Defining $\tilde{\boldsymbol{\sigma}}(\varepsilon) \in \Sigma(\Omega)$ in the following way:

$$\tilde{\sigma}_{33}(\varepsilon) = \sigma_{33}(\varepsilon), \qquad \tilde{\sigma}_{3\beta}(\varepsilon) = \varepsilon\sigma_{3\beta}(\varepsilon), \qquad \tilde{\sigma}_{\alpha\beta}(\varepsilon) = \varepsilon^2\sigma_{\alpha\beta}(\varepsilon),$$

then (26.99) is equivalent to

$$\int_\Omega \left[\frac{1+\nu}{E}\tilde{\sigma}_{ij}(\varepsilon) - \frac{\nu}{E}\tilde{\sigma}_{pp}\delta_{ij}\right]\tilde{\sigma}_{ij}(\varepsilon)\,\mathrm{d}\boldsymbol{x}$$
$$= -\alpha_T \int_\Omega \Theta\tilde{\sigma}_{pp}(\varepsilon)\,\mathrm{d}\boldsymbol{x} + \int_\Omega f_i u_i(\varepsilon)\,\mathrm{d}\boldsymbol{x} + \int_\Gamma g_i u_i(\varepsilon)\,\mathrm{d}a + \int_{\Gamma_L} h_i u_i(\varepsilon)\,\mathrm{d}a. \tag{26.100}$$

Using the positivity properties of coefficients E and ν (cf. (6.8)) one obtains

$$|\tilde{\boldsymbol{\sigma}}(\varepsilon)|^2_{0,\Omega} \leqslant C_1|\tilde{\boldsymbol{\sigma}}(\varepsilon)|_{0,\Omega} + C_2\|\boldsymbol{u}(\varepsilon)\|_{1,\Omega}, \tag{26.101}$$

where constant C_1 depends on Ω, E, ν, α_T and Θ, and where constant C_2 depends on Ω, E, $\nu, \boldsymbol{f}$ and $\boldsymbol{g}$.

On the other hand, for any $\boldsymbol{\tau} \in \Sigma(\Omega)$, from (26.49) we have

$$\int_\Omega e_{ij}(\boldsymbol{u}(\varepsilon))\tau_{ij}\,\mathrm{d}\boldsymbol{x} = a_0(\boldsymbol{\sigma}(\varepsilon),\boldsymbol{\tau}) + \varepsilon^2 a_2(\boldsymbol{\sigma}(\varepsilon),\boldsymbol{\tau}) + \varepsilon^4 a_4(\boldsymbol{\sigma}(\varepsilon),\boldsymbol{\tau})$$
$$- G_0(\boldsymbol{\tau}) - \varepsilon^2 G_2(\boldsymbol{\tau}), \tag{26.102}$$

from which we conclude for $0 < \varepsilon \leqslant 1$ that

$$\int_\Omega e_{ij}(\boldsymbol{u}(\varepsilon))\tau_{ij}\,\mathrm{d}\boldsymbol{x} \leqslant C_3|\tilde{\boldsymbol{\sigma}}(\varepsilon)|_{0,\Omega}|\boldsymbol{\tau}|_{0,\Omega} + C_4|\boldsymbol{\tau}|_{0,\Omega}. \tag{26.103}$$

Using now property (10.3) we obtain

$$\|\boldsymbol{u}(\varepsilon)\|_{1,\Omega} \leqslant C_5|\tilde{\boldsymbol{\sigma}}(\varepsilon)|_{0,\Omega} + C_6. \tag{26.104}$$

Finally estimations (10.4) and (10.5) are obtained combining (26.104) with (26.101) and using Young's inequality. □

We observe that in order to prove existence of $\boldsymbol{u}^0$ and in order to have the weak convergence $\boldsymbol{u}(\varepsilon) \rightharpoonup \boldsymbol{u}^0$ one does not need the temperature Θ to be of the form (26.66). However, we shall see that this condition allows us to obtain the strong convergence. In fact the following theorem (similar to Theorem 10.4), holds.

THEOREM 26.8. *Let $\boldsymbol{f}$, $\boldsymbol{g}$, $\boldsymbol{h}$ and Θ verify hypotheses* (26.65) *and* (26.66). *Assume that*

$$\boldsymbol{\Theta} = (\Theta_\alpha) \in W_2(\Omega). \tag{26.105}$$

Then, as $\varepsilon \to 0$, solution $(\boldsymbol{u}(\varepsilon), \boldsymbol{\sigma}(\varepsilon)) \in V(\Omega) \times \Sigma(\Omega)$ of the scaled problem (26.48)–(26.50) satisfies the following strong convergences:

$$\|\boldsymbol{u}(\varepsilon) - \boldsymbol{u}^0\|_{1,\Omega} \to 0, \tag{26.106}$$

$$|\sigma_{33}(\varepsilon) - \sigma_{33}^0|_{0,\Omega} \to 0, \tag{26.107}$$

$$\varepsilon|\sigma_{3\beta}(\varepsilon)|_{0,\Omega} \to 0, \qquad \varepsilon^2|\sigma_{\alpha\beta}(\varepsilon)|_{0,\Omega} \to 0. \tag{26.108}$$

PROOF. Using the same notation and ideas as in Theorem 10.4, from (26.48)–(26.55) one has

$$\begin{aligned}
&C\{|\sigma_{33}(\varepsilon) - \sigma_{33}^0|_{0,\Omega}^2 + \varepsilon^2|\sigma_{3\beta}(\varepsilon)|_{0,\Omega}^2 + \varepsilon^4|\sigma_{\alpha\beta}(\varepsilon)|_{0,\Omega}^2\} \\
&\quad \leqslant a_0(\boldsymbol{\sigma}(\varepsilon) - \tilde{\boldsymbol{\sigma}}^0, \boldsymbol{\sigma}(\varepsilon) - \tilde{\boldsymbol{\sigma}}^0) + \varepsilon^2 a_2(\boldsymbol{\sigma}(\varepsilon) - \tilde{\boldsymbol{\sigma}}^0, \boldsymbol{\sigma}(\varepsilon) - \tilde{\boldsymbol{\sigma}}^0) \\
&\qquad + \varepsilon^4 a_4(\boldsymbol{\sigma}(\varepsilon) - \tilde{\boldsymbol{\sigma}}^0, \boldsymbol{\sigma}(\varepsilon) - \tilde{\boldsymbol{\sigma}}^0) \\
&\quad = -b(\boldsymbol{\sigma}(\varepsilon) - \tilde{\boldsymbol{\sigma}}^0, \boldsymbol{u}(\varepsilon)) + \varepsilon^2 G_2(\boldsymbol{\sigma}(\varepsilon) - \tilde{\boldsymbol{\sigma}}^0) + b(\boldsymbol{\sigma}(\varepsilon) - \tilde{\boldsymbol{\sigma}}^0, \boldsymbol{u}^0) \\
&\qquad - \varepsilon^2 a_2(\tilde{\boldsymbol{\sigma}}^0, \boldsymbol{\sigma}(\varepsilon) - \tilde{\boldsymbol{\sigma}}^0) - \varepsilon^4 a_4(\tilde{\boldsymbol{\sigma}}^0, \boldsymbol{\sigma}(\varepsilon) - \tilde{\boldsymbol{\sigma}}^0) \\
&\quad = -b(\boldsymbol{\sigma}(\varepsilon), \boldsymbol{u}(\varepsilon) - \boldsymbol{u}^0) + b(\tilde{\boldsymbol{\sigma}}^0, \boldsymbol{u}(\varepsilon) - \boldsymbol{u}^0) \\
&\qquad + \varepsilon^2 G_2(\boldsymbol{\sigma}(\varepsilon) - \tilde{\boldsymbol{\sigma}}^0) - \varepsilon^2 a_2(\tilde{\boldsymbol{\sigma}}^0, \boldsymbol{\sigma}(\varepsilon) - \tilde{\boldsymbol{\sigma}}^0) \\
&\quad = \int_\Omega f_i(u_i(\varepsilon) - u_i^0)\,\mathrm{d}\boldsymbol{x} + \int_\Gamma g_i(u_i(\varepsilon) - u_i^0)\,\mathrm{d}a + \int_{\Gamma_L} h_i(u_i(\varepsilon) - u_i^0)\,\mathrm{d}a \\
&\qquad - \int_\Omega \sigma_{33}^0 e_{33}(\boldsymbol{u}(\varepsilon) - \boldsymbol{u}^0)\,\mathrm{d}\boldsymbol{x} \\
&\qquad - \varepsilon^2\alpha_T \int_\Omega \Theta\sigma_{\mu\mu}(\varepsilon)\,\mathrm{d}\boldsymbol{x} + \varepsilon^2\frac{\nu}{E}\int_\Omega \sigma_{33}^0\sigma_{\mu\mu}(\varepsilon)\,\mathrm{d}\boldsymbol{x}.
\end{aligned} \tag{26.109}$$

From the weak convergence results of the previous theorem we conclude that expression (26.109) converges to

$$\mathscr{L} = \int_\Omega \left(\frac{\nu}{E}\sigma_{33}^0 - \alpha_T\Theta\right)\Psi_{\mu\mu}\,\mathrm{d}\boldsymbol{x}. \tag{26.110}$$

Taking into account the form of σ_{33}^0 given by (26.72) together with properties (26.97) we conclude that $\mathscr{L}$ is of the form:

$$\mathscr{L} = -(1+\nu)\alpha_T \int_\Omega \Theta\Psi_{\mu\mu}\,\mathrm{d}\boldsymbol{x}. \tag{26.111}$$

If Θ is of the form (26.66) and verifies (26.105) then from property (26.96) we conclude that $\mathscr{L} = 0$ and, consequently, convergences (26.107) and (26.108) follow. The convergence of $\boldsymbol{u}(\varepsilon)$ to $\boldsymbol{u}^0$ in $V(\Omega)$ may be established as in Theorem 10.4. □

27. Asymptotic justification of an evolution linear thermoelastic model for rods

We consider that *all the notations and conventions defined in Section* 1, *with respect to beam* Ω^ε, *are now in force*. Moreover, the time derivatives $\partial u/\partial t$ will be denoted by u' and *the dependence on the time variable t of all the functions is implicitly assumed*, except otherwise stated.

We assume that the beam Ω^ε is made of an isotropic, nonhomogeneous, elastic material with Young's modulus $E^\varepsilon(\boldsymbol{x}^\varepsilon)$ and Poisson's ratio $\nu^\varepsilon(\boldsymbol{x}^\varepsilon)$. We suppose that the material has a thermal dilatation coefficient $\alpha_T^\varepsilon(\boldsymbol{x}^\varepsilon)$, a heat conductivity coefficient $k^\varepsilon(\boldsymbol{x}^\varepsilon)$, a specific heat coefficient $\beta^\varepsilon(\boldsymbol{x}^\varepsilon)$ and a specific mass $\rho^\varepsilon(\boldsymbol{x}^\varepsilon)$.

The beam Ω^ε is supposed to be clamped at both ends and kept at a uniform reference temperature T_0^ε at the ends. Other boundary conditions will be considered in the last part of this section. Moreover, it is submitted to body forces $\boldsymbol{f}^\varepsilon(\boldsymbol{x}^\varepsilon)$ applied in Ω^ε, surface forces $\boldsymbol{g}^\varepsilon(\boldsymbol{x}^\varepsilon)$ applied on the lateral surface Γ^ε and initial conditions over the displacement, velocity and temperature fields $(\overline{\boldsymbol{u}}^\varepsilon, \overline{\boldsymbol{v}}^\varepsilon, \overline{\Theta}^\varepsilon)$. Let $\boldsymbol{u}^\varepsilon = (u_i^\varepsilon)$ denote the displacement field, $\boldsymbol{\sigma}^\varepsilon = (\sigma_{ij}^\varepsilon)$ the stress tensor and $\Theta^\varepsilon = \xi^\varepsilon - T_0^\varepsilon$ the increment in the temperature field, where ξ^ε is the absolute temperature of the body and T_0^ε a reference temperature. Then, the dynamic problem corresponding to the elastic response of the beam Ω^ε is governed by following evolutive system of equations (cf. DUVAUT and LIONS [1972b], CARLSON [1972], NOWINSKI [1978], NOWACKI [1986]):

$$\rho^\varepsilon(\boldsymbol{u}_i^\varepsilon)'' - \partial_j^\varepsilon \sigma_{ij}^\varepsilon = f_i^\varepsilon \text{ in } \Omega^\varepsilon \times (0,T), \tag{27.1}$$

$$\beta^\varepsilon(\Theta^\varepsilon)' = \frac{1}{T_0^\varepsilon}\partial_j^\varepsilon(k^\varepsilon \partial_j^\varepsilon \Theta^\varepsilon) - \frac{E^\varepsilon \alpha_T^\varepsilon}{1-2\nu^\varepsilon} e_{kk}^\varepsilon((\boldsymbol{u}^\varepsilon)') \text{ in } \Omega^\varepsilon \times (0,T), \tag{27.2}$$

$$\boldsymbol{u}^\varepsilon = \boldsymbol{0} \text{ on } \Gamma_0^\varepsilon \cup \Gamma_L^\varepsilon \times (0,T), \tag{27.3}$$

$$\Theta^\varepsilon = 0 \text{ on } \Gamma_0^\varepsilon \cup \Gamma_L^\varepsilon \times (0,T), \tag{27.4}$$

$$\sigma_{i\alpha}^\varepsilon n_\alpha = g_i^\varepsilon \text{ on } \Gamma^\varepsilon \times (0,T), \tag{27.5}$$

$$k^\varepsilon \partial_\alpha^\varepsilon \Theta^\varepsilon n_\alpha = 0 \text{ on } \Gamma^\varepsilon \times (0,T), \tag{27.6}$$

$$\boldsymbol{u}^\varepsilon(0) = \overline{\boldsymbol{u}}^\varepsilon \text{ on } \Omega^\varepsilon, \tag{27.7}$$

$$(\boldsymbol{u}^\varepsilon)'(0) = \overline{\boldsymbol{v}}^\varepsilon \text{ on } \Omega^\varepsilon, \tag{27.8}$$

$$\Theta^\varepsilon(0) = \overline{\Theta}^\varepsilon \text{ on } \Omega^\varepsilon, \tag{27.9}$$

where the constitutive equation relating the stress tensor components σ_{ij}^ε to the linearized strain tensor and the temperature increment field Θ^ε is given by the linearized Duhamel–Neumann law (cf. (26.3)):

$$\sigma_{ij}^\varepsilon = \frac{E^\varepsilon}{1+\nu^\varepsilon} e_{ij}^\varepsilon(\boldsymbol{u}^\varepsilon) + \frac{E^\varepsilon \nu^\varepsilon}{(1+\nu^\varepsilon)(1-2\nu^\varepsilon)} e_{kk}^\varepsilon(\boldsymbol{u}^\varepsilon)\delta_{ij} - \frac{E^\varepsilon \alpha_T^\varepsilon}{(1-2\nu^\varepsilon)} \Theta^\varepsilon \delta_{ij}. \tag{27.10}$$

We define the functional spaces of feasible temperatures, displacements and stresses, respectively, as (cf. (1.10))

$$\begin{aligned} W(\Omega^\varepsilon) &= \{z^\varepsilon \in H^1(\Omega^\varepsilon) \colon z^\varepsilon = 0 \text{ on } \Gamma_0^\varepsilon \cup \Gamma_L^\varepsilon\}, \\ V(\Omega^\varepsilon) &= \{\boldsymbol{v}^\varepsilon \in [H^1(\Omega^\varepsilon)]^3 \colon \boldsymbol{v}^\varepsilon = \mathbf{0} \text{ on } \Gamma_0^\varepsilon \cup \Gamma_L^\varepsilon\}, \\ \Sigma(\Omega^\varepsilon) &= [L^2(\Omega^\varepsilon)]_s^9. \end{aligned} \tag{27.11}$$

Then, we can enunciate the following result of existence and uniqueness of a solution for the problem. The proof can be found in the works of MARSDEN and HUGHES [1983] or FRANCFORT [1983], where classical semigroup theory is used. A numerical approach of this problem can be found in VIAÑO [1981] and BERMÚDEZ and VIAÑO [1983].

THEOREM 27.1. *Under the following hypotheses*:

$$\begin{aligned} &\boldsymbol{f}^\varepsilon \in H^1(0,T;[L^2(\Omega^\varepsilon)]^3), \qquad \boldsymbol{g}^\varepsilon \in H^2(0,T;[L^2(\Gamma^\varepsilon)]^3), \\ &\overline{\boldsymbol{u}}^\varepsilon \in [H^2(\Omega^\varepsilon)]^3 \cap V(\Omega^\varepsilon), \qquad \overline{\boldsymbol{v}}^\varepsilon \in V(\Omega^\varepsilon), \qquad \overline{\Theta}^\varepsilon \in H^2(\Omega^\varepsilon) \cap W(\Omega^\varepsilon), \end{aligned} \tag{27.12}$$

there exists a unique solution $(\boldsymbol{u}^\varepsilon, \Theta^\varepsilon)$ *of the evolution problem* (27.1)–(27.10) *satisfying*

$$\begin{aligned} \boldsymbol{u}^\varepsilon &\in C^0([0,T];[H^2(\Omega^\varepsilon)]^3 \cap V(\Omega^\varepsilon)) \cap C^1([0,T];V(\Omega^\varepsilon)) \\ &\quad \cap C^2([0,T];[L^2(\Omega^\varepsilon)]^3), \\ \Theta^\varepsilon &\in C^0([0,T];H^2(\Omega^\varepsilon) \cap W(\Omega^\varepsilon)) \cap C^1([0,T];L^2(\Omega^\varepsilon)). \end{aligned} \tag{27.13}$$

In order to obtain a variational formulation for problem (27.1)–(27.10) we introduce the tensor $\boldsymbol{A}^\varepsilon \boldsymbol{\sigma}^\varepsilon$ defined as follows:

$$(\boldsymbol{A}^\varepsilon \boldsymbol{\sigma}^\varepsilon)_{ij} = \frac{1+\nu^\varepsilon}{E^\varepsilon}\sigma_{ij}^\varepsilon - \frac{\nu^\varepsilon}{E^\varepsilon}\sigma_{kk}^\varepsilon \delta_{ij} = e_{ij}^\varepsilon(\boldsymbol{u}^\varepsilon) - \alpha_T^\varepsilon \Theta^\varepsilon \delta_{ij}. \tag{27.14}$$

Thus, the problem takes the variational form

$$\begin{aligned} &\int_{\Omega^\varepsilon} (\boldsymbol{A}^\varepsilon \boldsymbol{\sigma}^\varepsilon)_{ij}\tau_{ij}^\varepsilon \,\mathrm{d}\boldsymbol{x}^\varepsilon - \int_{\Omega^\varepsilon} e_{ij}^\varepsilon(\boldsymbol{u}^\varepsilon)\tau_{ij}^\varepsilon \,\mathrm{d}\boldsymbol{x}^\varepsilon \\ &\qquad = -\int_{\Omega^\varepsilon} \alpha_T^\varepsilon \Theta^\varepsilon \tau_{kk}^\varepsilon \,\mathrm{d}\boldsymbol{x}^\varepsilon, \quad \text{for all } \boldsymbol{\tau}^\varepsilon \in \Sigma(\Omega^\varepsilon), \end{aligned} \tag{27.15}$$

$$\begin{aligned} &\int_{\Omega^\varepsilon} \rho^\varepsilon (u_i^\varepsilon)'' v_i^\varepsilon \,\mathrm{d}\boldsymbol{x}^\varepsilon + \int_{\Omega^\varepsilon} \sigma_{ij}^\varepsilon e_{ij}^\varepsilon(\boldsymbol{v}^\varepsilon)\,\mathrm{d}\boldsymbol{x}^\varepsilon \\ &\qquad = \int_{\Omega^\varepsilon} f_i^\varepsilon v_i^\varepsilon \,\mathrm{d}\boldsymbol{x}^\varepsilon + \int_{\Gamma^\varepsilon} g_i^\varepsilon v_i^\varepsilon \,\mathrm{d}a^\varepsilon, \quad \text{for all } \boldsymbol{v}^\varepsilon \in V(\Omega^\varepsilon), \end{aligned} \tag{27.16}$$

$$\int_{\Omega^\varepsilon} \beta^\varepsilon(\Theta^\varepsilon)' z^\varepsilon \, \mathrm{d}\boldsymbol{x}^\varepsilon + \int_{\Omega^\varepsilon} \frac{1}{T_0^\varepsilon} k^\varepsilon \partial_j^\varepsilon \Theta^\varepsilon \, \partial_j^\varepsilon z^\varepsilon \, \mathrm{d}\boldsymbol{x}^\varepsilon$$

$$+ \int_{\Omega^\varepsilon} \frac{E^\varepsilon \alpha_T^\varepsilon}{1-2\nu^\varepsilon} e_{kk}^\varepsilon((\boldsymbol{u}^\varepsilon)') z^\varepsilon \, \mathrm{d}\boldsymbol{x}^\varepsilon = 0, \quad \text{for all } z^\varepsilon \in W(\Omega^\varepsilon). \tag{27.17}$$

REMARK 27.1. Equation (27.15) can be expressed in the alternative form

$$\int_{\Omega^\varepsilon} (\boldsymbol{C}^\varepsilon \boldsymbol{u}^\varepsilon)_{ij} e_{ij}^\varepsilon(\boldsymbol{v}^\varepsilon) \, \mathrm{d}\boldsymbol{x}^\varepsilon - \int_{\Omega^\varepsilon} \sigma_{ij}^\varepsilon e_{ij}^\varepsilon(\boldsymbol{v}^\varepsilon) \, \mathrm{d}\boldsymbol{x}^\varepsilon$$

$$= \int_{\Omega^\varepsilon} \frac{E^\varepsilon}{1+\nu^\varepsilon} \alpha_T^\varepsilon \Theta^\varepsilon e_{kk}^\varepsilon(\boldsymbol{v}^\varepsilon) \, \mathrm{d}\boldsymbol{x}^\varepsilon, \quad \text{for all } \boldsymbol{v}^\varepsilon \in V(\Omega^\varepsilon), \tag{27.18}$$

where the new tensor $\boldsymbol{C}^\varepsilon \boldsymbol{u}^\varepsilon$ is defined as

$$\begin{aligned}(\boldsymbol{C}^\varepsilon \boldsymbol{u}^\varepsilon)_{ij} &= \frac{E^\varepsilon}{1+\nu^\varepsilon} e_{ij}^\varepsilon(\boldsymbol{u}^\varepsilon) + \frac{\nu^\varepsilon E^\varepsilon}{1+\nu^\varepsilon)(1-2\nu^\varepsilon)} e_{kk}^\varepsilon(\boldsymbol{u}^\varepsilon)\delta_{ij} \\ &= \sigma_{ij}^\varepsilon + \frac{E^\varepsilon}{1-2\nu^\varepsilon} \alpha_T^\varepsilon \Theta^\varepsilon \delta_{ij}.\end{aligned} \tag{27.19}$$

Following the technique of BERMÚDEZ and VIAÑO [1984] we define an equivalent problem posed on the fixed domain Ω. To this purpose, we make the following change of variable (cf. (2.4)):

$$\Pi^\varepsilon : \boldsymbol{x} = (x_1, x_2, x_3) \in \Omega \to \Pi^\varepsilon(\boldsymbol{x}) = \boldsymbol{x}^\varepsilon = (x_1^\varepsilon, x_2^\varepsilon, x_3^\varepsilon) = (\varepsilon x_1, \varepsilon x_2, x_3) \in \Omega^\varepsilon, \tag{27.20}$$

and a rescaling in the different fields. So, to each scalar field $z^\varepsilon(\boldsymbol{x}^\varepsilon) \in W(\Omega^\varepsilon)$, each vector field $\boldsymbol{v}^\varepsilon(\boldsymbol{x}^\varepsilon) \in V(\Omega^\varepsilon)$ and to each tensor field $\boldsymbol{\tau}^\varepsilon(\boldsymbol{x}^\varepsilon) \in \Sigma(\Omega^\varepsilon)$ we associate the fields $z(\varepsilon)(\boldsymbol{x}) \in W(\Omega)$, $\boldsymbol{v}(\varepsilon)(\boldsymbol{x}) \in V(\Omega)$ and $\boldsymbol{\tau}(\varepsilon)(\boldsymbol{x}) \in \Sigma(\Omega)$, respectively, where (cf. (2.6) and (2.21))

$$\begin{aligned}&z(\varepsilon)(\boldsymbol{x}) = z^\varepsilon(\boldsymbol{x}^\varepsilon), \\ &v_\alpha(\varepsilon)(\boldsymbol{x}) = \varepsilon v_\alpha^\varepsilon(\boldsymbol{x}^\varepsilon), \qquad v_3(\varepsilon)(\boldsymbol{x}) = v_3^\varepsilon(\boldsymbol{x}^\varepsilon), \\ &\tau_{\alpha\beta}(\varepsilon)(\boldsymbol{x}) = \varepsilon^{-2}\tau_{\alpha\beta}^\varepsilon(\boldsymbol{x}^\varepsilon), \qquad \tau_{\alpha 3}(\varepsilon)(\boldsymbol{x}) = \varepsilon^{-1}\tau_{\alpha 3}^\varepsilon(\boldsymbol{x}^\varepsilon), \qquad \tau_{33}(\varepsilon)(\boldsymbol{x}) = \tau_{33}^\varepsilon(\boldsymbol{x}^\varepsilon).\end{aligned} \tag{27.21}$$

Also, for the applied forces $\boldsymbol{f}^\varepsilon$ and $\boldsymbol{g}^\varepsilon$ we define the rescaled forces $\boldsymbol{f}(\varepsilon)$ and $\boldsymbol{g}(\varepsilon)$ given by (cf. (2.8), (2.9))

$$\begin{aligned}&f_\alpha^\varepsilon(\boldsymbol{x}^\varepsilon) = \varepsilon f_\alpha(\varepsilon)(\boldsymbol{x}), \qquad f_3^\varepsilon(\boldsymbol{x}^\varepsilon) = f_3(\varepsilon)(\boldsymbol{x}), \\ &g_\alpha^\varepsilon(\boldsymbol{x}^\varepsilon) = \varepsilon^2 g_\alpha(\varepsilon)(\boldsymbol{x}), \qquad g_3^\varepsilon(\boldsymbol{x}^\varepsilon) = \varepsilon g_3(\varepsilon)(\boldsymbol{x}).\end{aligned} \tag{27.22}$$

Finally, we asume that the coefficients of the thermoelastic material satisfy the following hypotheses, where E, ν, β, k, α, and ρ are functions defined in Ω independent of ε:

$$\begin{array}{lll} E^\varepsilon(\boldsymbol{x}^\varepsilon) = E(\boldsymbol{x}), \quad E > 0, & \nu^\varepsilon(\boldsymbol{x}^\varepsilon) = \nu(\boldsymbol{x}), \quad 0 < \nu < \frac{1}{2}, \\ \beta^\varepsilon(\boldsymbol{x}^\varepsilon) = \beta(\boldsymbol{x}), \quad \beta > 0, & k^\varepsilon(\boldsymbol{x}^\varepsilon) = k(\boldsymbol{x}), \quad k > 0, \\ \alpha_T^\varepsilon(\boldsymbol{x}^\varepsilon) = \alpha_T(\boldsymbol{x}), & \rho^\varepsilon(\boldsymbol{x}^\varepsilon) = \varepsilon^2\rho(\boldsymbol{x}), \quad \rho > 0, \\ T_0^\varepsilon = T_0. & \end{array} \tag{27.23}$$

REMARK 27.2. For the sake of simplicity, in what follows we assume that E, ν, β, k, α, and ρ are constants. However, we must remark that in the general case the technique herein used will still hold with the only extra assumption:

$$E,\ \nu,\ \beta,\ k,\ \alpha_T,\ \rho\ \in C^\infty(\overline{\Omega}).$$

The ε^2 dependence of ρ^ε is necessary in order to obtain a limit model sensitive to inertia effects (cf. RAOULT [1980]), and it is similar to other scalings appearing in several physical models (SANCHEZ-PALENCIA [1980], CIARLET and KESAVAN [1981]).

The ε dependence of body and surface forces is assumed in order to obtain a loading term in the limit problem.

Taking the new scalings into account, system (27.1)–(27.10) turns into

$$\begin{aligned} &\rho(u_\alpha(\varepsilon))'' - \partial_j\sigma_{\alpha j}(\varepsilon) = f_\alpha(\varepsilon) \quad \text{in } \Omega \times (0,T), \\ &\varepsilon^2\rho(u_3(\varepsilon))'' - \partial_j\sigma_{3j}(\varepsilon) = f_3(\varepsilon) \quad \text{in } \Omega \times (0,T), \\ &\beta(\Theta(\varepsilon))' = \frac{k}{T_0}[\varepsilon^{-2}\partial_{\alpha\alpha}\Theta(\varepsilon) + \partial_{33}\Theta(\varepsilon)] \\ &\qquad - \frac{E\alpha_T}{1-2\nu}[\varepsilon^{-2}e_{\gamma\gamma}((\boldsymbol{u}(\varepsilon))') + e_{33}((\boldsymbol{u}(\varepsilon))')] \quad \text{in } \Omega \times (0,T), \end{aligned} \tag{27.24}$$

$$\begin{aligned} &\boldsymbol{u}(\varepsilon) = \boldsymbol{0}, \quad \text{on } \Gamma_0 \cup \Gamma_L \times (0,T), \\ &\Theta(\varepsilon) = 0 \quad \text{on } \Gamma_0 \cup \Gamma_L \times (0,T), \\ &\sigma_{\alpha\beta}(\varepsilon)n_\beta = g_\alpha(\varepsilon) \quad \text{on } \Gamma \times (0,T), \\ &\sigma_{3\beta}(\varepsilon)n_\beta = g_3(\varepsilon) \quad \text{on } \Gamma \times (0,T), \\ &k\partial_\alpha\Theta(\varepsilon)n_\alpha = 0 \quad \text{on } \Gamma^\varepsilon \times (0,T), \end{aligned} \tag{27.25}$$

with initial conditions

$$\begin{aligned} &\boldsymbol{u}(\varepsilon)(0) = \overline{\boldsymbol{u}}(\varepsilon) \quad \text{in } \Omega, \\ &(\boldsymbol{u}(\varepsilon))'(0) = \overline{\boldsymbol{v}}(\varepsilon) \quad \text{in } \Omega, \\ &\Theta(\varepsilon)(0) = \overline{\Theta}(\varepsilon) \quad \text{in } \Omega, \end{aligned} \tag{27.26}$$

and where $\sigma_{ij}(\varepsilon)$ is given by *the rescaled constitutive law*:

$$\varepsilon^2\sigma_{\alpha\beta}(\varepsilon) = \varepsilon^{-2}\frac{E}{1+\nu}e_{\alpha\beta}(\boldsymbol{u}(\varepsilon))$$

$$
\begin{aligned}
&+\frac{E\nu}{(1+\nu)(1-2\nu)}[\varepsilon^{-2}e_{\gamma\gamma}(\boldsymbol{u}(\varepsilon))+e_{33}(\boldsymbol{u}(\varepsilon))]\delta_{\alpha\beta}\\
&-\frac{E\alpha_T}{(1-2\nu)}\Theta(\varepsilon)\delta_{\alpha\beta},
\end{aligned}
$$

$$
\varepsilon\sigma_{3\beta}(\varepsilon)=\varepsilon^{-1}\frac{E}{1+\nu}e_{3\beta}(\boldsymbol{u}(\varepsilon)), \tag{27.27}
$$

$$
\begin{aligned}
\sigma_{33}(\varepsilon)=\;&\frac{E}{1+\nu}e_{33}(\boldsymbol{u}(\varepsilon))+\frac{E\nu}{(1+\nu)(1-2\nu)}[\varepsilon^{-2}e_{\gamma\gamma}(\boldsymbol{u}(\varepsilon))+e_{33}(\boldsymbol{u}(\varepsilon))]\\
&-\frac{E\alpha_T}{(1-2\nu)}\Theta(\varepsilon).
\end{aligned}
$$

This system of evolutive equations can be written in the following variational formulation:

$$
a(\varepsilon)(\boldsymbol{\sigma}(\varepsilon),\boldsymbol{\tau})+b(\varepsilon)(\boldsymbol{u}(\varepsilon),\boldsymbol{\tau})=G(\varepsilon)(\Theta(\varepsilon),\boldsymbol{\tau}),\quad \text{for all } \boldsymbol{\tau}\in\Sigma(\Omega), \tag{27.28}
$$

$$
d(\varepsilon)((\boldsymbol{u}(\varepsilon))'',\boldsymbol{v})-b(\varepsilon)(\boldsymbol{v},\boldsymbol{\sigma}(\varepsilon))=F(\varepsilon)(\boldsymbol{v}),\quad \text{for all } \boldsymbol{v}\in V(\Omega), \tag{27.29}
$$

$$
e((\Theta(\varepsilon))',z)+l(\varepsilon)(\Theta(\varepsilon),z)+\tilde{G}(\varepsilon)(z,(\boldsymbol{u}(\varepsilon))')=0,\quad \text{for all } z\in W(\Omega), \tag{27.30}
$$

where

$$
\begin{aligned}
&a(\varepsilon)(\boldsymbol{\sigma},\boldsymbol{\tau})=a_0(\boldsymbol{\sigma},\boldsymbol{\tau})+\varepsilon^2a_2(\boldsymbol{\sigma},\boldsymbol{\tau})+\varepsilon^4a_4(\boldsymbol{\sigma},\boldsymbol{\tau}),\\
&G(\varepsilon)(z,\boldsymbol{\tau})=G_0(z,\boldsymbol{\tau})+\varepsilon^2G_2(z,\boldsymbol{\tau}),\\
&d(\varepsilon)(\boldsymbol{u},\boldsymbol{v})=d_0(\boldsymbol{u},\boldsymbol{v})+\varepsilon^2d_2(\boldsymbol{u},\boldsymbol{v}),\\
&F(\varepsilon)(\boldsymbol{v})=\int_\Omega f_i(\varepsilon)v_i\,\mathrm{d}\boldsymbol{x}+\int_\Gamma g_i(\varepsilon)v_i\,\mathrm{d}a,\\
&l(\varepsilon)(\Theta,z)=\varepsilon^{-2}l_2(\Theta,z)+l_0(\Theta,z),\\
&\tilde{G}(\varepsilon)(z,\boldsymbol{v})=\varepsilon^{-2}\tilde{G}_2(z,\boldsymbol{v})+\tilde{G}_0(z,\boldsymbol{v}),
\end{aligned} \tag{27.31}
$$

where for all $\boldsymbol{\sigma},\boldsymbol{\tau}\in\Sigma(\Omega)$, $\boldsymbol{u},\boldsymbol{v}\in V(\Omega)$, $\Theta,z\in W(\Omega)$,

$$
a_0(\boldsymbol{\sigma},\boldsymbol{\tau})=\int_\Omega\frac{1}{E}\sigma_{33}\tau_{33}\,\mathrm{d}\boldsymbol{x}, \tag{27.32}
$$

$$
a_2(\boldsymbol{\sigma},\boldsymbol{\tau})=\int_\Omega[\frac{2(1+\nu)}{E}\sigma_{3\beta}\tau_{3\beta}-\frac{\nu}{E}(\sigma_{33}\tau_{\mu\mu}+\sigma_{\alpha\alpha}\tau_{33})]\,\mathrm{d}\boldsymbol{x}, \tag{27.33}
$$

$$
a_4(\boldsymbol{\sigma},\boldsymbol{\tau})=\int_\Omega(\frac{1+\nu}{E}\sigma_{\alpha\beta}\tau_{\alpha\beta}-\frac{\nu}{E}\sigma_{\alpha\alpha}\tau_{\beta\beta})\,\mathrm{d}\boldsymbol{x}, \tag{27.34}
$$

$$b(\boldsymbol{\tau},\boldsymbol{u}) = -\int_\Omega e_{ij}(\boldsymbol{u})\tau_{ij}\,\mathrm{d}\boldsymbol{x}, \tag{27.35}$$

$$G_0(z,\boldsymbol{\tau}) = -\int_\Omega \alpha_T z\tau_{33}\,\mathrm{d}\boldsymbol{x}, \qquad G_2(z,\boldsymbol{\tau}) = -\int_\Omega \alpha_T z\tau_{\alpha\alpha}\,\mathrm{d}\boldsymbol{x}, \tag{27.36}$$

$$d_0(\boldsymbol{u},\boldsymbol{v}) = \int_\Omega \rho u_\alpha v_\alpha\,\mathrm{d}\boldsymbol{x}, \qquad d_2(\boldsymbol{u},\boldsymbol{v}) = \int_\Omega \rho u_3 v_3\,\mathrm{d}\boldsymbol{x}, \tag{27.37}$$

$$e(\Theta,z) = \int_\Omega \beta\Theta z\,\mathrm{d}\boldsymbol{x}, \tag{27.38}$$

$$l_2(\Theta,z) = \int_\Omega \frac{k}{T_0}\partial_\alpha\Theta\partial_\alpha z\,\mathrm{d}\boldsymbol{x}, \qquad l_0(\Theta,z) = \int_\Omega \frac{k}{T_0}\partial_3\Theta\partial_3 z\,\mathrm{d}\boldsymbol{x}, \tag{27.39}$$

$$\tilde{G}_2(z,\boldsymbol{v}) = \int_\Omega \frac{E\alpha_T}{1-2\nu}e_{\alpha\alpha}(\boldsymbol{v})z\,\mathrm{d}\boldsymbol{x}, \quad \tilde{G}_0(z,\boldsymbol{v}) = \int_\Omega \frac{E\alpha_T}{1-2\nu}e_{33}(\boldsymbol{v})z\,\mathrm{d}\boldsymbol{x}. \tag{27.40}$$

REMARK 27.3. Equation (27.28) can also be written in an alternative form as

$$c(\varepsilon)(\boldsymbol{u}(\varepsilon),\boldsymbol{v}) + b(\boldsymbol{v},\boldsymbol{\sigma}(\varepsilon)) = \tilde{G}(\varepsilon)(\Theta,\boldsymbol{v}), \quad \text{for all } \boldsymbol{v}\in V(\Omega), \tag{27.41}$$

where

$$c(\varepsilon)(\boldsymbol{u},\boldsymbol{v}) = c_0(\boldsymbol{u},\boldsymbol{v}) + \varepsilon^{-2}c_2(\boldsymbol{u},\boldsymbol{v}) + \varepsilon^{-4}c_4(\boldsymbol{u},\boldsymbol{v}), \tag{27.42}$$

and for all $\boldsymbol{u},\boldsymbol{v}\in V(\Omega)$

$$c_0(\boldsymbol{u},\boldsymbol{v}) = \int_\Omega \frac{E(1-\nu)}{(1+\nu)(1-2\nu)}e_{33}(\boldsymbol{u})e_{33}(\boldsymbol{v})\,\mathrm{d}\boldsymbol{x}, \tag{27.43}$$

$$\begin{aligned} c_2(\boldsymbol{u},\boldsymbol{v}) = \int_\Omega \{&\frac{E\nu}{(1+\nu)(1-2\nu)}[e_{33}(\boldsymbol{u})e_{\alpha\alpha}(\boldsymbol{v}) + e_{\alpha\alpha}(\boldsymbol{u})e_{33}(\boldsymbol{v})] \\ &+ \frac{E}{1+\nu}e_{3\alpha}(\boldsymbol{u})e_{3\alpha}(\boldsymbol{v})\,\mathrm{d}\boldsymbol{x}\}, \end{aligned} \tag{27.44}$$

$$c_4(\boldsymbol{u},\boldsymbol{v}) = \int_\Omega [\frac{E}{1+\nu}e_{\alpha\beta}(\boldsymbol{u})e_{\alpha\beta}(\boldsymbol{v}) + \frac{\nu E}{(1+\nu)(1-2\nu)}e_{\alpha\alpha}(\boldsymbol{u})e_{\beta\beta}(\boldsymbol{v})]\,\mathrm{d}\boldsymbol{x}.$$

Then, the results formulated in Theorem 27.1 become as follows

THEOREM 27.2. *Under the hypotheses*

$$\begin{aligned} &\boldsymbol{f}(\varepsilon)\in H^1(0,T;[L^2(\Omega)]^3), \qquad \boldsymbol{g}(\varepsilon)\in H^2(0,T;[L^2(\Gamma)]^3), \\ &\overline{\boldsymbol{u}}(\varepsilon)\in H^2(\Omega)]^3\cap V(\Omega), \qquad \overline{\boldsymbol{v}}(\varepsilon)\in V(\Omega), \qquad \overline{\Theta}(\varepsilon)\in H^2(\Omega)\cap W(\Omega), \end{aligned} \tag{27.45}$$

system (27.28)–(27.30), *with initial conditions* (27.26), *admits a unique solution* $(\boldsymbol{u}(\varepsilon), \Theta(\varepsilon))$ *verifying*

$$\begin{aligned}
&\boldsymbol{u}(\varepsilon) \in C^0([0,T];[H^2(\Omega)]^3 \cap V(\Omega)) \cap C^1([0,T];V(\Omega)) \cap C^2([0,T];[L^2(\Omega)]^3),\\
&\Theta(\varepsilon) \in C^0([0,T];H^2(\Omega) \cap W(\Omega)) \cap C^1([0,T];L^2(\Omega)).
\end{aligned} \tag{27.46}$$

Our aim now is to establish under some mild assumptions on the system of applied forces and initial data the convergence of the solution of problem (27.26)–(27.30) to the solution of a limit problem which we characterize. This result is a consequence of the following.

THEOREM 27.3. *If we assume the following convergences as* ε *goes to zero*:

$$\begin{aligned}
&\overline{\boldsymbol{u}}(\varepsilon) \rightharpoonup \overline{\boldsymbol{u}}^0 \quad \textit{weakly in } V(\Omega),\\
&\overline{\boldsymbol{v}}_\alpha(\varepsilon) \rightharpoonup \overline{\boldsymbol{v}}^0_\alpha, \qquad \varepsilon\overline{\boldsymbol{v}}_3(\varepsilon) \rightharpoonup \overline{\boldsymbol{v}}^0_3, \qquad \overline{\Theta}(\varepsilon) \rightharpoonup \overline{\Theta}^0, \quad \varepsilon^{-2}e_{\alpha\beta}(\overline{\boldsymbol{u}}(\varepsilon)) \rightharpoonup \overline{e}^0_{\alpha\beta},\\
&\varepsilon^{-1}e_{3\alpha}(\overline{\boldsymbol{u}}(\varepsilon)) \rightharpoonup \overline{e}^0_{3\alpha}, \qquad e_{33}(\overline{\boldsymbol{u}}(\varepsilon)) \rightharpoonup \overline{e}^0_{33}, \quad \textit{weakly in } L^2(\Omega),\\
&f_i(\varepsilon) \rightharpoonup f^0_i \quad \textit{weakly in } H^1(0,T;L^2(\Omega)),\\
&g_i(\varepsilon) \rightharpoonup g^0_i \quad \textit{weakly in } H^1(0,T;L^2(\Gamma)),
\end{aligned} \tag{27.47}$$

then, the following properties are satisfied independently of ε:

$$\begin{aligned}
&\boldsymbol{u}(\varepsilon) \quad \textit{is bounded in } L^\infty(0,T;V(\Omega)),\\
&(\boldsymbol{u}_\alpha(\varepsilon))', \qquad \varepsilon(\boldsymbol{u}_3(\varepsilon))', \qquad \varepsilon^{-2}e_{\alpha\beta}(\boldsymbol{u}(\varepsilon)), \qquad \varepsilon^{-1}e_{\alpha 3}(\boldsymbol{u}(\varepsilon)), \qquad e_{33}(\boldsymbol{u}(\varepsilon)),\\
&\varepsilon^2\sigma_{\alpha\beta}(\varepsilon), \qquad \varepsilon\sigma_{\alpha 3}(\varepsilon), \qquad \sigma_{33}(\varepsilon) \quad \textit{are bounded in } L^\infty(0,T;L^2(\Omega)),\\
&\Theta(\varepsilon) \quad \textit{is bounded in } L^\infty(0,T;L^2(\Omega)) \cap L^2(0,T;W(\Omega)),\\
&\varepsilon^{-1}\partial_\alpha\Theta(\varepsilon) \quad \textit{is bounded in } L^2(0,T;L^2(\Omega)).
\end{aligned} \tag{27.48}$$

PROOF. Taking $\boldsymbol{v} = (\boldsymbol{u}(\varepsilon))'$ and $z = \Theta(\varepsilon)$ in (27.29), (27.30) and (27.41), we obtain

$$\begin{aligned}
&c(\varepsilon)(\boldsymbol{u}(\varepsilon), (\boldsymbol{u}(\varepsilon))') + b((\boldsymbol{u}(\varepsilon))', \boldsymbol{\sigma}(\varepsilon)) = \tilde{G}(\varepsilon)(\Theta(\varepsilon), (\boldsymbol{u}(\varepsilon))'),\\
&d(\varepsilon)((\boldsymbol{u}(\varepsilon))'', (\boldsymbol{u}(\varepsilon))') - b((\boldsymbol{u}(\varepsilon))', \boldsymbol{\sigma}(\varepsilon)) = F(\varepsilon)((\boldsymbol{u}(\varepsilon))'),\\
&e((\Theta(\varepsilon))', \Theta(\varepsilon)) + l(\varepsilon)(\Theta(\varepsilon)), \Theta(\varepsilon)) + \tilde{G}(\varepsilon)(\Theta(\varepsilon), (\boldsymbol{u}(\varepsilon))') = 0.
\end{aligned}$$

Adding these three expressions and integrating the result over $[0,T]$ for $0 \leqslant$

$t \leqslant T$ we get

$$\int_\Omega \rho \left[\sum_{\alpha=1}^{2} |(u_\alpha(\varepsilon))'|^2 + |\varepsilon(u_3(\varepsilon))'|^2 \right] \mathrm{d}\boldsymbol{x} + c(\varepsilon)(\boldsymbol{u}(\varepsilon), \boldsymbol{u}(\varepsilon)) + \int_\Omega \beta |\Theta(\varepsilon)|^2 \,\mathrm{d}\boldsymbol{x}$$
$$+ 2 \int_0^t \int_\Omega \frac{k}{T_0} \left[\sum_{\alpha=1}^{2} |\varepsilon^{-1} \partial_\alpha \Theta(\varepsilon)|^2 + \partial_3 \Theta(\varepsilon)|^2 \right] \mathrm{d}\boldsymbol{x}$$
$$= \int_\Omega \rho \left[\sum_{\alpha=1}^{2} |\overline{v}_\alpha(\varepsilon)|^2 + |\varepsilon \overline{v}_3(\varepsilon)|^2 \right] \mathrm{d}\boldsymbol{x} + c(\varepsilon)(\overline{\boldsymbol{u}}(\varepsilon), \overline{\boldsymbol{u}}(\varepsilon)) + \int_\Omega \beta |\overline{\Theta}(\varepsilon)|^2 \,\mathrm{d}\boldsymbol{x}$$
$$+ 2 \left[\int_\Omega f_i(\varepsilon) u_i(\varepsilon) \,\mathrm{d}\boldsymbol{x} + \int_\Gamma g_i(\varepsilon) u_i(\varepsilon) \,\mathrm{d}a \right]_0^t$$
$$- 2 \int_0^t \left[\int_\Omega (f_i(\varepsilon))' u_i(\varepsilon) \,\mathrm{d}\boldsymbol{x} + \int_\Gamma (g_i(\varepsilon))' u_i(\varepsilon) \,\mathrm{d}a \right].$$

Due to the properties of the coefficients of the material and to hypothesis (27.47) we obtain the existence of a constant C_1, independent of ε, such that

$$\sum_{\alpha=1}^{2} \|(u_\alpha(\varepsilon))'\|^2_{L^\infty(0,T;L^2(\Omega))} + \|\varepsilon(u_3(\varepsilon))'\|^2_{L^\infty(0,T;L^2(\Omega))}$$
$$+ \sum_{\alpha,\beta=1}^{2} \|\varepsilon^{-2} e_{\alpha\beta}(\boldsymbol{u}(\varepsilon))\|^2_{L^\infty(0,T;L^2(\Omega))} + \sum_{\alpha=1}^{2} \|\varepsilon^{-1} e_{\alpha 3}(\boldsymbol{u}(\varepsilon))\|^2_{L^\infty(0,T;L^2(\Omega))}$$
$$+ \|e_{33}(\boldsymbol{u}(\varepsilon))\|^2_{L^\infty(0,T;L^2(\Omega))} + \|\Theta(\varepsilon)\|^2_{L^\infty(0,T;L^2(\Omega))}$$
$$+ \sum_{\alpha=1}^{2} \|\varepsilon^{-1} \partial_\alpha \Theta(\varepsilon)\|^2_{L^2(0,T;L^2(\Omega))} + \|\partial_3 \Theta(\varepsilon)\|^2_{L^2(0,T;L^2(\Omega))}$$
$$\leqslant C_1 \Bigg[\sum_{\alpha=1}^{2} \|\overline{v}_\alpha(\varepsilon)\|^2_{L^2(\Omega)} + \|\varepsilon \overline{v}_3(\varepsilon)\|^2_{L^2(\Omega)} + \sum_{\alpha,\beta=1}^{2} \|\varepsilon^{-2} e_{\alpha\beta}(\overline{\boldsymbol{u}}(\varepsilon))\|^2_{L^2(\Omega)}$$
$$+ \sum_{\alpha=1}^{2} \|\varepsilon^{-1} e_{\alpha 3}(\overline{\boldsymbol{u}}(\varepsilon))\|^2_{L^2(\Omega)} + \|e_{33}(\overline{\boldsymbol{u}}(\varepsilon))\|^2_{L^2(\Omega)} + \|\overline{\Theta}(\varepsilon)\|^2_{L^2(\Omega)}$$
$$+ \left(\|\boldsymbol{f}(\varepsilon)\|_{H^1(0,T;[L^2(\Omega)]^3)} + \|\boldsymbol{g}(\varepsilon)\|_{H^1(0,T;[L^2(\Gamma)]^3)}\right) \|\boldsymbol{u}(\varepsilon)\|_{L^\infty(0,T;V(\Omega))} \Bigg].$$

Using Poincaré's and Korn's inequalities, we have that for any $\varepsilon < 1$,

$$\|\boldsymbol{u}(\varepsilon)\|^2_{L^\infty(0,T;V(\Omega))} \leqslant C_2\Big[\sum_{\alpha,\beta=1}^{2} \|\varepsilon^{-2}e_{\alpha\beta}(\boldsymbol{u}(\varepsilon))\|^2_{L^\infty(0,T;L^2(\Omega))} + \sum_{\alpha=1}^{2}\|\varepsilon^{-1}e_{\alpha 3}(\boldsymbol{u}(\varepsilon))\|^2_{L^\infty(0,T;L^2(\Omega))} + \|e_{33}(\boldsymbol{u}(\varepsilon))\|_{L^\infty(0,T;L^2(\Omega))}\Big].$$

Thus, we obtain the desired estimates on $\boldsymbol{u}(\varepsilon)$, $\varepsilon^{-2}e_{\alpha\beta}(\boldsymbol{u}(\varepsilon))$, $\varepsilon^{-1}e_{\alpha 3}(\boldsymbol{u}(\varepsilon))$, $e_{33}(\boldsymbol{u}(\varepsilon))$, $(u_\alpha(\varepsilon))'$, $\varepsilon(u_3(\varepsilon))'$, $\Theta(\varepsilon)$ and $\varepsilon^{-1}\partial_\alpha\Theta(\varepsilon)$.

If we now take $\boldsymbol{\tau} = \boldsymbol{\sigma}(\varepsilon)$ in (27.28) we get

$$a(\varepsilon)(\boldsymbol{\sigma}(\varepsilon), \boldsymbol{\sigma}(\varepsilon)) + b(\boldsymbol{u}(\varepsilon), \boldsymbol{\sigma}(\varepsilon)) = G(\varepsilon)(\Theta(\varepsilon), \boldsymbol{\sigma}(\varepsilon)).$$

Thus,

$$\begin{aligned}
&\sum_{\alpha,\beta=1}^{2} \|\varepsilon^2\sigma_{\alpha\beta}(\varepsilon)\|^2_{L^\infty(0,T;L^2(\Omega))} + \sum_{\alpha=1}^{2}\|\varepsilon\sigma_{\alpha 3}(\varepsilon)\|^2_{L^\infty(0,T;L^2(\Omega))} + \|\sigma_{33}(\varepsilon)\|^2_{L^\infty(0,T;L^2(\Omega))} \\
&\quad\leqslant C_3\Big[\sum_{\alpha,\beta=1}^{2} \|\varepsilon^{-2}e_{\alpha\beta}(\boldsymbol{u}(\varepsilon))\|_{L^\infty(0,T;L^2(\Omega))} + \sum_{\alpha=1}^{2}\|\varepsilon^{-1}e_{\alpha 3}(\boldsymbol{u}(\varepsilon))\|_{L^\infty(0,T;L^2(\Omega))} + \|e_{33}(\boldsymbol{u}(\varepsilon))\|_{L^\infty(0,T;L^2(\Omega))} \\
&\qquad + \|\Theta(\varepsilon)\|_{L^\infty(0,T;L^2(\Omega))}\Big] \\
&\qquad\times\Big[\sum_{\alpha,\beta=1}^{2} \|\varepsilon^2\sigma_{\alpha\beta}(\varepsilon)\|_{L^\infty(0,T;L^2(\Omega))} + \sum_{\alpha=1}^{2}\|\varepsilon\sigma_{\alpha 3}(\varepsilon)\|_{L^\infty(0,T;L^2(\Omega))} + \|\sigma_{33}(\varepsilon)\|_{L^\infty(0,T;L^2(\Omega))}\Big].
\end{aligned}$$

In this way we obtain the estimates for $\varepsilon^2\sigma_{\alpha\beta}(\varepsilon)$, $\varepsilon\sigma_{\alpha 3}(\varepsilon)$ and $\sigma_{33}(\varepsilon)$. □

REMARK 27.4. The hypothesis required on the initial displacement $\overline{\boldsymbol{u}}(\varepsilon)$ implies that

$$e_{\alpha\beta}(\overline{\boldsymbol{u}}_0) = e_{\alpha 3}(\overline{\boldsymbol{u}}_0) = 0,$$

and then (see Theorem 4.3) $\overline{\boldsymbol{u}}_0 \in V_{\mathrm{BN}}(\Omega)$ where $V_{\mathrm{BN}}(\Omega)$ is the functional space defined by (4.13). Thus,

$$\begin{aligned}
&\overline{u}^0_\alpha \in H^2_0(0,L), \\
&\overline{u}^0_3 = \underline{\overline{u}}^0_3 - x_\alpha\partial_3\overline{u}^0_\alpha, \quad \text{where } \underline{\overline{u}}^0_3 \in H^1_0(0,L).
\end{aligned} \tag{27.49}$$

REMARK 27.5. It would be interesting to obtain error estimates on $(\boldsymbol{u}(\varepsilon))''$, but the lack of bounds for $\sigma_{\alpha 3}(\varepsilon)$ and $\sigma_{\alpha\beta}(\varepsilon)$ makes this fact impossible, because

$$\int_\Omega \rho(u_\alpha(\varepsilon))'' v_\alpha \, \mathrm{d}\boldsymbol{x} + \varepsilon^2 \int_\Omega \rho(u_3(\varepsilon))'' v_3 \, \mathrm{d}\boldsymbol{x} + \int_\Omega \sigma_{ij}(\varepsilon) e_{ij}(\boldsymbol{v}) \, \mathrm{d}\boldsymbol{x}$$
$$= \int_\Omega f_i(\varepsilon) v_i \, \mathrm{d}\boldsymbol{x} + \int_\Gamma g_i(\varepsilon) v_i \, \mathrm{d}a, \quad \text{for all } \boldsymbol{v} \in V(\Omega).$$

But if we restrict to $\boldsymbol{v} \in V_{\mathrm{BN}}(\Omega)$, this expression yields

$$\int_\Omega \rho(u_\alpha(\varepsilon))'' v_\alpha \, \mathrm{d}\boldsymbol{x} + \varepsilon^2 \int_\Omega \rho(u_3(\varepsilon))'' v_3 \, \mathrm{d}\boldsymbol{x} + \int_\Omega \sigma_{33}(\varepsilon) e_{33}(\boldsymbol{v}) \, \mathrm{d}\boldsymbol{x}$$
$$= \int_\Omega f_i(\varepsilon) v_i \, \mathrm{d}\boldsymbol{x} + \int_\Gamma g_i(\varepsilon) v_i \, \mathrm{d}a, \quad \text{for all } \boldsymbol{v} \in V_{\mathrm{BN}}(\Omega).$$

Consequently, from (27.47), we obtain that $((u_1(\varepsilon))'', (u_2(\varepsilon))'', \varepsilon^2(u_3(\varepsilon))'')$ is bounded in the space $L^\infty(0, T; (V_{\mathrm{BN}}(\Omega))')$ independently of ε.

COROLLARY 27.1. *Under hypothesis* (27.47) *we obtain the existence of at least a subsequence, still denoted in the same way, verifying the following convergences*:

$$\begin{aligned}
&\boldsymbol{u}(\varepsilon) \rightharpoonup \boldsymbol{u}^0 \quad \textit{weak - * in } L^\infty(0, T; V(\Omega)),\\
&(\boldsymbol{u}_\alpha(\varepsilon))' \rightharpoonup (u_\alpha^0)', \qquad \varepsilon(\boldsymbol{u}_3(\varepsilon))' \rightharpoonup 0,\\
&\varepsilon^{-2} e_{\alpha\beta}(\boldsymbol{u}(\varepsilon)) \rightharpoonup e_{\alpha\beta}^0, \qquad \varepsilon^{-1} e_{\alpha 3}(\boldsymbol{u}(\varepsilon)) \rightharpoonup e_{\alpha 3}^0, \qquad e_{33}(\boldsymbol{u}(\varepsilon)) \rightharpoonup e_{33}^0,\\
&\varepsilon^2 \sigma_{\alpha\beta}(\varepsilon) \rightharpoonup \sigma_{\alpha\beta}^0, \qquad \varepsilon \sigma_{\alpha 3}(\varepsilon) \rightharpoonup \sigma_{\alpha 3}^0,\\
&\sigma_{33}(\varepsilon) \rightharpoonup \sigma_{33}^0 \qquad \textit{weak - * in } L^\infty(0, T; L^2(\Omega)),\\
&\Theta(\varepsilon) \rightharpoonup \Theta^0 \quad \textit{weak - * in } L^\infty(0, T; L^2(\Omega)) \textit{ and weakly in } L^2(0, T; W(\Omega)),\\
&\varepsilon^{-1} \partial_\alpha \Theta(\varepsilon) \rightharpoonup r_\alpha^0 \quad \textit{weakly in } L^2(0, T; L^2(\Omega)).
\end{aligned} \tag{27.50}$$

REMARK 27.6. As a consequence of Corollary 27.1 we can characterize the linearized strain tensor corresponding to the limit displacement field $\boldsymbol{u}^0$. Firstly, from the weak - * convergence of $\boldsymbol{u}(\varepsilon)$ to $\boldsymbol{u}^0$ in $L^\infty(0, T; V(\Omega))$ we deduce that

$$e_{ij}(\boldsymbol{u}(\varepsilon)) \rightharpoonup e_{ij}(\boldsymbol{u}^0) \quad \text{weak - * in } L^\infty(0, T; L^2(\Omega)).$$

Bearing in mind that we have the weak - * convergence of $e_{33}(\boldsymbol{u}(\varepsilon))$ to e_{33}^0, we can conclude that $e_{33}(u^0) = e_{33}^0$.

Finally, because of the weak - * convergences of $\varepsilon^{-2} e_{\alpha\beta}(\boldsymbol{u}(\varepsilon))$ and $\varepsilon^{-1} e_{\alpha 3}(\boldsymbol{u}(\varepsilon))$, we deduce that $e_{\alpha\beta}(\boldsymbol{u}^0) = e_{\alpha 3}(\boldsymbol{u}^0) = 0$.

Thus,

$$\boldsymbol{u}^0 \in L^\infty(0,T;V_{\mathrm{BN}}(\Omega))$$

that is,

$$\begin{aligned} &u_\alpha^0 \in L^\infty(0,T;H_0^2(0,L)),\\ &u_3^0 = \underline{u}_3^0 - x_\alpha \partial_3 u_\alpha^0, \quad \text{where } \underline{u}_3^0 \in L^\infty(0,T;H_0^1(0,L)). \end{aligned} \tag{27.51}$$

On the other hand, the estimate $\varepsilon^{-1}\partial_\alpha \Theta(\varepsilon)$ assures that Θ^0 *is independent of* x_α.

Our main goal in this section is to characterize the limit problem solved by $(\boldsymbol{u}^0, \Theta^0)$, as well as to determine, as far as possible, the terms σ_{ij}^0, e_{ij}^0 and r_α^0.

Let $v \in W(\Omega)$, $\varphi \in C_0^\infty(0,T)$. Taking $\boldsymbol{w}(t,\boldsymbol{x}) = \varphi(t)(0,0,\varepsilon v(\boldsymbol{x}))$ as a test function in (27.29), integrating over the time interval $(0,T)$ and letting ε tend to zero we obtain that $\sigma_{\alpha 3}^0(t,x_3)$ verifies a.e. in $[0,T]\times[0,L]$:

$$\begin{aligned} \partial_\alpha \sigma_{\alpha 3}^0(t,x_3) &= 0 \quad \text{in } \omega,\\ \sigma_{\alpha 3}^0(t,x_3) n_\alpha &= 0 \quad \text{on } \gamma. \end{aligned} \tag{27.52}$$

The same procedure with $w(t,\boldsymbol{x}) = \varphi(t)(\varepsilon^2 v_1(\boldsymbol{x}), \varepsilon^2 v_2(\boldsymbol{x}), 0)$ where $(v_1,v_2) \in [W(\Omega)]^2$, $\varphi \in C_0^\infty(0,T)$ leads us to obtain from (27.29) that $\sigma_{\alpha\beta}^0(t,x_3)$ is a.e. in $[0,T]\times[0,L]$ a solution of the problem

$$\begin{aligned} \partial_\alpha \sigma_{\alpha\beta}^0(t,x_3) &= 0 \quad \text{in } \omega,\\ \sigma_{\alpha\beta}^0(t,x_3) n_\alpha &= 0 \quad \text{on } \gamma. \end{aligned} \tag{27.53}$$

Taking now $\boldsymbol{v} \in V_{\mathrm{BN}}(\Omega)$ in (27.29) and passing to the limit we have a.e. in $[0,T]$

$$\begin{aligned} &\int_\Omega \rho (u_\alpha^0)'' v_\alpha \,\mathrm{d}\boldsymbol{x} + \int_\Omega \sigma_{33}^0 \partial_3 v_3 \,\mathrm{d}\boldsymbol{x}\\ &\quad = \int_\Omega f_i^0 v_i \,\mathrm{d}\boldsymbol{x} + \int_\Gamma g_i^0 v_i \,\mathrm{d}a, \quad \text{for all } \boldsymbol{v} \in V_{\mathrm{BN}}(\Omega). \end{aligned} \tag{27.54}$$

Choosing $v_\alpha = 0$, we obtain from (27.54)

$$\partial_3\left(\int_\omega \sigma_{33}^0 \,\mathrm{d}\omega\right) + \int_\omega f_3^0 \,\mathrm{d}\omega + \int_\gamma g_3^0 \,\mathrm{d}\gamma = 0, \quad \text{a.e. in } [0,T]\times[0,L]. \tag{27.55}$$

In the same way, taking v_3 such that $\underline{v}_3 = 0$, we get from (27.54)

$$\int_\omega \rho(u_\alpha^0)''\,\mathrm{d}\omega - \partial_{33}\Big(\int_\omega x_\alpha \sigma_{33}^0\,\mathrm{d}\omega\Big)$$
$$= \int_\omega f_\alpha^0\,\mathrm{d}\omega + \int_\gamma g_\alpha^0\,\mathrm{d}\gamma$$
$$+ \partial_3\Big(\int_\omega x_\alpha f_3^0\,\mathrm{d}\omega + \int_\gamma x_\alpha g_3^0\,\mathrm{d}\gamma\Big) \quad \text{a.e. in } [0,T]\times[0,L]. \tag{27.56}$$

We conclude from (27.56) that $(\int_\omega \rho u_\alpha^0\,\mathrm{d}\omega)'' \in L^\infty(0,T;H^{-2}(0,L))$ and, consequently, $(\int_\omega \rho u_\alpha^0\,\mathrm{d}\omega)' \in C^0([0,T];H^{-2}(0,L))$. In order to evaluate the trace $(\int_\omega \rho u_\alpha^0\,\mathrm{d}\omega)'(0) \in H^{-2}(0,L)$ we take $\boldsymbol{v} \in V_{\mathrm{BN}}(\Omega)$, $\eta \in C_0^\infty([0,s))$ with $0 < s \leqslant t$, and choose $\boldsymbol{w}(t,\boldsymbol{x}) = \eta(t)\boldsymbol{v}(\boldsymbol{x})$ as test function in (27.29). After integrating over the interval $(0,s)$, using the integration by parts formula and letting ε tend to zero we obtain

$$\left(\int_\omega \rho u_\alpha^0\,\mathrm{d}\omega\right)'(0) = \int_\omega \rho \bar{v}_\alpha^0\,\mathrm{d}\omega. \tag{27.57}$$

On the other hand, as $u_\alpha(\varepsilon)$ converges weak - * to u_α^0 in $W^{1,\infty}(0,T;L^2(\Omega))$, $u_\alpha^0(\varepsilon)(0)$ converges to $u_\alpha^0(0)$ weakly in $L^2(\Omega)$, i.e.,

$$u_\alpha^0(0) = \bar{u}_\alpha^0. \tag{27.58}$$

In order to study tensor e_{ij}^0 we take $\boldsymbol{\psi} \in \Sigma(\Omega)$, $\varphi \in C_0^\infty(0,T)$ and choose $\boldsymbol{\tau}(t,\boldsymbol{x}) = \varphi(t)\boldsymbol{\psi}(\boldsymbol{x})$ as the test function in (27.28). Then, passing to the limit, we obtain

$$e_{33}^0 = \frac{1}{E}\sigma_{33}^0 - \frac{\nu}{E}\sigma_{\alpha\alpha}^0 + \alpha_T\Theta^0, \tag{27.59}$$

$$e_{\alpha 3}^0 = \frac{2(1+\nu)}{E}\sigma_{\alpha 3}^0, \tag{27.60}$$

$$e_{\alpha\beta}^0 = \frac{1+\nu}{E}\sigma_{\alpha\beta}^0 - \frac{\nu}{E}\sigma_{pp}^0\delta_{\alpha\beta} + \alpha_T\Theta^0\delta_{\alpha\beta}. \tag{27.61}$$

REMARK 27.7. It is possible to obtain an explicit expression for σ_{33}^0, which will not be the case for $\sigma_{\alpha 3}^0$ and $\sigma_{\alpha\beta}^0$, which we only know to be solutions of systems (27.52) and (27.53), respectively.

Considering suitable test functions it can be proved (see BERMÚDEZ and VIAÑO [1984] and (10.17)–(10.19) in Theorem 10.2) that

$$\int_\omega \sigma_{\alpha 3}^0\,\mathrm{d}\omega = 0, \qquad \int_\omega x_\alpha \sigma_{\alpha 3}^0\,\mathrm{d}\omega = 0,$$
$$\int_\omega \sigma_{\alpha\beta}^0\,\mathrm{d}\omega = 0, \qquad \int_\omega x_\mu \sigma_{\alpha\beta}^0\,\mathrm{d}\omega = 0. \tag{27.62}$$

These properties will be very useful in later computations.

Taking $\zeta \in W(\Omega)$, $\varphi \in C_0^\infty(0,T)$ and choosing $z(t,\boldsymbol{x}) = \varepsilon\varphi(t)\zeta(\boldsymbol{x})$ as test function in (27.30) we obtain that, a.e. in $[0,T] \times [0,L]$, $r_\alpha^0(t,x_3)$ is a solution of the problem

$$\begin{aligned} &\partial_\alpha(kr_\alpha^0(t,x_3)) = 0 \quad \text{in } \omega,\\ &kr_\alpha^0(t,x_3)n_\alpha = 0 \quad \text{on } \gamma. \end{aligned} \tag{27.63}$$

If we now take $z(t,\boldsymbol{x}) = \varphi(t)\zeta(\boldsymbol{x})$ with $\zeta \in H_0^1(0,L)$, as a test function, we get

$$\beta(\Theta^0)' = \frac{k}{T_0}\partial_{33}\Theta^0 - \int_\omega \frac{\alpha_T E}{1-2\nu}(e_{pp}^0)'\,\mathrm{d}\omega. \tag{27.64}$$

Then, if we regard expressions (27.59)–(27.61), Remark 27.7 and the fact that Θ^0 is independent of x_α (Remark 27.6), expression (27.64) yields

$$B(\Theta^0)' = \frac{k}{T_0}\partial_{33}\Theta^0 - \alpha_T E\int_\omega e_{33}((\boldsymbol{u}^0)')\,\mathrm{d}\omega, \tag{27.65}$$

where the effective coefficient B is given by

$$B = \beta + \frac{2\alpha_T^2 E(1+\nu)}{1-2\nu}. \tag{27.66}$$

This means that $\{B\Theta^0 + \alpha_T E\int_\omega e_{33}(\boldsymbol{u}^0)\,\mathrm{d}\omega\}$ belongs to $H^1(0,T;H^{-1}(0,L))$. In order to compute its trace at $t=0$ as an element of $H^{-1}(0,L)$, we use the same method as for the trace of $(\int_\omega \rho u_\alpha^0\,\mathrm{d}\omega)'$ at $t=0$. Thus, we obtain

$$\{B\Theta^0 + \alpha_T E\int_\omega e_{33}(\boldsymbol{u}^0)\,\mathrm{d}\omega\}(0) = \beta\int_\omega \overline{\Theta}^0\,\mathrm{d}\omega + \frac{\alpha_T E}{1-2\nu}\int_\omega \bar{e}_{pp}^0\,\mathrm{d}\omega. \tag{27.67}$$

On the other hand, since $u_3^0 = \underline{u}_3^0 - x_\alpha\partial_3 u_\alpha^0$, from (27.59) we obtain

$$\sigma_{33}^0 = E\partial_3\underline{u}_3^0 - Ex_\alpha\partial_{33}u_\alpha^0 - \alpha_T E\Theta^0 + \nu\sigma_{\mu\mu}^0, \tag{27.68}$$

and, as a consequence of the independence of $\underline{u}_3^0$ and Θ^0 with respect to x_α and of Remark 27.7, we have

$$\int_\omega \sigma_{33}^0\,\mathrm{d}\omega = E\partial_3\underline{u}_3^0 - \alpha_T E\Theta^0, \tag{27.69}$$

$$\int_\omega x_\alpha\sigma_{33}^0\,\mathrm{d}\omega = -EI_\alpha\partial_{33}u_\alpha^0 \quad \text{(no sum on } \alpha). \tag{27.70}$$

Finally, since

$$e_{33}(\boldsymbol{u}^0) = \partial_3 \underline{u}_3^0 - x_\alpha \partial_{33} u_\alpha^0, \tag{27.71}$$

we get

$$\int_\omega e_{33}((\boldsymbol{u}^0)')\,\mathrm{d}\omega = \partial_3(\underline{u}_3^0)'. \tag{27.72}$$

Summarizing, from Eqs. (27.56)–(27.58) and (27.70) we obtain that u_α^0 is a solution of the problem (see Theorem 27.5 for uniqueness)

$$\begin{aligned}
&\rho(u_\alpha^0)'' - EI_\alpha \partial_{3333} u_\alpha^0 \\
&\quad = \int_\omega f_\alpha^0\,\mathrm{d}\omega + \int_\gamma g_\alpha^0\,\mathrm{d}\gamma + \partial_3\left[\int_\omega x_\alpha f_3^0\,\mathrm{d}\omega + \int_\gamma x_\alpha g_3^0\,\mathrm{d}\gamma\right] = \mathscr{F}_\alpha^0, \\
&u_\alpha^0(t,0) = u_\alpha^0(t,L) = 0, \\
&\partial_3 u_\alpha^0(t,0) = \partial_3 u_\alpha^0(t,L) = 0, \\
&u_\alpha^0(0,x_3) = \overline{u}_\alpha^0(x_3), \\
&(u_\alpha^0)'(0,x_3) = \int_\omega \overline{v}_\alpha^0(x_3)\,\mathrm{d}\omega.
\end{aligned} \tag{27.73}$$

From Eqs. (27.55), (27.65), (27.67), (27.69) and (27.72), we obtain that $(\underline{u}_3^0, \Theta^0)$ is a solution of the problem (see Theorem 27.5 for uniqueness)

$$\begin{aligned}
&-E\partial_{33}\underline{u}_3^0 + \alpha_T E \partial_3 \Theta^0 = \int_\omega f_3^0\,\mathrm{d}\omega + \int_\gamma g_3^0\,\mathrm{d}\gamma = \mathscr{F}_3^0, \\
&B(\Theta^0)' = \frac{k}{T_0}\partial_{33}\Theta^0 - \alpha_T E \partial_3(\underline{u}_3^0)', \\
&\underline{u}_3^0(t,0) = \underline{u}_3^0(t,L) = 0, \\
&\Theta^0(t,0) = \Theta^0(t,L) = 0, \\
&B\Theta^0(0,x_3) = \beta \int_\omega \overline{\Theta}^0\,\mathrm{d}\omega + \frac{\alpha_T E}{1-2\nu}\int_\omega \overline{e}_{pp}^0\,\mathrm{d}\omega - \alpha_T E \int_\omega \partial_3 \underline{u}_3^0(0,x_3)\,\mathrm{d}\omega.
\end{aligned} \tag{27.74}$$

In this way, we have proved the following result.

THEOREM 27.4. *If hypotheses* (27.45) *and* (27.47) *hold true and if*

$$\partial_3 f_3^0 \in H^1(0,T;L^2(\Omega)), \qquad \partial_3 g_3^0 \in H^1(0,T;L^2(\Gamma)), \tag{27.75}$$

then, for the subsequences of Corollary 27.1, we deduce the following convergences:

$$
\begin{aligned}
&\boldsymbol{u}(\varepsilon) \rightharpoonup \boldsymbol{u}^0 \quad \textit{weak - * in } L^\infty(0,T;V(\Omega)),\\
&(\boldsymbol{u}_\alpha(\varepsilon))' \rightharpoonup (u_\alpha^0)' \qquad \varepsilon(\boldsymbol{u}_3(\varepsilon))' \rightharpoonup 0,\\
&\varepsilon^2\sigma_{\alpha\beta}(\varepsilon) \rightharpoonup \sigma_{\alpha\beta}^0, \qquad \varepsilon\sigma_{\alpha3}(\varepsilon) \rightharpoonup \sigma_{\alpha3}^0, \sigma_{33}(\varepsilon) \rightharpoonup \sigma_{33}^0,\\
&\varepsilon^{-2}e_{\alpha\beta}(\boldsymbol{u}(\varepsilon)) \rightharpoonup \frac{1+\nu}{E}\sigma_{\alpha\beta}^0 - \frac{\nu}{E}\sigma_{pp}^0\delta_{\alpha\beta} + \alpha\Theta^0\delta_{\alpha\beta},\\
&\varepsilon^{-1}e_{\alpha3}(\boldsymbol{u}(\varepsilon)) \rightharpoonup \frac{2(1+\nu)}{E}\sigma_{\alpha3}^0, \qquad \textit{weak - * in } L^\infty(0,T;L^2(\Omega)),\\
&\Theta(\varepsilon) \rightharpoonup \Theta^0 \quad \textit{weak - * in } L^\infty(0,T;L^2(\Omega)) \quad \textit{and weakly in } L^2(0,T;W(\Omega)),\\
&\varepsilon^{-1}\partial_\alpha\Theta(\varepsilon) \rightharpoonup r_\alpha^0 \quad \textit{weakly in } L^2(0,T;L^2(\Omega)),
\end{aligned}
\tag{27.76}
$$

where

(i)

$$\boldsymbol{u}^0 \in L^\infty(0,T;V_{\mathrm{BN}}(\Omega)), \qquad u_3^0 = \underline{u}_3^0 - x_\alpha\partial_3 u_\alpha^0,$$

u_α^0, $\underline{u}_3^0$ *and* Θ^0 *being solutions of problems* (27.73) *and* (27.74).

(ii) $\sigma_{\alpha\beta}^0$, $\sigma_{\alpha3}^0$ *and* r_α^0 *are solutions, respectively, of problems* (27.52), (27.53) *and* (27.63).

(iii) σ_{33}^0 *is given by expression* (27.68).

In the same way we can study the convergence of the bending moments $m_\alpha(\varepsilon) = \int_\omega x_\alpha\sigma_{33}(\varepsilon)\,\mathrm{d}\omega$, the axial force $q_3(\varepsilon) = \int_\omega \sigma_{33}(\varepsilon)\,\mathrm{d}\omega$ and the shear forces $q_\alpha(\varepsilon) = \int_\omega \sigma_{\alpha3}(\varepsilon)\,\mathrm{d}\omega$. Thus, the following result can be enunciated.

COROLLARY 27.2. *With the hypotheses of Theorem 27.4, we obtain, for the same subsequences, the following convergences as* ε *tends to zero*:

$$
\begin{aligned}
&q_3(\varepsilon) \rightharpoonup q_3^0, \qquad m_\alpha(\varepsilon) \rightharpoonup m_\alpha^0 \qquad \textit{weak - * in } L^\infty(0,T;L^2(0,L)),\\
&q_\alpha(\varepsilon) \rightharpoonup q_\alpha^0 \quad \textit{weakly in } L^2(0,T;L^2(0,L)),
\end{aligned}
\tag{27.77}
$$

where

$$
\begin{aligned}
&q_3^0 = E\partial_3\underline{u}_3^0 - E\alpha_T\Theta^0,\\
&m_\alpha^0 = -EI_\alpha\partial_{33}u_\alpha^0 \quad \textit{no sum on } \alpha,\\
&q_\alpha^0 = \int_\omega x_\alpha f_3^0\,\mathrm{d}\omega + \int_\gamma x_\alpha g_3^0\,\mathrm{d}\gamma + \partial_3 m_\alpha^0.
\end{aligned}
\tag{27.78}
$$

PROOF. The convergence of $q_3(\varepsilon)$ and $m_\alpha(\varepsilon)$ is a direct consequence of expressions (27.69) and (27.70). Let $v \in H_0^1(0,L)$, $\varphi \in C_0^\infty(0,T)$. Taking $\boldsymbol{w}(t,\boldsymbol{x}) =$

$\varphi(t)(0,0,x_\alpha v(x_3))$ as a test function in (27.29), integrating over $(0,T)$ and letting ε tend to zero we obtain the convergence of $q_\alpha(\varepsilon)$. □

Once secured the convergence of problem (27.1)–(27.10) to problem (27.73), (27.74), we are going to study the properties of the latter briefly.

The following theorem establishes the existence and uniqueness of solution for problem (27.73).

THEOREM 27.5. *If the system of applied loads possesses the following regularity:*

$$\begin{aligned} &f_i^0 \in L^2(0,T;L^2(\Omega)), \qquad g_i^0 \in L^2(0,T;L^2(\Gamma)),\\ &\partial_3 f_3^0 \in L^2(0,T;L^2(\Omega)), \qquad \partial_3 g_3^0 \in L^2(0,T;L^2(\Gamma)), \end{aligned} \tag{27.79}$$

and if the initial displacement and velocity verify, respectively,

$$\overline{\boldsymbol{u}}^0 \in V_{\mathrm{BN}}(\Omega), \qquad \overline{\boldsymbol{v}}^0 \in [L^2(\Omega)]^3, \tag{27.80}$$

then, problem (27.73) *admits a unique solution such that*

$$u_\alpha^0 \in C^0([0,T];H_0^2(0,L)) \cap C^1([0,T];L^2(0,L)). \tag{27.81}$$

PROOF. Due to the regularity required on the system of applied forces, the element

$$\mathcal{F}_\alpha^0 = \int_\omega f_\alpha^0\,\mathrm{d}\omega + \int_\gamma g_\alpha^0\,\mathrm{d}\gamma + \partial_3\left[\int_\omega x_\alpha f_3^0\,\mathrm{d}\omega + \int_\gamma x_\alpha g_3^0\,\mathrm{d}\gamma\right] \in L^2(0,T;L^2(0,L)).$$

On the other hand, we know that

$$\overline{u}_\alpha^0 \in H_0^2(0,L), \qquad \int_\omega \overline{v}_\alpha^0\,\mathrm{d}\omega \in L^2(0,L).$$

Then, in a classical way (cf. LIONS [1969], RAVIART and THOMAS [1983]) we conclude the existence and uniqueness of solution of problem (27.73) with the desired regularity. □

The second limit problem (27.74) consists of two coupled equations: a parabolic equation and a quasi-static one. For the study of the existence of solution we make use of the following result owed to Phillips and Lumer (cf. YOSIDA [1980]).

LEMMA 27.1. *Let H be a Hilbert space and let A be the infinitesimal generator of a strongly continuous contraction semigroup on H. If M is a self-adjoint isomorphism on H and if there exists a constant $\delta > 0$ such that*

$$(Mu,u)_H \geqslant \delta\|u\|_H^2, \quad \textit{for all } u \in H,$$

then, $M^{-1}A$ generates a strongly continuous semigroup on H.

Following the method introduced by BLANCHARD and FRANCFORT [1987], we obtain the following result.

THEOREM 27.6. *If the axial loads verify*

$$f_3^0 \in H^1(0,T;L^2(\Omega)), \qquad g_3^0 \in H^1(0,T;L^2(\Gamma)), \tag{27.82}$$

and the initial data are such that

$$\overline{\Theta}^0 \in L^2(\Omega), \qquad \overline{e}_{ij}^0 \in L^2(\Omega)), \tag{27.83}$$

then, problem (27.74) *admits a unique solution verifying*

$$(\underline{u}_3^0, \Theta^0) \in C^0([0,T];H_0^1(0,L) \times L^2(0,L)). \tag{27.84}$$

Besides, the initial value $\underline{u}_3^0(0,x_3)$ *is the unique solution of the problem*

$$\begin{aligned}
&\left(E + \frac{\alpha_T^2 E^2}{B}\right)\partial_{33}\underline{u}_3^0(0,x_3) \\
&\quad = \frac{\alpha_T E}{B}\partial_3 \int_\omega (\beta\overline{\Theta}^0 + \frac{\alpha_T E}{1-2\nu}\overline{e}_{pp}^0\, \mathrm{d}\omega) \\
&\qquad - \int_\omega f_3^0(0,x_3)\,\mathrm{d}\omega - \int_\gamma g_3^0(0,x_3)\,\mathrm{d}\gamma \quad \text{in } (0,L), \\
&\underline{u}_3^0(0,0) = \underline{u}_3^0(0,L) = 0.
\end{aligned} \tag{27.85}$$

PROOF. The positivity of coefficient E allows us to assure that the following mapping is an isomorphism:

$$\begin{aligned}
S : H_0^1(0,L) &\to H^{-1}(0,L), \\
v &\to S(v) = E\partial_{33}v.
\end{aligned}$$

In the same way it can be shown that the mapping

$$\begin{aligned}
L : L^2(0,L) &\to L^2(0,L) \\
\zeta &\to L(\zeta) = \alpha_T E\partial_3[S^{-1}(\alpha_T E\partial_3\zeta)]
\end{aligned}$$

is a self-adjoint, bounded, linear operator. Moreover, the following function belongs to $H^1(0,T;L^2(0,L))$:

$$r_0(t) = \alpha_T E\partial_3\left[S^{-1}\left(\int_\omega f_3^0(t)\,\mathrm{d}\omega + \int_\gamma g_3^0(t)\,\mathrm{d}\gamma\right)\right]$$

Then, problem (27.74) can be reformulated as

$$\underline{u}_3^0 = S^{-1}\left(-\int_\omega f_3^0\,\mathrm{d}\omega - \int_\gamma g_3^0\,\mathrm{d}\gamma + \alpha_T E\partial_3\Theta^0\right), \tag{27.86}$$

$$\begin{aligned}
&(BI + L)(\Theta^0)' = \frac{k}{T_0}\partial_{33}\Theta^0) + (r_0)',\\
&\Theta(t,0) = \Theta(t,L) = 0,\\
&(BI + L)\Theta^0(0,x_3) = \beta\int_\omega \overline{\Theta}^0\,\mathrm{d}\omega + \frac{\alpha_T E}{1-2\nu}\int_\omega \overline{e}_{pp}^0\,\mathrm{d}\omega + r_0(0).
\end{aligned} \tag{27.87}$$

Applying the previous lemma to problem (27.87) with

$$H = L^2(0,L), \qquad M = BI + L, \qquad \delta = B,$$

we obtain existence and unicity of Θ^0 in $C^0([0,T];L^2(0,L))$. Finally, due to Eq. (27.76), the existence and uniqueness of $\underline{u}_3^0$ in $C^0([0,T];H_0^1(0,L))$ can be derived.

Equation (27.85), which must be satisfied by $\underline{u}_3^0(0,x_3)$, is immediately obtained from Eqs. (27.74) written at time $t = 0$. □

REMARK 27.8. We remark that the required regularity in order to assure the weak convergence to the limit problem (27.73), (27.74) (Theorem 27.4) is sufficient to obtain the results of existence, uniqueness and regularity of the limit solutions (Theorems 27.5 and 27.6). Furthermore, in view of the regularity achieved for u_α^0 and $\underline{u}_3^0$, we obtain that the displacement field $(u_1^0, u_2^0, \underline{u}_3^0 - x_\alpha\partial_3 u_\alpha^0)$ belongs to the space $C^0([0,T];V_{\mathrm{BN}}(\Omega))$.

In Theorem 27.4 we proved that under certain hypotheses problem (27.26)–(27.30) converges weakly to the limit problem (27.73), (27.74). We shall now show that *under more restrictive assumptions on the system of applied loads and on the initial data, strong convergence can be obtained.*

The technique employed is also used in RAOULT [1980] and BLANCHARD and FRANCFORT [1987], and it is based on the convergence of the norms in the Hilbert space $L^2(0,T;L^2(\Omega))$. The following result holds.

THEOREM 27.7. *Let us assume that hypotheses* (27.45) *and* (27.75) *hold true, and that all the convergences in* (27.47) *are strong. Then, the following two assertions are equivalent*:

(i) *As ε tends to zero, the following strong convergences hold for the subsequences of Theorem 27.4 and Corollary 27.1:*

$$\begin{aligned}
&\boldsymbol{u}(\varepsilon) \rightharpoonup \boldsymbol{u}^0, \quad \textit{in } L^2(0,T;V(\Omega)),\\
&(u_\alpha(\varepsilon))' \rightharpoonup (u_\alpha^0)', \qquad \varepsilon(u_3(\varepsilon))' \rightharpoonup 0,\\
&\varepsilon^2\sigma_{\alpha\beta}(\varepsilon) \rightharpoonup \sigma_{\alpha\beta}^0, \qquad \varepsilon\sigma_{\alpha 3}(\varepsilon) \rightharpoonup \sigma_{\alpha 3}^0, \qquad \sigma_{33}(\varepsilon) \rightharpoonup \sigma_{33}^0,\\
&\varepsilon^{-2}e_{\alpha\beta}(\boldsymbol{u}(\varepsilon)) \rightharpoonup \frac{1+\nu}{E}\sigma_{\alpha\beta}^0 - \frac{\nu}{E}\sigma_{pp}^0\delta_{\alpha\beta} + \alpha_T\Theta^0\delta_{\alpha\beta},\\
&\varepsilon^{-1}e_{\alpha 3}(\boldsymbol{u}(\varepsilon)) \rightharpoonup \frac{2(1+\nu)}{E}\sigma_{\alpha 3}^0, \qquad \textit{in } L^2(0,T;L^2(\Omega)),\\
&\Theta(\varepsilon) \rightharpoonup \Theta^0, \qquad \sqrt{T-t}\,\partial_3\Theta(\varepsilon) \rightharpoonup \sqrt{T-t}\,\partial_3\Theta^0, \qquad \textit{in } L^2(0,T;L^2(\Omega)),\\
&\varepsilon^{-1}\sqrt{T-t}\,\partial_\alpha\Theta(\varepsilon) \rightharpoonup \sqrt{T-t}\,r_\alpha^0, \quad \textit{in } L^2(0,T;L^2(\Omega)).
\end{aligned} \tag{27.88}$$

(ii) *The following compatibility condition is verified by the initial data and the terms $\sigma_{\alpha i}^0$, r_α^0:*

$$\begin{aligned}
&T\Bigg\{\rho\int_0^L\left[\sum_{\alpha=1}^2\left(\int_\omega \overline{v}_\alpha^0\right)^2 \mathrm{d}\omega - \int_\omega\sum_{i=1}^3|\overline{v}_i^0|^2\,\mathrm{d}\omega\right]\mathrm{d}x_3 + EI_\alpha\int_0^L|\partial_{33}\overline{u}_\alpha^0|^2\,\mathrm{d}x_3\\
&\quad + E\int_0^L|\partial_3\underline{u}_3^0(0)|^2\,\mathrm{d}x_3 - c(\overline{e}^0,\overline{e}^0)\\
&\quad + \int_0^L\Big(B|\Theta^0(0)|^2 - \beta\int_\omega|\overline{\Theta}^0|^2\,\mathrm{d}\omega\Big)\mathrm{d}x_3 - 2\int_0^L\mathcal{F}_3^0(0)(\underline{u}_3^0(0) - \underline{u}_3^0)\,\mathrm{d}x_3\Bigg\}\\
&\quad + \frac{2k}{T_0}\int_0^T\int_0^t\int_0^L\int_\omega\sum_{\alpha=1}^2|r_\alpha^0|^2\,\mathrm{d}\omega\,\mathrm{d}s\,\mathrm{d}t\\
&\quad + \frac{8(1+\nu)}{E}\int_0^T\int_0^L\int_\omega\sum_{\alpha=1}^2|\sigma_{\alpha 3}^0|^2\,\mathrm{d}\omega\,\mathrm{d}x_3\,\mathrm{d}t\\
&\quad + \frac{1+\nu}{E}\int_0^T\int_0^L\int_\omega\sum_{\alpha,\beta=1}^2|\sigma_{\alpha\beta}^0|^2\,\mathrm{d}\omega\,\mathrm{d}x_3\,\mathrm{d}t\\
&\quad - \frac{\nu(1+\nu)}{E}\int_0^T\int_0^L\int_\omega|\sigma_{\mu\mu}^0|^2\,\mathrm{d}\omega\,\mathrm{d}x_3\,\mathrm{d}t = 0,
\end{aligned} \tag{27.89}$$

where $c(\cdot,\cdot)$ is the operator defined by

$$c(\boldsymbol{A},\boldsymbol{B}) = \int_\Omega \frac{E}{1+\nu}A_{ij}B_{ij}\,\mathrm{d}\boldsymbol{x} + \int_\Omega \frac{\nu E}{(1+\nu)(1-2\nu)}A_{ii}B_{jj}\,\mathrm{d}\boldsymbol{x}.$$

PROOF. We define the space

$$\mathscr{L} = \{\boldsymbol{G} = (G_1, G_2, G_3, G_4)\colon\ G_1 \in L^2(0,T;[L^2(\Omega)]^3), G_2 \in L^2(0,T;\Sigma(\Omega)), \\ G_3 \in L^2(0,T;L^2(\Omega)), \sqrt{T-t}\ G_4 \in L^2(0,T;[L^2(\Omega)]^3)\}.$$

We identify $\boldsymbol{G}(\varepsilon)$, $\boldsymbol{G}^0$, elements of $\mathscr{L}$, with

$$\begin{aligned}
G_1(\varepsilon) &= ((u_\alpha(\varepsilon))', \varepsilon(u_3(\varepsilon))'),\\
G_2(\varepsilon) &= (\varepsilon^{-2} e_{\alpha\beta}(\boldsymbol{u}(\varepsilon)), \varepsilon^{-1} e_{\alpha 3}(\boldsymbol{u}(\varepsilon)), e_{33}(\boldsymbol{u}(\varepsilon))),\\
G_3(\varepsilon) &= \Theta(\varepsilon),\\
G_4(\varepsilon) &= (\varepsilon^{-1}\partial_\alpha \Theta(\varepsilon), \partial_3 \Theta(\varepsilon)),\\
G_1^0 &= ((u_\alpha^0)', 0),\\
G_2^0 &= (e_{ij}^0),\\
G_3^0 &= \Theta^0,\\
G_4^0 &= (r_\alpha^0, \partial_3 \Theta^0).
\end{aligned}$$

Then, the strong convergences (27.88) are equivalent to the convergence $\boldsymbol{G}(\varepsilon) \to \boldsymbol{G}^0$ in $\mathscr{L}$ with respect to the usual product norm.

For any $\boldsymbol{G} \in \mathscr{L}$ we define the norm

$$|||\boldsymbol{G}||| = \left[\int_0^T \left\{ \int_\Omega \rho |G_1(t)|^2 \,\mathrm{d}\boldsymbol{x} + c(G_2(t), G_2(t)) + \int_\Omega \beta [G_3(t)]^2 \,\mathrm{d}\boldsymbol{x} \right.\right. \\ \left.\left. + \int_0^t \int_\Omega \frac{2k}{T_0} |G_4(s)|^2 \,\mathrm{d}\boldsymbol{x}\,\mathrm{d}s \right\} \mathrm{d}t \right]^{1/2},$$

where $|\cdot|$ denotes the Euclidean norm in $\mathbb{R}^3$.

Due to the positivity of all coefficients, $|||\cdot|||$ is equivalent to the usual norm in $\mathscr{L}$.

As a consequence of Corollary 27.1 we have the weak convergence $\boldsymbol{G}(\varepsilon) \to \boldsymbol{G}^0$ in $\mathscr{L}$. Then, the strong convergence is true if and only if

$$|||\boldsymbol{G}(\varepsilon)||| \to |||\boldsymbol{G}^0|||. \tag{27.90}$$

In order to evaluate $|||\boldsymbol{G}(\varepsilon)|||$ we integrate the equality

$$c(\varepsilon)(\boldsymbol{u}(\varepsilon), (\boldsymbol{u}(\varepsilon))') + d(\varepsilon)((\boldsymbol{u}(\varepsilon))'', (\boldsymbol{u}(\varepsilon))') + e((\Theta(\varepsilon))', \Theta(\varepsilon)) \\ + l(\varepsilon)(\Theta(\varepsilon), \Theta(\varepsilon)) = F(\varepsilon)((\boldsymbol{u}(\varepsilon))'),$$

over the interval $[0,t]$ for $0 \leqslant t \leqslant T$, and then again over the time interval $[0,T]$. So

$$\begin{aligned}|||\boldsymbol{G}(\varepsilon)|||^2 = T\Bigg[& c(\varepsilon)(\overline{\boldsymbol{u}}(\varepsilon),\overline{\boldsymbol{u}}(\varepsilon)) + d(\varepsilon)(\overline{\boldsymbol{v}}(\varepsilon),\overline{\boldsymbol{v}}(\varepsilon)) + e(\overline{\Theta}(\varepsilon),\overline{\Theta}(\varepsilon)) \\ & - 2\int_\Omega f_i(\varepsilon)(0)\overline{u}_i(\varepsilon)\,\mathrm{d}\boldsymbol{x} - 2\int_\Gamma g_i(\varepsilon)(0)\overline{u}_i(\varepsilon)\,\mathrm{d}a\Bigg] \\ & + 2\int_0^T\Bigg[\int_\Omega f_i(\varepsilon)u_i(\varepsilon)\,\mathrm{d}\boldsymbol{x} + \int_\Gamma g_i(\varepsilon)u_i(\varepsilon)\,\mathrm{d}a\Bigg]\mathrm{d}t \\ & - 2\int_0^T\Bigg[\int_\Omega (f_i(\varepsilon))'u_i(\varepsilon)\,\mathrm{d}\boldsymbol{x} + \int_\Gamma (g_i(\varepsilon))'u_i(\varepsilon)\,\mathrm{d}a\Bigg]\mathrm{d}s\,\mathrm{d}t. \qquad (27.91)\end{aligned}$$

Since convergences (27.47) are assumed to be strong by hypothesis, and since the weak convergence of $\boldsymbol{u}(\varepsilon)$ to $\boldsymbol{u}^0$ in $L^2(0,T;V(\Omega))$ is given by Corollary 27.1, we have

$$\begin{aligned}\lim_{\varepsilon\to 0}|||\boldsymbol{G}(\varepsilon)|||^2 = T\Bigg[& c(\overline{e}^0,\overline{e}^0) + \int_0^L \rho\int_\omega \sum_{i=1}^3 |\overline{v}_i^0|^2\,\mathrm{d}\omega\,\mathrm{d}x_3 + \int_0^L \beta\int_\omega |\overline{\Theta}^0|^2\,\mathrm{d}\omega\,\mathrm{d}x_3 \\ & - 2\int_0^L (\mathscr{F}_\alpha^0(0)\overline{u}_\alpha^0 + \mathscr{F}_3^0(0)\overline{\underline{u}}_3^0)\,\mathrm{d}x_3\Bigg] \\ & + 2\int_0^T\int_0^t (\mathscr{F}_\alpha^0 u_\alpha^0 + \mathscr{F}_3^0\underline{u}_3^0)\,\mathrm{d}x_3\,\mathrm{d}t \\ & - \int_0^T\int_0^t\int_0^L [(\mathscr{F}_\alpha^0)'u_\alpha^0 + (\mathscr{F}_3^0)'\underline{u}_3^0]\,\mathrm{d}x_3\,\mathrm{d}s\,\mathrm{d}t. \qquad (27.92)\end{aligned}$$

On the other hand, one has

$$\begin{aligned}|||\boldsymbol{G}^0|||^2 = \int_0^T\Bigg[& c(\boldsymbol{e}^0,\boldsymbol{e}^0) + \int_\Omega \rho\sum_{\alpha=1}^2 |(u_\alpha^0)'|^2\,\mathrm{d}\boldsymbol{x} + \beta\int_\Omega |\Theta^0|^2\,\mathrm{d}\boldsymbol{x} \\ & + 2\int_0^t \frac{k}{T_0}\int_\Omega\left(\sum_{\alpha=1}^2 |r_\alpha^0|^2 + |\partial_3\Theta^0|^2\right)\mathrm{d}\boldsymbol{x}\,\mathrm{d}s\Bigg]\mathrm{d}t. \qquad (27.93)\end{aligned}$$

From equalities (27.59), (27.61) and (27.68) we can obtain the following expression for $c(\boldsymbol{e}^0, \boldsymbol{e}^0)$:

$$c(\boldsymbol{e}^0, \boldsymbol{e}^0) = E\int_0^L [|\partial_3 \underline{u}_3^0|^2 + I_\alpha |\partial_{33} u_\alpha^0|^2]\,\mathrm{d}x_3 + \frac{2E\alpha_T^2(1+\nu)}{1-2\nu}\int_0^L |\Theta^0|^2\,\mathrm{d}x_3$$
$$+ \frac{8(1+\nu)}{E}\int_0^L\int_\omega \sum_{\alpha=1}^2 |\sigma_{\alpha 3}^0|^2\,\mathrm{d}\omega\,\mathrm{d}x_3$$
$$+ \frac{1+\nu}{E}\int_0^L\int_\omega \sum_{\alpha,\beta=1}^2 |\sigma_{\alpha\beta}^0|^2\,\mathrm{d}\omega\,\mathrm{d}x_3$$
$$- \frac{\nu(1+\nu)}{E}\int_0^L\int_\omega |\sigma_{\rho\rho}^0|^2\,\mathrm{d}\omega\,\mathrm{d}x_3 = 0. \tag{27.94}$$

Moreover, we know from (27.73), (27.74) that

$$\rho\int_0^L (u_\alpha^0)''(u_\alpha^0)'\,\mathrm{d}x_3 + EI_\alpha\int_0^L \partial_{33}u_\alpha^0\partial_{33}(u_\alpha^0)'\,\mathrm{d}x_3 = \int_0^L \mathscr{F}_\alpha^0(u_\alpha^0)'\,\mathrm{d}x_3, \tag{27.95}$$

$$E\int_0^L \partial_3\underline{u}_3^0\partial_3(\underline{u}_3^0)'\,\mathrm{d}x_3 + E\alpha_T\int_0^L \partial_3\Theta^0(\underline{u}_3^0)'\,\mathrm{d}x_3 = \int_0^L \mathscr{F}_3^0(\underline{u}_3^0)'\,\mathrm{d}x_3, \tag{27.96}$$

$$B\int_0^L (\Theta^0)'\Theta^0\,\mathrm{d}x_3 = -\int_0^L \frac{k}{T_0}\partial_3\Theta^0\partial_3\Theta^0\,\mathrm{d}x_3 + E\alpha_T\int_0^L \partial_3\Theta^0(\underline{u}_3^0)'\,\mathrm{d}x_3. \tag{27.97}$$

Adding together expressions (27.95)–(27.97), integrating the result over $[0, t]$ for $0 \leqslant t \leqslant T$, and again over $[0, T]$, substituting in (27.93) and taking (27.94) into account, we obtain

$$|||\boldsymbol{G}^0|||^2 = T[\rho\int_0^L \sum_{\alpha=1}^2 (\int_\omega \overline{v}_\alpha^0\,\mathrm{d}\omega)^2\,\mathrm{d}x_3 + E\int_0^L |\partial_3\underline{u}_3^0(0)|^2\,\mathrm{d}x_3$$
$$+ EI_\alpha\int_0^L |\partial_{33}\overline{u}_\alpha^0|^2\,\mathrm{d}x_3 + B\int_0^L |\Theta^0(0)|^2\,\mathrm{d}x_3]$$
$$- 2T\int_0^L [\mathscr{F}_\alpha^0(0)\overline{u}_\alpha^0 + \mathscr{F}_3^0(0)\overline{\underline{u}}_3^0(0)]\,\mathrm{d}x_3$$

$$+2\int_0^T\int_0^L(\mathcal{F}_\alpha^0 u_\alpha^0+\mathcal{F}_3^0\underline{u}_3^0)\,\mathrm{d}x_3\,\mathrm{d}t$$

$$-\int_0^T\int_0^t\int_0^L[(\mathcal{F}_\alpha^0)'u_\alpha^0+(\mathcal{F}_3^0)'\underline{u}_3^0]\,\mathrm{d}x_3\,\mathrm{d}s\,\mathrm{d}t$$

$$+\frac{2k}{T_0}\int_0^T\int_0^t\int_0^L\int_\omega\sum_{\alpha=1}^2|r_\alpha^0|^2\,\mathrm{d}\omega\,\mathrm{d}x_3\,\mathrm{d}s\,\mathrm{d}t$$

$$+\frac{8(1+\nu)}{E}\int_0^T\int_0^L\int_\omega\sum_{\alpha=1}^2|\sigma_{\alpha 3}^0|^2\,\mathrm{d}\omega\,\mathrm{d}x_3\,\mathrm{d}t$$

$$+\frac{1+\nu}{E}\int_0^T\int_0^L\int_\omega\sum_{\alpha,\beta=1}^2|\sigma_{\alpha\beta}^0|^2\,\mathrm{d}x_3\,\mathrm{d}t$$

$$-\frac{\nu(1+\nu)}{E}\int_0^T\int_0^L\int_\omega|\sigma_{\rho\rho}^0|^2\,\mathrm{d}\omega\,\mathrm{d}x_3\,\mathrm{d}t. \tag{27.98}$$

Thus, because of (27.92), the convergence (27.90) holds true if and only if the compatibility condition (27.89) is verified. □

REMARK 27.9. Since we have proved the uniqueness of $\boldsymbol{u}^0$ and Θ^0, strong convergence follows from Theorem 27.7, not only for the corresponding subsequence of $\boldsymbol{u}(\varepsilon)$ and $\Theta(\varepsilon)$, but also for the whole sequences.

REMARK 27.10. If we consider the following particular case:

$$\overline{v}_3^0=0,\qquad \overline{v}_\alpha^0,\overline{\Theta}^0\ \text{independent of}\ x_\alpha,\qquad \sigma_{\alpha 3}^0=\sigma_{\alpha\beta}^0=r_\alpha^0=0, \tag{27.99}$$

compatibility condition (27.89) turns into

$$EI_\alpha\int_0^L|\partial_{33}\overline{u}_\alpha^0|^2\,\mathrm{d}x_3+E\int_0^L|\partial_3\underline{u}_3^0(0)|^2\,\mathrm{d}x_3-c(\overline{e}^0,\overline{e}^0)$$

$$+\int_0^L[B|\Theta^0(0)|^2-\beta\int_\omega|\overline{\Theta}^0|^2\,\mathrm{d}\omega]\,\mathrm{d}x_3-2\int_0^L\mathcal{F}_3^0(0)[\underline{u}_3^0(0)-\overline{\underline{u}}_3^0]\,\mathrm{d}x_3=0. \tag{27.100}$$

Using (27.99), (27.59)–(27.61) and (27.68), here we also suppose that

$$\overline{e}_{\alpha 3}^0=0,\qquad \overline{e}_{\alpha\beta}^0=(-\nu\partial_3\overline{\underline{u}}_3^0+\nu x_\alpha\partial_{33}\overline{u}_\alpha^0+\alpha_T(1+\nu)\overline{\Theta}^0)\delta_{\alpha\beta},$$

then

$$c(\overline{e}^0,\overline{e}^0) = EI_\alpha \int_0^L |\partial_{33}\overline{u}_\alpha^0|^2\,\mathrm{d}x_3 + E\int_0^L |\partial_3\underline{\overline{u}}_3^0|^2\,\mathrm{d}x_3 + (B-\beta)\int_0^L |\overline{\Theta}^0|^2\,\mathrm{d}x_3,$$

and (27.100) becomes

$$E\int_0^L [|\partial_3\underline{u}_3^0(0)|^2 - |\partial_3\underline{\overline{u}}_3^0|^2]\,\mathrm{d}x_3 + B\int_0^L [|\Theta^0(0)|^2 - |\overline{\Theta}^0|^2]\,\mathrm{d}x_3$$
$$-2\int_0^L \mathcal{F}_3^0(0)[\underline{u}_3^0(0) - \underline{\overline{u}}_3^0]\,\mathrm{d}x_3 = 0, \tag{27.101}$$

which yields that (27.101) is true whenever fields $\underline{u}_3^0$ and Θ^0 verify

$$\underline{u}_3^0(0) = \underline{\overline{u}}_3^0, \qquad \Theta^0(0) = \overline{\Theta}^0.$$

In this case, a sufficient condition in order to obtain strong convergence seems to be that the initial conditions of $\underline{u}_3^0$ and Θ^0 do not undergo any changes. Similar conclusions have been obtained by BLANCHARD and FRANCFORT [1987], for the case of linearized thermoelastic plates. FRANCFORT [1983] also noticed that this change in the initial data can be avoided if we deal with the entropy field:

$$S^\varepsilon = \beta^\varepsilon\Theta^\varepsilon + \frac{E^\varepsilon\alpha_T^\varepsilon}{1-2\nu^\varepsilon}e_{\beta\beta}^\varepsilon(\boldsymbol{u}^\varepsilon),$$

instead of the temperature increment field Θ^ε. But this means working with a system with third order derivatives, which cannot be handled by the methods presented in this work.

The method developed in the previous sections for a linearized thermoelastic rod clamped at both ends can be used for the case of a beam submitted to *other boundary conditions*.

For example, we can consider the case of a beam clamped only at one of its ends Γ_0^ε, and we denote by $h^\varepsilon(\boldsymbol{x}^\varepsilon)$ the surface forces applied on the otherwise free end Γ_L^ε. Then, the problem so described corresponds to the system of Eqs. (27.1)–(27.10) to which we must add new boundary conditions over the free end:

$$\begin{aligned} &\sigma_{i3}^\varepsilon = h_i^\varepsilon \quad \text{on } \Gamma_L^\varepsilon\times(0,T),\\ &\Theta^\varepsilon = 0 \quad \text{on } \Gamma_L^\varepsilon\times(0,T). \end{aligned} \tag{27.102}$$

For this case, we define the functional space

$$V(\Omega^\varepsilon) = \{\boldsymbol{v}^\varepsilon \in [H^1(\Omega^\varepsilon)]^3\colon v_i^\varepsilon = 0 \text{ on } \Gamma_0^\varepsilon\}, \tag{27.103}$$

and, besides (27.12), we suppose

$$\boldsymbol{h}^\varepsilon \in H^2(0,T;[L^2(\Gamma_L^\varepsilon)]^3). \tag{27.104}$$

Then, there exists one and only one solution $(\boldsymbol{u}^\varepsilon, \Theta^\varepsilon)$ for the problem (27.1)–(27.10), (27.102) verifying the regularity results (27.13).

We add the following transformations to those defined in (27.12) (cf. (26.15)):

$$h_\alpha^\varepsilon(\boldsymbol{x}^\varepsilon) = \varepsilon h_\alpha(\varepsilon)(\boldsymbol{x}), \qquad h_3^\varepsilon(\boldsymbol{x}^\varepsilon) = h_3(\varepsilon)(\boldsymbol{x}). \tag{27.105}$$

Then, it can be shown that the variational formulation for the above problem is still (27.28)–(27.30) with the following definition for $F(\varepsilon)$:

$$F(\varepsilon) = \int_\Omega f_i(\varepsilon)v_i\,\mathrm{d}\boldsymbol{x} + \int_\Gamma g_i(\varepsilon)v_i\,\mathrm{d}a + \int_{\Gamma_L} h_i(\varepsilon)v_i\,\mathrm{d}a.$$

Now, besides the weak convergences (27.47) we also suppose

$$h_i(\varepsilon) \to h_i^0 \quad \text{weakly in } H^1(0,T;L^2(\Gamma_L)), \tag{27.106}$$

and then we obtain, in the same way, the weak convergences (27.50). Obviously, the characterization of the limit is slightly different. Consequently, we can enunciate the following result related to the limit problem:

THEOREM 27.8. *If hypotheses* (27.45), (27.47), (27.75), (27.104) *and* (27.106) *hold and if*

$$\partial_3 h_3^0 \in H^1(0,T;L^2(\Gamma_L)), \tag{27.107}$$

then the weak convergences (27.76) *are obtained, where*

(i) u_α^0 *is a solution of the problem*

$$\begin{aligned}
&\rho(u_\alpha^0)'' - EI_\alpha\partial_{3333}u_\alpha^0 \\
&\quad = \int_\omega f_\alpha^0\,\mathrm{d}\omega + \int_\gamma g_\alpha^0\,\mathrm{d}\gamma + \partial_3\Big[\int_\omega x_\alpha f_3^0\,\mathrm{d}\omega + \int_\gamma x_\alpha g_3^0\Big]\,\mathrm{d}\gamma = \mathcal{F}_\alpha^0, \\
&u_\alpha^0(t,0) = \partial_3 u_\alpha^0(t,0) = 0, \\
&\partial_{33}u_\alpha^0(t,L) = -\frac{1}{EI_\alpha}\int_\omega h_\alpha^0\,\mathrm{d}\omega, \\
&\partial_{333}u_\alpha^0(t,L) = -\frac{1}{EI_\alpha}\int_\omega x_\alpha h_3^0\,\mathrm{d}\omega, \\
&u_\alpha^0(0,x_3) = \overline{u}_\alpha^0(x_3), \\
&(u_\alpha^0)'(0,x_3) = \int_\omega \overline{v}_\alpha^0(x_3)\,\mathrm{d}\omega.
\end{aligned} \tag{27.108}$$

(ii) $u_3^0 = \underline{u}_3^0 - x_\alpha \partial_3 u_\alpha^0$, *where* $(\underline{u}_3^0, \Theta^0)$ *is a solution of the problem*

$$
\begin{aligned}
&-E\partial_{33}\underline{u}_3^0 + \alpha_T E \partial_3 \Theta^0 = \int_\omega f_3^0 \,\mathrm{d}\omega + \int_\gamma g_3^0 \,\mathrm{d}\gamma = \mathscr{F}_3^0,\\
&B(\Theta^0)' = \frac{k}{T_0}\partial_{33}\Theta^0 - \alpha_T E \partial_3 (\underline{u}_3^0)',\\
&\underline{u}_3^0(t,0) = 0,\\
&\partial_3 \underline{u}_3^0(t,L) = \frac{1}{E}\int_\omega h_3^0 \,\mathrm{d}\omega,\\
&\Theta^0(t,0) = \Theta^0(t,L) = 0,\\
&B\Theta^0(0,x_3) = \beta \int_\omega \overline{\Theta}^0 \,\mathrm{d}\omega + \frac{\alpha_T E}{1-2\nu}\int_\omega \overline{e}_{pp}^0 \,\mathrm{d}\omega - \alpha_T E \int_\omega \partial_3 \underline{u}_3^0(0,x_3)\,\mathrm{d}\omega,
\end{aligned}
\tag{27.109}
$$

$\underline{u}_3^0(0,x_3)$ *being the unique solution of the problem*

$$
\begin{aligned}
&\Big(E + \frac{\alpha_T^2 E^2}{B}\Big)\partial_{33}\underline{u}_3^0(0,x_3)\\
&\quad = \frac{\alpha_T E}{B}\partial_3 \int_\omega \Big(\beta\overline{\Theta}^0 + \frac{\alpha_T E}{1-2\nu}\overline{e}_{pp}^0\Big)\,\mathrm{d}\omega\\
&\qquad - \int_\omega f_3^0(0,x_3)\,\mathrm{d}\omega - \int_\gamma g_3^0(0,x_3)\,\mathrm{d}\gamma \quad \textit{in } (0,L),
\end{aligned}
\tag{27.110}
$$

$$
\begin{aligned}
&\underline{u}_3^0(0,0) = 0,\\
&\partial_3 \underline{u}_3^0(0,L) = \frac{1}{E}\int_\omega h_3^0(0)\,\mathrm{d}\omega.
\end{aligned}
$$

Other kind of boundary conditions for the heat equation can be studied with the same method. For example, instead of the Dirichlet homogeneous condition (27.6), we may take the Fourier condition (cf. VIAÑO [1983]):

$$
k^\varepsilon \partial_\alpha^\varepsilon \Theta^\varepsilon n_\alpha + \delta^\varepsilon \Theta^\varepsilon = j^\varepsilon \quad \text{on } \Gamma^\varepsilon \times (0,T), \tag{27.111}
$$

where δ^ε is the heat transfer coefficient depending on the material and j^ε is the heat flux on Γ^ε.

We assume

$$
\begin{aligned}
&\delta^\varepsilon(\boldsymbol{x}^\varepsilon) = \varepsilon\delta(\boldsymbol{x}), \quad \delta \geqslant 0,\\
&j^\varepsilon(\boldsymbol{x}^\varepsilon) = \varepsilon j(\varepsilon)(\boldsymbol{x}),
\end{aligned}
\tag{27.112}
$$

and, for simplicity, we consider δ independent of $\boldsymbol{x}$. Then, the variational formulation corresponding to the heat equation is given by

$$e((\Theta(\varepsilon))', z) + l(\varepsilon)(\Theta(\varepsilon), z) + \tilde{G}(\varepsilon)(z, (\boldsymbol{u}(\varepsilon))') + h(\Theta(\varepsilon), z) = 0,$$
$$\text{for all } z \in W(\Omega), \tag{27.113}$$

where

$$h(\Theta(\varepsilon), z) = \int_{\Gamma} \frac{1}{T_0}(\delta\Theta(\varepsilon) - j(\varepsilon))z \, \mathrm{d}a. \tag{27.114}$$

Consequently, if the following weak convergence is assumed together with (27.47):

$$j(\varepsilon) \rightharpoonup j^0 \text{ in } L^\infty(0, T; L^2(\Gamma)), \tag{27.115}$$

we obtain estimates (27.48) in the same way as in Theorem 27.3. In a similar way, the technique in Theorem 27.4 is still valid for this problem, but the equation corresponding to (27.74) is now the following one:

$$B(\Theta^0)' = \frac{k}{T_0}\partial_{33}\Theta^0 - E\alpha_T\partial_3(\underline{u}_3^0)' - \frac{|\gamma|}{T_0}\delta\Theta^0 + \frac{1}{T_0}\int_{\gamma} j^0 \, \mathrm{d}\gamma. \tag{27.116}$$

Introducing slight variations in the method, we can also study this case in the framework of the technique presented in this section.

Since $\boldsymbol{u}^0$, $\boldsymbol{\sigma}^0$, Θ^0, $\boldsymbol{m}^0$ and $\boldsymbol{q}^0$ are approximations of $\boldsymbol{u}(\varepsilon)$, $\boldsymbol{\sigma}(\varepsilon)$, $\Theta(\varepsilon)$, $\boldsymbol{m}(\varepsilon)$ and $\boldsymbol{q}(\varepsilon)$, respectively, as ε tends to zero, undoing the change of variables (27.20) and the scaling in the different fields, we get that the fields

$$\begin{aligned}
&\boldsymbol{u}^{0\varepsilon}(\boldsymbol{x}^\varepsilon) = (\varepsilon^{-1}u_\alpha^0(\boldsymbol{x}), u_3^0(\boldsymbol{x})),\\
&\boldsymbol{\sigma}^{0\varepsilon}(\boldsymbol{x}^\varepsilon) = (\varepsilon^2\sigma_{\alpha\beta}^0(\boldsymbol{x}), \varepsilon\sigma_{3\beta}^0(\boldsymbol{x}), \sigma_{33}^0(\boldsymbol{x})),\\
&\Theta^{0\varepsilon}(\boldsymbol{x}^\varepsilon) = \Theta^0(\boldsymbol{x}),\\
&\boldsymbol{m}^{0\varepsilon}(\boldsymbol{x}^\varepsilon) = (\varepsilon^3 m_\alpha^0(\boldsymbol{x})),\\
&\boldsymbol{q}^{0\varepsilon}(\boldsymbol{x}^\varepsilon) = (\varepsilon^3 q_\alpha^0(\boldsymbol{x}), \varepsilon^2 q_3^0(\boldsymbol{x})),
\end{aligned} \tag{27.117}$$

defined in Ω^ε, can be considered approximations of $\boldsymbol{u}^\varepsilon$, $\boldsymbol{\sigma}^\varepsilon$, Θ^ε, solutions of problem (27.1)–(27.9) and $m_\alpha^\varepsilon = \int_{\omega^\varepsilon} x_\alpha^\varepsilon \sigma_{33}^\varepsilon \, \mathrm{d}\omega^\varepsilon$, $q_i^\varepsilon = \int_{\omega^\varepsilon} \sigma_{i3}^\varepsilon \, \mathrm{d}\omega^\varepsilon$.

We can then enunciate the following result as an immediate consequence of Theorem 27.4 and Corollary 27.2.

THEOREM 27.9. *The approximations* $(\boldsymbol{u}^{0\varepsilon}, \sigma_{33}^{0\varepsilon}, \Theta^{0\varepsilon})$ *are uniquely characterized as follows:*

(i) $u_\alpha^{0\varepsilon}$ *is the unique solution of the following evolution problem:*

$$\begin{aligned}
&A^\varepsilon \rho^\varepsilon (u_\alpha^{0\varepsilon})'' - E^\varepsilon I_\alpha^\varepsilon \partial_{3333}^\varepsilon u_\alpha^{0\varepsilon} \\
&\quad = \int_{\omega^\varepsilon} f_\alpha^{0\varepsilon}\, \mathrm{d}\omega^\varepsilon + \int_{\gamma^\varepsilon} g_\alpha^{0\varepsilon}\, \mathrm{d}\gamma^\varepsilon + \partial_3^\varepsilon \Big[\int_{\omega^\varepsilon} x_\alpha^\varepsilon f_3^{0\varepsilon}\, \mathrm{d}\omega^\varepsilon + \int_{\gamma^\varepsilon} x_\alpha^\varepsilon g_3^{0\varepsilon}\, \mathrm{d}\gamma^\varepsilon\Big], \\
&u_\alpha^{0\varepsilon}(t,0) = u_\alpha^{0\varepsilon}(t,L) = 0, \\
&\partial_3^\varepsilon u_\alpha^{0\varepsilon}(t,0) = \partial_3^\varepsilon u_\alpha^{0\varepsilon}(t,L) = 0, \\
&u_\alpha^{0\varepsilon}(0,x_3^\varepsilon) = \overline{u}_\alpha^{0\varepsilon}(x_3^\varepsilon), \\
&A^\varepsilon (u_\alpha^{0\varepsilon})'(0,x_3^\varepsilon) = \int_{\omega^\varepsilon} \overline{v}_\alpha^{0\varepsilon}(x_3^\varepsilon)\, \mathrm{d}\omega^\varepsilon,
\end{aligned} \tag{27.118}$$

where

$$\begin{aligned}
&\boldsymbol{f}^{0\varepsilon}(\boldsymbol{x}^\varepsilon) = (\varepsilon f_\alpha^0(\boldsymbol{x}), f_3^0(\boldsymbol{x})), \\
&\boldsymbol{g}^{0\varepsilon}(\boldsymbol{x}^\varepsilon) = (\varepsilon^2 g_\alpha^0(\boldsymbol{x}), \varepsilon g_3^0(\boldsymbol{x})), \\
&\overline{u}_\alpha^{0\varepsilon}(\boldsymbol{x}^\varepsilon) = \varepsilon^{-1}\overline{u}_\alpha^0(\boldsymbol{x}), \\
&\overline{v}_\alpha^{0\varepsilon}(\boldsymbol{x}^\varepsilon) = \varepsilon^{-1}\overline{v}_\alpha^0(\boldsymbol{x}).
\end{aligned} \tag{27.119}$$

(ii) $u_3^{0\varepsilon} = \underline{u}_3^{0\varepsilon} - x_\alpha^\varepsilon \partial_3^\varepsilon u_\alpha^{0\varepsilon}$, *where* $(\underline{u}_3^{0\varepsilon}, \Theta^{0\varepsilon})$ *is the solution of the coupled system*

$$\begin{aligned}
&-A^\varepsilon E^\varepsilon \partial_{33}^\varepsilon \underline{u}_3^{0\varepsilon} + \alpha_T^\varepsilon E^\varepsilon \partial_3^\varepsilon \Theta^{0\varepsilon} = \int_{\omega^\varepsilon} f_3^{0\varepsilon}\, \mathrm{d}\omega^\varepsilon + \int_{\gamma^\varepsilon} g_3^{0\varepsilon}\, \mathrm{d}\gamma^\varepsilon, \\
&B^\varepsilon (\Theta^{0\varepsilon})' = \frac{k^\varepsilon}{T_0^\varepsilon}\partial_{33}^\varepsilon \Theta^{0\varepsilon} - \alpha_T^\varepsilon E^\varepsilon \partial_3^\varepsilon (\underline{u}_3^{0\varepsilon})', \\
&\underline{u}_3^{0\varepsilon}(t,0) = \underline{u}_3^{0\varepsilon}(t,L) = 0, \\
&\Theta^{0\varepsilon}(t,0) = \Theta^{0\varepsilon}(t,L) = 0, \\
&B^\varepsilon \Theta^{0\varepsilon}(0,x_3^\varepsilon) + \alpha_T^\varepsilon E^\varepsilon \int_{\omega^\varepsilon} \partial_3^\varepsilon \underline{u}_3^{0\varepsilon}(0,x_3^\varepsilon)\, \mathrm{d}\omega^\varepsilon \\
&\quad = \beta^\varepsilon \int_{\omega^\varepsilon} \overline{\Theta}^{0\varepsilon}\, \mathrm{d}\omega^\varepsilon + \frac{\alpha_T^\varepsilon E^\varepsilon}{1-2\nu}\Big[\int_{\omega^\varepsilon} \overline{\boldsymbol{e}}_{33}^{0\varepsilon}\, \mathrm{d}\omega^\varepsilon + \frac{1}{A^\varepsilon}\int_{\omega^\varepsilon} \overline{\boldsymbol{e}}_{\alpha\alpha}^{0\varepsilon}\, \mathrm{d}\omega^\varepsilon\Big],
\end{aligned} \tag{27.120}$$

where

$$\begin{aligned}
&\overline{\Theta}^{0\varepsilon}(\boldsymbol{x}^\varepsilon) = \overline{\Theta}^0(\boldsymbol{x}), \\
&\overline{\boldsymbol{e}}^{0\varepsilon}(\boldsymbol{x}^\varepsilon) = (\varepsilon^2 \overline{\boldsymbol{e}}_{\alpha\beta}^0(\boldsymbol{x}), \varepsilon \overline{\boldsymbol{e}}_{3\beta}^0(\boldsymbol{x}), \overline{\boldsymbol{e}}_{33}^0(\boldsymbol{x}), \\
&B^\varepsilon = \beta^\varepsilon + \frac{2E^\varepsilon(\alpha_T^\varepsilon)^2(1+\nu^\varepsilon)}{1-2\nu^\varepsilon}.
\end{aligned} \tag{27.121}$$

(iii) *The component* $\sigma_{33}^{0\varepsilon}$ *takes the form*

$$\sigma_{33}^{0\varepsilon} = E^\varepsilon \partial_3^\varepsilon \underline{u}_3^{0\varepsilon} - E^\varepsilon x_\alpha^\varepsilon \partial_{33}^\varepsilon u_\alpha^{0\varepsilon} - E^\varepsilon \alpha_T^\varepsilon \Theta^{0\varepsilon} + \frac{\nu^\varepsilon}{A^\varepsilon}\sigma_{\alpha\alpha}^{0\varepsilon}, \tag{27.122}$$

where $\sigma_{\alpha\beta}^{0\varepsilon}$ *is a solution of*

$$\begin{aligned} &\partial_\alpha^\varepsilon \sigma_{\alpha\beta}^{0\varepsilon} = 0 \quad \textit{in } \omega^\varepsilon, \\ &\sigma_{\alpha\beta}^{0\varepsilon} n_\alpha = 0 \quad \textit{on } \gamma^\varepsilon. \end{aligned} \tag{27.123}$$

(iv) *The normal force, bending moments and shear force components are given, respectively, by the expressions*

$$q_3^{0\varepsilon} = \int_{\omega^\varepsilon} \sigma_{33}^{0\varepsilon}\, \mathrm{d}\omega^\varepsilon = E^\varepsilon \partial_3^\varepsilon \underline{u}_3^{0\varepsilon} - E^\varepsilon \alpha_T^\varepsilon \Theta^{0\varepsilon}, \tag{27.124}$$

$$m_\alpha^{0\varepsilon} = \int_{\omega^\varepsilon} x_\alpha^\varepsilon \sigma_{33}^{0\varepsilon}\, \mathrm{d}\omega^\varepsilon = -E^\varepsilon I_\alpha^\varepsilon \partial_{33}^\varepsilon u_\alpha^{0\varepsilon}, \tag{27.125}$$

$$q_\alpha^{0\varepsilon} = \int_{\omega^\varepsilon} \sigma_{\alpha 3}^{0\varepsilon} = \int_{\omega^\varepsilon} x_\alpha^\varepsilon f_3^{0\varepsilon}\, \mathrm{d}\omega^\varepsilon + \int_{\gamma^\varepsilon} x_\alpha^\varepsilon g_3^{0\varepsilon}\, \mathrm{d}\gamma^\varepsilon + \partial_3^\varepsilon m_\alpha^{0\varepsilon}. \tag{27.126}$$

REMARK 27.11. If in problems (27.118) and (27.120) we neglect the time derivatives, we obtain the classical stretching–bending model for stationary thermoelastic rods (see BERMÚDEZ and VIAÑO [1984] and also (11.12)–(11.14), (11.16)–(11.19)):

$$-E^\varepsilon I_\alpha^\varepsilon \partial_{3333}^\varepsilon u_\alpha^{0\varepsilon} = \int_{\omega^\varepsilon} f_\alpha^{0\varepsilon}\, \mathrm{d}\omega^\varepsilon + \int_{\gamma^\varepsilon} g_\alpha^{0\varepsilon}\, \mathrm{d}\gamma^\varepsilon + \partial_3^\varepsilon \left[\int_{\omega^\varepsilon} x_\alpha^\varepsilon f_3^{0\varepsilon}\, \mathrm{d}\omega^\varepsilon + \int_{\gamma^\varepsilon} x_\alpha^\varepsilon g_3^{0\varepsilon}\, \mathrm{d}\gamma^\varepsilon \right], \tag{27.127}$$

$$\begin{aligned} &u_\alpha^{0\varepsilon}(0) = u_\alpha^{0\varepsilon}(L) = 0, \\ &\partial_3^\varepsilon u_\alpha^{0\varepsilon}(0) = \partial_3^\varepsilon u_\alpha^{0\varepsilon}(L) = 0, \\ &-A^\varepsilon E^\varepsilon \partial_{33}^\varepsilon \underline{u}_3^{0\varepsilon} = \int_{\omega^\varepsilon} f_3^{0\varepsilon}\, \mathrm{d}\omega^\varepsilon + \int_{\gamma^\varepsilon} g_3^{0\varepsilon}\, \mathrm{d}\gamma^\varepsilon, \\ &\underline{u}_3^{0\varepsilon}(0) = \underline{u}_3^{0\varepsilon}(L) = 0. \end{aligned} \tag{27.128}$$

For nonhomogeneous boundary conditions in the heat equation we obtain the general bending equation (cf. VINSON [1974], BERMÚDEZ and VIAÑO [1984]) together with Eq. (26.69):

$$-A^\varepsilon E^\varepsilon \partial_{33}^\varepsilon \underline{u}_3^{0\varepsilon} + \alpha_T^\varepsilon E^\varepsilon \partial_3^\varepsilon \Theta^{0\varepsilon} = \int_{\omega^\varepsilon} f_3^{0\varepsilon}\, \mathrm{d}\omega^\varepsilon + \int_{\gamma^\varepsilon} g_3^{0\varepsilon}\, \mathrm{d}\gamma^\varepsilon.$$

As in the classical theories, the term $(\nu^\varepsilon/A^\varepsilon)\sigma_{\alpha\alpha}^{0\varepsilon}$, for the axial stress $\sigma_{33}^{0\varepsilon}$, is not considered in practice.

CHAPTER VI

Asymptotic Method for Contact Problems in Rods and Rods with a Variable Cross Section

In this chapter the asymptotic expansion method is applied to two special cases: rods made of an homogeneous isotropic material in contact with a foundation and rods with a variable cross section.

In Section 28 we apply the asymptotic expansion method described in Chapter I to the *three-dimensional elasticity problem with unilateral frictionless contact conditions in one part of the boundary.* This allows us to obtain one-dimensional rod models in which the deformation is restricted either by the presence of a rigid foundation (Signorini's problem, cf. SIGNORINI [1943]) or by an elastic foundation (Winkler's model) (cf. KERR [1964], VILLAGIO [1967], BALL [1972], CIMATTI [1973], BIELAK and STEPHAN [1983]). The methodology is the same: variational formulation, change of variable and scaling of the unknowns, asymptotic expansion and identification of the first term in the expansion.

The major difference now, resides in the unilateral contact conditions in the three-dimensional elasticity problem which give rise to nonlinear problems and to variational inequalities, whose treatment is different from the one used so far.

We start from the case of a rigid foundation and the elastic case will be considered in the end. These results were obtained initially by VIAÑO [1985a] and ÁLVAREZ-DIOS [1986].

In Section 29 we adapt the asymptotic expansion method, to the case of elastic rods with a variable (both in shape and size) cross section along axis Ox_3. As a matter of fact this was the first case to be treated by the asymptotic expansion method by BERMÚDEZ and VIAÑO [1984]. A similar version for linear elastic plates of variable thickness is treated in VIAÑO [1983]. Thus we obtain from a mathematical point of view the justification and generalization of most of the well-known stretching, bending and torsion rod models in engineering literature.

Moreover, the difference in the calculations for the case of a variable cross section, when compared to the cylindrical case, resides in the change of variable because the Jacobian now depends, on the variable x_3. As an important consequence of this fact we obtain one dimensional models with variable coefficients for stretching, bending and torsion. We remark that the variations of the sections are assumed to occur "slowly". The case of rapidly variable cross sections could be considered by adapting the methods of KOHN and VOGELIUS [1984, 1985, 1986] and QUINTELA-ESTEVEZ [1989] for plates.

28. Asymptotic derivation of unilateral contact models for rods on a rigid or elastic foundation

The notations and hypotheses *which will not be explicitly introduced are assumed to be identical to those of Sections* 1–10. We consider a beam $\Omega^\varepsilon = \omega^\varepsilon \times (0, L)$ strongly clamped at its extremities, denoted by Γ_0^ε and Γ_L^ε and made of an homogeneous, isotropic material. Since we are going to restrict ourselves to the first term of the asymptotic expansion the result would have been the same had we considered a weak clamping condition (cf. (7.1)). We also assume that the boundary $\gamma^\varepsilon = \partial\omega^\varepsilon$ is divided into two nonempty disjoint parts denoted by γ_C^ε and γ_g^ε. Consequently, we denote

$$\Gamma_C^\varepsilon = \gamma_C^\varepsilon \times (0, L), \qquad \Gamma^\varepsilon = \gamma_g^\varepsilon \times (0, L). \tag{28.1}$$

We consider the beam subjected to the action of body forces, per unit volume, $\boldsymbol{f}^\varepsilon = (f_i^\varepsilon) : \Omega^\varepsilon \to \mathbb{R}^3$ and surface forces, per unit area, $\boldsymbol{g}^\varepsilon = (g_i^\varepsilon) : \Gamma^\varepsilon \to \mathbb{R}^3$ over the part Γ^ε of the lateral surface.

Moreover, we also assume that due to the action of this system of applied loads, the lateral surface Γ_C^ε *may come into contact, without friction, with a foundation.*

We introduce the function $s^\varepsilon : \Gamma_C^\varepsilon \to \mathbb{R}^+$ defining the initial gap, measured in the direction of the exterior normal vector $\boldsymbol{n}^\varepsilon$, between the foundation and the boundary Γ_C^ε. In the first place, we assume that *the foundation is rigid* and consequently the normal displacement component on Γ_C^ε, $u_n^\varepsilon = u_\alpha^\varepsilon n_\alpha^\varepsilon$, is less or equal than the initial gap s^ε.

This type of problems is studied in detail, for example in DUVAUT and LIONS [1972a], HLAVÁČEK, HASLINGER, NEČAS and LOVÍŠEK [1986], KIKUCHI and ODEN [1988], ODEN and PIRES [1983], PANAGIOTOPOULOS [1985], RODRIGUES [1987]. The equilibrium equations for this problem now take the form

$$-\partial_j^\varepsilon \sigma_{ij}^\varepsilon(\boldsymbol{u}^\varepsilon) = f_i^\varepsilon, \quad \text{in } \Omega^\varepsilon, \tag{28.2}$$

$$\sigma_{ij}^\varepsilon(\boldsymbol{u}^\varepsilon) n_j^\varepsilon = \sigma_{i\alpha}^\varepsilon(\boldsymbol{u}^\varepsilon) n_\alpha^\varepsilon = g_i^\varepsilon, \quad \text{on } \Gamma^\varepsilon, \tag{28.3}$$

$$u_i^\varepsilon = 0, \quad \text{on } \Gamma_0^\varepsilon \cup \Gamma_L^\varepsilon, \tag{28.4}$$

$$u_n^\varepsilon \leqslant s^\varepsilon, \qquad \sigma_n^\varepsilon \leqslant 0, \qquad \sigma_{ti}^\varepsilon = 0, \qquad \sigma_n^\varepsilon(u_n^\varepsilon - s^\varepsilon) = 0, \qquad \text{on } \Gamma_C^\varepsilon, \tag{28.5}$$

where, the following classical notation was used:

$$u_n^\varepsilon = u_i^\varepsilon n_i^\varepsilon, \qquad \sigma_n^\varepsilon = \sigma_{ij}^\varepsilon n_i^\varepsilon n_j^\varepsilon, \qquad \sigma_{ti}^\varepsilon = \sigma_{ij}^\varepsilon n_j^\varepsilon - \sigma_n^\varepsilon n_i^\varepsilon. \tag{28.6}$$

We consider that the beam is made of a homogeneous, isotropic linear elastic material, so that the constitutive equation is given by (cf. (1.9))

$$\sigma_{ij}^{\varepsilon}(\boldsymbol{u}^{\varepsilon}) = \frac{E^{\varepsilon}}{1+\nu^{\varepsilon}} e_{ij}^{\varepsilon}(\boldsymbol{u}^{\varepsilon}) + \frac{\nu^{\varepsilon} E^{\varepsilon}}{(1+\nu^{\varepsilon})(1-2\nu^{\varepsilon})} e_{pp}^{\varepsilon}(\boldsymbol{u}^{\varepsilon})\delta_{ij}. \tag{28.7}$$

We refer to BREZIS [1971], DUVAUT and LIONS [1972a], BREZZI, HAGER and RAVIART [1977], HASLINGER [1977, 1980, 1981], GLOWINSKI [1984], ODEN and KIM [1982], VIAÑO [1985b, 1986a,b] for variational formulation, existence and numerical approximation of general contact problems in elasticity.

Let us assume that

$$f_i^{\varepsilon} \in L^2(\Omega^{\varepsilon}), \qquad g_i^{\varepsilon} \in L^2(\Gamma^{\varepsilon}), \qquad s^{\varepsilon} \in L^2(\Gamma_C^{\varepsilon}), \tag{28.8}$$

and let $V(\Omega^{\varepsilon})$ denote the space defined in (1.10). Then, boundary conditions (28.4) and (28.5) lead us to look for the admissible displacements in the following nonempty, closed, convex subset of $V(\Omega^{\varepsilon})$:

$$K(\Omega^{\varepsilon}) = \{\boldsymbol{v}^{\varepsilon} = (v_i^{\varepsilon}) \in V(\Omega^{\varepsilon}) \colon v_n^{\varepsilon} = v_i^{\varepsilon} n_i^{\varepsilon} \leqslant s^{\varepsilon} \text{ a.e. on } \Gamma_C^{\varepsilon}\}. \tag{28.9}$$

In DUVAUT and LIONS [1972a] the equivalence between problem (28.2)–(28.6) and the following variational inequality is proved:

$$\begin{aligned} &\boldsymbol{u}^{\varepsilon} \in K(\Omega^{\varepsilon}), \\ &\int_{\Omega^{\varepsilon}} \sigma_{ij}^{\varepsilon}(\boldsymbol{u}^{\varepsilon}) e_{ij}^{\varepsilon}(\boldsymbol{v}^{\varepsilon} - \boldsymbol{u}^{\varepsilon})\, \mathrm{d}\boldsymbol{x}^{\varepsilon} \\ &\qquad \geqslant \int_{\Omega^{\varepsilon}} f_i^{\varepsilon}(v_i^{\varepsilon} - u_i^{\varepsilon})\, \mathrm{d}\boldsymbol{x}^{\varepsilon} + \int_{\Gamma^{\varepsilon}} g_i^{\varepsilon}(v_i^{\varepsilon} - u_i^{\varepsilon})\, \mathrm{d}a^{\varepsilon}, \quad \text{for all } \boldsymbol{v}^{\varepsilon} \in K(\Omega^{\varepsilon}). \end{aligned} \tag{28.10}$$

Since the bilinear form

$$B^{\varepsilon}(\boldsymbol{u}^{\varepsilon}, \boldsymbol{v}^{\varepsilon}) = \int_{\Omega^{\varepsilon}} \sigma_{ij}^{\varepsilon}(\boldsymbol{u}^{\varepsilon}) e_{ij}^{\varepsilon}(\boldsymbol{v}^{\varepsilon})\, \mathrm{d}\boldsymbol{x}^{\varepsilon}, \tag{28.11}$$

is symmetric, continuous and $V(\Omega^{\varepsilon})$-elliptic (cf. CIARLET [1988, Theorem 6.3-3]), we conclude that problem (28.10) possesses a unique solution, and moreover this solution is also the minimizer of the following problem (cf. EKELAND and TEMAM [1975], GLOWINSKI [1984]):

$$\boldsymbol{u}^{\varepsilon} \in K(\Omega^{\varepsilon}), \qquad J^{\varepsilon}(\boldsymbol{u}^{\varepsilon}) = \min_{\boldsymbol{v}^{\varepsilon} \in K(\Omega^{\varepsilon})} J^{\varepsilon}(\boldsymbol{v}^{\varepsilon}), \tag{28.12}$$

where

$$J^{\varepsilon}(\boldsymbol{v}^{\varepsilon}) = \tfrac{1}{2} \int_{\Omega^{\varepsilon}} \sigma_{ij}^{\varepsilon}(\boldsymbol{v}^{\varepsilon}) e_{ij}^{\varepsilon}(\boldsymbol{v}^{\varepsilon})\, \mathrm{d}\boldsymbol{x}^{\varepsilon} - \int_{\Omega^{\varepsilon}} f_i^{\varepsilon} v_i^{\varepsilon}\, \mathrm{d}\boldsymbol{x}^{\varepsilon} - \int_{\Gamma^{\varepsilon}} g_i^{\varepsilon} v_i^{\varepsilon}\, \mathrm{d}a^{\varepsilon}. \tag{28.13}$$

For the present case, it is also easier to use a mixed displacement–stress variational formulation. In order to do so one must invert the constitutive equation (28.7) and consider the displacement $\boldsymbol{u}^\varepsilon$ and the stress fields $\boldsymbol{\sigma}^\varepsilon$ as unknowns. One has (cf. (6.10)–(6.12))

$$\boldsymbol{u}^\varepsilon \in K(\Omega^\varepsilon), \qquad \boldsymbol{\sigma}^\varepsilon \in \Sigma(\Omega^\varepsilon), \tag{28.14}$$

$$\int_{\Omega^\varepsilon} [\frac{1+\nu^\varepsilon}{E^\varepsilon}\sigma_{ij}^\varepsilon - \frac{\nu^\varepsilon}{E^\varepsilon}\sigma_{pp}^\varepsilon\delta_{ij}]\tau_{ij}^\varepsilon \, \mathrm{d}\boldsymbol{x}^\varepsilon$$
$$= \int_{\Omega^\varepsilon} e_{ij}^\varepsilon(\boldsymbol{u}^\varepsilon)\tau_{ij}^\varepsilon \, \mathrm{d}\boldsymbol{x}^\varepsilon, \quad \text{for all } \boldsymbol{\tau}^\varepsilon \in \Sigma(\Omega^\varepsilon), \tag{28.15}$$

$$\int_{\Omega^\varepsilon} \sigma_{ij}^\varepsilon e_{ij}^\varepsilon(\boldsymbol{v}^\varepsilon - \boldsymbol{u}^\varepsilon) \, \mathrm{d}\boldsymbol{x}^\varepsilon$$
$$\geqslant \int_{\Omega^\varepsilon} f_i^\varepsilon(v_i^\varepsilon - u_i^\varepsilon) \, \mathrm{d}\boldsymbol{x}^\varepsilon + \int_{\Gamma^\varepsilon} g_i^\varepsilon(v_i^\varepsilon - u_i^\varepsilon) \, \mathrm{d}a^\varepsilon, \quad \text{for all } \boldsymbol{v}^\varepsilon \in K(\Omega^\varepsilon). \tag{28.16}$$

The main objective is now to study the behaviour of $(\boldsymbol{u}^\varepsilon, \boldsymbol{\sigma}^\varepsilon)$ as ε becomes very small. In order to do so we proceed as before: we introduce the change of variable (2.4) and the scalings on the unknowns defined by (2.6) and (6.15) in order to pass to an equivalent formulation written in the reference domain $\Omega = \omega \times (0, L)$.

We set

$$\Gamma_C = \Gamma_C^1, \qquad \Gamma = \Gamma^1, \tag{28.17}$$

and we consider that hypotheses (2.7)–(2.9) on the data hold. Moreover, we consider that the gap function $s^\varepsilon : \Gamma_C^\varepsilon \to \mathbb{R}^+$ is such that

$$s^\varepsilon(\boldsymbol{x}^\varepsilon) = \varepsilon^{-1}s(\boldsymbol{x}), \quad \text{for all } \boldsymbol{x}^\varepsilon = \Pi^\varepsilon(\boldsymbol{x}) \in \Gamma_C^\varepsilon, \ \boldsymbol{x} \in \Gamma_C, \tag{28.18}$$

where s is a function independent of ε. In the case of conditions (2.11)–(2.13) one should replace ε^{-1} by ε^{-1+r} in (28.18).

We define the following nonempty, closed, convex subset of $V(\Omega)$:

$$K(\Omega) = \{\boldsymbol{v} = (v_i) \in V(\Omega) \colon v_n = v_i n_i \leqslant s \text{ a.e. on } \Gamma_C\}. \tag{28.19}$$

Using notations (6.20)–(6.26) one immediately obtains the following theorem (cf. (6.25), (6.26)).

THEOREM 28.1. *The scaled displacement and stress pair $(\boldsymbol{u}(\varepsilon), \boldsymbol{\sigma}(\varepsilon))$ are the unique solution of the following problem:*

$$\boldsymbol{u}(\varepsilon) \in K(\Omega), \qquad \boldsymbol{\sigma}(\varepsilon) \in \Sigma(\Omega), \tag{28.20}$$

$$a_0(\boldsymbol{\sigma}(\varepsilon), \boldsymbol{\tau}) + \varepsilon^2 a_2(\boldsymbol{\sigma}(\varepsilon), \boldsymbol{\tau}) + \varepsilon^4 a_4(\boldsymbol{\sigma}(\varepsilon), \boldsymbol{\tau}) + b(\boldsymbol{\tau}, \boldsymbol{u}(\varepsilon)) = 0, \tag{28.21}$$
$$\textit{for all } \boldsymbol{\tau} \in \Sigma(\Omega),$$

$$b(\boldsymbol{\sigma}(\varepsilon), \boldsymbol{v} - \boldsymbol{u}(\varepsilon)) \leqslant F(\boldsymbol{v} - \boldsymbol{u}(\varepsilon)), \quad \textit{for all } \boldsymbol{v} \in K(\Omega). \tag{28.22}$$

Let us now assume that the following asymptotic expansion holds:

$$(\boldsymbol{u}(\varepsilon), \boldsymbol{\sigma}(\varepsilon)) = (\boldsymbol{u}^0, \boldsymbol{\sigma}^0) + \varepsilon^2(\boldsymbol{u}^2, \boldsymbol{\sigma}^2) + \varepsilon^4(\boldsymbol{u}^4, \boldsymbol{\sigma}^4) + \text{h.o.t.} \tag{28.23}$$

Substituting back into (28.20)–(28.22) one obtains that the zeroth order terms must be a solution of the following problem:

$$\begin{aligned} &\boldsymbol{u}^0 \in K(\Omega), \qquad \boldsymbol{\sigma}^0 \in \Sigma(\Omega), \\ &a_0(\boldsymbol{\sigma}^0, \boldsymbol{\tau}) + b(\boldsymbol{\tau}, \boldsymbol{u}^0) = 0, \quad \text{for all } \boldsymbol{\tau} \in \Sigma(\Omega), \\ &b(\boldsymbol{\sigma}^0, \boldsymbol{v} - \boldsymbol{u}^0) \leqslant F(\boldsymbol{v} - \boldsymbol{u}^0), \quad \text{for all } \boldsymbol{v} \in K(\Omega). \end{aligned} \tag{28.24}$$

In order to characterize elements $(\boldsymbol{u}^0, \boldsymbol{\sigma}^0)$ satisfying (28.24) we introduce the functions spaces $W_\alpha(\Omega)$ defined in (7.2) in such a way that $V(\Omega) = W_2(\Omega) \times W_1(\Omega)$. From this decomposition and considering that in $\Gamma_C \cup \Gamma$ the outward unit normal vector is of the form $(n_1, n_2, 0)$, we have

$$K(\Omega) = K_2(\Omega) \times W_1(\Omega), \tag{28.25}$$

where

$$K_2(\Omega) = \{(v_\beta) \in W_2(\Omega) \colon v_\alpha n_\alpha \leqslant s \text{ a.e. on } \Gamma_C\}. \tag{28.26}$$

Consequently, taking into account that $(0,0) \in K_2(\Omega)$ since $s \geqslant 0$, Eq. (28.22) is equivalent to

$$\begin{aligned} &(u_\beta(\varepsilon)) \in K_2(\Omega), \\ &\int_\Omega \sigma_{\alpha\beta}(\varepsilon)\partial_\alpha(v_\beta - u_\beta(\varepsilon))\,\mathrm{d}\boldsymbol{x} + \int_\Omega \sigma_{3\beta}(\varepsilon)\partial_3(v_\beta - u_\beta(\varepsilon))\,\mathrm{d}\boldsymbol{x} \\ &\quad \geqslant \int_\Omega f_\beta(v_\beta - u_\beta(\varepsilon))\,\mathrm{d}\boldsymbol{x} + \int_\Omega g_\beta(v_\beta - u_\beta(\varepsilon))\,\mathrm{d}a, \quad \text{for all } v_\beta \in K_2(\Omega), \end{aligned} \tag{28.27}$$

$$\begin{aligned} &u_3(\varepsilon) \in W_1(\Omega), \\ &\int_\Omega \sigma_{3\beta}(\varepsilon)\partial_\beta v_3\,\mathrm{d}\boldsymbol{x} + \int_\Omega \sigma_{33}(\varepsilon)\partial_3 v_3\,\mathrm{d}\boldsymbol{x} \\ &\quad = \int_\Omega f_3 v_3\,\mathrm{d}\boldsymbol{x} + \int_\Gamma g_3 v_3\,\mathrm{d}a, \quad \text{for all } v_3 \in W_1(\Omega). \end{aligned} \tag{28.28}$$

In a similar way, problem (28.24) may be written in the following equivalent form

$$(u_\alpha^0) \in K_2(\Omega), \qquad u_3^0 \in W_1(\Omega), \qquad \boldsymbol{\sigma}^0 \in \Sigma(\Omega), \tag{28.29}$$

$$\int_\Omega \frac{1}{E}\sigma_{33}^0 \tau_{33}\, \mathrm{d}\boldsymbol{x} - \int_\Omega \partial_3 u_3^0 \tau_{33}\, \mathrm{d}\boldsymbol{x} = 0, \quad \text{for all } \tau_{33} \in L^2(\Omega), \tag{28.30}$$

$$\int_\Omega (\partial_3 u_\beta^0 + \partial_\beta u_3^0)\tau_{3\beta}\, \mathrm{d}\boldsymbol{x} = 0, \quad \text{for all } (\tau_{3\beta}) \in [L^2(\Omega)]^2, \tag{28.31}$$

$$\tfrac{1}{2}\int_\Omega (\partial_\alpha u_\beta^0 + \partial_\beta u_\alpha^0)\tau_{\alpha\beta}\, \mathrm{d}\boldsymbol{x} = 0, \quad \text{for all } (\tau_{\alpha\beta}) \in [L^2(\Omega)]_s^4, \tag{28.32}$$

$$\int_\Omega (\sigma_{33}^0 \partial_3 v_3 + \sigma_{3\beta}^0 \partial_\beta v_3)\, \mathrm{d}\boldsymbol{x} = \int_\Omega f_3 v_3\, \mathrm{d}\boldsymbol{x} + \int_\Gamma g_3 v_3\, \mathrm{d}a, \quad \text{for all } v_3 \in W_1(\Omega), \tag{28.33}$$

$$\int_\Omega [\sigma_{\alpha\beta}^0 \partial_\alpha (v_\beta - u_\beta^0) + \sigma_{3\beta}^0 \partial_3 (v_\beta - u_\beta^0)]\, \mathrm{d}\boldsymbol{x} \geqslant \int_\Omega f_\beta (v_\beta - u_\beta^0)\, \mathrm{d}\boldsymbol{x} + \int_\Gamma g_\beta (v_\beta - u_\beta^0)\, \mathrm{d}a, \quad \text{for all } (v_\beta) \in K_2(\Omega). \tag{28.34}$$

The following result partially characterizes the solutions of (28.29)–(28.34). It is the equivalent of Theorems 6.3 and 23.1 to the present case (cf. VIAÑO [1985a], ÁLVAREZ-DIOS [1986]).

THEOREM 28.2. *Let the system of applied forces be such that*

$$\begin{aligned} &f_\alpha \in L^2(\Omega), \qquad f_3 \in H^1[0, L; L^2(\omega)], \\ &g_\alpha \in L^2(\Gamma), \qquad g_3 \in H^1[0, L; L^2(\gamma_g)]. \end{aligned} \tag{28.35}$$

Then, any solution of problem (28.29)–(28.34) *is of the following form*

(i) *Displacements u_α^0 depend only on variable x_3 and are given by*

$$\begin{aligned} &\boldsymbol{u}^0 \in V_{\mathrm{BN}}(\Omega) \cap K(\Omega), \qquad (u_\alpha^0) \in [H_0^2(0, L)]^2 \cap K_2(\Omega), \\ &u_3^0 = \underline{u}_3^0 - x_\alpha \partial_3 u_\alpha^0, \qquad \underline{u}_3^0 \in H_0^1(0, L). \end{aligned} \tag{28.36}$$

(ii) *Displacements (u_α^0) are the unique solution of the following variational inequality (no sum on α):*

$$(u_\alpha^0) \in [H_0^2(0, L)]^2 \cap K_2(\Omega) \quad \textit{and for all } (v_\alpha^0) \in [H_0^2(0, L)]^2 \cap K_2(\Omega):$$

$$\int_0^L EI_\alpha \partial_{33} u_\alpha^0 \partial_{33} (v_\alpha^0 - u_\alpha^0)\, \mathrm{d}x_3 \geqslant \int_0^L \left[\int_\omega f_\alpha\, \mathrm{d}\omega + \int_{\gamma_g} g_\alpha\, \mathrm{d}\gamma\right] (v_\alpha^0 - u_\alpha^0)\, \mathrm{d}x_3$$

$$-\int_0^L \left[\int_\omega x_\beta f_3 \, \mathrm{d}\omega + \int_{\gamma_g} x_\beta g_3 \, \mathrm{d}\gamma \right] \partial_3 (v_\alpha^0 - u_\alpha^0) \, \mathrm{d}x_3. \tag{28.37}$$

(iii) *The stretch component* $\underline{u}_3^0$ *is the unique solution of the following problem*:

$$\begin{aligned} &\underline{u}_3^0 \in H_0^1(0,L), \\ &\int_0^L EA\partial_3 \underline{u}_3^0 \partial_3 v_3^0 \, \mathrm{d}x_3 = \int_0^L \left[\int_\omega f_3 \, \mathrm{d}\omega + \int_{\gamma_g} g_3 \, \mathrm{d}\gamma \right] v_3^0 \, \mathrm{d}x_3, \\ &\quad \textit{for all } v_3^0 \in H_0^1(0,L). \end{aligned} \tag{28.38}$$

(iv) *The zeroth order axial stress component* σ_{33}^0 *is given by*

$$\sigma_{33}^0 = E\partial_3 u_3^0 = E(\partial_3 \underline{u}_3^0 - x_\alpha \partial_{33} u_\alpha^0), \tag{28.39}$$

and consequently the bending moment components m_β^0 *are given by*

$$m_\beta^0 = \int_\omega x_\beta \sigma_{33}^0 \, \mathrm{d}\omega = -EI_\beta \partial_{33} u_\beta^0 \quad (\textit{no sum on } \beta). \tag{28.40}$$

(v) *The zeroth order shear stress components* $\sigma_{3\beta}^0$ *are a solution of the following problem*:

$$\begin{aligned} &\sigma_{3\beta}^0 \in H^1[0,L;H^1(\omega)] \quad \textit{and a.e. in (0,L)}: \\ &-\partial_\beta \sigma_{3\beta}^0 = f_3 + \partial_3 \sigma_{33}^0, \quad \textit{in } \omega, \\ &\sigma_{3\beta}^0 n_\beta = g_3, \quad \textit{on } \gamma_g, \end{aligned} \tag{28.41}$$

therefore, the zeroth order shear force components q_β^0 *are given by*

$$q_\beta^0 = \int_\omega \sigma_{3\beta}^0 \, \mathrm{d}\omega = \partial_3 m_\beta^0 + \int_\omega x_\beta f_3 \, \mathrm{d}\omega + \int_{\gamma_g} x_\beta g_3 \, \mathrm{d}\gamma. \tag{28.42}$$

(vi) *The zeroth order plane stress components* $\sigma_{\alpha\beta}^0$ *are a solution of the following problem*:

$$\begin{aligned} &\sigma_{\alpha\beta}^0 \in L^2[0,L;H^1(\omega)] \quad \textit{and a.e. in (0,L)}: \\ &-\partial_\alpha \sigma_{\alpha\beta}^0 = f_\beta + \partial_3 \sigma_{3\beta}^0, \quad \textit{in } \omega, \\ &\sigma_{\alpha\beta}^0 n_\alpha = g_\beta, \quad \textit{on } \gamma_g, \\ &\sigma_{t\beta}^0 = 0, \qquad \sigma_n^0 \leqslant 0, \qquad \textit{on } \gamma_C. \end{aligned} \tag{28.43}$$

PROOF. In the proof we use the same ideas as in Theorems 6.3 and 23.1. We shall summarize them here. Equations (28.31) and (28.32) imply that $e_{\alpha\beta}(\boldsymbol{u}^0) = e_{3\beta}(\boldsymbol{u}^0) = 0$ and thus (28.36) follows. Expression (28.39) and consequently (28.40) follows from (28.30). In order to obtain (28.38) we substitute (28.39) into (28.33) and set $v_3 = v_3^0 \in H_0^1(0, L)$.

Equation (28.37) is obtained by choosing $v_\beta = v_\beta^0$ such that $(v_\beta^0) \in [H_0^2(0, L)]^2 \cap K_2(\Omega)$, in (28.34), and, in (28.33), $v_3 = x_\beta \partial_3(v_\beta^0 - u_\beta^0)$.

Equations (28.41) and (28.43) are a consequence of the strong interpretation of (28.33) and (28.34). Finally, expression (28.42) is obtained by setting $v_3 = x_\beta v_3^0$ in (28.33). □

REMARK 28.1. It is interesting to consider the following variational formulation equivalent to problems (28.37) and (28.38):

$$\begin{aligned}
&\boldsymbol{u}^0 \in V_{\mathrm{BN}}(\Omega) \cap K(\Omega),\\
&\int_\Omega E e_{33}(\boldsymbol{u}^0) e_{33}(\boldsymbol{v} - \boldsymbol{u}^0) \geqslant -F(\boldsymbol{v} - \boldsymbol{u}^0), \quad \text{for all } \boldsymbol{v} \in V_{\mathrm{BN}}(\Omega) \cap K(\Omega).
\end{aligned} \tag{28.44}$$

Among the new results, Eq. (28.37) is the most important one in this analysis: it represents a general bending model for a rod which may get in contact with an elastic foundation, written with respect to the reference domain Ω. In order to obtain the real model one should perform the change of variable (11.2), which we omit here in order to simplify the notation.

It is a model coupling both transverse bendings and although it is governed by one-dimensional equations, the convex set $K^0 = [H_0^2(0, L)]^2 \cap K^2(\Omega)$ of the admissible displacements still retains some three-dimensional information. We remark that K^0 may be written in the form

$$\begin{aligned}
K^0 &= [H_0^2(0, L)]^2 \cap K^2(\Omega)\\
&= \{\boldsymbol{\Psi} = (\Psi_\alpha) \in [H_0^2(0, L)]^2\colon\ \Psi_\alpha(x_3) n_\alpha(x_1, x_2) \leqslant s(x_1, x_2, x_3),\\
&\qquad \text{for all } x_3 \in (0, L), \text{ and a.e. } (x_1, x_2) \in \gamma_C\}.
\end{aligned} \tag{28.45}$$

Most of the well-known classical models may be obtained as particular cases of (28.37). Let us look at an example. Let us assume that the candidate contact surface Γ_C is plane and perpendicular to one of the inertia axes of the beam (Ox_1 for instance). Consequently the outward unit normal vector to Γ_C is constant and of the form $(\pm 1, 0, 0)$. In order to fix ideas let us consider that $\boldsymbol{n} = (-1, 0, 0)$. From (28.45) one deduces that the convex set of the admissible displacements K^0 is of the form

$$K^0 = U \times H_0^2(0, L),$$

where

$$U = \{\Phi \in H_0^2(0, L)\colon\ \Phi \geqslant -\hat{s} \text{ in } (0, L)\}, \tag{28.46}$$

with

$$\hat{s}(x_3) = \min_{(x_1,x_2)\in\gamma_C} s(x_1,x_2,x_3), \quad x_3 \in (0,L). \tag{28.47}$$

Setting successively, in (28.37), $(v_1^0, v_2^0) = (v_1^0, u_2^0)$, $v_1^0 \in U$, and $(v_1^0, v_2^0) = (u_1^0, v_2^0)$, $v_2^0 \in H_0^2(0,L)$, one sees that, in this case, (28.37) is equivalent to the following two problems:

$$\begin{aligned}
&u_1^0 \in U,\\
&\int_0^L EI_1 \partial_{33}u_1^0 \partial_{33}(v_1^0 - u_1^0)\,\mathrm{d}x_3\\
&\quad \geqslant \int_0^L \left[\int_\omega f_1\,\mathrm{d}\omega + \int_{\gamma_g} g_1\,\mathrm{d}\gamma\right](v_1^0 - u_1^0)\,\mathrm{d}x_3\\
&\quad - \int_0^L \left[\int_\omega x_1 f_3\,\mathrm{d}\omega + \int_{\gamma_g} x_1 g_3\,\mathrm{d}\gamma\right]\partial_3(v_1^0 - u_1^0)\,\mathrm{d}x_3, \quad \text{for all } v_1^0 \in U.
\end{aligned} \tag{28.48}$$

$$\begin{aligned}
&u_2^0 \in H_0^2(0,L),\\
&\int_0^L EI_2 \partial_{33}u_2^0 \partial_{33}v_2^0\,\mathrm{d}x_3\\
&\quad = \int_0^L \left[\int_\omega f_2\,\mathrm{d}\omega + \int_{\gamma_g} g_2\,\mathrm{d}\gamma\right] v_2^0\,\mathrm{d}x_3\\
&\quad - \int_0^L \left[\int_\omega x_2 f_3\,\mathrm{d}\omega + \int_{\gamma_g} x_2 g_3\,\mathrm{d}\gamma\right]\partial_3 v_2^0\,\mathrm{d}x_3, \quad \text{for all } v_2^0 \in H_0^2(0,L).
\end{aligned} \tag{28.49}$$

We observe that (28.49) is the usual bending model in the direction Ox_2 (cf. (7.59), (7.60)) and that (28.48) represents a classical *one-dimensional obstacle problem* (cf. CIMATTI [1973]). This type of variational inequality may be solved using the finite element method as can be seen in BERMÚDEZ and FERNÁNDEZ [1986]. As to the general model (28.37) its numerical treatment also using the finite element method for the approximation of variational inequalities may be seen in ÁLVAREZ-DIOS [1986].

REMARK 28.2. In order to complete this study, we should end up with the convergence analysis of $(\boldsymbol{u}(\varepsilon), \boldsymbol{\sigma}(\varepsilon))$ as ε goes to zero. This would allow us to justify the asymptotic expansion (28.27) together with problem (28.24). This can be done using a standard technique based on a priori estimates of the solution, weak convergence and strong convergence (with additional hypotheses if necessary). Although we shall not present it here, we remark that since $s \geqslant 0$ a.e. in Γ_C we may set $\boldsymbol{v} = \boldsymbol{0}$ in (28.22) and $\boldsymbol{\tau} = \boldsymbol{\sigma}(\varepsilon)$ in (28.21) so that we can write

$$a_0(\boldsymbol{\sigma}(\varepsilon), \boldsymbol{\sigma}(\varepsilon)) + \varepsilon^2 a_2(\boldsymbol{\sigma}(\varepsilon), \boldsymbol{\sigma}(\varepsilon)) + \varepsilon^4 a_4(\boldsymbol{\sigma}(\varepsilon), \boldsymbol{\sigma}(\varepsilon)) \\ = -b(\boldsymbol{\sigma}(\varepsilon), \boldsymbol{u}(\varepsilon)) \leqslant -F(\boldsymbol{u}(\varepsilon)).$$

Proceeding as in the proof of Theorem 10.1 the following a priori estimates are obtained:

$$|\sigma_{33}(\varepsilon)|_{0,\Omega} \leqslant C, \qquad \varepsilon|\sigma_{3\beta}(\varepsilon)|_{0,\Omega} \leqslant C, \qquad \varepsilon^2|\sigma_{\alpha\beta}(\varepsilon)|_{0,\Omega} \leqslant C, \tag{28.50}$$

$$\|\boldsymbol{u}(\varepsilon)\|_{1,\Omega} \leqslant C. \tag{28.51}$$

Consequently, one deduces the existence of at least a weakly convergent subsequence. The characterization of the limit of $\boldsymbol{u}(\varepsilon)$ and its identification with element $\boldsymbol{u}^0$ may be done as before if one takes into account some technical difficulties related to the form of the obstacle and of the candidate contact boundary Γ_C. Some hypotheses of the type

$$x_1 n_1 + x_2 n_2 \geqslant 0, \quad \text{a.e. in } \Gamma_C,$$

$$s(x_1, x_2, x_3) = \overline{u}_\alpha(x_3) n_\alpha(x_1, x_2), \quad \text{a.e. in } \Gamma_C,\ \overline{u}_\alpha \in H_0^2(0, L),$$

may greatly simplify the problem.

A closely related problem to the previous one is the case of the bending of a beam which may get in contact with a foundation made of an elastic material of the Winkler–Westergaard type. Assuming that penetration occurs in the direction of the outward unit normal and that the resisting force is proportional to it, we conclude that this type of material may be characterized by a distributed spring of rigidity k^ε acting normal to the contact boundary (cf. BIELAK and STEPHAN [1983] and respective bibliography). Constant k^ε depends only on the material the foundation is made of.

From what was said it is clear that now we may have $u_n^\varepsilon > s^\varepsilon$ in which case the foundation will react on the beam with a normal force proportional to the penetration. This means that boundary conditions (28.5) on the candidate contact surface Γ_C should be replaced by the following ones (cf. DUVAUT and LIONS [1972a]):

$$\begin{aligned} &\sigma_n^\varepsilon = 0, \quad \text{if } u_n^\varepsilon \leqslant s^\varepsilon, \\ &\sigma_n^\varepsilon = -k^\varepsilon(u_n^\varepsilon - s^\varepsilon), \quad \text{if } u_n^\varepsilon > s^\varepsilon,\ k^\varepsilon > 0, \\ &\sigma_{ti}^\varepsilon = 0. \end{aligned} \tag{28.52}$$

Thus, the problem under consideration is now described by Eqs. (28.2)–(28.4) and (28.52) with the notation of (28.6) together with the constitutive law (28.7).

We assume that the notations and hypotheses introduced for the case of a rigid foundation still hold, together with the standard notation $\varphi^+ = \max\{\varphi, 0\}$. Using a generalized Green formula, we may show that the three-dimensional elastic unilateral contact problem for a rod on an elastic foundation (28.1)–(28.4) and (28.52) is formally equivalent to the following nonlinear elliptic variational problem (cf. DUVAUT and LIONS [1972a]):

$$\begin{aligned}
&\boldsymbol{u}^\varepsilon \in V(\Omega^\varepsilon),\\
&\int_{\Omega^\varepsilon} \sigma_{ij}^\varepsilon(\boldsymbol{u}^\varepsilon) e_{ij}^\varepsilon(\boldsymbol{v}^\varepsilon)\,\mathrm{d}\boldsymbol{x}^\varepsilon + \int_{\Gamma_C^\varepsilon} k^\varepsilon (u_n^\varepsilon - s^\varepsilon)^+ v_n^\varepsilon\,\mathrm{d}a^\varepsilon \\
&\quad = \int_{\Omega^\varepsilon} f_i^\varepsilon v_i^\varepsilon\,\mathrm{d}\boldsymbol{x}^\varepsilon + \int_{\Gamma^\varepsilon} g_i^\varepsilon v_i^\varepsilon\,\mathrm{d}a^\varepsilon, \quad \text{for all } \boldsymbol{v}^\varepsilon \in V(\Omega^\varepsilon).
\end{aligned} \tag{28.53}$$

The existence, unicity and regularity of solutions to problems of this type can be seen in LIONS [1969], BREZIS, CRANDALL and PAZY [1970], BREZIS [1971], GLOWINSKI [1984].

The asymptotic expansion method and scalings considered in Section 4 may be applied to problem (28.53) without any major difficulty, except for the nonlinear contact term. As in the previous cases it is much easier to apply these techniques to a mixed displacement–stress formulation of problem (28.53).

In order to set up such a formulation we invert the constitutive law (28.7) and problem (28.53) becomes equivalent to the following mixed formulation:

$$\boldsymbol{u}^\varepsilon \in V(\Omega^\varepsilon), \qquad \boldsymbol{\sigma}^\varepsilon \in \Sigma(\Omega^\varepsilon): \tag{28.54}$$

$$\int_{\Omega^\varepsilon} \{\frac{1+\nu^\varepsilon}{E^\varepsilon}\sigma_{ij}^\varepsilon - \frac{\nu^\varepsilon}{E^\varepsilon}\sigma_{pp}^\varepsilon \delta_{ij}\}\tau_{ij}^\varepsilon\,\mathrm{d}\boldsymbol{x}^\varepsilon$$

$$= \int_{\Omega^\varepsilon} e_{ij}^\varepsilon(\boldsymbol{u}^\varepsilon)\tau_{ij}^\varepsilon\,\mathrm{d}\boldsymbol{x}^\varepsilon, \quad \text{for all } \boldsymbol{\tau}^\varepsilon \in \Sigma(\Omega^\varepsilon), \tag{28.55}$$

$$\begin{aligned}
&\int_{\Omega^\varepsilon} \sigma_{ij}^\varepsilon e_{ij}^\varepsilon(\boldsymbol{v}^\varepsilon)\,\mathrm{d}\boldsymbol{x}^\varepsilon + \int_{\Gamma_C^\varepsilon} k^\varepsilon (u_n^\varepsilon - s^\varepsilon)^+ v_n^\varepsilon\,\mathrm{d}a^\varepsilon \\
&\quad = \int_{\Omega^\varepsilon} f_i^\varepsilon v_i^\varepsilon\,\mathrm{d}\boldsymbol{x}^\varepsilon + \int_{\Gamma^\varepsilon} g_i^\varepsilon v_i^\varepsilon\,\mathrm{d}a^\varepsilon, \quad \text{for all } \boldsymbol{v}^\varepsilon \in V(\Omega^\varepsilon).
\end{aligned} \tag{28.56}$$

REMARK 28.3. Let $C(\Omega^\varepsilon)$ denote the following nonempty, closed, convex subset of $V(\Omega^\varepsilon) \times L^2(\Gamma_C^\varepsilon)$:

$$C(\Omega^\varepsilon) = \{(\boldsymbol{v}^\varepsilon, p^\varepsilon) \in V(\Omega^\varepsilon) \times L^2(\Gamma_C^\varepsilon) \colon p^\varepsilon + k^\varepsilon(v_n^\varepsilon - s^\varepsilon) \leqslant 0, \text{ a.e. in } \Gamma_C^\varepsilon\}. \tag{28.57}$$

Then, the previous problem is equivalent to

$$(\boldsymbol{u}^\varepsilon, \sigma_n^\varepsilon) \in C(\Omega^\varepsilon), \qquad \boldsymbol{\sigma}^\varepsilon \in \Sigma(\Omega^\varepsilon): \tag{28.58}$$

$$\int_{\Omega^\varepsilon} \{\frac{1+\nu^\varepsilon}{E^\varepsilon}\sigma_{ij}^\varepsilon - \frac{\nu^\varepsilon}{E^\varepsilon}\sigma_{pp}^\varepsilon\delta_{ij}\}\tau_{ij}^\varepsilon \, \mathrm{d}x^\varepsilon$$
$$= \int_{\Omega^\varepsilon} e_{ij}^\varepsilon(\boldsymbol{u}^\varepsilon)\tau_{ij}^\varepsilon \, \mathrm{d}x^\varepsilon, \quad \text{for all } \boldsymbol{\tau}^\varepsilon \in \Sigma(\Omega^\varepsilon), \tag{28.59}$$

$$\int_{\Omega^\varepsilon} \sigma_{ij}^\varepsilon e_{ij}^\varepsilon(\boldsymbol{v}^\varepsilon - \boldsymbol{u}^\varepsilon) \, \mathrm{d}x^\varepsilon + \int_{\Gamma_C^\varepsilon} \frac{1}{k^\varepsilon}\sigma_n^\varepsilon(p^\varepsilon - \sigma_n^\varepsilon) \, \mathrm{d}a^\varepsilon$$
$$\geqslant \int_{\Omega^\varepsilon} f_i^\varepsilon(v_i^\varepsilon - u_i^\varepsilon) \, \mathrm{d}x^\varepsilon \int_{\Gamma^\varepsilon} g_i^\varepsilon(v_i^\varepsilon - u_i^\varepsilon) \, \mathrm{d}a^\varepsilon, \quad \text{for all } (\boldsymbol{v}^\varepsilon, p^\varepsilon) \in C(\Omega^\varepsilon). \tag{28.60}$$

This formulation has the advantage of considering the normal contact pressure σ_n^ε as an unknown of the problem. The asymptotic expansion method may be adapted to this formulation also.

In order to study the behaviour of $(\boldsymbol{u}^\varepsilon, \boldsymbol{\sigma}^\varepsilon)$ as ε goes to zero, we make the change of variable (2.4) to the reference domain Ω and use the scalings in the unknowns (2.6) and (2.15). Moreover, besides the hypotheses on the data (2.7)–(2.9) and (28.18) we assume that the rigidity modulus of the foundation k^ε is such that

$$k^\varepsilon = \varepsilon^3 k, \tag{28.61}$$

where k is independent of ε.

REMARK 28.4. The veracity of this hypothesis depends on the material the elastic foundation is made of. If it holds then the contact effects show up in the first terms of the asymptotic expansion.

Using once again notation (6.20)–(6.26) together with a change of variable for integration it is possible to show the following result:

THEOREM 28.3. *The scaled displacement and stress fields* $(\boldsymbol{u}(\varepsilon), \boldsymbol{\sigma}(\varepsilon))$ *of problem* (28.54)–(28.56) *are the unique solution of the following problem*:

$$\boldsymbol{u}(\varepsilon) \in V(\Omega), \qquad \boldsymbol{\sigma}(\varepsilon) \in \Sigma(\Omega), \tag{28.62}$$
$$a_0(\boldsymbol{\sigma}(\varepsilon), \boldsymbol{\tau}) + \varepsilon^2 a_2(\boldsymbol{\sigma}(\varepsilon), \boldsymbol{\tau}) + \varepsilon^4 a_4(\boldsymbol{\sigma}(\varepsilon), \boldsymbol{\tau}) + b(\boldsymbol{\tau}, \boldsymbol{u}(\varepsilon)) = 0, \quad \textit{for all } \boldsymbol{\tau} \in \Sigma(\Omega), \tag{28.63}$$
$$b(\boldsymbol{\sigma}(\varepsilon), \boldsymbol{v}) - \int_{\Gamma_C} k(u_n(\varepsilon) - s)^+ v_n \, \mathrm{d}a = F(\boldsymbol{v}), \quad \textit{for all } \boldsymbol{v} \in V(\Omega). \tag{28.64}$$

Considering now a formal asymptotic expansion of the form (28.23) and substituting into the previous problem one concludes that $(\boldsymbol{u}^0, \boldsymbol{\sigma}^0)$ must be a solution of

$$
\begin{aligned}
&\boldsymbol{u}^0 \in V(\Omega), \qquad \boldsymbol{\sigma}^0 \in \Sigma(\Omega),\\
&a_0(\boldsymbol{\sigma}^0, \boldsymbol{\tau}) + b(\boldsymbol{\tau}, \boldsymbol{u}^0) = 0, \quad \text{for all } \boldsymbol{\tau} \in \Sigma(\Omega),\\
&b(\boldsymbol{\sigma}^0, \boldsymbol{v}) - \int_{\Gamma_C} k(u_n^0 - s)^+ v_n \,\mathrm{d}a = F(\boldsymbol{v}), \quad \text{for all } \boldsymbol{v} \in V(\Omega).
\end{aligned}
\tag{28.65}
$$

Writing these equations componentwise one obtains the following equivalent system:

$$
(u_\alpha^0) \in W_2(\Omega), \qquad u_3^0 \in W_1(\Omega), \qquad \boldsymbol{\sigma}^0 \in \Sigma(\Omega), \tag{28.66}
$$

$$
\int_\Omega \frac{1}{E}\sigma_{33}^0 \tau_{33}\,\mathrm{d}\boldsymbol{x} = \int_\Omega e_{ij}(\boldsymbol{u}^0)\tau_{ij}\,\mathrm{d}\boldsymbol{x}, \quad \text{for all } \boldsymbol{\tau} \in \Sigma(\Omega), \tag{28.67}
$$

$$
\begin{aligned}
&\int_\Omega \sigma_{\alpha\beta}^0 \partial_\alpha v_\beta \,\mathrm{d}\boldsymbol{x} + \int_\Omega \sigma_{3\beta}^0 \partial_3 v_\beta \,\mathrm{d}\boldsymbol{x} + \int_{\Gamma_C} k(u_n^0 - s)^+ v_n \,\mathrm{d}a\\
&\quad = \int_\Omega f_\alpha v_\alpha \,\mathrm{d}\boldsymbol{x} + \int_{\Gamma_C} g_\alpha v_\alpha \,\mathrm{d}a, \quad \text{for all } (v_\alpha) \in W_2(\Omega),
\end{aligned}
\tag{28.68}
$$

$$
\begin{aligned}
&\int_\Omega \sigma_{3\beta}^0 \partial_\beta v_3 \,\mathrm{d}\boldsymbol{x} + \int_\Omega \sigma_{33}^0 \partial_3 v_3 \,\mathrm{d}\boldsymbol{x}\\
&\quad = \int_\Omega f_3 v_3 \,\mathrm{d}\boldsymbol{x} + \int_{\Gamma_C} g_3 v_3 \,\mathrm{d}a, \quad \text{for all } v_3 \in W_1(\Omega).
\end{aligned}
\tag{28.69}
$$

With the same proof as in Theorem 28.2 it is possible to give the following partial characterization for the zeroth order displacement and stress fields $(\boldsymbol{u}^0, \sigma^0)$.

THEOREM 28.4. *Let us assume that the system of applied forces satisfies* (28.35). *Then, every solution of problem* (28.66)–(28.69) *is of the following form*

(i) *Displacements u_α^0 depend only on variable x_3 and are of the following form*

$$
\begin{aligned}
&\boldsymbol{u}^0 \in V_{\mathrm{BN}}(\Omega), \qquad (u_\alpha^0) \in [H_0^2(0, L)]^2,\\
&u_3^0 = \underline{u}_3^0 - x_\alpha \partial_3 u_\alpha^0, \qquad \underline{u}_3^0 \in H_0^1(0, L).
\end{aligned}
\tag{28.70}
$$

(ii) *Displacements (u_α^0) are the unique solution of the following problem (no sum on α):*

$(u_\alpha^0) \in [H_0^2(0, L)]^2$ *and for all* $(v_\alpha^0) \in [H_0^2(0, L)]^2$:

$$\int_0^L EI_\alpha \partial_{33} u_\alpha^0 \partial_{33} v_\alpha^0 \, \mathrm{d}x_3 + \int_0^L k[\int_{\gamma_C} (u_n^0 - s)^+ n_\alpha \, \mathrm{d}\gamma] v_\alpha \, \mathrm{d}x_3 \tag{28.71}$$

$$\int_0^L [\int_\omega f_\alpha \, \mathrm{d}\omega + \int_{\gamma_g} g_\alpha \, \mathrm{d}\gamma] v_\alpha^0 \, \mathrm{d}x_3 - \int_0^L [\int_\omega x_\alpha f_3 \, \mathrm{d}\omega + \int_{\gamma_g} x_\alpha g_3 \, \mathrm{d}\gamma] \partial_3 v_\alpha^0 \, \mathrm{d}x_3.$$

(iii) *The stretch component* $\underline{u}_3^0$ *is the unique solution of the following problem:*

$$\underline{u}_3^0 \in H_0^1(0, L),$$

$$\int_0^L EA \partial_3 \underline{u}_3^0 \partial_3 v_3^0 \, \mathrm{d}x_3 = \int_0^L [\int_\omega f_3 \, \mathrm{d}\omega + \int_{\gamma_g} g_3 \, \mathrm{d}\gamma] v_3^0 \, \mathrm{d}x_3, \tag{28.72}$$

for all $v_3^0 \in H_0^1(0, L)$.

(iv) *The zeroth order axial stress component* σ_{33}^0 *is given by*

$$\sigma_{33}^0 = E \partial_3 u_3^0 = E(\partial_3 \underline{u}_3^0 - x_\alpha \partial_{33} u_\alpha^0), \tag{28.73}$$

and consequently the bending moment components m_β^0 *are given by*

$$m_\beta^0 = \int_\omega x_\beta \sigma_{33}^0 \, \mathrm{d}\omega = -EI_\beta \partial_{33} u_\beta^0, \quad (\textit{no sum on } \beta). \tag{28.74}$$

(v) *The zeroth order shear stress components* $\sigma_{3\beta}^0$ *are a solution of the following problem:*

$$\begin{aligned} &\sigma_{3\beta}^0 \in H^1[0, L; H^1(\omega)] \quad \textit{and a.e. in } (0, L): \\ &-\partial_\beta \sigma_{3\beta}^0 = f_3 + \partial_3 \sigma_{33}^0, \quad \textit{in } \omega, \\ &\sigma_{3\beta}^0 n_\beta = g_3, \quad \textit{on } \gamma_g, \end{aligned} \tag{28.75}$$

thus, the the zeroth order shear force components q_β^0 *are given by*

$$q_\beta^0 = \int_\omega \sigma_{3\beta}^0 \, \mathrm{d}\omega = \partial_3 m_\beta^0 + \int_\omega x_\beta f_3 \, \mathrm{d}\omega + \int_{\gamma_g} x_\beta g_3 \, \mathrm{d}\gamma. \tag{28.76}$$

(vi) *The zeroth order plane stress components* $\sigma_{\alpha\beta}^0$ *are a solution of the following problem:*

$$\begin{aligned} &\sigma_{\alpha\beta}^0 \in L^2[0, L; H^1(\omega)] \quad \textit{and a.e. in } (0, L): \\ &-\partial_\alpha \sigma_{\alpha\beta}^0 = f_\beta + \partial_3 \sigma_{3\beta}^0, \quad \textit{in } \omega, \\ &\sigma_{\alpha\beta}^0 n_\alpha = g_\beta, \quad \textit{on } \gamma_g, \\ &\sigma_{t\beta}^0 = 0, \quad \sigma_n^0 + k(u_n^0 - s)^+ = 0, \quad \textit{on } \gamma_C. \end{aligned} \tag{28.77}$$

The most important result in this theorem is Eq. (28.71) because it constitutes a general Bernoulli–Navier bending model for beams that may get into contact with an elastic foundation. It generalizes most of the classical models used in engineering literature. Moreover, we observe that it is a model that couples both bending components through the nonlinear contact term, which in turn conserves three-dimensional information on γ_C.

Introducing the classical notation

$$F_\beta = \int_\omega f_\beta \, \mathrm{d}\omega + \int_\gamma g_\beta \, \mathrm{d}\gamma, \qquad M_\beta = \int_\omega x_\beta f_3 \, \mathrm{d}\omega + \int_\gamma x_\beta g_3 \, \mathrm{d}\gamma, \tag{28.78}$$

together with the *normal contact pressure* of the foundation along direction Ox_β:

$$p_\beta^0 = \int_{\gamma_C} k(u_n^0 - s)^+ n_\beta \, \mathrm{d}\gamma, \tag{28.79}$$

then, Eq. (28.71) possesses the following strong interpretation:

$$EI_\beta \partial_{3333} u_\beta^0 + p_\beta^0 = F_\beta + \partial_3 M_\beta, \quad \text{in } (0, L) \quad (\text{no sum on } \beta), \tag{28.80}$$

which coincides with the classical form adopted for these models (cf. for example BIELAK and STEPHAN [1983, Eq. (1.1)].

The most well-known classical models may be obtained as particular cases of (28.71). Let us assume, as an example, that *the candidate contact boundary Γ_C is plane and perpendicular to axis Ox_1* with an outward normal vector $(-1, 0, 0)$. Moreover, let us assume that the foundation is such that its distance s depends only on the variable x_3, i.e., $s(x_1, x_2, x_3) = \hat{s}(x_3)$. Then, in this case one has

$$p_1^0 = -k|\gamma_C|(-u_1^0 - s)^+, \qquad p_2^0 = 0, \tag{28.81}$$

or explicitly

$$p_1^0 = \begin{cases} 0 & \text{if } -u_1^0 < \hat{s}, \\ -k|\gamma_C|(-u_1^0 - s) & \text{if } -u_1^0 > \hat{s}. \end{cases} \tag{28.82}$$

In this particular case Eq. (28.71) is equivalent to the two following problems:

$$\begin{aligned} & u_1^0 \in H_0^2(0, L), \\ & \int_0^L EI_1 \partial_{33} u_1^0 \partial_{33} v \, \mathrm{d}x_3 - k|\gamma_C| \int_0^L (-u_n^0 - \hat{s})^+ v \, \mathrm{d}x_3 \\ & \quad = \int_0^L F_1 v \, \mathrm{d}x_3 - \int_0^L M_1 \partial_3 v \, \mathrm{d}x_3, \quad \text{for all } v \in H_0^2(0, L), \end{aligned} \tag{28.83}$$

$$u_2^0 \in H_0^2(0, L),$$
$$\int_0^L EI_2 \partial_{33} u_2^0 \partial_{33} v \, dx_3 = \int_0^L F_2 v \, dx_3 - \int_0^L M_2 \partial_3 v \, dx_3, \tag{28.84}$$
for all $v \in H_0^2(0, L)$.

Problem (28.84) is a classical bending model in direction Ox_2 (cf. (7.59), (7.60)) and problem (28.83) represents *a classical bending model of a beam on an elastic foundation, which was initially at a distance* $\hat{s}$ and whose strong interpretation is

$$EI_1 \partial_{3333} u_1^0 - k|\gamma_C|(-u_1^0 - s)^+ = F_1 + \partial_3 M_1, \quad \text{in } (0, L). \tag{28.85}$$

Problems of this type and their numerical treatment may be seen in CIMATTI [1973], BIELAK and STEPHAN [1983] and BERMÚDEZ and FERNÁNDEZ [1986].

REMARK 28.5. The method presented here should be justified with a convergence analysis which can be achieved essentially as in Section 10.

29. The asymptotic expansion method for rods with a variable cross section

We use all the conventions and notation of the previous sections unless otherwise stated. In what concerns the calculation of the first term of the asymptotic expansion we follow the work of BERMÚDEZ and VIAÑO [1984].

Starting with the reference domain $\omega \subset \mathbb{R}^2$ with a sufficiently smooth boundary $\gamma = \gamma_0 \cup \gamma_1 \cup \cdots \cup \gamma_p$, we define the rod under study as a linearly elastic solid occupying volume:

$$\Omega^\varepsilon = \{(x_1^\varepsilon, x_2^\varepsilon, x_3^\varepsilon) = (\varepsilon x_1 h(x_3), \varepsilon x_2 h(x_3), x_3)\colon (x_1, x_2) \in \omega, x_3 \in (0, L)\}. \tag{29.1}$$

It is thus a rod of length L, and the cross section at a distance x_3 of the origin of the system of axes is homothetic to the reference cross section ω with ratio $\varepsilon h(x_3)$. Denoting this cross section by $\omega^\varepsilon(x_3^\varepsilon)$ one has

$$\omega^\varepsilon(x_3^\varepsilon) = \varepsilon h(x_3)\omega = \{(\varepsilon x_1 h(x_3), \varepsilon x_2 h(x_3))\colon (x_1, x_2) \in \omega\}. \tag{29.2}$$

We assume that function h defining the change in the cross section satisfies the following conditions:

$$h : [0, L] \to \mathbb{R}^+, \qquad h \in W^{2,\infty}(0, L), \qquad h(x_3) \geqslant c > 0, \quad \text{for all } x_3 \in [0, L], \tag{29.3}$$

and whenever appropriate we denote by h' its derivative with respect to x_3.

Since we considered that system $Ox_1x_2x_3$ is a principal system of inertia for Ω we conclude that system $Ox_1^\varepsilon x_2^\varepsilon x_3^\varepsilon$ is also a principal system of inertia for Ω^ε. In particular this means that the longitudinal axis Ox_3^ε passes through the centroids of all the cross sections $\omega^\varepsilon(x_3^\varepsilon)$ and therefore

$$\int_{\omega^\varepsilon(x_3^\varepsilon)} x_\alpha^\varepsilon \, \mathrm{d}\omega^\varepsilon = \int_{\omega^\varepsilon(x_3^\varepsilon)} x_1^\varepsilon x_2^\varepsilon \, \mathrm{d}\omega^\varepsilon = 0, \quad x_3 \in (0, L).$$

We denote by $I_\alpha^\varepsilon(x_3)$ the second moments of area of the cross section $\omega^\varepsilon(x_3^\varepsilon)$, that is

$$I_\alpha^\varepsilon(x_3) = \int_{\omega^\varepsilon(x_3^\varepsilon)} (x_\alpha^\varepsilon)^2 \, \mathrm{d}\omega^\varepsilon, \tag{29.4}$$

and use the following additional notations:

$$\gamma^\varepsilon(x_3^\varepsilon) = \partial\omega^\varepsilon(x_3^\varepsilon) = \{(\varepsilon x_1 h(x_3), \varepsilon x_2 h(x_3)) \colon (x_1, x_2) \in \gamma\}, \tag{29.5}$$

$$\Gamma_0^\varepsilon = \omega^\varepsilon(0) \times \{0\}, \qquad \Gamma_L^\varepsilon = \omega^\varepsilon(L) \times \{L\}, \qquad \Gamma^\varepsilon = \bigcup_{x_3 \in (0,L)} \gamma^\varepsilon(x_3) \times \{x_3\}. \tag{29.6}$$

REMARK 29.1. Whenever there is no danger of confusion and in order to simplify the notation we will identify $\omega^\varepsilon(x_3) \times \{x_3\}$ with $\omega^\varepsilon(x_3)$ and $\gamma^\varepsilon(x_3) \times \{x_3\}$ with $\gamma^\varepsilon(x_3)$.

REMARK 29.2. In the work of BERMÚDEZ and VIAÑO [1984] a more general case is considered. In fact the beam under consideration there is of the form

$$\Omega^\varepsilon = \{(x_1^\varepsilon, x_2^\varepsilon, x_3^\varepsilon) = (\varepsilon x_1 h_1(x_3), \varepsilon x_2 h_2(x_3), x_3) \colon x_3 \in (0, L), (x_1, x_2) \in \omega\}, \tag{29.7}$$

where functions h_α satisfy conditions (29.3). The centroids are kept on the axial axis but the homotheties may be different along the tranverse directions. This case may be considered exactly in the same way as the one we are going to consider here for the geometry (29.1). However, since the calculation of the second order terms is very lengthy (more than one page for the torsion term alone) we decided to present a simpler case here, that possesses, nevertheless, all the features of the more general situation.

Since we are interested in evaluating the first terms of the asymptotic expansion we assume that the rod is weakly clamped at both extremities. As in the previous cases we also assume that the rod is subjected to a system of body forces per unit volume $\boldsymbol{f}^\varepsilon = (f_i^\varepsilon)$ in Ω^ε and to a system of surface tractions per unit area $\boldsymbol{g}^\varepsilon = (g_i^\varepsilon)$ acting on the lateral surface Γ_C. Moreover, we consider the rod made of a linearly, elastic, homogeneous isotropic material.

In order to avoid unnecessary repetitions with respect to the previous sections (cf. Sections 1–6) we start the study of this case directly with the mixed displacement–stress variational formulation.

Defining the space of admissible displacements,

$$V(\Omega^\varepsilon) = \{\boldsymbol{v}^\varepsilon = (v_i^\varepsilon) \in [H^1(\Omega^\varepsilon)]^3\colon \int_{\omega^\varepsilon(a)} v_i^\varepsilon \, \mathrm{d}\omega^\varepsilon = 0,$$
$$\int_{\omega^\varepsilon(a)} (x_i^\varepsilon v_j^\varepsilon - x_j^\varepsilon v_i^\varepsilon) \, \mathrm{d}\omega^\varepsilon = 0, \ a = 0, L\}, \tag{29.8}$$

the three-dimensional elasticity problem in Ω^ε may be written as (cf. (6.10)–(6.12)):

$$\boldsymbol{u}^\varepsilon \in V(\Omega^\varepsilon), \qquad \boldsymbol{\sigma}^\varepsilon \in \Sigma(\Omega^\varepsilon), \tag{29.9}$$

$$\int_{\Omega^\varepsilon} \left[\frac{1+\nu^\varepsilon}{E^\varepsilon}\sigma_{ij}^\varepsilon - \frac{\nu^\varepsilon}{E^\varepsilon}\sigma_{pp}^\varepsilon \delta_{ij}\right] \tau_{ij}^\varepsilon \, \mathrm{d}\boldsymbol{x}^\varepsilon$$
$$= \int_{\Omega^\varepsilon} e_{ij}^\varepsilon(\boldsymbol{u}^\varepsilon)\tau_{ij}^\varepsilon \, \mathrm{d}\boldsymbol{x}^\varepsilon, \quad \text{for all } \boldsymbol{\tau}^\varepsilon \in \Sigma(\Omega^\varepsilon), \tag{29.10}$$

$$\int_{\Omega^\varepsilon} \sigma_{ij}^\varepsilon e_{ij}^\varepsilon(\boldsymbol{v}^\varepsilon) \, \mathrm{d}\boldsymbol{x}^\varepsilon = \int_{\Omega^\varepsilon} f_i^\varepsilon v_i^\varepsilon \, \mathrm{d}\boldsymbol{x}^\varepsilon + \int_{\Gamma^\varepsilon} g_i^\varepsilon v_i^\varepsilon \, \mathrm{d}a^\varepsilon, \quad \text{for all } \boldsymbol{v}^\varepsilon \in V(\Omega^\varepsilon). \tag{29.11}$$

In order to study the behaviour of the solution of this problem when ε goes to zero, we use the same method: change of variable to a reference rod, scaling on the unknowns and on the data, asymptotic expansion and finally study of the convergence analysis.

The reference rod will be the same as before, that is,

$$\Omega = \omega \times (0, L). \tag{29.12}$$

We consider that all the notations and definitions concerning the reference beam and defined in Sections 1–10 are in force: lateral surface Γ, extremities Γ_0 and Γ_L, moments of area, functions and constants associated with the geometry, and so on.

The natural scaling used to transform problem (29.9)–(29.11) into an equivalent one, posed in Ω, is the following:

$$\Pi^\varepsilon : \boldsymbol{x} = (x_1, x_2, x_3) \in \overline{\Omega} \to \boldsymbol{x}^\varepsilon = \Pi^\varepsilon(\boldsymbol{x})$$
$$= (x_1^\varepsilon, x_2^\varepsilon, x_3^\varepsilon) = (\varepsilon x_1 h(x_3), \varepsilon x_2 h(x_3), x_3) \in \overline{\Omega}^\varepsilon. \tag{29.13}$$

One then has

$$\Omega^\varepsilon = \Pi^\varepsilon(\Omega), \qquad \Gamma^\varepsilon = \Pi^\varepsilon(\Gamma), \qquad \Gamma_0^\varepsilon = \Pi^\varepsilon(\Gamma_0), \qquad \Gamma_L^\varepsilon = \Pi^\varepsilon(\Gamma_L), \tag{29.14}$$

and the following result, which will be used later on, holds.

THEOREM 29.1. *Let* $\Phi^\varepsilon : \overline{\Omega}^\varepsilon \to \mathbb{R}$ *be a sufficiently regular function and let* $\Phi : \overline{\Omega} \to \mathbb{R}$ *be such that* $\Phi = \Phi^\varepsilon \circ \Pi^\varepsilon$, *that is,*

$$\Phi(\boldsymbol{x}) = \Phi^\varepsilon(\boldsymbol{x}^\varepsilon), \quad \textit{for all } \boldsymbol{x} \in \overline{\Omega}.$$

Then, one has

$$\int_{\Omega^\varepsilon} \Phi^\varepsilon \,\mathrm{d}\boldsymbol{x}^\varepsilon = \varepsilon^2 \int_{\Omega} h^2 \Phi \,\mathrm{d}\boldsymbol{x}, \tag{29.15}$$

$$\int_{\Gamma^\varepsilon} \Phi^\varepsilon \,\mathrm{d}a^\varepsilon = \varepsilon \int_{\Gamma} h h^*(\varepsilon) \Phi \,\mathrm{d}a, \tag{29.16}$$

where

$$\begin{aligned} &h^*(\varepsilon) = [1 + \varepsilon^2 (h')^2 (x_1 n_1 + x_2 n_2)^2]^{1/2}, \\ &\partial_\alpha^\varepsilon \Phi^\varepsilon(\boldsymbol{x}^\varepsilon) = \varepsilon^{-1} h^{-1}(x_3) \partial_\alpha \Phi(\boldsymbol{x}), \\ &\partial_3^\varepsilon \Phi^\varepsilon(\boldsymbol{x}^\varepsilon) = \partial_3 \Phi(\boldsymbol{x}) - h^{-1}(x_3) h'(x_3) x_\alpha \partial_\alpha \Phi(\boldsymbol{x}). \end{aligned} \tag{29.17}$$

Let us now assume that hypotheses (2.7)–(2.9) on the data hold with functions g_i such that g_i^* are independent of ε and where

$$g_i^* = h^*(\varepsilon) g_i. \tag{29.18}$$

More general hypotheses than (2.11)–(2.12) may be considered. On the reference spaces $V(\Omega)$ and $\Sigma(\Omega)$ we introduce the continuous bilinear forms $a_p(\cdot,\cdot) : \Sigma(\Omega) \times \Sigma(\Omega) \to \mathbb{R}$ $(p = 0, 2, 4)$, $b(\cdot,\cdot) : \Sigma(\Omega) \times V(\Omega) \to \mathbb{R}$ and the linear continuous form $F(\cdot) : V(\Omega) \to \mathbb{R}$ defined, for any element $\boldsymbol{\sigma}$, $\boldsymbol{\tau} \in \Sigma(\Omega)$, $\boldsymbol{v} \in V(\Omega)$, in the following way:

$$a_0(\boldsymbol{\sigma}, \boldsymbol{\tau}) = \int_\Omega \frac{1}{E} h^2 \sigma_{33} \tau_{33} \,\mathrm{d}\boldsymbol{x}, \tag{29.19}$$

$$a_2(\boldsymbol{\sigma}, \boldsymbol{\tau}) = \int_\Omega h^2 \left[\frac{2(1+\nu)}{E} \sigma_{3\beta} \tau_{3\beta} - \frac{\nu}{E} (\sigma_{33} \tau_{\mu\mu} + \sigma_{\mu\mu} \tau_{33}) \right] \mathrm{d}\boldsymbol{x}, \tag{29.20}$$

$$a_4(\boldsymbol{\sigma}, \boldsymbol{\tau}) = \int_\Omega h^2 \left[\frac{1+\nu}{E} \sigma_{\alpha\beta} - \frac{\nu}{E} \sigma_{\mu\mu} \delta_{\alpha\beta} \right] \tau_{\alpha\beta} \,\mathrm{d}\boldsymbol{x}, \tag{29.21}$$

$$b(\boldsymbol{\tau}, \boldsymbol{v}) = -\int_\Omega h^2 \tau_{ij} e_{ij}^*(\boldsymbol{v}) \,\mathrm{d}\boldsymbol{x}, \tag{29.22}$$

$$F(\boldsymbol{v}) = -\int_\Omega h^2 f_i v_i \,\mathrm{d}\boldsymbol{x} - \int_\Gamma h g_i^* v_i \,\mathrm{d}a, \tag{29.23}$$

where

$$\begin{aligned} e^*_{\alpha\beta}(\boldsymbol{v}) &= \tfrac{1}{2}h^{-1}(\partial_\alpha v_\beta + \partial_\beta v_\alpha),\\ e^*_{3\beta}(\boldsymbol{v}) &= e^*_{\beta 3}(\boldsymbol{v}) = \tfrac{1}{2}[h^{-1}\partial_\beta v_3 + \partial_3 v_\beta - h^{-1}h' x_\alpha \partial_\alpha v_\beta],\\ e^*_{33}(\boldsymbol{v}) &= \partial_3 v_3 - h^{-1}h' x_\alpha \partial_\alpha v_3. \end{aligned} \tag{29.24}$$

The next result will be used in what follows.

THEOREM 29.2. *The application*

$$\boldsymbol{v} \in V(\Omega) \to |\boldsymbol{e}^*(\boldsymbol{v})|_{0,\Omega},$$

is a norm in $V(\Omega)$ *equivalent to the* $\|\cdot\|_{1,\Omega}$ *norm.*

PROOF. Let

$$\Omega^* = \{(x_1 h(x_3), x_2 h(x_3), x_3) \in \mathbb{R}^3 \colon (x_1, x_2, x_3) \in \Omega\}. \tag{29.25}$$

Given $\boldsymbol{v} \in V(\Omega)$ let $\boldsymbol{v}^* \in V(\Omega^*)$ be the element defined by

$$\boldsymbol{v}^*(x_1 h(x_3), x_2 h(x_3), x_3) = \boldsymbol{v}(x_1, x_2, x_3). \tag{29.26}$$

Then, from Theorem 29.1, for $\varepsilon = 1$, one has

$$e^*_{ij}(\boldsymbol{v})(x_1, x_2, x_3) = e^*_{ij}(\boldsymbol{v}^*)(x_1 h(x_3), x_2 h(x_3), x_3).$$

Thus, using Korn's inequality in Ω^* we obtain

$$|\boldsymbol{e}^*(\boldsymbol{v})|_{0,\Omega} = |\boldsymbol{e}(\boldsymbol{v}) \circ \Pi^\varepsilon|_{0,\Omega} \geqslant C|\boldsymbol{e}^*(\boldsymbol{v})|_{0,\Omega^*} \geqslant C\|\boldsymbol{v}^*\|_{1,\Omega^*} \geqslant C\|\boldsymbol{v}\|_{1,\Omega}. \qquad \square$$

As a corollary the following Babuška–Brezzi type inequality is obtained.

THEOREM 29.3. *There exists a constant* $C > 0$ *such that*

$$\sup_{\boldsymbol{\tau} \in \Sigma(\Omega)} \frac{|b(\boldsymbol{\tau}, \boldsymbol{v})|}{|\boldsymbol{\tau}|_{0,\Omega}} \geqslant C\|\boldsymbol{v}\|_{1,\Omega}, \quad \textit{for all } \boldsymbol{v} \in V(\Omega).$$

PROOF. Given $\boldsymbol{v} \in V(\Omega)$, let $\boldsymbol{\tau}^* = e^*(\boldsymbol{v})$. From the previous theorem one has

$$\frac{|b(\boldsymbol{\tau}^*, \boldsymbol{v})|}{|\boldsymbol{\tau}^*|_{0,\Omega}} = \frac{\int_\Omega h^2 e^*_{ij}(\boldsymbol{v}) e^*_{ij}(\boldsymbol{v})\,\mathrm{d}\boldsymbol{x}}{|\boldsymbol{e}^*(\boldsymbol{v})|_{0,\Omega}} \geqslant C|\boldsymbol{e}^*(\boldsymbol{v})|_{0,\Omega} \geqslant C\|\boldsymbol{v}\|_{1,\Omega}. \qquad \square$$

Associated with hypotheses on the data (2.7)–(2.9) (or alternatively (2.12)–(2.13)) we consider the scaling on the unknowns and test functions given by (2.6) and (2.21) (or alternatively (2.11)). The following result is now a consequence of this scaling and of Theorem 29.1.

THEOREM 29.4. *The scaled displacement and stress fields* $(\boldsymbol{u}(\varepsilon), \boldsymbol{\sigma}(\varepsilon))$ *are the unique solution of the following mixed variational problem in* Ω:

$$\boldsymbol{u}(\varepsilon) \in V(\Omega), \qquad \boldsymbol{\sigma}(\varepsilon) \in \Sigma(\Omega), \tag{29.27}$$

$$a_0(\boldsymbol{\sigma}(\varepsilon), \boldsymbol{\tau}) + \varepsilon^2 a_2(\boldsymbol{\sigma}(\varepsilon), \boldsymbol{\tau}) + \varepsilon^4 a_4(\boldsymbol{\sigma}(\varepsilon), \boldsymbol{\tau}) + b(\boldsymbol{\tau}, \boldsymbol{u}(\varepsilon)) = 0, \quad \textit{for all } \boldsymbol{\tau} \in \Sigma(\Omega), \tag{29.28}$$

$$b(\boldsymbol{\sigma}(\varepsilon), \boldsymbol{v}) = F(\boldsymbol{v}), \quad \textit{for all } \boldsymbol{v} \in V(\Omega). \tag{29.29}$$

Following the standard asymptotic expansion technique, explained in the preceding sections, we consider a formal asymptotic expansion of the form

$$(\boldsymbol{u}(\varepsilon), \boldsymbol{\sigma}(\varepsilon)) = (\boldsymbol{u}^0, \boldsymbol{\sigma}^0) + \varepsilon^2(\boldsymbol{u}^2, \boldsymbol{\sigma}^2) + \varepsilon^4(\boldsymbol{u}^4, \boldsymbol{\sigma}^4) + \text{h.o.t.}, \tag{29.30}$$

and try to characterize the successive terms. Substituting into (29.28), (29.29) we conclude that functions $(\boldsymbol{u}^p, \boldsymbol{\sigma}^p)$ $(p = 0, 2, 4)$ must satisfy the following set of equations:

$$\int_\Omega \frac{1}{E} h^2 \sigma_{33}^0 \tau_{33} - \int_\Omega h^2 e_{33}^*(\boldsymbol{u}^0)\tau_{33} = 0, \quad \text{for all } \tau_{33} \in L^2(\Omega), \tag{29.31}$$

$$-\int_\Omega h^2 e_{3\beta}^*(\boldsymbol{u}^0)\tau_{3\beta}\, \mathrm{d}\boldsymbol{x} = 0, \quad \text{for all } (\tau_{3\beta}) \in [L^2(\Omega)]^2, \tag{29.32}$$

$$-\int_\Omega h^2 e_{\alpha\beta}^*(\boldsymbol{u}^0)\tau_{\alpha\beta}\, \mathrm{d}\boldsymbol{x} = 0, \quad \text{for all } (\tau_{\alpha\beta}) \in [L^2(\Omega)]_s^4, \tag{29.33}$$

$$\int_\Omega h^2[\sigma_{33}^0(\partial_3 v_3 - h^{-1}h' x_\alpha \partial_\alpha v_3) + h^{-1}\sigma_{3\beta}^0 \partial_\beta v_3]\, \mathrm{d}\boldsymbol{x} = \int_\Omega h^2 f_3 v_3\, \mathrm{d}\boldsymbol{x} + \int_\Gamma h g_3^* v_3\, \mathrm{d}a, \quad \text{for all } v_3 \in W_1(\Omega), \tag{29.34}$$

$$\int_\Omega h^2[h^{-1}\sigma_{\alpha\beta}^0 \partial_\alpha v_\beta + \sigma_{3\beta}^0(\partial_3 v_\beta - h^{-1}h' x_\alpha \partial_\alpha v_\beta]\, \mathrm{d}\boldsymbol{x} = \int_\Omega h^2 f_\beta v_\beta\, \mathrm{d}\boldsymbol{x} + \int_\Gamma h g_\beta^* v_\beta\, \mathrm{d}a, \quad \text{for all } (v_\beta) \in W_2(\Omega), \tag{29.35}$$

$$\int_\Omega \frac{1}{E} h^2 \sigma_{33}^2 \tau_{33}\, \mathrm{d}\boldsymbol{x} - \int_\Omega h^2 e_{33}^*(\boldsymbol{u}^2)\tau_{33}\, \mathrm{d}\boldsymbol{x} = \int_\Omega \frac{\nu}{E} h^2 \sigma_{\mu\mu}^0 \tau_{33}\, \mathrm{d}\boldsymbol{x}, \quad \text{for all } \tau_{33} \in L^2(\Omega), \tag{29.36}$$

$$-\int_\Omega 2h^2 e_{3\beta}^*(\boldsymbol{u}^2)\tau_{3\beta}\, \mathrm{d}\boldsymbol{x}$$

$$= -\int_\Omega \frac{2(1+\nu)}{E} h^2 \sigma^0_{3\beta} \tau_{3\beta}, \quad \text{for all } (\tau_{3\beta}) \in [L^2(\Omega)]^2, \tag{29.37}$$

$$-\int_\Omega h^2 e^*_{\alpha\beta}(\boldsymbol{u}^2) \tau_{\alpha\beta} \,\mathrm{d}\boldsymbol{x}$$

$$= \int_\Omega \frac{\nu}{E} h^2 \sigma^0_{33} \tau_{\alpha\beta} \delta_{\alpha\beta} \,\mathrm{d}\boldsymbol{x}, \quad \text{for all } (\tau_{\alpha\beta}) \in [L^2(\Omega)]^4_s, \tag{29.38}$$

$$\int_\Omega h^2 [\sigma^2_{33}(\partial_3 v_3 - h^{-1} h' x_\alpha \partial_\alpha v_3) + h^{-1} \sigma^2_{3\beta} \partial_\beta v_3] \,\mathrm{d}\boldsymbol{x} = 0,$$

$$\text{for all } v_3 \in W_1(\Omega), \tag{29.39}$$

$$\int_\Omega h^2 [h^{-1} \sigma^2_{\alpha\beta} \partial_\alpha v_\beta + \sigma^2_{3\beta}(\partial_3 v_\beta - h^{-1} h' x_\alpha \partial_\alpha v_\beta] \,\mathrm{d}\boldsymbol{x} = 0,$$

$$\text{for all } (v_\beta) \in W_2(\Omega), \tag{29.40}$$

$$\int_\Omega \frac{1}{E} h^2 \sigma^4_{33} \tau_{33} \,\mathrm{d}\boldsymbol{x} - \int_\Omega h^2 e^*_{33}(\boldsymbol{u}^4) \tau_{33} \,\mathrm{d}\boldsymbol{x}$$

$$= \int_\Omega \frac{\nu}{E} h^2 \sigma^2_{\mu\mu} \tau_{33} \,\mathrm{d}\boldsymbol{x}, \quad \text{for all } \tau_{33} \in L^2(\Omega), \tag{29.41}$$

$$-\int_\Omega 2h^2 e^*_{3\beta}(\boldsymbol{u}^4) \tau_{3\beta} \,\mathrm{d}\boldsymbol{x}$$

$$= -\int_\Omega \frac{2(1+\nu)}{E} h^2 \sigma^2_{3\beta} \tau_{3\beta}, \quad \text{for all } (\tau_{3\beta}) \in [L^2(\Omega)]^2, \tag{29.42}$$

$$-\int_\Omega h^2 e^*_{\alpha\beta}(\boldsymbol{u}^4) \tau_{\alpha\beta} \,\mathrm{d}\boldsymbol{x}$$

$$= \int_\Omega \frac{\nu}{E} h^2 \sigma^2_{33} \tau_{\alpha\beta} \delta_{\alpha\beta} \,\mathrm{d}\boldsymbol{x} - \int_\Omega h^2 \left[\frac{1+\nu}{E} \sigma^0_{\alpha\beta} - \frac{\nu}{E} \sigma_{\mu\mu} \delta_{\alpha\beta} \tau_{\alpha\beta} \right] \mathrm{d}\boldsymbol{x},$$

$$\text{for all } (\tau_{\alpha\beta}) \in [L^2(\Omega)]^4_s, \tag{29.43}$$

$$\int_\Omega h^2 [\sigma^4_{33}(\partial_3 v_3 - h^{-1} h' x_\alpha \partial_\alpha v_3) + h^{-1} \sigma^4_{3\beta} \partial_\beta v_3] \,\mathrm{d}\boldsymbol{x} = 0,$$

$$\text{for all } v_3 \in W_1(\Omega), \tag{29.44}$$

$$\int_\Omega h^2 [h^{-1} \sigma^4_{\alpha\beta} \partial_\alpha v_\beta + \sigma^4_{3\beta}(\partial_3 v_\beta - h^{-1} h' x_\alpha \partial_\alpha v_\beta)] \,\mathrm{d}\boldsymbol{x} = 0,$$

$$\text{for all } (v_\beta) \in W_2(\Omega), \tag{29.45}$$

Equations (29.31)–(29.45) are analogous to Eqs. (7.6)–(7.20) if one introduces the variable coefficients h^2 and $h^{-1}h'$ and if one substitutes e_{ij} by e^*_{ij}. In BERMÚDEZ and VIAÑO [1984] a partial characterization of the solutions of the system (29.31)–(29.35) is made and it is shown that the method used in Theorem 7.1 for (7.6)–(7.10) is also valid for this more general case. For this reason, in the remaining part of this section we shall state the analogous results of Theorems 7.1 and 9.1, characterizing the solutions but *we will insist only on those aspects that represent an essential difference with respect to the previous sections. Moreover, in order to avoid extremely long computations we omit the expressions for* $\sigma^0_{\alpha\beta}$, *for* $\underline{u}^2_3$ *and for* z^2_α. In this way, *we will only show the model for the torsion problem of a rod with a variable cross section.*

Equations (29.32) and (29.33) lead us in a natural way to the following Bernoulli–Navier displacement space:

$$V^*_{\mathrm{BN}}(\Omega) = \{\boldsymbol{v} = (v_i) \in V(\Omega)\colon v_\alpha \in H^2_0(0,L),\ v_3 = \underline{v}_3 - hx_\alpha \partial_3 v_\alpha,\ \underline{v}_3 \in H^1_0(0,L)\}, \tag{29.46}$$

equipped with the $\|\cdot\|_{\mathrm{BN}}$ norm defined in (4.15) which is equivalent to the $\|\cdot\|_{1,\Omega}$ norm. In fact the following result owed to BERMÚDEZ and VIAÑO [1984] holds.

THEOREM 29.5. *Let* $\boldsymbol{v} \in V(\Omega)$. *Then, the following conditions are equivalent*:

(i) $e^*_{\alpha\beta}(\boldsymbol{v}) = e^*_{3\beta}(\boldsymbol{v}) = 0$,

(ii) $\boldsymbol{v} \in V^*_{\mathrm{BN}}(\Omega)$.

PROOF. It is clear that every element of $V^*_{\mathrm{BN}}(\Omega)$ satisfies (i). Conversely, if $\boldsymbol{v} \in V(\Omega)$ satisfies (i) then condition $e^*_{\alpha\beta}(\boldsymbol{v}) = 0$ implies that

$$\partial_1 v_1 = 0, \qquad \partial_2 v_2 = 0, \qquad \partial_1 v_2 + \partial_2 v_1 = 0.$$

Reasoning as in the proof of Theorem 4.3 one shows that there exist functions $z, z_\alpha \in H^1_0(0,L)$ such that

$$v_1 = z_1 + x_2 hz, \qquad v_2 = z_2 - x_1 hz. \tag{29.47}$$

Now using condition $e^*_{3\beta}(\boldsymbol{v}) = 0$ one obtains

$$\begin{aligned} h^{-1}\partial_1 v_3 &= -\partial_3 v_1 + h^{-1}h' x_\alpha \partial_\alpha v_1 = -\partial_3 z_1 - x_2 h \partial_3 z, \\ h^{-1}\partial_2 v_3 &= -\partial_3 v_2 + h^{-1}h' x_\alpha \partial_\alpha v_2 = -\partial_3 z_2 + x_1 h \partial_3 z, \end{aligned} \tag{29.48}$$

from which we conclude that

$$\begin{aligned} &\partial_{12} v_3 = -h^2 \partial_3 z = \partial_{21} v_3 = h^2 \partial_3 z = 0, \\ &\partial_{11} v_3 = \partial_{22} v_3 = 0. \end{aligned}$$

Consequently, as $h^2 > 0$ one has $\partial_3 z = 0$, and as $z \in H_0^1(0, L)$ we conclude that $z = 0$.

From (29.47) and (29.48) one now obtains

$$v_\alpha \in H_0^1(0, L), \qquad \partial_\alpha v_3 = -h\partial_3 v_\alpha,$$

from which we conclude the existence of $\underline{v}_3$ depending only on x_3 and such that

$$v_3 = \underline{v}_3 - hx_\alpha \partial_3 v_\alpha.$$

The fact that $v_3 \in W_1(\Omega)$ implies that $\underline{v}_3 \in H_0^1(0, L)$ and that $\partial_3 v_\alpha \in H_0^1(0, L)$ which proves that $v_\alpha \in H_0^2(0, L)$ and this completes the proof. □

Besides using functions and constants depending on the geometry of the cross section and introduced in Section 7, for the case under study, we also consider functions λ_α defined as the unique solution of the following problem:

$$\begin{aligned} &\lambda_\alpha \in H^1(\omega), \\ &\int_\omega \partial_\beta \lambda_\alpha \partial_\beta \varphi \, \mathrm{d}\omega = -\int_\gamma x_\alpha (x_1 n_1 + x_2 n_2) \varphi \, \mathrm{d}\gamma, \quad \text{for all } \varphi \in H^1(\omega), \\ &\int_\omega \lambda_\alpha \, \mathrm{d}\omega = 0. \end{aligned} \tag{29.49}$$

Associated with these functions we define the following constants:

$$L_{\alpha\beta}^\lambda = \int_\omega x_\beta \lambda_\alpha \, \mathrm{d}\omega. \tag{29.50}$$

The following property of functions λ_α is very often used:

$$\int_\omega \partial_\alpha \lambda_\beta \, \mathrm{d}\omega = -4 \int_\omega x_\alpha x_\beta. \tag{29.51}$$

The following theorem (partially) characterizes terms $\boldsymbol{u}^0$, $\boldsymbol{u}^2$ and $\boldsymbol{\sigma}^0$ showing up in the system of equations (29.31)–(29.45). It is a result of the same type as Theorems 7.1, 23.1, 24.2 and 26.6. The equations for the zeroth order displacement components were derived by BERMÚDEZ and VIAÑO [1984].

THEOREM 29.6. *Let the system of applied forces be such that*

$$\begin{aligned} &f_\alpha \in L^2(\Omega), \qquad g_\alpha^* \in L^2(\Gamma), \\ &f_3 \in H^1[0, L; L^2(\omega)], \qquad g_3^* \in H^1[0, L; L^2(\gamma)]. \end{aligned} \tag{29.52}$$

Then, Eqs. (29.31)–(29.45) *determine elements* $(\boldsymbol{u}^0, \boldsymbol{\sigma}^0) \in V(\Omega) \times \Sigma(\Omega)$, $\boldsymbol{u}^2 \in V(\Omega)$ *and* $\sigma_{33}^2 \in L^2(\Omega)$ *in a unique way, through the following characterization:*

(i) *Displacements* $\boldsymbol{u}^0$ *verify*

$$\boldsymbol{u}^0 \in V_{\mathrm{BN}}^*(\Omega), \qquad u_3^0 = \underline{u}_3^0 - hx_\alpha \partial_3 u_\alpha^0, \quad \underline{u}_3^0 \in H_0^1(0,L), \quad u_\alpha^0 \in H_0^2(0,L), \tag{29.53}$$

where functions u_α^0 *and* $\underline{u}_3^0$ *are the unique solutions of the following variational problems, respectively:*

$$u_\beta^0 \in H_0^2(0,L), \quad \textit{and for all } v \in H_0^2(0,L) \quad (\textit{no sum on } \beta):$$

$$\begin{aligned}\int_0^L & EI_\beta h^4 \partial_{33} u_\beta^0 \partial_{33} v \, \mathrm{d}x_3 \\ &= \int_0^L \left[h^2 \int_\omega f_\beta \, \mathrm{d}\omega + h \int_\gamma g_\beta^* \, \mathrm{d}\gamma \right] v \, \mathrm{d}x_3 \\ &\quad - \int_0^L \left[h^3 \int_\omega x_\beta f_3 \, \mathrm{d}\omega + h^2 \int_\gamma x_\beta g_3^* \, \mathrm{d}\gamma \right] \partial_3 v \, \mathrm{d}x_3,\end{aligned} \tag{29.54}$$

$$\underline{u}_3^0 \in H_0^1(0,L) \quad \textit{and for all } v \in H_0^1(0,L):$$

$$\int_0^L EAh^2 \partial_3 \underline{u}_3^0 \partial_3 v \, \mathrm{d}x_3 = \int_0^L \left[h^2 \int_\omega f_3 \, \mathrm{d}\omega + h \int_\gamma g_3^* \, \mathrm{d}\gamma \right] v \, \mathrm{d}x_3. \tag{29.55}$$

(ii) *The stress component* σ_{33}^0 *is a function of* $\boldsymbol{u}^0$, *given by*

$$\sigma_{33}^0 = E\partial_3 u_3^0 = E(\partial_3 \underline{u}_3^0 - hx_\alpha \partial_{33} u_\alpha^0). \tag{29.56}$$

and the respective bending moment component is

$$m_\beta^0 = \int_\omega x_\beta \sigma_{33}^0 = -EI_\beta h \partial_{33} u_\beta^0, \quad \textit{no sum on } \beta. \tag{29.57}$$

(iii) *The second order displacements* $\boldsymbol{u}^2 \in V(\Omega)$ *are of the following form*

$$u_1^2 = z_1^2 + hx_2 z^2 - \nu(hx_1 \partial_3 \underline{u}_3^0 - h^2 \Phi_{1\beta} \partial_{33} u_\beta^0), \tag{29.58}$$

$$u_2^2 = z_2^2 - hx_1 z^2 - \nu(hx_2 \partial_3 \underline{u}_3^0 - h^2 \Phi_{2\beta} \partial_{33} u_\beta^0), \tag{29.59}$$

$$\begin{aligned} u_3^2 &= \underline{u}_3^2 - hx_\alpha \partial_3 z_\alpha^2 - h^2 w \partial_3 z^2 - \nu h^2 \left[\tfrac{1}{2}(x_1^2 + x_2^2) - \frac{1}{2A}(I_1 + I_2) \right] \partial_{33} \underline{u}_3^0 \\ &\quad + 2(1+\nu) hh' \left[\tfrac{1}{2}(x_1^2 + x_2^2) - \frac{1}{2A}(I_1 + I_2) \right] \partial_3 \underline{u}_3^0 + h^3 [(1+\nu)\eta_\alpha + \nu \theta_\alpha] \partial_{333} u_\alpha^0 \\ &\quad + 2(1+\nu) h^2 h' \lambda_\alpha \partial_{33} u_\alpha + \frac{2(1+\nu)}{E} w^0, \end{aligned} \tag{29.60}$$

where functions z^2, w^0, $\underline{u}_3^2$ *and* z_α^2 *are uniquely determined from the data and from functions* $\boldsymbol{u}^0$ *through the following problems* (a)–(d):

(a) *The angle of twist* z^2 *depends only on* x_3 *and is the unique solution of the following variational problem:*

$$
\begin{aligned}
& z^2 \in H^1(0,L),\\
& \int_0^L \frac{EJ}{2(1+\nu)} h^4 \partial_3 z^2 \partial_3 v \, \mathrm{d}x_3 = M_3^2(v), \quad \text{for all } v \in H_0^1(0,L),\\
& z^2(a) = h(a)\frac{\nu}{(I_1+I_2)}[H_2\partial_{33}u_1^0(a) - H_1\partial_{33}u_2^0(a)] \quad \text{at } a = 0, L,
\end{aligned}
\tag{29.61}
$$

where

$$
\begin{aligned}
M_3^2(v) = {} & \int_0^L \left[h^3 \int_\omega (x_2 f_1 - x_1 f_2)\,\mathrm{d}\omega + h^2 \int_\gamma (x_2 g_1^* - x_1 g_2^*)\,\mathrm{d}\gamma \right] v \,\mathrm{d}x_3\\
& + \int_0^L \left[h^4 \int_\omega w f_3 \,\mathrm{d}\omega + h^3 \int_\gamma w g_3^* \,\mathrm{d}\gamma \right] \partial_3 v \,\mathrm{d}x_3\\
& + \int_0^L \frac{E}{2(1+\nu)} \{ h^5[(1+\nu)I_\alpha^w + \nu I_\alpha^\Psi]\partial_{333}u_\alpha^0\\
& \qquad + 4(1+\nu)h^4 h'(I_\alpha^w + I_\alpha^\Psi)\partial_{33}u_\alpha^0 \}\partial_3 v \,\mathrm{d}x_3.
\end{aligned}
\tag{29.62}
$$

(b) *The additional warping* w^0 *is the unique solution of the following variational problem:*

$$
\begin{aligned}
& w^0 \in H^1[0,L;H^1(\omega)] \quad \text{and for all } x_3 \in [0,L],\\
& \int_\omega \partial_\beta w^0(x_3)\partial_\beta \varphi \,\mathrm{d}\omega\\
& \quad = h^2 \int_\omega f_3(x_3)\varphi \,\mathrm{d}\omega + h \int_\gamma g_3^*(x_3)\varphi \,\mathrm{d}\gamma\\
& \quad - \frac{1}{A}\left[h^2 \int_\omega f_3(x_3)\,\mathrm{d}\omega + h \int_\gamma g_3^*(x_3)\,\mathrm{d}\gamma \right] \int_\omega \varphi \,\mathrm{d}\omega, \quad \text{for all } \varphi \in H^1(\omega),\\
& \int_\omega w^0(x_3)\,\mathrm{d}\omega = 0.
\end{aligned}
\tag{29.63}
$$

(c) *The stretching component* $\underline{u}_3^2$ *depends only on* x_3 *and uniquely solves the following problem*:

$$\underline{u}_3^2 \in H_0^1(0,L),$$

$$\int_0^L EAh^2\partial_3\underline{u}_3^2\partial_3 v \,\mathrm{d}x_3 = \int_0^L G_3^2\partial_3 v\,\mathrm{d}x_3, \quad \textit{for all } v \in H_0^1(0,L), \tag{29.64}$$

where

$$G_3^2 = -\int_\omega h^2(\partial_3 U_3^2 - h^{-1}h'x_\alpha\partial_\alpha U_3^2 + \nu\sigma_{\mu\mu}^0)\,\mathrm{d}\omega, \tag{29.65}$$

$$U_3^2 = u_3^2 - \underline{u}_3^2 + hx_\alpha\partial_3 z_\alpha^2. \tag{29.66}$$

(d) *The bending component* z_α^2 *depends only on the variable* x_3 *and is the unique solution of the following variational problem* (*no sum on* α, $\beta \neq \alpha$):

$$z_\alpha^2 \in H^2(0,L) \quad \textit{and for all } v \in H_0^2(0,L)\textit{:}$$

$$\int_0^L EI_\alpha h^4\partial_{33}z_\alpha^2\partial_{33}v\,\mathrm{d}x_3 = \int_0^L M_\alpha^2\partial_{33}v\,\mathrm{d}x_3,$$

$$z_\alpha^2(a) = -\tfrac{1}{2}\nu(I_\alpha - I_\beta)h^2(a)\partial_{33}u_\alpha^0(a) \quad \textit{at } a = 0, L,$$

$$\begin{aligned}\partial_3 z_\alpha^2(a) = \frac{1}{h(a)I_\alpha}\Bigg\{&\frac{2(1+\nu)}{E}\int_\omega x_\alpha w^0(a)\,\mathrm{d}\omega - \tfrac{1}{2}I_\alpha^w h^2(a)\partial_3 z^2(a)\\ &+\nu H_\alpha h^2(a)\partial_{33}\underline{u}_3^0(a) + 2(1+\nu)H_\alpha h(a)h'(a)\partial_3\underline{u}_3^0(a)\\ &+[(1+\nu)L_{\alpha\beta}^\eta + \nu L_{\alpha\beta}^\theta]h^3(a)\partial_{333}u_\beta^0(a)\\ &+2(1+\nu)L_{\alpha\beta}^\lambda h^2(a)h'(a)\partial_{33}u_\beta^0(a)\Bigg\} \quad \textit{at } a = 0, L,\end{aligned} \tag{29.67}$$

where

$$M_\alpha^2 = \int_\omega Eh^3x_\alpha(\partial_3 U_3^2 - h^{-1}h'x_\beta\partial_\beta U_3^2 + \nu\sigma_{\mu\mu}^0)\,\mathrm{d}\omega. \tag{29.68}$$

(iv) *The shear stress components* $\sigma_{3\beta}^0$ *are uniquely determined by*

$$\begin{aligned}\sigma_{31}^0 = \frac{E}{2(1+\nu)}\{&-\partial_2\Psi h\partial_3 z^2 + h^2[(1+\nu)\partial_1\eta_\beta + \nu(\partial_1\theta_\beta + \Phi_{1\beta})]\partial_{333}u_\beta^0\\ &+2(1+\nu)h'x_1\partial_3\underline{u}_3^0 + 2(1+\nu)hh'\partial_1\lambda_\alpha\partial_{33}u_\alpha^0\} + h^{-1}\partial_1 w^0,\end{aligned} \tag{29.69}$$

$$\begin{aligned}\sigma_{32}^0 = \frac{E}{2(1+\nu)}\{&\partial_1\Psi h\partial_3 z^2 + h^2[(1+\nu)\partial_2\eta_\beta + \nu(\partial_2\theta_\beta + \Phi_{2\beta})]\partial_{333}u_\beta^0\\ &+2(1+\nu)h'x_2\partial_3\underline{u}_3^0 + 2(1+\nu)hh'\partial_2\lambda_\alpha\partial_{33}u_\alpha^0\} + h^{-1}\partial_2 w^0,\end{aligned} \tag{29.70}$$

and the shear force components $q^0_\beta = \int_\omega \sigma^0_{3\beta}$ *are given by*

$$q^0_\beta = -EI_\beta h^2 \partial_{333} u^0_\beta - 4EI_\beta h h' \partial_{33} u^0_\beta + \int_\omega h x_\beta f_3 \,\mathrm{d}\omega + \int_\gamma x_\beta g^*_3 \,\mathrm{d}\gamma \quad (\textit{no sum on } \beta). \tag{29.71}$$

(v) *The plane stress components* $\sigma^0_{\alpha\beta}$ *are given by*

$$\sigma^0_{\alpha\beta} = S^*_{\alpha\beta}(\underline{\boldsymbol{u}}^4) + S^0_{\alpha\beta}, \tag{29.72}$$

where

$$\begin{aligned} S^0_{\alpha\beta} &= \frac{\nu E}{(1+\nu)(1-2\nu)} \partial_3 (U^2_3 - h^{-1} h' x_\mu \partial_\mu U^2_3) \delta_{\alpha\beta}, \\ S^*_{\alpha\beta}(\underline{\boldsymbol{u}}^4) &= \frac{E}{(1+\nu)} e^*_{\alpha\beta}(\underline{\boldsymbol{u}}^4) + \frac{\nu E}{(1+\nu)(1-2\nu)} e^*_{\mu\mu}(\underline{\boldsymbol{u}}^4) \delta_{\alpha\beta}, \end{aligned} \tag{29.73}$$

and $\underline{\boldsymbol{u}}^4 = (\underline{u}^4_\alpha)$ *is the unique solution of the following plane elasticity problem*:

$$\begin{aligned} &\underline{\boldsymbol{u}}^4 \in L^2[0, L; (H^1(\omega))^2], \quad \textit{and a.e. in } (0, L), \textit{ for all } \boldsymbol{\varphi} \in [H^1(\omega)]^2: \\ &\int_\omega h^2 S^*_{\alpha\beta}(\underline{\boldsymbol{u}}^4) e^*_{\alpha\beta}(\boldsymbol{\varphi}) \,\mathrm{d}\omega \\ &\quad = h^2 \int_\omega f_\beta \varphi_\beta \,\mathrm{d}\omega + h \int_\gamma g^*_\beta \varphi_\beta \,\mathrm{d}\gamma + \int_\omega \partial_3 (h^2 \sigma^0_{3\beta}) \varphi_\beta \,\mathrm{d}\omega \\ &\qquad + \int_\omega h h' \sigma^0_{3\beta} x_\mu \partial_\mu \varphi_\beta \,\mathrm{d}\omega - \int_\omega h^2 S^0_{\alpha\beta} e^*_{\alpha\beta}(\boldsymbol{\varphi}) \,\mathrm{d}\omega, \\ &\int_\omega \underline{u}^4_\alpha \,\mathrm{d}\omega = \int_\omega [x_2 \underline{u}^4_1 - x_1 \underline{u}^4_2] \,\mathrm{d}\omega = 0. \end{aligned} \tag{29.74}$$

(vi) *The axial stress* σ^2_{33} *and the second order bending moment components are given by*

$$\begin{aligned} \sigma^2_{33} &= E e^*_{33}(\boldsymbol{u}^2) + \nu \sigma^0_{\mu\mu} \\ &= E \partial_3 \underline{u}^2_3 - E h x_\alpha \partial_{33} z^2_\alpha + E(\partial_3 U^2_3 - h^{-1} h' x_\beta \partial_\beta U^2_3) + \nu \sigma^0_{\mu\mu}, \end{aligned} \tag{29.75}$$

$$m^2_\beta = -EI_\beta h \partial_{33} z^2_\beta + M^2_\beta \quad (\textit{no sum on } \beta). \tag{29.76}$$

PROOF. The proof of Theorem 29.6 is similar to the proof of Theorem 7.1 and the most important differences are due to the fact that h' does not vanish.

Therefore we indicate only the major steps of the proof without detailing the calculations.

Step 1. Equations (29.32) and (29.33) imply that $e^*_{\alpha\beta}(\boldsymbol{u}^0) = e^*_{3\beta}(\boldsymbol{u}^0) = 0$ and consequently, from Theorem 29.5, $\boldsymbol{u}^0$ is of the form (29.53). Equations (29.56) and (29.57) are now a consequence of (29.31). Substituting σ^0_{33} into Eq. (29.34) and taking $v_3 = v \in H^1_0(0, L)$ one obtains (29.55). Equations for u^0_α are obtained by the same method setting in (29.34) and (29.35) $v_3 = hx_\alpha\partial_3 v_\alpha$, with $v_\alpha \in H^2_0(0, L)$.

Step 2. The form for the displacement components u^2_α (cf. (29.58), (29.59)) is obtained from a direct integration of the following equations which in turn may be deduced from (29.38):

$$\partial_1 u^2_1 = \partial_2 u^2_2 = -\nu(h\partial_3 \underline{u}^0_3 - h^2 x_\alpha \partial_{33} u^0_\alpha),$$
$$\partial_1 u^2_2 + \partial_2 u^2_1 = 0.$$

From Eq. (29.37) one obtains:

$$\sigma^0_{3\beta} = \frac{E}{2(1+\nu)}(h^{-1}\partial_\beta u^2_3 + \partial_3 u^2_\beta - h^{-1}h' x_\alpha \partial_\alpha u^2_\beta). \tag{29.77}$$

Substituting this expression in (29.34) we conclude that u^2_3 satisfies the following problem in $(0, L)$:

$$\begin{aligned}
&\frac{E}{2(1+\nu)}\int_\omega \partial_\beta u^2_3 \partial_\beta \varphi \, \mathrm{d}\omega \\
&\quad = -\frac{E}{2(1+\nu)}\left[\int_\omega h\partial_3 u^2_\beta \partial_\beta \varphi \, \mathrm{d}\omega - h' \int_\omega x_\alpha \partial_\alpha u^2_\beta \partial_\beta \varphi \, \mathrm{d}\omega\right] \\
&\qquad + h^2 \int_\omega \partial_3(h^2 \sigma^0_{33})\varphi \, \mathrm{d}\omega + hh' \int_\omega \sigma^0_{33} x_\alpha \partial_\alpha \varphi \, \mathrm{d}\omega \\
&\qquad + h^2 \int_\omega f_3 \varphi \, \mathrm{d}\omega + h \int_\gamma g^*_3 \varphi \, \mathrm{d}\gamma, \quad \text{for all } \varphi \in H^1(\omega).
\end{aligned}$$

Considering now the form of σ^0_{33} and of u^2_β this problem reduces to the following:

$$\begin{aligned}
&\frac{E}{2(1+\nu)}\int_\omega \partial_\beta u^2_3 \partial_\beta \varphi \, \mathrm{d}\omega \\
&\quad = -\frac{E}{2(1+\nu)}\left\{h\partial_3 z^2_\beta \int_\omega \partial_\beta \varphi \, \mathrm{d}\omega + h^2 \partial_3 z^2 \int_\omega (x_2 \partial_1 \varphi - x_1 \partial_2 \varphi) \, \mathrm{d}\omega\right.
\end{aligned}$$

$$- [\nu h^2 \partial_{33} \underline{u}_3^0 + 2(1+\nu) h h' \partial_3 \underline{u}_3^0] \int_\omega x_\beta \partial_\beta \varphi \, d\omega$$

$$+ h^3 \partial_{333} u_\alpha^0 \int_\omega [2(1+\nu) x_\alpha \varphi + \nu \Phi_{\beta\alpha} \partial_\beta \varphi] \, d\omega$$

$$+ 2(1+\nu) h^2 h' \partial_{33} u_\alpha^0 \int_\omega (3 x_\alpha \varphi + x_\alpha x_\beta \partial_\beta \varphi) \, d\omega \Bigg\}$$

$$+ h^2 \int_\omega f_3 \varphi \, d\omega + h \int_\gamma g_3^* \varphi \, d\gamma - \frac{1}{A} \left(h^2 \int_\omega f_3 \, d\omega + h \int_\gamma g_3^* \, d\gamma \right) \int_\omega \varphi \, d\omega,$$

for all $\varphi \in H^1(\omega)$. (29.78)

Equation (29.60) for u_3^2 is directly obtained from (29.78) after introducing the definitions of functions w, η_α, θ_α, λ_α and w^0.

Step 3. Expressions for $\underline{u}_3^2(a)$, $z^2(a)$, $z_\alpha^2(a)$ and $\partial_3 z_\alpha^2(a)$ $(a = 0, L)$ are obtained as necessary conditions in order to have $\boldsymbol{u}^2 \in V(\Omega)$.

Step 4. Expressions (29.69) and (29.70) for the shear stress components $\sigma_{3\beta}^0$ are a consequence of the form of the displacement field $\boldsymbol{u}^2$ and of (29.77). In order to obtain expression (29.71) for the shear force components one may set $v_3 = h x_\alpha v_\alpha$, with $v_\alpha \in H_0^1(0, L)$, in (29.34), and employ expression (29.57).

Step 5. Setting $v_1 = x_2 h v$ and $v_2 = -x_1 h v$, for any $v \in H_0^1(0, L)$, in Eq. (29.35), one obtains

$$\int_0^L h^3 \left[\int_\omega (x_2 \sigma_{31}^0 - x_1 \sigma_{32}^0) \, d\omega \right] \partial_3 v \, dx_3$$

$$= \int_0^L \left[h^3 \int_\omega (x_2 f_1 - x_1 f_2) \, d\omega + h^2 \int_\gamma (x_2 g_1^* - x_1 g_2^*) \, d\gamma \right] v \, dx_3.$$

On the other hand, from (29.69) and (29.70) it is possible to show that

$$\int_\omega (x_2 \sigma_{31}^0 - x_1 \sigma_{32}^0) \, d\omega$$

$$= \frac{E}{2(1+\nu)} \{ J h \partial_3 z^2 - h^2 [(1+\nu) I_\alpha^w + \nu I_\alpha^\Psi] \partial_{333} u_\alpha^0$$

$$- 4(1+\nu) h h' (I_\alpha^w + I_\alpha^\Psi) \partial_{33} u_\alpha^0 \} + h \int_\omega w f_3 \, d\omega + \int_\gamma w g_3^* \, d\gamma.$$

Problem (29.61), (29.62) is now obtained after substituting this expression into the previous one.

Step 6. With the same arguments as in Theorem 7.1, from Eqs. (29.36) and (29.43), one has

$$\sigma^0_{\alpha\beta} = S^*_{\alpha\beta}(\boldsymbol{u}^4) + \frac{\nu E}{(1+\nu)(1-2\nu)} e^*_{33}(\boldsymbol{u}^2)\delta_{\alpha\beta}$$
$$= S^*_{\alpha\beta}(\boldsymbol{u}^4) + \frac{\nu E}{(1+\nu)(1-2\nu)} e^*_{33}(\underline{u}^2_3 - hx_\mu \partial_3 z^2_\mu)\delta_{\alpha\beta} + S^0_{\alpha\beta}.$$

Let us consider the equation

$$S^*_{\alpha\beta}(\boldsymbol{\xi}^4) = -\frac{\nu E}{(1+\nu)(1-2\nu)} e^*_{33}(\underline{u}^2_3 - hx_\mu \partial_3 z^2_\mu),$$

which is equivalent to

$$\partial_1 \xi^4_1 = \partial_2 \xi^4_2 = -\nu(h\partial_3 \underline{u}^2_3 - h^2 x_\mu \partial_{33} z^2_\mu),$$
$$\partial_1 \xi^4_2 + \partial_2 \xi^4_1 = 0,$$

which may be integrated directly as in Step 2. Setting $\underline{\boldsymbol{u}}^4 = \boldsymbol{u}^4 - \boldsymbol{\xi}^4$, (29.72) is obtained. Equation (29.74) for $\underline{\boldsymbol{u}}^4$ is a consequence of (29.35) and of the condition $(u^4_\alpha) \in W_2(\Omega)$.

Step 7. Expressions (29.75) and (29.76) for σ^2_{33} and for m^2_β are a direct consequence of Eq. (29.36). Using these expressions into (29.39) and (29.40) with the same test functions as in Step 1, one obtains Eqs. (29.64) and (29.65) for $\underline{u}^2_3$ and (29.67), (29.68) for z^2_α. This concludes the summary of the proof. □

As for the case of a constant cross section it is also possible to obtain a convergence result which justifies the asymptotic expansion (29.30), albeit partially. It is clear that from the properties assumed in (29.3) for the function h, and from Theorems 29.2 and 29.3, it is possible to obtain a priori estimates (10.4) and (10.5) for the solution $(\boldsymbol{u}(\varepsilon), \boldsymbol{\sigma}(\varepsilon))$ of the scaled problem (29.27)–(29.29). From this estimate similar results as the ones established in Theorems 10.2 and 10.3 may also be obtained. Thus, we omit the proof of the following result which can be done as in Section 10 or as in BERMÚDEZ and VIAÑO [1984].

THEOREM 29.7. *Let* $(\boldsymbol{u}(\varepsilon), \boldsymbol{\sigma}(\varepsilon)) \in V(\Omega) \times \Sigma(\Omega)$ *be the unique solution of the scaled problem* (29.27)–(29.29). *Let* $\boldsymbol{u}^0 \in V(\Omega)$, $\sigma^0_{33} \in L^2(\Omega)$, m^0_β, $q^0_\beta \in L^2(0, L)$ *be the elements defined in Theorem* 29.6. *Then, as* $\varepsilon \to 0$, *one has*

$$\boldsymbol{u}(\varepsilon) \to \boldsymbol{u}^0 \quad \text{in } V(\Omega), \tag{29.79}$$
$$\sigma_{33}(\varepsilon) \to \sigma^0_{33}, \qquad \varepsilon\sigma_{3\beta}(\varepsilon) \to 0, \qquad \varepsilon^2\sigma_{\alpha\beta}(\varepsilon) \to 0 \quad \text{in } L^2(\Omega), \tag{29.80}$$
$$m_\beta(\varepsilon) \to m^0_\beta, \qquad q_\beta(\varepsilon) \rightharpoonup q^0_\beta \qquad \text{in } L^2(0, L). \tag{29.81}$$

Finally, we end this section by studying the models obtained but written with respect to the original domain. Descaling the unknowns, one has (cf. (11.2))

$$\begin{aligned}
&u_\alpha^{0\varepsilon}(\boldsymbol{x}^\varepsilon) = \varepsilon^{-1}u_\alpha^0(\boldsymbol{x}), \qquad u_3^{0\varepsilon}(\boldsymbol{x}^\varepsilon) = u_3^0(\boldsymbol{x}),\\
&u_\alpha^{2\varepsilon}(\boldsymbol{x}^\varepsilon) = \varepsilon u_\alpha^2(\boldsymbol{x}), \qquad u_3^{2\varepsilon}(\boldsymbol{x}^\varepsilon) = \varepsilon^2 u_3^2(\boldsymbol{x}),\\
&\sigma_{\alpha\beta}^{0\varepsilon}(\boldsymbol{x}^\varepsilon) = \varepsilon^2\sigma_{\alpha\beta}^0(\boldsymbol{x}), \qquad \sigma_{3\beta}^{0\varepsilon}(\boldsymbol{x}^\varepsilon) = \varepsilon\sigma_{3\beta}^0(\boldsymbol{x}), \qquad \sigma_{33}^{0\varepsilon}(\boldsymbol{x}^\varepsilon) = \sigma_{33}^0(\boldsymbol{x}).
\end{aligned} \tag{29.82}$$

The first order models are obtained considering only the displacement component $\boldsymbol{u}^{0\varepsilon}$ and the second order models are obtained considering the displacement field given by $\hat{\boldsymbol{u}}^\varepsilon = \boldsymbol{u}^{0\varepsilon} + \boldsymbol{u}^{2\varepsilon}$.

We denote by $\Phi_{\alpha\beta}^\varepsilon(x_3^\varepsilon)$, $w^\varepsilon(x_3^\varepsilon)$, $\Psi^\varepsilon(x_3^\varepsilon)$, $\eta_\alpha^\varepsilon(x_3^\varepsilon)$, $\theta_\alpha^\varepsilon(x_3^\varepsilon)$, $\lambda_\alpha^\varepsilon(x_3^\varepsilon)$ the functions depending on the geometry of the cross section $\omega^\varepsilon(x_3^\varepsilon)$. Moreover, we denote also the corresponding geometric constants by $I_\beta^\varepsilon(x_3^\varepsilon)$, $I_\alpha^{w^\varepsilon}(x_3^\varepsilon)$, $I_\alpha^{\Psi^\varepsilon}(x_3^\varepsilon)$, $J^\varepsilon(x_3^\varepsilon)$, $L_{\alpha\beta}^{\eta^\varepsilon}(x_3^\varepsilon)$, $L_{\alpha\beta}^{\theta^\varepsilon}(x_3^\varepsilon)$ and $L_{\alpha\beta}^{\lambda^\varepsilon}(x_3^\varepsilon)$. Then, the following properties hold:

$$\begin{aligned}
&\Phi_{\alpha\beta}^\varepsilon(x_3^\varepsilon)(x_1^\varepsilon, x_2^\varepsilon) = \varepsilon^2 h^2(x_3)\Phi_{\alpha\beta}(x_1, x_2),\\
&w^\varepsilon(x_3^\varepsilon)(x_1^\varepsilon, x_2^\varepsilon) = \varepsilon^2 h^2(x_3) w(x_1, x_2),\\
&\Psi^\varepsilon(x_3^\varepsilon)(x_1^\varepsilon, x_2^\varepsilon) = \varepsilon^2 h^2(x_3)\Psi(x_1, x_2),
\end{aligned} \tag{29.83}$$

$$\begin{aligned}
&\eta_\alpha^\varepsilon(x_3^\varepsilon)(x_1^\varepsilon, x_2^\varepsilon) = \varepsilon^3 h^3(x_3)\eta_\alpha(x_1, x_2),\\
&\theta_\alpha^\varepsilon(x_3^\varepsilon)(x_1^\varepsilon, x_2^\varepsilon) = \varepsilon^3 h^3(x_3)\theta_\alpha(x_1, x_2),\\
&\lambda_\alpha^\varepsilon(x_3^\varepsilon)(x_1^\varepsilon, x_2^\varepsilon) = \varepsilon^3 h^3(x_3)\lambda_\alpha(x_1, x_2).
\end{aligned} \tag{29.84}$$

From these relations the corresponding relations for the associated constants follow, that is,

$$\begin{aligned}
&A^\varepsilon(x_3^\varepsilon) = \varepsilon^2 h^2(x_3) A,\\
&H_\alpha^\varepsilon(x_3^\varepsilon) = \varepsilon^5 h^5(x_3) H_\alpha,\\
&I_\beta^\varepsilon(x_3^\varepsilon) = \varepsilon^4 h^4(x_3) I_\beta,\\
&I_\beta^{\Psi^\varepsilon}(x_3^\varepsilon) = \varepsilon^5 h^5(x_3) I_\beta^\Psi,\\
&J^\varepsilon(x_3^\varepsilon) = \varepsilon^4 h^4(x_3) J,\\
&I_\beta^{w^\varepsilon}(x_3^\varepsilon) = \varepsilon^5 h^5(x_3) I_\beta^w,\\
&L_{\alpha\beta}^{\eta^\varepsilon}(x_3^\varepsilon) = \varepsilon^6 h^6(x_3) L_{\alpha\beta}^\eta,\\
&L_{\alpha\beta}^{\theta^\varepsilon}(x_3^\varepsilon) = \varepsilon^6 h^6(x_3) L_{\alpha\beta}^\theta,\\
&L_{\alpha\beta}^{\lambda^\varepsilon}(x_3^\varepsilon) = \varepsilon^6 h^6(x_3) L_{\alpha\beta}^\lambda.
\end{aligned} \tag{29.85}$$

Descaling the data and taking Eqs. (29.82)–(29.85) into account, from Theorem 29.6 the following result follows.

THEOREM 29.8. *The displacement* $\boldsymbol{u}^{0\varepsilon}$ *and* $\boldsymbol{u}^{2\varepsilon}$ *and the stress* $\boldsymbol{\sigma}^{0\varepsilon}$ *fields, defined in* (29.82), *corresponding to the original rod are uniquely determined in the following way:*

(i) *The displacement components $u_\beta^{0\varepsilon}$ depend only on the variable x_3^ε and are the unique solution of the following problem*:

$$u_\alpha^{0\varepsilon} \in H_0^2(0,L),$$

$$\int_0^L E^\varepsilon I_\alpha^\varepsilon \partial_{33}^\varepsilon u_\alpha^{0\varepsilon} \partial_{33} v^0 \, dx_3^\varepsilon$$

$$= \int_0^L \left[\int_{\omega^\varepsilon(x_3^\varepsilon)} f_\alpha^\varepsilon \, d\omega^\varepsilon + \int_{\gamma^\varepsilon(x_3^\varepsilon)} g_\alpha^\varepsilon \, d\gamma^\varepsilon \right] v^0 \, dx_3^\varepsilon \tag{29.86}$$

$$- \int_0^L \left[\int_{\omega^\varepsilon(x_3^\varepsilon)} x_\alpha^\varepsilon f_3^\varepsilon \, d\omega^\varepsilon + \int_{\gamma^\varepsilon(x_3^\varepsilon)} x_\alpha^\varepsilon g_3^\varepsilon \, d\gamma^\varepsilon \right] \partial_3^\varepsilon v^0 \, dx_3^\varepsilon,$$

for all $v^0 \in H_0^2(0,L)$.

(ii) *The displacement component $u_3^{0\varepsilon}$ is of the form*

$$u_3^{0\varepsilon} = \underline{u}_3^{0\varepsilon} - x_\alpha^\varepsilon \partial_{33} u_\alpha^{0\varepsilon}, \tag{29.87}$$

where $\underline{u}_3^{0\varepsilon}$ depends only on variable x_3^ε and is the unique solution of the following problem:

$$\underline{u}_3^{0\varepsilon} \in H_0^1(0,L),$$

$$\int_0^L E^\varepsilon A^\varepsilon \partial_3 \underline{u}_3^{0\varepsilon} \partial_3^\varepsilon \underline{v}^0 \, dx_3^\varepsilon = \int_0^L \left[\int_{\omega^\varepsilon(x_3^\varepsilon)} f_3^\varepsilon \, d\omega^\varepsilon + \int_{\gamma^\varepsilon(x_3^\varepsilon)} g_3^\varepsilon \, d\gamma^\varepsilon \right] \underline{v}^0 \, dx_3^\varepsilon, \tag{29.88}$$

for all $\underline{v}^0 \in H_0^1(0,L)$.

(iii) *Displacements $u_i^{2\varepsilon} \in V(\Omega^\varepsilon)$ are of the form*

$$u_1^{2\varepsilon} = z_1^{2\varepsilon} + x_2^\varepsilon z^{2\varepsilon} - \nu^\varepsilon (x_1^\varepsilon \partial_3^\varepsilon \underline{u}_3^{0\varepsilon} - \Phi_{1\beta}^\varepsilon \partial_{33}^\varepsilon u_\beta^{0\varepsilon}), \tag{29.89}$$

$$u_2^{2\varepsilon} = z_2^{2\varepsilon} - x_1^\varepsilon z^{2\varepsilon} - \nu^\varepsilon (x_2^\varepsilon \partial_3^\varepsilon \underline{u}_3^{0\varepsilon} - \Phi_{2\beta}^\varepsilon \partial_{33}^\varepsilon u_\beta^{0\varepsilon}), \tag{29.90}$$

$$\begin{aligned} u_3^{2\varepsilon} = {} & \underline{u}_3^{2\varepsilon} - x_\alpha^\varepsilon \partial_3^\varepsilon z_\alpha^{2\varepsilon} - w^\varepsilon \partial_3^\varepsilon z^{2\varepsilon} \\ & + \nu^\varepsilon \left\{ \tfrac{1}{2}[(x_1^\varepsilon)^2 + (x_2^\varepsilon)^2] - \frac{1}{2A^\varepsilon}(I_1^\varepsilon + I_2^\varepsilon) \right\} \partial_{33}^\varepsilon \underline{u}_3^{0\varepsilon} \\ & + [(1+\nu^\varepsilon)\eta_\alpha^\varepsilon + \nu^\varepsilon \theta_\alpha^\varepsilon] \partial_{333}^\varepsilon u_\alpha^{0\varepsilon} + 2(1+\nu^\varepsilon) h^{-1} h' \lambda_\alpha^\varepsilon \partial_{33} u_\alpha^{0\varepsilon} \\ & + \frac{2(1+\nu^\varepsilon)}{E^\varepsilon} w^{0\varepsilon}, \end{aligned} \tag{29.91}$$

where functions $z^{2\varepsilon}$, $w^{0\varepsilon}$, $\underline{u}_3^{2\varepsilon}$ and $z_\alpha^{2\varepsilon}$ are characterized in a unique way from the data and the zeroth order displacement components through the following problems:

(a) *The angle of twist $z^{2\varepsilon}$ depends only on x_3^ε and is the unique solution of the following one dimensional problem*:

$$
\begin{aligned}
& z^{2\varepsilon} \in H^1(0,L),\\
& \int_0^L \frac{E^\varepsilon J^\varepsilon}{2(1+\nu^\varepsilon)} \partial_3^\varepsilon z^{2\varepsilon} \partial_3^\varepsilon v \,\mathrm{d}x_3^\varepsilon = M_3^{2\varepsilon}(v)\,\mathrm{d}x_3^\varepsilon, \quad \textit{for all } v \in H_0^1(0,L),\\
& z^{2\varepsilon}(a) = \frac{\nu^\varepsilon}{(I_1^\varepsilon(a)+I_2^\varepsilon(a))}[H_2^\varepsilon(a)\partial_{33}^\varepsilon u_1^{0\varepsilon}(a) - H_1^\varepsilon(a)\partial_{33}^\varepsilon u_2^{0\varepsilon}(a)]\\
& \quad \textit{at } a = 0, L,
\end{aligned}
\tag{29.92}
$$

where

$$
\begin{aligned}
M_3^{2\varepsilon}(v) = {} & \int_0^L \left[\int_{\omega^\varepsilon(x_3^\varepsilon)} (x_2^\varepsilon f_1^\varepsilon - x_1^\varepsilon f_2^\varepsilon)\,\mathrm{d}\omega^\varepsilon + \int_{\gamma^\varepsilon(x_3^\varepsilon)} (x_2^\varepsilon g_1^\varepsilon - x_1^\varepsilon g_2^\varepsilon)\,\mathrm{d}\gamma^\varepsilon \right] v \,\mathrm{d}x_3^\varepsilon\\
& + \int_0^L \left[\int_{\omega^\varepsilon(x_3^\varepsilon)} w^\varepsilon f_3^\varepsilon \,\mathrm{d}\omega^\varepsilon + \int_{\gamma^\varepsilon(x_3^\varepsilon)} w^\varepsilon g_3^\varepsilon \,\mathrm{d}\gamma^\varepsilon \right] \partial_3^\varepsilon v \,\mathrm{d}x_3^\varepsilon\\
& + \int_0^L \frac{E^\varepsilon}{2(1+\nu^\varepsilon)} \{[(1+\nu^\varepsilon) I_\alpha^{w^\varepsilon} + \nu^\varepsilon I_\alpha^{\Psi^\varepsilon}]\partial_{333}^\varepsilon u_\alpha^{0\varepsilon}\\
& \qquad + 4(1+\nu^\varepsilon) h^{-1} h' (I_\alpha^{w^\varepsilon} + I_\alpha^{\Psi^\varepsilon}) \partial_{33}^\varepsilon u_\alpha^{0\varepsilon}\} \partial_3^\varepsilon v \,\mathrm{d}x_3^\varepsilon.
\end{aligned}
\tag{29.93}
$$

(b) *The additional warping $w^{0\varepsilon}$ is the unique solution of the following problem*:

$$
\begin{aligned}
& w^{0\varepsilon} \in H^1[0,L;H^1(\omega^\varepsilon)] \quad \textit{and for all } x_3^\varepsilon \in [0,L] \textit{ and all } \varphi^\varepsilon \in H^1(\omega^\varepsilon)\textit{:}\\
& \int_{\omega^\varepsilon(x_3^\varepsilon)} \partial_\beta^\varepsilon w^{0\varepsilon}(x_3^\varepsilon) \partial_\beta^\varepsilon \varphi^\varepsilon \,\mathrm{d}\omega^\varepsilon\\
& \quad = \int_{\omega^\varepsilon(x_3^\varepsilon)} f_3^\varepsilon(x_3^\varepsilon)\varphi^\varepsilon \,\mathrm{d}\omega^\varepsilon + \int_{\gamma^\varepsilon(x_3^\varepsilon)} g_3^\varepsilon(x_3^\varepsilon)\varphi^\varepsilon \,\mathrm{d}\gamma^\varepsilon\\
& \quad - \frac{1}{A(x_3^\varepsilon)} \left[\int_{\omega^\varepsilon(x_3^\varepsilon)} f_3^\varepsilon(x_3^\varepsilon)\,\mathrm{d}\omega^\varepsilon + \int_{\gamma^\varepsilon(x_3^\varepsilon)} g_3^\varepsilon(x_3^\varepsilon)\,\mathrm{d}\gamma^\varepsilon \right] \int_\omega^\varepsilon \varphi^\varepsilon \,\mathrm{d}\omega^\varepsilon,\\
& \int_{\omega^\varepsilon(x_3^\varepsilon)} w^{0\varepsilon}(x_3^\varepsilon)\,\mathrm{d}\omega^\varepsilon = 0.
\end{aligned}
\tag{29.94}
$$

(c) *The stretching component* $\underline{u}_3^{2\varepsilon}$ *depends only on* x_3^ε *and uniquely solves the following problem*:

$$\underline{u}_3^{2\varepsilon} \in H_0^1(0,L),$$

$$\int_0^L E^\varepsilon A^\varepsilon \partial_3^\varepsilon \underline{u}_3^{2\varepsilon} \partial_3^\varepsilon v \,\mathrm{d}x_3^\varepsilon = \int_0^L G_3^{2\varepsilon} \partial_3^\varepsilon v \,\mathrm{d}x_3^\varepsilon, \quad \textit{for all } v \in H_0^1(0,L), \tag{29.95}$$

where

$$G_3^{2\varepsilon} = -\int_{\omega^\varepsilon(x_3^\varepsilon)} (\partial_3^\varepsilon U_3^{2\varepsilon} + \nu^\varepsilon \sigma_{\mu\mu}^{0\varepsilon}) \,\mathrm{d}\omega^\varepsilon, \tag{29.96}$$

$$U_3^{2\varepsilon} = u_3^{2\varepsilon} - \underline{u}_3^{2\varepsilon} - x_\alpha^\varepsilon \partial_3^\varepsilon z_\alpha^{2\varepsilon}. \tag{29.97}$$

(d) *The bending component* $z_\alpha^{2\varepsilon}$ *depends only on the variable* x_3^ε *and is the unique solution of the following variational problem (no sum on* α, $\beta \neq \alpha$):

$$z_\alpha^{2\varepsilon} \in H^2(0,L),$$

$$\int_0^L E^\varepsilon I_\alpha^\varepsilon \partial_{33}^\varepsilon z_\alpha^{2\varepsilon} \partial_{33}^\varepsilon v \,\mathrm{d}x_3^\varepsilon = \int_0^L M_\alpha^{2\varepsilon} \partial_{33}^\varepsilon v \,\mathrm{d}x_3^\varepsilon, \quad \textit{for all } v \in H_0^2(0,L), \tag{29.98}$$

$$z_\alpha^{2\varepsilon}(a) = -\frac{1}{2A^\varepsilon(a)} \nu^\varepsilon (I_\alpha^\varepsilon(a) - I_\beta^\varepsilon(a)) \partial_{33}^\varepsilon u_\alpha^{0\varepsilon}(a) \quad \textit{at } a = 0, L, \tag{29.99}$$

$$\begin{aligned}
\partial_3^\varepsilon z_\alpha^{2\varepsilon}(a) = \frac{1}{I_\alpha^\varepsilon(a)} \Bigg\{ &\frac{2(1+\nu^\varepsilon)}{E^\varepsilon} \int_{\omega^\varepsilon(a)} x_\alpha^\varepsilon w^{0\varepsilon}(a) \,\mathrm{d}\omega^\varepsilon - \tfrac{1}{2} I_\alpha^{w^\varepsilon}(a) \partial_3^\varepsilon z^{2\varepsilon}(a) \\
&+ \nu^\varepsilon H_\alpha^\varepsilon(a) \partial_{33}^\varepsilon \underline{u}_3^{0\varepsilon}(a) + [(1+\nu^\varepsilon) L_{\alpha\beta}^{\eta^\varepsilon}(a) + \nu^\varepsilon L_{\alpha\beta}^{\theta^\varepsilon}(a)] \partial_{333}^\varepsilon u_\beta^{0\varepsilon}(a) \\
&+ 2(1+\nu^\varepsilon) h^{-1}(a) h'(a) H_\alpha^\varepsilon(a) \partial_3^\varepsilon \underline{u}_3^{0\varepsilon}(a) \\
&+ 2(1+\nu^\varepsilon) h^{-1}(a) h'(a) L_{\alpha\beta}^{\lambda^\varepsilon}(a) \partial_{33}^\varepsilon u_\beta^{0\varepsilon}(a) \Bigg\}, \quad \textit{at } a = 0, L,
\end{aligned} \tag{29.100}$$

where

$$M_\alpha^{2\varepsilon} = E^\varepsilon \int_{\omega^\varepsilon(x_3^\varepsilon)} x_\alpha^\varepsilon (\partial_3^\varepsilon U_3^{2\varepsilon} + \nu^\varepsilon \sigma_{\mu\mu}^{0\varepsilon}) \,\mathrm{d}\omega^\varepsilon. \tag{29.101}$$

(iv) *The stress components are of the following form*

$$\sigma_{33}^{0\varepsilon} = E^\varepsilon (\partial_3^\varepsilon \underline{u}_3^{0\varepsilon} - x_\alpha^\varepsilon \partial_{33}^\varepsilon u_\alpha^{0\varepsilon}). \tag{29.102}$$

$$\sigma_{31}^{0\varepsilon} = \frac{E^\varepsilon}{2(1+\nu^\varepsilon)}\{-\partial_2^\varepsilon \Psi^\varepsilon \partial_3^\varepsilon z^{2\varepsilon} + [(1+\nu^\varepsilon)\partial_1^\varepsilon \eta_\beta^\varepsilon + \nu^\varepsilon(\partial_1^\varepsilon \theta_\beta^\varepsilon + \Phi_{1\beta}^\varepsilon)]\partial_{333}^\varepsilon u_\beta^{0\varepsilon}$$
$$+ 2(1+\nu^\varepsilon)h^{-1}h'(x_1^\varepsilon \partial_3^\varepsilon \underline{u}_3^\varepsilon + \partial_1^\varepsilon \lambda_\alpha^\varepsilon \partial_{33} u_\alpha^{0\varepsilon})\} + \partial_1^\varepsilon w^0, \quad (29.103)$$

$$\sigma_{32}^{0\varepsilon} = \frac{E^\varepsilon}{2(1+\nu^\varepsilon)}\{\partial_1^\varepsilon \Psi^\varepsilon \partial_3^\varepsilon z^{2\varepsilon} + [(1+\nu^\varepsilon)\partial_2^\varepsilon \eta_\beta^\varepsilon + \nu^\varepsilon(\partial_2^\varepsilon \theta_\beta^\varepsilon + \Phi_{2\beta}^\varepsilon)]\partial_{333}^\varepsilon u_\beta^{0\varepsilon}$$
$$+ 2(1+\nu^\varepsilon)h^{-1}h'(x_2^\varepsilon \partial_3^\varepsilon \underline{u}_3^\varepsilon + \partial_2^\varepsilon \lambda_\alpha^\varepsilon \partial_{33} u_\alpha^\varepsilon)\} + \partial_2^\varepsilon w^0, \quad (29.104)$$

$$\sigma_{\alpha\beta}^{0\varepsilon} = S_{\alpha\beta}^\varepsilon(\underline{u}^{4\varepsilon}) + S_{\alpha\beta}^{0\varepsilon}, \quad (29.105)$$

where

$$S_{\alpha\beta}^{0\varepsilon} = \frac{\nu^\varepsilon E^\varepsilon}{(1+\nu^\varepsilon)(1-2\nu^\varepsilon)}\partial_3^\varepsilon U_3^{2\varepsilon}\delta_{\alpha\beta}, \quad (29.106)$$

$$S_{\alpha\beta}^\varepsilon(\underline{u}^{4\varepsilon}) = \frac{E^\varepsilon}{(1+\nu^\varepsilon)}e_{\alpha\beta}^\varepsilon(\underline{u}^{4\varepsilon}) + \frac{\nu^\varepsilon E^\varepsilon}{(1+\nu^\varepsilon)(1-2\nu^\varepsilon)}e_{\mu\mu}^\varepsilon(\underline{u}^{4\varepsilon})\delta_{\alpha\beta}, \quad (29.107)$$

and $\underline{u}^{4\varepsilon} = (\underline{u}_\alpha^{4\varepsilon})$ *is the unique solution of the following plane elasticity problem:*

$$\underline{u}^{4\varepsilon} \in L^2[0,L;(H^1(\omega^\varepsilon))^2], \quad \text{and a.e. in } (0,L), \text{ for all } \varphi^\varepsilon \in [H^1(\omega^\varepsilon)]^2:$$
$$\int_{\omega^\varepsilon(x_3^\varepsilon)} S_{\alpha\beta}^\varepsilon(\underline{u}^{4\varepsilon})e_{\alpha\beta}^\varepsilon(\varphi^\varepsilon)\,d\omega^\varepsilon$$
$$= \int_{\omega^\varepsilon(x_3^\varepsilon)} f_\beta^\varepsilon \varphi_\beta^\varepsilon\,d\omega^\varepsilon + \int_{\gamma^\varepsilon(x_3^\varepsilon)} g_\beta^\varepsilon \varphi_\beta^\varepsilon\,d\gamma^\varepsilon$$
$$+ \int_{\omega^\varepsilon(x_3^\varepsilon)} \partial_3^\varepsilon \sigma_{3\beta}^{0\varepsilon}\varphi_\beta^\varepsilon\,d\omega^\varepsilon - \int_{\omega^\varepsilon(x_3^\varepsilon)} S_{\alpha\beta}^{0\varepsilon}e_{\alpha\beta}^\varepsilon(\varphi^\varepsilon)\,d\omega^\varepsilon \quad (29.108)$$
$$+ \int_{\omega^\varepsilon(x_3^\varepsilon)} h^{-1}h'(x_\alpha^\varepsilon \partial_\alpha^\varepsilon \sigma_{3\beta}^{0\varepsilon} + 2\sigma_{3\beta}^{0\varepsilon})\varphi_\beta\,d\omega^\varepsilon,$$
$$\int_{\omega^\varepsilon(x_3^\varepsilon)} \underline{u}_\alpha^{4\varepsilon}\,d\omega^\varepsilon = \int_{\omega^\varepsilon(x_3^\varepsilon)} [x_2^\varepsilon \underline{u}_1^{4\varepsilon} - x_1^\varepsilon \underline{u}_2^{4\varepsilon}]\,d\omega^\varepsilon = 0. \qquad \square$$

REMARK 29.3. It is important to observe that Eqs. (29.86)–(29.108) are analogous to those of the case of a constant cross section, obtained in (11.7), (11.8) and (12.2)–(12.17). In (29.86)–(29.108) one must take into account the fact that the functions and constants depending on the geometry of the cross section now depend on the variable x_3^ε also. Moreover, some additional terms are also obtained from the fact that h' does not vanish, that is, the way the cross section changes introduces additional terms in the classical theories.

The first order models obtained may be written in the classical form as

$$\frac{\mathrm{d}^2}{\mathrm{d}(x_3^\varepsilon)^2}\left(E^\varepsilon I_\beta^\varepsilon \frac{\mathrm{d}^2 u_\beta^{0\varepsilon}}{\mathrm{d}(x_3^\varepsilon)^2}\right) = \int_{\omega^\varepsilon(x_3^\varepsilon)} f_\beta^\varepsilon \,\mathrm{d}\omega^\varepsilon + \int_{\gamma^\varepsilon(x_3^\varepsilon)} g_\beta^\varepsilon \,\mathrm{d}\gamma^\varepsilon + \frac{\mathrm{d}}{\mathrm{d}x_3^\varepsilon}\left[\int_{\omega^\varepsilon(x_3)} x_\beta^\varepsilon f_3^\varepsilon \,\mathrm{d}\omega^\varepsilon + \int_{\gamma^\varepsilon(x_3)} x_\beta^\varepsilon g_3^\varepsilon \,\mathrm{d}\gamma^\varepsilon\right] \quad \text{in } (0,L), \tag{29.109}$$

$$u_\beta^{0\varepsilon}(0) = u_\beta^{0\varepsilon}(L) = 0,$$

$$\frac{\mathrm{d}u_\beta^{0\varepsilon}}{\mathrm{d}x_3}(0) = \frac{\mathrm{d}u_\beta^{0\varepsilon}}{\mathrm{d}x_3^\varepsilon}(L) = 0.$$

$$-\frac{\mathrm{d}}{\mathrm{d}x_3^\varepsilon}\left(E^\varepsilon A^\varepsilon \frac{\mathrm{d}\underline{u}_3^{0\varepsilon}}{\mathrm{d}x_3^\varepsilon}\right) = \int_{\omega^\varepsilon(x_3^\varepsilon)} f_3^\varepsilon \,\mathrm{d}\omega^\varepsilon + \int_{\gamma^\varepsilon(x_3^\varepsilon)} g_3^\varepsilon \,\mathrm{d}\gamma^\varepsilon \quad \text{in } (0,L), \tag{29.110}$$

$$\underline{u}_3^{0\varepsilon}(0) = \underline{u}_3^{0\varepsilon}(L) = 0.$$

REMARK 29.4. The dependence on variable x_3^ε, of the functions and constants of the geometry, is known explicitly (cf. (29.83)–(29.85)). Therefore, it is enough to know these quantities for the reference cross section ω. This is a consequence of the fact that the variation with respect to ω is very simple: an homothety of ratio $\varepsilon h(x_3)$.

Although not mathematically justified, models (29.109) and (29.110) may be considered valid also for the case of more general geometries of the form $\Omega^\varepsilon = \bigcup_{x_3^\varepsilon \in (0,L)} \omega^\varepsilon(x_3^\varepsilon)$, where $\omega^\varepsilon(x_3^\varepsilon)$ is not necessarily obtained from ω through an homothety, but in such a way that axis Ox_3^ε is still a principal axis of inertia.

An interesting case, studied by VEIGA [1993b,c], where the asymptotic expansion method may be applied in order to justify and to obtain general models, is the one where the cross section $\omega^\varepsilon(x_3^\varepsilon)$ is obtained from $\omega^\varepsilon = \varepsilon\omega$ through a rotation of an angle $\xi(x_3^\varepsilon)$, that is,

$$\omega^\varepsilon(x_3) = \{(x_1^\varepsilon, x_2^\varepsilon) \in \mathbb{R}^2 \colon x_1^\varepsilon = \varepsilon(x_1 \cos\xi(x_3) + x_2 \sin\xi(x_3)),$$
$$x_2^\varepsilon = \varepsilon(-x_1 \sin\xi(x_3) + x_2 \cos\xi(x_3)) \colon (x_1, x_2) \in \omega\}.$$

In this case the stretching, bending and torsion components are completely coupled.

CHAPTER VII

Rods with a Multicellular Cross Section. Some Homogenization and Asymptotic Results

In this chapter we consider a rod made of a linearly elastic material with a multiply connected cross section. The results of Sections 1–12 immediatly apply to this case. However, if the structure is finely perforated and if one tries to compute the corresponding displacement and stress fields, given by the asymptotic expansion method, one is faced with a problem that may lead to numerical instabilities and to large computing times. Therefore, it is desirable to have an alternative reliable model giving the displacement and the stress fields within a certain accuracy.

Let us describe the major difficulty in this type of analysis. Consider the cross section made of a periodic repetition of cells whose characteristic dimension δ is much smaller than ε, a measure of the size of the cross section, which in turn is much smaller than the beam's length L.

Considering the three-dimensional elasticity problem for this case one obtains the displacement and the stress fields as a function of these two parameters, which we denote by $\boldsymbol{u}^{\delta,\varepsilon}$ and $\boldsymbol{\sigma}^{\delta,\varepsilon}$.

If one whishes to replace the three-dimensional model by the corresponding generalizations of the classical theories as given by the asymptotic expansion method, two main approaches may now be followed:

(i) ε is held fixed and δ is taken to zero, i.e., we obtain the three-dimensional homogenized elasticity equations for a rod. The homogenized material is geometrically anisotropic and we may apply the results of Section 25 in order to take now the limit as ε goes to zero.

(ii) δ is held fixed and we first take the limit as ε goes to zero. We obtain the one-dimensional generalized beam equations and then take the limit as δ goes to zero.

A major question is whether these two limit processes interchange, i.e., is the model obtained the same in either approach? Equivalently,

$$\lim_{\delta\to 0}\lim_{\varepsilon\to 0}\boldsymbol{u}^{\delta,\varepsilon} = \lim_{\varepsilon\to 0}\lim_{\delta\to 0}\boldsymbol{u}^{\delta,\varepsilon}?$$

$$\lim_{\delta\to 0}\lim_{\varepsilon\to 0}\boldsymbol{\sigma}^{\delta,\varepsilon} = \lim_{\varepsilon\to 0}\lim_{\delta\to 0}\boldsymbol{\sigma}^{\delta,\varepsilon}?$$

The answer to this question is: it depends. We refer to GEYMONAT, KRASUCKI and MARIGO [1987] and TUTEK [1987] for some of the details.

In Sections 30 and 31 we shall be concerned mainly with the obtainment of the homogenized equations corresponding to the classical Bernoulli–Navier, Saint Venant, Timoshenko and Vlassov beam theories given through the general asymptotic model. These results were first obtained by GEYMONAT, KRASUCKI and MARIGO [1987], TUTEK [1987] for the zeroth order term and by MASCARENHAS and TRABUCHO [1990] for the second order terms.

We shall give the homogenized behaviour of all the functions and constants showing up in the generalizations of Saint Venant's torsion theory and Timoshenko's beam theory. In order to preserve the relationship between the torsion and the warping functions, the derivation of the homogenized torsion problem presented here is different from the one contained in CIORANESCU and SAINT JEAN PAULIN [1979]. Nevertheless it can be proved that the two limit problems are the same.

We indicate the relationship between the coefficients of the so obtained limit problems and those of the homogenized three-dimensional elasticity model. We consider the homogenized behaviour of the shear and plane stress fields which takes into account the deformation of the cross section on its own plane.

In Section 32 we follow the inverse process. First, we calculate the three-dimensional homogenized model ($\delta \to 0$) applying the asymptotic method ($\varepsilon \to 0$) afterwards. In this way, we obtain the first order homogenized theory which coincides with the model obtained in Section 31. The results in this chapter are due to MASCARENHAS and TRABUCHO [1990].

30. The asymptotic model for rods with a multicellular cross section

We start by adapting some of the notations to standard ones in homogenization literature (cf. BENSSOUSSAN, LIONS and PAPANICOLOAU [1978], CIORANESCU and SAINT JEAN PAULIN [1979], SANCHEZ-PALENCIA [1980]), through the description of the periodic structure of the cross section that we are going to consider.

With the notation of Section 2 in mind, namely Eqs. (2.1)–(2.5), we associate with each ω_i ($i = 1, \ldots, N$) a hole periodically distributed in ω. More specifically consider $Y = [0,1] \times [0,1]$ and let T be an open subset of Y, with a regular boundary ∂T, such that $\overline{T} \subset \operatorname{int} Y$. Define $Y^* = Y - T$ and let χ be its characteristic function, periodically extended to all the planes generated by (e_α). Let Θ be the area of Y^*, that is, $\Theta = |Y^*|$. Given $\delta \in \mathbb{R}, \delta > 0$, consider χ^δ given by $\chi^\delta(\boldsymbol{x}) = \chi(\boldsymbol{x}/\delta)$, for all $\boldsymbol{x} \in \mathbb{R}^2$. χ^δ defines a subset of $\mathbb{R}^2$ with a periodic structure having δY^* as its basic cell. Suppose δ is a small parameter. If χ_ω represents the characteristic function of ω, $\chi^{\delta,1} = \chi_\omega \chi^\delta$ defines a subset $\omega^{\delta,1}$, of ω, obtained by perforating ω with holes δT, with a periodicity δY. To avoid a nonregular boundary, only holes whose closure does not touch γ are to be considered, i.e., whenever the boundary of a hole intersects γ we will consider $\chi^{\delta,1} = 1$ in all the respective cells. This adjusted function $\chi^{\delta,1}$ defines a set that will be represented by $\omega^{\delta,1}$, with boundary $\gamma^{\delta,1}$. Define

$$\Omega^{\delta,1} = \omega^{\delta,1} \times (0, L), \qquad \Gamma_0^{\delta,1} = \omega^{\delta,1} \times \{0\},$$
$$\Gamma_L^{\delta,1} = \omega^{\delta,1} \times \{L\}, \qquad \Gamma^{\delta,1} = \gamma^{\delta,1} \times (0, L).$$

Representing by $\partial T^{\delta,1} = \bigcup_{k=1}^{N(\delta)} \partial T_k^{\delta,1}$ the union of the boundaries of the $N(\delta)$ holes periodically made in ω, the following disjoint union holds: $\gamma^{\delta,1} = \gamma \cup \partial T^{\delta,1}$.

As in Section 2 (cf. (2.4)) for every $\varepsilon \in \mathbb{R}$, $\varepsilon > 0$, consider the homothety $\Pi^\varepsilon(x_1, x_2) = (\varepsilon x_1, \varepsilon x_2) = (x_1^\varepsilon, x_2^\varepsilon)$, for all $(x_1, x_2) \in \mathbb{R}^2$ and denote

$$\omega^\varepsilon = \Pi^\varepsilon(\omega), \qquad \gamma^\varepsilon = \Pi^\varepsilon(\gamma), \qquad \omega^{\delta,\varepsilon} = \Pi^\varepsilon(\omega^{\delta,1}),$$
$$\partial T^{\delta,\varepsilon} = \Pi^\varepsilon(\partial T^{\delta,1}), \qquad \gamma^{\delta,\varepsilon} = \Pi^\varepsilon(\gamma^{\delta,1}),$$
$$\Omega^\varepsilon = \omega^\varepsilon \times (0, L), \qquad \Gamma_0^\varepsilon = \omega^\varepsilon \times \{0\},$$
$$\Gamma_L^\varepsilon = \omega^\varepsilon \times \{L\}, \qquad \Gamma^\varepsilon = \gamma^\varepsilon \times (0, L),$$
$$\Omega^{\delta,\varepsilon} = \omega^{\delta,\varepsilon} \times (0, L), \qquad \Gamma_0^{\delta,\varepsilon} = \omega^{\delta,\varepsilon} \times \{0\},$$
$$\Gamma_L^{\delta,\varepsilon} = \omega^{\delta,\varepsilon} \times \{L\}, \qquad \Gamma^{\delta,\varepsilon} = \gamma^{\delta,\varepsilon} \times (0, L).$$

The characteristic function of the set $\omega^{\delta,\varepsilon}$ will be represented by $\chi^{\delta,\varepsilon}$.

In what follows δ and ε will be considered two small and independent parameters. Parameter ε is a measure of the size of section $\omega^{\delta,\varepsilon}$: in fact, since the area of ω^ε is precisely ε^2, the area of $\omega^{\delta,\varepsilon}$ may be considered to be $\Theta\varepsilon^2$ and so, independent of δ.

Parameter $\eta = \varepsilon\delta$ measures the periodicity of perforation in section $\omega^{\delta,\varepsilon}$: in fact, representing by $\chi_{\omega^\varepsilon}$ the characteristic function of ω^ε, one has $\chi^{\delta,\varepsilon}(\boldsymbol{x}^\varepsilon) = \chi_{\omega^\varepsilon}(\boldsymbol{x}^\varepsilon)\chi(\boldsymbol{x}^\varepsilon/\varepsilon\delta)$, except for the holes that touch the boundary γ^ε.

System (e_i) is still a principal system of inertia associated with Ω^ε. Since δ is a small parameter we shall admit, for the sake of simplicity, that the same principal system of inertia with respect to $\Omega^{\delta,\varepsilon}$ is kept independently of δ. This actually happens if ω admits two orthogonal axes of symmetry.

As in Sections 2–5 it is possible to give a displacement approach formulation for this problem and to proceed in order to show that the scaled stress field may be expanded in positive powers of ε only. However, since the calculations in the displacement–stress approach formulation are rather simpler to present, we shall adopt it here. Consequently, we consider that an equivalent result to those of Theorems 5.1, 5.3, 5.5 and Corollary 5.3 has been proved for the present case and which we shall omit here because the proof can be done exactly as for those cases, except for the notation.

Let the beam $\Omega^{\delta,\varepsilon}$ be made of a homogeneous and isotropic linearly elastic material of modulus of elasticity E and Poisson's ratio ν, both independent of δ and ε.

Let $f_i^\varepsilon \in L^2(\Omega^{\delta,\varepsilon})$ and $g_i^\varepsilon \in L^2(\Gamma^{\delta,\varepsilon})$ be the ith components of the volume density of applied body forces and surface density of applied surface tractions, respectively. To simplify we have admitted that the system of applied forces does not depend on the parameter δ. We also admit that $g_i^\varepsilon(\boldsymbol{x}^\varepsilon) = 0$ if $\boldsymbol{x}^\varepsilon \in \partial T^{\delta,\varepsilon} \times (0, L)$.

Consider the spaces

$$V(\Omega^{\delta,\varepsilon}) = \{\boldsymbol{v} = (v_i) \in [H^1(\Omega^{\delta,\varepsilon})]^3;$$
$$\int_{\Gamma_0^{\delta,\varepsilon} \cup \Gamma_L^{\delta,\varepsilon}} \boldsymbol{v} \, \mathrm{d}\omega^{\delta,\varepsilon} = \int_{\Gamma_0^{\delta,\varepsilon} \cup \Gamma_L^{\delta,\varepsilon}} \boldsymbol{x} \times \boldsymbol{v} \, \mathrm{d}\omega^{\delta,\varepsilon} = \boldsymbol{0}\}$$
$$\Sigma(\Omega^{\delta,\varepsilon}) = [L^2(\Omega^{\delta,\varepsilon})]_s^9 = \{\boldsymbol{\tau} = (\tau_{ij}) \in [L^2(\Omega^{\delta,\varepsilon})]^9; \tau_{ij} = \tau_{ji}\}$$

equipped with the usual norms. Then the three-dimensional elasticity problem in a displacement–stress approach may be written as

$$\boldsymbol{u}^{\delta,\varepsilon} \in V(\Omega^{\delta,\varepsilon}):$$
$$\int_{\Omega^{\delta,\varepsilon}} \sigma_{ij}^{\delta,\varepsilon} e_{ij}^\varepsilon(\boldsymbol{v}) \, \mathrm{d}\boldsymbol{x}^\varepsilon = \int_{\Omega^{\delta,\varepsilon}} f_i^\varepsilon v_i \, \mathrm{d}\boldsymbol{x}^\varepsilon + \int_{\Gamma^{\delta,\varepsilon}} g_i^\varepsilon v_i \, \mathrm{d}a^\varepsilon, \quad \text{for all } \boldsymbol{v} \in V(\Omega^{\delta,\varepsilon}),$$
$$\boldsymbol{\sigma}^{\delta,\varepsilon} \in \Sigma(\Omega^{\delta,\varepsilon}): \boldsymbol{\sigma}^{\delta,\varepsilon} = 2\mu \boldsymbol{e}^\varepsilon(\boldsymbol{u}^{\delta,\varepsilon}) + \lambda \operatorname{tr} \boldsymbol{e}^\varepsilon(\boldsymbol{u}^{\delta,\varepsilon}) \, \mathbf{Id},$$

where the same notation as in the previous sections was used.

As in Section 6 (cf. (6.10)–(6.12)) it is a classical result that this problem has a unique solution. We are interested in studying the behaviour of the displacement $\boldsymbol{u}^{\delta,\varepsilon}$ and the stress $\boldsymbol{\sigma}^{\delta,\varepsilon}$ fields, when δ and ε are very small parameters.

Considering $\varepsilon \to 0$ while $\delta > 0$ is fixed, the number of holes and its relative distribution in ω^ε are kept and a zoom is made over the entire section. The asymptotic expansion technique for beam models (see Section 7) may be performed to obtain a generalization of the classical theory of Bernoulli–Navier. Moreover considering higher order terms as in Sections 8 and 9 it is possible to obtain generalizations of the classical torsion theory of Saint Venant and of Timoshenko and Vlassov's beams theories. If we then consider the limit as $\delta \to 0$, the behaviour of the homogenized theories is obtained.

The case of Bernoulli–Navier's beam theory, for a composite material, has been considered in GEYMONAT, KRASUKI and MARIGO [1987a,b]. The axially perforated case was considered in MASCARENHAS and TRABUCHO [1990] where not only Bernoulli–Navier's but also Saint Venant's, Timoshenko's and Vlassov's beam theories were considered, through a generalized asymptotic model.

Moreover, for the case where the coefficients depend only on the axial component x_3^ε, it is also known that the limit as $\delta \to 0$ and as $\varepsilon \to 0$ may be interchanged. These results also apply to the present case and the proof can be made with minor modifications.

REMARK 30.1. Since we now have two small parameters, δ and ε, in this section whenever we refer to Eqs. (11.6)–(11.10) and (12.1)–(12.20) or even to those characterizing the geometry of the cross section (7.21)–(7.56) we shall always assume that they are written with respect to the actual domain $\Omega^{\delta,\varepsilon}$ or with respect to the reference domain $\Omega^{\delta,1}$ for the case $\varepsilon = 1$. We shall denote this dependence by using superscript δ, ε or $\delta, 1$ respectively. Moreover for the zeroth-, second- and higher order displacement and stress components we use the superscript $\delta(0)\varepsilon$ instead of 0ε and $\delta(2), 1$ instead of 2 for example.

In order to make all this clear and in order to make this section self contained we summarize Theorems 11.1 and 12.1 here with this new notation. Let us first write the constants and functions characterizing the geometry of the cross section $\omega^{\delta,\varepsilon}$ with the new notation.

(a) Constants $H_i^{\delta,\varepsilon}$ and $I_\alpha^{\delta,\varepsilon}$.

$$\begin{aligned} H_\alpha^{\delta,\varepsilon} &= \int_{\omega^{\delta,\varepsilon}} \tfrac{1}{2} x_\alpha^\varepsilon [(x_1^\varepsilon)^2 + (x_2^\varepsilon)^2] \, d\omega^{\delta,\varepsilon}, \\ H_3^{\delta,\varepsilon} &= \int_{\omega^{\delta,\varepsilon}} \tfrac{1}{4} [(x_1^\varepsilon)^2 + (x_2^\varepsilon)^2]^2 \, d\omega^{\delta,\varepsilon}, \\ I_\alpha^{\delta,\varepsilon} &= \int_{\omega^{\delta,\varepsilon}} (x_\alpha^\varepsilon)^2 \, d\omega^{\delta,\varepsilon}, \end{aligned} \tag{30.1}$$

where $I_\alpha^{\delta,\varepsilon}$ represents the second moment of area with respect to axis e_β $(\alpha \neq \beta)$.

(b) The warping function $w^{\delta,\varepsilon}$ is the unique solution of the problem

$$\begin{aligned} &-\partial_{\alpha\alpha}^\varepsilon w^{\delta,\varepsilon} = 0, \quad \text{in } \omega^{\delta,\varepsilon}, \\ &\partial_n^\varepsilon w^{\delta,\varepsilon} = x_2^\varepsilon n_1^\varepsilon - x_1^\varepsilon n_2^\varepsilon, \quad \text{on } \gamma^{\delta,\varepsilon}, \\ &\int_{\omega^{\delta,\varepsilon}} w^{\delta,\varepsilon} \, d\omega^{\delta,\varepsilon} = 0. \end{aligned} \tag{30.2}$$

(c) Constants $I_\beta^{w^{\delta,\varepsilon}}$ are given by

$$I_\beta^{w^{\delta,\varepsilon}} = 2 \int_{\omega^{\delta,\varepsilon}} x_\beta^{\delta,\varepsilon} w^{\delta,\varepsilon} \, d\omega^{\delta,\varepsilon}. \tag{30.3}$$

(d) Saint Venant's torsion function $\Psi^{\delta,\varepsilon}$ is the unique solution of problem:

$$\begin{aligned} &-\partial_{\alpha\alpha}^\varepsilon \Psi^{\delta,\varepsilon} = 2, \quad \text{in } \omega^{\delta,\varepsilon}, \\ &\Psi^{\delta,\varepsilon} = 0, \quad \text{on } \gamma^\varepsilon, \\ &\partial_\tau^\varepsilon \Psi^{\delta,\varepsilon} = 0, \quad \text{on } \partial T^{\delta,\varepsilon}, \\ &\int_{\partial T_k^{\delta,\varepsilon}} \partial_n^\varepsilon \Psi^{\delta,\varepsilon} \, dT_k^{\delta,\varepsilon} = 2|T_k^{\delta,\varepsilon}|, \quad k = 1, \ldots, N(\delta). \end{aligned} \tag{30.4}$$

It is possible to prove that function $\Psi^{\delta,\varepsilon}$ is constant on the boundary of each hole $T_k^{\delta,\varepsilon}$, $k=1,2,\dots,N(\delta)$. For each $k=1,2,\dots,N(\delta)$ let $c_k^{\delta,\varepsilon}$ denote the trace of $\Psi^{\delta,\varepsilon}$ on $\partial T_k^{\delta,\varepsilon}$.

(e) Constants $I_\alpha^{\Psi^{\delta,\varepsilon}}$ are given by

$$\begin{aligned} I_1^{\Psi^{\delta,\varepsilon}} &= -\int_{\omega^{\delta,\varepsilon}} (x_2^\varepsilon)^2 \partial_2^\varepsilon \Psi^{\delta,\varepsilon}\,\mathrm{d}\omega^{\delta,\varepsilon} \\ &= 2\int_{\omega^{\delta,\varepsilon}} x_2^\varepsilon \Psi^{\delta,\varepsilon}\,\mathrm{d}\omega^{\delta,\varepsilon} + 2\sum_{k=1}^{N(\delta)} c_k^{\delta,\varepsilon} \int_{T_k^{\delta,\varepsilon}} x_2^\varepsilon\,\mathrm{d}T_k^{\delta,\varepsilon}, \\ I_2^{\Psi^{\delta,\varepsilon}} &= \int_{\omega^{\delta,\varepsilon}} (x_1^\varepsilon)^2 \partial_1^\varepsilon \Psi^{\delta,\varepsilon}\,\mathrm{d}\omega^{\delta,\varepsilon} \\ &= -2\int_{\omega^{\delta,\varepsilon}} x_1^\varepsilon \Psi^{\delta,\varepsilon}\,\mathrm{d}\omega^{\delta,\varepsilon} - 2\sum_{k=1}^{N(\delta)} c_k^{\delta,\varepsilon} \int_{T_k^{\delta,\varepsilon}} x_1^\varepsilon\,\mathrm{d}T_k^{\delta,\varepsilon}. \end{aligned} \tag{30.5}$$

(f) The torsion constant is

$$J^{\delta,\varepsilon} = -\int_{\omega^{\delta,\varepsilon}} x_\alpha^\varepsilon \partial_\alpha^\varepsilon \Psi^{\delta,\varepsilon}\,\mathrm{d}\omega^{\delta,\varepsilon} = 2\int_{\omega^{\delta,\varepsilon}} \Psi^{\delta,\varepsilon}\,\mathrm{d}\omega^{\delta,\varepsilon} + 2\sum_{k=1}^{N(\delta)} c_k^{\delta,\varepsilon} |T_k^{\delta,\varepsilon}|. \tag{30.6}$$

(g) Functions $\eta_\beta^{\delta,\varepsilon}$ are the unique solution of problem

$$\begin{aligned} &-\partial_{\alpha\alpha}^\varepsilon \eta_\beta^{\delta,\varepsilon} = -2x_\beta^\varepsilon, \quad \text{in } \omega^{\delta,\varepsilon}, \\ &\partial_n^\varepsilon \eta_\beta^{\delta,\varepsilon} = 0, \quad \text{on } \gamma^{\delta,\varepsilon}, \\ &\int_{\omega^{\delta,\varepsilon}} \eta_\beta^{\delta,\varepsilon}\,\mathrm{d}\omega^{\delta,\varepsilon} = 0. \end{aligned} \tag{30.7}$$

(h) Functions $\theta_\beta^{\delta,\varepsilon}$ are the unique solution of problem

$$\begin{aligned} &-\partial_{\alpha\alpha}^\varepsilon \theta_\beta^{\delta,\varepsilon} = 2x_\beta^\varepsilon, \quad \text{in } \omega^{\delta,\varepsilon}, \\ &\partial_n^\varepsilon \theta_\beta^{\delta,\varepsilon} = -\Phi_{\alpha\beta}^\varepsilon n_\alpha^\varepsilon, \quad \text{on } \gamma^{\delta,\varepsilon}, \\ &\int_{\omega^{\delta,\varepsilon}} \theta_\beta^{\delta,\varepsilon}\,\mathrm{d}\omega^{\delta,\varepsilon} = 0. \end{aligned} \tag{30.8}$$

where, for each $(x_1^\varepsilon, x_2^\varepsilon) \in \omega^\varepsilon$ we defined

$$\boldsymbol{\Phi}^\varepsilon(x_1^\varepsilon, x_2^\varepsilon) = (\Phi_{\alpha\beta}^\varepsilon)(x_1^\varepsilon, x_2^\varepsilon) = \begin{bmatrix} \frac{1}{2}[(x_1^\varepsilon)^2 - (x_2^\varepsilon)^2] & x_1^\varepsilon x_2^\varepsilon \\ x_1^\varepsilon x_2^\varepsilon & \frac{1}{2}[(x_2^\varepsilon)^2 - (x_1^\varepsilon)^2] \end{bmatrix}. \tag{30.9}$$

(i) Constants $L^{\eta^{\delta,\varepsilon}}_{\alpha\beta}$, $L^{\theta^{\delta,\varepsilon}}_{\alpha\beta}$, $K^{\eta^{\delta,\varepsilon}}_{\alpha\beta}$, and $K^{\theta^{\delta,\varepsilon}}_{\alpha\beta}$ are defined by

$$L^{\eta^{\delta,\varepsilon}}_{\alpha\beta} = \int_{\omega^{\delta,\varepsilon}} x^\varepsilon_\alpha \eta^{\delta,\varepsilon}_\beta \, d\omega^{\delta,\varepsilon}, \qquad L^{\theta^{\delta,\varepsilon}}_{\alpha\beta} = \int_{\omega^{\delta,\varepsilon}} x^\varepsilon_\alpha \theta^{\delta,\varepsilon}_\beta \, d\omega^{\delta,\varepsilon}, \tag{30.10}$$

$$K^{\eta^{\delta,\varepsilon}}_{\alpha\beta} = \int_{\omega^{\delta,\varepsilon}} \Phi^\varepsilon_{\alpha\mu} \partial^\varepsilon_\mu \eta^{\delta,\varepsilon} \, d\omega^{\delta,\varepsilon}, \qquad K^{\theta^{\delta,\varepsilon}}_{\alpha\beta} = \int_{\omega^{\delta,\varepsilon}} \Phi^\varepsilon_{\alpha\mu} \partial^\varepsilon_\mu \theta^{\delta,\varepsilon} \, d\omega^{\delta,\varepsilon}. \tag{30.11}$$

We now briefly summarize the results of Theorems 11.1 and 12.2 adapted to the present case, in the following theorem.

THEOREM 30.1. *Let the system of applied forces be such that*

$$f^\varepsilon_i \in L^2(\Omega^{\delta,\varepsilon}), \qquad g^\varepsilon_i \in L^2(\Gamma^{\delta,\varepsilon}),$$
$$g^\varepsilon_i(\boldsymbol{x}^\varepsilon) = 0 \quad \text{if } \boldsymbol{x}^\varepsilon \in \partial T^{\delta,\varepsilon} \times (0, L).$$

Then, the zeroth order terms $\boldsymbol{u}^{\delta(0)\varepsilon}$, $\sigma^{\delta(0)\varepsilon}_{33}$, $m^{\delta(0)\varepsilon}_\beta$, $q^{\delta(0)\varepsilon}_\beta$, *are uniquely characterized as follows:*

(i) *The displacement field* $\boldsymbol{u}^{\delta(0)\varepsilon}$ *is of the Bernoulli–Navier type:*

$$\boldsymbol{u}^{\delta(0)\varepsilon} \in V_{\mathrm{BN}}(\Omega^{\delta,\varepsilon}): \quad \begin{aligned} &u^{\delta(0)\varepsilon}_3 = \underline{u}^{\delta(0)\varepsilon}_3 - x^\varepsilon_\alpha \partial^\varepsilon_3 u^{\delta(0)\varepsilon}_\alpha, \\ &\underline{u}^{\delta(0)\varepsilon}_3 \in H^1_0(0, L), \ \ u^{\delta(0)\varepsilon}_\alpha \in H^2_0(0, L), \end{aligned} \tag{30.12}$$

where the axial extension $\underline{u}^{\delta(0)\varepsilon}_3$ *and the bending* $u^{\delta(0)\varepsilon}_\alpha$ *components are the unique solutions of the following one-dimensional problems*

$$u^{\delta(0)\varepsilon}_\alpha \in H^2_0(0, L),$$

$$\begin{aligned} &\int_0^L E^\varepsilon I^{\delta,\varepsilon}_\alpha \partial^\varepsilon_{33} u^{\delta(0)\varepsilon}_\alpha \partial_{33} v^0_\alpha \, dx^\varepsilon_3 \\ &\quad = \int_0^L \left[\int_{\omega^{\delta,\varepsilon}} f^\varepsilon_\alpha \, d\omega^{\delta,\varepsilon} + \int_{\gamma^\varepsilon} g^\varepsilon_\alpha \, d\gamma^\varepsilon \right] v^0 \, dx^\varepsilon_3 \\ &\quad - \int_0^L \left[\int_{\omega^{\delta,\varepsilon}} x^\varepsilon_\alpha f^\varepsilon_3 \, d\omega^{\delta,\varepsilon} + \int_{\gamma^\varepsilon} x^\varepsilon_\alpha g^\varepsilon_3 \, d\gamma^\varepsilon \right] \partial^\varepsilon_3 v^0 \, dx^\varepsilon_3, \end{aligned} \tag{30.13}$$

for all $\boldsymbol{v}^0 \in H^2_0(0, L)$.

$$\underline{u}^{\delta(0)\varepsilon}_3 \in H^1_0(0, L),$$

$$\int_0^L E^\varepsilon |\omega^{\delta,\varepsilon}| \partial_3 \underline{u}^{\delta(0)\varepsilon}_3 \partial^\varepsilon_3 \underline{v}^0 \, dx^\varepsilon_3 = \int_0^L \left[\int_{\omega^{\delta,\varepsilon}} f^\varepsilon_3 \, d\omega^{\delta,\varepsilon} + \int_{\gamma^\varepsilon} g^\varepsilon_3 \, d\gamma^\varepsilon \right] \underline{v}^0, \tag{30.14}$$

for all $\underline{v}^0 \in H^1_0(0, L)$.

(ii) *The axial stress component* $\sigma_{33}^{\delta(0)\varepsilon}$ *is obtained from* $u_i^{\delta(0)\varepsilon}$ *in the following way:*

$$\sigma_{33}^{\delta(0)\varepsilon} = E^\varepsilon(\partial_3^\varepsilon \underline{u}_3^{\delta(0)\varepsilon} - x_\alpha^\varepsilon \partial_{33}^\varepsilon u_\alpha^{\delta(0)\varepsilon}). \tag{30.15}$$

(iii) *The bending moment* $m_\beta^{\delta(0)\varepsilon}$ *and shear force* $q_\beta^{\delta(0)\varepsilon}$ *components are given by*

$$m_\beta^{\delta(0)\varepsilon} = \int_{\omega^{\delta,\varepsilon}} x_\beta^\varepsilon \sigma_{33}^{\delta(0)\varepsilon} \,\mathrm{d}\omega^{\delta,\varepsilon} = -E^\varepsilon I_\beta^{\delta,\varepsilon} \partial_{33}^\varepsilon u_\beta^{\delta(0)\varepsilon} \quad (\textit{no sum on } \beta), \tag{30.16}$$

$$\begin{aligned} q_\beta^{\delta(0)\varepsilon} &= \int_{\omega^{\delta,\varepsilon}} \sigma_{3\beta}^{\delta(0)\varepsilon} \,\mathrm{d}\omega^{\delta,\varepsilon} \\ &= -E^\varepsilon I_\beta^{\delta,\varepsilon} \partial_{333}^\varepsilon u_\beta^{\delta(0)\varepsilon} + \int_{\omega^{\delta,\varepsilon}} x_\beta^\varepsilon f_3^\varepsilon \,\mathrm{d}\omega^{\delta,\varepsilon} + \int_{\gamma^\varepsilon} x_\beta^\varepsilon g_3^\varepsilon \,\mathrm{d}\gamma^\varepsilon \quad (\textit{no sum on } \beta), \end{aligned} \tag{30.17}$$

Moreover, let the system of applied forces be such that

$$\begin{aligned} &f_\alpha^\varepsilon \in L^2(\Omega^{\delta,\varepsilon}), \qquad f_3^\varepsilon \in H^1[0,L;L^2(\omega^{\delta,\varepsilon})], \\ &g_\alpha^\varepsilon \in L^2(\Gamma^{\delta,\varepsilon}), \qquad g_3^\varepsilon \in H^1[0,L;L^2(\gamma^{\delta,\varepsilon})]. \end{aligned} \tag{30.18}$$

Then, besides the previous expressions characterizing $\boldsymbol{u}^{\delta(0)\varepsilon}$, $\sigma_{33}^{\delta(0)\varepsilon}$, $m_\beta^{\delta(0)\varepsilon}$ *and* $q_\beta^{\delta(0)\varepsilon}$, *we have the following characterization of* $\boldsymbol{u}^{\delta(2)\varepsilon}$, $\sigma_{33}^{\delta(2)\varepsilon}$, $m_\beta^{\delta(2)\varepsilon}$, $q_\beta^{\delta(2)\varepsilon}$ *and* $\sigma_{i\beta}^{\delta(0)\varepsilon}$:

(iv) *Displacements* $u_i^{\delta(2)\varepsilon}$ *are of the form:*

$$u_1^{\delta(2)\varepsilon} = z_1^{\delta(2)\varepsilon} + x_2^\varepsilon z^{\delta(2)\varepsilon} - \nu^\varepsilon(x_1^\varepsilon \partial_3^\varepsilon \underline{u}_3^{\delta(0)\varepsilon} - \Phi_{1\beta}^\varepsilon \partial_{33}^\varepsilon u_\beta^{\delta(0)\varepsilon}), \tag{30.19}$$

$$u_2^{\delta(2)\varepsilon} = z_2^{\delta(2)\varepsilon} - x_1^\varepsilon z^{\delta(2)\varepsilon} - \nu^\varepsilon(x_2^\varepsilon \partial_3^\varepsilon \underline{u}_3^{\delta(0)\varepsilon} - \Phi_{2\beta}^\varepsilon \partial_{33}^\varepsilon u_\beta^{\delta(0)\varepsilon}), \tag{30.20}$$

$$\begin{aligned} u_3^{\delta(2)\varepsilon} &= \underline{u}_3^{\delta(2)\varepsilon} - x_\alpha^\varepsilon \partial_3^\varepsilon z_\alpha^{\delta(2)\varepsilon} - w^{\delta,\varepsilon} \partial_3^\varepsilon z^{\delta(2)\varepsilon} \\ &\quad - \nu^\varepsilon[\tfrac{1}{2}((x_1^\varepsilon)^2 + (x_2^\varepsilon)^2) - \frac{1}{2|\omega^{\delta,\varepsilon}|}(I_1^{\delta,\varepsilon} + I_2^{\delta,\varepsilon})]\partial_{33}^\varepsilon \underline{u}_3^{\delta(0)\varepsilon} \\ &\quad + [(1+\nu^\varepsilon)\eta_\alpha^{\delta,\varepsilon} + \nu^\varepsilon \theta_\alpha^{\delta,\varepsilon}]\partial_{333}^\varepsilon u_\alpha^{\delta(0)\varepsilon} + \frac{2(1+\nu^\varepsilon)}{E^\varepsilon} w^{\delta(0)\varepsilon}, \end{aligned} \tag{30.21}$$

where functions $z^{\delta(2)\varepsilon}$, $w^{\delta(0)\varepsilon}$, $\underline{u}_3^{\delta(2)\varepsilon}$ *and* $z_\alpha^{\delta(2)\varepsilon}$ *are characterized in the following way:*

(a) *The angle of twist* $z^{\delta(2)\varepsilon}$ *depends only on* x_3^ε *and is the unique solution of the following one dimensional problem*:

$$z^{\delta(2)\varepsilon} \in H^1(0,L),$$

$$\int_0^L \frac{E^\varepsilon J^{\delta,\varepsilon}}{2(1+\nu^\varepsilon)} \partial_3^\varepsilon z^{\delta(2)\varepsilon} \partial_3^\varepsilon v \, \mathrm{d}x_3^\varepsilon = \int_0^L M_3^{\delta(2)\varepsilon} v \, \mathrm{d}x_3^\varepsilon \quad \textit{for all } v \in H_0^1(0,L),$$

$$z^{\delta(2)\varepsilon}(a) = \frac{\nu^\varepsilon}{(I_1^{\delta,\varepsilon} + I_2^{\delta,\varepsilon})} [H_2^{\delta,\varepsilon} \partial_{33}^\varepsilon u_1^{\delta(0)\varepsilon}(a) - H_1^{\delta,\varepsilon} \partial_{33}^\varepsilon u_2^{\delta(0)\varepsilon}(a)] \quad \textit{at } a = 0, L, \tag{30.22}$$

where

$$\begin{aligned} M_3^{\delta(2)\varepsilon} = &\int_{\omega^{\delta,\varepsilon}} (x_2^\varepsilon f_1^\varepsilon - x_1^\varepsilon f_2^\varepsilon) \, \mathrm{d}\omega^{\delta,\varepsilon} + \int_{\gamma^\varepsilon} (x_2^\varepsilon g_1^\varepsilon - x_1^\varepsilon g_2^\varepsilon) \, \mathrm{d}\gamma^\varepsilon \\ &+ \int_{\omega^{\delta,\varepsilon}} w^{\delta,\varepsilon} \partial_3^\varepsilon f_3^\varepsilon \, \mathrm{d}\omega^{\delta,\varepsilon} + \int_{\gamma^\varepsilon} w^{\delta,\varepsilon} \partial_3^\varepsilon g_3^\varepsilon \, \mathrm{d}\gamma^\varepsilon \\ &- \frac{E^\varepsilon}{2(1+\nu^\varepsilon)} [(1+\nu^\varepsilon) I_\alpha^{w^{\delta,\varepsilon}} + \nu^\varepsilon I_\alpha^{\Psi^{\delta,\varepsilon}}] \partial_{3333}^\varepsilon u_\alpha^{\delta(0)\varepsilon}. \end{aligned} \tag{30.23}$$

(b) *The additional warping* $w^{\delta(0)\varepsilon}$ *is the unique solution of the following problem*:

$$w^{\delta(0)\varepsilon} \in H^1[0,L; H^1(\omega^{\delta,\varepsilon})] \quad \textit{and for all } x_3^\varepsilon \in [0,L],$$

$$\begin{aligned} &\int_{\omega^{\delta,\varepsilon}} \partial_\beta^\varepsilon w^{\delta(0)\varepsilon}(x_3^\varepsilon) \partial_\beta^\varepsilon \varphi^\varepsilon \, \mathrm{d}\omega^{\delta,\varepsilon} \\ &= \int_{\omega^{\delta,\varepsilon}} f_3^\varepsilon(x_3^\varepsilon) \varphi^\varepsilon \, \mathrm{d}\omega^{\delta,\varepsilon} + \int_{\gamma^\varepsilon} g_3^\varepsilon(x_3^\varepsilon) \varphi^\varepsilon \, \mathrm{d}\gamma^\varepsilon - \frac{1}{|\omega^{\delta,\varepsilon}|} \Bigg[\int_{\omega^{\delta,\varepsilon}} f_3^\varepsilon(x_3^\varepsilon) \, \mathrm{d}\omega^{\delta,\varepsilon} \\ &\quad + \int_{\gamma^\varepsilon} g_3^\varepsilon(x_3^\varepsilon) \, \mathrm{d}\gamma^\varepsilon \Bigg] \int_{\omega^{\delta,\varepsilon}} \varphi^\varepsilon \, \mathrm{d}\omega^{\delta,\varepsilon}, \quad \textit{for all } \varphi^\varepsilon \in H^1(\omega^{\delta,\varepsilon}), \\ &\int_{\omega^{\delta,\varepsilon}} w^{\delta(0)\varepsilon}(x_3^\varepsilon) \, \mathrm{d}\omega^{\delta,\varepsilon} = 0. \end{aligned} \tag{30.24}$$

(c) *The stretching component* $\underline{u}_3^{\delta(2)\varepsilon}$ *depends only on* x_3^ε *and uniquely solves the following problem*:

$$\underline{u}_3^{\delta(2)\varepsilon} \in H_0^1(0,L),$$

$$\int_0^L E^\varepsilon |\omega^{\delta,\varepsilon}| \partial_3^\varepsilon \underline{u}_3^{\delta(2)\varepsilon} \partial_3^\varepsilon v \, \mathrm{d}x_3^\varepsilon = \int_0^L \nu^\varepsilon G_3^{\delta(2)\varepsilon} \partial_3^\varepsilon v \, \mathrm{d}x_3^\varepsilon, \tag{30.25}$$

for all $v \in H_0^1(0,L)$,

where

$$
\begin{aligned}
G_3^{\delta(2)\varepsilon} = & -\tfrac{1}{2}E^\varepsilon(I_1^{\delta,\varepsilon}+I_2^{\delta,\varepsilon})\partial^\varepsilon_{333}\underline{u}_3^{\delta(0)\varepsilon} + E^\varepsilon H_\alpha^{\delta,\varepsilon}\partial^\varepsilon_{3333}u_\alpha^{\delta(0)\varepsilon} \\
& - \int_{\omega^{\delta,\varepsilon}} x_\alpha^\varepsilon f_\alpha^\varepsilon \, d\omega^{\delta,\varepsilon} - \int_{\gamma^\varepsilon} x_\alpha^\varepsilon g_\alpha^\varepsilon \, d\gamma^\varepsilon \\
& - \int_{\omega^{\delta,\varepsilon}} \tfrac{1}{2}[(x_1^\varepsilon)^2+(x_2^\varepsilon)^2]\partial_3^\varepsilon f_3^\varepsilon \, d\omega^{\delta,\varepsilon} - \int_{\gamma^\varepsilon} \tfrac{1}{2}[(x_1^\varepsilon)^2+x_2^\varepsilon)^2]\partial_3^\varepsilon g_3^\varepsilon \, d\gamma^\varepsilon .
\end{aligned}
\tag{30.26}
$$

(d) *The bending component* $z_\alpha^{\delta(2)\varepsilon}$ *depends only on the variable* x_3^ε *and is the unique solution of the following variational problem (no sum on* α, $\beta \neq \alpha$*):*

$$
\begin{aligned}
& z_\alpha^{\delta(2)\varepsilon} \in H^2(0,L) \quad \textit{and for all } v \in H_0^2(0,L)\textit{:} \\
& \int_0^L E^\varepsilon I_\alpha^{\delta,\varepsilon}\partial^\varepsilon_{33} z_\alpha^{\delta(2)\varepsilon}\partial^\varepsilon_{33} v \, dx_3^\varepsilon = \int_0^L M_\alpha^{\delta(2)\varepsilon}\partial^\varepsilon_{33} v \, dx_3^\varepsilon, \\
& z_\alpha^{\delta(2)\varepsilon}(a) = -\tfrac{1}{2}\frac{\nu^\varepsilon}{|\omega^{\delta,\varepsilon}|}(I_\alpha^{\delta,\varepsilon}-I_\beta^{\delta,\varepsilon})\partial^\varepsilon_{33}u_\alpha^{\delta(0)\varepsilon}(a) \quad \textit{at } a=0,L, \\
& \partial_3^\varepsilon z_\alpha^{\delta(2)\varepsilon}(a) = \frac{1}{I_\alpha^{\delta,\varepsilon}}\left\{ \frac{2(1+\nu^\varepsilon)}{E^\varepsilon}\int_{\omega^{\delta,\varepsilon}} x_\alpha^\varepsilon w^{\delta(0)\varepsilon}(a)\, d\omega^{\delta,\varepsilon} \right. \\
& \qquad - \tfrac{1}{2}I_\alpha^{w^{\delta,\varepsilon}}\partial_3^\varepsilon z^{\delta(2)\varepsilon}(a) + \nu^\varepsilon H_\alpha^{\delta,\varepsilon}\partial^\varepsilon_{33}\underline{u}_3^{\delta(0)\varepsilon}(a) \\
& \qquad \left. + [(1+\nu^\varepsilon)L_{\alpha\beta}^{\eta^{\delta,\varepsilon}} + \nu^\varepsilon L_{\alpha\beta}^{\theta^{\delta,\varepsilon}}]\partial^\varepsilon_{333}u_\beta^{0^\varepsilon}(a) \right\} \quad \textit{at } a=0,L,
\end{aligned}
\tag{30.27}
$$

where

$$
\begin{aligned}
M_\alpha^{\delta(2)\varepsilon} = & E^\varepsilon\{(1+\nu^\varepsilon)L_{\alpha\beta}^{\eta^{\delta,\varepsilon}} + \nu^\varepsilon L_{\alpha\beta}^{\theta^{\delta,\varepsilon}} + \tfrac{1}{2}\nu^\varepsilon K_{\alpha\beta}^{\eta^{\delta,\varepsilon}} \\
& + \frac{(\nu^\varepsilon)^2}{2(1+\nu^\varepsilon)}(K_{\alpha\beta}^{\theta^{\delta,\varepsilon}} + H_3^{\delta,\varepsilon}\delta_{\alpha\beta})\}\partial^\varepsilon_{3333}u_\beta^{\delta(0)\varepsilon} \\
& + \nu^\varepsilon E^\varepsilon H_\alpha^{\delta,\varepsilon}\partial^\varepsilon_{333}\underline{u}_3^{\delta(0)\varepsilon} - \frac{E^\varepsilon}{2(1+\nu^\varepsilon)}[(1+\nu^\varepsilon)I_\alpha^{w^{\delta,\varepsilon}} + \nu^\varepsilon I_\alpha^{\Psi^{\delta,\varepsilon}}]\partial^\varepsilon_{33}z^{\delta(2)\varepsilon} \\
& - \int_{\omega^{\delta,\varepsilon}} [(1+\nu^\varepsilon)\eta_\alpha^{\delta,\varepsilon} + \nu^\varepsilon\theta_\alpha^{\delta,\varepsilon}]\partial_3^\varepsilon f_3^\varepsilon \, d\omega^{\delta,\varepsilon} \\
& - \int_{\gamma^\varepsilon} [(1+\nu^\varepsilon)\eta_\alpha^{\delta,\varepsilon} + \nu^\varepsilon\theta_\alpha^{\delta,\varepsilon}]\partial_3^\varepsilon g_3^\varepsilon \, d\gamma^\varepsilon \\
& + \int_{\omega^{\delta,\varepsilon}} \nu^\varepsilon \Phi^\varepsilon_{\alpha\beta} f_\beta^\varepsilon \, d\omega^\varepsilon + \int_{\gamma^\varepsilon} \nu^\varepsilon \Phi^\varepsilon_{\alpha\beta} g_\beta^\varepsilon \, d\gamma^\varepsilon .
\end{aligned}
\tag{30.28}
$$

(v) *The shear stress components* $\sigma_{3\beta}^{\delta(0)\varepsilon}$ *are uniquely determined by*

$$\sigma_{31}^{\delta(0)\varepsilon} = \frac{E^\varepsilon}{2(1+\nu^\varepsilon)}\{-\partial_2^\varepsilon \Psi^{\delta,\varepsilon}\partial_3^\varepsilon z^{\delta(2)\varepsilon} + [(1+\nu^\varepsilon)\partial_1^\varepsilon \eta_\beta^{\delta,\varepsilon} + \nu^\varepsilon(\partial_1^\varepsilon \theta_\beta^{\delta,\varepsilon} + \Phi_{1\beta}^\varepsilon)]\partial_{333}^\varepsilon u_\beta^{\delta(0)\varepsilon}\} + \partial_1^\varepsilon w^{\delta(0)\varepsilon}, \tag{30.29}$$

$$\sigma_{32}^{\delta(0)\varepsilon} = \frac{E^\varepsilon}{2(1+\nu^\varepsilon)}\{\partial_1^\varepsilon \Psi^{\delta,\varepsilon}\partial_3^\varepsilon z^{\delta(2)\varepsilon} + [(1+\nu^\varepsilon)\partial_2^\varepsilon \eta_\beta^{\delta,\varepsilon} + \nu^\varepsilon(\partial_2^\varepsilon \theta_\beta^{\delta,\varepsilon} + \Phi_{2\beta}^\varepsilon)]\partial_{333}^\varepsilon u_\beta^{\delta(0)\varepsilon}\} + \partial_2^\varepsilon w^{\delta(0)\varepsilon}, \tag{30.30}$$

(vi) *The plane stress components* $\sigma_{\alpha\beta}^{\delta(0)\varepsilon}$ *are given by*

$$\sigma_{\alpha\beta}^{\delta(0)\varepsilon} = S_{\alpha\beta}^{\delta,\varepsilon}(\underline{\boldsymbol{u}}^4) + S_{\alpha\beta}^{\delta(0)\varepsilon}, \tag{30.31}$$

where

$$S_{\alpha\beta}^{\delta(0)\varepsilon} = \frac{\nu^\varepsilon E^\varepsilon}{(1+\nu^\varepsilon)(1-2\nu^\varepsilon)}\partial_3^\varepsilon(u_3^{\delta(2)\varepsilon} - \underline{u}_3^{\delta(2)\varepsilon} + x_\mu^\varepsilon \partial_3^\varepsilon z_\mu^{\delta(2)\varepsilon})\delta_{\alpha\beta}, \tag{30.32}$$

$$S_{\alpha\beta}^{\delta,\varepsilon}(\underline{\boldsymbol{u}}^4) = \frac{E^\varepsilon}{(1+\nu^\varepsilon)}e_{\alpha\beta}^\varepsilon(\underline{\boldsymbol{u}}^{\delta(4)\varepsilon}) + \frac{\nu^\varepsilon E^\varepsilon}{(1+\nu^\varepsilon)(1-2\nu^\varepsilon)}e_{\mu\mu}^\varepsilon(\underline{\boldsymbol{u}}^{\delta(4)\varepsilon})\delta_{\alpha\beta}, \tag{30.33}$$

and $\underline{\boldsymbol{u}}^{\delta(4)\varepsilon} = (\underline{u}_\alpha^{\delta(4)\varepsilon})$ *is the unique solution of the following plane elasticity problem*:

$$\begin{aligned}
&\underline{\boldsymbol{u}}^{\delta(4)\varepsilon} \in L^2[0,L;(H^1(\omega^{\delta,\varepsilon}))^2], \quad \textit{and a.e. in } (0,L), \textit{ for all } \boldsymbol{\varphi}^\varepsilon \in [H^1(\omega^{\delta,\varepsilon})]^2: \\
&\int_\omega S_{\alpha\beta}^{\delta,\varepsilon}(\underline{\boldsymbol{u}}^{\delta(4)\varepsilon})e_{\alpha\beta}^\varepsilon(\boldsymbol{\varphi}^\varepsilon)\,\mathrm{d}\omega^{\delta,\varepsilon} \\
&\quad = \int_{\omega^{\delta,\varepsilon}} f_\beta^\varepsilon \varphi_\beta^\varepsilon\,\mathrm{d}\omega^{\delta,\varepsilon} + \int_{\gamma^\varepsilon} g_\beta^\varepsilon \varphi_\beta^\varepsilon\,\mathrm{d}\gamma^\varepsilon \\
&\quad + \int_{\omega^{\delta,\varepsilon}} \partial_3^\varepsilon \sigma_{3\beta}^{\delta(0)\varepsilon}\varphi_\beta^\varepsilon\,\mathrm{d}\omega^{\delta,\varepsilon} - \int_{\omega^{\delta,\varepsilon}} S_{\alpha\beta}^{\delta(0)\varepsilon}e_{\alpha\beta}^\varepsilon(\varphi^\varepsilon)\,\mathrm{d}\omega^{\delta,\varepsilon}, \\
&\int_{\omega^{\delta,\varepsilon}} \underline{u}_\alpha^{\delta(4)\varepsilon}\,\mathrm{d}\omega^{\delta,\varepsilon} = \int_{\omega^{\delta,\varepsilon}} [x_2^\varepsilon \underline{u}_1^{\delta(4)\varepsilon} - x_1^\varepsilon \underline{u}_2^{\delta(4)\varepsilon}]\,\mathrm{d}\omega^{\delta,\varepsilon} = 0.
\end{aligned} \tag{30.34}$$

(vii) *The axial stress* $\sigma_{33}^{\delta(2)\varepsilon}$ *and the second order bending moment components are given by*

$$
\begin{aligned}
\sigma_{33}^{\delta(2)\varepsilon} &= E^\varepsilon \partial_3^\varepsilon u_3^{\delta(2)\varepsilon} + \nu^\varepsilon \sigma_{\mu\mu}^{\delta(0)\varepsilon} \\
&= E^\varepsilon \{\partial_3^\varepsilon \underline{u}_3^{\delta(2)\varepsilon} - x_\alpha^\varepsilon \partial_{33}^\varepsilon z_\alpha^{\delta(2)\varepsilon} - w^{\delta,\varepsilon} \partial_{33}^\varepsilon z^{\delta(2)\varepsilon} \\
&\quad - \nu^\varepsilon [\tfrac{1}{2}[(x_1^\varepsilon)^2 + (x_2^\varepsilon)^2] - \tfrac{1}{2}(I_1^{\delta,\varepsilon} + I_2^{\delta,\varepsilon})] \partial_{333}^\varepsilon \underline{u}_3^{\delta(0)\varepsilon} \\
&\quad + [(1+\nu^\varepsilon)\eta_\alpha^{\delta,\varepsilon} + \nu^\varepsilon \theta_\alpha^{\delta,\varepsilon}] \partial_{3333}^\varepsilon u_\alpha^{\delta(0)\varepsilon}\} + 2(1+\nu^\varepsilon)\partial_3^\varepsilon w^{\delta(0)\varepsilon} + \nu^\varepsilon \sigma_{\mu\mu}^{\delta(0)\varepsilon}.
\end{aligned}
\tag{30.35}
$$

(viii) *The bending moment* $m_\beta^{\delta(2)\varepsilon} = \int_{\omega^{\delta,\varepsilon}} x_\beta^\varepsilon \sigma_{33}^{\delta(2)\varepsilon}\, d\omega^{\delta,\varepsilon}$ *and the shear force* $q_\beta^{\delta(2)\varepsilon} = \int_{\omega^{\delta,\varepsilon}} \sigma_{3\beta}^{\delta(2)\varepsilon}\, d\omega^{\delta,\varepsilon}$ *components, are given by the following expressions:*

$$
m_\beta^{\delta(2)\varepsilon} = -E^\varepsilon I_\beta^{\delta,\varepsilon} \partial_{33}^\varepsilon z_\beta^{\delta(2)\varepsilon} + M_\beta^{\delta(2)\varepsilon} \quad (\textit{no sum on } \beta). \tag{30.36}
$$

$$
q_\beta^{\delta(2)\varepsilon} = \partial_3^\varepsilon m_\beta^{\delta(2)\varepsilon} = -E^\varepsilon I_\alpha^{\delta,\varepsilon} \partial_{333}^\varepsilon z_\beta^{\delta(2)\varepsilon} + \partial_3^\varepsilon M_\beta^{\delta(2)\varepsilon} \quad (\textit{no sum on } \beta). \tag{30.37}
$$

31. Homogenization of the generalized beam theories

This section is concerned with the homogenized behaviour of the generalized beam theories, described by Eqs. (30.12)–(30.37) of Theorem 30.1, when δ goes to zero. The first step in this direction is the study of the limits of the constants and functions determined by the geometry of the cross section and defined by (30.1)–(30.11). This will be the object of Theorem 31.1 and Corollary 31.1. The second step is to pass to the limit in Eqs. (30.12)–(30.37) themselves, the result of which will be stated in Theorem 31.2.

For the general periodic homogenization methods used we refer to BENSSOUSSAN, LIONS and PAPANICOLAOU [1978], CIORANESCU and SAINT JEAN PAULIN [1979], SANCHEZ-PALENCIA [1980] and LÉNÉ [1984]. In order to study the nonhomogeneous Neumann problems some homogenization techniques are adapted from H-convergence theory, due to MURAT [1977/8] and TARTAR [1977], and partially contained in MURAT [1977/8]. The results presented here follow MASCARENHAS and TRABUCHO [1990]. The following lemmas are used in the sequel.

LEMMA 31.1. *Let* $Q \subset \mathbb{R}^m$ $(m \in \mathbb{N})$ *be an m-cube and* $f \in L^p_{\text{loc}}(\mathbb{R}^m)$ $(1 \leqslant p \leqslant +\infty)$ *be Q-periodic. Define, for each* $\delta > 0, \boldsymbol{x} \in \mathbb{R}^m, f^\delta(\boldsymbol{x}) = f(\boldsymbol{x}/\delta)$. *Then, as* $\delta \to 0, f^\delta \to (1/|Q|)\int_Q f(y)\,dy$, *weakly in* $L^p(\Omega)$, *if* $p < +\infty$, *and weakly* in* $L^\infty(\Omega)$, *if* $p = +\infty$, *for any open bounded subset* $\Omega \subset \mathbb{R}^m$.

PROOF. See BALL and MURAT [1984, p. 249]. □

LEMMA 31.2. *Let* Ω *be an arbitrary open subset in* $\mathbb{R}^m$. *Let* (v_n) *and* (w_n) *be two sequences, respectively in* $[L^2(\Omega)]^m$ *and* $H^1(\Omega)$, *satisfying* $v_n \to v$, *weakly in*

$[L^2(\Omega)]^m$; $w_n \to w$, weakly in $H^1(\Omega)$; $\operatorname{div} v_n \to \operatorname{div} v$, strongly in $H^{-1}(\Omega)$, then representing by $\langle \cdot, \cdot \rangle$ the inner product in $\mathbb{R}^m$, $\langle v_n, \nabla w_n \rangle \to \langle v, \nabla w \rangle$, in $\mathcal{D}'(\Omega)$.

PROOF. See MURAT [1977/78]. □

This lemma is a particular case of a result owed to Murat and Tartar (see MURAT [1978]) known as the divergence–curl compactness theorem. However, this particular case has a simple direct proof, by integrating by parts and passing to the limit, with the help of the compact inclusion of $H^1(\Omega)$ in $L^2(\Omega)$.

LEMMA 31.3. *Let $\omega^{\delta,1}$ be defined as in Section 21. There exists a linear continuous extension operator P^δ from $H^1(\omega^{\delta,1})$ to $H^1(\omega)$, satisfying*

$$P^\delta \boldsymbol{v} = \boldsymbol{v}, \quad \text{a.e. in } \omega^{\delta,1}, \text{ for all } \boldsymbol{v} \in H^1(\omega^{\delta,1}),$$

$$\|\nabla(P^\delta \boldsymbol{v})\|_{L^2(\omega)} \leqslant C \|\nabla \boldsymbol{v}\|_{L^2(\omega^{\delta,1})}, \quad \text{for all } \boldsymbol{v} \in H^1(\omega^{\delta,1}),$$

where C is a constant independent of δ.

PROOF. See CIORANESCU and SAINT JEAN PAULIN [1979, p. 603], TARTAR [1977]). □

LEMMA 31.4. *Let $\omega^{\delta,1}$ be defined as in Section 29. There exists a linear continuous extension operator Q^δ from $[H^1(\omega^{\delta,1})]^2$ to $[H^1(\omega)]^2$, satisfying*

$$Q^\delta \boldsymbol{v} = \boldsymbol{v}, \quad \text{a.e. in } \omega^{\delta,1}, \text{ for all } \boldsymbol{v} \in [H^1(\omega^{\delta,1})]^2,$$

$$\|e(Q^\delta \boldsymbol{v})\|_{L^2(\omega)} \leqslant C \|e(\boldsymbol{v})\|_{L^2(\omega^{\delta,1})}, \text{ for all } \boldsymbol{v} \in [H^1(\omega^{\delta,1})]^2,$$

where C is a constant independent of δ.

PROOF. See LÉNÉ [1984, Chapter 2, Section 3.1]. □

The following result concerning the cross section $\omega^{\delta,1}$ defined in Section 30 is due to MASCARENHAS and TRABUCHO [1990, Lemma 3.3].

LEMMA 31.5. *Let $\omega^{\delta,1}$ be defined as in Section 30. There exists a constant C, independent of δ, such that, for all $v \in H^1(\omega^{\delta,1})$, satisfying $\int_{\omega^{\delta,1}} v \, \mathrm{d}\omega^{\delta,1} = 0$,*

$$\|v\|_{H^1(\omega^{\delta,1})} \leqslant C \|\nabla v\|_{L^2(\omega^{\delta,1})}. \tag{31.1}$$

PROOF. From Lemma 31.3, for each element δ of the sequence $\delta \to 0$, let P^δ be a linear extension operator from $H^1(\omega^{\delta,1})$ to $H^1(\omega)$, satisfying

$$\text{(i)} \ \|P^\delta v\|_{L^2(\omega)} \leqslant \overline{C} \|v\|_{L^2(\omega^{\delta,1})}, \tag{31.2}$$
$$\text{(ii)} \ \|\nabla P^\delta v\|_{L^2(\omega)} \leqslant \overline{C} \|\nabla v\|_{L^2(\omega^{\delta,1})},$$

for a constant $\overline{C}$ independent of δ and for all $v \in H^1(\omega^{\delta,1})$. We prove, by contradiction, that

$$\hat{C} = \sup\left\{ \frac{\|P^{\delta}v\|_{H^1(\omega)}}{\|\nabla(P^{\delta}v)\|_{L^2(\omega)}} : v \in H^1(\omega^{\delta,1}),\ \int_{\omega^{\delta,1}} v \,\mathrm{d}\omega^{\delta,1} = 0,\ v \neq 0 \right\} < \infty. \tag{31.3}$$

Suppose that $\hat{C} = \infty$. For each $n \in \mathbb{N}$ there exists a δ_n, element of the sequence $\delta \to 0$, and a $v_n \in H^1(\omega^{\delta_n,1})$, $v_n \neq 0$, $\int_{\omega^{\delta_n,1}} v_n = 0$, such that $\|P^{\delta_n}v_n\|_{H^1(\omega)} = 1$, $\|\nabla(P^{\delta_n}v_n)\|_{L^2(\omega)} < 1/n$.

Since (δ_n) is a sequence of elements $\delta \to 0$, we may suppose, up to a subsequence, that $\delta_n \to 0$ or that $\delta_n = \delta^*$ for n big enough.

As a consequence of $\|P^{\delta_n}v_n\|_{H^1(\omega)} = 1$ there exists a sequence n_k and a $v^* \in H^1(\omega)$, satisfying $P^{\delta_{nk}}v_{n_k} \to v^*$, weakly in $H^1(\omega)$ and, by the compact injection of $H^1(\omega)$ in $L^2(\omega)$, strongly in $L^2(\omega)$. Since $\nabla(P^{\delta_{nk}}v_{n_k}) \to \nabla v^*$, weakly in $[L^2(\omega)]^2$, $\|\nabla(P^{\delta_{nk}}v_{n_k})\|_{L^2(\omega)} \to 0$ and since ω is a connected set, one concludes that v^* is a constant in (ω).

If $\delta_n \to 0$, using Lemma 31.1, we obtain that the characteristic function $\chi^{\delta_n,1}$ of $\omega^{\delta_n,1}$ converges, weakly in $L^2(\omega)$, to its mean Θ. So $0 = \int_{\omega^{\delta_{n_k},1}} v_{n_k} \,\mathrm{d}\omega^{\delta_{n_k},1} = \int_\omega \chi^{\delta_{n_k},1} P^{\delta_{n_k}} v_{n_k} \to \Theta \int_\omega v^* \,\mathrm{d}\omega = 0$. Since v^* is constant, $v^* = 0$ a.e. in ω, passing to the limit, as $k \to +\infty$, in

$$1 = \|P^{\delta_{n_k}}v_{n_k}\|_{H^1(\omega)} = \|P^{\delta_{n_k}}v_{n_k}\|_{L^2(\omega)} + \|\nabla(P^{\delta_{n_k}}v_{n_k})\|_{L^2(\omega)},$$

one obtains that $\|v^*\|_{L^2(\omega)} = 1$, which contradicts $v^* = 0$. If $\delta_n = \delta^*$, for n big enough, the same argument holds since we have, for this case,

$$\begin{aligned} 0 &= \int_{\omega^{\delta^*,1}} v_{n_k} \,\mathrm{d}\omega^{\delta^*,1} \\ &= \int_{\omega^{\delta^*,1}} P^{\delta_{n_k}} v_{n_k} \,\mathrm{d}\omega^{\delta^*,1} \to \int_{\omega^{\delta^*,1}} v^* \,\mathrm{d}\omega^{\delta^*,1} = |\omega^{\delta^*,1}| v^* = 0. \end{aligned}$$

Then, from (31.2) and (31.3) one has, for all $v \in H^1(\omega^{\delta,1})$ such that $\int_{\omega^{\delta,1}} v \,\mathrm{d}\omega^{\delta,1} = 0$, and for all δ,

$$\|v\|_{H^1(\omega^{\delta,1})} \leqslant \|P^{\delta}v\|_{H^1(\omega)} \leqslant \hat{C}\|\nabla(P^{\delta}v)\|_{L^2(\omega)} \leqslant \hat{C}\overline{C}\|v\|_{L^2(\omega^{\delta,1})},$$

which proves the result, with $C = \hat{C}\overline{C}$. □

With an analogous proof, adapted to the vectorial case, where the gradient is replaced by the tensor $\boldsymbol{e}(\boldsymbol{u})$ and constant functions by rigid displacements, with the help of Korn's inequality and Lemma 31.4, the following result holds (cf. MASCARENHAS and TRABUCHO [1990, Lemma 3.4]).

LEMMA 31.6. *There exists a constant C, independent of δ, such that, for all $\boldsymbol{v} \in [H^1(\omega^{\delta,1})]^2$, satisfying $\int_{\omega^{\delta,1}} v_\alpha \, d\omega^{\delta,1} = \int_{\omega^{\delta,1}} (x_2 v_1 - x_1 v_2) \, d\omega^{\delta,1} = 0$, one has*

$$\|\boldsymbol{v}\|_{H^1(\omega^{\delta,1})} \leqslant C\|\boldsymbol{e}(\boldsymbol{v})\|_{L^2(\omega^{\delta,1})}.$$

Through the following result, we now define the limit problems corresponding to the functions defined by (30.2), (30.4), (30.7) and (30.8).

THEOREM 31.1. *Solutions $w^{\delta,\varepsilon}$, $\Psi^{\delta,\varepsilon}$, $\eta_\beta^{\delta,\varepsilon}$, $\theta_\beta^{\delta,\varepsilon}$, of problems* (30.2), (30.4), (30.7) *and* (30.8), *respectively, admit extensions in $H^1(\omega^\varepsilon)$, bounded independently of δ. If $\hat{w}^{\delta,\varepsilon}$, $\hat{\Psi}^{\delta,\varepsilon}$, $\hat{\eta}_\beta^{\delta,\varepsilon}$, $\hat{\theta}_\beta^{\delta,\varepsilon}$, are any extensions of $w^{\delta,\varepsilon}$, $\Psi^{\delta,\varepsilon}$, $\eta_\beta^{\delta,\varepsilon}$, $\theta_\beta^{\delta,\varepsilon}$, respectively, bounded in $H^1(\omega^\varepsilon)$, independently of δ, then, for each fixed ε and as $\delta \to 0$, the following convergences hold, for the weak topology of $H^1(\omega^\varepsilon)$:*

$$\hat{w}^{\delta,\varepsilon} \rightharpoonup w^{*,\varepsilon}, \qquad \hat{\Psi}^{\delta,\varepsilon} \rightharpoonup \Psi^{*,\varepsilon}, \qquad \hat{\eta}_\beta^{\delta,\varepsilon} \rightharpoonup \eta_\beta^{*,\varepsilon}, \qquad \hat{\theta}_\beta^{\delta,\varepsilon} \rightharpoonup \theta_\beta^{*,\varepsilon},$$

where $w^{*,\varepsilon}$, $\Psi^{*,\varepsilon}$, $\eta_\beta^{*,\varepsilon}$, $\theta_\beta^{*,\varepsilon}$, are respectively, the unique solutions of the following limit problems:

$$\begin{aligned} &-\operatorname{div}^\varepsilon(A^*\nabla^\varepsilon w^{*,\varepsilon}) = 0, \quad \text{in } \omega^\varepsilon,\\ &(A^*\nabla^\varepsilon w^{*,\varepsilon})\cdot\boldsymbol{n}^\varepsilon = A^*(x_2^\varepsilon, -x_1^\varepsilon)\cdot\boldsymbol{n}^\varepsilon, \quad \text{on } \gamma^\varepsilon,\\ &\int_{\omega^\varepsilon} w^{*,\varepsilon}\, d\omega^\varepsilon = 0, \end{aligned} \tag{31.4}$$

$$\begin{aligned} &-\operatorname{div}^\varepsilon(A^*\nabla^\varepsilon \Psi^{*,\varepsilon}) = 2\det A^*, \quad \text{in } \omega^\varepsilon,\\ &\Psi^{*,\varepsilon} = 0, \quad \text{on } \gamma^\varepsilon, \end{aligned} \tag{31.5}$$

$$\begin{aligned} &-\operatorname{div}^\varepsilon(A^*\nabla^\varepsilon \eta_\beta^{*,\varepsilon}) = -2\Theta x_\beta^\varepsilon, \quad \text{in } \omega^\varepsilon,\\ &(A^*\nabla^\varepsilon \eta_\beta^{*,\varepsilon}).\boldsymbol{n}^\varepsilon = 0, \quad \text{on } \gamma^\varepsilon,\\ &\int_{\omega^\varepsilon} \eta_\beta^{*,\varepsilon} = 0, \end{aligned} \tag{31.6}$$

$$\begin{aligned} &-\operatorname{div}^\varepsilon(A^*\nabla^\varepsilon \theta_\beta^{*,\varepsilon}) = x_\beta^\varepsilon\ (\operatorname{tr} A^*), \quad \text{in } \omega^\varepsilon,\\ &(A^*\nabla^\varepsilon \theta_\beta^{*,\varepsilon})\cdot\boldsymbol{n}^\varepsilon = -A^*(\Phi_{\beta 1}^\varepsilon, \Phi_{\beta 2}^\varepsilon)\cdot\boldsymbol{n}^\varepsilon, \quad \text{on } \gamma^\varepsilon,\\ &\int_{\omega^\varepsilon} \theta_\beta^{*,\varepsilon}\, d\omega^\varepsilon = 0, \end{aligned} \tag{31.7}$$

where $A^* = (a^*_{\alpha\beta})$ is a 2×2 symmetric and positive definite matrix, its coefficients $a^*_{\alpha\beta}$ are defined by

$$a^*_{\alpha\beta} = \Theta\delta_{\alpha\beta} + \int_{Y^*} \partial_\alpha \chi_\beta \, \mathrm{d}Y^* = \int_{Y^*} \nabla(\chi_\alpha + y_\alpha) \cdot \nabla(\chi_\beta + y_\beta) \, \mathrm{d}Y^*, \tag{31.8}$$

and where χ_β is the unique solution of the following problem defined in the cell Y^*:

$$\begin{aligned} &\chi_\beta \quad Y\text{-periodic}, \\ &-\Delta\chi_\beta = 0, \quad \text{in } Y^*, \\ &\partial_n \chi_\beta = -n_\beta, \quad \text{on } \partial T. \\ &\int_{Y^*} \chi_\beta \, \mathrm{d}Y^* = 0. \end{aligned} \tag{31.9}$$

REMARK 31.1. (i) Problem (31.9) is equivalent to solving:

$$\chi_\beta \in V(Y^*) = \{\boldsymbol{v} = (v_\beta) \in H^1(Y^*)\text{: } \boldsymbol{v}\ Y\text{-periodic and } \int_{Y^*} \boldsymbol{v} \, \mathrm{d}Y^* = \boldsymbol{0}\},$$

$$\int_{Y^*} \nabla(\chi_\beta + y_\beta) \cdot \nabla\varphi \, \mathrm{d}Y^* = 0, \quad \text{for all } \varphi \in V(Y^*).$$

(ii) It is important to remark that the coefficients $a^*_{\alpha\beta}$ of matrix A^* are constants, independent of ε, and depending only on the geometry of the unit cell Y^* (cf. (31.8) and (31.9)). Consequently, the limit functions defined by (31.3)–(31.7) are independent of the extensions used to approach them.

PROOF. With no loss of generality, and in order to simplify the notations, we shall prove the theorem for $\varepsilon = 1$.

(A) We begin with the homogenization of the warping function. Consider

$$\begin{aligned} &-\partial_{\alpha\alpha} w^{\delta,1} = 0, \quad \text{in } \omega^{\delta,1}, \\ &\partial_n w^{\delta,1} = x_2 n_1 - x_1 n_2, \quad \text{on } \gamma^{\delta,1}, \\ &\int_{\omega^{\delta,1}} w^{\delta,1} \, \mathrm{d}\omega^{\delta,1} = 0. \end{aligned} \tag{31.10}$$

For each δ, existence and uniqueness of a solution $w^{\delta,1} \in H^1(\omega^{\delta,1})$ of this problem is quite immediate. From its variational setting in $H^1(\omega^{\delta,1})$, using the divergence theorem and the Cauchy–Schwarz inequality (in $\mathbb{R}^2$ and in $L^2(\omega^{\delta,1})$) we conclude the existence of a constant C, independent of δ, such that

$$\|\nabla w^{\delta,1}\|_{L^2(\omega^{\delta,1})} \leqslant C. \tag{31.11}$$

For each δ consider the classical extension operator P^δ given by Lemma 31.3. Then, there exists a constant C, independent of δ such that, for all $\boldsymbol{u} \in H^1(\omega^{\delta,1})$

$$
\begin{aligned}
&(P^\delta \boldsymbol{u})(\boldsymbol{x}) = \boldsymbol{u}(\boldsymbol{x}), \quad \text{a.e. } \boldsymbol{x} \in \omega^{\delta,1},\\
&\|P^\delta \boldsymbol{u}\|_{L^2(\omega)} \leqslant C\|\boldsymbol{u}\|_{L^2(\omega^{\delta,1})},\\
&\|\nabla P^\delta \boldsymbol{u}\|_{L^2(\omega)} \leqslant C\|\nabla \boldsymbol{u}\|_{L^2(\omega^{\delta,1})}.
\end{aligned}
$$

Since $\int_{\omega^{\delta,1}} w^{\delta,1}\,\mathrm{d}\omega^{\delta,1} = 0$, for all δ, we conclude, from Lemma 31.5, that $P^\delta w^{\delta,1}$ is bounded in $H^1(\omega)$, independently of δ. Then, there exists a subsequence δ' of δ and a function $w^{*,1}$, such that, if $P^{\delta'} w^{\delta',1} = \hat{w}^{\delta',1}$ is such an extension, we have as $\delta \to 0$,

$$\hat{w}^{\delta',1} \to w^{*,1}, \quad \text{weakly in } H^1(\omega), \quad \text{strongly in } L^2(\omega). \tag{31.12}$$

Consider the following zero extensions:

$$\tilde{w}^{\delta,1} = w^{\delta,1}, \quad \text{on } \omega^{\delta,1}; \qquad \tilde{w}^{\delta,1} = 0, \quad \text{on } \omega - \omega^{\delta,1}, \tag{31.13}$$

$$[\nabla w^{\delta,1}]^\sim = [\nabla w^{\delta,1}], \quad \text{on } \omega^{\delta,1}; \qquad [\nabla w^{\delta,1}]^\sim = 0, \quad \text{on } \omega - \omega^{\delta,1}. \tag{31.14}$$

Both $\|\tilde{w}^{\delta,1}\|_{L^2(\omega)}$ and $\|[\nabla w^{\delta,1}]^\sim\|_{L^2(\omega)}$ are bounded independently of δ and so there exists a subsequence, still denoted by δ', and functions $\tilde{w}$ and $\sigma_w^{*,1}$, such that, as $\delta \to 0$,

$$
\begin{aligned}
&\tilde{w}^{\delta',1} \to \tilde{w}, \quad \text{weakly in } L^2(\omega),\\
&[\nabla w^{\delta',1}]^\sim \to \sigma_w^{*,1}, \quad \text{weakly in } [L^2(\omega)]^2.
\end{aligned}
\tag{31.15}
$$

Since $\chi^{\delta,1} \to \Theta$, weakly* in $L^\infty(\omega)$ (Lemma 31.1), $\hat{w}^{\delta',1} \to w^{*,1}$, strongly in $L^2(\omega)$ and $\chi^{\delta,1}\hat{w}^{\delta,1} = \tilde{w}^{\delta,1}$, one has $\Theta w^{*,1} = \tilde{w}$ and, consequently, $\int_\omega w^{*,1}\,\mathrm{d}\omega = 0$.

In particular this implies that any other converging sequence of extensions of $w^{\delta,1}$ must have the same limit $w^{*,1}$.

Passing to the limit in the variational formulation of (31.10), one has

$$\int_\omega [\nabla w^{\delta,1}]^\sim \cdot \nabla\varphi\,\mathrm{d}\omega = \int_\omega \chi^{\delta,1}(\partial_1\varphi x_2 - \partial_2\varphi x_1)\,\mathrm{d}\omega, \quad \text{for all } \varphi \in H^1(\omega).$$

Passing to the limit as $\delta \to 0$ one obtains that

$$-\operatorname{div}\sigma_w^{*,1} = 0, \quad \text{in } \omega; \qquad \sigma_w^{*,1}\cdot\boldsymbol{n} = \Theta(x_2 n_1 - x_1 n_2), \quad \text{on } \gamma. \tag{31.16}$$

In order to identify the relationship between $\sigma_w^{*,1}$ and $w^{*,1}$ we shall use Tartar's energy method (see TARTAR [1977]). Consider $\hat{\chi}_\beta \in H^1_{\mathrm{loc}}(\mathbb{R}^2)$, Y-periodic, as the solution of problem (31.9) extended to the entire cell Y and periodically defined in all $\mathbb{R}^2$. For each $\delta > 0$, define $\Psi_\beta^\delta(\boldsymbol{x}) = x_\beta + \delta\hat{\chi}_\beta(\boldsymbol{x}/\delta)$. One will have $(\nabla \Psi_\beta^\delta)(\boldsymbol{x}) = e_\beta + (\nabla \hat{\chi}_\beta)(\boldsymbol{x}/\delta)$ and by Lemma 31.1, $\Psi_\beta^\delta \rightharpoonup x_\beta$, weakly in $H^1_{\mathrm{loc}}(\mathbb{R}^2)$.

Representing by $\langle \cdot, \cdot \rangle$ the inner product of $\mathbb{R}^2$, the following equality holds

$$\begin{aligned}&\langle [\nabla w^{\delta,1}]^\sim - \chi^{\delta,1}(x_2, -x_1), \nabla \Psi_\beta^\delta \rangle \\ &\quad = \langle \nabla(\hat{w}^{\delta,1}), (\nabla \Psi_\beta^\delta)\chi^{\delta,1} \rangle - \langle (x_2, -x_1), (\nabla \Psi_\beta^\delta)\chi^{\delta,1} \rangle, \quad \text{a.e. in } \omega. \qquad (31.17)\end{aligned}$$

From Lemma 31.1, one has $(\nabla \Psi_\beta^\delta)\chi^{\delta,1} \rightharpoonup \Theta e_\beta + \int_{Y^*} \nabla \chi_\beta \, \mathrm{d}Y^* = A^* e_\beta$ weakly in $[L^2(\omega)]^2$. Since $\operatorname{div}\{[\nabla w^{\delta,1}]^\sim - \chi^{\delta,1}(x_2, -x_1)\} = 0$, in ω and $\operatorname{div}[(\nabla \Psi_\beta^\delta)\chi^{\delta,1}] = 0$, in ω, Lemma 3.2 allows us to pass to the limit, as $\delta \to 0$, in both sides of (31.17). One obtains

$$\langle \sigma_w^{*,1} - \Theta(x_2, -x_1), e_\beta \rangle = \langle \nabla w^{*,1}, A^* e_\beta \rangle - \langle (x_2, -x_1), A^* e_\beta \rangle, \quad \text{a.e. in } \omega. \qquad (31.18)$$

Since matrix A^* is symmetric (see CIORANESCU and SAINT JEAN PAULIN [1979]), expression (31.18) leads to

$$\sigma_w^{*,1} - \Theta(x_2, -x_1) = A^* \nabla w^{*,1} - A^*(x_2, -x_1), \quad \text{a.e. in } \omega. \qquad (31.19)$$

The symmetry of A^* implies that $\operatorname{div}[A^*(x_2, -x_1)] = 0$ and the limit problem (31.4), with $\varepsilon = 1$, follows from (31.16) and (31.19). The fact that A^* is positive definite (see BENSSOUSSAN, LIONS and PAPANICOLAOU [1978, p. 17]) guarantees the uniqueness of the solution $w^{*,1}$ of problem (31.4) and consequently the convergence of the entire sequence $\hat{w}^{\delta,1}$ towards $w^{*,1}$.

(B) In what concerns problems (30.7) and (30.8) the same technique holds. Using the previous notations we obtain the existence of extensions

$$\hat{\eta}_\beta^{\delta,1} \rightharpoonup \eta_\beta^{*,1}, \qquad \hat{\theta}_\beta^{\delta,1} \rightharpoonup \theta_\beta^{*,1}, \qquad \text{weakly in } H^1(\omega), \text{ strongly in } L^2(\omega), \qquad (31.20)$$

$$[\nabla \eta_\beta^{\delta,1}]^\sim \rightharpoonup \sigma_{\eta_\beta}^{*,1}, \qquad [\nabla \theta_\beta^{\delta,1}]^\sim \rightharpoonup \sigma_{\theta_\beta}^{*,1}, \quad \text{weakly in } [L^2(\omega)]^2. \qquad (31.21)$$

Using the same test functions as in part (A), we identify the relationship between $\sigma_{\eta_\beta}^{*,1}$, $\sigma_{\theta_\beta}^{*,1}$ and $\eta_\beta^{*,1}$, $\theta_\beta^{*,1}$, respectively, obtaining

$$\sigma_{\eta_\beta}^{*,1} = A^* \nabla \eta_\beta^{*,1}, \quad \text{a.e. in } \omega, \qquad (31.22)$$

$$\sigma_{\theta_\beta}^{*,1} + \Theta(\Phi_{\beta 1}, \Phi_{\beta 2}) = A^* \nabla \theta_\beta^{*,1} + A^*(\Phi_{\beta 1}, \Phi_{\beta 2}), \quad \text{a.e. in } \omega. \qquad (31.23)$$

From these results (31.6) and (31.7) follow.

(C) The study of the homogenization problem concerning the torsion function is classical and presented in CIORANESCU and SAINT JEAN PAULIN [1979]. However, an important relation between $\Psi^{\delta,1}$ and the warping function $w^{\delta,1}$

suggests a direct proof of (31.5), where the matrix A^* is exactly the same as in the homogenized warping problem. Following GIRAULT and RAVIART [1981, Corollary 3.1, p. 26], since

$$\begin{aligned} &\operatorname{div}[(\partial_1 w^{\delta,1} - x_2), (\partial_2 w^{\delta,1} + x_1)] = 0, && \text{in } \omega^{\delta,1}, \\ &(\partial_1 w^{\delta,1} - x_2)n_1 + (\partial_2 w^{\delta,1} + x_1)n_2 = 0, && \text{on } \gamma^{\delta,1}, \end{aligned}$$

there exists a unique function $\overline{\Psi}^{\delta,1} \in H^1(\omega^{\delta,1})$ such that $\overline{\Psi}^{\delta,1} = 0$, on γ, satisfying

$$\begin{aligned} \partial_1 w^{\delta,1} - x_2 &= \partial_2 \overline{\Psi}^{\delta,1}, && \text{a.e. in } \omega^{\delta,1}, \\ \partial_2 w^{\delta,1} + x_1 &= -\partial_1 \overline{\Psi}^{\delta,1}, && \text{a.e. in } \omega^{\delta,1}. \end{aligned} \tag{31.24}$$

Moreover, this function $\overline{\Psi}^{\delta,1}$ is the solution of (30.4) and, consequently, $\overline{\Psi}^{\delta,1} = \Psi^{\delta,1}$, a.e. in $\omega^{\delta,1}$. Since the trace of $\Psi^{\delta,1}$ on each connex component of the boundary $\partial T_k^{\delta,1}$ is a constant, denoted $C_k^{\delta,1}$, $\Psi^{\delta,1}$ has a natural extension $\hat{\Psi}^{\delta,1}$, in $H^1(\omega)$, defined by

$$\begin{aligned} &\hat{\Psi}^{\delta,1}(\boldsymbol{x}) = \Psi^{\delta,1}(\boldsymbol{x}), && \text{for all } \boldsymbol{x} \in \omega^{\delta,1}, \\ &\hat{\Psi}^{\delta,1}(\boldsymbol{x}) = C_k^{\delta,1}, && \text{for all } \boldsymbol{x} \in T_k^{\delta,1}, \ k = 1, 2, \ldots, N(\delta). \end{aligned} \tag{31.25}$$

Representing by $[\boldsymbol{\nabla} \Psi^{\delta,1}]^{\sim}$ the zero extension of $\boldsymbol{\nabla} \Psi^{\delta,1}$ to all ω one has

$$[\boldsymbol{\nabla} \Psi^{\delta,1}]^{\sim} = \boldsymbol{\nabla} \hat{\Psi}^{\delta,1}. \tag{31.26}$$

From (31.24), (31.11), and (31.26), with a constant C, independent of δ, the following estimate holds $\|\boldsymbol{\nabla} \Psi^{\delta,1}\|_{L^2(\omega^{\delta,1})} \leqslant C$, which implies, by Poincaré's inequality, that $\hat{\Psi}^{\delta,1}$ is bounded in $H^1(\omega)$, independently of δ. There exists then a subsequence δ' of δ, such that

$$\hat{\Psi}^{\delta',1} \rightharpoonup \Psi^{*,1}, \quad \text{weakly in } H^1(\omega), \ \text{strongly in } L^2(\omega). \tag{31.27}$$

Using (31.24) and (31.26) one has

$$\begin{aligned} -\partial_1 \hat{\Psi}^{\delta,1} &= [\partial_2 w^{\delta,1}]^{\sim} + x_1 \chi^{\delta,1}, && \text{a.e. in } \omega, \\ \partial_2 \hat{\Psi}^{\delta,1} &= [\partial_1 w^{\delta,1}]^{\sim} - x_2 \chi^{\delta,1}, && \text{a.e. in } \omega. \end{aligned} \tag{31.28}$$

By convergences (31.27) and (31.15), as $\delta \to 0$, one obtains from (31.28)

$$\begin{aligned} -\partial_1 \Psi^{*,1} &= (\sigma_w^{*,1})_2 + x_1 \Theta, && \text{a.e. in } \omega, \\ \partial_2 \Psi^{*,1} &= (\sigma_w^{*,1})_1 - x_2 \Theta, && \text{a.e. in } \omega. \end{aligned} \tag{31.29}$$

As a consequence of (31.19) and (31.29) we conclude that $\Psi^{*,1}$ is the unique solution of problem (31.5), with $\varepsilon = 1$, which completes the proof. □

COROLLARY 31.1. *For constants* $I_\beta^{\delta,\varepsilon}$, $H_i^{\delta,\varepsilon}$, $I_\beta^{w^{\delta,\varepsilon}}$, $I_\beta^{\Psi^{\delta,\varepsilon}}$, $J^{\delta,\varepsilon}$, $L_{\alpha\beta}^{\eta^{\delta,\varepsilon}}$, $L_{\alpha\beta}^{\theta^{\delta,\varepsilon}}$, $K_{\alpha\beta}^{\eta^{\delta,\varepsilon}}$, $K_{\alpha\beta}^{\theta^{\delta,\varepsilon}}$, *defined by expressions* (30.1), (30.3), (30.5), (30.6), (30.10) *and* (30.11), *respectively, the following convergences hold, for each fixed* ε *and as* $\delta \to 0$,

$$I_\beta^{\delta,\varepsilon} \to \Theta I_\beta^\varepsilon, \quad \textit{with } I_\beta^\varepsilon = \int_{\omega^\varepsilon} (x_\beta)^2 \, \mathrm{d}\omega^\varepsilon, \tag{31.30}$$

$$H_i^{\delta,\varepsilon} \to \Theta H_i^\varepsilon, \quad \textit{with} \quad \begin{aligned} H_\beta^\varepsilon &= \int_{\omega^\varepsilon} \frac{x_\beta^\varepsilon}{2}[(x_1^\varepsilon)^2 + (x_2^\varepsilon)^2] \, \mathrm{d}\omega^\varepsilon, \\ H_3^\varepsilon &= \int_{\omega^\varepsilon} \tfrac{1}{4}[(x_1^\varepsilon)^2 + (x_2^\varepsilon)^2]^2 \, \mathrm{d}\omega^\varepsilon, \end{aligned} \tag{31.31}$$

$$I_\beta^{w^{\delta,\varepsilon}} \to \Theta I_\beta^{w^{*,\varepsilon}}, \quad \textit{with} \quad I_\beta^{w^{*,\varepsilon}} = 2\int_{\omega^\varepsilon} x_\beta w^{*,\varepsilon} \, \mathrm{d}\omega^\varepsilon, \tag{31.32}$$

$$I_\beta^{\Psi^{\delta,\varepsilon}} \to I_\beta^{\Psi^{*,\varepsilon}}, \quad \textit{with} \quad \begin{aligned} I_1^{\Psi^{\delta,\varepsilon}} &= 2\int_{\omega^\varepsilon} x_2 \Psi^{*,\varepsilon} \, \mathrm{d}\omega^\varepsilon, \\ I_2^{\Psi^{\delta,\varepsilon}} &= -2\int_{\omega^\varepsilon} x_1 \Psi^{*,\varepsilon} \, \mathrm{d}\omega^\varepsilon, \end{aligned} \tag{31.33}$$

$$J^{\delta,\varepsilon} \to J^{*,\varepsilon}, \quad \textit{with} \quad J^{*,\varepsilon} = 2\int_{\omega^\varepsilon} \Psi^{*,\varepsilon} \, \mathrm{d}\omega^\varepsilon, \tag{31.34}$$

$$L_{\alpha\beta}^{\eta^{\delta,\varepsilon}} \to \Theta L_{\alpha\beta}^{\eta^{*,\varepsilon}}, \quad \textit{with} \quad L_{\alpha\beta}^{\eta^{*,\varepsilon}} = \int_{\omega^\varepsilon} x_\alpha \eta_\beta^{*,\varepsilon} \, \mathrm{d}\omega^\varepsilon, \tag{31.35}$$

$$L_{\alpha\beta}^{\theta^{\delta,\varepsilon}} \to \Theta L_{\alpha\beta}^{\theta^{*,\varepsilon}}, \quad \textit{with} \quad L_{\alpha\beta}^{\theta^{*,\varepsilon}} = \int_{\omega^\varepsilon} x_\alpha \theta_\beta^{*,\varepsilon} \, \mathrm{d}\omega^\varepsilon, \tag{31.36}$$

$$K_{\alpha\beta}^{\eta^{\delta,\varepsilon}} \to K_{\alpha\beta}^{\eta^{*,\varepsilon}}, \quad \textit{with} \quad K_{\alpha\beta}^{\eta^{*,\varepsilon}} = \int_{\omega^\varepsilon} \Phi_{\alpha\mu}^\varepsilon a_{\mu\lambda}^* \partial_\lambda^\varepsilon \eta_\beta^{*,\varepsilon} \, \mathrm{d}\omega^\varepsilon, \tag{31.37}$$

$$K_{\alpha\beta}^{\theta^{\delta,\varepsilon}} \to K_{\alpha\beta}^{\theta^{*,\varepsilon}} + N_{\alpha\beta}^{*,\varepsilon} - \Theta H_3^\varepsilon \delta_{\alpha\beta}, \quad \textit{with} \quad \begin{aligned} K_{\alpha\beta}^{\theta^{*,\varepsilon}} &= \int_{\omega^\varepsilon} \Phi_{\alpha\mu}^\varepsilon a_{\mu\lambda}^* \partial_\lambda^\varepsilon \theta_\beta^{*,\varepsilon} \, \mathrm{d}\omega^\varepsilon, \\ N_{\alpha\beta}^{*,\varepsilon} &= \int_{\omega^\varepsilon} \Phi_{\alpha\mu}^\varepsilon a_{\mu\lambda}^* \Phi_{\beta\mu}^\varepsilon \, \mathrm{d}\omega^\varepsilon. \end{aligned} \tag{31.38}$$

PROOF. Since $I_\beta^{\delta,\varepsilon} = \int_{\omega^{\delta,\varepsilon}} (x_\beta^\varepsilon)^2 \, d\omega^{\delta,\varepsilon} = \int_{\omega^\varepsilon} x_\beta^\varepsilon \chi^{\delta,\varepsilon} \, d\omega^\varepsilon$ and, by Lemma 31.1, $\chi^{\delta,\varepsilon} \to \Theta$, weakly in $L^2(\omega^\varepsilon)$, as $\delta \to 0$, convergence (31.30) holds. The same argument proves (31.31). Since $I_\beta^{w^{\delta,\varepsilon}} = 2\int_{\omega^{\delta,\varepsilon}} x_\beta^\varepsilon w^{\delta,\varepsilon} \, d\omega^{\delta,\varepsilon} = 2\int_{\omega^{\delta,\varepsilon}} x_\beta^\varepsilon \tilde{w}^{\delta,\varepsilon} \, d\omega^{\delta,\varepsilon}$, and, by (31.15), $\tilde{w}^{\delta,\varepsilon} \to \tilde{w}^\varepsilon = \Theta w^{*,\varepsilon}$, weakly in $L^2(\omega^\varepsilon)$, as $\delta \to 0$, convergence (31.32) holds. The same argument proves (31.35) and (31.36). For the quantities depending on the torsion constant one has, integrating by parts, $J^{\delta,\varepsilon} = 2\int_{\omega^{\delta,\varepsilon}} \Psi^{\delta,\varepsilon} \, d\omega^{\delta,\varepsilon} + 2c_k^{\delta,\varepsilon} \mid T_k^{\delta,\varepsilon} \mid = 2\int_{\omega^\varepsilon} \hat{\Psi}^{\delta,\varepsilon} \, d\omega^\varepsilon$, $I_1^{\Psi^{\delta,\varepsilon}} = 2\int_{\omega^\varepsilon} x_2^\varepsilon \hat{\Psi}^{\delta,\varepsilon} \, d\omega^\varepsilon$ and $I_2^{\Psi^{\delta,\varepsilon}} = -2\int_{\omega^\varepsilon} x_1^\varepsilon \hat{\Psi}^{\delta,\varepsilon} \, d\omega^\varepsilon$. In view of (31.27) one has $\int_{\omega^\varepsilon} \hat{\Psi}^{\delta,\varepsilon} \, d\omega^\varepsilon \to \int_{\omega^\varepsilon} \Psi^{*,\varepsilon} \, d\omega^\varepsilon$, which proves (31.33) and (31.34). Finally, consider $K_{\alpha\beta}^{\eta^{\delta,\varepsilon}}$ and $K_{\alpha\beta}^{\theta^{\delta,\varepsilon}}$ given by (30.11). From (31.21) the following convergences hold: $K_{\alpha\beta}^{\eta^{\delta,\varepsilon}} \to \int_{\omega^\varepsilon} \Phi_{\alpha\mu}^\varepsilon (\sigma_{\eta_\beta}^{*,\varepsilon})_\mu \, d\omega^\varepsilon$ and $K_{\alpha\beta}^{\theta^{\delta,\varepsilon}} \to \int_{\omega^\varepsilon} \Phi_{\alpha\mu}^\varepsilon (\sigma_{\theta_\beta}^{*,\varepsilon})_\mu \, d\omega^\varepsilon$. Expressions (31.37) and (31.38) follow from (31.22) and (31.23), respectively, which completes the proof. □

REMARK 31.2. It is interesting to compare the behaviour of the limit functions and constants given by Theorem 31.1 and Corollary 31.1 with the corresponding functions and constants for the isotropic and simply connected case. Macroscopically, the multicellular cross section behaves as a simply connected but geometrically anisotropic domain. (cf. Section 25 and also ÁLVAREZ-DIOS [1992], ÁLVAREZ-DIOS and VIAÑO [1992a,b]).

We also remark that replacing A^* by the identity, the functions and constants given by Theorem 31.1 and Corollary 31.1 just reduce to the corresponding ones for the isotropic and simply connected cases.

In the following, let $\langle \cdot, \cdot \rangle_H$ represent the inner product in any Hilbert space H.

THEOREM 31.2. *Consider the functions and constants defined by* (31.4)–(31.9) *and* (31.30)–(31.38). *Suppose that the system of applied loads satisfies*

$$\begin{aligned} &f_\alpha^\varepsilon \in L^2(\Omega^\varepsilon), \qquad g_\alpha^\varepsilon \in L^2(\Gamma^\varepsilon), \\ &f_3^\varepsilon \in H^1[0,L;L^2(\omega^\varepsilon)], \qquad g_3^\varepsilon \in H^1[0,L;L^2(\gamma^\varepsilon)], \end{aligned} \tag{31.39}$$

then, for elements $\boldsymbol{u}^{\delta(0)\varepsilon}$, $\boldsymbol{u}^{\delta(2)\varepsilon}$, $\boldsymbol{\sigma}^{\delta(0)\varepsilon}$ and $\boldsymbol{\sigma}^{\delta(2)\varepsilon}$ characterized in Theorem 30.1, where the volumic forces are the restriction of $\boldsymbol{f}^\varepsilon$ to $\Omega^{\delta,\varepsilon}$ and where the surface forces coincide with $\boldsymbol{g}^\varepsilon$ on Γ^ε, the following convergences hold, for each fixed ε, as $\delta \to 0$:

$$\text{(a) } u_\beta^{\delta(0)\varepsilon} \to u_\beta^{*(0)\varepsilon}, \quad \textit{strongly in } H^4(0,L), \tag{31.40}$$

where $u_\beta^{*(0)\varepsilon}$ is the unique solution of

$$\begin{aligned} &u_\beta^{*(0)\varepsilon} \in H_0^2(0,L) \cap H^4(0,L), \\ &\Theta E^\varepsilon I_\beta^\varepsilon \partial_{3333}^\varepsilon u_\beta^{*(0)\varepsilon} = F_\beta^{*(0)\varepsilon} - \partial_3^\varepsilon M_\beta^{*(0)\varepsilon}, \quad \textit{in } (0,L) \textit{ (no sum on } \beta), \end{aligned} \tag{31.41}$$

where

$$F_\beta^{*(0)\varepsilon} = \Theta \int_{\omega^\varepsilon} f_\beta^\varepsilon \, d\omega^\varepsilon + \int_{\gamma^\varepsilon} g_\beta^\varepsilon \, d\gamma^\varepsilon,$$
$$M_\beta^{*(0)\varepsilon} = \Theta \int_{\omega^\varepsilon} x_\beta^\varepsilon f_3^\varepsilon \, d\omega^\varepsilon + \int_{\gamma^\varepsilon} x_\beta^\varepsilon g_3^\varepsilon \, d\gamma^\varepsilon, \tag{31.42}$$

(b) *The stretching component* $\hat{u}_3^{\delta(0)\varepsilon}$ *of* $u_3^{*(0)\varepsilon}$, *to* Ω^ε, *defined by*:

$$\hat{u}_3^{\delta(0)\varepsilon} = \underline{u}_3^{\delta(0)\varepsilon} - x_\alpha^\varepsilon \partial_3^\varepsilon u_\alpha^{\delta(0)\varepsilon}, \tag{31.43}$$

satisfies

$$\hat{u}_3^{\delta(0)\varepsilon} \to u_3^{*(0)\varepsilon}, \quad \textit{strongly in } H^3[0,L;C^m(\omega^\varepsilon)], \textit{ for all } m \in \mathbb{N}_0, \tag{31.44}$$

with

$$u_3^{*(0)\varepsilon} = \underline{u}_3^{*(0)\varepsilon} - x_\alpha^\varepsilon \partial_3^\varepsilon u_\alpha^{*(0)\varepsilon}, \tag{31.45}$$

where the stretching component $\underline{u}_3^{*(0)\varepsilon}$ *is the unique solution of the following problem*:

$$\underline{u}_3^{*(0)\varepsilon} \in H_0^1(0,L) \cap H^3(0,L),$$
$$-\Theta E^\varepsilon |\omega^\varepsilon| \partial_{33}^\varepsilon \underline{u}_3^{*(0)\varepsilon} = F_3^{*(0)\varepsilon}, \quad \textit{in } (0,L), \tag{31.46}$$

with $F_3^{*(0)\varepsilon}$ *given by*

$$F_3^{*(0)\varepsilon} = \Theta \int_{\omega^\varepsilon} f_3^\varepsilon \, d\omega^\varepsilon + \int_{\gamma^\varepsilon} g_3^\varepsilon \, d\gamma^\varepsilon. \tag{31.47}$$

(c) *The zero extension* $\tilde{\sigma}_{33}^{\delta(0)\varepsilon}$ *of* $\sigma_{33}^{\delta(0)\varepsilon}$, *to* Ω^ε, *satisfies*

$$\langle \tilde{\sigma}_{33}^{\delta(0)\varepsilon}, \varphi \rangle_{L^2(\omega^\varepsilon)} \to \langle \sigma_{33}^{*(0)\varepsilon}, \varphi \rangle_{L^2(\omega^\varepsilon)}, \quad \textit{strongly in } H^2(0,L), \textit{ for all } \varphi \in L^2(\omega^\varepsilon), \tag{31.48}$$

where

$$\sigma_{33}^{*(0)\varepsilon} = \Theta E^\varepsilon \partial_3^\varepsilon u_3^{*(0)\varepsilon} = \Theta E^\varepsilon (\partial_3^\varepsilon \underline{u}_3^{*(0)\varepsilon} - x_\alpha^\varepsilon \partial_{33}^\varepsilon u_\alpha^{*(0)\varepsilon}). \tag{31.49}$$

(d) *For extensions* $\hat{u}_i^{\delta(2)\varepsilon}$ *of* $u_i^{\delta(2)\varepsilon}$, *to* Ω^ε, *defined by*

$$\hat{u}_1^{\delta(2)\varepsilon} = z_1^{\delta(2)\varepsilon} + x_2^\varepsilon z^{\delta(2)\varepsilon} - \nu^\varepsilon (x_1^\varepsilon \partial_3^\varepsilon \underline{u}_3^{\delta(0)\varepsilon} - \Phi_{1\beta}^\varepsilon \partial_{33}^\varepsilon u_\beta^{\delta(0)\varepsilon}), \tag{31.50}$$

$$\hat{u}_2^{\delta(2)\varepsilon} = z_2^{\delta(2)\varepsilon} - x_1^\varepsilon z^{\delta(2)\varepsilon} - \nu^\varepsilon (x_2^\varepsilon \partial_3^\varepsilon \underline{u}_3^{\delta(0)\varepsilon} - \Phi_{2\beta}^\varepsilon \partial_{33}^\varepsilon u_\beta^{\delta(0)\varepsilon}), \tag{31.51}$$

$$\begin{aligned}\hat{u}_3^{\delta(2)\varepsilon} &= \underline{u}_3^{\delta(2)\varepsilon} - x_\alpha^\varepsilon \partial_\alpha^\varepsilon z_\alpha^{\delta(2)\varepsilon} - \hat{w}^{\delta,\varepsilon} \partial_3^\varepsilon z^{\delta(2)\varepsilon} + [(1+\nu^\varepsilon)\hat{\eta}_\alpha^{\delta,\varepsilon} + \nu^\varepsilon \hat{\theta}_\alpha^{\delta,\varepsilon}] \partial_{333}^\varepsilon u_\alpha^{\delta(0)\varepsilon} \\ &\quad + \nu^\varepsilon \left\{ \tfrac{1}{2}[(x_1^\varepsilon)^2 + (x_2^\varepsilon)^2] - \frac{1}{2|\omega^{\delta,\varepsilon}|}(I_1^{\delta,\varepsilon} + I_2^{\delta,\varepsilon}) \right\} \partial_{33}^\varepsilon \underline{u}_3^{\delta(0)\varepsilon} \\ &\quad + \frac{2(1+\nu^\varepsilon)}{E^\varepsilon} \hat{w}^{\delta(0)\varepsilon},\end{aligned} \tag{31.52}$$

where $\hat{w}^{\delta,\varepsilon}$, $\hat{\eta}_\alpha^{\delta,\varepsilon}$, $\hat{\theta}_\alpha^{\delta,\varepsilon}$ and $\hat{w}^{\delta(0)\varepsilon}$, for each $x_3^\varepsilon \in [0, L]$, are the classical extensions of $w^{\delta,\varepsilon}$, $\eta_\alpha^{\delta,\varepsilon}$, $\theta_\alpha^{\delta,\varepsilon}$ and $w^{\delta(0)\varepsilon}$, respectively, the following convergences hold

$$\hat{u}_\alpha^{\delta(2)\varepsilon} \to u_\alpha^{*(2)\varepsilon}, \quad \text{strongly in } H^2[0, L; C^m(\omega^\varepsilon)], \text{ for all } m \in \mathbb{N}_0, \tag{31.53}$$

$$\langle \hat{u}_3^{\delta(2)\varepsilon}, \varphi\rangle_{H^1(\omega^\varepsilon)} \to \langle u_3^{*(2)\varepsilon}, \varphi\rangle_{H^1(\omega^\varepsilon)}, \quad \text{strongly in } H^1(0, L), \tag{31.54}$$

valid for all $\varphi \in H^1(\omega^\varepsilon)$, and with

$$u_1^{*(2)\varepsilon} = z_1^{*(2)\varepsilon} + x_2^\varepsilon z^{*(2)\varepsilon} - \nu^\varepsilon (x_1^\varepsilon \partial_3^\varepsilon \underline{u}_3^{*(0)\varepsilon} - \Phi_{1\beta}^\varepsilon \partial_{33}^\varepsilon u_\beta^{*(0)\varepsilon}), \tag{31.55}$$

$$u_2^{*(2)\varepsilon} = z_2^{*(2)\varepsilon} - x_1^\varepsilon z^{*(2)\varepsilon} - \nu^\varepsilon (x_2^\varepsilon \partial_3^\varepsilon \underline{u}_3^{*(0)\varepsilon} - \Phi_{2\beta}^\varepsilon \partial_{33}^\varepsilon u_\beta^{*(0)\varepsilon}), \tag{31.56}$$

$$\begin{aligned} u_3^{*(2)\varepsilon} =& \underline{u}_3^{*(2)\varepsilon} - x_\alpha^\varepsilon \partial_\alpha^\varepsilon z_\alpha^{*(2)\varepsilon} - w^{*,\varepsilon} \partial_3^\varepsilon z^{*(2)\varepsilon} + [(1+\nu^\varepsilon)\eta_\alpha^{*,\varepsilon} + \nu^\varepsilon \theta_\alpha^{*,\varepsilon}] \partial_{333}^\varepsilon u_\alpha^{*(0)\varepsilon} \\ &+ \nu^\varepsilon \left\{ \tfrac{1}{2}[(x_1^\varepsilon)^2 + (x_2^\varepsilon)^2] - \frac{1}{2|\omega^\varepsilon|}(I_1^\varepsilon + I_2^\varepsilon) \right\} \partial_{33}^\varepsilon \underline{u}_3^{*(0)\varepsilon} + \frac{2(1+\nu^\varepsilon)}{E^\varepsilon} w^{*(0)\varepsilon}, \end{aligned} \tag{31.57}$$

where functions $z^{*(2)\varepsilon}$, $w^{*(0)\varepsilon}$, $\underline{u}_3^{*(2)\varepsilon}$ and $z_\alpha^{*(2)\varepsilon}$ are the unique solutions of the following problems:

(d1) *Homogenized angle of twist.*

$$\begin{aligned} & z^{*(2)\varepsilon} \in H^2(0, L), \\ & -\frac{EJ^{*,\varepsilon}}{2(1+\nu^\varepsilon)} \partial_{33}^\varepsilon z^{*(2)\varepsilon} = M_3^{*(2)\varepsilon}, \quad \text{in } (0, L), \\ & z^{*(2)\varepsilon}(a) = \frac{\nu^\varepsilon}{I_1^\varepsilon + I_2^\varepsilon} [H_2^\varepsilon \partial_{33}^\varepsilon u_1^{*(0)\varepsilon}(a) - H_1^\varepsilon \partial_{33}^\varepsilon u_2^{*(0)\varepsilon}(a)], \quad \text{at } a = 0, L, \end{aligned} \tag{31.58}$$

where

$$\begin{aligned} M_3^{*(2)\varepsilon} =& \Theta \int_{\omega^\varepsilon} (x_2^\varepsilon f_1^\varepsilon - x_1^\varepsilon f_2^\varepsilon)\, d\omega^\varepsilon + \int_{\gamma^\varepsilon} (x_2^\varepsilon g_1^\varepsilon - x_1^\varepsilon g_2^\varepsilon)\, d\gamma^\varepsilon \\ &+ \Theta \int_{\omega^\varepsilon} w^{*,\varepsilon} \partial_3^\varepsilon f_3^\varepsilon \, d\omega^\varepsilon + \int_{\gamma^\varepsilon} w^{*,\varepsilon} \partial_3^\varepsilon g_3^\varepsilon \, d\gamma^\varepsilon \\ &- \frac{E^\varepsilon}{2(1+\nu^\varepsilon)} [(1+\nu^\varepsilon)\Theta I_\alpha^{w^{*,\varepsilon}} + \nu^\varepsilon I_\alpha^{\Psi^{*,\varepsilon}}] \partial_{3333}^\varepsilon u_\alpha^{*(0)\varepsilon}. \end{aligned} \tag{31.59}$$

(d2) *Homogenized additional warping.*

$$\begin{aligned} & w^{*(0)\varepsilon} \in H^1[0, L; H^1(\omega^{*,\varepsilon})] \quad \text{such that for all } x_3^\varepsilon \in (0, L), \\ & -\operatorname{div}^\varepsilon(A^* \nabla^\varepsilon w^{*(0)\varepsilon}(x_3^\varepsilon)) \\ & \quad = \Theta f_3^\varepsilon(x_3^\varepsilon) - \frac{1}{|\omega^\varepsilon|} \left[\Theta \int_{\omega^\varepsilon} f_3^\varepsilon(x_3^\varepsilon)\, d\omega^\varepsilon + \int_{\gamma^\varepsilon} g_3^\varepsilon(x_3^\varepsilon)\, d\gamma^\varepsilon \right], \quad \text{in } \omega^\varepsilon, \\ & (A^* \nabla^\varepsilon w^{*(0)\varepsilon}(x_3^\varepsilon)) \cdot \boldsymbol{n}^\varepsilon = g_3^\varepsilon(x_3^\varepsilon), \quad \text{on } \gamma^\varepsilon, \\ & \int_{\omega^\varepsilon} w^{*(0)\varepsilon}(x_3^\varepsilon)\, d\omega^\varepsilon = 0. \end{aligned} \tag{31.60}$$

the coefficients of A^* *being given by* (31.8), (31.9).

(d3) *Homogenized stretching component* $\underline{u}_3^{*(2)\varepsilon}$.

$$\begin{aligned} &\underline{u}_3^{*(2)\varepsilon} \in H_0^1(0,L), \\ &-\Theta E|\omega^\varepsilon|\partial_{33}^\varepsilon \underline{u}_3^{*(2)\varepsilon} = -\nu^\varepsilon \partial_3^\varepsilon G_3^{*(2)\varepsilon}, \end{aligned} \tag{31.61}$$

where

$$\begin{aligned} G_3^{*(2)\varepsilon} = &-[\tfrac{1}{2}\Theta E(I_1^\varepsilon + I_2^\varepsilon)]\partial_{333}^\varepsilon \underline{u}_3^{*(0)\varepsilon} + \Theta E^\varepsilon H_\alpha^\varepsilon \partial_{3333}^\varepsilon u_\alpha^{*(0)\varepsilon} \\ &- \Theta \int_{\omega^\varepsilon} \{\tfrac{1}{2}[(x_1^\varepsilon)^2 + (x_2^\varepsilon)^2]\}\partial_3^\varepsilon f_3^\varepsilon \,\mathrm{d}\omega^\varepsilon - \int_{\gamma^\varepsilon} \{\tfrac{1}{2}[(x_1^\varepsilon)^2 + (x_2^\varepsilon)^2]\}\partial_3^\varepsilon g_3^\varepsilon \,\mathrm{d}\gamma^\varepsilon \\ &- \Theta \int_{\omega^\varepsilon} x_\alpha^\varepsilon f_\alpha^\varepsilon \,\mathrm{d}\omega^\varepsilon - \int_{\gamma^\varepsilon} x_\alpha^\varepsilon g_\alpha^\varepsilon \,\mathrm{d}\gamma^\varepsilon. \end{aligned} \tag{31.62}$$

(d4) *The homogenized bending component* $z_\alpha^{*(2)\varepsilon}$ *depends only on variable* x_3^ε *and is the unique solution of (no sum on* α, $\beta \neq \alpha$*)*

$$\begin{aligned} &z_\alpha^{*(2)\varepsilon} \in H^2(0,L), \\ &\Theta E I_\alpha^\varepsilon \partial_{3333}^\varepsilon z_\alpha^{*(2)\varepsilon} = \partial_{33}^\varepsilon M_\alpha^{*(2)\varepsilon}, \quad \textit{in } (0,L), \\ &z_\alpha^{*(2)\varepsilon}(a) = -\frac{\nu^\varepsilon(I_\alpha^\varepsilon - I_\beta^\varepsilon)}{2\,|\,\omega^\varepsilon\,|}\partial_{33}^\varepsilon u_\alpha^{*(0)\varepsilon}(a) \quad \textit{at } a = 0, L, \\ &\partial_3^\varepsilon z_\alpha^{*(2)\varepsilon}(a) = \frac{1}{I_\alpha^\varepsilon}\left\{\frac{2(1+\nu^\varepsilon)}{E}\int_{\omega^\varepsilon} x_\alpha^\varepsilon w^{*(0)\varepsilon}(a)\,\mathrm{d}\omega^\varepsilon - \frac{I_\alpha^{w^*,\varepsilon}}{2}\partial_3^\varepsilon z^{*(2)\varepsilon}(a)\right. \\ &\qquad \left. + \nu^\varepsilon H_\alpha^\varepsilon \partial_{33}^\varepsilon \underline{u}_3^{*(0)\varepsilon}(a) + [(1+\nu^\varepsilon)L_{\alpha\beta}^{\eta^*,\varepsilon} + \nu^\varepsilon L_{\alpha\beta}^{\theta^*,\varepsilon}]\partial_{333}^\varepsilon u_\beta^{*(0)\varepsilon}(a)\right\} \\ &\qquad \textit{at } a = 0 \textit{ and } L, \end{aligned} \tag{31.63}$$

where

$$\begin{aligned} M_\alpha^{*(2)\varepsilon} = &E^\varepsilon\left\{(1+\nu^\varepsilon)\Theta L_{\alpha\beta}^{\eta^*,\varepsilon} + \nu^\varepsilon \Theta L_{\alpha\beta}^{\theta^*,\varepsilon}\right. \\ &\left. + \frac{\nu^\varepsilon}{2}K_{\alpha\beta}^{\eta^*,\varepsilon} + \frac{(\nu^\varepsilon)^2}{2(1+\nu^\varepsilon)}(K_{\alpha\beta}^{\theta^*,\varepsilon} + N_{\alpha\beta}^{*,\varepsilon})\right\}\partial_{3333}^\varepsilon u_\beta^{*(0)\varepsilon} \\ &+ E^\varepsilon \nu^\varepsilon \Theta H_\alpha^\varepsilon \partial_{333}^\varepsilon \underline{u}_3^{*(0)\varepsilon} - [(1+\nu^\varepsilon)\Theta I_\alpha^{w^*,\varepsilon} + \nu^\varepsilon I_\alpha^{\Psi^*,\varepsilon}]\frac{E^\varepsilon}{2(1+\nu^\varepsilon)}\partial_{33}^\varepsilon z^{*(2)\varepsilon} \\ &- \Theta \int_{\omega^\varepsilon} [(1+\nu^\varepsilon)\eta_\alpha^{*,\varepsilon} + \nu^\varepsilon \theta_\alpha^{*,\varepsilon}]\partial_3^\varepsilon f_3^\varepsilon \,\mathrm{d}\omega^\varepsilon \\ &- \int_{\gamma^\varepsilon} [(1+\nu^\varepsilon)\eta_\alpha^{*,\varepsilon} + \nu^\varepsilon \theta_\alpha^{*,\varepsilon}]\partial_3^\varepsilon g_3^\varepsilon \,\mathrm{d}\gamma^\varepsilon \\ &+ \nu^\varepsilon \Theta \int_{\omega^\varepsilon} \Phi_{\alpha\beta}^\varepsilon f_\beta^\varepsilon \,\mathrm{d}\omega^\varepsilon + \nu^\varepsilon \int_{\gamma^\varepsilon} \Phi_{\alpha\beta}^\varepsilon g_\beta^\varepsilon \,\mathrm{d}\gamma^\varepsilon, \end{aligned} \tag{31.64}$$

(e) *The zero extensions* $\tilde{\sigma}_{3\beta}^{\delta(0)\varepsilon}$ *of* $\sigma_{3\beta}^{\delta(0)\varepsilon}$, *to* Ω^ε, *satisfy*

$$\begin{aligned}&\langle\tilde{\sigma}_{3\beta}^{\delta(0)\varepsilon},\varphi\rangle_{L^2(\omega^\varepsilon)}\to\langle\sigma_{3\beta}^{*(0)\varepsilon},\varphi\rangle_{L^2(\omega^\varepsilon)},\\&\quad \text{strongly in } H^1(0,L), \text{ for all } \varphi\in L^2(\omega^\varepsilon),\end{aligned}\tag{31.65}$$

with

$$\sigma_{31}^{*(0)\varepsilon}=\frac{E^\varepsilon}{2(1+\nu^\varepsilon)}(-\partial_2^\varepsilon\Psi^{*,\varepsilon}\partial_3^\varepsilon z^{*(2)\varepsilon})+\underline{\sigma}_{31}^{*(0)\varepsilon},\tag{31.66}$$

$$\sigma_{32}^{*(0)\varepsilon}=\frac{E^\varepsilon}{2(1+\nu^\varepsilon)}(\partial_1^\varepsilon\Psi^{*,\varepsilon}\partial_3^\varepsilon z^{*(2)\varepsilon})+\underline{\sigma}_{32}^{*(0)\varepsilon},\tag{31.67}$$

where

$$\begin{aligned}\underline{\sigma}_{3\gamma}^{*(0)\varepsilon}=a_{\gamma\alpha}^*\bigg\{&\frac{E^\varepsilon}{2(1+\nu^\varepsilon)}[(1+\nu^\varepsilon)\partial_\alpha^\varepsilon\eta_\beta^{*,\varepsilon}\\&+\nu^\varepsilon(\partial_\alpha^\varepsilon\theta_\beta^{*,\varepsilon}+\Phi_{\alpha\beta}^\varepsilon)]\partial_{333}^\varepsilon u_\beta^{*(0)\varepsilon}+\partial_\alpha^\varepsilon w^{*(0)\varepsilon}\bigg\}.\end{aligned}\tag{31.68}$$

(f) *The bending moment* $m_\beta^{\delta(0)\varepsilon}=\int_{\omega^{\delta,\varepsilon}}x_\beta^\varepsilon\sigma_{33}^{\delta(0)\varepsilon}\,\mathrm{d}\omega^{\delta,\varepsilon}$ *and shear force* $q_\beta^{\delta(0)\varepsilon}=\int_{\omega^{\delta,\varepsilon}}\sigma_{3\beta}^{\delta(0)\varepsilon}\,\mathrm{d}\omega^{\delta,\varepsilon}$ *components converge, respectively, to* $m_\beta^{*(0)\varepsilon}=\int_{\omega^\varepsilon}x_\beta^\varepsilon\sigma_{33}^{*(0)\varepsilon}\,\mathrm{d}\omega^\varepsilon$, *strongly in* $H^2(0,L)$, *and to* $q_\beta^{*(0)\varepsilon}=\int_{\omega^\varepsilon}\sigma_{3\beta}^{*(0)\varepsilon}\,\mathrm{d}\omega^\varepsilon$, *strongly in* $H^1(0,L)$, *satisfying (no sum on* β*),*

$$m_\beta^{*(0)\varepsilon}=-\Theta E^\varepsilon I_\beta^\varepsilon\partial_{33}u_\beta^{*(0)\varepsilon},\tag{31.69}$$

$$q_\beta^{*(0)\varepsilon}=-\Theta E^\varepsilon I_\beta^\varepsilon\partial_{333}u_\beta^{*(0)\varepsilon}+\Theta\int_{\omega^\varepsilon}x_\beta^\varepsilon f_3^\varepsilon\,\mathrm{d}\omega^\varepsilon+\int_{\gamma^\varepsilon}x_\beta^\varepsilon g_3^\varepsilon\,\mathrm{d}\gamma^\varepsilon.\tag{31.70}$$

(g) *The zero extensions* $\tilde{\sigma}_{\alpha\beta}^{\delta(0)\varepsilon}$ *of* $\sigma_{\alpha\beta}^{\delta(0)\varepsilon}$, *to* Ω^ε, *satisfy*

$$\langle\tilde{\sigma}_{\alpha\beta}^{\delta(0)\varepsilon},\varphi\rangle_{L^2(\omega^\varepsilon)}\to\langle\sigma_{\alpha\beta}^{*(0)\varepsilon},\varphi\rangle_{L^2(\omega^\varepsilon)},\quad\text{strongly in } L^2(0,L),\ \text{for all } \varphi\in L^2(\omega^\varepsilon),\tag{31.71}$$

with

$$\begin{aligned}\sigma_{\alpha\beta}^{*(0)\varepsilon}&=S_{\alpha\beta}^{*\varepsilon}(\underline{\boldsymbol{u}}^{*(4)\varepsilon})+S_{\alpha\beta}^{*(0)\varepsilon},\\S_{\alpha\beta}^{*\varepsilon}(\underline{\boldsymbol{u}}^{*(4)\varepsilon})&=\frac{E}{(1+\nu^\varepsilon)}[\Theta e_{\alpha\beta}^\varepsilon(\underline{\boldsymbol{u}}^{*(4)\varepsilon})+b_{\alpha\beta\zeta\rho}^*e_{\zeta\rho}^\varepsilon(\underline{\boldsymbol{u}}^{*(4)\varepsilon})]\\&\quad+\frac{\nu^\varepsilon E^\varepsilon}{(1+\nu^\varepsilon)(1-2\nu^\varepsilon)}[\Theta e_{\mu\mu}^\varepsilon(\underline{\boldsymbol{u}}^{*(4)\varepsilon})+b_{\mu\mu\zeta\rho}^*e_{\zeta\rho}^\varepsilon(\underline{\boldsymbol{u}}^{*(4)\varepsilon})]\delta_{\alpha\beta},\\S_{\alpha\beta}^{*(0)\varepsilon}&=\frac{\nu^\varepsilon E^\varepsilon}{(1+\nu^\varepsilon)(1-2\nu^\varepsilon)}\partial_3^\varepsilon(u_3^{*(2)\varepsilon}-\underline{u}_3^{*(2)\varepsilon}+x_\mu^\varepsilon\partial_3^\varepsilon z_\mu^{*(2)\varepsilon})\delta_{\alpha\beta},\end{aligned}\tag{31.72}$$

where $\underline{\boldsymbol{u}}^{*(4)\varepsilon} = (\underline{\boldsymbol{u}}_\alpha^{*(4)\varepsilon})$ *is the unique solution of the following homogenized plane elasticity problem:*

$$\underline{\boldsymbol{u}}^{*(4)\varepsilon} \in L^2[0, L; (H^1(\omega^\varepsilon))^2], \quad \textit{and a.e. in } (0, L), \textit{ for all } \boldsymbol{\varphi} \in [H^1(\omega^\varepsilon)]^2:$$

$$\begin{aligned}
&\int_{\omega^\varepsilon} S_{\alpha\beta}^{*\varepsilon}(\underline{\boldsymbol{u}}^{*(4)\varepsilon}) e_{\alpha\beta}^\varepsilon(\boldsymbol{\varphi})\,\mathrm{d}\omega^\varepsilon \\
&\quad = \Theta \int_{\omega^\varepsilon} f_\beta^\varepsilon \varphi_\beta\,\mathrm{d}\omega^\varepsilon + \int_{\gamma^\varepsilon} g_\beta^\varepsilon \varphi_\beta\,\mathrm{d}\gamma^\varepsilon \\
&\qquad + \int_{\omega^\varepsilon} \partial_3^\varepsilon \sigma_{3\beta}^{*(0)\varepsilon} \varphi_\beta\,\mathrm{d}\omega^\varepsilon - \int_{\omega^\varepsilon} S_{\alpha\beta}^{*(0)\varepsilon} e_{\alpha\beta}(\boldsymbol{\varphi})\,\mathrm{d}\omega^\varepsilon,
\end{aligned} \tag{31.73}$$

$$\int_{\omega^{*,\varepsilon}} \underline{\boldsymbol{u}}_\alpha^{*(4)\varepsilon}\,\mathrm{d}\omega^\varepsilon = \int_{\omega^{*,\varepsilon}} [x_2^\varepsilon \underline{\boldsymbol{u}}_1^{*(4)\varepsilon} - x_1^\varepsilon \underline{\boldsymbol{u}}_2^{*(4)\varepsilon}]\,\mathrm{d}\omega^\varepsilon = 0,$$

the coefficients $b^*_{\alpha\beta\zeta\rho}$ *being constants independent of* ε, *given by*

$$b^*_{\alpha\beta\zeta\rho} = \int_{Y^*} e_{\alpha\beta}(\boldsymbol{\chi}^{\zeta\rho})\,\mathrm{d}Y^*, \tag{31.74}$$

where $\boldsymbol{\chi}^{\zeta\rho} = (\chi_\alpha^{\zeta\rho})$ *is the unique solution, defined up to a constant vector, of the following problem in the cell* Y^*:

$$\boldsymbol{\chi}^{\zeta\rho} \in [H^1(Y^*)]^2, \quad \boldsymbol{\chi}^{\zeta\rho}\ Y\textit{-periodic, and for all } \boldsymbol{\varphi} \in [H^1(Y^*)]^2,\ Y\textit{-periodic:}$$

$$\begin{aligned}
&\int_{Y^*} [2\mu e_{\alpha\beta}(\boldsymbol{\chi}^{\zeta\rho}) + \lambda e_{\mu\mu}(\boldsymbol{\chi}^{\zeta\rho})\delta_{\alpha\beta}] e_{\alpha\beta}(\boldsymbol{\varphi})\,\mathrm{d}Y^* \\
&\quad = -\int_{Y^*} [2\mu e_{\zeta\rho}(\boldsymbol{\varphi}) + \lambda e_{\mu\mu}(\boldsymbol{\varphi})\delta_{\zeta\rho}]\,\mathrm{d}Y^*.
\end{aligned} \tag{31.75}$$

(h) *The zero extension* $\tilde{\sigma}_{33}^{\delta(2)\varepsilon}$ *of the axial stress component* $\sigma_{33}^{\delta(2)\varepsilon}$, *to* Ω^ε, *satisfies*

$$\begin{aligned}
&\langle \tilde{\sigma}_{33}^{\delta(2)\varepsilon}, \varphi\rangle_{L^2(\omega^\varepsilon)} \to \langle \sigma_{33}^{*(2)\varepsilon}, \varphi\rangle_{L^2(\omega^\varepsilon)}, \\
&\quad \textit{strongly in } L^2(0, L), \textit{ for all } \varphi \in L^2(\omega^\varepsilon),
\end{aligned} \tag{31.76}$$

with

$$\begin{aligned}
\sigma_{33}^{*(2)\varepsilon} &= \Theta E^\varepsilon \partial_3^\varepsilon u_3^{*(2)\varepsilon} + \nu^\varepsilon \sigma_{\mu\mu}^{*(0)\varepsilon} \\
&= \Theta E^\varepsilon \{\partial_3^\varepsilon \underline{u}_3^{*(2)\varepsilon} - x_\alpha^\varepsilon \partial_{33}^\varepsilon z_\alpha^{*(2)\varepsilon} - w^{*,\varepsilon} \partial_{33}^\varepsilon z^{*(2)\varepsilon} \\
&\quad + \nu^\varepsilon [\tfrac{1}{2}[(x_1^\varepsilon)^2 + (x_2^\varepsilon)^2] - \frac{1}{2|\omega^\varepsilon|}(I_1^\varepsilon + I_2^\varepsilon)] \partial_{333}^\varepsilon \underline{u}_3^{*(0)\varepsilon} \\
&\quad + [(1 + \nu^\varepsilon)\eta_\alpha^{*,\varepsilon} + \nu^\varepsilon \theta_\alpha^{*,\varepsilon}] \partial_{3333}^\varepsilon u_\alpha^{*(0)\varepsilon} + \frac{2(1 + \nu^\varepsilon)}{E^\varepsilon} \partial_3^\varepsilon w^{*(0)\varepsilon}\} + \nu^\varepsilon \sigma_{\mu\mu}^{*(0)\varepsilon}.
\end{aligned} \tag{31.77}$$

(i) *The second order bending moment components* $m_\beta^{\delta(2)\varepsilon} = \int_{\omega^{\delta,\varepsilon}} x_\beta^\varepsilon \sigma_{33}^{\delta(2)\varepsilon} \, d\omega^{\delta,\varepsilon}$ *converge, strongly in* $L^2(0,L)$, *to* $m_\beta^{*(2)\varepsilon} = \int_{\omega^\varepsilon} x_\beta^\varepsilon \sigma_{33}^{*(2)\varepsilon} \, d\omega^\varepsilon$, *satisfying*

$$m_\beta^{*(2)\varepsilon} = -\Theta E^\varepsilon I_\beta^\varepsilon \partial_{33}^\varepsilon z_\beta^{*(2)\varepsilon} + M_\beta^{*(2)\varepsilon}, \quad (\textit{no sum on } \beta). \tag{31.78}$$

PROOF. *Step* 1. *Proof of* (a), (b) *and* (c). Using convergences (31.30) and the fact that by Lemma 31.1, $\chi^{\delta,\varepsilon} \to \Theta$, weakly in $L^2(\omega^{\delta,\varepsilon})$, as $\delta \to 0$, one concludes from (30.35) and (30.38) that

$$\frac{1}{E^\varepsilon I_\beta^{\delta,\varepsilon}} F_\beta^{\delta(0)\varepsilon} \to \frac{1}{\Theta E^\varepsilon I_\beta^\varepsilon} F_\beta^{*(0)\varepsilon}, \quad \text{strongly in } L^2(0,L), \tag{31.79}$$

$$\frac{1}{E^\varepsilon |\omega^{\delta,\varepsilon}|} F_3^{\delta(0)\varepsilon} \to \frac{1}{\Theta E^\varepsilon |\omega^{\delta,\varepsilon}|} F_3^{*(0)\varepsilon}, \quad \text{strongly in } H^1(0,L), \tag{31.80}$$

which implies convergence (31.40), by the regular dependence solution–data, and also

$$\underline{u}_3^{\delta(0)\varepsilon} \to \underline{u}_3^{*(0)\varepsilon}, \quad \text{strongly in } H^3(0,L), \tag{31.81}$$

for $\underline{u}_3^{*(0)\varepsilon}$ satisfying (31.46) and (31.47). Passing to the limit in (31.43), as $\delta \to 0$, one obtains (31.44), (31.45) and since, by (30.39),

$$\tilde{\sigma}_{33}^{\delta(0)\varepsilon} = \chi^{\delta,\varepsilon} E^\varepsilon \partial_3^\varepsilon u_3^{\delta(0)\varepsilon}, \quad \text{in } \Omega^\varepsilon, \tag{31.82}$$

convergence (31.48) and expression (31.49) hold.

Step 2. *Proof of* (d), (e) *and* (f). If we consider convergences (31.40), (31.81) and those given by Theorem 31.1 and Corollary 31.1, the same regularity argument used above holds and we conclude that the solution $z^{\delta(2)\varepsilon}$ of problem (30.22), (30.23) satisfies

$$z^{\delta(2)\varepsilon} \to z^{*(2)\varepsilon}, \quad \text{strongly in } H^2(0,L), \tag{31.83}$$

where $z^{*(2)\varepsilon}$ is the solution of (31.58), (31.59), and the solution $\underline{u}_3^{\delta(2)\varepsilon}$ of problem (30.25), (30.26) satisfies

$$\underline{u}_3^{\delta(2)\varepsilon} \to \underline{u}_3^{*(2)\varepsilon}, \quad \text{strongly in } H^1(0,L), \tag{31.84}$$

where $\underline{u}_3^{*(2)\varepsilon}$ is the solution of (31.61), (31.62).

Consider now $w^{\delta(0)\varepsilon}$, solution of problem (30.24). From the linearity of the problem and the regularity of the data one sees not only that $w^{\delta(0)\varepsilon} \in$

$H^1[0, L; H^1(\omega^\varepsilon)]$, but also that a.e. in $x_3^\varepsilon \in (0, L)$, $\partial_3^\varepsilon w^{\delta(0)\varepsilon}(x_3^\varepsilon)$ is the unique solution of

$$
\begin{aligned}
& -\Delta \partial_3^\varepsilon w^{\delta(0)\varepsilon}(x_3^\varepsilon) \\
& \quad = \partial_3^\varepsilon f_3^\varepsilon(x_3^\varepsilon) - \frac{1}{|\omega^{\delta,\varepsilon}|}\left[\int_{\omega^{\delta,\varepsilon}} \partial_3^\varepsilon f_3^\varepsilon(x_3^\varepsilon)\ \mathrm{d}\omega^{\delta,\varepsilon} + \int_{\gamma^\varepsilon} \partial_3^\varepsilon g_3^\varepsilon(x_3^\varepsilon)\,\mathrm{d}\gamma^\varepsilon\right], \quad \text{in } \omega^{\delta,\varepsilon}, \\
& \partial_n^\varepsilon \partial_3^\varepsilon w^{\delta(0)\varepsilon}(x_3^\varepsilon) = \partial_3^\varepsilon g_3^\varepsilon(x_3^\varepsilon) \quad \text{on } \gamma^\varepsilon, \\
& \int_{\omega^{\delta,\varepsilon}} \partial_3^\varepsilon w^{\delta(0)\varepsilon}(x_3^\varepsilon)\,\mathrm{d}\omega^{\delta,\varepsilon} = 0.
\end{aligned}
\tag{31.85}
$$

Using a classical linear extension operator define, for each $x_3^\varepsilon \in [0, L]$, $\hat{w}^{\delta(0)\varepsilon}(x_3^\varepsilon) \in H^1(\omega^\varepsilon)$, such that

$$
\begin{aligned}
& \|\hat{w}^{\delta(0)\varepsilon}(x_3^\varepsilon)\|_{L^2(\omega^\varepsilon)} \leqslant C\|w^{\delta(0)\varepsilon}(x_3^\varepsilon)\|_{L^2(\omega^{\delta,\varepsilon})}, \\
& \|\nabla^\varepsilon \hat{w}^{\delta(0)\varepsilon}(x_3^\varepsilon)\|_{L^2(\omega^\varepsilon)} \leqslant C\|\nabla^\varepsilon w^{\delta(0)\varepsilon}(x_3^\varepsilon)\|_{L^2(\omega^{\delta,\varepsilon})},
\end{aligned}
\tag{31.86}
$$

for C independent of δ and x_3^ε. Since application $x_3^\varepsilon \in [0, L] \to w^{\delta(0)\varepsilon}(x_3^\varepsilon) \in H^1(\omega^{\delta,\varepsilon})$ is absolutely continuous, with derivative in $L^2[0, L; H^1(\omega^{\delta,\varepsilon})]$, and the extension operator is linear and continuous, the same holds for $x_3^\varepsilon \in [0, L] \to \hat{w}^{\delta(0)\varepsilon}(x_3^\varepsilon) \in H^1(\omega^\varepsilon)$ and we then conclude that $\hat{w}^{\delta(0)\varepsilon} \in H^1[0, L; H^1(\omega^\varepsilon)]$ and that $\partial_3^\varepsilon \hat{w}^{\delta(0)\varepsilon}(x_3^\varepsilon)$ is an extension of $\partial_3^\varepsilon w^{\delta(0)\varepsilon}(x_3^\varepsilon)$, satisfying the following estimates a.e. in $x_3^\varepsilon \in (0, L)$:

$$
\begin{aligned}
& \|\partial_3^\varepsilon \hat{w}^{\delta(0)\varepsilon}(x_3^\varepsilon)\|_{L^2(\omega^\varepsilon)} \leqslant C\|\partial_3^\varepsilon w^{\delta(0)\varepsilon}(x_3^\varepsilon)\|_{L^2(\omega^{\delta,\varepsilon})}, \\
& \|\nabla^\varepsilon \partial_3^\varepsilon \hat{w}^{\delta(0)\varepsilon}(x_3^\varepsilon)\|_{L^2(\omega^\varepsilon)} \leqslant C\|\nabla^\varepsilon \partial_3^\varepsilon w^{\delta(0)\varepsilon}(x_3^\varepsilon)\|_{L^2(\omega^{\delta,\varepsilon})}.
\end{aligned}
\tag{31.87}
$$

From the variational setting of problems (30.25) and (31.85), using Lemma 31.3, the properties of the extension operator and conditions (31.39), one obtains respectively,

$$
\|\hat{w}^{\delta(0)\varepsilon}(x_3^\varepsilon)\|_{H^1(\omega^\varepsilon)} \leqslant h_1(x_3^\varepsilon), \tag{31.88}
$$

$$
\|\partial_3^\varepsilon \hat{w}^{\delta(0)\varepsilon}(x_3^\varepsilon)\|_{H^1(\omega^\varepsilon)} \leqslant h_2(x_3^\varepsilon), \tag{31.89}
$$

with $h_1 \in L^2(0, L)$ and $h_2 \in L^2(0, L)$ being independent of δ. Following the same method and notation as in the proof of Theorem 31.1, one obtains

$$
\begin{aligned}
& \hat{w}^{\delta(0)\varepsilon}(x_3^\varepsilon) \rightharpoonup w^{*(0)\varepsilon}(x_3^\varepsilon), \quad \text{weakly in } H^1(\omega^\varepsilon), \text{ for all } x_3^\varepsilon \in [0, L], \\
& \partial_3^\varepsilon \hat{w}^{\delta(0)\varepsilon}(x_3^\varepsilon) \rightharpoonup \zeta^{*(0)\varepsilon}(x_3^\varepsilon), \quad \text{weakly in } H^1(\omega^\varepsilon), \text{ a.e. } x_3^\varepsilon \in (0, L), \\
& [\nabla^\varepsilon w^{\delta(0)\varepsilon}(x_3^\varepsilon)]^{\sim} \rightharpoonup \sigma_w^{*(0)\varepsilon}(x_3^\varepsilon), \quad \text{weakly in } [L^2(\omega^\varepsilon)]^2, \text{ for all } x_3^\varepsilon \in [0, L], \\
& [\nabla^\varepsilon \partial_3^\varepsilon w^{\delta(0)\varepsilon}(x_3^\varepsilon)]^{\sim} \rightharpoonup \sigma_\zeta^{*(0)\varepsilon}(x_3^\varepsilon), \quad \text{weakly in } [L^2(\omega^\varepsilon)]^2, \text{ a.e. } x_3^\varepsilon \in (0, L),
\end{aligned}
$$

satisfying, for all $x_3^\varepsilon \in [0, L]$,

$$\begin{aligned}
&- \operatorname{div}^\varepsilon \sigma_w^{*(0)\varepsilon}(x_3^\varepsilon) \\
&\quad = \Theta f_3^\varepsilon(x_3^\varepsilon) - \frac{1}{|\omega^\varepsilon|}\left[\Theta \int_{\omega^\varepsilon} f_3^\varepsilon(x_3^\varepsilon)\,\mathrm{d}\omega^\varepsilon + \int_{\gamma^\varepsilon} g_3^\varepsilon(x_3^\varepsilon)\,\mathrm{d}\gamma^\varepsilon\right], \quad \text{in } \omega^\varepsilon, \\
&\sigma_w^{*(0)\varepsilon}(x_3^\varepsilon)\cdot \boldsymbol{n}^\varepsilon = g_3^\varepsilon(x_3^\varepsilon), \quad \text{on } \gamma^\varepsilon, \\
&\sigma_w^{*(0)\varepsilon}(x_3^\varepsilon) = A^*\nabla^\varepsilon w^{*(0)\varepsilon}(x_3^\varepsilon), \quad \text{in } \omega^\varepsilon, \\
&\int_{\omega^\varepsilon} w^{*(0)\varepsilon}(x_3^\varepsilon)\,\mathrm{d}\omega^\varepsilon = 0,
\end{aligned} \tag{31.90}$$

which is equivalent to problem (31.60) and a.e. in $x_3^\varepsilon \in (0, L)$,

$$\begin{aligned}
&- \operatorname{div}^\varepsilon \sigma_\zeta^{*(0)\varepsilon}(x_3^\varepsilon) \\
&\quad = \Theta \partial_3^\varepsilon f_3^\varepsilon(x_3^\varepsilon) - \frac{1}{|\omega^\varepsilon|}\left[\Theta \int_{\omega^\varepsilon} \partial_3^\varepsilon f_3^\varepsilon(x_3^\varepsilon)\,\mathrm{d}\omega^\varepsilon + \int_{\gamma^\varepsilon} \partial_3^\varepsilon g_3^\varepsilon(x_3^\varepsilon)\,\mathrm{d}\gamma^\varepsilon\right], \quad \text{in } \omega^\varepsilon, \\
&\sigma_\zeta^{*(0)\varepsilon}(x_3^\varepsilon)\cdot \boldsymbol{n}^\varepsilon = \partial_3^\varepsilon g_3^\varepsilon(x_3^\varepsilon), \quad \text{on } \gamma^\varepsilon, \\
&\sigma_\zeta^{*(0)\varepsilon}(x_3^\varepsilon) = A^*\nabla^\varepsilon \zeta^{*(0)\varepsilon}(x_3^\varepsilon), \quad \text{in } \omega^\varepsilon, \\
&\int_{\omega^\varepsilon} \zeta^{*(0)\varepsilon}(x_3^\varepsilon)\,\mathrm{d}\omega^\varepsilon = 0.
\end{aligned} \tag{31.91}$$

From (31.90), (31.91), the linearity of the problems and the regularity of the respective data one concludes that $w^{*(0)\varepsilon} \in H^1[0, L; H^1(\omega^\varepsilon)]$ and

$$\partial_3^\varepsilon w^{*(0)\varepsilon}(x_3^\varepsilon) = \zeta^{*(0)\varepsilon}(x_3^\varepsilon), \quad \text{a.e. } x_3^\varepsilon \in (0, L).$$

Summarizing: there exists an extension $\hat{w}^{\delta(0)\varepsilon} \in H^1[0, L; H^1(\omega^\varepsilon)]$ of $w^{\delta(0)\varepsilon}$, solution of problem (30.45) and such that the following weak convergences hold:

$$\hat{w}^{\delta(0)\varepsilon}(x_3^\varepsilon) \rightharpoonup w^{*(0)\varepsilon}(x_3^\varepsilon), \quad \text{in } H^1(\omega^\varepsilon), \text{ for all } x_3^\varepsilon \in [0, L], \tag{31.92}$$

$$\partial_3^\varepsilon \hat{w}^{\delta(0)\varepsilon}(x_3^\varepsilon) \rightharpoonup \partial_3^\varepsilon w^{*(0)\varepsilon}(x_3^\varepsilon), \quad \text{in } H^1(\omega^\varepsilon), \text{ a.e. } x_3^\varepsilon \in (0, L), \tag{31.93}$$

$$[\nabla^\varepsilon w^{\delta(0)\varepsilon}(x_3^\varepsilon)]^\sim \rightharpoonup A^*\nabla^\varepsilon w^{*(0)\varepsilon}(x_3^\varepsilon), \quad \text{in } [L^2(\omega^\varepsilon)]^2, \text{ for all } x_3^\varepsilon \in [0, L], \tag{31.94}$$

$$\begin{aligned}
&[\nabla^\varepsilon \partial_3^\varepsilon w^{\delta(0)\varepsilon}(x_3^\varepsilon)]^\sim \rightharpoonup A^*\nabla^\varepsilon \partial_3^\varepsilon w^{*(0)\varepsilon}(x_3^\varepsilon), \\
&\quad \text{in } [L^2(\omega^\varepsilon)]^2, \text{ a.e. } x_3^\varepsilon \in (0, L),
\end{aligned} \tag{31.95}$$

where $w^{*(0)\varepsilon}$ is the solution of problem (31.60). Furthermore, in view of (31.88) and (31.89), one has, by Lebesgue's theorem, and for all $\varphi \in H^1(\omega^\varepsilon)$,

$$\langle \hat{w}^{\delta(0)\varepsilon}(x_3^\varepsilon), \varphi\rangle_{H^1(\omega^\varepsilon)} \to \langle w^{*(0)\varepsilon}(x_3^\varepsilon), \varphi\rangle_{H^1(\omega^\varepsilon)}, \quad \text{strongly in } H^1(0,L). \tag{31.96}$$

From convergences (31.40), (31.81), (31.83), (31.92) and those of Theorem 31.1 and Corollary 31.1 it follows that the data of problem (30.27), (30.28) converges, in the appropriate spaces, to the data of problems (31.63), (31.64) and so,

$$z_\alpha^{\delta(2)\varepsilon} \to z_\alpha^{*(2)\varepsilon}, \quad \text{strongly in } H^2(0,L). \tag{31.97}$$

Convergences (31.97), (31.83), (31.81) and (31.40) imply convergence (31.53) and, together with (31.84), (31.96) and the convergences given in Theorem 31.1 and Corollary 31.1, one obtains (31.54).

Convergence (31.65) follows from (31.26) and (31.27), (31.83), (31.21) and (31.22), (31.23), (31.40), (31.95) and (31.96).

Finally, the convergence of $m_\beta^{\delta(0)\varepsilon}$ and of $q_\beta^{\delta(0)\varepsilon}$ are particular cases of convergences (31.48) and (31.65) with $\varphi = x_\beta^\varepsilon$ and $\varphi = 1$, respectively.

Step 3. *Proof of* (g), (h) *and* (i). The passage to the limit in the elasticity problem (30.34) applies essentially the same homogenization technique used in Theorem 31.1 and in Step 2, for problem (30.24).

From Lemma 31.4, there exists an extension operator $Q^{\delta,\varepsilon} : [H^1(\omega^{\delta,\varepsilon})]^2 \to [H^1(\omega^\varepsilon)]^2$ satisfying, for a constant C independent of δ and ε,

$$(Q^{\delta,\varepsilon}\boldsymbol{u})(\boldsymbol{x}^\varepsilon) = \boldsymbol{u}(\boldsymbol{x}^\varepsilon), \quad \text{a.e. } \boldsymbol{x}^\varepsilon \in \omega^{\delta,\varepsilon},$$
$$\|Q^{\delta,\varepsilon}\boldsymbol{u}\|_{L^2(\omega^\varepsilon)} \leqslant \|\boldsymbol{u}\|_{L^2(\omega^{\delta,\varepsilon})},$$
$$\|\boldsymbol{e}(Q^{\delta,\varepsilon}\boldsymbol{u})\|_{L^2(\omega^\varepsilon)} \leqslant \|\boldsymbol{e}(\boldsymbol{u})\|_{L^2(\omega^{\delta,\varepsilon})},$$

for all $\boldsymbol{u} \in [H^1(\omega^{\delta,\varepsilon})]^2$. For a.e. $x_3^\varepsilon \in (0,L)$, let $\underline{\hat{\boldsymbol{u}}}^{\delta(4)\varepsilon}(x_3^\varepsilon) = Q^{\delta,\varepsilon}(\underline{\boldsymbol{u}}^{\delta(4)\varepsilon}(x_3^\varepsilon))$; obviously $\underline{\hat{\boldsymbol{u}}}^{\delta(4)\varepsilon} \in L^2[0,L;H^1(\omega^\varepsilon)]$. From the variational setting of (30.34), from the properties of the extension operator and using Lemma 31.6 one obtains, a.e. $x_3^\varepsilon \in (0,L)$, the following estimate:

$$\begin{aligned}\|\underline{\hat{\boldsymbol{u}}}^{\delta(4)\varepsilon}(x_3^\varepsilon)\|_{H^1(\omega^\varepsilon)} \leqslant \overline{C}(&\|f_\beta^\varepsilon(x_3^\varepsilon)\|_{L^2(\omega^\varepsilon)} + \|g_\beta^\varepsilon(x_3^\varepsilon)\|_{L^2(\gamma^\varepsilon)} \\ &+ \|\partial_3^\varepsilon \tilde{\sigma}_{3\beta}^{\delta(0)\varepsilon}(x_3^\varepsilon)\|_{L^2(\omega^\varepsilon)} + \|\tilde{S}_{\alpha\beta}^{\delta(0)\varepsilon}(x_3^\varepsilon)\|_{L^2(\omega^\varepsilon)}),\end{aligned}$$

where $\overline{C}$ is a constant independent of δ. From the expressions of $\sigma_{3\beta}^{\delta(0)\varepsilon}$ and $S_{\alpha\beta}^{\delta(0)\varepsilon}$, from the definitions (30.13) of $u_\beta^{\delta(0)\varepsilon}$, from the definition (30.22), (30.23) of $z^{\delta(2)\varepsilon}$, from (31.88) and (31.89) one sees that $\|\partial_3^\varepsilon \tilde{\sigma}_{3\beta}^{\delta(0)\varepsilon}(x_3^\varepsilon)\|_{L^2(\omega^\varepsilon)}$ and that

$\|\tilde{S}_{\alpha\beta}^{\delta(0)\varepsilon}(x_3^\varepsilon)\|_{L^2(\omega^\varepsilon)}$ are bounded as $L^2(0,L)$ functions, independently of δ. Consequently, the following estimate holds, a.e. in $x_3^\varepsilon \in (0,L)$,

$$\|\underline{\hat{\boldsymbol{u}}}^{\delta(4)\varepsilon}\|_{H^1(\omega^\varepsilon)} \leqslant h(x_3^\varepsilon), \quad h \in L^2(0,L), \tag{31.98}$$

h being independent of δ. Up to a subsequence of δ one has, as $\delta \to 0$, a.e. in $x_3^\varepsilon \in (0,L)$ and for the weak topologies of $[H^1(\omega^\varepsilon)]^2$ and $L^2(\omega^\varepsilon)$, respectively,

$$\underline{\hat{\boldsymbol{u}}}^{\delta(4)\varepsilon}(x_3^\varepsilon) \rightharpoonup \underline{\boldsymbol{u}}^{*(4)\varepsilon}(x_3^\varepsilon), \tag{31.99}$$

$$\tilde{\sigma}_{\alpha\beta}^{\delta(0)\varepsilon}(x_3^\varepsilon) \rightharpoonup \sigma_{\alpha\beta}^{*(0)\varepsilon}(x_3^\varepsilon) = S_{\alpha\beta}^{*\varepsilon}(x_3^\varepsilon) + S_{\alpha\beta}^{*(0)\varepsilon}(x_3^\varepsilon), \tag{31.100}$$

with $S_{\alpha\beta}^{*(0)\varepsilon}(x_3^\varepsilon)$ given by (31.72) and $S_{\alpha\beta}^{*\varepsilon}(x_3^\varepsilon)$ satisfying a.e. in $x_3^\varepsilon \in (0,L)$, and for all $\boldsymbol{\varphi} \in [H^1(\omega^\varepsilon)]^2$,

$$\begin{aligned}
&\int_{\omega^\varepsilon} S_{\alpha\beta}^{*\varepsilon}(x_3^\varepsilon) e_{\alpha\beta}^\varepsilon(\boldsymbol{\varphi})\,\mathrm{d}\omega^\varepsilon \\
&\quad = \Theta \int_{\omega^\varepsilon} f_\beta^\varepsilon(x_3^\varepsilon)\varphi_\beta\,\mathrm{d}\omega^\varepsilon + \int_{\gamma^\varepsilon} g_\beta^\varepsilon(x_3^\varepsilon)\varphi_\beta\,\mathrm{d}\gamma^\varepsilon \\
&\qquad + \int_{\omega^\varepsilon} \partial_3^\varepsilon \sigma_{3\beta}^{*(0)\varepsilon}(x_3^\varepsilon)\varphi_\beta\,\mathrm{d}\omega^\varepsilon - \int_{\omega^\varepsilon} S_{\alpha\beta}^{*(0)\varepsilon}(x_3^\varepsilon) e_{\alpha\beta}^\varepsilon(\boldsymbol{\varphi})\,\mathrm{d}\omega^\varepsilon,
\end{aligned} \tag{31.101}$$

$$\int_{\omega^\varepsilon} \underline{u}_\alpha^{*(4)\varepsilon}(x_3^\varepsilon)\,\mathrm{d}\omega^\varepsilon = \int_{\omega^\varepsilon} [x_2^\varepsilon \underline{u}_1^{*(4)\varepsilon}(x_3^\varepsilon) - x_1^\varepsilon \underline{u}_2^{*(4)\varepsilon}(x_3^\varepsilon)]\,\mathrm{d}\omega^\varepsilon = 0.$$

To identify the relationship between $S_{\alpha\beta}^{*\varepsilon}(x_3^\varepsilon)$ and $\underline{\boldsymbol{u}}^{*(4)\varepsilon}(x_3^\varepsilon)$, consider $\hat{\chi}^{\zeta\rho} \in [H^1_{\mathrm{loc}}(\mathbb{R}^2)]^2$, Y-periodic, as the solution of problem (31.75) extended to the entire cell Y and periodically defined in all $\mathbb{R}^2$. For each δ and ε define

$$\begin{aligned}
&\Psi_{11}^{\delta,\varepsilon}(\boldsymbol{x}^\varepsilon) = (x_1^\varepsilon, 0) + \delta\varepsilon\hat{\chi}^{11}\Big(\frac{\boldsymbol{x}^\varepsilon}{\delta\varepsilon}\Big), \qquad &&\Psi_{22}^{\delta,\varepsilon}(\boldsymbol{x}^\varepsilon) = (0, x_2^\varepsilon) + \delta\varepsilon\hat{\chi}^{22}\Big(\frac{\boldsymbol{x}^\varepsilon}{\delta\varepsilon}\Big),\\
&\Psi_{12}^{\delta,\varepsilon}(\boldsymbol{x}^\varepsilon) = (x_2^\varepsilon, 0) + \delta\varepsilon\hat{\chi}^{12}\Big(\frac{\boldsymbol{x}^\varepsilon}{\delta\varepsilon}\Big), &&\Psi_{21}^{\delta,\varepsilon}(\boldsymbol{x}^\varepsilon) = (0, x_1^\varepsilon) + \delta\varepsilon\hat{\chi}^{21}\Big(\frac{\boldsymbol{x}^\varepsilon}{\delta\varepsilon}\Big).
\end{aligned}$$

By Lemma 31.1 it follows, that for each fixed ε and as $\delta \to 0$,

$$\left.\begin{aligned}
&\Psi_{11}^{\delta,\varepsilon}(\boldsymbol{x}^\varepsilon) \rightharpoonup (x_1^\varepsilon, 0), \qquad &&\Psi_{22}^{\delta,\varepsilon}(\boldsymbol{x}^\varepsilon) \rightharpoonup (0, x_2^\varepsilon),\\
&\Psi_{12}^{\delta,\varepsilon}(\boldsymbol{x}^\varepsilon) \rightharpoonup (x_2^\varepsilon, 0), &&\Psi_{21}^{\delta,\varepsilon}(\boldsymbol{x}^\varepsilon) \rightharpoonup (0, x_1^\varepsilon),
\end{aligned}\right\} \quad \text{weakly in } [H^1_{\mathrm{loc}}(\mathbb{R}^2)]^2. \tag{31.102}$$

Let $a_{\alpha\beta\zeta\rho} = \mu(\delta_{\alpha\zeta}\delta_{\beta\rho} + \delta_{\alpha\rho}\delta_{\beta\zeta}) + \lambda\delta_{\alpha\beta}\delta_{\zeta\rho}$ represent the coefficients of the operator $S_{\alpha\beta}^{\delta,\varepsilon}$, defined in (30.31)–(30.33), i.e., $S_{\alpha\beta}^{\delta,\varepsilon}(\boldsymbol{v}) = a_{\alpha\beta\zeta\rho}e_{\zeta\rho}^\varepsilon(\boldsymbol{v})$. One has

$$a_{\alpha\beta\zeta\rho} = a_{\zeta\rho\alpha\beta} = a_{\beta\alpha\zeta\rho}, \tag{31.103}$$

and the second expression in (31.75) reads

$$\int_{Y^*} a_{\alpha\beta\mu\lambda}[e_{\mu\lambda}(\boldsymbol{\chi}^{\zeta\rho}) + \delta_{\mu\rho}\delta_{\lambda\zeta}]\partial_\alpha \varphi_\beta \, \mathrm{d}Y^* = 0, \tag{31.104}$$

for all $\boldsymbol{\varphi} \in [H^1(Y^*)]^2$, $\boldsymbol{\varphi}$ Y–periodic, which implies that $\partial_\alpha\{\chi a_{\alpha\beta\mu\lambda}[e_{\mu\lambda}(\hat{\boldsymbol{\chi}}^{\zeta\rho}) + \delta_{\mu\rho}\delta_{\lambda\zeta}]\} = 0$ in ω, and consequently,

$$\partial^\varepsilon_\alpha[\chi^{\delta,\varepsilon} S^{\delta,\varepsilon}_{\alpha\beta}(\boldsymbol{\Psi}^{\delta,\varepsilon}_{\zeta\rho})] = \partial^\varepsilon_\alpha\{\chi^{\delta,\varepsilon} a_{\alpha\beta\mu\lambda}[e_{\mu\lambda}(\hat{\boldsymbol{\chi}}^{\zeta\rho})(\frac{\boldsymbol{x}^\varepsilon}{\delta\varepsilon}) + \delta_{\mu\rho}\delta_{\lambda\zeta}]\} = 0, \quad \text{in } \omega^\varepsilon. \tag{31.105}$$

From Lemma 31.1 the following convergence holds:

$$\chi^{\delta,\varepsilon} S^{\delta,\varepsilon}_{\alpha\beta}(\boldsymbol{\Psi}^{\delta,\varepsilon}_{\zeta\rho}) \rightarrow a^h_{\beta\alpha\zeta\rho}, \quad \text{weakly in } L^2(\omega^\varepsilon),$$

with

$$a^h_{\beta\alpha\zeta\rho} = a_{\alpha\beta\mu\lambda} \int_{Y^*} e_{\mu\lambda}(\hat{\boldsymbol{\chi}}^{\zeta\rho}) \, \mathrm{d}Y^* + a_{\beta\alpha\zeta\rho}\Theta. \tag{31.106}$$

From (30.31) and (31.103) one obtains, a.e. in $x^\varepsilon_3 \in (0, L)$,

$$[\tilde{\sigma}^{\delta(0)\varepsilon}_{\alpha\beta}(x^\varepsilon_3) - \tilde{S}^{\delta(0)\varepsilon}_{\alpha\beta}(x^\varepsilon_3)]\partial^\varepsilon_\alpha(\boldsymbol{\Psi}^{\delta,\varepsilon}_{\zeta\rho})_\beta = \partial^\varepsilon_\alpha \underline{\hat{u}}^{\delta(4)\varepsilon}_\beta \chi^{\delta,\varepsilon} S^{\delta,\varepsilon}_{\alpha\beta}(\boldsymbol{\Psi}^{\delta,\varepsilon}_{\zeta\rho}), \quad \text{a.e. in } \omega^\varepsilon, \tag{31.107}$$

and also from (30.31), (30.34) and (31.65)

$$\begin{aligned} &-\partial^\varepsilon_\alpha[\tilde{\sigma}^{\delta(0)\varepsilon}_{\alpha\beta}(x^\varepsilon_3) - \tilde{S}^{\delta(0)\varepsilon}_{\alpha\beta}(x^\varepsilon_3)] \\ &\quad = \chi^{\delta,\varepsilon} f^\varepsilon_\beta(x^\varepsilon_3) + \partial^\varepsilon_3 \tilde{\sigma}^{\delta(0)\varepsilon}_{3\beta}(x^\varepsilon_3) \rightarrow \Theta f^\varepsilon_\beta(x^\varepsilon_3) + \partial^\varepsilon_3 \tilde{\sigma}^{*(0)\varepsilon}_{3\beta}(x^\varepsilon_3), \end{aligned} \tag{31.108}$$

weakly in $L^2(\omega^\varepsilon)$ and, consequently, strongly in $H^{-1}(\omega^\varepsilon)$, as $\delta \rightarrow 0$.

Since by (31.105) and (31.108), the conditions of Lemma 31.2 are satisfied, passing to the limit in (31.107), one obtains, in view of (31.99) and (31.102),

$$\sigma^{*(0)\varepsilon}_{\zeta\rho}(x^\varepsilon_3) - S^{*(0)\varepsilon}_{\zeta\rho}(x^\varepsilon_3) = \partial^\varepsilon_\alpha \underline{u}^{*(4)\varepsilon}_\beta(x^\varepsilon_3) a^h_{\alpha\beta\zeta\rho}. \tag{31.109}$$

The homogenized tensor $(a^h_{\alpha\beta\zeta\rho})$ is symmetric and positive definite. In fact, if we consider $\varphi = \boldsymbol{\chi}^{\overline{\zeta\rho}}$ in the variational formulation (31.104), the following equality holds:

$$\int_{Y^*} a_{\alpha\beta\lambda\mu}[e_{\lambda\mu}(\boldsymbol{\chi}^{\zeta\rho}) + \delta_{\mu\rho}\delta_{\lambda\zeta}][e_{\alpha\beta}(\boldsymbol{\chi}^{\overline{\zeta\rho}}) + \delta_{\alpha\bar{\rho}}\delta_{\beta\bar{\zeta}}] \, \mathrm{d}Y^* = a^h_{\bar{\rho}\bar{\zeta}\rho\zeta}, \tag{31.110}$$

which implies that

$$a^h_{\alpha\beta\zeta\rho} = a^h_{\zeta\rho\alpha\beta} = a^h_{\beta\alpha\zeta\rho}. \tag{31.111}$$

From (31.110) one also obtains that

$$a^h_{\overline{\rho\zeta\rho\zeta}}\tau_{\overline{\rho\zeta}}\tau_{\rho\zeta} = \int_{Y^*} a_{\alpha\beta\lambda\mu}[e_{\mu\lambda}(\tau_{\zeta\rho}\chi^{\zeta\rho}) + \tau_{\mu\lambda}][e_{\alpha\beta}(\tau_{\zeta\rho}\chi^{\zeta\rho}) + \tau_{\alpha\beta}]\,\mathrm{d}Y^*, \tag{31.112}$$

for $\boldsymbol{\tau} = (\tau_{\alpha\beta})$ symmetric. From (31.112) and the fact that $a_{\alpha\beta\mu\lambda}\tau_{\alpha\beta}\tau_{\mu\lambda} \geqslant 0$, for all $\boldsymbol{\tau}$ symmetric, and $a_{\alpha\beta\mu\lambda}\tau_{\alpha\beta}\tau_{\mu\lambda} = 0$ if and only if $\boldsymbol{\tau} = \mathbf{0}$, the same property holds for $a^h_{\overline{\rho\zeta\rho\zeta}}$. In fact, if $a^h_{\overline{\rho\zeta\rho\zeta}}\tau_{\overline{\rho\zeta}}\tau_{\rho\zeta} = 0$ one has $\boldsymbol{e}[\tau_{\zeta\rho}(\boldsymbol{\chi}^{\zeta\rho} + \boldsymbol{P}^{\zeta\rho})] = (e_{\alpha\beta}(\tau_{\zeta\rho}\boldsymbol{\chi}^{\zeta\rho}) + \tau_{\alpha\beta}) = \mathbf{0}$, in Y^*, meaning by $\boldsymbol{P}^{\zeta\rho}$ the vector whose α-component is $P^{\zeta\rho}_\alpha = y_\rho\delta_{\zeta\alpha}$. The fact that $\boldsymbol{e}[\tau_{\zeta\rho}(\boldsymbol{\chi}^{\zeta\rho} + \boldsymbol{P}^{\zeta\rho})] = \mathbf{0}$ implies that $\tau_{\zeta\rho}\boldsymbol{\chi}^{\zeta\rho} = -\tau_{\zeta\rho}\boldsymbol{P}^{\zeta\rho} + \mathcal{R}$, where $\mathcal{R}$ is a rigid body displacement. Using the periodicity of the first member of the last equality one obtains $\boldsymbol{\tau} = \mathbf{0}$.

Expression (31.72) follows from (31.111), (31.106) and definition (31.74). The coercivity of coefficients $a^h_{\alpha\beta\mu\lambda}$ guarantee the uniqueness of the solution $\underline{\boldsymbol{u}}^{*(4)\varepsilon}$ of problem (31.73), which implies the convergence of all the sequence δ, in (31.99) and (31.100).

Convergence (31.71) follows from (31.100), (30.31) and (31.98), with the help of Lebesgue's theorem.

Convergence (31.76) is an immediate consequence of (31.71) and (31.54), and it implies the convergence of the second order bending moment, stated in (i). □

32. Generalized beam theory of the homogenized material

This section is devoted to the behaviour of the homogenized material concerning the first order stretching–bending theories. First, the homogenized material is determined (Theorem 32.1). This corresponds to taking the limit as $\delta \to 0$. The homogenized material is independent of the domain and therefore it is independent of ε. Afterwards, the three-dimensional homogenized problem in Ω^ε is considered and, using the change of variables (2.11)–(2.13), transposed to the fixed domain Ω. Letting $\varepsilon \to 0$ and using the inverse scalings to return to Ω^ε, one obtains characterizations (31.41), (31.42), (31.45), (31.46), (31.47) and (31.49). The results presented in this section are owed to MASCARENHAS and TRABUCHO [1990].

Consider the three-dimensional linearized elasticity problem (1.9), (1.11) with the weakly clamping boundary condition (5.1), instead of (1.10). In order to present the corresponding homogenized problem (Theorem 32.1) one needs the following lemmas generalizing to the three-dimensional case Lemmas 31.4 and 31.6, respectively.

LEMMA 32.1. *Let Ω^ε and $\Omega^{\delta,\varepsilon}$ be defined as in Sections* 1 *and* 29, *respectively. For each δ and ε there exists a linear operator $\Pi^{\delta,\varepsilon}$, from $[H^1(\Omega^{\delta,\varepsilon})]^3$ into $H^1(\Omega^\varepsilon)]^3$ and a constant C, independent of δ and ε, such that, for all $\boldsymbol{u} \in [H^1(\Omega^{\delta,\varepsilon})]^3$, one has*

$$\Pi^{\delta,\varepsilon}\boldsymbol{u} = \boldsymbol{u}, \quad \text{a.e. in } \Omega^{\delta,\varepsilon}, \tag{32.1}$$
$$\|\Pi^{\delta,\varepsilon}\boldsymbol{u}\|_{L^2(\Omega^\varepsilon)} \leqslant C\|\boldsymbol{u}\|_{L^2(\Omega^{\delta,\varepsilon})}, \tag{32.2}$$
$$\|\boldsymbol{e}^\varepsilon(\Pi^{\delta,\varepsilon}\boldsymbol{u})\|_{L^2(\Omega^\varepsilon)} \leqslant C\|\boldsymbol{e}^\varepsilon(\boldsymbol{u})\|_{L^2(\Omega^{\delta,\varepsilon})}. \tag{32.3}$$

PROOF. *Step* 1. Consider the unit cells $Z = Y \times [0,1]$ and $Z^* = Y^* \times [0,1]$, where Y and Y^* are defined in Section 31. In this step we prove the existence of a linear operator $\Pi : [H^1(Z^*)]^3 \to [H^1(Z)]^3$ and of a constant C, satisfying

$$\Pi\boldsymbol{u} = \boldsymbol{u}, \quad \text{a.e. in } Z^*, \tag{32.4}$$
$$\|\Pi\boldsymbol{u}\|_{L^2(Z)} \leqslant C\|\boldsymbol{u}\|_{L^2(Z^*)}, \tag{32.5}$$
$$\|\boldsymbol{e}(\Pi\boldsymbol{u})\|_{L^2(Z)} \leqslant C\|\boldsymbol{e}(\boldsymbol{u})\|_{L^2(Z^*)}. \tag{32.6}$$

If $\boldsymbol{u}(\boldsymbol{y},0) = \boldsymbol{v}(\boldsymbol{y},1)$, a.e. in $\boldsymbol{y} \in Y^*$ then

$$(\Pi\boldsymbol{u})(\boldsymbol{y},0) = (\Pi\boldsymbol{v})(\boldsymbol{y},1), \quad \text{a.e. in } \boldsymbol{y} \in Y^*, \text{ for all } \boldsymbol{u} \in [H^1(Z^*)]^3. \tag{32.7}$$

Consider a linear operator $Q : [H^1(Y^*)]^2 \to [H^1(Y)]^2$, satisfying, for a constant C_1 and for all $\boldsymbol{v} \in [H^1(Y^*)]^2$,

$$Q\boldsymbol{v} = \boldsymbol{v}, \quad \text{a.e. in } Y^*, \tag{32.8}$$
$$\|Q\boldsymbol{v}\|_{L^2(Y)} \leqslant C_1\|\boldsymbol{v}\|_{L^2(Y^*)}, \tag{32.9}$$
$$\|\boldsymbol{e}(Q\boldsymbol{v})\|_{L^2(Y)} \leqslant C_1\|\boldsymbol{e}(\boldsymbol{v})\|_{L^2(Y^*)}. \tag{32.10}$$

(For the construction of such an operator see Lemma 31.4.) Consider a linear operator $P : H^1(Y^*) \to H^1(Y)$, satisfying, for all $w \in H^1(Y^*)$ and for a constant C_2,

$$Pw = w, \quad \text{a.e. in } Y^*, \tag{32.11}$$
$$\|Pw\|_{L^2(Y)} \leqslant C_2\|w\|_{L^2(Y^*)}, \tag{32.12}$$
$$\|Pw\|_{H^1(Y)} \leqslant C_2\|w\|_{H^1(Y^*)}. \tag{32.13}$$
$$P(a + by_1 + cy_2) = a + by_1 + cy_2, \quad \text{with } a,\, b,\, c \in \mathbb{R}. \tag{32.14}$$

We construct such an operator by taking an extension operator P_0 satisfying (32.11)–(32.13) (see BREZIS [1983]); by considering for each $w \in H^1(Y^*)$, its L^2-projection w_F, in the finite dimensional subspace F of all functions of the form $a + by_1 + cy_2$ and by defining

$$P(w) = P_0(w - w_F) + w_F. \tag{32.15}$$

For each $\boldsymbol{u} = (u_1, u_2, u_3) \in [H^1(Z^*)]^3$ and $\lambda \in [0,1]$ let $w_\lambda(\boldsymbol{y}) = u_3(\boldsymbol{y}, \lambda)$ and $\boldsymbol{v}_\lambda(\boldsymbol{y}) = (u_1(\boldsymbol{y}, \lambda), u_2(\boldsymbol{y}, \lambda))$. Since $w_\lambda \in H^1(Y^*)$ and $\boldsymbol{v}_\lambda \in [H^1(Y^*)]^2$, define

$$(\Pi \boldsymbol{u})(\boldsymbol{y}, \lambda) = [(Q\boldsymbol{v}_\lambda)(\boldsymbol{y}), (Pw_\lambda)(\boldsymbol{y})]. \tag{32.16}$$

From (32.9), (32.10), (32.12) and (32.13) one has

$$\|(\Pi \boldsymbol{u})(\cdot, \lambda)\|_{L^2(Y)} \leqslant C_3 \|\boldsymbol{u}(\cdot, \lambda)\|_{L^2(Y^*)}, \tag{32.17}$$

$$\|(\Pi \boldsymbol{u})(\cdot, \lambda)\|_{H^1(Y)} \leqslant C_3 \|\boldsymbol{u}(\cdot, \lambda)\|_{H^1(Y^*)}. \tag{32.18}$$

By integrating (32.18), in λ, one obtains $\Pi\boldsymbol{u} \in L^2[0,1; [H^1(Y)]^3]$. Since $\boldsymbol{u}$ is absolutely continuous in λ, with values in $[L^2(Y^*)]^3$, from (32.16), (32.17) and the linearity of Q and P, one concludes that the same holds for $\Pi\boldsymbol{u}$ and, consequently, that $\Pi\boldsymbol{u} \in H^1[0,1; [L^2(Y)]^3]$, with

$$\|\partial_\lambda(\Pi \boldsymbol{u})(\cdot, \lambda)\|_{L^2(Y)} \leqslant C_3 \|\partial_\lambda \boldsymbol{u}(\cdot, \lambda)\|_{L^2(Y^*)}. \tag{32.19}$$

Then the operator Π is linear and continuous from $[H^1(Z^*)]^3$ into $[H^1(Z)]^3$ and satisfies (32.4), (32.5) and (32.7). Π also satisfies (32.6) since by definition the image of a rigid displacement is a rigid displacement (see (32.16), (32.10), (32.15)); in fact, representing by $\boldsymbol{u}_R$ the L^2-projection of $\boldsymbol{u}$ on the space of rigid displacements, one has, for a constant C_4,

$$\begin{aligned}\|\boldsymbol{e}(\Pi\boldsymbol{u})\|_{L^2(Z)} &= \|\boldsymbol{e}(\Pi(\boldsymbol{u} - \boldsymbol{u}_R))\|_{L^2(Z)} \\ &\leqslant \|\Pi(\boldsymbol{u} - \boldsymbol{u}_R)\|_{H^1(Z)} \leqslant C_4 \|\boldsymbol{u} - \boldsymbol{u}_R\|_{H^1(Z^*)},\end{aligned}$$

and, using Korn's inequality to estimate the last member,

$$\|\boldsymbol{u} - \boldsymbol{u}_R\|_{H^1(Z^*)} \leqslant C_5 \|\boldsymbol{e}(\boldsymbol{u})\|_{L^2(Z^*)},$$

for a constant C_5, which implies (32.6).

Step 2. Let $\eta = \delta\varepsilon$. To simplify we consider that $\Omega^{\delta,\varepsilon}$ is the union of cells $\eta Z^* + C_k^\eta$, $k = 1, \ldots, N(\eta)$, obtained by homothety and translation of the unit cell Z^*. Let $\boldsymbol{u} \in [H^1(\Omega^{\delta,\varepsilon})]^3$ and consider, for each k, $\boldsymbol{v} \in [H^1(Z^*)]^3$ given by $\boldsymbol{v}(\boldsymbol{y}) = \boldsymbol{u}(\eta\boldsymbol{y} + C_k^\eta)$. Define

$$(\Pi^{\delta,\varepsilon}\boldsymbol{u})(\boldsymbol{x}^\varepsilon) = (\Pi\boldsymbol{v})\left(\frac{1}{\eta}(\boldsymbol{x}^\varepsilon - C_k^\eta)\right), \quad \text{for } \boldsymbol{x}^\varepsilon \in \eta Z + C_k^\eta.$$

Since Ω^ε is the union of all cells $\eta Z + C_k^\eta$, $k = 1, \ldots, N(\eta)$, and Π satisfies (32.7), $\Pi^{\delta,\varepsilon} \in [H^1(\Omega^\varepsilon)]^3$ and satisfies (32.1). On the other hand,

$$\begin{aligned}\|\boldsymbol{e}^\varepsilon(\Pi^{\delta,\varepsilon}(\boldsymbol{u}))\|^2_{L^2(\Omega^\varepsilon)} &= \sum_{k=1}^{N(\eta)} \int_{\eta Z + C_k^\eta} \sum_{ij} |e_{ij}^\varepsilon(\Pi^{\delta,\varepsilon}\boldsymbol{u})|^2 \, \mathrm{d}\boldsymbol{x}^\varepsilon \\ &= \sum_{k=1}^{N(\eta)} \eta^{-2} \int_{\eta Z + C_k^\eta} \sum_{ij} |e_{ij}^\varepsilon(\Pi\boldsymbol{v})|^2 \, \mathrm{d}\boldsymbol{x}^\varepsilon \\ &= \sum_{k=1}^{N(\eta)} \eta^{-2}\eta^3 \int_Z \sum_{ij} |e_{ij}^\varepsilon(\Pi\boldsymbol{v})|^2 \, \mathrm{d}\boldsymbol{y}.\end{aligned} \tag{32.20}$$

Using estimates (32.6) one obtains, from (32.20),

$$\|e^{\varepsilon}(\Pi^{\delta,\varepsilon}u)\|^2_{L^2(\Omega^\varepsilon)} \leqslant \sum_{k=1}^{N(\eta)} \eta^{-2}\eta^3 C^2 \int_{Z^*} \sum_{ij} |e_{ij}(v)|^2\, dy$$
$$= C^2 \|e^{\varepsilon}(u)\|^2_{L^2(\Omega^{\delta,\varepsilon})},$$

i.e., estimate (32.3). Analogously one proves (32.2). □

LEMMA 32.2. *Let $\Omega^{\delta,\varepsilon}$, $\Gamma_0^{\delta,\varepsilon}$, $\Gamma_L^{\delta,\varepsilon}$, $V(\Omega^\varepsilon)$ and $V(\Omega^{\delta,\varepsilon})$ be defined as in Section 30. There exists a constant C, independent of δ, such that, for all $v \in V(\Omega^{\delta,\varepsilon})$,*

$$\|v\|_{H^1(\Omega^{\delta,\varepsilon})} \leqslant C\|e^{\varepsilon}(v)\|_{L^2(\Omega^{\delta,\varepsilon})}. \tag{32.21}$$

The proof of this lemma is quite similar to those of Lemma 31.3 and Lemma 31.4.

LEMMA 32.3. *Consider the spaces $V(\Omega^\varepsilon)$ and $V(\Omega^{\delta,\varepsilon})$ Then, for all $v \in V(\Omega^\varepsilon)$, there exists a sequence $(v^\delta)_{\delta\to 0}$, with $v^\delta \in V(\Omega^{\delta,\varepsilon})$ converging to v in $[H^1(\Omega^\varepsilon)]^3$.*

PROOF. Defining

$$a^{\delta,\varepsilon}(x_3^\varepsilon) = \frac{1}{L|\omega^{\delta,\varepsilon}|}\left[(L - x_3^\varepsilon)\int_{\omega^{\delta,\varepsilon}} v(0)\, d\omega^{\delta,\varepsilon} + x_3^\varepsilon \int_{\omega^{\delta,\varepsilon}} v(L)\, d\omega^{\delta,\varepsilon}\right],$$

$$b_1^{\delta,\varepsilon}(x_3^\varepsilon) = \frac{1}{LI_2^{\delta,\varepsilon}}\left[(L - x_3^\varepsilon)\int_{\omega^{\delta,\varepsilon}} v_3(0)x_2^\varepsilon\, d\omega^{\delta,\varepsilon} + x_3^\varepsilon \int_{\omega^{\delta,\varepsilon}} v_3(L)x_2^\varepsilon\, d\omega^{\delta,\varepsilon}\right],$$

$$b_2^{\delta,\varepsilon}(x_3^\varepsilon) = -\frac{1}{LI_1^{\delta,\varepsilon}}\left[(L - x_3^\varepsilon)\int_{\omega^{\delta,\varepsilon}} v_3(0)x_1^\varepsilon\, d\omega^{\delta,\varepsilon} + x_3^\varepsilon \int_{\omega^{\delta,\varepsilon}} v_3(L)x_1^\varepsilon\, d\omega^{\delta,\varepsilon}\right],$$

$$b_3^{\delta,\varepsilon}(x_3^\varepsilon) = -\left[(L - x_3^\varepsilon)\int_{\omega^{\delta,\varepsilon}} (v_1(0)x_2^\varepsilon - v_2(0)x_1^\varepsilon)\, d\omega^{\delta,\varepsilon}\right.$$
$$\left. + \frac{1}{L(I_1^{\delta,\varepsilon} + I_2^{\delta,\varepsilon})} x_3^\varepsilon \int_{\omega^{\delta,\varepsilon}} (v_1(L)x_2^\varepsilon - v_2(L)x_1^\varepsilon)\, d\omega^{\delta,\varepsilon}\right],$$

$$c^{\delta,\varepsilon}(x_3^\varepsilon) = (-b_3^{\delta,\varepsilon}(x_3^\varepsilon)x_2^\varepsilon, b_3^{\delta,\varepsilon}(x_3^\varepsilon)x_1^\varepsilon, b_1^{\delta,\varepsilon}(x_3^\varepsilon)x_2^\varepsilon - b_2^{\delta,\varepsilon}(x_3^\varepsilon)x_1^\varepsilon),$$

and

$$v^{\delta,\varepsilon} = v - (a^{\delta,\varepsilon} + c^{\delta,\varepsilon}),$$

one obtains that $\boldsymbol{v}^{\delta,\varepsilon} \in V(\Omega^{\delta,\varepsilon})$ and since $\boldsymbol{v} \in V(\Omega^{\varepsilon})$, $\boldsymbol{a}^{\delta,\varepsilon}$, $\boldsymbol{b}^{\delta,\varepsilon}$ converge to zero in $H^1(0,L)$, as $\delta \to 0$, which implies the convergence of $\boldsymbol{v}^{\delta,\varepsilon}$ to $\boldsymbol{v}$ in $[H^1(\Omega^{\varepsilon})]^3$ as $\delta \to 0$. □

THEOREM 32.1. *Let $\boldsymbol{f}^{\varepsilon} \in [L^2(\Omega^{\varepsilon})]^3$ and $\boldsymbol{g}^{\varepsilon} \in [L^2(\Gamma^{\varepsilon})]^3$. The solution $\boldsymbol{u}^{\delta,\varepsilon} \in V(\Omega^{\delta,\varepsilon})$ of problem* (30.1) *where the volumic forces are the restrictions of $\boldsymbol{f}^{\varepsilon}$ to $\Omega^{\delta,\varepsilon}$ and where the surface forces coincide in Γ^{ε} with $\boldsymbol{g}^{\varepsilon}$, admits an extension in $[H^1(\Omega^{\varepsilon})]^3$, bounded independently of δ. If $\hat{\boldsymbol{u}}^{\delta,\varepsilon}$ is any extension of $\boldsymbol{u}^{\delta,\varepsilon}$, bounded in $[H^1(\Omega^{\varepsilon})]^3$, independently of δ, and $\hat{\boldsymbol{\sigma}}^{\delta,\varepsilon}$ is the zero extension to Ω^{ε} of $\boldsymbol{\sigma}^{\delta,\varepsilon} \in \Sigma^{\delta,\varepsilon}$, then, for each fixed ε and as $\delta \to 0$, the following weak convergences hold in $[H^1(\Omega^{\varepsilon})]^3$ and $[L^2(\Omega^{\varepsilon})]^9$, respectively:*

$$\begin{aligned} &\hat{\boldsymbol{u}}^{\delta,\varepsilon} \rightharpoonup \boldsymbol{u}^{*,\varepsilon}, \\ &\hat{\boldsymbol{\sigma}}^{\delta,\varepsilon} \rightharpoonup \boldsymbol{\sigma}^{*,\varepsilon}, \end{aligned} \tag{32.22}$$

where $\boldsymbol{u}^{*,\varepsilon}$ is the unique solution of the following problem:

$$\boldsymbol{u}^{*,\varepsilon} \in V(\Omega^{\varepsilon}), \tag{32.23}$$

$$\int_{\Omega^{\varepsilon}} \sigma_{ij}^{*,\varepsilon} e_{ij}^{\varepsilon}(\boldsymbol{v})\, \mathrm{d}\boldsymbol{x}^{\varepsilon} = \Theta \int_{\Omega^{\varepsilon}} f_i^{\varepsilon} v_i \, \mathrm{d}\boldsymbol{x}^{\varepsilon} + \int_{\Gamma^{\varepsilon}} g_i^{\varepsilon} v_i \, \mathrm{d}a^{\varepsilon}, \tag{32.24}$$

for all $\boldsymbol{v} \in V(\Omega^{\varepsilon})$,

where $\boldsymbol{\sigma}_{ij}^{*,\varepsilon} \in \Sigma^{\varepsilon} = \{\boldsymbol{\tau} = (\tau_{ij}) \in [L^2(\Omega^{\varepsilon})]^9\colon\ \tau_{ij} = \tau_{ji}\}$ is given by

$$\begin{aligned} \boldsymbol{\sigma}_{ij}^{*,\varepsilon} = {}&2\mu[\Theta e_{ij}^{\varepsilon}(\boldsymbol{u}^{*,\varepsilon}) + q_{ijkl}^{*} e_{kl}(\boldsymbol{u}^{*,\varepsilon})] \\ &+ \lambda[\Theta e_{pp}^{\varepsilon}(\boldsymbol{u}^{*,\varepsilon}) + q_{ppkl}^{*} e_{kl}(\boldsymbol{u}^{*,\varepsilon})]\delta_{ij}, \end{aligned} \tag{32.25}$$

the coefficients q_{ijkl}^{*} being the constants, independent of ε, given by

$$q_{ijkl}^{*} = \int_{Y^*} e_{ij}(\underline{\boldsymbol{\chi}}^{kl}), \tag{32.26}$$

where $\underline{\boldsymbol{\chi}}^{kl} = \underline{\boldsymbol{\chi}}^{lk}$; k, $l \in \{1,2,3\}$, are defined in Y^* as follows:

(1) $\underline{\boldsymbol{\chi}}^{\alpha\beta}(\boldsymbol{y}) = (\chi_1^{\alpha\beta}(y_1,y_2), \chi_2^{\alpha\beta}(y_1,y_2), 0)$, *where $\boldsymbol{\chi}^{\alpha\beta} = (\chi_1^{\alpha\beta}, \chi_2^{\alpha\beta})$ is the solution, defined up to a constant vector, of problem* (31.75).

(2) $\underline{\boldsymbol{\chi}}^{33}(\boldsymbol{y}) = (\chi_1^{33}(y_1,y_2), \chi_2^{33}(y_1,y_2), 0)$, *where $\boldsymbol{\chi}^{33} = (\chi_1^{33}, \chi_2^{33})$ is the solution, defined up to a constant vector, of*

$$\boldsymbol{\chi}^{33} \in [H^1(Y^*)]^2,\quad Y\text{-periodic, and for all } \boldsymbol{\psi} \in [H^1(Y^*)]^2,\ Y\text{-periodic:}$$

$$\int_{Y^*} [2\mu e_{\alpha\beta}(\boldsymbol{\chi}^{33}) e_{\alpha\beta}(\boldsymbol{\psi}) + \lambda e_{\mu\mu}(\boldsymbol{\chi}^{33}) e_{\lambda\lambda}(\boldsymbol{\psi})]\, \mathrm{d}Y^* = -\lambda \int_{Y^*} e_{\rho\rho}(\boldsymbol{\psi})\, \mathrm{d}Y^*. \tag{32.27}$$

(3) $\underline{\boldsymbol{\chi}}^{3\beta}(y) = \underline{\boldsymbol{\chi}}^{\beta 3}(y) = (0, 0, \chi_\beta(y_1, y_2))$, *where* χ_β *is the solution of problem* (31.9).

REMARK 32.1. The homogenized coefficients

$$a^h_{ijkl} = \Theta a_{ijkl} + a_{ijmn} q^*_{mnkl}, \tag{32.28}$$

with

$$a_{ijkl} = \mu(\delta_{ik}\delta_{jl} + \delta_{il}\delta_{jk}) + \lambda \delta_{ij}\delta_{kl}, \tag{32.29}$$

satisfy

$$\begin{aligned} &a^h_{ijkl} = a^h_{jikl} = a^h_{klij}, \\ &a^h_{ijkl}\tau_{ij}\tau_{kl} \geqslant M \tau_{ij}\tau_{ij}, \end{aligned} \tag{32.30}$$

for a positive constant M and for all $\boldsymbol{\tau} = (\tau_{ij})$ symmetric.

Coefficients q^*_{ijkl} satisfy the following equalities (cf. (31.8) and (31.74)):

$$\begin{aligned} &q^*_{\alpha\beta\zeta\rho} = b^*_{\alpha\beta\zeta\rho}, \\ &q^*_{3\alpha 3\beta} = q^*_{3\beta 3\alpha} = q^*_{\alpha 33\beta} = q^*_{3\alpha\beta 3} = \tfrac{1}{2}(a^*_{\alpha\beta} - \Theta\delta_{\alpha\beta}), \end{aligned} \tag{32.31}$$

and exception made to

$$q^*_{\alpha\beta 33} = \left(\frac{\lambda}{2\mu}\right) b^*_{\rho\rho\alpha\beta} - \left[\frac{\lambda^2}{4\mu(\lambda + \mu)}\right] b^*_{\rho\rho\zeta\zeta}\delta_{\alpha\beta},$$

all the coefficients vanish.

PROOF. The homogenization process of the three-dimensional elasticity problem posed in $\Omega^{\delta,\varepsilon}$ is essentially the one explained in the proof of Theorem 31.2 Step 3, for the two dimensional elasticity problem (31.73). Consequently, we shall detail the technical differences only.

From the variational setting (30.1) one obtains the following estimate:

$$\|\boldsymbol{e}^\varepsilon(\boldsymbol{u}^{\delta,\varepsilon})\|_{L^2(\Omega^{\delta,\varepsilon})} \leqslant \overline{C}\|\boldsymbol{u}^{\delta,\varepsilon}\|_{H^1(\Omega^{\delta,\varepsilon})}, \tag{32.32}$$

where $\overline{C}$ is a constant independent of δ. Taking the boundary conditions of $\boldsymbol{u}^{\delta,\varepsilon}$ on $\Gamma_0^{\delta,\varepsilon} \cup \Gamma_L^{\delta,\varepsilon}$ into account, it follows from (32.32), from Lemma 32.2 and from the properties of the extension operator defined in Lemma 32.1, that $\Pi^{\delta,\varepsilon}\boldsymbol{u}^{\delta,\varepsilon}$ is bounded in $[H^1(\Omega^\varepsilon)]^3$, independently of δ.

From Hooke's law it follows that the zero extension $\tilde{\boldsymbol{\sigma}}^{\delta,\varepsilon}$ of $\boldsymbol{\sigma}^{\delta,\varepsilon}$ to $\Omega^{\delta,\varepsilon}$ is also bounded in $[L^2(\Omega^\varepsilon)]^9$, independently of δ. So, up to a subsequence of δ and for each fixed ε, one has

$$\begin{aligned} &\Pi^{\delta,\varepsilon}\boldsymbol{u}^{\delta,\varepsilon} \rightharpoonup \boldsymbol{u}^{*,\varepsilon} \in V(\Omega^\varepsilon), \quad \text{weakly in } [H^1(\Omega^\varepsilon)]^3, \\ &\tilde{\boldsymbol{\sigma}}^{\delta,\varepsilon} \rightharpoonup \boldsymbol{\sigma}^{*,\varepsilon} \in \boldsymbol{\Sigma}(\Omega^\varepsilon), \quad \text{weakly in } [L^2(\Omega^\varepsilon)]^9. \end{aligned} \tag{32.33}$$

For each $\boldsymbol{v}^\varepsilon \in V(\Omega^\varepsilon)$, let $\boldsymbol{v}^{\delta,\varepsilon} \in V(\Omega^{\delta,\varepsilon})$ be such that $\boldsymbol{v}^{\delta,\varepsilon} \to \boldsymbol{v}^\varepsilon$, strongly in $[H^1(\Omega^\varepsilon)]^3$ (see Lemma 32.3). From the variational formulation of the three-dimensional elasticity problem in $\Omega^{\delta,\varepsilon}$ one has, for each δ and ε

$$\int_{\Omega^\varepsilon} \tilde{\sigma}_{ij}^{\delta,\varepsilon} e_{ij}^\varepsilon(\boldsymbol{v}^{\delta,\varepsilon})\,\mathrm{d}\boldsymbol{x}^\varepsilon = \int_{\Omega^\varepsilon} \chi^{\delta,\varepsilon} f_i^\varepsilon v_i^{\delta,\varepsilon}\,\mathrm{d}\boldsymbol{x}^\varepsilon + \int_{\Gamma^\varepsilon} g_i^\varepsilon v_i^{\delta,\varepsilon}\,\mathrm{d}a^\varepsilon. \tag{32.34}$$

Passing to the limit, as $\delta \to 0$, one obtains

$$\int_{\Omega^\varepsilon} \sigma_{ij}^{*,\varepsilon} e_{ij}^\varepsilon(\boldsymbol{v}^\varepsilon)\,\mathrm{d}\boldsymbol{x}^\varepsilon = \Theta \int_{\Omega^\varepsilon} f_i^\varepsilon v_i^\varepsilon\,\mathrm{d}\boldsymbol{x}^\varepsilon + \int_{\Gamma^\varepsilon} g_i^\varepsilon v_i^\varepsilon\,\mathrm{d}a^\varepsilon, \tag{32.35}$$

valid for any $\boldsymbol{v}^\varepsilon \in V(\Omega^\varepsilon)$.

In order to identify the relationship between $\boldsymbol{\sigma}^{*,\varepsilon}$ and $\boldsymbol{u}^{*,\varepsilon}$ we proceed exactly as in the proof of Theorem 31.1 Step 3, but now for the three-dimensional case, and first find that (32.25), (32.26) hold for $\underline{\boldsymbol{\chi}}^{kl}$ solution of the following problem, defined up to a constant vector:

$$\underline{\boldsymbol{\chi}}^{kl} \in H^1(Q^*),\quad Q\text{-periodic},\quad \text{and for all } \Psi \in [H^1(Q^*)]^3,\ Q\text{-periodic},$$

$$\int_{Q^*} a_{ijmn}[e_{mn}(\underline{\boldsymbol{\chi}}^{kl}) + \delta_{ml}\delta_{nk}]\partial_i \Psi_j\,\mathrm{d}Q^* = 0, \tag{32.36}$$

where a_{ijmn} are given by (4.29) and $Q = Y \times [0,1]$, $Q^* = Y^* \times [0,1]$. It can be shown that the solution of problem (32.36), with $Q = Y \times [0,T]$, $Q^* = Y^* \times [0,T]$, for any $0 < T < 1$, coincides with $\underline{\boldsymbol{\chi}}^{kl}$, which implies that $\underline{\boldsymbol{\chi}}^{kl}$ is independent of y_3 and Y-periodic in the variable (y_1, y_2). Functions $\underline{\boldsymbol{\chi}}^{kl}$ can also be specified as stated in (1), (2) and (3) of Theorem 32.1 (for the proof see LÉNÉ [1984, Chapter 4, Theorem 4.1]). □

The present aim is to study the behaviour of the displacement field $\boldsymbol{u}^{*,\varepsilon}$ and of the stress field $\boldsymbol{\sigma}^{*,\varepsilon}$, defined by (32.23)–(32.25), as $\varepsilon \to 0$. As in Section 30 we use the change of variable (2.11)–(2.13) in order to pass to the fixed domain Ω.

Let $\boldsymbol{u}^*(\varepsilon) \in V(\Omega)$, $\boldsymbol{\sigma}^*(\varepsilon) \in \Sigma(\Omega)$, $\boldsymbol{f}(\varepsilon) \in [L^2(\Omega)]^3$ and $\boldsymbol{g}(\varepsilon) \in [L^2(\Gamma)]^3$ be obtained from $\boldsymbol{u}^{*,\varepsilon} \in V(\Omega^\varepsilon)$, $\boldsymbol{\sigma}^{*,\varepsilon} \in \boldsymbol{\Sigma}(\Omega^\varepsilon)$, $\boldsymbol{f}^\varepsilon \in [L^2(\Omega^\varepsilon)]^3$ and $\boldsymbol{g}^\varepsilon \in [L^2(\Gamma^\varepsilon)]^3$, respectively, satisfying Eqs. (32.24), (32.25), through transformations (2.11)–(2.13). Then $\boldsymbol{u}^*(\varepsilon)$, $\boldsymbol{\sigma}^*(\varepsilon)$, $\boldsymbol{f}(\varepsilon)$ and $\boldsymbol{g}(\varepsilon)$ satisfy the following equations:

$$\int_\Omega \sigma_{ij}^*(\varepsilon) e_{ij}(\boldsymbol{v})\,\mathrm{d}\boldsymbol{x} = \Theta \int_\Omega f_i(\varepsilon) v_i\,\mathrm{d}\boldsymbol{x} + \int_\Gamma g_i(\varepsilon) v_i\,\mathrm{d}a,$$

$$\text{for all } \boldsymbol{v} \in V(\Omega), \tag{32.37}$$

$$\sigma^*_{33}(\varepsilon) = [\Theta(\lambda+2\mu) + \lambda q^*_{\gamma\gamma 33}]e_{33}(\boldsymbol{u}^*(\varepsilon)) + \varepsilon^{-2}\lambda[\Theta\delta_{\alpha\beta} + q^*_{\gamma\gamma\alpha\beta}]e_{\alpha\beta}(\boldsymbol{u}^*(\varepsilon)), \tag{32.38}$$

$$\sigma^*_{3\beta}(\varepsilon) = \varepsilon^{-2}2\mu(\Theta\delta_{\alpha\beta} + 2q^*_{3\beta 3\alpha})e_{3\alpha}(\boldsymbol{u}^*(\varepsilon)), \tag{32.39}$$

$$\begin{aligned}\sigma^*_{\alpha\beta}(\varepsilon) &= \varepsilon^{-2}(\lambda\Theta\delta_{\alpha\beta} + \lambda q^*_{\gamma\gamma 33}\delta_{\alpha\beta} + 2\mu q^*_{\alpha\beta 33})e_{33}(\boldsymbol{u}^*(\varepsilon)) \\ &\quad + \varepsilon^{-4}[2\mu\Theta e_{\alpha\beta}(\boldsymbol{u}^*(\varepsilon)) + 2\mu q^*_{\alpha\beta\zeta\rho}e_{\zeta\rho}(\boldsymbol{u}^*(\varepsilon)) + \lambda\Theta e_{\gamma\gamma}(\boldsymbol{u}^*(\varepsilon))\delta_{\alpha\beta} \\ &\quad + \lambda q^*_{\gamma\gamma\zeta\rho}e_{\zeta\rho}(\boldsymbol{u}^*(\varepsilon))\delta_{\alpha\beta}].\end{aligned} \tag{32.40}$$

THEOREM 32.2. *Let $\boldsymbol{f}(\varepsilon)$ and $\mathbf{g}(\varepsilon)$ be bounded, independently of ε, in $[L^2(\Omega)]^3$ and in $[L^2(\Gamma)]^3$, respectively, then, up to a subsequence of ε, the following convergences hold:*

$$\boldsymbol{u}^*(\varepsilon) \rightharpoonup \boldsymbol{u}^*, \quad \textit{weakly in } [H^1(\Omega)]^3, \tag{32.41}$$

where $\boldsymbol{u}^* \in V_{\mathrm{BN}}(\Omega)$,

$$\varepsilon^{-1}e_{3\beta}(\boldsymbol{u}^*(\varepsilon)) \rightharpoonup \tau_{3\beta}, \quad \textit{weakly in } L^2(\Omega), \tag{32.42}$$

$$\varepsilon^{-2}e_{\alpha\beta}(\boldsymbol{u}^*(\varepsilon)) \rightharpoonup \tau_{\alpha\beta} = e_{\alpha\beta}(\boldsymbol{w}), \quad \textit{weakly in } L^2(\Omega), \tag{32.43}$$

where $\boldsymbol{w} \in L^2[0,L;(H^1(\omega))^3]$.

PROOF. From (32.37)–(32.40) one obtains

$$\int_\Omega \sigma^*_{ij}(\varepsilon)e_{ij}(\boldsymbol{u}^*(\varepsilon)) = \int_\Omega 2\mu[\Theta\tau_{ij}(\varepsilon)\tau_{ij}(\varepsilon) + q^*_{ijkl}\tau_{kl}(\varepsilon)\tau_{ij}(\varepsilon)]\,\mathrm{d}\boldsymbol{x} + \int_\Omega \lambda[\Theta\tau_{pp}(\varepsilon)\tau_{rr}(\varepsilon) + q^*_{ppkl}\tau_{kl}(\varepsilon)\tau_{rr}(\varepsilon)]\,\mathrm{d}\boldsymbol{x}, \tag{32.44}$$

where $\tau_{\alpha\beta}(\varepsilon) = \varepsilon^{-2}e_{\alpha\beta}(\boldsymbol{u}^*(\varepsilon))$, $\tau_{3\beta}(\varepsilon) = \varepsilon^{-1}e_{3\beta}(\boldsymbol{u}^*(\varepsilon))$, and $\tau_{33}(\varepsilon) = e_{33}(\boldsymbol{u}^*(\varepsilon))$. From (32.37), (32.44), the coercivity of the homogenized coefficients (see Remark 32.1) and Korn's inequality, one obtains, for a constant C, independent of ε,

$$\begin{aligned}&\varepsilon^{-2}\|e_{\alpha\beta}(\boldsymbol{u}^*(\varepsilon))\|_{L^2(\Omega)} \leqslant C, \\ &\varepsilon^{-1}\|e_{3\beta}(\boldsymbol{u}^*(\varepsilon))\|_{L^2(\Omega)} \leqslant C, \\ &\|e_{33}(\boldsymbol{u}^*(\varepsilon))\|_{L^2(\Omega)})^2 \leqslant C.\end{aligned} \tag{32.45}$$

As a consequence there exists a subsequence of ε, $\boldsymbol{u}^* \in V(\Omega)$, and $\tau_{3\beta}, \tau_{\alpha\beta} \in L^2(\Omega)$, such that

$$\begin{aligned}&\boldsymbol{u}^*(\varepsilon) \rightharpoonup \boldsymbol{u}^*, \quad \text{weakly in } [H^1(\Omega)]^3, \\ &\varepsilon^{-1}e_{3\beta}(\boldsymbol{u}^*(\varepsilon)) \rightharpoonup \tau_{3\beta}, \quad \text{weakly in } L^2(\Omega), \\ &\varepsilon^{-2}e_{\alpha\beta}(\boldsymbol{u}^*(\varepsilon)) \rightharpoonup \tau_{\alpha\beta}, \quad \text{weakly in } L^2(\Omega).\end{aligned}$$

Since $e_{\alpha\beta}(\boldsymbol{u}^*) = e_{3\beta}(\boldsymbol{u}^*) = 0$, one has $\boldsymbol{u}^* \in V_{\mathrm{BN}}(\Omega)$. Since there exists $\boldsymbol{w}(\varepsilon) \in L^2[0, L; (H^1(\omega))^3]$, $\boldsymbol{w}(\varepsilon) = (w_1(\varepsilon), w_2(\varepsilon), 0)$, $e_{\alpha\beta}(\boldsymbol{w}(\varepsilon)) = e_{\alpha\beta}(\varepsilon^{-2}\boldsymbol{u}^*(\varepsilon))$ and $\int_\omega w_\alpha(\varepsilon)\,\mathrm{d}\omega = \int_\omega (x_2 w_1(\varepsilon) - x_1 w_2(\varepsilon))\,\mathrm{d}\omega = 0$, a.e. in $x_3 \in (0, L)$, it can be proved that $\tau_{\alpha\beta} = e_{\alpha\beta}(\boldsymbol{w})$, for some $\boldsymbol{w} \in L^2[0, L; (H^1(\omega))^3]$, $\boldsymbol{w} = (w_1, w_2, 0)$. □

Convergences (32.41)–(32.43) still hold for the strong topology. In order to prove it we follow GEYMONAT, KRASUCKI and MARIGO [1987a,b]. Let

$$C_{-4}(\boldsymbol{u}, \boldsymbol{v}) = \int_\Omega 2\mu[\Theta e_{\alpha\beta}(\boldsymbol{u}) e_{\alpha\beta}(\boldsymbol{v}) + q^*_{\alpha\beta\zeta\rho} e_{\zeta\rho}(\boldsymbol{u}) e_{\alpha\beta}(\boldsymbol{v})]\,\mathrm{d}\boldsymbol{x} + \int_\Omega \lambda[\Theta e_{\alpha\alpha}(\boldsymbol{u}) e_{\gamma\gamma}(\boldsymbol{v}) + q^*_{\alpha\alpha\zeta\rho} e_{\zeta\rho}(\boldsymbol{u}) e_{\gamma\gamma}(\boldsymbol{v})]\,\mathrm{d}\boldsymbol{x},$$

$$C^0_{-2}(\boldsymbol{u}, \boldsymbol{v}) = \int_\Omega \lambda(\Theta\delta_{\alpha\beta} + q^*_{\gamma\gamma\alpha\beta}) e_{\alpha\beta}(\boldsymbol{u}) e_{33}(\boldsymbol{v})\,\mathrm{d}\boldsymbol{x},$$

$$C^1_{-2}(\boldsymbol{u}, \boldsymbol{v}) = \int_\Omega (\lambda\Theta\delta_{\alpha\beta} + \lambda q^*_{\gamma\gamma 33}\delta_{\alpha\beta} + 2\mu q^*_{\alpha\beta 33}) e_{33}(\boldsymbol{u}) e_{\alpha\beta}(\boldsymbol{v})\,\mathrm{d}\boldsymbol{x},$$

$$C^2_{-2}(\boldsymbol{u}, \boldsymbol{v}) = \int_\Omega 4\mu(\Theta\delta_{\alpha\beta} + 2q^*_{3\beta 3\alpha}) e_{3\alpha}(\boldsymbol{u}) e_{3\beta}(\boldsymbol{v})\,\mathrm{d}\boldsymbol{x},$$

$$C_0(\boldsymbol{u}, \boldsymbol{v}) = \int_\Omega [\Theta(\lambda + 2\mu) + \lambda q^*_{\gamma\gamma 33}] e_{33}(\boldsymbol{u}) e_{33}(\boldsymbol{v})\,\mathrm{d}\boldsymbol{x},$$

and

$$F(\boldsymbol{v}) = \Theta \int_\Omega f_i(\varepsilon) v_i\,\mathrm{d}\boldsymbol{x} + \int_\Gamma g_i(\varepsilon) v_i\,\mathrm{d}a.$$

Equation (32.47) reads, for all $\boldsymbol{v} \in V(\Omega)$,

$$C^\varepsilon(\boldsymbol{u}^*(\varepsilon), \boldsymbol{v}) = \varepsilon^{-4} C_{-4}(\boldsymbol{u}^*(\varepsilon), \boldsymbol{v}) + \varepsilon^{-2} C^0_{-2}(\boldsymbol{u}^*(\varepsilon), \boldsymbol{v}) + \varepsilon^{-2} C^1_{-2}(\boldsymbol{u}^*(\varepsilon), \boldsymbol{v}) + \varepsilon^{-2} C^2_{-2}(\boldsymbol{u}^*(\varepsilon), \boldsymbol{v}) + C_0(\boldsymbol{u}^*(\varepsilon), \boldsymbol{v}) = F(\boldsymbol{v}). \tag{32.46}$$

LEMMA 32.4. *The following equalities hold*:

$$C^0_{-2}(\boldsymbol{w}, \boldsymbol{v}) + C_0(\boldsymbol{u}^*, \boldsymbol{v}) = F(\boldsymbol{v}), \quad \textit{for all } \boldsymbol{v} \in V_{\mathrm{BN}}(\Omega), \tag{32.47}$$

$$C^1_{-2}(\boldsymbol{u}^*, \boldsymbol{v}) + C_{-4}(\boldsymbol{w}, \boldsymbol{v}) = 0, \quad \textit{for all } \boldsymbol{v} \in V(\Omega) \textit{ and for } \boldsymbol{v} = \boldsymbol{w}. \tag{32.48}$$

PROOF. One obtains (32.47) by considering $\boldsymbol{v} \in V_{\mathrm{BN}}(\Omega)$, in (32.46), and taking the limit as $\varepsilon \to 0$. One obtains (32.48) for all $\boldsymbol{v} \in V(\Omega)$ by multiplying (32.46) by ε^2 and taking the limit as $\varepsilon \to 0$. One obtains (32.48) for $\boldsymbol{v} = \boldsymbol{w}$, by considering $\boldsymbol{v} = (\varepsilon^{-2} u_1^*(\varepsilon), \varepsilon^{-2} u_2^*(\varepsilon), 0)$ in (32.48) and taking the limit as $\varepsilon \to 0$. Convergences (32.43) are used. □

THEOREM 32.3. *Under the same hypotheses of Theorem 32.2 the following convergences hold, for a subsequence of ε, and strongly in $L^2(\Omega)$,*

$$\begin{aligned} &e_{33}(\boldsymbol{u}^*(\varepsilon)) \to e_{33}(\boldsymbol{u}^*),\\ &\varepsilon^{-1}e_{3\beta}(\boldsymbol{u}^*(\varepsilon)) \to 0,\\ &\varepsilon^{-2}e_{\alpha\beta}(\boldsymbol{u}^*(\varepsilon)) \to e_{\alpha\beta}(\boldsymbol{w}). \end{aligned} \tag{32.49}$$

PROOF. By the coercivity of the homogenized coefficients (see Remark 32.1), one has, for a constant C, and all $\boldsymbol{\gamma} \in \boldsymbol{\Sigma}(\Omega)$,

$$\begin{aligned} &\int_\Omega 2\mu[\Theta\gamma_{ij}\gamma_{ij} + q^*_{ijkl}\gamma_{kl}\gamma_{ij}]\,\mathrm{d}\boldsymbol{x} + \int_\Omega \lambda[\Theta\gamma_{pp}\gamma_{rr} + q^*_{ppkl}\gamma_{kl}\gamma_{rs}]\,\mathrm{d}\boldsymbol{x}\\ &\quad \geqslant C\|\boldsymbol{\gamma}\|_{L^2(\Omega)}. \end{aligned} \tag{32.50}$$

Replacing in this equation $\boldsymbol{\gamma}$ by $\boldsymbol{\tau}(\varepsilon) - \boldsymbol{\tau}$, where

$$\begin{aligned} &\tau_{\alpha\beta}(\varepsilon) = \varepsilon^{-2}e_{\alpha\beta}(\boldsymbol{u}^*(\varepsilon)), \qquad \tau_{3\beta}(\varepsilon) = \varepsilon^{-1}e_{3\beta}(\boldsymbol{u}^*(\varepsilon)), \qquad \tau_{33}(\varepsilon) = e_{33}(\boldsymbol{u}^*(\varepsilon)),\\ &\tau_{\alpha\beta} = e_{\alpha\beta}(\boldsymbol{w}), \qquad \tau_{3\beta} = 0, \qquad \tau_{33} = e_{33}(\boldsymbol{u}^*), \end{aligned} \tag{32.51}$$

inequality (32.50) reads

$$\begin{aligned} &F(\boldsymbol{u}^*(\varepsilon)) + C_0(\boldsymbol{u}^*, \boldsymbol{u}^*) - 2C_0(\boldsymbol{u}^*(\varepsilon), \boldsymbol{u}^*)\\ &\quad + C^0_{-2}(\boldsymbol{w}, \boldsymbol{u}^*) - C^0_{-2}(\varepsilon^{-2}\boldsymbol{u}^*(\varepsilon), \boldsymbol{u}^*) - C^0_{-2}(\boldsymbol{w}, \boldsymbol{u}^*(\varepsilon))\\ &\quad + C^1_{-2}(\boldsymbol{u}^*, \boldsymbol{w}) - C^1_{-2}(\boldsymbol{u}^*, \varepsilon^{-2}\boldsymbol{u}^*(\varepsilon)) - C^1_{-2}(\boldsymbol{u}^*(\varepsilon), \boldsymbol{w})\\ &\quad + C_{-4}(\boldsymbol{w}, \boldsymbol{w}) - 2C_{-4}(\varepsilon^{-2}\boldsymbol{u}^*(\varepsilon), \boldsymbol{w})\\ &\quad \geqslant C(\|\varepsilon^{-2}e_{\alpha\beta}(\boldsymbol{u}^*(\varepsilon)) - e_{\alpha\beta}(\boldsymbol{w})\|^2_{L^2(\Omega)} + 2\|\varepsilon^{-1}e_{3\beta}(\boldsymbol{u}^*(\varepsilon))\|^2_{L^2(\Omega)}\\ &\quad + \|e_{33}(\boldsymbol{u}^*(\varepsilon)) - e_{33}(\boldsymbol{u}^*)\|^2_{L^2(\Omega)}). \end{aligned} \tag{32.52}$$

Taking the limit, as $\varepsilon \to 0$, and using Lemma 32.4, one sees that the left-hand side of inequality (32.52) tends to zero, which implies convergences (32.49). □

As for the isotropic case, and using similar techniques, we prove the following result.

THEOREM 32.4. *If $\boldsymbol{f}(\varepsilon) \to \boldsymbol{f}$, weakly in $L^2(\Omega)$ and $\boldsymbol{g}(\varepsilon) \to \boldsymbol{g}$, weakly in $L^2(\Gamma)$, then*

$$\boldsymbol{u}^*(\varepsilon) \to \boldsymbol{u}^*, \quad \text{strongly in } [H^1(\Omega)]^3, \tag{32.53}$$

$$\varepsilon^2\sigma^*_{\alpha\beta}(\varepsilon) \to 0, \qquad \varepsilon\sigma^*_{3\beta}(\varepsilon) \to 0, \qquad \sigma^*_{33}(\varepsilon) \to \sigma^*_{33}, \tag{32.54}$$

strongly in $L^2(\Omega)$,

the limits $\boldsymbol{u}^*$ and σ_{33}^* being uniquely determined by

$$u_3^* = \underline{u}_3^* - x_\alpha \partial_3 u_\alpha^*, \tag{32.55}$$

where $\underline{u}_3^*$ and u_β^* depend only on x_3 and are the unique solutions of, respectively,

$$\begin{aligned} &\underline{u}_3^* \in H_0^1(0,L) \cap H^2(0,L), \\ &-\Theta E|\omega|\partial_{33}\underline{u}_3^* = \Theta \int_\omega f_3 \,\mathrm{d}\omega + \int_\gamma g_3 \,\mathrm{d}\gamma, \quad \textit{in } (0,L), \end{aligned} \tag{32.56}$$

and, with no sum on β,

$$\begin{aligned} &u_\beta^* \in H_0^2(0,L) \cap H^4(0,L), \\ &\Theta E I_\beta \partial_{3333} u_\beta^* = \Theta \int_\omega f_\beta \,\mathrm{d}\omega + \int_\gamma g_\beta \,\mathrm{d}\gamma \\ &\qquad + \Theta \int_\omega x_\beta \partial_3 f_3 \,\mathrm{d}\omega + \int_\gamma x_\beta \partial_3 g_3 \,\mathrm{d}\gamma, \quad \textit{in } (0,L), \end{aligned} \tag{32.57}$$

and

$$\sigma_{33}^* = \Theta E \partial_3 u_3^*. \tag{32.58}$$

PROOF. From (32.49) one concludes that $\boldsymbol{e}(\boldsymbol{u}^*(\varepsilon)) \to \boldsymbol{e}(\boldsymbol{u}^*)$, strongly in $[L^2(\Omega)]^9$ and, by Korn's inequality, (32.53) holds. Since $\boldsymbol{u}^* \in V_{\mathrm{BN}}(\Omega)$, it is necessarily of the form (32.55), with $\underline{u}_3^* \in H_0^1(0,L)$ and $u_\beta^* \in H_0^2(0,L)$. From (32.38)–(32.40) and (32.49) one has, as $\varepsilon \to 0$,

$$\sigma_{33}^*(\varepsilon) \to \xi_{33} = [\Theta(\lambda + 2\mu) + \lambda q_{\gamma\gamma 33}^*] e_{33}(\boldsymbol{u}^*) + \lambda(\Theta\delta_{\alpha\beta} + q_{\gamma\gamma\alpha\beta}^*) e_{\alpha\beta}(\boldsymbol{w}), \tag{32.59}$$

$$\varepsilon\sigma_{3\beta}^*(\varepsilon) \to \xi_{3\beta} = 0, \tag{32.60}$$

$$\begin{aligned} \varepsilon^2\sigma_{\alpha\beta}^*(\varepsilon) \to \xi_{\alpha\beta} &= (\Theta\lambda\delta_{\alpha\beta} + \lambda q_{\gamma\gamma 33}^*\delta_{\alpha\beta} + 2\mu q_{\alpha\beta 33}^*) e_{33}(\boldsymbol{u}^*) \\ &\quad + 2\mu[\Theta e_{\alpha\beta}(\boldsymbol{w}) + q_{\alpha\beta\zeta\rho}^* e_{\zeta\rho}(\boldsymbol{w})] \\ &\quad + \lambda[\Theta e_{\gamma\gamma}(\boldsymbol{w})\delta_{\alpha\beta} + \lambda q_{\gamma\gamma\zeta\rho}^* e_{\zeta\rho}(\boldsymbol{w})\delta_{\alpha\beta}], \end{aligned} \tag{32.61}$$

strongly in $L^2(\Omega)$. Expression (32.48) is equivalent to

$$\int_\Omega \xi_{\alpha\beta} e_{\alpha\beta}(\boldsymbol{v}) \,\mathrm{d}\boldsymbol{x} = 0, \quad \text{for all } \boldsymbol{v} \in V(\Omega),$$

which implies (for $v_\beta = x_\alpha\varphi(x_3)$, $v_\beta = \frac{1}{2}x_\gamma^2\varphi(x_3)$, $v_\beta = x_1x_2\varphi(x_3)$, $\varphi \in H_0^1(0,L)$) that

$$\int_\omega \xi_{\alpha\beta} \,\mathrm{d}\omega = \int_\omega x_\gamma \xi_{\alpha\beta} \,\mathrm{d}\omega = 0. \tag{32.62}$$

Since

$$\xi_{11} + \xi_{22} = [\Theta 2\lambda + 2(\lambda + \mu)q^*_{\gamma\gamma 33}]e_{33}(\boldsymbol{u}^*) + 2(\lambda + \mu)[\Theta e_{\gamma\gamma}(\boldsymbol{w}) + q^*_{\gamma\gamma\zeta\rho}e_{\zeta\rho}(\boldsymbol{w})],$$

one obtains, for the expression of ξ_{33} (see (32.59) and (6.8)),

$$\xi_{33} = \Theta E\partial_3 u^*_3 + \nu\xi_{\alpha\alpha} = \Theta E(\partial_3 \underline{u}^*_3 - x_\beta\partial_{33}u^*_\beta) + \nu\xi_{\alpha\alpha}. \tag{32.63}$$

From (32.37), with $\boldsymbol{v} = (0, 0, \varphi)$, $\varphi \in H^1_0(0, L)$, one obtains

$$\int_0^L \left[\int_\omega \sigma^*_{33}(\varepsilon)\,\mathrm{d}\omega\right] \partial_3\varphi\,\mathrm{d}x_3$$
$$= \int_0^L \left[\Theta\int_\omega f_3(\varepsilon)\,\mathrm{d}\omega + \int_\gamma g_3(\varepsilon)\,\mathrm{d}\gamma\right] \varphi\,\mathrm{d}x_3, \quad \text{for all } \varphi \in H^1_0(0, L). \tag{32.64}$$

Taking the limit when $\varepsilon \to 0$ in (32.64) and using (32.59), (32.63) and (32.62) one obtains the variational setting of problem (32.56) for $\underline{u}^*_3$. From (32.37) with $\boldsymbol{v} = (\varphi, 0, -x_\beta\partial_3\varphi)$, one obtains, for all $\varphi \in H^2_0(0, L)$,

$$-\int_0^L \left[\int_\omega x_\beta\sigma^*_{33}(\varepsilon)\,\mathrm{d}\omega\right] \partial_{33}\varphi\,\mathrm{d}x_3$$
$$= \int_0^L \left[\Theta\int_\omega f_\beta(\varepsilon)\,\mathrm{d}\omega + \int_\gamma g_\beta(\varepsilon)\,\mathrm{d}\gamma\right] \varphi\,\mathrm{d}x_3$$
$$+ \int_0^L \left[\Theta\int_\omega x_\beta\partial_3 f_3(\varepsilon)\,\mathrm{d}\omega + \int_\gamma x_\beta\partial_3 g_3(\varepsilon)\,\mathrm{d}\gamma\right] \varphi\,\mathrm{d}x_3. \tag{32.65}$$

Taking the limit as $\varepsilon \to 0$ and using (32.59), (32.63) and (32.62) one obtains the variational setting of problem (32.57) for u^*_β.

Finally, we prove (32.54) and (32.58). Since the homogenized tensor is positive definite (see Remark 32.1) we may invert the relation

$$\sigma_{ij} = 2\mu(\Theta\gamma_{ij} + q^*_{ijkl}\gamma_{kl}) + \lambda(\Theta\gamma_{pp} + q^*_{ppkl}\gamma_{kl})\delta_{ij},$$

and obtain

$$\gamma_{\alpha\beta} = \frac{1+\nu}{E}k_{\alpha\beta\zeta\rho}\sigma_{\zeta\rho} - \frac{\nu}{E}k_{\alpha\beta\zeta\rho}\left(\delta_{\zeta\rho} - \frac{1}{\Theta}q^*_{\zeta\rho 33}\right)\sigma_{\gamma\gamma}$$
$$- \frac{1}{E}k_{\alpha\beta\zeta\rho}\left(\nu\delta_{\zeta\rho} + \frac{1}{\Theta}q^*_{\zeta\rho 33}\right)\sigma_{33},$$
$$\gamma_{3\beta} = \frac{1+\nu}{E}k_{\alpha\beta}\sigma_{3\alpha},$$
$$\gamma_{33} = \frac{1}{\Theta E}(\sigma_{33} - \nu\sigma_{\gamma\gamma}),$$

where $(k_{\alpha\beta\zeta\rho})$ and $(k_{\alpha\beta})$ are the inverse tensor components of $(\frac{1}{2}\Theta[\delta_{\alpha\zeta}\delta_{\beta\rho} + \delta_{\alpha\rho}\delta_{\beta\zeta}] + q^*_{\alpha\beta\zeta\rho})$ and $(a^*_{\alpha\beta})$, respectively (expression (32.31) relates these components with the components of the positive definite homogenized tensors defined in Section 31, in the proofs of Theorem 31.2 Step 3, and Theorem 31.1, respectively).

Defining, for all $\boldsymbol{\sigma}, \boldsymbol{\tau} \in \Sigma(\Omega)$,

$$a_0(\boldsymbol{\sigma}, \boldsymbol{\tau}) = \int_\Omega \frac{1}{\Theta E}\sigma_{33}\tau_{33}\,\mathrm{d}\boldsymbol{x},$$

$$a_2(\boldsymbol{\sigma}, \boldsymbol{\tau}) = \int_\Omega \frac{2(1+\nu)}{E}k_{\alpha\beta}\sigma_{3\alpha}\tau_{3\beta}\,\mathrm{d}\boldsymbol{x} - \int_\Omega \frac{1}{E}k_{\alpha\beta\zeta\rho}\left(\nu\delta_{\zeta\rho} + \frac{1}{\Theta}q^*_{\zeta\rho33}\right)\sigma_{33}\tau_{\alpha\beta}\,\mathrm{d}\boldsymbol{x}$$
$$- \int_\Omega \frac{\nu}{\Theta E}\sigma_{\gamma\gamma}\tau_{33}\,\mathrm{d}\boldsymbol{x},$$

$$a_4(\boldsymbol{\sigma}, \boldsymbol{\tau}) = \int_\Omega \frac{(1+\nu)}{E}k_{\alpha\beta\zeta\rho}\sigma_{\zeta\rho}\tau_{\alpha\beta}\,\mathrm{d}\boldsymbol{x} - \int_\Omega \frac{\nu}{E}k_{\alpha\beta\zeta\rho}\left(\delta_{\zeta\rho} - \frac{1}{\Theta}q^*_{\zeta\rho33}\right)\sigma_{\gamma\gamma}\tau_{\alpha\beta},$$

one has, for all $\boldsymbol{\tau} \in \Sigma(\Omega)$,

$$\int_\Omega \gamma_{ij}\tau_{ij}\,\mathrm{d}\boldsymbol{x} = a_0(\boldsymbol{\sigma}, \boldsymbol{\tau}) + a_2(\boldsymbol{\sigma}, \boldsymbol{\tau}) + a_4(\boldsymbol{\sigma}, \boldsymbol{\tau}),$$

and, by a coercivity argument, there exists a positive constant C, such that, for all $\boldsymbol{\sigma} \in \Sigma(\Omega)$,

$$a_0(\boldsymbol{\sigma}, \boldsymbol{\sigma}) + a_2(\boldsymbol{\sigma}, \boldsymbol{\sigma}) + a_4(\boldsymbol{\sigma}, \boldsymbol{\sigma}) \geqslant C\|\boldsymbol{\sigma}\|^2_{L^2(\Omega)}. \tag{32.66}$$

Replacing $\boldsymbol{\sigma}$ by $\boldsymbol{S}(\varepsilon) - \boldsymbol{S}$, where

$$S_{\alpha\beta}(\varepsilon) = \varepsilon^2\sigma^*_{\alpha\beta}(\varepsilon), \qquad S_{3\beta}(\varepsilon) = \varepsilon\sigma^*_{3\beta}(\varepsilon), \qquad S_{33}(\varepsilon) = \sigma^*_{33}(\varepsilon),$$
$$S_{\alpha\beta} = 0, \qquad S_{3\beta} = 0, \qquad S_{33} = \Theta E\partial_3 u^*_3,$$

inequality (32.66) reads

$$\begin{aligned} &a_0(\boldsymbol{\sigma}^*(\varepsilon) - \boldsymbol{S}, \boldsymbol{\sigma}^*(\varepsilon) - \boldsymbol{S}) + \varepsilon^2 a_2(\boldsymbol{\sigma}^*(\varepsilon) - \boldsymbol{S}, \boldsymbol{\sigma}^*(\varepsilon) - \boldsymbol{S}) \\ &\quad + \varepsilon^4 a_4(\boldsymbol{\sigma}^*(\varepsilon) - \boldsymbol{S}, \boldsymbol{\sigma}^*(\varepsilon) - \boldsymbol{S}) \\ &\quad \geqslant C(\|\varepsilon^2\sigma^*_{\alpha\beta}(\varepsilon)\|^2_{L^2(\Omega)} + 2\|\varepsilon\sigma^*_{3\beta}(\varepsilon)\|^2_{L^2(\Omega)} + \|\sigma^*_{33}(\varepsilon) - \Theta E\partial_3 u^*_3\|^2_{L^2(\Omega)}. \end{aligned} \tag{32.67}$$

We prove that the left-hand side of this inequality tends to zero, as $\varepsilon \to 0$, and consequently, (32.54) and (32.58) hold. In fact, since for all $\boldsymbol{\tau} \in \Sigma(\Omega)$ (see (32.25)),

$$\int_\Omega \boldsymbol{e}(\boldsymbol{u}^*)\boldsymbol{\tau}\,\mathrm{d}\boldsymbol{x} = a_0(\boldsymbol{\sigma}^*(\varepsilon), \boldsymbol{\tau}) + \varepsilon^2 a_2(\boldsymbol{\sigma}^*(\varepsilon), \boldsymbol{\tau}) + \varepsilon^4 a_4(\boldsymbol{\sigma}^*(\varepsilon), \boldsymbol{\tau}),$$

one has

$$
\begin{aligned}
&a_0(\boldsymbol{\sigma}^*(\varepsilon)-\boldsymbol{S},\boldsymbol{\sigma}^*(\varepsilon)-\boldsymbol{S})+\varepsilon^2 a_2(\boldsymbol{\sigma}^*(\varepsilon)-\boldsymbol{S},\boldsymbol{\sigma}^*(\varepsilon)-\boldsymbol{S})\\
&\quad+\varepsilon^4 a_4(\boldsymbol{\sigma}^*(\varepsilon)-\boldsymbol{S},\boldsymbol{\sigma}^*(\varepsilon)-\boldsymbol{S})\\
&\quad=\int_\Omega \boldsymbol{e}(\boldsymbol{u}^*)(\boldsymbol{\sigma}^*(\varepsilon)-\boldsymbol{S})\,\mathrm{d}\boldsymbol{x}-a_0(\boldsymbol{S},\boldsymbol{\sigma}^*(\varepsilon)-\boldsymbol{S})-\varepsilon^2 a_2(\boldsymbol{S},\boldsymbol{\sigma}^*(\varepsilon)-\boldsymbol{S}).
\end{aligned}
$$

From (32.37), (32.53)–(32.57) and the assumed convergences of $f_i(\varepsilon)$ and $g_i(\varepsilon)$, one has, as $\varepsilon\to 0$,

$$
\begin{aligned}
&\int_\Omega \boldsymbol{e}(\boldsymbol{u}^*)(\boldsymbol{\sigma}^*(\varepsilon)-\boldsymbol{S})\,\mathrm{d}\boldsymbol{x}\\
&\quad=\Theta\int_\Omega f_i(\varepsilon)u_i^*(\varepsilon)\,\mathrm{d}\boldsymbol{x}+\int_\Gamma g_i(\varepsilon)u_i^*(\varepsilon)\,\mathrm{d}a-\Theta\int_\Omega E\partial_3u_3^*(\varepsilon)\partial_3u_3^*\,\mathrm{d}\boldsymbol{x}\\
&\quad\to\Theta\int_\Omega f_iu_i^*\,\mathrm{d}\boldsymbol{x}+\int_\Gamma g_iu_i^*\,\mathrm{d}a-\Theta\int_\Omega E\partial_3u_3^*\partial_3u_3^*\,\mathrm{d}\boldsymbol{x}=0.
\end{aligned}
$$

From (32.53), (32.59), (32.63), (32.55) and (32.62), one has, as $\varepsilon\to 0$,

$$
\begin{aligned}
a_0(\boldsymbol{S},\boldsymbol{\sigma}^*(\varepsilon)-\boldsymbol{S})&=\Theta\int_\Omega E\partial_3u_3^*(\sigma_{33}^*(\varepsilon)-\Theta E\partial_3u_3^*)\,\mathrm{d}\boldsymbol{x}\\
&\to\Theta E\int_\Omega \partial_3u_3^*\xi_{\alpha\alpha}\,\mathrm{d}\boldsymbol{x}=0.
\end{aligned}
$$

Finally, using (32.61), (32.55) and (32.62), one obtains, as $\varepsilon\to 0$,

$$
\varepsilon^2 a_2(\boldsymbol{S},\boldsymbol{\sigma}^*(\varepsilon)-\boldsymbol{S})\to-\Theta k_{\alpha\beta\zeta\rho}\Big(\nu\delta_{\zeta\rho}+\frac{1}{\Theta}q_{\zeta\rho33}^*\Big)\int_\Omega \partial_3u_3^*\xi_{\alpha\beta}\,\mathrm{d}\boldsymbol{x}=0,
$$

which completes the proof. □

REMARK 32.2. Using the inverse scalings to return to the domain Ω^ε, from $\boldsymbol{u}^*$ and σ_{33}^*, one obtains, respectively, the limits $\boldsymbol{u}^{*(0)\varepsilon}$ and $\sigma_{33}^{*(0)\varepsilon}$ described in (a), (b) and (c) of Theorem 31.2, which constitute the homogenized displacement and stress fields of the homogenized Bernoulli–Navier beam model.

CHAPTER VIII

A Galerkin Type Method for Linear Elastic Rods

In Sections 1–5 we observed that in order to evaluate higher order terms in the asymptotic expansion, we had to consider different types of boundary conditions in order to comply with a boundary layer effect. It was also this problem that prevented the obtainment of an H^1 estimate for second- and higher order terms.

One could ask the question whether it would still be possible to find an expansion in powers of ε for displacements and stress and at the same time have an estimate in the H^1 norm for the successive powers of ε.

From the previous sections it is clear that if this is possible then one possibly has to abandon the identification between the different terms in the asymptotic expansion and the different rod theories.

In other words, this question may be addressed as follows. The natural space for the three-dimensional linearized elasticity problem is $H^1(\Omega)$. From the classical rod theories we know that the displacement and stress fields are given as sums of products of functions of x_3^ε times functions of $(x_1^\varepsilon, x_2^\varepsilon)$ (which is also true for the case of plates and shells). Then it is natural to ask what is the "best" approximation, of the three-dimensional solution, of this form and in the H^1 norm.

The results for straight rods presented in Section 33 are owed to MIARA and TRABUCHO [1990, 1992], the multicellular and homogenization cases, presented in Sections 34, 35 and 36 are owed to MASCARENHAS and TRABUCHO [1991, 1992], and finally in Sections 37, 38 and 39 the case of a three-dimensional shallow curved rod is considered following the work of FIGUEIREDO and TRABUCHO [1992b].

33. A Galerkin type approximation in linearized beam theory

In Sections 1–5 it was established that the scaled displacement field $\boldsymbol{u}(\varepsilon)$, for a linearized elastic weakly clamped rod, is the unique solution of a variational problem of the following form (cf. (2.14), (2.15)):

$$\begin{aligned} &\boldsymbol{u}(\varepsilon) \in V(\Omega), \\ &B(\varepsilon)(\boldsymbol{u}(\varepsilon), \boldsymbol{v}) = F(\boldsymbol{v}), \quad \text{for all } \boldsymbol{v} \in V(\Omega), \end{aligned} \tag{33.1}$$

The idea of the Galerkin approximation is to replace the solution of this three-dimensional problem by a finite number of problems posed on lower dimensional spaces by applying a Galerkin type method. For each $N \geqslant 0$, the Galerkin approximation $\boldsymbol{u}_N(\varepsilon)$ of $\boldsymbol{u}(\varepsilon)$ is, by definition, the projection of $\boldsymbol{u}(\varepsilon)$ onto an approximation space $V_N(\Omega)$ of $V(\Omega)$, of the form (no sum on i):

$$V_N(\Omega) = \{\boldsymbol{v} \in V(\Omega)\colon \boldsymbol{v} = \left(\sum_{k=0}^{N} T_i^k(x_1,x_2)v_i^k(x_3)\right)_{1\leqslant i\leqslant 3}, \\ (x_1,x_2) \in \omega,\ x_3 \in [0,L]\}. \tag{33.2}$$

Thus, if the basis functions T_i^k are known, coefficients $u_i^k(\varepsilon)$ of $\boldsymbol{u}_N(\varepsilon) = [\sum_{k=0}^{N} T_i^k(x_1,x_2)u_i^k(x_3)]_{1\leqslant i\leqslant 3}$, are solutions of one-dimensional problems. This significantly simplifies the computation of $\boldsymbol{u}(\varepsilon)$. Moreover, using an argument from VOGELIUS and BABUŠKA [1981], we show that, for special loadings, the basis functions T_i^k can be chosen so as to minimize the approximation error $\|\boldsymbol{u}(\varepsilon) - \boldsymbol{u}_N(\varepsilon)\|_{H^1(\Omega)}$ with respect to parameter ε. We also give an example of how to compute the coefficients $u_i^k(\varepsilon)$ of $\boldsymbol{u}_N(\varepsilon)$. For thin clamped plates and shallow shells the same kind of results have already been obtained by MIARA [1989] and by FIGUEIREDO and TRABUCHO [1992a], respectively.

Associating with the displacement fields $\boldsymbol{u}^\varepsilon : \overline{\Omega}^\varepsilon \to \mathbb{R}^3$ and $\boldsymbol{v}^\varepsilon \in V(\Omega^\varepsilon)$, functions $\boldsymbol{u}(\varepsilon) = (u_i(\varepsilon)) : \overline{\Omega} \to \mathbb{R}^3$ and $\boldsymbol{v} = (v_i) : \overline{\Omega} \to \mathbb{R}^3$, respectively defined by the scalings:

$$\begin{aligned} u_\alpha^\varepsilon(\boldsymbol{x}^\varepsilon) &= \varepsilon u_\alpha(\varepsilon)(\boldsymbol{x}), \quad \text{for all } \boldsymbol{x}^\varepsilon \in \overline{\Omega}^\varepsilon, \\ u_3^\varepsilon(\boldsymbol{x}^\varepsilon) &= \varepsilon^2 u_3(\varepsilon)(\boldsymbol{x}), \quad \text{for all } \boldsymbol{x}^\varepsilon \in \overline{\Omega}^\varepsilon, \\ v_\alpha^\varepsilon(\boldsymbol{x}^\varepsilon) &= \varepsilon v_\alpha(\boldsymbol{x}), \quad \text{for all } \boldsymbol{x}^\varepsilon \in \overline{\Omega}^\varepsilon, \\ v_3^\varepsilon(\boldsymbol{x}^\varepsilon) &= \varepsilon^2 v_3(\boldsymbol{x}), \quad \text{for all } \boldsymbol{x}^\varepsilon \in \overline{\Omega}^\varepsilon, \end{aligned} \tag{33.3}$$

and assuming that there exist functions $\boldsymbol{f} \in [L^2(\Omega)]^3$ and $\boldsymbol{g} \in [L^2(\Gamma)]^3$ independent of ε such that

$$\begin{aligned} f_\alpha^\varepsilon(\boldsymbol{x}^\varepsilon) &= \varepsilon^3 f_\alpha(\varepsilon)(\boldsymbol{x}), \quad f_3^\varepsilon(\boldsymbol{x}^\varepsilon) = \varepsilon^2 f_3(\varepsilon)(\boldsymbol{x}), \quad \text{for all } \boldsymbol{x}^\varepsilon \in \overline{\Omega}^\varepsilon, \\ g_\alpha^\varepsilon(\boldsymbol{x}^\varepsilon) &= \varepsilon^4 g_\alpha(\varepsilon)(\boldsymbol{x}), \quad g_3^\varepsilon(\boldsymbol{x}^\varepsilon) = \varepsilon^3 g_3(\varepsilon)(\boldsymbol{x}), \quad \text{for all } \boldsymbol{x}^\varepsilon \in \Gamma^\varepsilon, \end{aligned} \tag{33.4}$$

we can thus reformulate the three-dimensional elasticity problem, in the displacement formulation (cf. (1.7)–(1.11)), into an equivalent form. Consider the Hilbert space $V = \{\boldsymbol{v} = (v_i) \in [H^1(\Omega)]^3\colon \boldsymbol{v} = \boldsymbol{0} \text{ on } \Gamma_0 \cup \Gamma_L\}$, equipped with the H^1-norm. Let $B(\varepsilon)(\cdot,\cdot) : V \times V \to \mathbb{R}$ denote the continuous, symmetric, bilinear form defined by

$$B(\varepsilon)(\boldsymbol{u}(\varepsilon),(\boldsymbol{v})) = \varepsilon^2 \int_\Omega [\lambda\, e_{\alpha\alpha}(\boldsymbol{u}(\varepsilon))\, e_{\beta\beta}(\boldsymbol{v}) + 2\mu e_{\alpha\beta}(\boldsymbol{u}(\varepsilon))\, e_{\alpha\beta}(\boldsymbol{v})]\, \mathrm{d}\boldsymbol{x}$$

$$+\,\varepsilon^4 \int_\Omega \lambda\,[e_{33}(\boldsymbol{u}(\varepsilon))\,e_{\beta\beta}(\boldsymbol{v}) + e_{\beta\beta}(\boldsymbol{u}(\varepsilon))\,e_{33}(\boldsymbol{v})]\,\mathrm{d}\boldsymbol{x}$$

$$+\,\varepsilon^4 \int_\Omega 4\mu e_{3\alpha}(\boldsymbol{u}(\varepsilon))\,e_{3\alpha}(\boldsymbol{v})\}\,\mathrm{d}\boldsymbol{x}$$

$$+\,\varepsilon^6 \int_\Omega [(\lambda+2\mu)\,e_{33}(\boldsymbol{u}(\varepsilon))\,e_{33}(\boldsymbol{v})]\,\mathrm{d}\boldsymbol{x}, \tag{33.5}$$

and let $F(\cdot): V \to \mathbb{R}$ denote the continuous, linear form, defined by

$$F(\boldsymbol{v}) = \varepsilon^6 \int_\Omega f_i\, v_i\,\mathrm{d}\boldsymbol{x} + \varepsilon^6 \int_\Gamma g_i\, v_i\,\mathrm{d}a. \tag{33.6}$$

The scaled displacement field $\boldsymbol{u}(\varepsilon)$ is then the solution of the variational problem

$$B(\varepsilon)(\boldsymbol{u}(\varepsilon), \boldsymbol{v}) = F(\boldsymbol{v}), \quad \text{for all } \boldsymbol{v} \in V. \tag{33.7}$$

THEOREM 33.1. *For each $\varepsilon > 0$ the bilinear form $B(\varepsilon)(\cdot,\cdot)$ is continuous on $V \times V$ and V-elliptic. This means that there exist two constants m and M (independent of ε) such that*

$$\begin{aligned} &B(\varepsilon)(\boldsymbol{u},\boldsymbol{v}) \leqslant \varepsilon^2\, M\, \|\boldsymbol{u}\|_{H^1(\Omega)}\, \|\boldsymbol{v}\|_{H^1(\Omega)}, \quad \text{for all } \boldsymbol{u},\boldsymbol{v} \in V, \\ &B(\varepsilon)(\boldsymbol{v},\boldsymbol{v}) \geqslant \varepsilon^6\, m\, \|\boldsymbol{v}\|^2_{H^1(\Omega)}, \quad \text{for all } \boldsymbol{u},\boldsymbol{v} \in V. \end{aligned} \tag{33.8}$$

REMARK 33.1. The above scalings amount to multiplying both sides of (2.6) by ε^{-2} and to multiplying both sides of (2.8), (2.9) by ε^3. The scaled variational problem (33.7), may also be obtained from (2.15) by multiplying both sides by ε^6. This does not change the nature of the problem and makes the presentation simpler.

A Galerkin method consists in seeking the solution of a variational problem in terms of a truncated series of known smooth functions (for example, polynomials, trigonometric functions) taken from an approximation space $V_N(\Omega)$. More precisely, let $V_N(\Omega) \subset V(\Omega)$ be the approximation space and $\boldsymbol{u}_N(\varepsilon) \in V_N(\Omega)$ be the Galerkin approximation to the unique solution $\boldsymbol{u}(\varepsilon) \in V$ to the variational problem (33.7), then by definition, the approximation $\boldsymbol{u}_N(\varepsilon)$ is the unique solution of the problem

$$B(\varepsilon)(\boldsymbol{u}_N(\varepsilon), \boldsymbol{v}) = F(\boldsymbol{v}), \quad \text{for all } \boldsymbol{v} \in V_N(\Omega), \tag{33.9}$$

which can also be characterized as the unique solution of the minimization problem

$$\begin{aligned} &B(\varepsilon)(\boldsymbol{u}(\varepsilon) - \boldsymbol{u}_N(\varepsilon), \boldsymbol{u}(\varepsilon) - \boldsymbol{u}_N(\varepsilon)) \\ &\quad = \inf_{z_N \in V_N(\Omega)} \{B(\varepsilon)(\boldsymbol{u}(\varepsilon) - z_N, \boldsymbol{u}(\varepsilon) - z_N)\}, \end{aligned} \tag{33.10}$$

thus, $\boldsymbol{u}_N(\varepsilon)$ is the projection of $\boldsymbol{u}(\varepsilon)$ onto the approximation space V_N with respect to the inner product associated with the quadratic form $B(\varepsilon)(\cdot,\cdot)$.

In this section we give the structure of a possible approximation space $V_N(\Omega)$ and we state a convergence theorem when N goes to infinity.

As will be seen later, it is convenient to split the bilinear form $B(\varepsilon)$ into an "horizontal part" $B_H(\varepsilon)$ (this means that $B_H(\varepsilon)(\boldsymbol{u}(\varepsilon),\cdot)$ acts only on the "horizontal component" $\boldsymbol{v}_H=(v_\alpha)$ of any test function $\boldsymbol{v}=(v_i)=(\boldsymbol{v}_H,v_3)$ and an "axial part" $B_3(\varepsilon)$ (this means that $B_3(\varepsilon)(\boldsymbol{u}(\varepsilon),\cdot)$ acts only on the "axial component"). More specifically, we write,

$$B(\varepsilon)(\boldsymbol{u}(\varepsilon),\boldsymbol{v})=B_H(\varepsilon)(\boldsymbol{u}(\varepsilon),\boldsymbol{v}_H)+B_3(\varepsilon)(\boldsymbol{u}(\varepsilon),v_3),\quad \text{for all } \boldsymbol{v}=(\boldsymbol{v}_H,v_3)\in V(\Omega), \tag{33.11}$$

with the following explicit expressions for $B_H(\varepsilon)(\cdot,\cdot)$ and $B_3(\varepsilon)(\cdot,\cdot)$:

$$\begin{aligned}B_H(\varepsilon)(\boldsymbol{u}(\varepsilon),\boldsymbol{v}_H)=\;&\varepsilon^2\int_\Omega[\lambda\, e_{\alpha\alpha}(\boldsymbol{u}(\varepsilon))\; e_{\beta\beta}(\boldsymbol{v})+2\mu e_{\alpha\beta}(\boldsymbol{u}(\varepsilon))\; e_{\alpha\beta}(\boldsymbol{v})]\,\mathrm{d}\boldsymbol{x}\\&+\varepsilon^4\int_\Omega[\lambda\, e_{33}(\boldsymbol{u}(\varepsilon))\; e_{\beta\beta}(\boldsymbol{v})+2\mu e_{3\alpha}(\boldsymbol{u}(\varepsilon))\;\partial_3\, v_\alpha]\,\mathrm{d}\boldsymbol{x},\end{aligned} \tag{33.12}$$

$$\begin{aligned}B_3(\varepsilon)(\boldsymbol{u}(\varepsilon),v_3)=\;&\varepsilon^4\int_\Omega[\lambda\, e_{\alpha\alpha}(\boldsymbol{u}(\varepsilon))\; e_{33}(\boldsymbol{v})+2\mu e_{3\alpha}(\boldsymbol{u}(\varepsilon))\;\partial_\alpha\, v_3]\,\mathrm{d}\boldsymbol{x}\\&+\varepsilon^6\int_\Omega[(\lambda+2\mu)\; e_{33}(\boldsymbol{u}(\varepsilon))\; e_{33}(\boldsymbol{v})]\,\mathrm{d}\boldsymbol{x}.\end{aligned} \tag{33.13}$$

Similarly, the linear form $F(\cdot)$ can be written, for all $\boldsymbol{v}=(\boldsymbol{v}_H,v_3)\in V(\Omega)$, as

$$F(\boldsymbol{v})=F_H(\varepsilon)(\boldsymbol{v}_H)+F_3(\varepsilon)(v_3), \tag{33.14}$$

$$F_H(\boldsymbol{v}_H)=\varepsilon^6\left[\int_\Omega f_\alpha v_\alpha\,\mathrm{d}\boldsymbol{x}+\int_\Gamma g_\alpha v_\alpha\,\mathrm{d}a\right], \tag{33.15}$$

$$F_3(v_3)=\varepsilon^6\left[\int_\Omega f_3 v_3\,\mathrm{d}\boldsymbol{x}+\int_\Gamma g_3 v_3\,\mathrm{d}a\right]. \tag{33.16}$$

Consequently, the scaled displacement field $\boldsymbol{u}(\varepsilon)$ solves the system

$$\begin{aligned}&B_H(\varepsilon)(\boldsymbol{u}(\varepsilon),\boldsymbol{v}_H)=F_H(\varepsilon)(\boldsymbol{v}_H),\\&B_3(\varepsilon)(\boldsymbol{u}(\varepsilon),v_3)=F_3(\varepsilon)(v_3),\quad \text{for all } \boldsymbol{v}=(\boldsymbol{v}_H,v_3)\in V(\Omega).\end{aligned} \tag{33.17}$$

We also distinguish between the "horizontal part" (P_α^k) and the "axial part" Q^k of the basis functions $\boldsymbol{T}^k$ of (33.2), that is $\boldsymbol{T}^k = (P_1^k, P_2^k, Q^k)$ for $k \geqslant 0$. The approximation space $V_N(\Omega)$ of the space $V(\Omega)$, is therefore defined by

$$\begin{aligned} V_N(\Omega) = \{\boldsymbol{v} = (v_i) \in V\colon v_\alpha = \sum_{k=0}^{N} P_\alpha^k(x_1,x_2)\, v_\alpha^k(x_3), \\ v_\alpha^k \in H_0^1(0,L),\ P_\alpha^k \in H^1(\omega), v_3 = \sum_{k=0}^{N} Q^k(x_1,x_2)\, v_3^k(x_3), \\ v_3^k \in H_0^1(0,L),\ Q^k \in H^1(\omega),\ \text{for all } 0 \leqslant k \leqslant N,\ (\text{no sum on } \alpha)\}. \end{aligned} \tag{33.18}$$

Following the results obtained for plates by MIARA [1989], one can choose a particular element $\boldsymbol{\Psi}_N \in V_N$ that simplifies both the expressions of $B_H(\varepsilon)(\boldsymbol{\Psi}_N, \boldsymbol{v}_H)$ and $B_3(\varepsilon)(\boldsymbol{\Psi}_N, v_3)$ for all $\boldsymbol{v} = (\boldsymbol{v}_H, v_3) \in V$.

THEOREM 33.2. *For $N \geqslant 0$ let $\Psi_1^0 = \Psi_2^0 \in H_0^{2N+2}(0,L)$ and let $\boldsymbol{\Psi}^k = (\Psi_1^k, \Psi_2^k, \Psi_3^k)$, $k \geqslant 0$, be defined by*

$$\begin{aligned} &\Psi_3^0(x_3) = \partial_3\ \Psi_1^0(x_3) = \partial_3\ \Psi_2^0(x_3), \qquad \Psi_3^k(x_3) = \partial_{33}\ \Psi_3^{k-1}(x_3), \quad k \geqslant 1, \\ &\Psi_1^k(x_3) = \Psi_2^k(x_3) = \partial_3\ \Psi_3^{k-1}(x_3) = \partial_{33}\ \Psi_1^{k-1}(x_3) = \partial_{33}\ \Psi_2^{k-1}(x_3), \quad k \geqslant 1. \end{aligned}$$

Then, the element $\boldsymbol{\Psi}_N = \sum_{k=0}^N \varepsilon^{2k}(\Psi_1^k P_1^k, \Psi_2^k P_2^k, \Psi_3^k Q^k)$ belongs to $V_N(\Omega)$ and solves the following equations, valid for all $\boldsymbol{v} = (\boldsymbol{v}_H, v_3) \in V(\Omega)$:

$$\begin{aligned} B_H(\varepsilon)(\boldsymbol{\Psi}_N, \boldsymbol{v}_H) = \varepsilon^{2N+4} \int_\omega [\lambda\ Q^N\ e_{\alpha\alpha}(\boldsymbol{s}^{N+1}) - \mu\ (P_\alpha^N + \partial_\alpha\ Q^N)\ s_\alpha^{N+1}]\,\mathrm{d}\omega \\ + \sum_{k=0}^{N} \varepsilon^{2k+2} \int_\omega [\lambda\ e_{\alpha\alpha}(\boldsymbol{P}^k)\ e_{\beta\beta}(\boldsymbol{s}^k) + 2\mu e_{\alpha\beta}(\boldsymbol{P}^k)\ \partial_\alpha\ s_\beta^k]\,\mathrm{d}\omega \\ + \sum_{k=0}^{N} \varepsilon^{2k+2} \int_\omega [\lambda\ Q^{k-1}\ e_{\alpha\alpha}(\boldsymbol{s}^k) - \mu\ (\partial_\alpha\ Q^{k-1} + P_\alpha^{k-1})\ s_\alpha^k]\,\mathrm{d}\omega, \end{aligned}$$

where (with no sum on α)

$$s_\alpha^k(x_1,x_2) = \int_0^L \Psi_\alpha^k(x_3)\ v_\alpha(x_1,x_2,x_3)\,\mathrm{d}x_3 \in H^1(\omega), \quad 0 \leqslant k \leqslant N,$$

$$s_\alpha^{N+1}(x_1,x_2) = \int_0^L \partial_3\ \Psi_3^N(x_3)\ v_\alpha(x_1,x_2,x_3)\,\mathrm{d}x_3 \in H^1(\omega),$$

and

$$B_3(\varepsilon)(\boldsymbol{\Psi}_N, v_3) = \varepsilon^{2N+6} \int\limits_\omega (\lambda + 2\mu)\, Q^N\, r^{N+1}\, \mathrm{d}\omega$$
$$+ \sum_{k=0}^{N} \varepsilon^{2k+4} \int\limits_\omega [\mu(\partial_\alpha\, Q^k + P_\alpha^k)\, \partial_\alpha\, r^k]\, \mathrm{d}\omega$$
$$- \sum_{k=0}^{N} \varepsilon^{2k+4} \int\limits_\omega [(\lambda + 2\mu)\, Q^{k-1} + \lambda\, \partial_\alpha\, P_\alpha^k]\, r^k\, \mathrm{d}\omega,$$

where

$$r^k(x_1, x_2) = \int_0^L \Psi_3^k(x_3)\, v_3(x_1, x_2, x_3)\, \mathrm{d}x_3 \in H^1(\omega), \quad 0 \leqslant k \leqslant N,$$

$$r^{N+1}(x_1, x_2) = \int_0^L \partial_3\, \Psi_3^N(x_3)\, \partial_3\, v_3(x_1, x_2, x_3)\, \mathrm{d}x_3 \in H^1(\omega),$$

and where, by convention, $\boldsymbol{P}^{-1} = \mathbf{0}$, $Q^{-1} = 0$.

PROOF. It is enough to replace $\boldsymbol{u}(\varepsilon)$ by $\boldsymbol{\Psi}_N$ in system (33.17). □

The idea now is to select the basis functions $(\boldsymbol{P}^k, Q^k)$ so that $B(\varepsilon)(\boldsymbol{u}(\varepsilon) - \boldsymbol{\Psi}_N, \boldsymbol{v})$ be as small as possible, with respect to ε, for all $\boldsymbol{v} \in V(\Omega)$. For simplicity, we suppose hereafter, and with no loss of generality (see Remark 33.2 ahead), that all the forces vanish except g_3 (the axial component of the surface force), and that it depends only on x_3. Accordingly, the construction of the basis functions is achieved by the following scheme:

(i) in $B(\varepsilon)(\boldsymbol{\Psi}_N, \boldsymbol{v})$, cancel all the coefficients of ε^{2k+2}, $k \leqslant N+1$, except that of ε^6,

(ii) in $B(\varepsilon)(\boldsymbol{\Psi}_N, \boldsymbol{v})$, set the coefficient of ε^6 equal to $\int_\Gamma g_3\, v_3\, \mathrm{d}a$.

This yields

$$B(\boldsymbol{u}(\varepsilon) - \boldsymbol{\Psi}_N, \boldsymbol{v}) = -\,\varepsilon^{2N+4} \int\limits_\omega \int_0^L [\lambda\, Q^N\, \partial_3 \boldsymbol{\Psi}_3^N\, \partial_\beta v_\beta]\, \mathrm{d}x_3\, \mathrm{d}\omega$$
$$+\,\varepsilon^{2N+4} \int\limits_\omega \int_0^L [\mu(P_\beta^N\, \partial_3 \Psi_\beta^N + \partial_\beta\, Q^N\, \Psi_3^N)\, \partial_3 v_\beta]\, \mathrm{d}x_3\, \mathrm{d}\omega$$
$$-\,\varepsilon^{2N+6} \int\limits_\omega \int_0^L [(\lambda + 2\mu)\, Q^N\, \partial_3 \Psi_3^N\, \partial_3 v_3]\, \mathrm{d}x_3\, \mathrm{d}\omega, \quad (33.19)$$

and as a consequence the basis functions $(\boldsymbol{P}^k, Q^k)$ are given by the following recursion formulas for $k \geqslant 0$, and with $\boldsymbol{P}^{-1} = \boldsymbol{0}$, $Q^{-1} = 0$:

$$\int_\omega [\lambda\, e_{\alpha\alpha}(\boldsymbol{P}^k)\, e_{\beta\beta}(\boldsymbol{s}) + 2\mu e_{\alpha\beta}(\boldsymbol{P}^k)\, \partial_\alpha s_\beta]\, \mathrm{d}\omega$$

$$+ \int_\omega [\lambda\, Q^{k-1}\, e_{\alpha\alpha}(\boldsymbol{s}) - \mu(\partial_\alpha\, Q^{k-1} + P_\alpha^{k-1}) s_\alpha]\, \mathrm{d}\omega = 0,$$

$$\text{for all } \boldsymbol{s} \in [H^1(\omega)]^2, \tag{33.20}$$

$$\int_\omega \mu(\partial_\alpha Q^k + P_\alpha^k)\, \partial_\alpha r\, \mathrm{d}\omega - \int_\omega [(\lambda + 2\mu) Q^{k-1} + \lambda \partial_\alpha P_\alpha^k] r\, \mathrm{d}\omega$$

$$= \delta_1^k \int_\gamma r\, \mathrm{d}\gamma, \quad \text{for all } r \in H^1(\omega). \tag{33.21}$$

We now give the interpretation of these equations. The vector valued function $\boldsymbol{P}^k = (P_1^k, P_2^k)$ can be interpreted as a weak solution of a plane deformation problem of linearized elasticity in ω. The two-dimensional displacement field $\boldsymbol{P}^k$ expresses the deformation of a body with Lamé constants λ and μ subjected to volume forces $(\lambda + \mu)\partial_\alpha Q^{k-1} + \mu P_\alpha^{k-1}$ in ω and to surface forces $-\lambda Q^{k-1} n_\alpha$ on γ where $\boldsymbol{n} = (n_\alpha)$ is the outward unit normal along the boundary γ. This effect is not present in the classical engineering beam theories. The compatibility conditions for Eq. (33.20) express the fact that these applied forces are in equilibrium, namely the resultant and the total moment vanish:

$$\int_\omega (\partial_\alpha Q^{k-1} + P_\alpha^{k-1})\, \mathrm{d}\omega = 0, \tag{33.22}$$

$$\int_\omega [(\partial_1 Q^{k-1} + P_1^{k-1})\, x_2 - (\partial_2 Q^{k-1} + P_2^{k-1})\, x_1]\, \mathrm{d}\omega = 0. \tag{33.23}$$

Therefore, if these conditions are satisfied, there exists a function $\boldsymbol{P}^k$, unique up to a planar infinitesimal rigid displacement, that solves variational Eqs. (33.20).

The function Q^k can be interpreted as a weak solution of a plane membrane or torsion problem of linearized elasticity in ω. This takes into account the warping of the cross section associated with Saint Venant's torsion theory. The two-dimensional body with shear modulus μ and cross section ω is subjected to volume forces $(\lambda + 2\mu)Q^{k-1} + (\lambda + \mu)\partial_\alpha P_\alpha^k$ in ω and to surface forces $-\mu P_\alpha^k n_\alpha + \delta_1^k$ on γ. The compatibility conditions for Eq. (33.21) express the fact that the resultant of the applied forces vanishes:

$$\int_\omega [(\lambda + 2\mu)\, Q^{k-1} + \lambda \partial_\alpha P_\alpha^k]\, \mathrm{d}\boldsymbol{x} = \delta_1^k |\gamma|. \tag{33.24}$$

Therefore, if this condition is satisfied, there exists a solution Q^k of Eqs. (33.21), defined up to an additive constant.

REMARK 33.2 (*Extensions to other loadings*). We considered that the system of applied forces reduces to $g_3(x_1, x_2, x_3) = g_3(x_3)$. This is not a restriction since other types of loadings are possible. For example:

(i) If $f_3(x_1, x_2, x_3) = f_3(x_3)$ then we must replace the right-hand side of (33.21) by $\delta_1^k \int_\omega r \, d\omega$ and the right-hand side of (33.20) by zero.

(ii) If $f_3(x_1, x_2, x_3) = f_3^1(x_1, x_2)\, f_3^2(x_3)$ (respectively, $g_3(x_1, x_2, x_3) = g_3^1(x_1, x_2) \times g_3^2(x_3)$), we must replace the right-hand side of (33.21) by $\delta_1^k \int_\omega f_3^1(x_1, x_2) r \, d\omega$, (respectively, $\delta_1^k \int_\gamma g_3^1(x_1, x_2) r \, d\gamma$), and the right-hand side of (33.20) by zero.

(iii) If $f_\alpha(x_1, x_2, x_3) = f_\alpha(x_3)$ (respectively, $g_\alpha(x_1, x_2, x_3) = g_\alpha(x_3)$), then we must replace the right-hand side of (33.20) by $\delta_2^k \int_\omega (s_1 + s_2) \, d\omega$, (respectively, $\delta_2^k \int_\gamma (s_1 + s_2) \, d\gamma$), and the right-hand side of (33.21) by zero. Moreover, the following compatibility condition must hold: $f_1(x_1, x_2, x_3) = f_2(x_1, x_2, x_3)$ (respectively, $g_1(x_1, x_2, x_3) = g_2(x_1, x_2, x_3)$).

(iv) In the case where $f_\alpha(x_1, x_2, x_3) = f_\alpha^1(x_1, x_2) f_\alpha^2(x_3)$ (with no sum on α) (respectively, $g_\alpha(x_1, x_2, x_3) = g_\alpha^1(x_1, x_2) g_\alpha^2(x_3)$), we must put $\delta_2^k \int_\omega f_\alpha^1(x_1, x_2) s_\alpha \, d\omega$ (respectively, $\delta_2^k \int_\gamma g_\alpha^1(x_1, x_2) s_\alpha \, d\gamma$), in the right-hand side of (33.20) and replace the right-hand side of (33.21) by zero. Moreover, the following compatibility condition must hold: $f_1(x_1, x_2, x_3) = f_2(x_1, x_2, x_3)$ (respectively, $g_1(x_1, x_2, x_3) = g_2(x_1, x_2, x_3)$).

It is possible to show (cf. MIARA and TRABUCHO [1992]) that problems (33.20), (33.21) defining the basis functions are well posed. Since we are going to consider the more general multicellular and homogenization cases, we omit the proof here and refer the reader to the next section.

It is possible to show that the basis functions may be given as linear combinations of the functions and associated constants characterizing the cross section and defined in (7.21)–(7.46). In fact the following result holds.

THEOREM 33.3. *In addition to functions and constants defined in* (7.21)–(7.46), *let H be the solution of the following problem*:

$$\begin{aligned} &-\partial_{\alpha\alpha} H = -\frac{|\gamma|}{|\omega|}, \quad \text{in } \omega, \\ &\partial_n H = 1, \quad \text{on } \gamma, \\ &\int_\omega H \, d\omega = 0. \end{aligned} \tag{33.25}$$

Then, Eqs. (33.20) *and* (33.21), *for* $0 \leqslant k \leqslant j + 2$ *uniquely define the basis func-*

tions $(\boldsymbol{P}^k, Q^k)_{0\leqslant k\leqslant j}$. *Moreover,*

$$P_\alpha^0 = \frac{1}{\mu} C_\alpha^0, \qquad Q^0 = \frac{1}{\mu}(D^0 - C_\alpha^0\, x_\alpha), \tag{33.26}$$

where the constants C_α^0 *and* D^0 *are defined by*

$$C_\alpha^0 = \frac{\lambda+\mu}{3\lambda+2\mu}\frac{\int_\gamma x_\alpha\, d\gamma}{I_\alpha} \quad \textit{(no sum on } \alpha\textit{)}, \tag{33.27}$$

$$D^0 = -\frac{\lambda+\mu}{3\lambda+2\mu}\frac{|\gamma|}{|\omega|}, \tag{33.28}$$

$$\begin{aligned} P_1^1 &= \frac{\lambda}{2\mu(\lambda+\mu)}(-D^0\, x_1 + C_\alpha^0\, \Phi_{1\alpha}) + \frac{\lambda}{\mu^2}(-K^1\, x_2 + C_1^1), \\ P_2^1 &= \frac{\lambda}{2\mu(\lambda+\mu)}(-D^0\, x_2 + C_\alpha^0\, \Phi_{2\alpha}) + \frac{\lambda}{\mu^2}(K^1\, x_1 + C_2^1), \\ Q^1 &= \frac{\lambda}{\mu^2}K^1\, w + \frac{3\lambda+2\mu}{2\mu(\lambda+\mu)}C_\alpha^0\, \eta_\alpha + \frac{\lambda}{2\mu(\lambda+\mu)}C_\alpha^0\theta_\alpha \\ &\quad - \frac{\lambda}{\mu^2}(C_\alpha^1 x_\alpha - D^1) + \frac{\lambda}{2\mu(\lambda+\mu)}D^0\left[\frac{1}{2}(x_1^2+x_2^2) - \frac{(I_1+I_2)}{2|\omega|}\right] + \frac{1}{\mu}H, \end{aligned} \tag{33.29}$$

where (with sum on α*), constants* K^1 *and* D^1 *are given by*

$$K^1 = \frac{\mu}{\lambda J}\left[-\frac{3\lambda+2\mu}{2(\lambda+\mu)}C_\alpha^0 I_\alpha^w - \frac{\lambda}{2(\lambda+\mu)}C_\alpha^0 I_\alpha^\Psi + \int_\gamma w\, d\gamma\right], \tag{33.30}$$

$$D^1 = \frac{\mu}{4\,(3\lambda+2\mu)|\omega|}\left[\frac{2\,H_\alpha}{I_\alpha}\int_\gamma x_\alpha\, d\gamma - \int_\gamma (x_1^2+x_2^2)\, d\gamma + \frac{|\gamma|}{|\omega|}(I_1+I_2)\right]. \tag{33.31}$$

Moreover,

$$\begin{aligned} P_2^1 &= \frac{\lambda^2}{\mu^2}\left[\frac{1}{2(\lambda+\mu)}(-D^1 x_1 + C_\alpha^1\Phi_{1\alpha}) + \frac{1}{\mu}(-K^2 x_2 + C_1^2)\right] + \overline{P}_\alpha^2, \\ P_2^2 &= \frac{\lambda^2}{\mu^2}\left[\frac{1}{2(\lambda+\mu)}(-D^1 x_2 + C_\alpha^1\Phi_{2\alpha}) + \frac{1}{\mu}(-K^2 x_2 + C_2^2)\right] + \overline{P}_\alpha^2, \end{aligned} \tag{33.32}$$

where components $\overline{P}^2_\alpha$ *are defined as follows:*

$$\overline{P}^2_\alpha = \frac{\lambda^2}{\mu^2}K^1 P^{2,1}_\alpha + \frac{\lambda}{2\mu}P^{2,2}_\alpha + \frac{\lambda^2}{2\mu(3\lambda+2\mu)}P^{2,3}_\alpha - \frac{\lambda^2}{2\mu(3\lambda+2\mu)}\frac{|\gamma|}{|\omega|}P^{2,4}_\alpha + \frac{\lambda}{\mu}P^{2,5}_\alpha + \mu P^{2,6}_\alpha, \tag{33.33}$$

with $P^{2,k}_\alpha$, $1 \leqslant k \leqslant 6$, *the unique solutions of the following plane deformation elasticity problems of the form:*

$$\begin{aligned} &-(\lambda+\mu)\,\partial_{\beta\alpha}\,P^{2,k}_\alpha - \mu\partial_{\alpha\alpha}\,P^{2,k}_\beta = F^k_\beta \quad \text{in } \omega, \\ &\lambda\,\partial_\alpha P^{2,k}_\alpha\,n_\beta + \mu(\partial_n P^{2,k}_\beta + \partial_\beta P^{2,k}_\alpha\,n_\alpha) = G^k_\beta \quad \text{on } \gamma, \\ &\int_\omega P^{2,k}_\alpha\,\mathrm{d}\omega = \int_\omega (P^{2,k}_1\,x_2 - P^{2,k}_2\,x_1)\,\mathrm{d}\omega = 0, \end{aligned} \tag{33.34}$$

and the data F^k_β, G^k_β *associated with* P^{2k}_β, $1 \leqslant k \leqslant 6$, *are given by (with sum on* α*):*

$$\begin{aligned} F^1_\beta &= \partial_\beta w, \\ G^1_\beta &= -w\,n_\beta, \end{aligned} \tag{33.35}$$

$$\begin{aligned} F^2_\beta &= \frac{\int_\gamma x_\alpha\,\mathrm{d}\gamma}{I_\alpha}\,\partial_\beta\eta_\alpha, \\ G^2_\beta &= -\frac{\int_\gamma x_\alpha\,\mathrm{d}\gamma}{I_\alpha}\eta_\alpha\,n_\beta, \end{aligned} \tag{33.36}$$

$$\begin{aligned} F^3_\beta &= \frac{\int_\gamma x_\alpha\,\mathrm{d}\gamma}{I_\alpha}\partial_\beta\theta_\alpha, \\ G^3_\beta &= -\frac{\int_\gamma x_\alpha\,\mathrm{d}\gamma}{I_\alpha}\theta_\alpha n_\beta, \end{aligned} \tag{33.37}$$

$$\begin{aligned} F^4_\beta &= x_\beta, \\ G^4_\beta &= -\left[\frac{(x_1^2+x_2^2)}{2} - \frac{(I_1+I_2)}{|\omega|}\right]n_\beta, \end{aligned} \tag{33.38}$$

$$\begin{aligned} F^5_\beta &= \partial_\beta H, \\ G^5_\beta &= -Hn_\beta, \end{aligned} \tag{33.39}$$

$$\begin{aligned} F^6_1 &= \frac{\lambda}{2\mu(\lambda+\mu)}C^0_\alpha(\partial_1\theta_\alpha + \Phi_{1\alpha}) + \frac{(3\lambda+2\mu)}{2\mu(\lambda+\mu)}C^0_\alpha\partial_1\eta_\alpha + \frac{1}{\mu}\partial_1 H + \frac{\lambda}{\mu^2}K^1\partial_2\Psi, \\ F^6_2 &= \frac{\lambda}{2\mu(\lambda+\mu)}C^0_\alpha(\partial_2\theta_\alpha + \Phi_{2\alpha}) + \frac{(3\lambda+2\mu)}{2\mu(\lambda+\mu)}C^0_\alpha\partial_2\eta_\alpha + \frac{1}{\mu}\partial_2 H - \frac{\lambda}{\mu^2}K^1\partial_1\Psi, \\ G^6_\beta &= 0. \end{aligned} \tag{33.40}$$

Constants C^1_α, C^2_α, which can be determined from (33.20) and (33.21) with $k = 2, 3, 4$, and the expression for Q^2, which is defined by (33.21) with $k = 2$, are omitted because they are too lengthy in the general case. We shall nevertheless write them down for the simpler case of a circular cross section of radius R.

For this case the basis functions are polynomials since the elementary functions w, Ψ, η_α, θ_α, and H, introduced previously, are polynomials. In fact, we have

$$w = 0, \qquad \Psi = \tfrac{1}{2}[R^2 - (x_1^2 + x_2^2)], \qquad \int_\gamma x_\alpha \, \mathrm{d}\gamma = 0, \tag{33.41}$$

$$\begin{aligned} \eta_\alpha &= \tfrac{1}{4}(x_1^2 + x_2^2 - 3R^2)\, x_\alpha, \\ \theta_\alpha &= -\tfrac{1}{4}(x_1^2 + x_2^2 - R^2) x_\alpha, \\ H &= \frac{1}{2R}(x_1^2 + x_2^2 - \tfrac{1}{2}R^2), \end{aligned} \tag{33.42}$$

$$\begin{aligned} &I_\alpha = \tfrac{1}{4}\pi R^4, \qquad J = \tfrac{1}{2}\pi R^4, \qquad I_\alpha^w = 0, \qquad I_\alpha^\Psi = 0, \\ &H_\alpha = 0, \qquad H_3 = \tfrac{1}{12}\pi R^6, \end{aligned} \tag{33.43}$$

and we get the following expressions for P^0_α, P^1_α, Q^0 and Q^1:

$$\begin{aligned} P^0_\beta &= 0, \\ Q^0 &= -\frac{2(\lambda + \mu)}{\mu(3\lambda + 2\mu)R}, \end{aligned} \tag{33.44}$$

$$\begin{aligned} P^1_\beta &= \frac{\lambda}{\mu(3\lambda + 2\mu)} \frac{x_\beta}{R}, \\ Q^1 &= -\frac{R}{4\mu} + \frac{(\lambda + \mu)}{\mu(3\lambda + 2\mu)} \frac{(x_1^2 + x_2^2)}{R}, \end{aligned} \tag{33.45}$$

and for $\boldsymbol{P}^2$, Q^2 (which are defined up to additive constants which can be determined using higher order terms):

$$\begin{aligned} P^2_\beta &= \frac{3(\lambda + \mu)}{8\mu(3\lambda + 2\mu)} R x_\beta + \frac{\lambda^2}{\mu^3} C^2_\beta - \frac{2\lambda + \mu}{8\mu(3\lambda + 2\mu)} (x_1^2 + x_2^2) \frac{x_\beta}{R}, \\ Q^2 &= \frac{2\lambda + \mu}{16\mu(3\lambda + 2\mu} (x_1^2 + x_2^2) R + \frac{\lambda^2}{32\mu(3\lambda + 2\mu)(\lambda + \mu)} R^3 \\ &\quad - \frac{3(\lambda + \mu)}{32\mu(3\lambda + 2\mu)} (x_1^2 + x_2^2)^2 \frac{1}{R} + \lambda^2 \mu^3 (D^2 - C^2_\alpha x_\alpha). \end{aligned} \tag{33.46}$$

REMARK 33.3. It is possible to show that, for the homogeneous isotropic case, the basis functions for the circular cross section are polynomials.

For a sufficiently smooth data g_3 (all other forces vanish by assumption), the Galerkin approximation $\boldsymbol{u}_N(\varepsilon)$ gives a "good approximation" of the three-dimensional solution $\boldsymbol{u}(\varepsilon)$ when ε goes to zero. This is the result we shall state next.

Let $G_3 = \{g_3 \colon x \in (0,L) \to g_3(x) = \partial_3 h, h \in H_0^4(0,L)\}$, then the following result holds.

THEOREM 33.4. *If the system of applied forces is such that* $f_i = 0$, $g_\alpha = 0$ *and the component* $g_3 = g_3(x_3) \in G_3 \cap H_0^{2N-1}(0,L)$ *for* $N \geqslant 1$, *there exists a constant* C_N, *independent of* ε, *such that*

$$\|\boldsymbol{u}(\varepsilon) - \boldsymbol{u}_N(\varepsilon)\|_{H^1(\Omega)} \leqslant C_N \varepsilon^{2N-2}, \qquad \|\boldsymbol{u}^\varepsilon - \boldsymbol{u}_N^\varepsilon\|_{H^1(\Omega^\varepsilon)} \leqslant C_N \varepsilon^{2N-1}. \tag{33.47}$$

PROOF. By assumption the element

$$\tilde{\Psi}_N = \sum_{k=0}^{N} \varepsilon^{2k} \big(P_1^k(x_1,x_2)\partial_3^{2k-3} g_3(x_3),\ P_2^k(x_1,x_2)\partial_3^{2k-3} g_3(x_3), Q^k(x_1,x_2)\partial_3^{2k-2} g_3(x_3)\big)$$

belongs to $V_N(\Omega)$ and satisfies Theorem 33.2 (this element is obtained choosing $\Psi_1^3(x_3) = g_3(x_3)$). Then, using the definition of the basis functions, we get for all $\boldsymbol{v} \in V(\Omega)$,

$$\begin{aligned} B(\varepsilon)(\boldsymbol{u}(\varepsilon) - \tilde{\Psi}_N, \boldsymbol{v}) = &- \varepsilon^{2N+4} \int_0^L \int_\omega [\lambda Q^N \partial_3^{2N-1} g_3 \partial_\beta v_\beta \\ &+ \mu(P_\beta^N \partial_3^{2N-2} g_3 + \partial_\beta Q^N \partial_3^{2N-2} g_3)\partial_3 v_\beta] \,\mathrm{d}\omega\,\mathrm{d}x_3 \\ &- \varepsilon^{2N+6} \int_0^L \int_\omega [(\lambda + 2\mu) Q^N \partial_3^{2N-1} g_3 \partial_3 v_3] \,\mathrm{d}\omega\,\mathrm{d}x_3, \end{aligned}$$

therefore, $|B(\varepsilon)(\boldsymbol{u}(\varepsilon) - \tilde{\Psi}_N, \boldsymbol{v})| \leqslant C_N \varepsilon^{2N+4} \|\boldsymbol{v}\|_{H^1(\Omega)}$. Since $\boldsymbol{u}_N(\varepsilon)$ is the Galerkin approximation of $\boldsymbol{u}(\varepsilon)$ the result is a consequence of the coerciveness of $B(\varepsilon)(\cdot,\cdot)$ and of the scalings defined by (33.3). □

REMARK 33.4 (*Extension to other loadings*). With reference to Remark 33.2 we have, in addition

(i) If $f_3(x_1,x_2,x_3) = f_3(x_3)$, we set $\Psi_3^1(x_3) = f_3(x_3)$.

(ii) If $f_3(x_1,x_2,x_3) = f_3^1(x_1,x_2) f_3^2(x_3)$, (respectively $g_3(x_1,x_2,x_3) = g_3^1(x_1,x_2) \times g_3^2(x_3)$), we set $\Psi_3^1(x_3) = f_3(x_3)$ (respectively, $\Psi_3^1(x_3) = g_3(x_3)$).

(iii) If $f_\alpha(x_1,x_2,x_3) = f_\alpha(x_3)$ (respectively $g_\alpha(x_1,x_2,x_3) = g_\alpha(x_3)$), then we choose $\Psi_\alpha^2(x_3) = f_\alpha(x_3)$ (respectively $\Psi_\alpha^2(x_3) = g_\alpha(x_3)$).

(iv) In the case where $f_\alpha(x_1,x_2,x_3) = f_\alpha^1(x_1,x_2) f_\alpha^2(x_3)$ (with no sum on α) (respectively, $g_\alpha(x_1,x_2,x_3) = g_\alpha^1(x_1,x_2) g_\alpha^2(x_3)$), we choose $\Psi_\alpha^2(x_3) = f_\alpha(x_3)$ (respectively $\Psi_\alpha^2(x_3) = g_\alpha(x_3)$).

We finally remark that in cases (ii) and (iv), the basis functions for the circular case are not necessarily polynomials.

REMARK 33.5. For ε sufficiently small, MASCARENHAS and TRABUCHO [1991, 1992] have shown that $C_N\ \varepsilon^{2N-2}$ goes to zero as N goes to infinity (cf. next section also).

We finish this section with some examples of the calculation of the components u_i^k, $1 \leqslant i \leqslant 3$, $1 \leqslant k \leqslant N$.

The approximation $\boldsymbol{u}_1(\varepsilon)$, as defined, is expressed in terms of Q^0, Q^1, $\boldsymbol{P}^0$, $\boldsymbol{P}^1$ by

$$\boldsymbol{u}_1(\varepsilon) = \begin{Bmatrix} u_1^0 P_1^0 + \varepsilon^2 u_1^1 P_1^1 \\ u_2^0 P_2^0 + \varepsilon^2 u_2^1 P_2^1 \\ u_3^0 Q^0 + \varepsilon^2 u_3^1 Q^1 \end{Bmatrix} \tag{33.48}$$

and it is a solution of the following variational equations:

$$B(\varepsilon)(\boldsymbol{u}_1(\varepsilon), \boldsymbol{v}) = F(\varepsilon)(\boldsymbol{v}), \quad \text{for all } \boldsymbol{v} \in V_1. \tag{33.49}$$

In the general case, if $P_\alpha^0 \neq 0$, $P_\alpha^1 \neq 0$, $Q^0 \neq 0$, $Q^1 \neq 0$, we have for all $v \in H_0^1(0, L)$,

$$\begin{aligned} &\int_0^L \mu[\partial_3 u_1^0 (P_1^0, P_1^0) + u_3^0 (P_1^0, \partial_1 Q^0)] \partial_3 v \, \mathrm{d}x_3 \\ &\quad + \varepsilon^2 \int_0^L \mu[\partial_3 u_1^1 (P_1^0, P_1^1) + u_3^1 (P_1^0, \partial_1 Q^1)] \partial_3 v \, \mathrm{d}x_3 = \varepsilon^2 \int_0^L F_1^0 v \, \mathrm{d}x_3, \end{aligned} \tag{33.50}$$

$$\begin{aligned} &\int_0^L \{(\lambda + 2\mu)(\partial_1 P_1^1, \partial_1 P_1^1) u_1^1 v + \lambda (\partial_1 P_1^1, \partial_2 P_2^1) u_2^1 v \\ &\quad + \mu[(\partial_2 P_1^1, \partial_2 P_1^1) u_1^1 + (\partial_1 P_2^1, \partial_2 P_1^1) u_2^1] v \\ &\quad + \lambda (\partial_1 P_1^1, Q^0) \partial_3 u_3^0 v + \mu[(P_1^1, P_1^0) \partial_3 u_1^0 + (P_1^1, \partial_1 Q^0) u_3^0] \partial_3 v\} \, \mathrm{d}x_3 \\ &\quad + \varepsilon^2 \int_0^L \{\lambda (\partial_1 P_1^1, Q^1) \partial_3 u_3^1 v + \mu[(P_1^1, P_1^1) \partial_3 u_1^1 + (P_1^1, \partial_1 Q^1) u_3^1] \partial_3 v\} \, \mathrm{d}x_3 \\ &= \varepsilon^2 \int_0^L F_1^1 v \, \mathrm{d}x_3, \end{aligned} \tag{33.51}$$

$$\int_0^L \mu[\partial_3 u_2^0(P_2^0, P_2^0) + u_3^0(P_2^0, \partial_2 Q^0)]\partial_3 v \, \mathrm{d}x_3$$

$$+ \varepsilon^2 \int_0^L \mu[\partial_3 u_2^1(P_2^0, P_2^1) + u_3^1(P_2^0, \partial_2 Q^1)]\partial_3 v \, \mathrm{d}x_3 = \varepsilon^2 \int_0^L F_2^0 v \, \mathrm{d}x_3, \tag{33.52}$$

$$\int_0^L \{(\lambda + 2\mu)(\partial_2 P_2^1, \partial_2 P_2^1)u_2^1 v + \lambda(\partial_1 P_1^1, \partial_2 P_2^1)u_1^1 v$$

$$+ \mu[(\partial_2 P_1^1, \partial_1 P_2^1)u_1^1 + (\partial_1 P_2^1, \partial_1 P_2^1)u_2^1]v$$

$$+ \lambda(\partial_2 P_2^1, Q^0)\partial_3 u_3^0 v + \mu[(P_2^1, P_2^0)\partial_3 u_2^0 + (P_2^1, \partial_2 Q^0)u_3^0]\partial_3 v\} \, \mathrm{d}x_3$$

$$+ \varepsilon^2 \int_0^L \{\lambda(\partial_2 P_2^1, Q^1)\partial_3 u_3^1 v + \mu[(P_2^1, P_2^1)\partial_3 u_2^1 + (P_2^1, \partial_2 Q^1)u_3^1]\partial_3 v\} \, \mathrm{d}x_3$$

$$= \varepsilon^2 \int_0^L F_2^1 v \, \mathrm{d}x_3, \tag{33.53}$$

$$\int_0^L \mu[(\partial_1 Q^0, P_1^0)\partial_3 u_1^0 + (\partial_2 Q^0, P_2^0)\partial_3 u_2^0 + (\partial_\beta Q^0, \partial_\beta Q^0)u_3^0]v \, \mathrm{d}x_3$$

$$+ \varepsilon^2 \int_0^L \{(\lambda + 2\mu)(Q^0, Q^0)\partial_3 u_3^0 \partial_3 v + \lambda[(Q^0, \partial_1 P_1^1)u_1^1 + (Q^0, \partial_2 P_2^1)u_2^1]\partial_3 v$$

$$+ \mu[(\partial_1 Q^0, P_1^1)\partial_3 u_1^1 + (\partial_2 Q^0, P_2^1)\partial_3 u_2^1 + (\partial_\beta Q^0, \partial_\beta Q^1)u_3^1]v\} \, \mathrm{d}x_3$$

$$+ \varepsilon^4 \int_0^L (\lambda + 2\mu)(Q^0, Q^1)\partial_3 u_3^1 \partial_3 v \, \mathrm{d}x_3 = \varepsilon^2 \int_0^L F_3^0 v \, \mathrm{d}x_3, \tag{33.54}$$

$$\int_0^L \mu[(\partial_1 Q^1, P_1^0)\partial_3 u_1^0 + (\partial_2 Q^1, P_2^0)\partial_3 u_2^0 + (\partial_\beta Q^1, \partial_\beta Q^0)u_3^0]v \, \mathrm{d}x_3$$

$$+\varepsilon^2 \int_0^L \{(\lambda+2\mu)(Q^1,Q^0)\partial_3 u_3^0 \partial_3 v + \lambda[(Q^1,\partial_1 P_1^1)u_1^1 + (Q^1,\partial_2 P_2^1)u_2^1]\partial_3 v$$

$$+ \mu[(\partial_1 Q^1, P_1^1)\partial_3 u_1^1 + (\partial_2 Q^1, P_2^1)\partial_3 u_2^1 + (\partial_\beta Q^1, \partial_\beta Q^1)u_3^1]v\}\,\mathrm{d}x_3$$

$$+\varepsilon^4 \int_0^L (\lambda+2\mu)(Q^1,Q^1)\partial_3 u_3^1 \partial_3 v\,\mathrm{d}x_3 = \varepsilon^2 \int_0^L F_3^1 v\,\mathrm{d}x_3, \tag{33.55}$$

where (with no sum on α; $m = 0,1$), $F_\alpha^m = \int_\omega f_\alpha P_\alpha^m\,\mathrm{d}\omega + \int_\gamma g_\alpha P_\alpha^m\,\mathrm{d}\gamma$; $F_3^m = \int_\omega f_3 Q^m\,\mathrm{d}\omega + \int_\gamma g_3 Q^m\,\mathrm{d}\gamma$, and where $(\cdot,\cdot)$ denotes the $L^2(\omega)$ inner product.

We shall now consider the problem of existence and uniqueness of the solution of system (33.50)–(33.55). To this end, we substitute this problem by an equivalent one obtained by considering the following linear combinations of (33.50)–(33.55):

$$\frac{(P_1^1,P_1^1)}{(P_1^0,P_1^1)} \times [\text{Eq. (33.50)}] - [\text{Eq. (33.51)}], \tag{33.56}$$

$$-\frac{(P_1^1,P_1^0)}{(P_1^0,P_1^0)} \times [\text{Eq. (33.50)}] + [\text{Eq. (33.51)}], \tag{33.57}$$

$$\frac{(P_2^1,P_2^1)}{(P_2^0,P_2^1)} \times [\text{Eq. (33.52)}] - [\text{Eq. (33.53)}], \tag{33.58}$$

$$-\frac{(P_2^1,P_2^0)}{(P_2^0,P_2^0)} \times [\text{Eq. (33.52)}] + [\text{Eq. (33.53)}], \tag{33.59}$$

$$\frac{(Q^1,Q^1)}{(Q^0,Q^1)} \times [\text{Eq. (33.54)}] - [\text{Eq. (33.55)}], \tag{33.60}$$

$$-\frac{(Q^1,Q^0)}{(Q^0,Q^0)} \times [\text{Eq. (33.54)}] + [\text{Eq. (33.55)}]. \tag{33.61}$$

Let W denote the Sobolev space $W = \{\boldsymbol{w} = (w_\alpha^0, w_3^0, w_\alpha^1, w_3^1) \in [H_0^1(0,L)]^6\}$. Let $C(\cdot,\cdot) : W \times W \to \mathbb{R}$ denote the bilinear form associated with the variational formulation of problem (33.56)–(33.61), and let $M(\cdot) : W \to \mathbb{R}$ denote the linear form associated with the variational formulation of the same system. We then have

THEOREM 33.5. *The bilinear form $C(\cdot,\cdot)$ is continuous on W and satisfies the following inequality of the Garding type on W: there exist six positive constants* *(independent of* ε*)* A_k, $1 \leqslant k \leqslant 6$ *such that for any* $\boldsymbol{w} \in W$:

$$\begin{aligned} C(\boldsymbol{w}, \boldsymbol{w}) \geqslant\ & A_1(|\partial_3 w_1^0|^2 + |\partial_3 w_2^0|^2 + |w_1^1|^2 + |w_2^1|^2 + |w_3^0|^2) \\ & + \varepsilon^2 A_2(|\partial_3 w_1^0|^2 + |\partial_3 w_2^0|^2 + |\partial_3 w_1^1|^2 + |\partial_3 w_2^1|^2 + |\partial_3 w_3^0|^2 + |w_3^1|^2) \\ & + \varepsilon^4 A_3 |\partial_3 w_3^1|^2 \\ & - A_4(|w_1^0|^2 + |w_2^0|^2 + |w_3^0|^2 + |w_1^1|^2 + |w_2^1|^2 + |w_3^1|^2) \\ & - \varepsilon^2 A_5(|w_3^0|^2 + |w_1^1|^2 + |w_2^1|^2 + |w_3^1|^2) \\ & - \frac{1}{\varepsilon^2} A_6(|w_1^0|^2 + |w_2^0|^2 + |w_1^1|^2 + |w_2^1|^2). \end{aligned}$$

PROOF. An application of Young's inequality. □

Since the linear form $M(\cdot)$ is continuous on W, we then have the following.

THEOREM 33.6. *For any* $\varepsilon \neq 0$, *if* 0 *is not an eigenvalue associated with the bilinear form* $C(\cdot,\cdot)$, *system* (33.56)–(33.61) *and consequently system* (33.50)–(33.55) *have a unique solution on* W.

PROOF. See NEČAS [1967, p. 53]. □

When the beam's cross section is circular (of radius R), the first basis function $\boldsymbol{P}^0$ vanishes and the previous equations (33.50) and (33.52) are identically satisfied. One can show that when $\varepsilon = 0$ the system of three equations (33.51), (33.53), (33.54) gives a Bernoulli–Navier type displacement field and the governing equations coincide with those associated with the zeroth order terms obtained via the asymptotic expansion method.

Generalization of this method to homogeneous anisotropic elastic rods has been done in VEIGA [1993a].

34. A Galerkin type approximation for the multicellular case

In this section we combine the approaches of Sections 30, 31 and 33. Specifically, let the notations of Sections 30 and 31 hold and consider the three-dimensional elasticity problem in $\Omega^{\delta,\varepsilon}$. As previously, we consider that the beam $\Omega^{\delta,\varepsilon}$ is made of a homogeneous isotropic linearly elastic material with Lamé's constants λ and μ, both independent of δ and ε. Let $f_i^\varepsilon \in L^2(\Omega^{\delta,\varepsilon})$ and $g_i^\varepsilon \in L^2(\Gamma^{\delta,\varepsilon})$ be the ith components of the volume density of applied body forces and surface density of applied surface tractions, respectively. To simplify we consider that the system of applied forces does not depend on the parameter δ. However, for certain kinds of oscillations in δ the same type of results are obtained, for instance, if the sequence $(f_i^{\delta\varepsilon})_\delta$ is compact in $L^2(\Omega^{\delta,\varepsilon})$ (respectively $(g_i^{\delta\varepsilon})_\delta$ is compact in $L^2(\Gamma^{\delta,\varepsilon})$). We also admit that $g_i^\varepsilon(\boldsymbol{x}^\varepsilon) = 0$ if $\boldsymbol{x}^\varepsilon \in \partial T^{\delta,\varepsilon} \times (0, L)$.

Consider the space of admissible displacements $V(\Omega^{\delta,\varepsilon}) = \{\boldsymbol{v} = (v_i) \in [H^1(\Omega^{\delta,\varepsilon})]^3\colon \boldsymbol{v} = 0 \text{ on } \Gamma_0^{\delta,\varepsilon} \cup \Gamma_L^{\delta,\varepsilon}\}$, equipped with the usual H^1-norm and consider the classical three-dimensional, linearized, elasticity problem in $\Omega^{\delta,\varepsilon}$

$$\begin{aligned}
&\boldsymbol{u}^{\delta,\varepsilon} \in V(\Omega^{\delta,\varepsilon}),\\
&\int_{\Omega^{\delta,\varepsilon}} [2\mu e^{\varepsilon}(\boldsymbol{u}^{\delta,\varepsilon}) + \lambda \operatorname{tr} e^{\varepsilon}(\boldsymbol{u}^{\delta,\varepsilon}) \operatorname{Id}] e^{\varepsilon}(\boldsymbol{v}^{\delta,\varepsilon}) \, \mathrm{d}\boldsymbol{x}^{\delta,\varepsilon}\\
&\quad = \int_{\Omega^{\delta,\varepsilon}} \boldsymbol{f}^{\varepsilon} \cdot \boldsymbol{v} \, \mathrm{d}\boldsymbol{x}^{\delta,\varepsilon} + \int_{\Gamma^{\delta,\varepsilon}} \boldsymbol{g}^{\varepsilon} \cdot \boldsymbol{v} \, \mathrm{d}a^{\delta,\varepsilon}, \quad \text{for all } \boldsymbol{v} \in V(\Omega^{\delta,\varepsilon}).
\end{aligned} \tag{34.1}$$

It is a classical result that problem (34.1) has a unique solution. We are interested in the behaviour of the displacement field $\boldsymbol{u}^{\delta,\varepsilon}$, when δ and ε tend to zero. This type of study has already been done, by means of the asymptotic expansion method, by GEYMONAT, KRASUCKI and MARIGO [1987], by TUTEK [1987], for the case of a composite material and concerning the first order Bernoulli–Navier bending–extension theory, and by MASCARENHAS and TRABUCHO [1990], for the second order theories of Saint Venant and Timoshenko.

The asymptotic expansion method, however, does not allow good error estimates for the second order terms, where important beam theories also show up (see TRABUCHO and VIAÑO [1989, 1990] and Chapter II also). Therefore, in this section, we combine homogenization techniques and a Galerkin approach, developed in MIARA and TRABUCHO [1992] (cf. Section 33), and which has the advantage of giving suitable error estimates. These results were obtained by MASCARENHAS and TRABUCHO [1991, 1992].

Specifically, an equivalent formulation of the three-dimensional elasticity problem (34.1) posed now in a fixed (with respect to ε) domain $\Omega^{\delta,1}$ is considered. This is achieved using the same type of transformation as in Section 33 (cf. (33.3), (33.4) and also Remark 33.1). Then, the transformed displacement field $\boldsymbol{u}^{\delta}(\varepsilon) : \overline{\Omega}^{\delta,1} \to \mathbb{R}^3$ is obtained as the unique solution to an elliptic variational problem of the type

$$\boldsymbol{u}^{\delta}(\varepsilon) \in V(\Omega^{\delta,1}), \qquad B(\varepsilon)(\boldsymbol{u}^{\delta}(\varepsilon), \boldsymbol{v}) = F(\varepsilon)(\boldsymbol{v}), \quad \text{for all } \boldsymbol{v} \in V(\Omega^{\delta,1}). \tag{34.2}$$

We then seek an approximation $\boldsymbol{u}_N^{\delta}(\varepsilon)$ of $\boldsymbol{u}^{\delta}(\varepsilon)$ as the projection of $\boldsymbol{u}^{\delta}(\varepsilon)$ onto the approximation space $V_N(\Omega^{\delta,1})$ of $V(\Omega^{\delta,1})$, of the form (no sum on i):

$$\begin{aligned}
V_N(\Omega^{\delta,1}) = \{\boldsymbol{v} \in V(\Omega^{\delta,1})\colon \boldsymbol{v} = \left(\sum_{k=0}^{N} R_i^{k\delta}(x_1, x_2) v_i^k(x_3) \right)_{1 \leqslant i \leqslant 3},\\
(x_1, x_2) \in \omega^{\delta,1}, x_3 \in [0, L]\},
\end{aligned} \tag{34.3}$$

where $R_i^{k\delta}$ are some fixed, and conveniently chosen functions in $H^1(\omega^{\delta,1})$, called "basis functions", and v_i^k are coefficient functions in $H_0^1(0, L)$.

We proceed in order to study the behaviour of approximations $\boldsymbol{u}_N^{\delta}(\varepsilon)$, as δ becomes very small. This means studying the limit behaviour of the basis functions $R_i^{k\delta}$ and of the corresponding coefficient functions, which is done in the next section, thus obtaining the homogenized basis functions R_i^{k*}. In Section 36 we prove that the limits R_i^{k*} of $R_i^{k\delta}$, are exactly the basis functions associated with the Galerkin approximation of the homogenized three-dimensional problem, but the same does not happen with the limit of $\boldsymbol{u}_N^{\delta}(\varepsilon)$, as δ goes to zero.

As in Section 33 we associate with the displacement fields $\boldsymbol{u}^{\delta,\varepsilon} : \overline{\Omega}^{\delta,\varepsilon} \to \mathbb{R}^3$ and $\boldsymbol{v}^{\delta,\varepsilon} \in V(\Omega^{\delta,\varepsilon})$, functions $\boldsymbol{u}^{\delta}(\varepsilon) = (u_i^{\delta}(\varepsilon)) : \overline{\Omega}^{\delta,1} \to \mathbb{R}^3$ and $\boldsymbol{v} = (v_i) : \overline{\Omega}^{\delta,1} \to \mathbb{R}^3$, respectively, defined by the scalings

$$\begin{aligned} u_\alpha^{\delta,\varepsilon}(\boldsymbol{x}^\varepsilon) &= \varepsilon u_\alpha^{\delta}(\varepsilon)(\boldsymbol{x}), && \text{for all } \boldsymbol{x}^\varepsilon \in \overline{\Omega}^{\delta,\varepsilon}, \\ u_3^{\delta,\varepsilon}(\boldsymbol{x}^\varepsilon) &= \varepsilon^2 u_3^{\delta}(\varepsilon)(\boldsymbol{x}), && \text{for all } \boldsymbol{x}^\varepsilon \in \overline{\Omega}^{\delta,\varepsilon}, \\ v_\alpha^{\delta,\varepsilon}(\boldsymbol{x}^\varepsilon) &= \varepsilon v_\alpha(\boldsymbol{x}), && \text{for all } \boldsymbol{x}^\varepsilon \in \overline{\Omega}^{\delta,\varepsilon}, \\ v_3^{\delta,\varepsilon}(\boldsymbol{x}^\varepsilon) &= \varepsilon^2 v_3(\boldsymbol{x}), && \text{for all } \boldsymbol{x}^\varepsilon \in \overline{\Omega}^{\delta,\varepsilon}. \end{aligned} \tag{34.4}$$

We also assume that there exist functions $\boldsymbol{f}^{\varepsilon} \in [L^2(\Omega^{\delta,1})]^3$ and $\boldsymbol{g}^{\varepsilon} \in [L^2(\Gamma^{\delta,1})]^3$ independent of ε such that

$$\begin{aligned} f_\alpha^\varepsilon(\boldsymbol{x}^\varepsilon) &= \varepsilon^3 f_\alpha(\boldsymbol{x}), & f_3^\varepsilon(\boldsymbol{x}^\varepsilon) &= \varepsilon^2 f_\alpha(\boldsymbol{x}), && \text{for all } \boldsymbol{x}^\varepsilon \in \Omega^{\delta,\varepsilon}, \\ g_\alpha^\varepsilon(\boldsymbol{x}^\varepsilon) &= \varepsilon^4 g_\alpha(\boldsymbol{x}), & g_3^\varepsilon(\boldsymbol{x}^\varepsilon) &= \varepsilon^3 g_\alpha(\boldsymbol{x}), && \text{for all } \boldsymbol{x}^\varepsilon \in \Omega^{\delta,\varepsilon}, \end{aligned} \tag{34.5}$$

then the following proposition is an immediate consequence of these transformations and of Lax Milgram's lemma.

THEOREM 34.1. *Let $\boldsymbol{u}^{\delta}(\varepsilon) \in V(\Omega^{\delta,1})$ be the element obtained from solution $\boldsymbol{u}^{\delta,\varepsilon}$ of* (34.1) *through the transformations* (34.4). *Then, $\boldsymbol{u}^{\delta}(\varepsilon)$, is the unique solution to the following variational problem*:

$$\boldsymbol{u}^{\delta}(\varepsilon) \in V(\Omega^{\delta,1}), \qquad B(\varepsilon)(\boldsymbol{u}^{\delta}(\varepsilon), \boldsymbol{v}) = F(\varepsilon)(\boldsymbol{v}), \quad \textit{for all } \boldsymbol{v} \in V(\Omega^{\delta,1}), \tag{34.6}$$

where $B(\varepsilon)(\cdot,\cdot) : V(\Omega^{\delta,1}) \times V(\Omega^{\delta,1}) \to \mathbb{R}$ denotes the symmetric bilinear form defined by

$$\begin{aligned} B(\varepsilon)(\boldsymbol{w}, \boldsymbol{v}) = {} & \varepsilon^2 \int_{\Omega^{\delta,1}} [\lambda e_{\alpha\alpha}(\boldsymbol{w}) e_{\beta\beta}(\boldsymbol{v}) + 2\mu e_{\alpha\beta}(\boldsymbol{w}) e_{\alpha\beta}(\boldsymbol{v})] \, \mathrm{d}\boldsymbol{x}^{\delta,1} \\ & + \varepsilon^4 \int_{\Omega^{\delta,1}} \{\lambda [e_{33}(\boldsymbol{w}) e_{\beta\beta}(\boldsymbol{v}) + e_{\alpha\alpha}(\boldsymbol{w}) e_{33}(\boldsymbol{v})] \\ & \qquad + 4\mu e_{3\alpha}(\boldsymbol{w}) e_{3\alpha}(\boldsymbol{v})\} \, \mathrm{d}\boldsymbol{x}^{\delta,1} \\ & + \varepsilon^6 \int_{\Omega^{\delta,1}} [(\lambda + 2\mu) e_{33}(\boldsymbol{w}) e_{33}(\boldsymbol{v})] \, \mathrm{d}\boldsymbol{x}^{\delta,1}, \end{aligned} \tag{34.7}$$

and where $F(\varepsilon)(\cdot) : V(\Omega^{\delta,1}) \to \mathbb{R}$ *denotes the continuous, linear form, defined by*

$$F(\varepsilon)(\boldsymbol{v}) = \varepsilon^6 \left[\int_{\Omega^{\delta,1}} \boldsymbol{f} \cdot \boldsymbol{v}\, \mathrm{d}x^{\delta,1} + \int_{\Gamma^{\delta,1}} \boldsymbol{g} \cdot \boldsymbol{v}\, \mathrm{d}a^{\delta,1} \right]. \tag{34.8}$$

Let $V_N(\Omega^{\delta,1}) \subset V(\Omega^{\delta,1})$ be the approximation space and $\boldsymbol{u}_N^{\delta}(\varepsilon) \in V_N(\Omega^{\delta,1})$ the Galerkin approximation of the unique solution $\boldsymbol{u}^{\delta}(\varepsilon) \in V(\Omega^{\delta,1})$ to the variational problem (34.6)–(34.8). By definition $\boldsymbol{u}_N^{\delta}(\varepsilon)$ is the unique solution to the following variational problem:

$$\boldsymbol{u}_N^{\delta}(\varepsilon) \in V_N(\Omega^{\delta,1}), \qquad B(\varepsilon)(\boldsymbol{u}_N^{\delta}(\varepsilon), \boldsymbol{v}) = F(\varepsilon)(\boldsymbol{v}), \quad \text{for all } \boldsymbol{v} \in V_N(\Omega^{\delta,1}). \tag{34.9}$$

In order to simplify the notation, and since we are just interested in the homogenization process, we shall consider that all the applied forces vanish with the exception of the axial component g_3. We refer to MIARA and TRABUCHO [1990], or to Remarks 33.2 and 33.4, for the application of the method to other types of loads. The following result holds.

THEOREM 34.2. *Let* $f_i(\boldsymbol{x}) = 0$ *if* $\boldsymbol{x} \in \Omega^{\delta,1}$, $g_\alpha(\boldsymbol{x}) = 0$ *if* $\boldsymbol{x} \in \Gamma^{\delta,1}$, *and* $g_3(\boldsymbol{x}) = 0$ *if* $\boldsymbol{x} \in \partial T^{\delta,1} \times (0, L)$. *Let* $G_3 = \{g_3\colon \boldsymbol{x} \in (0, L) \to g_3(\boldsymbol{x}) = \partial_3 h, h \in H_0^4((0, L))\}$, *then, if the force component* $g_3 \in G_3 \cap H_0^{2N-1}((0, L))$ *for* $N \geqslant 1$, *there exist constants C and K, independent of* ε, δ *and N, such that*

$$\begin{aligned} \|\boldsymbol{u}^{\delta}(\varepsilon) - \boldsymbol{u}_N^{\delta}(\varepsilon)\|_{H^1(\Omega^{\delta,1})} &\leqslant CNK^{N-1}\varepsilon^{2N-2}, \\ \|\boldsymbol{u}^{\delta,\varepsilon} - \boldsymbol{u}_N^{\delta,\varepsilon)}\|_{H^1(\Omega^{\delta,\varepsilon})} &\leqslant CNK^{N-1}\varepsilon^{2N-1}, \end{aligned} \tag{34.10}$$

where $\boldsymbol{u}^{\delta,\varepsilon}$ *is the transformed of* $\boldsymbol{u}^{\delta}(\varepsilon)$ *through the inverse transformations of* (33.4), (33.5). *The approximation space* $V_N(\Omega^{\delta,1})$ *is given by*

$$\begin{aligned} V_N(\Omega^{\delta,1}) = \{ \boldsymbol{v} = (v_i) \in V(\Omega^{\delta,1})\colon & \\ & v_\alpha = \sum_{k=0}^{N} P_\alpha^{k\delta}(x_1, x_2) v_\alpha^k(x_3), \\ & v_\alpha^k \in H_0^1(0, L), P_\alpha^{k\delta} \in H^1(\omega^{\delta,1}), \\ & v_3 = \sum_{k=0}^{N} Q^{k\delta}(x_1, x_2) v_3^k(x_3), v_3^k \in H_0^1(0, L), \\ & Q^{k\delta} \in H^1(\omega^{\delta,1}), \quad \textit{for all } 0 \leqslant k \leqslant N \ (\textit{no sum on } \delta) \} \end{aligned} \tag{34.11}$$

and the basis functions are defined, by recurrence on k, as the solution to the following boundary value problems:

$$\int_{\omega^{\delta,1}} [\lambda e_{\alpha\alpha}(\boldsymbol{P}^{k\delta}) e_{\beta\beta}(\boldsymbol{s}) + 2\mu e_{\alpha\beta}(\boldsymbol{P}^{k\delta})\partial_\alpha s_\beta]\, \mathrm{d}\omega^{\delta,1} + \int_{\omega^{\delta,1}} [\lambda Q^{(k-1)\delta} e_{\alpha\alpha}(\boldsymbol{s}) - \mu(\partial_\alpha Q^{(k-1)\delta} + P_\alpha^{(k-1)\delta}) s_\alpha]\, \mathrm{d}\omega^{\delta,1} = 0, \tag{34.12}$$

for all $\boldsymbol{s} \in [H^1(\omega^{\delta,1})]^2$,

$$\int_{\omega^{\delta,1}} \mu(\partial_\alpha Q^{k\delta} + P_\alpha^{k\delta})\partial_\alpha r\, \mathrm{d}\omega^{\delta,1} - \int_{\omega^{\delta,1}} [(\lambda + 2\mu) Q^{(k-1)\delta} + \lambda \partial_\alpha P_\alpha^{k\delta}] r\, \mathrm{d}\omega^{\delta,1} = \delta_{k1} \int_\gamma r\, \mathrm{d}\gamma, \tag{34.13}$$

for all $r \in H^1(\omega^{\delta,1})$,

with $P_\alpha^{-1\delta} = 0$ *and* $Q^{-1\delta} = 0$.

Before proving Theorem 34.2 we give an equivalent formulation to problems (34.12) and (34.13).

THEOREM 34.3. *Let* $\boldsymbol{P}^{k\delta}$ *and* $Q^{k\delta}$, *for* $k \in Z$, $k \geqslant -1$, *satisfy expressions* (34.12) *and* (34.13). *Then they are uniquely determined, by recurrence, as follows*:

$$\boldsymbol{P}^{-1\delta} = \boldsymbol{0}, \qquad Q^{-1\delta} = 0 \qquad (k = -1); \tag{34.14}$$

$$\boldsymbol{P}^{k\delta} = \overline{\boldsymbol{P}}^{k\delta} + {}^R\boldsymbol{P}^{k\delta}, \qquad Q^{k\delta} = \overline{Q}^{k\delta} + {}^R Q^{k\delta} \qquad (k \geqslant 0), \tag{34.15}$$

where $\overline{\boldsymbol{P}}^{k\delta}$ *is the solution of the following plane deformation elasticity problem in* $\omega^{\delta,1}$:

$$\begin{aligned} & -\partial_\alpha S_{\alpha\beta}^{k\delta} = \mu(P_\beta^{(k-1)\delta} + \partial_\beta Q^{(k-1)\delta}) + \lambda \partial_\beta Q^{(k-1)\delta}, \quad \text{in } \omega^{\delta,1}, \\ & S_{\alpha\beta}^{k\delta} n_\alpha = -\lambda Q^{(k-1)\delta} n_\beta, \quad \text{on } \gamma^{\delta,1}, \\ & \int_{\omega^{\delta,1}} \overline{P}_\alpha^{k\delta}\, \mathrm{d}\omega^{\delta,1} = \int_{\omega^{\delta,1}} (\overline{P}_1^{k\delta} x_2 - \overline{P}_2^{k\delta} x_1)\, \mathrm{d}\omega^{\delta,1} = 0, \end{aligned} \tag{34.16}$$

with $S_{\alpha\beta}^{k\delta}$ *given by* $S_{\alpha\beta}^{k\delta} = 2\mu e_{\alpha\beta}(\overline{\boldsymbol{P}}^{k\delta}) + \lambda e_{\tau\tau}(\overline{\boldsymbol{P}}^{k\delta})\delta_{\alpha\beta}$, *that is,*

$$S_{\alpha\beta}^{k\delta} = S_{\alpha\beta\zeta\rho} e_{\zeta\rho}(\overline{\boldsymbol{P}}^{k\delta}), \qquad S_{\alpha\beta\zeta\rho} = \mu(\delta_{\alpha\zeta}\delta_{\beta\rho} + \delta_{\alpha\rho}\delta_{\beta\zeta}) + \lambda \delta_{\alpha\beta}\delta_{\zeta\rho}; \tag{34.17}$$

$\overline{Q}^{k\delta}$ *is the solution of the following membrane or torsion type problem in* $\omega^{\delta,1}$:

$$
\begin{aligned}
&-\mu\partial_{\beta\beta}\overline{Q}^{k\delta} = (\lambda+\mu)\partial_\alpha\overline{P}^{k\delta}_\alpha + (\lambda+2\mu)Q^{(k-1)\delta}, \quad \textit{in } \omega^{\delta,1},\\
&\partial_n\overline{Q}^{k\delta} = -\overline{\boldsymbol{P}}^{k\delta}\cdot\boldsymbol{n} + \frac{1}{\mu}\delta_{k1}, \quad \textit{on } \gamma,\\
&\partial_n\overline{Q}^{k\delta} = -\overline{\boldsymbol{P}}^{k\delta}\cdot\boldsymbol{n}, \quad \textit{on } \partial T^{\delta,1},\\
&\int_{\omega^{\delta,1}} \overline{Q}^{k\delta}\,\mathrm{d}\omega^{\delta,1} = 0;
\end{aligned}
\tag{34.18}
$$

$$
{}^RP_1^{k\delta} = a_1^{k\delta} + b^{k\delta}x_2, \qquad {}^RP_2^{k\delta} = a_2^{k\delta} - b^{k\delta}x_1 \qquad \textit{(rigid displacement)} \tag{34.19}
$$

$$
{}^RQ^{k\delta} = -b^{k\delta}w^\delta - a_1^{k\delta}x_1 - a_2^{k\delta}x_2 + c^{k\delta}, \tag{34.20}
$$

with constants $a_\alpha^{k\delta}$, $b^{k\delta}$ *and* $c^{k\delta}$ *defined by* (*no sum on* α, $\tau \neq \alpha$),

$$
\begin{aligned}
a_\alpha^{k\delta} = {} & \frac{1}{EI_\alpha^\delta}\int_\gamma x_\alpha\,\mathrm{d}\gamma\,\delta_{k0} + \frac{1}{I_\alpha^\delta}\int_{\omega^{\delta,1}} \overline{Q}^{k\delta}x_\alpha\,\mathrm{d}\omega^{\delta,1}\\
&+ \frac{\nu\mu}{EI_\alpha^\delta}\int_{\omega^{\delta,1}} \left(\overline{P}_\beta^{k\delta} + \partial_\beta\overline{Q}^{k\delta}\right)\Phi_{\alpha\beta}\,\mathrm{d}\omega^{\delta,1}\\
&- \frac{1}{2I_\alpha^\delta}b^{k\delta}I_\alpha^{w^\delta} - \frac{\nu\mu b^{k\delta}}{EI_\alpha^\delta}\left(H_\tau^\delta + \int_{\omega^{\delta,1}} \partial_\beta w^\delta\phi_{\alpha\beta}\,\mathrm{d}\omega^{\delta,1}\right),
\end{aligned}
\tag{34.21}
$$

$$
b^{k\delta} = \frac{1}{J^\delta}\int_{\omega^{\delta,1}} \left(\partial_2\overline{Q}^{k\delta}x_1 - \partial_1\overline{Q}^{k\delta}x_2\right)\mathrm{d}\omega^{\delta,1}, \tag{34.22}
$$

$$
c^{k\delta} = -\frac{1}{E|\omega^{\delta,1}|}\left[|\gamma|\delta_{k0} + \nu\mu\int_{\omega^{\delta,1}} (\overline{P}_\beta^{k\delta} + \partial_\beta\overline{Q}^{k\delta})x_\beta\,\mathrm{d}\omega^{\delta,1}\right], \tag{34.23}
$$

where, with the same notation as before, E *and* ν *are, respectively, the modulus of elasticity and Poisson's ratio, related with Lamé's constants* λ *and* μ *by* $E = \mu(3\lambda+2\mu)/(\lambda+\mu)$, *and* $\nu = \lambda/[2(\lambda+\mu)]$. *Moreover, the warping function* w^δ *is given as the unique solution of the following boundary value problem*:

$$
\begin{aligned}
&-\partial_{\alpha\alpha}w^\delta = 0, \quad \textit{in } \omega^{\delta,1},\\
&\partial_n w^\delta = x_2n_1 - x_1n_2, \quad \textit{on } \gamma^{\delta,1},\\
&\int_{\omega^{\delta,1}} w^\delta\,\mathrm{d}\omega^{\delta,1} = 0,
\end{aligned}
\tag{34.24}
$$

and the remaining constants and functions are given by

$$I_\alpha^\delta = \int_{\omega^{\delta,1}} (x_\alpha)^2 \,\mathrm{d}\omega^{\delta,1} \quad \textit{(second moment of area with respect to } e_\beta\ (\alpha \neq \beta)\textit{)}, \tag{34.25}$$

$$\Phi_{\alpha\beta}(x_1, x_2) = \begin{pmatrix} \frac{1}{2}[(x_1)^2 - (x_2)^2] & x_1x_2 \\ x_1x_2 & \frac{1}{2}[(x_2)^2 - (x_1)^2] \end{pmatrix}, \tag{34.26}$$

$$H_\alpha^\delta = \frac{1}{2} \int_{\omega^{\delta,1}} x_\alpha[(x_1)^2 + (x_2)^2] \,\mathrm{d}\omega^{\delta,1}, \tag{34.27}$$

$$J^\delta = I_1^\delta + I_2^\delta + \int_{\omega^{\delta,1}} (\partial_2 w^\delta x_1 - \partial_1 w^\delta x_2) \,\mathrm{d}\omega^{\delta,1} \quad \textit{(torsional constant)}, \tag{34.28}$$

$$I_\beta^{w^\delta} = 2 \int_{\omega^{\delta,1}} x_\beta w^\delta \,\mathrm{d}\omega^{\delta,1}. \tag{34.29}$$

Conversely, the functions defined by (34.14)–(34.29) *satisfy* (34.12) *and* (34.13).

PROOF. The equivalence between the variational formulation (34.12) and problem $(34.15)_1$–(34.17) and (34.19) is clear. In what concerns the variational formulation (34.13) it is also clear that its solution is given by the sum of the solution of problem (34.18) with the solution, defined up to an additive constant, denoted by $c^{k\delta}$, of the following problem:

$$\begin{aligned} &-\mu\partial_{\alpha\alpha}{}^R Q^{k\delta} = (\lambda + \mu)\partial_\alpha{}^R P_\alpha^{k\delta}, \quad \text{in } \omega^{\delta,1}, \\ &\partial_n{}^R Q^{k\delta} = -{}^R P_\alpha^{k\delta} \cdot \boldsymbol{n}, \quad \text{in } \gamma^{\delta,1}, \end{aligned} \tag{34.30}$$

which is given by (34.20) (see (34.24)), since ${}^R\boldsymbol{P}^{k\delta}$ has the particular form of the infinitesimal rigid displacement (34.19).

To determine constants $a_\alpha^{k\delta}$, $b^{k\delta}$ and $c^{k\delta}$ it is enough to consider the compatibility conditions associated to problems (34.16)–(34.18) and consequently, included in the variational formulation (34.12), (34.13). These conditions are

$$\int_{\omega^{\delta,1}} \left(P_\beta^{(k-1)\delta} + \partial_\beta Q^{(k-1)\delta}\right) \mathrm{d}\omega^{\delta,1} = 0, \tag{34.31}$$

$$\int_{\omega^{\delta,1}} \left[\left(P_1^{(k-1)\delta} + \partial_1 Q^{(k-1)\delta}\right) x_2 - \left(P_2^{(k-1)\delta} + \partial_2 Q^{(k-1)\delta}\right) x_1\right] \mathrm{d}\omega^{\delta,1} = 0, \tag{34.32}$$

$$\int_{\omega^{\delta,1}} \left[\lambda\partial_\alpha \overline{P}_\alpha^{k\delta} + (\lambda + 2\mu) Q^{(k-1)\delta}\right] \mathrm{d}\omega^{\delta,1} + |\gamma|\delta_{k1} = 0. \tag{34.33}$$

Condition (34.32), for the order $k+1$, enables us to determine constant $b^{k\delta}$. In fact, using decompositions $\boldsymbol{P}^{k\delta} = \overline{\boldsymbol{P}}^{k\delta} + {}^R\boldsymbol{P}^{k\delta}$, with ${}^R\boldsymbol{P}^{k\delta}$ given by (34.19), $Q^{k\delta} = \overline{Q}^{k\delta} + {}^RQ^{k\delta}$, with ${}^RQ^{k\delta}$ given by (34.20), and the expression for J^δ, given by (34.28), we obtain expression (34.22), from (34.32).

Condition (34.33), for the order $k+1$, enables us to determine constant $c^{k\delta}$: using the previous decomposition of $Q^{k\delta}$ (see (34.15), (34.20) and (34.24)), condition (34.33), for the order $k+1$, reads

$$\int_{\omega^{\delta,1}} \lambda \partial_\alpha \overline{P}_\alpha^{(k+1)\delta} \, \mathrm{d}\omega^{\delta,1} + (\lambda + 2\mu)|\omega^{\delta,1}| c^{k\delta} + |\gamma| \delta_{k0} = 0. \tag{34.34}$$

From the variational formulation of $\overline{\boldsymbol{P}}^{(k+1)\delta}$, using as test functions $v_\beta = x_\beta$ and the explicit formulas of ${}^R\boldsymbol{P}^{k\delta}$ and ${}^RQ^{k\delta}$, respectively (34.19) and (34.20), one obtains

$$\begin{aligned} 2(\lambda+\mu) \int_{\omega^{\delta,1}} \partial_\alpha \overline{P}_\alpha^{(k+1)\delta} \, \mathrm{d}\omega^{\delta,1} + 2\lambda |\omega^{\delta,1}| c^{k\delta} \\ - \mu \int_{\omega^{\delta,1}} (\overline{P}_\beta^{k\delta} + \partial_\beta \overline{Q}^{k\delta}) x_\beta \, \mathrm{d}\omega^{\delta,1} \\ = -\mu b^{k\delta} \int_{\omega^{\delta,1}} (\partial_1 w^\delta x_1 + \partial_2 w^\delta x_2) \, \mathrm{d}\omega^{\delta,1}. \end{aligned} \tag{34.35}$$

From the variational formulation of w^δ (see (34.24)), using $\varphi = \frac{1}{2}(x_1)^2 + (x_2)^2$ as a test function, one proves that the second member of (34.35) vanishes. This last equation, together with (34.34), allows us to determine expression (34.23) for $c^{k\delta}$.

Finally, condition (34.31), for the order $k+2$, enables us to determine constant $a_\alpha^{k\delta}$. From the variational formulation of $Q^{(k+1)\delta}$ (see (34.13)), using $v = x_1$ as test function, and from condition (34.31), one obtains

$$\lambda \int_{\omega^{\delta,1}} \partial_\beta P_\beta^{(k+1)\delta} x_1 \, \mathrm{d}\omega^{\delta,1} + (\lambda + 2\mu) \int_{\omega^{\delta,1}} Q^{k\delta} x_1 \, \mathrm{d}\omega^{\delta,1} + \delta_{k0} \int_\gamma x_1 \, \mathrm{d}\gamma = 0; \tag{34.36}$$

from the variational formulation of $P_\beta^{(k+1)\delta}$ (see (34.12)), using $v_1 = \Phi_{11}$, $v_2 = \Phi_{12}$, as test functions, one obtains

$$\begin{aligned} \lambda \int_{\omega^{\delta,1}} \partial_\beta P_\beta^{(k+1)\delta} x_1 \, \mathrm{d}\omega^{\delta,1} \\ = \frac{1}{2(\lambda+\mu)} \left[-2\lambda^2 \int_{\omega^{\delta,1}} Q^{k\delta} x_1 \, \mathrm{d}\omega^{\delta,1} + \lambda\mu \int_{\omega^{\delta,1}} (P_\beta^{k\delta} + \partial_\beta Q^{k\delta}) \Phi_{1\beta} \, \mathrm{d}\omega^{\delta,1} \right]. \end{aligned} \tag{34.37}$$

Both expressions (34.36) and (34.37), together with the decomposition of $\boldsymbol{P}^{k\delta}$ and $Q^{k\delta}$, enable us to determine $a_1^{k\delta}$ and analogously $a_2^{k\delta}$.

Conversely, it is clear that the solutions $\boldsymbol{P}^{k\delta}$ and $Q^{k\delta}$, obtained from (34.14)–(34.29) satisfy equalities (34.12) and (34.13), since the expressions chosen for the constants $a_\alpha^{k\delta}$, $b^{k\delta}$ and $c^{k\delta}$ also imply the compatibility conditions (34.31), for order $k+2$; (34.32), for order $k+1$; and (34.33), for order $k+1$, respectively. □

PROOF (of Theorem 34.2). As in MIARA and TRABUCHO [1992] one obtains the following estimates:

$$\begin{aligned} &\|\boldsymbol{u}^{\delta}(\varepsilon)-\boldsymbol{u}_N^{\delta}(\varepsilon)\|_{H^1(\Omega^{\delta,1})} \leqslant C_N^{\delta}\varepsilon^{2N-2}, \\ &\|\boldsymbol{u}^{\delta,\varepsilon}-\boldsymbol{u}_N^{\delta,\varepsilon}\|_{H^1(\Omega^{\delta,\varepsilon})} \leqslant C_N^{\delta}\varepsilon^{2N-1}, \end{aligned} \tag{34.38}$$

where $C_N^{\delta}=\max\{\|Q^{N\delta}\|_{H^1(\omega^{\delta,1})}, \|\boldsymbol{P}^{N\delta}\|_{L^2(\omega^{\delta,1})}\}$. From decomposition (34.15) we write

$$\|Q^{N\delta}\|_{H^1(\omega^{\delta,1})} \leqslant \|\overline{Q}^{N\delta}\|_{H^1(\omega^{\delta,1})} + \|{}^R Q^{N\delta}\|_{H^1(\omega^{\delta,1})},$$

$$\|\boldsymbol{P}^{N\delta}\|_{H^1(\omega^{\delta,1})} \leqslant \|\overline{\boldsymbol{P}}^{N\delta}\|_{H^1(\omega^{\delta,1})} + \|{}^R \boldsymbol{P}^{N\delta}\|_{H^1(\omega^{\delta,1})},$$

and we evaluate these norms by recurrence. In what follows, C will denote different constants, independent of δ. From the variational formulation of problem (34.16), (34.17) and Korn's inequality we conclude that there exists a constant C, independent of δ, such that

$$\|\overline{\boldsymbol{P}}^{k\delta}\|_{H^1(\omega^{\delta,1})} \leqslant C(\|Q^{(k-1)\delta}\|_{H^1(\omega^{\delta,1})} + \|\boldsymbol{P}^{(k-1)\delta}\|_{L^2(\omega^{\delta,1})}).$$

In a similar way, from the variational formulation of problem (34.18) and the Poincaré–Wirtinger inequality we conclude that there exists a constant C, independent of δ, such that

$$\|\overline{Q}^{k\delta}\|_{H^1(\omega^{\delta,1})} \leqslant C(\|Q^{(k-1)\delta}\|_{H^1(\omega^{\delta,1})} + \|\boldsymbol{P}^{(k-1)\delta}\|_{L^2(\omega^{\delta,1})} + 1).$$

From definitions (34.19) and (34.20) we obtain

$$\begin{aligned} \|{}^R Q^{k\delta}\|^2_{L^2(\omega^{\delta,1})} &= (c^{k\delta})^2|\omega^{\delta,1}| + (a_1^{k\delta})^2 I_1 + (a_2^{k\delta})^2 I_2 \\ &\quad - \tfrac{1}{2}b^{k\delta}(a_1^{k\delta}I_1^{w^\delta} + a_2^{k\delta}I_2^{w^\delta}) + (b^{k\delta})^2\|w\|_{L^2(\omega^{\delta,1})}, \end{aligned}$$

$$\|{}^R \boldsymbol{p}^{k\delta}\|^2_{L^2(\omega^{\delta,1})} = (b^{k\delta})^2(I_1+I_2) + [(a_1^{k\delta})^2 + a_2^{k\delta})^2)]|\omega^{\delta,1}|.$$

From Eqs. (34.21)–(34.23) we obtain that there exists a constant C, independent of δ, such that

$$(a_\alpha^{k\delta})^2 \leqslant C(\|\overline{Q}^{k\delta}\|^2_{H^1(\omega^{\delta,1})} + \|\overline{\boldsymbol{P}}^{k\delta}\|^2_{L^2(\omega^{\delta,1})} + 1),$$

$$(b^{k\delta})^2 \leqslant C(\|\overline{Q}^{k\delta}\|^2_{H^1(\omega^{\delta,1})} + \|\overline{\boldsymbol{P}}^{k\delta}\|^2_{L^2(\omega^{\delta,1})}),$$

$$(c^{k\delta})^2 \leqslant C(\|\overline{Q}^{k\delta}\|^2_{H^1(\omega^{\delta,1})} + \|\overline{\boldsymbol{P}}^{k\delta}\|^2_{L^2(\omega^{\delta,1})} + 1),$$

and consequently,

$$\|{}^R Q^{k\delta}\|^2_{L^2(\omega^{\delta,1})} \leqslant C(\|\overline{Q}^{k\delta}\|^2_{H^1(\omega^{\delta,1})} + \|\overline{\boldsymbol{P}}^{k\delta}\|^2_{L^2(\omega^{\delta,1})} + 1),$$

$$\|{}^R \boldsymbol{P}^{k\delta}\|^2_{L^2(\omega^{\delta,1})} \leqslant C(\|\overline{Q}^{k\delta}\|^2_{H^1(\omega^{\delta,1})} + \|\overline{\boldsymbol{P}}^{k\delta}\|^2_{L^2(\omega^{\delta,1})} + 1).$$

We then conclude that

$$\|Q^{k\delta}\|^2_{L^2(\omega^{\delta,1})} \leqslant C(\|\overline{Q}^{k\delta}\|^2_{H^1(\omega^{\delta,1})} + \|\overline{\boldsymbol{P}}^{k\delta}\|^2_{L^2(\omega^{\delta,1})} + 1),$$

$$\|\boldsymbol{P}^{k\delta}\|^2_{L^2(\omega^{\delta,1})} \leqslant C(\|\overline{Q}^{k\delta}\|^2_{H^1(\omega^{\delta,1})} + \|\overline{\boldsymbol{P}}^{k\delta}\|^2_{L^2(\omega^{\delta,1})} + 1).$$

In a similar way, from (34.19)–(34.23) and the variational formulation of problem (34.30), we conclude the existence of a constant C, independent of δ, such that

$$\|{}^R Q^{k\delta}\|^2_{H^1(\omega^{\delta,1})} \leqslant C(\|\overline{Q}^{k\delta}\|^2_{H^1(\omega^{\delta,1})} + \|\overline{\boldsymbol{P}}^{k\delta}\|^2_{L^2(\omega^{\delta,1})} + 1),$$

$$\|{}^R \boldsymbol{P}^{k\delta}\|^2_{H^1(\omega^{\delta,1})} \leqslant C(\|\overline{Q}^{k\delta}\|^2_{H^1(\omega^{\delta,1})} + \|\overline{\boldsymbol{P}}^{k\delta}\|^2_{L^2(\omega^{\delta,1})} + 1).$$

Putting all these results together, we obtain

$$\|Q^{k\delta}\|^2_{H^1(\omega^{\delta,1})} \leqslant C(\|\overline{Q}^{(k-1)\delta}\|^2_{H^1(\omega^{\delta,1})} + \|\overline{\boldsymbol{P}}^{(k-1)\delta}\|^2_{L^2(\omega^{\delta,1})} + 1),$$

$$\|\boldsymbol{P}^{k\delta}\|^2_{H^1(\omega^{\delta,1})} \leqslant C(\|\overline{Q}^{(k-1)\delta}\|^2_{H^1(\omega^{\delta,1})} + \|\overline{\boldsymbol{P}}^{(k-1)\delta}\|^2_{L^2(\omega^{\delta,1})} + 1)$$

from which we conclude that

$$C_N^\delta \leqslant C(\|\overline{Q}^{(N-1)\delta}\|^2_{H^1(\omega^{\delta,1})} + \|\overline{\boldsymbol{P}}^{(N-1)\delta}\|^2_{L^2(\omega^{\delta,1})} + 1).$$

Applying this expression recursively and taking into account that $\overline{Q}^{0\delta} = 0$ and $\overline{\boldsymbol{P}}^{0\delta} = 0$, one has $C_N \leqslant \sum_{k=1}^N 2^{k-1} C^k$, and the result follows with $K = 2C$.

Moreover, if $\varepsilon^2 < 1/2C$ then $\|\boldsymbol{u}^\delta(\varepsilon) - \boldsymbol{u}_N^\delta(\varepsilon)\|_{H^1(\Omega^{\delta,1})}$ goes to zero when N goes to infinity.

The second inequality in (34.10) is just a consequence of the transformations (34.4) applied to this result. □

35. Homogenization of the basis functions

In this section we pass to the limit as δ goes to zero, in the recurrence problems used to define the basis functions $\boldsymbol{P}^{k\delta}$ and $Q^{k\delta}$ and find the limit functions $\boldsymbol{P}^{k*}$ and Q^{k*}.

We recall the following result issued from Theorem 31.1.

THEOREM 35.1 (MASCARENHAS and TRABUCHO [1990, Proposition 3.1]). *Let w^δ, be the solution of problem (34.24), then there exists an extension $\hat{w}^\delta$ of w^δ, bounded in $H^1(\omega)$, independently of δ. Any other extension $\hat{w}^\delta$ of w^δ, bounded in $H^1(\omega)$, independently of δ, satisfies $\hat{w}^\delta \to w^*$, weakly in $H^1(\omega)$, where w^* is the unique solution of the following problem:*

$$\begin{aligned} &-\operatorname{div}(A^*\nabla w^*) = 0, \quad \text{in } \omega,\\ &(A^*\nabla w^*)\cdot \boldsymbol{n} = A^*(x_2, -x_1)\cdot \boldsymbol{n}, \quad \text{on } \gamma,\\ &\int_\omega w^*\,\mathrm{d}\omega = 0, \end{aligned} \tag{35.1}$$

where $A^* = (a^*_{\alpha\beta})$ is a 2×2 symmetric and positive definite matrix, whose coefficients $a^*_{\alpha\beta}$ are defined by

$$a^*_{\alpha\beta} = \Theta\delta_{\alpha\beta} + \int_{Y^*} \partial_\alpha \chi_\beta\,\mathrm{d}Y^* = \int_{Y^*} \nabla(\chi_\alpha + y_\alpha)\cdot\nabla(\chi_\beta + y_\beta)\,\mathrm{d}Y^*, \tag{35.2}$$

and where χ_β is the unique solution of the following problem defined in the cell Y^*:

$$\begin{aligned} &\chi_\beta\ Y\text{-periodic}, \qquad \int_{Y^*}\chi_\beta\,\mathrm{d}Y^* = 0,\\ &-\Delta\chi_\beta = 0, \quad \text{in } Y^*,\\ &\partial_n\chi_\beta = -n_\beta, \quad \text{on } \partial T. \end{aligned} \tag{35.3}$$

Moreover, representing by $[\nabla w^\delta]^\sim$ the zero extension of ∇w^δ to the entire ω, the following convergence holds:

$$[\nabla w^\delta]^\sim \to A^*\nabla w^* + (A^* - \Theta\,\mathrm{Id})(-x_2, x_1), \quad \text{weakly in } [L^2(\omega)]^2. \tag{35.4}$$

PROOF. See the proof of Theorem 31.1. □

We now state and prove the homogenization result, concerning the basis functions. This result is also owed to MASCARENHAS and TRABUCHO [1991, Proposition 4.8].

THEOREM 35.2. *Let $\boldsymbol{P}^{k\delta}$ and $Q^{k\delta}$ be the basis functions defined by* (34.12), (34.13). *For each k there exist extensions $\hat{\boldsymbol{P}}^{k\delta}$ and $\hat{Q}^{k\delta}$ of $\boldsymbol{P}^{k\delta}$ and $Q^{k\delta}$, respectively, which are bounded, in $[H^1(\omega)]^2$ and $H^1(\omega)$, respectively, independent of δ. Any other extensions $\hat{\boldsymbol{P}}^{k\delta}$ and $\hat{Q}^{k\delta}$, bounded independently of δ, satisfy the following convergences:*

$$\hat{\boldsymbol{P}}^{k\delta} \rightarrow \boldsymbol{P}^{k*}, \quad \textit{weakly in } [H^1(\omega)]^2, \tag{35.5}$$

$$\hat{Q}^{k\delta} \rightarrow Q^{k*}, \quad \textit{weakly in } H^1(\omega), \tag{35.6}$$

where $\boldsymbol{P}^{k*}$ and Q^{k*} are uniquely defined by the following recurrence formulas:

$$\boldsymbol{P}^{-1*} = 0, \qquad Q^{-1*} = 0 \qquad (k = -1), \tag{35.7}$$

$$\boldsymbol{P}^{k*} = \overline{\boldsymbol{P}}^{k*} + {}^R\boldsymbol{P}^{k*}, \qquad Q^{k*} = \overline{Q}^{k*} + {}^R Q^{k*} \qquad (k \geqslant 0), \tag{35.8}$$

and where $\overline{\boldsymbol{P}}^{k*}$ is the solution of the following plane deformation elasticity problem in ω:

$$\begin{aligned} &-\partial_\alpha S^{k*}_{\alpha\beta} = \mu a^*_{\alpha\beta}(P^{(k-1)*}_\alpha + \partial_\alpha Q^{(k-1)*}) + \lambda(\Theta\delta_{\alpha\beta} + s^*_{\rho\rho\alpha\beta})\partial_\alpha Q^{(k-1)*}, \quad \textit{in } \omega, \\ &S^{k*}_{\alpha\beta} n_\alpha = -\lambda(\Theta\delta_{\alpha\beta} + s^*_{\rho\rho\alpha\beta}) Q^{(k-1)*} n_\alpha, \quad \textit{on } \gamma, \\ &\int_\omega \overline{P}^{k*}_\alpha \,\mathrm{d}\omega = \int_\omega (\overline{P}^{k*}_1 x_2 - \overline{P}^{k*}_2 x_1)\,\mathrm{d}\omega = 0, \end{aligned} \tag{35.9}$$

with $S^{k*}_{\alpha\beta}$ given by

$$S^{k*}_{\alpha\beta} = S^*_{\alpha\beta\zeta\rho} e_{\zeta\rho}(\overline{\boldsymbol{P}}^{k*}), \qquad S^*_{\alpha\beta\zeta\rho} = \Theta S_{\alpha\beta\zeta\rho} + S_{\alpha\beta\tau\sigma} s^*_{\tau\sigma\zeta\rho}, \tag{35.10}$$

$$s^*_{\alpha\beta\zeta\rho} = \int_{Y^*} e_{\alpha\beta}(\boldsymbol{\chi}^{\zeta\rho})\,\mathrm{d}Y^*, \tag{35.11}$$

for $\boldsymbol{\chi}^{\zeta\rho} = (\chi^{\zeta\rho}_\alpha)$ defined, up to a constant vector, as the solution of the following problem, in the unit cell Y^* and valid for all $\boldsymbol{\varphi} \in [H^1(Y^*)]^2$, Y-periodic:

$$\begin{aligned} &\boldsymbol{\chi}^{\zeta\rho} \in [H^1(Y^*)]^2, \qquad \boldsymbol{\chi}^{\zeta\rho} \ \textit{Y-periodic}, \\ &\int_{Y^*} [2\mu e_{\alpha\beta}(\boldsymbol{\chi}^{\zeta\rho}) + \lambda e_{\mu\mu}(\boldsymbol{\chi}^{\zeta\rho})\delta_{\alpha\beta}] e_{\alpha\beta}(\boldsymbol{\varphi})\,\mathrm{d}Y^* \\ &\quad = -\int_{Y^*} [2\mu e_{\zeta\rho}(\boldsymbol{\varphi}) + \lambda e_{\mu\mu}(\boldsymbol{\varphi})\delta_{\zeta\rho}]\,\mathrm{d}Y^*, \end{aligned} \tag{35.12}$$

the elasticity tensor $(S^*_{\alpha\beta\zeta\rho})$ being positive definite, satisfying $S^*_{\alpha\beta\zeta\rho} = S^*_{\beta\alpha\zeta\rho} = S^*_{\zeta\rho\alpha\beta}$, and $(a^*_{\alpha\beta}) = A^*$ defined as in Theorem 35.1. Moreover, $\overline{Q}^{k*}$ is the solution *of the following membrane or (anisotropic) torsion type problem in linearized elasticity, for a two-dimensional body occupying volume ω:*

$$\begin{aligned}
&-\mu\,\mathrm{div}(A^*\nabla\overline{Q}^{k*})\\
&\quad=\mu\,\mathrm{div}(A^*\overline{\boldsymbol{P}}^{k*})+\lambda[\Theta e_{\beta\beta}(\overline{\boldsymbol{P}}^{k*})+s^*_{\beta\beta\zeta\rho}e_{\zeta\rho}(\overline{\boldsymbol{P}}^{k*})]\\
&\qquad+\frac{\lambda^2}{2(\lambda+\mu)}s^*_{\mu\mu\beta\beta}Q^{(k-1)*}+(\lambda+2\mu)\Theta Q^{(k-1)*},\quad \text{in } \omega,\\
&(A^*\nabla\overline{Q}^{k*})\cdot\boldsymbol{n}=-(A^*\overline{\boldsymbol{P}}^{k*})\cdot\boldsymbol{n}+\frac{1}{\mu}\delta_{k1},\quad \text{on } \gamma,\\
&\int_\omega \overline{Q}^{k*}\,\mathrm{d}\omega=0,
\end{aligned}\tag{35.13}$$

with

$${}^RP_1^{k*}=a_1^{k*}+b^{k*}x_2,\qquad {}^RP_2^{k*}=a_2^{k*}-b^{k*}x_2,\tag{35.14}$$

$${}^RQ^{k*}=-b^{k*}w^*-a_1^{k*}x_1-a_2^{k*}x_2+c^{k*},\tag{35.15}$$

where w^* is the solution of problem (35.1) *and where constants a_α^{k*}, b^{k*} and c^{k*} are given by*

$$\begin{aligned}
a_\alpha^{k*}={}&\frac{\delta_{k0}}{E\Theta I_\alpha}\int_\gamma x_\alpha\,\mathrm{d}\gamma+\frac{1}{I_\alpha}\int_\omega \overline{Q}^{k*}x_\alpha\,\mathrm{d}\omega\\
&+\frac{\nu\mu}{E\Theta I_\alpha}\int_\omega[(a^*_{\beta\zeta}(\overline{P}_\zeta^{k*}+\partial_\zeta\overline{Q}^{k*})]\Phi_{\alpha\beta}\,\mathrm{d}\omega\\
&-\frac{1}{2I_\alpha}b^{k*}I_\alpha^{w^*}-\frac{\nu\mu b^{k*}}{E\Theta I_\alpha}\int_\omega[a^*_{\beta\zeta}(\partial_\zeta w^*-(x_2,-x_1))]\phi_{\alpha\beta}\,\mathrm{d}\omega,
\end{aligned}\tag{35.16}$$

$$b^{k*}=\frac{1}{J^*}\int_\omega[A^*(\overline{\boldsymbol{P}}^{k*}+\nabla\overline{Q}^{k*})\cdot(-x_2,x_1)]\,\mathrm{d}\omega,\tag{35.17}$$

$$c^{k*}=-\frac{1}{E\Theta|\omega|}[|\gamma|\delta_{k0}+\nu\mu\int_\omega[A^*(\overline{\boldsymbol{P}}^{k*}+\nabla\overline{Q}^{k*})\cdot\boldsymbol{x}]\,\mathrm{d}\omega.\tag{35.18}$$

Matrix $\Phi_{\alpha\beta}$ is defined by (34.26) *and I_α (second moment of area with respect to axis $e_\beta(\alpha\neq\beta)$), J^* (anisotropic torsion constant) and $I_\alpha^{w^*}$ stand for*

$$I_\alpha=\int_\omega(x_\alpha)^2\,\mathrm{d}\omega,\tag{35.19}$$

$$J^*=a^*_{11}I_2+a^*_{22}I_1-\int_\omega(A^*\nabla w^*)\cdot(-x_2,x_1)\,\mathrm{d}\omega,\tag{35.20}$$

$$I_\alpha^{w^*}=2\int_\omega x_\alpha w^*\,\mathrm{d}\omega.\tag{35.21}$$

PROOF. The passage to the limit in Eqs. (34.16)–(34.18) consists, essentially, in three steps: (1) extensions and estimates independent of δ; (2) passage to the limit in the equilibrium equation; (3) identification of the constitutive law of the homogenized material.

Since we are dealing with a recurrence process we first analyze the case $k = -1$: our thesis already holds. In fact, if, for instance, $\hat{Q}^{-1\delta}$ is an extension of $Q^{-1\delta}$ to all ω, converging weakly in $H^1(\omega)$ towards Q^{-1*}, one has that $\hat{Q}^{-1\delta} \to Q^{-1*}$ strongly in $L^2(\omega)$. By Lemma 31.1 the characteristic function $\chi^{\delta,1}$ converges weakly* $L^\infty(\omega)$ to its mean value Θ and, consequently, $\hat{Q}^{-1\delta} \to \Theta Q^{-1*}$, weakly in $L^2(\omega)$. Since $\chi^{\delta,1}\hat{Q}^{-1\delta} = 0$ in ω, one concludes that $Q^{-1*} = 0$, in ω. Let us assume that the result holds for the functions $\boldsymbol{P}^{(k-1)\delta}$ and $Q^{(k-1)\delta}$ and, additionally, that, if we represent by $[\boldsymbol{\nabla}\overline{Q}^{(k-1)\delta}]^{\sim}$ the zero extension of $\boldsymbol{\nabla}\overline{Q}^{(k-1)\delta}$, to all ω, the following convergence also takes place:

$$[\boldsymbol{\nabla}\overline{Q}^{(k-1)\delta}]^{\sim} \to A^*\boldsymbol{\nabla}\overline{Q}^{(k-1)*} + (A^* - \Theta\,\mathrm{Id})\overline{\boldsymbol{P}}^{(k-1)*}, \quad \text{weakly in } [L^2(\omega)]^2, \tag{35.22}$$

which is clearly satisfied at order $k = -1$.

Consider now problem (34.16), (34.17) which defines $\overline{\boldsymbol{P}}^{k\delta}$.

(1) Using $\overline{\boldsymbol{P}}^{k\delta}$ as test function one obtains, from its variational formulation and from the recurrence hypothesis, that $\|e(\overline{\boldsymbol{P}}^{k\delta})\|_{L^2(\omega^{\delta,1})}$ and $\|\boldsymbol{S}^{k\delta}\|_{L^2(\omega^{\delta,1})}$ are bounded, independently of δ. Lemma 31.4, together with Lemma 31.6, guarantee the existence of an extension $\hat{\overline{\boldsymbol{P}}}^{k\delta}$ of $\overline{\boldsymbol{P}}^{k\delta}$, to all ω, bounded in $[H^1(\omega)]^2$, independently of δ. Representing by $\tilde{\boldsymbol{S}}^{k\delta}$ the zero extension of $\boldsymbol{S}^{k\delta}$ to all ω, one clearly has $\tilde{\boldsymbol{S}}^{k\delta}$ bounded in $[L^2(\omega)]^4$, independently of δ. Up to a subsequence of δ, the following convergences hold:

$$\hat{\overline{\boldsymbol{P}}}^{k\delta} \to \overline{\boldsymbol{P}}^{k*}, \quad \text{weakly in } [H^1(\omega)]^2, \text{ strongly in } [L^2(\omega)]^2, \tag{35.23}$$

$$\tilde{\boldsymbol{S}}^{k\delta} \to \boldsymbol{S}^{k\delta}, \quad \text{weakly in } [L^2(\omega)]^4. \tag{35.24}$$

Passing to the limit in $\int_{\omega^{\delta,1}} \overline{P}_\alpha^{k\delta}\,\mathrm{d}\omega = \int_\omega \chi^{\delta,1}\hat{\overline{\boldsymbol{P}}}^{k\delta}\,\mathrm{d}\omega = 0$, and in $\int_{\omega^{\delta,1}} (\overline{P}_1^{k\delta} x_2 - \overline{P}_2^{k\delta} x_1)\,\mathrm{d}\omega = \int_\omega \chi^{\delta,1}(\hat{\overline{P}}_1^{k\delta} x_2 - \hat{\overline{P}}_2^{k\delta} x_1)\,\mathrm{d}\omega = 0$, one obtains, $\int_\omega \overline{P}_\alpha^{k*}\,\mathrm{d}\omega = \int_\omega (\overline{P}_1^{k*} x_2 - \overline{P}_2^{k*} x_1)\,\mathrm{d}\omega = 0$, since $\chi^{\delta,1} \to \Theta \neq 0$, weakly* in $L^\infty(\omega)$.

(2) From the variational formulation of problem (34.16), (34.17), using the extensions we have just defined, the extensions obtained by the recurrence hypothesis and the characteristic function $\chi^{\delta,1}$, we have, for all $v_\beta \in H^1(\omega)$,

$$\int_\omega \tilde{S}_{\alpha\beta}^{k\delta}\partial_\alpha v_\beta\,\mathrm{d}\omega = \mu \int_\omega \chi^{\delta,1}(\hat{P}_\beta^{(k-1)\delta} + \partial_\beta \hat{Q}^{(k-1)\delta})v_\beta\,\mathrm{d}\omega - \lambda \int_\omega \chi^{\delta,1}\hat{Q}^{(k-1)\delta}\partial_\beta v_\beta\,\mathrm{d}\omega. \tag{35.25}$$

Using convergences (35.23), (35.5) and (35.6) for $k-1$, and (35.22), in order to pass to the limit, as δ goes to zero, one obtains, for all $v_\beta \in H^1(\omega)$,

$$\int_\omega S^{k\delta}_{\alpha\beta}\partial_\alpha v_\beta \, d\omega = \mu \int_\omega a^*_{\beta\zeta}(P^{(k-1)*}_\zeta + \partial_\zeta Q^{(k-1)*}) v_\beta \, d\omega - \lambda\Theta \int_\omega Q^{(k-1)*}\partial_\beta v_\beta \, d\omega. \tag{35.26}$$

(3) In order to identify the relationship between $\boldsymbol{S}^k$ and $\overline{\boldsymbol{P}}^{k*}$ we will use Lemma 31.2. We need convenient test functions $\boldsymbol{\Psi}^\delta_{\zeta\rho}$, which we define as follows:

$$\boldsymbol{\Psi}^\delta_{\zeta\rho}(x_1,x_2) = x_\rho e_\zeta + \delta\hat{\boldsymbol{\chi}}^{\zeta\rho}\left(\frac{x}{\delta}\right), \tag{35.27}$$

$\hat{\boldsymbol{\chi}}^{\zeta\rho}$ being an $[H^1(Y)]^2$-extension of the solution $\boldsymbol{\chi}^{\zeta\rho}$ of problem (35.12), periodically extended to all $\mathbb{R}^2$. By Lemma 31.1 one has

$$\boldsymbol{\Psi}^\delta_{\zeta\rho} \rightharpoonup x_\rho e_\zeta, \quad \text{weakly in } [H^1_{\text{loc}}(\mathbb{R}^2)]^2. \tag{35.28}$$

Representing by $\langle\cdot,\cdot\rangle$ the $\mathbb{R}^2$ inner product one has, using the definition of $S^{k\delta}_{\alpha\beta}$, and for each fixed ζ, ρ and β (with no sum on β),

$$\begin{aligned}&\langle \tilde{S}^{k\delta}_{\alpha\beta} + \lambda\chi^{\delta,1}\hat{Q}^{(k-1)\delta}\delta_{\alpha\beta}, \partial_\alpha(\boldsymbol{\Psi}^\delta_{\zeta\rho})_\beta\rangle \\ &\quad = \langle \partial_\alpha \hat{P}^{k\delta}_\beta, \chi^{\delta,1}[2\mu e_{\alpha\beta}(\boldsymbol{\Psi}^\delta_{\zeta\rho}) + \lambda e_{\tau\tau}(\boldsymbol{\Psi}^\delta_{\zeta\rho})\delta_{\alpha\beta}]\rangle + \lambda\hat{Q}^{(k-1)\delta}\chi^{\delta,1}e_{\beta\beta}(\boldsymbol{\Psi}^\delta_{\zeta\rho}).\end{aligned} \tag{35.29}$$

From (35.25) one sees that, for each fixed β, $\partial_\alpha[\tilde{S}^{k\delta}_{\alpha\beta} + \lambda\chi^{\delta,1}\hat{Q}^{(k-1)\delta}\delta_{\alpha\beta}]$ is compact in $H^{-1}(\omega)$. From the definition of $\boldsymbol{\Psi}^\delta_{\zeta\rho}$ and the variational formulation of the function $\boldsymbol{\chi}^{\zeta\rho}$ one sees that, for each β, $\partial_\alpha\{\chi^{\delta,1}[2\mu e_{\alpha\beta}(\boldsymbol{\Psi}^\delta_{\zeta\rho}) + \lambda e_{\tau\tau}(\boldsymbol{\Psi}^\delta_{\zeta\rho})\delta_{\alpha\beta}]\}$ is compact in $H^{-1}(\omega)$. Lemma 31.1 and Lemma 31.2 allow us to pass to the limit, in the sense of distributions, in both sides of (35.29) and obtain the following identity, almost everywhere in ω:

$$\begin{aligned}&\langle S^k_{\alpha\beta} + \lambda\Theta Q^{(k-1)*}\delta_{\alpha\beta}, \delta_{\alpha\rho}\delta_{\beta\zeta}\rangle \\ &\quad = \langle \partial_\alpha\overline{P}^{k*}_\beta, \mu[(\delta_{\alpha\zeta}\delta_{\beta\rho} + \delta_{\alpha\rho}\delta_{\beta\zeta})\Theta + 2s^*_{\alpha\beta\zeta\rho}] + \lambda(\delta_{\alpha\beta}\delta_{\zeta\rho}\Theta + s^*_{\tau\tau\zeta\rho}\delta_{\alpha\beta})\rangle \\ &\qquad + \lambda Q^{(k-1)*}(\delta_{\beta\zeta}\delta_{\beta\rho}\Theta + s^*_{\beta\beta\zeta\rho}) \\ &\quad = \langle \partial_\alpha\overline{P}^{k*}_\beta, S^*_{\alpha\beta\tau\sigma}\delta_{\tau\zeta}\delta_{\sigma\rho}\rangle + \lambda Q^{(k-1)*}(\delta_{\beta\zeta}\delta_{\beta\rho}\Theta + s^*_{\beta\beta\zeta\rho}) \quad \text{(no sum on } \beta).\end{aligned} \tag{35.30}$$

Since this expression holds for every ζ, ρ and β, summing in β on each side of (35.30) and using the symmetric properties of $(S^*_{\alpha\beta\tau\sigma})$ one obtains, almost everywhere in ω,

$$S^k_{\alpha\beta} + \lambda\Theta Q^{(k-1)*}\delta_{\alpha\beta} = S^*_{\alpha\beta\tau\sigma}e_{\tau\sigma}(\overline{\boldsymbol{P}}^{k*}) + \lambda Q^{(k-1)*}(\Theta\delta_{\alpha\beta} + s^*_{\mu\mu\alpha\beta}), \tag{35.31}$$

which together with (35.26) is the variational formulation of problem (35.9), (35.10). The uniqueness of $\overline{\boldsymbol{P}}^{k*}$ as a solution of problem (35.9)–(35.11) determines that all the sequence $\hat{\overline{\boldsymbol{P}}}^{k\delta}$ converges to $\overline{\boldsymbol{P}}^{k*}$.

For the proof of the elliptic and symmetric properties of the homogenized coefficients $(S^*_{\alpha\beta\tau\sigma})$ we refer to LÉNÉ [1984], where an adaptation of the classical proof for the conductivity problem (BENSSOUSSAN, LIONS and PAPANICOLAOU [1978]), is presented.

Since we shall need it, so as to pass to the limit in the problem defining $\overline{Q}^{k\delta}$, we compute the weak limit of the zero extension $[\partial_\beta \overline{P}^{k\delta}_\beta]^\sim$, for all ω, of $\partial_\beta \overline{P}^{k\delta}_\beta$. From (34.17) and (35.24) one has

$$[\partial_\beta \overline{P}^{k\delta}_\beta]^\sim = \frac{1}{2(\lambda+\mu)}\tilde{S}^{k\delta}_{\beta\beta} \to \frac{1}{2(\lambda+\mu)} S^k_{\beta\beta}, \quad \text{weakly in } L^2(\omega).$$

From (35.31) we have,

$$\begin{aligned} S^k_{\beta\beta} &= S^*_{\beta\beta\zeta\rho} e_{\zeta\rho}(\overline{\boldsymbol{P}}^{k*}) + \lambda s^*_{\mu\mu\beta\beta} Q^{(k-1)*} \\ &= 2(\lambda+\mu)[\Theta e_{\beta\beta}(\overline{\boldsymbol{P}}^{k*}) + s^*_{\beta\beta\zeta\rho} e_{\zeta\rho}(\overline{\boldsymbol{P}}^{k*})] + \lambda s^*_{\mu\mu\beta\beta} Q^{(k-1)*}, \end{aligned}$$

and, consequently,

$$[\partial_\beta \overline{P}^{k\delta}_\beta]^\sim \to [\Theta e_{\beta\beta}(\overline{\boldsymbol{P}}^{k*}) + s^*_{\beta\beta\zeta\rho} e_{\zeta\rho}(\overline{\boldsymbol{P}}^{k*})] + \frac{\lambda}{2(\lambda+\mu)} s^*_{\mu\mu\beta\beta} Q^{(k-1)*} \tag{35.32}$$

weakly in $L^2(\omega)$. Consider now problem (35.13) defining $\overline{Q}^{k\delta}$.

(1) Using $\overline{Q}^{k\delta}$ as a test function in its own variational formulation, with help of Lemma 31.3, Lemma 31.5, the recurrence hypothesis and (35.32), one obtains the existence of an extension $\hat{Q}^{k\delta}$ to all ω, such that $\|\nabla \overline{Q}^{k\delta}\|_{L^2(\omega^{\delta,1})}$ and $\|\hat{\overline{Q}}^{k\delta}\|_{H^1(\omega)}$ are bounded, independently of δ. Representing by $[\nabla \overline{Q}^{k\delta}]^\sim$ the zero extension of $\nabla \hat{Q}^{k\delta}$ to all ω, one also has that $[\nabla \overline{Q}^{k\delta}]^\sim$ is bounded in $[L^2(\omega)]^2$, independently of δ. Up to a subsequence, the following convergences hold:

$$\hat{Q}^{k\delta} \to \overline{Q}^{k*}, \quad \text{weakly in } H^1(\omega), \ \text{strongly in } L^2(\omega), \tag{35.33}$$

$$[\nabla \overline{Q}^{k\delta}]^\sim \to \boldsymbol{\sigma}^k, \quad \text{weakly in } [L^2(\omega)]^2. \tag{35.34}$$

Passing to the limit in $\int_{\omega^{\delta,1}} \overline{Q}^{k\delta}\, d\omega = 0$ and since $\chi^{\delta,1} \to \Theta \neq 0$, weakly* in $L^\infty(\omega)$, one obtains that $\int_\omega \overline{Q}^{k*}\, d\omega = 0$.

(2) From the variational formulation of problem (35.13), using the extensions already defined and the characteristic function $\chi^{\delta,1}$, defined in Section 30, we

have, for all $v \in H^1(\omega)$,

$$\mu \int_\omega \left(\left[\nabla \overline{Q}^{k\delta}\right]^{\sim} + \chi^{\delta,1}\hat{\overline{\boldsymbol{P}}}^{k\delta}\right) \cdot \nabla v \, \mathrm{d}\omega$$

$$= \lambda \int_\omega \left[\partial_\beta \overline{P}_\beta^{k\delta}\right]^{\sim} v \, \mathrm{d}\omega + (\lambda + 2\mu) \int_\omega \chi^{\delta,1} \hat{Q}^{(k-1)\delta} v \, \mathrm{d}\omega + \delta_{k1} \int_\gamma v \, \mathrm{d}\gamma. \quad (35.35)$$

Using convergences (35.34), (35.23), (35.32) and (35.6), for $k-1$, in order to pass to the limit, as δ goes to zero, one obtains, for all $v \in H^1(\omega)$,

$$\mu \int_\omega (\boldsymbol{\sigma}^k + \Theta \overline{\boldsymbol{P}}^{k*}) \cdot \nabla v \, \mathrm{d}\omega$$

$$= \lambda \int_\omega [\Theta e_{\beta\beta}(\overline{\boldsymbol{P}}^{k*}) + s^*_{\beta\beta\zeta\rho} e_{\zeta\rho}(\overline{\boldsymbol{P}}^{k*})] v \, \mathrm{d}\omega$$

$$+ \frac{\lambda^2}{2(\lambda + \mu)} \int_\omega s^*_{\mu\mu\beta\beta} Q^{(k-1)*} v \, \mathrm{d}\omega + (\lambda + 2\mu)\Theta \int_\omega Q^{(k-1)*} v \, \mathrm{d}\omega + \delta_{k1} \int_\gamma v \, \mathrm{d}\gamma.$$

(35.36)

(3) For the identification of the relationship between $\boldsymbol{\sigma}^k$ and $\overline{Q}^{k*}$ we use test functions Ψ_β^δ defined as follows

$$\Psi_\beta^\delta(x_1, x_2) = x_\beta + \delta \hat{\chi}_\beta \left(\frac{\boldsymbol{x}}{\delta}\right), \quad (35.37)$$

where $\hat{\chi}_\beta$ is an $[H^1(Y)]$-extension of the solution χ_β of problem (35.3), periodically repeated to all $\mathbb{R}^2$. By Lemma 31.1 one has

$$\Psi_\beta^\delta \to x_\beta, \quad \text{weakly in } H^1_{\mathrm{loc}}(\mathbb{R}^2). \quad (35.38)$$

From the definition of Ψ_β^δ one sees that $\operatorname{div}(\chi^{\delta,1} \nabla \Psi_\beta^\delta)$ is compact in $H^{-1}(\omega)$. From (35.35), $\operatorname{div}([\nabla \overline{Q}^{k\delta}]^{\sim} + \chi^{\delta,1}\hat{\overline{\boldsymbol{P}}}^{k\delta})$ is compact in $H^{-1}(\omega)$. Using Lemma 31.1 and Lemma 31.2 we pass to the limit, in the sense of distributions, in both sides of the following equality:

$$\langle [\nabla \overline{Q}^{k\delta}]^{\sim} + \chi^{\delta,1}\hat{\overline{\boldsymbol{P}}}^{k\delta}, \nabla \Psi_\beta^\delta \rangle = \langle \nabla \hat{\overline{Q}}^{k\delta}, \chi^{\delta,1} \nabla \Psi_\beta^\delta \rangle + \langle \hat{\overline{\boldsymbol{P}}}^{k\delta}, \chi^{\delta,1} \nabla \Psi_\beta^\delta \rangle, \quad (35.39)$$

and obtain, using (35.2),

$$\langle \boldsymbol{\sigma}^k + \Theta \overline{\boldsymbol{P}}^{k*}, e_\beta \rangle = \langle \nabla \overline{Q}^{k*}, A^* e_\beta \rangle + \langle \overline{\boldsymbol{P}}^{k*}, A^* e_\beta \rangle,$$

almost everywhere in ω, which implies, by the symmetry of A^* and the arbitrarity of β, that

$$\boldsymbol{\sigma}^k = A^* \nabla \overline{Q}^{k*} + (A^* - \Theta \,\mathrm{Id})\overline{\boldsymbol{P}}^{k*}, \quad \text{a.e. in } \omega. \tag{35.40}$$

Expression (35.22), together with (35.36) constitute the variational formulation of problem (35.13). The uniqueness of $\overline{Q}^{k*}$ as a solution of problem (35.13) determines that all the sequences $\hat{\overline{Q}}^{k\delta}$ converges to $\overline{Q}^{k*}$.

To complete the proof we consider the natural extensions of ${}^R\boldsymbol{P}^{k\delta}$ and ${}^RQ^{k\delta}$ to all ω and pass to the limit, using Theorem 35.1 together with all the precedent results about $\overline{\boldsymbol{P}}^{k\delta}$ and $\overline{Q}^{k\delta}$. For any other bounded extensions of ${}^R\boldsymbol{P}^{k\delta}$ and ${}^RQ^{k\delta}$ the same limits are found since the characteristic function of $\omega^{\delta,1}$, $\chi^{\delta,1}$, converges, in the weak sense, to $\Theta \neq 0$.

Expression (35.32), for order k, is a consequence of (35.40) and convergence (35.34), together with (34.19), (34.20) and the convergence (35.4). □

36. A Galerkin approximation of the homogenized three-dimensional equations

This section is concerned with the limit, as δ goes to zero, of the Galerkin approximation $\boldsymbol{u}_N^\delta(\varepsilon)$ and its relationship with the same order Galerkin approximation, related to the three-dimensional homogenized problem.

The following classical homogenization result holds (cf. LÉNÉ [1984], MASCARENHAS and TRABUCHO [1990], and Theorem 32.1).

THEOREM 36.1. *Let $\boldsymbol{f}^\varepsilon \in [L^2(\Omega^\varepsilon)]^3$ and $\boldsymbol{g}^\varepsilon \in [L^2(\Gamma^\varepsilon)]^3$. The solution $\boldsymbol{u}^{\delta,\varepsilon} \in V^{\delta,\varepsilon}$ of problem* (34.1)*, where the volumic forces are the restrictions of $\boldsymbol{f}^\varepsilon$ to $\Omega^{\delta,\varepsilon}$ and where the surface forces coincide in Γ^ε with $\boldsymbol{g}^\varepsilon$, admits an extension in $[H^1(\Omega^\varepsilon)]^3$, bounded independently of δ. If $\hat{\boldsymbol{u}}^{\delta,\varepsilon}$ is any extension of $\boldsymbol{u}^{\delta,\varepsilon}$, bounded in $[H^1(\Omega^\varepsilon)]^3$, independently of δ, then, for each fixed ε and as $\delta \to 0$, $\hat{\boldsymbol{u}}^{\delta,\varepsilon}$ converges to $\boldsymbol{u}^{*,\varepsilon}$, weakly in $[H^1(\Omega^\varepsilon)]^3$, where $\boldsymbol{u}^{*,\varepsilon}$ is the unique solution of the following problem*:

$$\boldsymbol{u}^{*,\varepsilon} \in V(\Omega^\varepsilon) = \{\boldsymbol{v} = (v_i) \in [H^1(\Omega^\varepsilon)]^3 \colon \boldsymbol{v} = \boldsymbol{0} \quad \text{on } \Gamma_0^\varepsilon \cup \Gamma_L^\varepsilon\}, \tag{36.1}$$

$$\int_{\Omega^\varepsilon} \sigma_{ij}^{*,\varepsilon} e_{ij}^\varepsilon(\boldsymbol{v}) \, \mathrm{d}\boldsymbol{x}^\varepsilon = \Theta \int_{\Omega^\varepsilon} f_i^\varepsilon v_i \, \mathrm{d}\boldsymbol{x}^\varepsilon + \int_{\Gamma^\varepsilon} g_i^\varepsilon v_i \, \mathrm{d}a^\varepsilon, \quad \text{for all } \boldsymbol{v} \in V(\Omega^\varepsilon), \tag{36.2}$$

$$\sigma_{ij}^{*,\varepsilon} = 2\mu[\Theta e_{ij}^\varepsilon(\boldsymbol{u}^{*,\varepsilon}) + q_{ijkl}^* e_{kl}^\varepsilon(\boldsymbol{u}^{*,\varepsilon})] + \lambda[\Theta e_{pp}^\varepsilon(\boldsymbol{u}^{*,\varepsilon}) + q_{ppkl}^* e_{kl}^\varepsilon(\boldsymbol{u}^{*,\varepsilon})]\delta_{ij}, \tag{36.3}$$

where the coefficients q_{ijkl}^* satisfy

$$q_{\alpha\beta\zeta\rho}^* = s_{\alpha\beta\zeta\rho}^*,$$

$$q^*_{3\alpha 3\beta} = q^*_{3\beta 3\alpha} = q^*_{\alpha 33\beta} = q^*_{3\alpha\beta 3} = \tfrac{1}{2}(a^*_{\alpha\beta} - \Theta\delta_{\alpha\beta}), \tag{36.4}$$
$$q^*_{\alpha\beta 33} = \frac{\lambda}{2\mu}s^*_{\rho\rho\alpha\beta} - \frac{\lambda^2}{4\mu(\lambda+\mu)}s^*_{\rho\rho\zeta\zeta}\delta_{\alpha\beta},$$

and all the others vanish, (see (35.2), (35.3), (35.11) *and* (35.12)*).*

THEOREM 36.2. *Functions P^{k*}_α, Q^{k*}, $k \geqslant -1$, $\alpha = 1, 2$, defined in Theorem 35.2, are the Galerkin basis functions corresponding to the homogenized problem* (36.1)–(36.3).

PROOF. For the constitutive law (36.3), following the method used in MIARA and TRABUCHO [1990] for the isotropic case (cf. Section 33), we find that the basis functions corresponding to the homogenized problem must satisfy $\boldsymbol{P}^{-1} = \boldsymbol{0}$, $Q^{-1} = 0$, and for $k \geqslant 0$,

$$\int_\omega \{2\mu[\Theta e_{\alpha\beta}(\boldsymbol{P}^k) + q^*_{\alpha\beta\zeta\rho}e_{\zeta\rho}(\boldsymbol{P}^k)]e_{\alpha\beta}(\boldsymbol{v})$$
$$+ \lambda[\Theta e_{\mu\mu}(\boldsymbol{P}^k) + q^*_{\mu\mu\zeta\rho}e_{\zeta\rho}(\boldsymbol{P}^k)]e_{\beta\beta}(\boldsymbol{v})$$
$$+ [2\mu q^*_{\alpha\beta 33} + \lambda(\Theta + q^*_{\mu\mu 33})\delta_{\alpha\beta}]Q^{(k-1)}e_{\alpha\beta}(\boldsymbol{v})$$
$$- \mu[\Theta(P^{(k-1)}_\alpha + \partial_\alpha Q^{(k-1)}) + 2q^*_{3\alpha 3\rho}(P^{(k-1)}_\rho + \partial_\rho Q^{(k-1)})]v_\alpha\}\,\mathrm{d}\omega = 0, \tag{36.5}$$

for all $\boldsymbol{v} \in [H^1(\omega)]^2$, and

$$\int_\omega \{\mu[\Theta(P^k_\alpha + \partial_\alpha Q^k) + 2q^*_{3\alpha 3\rho}(P^k_\rho + \partial_\rho Q^k)]\partial_\alpha v$$
$$- \lambda[\Theta e_{\mu\mu}(\boldsymbol{P}^k) + q^*_{\mu\mu\zeta\rho}e_{\zeta\rho}(\boldsymbol{P}^k)]v$$
$$- [(\lambda + 2\mu)\Theta + \lambda q^*_{\mu\mu 33}]Q^{(k-1)}v\}\,\mathrm{d}\omega = \int_\gamma \delta_{1k}v\,\mathrm{d}\gamma, \quad \text{for all } v \in H^1(\omega). \tag{36.6}$$

Using relations (36.4) one sees that (36.5) and (36.6) are equivalent to, respectively,

$$\int_\omega \{2\mu[\Theta e_{\alpha\beta}(\boldsymbol{P}^k) + s^*_{\alpha\beta\zeta\rho}e_{\zeta\rho}(\boldsymbol{P}^k)]e_{\alpha\beta}(\boldsymbol{v})$$
$$+ \lambda[\Theta e_{\mu\mu}(\boldsymbol{P}^k) + s^*_{\mu\mu\zeta\rho}e_{\zeta\rho}(\boldsymbol{P}^k)]e_{\beta\beta}(\boldsymbol{v}) + \lambda(\Theta\delta_{\alpha\beta} + s^*_{\mu\mu\zeta\rho})Q^{(k-1)}e_{\alpha\beta}(\boldsymbol{v})$$
$$- \mu a^*_{\alpha\rho}(P^{(k-1)}_\rho + \partial_\rho Q^{(k-1)})]v_\alpha\}\,\mathrm{d}\omega = 0, \quad \text{for all } \boldsymbol{v} \in [H^1(\omega)]^2, \tag{36.7}$$

$$\int_\omega \{\mu a^*_{\alpha\rho}(P^k_\rho + \partial_\rho Q^k)\partial_\alpha v - \lambda[\Theta e_{\mu\mu}(\boldsymbol{P}^k) + s^*_{\mu\mu\zeta\rho}e_{\zeta\rho}(\boldsymbol{P}^k)]v$$

$$- [(\lambda+2\mu)\Theta + \frac{\lambda^2}{2(\lambda+\mu)}s^*_{\mu\mu\rho\rho}]Q^{(k-1)}v\}\,\mathrm{d}\omega$$

$$= \int_\gamma \delta_{1k}v\,\mathrm{d}\gamma, \quad \text{for all } v \in H^1(\omega). \tag{36.8}$$

Reasoning as in Theorem 34.3 one sees that these variational formulations are equivalent to problems (35.7)–(35.21), which, by an unicity argument, proves the theorem. □

Recalling Theorem 34.1, consider $\boldsymbol{u}^\delta(\varepsilon) \in V(\Omega^{\delta,1})$, transformed of $\boldsymbol{u}^{\delta,\varepsilon}$, and solution of

$$B(\varepsilon)(\boldsymbol{u}^\delta(\varepsilon), \boldsymbol{v}) = F(\varepsilon)(\boldsymbol{v}), \quad \text{for all } \boldsymbol{v} \in V(\Omega^{\delta,1}).$$

Using standard homogenization techniques, one finds that any extension of $\boldsymbol{u}^\delta(\varepsilon)$, bounded in $[H^1(\Omega)]^3$ independently of δ, converges weakly in $[H^1(\Omega)]^3$ to $\boldsymbol{u}^*(\varepsilon)$, solution of

$$B^*(\varepsilon)(\boldsymbol{u}^*(\varepsilon), \boldsymbol{v}) = F^*(\varepsilon)(\boldsymbol{v}), \quad \text{for all } \boldsymbol{v} \in V(\Omega), \tag{36.9}$$

where for all $\boldsymbol{w}, \boldsymbol{v} \in V(\Omega)$,

$$B^*(\varepsilon)(\boldsymbol{w}, \boldsymbol{v})$$

$$= \int_\Omega \{\Theta[\lambda e_{\mu\mu}(\boldsymbol{w})e_{\beta\beta}(\boldsymbol{v}) + 2\mu e_{\alpha\beta}(\boldsymbol{w})e_{\alpha\beta}(\boldsymbol{v})]$$

$$+ [\lambda q^*_{\mu\mu\zeta\rho}e_{\zeta\rho}(\boldsymbol{w})e_{\beta\beta}(\boldsymbol{v}) + 2\mu q^*_{\alpha\beta\zeta\rho}e_{\zeta\rho}(\boldsymbol{w})e_{\alpha\beta}(\boldsymbol{v})]\}\,\mathrm{d}\boldsymbol{x}$$

$$+ \varepsilon^2 \int_\Omega \{\Theta[\lambda e_{33}(\boldsymbol{w})e_{\beta\beta}(\boldsymbol{v}) + \lambda e_{\alpha\alpha}(\boldsymbol{w})e_{33}(\boldsymbol{v}) + 4\mu e_{3\beta}(\boldsymbol{w})e_{3\beta}(\boldsymbol{v})]$$

$$+ [\lambda q^*_{\mu\mu 33}e_{33}(\boldsymbol{w})e_{\beta\beta}(\boldsymbol{v}) + 2\mu q^*_{\alpha\beta 33}e_{33}(\boldsymbol{w})e_{\alpha\beta}(\boldsymbol{v})]$$

$$+ [\lambda q^*_{\mu\mu\zeta\rho}e_{\zeta\rho}(\boldsymbol{w})e_{33}(\boldsymbol{v}) + 8\mu q^*_{3\beta 3\alpha}e_{3\beta}(\boldsymbol{w})e_{3\alpha}(\boldsymbol{v})]\}\,\mathrm{d}\boldsymbol{x}$$

$$+ \varepsilon^4 \int_\Omega \Theta[(\lambda+2\mu)e_{33}(\boldsymbol{w})e_{33}(\boldsymbol{v}) + \lambda q^*_{\alpha\alpha 33}e_{33}(\boldsymbol{w})e_{33}(\boldsymbol{v})]\,\mathrm{d}\boldsymbol{x}, \tag{36.10}$$

and

$$F^*(\varepsilon)(\boldsymbol{v}) = \Theta \int_\Omega f_i v_i \, \mathrm{d}\boldsymbol{x} + \int_\Gamma g_i v_i \, \mathrm{d}a. \tag{36.11}$$

One easily verifies that the solution $\boldsymbol{u}^*(\varepsilon)$ of (36.9) is the transformed of $\boldsymbol{u}^{*,\varepsilon}$, solution of problem (36.1)–(36.3), by the scalings (34.4), (34.5).

The same does not happen, however, with the projections $\boldsymbol{u}_N^\delta(\varepsilon)$ of $\boldsymbol{u}^\delta(\varepsilon)$, in the approximation spaces $V_N(\Omega^{\delta,1})$. In fact, for fixed N, let $\boldsymbol{u}_N^\delta(\varepsilon) \in V_N(\Omega^{\delta,1})$ be the solution of

$$B(\varepsilon)(\boldsymbol{u}_N^\delta(\varepsilon), \boldsymbol{v}) = F(\varepsilon)(\boldsymbol{v}), \quad \text{for all } \boldsymbol{v} \in V_N(\Omega^{\delta,1}), \tag{36.12}$$

and let $\boldsymbol{u}_N^*(\varepsilon) \in V_N^*(\Omega)$ be the solution of

$$B^*(\varepsilon)(\boldsymbol{u}_N^*(\varepsilon), \boldsymbol{v}) = F^*(\varepsilon)(\boldsymbol{v}), \quad \text{for all } \boldsymbol{v} \in V_N^*(\Omega), \tag{36.13}$$

where $V_N^*(\Omega)$ stands for the space

$$\begin{aligned} V_N^*(\Omega) = \{\boldsymbol{v} = (v_i) \in V(\Omega)\colon v_\alpha &= \sum_{k=0}^N P_\alpha^{k*}(x_1,x_2) v_\alpha^k(x_3), \\ & v_\alpha^k \in H_0^1(0,L),\ P_\alpha^{k*} \in H^1(\omega), \\ v_3 &= \sum_{k=0}^N Q^{k*}(x_1,x_2) v_3^k(x_3), \\ & v_3^k \in H_0^1(0,L), Q^{k*} \in H^1(\omega), \quad \text{for all } 0 \leqslant k \leqslant N\} \end{aligned}$$

one has (cf. MIARA and TRABUCHO [1990])

$$\boldsymbol{u}_N^\delta(\varepsilon) = \sum_{k=0}^N \varepsilon^{2k} (u_1^{k\delta} P_1^{k\delta}, u_2^{k\delta} P_2^{k\delta}, u_3^{k\delta} Q^{k\delta}), \tag{36.14}$$

$$\boldsymbol{u}_N^*(\varepsilon) = \sum_{k=0}^N \varepsilon^{2k} (u_1^{k*} P_1^{k*}, u_2^{k*} P_2^{k*}, u_3^{k*} Q^{k*}), \tag{36.15}$$

where the coefficient functions $u_i^{k\delta}$ (respectively u_i^{k*}) ($i = 1,2,3, k = 0,1,\ldots,N$) are the solutions, in $H_0^1(0,L)$, of a linear system, of ordinary differential equations with constant coefficients, depending on the basis functions $P_1^{k\delta}, P_2^{k\delta}, Q^{k\delta}$ (respectively $(P_1^{k*}, P_2^{k*}, Q^{k*})$).

Since the coefficients of the system defining $u_i^{k\delta}$ are bounded, independently of δ, its solution converges, up to a subsequence, strongly in $H_0^1(0,L)$, to some functions u_i^k. These functions are not, in general, equal to u_i^{k*}. Consequently, the weak limit of an extension $\hat{\boldsymbol{u}}_N^\delta(\varepsilon)$ of $\boldsymbol{u}_N^\delta(\varepsilon)$ is not, in general, equal to $\boldsymbol{u}_N^*(\varepsilon)$.

As an example, we give the computations for the simplest case: $N = 0$.

Replacing $\boldsymbol{u}^{\delta}_{N=0}(\varepsilon) = (u_1^{0\delta}P_1^{0\delta}, u_2^{0\delta}P_2^{0\delta}, u_3^{0\delta}Q^{0\delta})$ in (36.12) (see also (34.7)–(34.8)) and making $\boldsymbol{v}$ successively equal to $(\varphi P_1^{0\delta}, 0, 0)$, $(0, \varphi P_2^{0\delta}, 0)$ and $(0, 0, \varphi Q^{0\delta})$, one concludes that $u_i^{0\delta}$ satisfies, for all $\varphi \in H_0^1(0, L)$, the following system (with sum on α):

$$\int_0^L \mu \left[\partial_3 u_1^{0\delta} \int_{\omega^{\delta,1}} (P_1^{0\delta} P_1^{0\delta}) \,\mathrm{d}\omega^{\delta,1} + u_3^{0\delta} \int_{\omega^{\delta,1}} (P_1^{0\delta} \partial_1 Q^{0\delta}) \,\mathrm{d}\omega^{\delta,1} \right] \partial_3 \varphi \,\mathrm{d}x_3 = 0,$$

$$\int_0^L \mu \left[\partial_3 u_2^{0\delta} \int_{\omega^{\delta,1}} (P_2^{0\delta} P_2^{0\delta}) \,\mathrm{d}\omega^{\delta,1} + u_3^{0\delta} \int_{\omega^{\delta,1}} (P_2^{0\delta} \partial_2 Q^{0\delta}) \,\mathrm{d}\omega^{\delta,1} \right] \partial_3 \varphi \,\mathrm{d}x_3 = 0,$$

$$\begin{aligned} &\int_0^L \mu \left[\partial_3 u_\alpha^{0\delta} \int_{\omega^{\delta,1}} (P_\alpha^{0\delta} \partial_\alpha Q^{0\delta}) \,\mathrm{d}\omega^{\delta,1} + u_3^{0\delta} \int_{\omega^{\delta,1}} (\partial_\beta Q^{0\delta} \partial_\beta Q^{0\delta}) \,\mathrm{d}\omega^{\delta,1} \right] \varphi \,\mathrm{d}x_3 \\ &\quad + \varepsilon^2 \int_0^L \left[(\lambda + 2\mu) \partial_3 u_3^{0\delta} \int_{\omega^{\delta,1}} (Q^{0\delta} Q^{0\delta}) \,\mathrm{d}\omega^{\delta,1} \right] \partial_3 \varphi \,\mathrm{d}x_3 \\ &\quad = \varepsilon^2 \int_0^L \int_\gamma g_3 Q^{0\delta} \varphi \,\mathrm{d}\gamma^{\delta,1} \,\mathrm{d}x_3. \end{aligned} \tag{36.16}$$

Since $P_\alpha^{0\delta} = a_\alpha^{0\delta}$, $Q^{0\delta} = -a_\alpha^{0\delta} x_\alpha + C^{0\delta}$, $P_\alpha^{0*} = a_\alpha^{0*}$, $Q^{0*} = -a_\alpha^{0*} x_\alpha + C^{0*}$, $a_\alpha^{0\delta} \to a_\alpha^{0*}$ and $C^{0\delta} \to C^{0*}$, one concludes the convergence of $u_i^{0\delta}$, as δ goes to zero, to the solution u_i^0 of the following system:

$$\int_0^L \Theta\mu \left[\partial_3 u_1^0 \int_\omega (P_1^{0*} P_1^{0*}) \,\mathrm{d}\omega + u_3^0 \int_\omega (P_1^{0*} \partial_1 Q^{0*}) \,\mathrm{d}\omega \right] \partial_3 \varphi \,\mathrm{d}x_3 = 0,$$

$$\int_0^L \Theta\mu \left[\partial_3 u_2^0 \int_\omega (P_2^{0*} P_2^{0*}) \,\mathrm{d}\omega + u_3^0 \int_\omega (P_2^{0*} \partial_2 Q^{0*}) \,\mathrm{d}\omega \right] \partial_3 \varphi \,\mathrm{d}x_3 = 0,$$

$$\begin{aligned} &\int_0^L \Theta\mu \left[\partial_3 u_\alpha^0 \int_\omega (P_\alpha^{0*} \partial_\alpha Q^{0*}) \,\mathrm{d}\omega + u_3^0 \int_\omega (\partial_\beta Q^{0*} \partial_\beta Q^{0*}) \,\mathrm{d}\omega \right] \varphi \,\mathrm{d}x_3 \\ &\quad + \varepsilon^2 \int_0^L \left[(\lambda + 2\mu) \partial_3 u_3^0 \int_\omega (Q^{0*} Q^{0*}) \,\mathrm{d}\omega \right] \partial_3 \varphi \,\mathrm{d}x_3 = \varepsilon^2 \int_0^L \int_\gamma g_3 Q^{0*} \varphi \,\mathrm{d}\gamma \,\mathrm{d}x_3. \end{aligned} \tag{36.17}$$

On the other hand, replacing $\boldsymbol{u}^*_{N=0}(\varepsilon) = (u_1^{0*}P_1^{0*}, u_2^{0*}P_2^{0*}, u_3^{0*}Q^{0*})$ in (36.13) (see also (36.10), (36.11)) and taking $\boldsymbol{v}$ successively equal to $(\varphi P_1^{0*}, 0, 0)$, $(0, \varphi P_2^{0*}, 0)$ and $(0, 0, \varphi Q^{0*})$, one concludes that u_i^{0*} satisfies, for all $\varphi \in H_0^1(0, L)$, the following system (with sum on α):

$$\begin{aligned}
&\int_0^L \mu(\Theta\delta_{1\alpha} + 2q^*_{313\alpha})\left[\partial_3 u_\alpha^{0*}\int_\omega (P_\alpha^{0*}P_1^{0*})\,\mathrm{d}\omega\right.\\
&\qquad\left. + u_3^{0*}\int_\omega (P_1^{0*}\partial_\alpha Q^{0*})\,\mathrm{d}\omega\right]\partial_3\varphi\,\mathrm{d}x_3 = 0,\\
&\int_0^L \mu(\Theta\delta_{2\alpha} + 2q^*_{323\alpha})\left[\partial_3 u_\alpha^{0*}\int_\omega (P_\alpha^{0*}P_2^{0*})\,\mathrm{d}\omega\right.\\
&\qquad\left. + u_3^{0*}\int_\omega (P_2^{0*}\partial_\alpha Q^{0*})\,\mathrm{d}\omega\right]\partial_3\varphi\,\mathrm{d}x_3 = 0,\\
&\int_0^L \mu(\Theta\delta_{\alpha\beta} + 2q^*_{3\alpha 3\beta})\left[\partial_3 u_\alpha^{0*}\int_\omega (P_\alpha^{0*}\partial_\beta Q^{0*})\,\mathrm{d}\omega\right.\\
&\qquad\left. + u_3^{0*}\int_\omega (\partial_\alpha Q^{0*}\partial_\beta Q^{0*})\,\mathrm{d}\omega\right]\varphi\,\mathrm{d}x_3\\
&\qquad + \varepsilon^2\int_0^L \left\{[(\lambda+2\mu)\Theta + \lambda q^*_{\alpha\alpha 33})]\partial_3 u_3^{0*}\int_\omega (Q^{0*}Q^{0*})\,\mathrm{d}\omega\right\}\partial_3\varphi\,\mathrm{d}x_3\\
&\quad = \varepsilon^2\int_0^L\int_\gamma g_3 Q^{0*}\varphi\,\mathrm{d}\gamma\,\mathrm{d}x_3.
\end{aligned} \tag{36.18}$$

Systems (36.17) and (36.18) are not equivalent.

REMARK 36.1. Although in general the weak limit $\boldsymbol{u}_N(\varepsilon)$ of the extended projections $\hat{\boldsymbol{u}}_N^\delta(\varepsilon)$ is different from $\boldsymbol{u}_N^*(\varepsilon)$, they both represent an approximation of the same order of the homogenized solution $\boldsymbol{u}^*(\varepsilon)$. To prove this we consider the extension operator Π^δ, from $[H^1(\Omega^{\delta,1})]^3$ into $[H^1(\Omega)]^3$, defined as follows: for each $\boldsymbol{u} = (u_1, u_2, u_3) \in [H^1(\Omega^{\delta,1})]^3$, $\lambda \in [0, L]$, $\boldsymbol{x} \in \omega^{\delta,1}$, set $\boldsymbol{v}_\lambda(\boldsymbol{x}) = (u_1(\boldsymbol{x}, \lambda), u_2(\boldsymbol{x}, \lambda))$, $w_\lambda(\boldsymbol{x}) = u_3(\boldsymbol{x}, \lambda)$ and define $(\Pi^\delta\boldsymbol{u})(\boldsymbol{x}, \lambda) = [(\Pi_2^\delta\boldsymbol{v}_\lambda)(\boldsymbol{x}), (\Pi_1^\delta w_\lambda)(\boldsymbol{x})]$, where Π_2^δ and Π_1^δ are the extension operators defined in Lemma 31.4 and Lemma 31.3, respectively. Operator Π^δ is linear, continuous and satisfies, for a constant $\tilde{c}$, independent of δ, and for all $\boldsymbol{u} \in [H^1(\Omega^{\delta,1})]^3$,

$$\|\Pi^\delta\boldsymbol{u}\|_{H^1(\Omega)} \leqslant \tilde{c}\|\boldsymbol{u}\|_{H^1(\Omega^{\delta,1})}. \tag{36.19}$$

If we consider $\boldsymbol{u}_N^\delta(\varepsilon)$ given by (36.14) then

$$\Pi^\delta \boldsymbol{u}_N^\delta(\varepsilon) = \sum_{k=0}^{N} \varepsilon^{2k}(u_1^{k\delta}\hat{P}_1^{k\delta}, u_2^{k\delta}\hat{P}_2^{k\delta}, u_3^{k\delta}\hat{Q}^{k\delta}), \tag{36.20}$$

where $\hat{\boldsymbol{P}}^{k\delta} = \Pi_2^\delta \boldsymbol{P}^{k\delta}$ and $\hat{Q}^{k\delta} = \Pi_1^\delta Q^{k\delta}$, converges weakly to

$$\boldsymbol{u}_N(\varepsilon) = \sum_{k=0}^{N} \varepsilon^{2k}(u_1^k P_1^{k*}, u_2^k P_2^{k*}, u_3^k Q^{k*}). \tag{36.21}$$

THEOREM 36.3. *Under the same hypotheses of Theorem 34.2 and Theorem 36.1, and using the previous notation, there exist constants $\overline{C}$ and $\overline{K}$, C^* and K^*, independent of ε, δ and N, such that*

$$\begin{aligned} \|\boldsymbol{u}^*(\varepsilon) - \boldsymbol{u}_N(\varepsilon)\|_{H^1(\Omega)} &\leqslant \overline{C}N\overline{K}^{N-1}\varepsilon^{2N-2}, \\ \|\boldsymbol{u}^{*,\varepsilon} - \boldsymbol{u}_N^\varepsilon\|_{H^1(\Omega^\varepsilon)} &\leqslant \overline{C}N\overline{K}^{N-1}\varepsilon^{2N-1}, \end{aligned} \tag{36.22}$$

$$\begin{aligned} \|\boldsymbol{u}^*(\varepsilon) - \boldsymbol{u}_N^*(\varepsilon)\|_{H^1(\Omega)} &\leqslant C^*N(K^*)^{N-1}\varepsilon^{2N-2}, \\ \|\boldsymbol{u}^{*,\varepsilon} - \boldsymbol{u}_N^{*,\varepsilon}\|_{H^1(\Omega^\varepsilon)} &\leqslant C^*N(K^*)^{N-1}\varepsilon^{2N-1}. \end{aligned} \tag{36.23}$$

PROOF. In order to prove the first inequality of (36.22) we note that, by (36.19),

$$\begin{aligned} \|\Pi^\delta \boldsymbol{u}^\delta(\varepsilon) - \Pi^\delta \boldsymbol{u}_N^\delta(\varepsilon)\|_{H^1(\Omega)} &= \|\Pi^\delta(\boldsymbol{u}^\delta(\varepsilon) - \boldsymbol{u}_N^\delta(\varepsilon))\|_{H^1(\Omega)} \\ &\leqslant \tilde{c}\|\boldsymbol{u}^\delta(\varepsilon) - \boldsymbol{u}_N^\delta(\varepsilon)\|_{H^1(\Omega^{\delta,1})} \end{aligned}$$

and, using the first inequality of (34.10), we establish

$$\|\Pi^\delta \boldsymbol{u}^\delta(\varepsilon) - \Pi^\delta \boldsymbol{u}_N^\delta(\varepsilon)\|_{H^1(\Omega)} \leqslant \tilde{c}CNk^{N-1}\varepsilon^{2N-2}.$$

Since, by (36.19), $\Pi^\delta \boldsymbol{u}^\delta(\varepsilon)$ is bounded in $H^1(\Omega)$, independently of δ, it converges weakly, as δ goes to zero, to $\boldsymbol{u}^*(\varepsilon)$ and, by the lower semi-continuity of the norm, with respect to the weak topology, we obtain

$$\|\boldsymbol{u}^*(\varepsilon) - \boldsymbol{u}_N(\varepsilon)\|_{H^1(\Omega)} \leqslant \tilde{c}CNk^{N-1}\varepsilon^{2N-2}.$$

The second inequality of (36.22) is obtained using scalings (34.4). For the homogenized case, the proof of (36.23) follows the same method as presented in the proof of Theorem 34.2. □

REMARK 36.2. In Sections 34, 35 and 36 we studied an approximation of the three-dimensional elasticity problem for a beam with a multiply connected cross section. This was done using both a Galerkin approximation technique and homogenization methods.

The homogenization theory was used to build an "equivalent" model when the beam is finely perforated.

The Galerkin method was employed in order to replace the full three-dimensional computation of the displacement field by a series of two- and one-dimensional problems.

In both cases, and for sufficiently smooth loads, this was achieved within an appropriate error estimate as stated in Theorems 34.2 and 36.5. However, the homogenization and Galerkin techniques do not commute. It is thus possible to have "different" approximations of the three-dimensional solution within the required estimate as stated in Theorem 36.5.

Nevertheless, when $\varepsilon \to 0$ and $N \to \infty$, the limit of all approximations is the same and coincides with the classical Bernoulli–Navier displacement field. It is only at the level of Saint Venant's torsion theory and Timoshenko's beam theory that the differences show up.

The method presented here is also applicable to other types of loading and boundary conditions (see MIARA and TRABUCHO [1990] or Remarks 33.2 and 33.4).

In the case of two materials, one occupying $\Omega^{\delta,\varepsilon}$ with Lamé's constants λ_1 and μ_1 and the other occupying $\Omega^\varepsilon \setminus \Omega^{\delta,\varepsilon}$, with λ_2 and μ_2 the same type of results hold.

More complex is the case of materials with smoothly oscillating coefficients λ and μ. The first difficulty appears when defining the basis functions: if we neglect the dependence on δ and consider $\lambda = \lambda(x_\alpha)$ and $\mu = \mu(x_\alpha)$ in the constant domain ω, some restrictive conditions must be satisfied by λ and μ, otherwise problems (34.12) and (34.13) may have no solution. In fact let $\boldsymbol{P}^0 = (a_1^0, a_2^0)$ and $Q^0 = -a_\alpha x_\alpha + c^0$ be solutions of (34.12) and (34.13) respectively, for $k = 0$. The compatibility conditions for problem (34.12) with $k = 1$ are satisfied. Let $\overline{\boldsymbol{P}}^1$ be the corresponding solution such that $\int_\omega \overline{P}^1_\alpha \,\mathrm{d}\omega = \int_\omega (x_2\overline{P}^1_1 - x_1\overline{P}^1_2)\,\mathrm{d}\omega = 0$. One has

$$\overline{\boldsymbol{P}}^1 = \boldsymbol{M}^1_\zeta a^0_\zeta + \boldsymbol{N}^1 c^0, \tag{36.24}$$

where $\boldsymbol{M}^1_\zeta$ and $\boldsymbol{N}^1$ are defined as the solutions of the following problems:

$$\begin{aligned} &\int_\omega [2\mu e_{\alpha\beta}(\boldsymbol{M}^1_\zeta) e_{\alpha\beta}(\boldsymbol{v}) + \lambda e_{\mu\mu}(\boldsymbol{M}^1_\zeta) e_{\tau\tau}(\boldsymbol{v})]\,\mathrm{d}\omega \\ &\quad = \int_\omega \lambda x_\zeta \partial_\beta v_\beta \,\mathrm{d}\omega, \quad \text{for all } v_\beta \in H^1(\omega), \\ &\int_\omega (\boldsymbol{M}^1_\alpha)_\zeta \,\mathrm{d}\omega = \int_\omega [x_2(\boldsymbol{M}^1_1)_\zeta - x_1(\boldsymbol{M}^1_2)_\zeta]\,\mathrm{d}\omega = 0, \end{aligned} \tag{36.25}$$

$$\int_{\omega} [2\mu e_{\alpha\beta}(\boldsymbol{N}^1)e_{\alpha\beta}(\boldsymbol{v}) + \lambda e_{\mu\mu}(\boldsymbol{N}^1)e_{\tau\tau}(\boldsymbol{v})]\,\mathrm{d}\omega$$

$$= -\int_{\omega} \lambda \partial_\beta v_\beta \,\mathrm{d}\omega, \quad \text{for all } v_\beta \in H^1(\omega), \tag{36.26}$$

$$\int_{\omega} N^1_\alpha \,\mathrm{d}\omega = \int_{\omega} (x_2 N^1_1 - x_1 N^1_2)\,\mathrm{d}\omega = 0.$$

Equation (34.13), with $k = 1$ and r equal to a constant implies that

$$\int_{\omega} \lambda \partial_\alpha \overline{P}^1_\alpha \,\mathrm{d}\omega - \left[\int_{\omega} (\lambda + 2\mu)x_\alpha \,\mathrm{d}\omega\right] a^0_\alpha + \left[\int_{\omega} (\lambda + 2\mu)\,\mathrm{d}\omega\right] c^0 + |\gamma| = 0. \tag{36.27}$$

Equation (34.12), with $k = 2$ and s equal to the canonical basis vector in $\mathbb{R}^2$, together with Eq. (34.13), with $k = 1$ and $r = x_\alpha$, implies that

$$\int_{\omega} \lambda \partial_\alpha \overline{P}^1_\alpha x_1 \,\mathrm{d}\omega - \left[\int_{\omega} (\lambda + 2\mu)x_1^2 \,\mathrm{d}\omega\right] a^0_1 - \left[\int_{\omega} (\lambda + 2\mu)x_1 x_2 \,\mathrm{d}\omega\right] a^0_2$$

$$+ \left[\int_{\omega} (\lambda + 2\mu)x_1 \,\mathrm{d}\omega\right] c^0 + \int_{\gamma} x_1 \,\mathrm{d}\gamma = 0. \tag{36.28}$$

$$\int_{\omega} \lambda \partial_\alpha \overline{P}^1_\alpha x_2 \,\mathrm{d}\omega - \left[\int_{\omega} (\lambda + 2\mu)x_1 x_2 \,\mathrm{d}\omega\right] a^0_1 - \left[\int_{\omega} (\lambda + 2\mu)x_2^2 \,\mathrm{d}\omega\right] a^0_2$$

$$+ \left[\int_{\omega} (\lambda + 2\mu)x_2 \,\mathrm{d}\omega\right] c^0 + \int_{\gamma} x_2 \,\mathrm{d}\gamma = 0. \tag{36.29}$$

In view of (36.24) and (36.27)–(36.29), constants a^0_α and c^0 satisfy a nonhomogeneous linear system of the matrix

$$\begin{pmatrix} \int_\omega A_1 \,\mathrm{d}\omega & \int_\omega A_2 \,\mathrm{d}\omega & \int_\omega A_3 \,\mathrm{d}\omega \\ \int_\omega A_1 x_1 \,\mathrm{d}\omega & \int_\omega A_2 x_1 \,\mathrm{d}\omega & \int_\omega A_3 x_1 \,\mathrm{d}\omega \\ \int_\omega A_1 x_2 \,\mathrm{d}\omega & \int_\omega A_2 x_2 \,\mathrm{d}\omega & \int_\omega A_3 x_2 \,\mathrm{d}\omega \end{pmatrix}$$

whose determinant must not vanish in order to obtain compatibility for system (34.12), (34.13), and where

$$A_1 = \lambda \partial_\alpha (M^1_1)_\alpha - (\lambda + 2\mu)x_1,$$

$$A_2 = \lambda \partial_\alpha (M^1_2)_\alpha - (\lambda + 2\mu)x_2,$$

$$A_3 = \lambda \partial_\alpha N^1_\alpha + \lambda + 2\mu.$$

This condition is clearly satisfied in the case of λ and μ constants (see Theorem 34.3).

37. The three-dimensional equations for a linearly elastic curved rod

In this section we apply a Galerkin method to the equations of linearized three-dimensional elasticity for a curved rod. We explicitly construct an approximate solution of the three-dimensional elasticity problem as a minimizer of the H^1 norm with respect to a geometric parameter. We prove that the approximation error, in the H^1 norm, is a function of the loading, the geometry, the order of approximation and the curvature and torsion tensors associated with the definition of the beam's axis.

As before, let ε and L be two positive scalars and let ω^ε denote an open, bounded, simply connected subset of $\mathbb{R}^2$, with a Lipschitz continuous boundary γ^ε having area $A(\omega^\varepsilon) = \varepsilon^2$. We define the beam's axis as the set

$$C^\varepsilon = \left\{ \left(\theta_1^\varepsilon(x_3^\varepsilon), \theta_2^\varepsilon(x_3^\varepsilon), \theta_3^\varepsilon(x_3^\varepsilon)\right) \in \mathbb{R}^3 \colon x_3^\varepsilon \in [0, L] \right\}, \tag{37.1}$$

where, for each $\varepsilon > 0$, $\theta_i^\varepsilon : [0, L] \to \mathbb{R}$ are sufficiently smooth functions (one may assume that θ_i^ε are C^∞ functions).

At each point of the beam's axis C^ε, consider the Frenet basis $\{\boldsymbol{t}^\varepsilon, \boldsymbol{n}^\varepsilon, \boldsymbol{b}^\varepsilon\}$, whose unitary components, with respect to the canonical basis $\{\boldsymbol{e}_i\}$ $(i = 1, 2, 3)$, of the Euclidean space $\mathbb{R}^3$, are

$$\boldsymbol{t}^\varepsilon = \begin{pmatrix} t_1^\varepsilon \\ t_2^\varepsilon \\ t_3^\varepsilon \end{pmatrix} = \frac{1}{\sqrt{\alpha^\varepsilon}} \begin{pmatrix} \partial_3^\varepsilon \theta_1^\varepsilon \\ \partial_3^\varepsilon \theta_2^\varepsilon \\ \partial_3^\varepsilon \theta_3^\varepsilon \end{pmatrix}, \tag{37.2}$$

$$\boldsymbol{n}^\varepsilon = \begin{pmatrix} n_1^\varepsilon \\ n_2^\varepsilon \\ n_3^\varepsilon \end{pmatrix} = \frac{1}{\sqrt{\alpha^\varepsilon \beta^\varepsilon}} \begin{pmatrix} \alpha^\varepsilon \partial_{33}^\varepsilon \theta_1^\varepsilon - \partial_3^\varepsilon \theta_1^\varepsilon \partial_3^\varepsilon \theta_j^\varepsilon \partial_{33}^\varepsilon \theta_j^\varepsilon \\ \alpha^\varepsilon \partial_{33}^\varepsilon \theta_2^\varepsilon - \partial_3^\varepsilon \theta_2^\varepsilon \partial_3^\varepsilon \theta_j^\varepsilon \partial_{33}^\varepsilon \theta_j^\varepsilon \\ \alpha^\varepsilon \partial_{33}^\varepsilon \theta_3^\varepsilon - \partial_3^\varepsilon \theta_3^\varepsilon \partial_3^\varepsilon \theta_j^\varepsilon \partial_{33}^\varepsilon \theta_j^\varepsilon \end{pmatrix}, \tag{37.3}$$

$$\boldsymbol{b}^\varepsilon = \begin{pmatrix} b_1^\varepsilon \\ b_2^\varepsilon \\ b_3^\varepsilon \end{pmatrix} = \frac{1}{\sqrt{\beta^\varepsilon}} \begin{pmatrix} \partial_3^\varepsilon \theta_2^\varepsilon \partial_{33}^\varepsilon \theta_3^\varepsilon - \partial_3^\varepsilon \theta_3^\varepsilon \partial_{33}^\varepsilon \theta_2^\varepsilon \\ \partial_3^\varepsilon \theta_3^\varepsilon \partial_{33}^\varepsilon \theta_1^\varepsilon - \partial_3^\varepsilon \theta_1^\varepsilon \partial_{33}^\varepsilon \theta_3^\varepsilon \\ \partial_3^\varepsilon \theta_1^\varepsilon \partial_{33}^\varepsilon \theta_2^\varepsilon - \partial_3^\varepsilon \theta_2^\varepsilon \partial_{33}^\varepsilon \theta_1^\varepsilon \end{pmatrix}, \tag{37.4}$$

where

$$\begin{aligned} \alpha^\varepsilon &= \partial_3^\varepsilon \theta_i^\varepsilon \partial_3^\varepsilon \theta_i^\varepsilon, \\ \beta^\varepsilon &= \partial_3^\varepsilon \theta_i^\varepsilon \partial_3^\varepsilon \theta_i^\varepsilon \partial_{33}^\varepsilon \theta_j^\varepsilon \partial_{33}^\varepsilon \theta_j^\varepsilon - \partial_3^\varepsilon \theta_i^\varepsilon \partial_{33}^\varepsilon \theta_i^\varepsilon \partial_3^\varepsilon \theta_j^\varepsilon \partial_{33}^\varepsilon \theta_j^\varepsilon. \end{aligned} \tag{37.5}$$

The radius of curvature R^ε, and the radius of torsion T^ε, of the curve C^ε, are defined by

$$R^\varepsilon = \frac{(\alpha^\varepsilon)^{3/2}}{\sqrt{\beta^\varepsilon}}, \qquad T^\varepsilon = \frac{\sqrt{\alpha^\varepsilon}}{|\partial_3^\varepsilon \boldsymbol{b}^\varepsilon|}, \tag{37.6}$$

where $|\cdot|$ denotes the usual Euclidean norm in $\mathbb{R}^3$.

The beam under study is now identified with the three-dimensional body occupying the volume $\overline{\tilde{\Omega}^\varepsilon}$, where

$$\tilde{\Omega}^\varepsilon = \boldsymbol{\varphi}^\varepsilon(\Omega^\varepsilon), \tag{37.7}$$

and where the mapping $\boldsymbol{\varphi}^\varepsilon$ is defined by letting for all $\boldsymbol{x}^\varepsilon = (x_1^\varepsilon, x_2^\varepsilon, x_3^\varepsilon) \in \overline{\omega^\varepsilon} \times [0, L]$

$$\begin{aligned} &\boldsymbol{\varphi}^\varepsilon(\boldsymbol{x}^\varepsilon) = \left(\varphi_1^\varepsilon(\boldsymbol{x}^\varepsilon), \varphi_2^\varepsilon(\boldsymbol{x}^\varepsilon), \varphi_3^\varepsilon(\boldsymbol{x}^\varepsilon)\right), \\ &\varphi_i^\varepsilon(x_1^\varepsilon, x_2^\varepsilon, x_3^\varepsilon) = \theta_i^\varepsilon(x_3^\varepsilon) + x_1^\varepsilon n_i^\varepsilon(x_3^\varepsilon) + x_2^\varepsilon b_i^\varepsilon(x_3^\varepsilon). \end{aligned} \tag{37.8}$$

If we particularize the mappings θ_i^ε we obtain different types of beams, including the cases where the beam's axis exhibits pure bending ($T^\varepsilon = +\infty$), or pure torsion ($R^\varepsilon = +\infty$), and also the straight beam case ($T^\varepsilon = +\infty$ and $R^\varepsilon = +\infty$).

The boundary of $\tilde{\Omega}^\varepsilon$, $\partial\tilde{\Omega}^\varepsilon$, is the union of the end faces $\tilde{\Gamma}_0^\varepsilon = \boldsymbol{\varphi}^\varepsilon(\omega^\varepsilon \times \{0\})$, $\tilde{\Gamma}_L^\varepsilon = \boldsymbol{\varphi}^\varepsilon(\omega^\varepsilon \times \{L\})$ and of the lateral surface $\tilde{\Gamma}^\varepsilon = \boldsymbol{\varphi}^\varepsilon(\gamma^\varepsilon \times (0, L))$. Let $\tilde{\boldsymbol{x}}^\varepsilon = (\tilde{x}_1^\varepsilon, \tilde{x}_2^\varepsilon, \tilde{x}_3^\varepsilon)$ denote a generic point in $\overline{\tilde{\Omega}^\varepsilon}$ and let

$$\tilde{\partial}_i^\varepsilon = \frac{\partial}{\partial \tilde{x}_i^\varepsilon}. \tag{37.9}$$

Let us assume that the beam $\tilde{\Omega}^\varepsilon$ is clamped at both ends and that it is subjected to body forces $\tilde{\boldsymbol{f}}^\varepsilon = (\tilde{f}_i^\varepsilon) : \tilde{\Omega}^\varepsilon \to \mathbb{R}^3$ and to surface tractions $\tilde{\boldsymbol{g}}^\varepsilon = (\tilde{g}_i^\varepsilon) : \tilde{\Gamma}^\varepsilon \to \mathbb{R}^3$. Then, the variational equilibrium problem for the clamped rod $\overline{\tilde{\Omega}^\varepsilon}$, made of an isotropic, homogeneous and linear elastic material is the following:

$$\begin{aligned} &\text{Find } \tilde{\boldsymbol{u}}^\varepsilon \in \boldsymbol{V}(\tilde{\Omega}^\varepsilon), \text{such that} \\ &\tilde{B}^\varepsilon(\tilde{\boldsymbol{u}}^\varepsilon, \tilde{\boldsymbol{v}}^\varepsilon) = \tilde{F}^\varepsilon(\tilde{\boldsymbol{v}}^\varepsilon), \quad \text{for all } \tilde{\boldsymbol{v}}^\varepsilon \in \boldsymbol{V}(\tilde{\Omega}^\varepsilon). \end{aligned} \tag{37.10}$$

The space of admissible displacements $\boldsymbol{V}(\tilde{\Omega}^\varepsilon)$ is defined by

$$\boldsymbol{V}(\tilde{\Omega}^\varepsilon) = \left\{\tilde{\boldsymbol{v}}^\varepsilon = (\tilde{v}_i^\varepsilon) \in [H^1(\tilde{\Omega}^\varepsilon)]^3 \colon \tilde{\boldsymbol{v}}^\varepsilon = \boldsymbol{0} \text{ on } \tilde{\Gamma}_0^\varepsilon \cup \tilde{\Gamma}_L^\varepsilon\right\}, \tag{37.11}$$

and it is equipped with the norm induced by the norm of H^1, that is,

$$\|\tilde{\boldsymbol{v}}^\varepsilon\|_{\boldsymbol{V}(\tilde{\Omega}^\varepsilon)} = \left[\sum_{i=1}^3 \|\tilde{v}_i^\varepsilon\|^2_{H^1(\tilde{\Omega}^\varepsilon)}\right]^{1/2}. \tag{37.12}$$

If $\tilde{f}_i^\varepsilon \in L^2(\tilde{\Omega}^\varepsilon)$ and $\tilde{g}_i^\varepsilon \in L^2(\tilde{\Gamma}^\varepsilon)$ then, the continuous linear form $\tilde{F}^\varepsilon : \boldsymbol{V}(\tilde{\Omega}^\varepsilon) \to \mathbb{R}$ is given by

$$\tilde{F}^\varepsilon(\tilde{\boldsymbol{v}}^\varepsilon) = \int_{\tilde{\Omega}^\varepsilon} \tilde{f}_i^\varepsilon \tilde{v}_i^\varepsilon \, \mathrm{d}\tilde{\boldsymbol{x}}^\varepsilon + \int_{\tilde{\Gamma}^\varepsilon} \tilde{g}_i^\varepsilon \tilde{v}_i^\varepsilon \, \mathrm{d}\tilde{a}^\varepsilon, \quad \text{for all } \tilde{\boldsymbol{v}}^\varepsilon \in \boldsymbol{V}(\tilde{\Omega}^\varepsilon), \tag{37.13}$$

and the continuous bilinear and symmetric form $\tilde{B}^\varepsilon(\cdot,\cdot) : \boldsymbol{V}(\tilde{\Omega}^\varepsilon) \times \boldsymbol{V}(\tilde{\Omega}^\varepsilon) \to \mathbb{R}$ is defined as

$$\tilde{B}^\varepsilon(\tilde{\boldsymbol{u}}^\varepsilon, \tilde{\boldsymbol{v}}^\varepsilon) = \int_{\tilde{\Omega}^\varepsilon} \{\lambda \tilde{e}^\varepsilon_{pp}(\tilde{\boldsymbol{u}}^\varepsilon)\tilde{e}^\varepsilon_{qq}(\tilde{\boldsymbol{v}}^\varepsilon) + 2\mu \tilde{e}^\varepsilon_{ij}(\tilde{\boldsymbol{u}}^\varepsilon)\tilde{e}^\varepsilon_{ij}(\tilde{\boldsymbol{v}}^\varepsilon)\} \, d\tilde{\boldsymbol{x}}^\varepsilon. \tag{37.14}$$

In (37.14) the Lamé constants λ and μ of the beam $\tilde{\Omega}^\varepsilon$ are assumed to be independent of ε and $\tilde{e}^\varepsilon_{ij}$ are the components of the linearized strain tensor related to the displacement field $\tilde{\boldsymbol{v}}^\varepsilon$ by the formula

$$\tilde{e}^\varepsilon_{ij}(\tilde{\boldsymbol{v}}^\varepsilon) = \tfrac{1}{2}(\tilde{\partial}^\varepsilon_i \tilde{v}^\varepsilon_j + \tilde{\partial}^\varepsilon_j \tilde{v}^\varepsilon_i). \tag{37.15}$$

The existence and uniqueness of the solution of problem (37.10) relies on Korn's inequality and the Lax–Milgram theorem (cf., e.g., CIARLET [1988], pp. 288–292).

THEOREM 37.1. *Let* $\tilde{\boldsymbol{f}}^\varepsilon \in [L^2(\tilde{\Omega}^\varepsilon)]^3$ *and* $\tilde{\boldsymbol{g}}^\varepsilon \in [L^2(\tilde{\Gamma}^\varepsilon)]^3$, *then there exists a unique displacement field* $\tilde{\boldsymbol{u}}^\varepsilon$ *in* $\boldsymbol{V}(\tilde{\Omega}^\varepsilon)$ *that solves the variational equation* (37.10).

We shall now give a brief outline for Sections 38 and 39. In the next section we rewrite the three-dimensional elasticity problem (37.10) in $\Omega^\varepsilon = \omega^\varepsilon \times (0, L)$. Moreover, transforming the curvature and torsion tensors associated with the beam's axis and making some suitable hypotheses on functions θ^ε_i and on the loading, we can reformulate problem (37.10) in a set $\Omega = \omega \times (0, L)$ independent of ε. We then observe that one part of the so transformed bilinear form $\tilde{B}^\varepsilon(\cdot,\cdot)$, is formally equivalent to a three-dimensional elasticity problem, written with respect to a "straight beam" $\overline{\tilde{\Omega}^\varepsilon_{\text{sb}}} = \overline{\omega^\varepsilon} \times [a, b]$, where $a = \theta^\varepsilon_3(0)$ and $b = \theta^\varepsilon_3(L)$.

We may then apply, to this term, the results of MIARA and TRABUCHO [1990, 1992], for straight beams (cf. also Section 33), and estimate the remaining terms, as functions of the geometry, the loading, the Lamé constants, the curvature and the torsion tensors, which is done in Section 39.

38. The curved rod problem posed in a fixed domain

We shall first give another formulation of the equilibrium problem (37.10), posed over the domain $\Omega^\varepsilon = \omega^\varepsilon \times (0, L)$.

We suppose that the mapping $\boldsymbol{\varphi}^\varepsilon$ (cf. (37.8)) is a C^1–diffeomorphism and consequently induces a continuous bijection between the spaces $H^1(\tilde{\Omega}^\varepsilon)$ and $H^1(\Omega^\varepsilon)$ (cf. ADAMS [1975], p. 64) defined by

$$\tilde{v}^\varepsilon \in H^1(\tilde{\Omega}^\varepsilon) \longrightarrow v^\varepsilon = \tilde{v}^\varepsilon \circ \boldsymbol{\varphi}^\varepsilon \in H^1(\Omega^\varepsilon). \tag{38.1}$$

Let $D\boldsymbol{\varphi}^\varepsilon(\boldsymbol{x}^\varepsilon)$ be the matrix representing the derivative of the mapping $\boldsymbol{\varphi}^\varepsilon$ at the point $\boldsymbol{x}^\varepsilon$ of Ω^ε. It is a matter of computation to verify that the components

$b_{ij}^{\varepsilon}(\boldsymbol{x}^{\varepsilon})$ of the inverse matrix of $D\boldsymbol{\varphi}^{\varepsilon}(\boldsymbol{x}^{\varepsilon})$ are

$$\begin{aligned}
b_{1j}^{\varepsilon} &= \frac{1}{d^{\varepsilon}}\left[\sqrt{\alpha^{\varepsilon}}\left(1-\frac{1}{R^{\varepsilon}}x_1^{\varepsilon}\right)n_j^{\varepsilon}+\frac{1}{T^{\varepsilon}}x_2^{\varepsilon}\partial_3^{\varepsilon}\theta_j^{\varepsilon}\right],\\
b_{2j}^{\varepsilon} &= \frac{1}{d^{\varepsilon}}\left[\sqrt{\alpha^{\varepsilon}}\left(1-\frac{1}{R^{\varepsilon}}x_1^{\varepsilon}\right)b_j^{\varepsilon}-\frac{1}{T^{\varepsilon}}x_1^{\varepsilon}\partial_3^{\varepsilon}\theta_j^{\varepsilon}\right],\\
b_{3j}^{\varepsilon} &= \frac{1}{d^{\varepsilon}\sqrt{\alpha^{\varepsilon}}}\partial_3^{\varepsilon}\theta_j^{\varepsilon},
\end{aligned}\tag{38.2}$$

where d^{ε} is the determinant of matrix $D\boldsymbol{\varphi}^{\varepsilon}$, that is,

$$d^{\varepsilon} = \sqrt{\alpha^{\varepsilon}}(1-\frac{1}{R^{\varepsilon}}x_1^{\varepsilon})+\frac{1}{T^{\varepsilon}}\left(x_1^{\varepsilon}b_i^{\varepsilon}-x_2^{\varepsilon}n_i^{\varepsilon}\right)\partial_3^{\varepsilon}\theta_i^{\varepsilon}. \tag{38.3}$$

Using now the change of variable formula, we obtain the following result.

THEOREM 38.1. *Let $\boldsymbol{\varphi}^{\varepsilon}:\overline{\Omega^{\varepsilon}}\to\overline{\tilde{\Omega}^{\varepsilon}}$ be a C^1-diffeomorphism satisfying the orientation preserving condition*

$$d^{\varepsilon}(\boldsymbol{x}^{\varepsilon}) = \det D\boldsymbol{\varphi}^{\varepsilon}(\boldsymbol{x}^{\varepsilon}) > 0, \quad \textit{for all } \boldsymbol{x}^{\varepsilon}\in\Omega^{\varepsilon}, \tag{38.4}$$

then problem (37.10) *is equivalent to the following problem:*

$$\begin{aligned}
&\textit{Find } \boldsymbol{u}^{\varepsilon}\in \boldsymbol{V}(\Omega^{\varepsilon}), \quad \textit{such that}\\
&B^{\varepsilon}(\boldsymbol{u}^{\varepsilon},\boldsymbol{v}^{\varepsilon}) = F^{\varepsilon}(\boldsymbol{v}^{\varepsilon}), \quad \textit{for all } \boldsymbol{v}^{\varepsilon}\in \boldsymbol{V}(\Omega^{\varepsilon}).
\end{aligned}\tag{38.5}$$

where the space of admissible displacements $\boldsymbol{V}(\Omega^{\varepsilon})$ is defined by

$$\boldsymbol{V}(\Omega^{\varepsilon}) = \left\{\boldsymbol{v}^{\varepsilon}=(v_i^{\varepsilon})\in[H^1(\Omega^{\varepsilon})]^3\colon \boldsymbol{v}^{\varepsilon}=\boldsymbol{0} \textit{ on } \Gamma_0^{\varepsilon}\cup\Gamma_L^{\varepsilon}=\omega^{\varepsilon}\times\{0,L\}\right\},$$

and is equipped with the norm induced by the norm of $H^1(\Omega^{\varepsilon})$, that is,

$$\|\boldsymbol{v}^{\varepsilon}\|_{\boldsymbol{V}(\Omega^{\varepsilon})} = \left[\sum_{i=1}^{3}\|v_i^{\varepsilon}\|^2_{H^1(\Omega^{\varepsilon})}\right]^{1/2}.$$

The continuous and linear form $F^{\varepsilon}:\boldsymbol{V}(\Omega^{\varepsilon})\to\mathbb{R}$ is given by

$$F^{\varepsilon}(\boldsymbol{v}^{\varepsilon}) = \int_{\Omega^{\varepsilon}} f_i^{\varepsilon}v_i^{\varepsilon}d^{\varepsilon}\,\mathrm{d}\boldsymbol{x}^{\varepsilon}+\int_{\Gamma^{\varepsilon}} g_i^{\varepsilon}v_i^{\varepsilon}d^{\varepsilon}\left\{\sum_{j=1}^{3}(b_{1j}^{\varepsilon}r_1^{\varepsilon}+b_{2j}^{\varepsilon}r_2^{\varepsilon})^2\right\}^{1/2}\mathrm{d}a^{\varepsilon}, \tag{38.6}$$

where $\Gamma^\varepsilon = \gamma^\varepsilon \times (0, L)$ *is the lateral surface of beam* Ω^ε; $\boldsymbol{f}^\varepsilon = \tilde{\boldsymbol{f}}^\varepsilon \circ \boldsymbol{\varphi}^\varepsilon$; $\boldsymbol{g}^\varepsilon = \tilde{\boldsymbol{g}}^\varepsilon \circ \boldsymbol{\varphi}^\varepsilon$; $\boldsymbol{r}^\varepsilon = r^\varepsilon_\alpha \boldsymbol{e}_\alpha$ *is the outward unit normal to surface* Γ^ε *and* $B^\varepsilon(\cdot,\cdot) : \boldsymbol{V}(\Omega^\varepsilon) \times \boldsymbol{V}(\Omega^\varepsilon) \to \mathbb{R}$ *stands for the continuous, symmetric bilinear form*

$$B^\varepsilon(\boldsymbol{u}^\varepsilon, \boldsymbol{v}^\varepsilon) = \int_{\Omega^\varepsilon} \lambda \partial^\varepsilon_j u^\varepsilon_k b^\varepsilon_{jk} \partial^\varepsilon_i v^\varepsilon_p b^\varepsilon_{ip} \, d^\varepsilon \, \mathrm{d}\boldsymbol{x}^\varepsilon + \int_{\Omega^\varepsilon} \mu (\partial^\varepsilon_k u^\varepsilon_j b^\varepsilon_{ki} \partial^\varepsilon_l v^\varepsilon_j b^\varepsilon_{li} + \partial^\varepsilon_p u^\varepsilon_j b^\varepsilon_{pi} \partial^\varepsilon_m v^\varepsilon_i b^\varepsilon_{mj}) \, d^\varepsilon \, \mathrm{d}\boldsymbol{x}^\varepsilon. \tag{38.7}$$

Let us now assume that the mappings θ^ε_i (cf. (37.1)), which define the beam's axis, satisfy, for all $x^\varepsilon_3 \in [0, L]$,

$$\theta^\varepsilon_1(x^\varepsilon_3) = \varepsilon^{k_1}\theta_1(x^\varepsilon_3), \qquad \theta^\varepsilon_2(x^\varepsilon_3) = \varepsilon^{k_2}\theta_2(x^\varepsilon_3), \qquad \theta^\varepsilon_3(x^\varepsilon_3) = \theta_3(x^\varepsilon_3), \tag{38.8}$$

where functions θ_i are independent of ε, smooth enough and where the exponents k_1, k_2, are positive real numbers, not both zero, verifying, with no loss of generality

$$k_1 \leqslant k_2. \tag{38.9}$$

Assuming that the cross section verifies (38.14), we obtain the following result.

THEOREM 38.2. *If hypotheses* (38.8) *and* (38.9) *are satisfied and if for all* $x^\varepsilon_3 \in (0, L)$,

$$\begin{aligned} &(\partial^\varepsilon_3\theta_3\partial^\varepsilon_{33}\theta_1 - \partial^\varepsilon_3\theta_1\partial^\varepsilon_{33}\theta_3)(x^\varepsilon_3) > 0, \\ &\partial^\varepsilon_3\theta_3(x^\varepsilon_3) > 0, \end{aligned} \tag{38.10}$$

then the components of the binormal and normal vectors, the curvature and the torsion of the beam's axis, verify

$$\begin{aligned} b^\varepsilon_1 &= \mathrm{O}(\varepsilon^{k_2-k_1}), & n^\varepsilon_1 &= 1 + \mathrm{O}(\varepsilon^{2\sigma}), \\ b^\varepsilon_2 &= 1 + \mathrm{O}(\varepsilon^{2(k_2-k_1)}), & n^\varepsilon_2 &= \mathrm{O}(\varepsilon^{k_2-k_1}), \\ b^\varepsilon_3 &= \mathrm{O}(\varepsilon^{k_2}), & n^\varepsilon_3 &= \mathrm{O}(\varepsilon^{k_1}), \end{aligned} \tag{38.11}$$

where $\sigma = \min\{k_1,\ k_2 - k_1\}$ *and*

$$\frac{1}{R^\varepsilon} = \mathrm{O}(\varepsilon^{k_1}), \qquad \frac{1}{T^\varepsilon} = \mathrm{O}(\varepsilon^{k_2-k_1}). \tag{38.12}$$

Moreover the formulas for the determinant d^ε *and its inverse, and for the elements*

b^ε_{ij}, of matrix $(D\boldsymbol{\varphi}^\varepsilon)^{-1}$, are the following:

$$
\begin{aligned}
&d^\varepsilon = \partial_3^\varepsilon \theta_3 + \mathrm{O}(\varepsilon^p), \quad p = \min\{2k_1,\ k_1+1\},\\
&\frac{1}{d^\varepsilon} = \frac{1}{\partial_3^\varepsilon \theta_3} + \mathrm{O}(\varepsilon^p),\\
&b^\varepsilon_{\alpha\alpha} = 1 + \mathrm{O}(\varepsilon^q), \quad q = \min\{2\sigma,\ p\},\\
&b^\varepsilon_{\alpha\beta} = \mathrm{O}(\varepsilon^{k_2-k_1}), \quad \alpha \neq \beta,\\
&b^\varepsilon_{13} = \mathrm{O}(\varepsilon^r), \quad r = \min\{k_1,\ k_2-k_1+1\},\\
&b^\varepsilon_{23} = \mathrm{O}(\varepsilon^z), \quad z = \min\{k_2,\ k_2-k_1+1\},\\
&b^\varepsilon_{31} = \mathrm{O}(\varepsilon^{k_1}),\\
&b^\varepsilon_{32} = \mathrm{O}(\varepsilon^{k_2}),\\
&b^\varepsilon_{33} = \frac{1}{\partial_3^\varepsilon \theta_3} + \mathrm{O}(\varepsilon^p).
\end{aligned}
\tag{38.13}
$$

PROOF. The proof involves many calculations, so we shall only give a brief outline. We first replace θ_i^ε by $\varepsilon^{k_i}\theta_i$ (where $k_3 = 0$) in the definitions of n_i^ε, b_i^ε, $1/R^\varepsilon$, $1/T^\varepsilon$. We then get new functions of the parameter ε. As ε is very small it is possible to express these functions by their Taylor expansions in powers of ε, which leads to formulas (38.11)–(38.12). Formulas (38.13) are a direct consequence of (38.11)–(38.12), because by definition (cf. (38.2)–(38.3)), functions b^ε_{ij} and determinant d^ε are linear combinations of n_i^ε, b_i^ε, $1/R^\varepsilon$, $1/T^\varepsilon$. □

We shall now define a problem equivalent to problem (38.5) (and to problem (37.10)), but posed over a domain that does not depend on ε. Let us define the "reference cross section" ω, obtained from ω^ε, by

$$\omega = \{(x_1, x_2) \in \mathbb{R}^2\colon (x_1, x_2) = \varepsilon^{-1}(x_1^\varepsilon, x_2^\varepsilon), \text{for all } (x_1^\varepsilon, x_2^\varepsilon) \in \omega^\varepsilon\}, \tag{38.14}$$

such that the area of ω, $A(\omega) = |\omega| = 1$. We also define

$$
\begin{aligned}
&\Omega = \omega \times (0, L), \qquad \Gamma_0 = \omega \times \{0\},\\
&\Gamma_L = \omega \times \{L\}, \qquad \Gamma = \gamma \times (0, L) \qquad (\gamma = \partial\omega).
\end{aligned}
\tag{38.15}
$$

In the following, we shall set $\partial_i = \partial/\partial x_i$. With each point $\boldsymbol{x} = (x_1, x_2, x_3) \in \overline{\Omega}$ we associate the point $\boldsymbol{x}^\varepsilon \in \overline{\Omega^\varepsilon}$, through the bijection

$$\boldsymbol{T}^\varepsilon : \overline{\Omega} \longrightarrow \overline{\Omega^\varepsilon}, \boldsymbol{x} = (x_1, x_2, x_3) \longrightarrow \boldsymbol{x}^\varepsilon = (\varepsilon x_1, \varepsilon x_2, x_3). \tag{38.16}$$

Following (33.3) and (33.4) we associate with the displacement fields $\boldsymbol{u}^\varepsilon$, $\boldsymbol{v}^\varepsilon \in \boldsymbol{V}(\Omega^\varepsilon)$, the functions $\boldsymbol{u}(\varepsilon)$ and $\boldsymbol{v}$, defined in Ω by the scalings

$$\begin{aligned}
u_\alpha(\varepsilon)(\boldsymbol{x}) &= \varepsilon^{-1}u_\alpha^\varepsilon \circ \boldsymbol{T}^\varepsilon(\boldsymbol{x}) = \varepsilon^{-1}\tilde{u}_\alpha^\varepsilon \circ \boldsymbol{\varphi}^\varepsilon \circ \boldsymbol{T}^\varepsilon(\boldsymbol{x}), && \text{for all } \boldsymbol{x} \in \overline{\Omega},\\
u_3(\varepsilon)(\boldsymbol{x}) &= \varepsilon^{-2}\, u_3^\varepsilon \circ \boldsymbol{T}^\varepsilon(\boldsymbol{x}) = \varepsilon^{-2}\tilde{u}_3^\varepsilon \circ \boldsymbol{\varphi}^\varepsilon \circ \boldsymbol{T}^\varepsilon(\boldsymbol{x}), && \text{for all } \boldsymbol{x} \in \overline{\Omega},\\
v_\alpha(\boldsymbol{x}) &= \varepsilon^{-1}\, v_\alpha^\varepsilon \circ \boldsymbol{T}^\varepsilon(\boldsymbol{x}) = \varepsilon^{-1}\tilde{v}_\alpha^\varepsilon \circ \boldsymbol{\varphi}^\varepsilon \circ \boldsymbol{T}^\varepsilon(\boldsymbol{x}), && \text{for all } \boldsymbol{x} \in \overline{\Omega},\\
v_3(\boldsymbol{x}) &= \varepsilon^{-2}v_3^\varepsilon \circ \boldsymbol{T}^\varepsilon(\boldsymbol{x}) = \varepsilon^{-2}\tilde{v}_3^\varepsilon \circ \boldsymbol{\varphi}^\varepsilon \circ \boldsymbol{T}^\varepsilon(\boldsymbol{x}), && \text{for all } \boldsymbol{x} \in \overline{\Omega}.
\end{aligned} \tag{38.17}$$

We also assume that there exist functions $\boldsymbol{f} \in [L^2(\Omega)]^3$ and $\boldsymbol{g} \in [L^2(\Gamma)]^3$, independent of ε, such that for all $\boldsymbol{x}^\varepsilon$ in $\overline{\Omega^\varepsilon}$,

$$\begin{aligned}
f_\alpha^\varepsilon(\boldsymbol{x}^\varepsilon) &= \varepsilon^3 f_\alpha \circ (\boldsymbol{T}^\varepsilon)^{-1}(\boldsymbol{x}^\varepsilon), & g_\alpha^\varepsilon(\boldsymbol{x}^\varepsilon) &= \varepsilon^4 g_\alpha \circ (\boldsymbol{T}^\varepsilon)^{-1}(\boldsymbol{x}^\varepsilon),\\
f_3^\varepsilon(\boldsymbol{x}^\varepsilon) &= \varepsilon^2 f_3 \circ (\boldsymbol{T}^\varepsilon)^{-1}(\boldsymbol{x}^\varepsilon), & g_3^\varepsilon(\boldsymbol{x}^\varepsilon) &= \varepsilon^3 g_3 \circ (\boldsymbol{T}^\varepsilon)^{-1}(\boldsymbol{x}^\varepsilon).
\end{aligned} \tag{38.18}$$

Let us consider the Hilbert space

$$\boldsymbol{V}(\Omega) = \left\{\boldsymbol{v} = (v_i) \in [H^1(\Omega)]^3\colon \boldsymbol{v} = \boldsymbol{0} \text{ on } \Gamma_0 \cup \Gamma_L\right\}, \tag{38.19}$$

equipped with the norm induced by the norm of $H^1(\Omega)$, that is,

$$\|\boldsymbol{v}\|_{\boldsymbol{V}(\Omega)} = \left[\sum_{i=1}^{3} \|v_i\|^2_{H^1(\Omega)}\right]^{1/2}, \tag{38.20}$$

then, the equivalent formulation of problem (1.10) (also (38.5)), posed over the domain $\overline{\Omega} = \overline{\omega} \times [0, L]$, is the following.

THEOREM 38.3. *Let the applied force densities $\tilde{\boldsymbol{f}}^\varepsilon$ and $\tilde{\boldsymbol{g}}^\varepsilon$ be such that the associated forces $\boldsymbol{f}^\varepsilon$ and $\boldsymbol{g}^\varepsilon$ satisfy* (38.18), *and let the mapping $\boldsymbol{\varphi}^\varepsilon$ be a C^1-diffeomorphism, such that $d^\varepsilon(\boldsymbol{x}^\varepsilon) > 0$ for all $\boldsymbol{x}^\varepsilon \in \Omega^\varepsilon$. Then, if $\boldsymbol{u}(\varepsilon) \in \boldsymbol{V}(\Omega)$ is the displacement field associated to the unique solution $\tilde{\boldsymbol{u}}^\varepsilon \in \boldsymbol{V}(\tilde{\Omega}^\varepsilon)$ of problem* (37.10), *by formulas* (38.17), *$\boldsymbol{u}(\varepsilon)$ is the unique solution of the following variational equation:*

$$B(\varepsilon)(\boldsymbol{u}(\varepsilon), \boldsymbol{v}) = F(\varepsilon)(\boldsymbol{v}), \quad \textit{for all } \boldsymbol{v} \in \boldsymbol{V}(\Omega). \tag{38.21}$$

The forms $F(\varepsilon)(\cdot)$ and $B(\varepsilon)(\cdot,\cdot)$ depend explicitly on ε and verify:

(i) *$F(\varepsilon) : \boldsymbol{V}(\Omega) \to \mathbb{R}$ is a continuous linear form given by*

$$F(\varepsilon)(\boldsymbol{v}) = \varepsilon^6 \int_\Omega f_i v_i (\partial_3\theta_3 + \mathrm{O}(\varepsilon^p))\,\mathrm{d}\boldsymbol{x} + \varepsilon^6 \int_\Gamma (g_i + \varepsilon^m h_i^\varepsilon) v_i \partial_3\theta_3\,\mathrm{d}a, \tag{38.22}$$

where p is the same as in (38.13) *and*

$$\begin{aligned}
m &= \min\{k_2 - k_1,\ 2k_1,\ k_1 + 1\},\\
h_i^\varepsilon &= \mathcal{F}(g_j, \partial_3\theta_k) + \mathrm{O}(\varepsilon^{p-m}),
\end{aligned} \tag{38.23}$$

and where $\mathcal{F}(g_j, \partial_3\theta_k)$ is a function depending on g_j and on $\partial_3\theta_k$, but independent of ε.

(ii) $B(\varepsilon)(\cdot,\cdot) : V(\Omega) \times V(\Omega) \to \mathbb{R}$ *is a continuous, bilinear and symmetric form defined by*

$$B(\varepsilon)(u,v) = C(\varepsilon)(u,v) + \varepsilon^{\sigma+2} D(\varepsilon)(u,v), \tag{38.24}$$

where

$$\sigma = \min\{k_1,\ k_2 - k_1\}, \tag{38.25}$$

and for arbitrary elements $u, v \in V(\Omega)$,

$$\begin{aligned} C(\varepsilon)(u,v) = {} & \varepsilon^2 \int_\Omega \Big[\lambda e_{\alpha\alpha}(u) e_{\beta\beta}(v) + 2\mu e_{\alpha\beta}(u) e_{\alpha\beta}(v)\Big] \partial_3\theta_3 \, dx \\ & + \varepsilon^4 \int_\Omega \Big[\lambda\big(e_{\alpha\alpha}(u)e_{33}(v) + e_{\beta\beta}(v)e_{33}(u)\big) + \mu\Big(\partial_3 u_\beta \partial_3 v_\beta \frac{1}{(\partial_3\theta_3)^2} \\ & + \partial_\beta u_3 \partial_\beta v_3 + \partial_3 u_\beta \partial_\beta v_3 \frac{1}{\partial_3\theta_3} + \partial_\beta u_3 \partial_3 v_\beta \frac{1}{\partial_3\theta_3}\Big)\partial_3\theta_3\Big] \, dx \\ & + \varepsilon^6 \int_\Omega (\lambda + 2\mu) e_{33}(u) e_{33}(v) \frac{1}{\partial_3\theta_3} \, dx, \end{aligned} \tag{38.26}$$

and there exists a constant C independent of ε *and depending on* $\partial_3\theta_i$, *such that*

$$|\varepsilon^{\sigma+2} D(\varepsilon)(u,v)| \leqslant \varepsilon^{\sigma+2} C \|u\|_{V(\Omega)} \|v\|_{V(\Omega)}, \quad \textit{for all } (u,v) \in V(\Omega) \times V(\Omega). \tag{38.27}$$

REMARK 38.1. In (38.26) we have set for any $v \in V(\Omega)$

$$e_{ij}(v) = \tfrac{1}{2}(\partial_i u_j + \partial_j u_i). \tag{38.28}$$

The proof of Theorem 2.3 is rather technical. In order to obtain the results of this theorem we must introduce in the definition of the forms B^ε and F^ε (cf. (38.6)–(38.7)), the expressions (38.13) for the functions d^ε, $(d^\varepsilon)^{-1}$, b^ε_{ij} and also to take the scalings (38.17)–(38.18) into account.

It is important to remark that the bilinear form $C(\varepsilon)(\cdot,\cdot)$ corresponds exactly (after transforming the displacement fields by scalings of the type (38.17)) to the bilinear form associated to the linear elastic equilibrium problem for the "straight beam" $\overline{\Omega^\varepsilon_{\text{sb}}} = \overline{\omega^\varepsilon} \times [a,b]$, ("sb" stands for "straight beam"), whose cross section is ω^ε (the same cross section of beam $\tilde{\Omega}^\varepsilon$) and whose axis is the line segment $[a,b]$, where

$$a = \theta_3(0), \qquad b = \theta_3(L). \tag{38.29}$$

In order to verify this statement let us rewrite the bilinear form $C(\varepsilon)(\cdot,\cdot)$ in the domain $\tilde{\Omega}^\varepsilon_{\rm sb}$. Let $\boldsymbol{\psi}^\varepsilon$ be the mapping

$$\begin{aligned}&\boldsymbol{\psi}^\varepsilon : \overline{\Omega^\varepsilon} = \overline{\omega^\varepsilon} \times [0,L] \longrightarrow \overline{\tilde{\Omega}^\varepsilon_{\rm sb}} = \overline{\omega^\varepsilon} \times [a,b],\\ &(x_1^\varepsilon, x_2^\varepsilon, x_3^\varepsilon) \longrightarrow \big(x_1^\varepsilon, x_2^\varepsilon, \theta_3(x_3^\varepsilon)\big),\end{aligned} \tag{38.30}$$

and $\tilde{\boldsymbol{y}}^\varepsilon = (\tilde{y}_1^\varepsilon, \tilde{y}_2^\varepsilon, \tilde{y}_3^\varepsilon)$ be an arbitrary point of $\overline{\tilde{\Omega}^\varepsilon_{\rm sb}}$. Let $\boldsymbol{V}(\tilde{\Omega}^\varepsilon_{\rm sb})$ be the Hilbert space defined by

$$\boldsymbol{V}(\tilde{\Omega}^\varepsilon_{\rm sb}) = \Big\{\tilde{\boldsymbol{v}}^\varepsilon_{\rm sb} = (\tilde{v}^\varepsilon_{{\rm sb}i}) \in [H^1(\tilde{\Omega}^\varepsilon_{\rm sb})]^3 \colon \tilde{\boldsymbol{v}}^\varepsilon_{\rm sb} = \boldsymbol{0} \text{ on } \tilde{\Gamma}^\varepsilon_{\rm sb0} \cup \tilde{\Gamma}^\varepsilon_{{\rm sb}L} = \omega^\varepsilon \times \{a,b\}\Big\}, \tag{38.31}$$

and equipped with the norm induced by the norm of $[H^1(\tilde{\Omega}^\varepsilon_{\rm sb})]^3$, that is,

$$\|\tilde{\boldsymbol{v}}^\varepsilon_{\rm sb}\|_{\boldsymbol{V}(\tilde{\Omega}^\varepsilon_{\rm sb})} = \left[\sum_{i=1}^3 \|\tilde{v}^\varepsilon_{{\rm sb}i}\|^2_{H^1(\tilde{\Omega}^\varepsilon_{\rm sb})}\right]^{1/2}. \tag{38.32}$$

For any $\tilde{\boldsymbol{v}}^\varepsilon_{\rm sb}$ in $\boldsymbol{V}(\tilde{\Omega}^\varepsilon_{\rm sb})$ we define $\boldsymbol{v}(\varepsilon)$ in $\boldsymbol{V}(\Omega)$ as follows:

$$\begin{aligned}v_\alpha(\varepsilon)(\boldsymbol{x}) &= \varepsilon^{-1}\tilde{v}^\varepsilon_{{\rm sb}\alpha} \circ \boldsymbol{\psi}^\varepsilon \circ \boldsymbol{T}^\varepsilon(\boldsymbol{x}), \quad \text{for all } \boldsymbol{x} \in \overline{\Omega},\\ v_3(\varepsilon)(\boldsymbol{x}) &= \varepsilon^{-2}\tilde{v}^\varepsilon_{{\rm sb}3} \circ \boldsymbol{\psi}^\varepsilon \circ \boldsymbol{T}^\varepsilon(\boldsymbol{x}), \quad \text{for all } \boldsymbol{x} \in \overline{\Omega}.\end{aligned} \tag{38.33}$$

Then, using the formula of change of variable in multiple integrals, we conclude that

$$C(\varepsilon)(\boldsymbol{u}(\varepsilon), \boldsymbol{v}(\varepsilon)) = \tilde{C}^\varepsilon(\tilde{\boldsymbol{u}}^\varepsilon_{\rm sb}, \tilde{\boldsymbol{v}}^\varepsilon_{\rm sb}), \tag{38.34}$$

where $\tilde{C}^\varepsilon(\cdot,\cdot)$ is a continuous, symmetric and bilinear form defined on $\boldsymbol{V}(\tilde{\Omega}^\varepsilon_{\rm sb}) \times \boldsymbol{V}(\tilde{\Omega}^\varepsilon_{\rm sb})$ with values on $\mathbb{R}$, such that for any $(\tilde{\boldsymbol{u}}^\varepsilon_{\rm sb}, \tilde{\boldsymbol{v}}^\varepsilon_{\rm sb}) \in \boldsymbol{V}(\tilde{\Omega}^\varepsilon_{\rm sb}) \times \boldsymbol{V}(\tilde{\Omega}^\varepsilon_{\rm sb})$,

$$\tilde{C}^\varepsilon(\tilde{\boldsymbol{u}}^\varepsilon_{\rm sb}, \tilde{\boldsymbol{v}}^\varepsilon_{\rm sb}) = \int_{\tilde{\Omega}^\varepsilon_{\rm sb}} \Big[\lambda \tilde{e}^\varepsilon_{pp}(\tilde{\boldsymbol{u}}^\varepsilon_{\rm sb})\tilde{e}^\varepsilon_{qq}(\tilde{\boldsymbol{v}}^\varepsilon_{\rm sb}) + 2\mu \tilde{e}^\varepsilon_{ij}(\tilde{\boldsymbol{u}}^\varepsilon_{\rm sb})\tilde{e}^\varepsilon_{ij}(\tilde{\boldsymbol{v}}^\varepsilon_{\rm sb})\Big]\,{\rm d}\tilde{\Omega}^\varepsilon_{\rm sb}, \tag{38.35}$$

where we set

$$\tilde{e}^\varepsilon_{ij}(\tilde{\boldsymbol{u}}^\varepsilon_{\rm sb}) = \frac{1}{2}\left(\frac{\partial \tilde{u}^\varepsilon_{{\rm sb}i}}{\partial \tilde{y}^\varepsilon_j} + \frac{\partial \tilde{u}^\varepsilon_{{\rm sb}j}}{\partial \tilde{y}^\varepsilon_i}\right). \tag{38.36}$$

Therefore, the statement is proved because $\tilde{C}^\varepsilon(\cdot,\cdot)$ is exactly (cf. MIARA and TRABUCHO [1990, 1992] or Section 33 or else (1.9), (1.12)) the bilinear form that is associated to the three-dimensional linearized elasticity problem for a straight beam, made of a homogeneous linear elastic material and whose reference configuration $\overline{\tilde{\Omega}^\varepsilon_{\rm sb}}$ is a natural state.

Before finishing this section let us remark that the bilinear forms $B(\varepsilon)(\cdot,\cdot)$ and $C(\varepsilon)(\cdot,\cdot)$ associated, respectively, with the domains $\tilde{\Omega}^\varepsilon$ and $\tilde{\Omega}^\varepsilon_{\rm sb}$, are $\boldsymbol{V}(\Omega)$-elliptic. Indeed the following result holds.

THEOREM 38.4. *For each $\varepsilon > 0$, there exist positive constants C_1, C_2 independent of ε, such that for any $\boldsymbol{v}$ in $\boldsymbol{V}(\Omega)$,*

$$B(\varepsilon)(\boldsymbol{v},\boldsymbol{v}) \geqslant \varepsilon^6 C_1 C(\tilde{\Omega}^\varepsilon)\|\boldsymbol{v}\|^2_{\boldsymbol{V}(\Omega)}, \tag{38.37}$$

$$C(\varepsilon)(\boldsymbol{v},\boldsymbol{v}) \geqslant \varepsilon^6 C_2\|\boldsymbol{v}\|^2_{\boldsymbol{V}(\Omega)}, \tag{38.38}$$

where $C(\tilde{\Omega}^\varepsilon)$ is a positive constant, less than 1, depending on the geometry of $\tilde{\Omega}^\varepsilon$.

PROOF. As a consequence of the three-dimensional Korn's inequality, the bilinear form $B(\varepsilon)(\cdot,\cdot)$ satisfies, for all $\boldsymbol{v} \in \boldsymbol{V}(\Omega)$ (verifying (38.17))

$$B(\varepsilon)(\boldsymbol{v},\boldsymbol{v}) = \tilde{B}^\varepsilon(\tilde{\boldsymbol{v}}^\varepsilon,\tilde{\boldsymbol{v}}^\varepsilon) \geqslant C(\tilde{\Omega}^\varepsilon)2\mu\|\tilde{\boldsymbol{v}}^\varepsilon\|^2_{\boldsymbol{V}(\tilde{\Omega}^\varepsilon)}, \tag{38.39}$$

where the positive constant $C(\tilde{\Omega}^\varepsilon)$ depends on the domain $\tilde{\Omega}^\varepsilon$ and is less than 1. On the other hand if $\boldsymbol{v} \in \boldsymbol{V}(\Omega)$ is related to $\tilde{\boldsymbol{v}}^\varepsilon \in \boldsymbol{V}(\tilde{\Omega}^\varepsilon)$ by (38.17) then

$$\exists C_3 > 0\colon\ \|\boldsymbol{v}\|^2_{\boldsymbol{V}(\Omega)} \leqslant \varepsilon^{-6} C_3\|\tilde{\boldsymbol{v}}^\varepsilon\|^2_{\boldsymbol{V}(\tilde{\Omega}^\varepsilon)}, \tag{38.40}$$

where constant C_3 is independent of ε. Hence (38.37) follows.

For the bilinear form $C(\varepsilon)(\cdot,\cdot)$, a simple calculation shows that

$$C(\varepsilon)(\boldsymbol{v},\boldsymbol{v}) \geqslant \varepsilon^6 C(\theta_3)\|\boldsymbol{v}\|^2_{\boldsymbol{V}(\Omega)}, \tag{38.41}$$

where the positive constant $C(\theta_3)$ depends on θ_3. □

REMARK 38.2. As $B(\varepsilon)(\boldsymbol{v},\boldsymbol{v}) = C(\varepsilon)(\boldsymbol{v},\boldsymbol{v}) + \varepsilon^{\sigma+2}D(\varepsilon)(\boldsymbol{v},\boldsymbol{v})$, for all $\boldsymbol{v} \in \boldsymbol{V}(\Omega)$ (cf. (38.24)), if we suppose that $\sigma + 2 > 6$, then, we conclude from (38.41) and (38.27) that

$$B(\varepsilon)(\boldsymbol{v},\boldsymbol{v}) \geqslant \varepsilon^6[C_2 - \varepsilon^{\sigma-4}C_3]\|\boldsymbol{v}\|^2_{\boldsymbol{V}(\Omega)}, \tag{38.42}$$

where C_2 and C_3 are positive constants independent of ε. So, if ε is small enough, we obtain from (38.42)

$$B(\varepsilon)(\boldsymbol{v},\boldsymbol{v}) \geqslant \varepsilon^6 C\|\boldsymbol{v}\|^2_{\boldsymbol{V}(\Omega)}, \quad \text{for all } \boldsymbol{v} \in \boldsymbol{V}(\Omega), \tag{38.43}$$

where now C is another positive constant which does not depend on ε.

39. The Galerkin method for curved rods

39.1. *The Galerkin method for the straight beam* $\tilde{\Omega}^\varepsilon_{\rm sb}$

As in the previous sections we suppose that all the forces acting on the beam $\tilde{\Omega}^\varepsilon$ vanish, except $\tilde{g}^\varepsilon_3$, which is not a restrictive hypothesis. We then have (cf. (38.22))

$$F(\varepsilon)(\boldsymbol{v}) = \varepsilon^6 \int_\Gamma G_3(\varepsilon) v_3 \partial_3 \theta_3 \, {\rm d}a, \quad \text{for all } \boldsymbol{v} \in \boldsymbol{V}(\Omega), \tag{39.1}$$

where

$$G_3(\varepsilon) = g_3 + \varepsilon^m h^\varepsilon_3. \tag{39.2}$$

Let us now assume that the "straight beam" $\tilde{\Omega}^\varepsilon_{\rm sb}$ is subjected to the action of the surface force $\tilde{G}^\varepsilon_3 : \tilde{\Gamma}^\varepsilon_{\rm sb1} \to \mathbb{R}$ where,

$$\begin{aligned} &\tilde{\Gamma}^\varepsilon_{\rm sb1} = \gamma^\varepsilon \times]a, b[, \\ &G_3(\varepsilon)(\boldsymbol{x}) = \varepsilon^{-3} \tilde{G}^\varepsilon_3 \circ \boldsymbol{\psi}^\varepsilon \circ \boldsymbol{T}^\varepsilon(\boldsymbol{x}), \quad \text{for all } \boldsymbol{x} \in \Gamma_1. \end{aligned} \tag{39.3}$$

Let us set

$$\Omega_{\rm sb} = \omega \times (a, b), \qquad \Gamma_{\rm sb0} \cup \Gamma_{{\rm sb}L} = \omega \times \{a, b\}, \qquad \Gamma_{\rm sb} = \gamma \times (a, b), \tag{39.4}$$

and let $\boldsymbol{y} = (y_1, y_2, y_3) = (x_1, x_2, y_3) = (x_1, x_2, \theta_3(x_3))$ denote a generic point of $\overline{\Omega_{\rm sb}}$. We also define the mapping

$$\begin{aligned} \boldsymbol{S} : \overline{\Omega} = \overline{\omega} \times [0, L] &\longrightarrow \overline{\Omega_{\rm sb}} = \overline{\omega} \times [a, b], \\ (x_1, x_2, x_3) &\longrightarrow (x_1, x_2, \theta_3(x_3)) \end{aligned} \tag{39.5}$$

and the space of admissible displacements

$$\boldsymbol{V}(\Omega_{\rm sb}) = \left\{ \boldsymbol{v}_{\rm sb} = (v_{{\rm sb}i}) \in [H^1(\Omega_{\rm sb})]^3 : \boldsymbol{v}_{\rm sb} = \boldsymbol{0} \text{ on } \Gamma_{\rm sb0} \cup \Gamma_{{\rm sb}L} \right\}, \tag{39.6}$$

equipped with the norm induced by the norm of $H^1(\Omega_{\rm sb})$. If $\boldsymbol{v}_{\rm sb} \in \boldsymbol{V}(\Omega_{\rm sb})$ we define $\tilde{\boldsymbol{v}}^\varepsilon_{\rm sb} \in \boldsymbol{V}(\tilde{\Omega}^\varepsilon_{\rm sb})$ and $\boldsymbol{v}(\varepsilon) \in \boldsymbol{V}(\Omega_{\rm sb})$ as follows (cf. (38.33)):

$$\begin{aligned} &\varepsilon\, v_{{\rm sb}\alpha} \circ \boldsymbol{S}(\boldsymbol{x}) = \tilde{v}^\varepsilon_{{\rm sb}\alpha} \circ \boldsymbol{\psi}^\varepsilon \circ \boldsymbol{T}^\varepsilon(\boldsymbol{x}) = \varepsilon\, v_\alpha(\varepsilon)(\boldsymbol{x}), \\ &\varepsilon^2\, v_{{\rm sb}3} \circ \boldsymbol{S}(\boldsymbol{x}) = \tilde{v}^\varepsilon_{{\rm sb}3} \circ \boldsymbol{\psi}^\varepsilon \circ \boldsymbol{T}^\varepsilon(\boldsymbol{x}) = \varepsilon^2\, v_3(\varepsilon)(\boldsymbol{x}), \end{aligned} \tag{39.7}$$

for all $\boldsymbol{x} \in \overline{\Omega}$. Therefore, we can define a new bilinear form $C^\varepsilon(\cdot,\cdot)$ in $\boldsymbol{V}(\Omega_{\rm sb}) \times \boldsymbol{V}(\Omega_{\rm sb})$ associated to the bilinear forms $C(\varepsilon)(\cdot,\cdot)$ and $\tilde{C}^\varepsilon(\cdot,\cdot)$ in the following way (see MIARA and TRABUCHO [1990, 1992] or Section 33):

$$C^\varepsilon(\boldsymbol{u}_{\rm sb}, \boldsymbol{v}_{\rm sb}) = C(\varepsilon)(\boldsymbol{u}(\varepsilon), \boldsymbol{v}(\varepsilon)) = \tilde{C}^\varepsilon(\tilde{\boldsymbol{u}}^\varepsilon_{\rm sb}, \tilde{\boldsymbol{v}}^\varepsilon_{\rm sb}), \tag{39.8}$$

where, for all $\boldsymbol{u}_{\text{sb}}, \boldsymbol{v}_{\text{sb}} \in \boldsymbol{V}(\Omega_{\text{sb}})$,

$$\begin{aligned} C^{\varepsilon}(\boldsymbol{u}_{\text{sb}}, \boldsymbol{v}_{\text{sb}}) = \varepsilon^2 \int_{\Omega_{\text{sb}}} & \left[\lambda e_{\alpha\alpha}(\boldsymbol{u}_{\text{sb}}) e_{\beta\beta}(\boldsymbol{v}_{\text{sb}}) + 2\mu e_{\alpha\beta}(\boldsymbol{u}_{\text{sb}}) e_{\alpha\beta}(\boldsymbol{v}_{\text{sb}})\right] \mathrm{d}\Omega_{\text{sb}} \\ + \varepsilon^4 \int_{\Omega_{\text{sb}}} & \lambda \left[e_{33}(\boldsymbol{u}_{\text{sb}}) e_{\beta\beta}(\boldsymbol{v}_{\text{sb}}) + e_{33}(\boldsymbol{v}_{\text{sb}})\, e_{\beta\beta}(\boldsymbol{u}_{\text{sb}})\right] \mathrm{d}\Omega_{\text{sb}} \\ + \varepsilon^4 \int_{\Omega_{\text{sb}}} & 4\mu e_{3\alpha}(\boldsymbol{u}_{\text{sb}}) e_{3\alpha}(\boldsymbol{v}_{\text{sb}})\, \mathrm{d}\Omega_{\text{sb}} \\ + \varepsilon^6 \int_{\Omega_{\text{sb}}} & (\lambda + 2\mu) e_{33}(\boldsymbol{u}_{\text{sb}}) e_{33}(\boldsymbol{v}_{\text{sb}})\, \mathrm{d}\Omega_{\text{sb}}, \end{aligned} \tag{39.9}$$

with

$$e_{ij}(\boldsymbol{u}_{\text{sb}}) = \tfrac{1}{2}\left(\frac{\partial u_{\text{sb}i}}{\partial y_j} + \frac{\partial u_{\text{sb}j}}{\partial y_i}\right), \tag{39.10}$$

and

$$\mathrm{d}\Omega_{\text{sb}} = \partial_3 \theta_3\, \mathrm{d}\boldsymbol{x}. \tag{39.11}$$

Thus, the linear elastic equilibrium problem for the "straight beam" $\tilde{\Omega}^{\varepsilon}_{\text{sb}}$, subjected to the action of the surface force $\tilde{G}^{\varepsilon}_3$, and clamped at both ends $\tilde{\Gamma}^{\varepsilon}_{\text{sb}0}$, can be written in an equivalent form, on the set Ω_{sb}, as

Find $\boldsymbol{u}^{\varepsilon}_{\text{sb}} \in \boldsymbol{V}(\Omega_{\text{sb}})$ such that,

$$C^{\varepsilon}(\boldsymbol{u}^{\varepsilon}_{\text{sb}}, \boldsymbol{v}_{\text{sb}}) = \varepsilon^6 \int_{\Gamma_{\text{sb}}} G^{\varepsilon}_3 v_{\text{sb}3}\, \mathrm{d}\Gamma_{\text{sb}}, \quad \text{for all } \boldsymbol{v}_{\text{sb}} \in \boldsymbol{V}(\Omega_{\text{sb}}), \tag{39.12}$$

where $G^{\varepsilon}_3 \circ \boldsymbol{S}(\boldsymbol{x}) = G_3(\varepsilon)(\boldsymbol{x})$, for all $\boldsymbol{x} \in \Gamma$. Following the results of MIARA and TRABUCHO [1990, 1992] (cf. Section 33), we choose the approximation space for the straight beam as

$$\begin{aligned} \boldsymbol{V}^N(\Omega_{\text{sb}}) = \Big\{ & \boldsymbol{v}_{\text{sb}} = (v_{\text{sb}i}) \in \boldsymbol{V}(\Omega_{\text{sb}}) \colon v_{\text{sb}\alpha}(\boldsymbol{y}) = \sum_{k=0}^{N} P^k_{\alpha}(x_1, x_2) v^k_{\text{sb}\alpha}(y_3), \\ & v^k_{\text{sb}\alpha} \in H^1_0(a,b),\ P^k_{\alpha} \in H^1(\omega), \text{for all } 0 \leqslant k \leqslant N, \\ & v_{\text{sb}3}(\boldsymbol{y}) = \sum_{k=0}^{N} Q^k(x_1, x_2) v^k_{\text{sb}3}(y_3), \\ & v^k_{\text{sb}3} \in H^1_0(a,b), Q^k \in H^1(\omega), \text{for all } 0 \leqslant k \leqslant N \Big\}, \end{aligned} \tag{39.13}$$

where functions $\boldsymbol{P}^k = (P_1^k, P_2^k)$ and Q^k for $k = 0, 1, \ldots, N$, are exactly the same functions given by (33.20) and (33.21). Then, provided that $G_3^\varepsilon = G_3^\varepsilon(y_3) \in H_0^{2N}(a,b)$, for all $N \geqslant 1$ and $G_3^\varepsilon = \partial s^\varepsilon/\partial y_3$, $s^\varepsilon \in H_0^4(a,b)$ it is possible to determine a function $\boldsymbol{\Psi}_{\mathrm{sb}}^{\varepsilon N}$ belonging to $\boldsymbol{V}^N(\Omega_{\mathrm{sb}})$, depending on force G_3^ε, given by

$$\boldsymbol{\Psi}_{\mathrm{sb}}^{\varepsilon N}(x_1,x_2,y_3) = \left(\Psi_{\mathrm{sb1}}^{\varepsilon N}(x_1,x_2,y_3), \Psi_{\mathrm{sb2}}^{\varepsilon N}(x_1,x_2,y_3), \Psi_{\mathrm{sb3}}^{\varepsilon N}(x_1,x_2,y_3)\right),$$

$$\Psi_{\mathrm{sb}\alpha}^{\varepsilon N}(x_1,x_2,y_3) = \sum_{k=0}^{N} \varepsilon^{2k} P_\alpha^k(x_1,x_2) \frac{\partial^{2k-3} G_3^\varepsilon}{\partial (y_3)^{2k-3}}(y_3), \quad \alpha = 1,2, \tag{39.14}$$

$$\Psi_{\mathrm{sb3}}^{\varepsilon N}(x_1,x_2,y_3) = \sum_{k=0}^{N} \varepsilon^{2k} Q^k(x_1,x_2) \frac{\partial^{2k-2} G_3^\varepsilon}{\partial (y_3)^{2k-2}}(y_3),$$

and such that

$$C^\varepsilon(\boldsymbol{\Psi}_{\mathrm{sb}}^{\varepsilon N}, \boldsymbol{v}_{\mathrm{sb}}) = \varepsilon^6 \int_{\Gamma_{\mathrm{sb}}} G_3^\varepsilon v_{\mathrm{sb3}} \, \mathrm{d}\Gamma_{\mathrm{sb}} + \varepsilon^{2N+4} J(\boldsymbol{\Psi}_{\mathrm{sb}}^{\varepsilon N}, \boldsymbol{v}_{\mathrm{sb}}), \tag{39.15}$$

where the bilinear form $J(\cdot,\cdot)$ satisfies

$$|J(\boldsymbol{\Psi}_{\mathrm{sb}}^{\varepsilon N}, \boldsymbol{v}_{\mathrm{sb}})| \leqslant C_N^\varepsilon \|\boldsymbol{v}_{\mathrm{sb}}\|_{\boldsymbol{V}(\Omega_{\mathrm{sb}})}, \tag{39.16}$$

with C_N^ε a constant depending on N and ε, verifying

$$C_N^\varepsilon \leqslant C \max_{y_3 \in [a,b]} \left| \frac{\partial^{2N-1} G_3^\varepsilon}{\partial y_3^{2N-1}}(y_3) \right| \max\left\{ \|\boldsymbol{P}^N\|_{H^1(\omega)}, \|Q^N\|_{H^1(\omega)} \right\}, \tag{39.17}$$

where C is independent of ε and N but depends on the Lamé constants λ and μ.

It is also possible to conclude that there exists another constant C, independent of ε and depending on θ_3, such that

$$\|\boldsymbol{v}_{\mathrm{sb}}\|_{\boldsymbol{V}(\Omega_{\mathrm{sb}})} \leqslant C \|\boldsymbol{v}\|_{\boldsymbol{V}(\Omega)}, \tag{39.18}$$

if the vector fields $\boldsymbol{v}_{\mathrm{sb}}$ and $\boldsymbol{v}$ satisfy (cf. (39.7))

$$\boldsymbol{v}_{\mathrm{sb}} \circ \boldsymbol{S}(\boldsymbol{x}) = \boldsymbol{v}(\boldsymbol{x}), \quad \text{for all } \boldsymbol{x} \in \Omega. \tag{39.19}$$

Now if $\boldsymbol{u}_{\mathrm{sb}}^\varepsilon$ is the unique solution of (39.12) and $\boldsymbol{u}_{\mathrm{sb}}^{\varepsilon N}$ is the orthogonal projection of $\boldsymbol{u}_{\mathrm{sb}}^\varepsilon$ in the space $\boldsymbol{V}^N(\Omega_{\mathrm{sb}})$ (for the inner product induced by $C^\varepsilon(\cdot,\cdot)$ in $\boldsymbol{V}(\Omega_{\mathrm{sb}})$) and if we set

$$\boldsymbol{u}_{\mathrm{sb}}(\varepsilon) = \boldsymbol{u}_{\mathrm{sb}}^\varepsilon \circ \boldsymbol{S}, \qquad \boldsymbol{u}_{\mathrm{sb}}^N(\varepsilon) = \boldsymbol{u}_{\mathrm{sb}}^{\varepsilon N} \circ \boldsymbol{S}, \qquad \boldsymbol{\Psi}_{\mathrm{sb}}^N(\varepsilon) = \boldsymbol{\Psi}_{\mathrm{sb}}^{\varepsilon N} \circ \boldsymbol{S}, \tag{39.20}$$

then $\boldsymbol{u}_{\text{sb}}(\varepsilon)$, $\boldsymbol{u}^N_{\text{sb}}(\varepsilon)$, $\boldsymbol{\Psi}^N_{\text{sb}}(\varepsilon)$ are in $\boldsymbol{V}(\Omega)$ and we obtain from (39.8), (39.12), (39.15), (39.16), (39.18) and MIARA and TRABUCHO [1990, 1992],

$$\begin{aligned} C(\varepsilon)(\boldsymbol{u}_{\text{sb}}(\varepsilon) - \boldsymbol{\Psi}^N_{\text{sb}}(\varepsilon), \boldsymbol{v}) &\leqslant C^\varepsilon_N \varepsilon^{2N+4} C\|\boldsymbol{v}\|_{V(\Omega)}, \\ C(\varepsilon)(\boldsymbol{u}_{\text{sb}}(\varepsilon) - \boldsymbol{u}^N_{\text{sb}}(\varepsilon), \boldsymbol{v}) &\leqslant C^\varepsilon_N \varepsilon^{2N+4} C\|\boldsymbol{v}\|_{V(\Omega)}, \end{aligned} \tag{39.21}$$

for all $\boldsymbol{v}$ in $\boldsymbol{V}(\Omega)$ where C is a constant independent of ε and N. The inequalities in (39.21) together with the $\boldsymbol{V}(\Omega)$-ellipticity of the bilinear form $C(\varepsilon)(\cdot,\cdot)$ (cf. (38.38)) imply that

$$\begin{aligned} \|\boldsymbol{u}_{\text{sb}}(\varepsilon) - \boldsymbol{\Psi}^N_{\text{sb}}(\varepsilon)\|_{V(\Omega)} &\leqslant C^\varepsilon_N \varepsilon^{2N-2} C, \\ \|\boldsymbol{u}_{\text{sb}}(\varepsilon) - \boldsymbol{u}^N_{\text{sb}}(\varepsilon)\|_{V(\Omega)} &\leqslant C^\varepsilon_N \varepsilon^{2N-2} C. \end{aligned} \tag{39.22}$$

Constant C is independent of ε and N, but depends on $\partial_3\theta_3$ and on the Lamé constants. Constant C^ε_N can be estimated as in Theorem 34.2 and the following result holds (cf. MASCARENHAS and TRABUCHO [1991, Theorem 3.2]):

THEOREM 39.1. *Let $G^\varepsilon_3 = G^\varepsilon_3(y_3) \in H^{2N}_0(a,b)$ for all $N \geqslant 1$, $G^\varepsilon_3 = \partial s^\varepsilon/\partial y_3$, $s^\varepsilon \in H^4_0(a,b)$ and $g_3 = g_3(x_3) \in H^{2N}_0(0,L)$ for all $N \geqslant 1$. Then there exist constants C_1 and C_2 independent of ε and N such that for ε sufficiently small, one has*

$$C^\varepsilon_N \leqslant C_1 N C_2^{N-1}. \tag{39.23}$$

Combining (39.22) and (39.23), we conclude that the application of the method to the "straight beam" $\tilde{\Omega}^\varepsilon_{\text{sb}}$, subjected to the axial force $\tilde{G}^\varepsilon_3$, gives the following result.

THEOREM 39.2. *Let the assumptions of Theorem 39.1 hold. Then for a fixed small enough ε, $\|\boldsymbol{u}^N_{\text{sb}}(\varepsilon) - \boldsymbol{u}_{\text{sb}}(\varepsilon)\|_{V(\Omega)} \to 0$ as $N \to +\infty$ and there exist constants C_1, C_2, independent of ε and N, such that*

$$\|\boldsymbol{u}_{\text{sb}}(\varepsilon) - \boldsymbol{u}^N_{\text{sb}}(\varepsilon)\|_{V(\Omega)} \leqslant \varepsilon^{2N-2} C_1 N C_2^{N-1}. \tag{39.24}$$

REMARK 39.1. From (39.22) we conclude that the second member of (39.24) is also an upper bound for the difference $\|\boldsymbol{u}_{\text{sb}}(\varepsilon) - \boldsymbol{\Psi}^N_{\text{sb}}(\varepsilon)\|_{V(\Omega)}$.

39.2. The Galerkin method for the beam $\tilde{\Omega}^\varepsilon$

The structure of the bilinear form $B(\varepsilon)$, which may be split into a bilinear form associated to the "straight beam" $\tilde{\Omega}^\varepsilon_{\text{sb}}$ and another bilinear form (cf. (38.24)), suggests choosing the analog of the approximation space $\boldsymbol{V}^N(\Omega_{\text{sb}})$ for the approximation space $\boldsymbol{V}^N(\Omega)$. Therefore, we define $\boldsymbol{V}^N(\Omega)$ as follows:

$$\boldsymbol{V}^N(\Omega) = \Big\{ \boldsymbol{v} = (v_i) \in \boldsymbol{V}(\Omega)\colon v_\alpha(\boldsymbol{x}) = \sum_{k=0}^{N} P^k_\alpha(x_1,x_2) v^k_\alpha(x_3),$$

$$v_\alpha^k \in H_0^1(0,L),\ P_\alpha^k \in H^1(\omega), \text{for all } 0 \leqslant k \leqslant N,$$

$$v_3(\boldsymbol{x}) = \sum_{k=0}^{N} Q^k(x_1,x_2)v_3^k(x_3),$$

$$v_3^k \in H_0^1(0,L),\ Q^k \in H^1(\omega), \text{for all } 0 \leqslant k \leqslant N\Big\}, \tag{39.25}$$

where $\boldsymbol{P}^k$ and Q^k are the same functions as in the space $\boldsymbol{V}^N(\Omega_{\mathrm{sb}})$ (cf. (39.13)).

Let $\boldsymbol{u}^N(\varepsilon)$ be the orthogonal projection in the space $\boldsymbol{V}^N(\Omega)$ (associated to the inner product $B(\varepsilon)(\cdot,\cdot)$) of the unique solution $\boldsymbol{u}(\varepsilon)$ of problem (38.21). In order to determine an estimate for the norm difference $\|\boldsymbol{u}(\varepsilon) - \boldsymbol{u}^N(\varepsilon)\|_{V(\Omega)}$ we first establish the following result.

THEOREM 39.3. *There exist constants C_1 and C_2 independent of ε and N, but dependent on $\partial_3\theta_i$, such that for any $\boldsymbol{v} \in \boldsymbol{V}(\Omega)$,*

$$|B(\varepsilon)(\boldsymbol{u}(\varepsilon) - \boldsymbol{\Psi}_{\mathrm{sb}}^N(\varepsilon), \boldsymbol{v})| \leqslant \Big\{C_N^\varepsilon \varepsilon^{2N+4} C_1 + \varepsilon^{\sigma+2} C_2 \|\boldsymbol{\Psi}_{\mathrm{sb}}^N(\varepsilon)\|_{V(\Omega)}\Big\}\|\boldsymbol{v}\|_{V(\Omega)}, \tag{39.26}$$

where C_N^ε is the constant defined in (39.16).

PROOF. The bilinear form $B(\varepsilon)(\cdot,\cdot)$ verifies

$$\begin{aligned} &B(\varepsilon)(\boldsymbol{u}(\varepsilon),\boldsymbol{v}) = \varepsilon^6 \int_\Gamma G_3(\varepsilon) v_3 \partial\theta_3 \,\mathrm{d}a, \\ &B(\varepsilon)(\boldsymbol{\Psi}_{\mathrm{sb}}^N(\varepsilon),\boldsymbol{v}) = C(\varepsilon)(\boldsymbol{\Psi}_{\mathrm{sb}}^N(\varepsilon),\boldsymbol{v}) + \varepsilon^{\sigma+2} D(\varepsilon)(\boldsymbol{\Psi}_{\mathrm{sb}}^N(\varepsilon),\boldsymbol{v}), \end{aligned} \tag{39.27}$$

for all $\boldsymbol{v} \in \boldsymbol{V}(\Omega)$. So because of (39.8) and (39.15) we get

$$\begin{aligned} &B(\varepsilon)(\boldsymbol{u}(\varepsilon) - \boldsymbol{\Psi}_{\mathrm{sb}}^N(\varepsilon),\boldsymbol{v}) \\ &\quad = B(\varepsilon)(\boldsymbol{u}(\varepsilon),\boldsymbol{v}) - \varepsilon^6 \int_\Gamma G_3(\varepsilon) v_3 \partial_3\theta_3 \,\mathrm{d}a \\ &\quad\quad - \varepsilon^{2N+4} J(\boldsymbol{\Psi}_{\mathrm{sb}}^{\varepsilon N}, \boldsymbol{v}_{\mathrm{sb}}) - \varepsilon^{\sigma+2} D(\varepsilon)(\boldsymbol{\Psi}_{\mathrm{sb}}^N(\varepsilon),\boldsymbol{v}). \end{aligned} \tag{39.28}$$

But the first two terms in the second member of (39.28) cancel and then the result is just a consequence of (39.16), (39.18) and the continuity of $D(\varepsilon)$ (cf. (38.27)). □

The next result will be used in the calculation of an estimate for $\|\boldsymbol{\Psi}_{\mathrm{sb}}^N(\varepsilon)\|_{V(\Omega)}$.

THEOREM 39.4. *Let $\varepsilon > 0$ be a sufficiently small parameter. Then, if G_3^ε satisfies the assumptions of Theorem* 39.1

$$\int_{\Gamma_{\mathrm{sb}}} G_3^\varepsilon \Psi_{\mathrm{sb}3}^{\varepsilon N} \,\mathrm{d}\Gamma_{\mathrm{sb}} \leqslant 0. \tag{39.29}$$

PROOF. By definition we have (cf. (39.14))

$$\Psi_{\mathrm{sb}3}^{\varepsilon N}(\boldsymbol{y}) = \sum_{k=0}^{N} \varepsilon^{2k} \Psi_3^k(y_3) Q^k(x_1, x_2), \tag{39.30}$$

for all $\boldsymbol{y} = (x_1, x_2, y_3) \in \Omega_{\mathrm{sb}} = \omega \times (a, b)$ and (cf. (33.26), (33.28))

$$\begin{aligned}
&\Psi_3^k(y_3) = \partial_3^{2k-2} G_3^\varepsilon(y_3),\\
&G_3^\varepsilon(y_3) = \Psi_3^1(y_3) = \partial_3^2 \Psi_3^0(y_3),\\
&Q^0(x_1, x_2) = \frac{1}{\mu}(D^0 - C_\alpha^0 x_\alpha),\\
&D^0 = -\frac{\lambda + \mu}{3\lambda + 2\mu} \frac{|\gamma|}{|\omega|},\\
&C_\alpha^0 = \frac{\lambda + \mu}{3\lambda + 2\mu} \int_\gamma \frac{x_\alpha}{I_\alpha}\, \mathrm{d}\gamma,\\
&I_\alpha = \int_\omega (x_\alpha)^2\, \mathrm{d}\omega,
\end{aligned} \tag{39.31}$$

where we have set $\partial_3 = \partial/\partial y_3$ to simplify the notations, and $|\gamma|$, $|\omega|$ are the measures of γ and ω respectively. Consequently, we obtain

$$\begin{aligned}
\int_{\Gamma_{\mathrm{sb}}} G_3^\varepsilon \Psi_{\mathrm{sb}3}^{\varepsilon N}\, \mathrm{d}\Gamma_{\mathrm{sb}} &= \sum_{k=0}^{N} \varepsilon^{2k} \int_\gamma Q^k(x_1, x_2)\, \mathrm{d}\gamma \int_a^b G_3^\varepsilon(y_3) \Psi_3^k(y_3)\, \mathrm{d}y_3\\
&= \int_\gamma Q^0(x_1, x_2)\, \mathrm{d}\gamma \int_a^b |\partial_3 \Psi_3^0(y_3)|^2\, \mathrm{d}y_3\\
&\quad + \sum_{k=1}^{N} \varepsilon^{2k} \int_\gamma Q^k(x_1, x_2)\, \mathrm{d}\gamma \int_a^b |\partial_3^{k-1} G_3^\varepsilon(y_3)|^2 \mathrm{d}y_3.
\end{aligned} \tag{39.32}$$

But by definition of Ψ_3^0 (cf. (39.31)) and the formula of change of variable, we get

$$\partial_3 \Psi_3^0(y_3) = \int_a^{y_3} G_3^\varepsilon(t)\, \mathrm{d}t = \int_0^{x_3} [g_3(t) + \mathrm{O}(\varepsilon^m)] \partial_3 \theta_3(t)\, \mathrm{d}t. \tag{39.33}$$

On the other hand, by definition of $Q^0(x_1, x_2)$ (cf. (39.31)), we also obtain

$$\int_\gamma Q^0(x_1, x_2)\, \mathrm{d}\gamma = -\frac{\lambda + \mu}{\mu(3\lambda + 2\mu)} \int_\gamma \left\{ \frac{|\gamma|^2}{|\omega|} + \frac{[\int_\gamma x_\alpha\, \mathrm{d}\gamma]^2}{I_\alpha} \right\} \mathrm{d}\gamma \leqslant 0. \tag{39.34}$$

Consequently, we have

$$\|\boldsymbol{u}_{\rm sb}^{\varepsilon N}\|_{\rm sb}^2 \leqslant 2(\boldsymbol{\Psi}_{\rm sb}^{\varepsilon N}, \boldsymbol{u}_{\rm sb}^{\varepsilon})_{\rm sb} + \left\| \boldsymbol{\Psi}_{\rm sb}^{\varepsilon N} - \frac{\boldsymbol{u}_{\rm sb}^{\varepsilon N} + \boldsymbol{u}_{\rm sb}^{\varepsilon}}{2} \right\|_{\rm sb}^2. \tag{39.41}$$

By definition of $(\boldsymbol{\Psi}_{\rm sb}^{\varepsilon N}, \boldsymbol{u}_{\rm sb}^{\varepsilon})_{\rm sb}$ and because of Theorem 39.4 one gets

$$(\boldsymbol{\Psi}_{\rm sb}^{\varepsilon N}, \boldsymbol{u}_{\rm sb}^{\varepsilon})_{\rm sb} = C^{\varepsilon}(\boldsymbol{\Psi}_{\rm sb}^{\varepsilon N}, \boldsymbol{u}_{\rm sb}^{\varepsilon}) = C^{\varepsilon}(\boldsymbol{\Psi}_{\rm sb}^{\varepsilon N}, \boldsymbol{u}_{\rm sb}^{\varepsilon N}) = \varepsilon^6 \int_{\Gamma_{\rm sb}} G_3^{\varepsilon} \Psi_{\rm sb3}^{\varepsilon N}\, \mathrm{d}\Gamma_{\rm sb} \leqslant 0, \tag{39.42}$$

and Eqs. (39.41), (39.42) imply that

$$\begin{aligned}\|\boldsymbol{u}_{\rm sb}^{\varepsilon N}\|_{\rm sb}^2 &\leqslant \| \boldsymbol{\Psi}_{\rm sb}^{\varepsilon N} - \frac{\boldsymbol{u}_{\rm sb}^{\varepsilon N} + \boldsymbol{u}_{\rm sb}^{\varepsilon}}{2} \|_{\rm sb}^2 \\ &\leqslant \tfrac{1}{2}\| \boldsymbol{\Psi}_{\rm sb}^{\varepsilon N} - \boldsymbol{u}_{\rm sb}^{\varepsilon N}\|_{\rm sb}^2 + \tfrac{1}{2}\| \boldsymbol{\Psi}_{\rm sb}^{\varepsilon N} - \boldsymbol{u}_{\rm sb}^{\varepsilon}\|_{\rm sb}^2.\end{aligned} \tag{39.43}$$

But we also know (cf. (39.21), (39.22)) that

$$\begin{aligned}\| \boldsymbol{\Psi}_{\rm sb}^{\varepsilon N} - \boldsymbol{u}_{\rm sb}^{\varepsilon}\|_{\rm sb}^2 &= C(\varepsilon)(\boldsymbol{\Psi}_{\rm sb}^{N}(\varepsilon) - \boldsymbol{u}_{\rm sb}(\varepsilon), \boldsymbol{\Psi}_{\rm sb}^{N}(\varepsilon) - \boldsymbol{u}_{\rm sb}(\varepsilon)) \\ &\leqslant C_N^{\varepsilon} \varepsilon^{2N+4} C \| \boldsymbol{\Psi}_{\rm sb}^{N}(\varepsilon) - \boldsymbol{u}_{\rm sb}(\varepsilon)\|_{V(\Omega)} \\ &\leqslant (C_N^{\varepsilon})^2 \varepsilon^{4N+2} C^2,\end{aligned} \tag{39.44}$$

and also

$$\begin{aligned}\| \boldsymbol{\Psi}_{\rm sb}^{\varepsilon N} - \boldsymbol{u}_{\rm sb}^{\varepsilon N}\|_{\rm sb}^2 &\leqslant 2\| \boldsymbol{\Psi}_{\rm sb}^{\varepsilon N} - \boldsymbol{u}_{\rm sb}^{\varepsilon}\|_{\rm sb}^2 + 2\|\boldsymbol{u}_{\rm sb}^{\varepsilon} - \boldsymbol{u}_{\rm sb}^{\varepsilon N}\|_{\rm sb}^2 \\ &\leqslant (C_N^{\varepsilon})^2 \varepsilon^{4N+2} C^2,\end{aligned} \tag{39.45}$$

so we deduce from (39.43)–(39.45) and (39.39) that

$$\|\boldsymbol{u}_{\rm sb}^{N}(\varepsilon)\|_{V(\Omega)} \leqslant \varepsilon^{2N-2} C_N^{\varepsilon} C, \tag{39.46}$$

and the proof is complete. □

Finally, as a consequence of Theorems 39.3 and 39.5, the following estimate holds.

THEOREM 39.6. *If ε is a small positive parameter, $G_3^{\varepsilon} = G_3^{\varepsilon}(y_3) \in H_0^{2N}(a,b)$, for all $N \geqslant 1$, $G_3^{\varepsilon} = \partial s^{\varepsilon}/\partial y_3$, $s^{\varepsilon} \in H_0^4(a^{\varepsilon}, b^{\varepsilon})$, $g_3 = g_3(x_3) \in H_0^{2N}(0,L)$ for all $N \geqslant 1$ then, for a fixed small enough ε, $\|\boldsymbol{u}(\varepsilon) - \boldsymbol{u}^N(\varepsilon)\|_{V(\Omega)} \to 0$ as $N \to +\infty$ and*

$$\|\boldsymbol{u}(\varepsilon) - \boldsymbol{u}^N(\varepsilon)\|_{V(\Omega)} \leqslant \varepsilon^{2N-2} C_1 N C_2^{N-1} [C(\tilde{\Omega}^{\varepsilon})]^{-1} \{1 + \varepsilon^{\sigma-4}\}, \tag{39.47}$$

where C_1, C_2 are constants independent of ε and N, $C(\tilde{\Omega}^\varepsilon)$ is a positive constant defined in (38.37) *and the exponent σ is defined in* (38.25).

PROOF. From (39.26) and (39.37) a simple calculation leads to

$$|B(\boldsymbol{u}(\varepsilon) - \boldsymbol{\Psi}_{\mathrm{sb}}^N(\varepsilon), \boldsymbol{v})| \leqslant C_N^\varepsilon C \|\boldsymbol{v}\|_{V(\Omega)} \{\varepsilon^{2N+4} + \varepsilon^{\sigma+2N}\}.$$

Thus, estimate (39.47) is a just a consequence of the last inequality, the estimate (39.23) for C_N^ε, the definition of $\boldsymbol{u}^N(\varepsilon)$ and the $\boldsymbol{V}(\Omega)$-ellipticity of $B(\varepsilon)(\cdot,\cdot)$ (cf. (38.37)). □

REMARK 39.2. (i) The hypothesis that all the forces vanish, except $\tilde{g}_3$, is not restrictive, since for the straight beam $\tilde{\Omega}_{\mathrm{sb}}^\varepsilon$ the estimate of the Galerkin approximation can be extended to other loadings (cf. MIARA and TRABUCHO [1992] or Remarks 33.2 and 33.4).

(ii) The fact that the force $G_3(\varepsilon)$ depends on ε does not change the computation of the basis functions $\boldsymbol{P}^k$ and Q^k, for the "straight beam" $\tilde{\Omega}_{\mathrm{sb}}^\varepsilon$. The calculation is precisely the same as in MIARA and TRABUCHO [1990, 1992] (cf. (33.25)–(33.40) also).

(iii) With respect to the constant $C(\tilde{\Omega}^\varepsilon)$ we can only say that it is less than one and depends on the geometry of $\tilde{\Omega}^\varepsilon$. But for some special geometries it is possible to specify this dependence (cf. Remark 38.2 and also KONDRATIEV and OLEINIK [1989]).

(iv) Estimate (39.47) is valid for a general beam. Therefore, if we particularize the values of the exponents k_1, k_2, or the mappings θ_i, we shall get estimates for beams whose axis undergoes no torsion and also for straight beams. More precisely, if $k_2 = +\infty$, or for example $\theta_2 = 0$ and $\theta_3(x_3) = x_3$, then the torsion of the beam's axis is $1/T^\varepsilon = 0$, and estimate $\|\boldsymbol{u}(\varepsilon) - \boldsymbol{u}^N(\varepsilon)\|_{V(\Omega)}$ in Ω is given by (39.47), with $\sigma = k_1$. If the torsion of the axis is zero and $k_1 = +\infty$, or for example $\theta_1 = \theta_2 = 0$ and $\theta_3(x_3) = x_3$, then the curvature is $1/R^\varepsilon = 0$, the beam $\tilde{\Omega}^\varepsilon$ is straight and estimate (39.47) becomes

$$\|\boldsymbol{u}(\varepsilon) - \boldsymbol{u}^N(\varepsilon)\|_{V(\Omega)} \leqslant \varepsilon^{2N-2} C_1 N C_2^{N-1},$$

which coincides with (33.47), since $C_N \leqslant C_1 N C_2^{N-1}$ (cf. Theorem 34.2).

(v) If we interchange axis x_1^ε and x_2^ε, that is if the beam $\tilde{\Omega}^\varepsilon$ is defined by the mapping $\boldsymbol{\varphi}^\varepsilon$, such that (compare with (37.8)),

$$\varphi_i^\varepsilon(x_1^\varepsilon, x_2^\varepsilon, x_3^\varepsilon) = \theta_i^\varepsilon(x_3^\varepsilon) + x_2^\varepsilon\, n_i^\varepsilon(x_3^\varepsilon) + x_1^\varepsilon\, b_i^\varepsilon(x_3^\varepsilon),$$

and if exponents k_i (with $k_3 = 0$) and functions $\theta_i^\varepsilon = \varepsilon^{k_i}\theta_i$ satisfy the relations (compare with (38.9) and (38.10))

$$\begin{aligned}
&k_2 \leqslant k_1 (k_2 \text{ and } k_1 \text{ not both zero}),\\
&\partial_3^\varepsilon\theta_2\partial_{33}^\varepsilon\theta_3 - \partial_3^\varepsilon\theta_3\partial_{33}^\varepsilon\theta_2 > 0, \quad \text{in } (0, L),\\
&\partial_3^\varepsilon\theta_3 > 0, \quad \text{in } (0, L),
\end{aligned}$$

we still obtain estimate (39.47), for the norm difference $\|\boldsymbol{u}(\varepsilon) - \boldsymbol{u}^N(\varepsilon)\|_{V(\Omega)}$, but the position of k_1 is interchanged with the position of k_2 in the definition of the exponent σ.

(vi) Finally, estimate (39.47) implies that for N fixed and special geometries of the beam $\tilde{\Omega}^\varepsilon$ (cf. Remark 38.2 and (iii) of Remark 39.2)

$$\lim_{\varepsilon \to 0} \|\boldsymbol{u}(\varepsilon) - \boldsymbol{u}^N(\varepsilon)\|_{V(\Omega)} = 0.$$

we will again estimate $\delta M/M$ for the norm difference $||x[1] - x[2]||_{\infty}$, ... but the position $x[1]$ is interchanged with the position $x[2]$ in the definition of the exponent α.

Finally, equation (9.37) implies that for A ... and special conditions ... on mechanism ... for Theorem 9.6.3 and (iii) of Theorem 9.6.3.

$$\lim_{n \to \infty} ||x[n] - x||_{\infty} = 0.$$

CHAPTER IX

Asymptotic Method for Nonlinear Elastic Rods. Stationary Case

In this chapter we initiate the asymptotic study of the geometrically nonlinear three-dimensional elasticity model through the analysis of the stationary case. The dynamic case is the object of study of Chapter X.

The main results of this chapter were obatined first by CIMETIÈRE, GEYMONAT, LE DRET, RAOULT and TUTEK [1986, 1988] whose work is at the origin of the technique to evaluate the zeroth order stress components through the compatibility with higher order terms. However, here we present a displacement approach for this case (Sections 40–44), motivated by CIARLET [1990] for the plate case. In Section 45 the classical displacement–stress approach is also given and the results compared with the classical nonlinear rod theories.

40. The three-dimensional equations of a nonlinearly elastic clamped rod

With the same notation as before (Sections 1–10), we now assume that for each $\varepsilon > 0$, the set $\overline{\Omega}^{\varepsilon}$ is the *reference configuration* occupied by the body in the absence of applied forces. As before, for each $\varepsilon > 0$, we consider that the beam is subjected to two kinds of *applied forces*: *applied body forces* acting in its interior, of density $\boldsymbol{f}^{\varepsilon} = (f_i^{\varepsilon}) : \Omega^{\varepsilon} \to \mathbb{R}^3$, and *applied surface tractions* acting on the lateral surface, of density $\boldsymbol{g}^{\varepsilon} = (g_i^{\varepsilon}) : \Gamma^{\varepsilon} \to \mathbb{R}^3$. Since these densities do not depend on the unknown, the applied forces considered here are thus *dead loads* (cf. CIARLET [1988, Section 2.7]).

We assume that

$$f_i^{\varepsilon} \in L^2(\Omega^{\varepsilon}), \qquad g_i^{\varepsilon} \in L^2(\Gamma^{\varepsilon}).$$

As in the previous sections the unknown of the problem is the *displacement vector field* $\boldsymbol{u}^{\varepsilon} = (u_i^{\varepsilon}) : \overline{\Omega}^{\varepsilon} \to \mathbb{R}^3$, i.e., the vector $\boldsymbol{u}^{\varepsilon}(\boldsymbol{x}^{\varepsilon})$ represents the displacement field that each point $\boldsymbol{x}^{\varepsilon} \in \overline{\Omega}^{\varepsilon}$ undergoes by the action of the applied forces.

The set $\boldsymbol{\varphi}^{\varepsilon}(\overline{\Omega}^{\varepsilon})$, where $\boldsymbol{\varphi}^{\varepsilon} = \mathbf{id} + \boldsymbol{u}^{\varepsilon}$, is then called a *deformed configuration* and the mapping $\boldsymbol{\varphi}^{\varepsilon}$ is a *deformation.* Since the approach in this section is formal, we assume that the requirement that the deformation $\boldsymbol{\varphi}^\varepsilon$ should satisfy in order to be physically admissible (orientation-preserving character and injectivity) are satisfied for all the values of ε that are considered.

We assume that the rod is *clamped* at both ends $\Gamma_0^\varepsilon \cup \Gamma_L^\varepsilon$, in the sense that the following *boundary condition of place*

$$\boldsymbol{u}^\varepsilon = \mathbf{0}, \quad \text{on } \Gamma_0^\varepsilon \cup \Gamma_L^\varepsilon, \tag{40.1}$$

is imposed on the displacement components. Later on we shall weaken this condition in order to be able to evaluate the second order terms in an asymptotic expansion of the displacement field, but all the same the zeroth order terms still verify a Dirichlet type boundary condition.

Using the same notations as in Section 1 the following *equations of equilibrium* in *reference configuration* (cf. CIARLET [1988, Section 2.6]) are satisfied:

$$\begin{aligned} -\partial_j^\varepsilon(\sigma_{ij}^\varepsilon + \sigma_{kj}^\varepsilon \partial_k^\varepsilon u_i^\varepsilon) &= f_i^\varepsilon, \quad \text{in } \Omega^\varepsilon, \\ (\sigma_{ij}^\varepsilon + \sigma_{kj}^\varepsilon \partial_k^\varepsilon u_i^\varepsilon) n_j^\varepsilon &= g_i^\varepsilon, \quad \text{on } \Gamma^\varepsilon, \end{aligned} \tag{40.2}$$

where $\boldsymbol{\Sigma}^\varepsilon = (\sigma_{ij}^\varepsilon) : \overline{\Omega}^\varepsilon \to S^3$ denotes the *second Piola–Kirchhoff stress tensor* and S^3 denotes the set of all symmetric matrices of order 3. For simplicity and because we shall not use any other stress tensor in this chapter we shall often call *stress* the components $\sigma_{ij}^\varepsilon : \overline{\Omega}^\varepsilon \to \mathbb{R}$ of the second Piola–Kirchhoff stress tensor field.

The boundary condition on Γ^ε is referred to as *boundary condition of traction*.

Let $V(\Omega^\varepsilon)$ denote a space of "sufficiently smooth" vector-valued functions $\boldsymbol{v}^\varepsilon = (v_i^\varepsilon) : \overline{\Omega}^\varepsilon \to \mathbb{R}^3$ that vanish on $\Gamma_0^\varepsilon \cup \Gamma_L^\varepsilon$, and whose smoothness will be specified later. Then, if the unknown displacement $\boldsymbol{u}^\varepsilon$ belongs to the space $V(\Omega^\varepsilon)$, it can be shown that *the equations of equilibrium* (4.2) *are formally equivalent to the principle of virtual work*, which states that

$$\begin{aligned} &\int_{\Omega^\varepsilon} (\sigma_{ij}^\varepsilon \partial_j^\varepsilon v_i^\varepsilon + \sigma_{kj}^\varepsilon \partial_k^\varepsilon u_i^\varepsilon \partial_j^\varepsilon v_i^\varepsilon)\, \mathrm{d}\boldsymbol{x}^\varepsilon \\ &\quad = \int_{\Omega^\varepsilon} f_i^\varepsilon v_i^\varepsilon \, \mathrm{d}\boldsymbol{x}^\varepsilon + \int_{\Gamma^\varepsilon} g_i^\varepsilon v_i^\varepsilon \, \mathrm{d}a^\varepsilon, \quad \text{for all } \boldsymbol{v}^\varepsilon \in V(\Omega^\varepsilon). \end{aligned} \tag{40.3}$$

We remark that the principle of virtual work is nothing but the *weak* or *variational* form of the equations of equilibrium, and that functions $\boldsymbol{v}^\varepsilon \in V(\Omega^\varepsilon)$ may be viewed as "variations" around the actual configuration $\mathbf{id} + \boldsymbol{u}^\varepsilon$. As in the previous chapters it is this variational form that will be our starting point.

We finally assume that, for each $\varepsilon > 0$, the material constituting the beam is *elastic, homogeneous and isotropic*, and that its reference configuration $\overline{\Omega^\varepsilon}$ is a natural state (see CIARLET [1988, Chapter 3]). Later on we shall make some remarks concerning more general materials. For the moment we consider a particular class of such elastic materials, which allows for an easier exposition, while retaining the essential features of the method we wish to describe in this chapter.

More specifically, let

$$E_{ij}^{\varepsilon}(\boldsymbol{u}^{\varepsilon}) = \tfrac{1}{2}(\partial_i^{\varepsilon} u_j^{\varepsilon} + \partial_j^{\varepsilon} u_i^{\varepsilon} + \partial_i^{\varepsilon} u_k^{\varepsilon} \partial_j^{\varepsilon} u_k^{\varepsilon}), \tag{40.4}$$

denote the components of the *Green–Saint Venant strain tensor* $\boldsymbol{E}(\boldsymbol{u}^{\varepsilon}) = (E_{ij}^{\varepsilon}(\boldsymbol{u}^{\varepsilon}))$. We assume that, for each $\varepsilon > 0$, the material constituting the rod is a Saint Venant–Kirchhoff material. This means (see CIARLET [1988, Section 3.9]) that for each $\varepsilon > 0$, the second Piola–Kirchhoff stress tensor is expressed in terms of the Green–Saint Venant strain tensor through a *constitutive equation* of the form

$$\boldsymbol{\Sigma}^{\varepsilon} = \hat{\boldsymbol{\Sigma}}^{\varepsilon}(\boldsymbol{\nabla}^{\varepsilon} \boldsymbol{u}^{\varepsilon}) = \lambda^{\varepsilon}(\operatorname{tr}(\boldsymbol{E}^{\varepsilon}(\boldsymbol{u}^{\varepsilon}))\boldsymbol{I} + 2\mu^{\varepsilon} \boldsymbol{E}^{\varepsilon}(\boldsymbol{u}^{\varepsilon}),$$

where λ^{ε} and μ^{ε} are the *Lamé constants* of the material and matrix $\boldsymbol{\nabla}^{\varepsilon} \boldsymbol{u}^{\varepsilon} = (\partial_i^{\varepsilon} u_j^{\varepsilon})$ is the *displacement gradient.* Componentwise the constitutive equation thus reads

$$\sigma_{ij}^{\varepsilon} = \hat{\sigma}_{ij}^{\varepsilon}(\boldsymbol{\nabla}^{\varepsilon} \boldsymbol{u}^{\varepsilon}) = \lambda^{\varepsilon} E_{pp}^{\varepsilon}(\boldsymbol{u}^{\varepsilon})\delta_{ij} + 2\mu^{\varepsilon} E_{ij}^{\varepsilon}(\boldsymbol{u}^{\varepsilon}). \tag{40.5}$$

Consequently, in the present case, in order for the integrals appearing in the left-hand side of the principle of virtual work to make sense we are naturally led to the *Sobolev space*

$$\boldsymbol{W}^{1,4}(\Omega^{\varepsilon}) = \{\boldsymbol{v}^{\varepsilon} = (v_i^{\varepsilon}) \colon v_i^{\varepsilon} \in W^{1,4}(\Omega^{\varepsilon})\}.$$

Hence $V(\Omega^{\varepsilon})$ may be defined in the present case as

$$V(\Omega^{\varepsilon}) = \{\boldsymbol{v}^{\varepsilon} \in \boldsymbol{W}^{1,4}(\Omega^{\varepsilon}) \colon \boldsymbol{v}^{\varepsilon} = \boldsymbol{0} \text{ on } \Gamma_0^{\varepsilon} \cup \Gamma_L^{\varepsilon}\}. \tag{40.6}$$

If we assume that $\boldsymbol{u}^{\varepsilon} \in \boldsymbol{W}^{1,4}(\Omega^{\varepsilon})$, we also have

$$\boldsymbol{\Sigma}^{\varepsilon} = \lambda^{\varepsilon}(\operatorname{tr}(\boldsymbol{E}^{\varepsilon}(\boldsymbol{u}^{\varepsilon}))I + 2\mu^{\varepsilon} \boldsymbol{E}^{\varepsilon}(\boldsymbol{u}^{\varepsilon}) \in \boldsymbol{L}_s^2(\Omega^{\varepsilon}) \tag{40.7}$$

where,

$$\boldsymbol{L}_s^2(\Omega^{\varepsilon}) = \{\boldsymbol{\tau}^{\varepsilon} = (\tau_{ij}^{\varepsilon}) \colon \tau_{ij}^{\varepsilon} = \tau_{ji}^{\varepsilon},\ \tau_{ij}^{\varepsilon} \in L^2(\Omega^{\varepsilon})\}. \tag{40.8}$$

To sum up, *the displacement field* $\boldsymbol{u}^{\varepsilon} = (u_i^{\varepsilon})$ *solves the following problem*:

$$\begin{aligned} &\boldsymbol{u}^{\varepsilon} \in V(\Omega^{\varepsilon}), \\ &\int_{\Omega^{\varepsilon}} \{\hat{\sigma}_{ij}^{\varepsilon}(\boldsymbol{\nabla}^{\varepsilon} \boldsymbol{u}^{\varepsilon})\partial_j^{\varepsilon} v_i^{\varepsilon} + \hat{\sigma}_{kj}^{\varepsilon}(\boldsymbol{\nabla}^{\varepsilon} \boldsymbol{u}^{\varepsilon})\partial_k^{\varepsilon} u_i^{\varepsilon} \partial_j^{\varepsilon} v_i^{\varepsilon}\} \,\mathrm{d}\boldsymbol{x}^{\varepsilon} \\ &\quad = \int_{\Omega^{\varepsilon}} f_i^{\varepsilon} v_i^{\varepsilon} \,\mathrm{d}\boldsymbol{x}^{\varepsilon} + \int_{\Gamma^{\varepsilon}} g_i^{\varepsilon} v_i^{\varepsilon} \,\mathrm{d}a^{\varepsilon}, \quad \text{for all } \boldsymbol{v}^{\varepsilon} \in V(\Omega^{\varepsilon}), \end{aligned} \tag{40.9}$$

where

$$\hat{\sigma}_{ij}^{\varepsilon}(\nabla^{\varepsilon}\boldsymbol{u}^{\varepsilon}) = \lambda^{\varepsilon}E_{pp}^{\varepsilon}(\boldsymbol{u}^{\varepsilon})\delta_{ij} + 2\mu^{\varepsilon}E_{ij}^{\varepsilon}(\boldsymbol{u}^{\varepsilon}), \tag{40.10}$$

$$2E_{ij}^{\varepsilon}(\boldsymbol{u}^{\varepsilon}) = (\partial_i^{\varepsilon}u_j^{\varepsilon} + \partial_j^{\varepsilon}u_i^{\varepsilon} + \partial_i^{\varepsilon}u_k^{\varepsilon}\partial_j^{\varepsilon}u_k^{\varepsilon}). \tag{40.11}$$

Problem (40.9)–(40.11) constitutes the three-dimensional problem of a (geometrically) nonlinear elastic clamped rod (made of Saint Venant–Kirchhoff material).

We remark that, as of now, *there is no result guaranteeing the existence of a solution* $\boldsymbol{u}^{\varepsilon}$ of problem (40.9)–(40.11). The only available existence result valid for Saint Venant–Kirchhoff materials is based on the implicit function theorem, and for this reason, is restricted to a special class of boundary conditions which does not include those considered here (see CIARLET [1988, Section 6.7] for a discussion of this issue).

The more powerful existence theory of BALL [1977] does include boundary conditions of the type considered here but, even within the class of elastic materials to which it applies (which does not include Saint Venant–Kirchhoff materials), it does not provide the existence of a solution to the corresponding problem (40.9)–(40.11) (see CIARLET [1988, Section 7.10]).

41. The fundamental scalings on the unknowns and assumptions on the data: the displacement approach

In order to carry out the asymptotic treatment of problem (40.9)–(40.11) by considering ε as a small parameter, we must:

(i) *specify the way the displacement field* (u_i^{ε}) *and more generally the functions* (v_i^{ε}) *of the space* $V(\Omega^{\varepsilon})$, *are mapped into vector-valued functions defined over the set* $\overline{\Omega}$;

(ii) *control the way the Lamé constants and the applied forces depend on* ε.

Accordingly, we first *set* the following *correspondences between the displacement fields* (cf. Section 2 for the notation).

With the displacement field $\boldsymbol{u}^{\varepsilon} = (u_i^{\varepsilon}) : \overline{\Omega}^{\varepsilon} \to \mathbb{R}^3$, we associate the *scaled displacement field* $\boldsymbol{u}(\varepsilon) = (u_i(\varepsilon)) : \overline{\Omega} \to \mathbb{R}^3$, defined by the following *scalings*, introduced firstly by CIMETIÈRE, GEYMONAT, LE DRET, RAOULT, and TUTEK [1986, 1988]:

$$\left.\begin{aligned} u_{\alpha}^{\varepsilon}(\boldsymbol{x}^{\varepsilon}) &= \varepsilon u_{\alpha}(\varepsilon)(\boldsymbol{x}), \\ u_3^{\varepsilon}(\boldsymbol{x}^{\varepsilon}) &= \varepsilon^2 u_3(\varepsilon)(\boldsymbol{x}), \end{aligned}\right\} \quad \text{for all } \boldsymbol{x}^{\varepsilon} = \Pi^{\varepsilon}\boldsymbol{x} \in \overline{\Omega}^{\varepsilon}, \tag{41.1}$$

and, as before, we call *scaled displacements* the functions $\boldsymbol{u}_i(\varepsilon) : \overline{\Omega} \to \mathbb{R}^3$.

Likewise, we associate with any function $\boldsymbol{v}^{\varepsilon} = (v_i^{\varepsilon}) \in V(\Omega^{\varepsilon})$ the *scaled function* $\boldsymbol{v} = (v_i) : \overline{\Omega} \to \mathbb{R}^3$ defined by the *scalings*:

$$\left.\begin{aligned} v_{\alpha}^{\varepsilon}(\boldsymbol{x}^{\varepsilon}) &= \varepsilon v_{\alpha}(\boldsymbol{x}), \\ v_3^{\varepsilon}(\boldsymbol{x}^{\varepsilon}) &= \varepsilon^2 v_3(\boldsymbol{x}), \end{aligned}\right\} \quad \text{for all } \boldsymbol{x}^{\varepsilon} = \Pi^{\varepsilon}\boldsymbol{x} \in \overline{\Omega}^{\varepsilon}, \tag{41.2}$$

Hence, both the scaled displacement $\boldsymbol{u}(\varepsilon)$ and the scaled function $\boldsymbol{v}$ belong to the space

$$V(\Omega) = \{\boldsymbol{v} = (v_i) \in \boldsymbol{W}^{1,4}(\Omega) \colon \boldsymbol{v} = \boldsymbol{0} \text{ on } \Gamma_0 \cup \Gamma_L\}. \tag{41.3}$$

Next, we make the following *assumptions on the data*. We *require* that *the Lamé constants*, the *applied body forces*, and the *applied surface tractions*, should be of the following form:

$$\lambda^\varepsilon = \lambda, \qquad \mu^\varepsilon = \mu, \tag{41.4}$$

or equivalently,

$$E^\varepsilon = E, \qquad \nu^\varepsilon = \nu, \tag{41.5}$$

$$\left.\begin{aligned} f_\alpha^\varepsilon(\boldsymbol{x}^\varepsilon) &= \varepsilon^3 f_\alpha(\boldsymbol{x}), \\ f_3^\varepsilon(\boldsymbol{x}^\varepsilon) &= \varepsilon^2 f_3(\boldsymbol{x}), \end{aligned}\right\} \quad \text{for all } \boldsymbol{x}^\varepsilon = \Pi^\varepsilon \boldsymbol{x} \in \overline{\Omega}^\varepsilon, \tag{41.6}$$

$$\left.\begin{aligned} g_\alpha^\varepsilon(\boldsymbol{x}^\varepsilon) &= \varepsilon^4 g_\alpha(\boldsymbol{x}), \\ g_3^\varepsilon(\boldsymbol{x}^\varepsilon) &= \varepsilon^3 g_3(\boldsymbol{x}), \end{aligned}\right\} \quad \text{for all } \boldsymbol{x}^\varepsilon = \Pi^\varepsilon \boldsymbol{x} \in \overline{\Omega}^\varepsilon, \tag{41.7}$$

where constants $\lambda > 0$, $\mu > 0$, functions $f_i \in L^2(\Omega)$ and $g_i \in L^2(\Gamma)$ are independent of ε.

Using these scalings on the unknowns, the assumptions on the data, and simple computations, in the following theorem we reformulate the three-dimensional problem (40.9)–(40.11) as an equivalent problem posed in the reference domain $\overline{\Omega}$.

THEOREM 41.1. *Assume that* $\boldsymbol{u}^\varepsilon \in \boldsymbol{W}^{1,4}(\Omega^\varepsilon)$. *Then, the scaled displacement* $\boldsymbol{u}(\varepsilon) = (u_i(\varepsilon))$ *solves the following problem*:

$$\boldsymbol{u}(\varepsilon) \in V(\Omega),$$

$$\begin{aligned} &\int_\Omega \hat{\sigma}_{ij}(\varepsilon; \nabla \boldsymbol{u}(\varepsilon)) \partial_j v_i \,\mathrm{d}\boldsymbol{x} + \int_\Omega \hat{\sigma}_{ij}(\varepsilon; \nabla \boldsymbol{u}(\varepsilon)) \partial_i u_\gamma(\varepsilon) \partial_j v_\gamma \,\mathrm{d}\boldsymbol{x} \\ &\qquad + \varepsilon^2 \int_\Omega \hat{\sigma}_{ij}(\varepsilon; \nabla \boldsymbol{u}(\varepsilon)) \partial_i u_3(\varepsilon) \partial_j v_3 \,\mathrm{d}\boldsymbol{x} \\ &\quad = \int_\Omega f_i v_i \,\mathrm{d}\boldsymbol{x} + \int_\Gamma g_i v_i \,\mathrm{d}a, \quad \textit{for all } \boldsymbol{v} \in V(\Omega), \end{aligned} \tag{41.8}$$

where

$$\begin{aligned} \hat{\sigma}_{\alpha\beta}(\varepsilon; \nabla \boldsymbol{u}(\varepsilon)) &= \varepsilon^{-4}[\lambda E^0_{\gamma\gamma}(\boldsymbol{u}(\varepsilon))\delta_{\alpha\beta} + 2\mu E^0_{\alpha\beta}(\boldsymbol{u}(\varepsilon))] \\ &\quad + \varepsilon^{-2}\{\lambda[E^2_{\gamma\gamma}(\boldsymbol{u}(\varepsilon)) + E^0_{33}(\boldsymbol{u}(\varepsilon))]\delta_{\alpha\beta} + 2\mu E^2_{\alpha\beta}(\boldsymbol{u}(\varepsilon))\} \\ &\quad + \lambda E^2_{33}(\boldsymbol{u}(\varepsilon))\delta_{\alpha\beta}, \end{aligned} \tag{41.9}$$

$$\hat{\sigma}_{3\beta}(\varepsilon; \nabla \boldsymbol{u}(\varepsilon)) = \varepsilon^{-2} 2\mu E^0_{3\beta}(\boldsymbol{u}(\varepsilon)) + 2\mu E^2_{3\beta}(\boldsymbol{u}(\varepsilon)), \tag{41.10}$$

$$\begin{aligned} \hat{\sigma}_{33}(\varepsilon; \nabla \boldsymbol{u}(\varepsilon)) &= \varepsilon^{-2}\lambda E^0_{\alpha\alpha}(\boldsymbol{u}(\varepsilon)) + (\lambda + 2\mu)E^0_{33}(\boldsymbol{u}(\varepsilon)) \\ &\quad + \lambda E^2_{\alpha\alpha}(\boldsymbol{u}(\varepsilon)) + \varepsilon^2(\lambda + 2\mu)E^2_{33}(\boldsymbol{u}(\varepsilon)), \end{aligned} \tag{41.11}$$

and where

$$2E^0_{ij}(\boldsymbol{u}(\varepsilon)) = \partial_i u_j(\varepsilon) + \partial_j u_i(\varepsilon) + \partial_i u_\gamma(\varepsilon)\partial_j u_\gamma(\varepsilon), \tag{41.12}$$

$$2E^2_{ij}(\boldsymbol{u}(\varepsilon)) = \partial_i u_3(\varepsilon)\partial_j u_3(\varepsilon). \tag{41.13}$$

REMARK 41.1. As a consequence of the scalings on the displacement field (cf. (41.1)), the Green–Saint Venant strain tensor $\boldsymbol{E}^\varepsilon(\boldsymbol{u}^\varepsilon)$ is now "scaled" as follows:

$$\begin{aligned} E^\varepsilon_{33}(\boldsymbol{u}^\varepsilon) &= \varepsilon^2[E^0_{33}(\boldsymbol{u}(\varepsilon)) + \varepsilon^2 E^2_{33}(\boldsymbol{u}(\varepsilon))],\\ E^\varepsilon_{3\beta}(\boldsymbol{u}^\varepsilon) &= \varepsilon[E^0_{3\beta}(\boldsymbol{u}(\varepsilon)) + \varepsilon^2 E^2_{3\beta}(\boldsymbol{u}(\varepsilon))],\\ E^\varepsilon_{\alpha\beta}(\boldsymbol{u}^\varepsilon) &= E^0_{\alpha\beta}(\boldsymbol{u}(\varepsilon)) + \varepsilon^2 E^2_{\alpha\beta}(\boldsymbol{u}(\varepsilon)). \end{aligned} \tag{41.14}$$

As in the linearized case, the "passage from Ω^ε to Ω" has the effect of setting up a *polynomial dependence on parameter* ε. More specifically, problem (41.8)–(41.13) reads

$$\begin{aligned} &\mathcal{B}(\hat{\boldsymbol{\Sigma}}(\varepsilon; \nabla\boldsymbol{u}(\varepsilon)), \nabla\boldsymbol{v}) + \mathcal{T}^0(\hat{\boldsymbol{\Sigma}}(\varepsilon; \nabla\boldsymbol{u}(\varepsilon)), \nabla\boldsymbol{u}(\varepsilon), \nabla\boldsymbol{v})\\ &\quad + \varepsilon^2\mathcal{T}^2(\hat{\boldsymbol{\Sigma}}(\varepsilon; \nabla\boldsymbol{u}(\varepsilon)), \nabla\boldsymbol{u}(\varepsilon), \nabla\boldsymbol{v}) = \mathcal{L}(\boldsymbol{v}), \quad \text{for all } \boldsymbol{v} \in V(\Omega), \end{aligned} \tag{41.15}$$

where the tensor

$$\hat{\boldsymbol{\Sigma}}(\varepsilon; \nabla\boldsymbol{u}(\varepsilon)) = (\hat{\sigma}_{ij}(\varepsilon; \nabla\boldsymbol{u}(\varepsilon))), \tag{41.16}$$

is of the form

$$\begin{aligned} \hat{\boldsymbol{\Sigma}}(\varepsilon; \nabla\boldsymbol{u}(\varepsilon)) &= \varepsilon^{-4}\hat{\boldsymbol{S}}^{-4}(\nabla\boldsymbol{u}(\varepsilon)) + \varepsilon^{-2}\hat{\boldsymbol{S}}^{-2}(\nabla\boldsymbol{u}(\varepsilon))\\ &\quad + \hat{\boldsymbol{S}}^0(\nabla\boldsymbol{u}(\varepsilon)) + \varepsilon^2\hat{\boldsymbol{S}}^2(\nabla\boldsymbol{u}(\varepsilon)), \end{aligned} \tag{41.17}$$

and where the linear form $\mathcal{L}$, the bilinear form $\mathcal{B}$, the trilinear forms $\mathcal{T}^0$, $\mathcal{T}^2$ and the tensor-valued mappings $\hat{\boldsymbol{S}}^{-4}$, $\hat{\boldsymbol{S}}^{-2}$, $\hat{\boldsymbol{S}}^0$ and $\hat{\boldsymbol{S}}^2$ are all independent of ε and given by

$$\mathcal{L}(\boldsymbol{v}) = \int_\Omega f_i v_i \,\mathrm{d}\boldsymbol{x} + \int_\Gamma g_i v_i \,\mathrm{d}a, \tag{41.18}$$

$$\mathcal{B}(\boldsymbol{\Sigma}, \nabla\boldsymbol{v}) = \int_\Omega \sigma_{ij}\partial_j v_i \,\mathrm{d}\boldsymbol{x}, \tag{41.19}$$

$$\mathcal{T}^0(\boldsymbol{\Sigma}, \nabla\boldsymbol{w}, \nabla\boldsymbol{v}) = \int_\Omega \sigma_{ij}\partial_i w_\gamma \partial_j v_\gamma \,\mathrm{d}\boldsymbol{x}, \tag{41.20}$$

$$\mathcal{T}^2(\boldsymbol{\Sigma}, \nabla\boldsymbol{w}, \nabla\boldsymbol{v}) = \int_\Omega \sigma_{ij}\partial_i w_3 \partial_j v_3 \,\mathrm{d}\boldsymbol{x}, \tag{41.21}$$

$$\hat{\boldsymbol{S}}^{-4}(\boldsymbol{\nabla v}) = \begin{pmatrix} \lambda E^0_{\gamma\gamma}(\boldsymbol{v}) + 2\mu E^0_{11}(\boldsymbol{v}) & 2\mu E^0_{12}(\boldsymbol{v}) & 0 \\ 2\mu E^0_{21}(\boldsymbol{v}) & \lambda E^0_{\gamma\gamma}(\boldsymbol{v}) + 2\mu E^0_{22}(\boldsymbol{v}) & 0 \\ 0 & 0 & 0 \end{pmatrix}, \tag{41.22}$$

$$\hat{\boldsymbol{S}}^{-2}(\boldsymbol{\nabla v}) = \begin{pmatrix} \lambda[E^2_{\gamma\gamma}(\boldsymbol{v}) + E^0_{33}(\boldsymbol{v})] + 2\mu E^2_{11}(\boldsymbol{v}) & 2\mu E^2_{12}(\boldsymbol{v}) & 2\mu E^0_{13}(\boldsymbol{v}) \\ 2\mu E^2_{21}(\boldsymbol{v}) & \lambda[E^2_{\gamma\gamma}(\boldsymbol{v}) + E^0_{33}(\boldsymbol{v})] + 2\mu E^2_{22}(\boldsymbol{v}) & 2\mu E^0_{32}(\boldsymbol{v}) \\ 2\mu E^0_{31}(\boldsymbol{v}) & 2\mu E^0_{32}(\boldsymbol{v}) & \lambda E^0_{\gamma\gamma}(\boldsymbol{v}) \end{pmatrix}, \tag{41.23}$$

$$\hat{\boldsymbol{S}}^{0}(\boldsymbol{\nabla v}) = \begin{pmatrix} \lambda E^2_{33}(\boldsymbol{v}) & 0 & 2\mu E^2_{33}(\boldsymbol{v}) \\ 0 & \lambda E^2_{33}(\boldsymbol{v}) & 2\mu E^2_{33}(\boldsymbol{v}) \\ 2\mu E^2_{31}(\boldsymbol{v}) & 2\mu E^2_{32}(\boldsymbol{v}) & (\lambda + 2\mu)E^0_{33}(\boldsymbol{v}) + \lambda E^2_{\alpha\alpha}(\boldsymbol{v}) \end{pmatrix}, \tag{41.24}$$

$$\hat{\boldsymbol{S}}^{2}(\boldsymbol{\nabla v}) = \begin{pmatrix} 0 & 0 & 0 \\ 0 & 0 & 0 \\ 0 & 0 & (\lambda + 2\mu)E^2_{33}(\boldsymbol{v}) \end{pmatrix}. \tag{41.25}$$

This polynomial dependence with respect to parameter ε, and the fact that ε is thought of as a "small" parameter, led us to adapt to the present case the asymptotic expansion technique employed in Section 4 for the linear case.

Before we proceed to actually compute the first terms of the asymptotic expansion of $\boldsymbol{u}(\varepsilon)$, let us record an equivalent formulation of Theorem 41.1, which, together with the results of the next section, will be the basis to another application of the method of asymptotic expansions: *the displacement stress approach.* More specifically, instead of scaling problem (41.8)–(41.13), we now scale both Eqs. (40.3) and (40.5) of Section 40, in the form of a *scaled principle of virtual work* and a *scaled constitutive equation.*

THEOREM 41.2. (a) *Assume* $\boldsymbol{u}^\varepsilon \in \boldsymbol{W}^{1,4}(\Omega^\varepsilon)$. *The scaled displacement* $\boldsymbol{u}(\varepsilon) = (u_i(\varepsilon)) \in V(\Omega)$ *solves the following equations*:

$$\int_\Omega \sigma_{ij}(\varepsilon)\partial_j v_i \,\mathrm{d}\boldsymbol{x} + \int_\Omega \sigma_{ij}(\varepsilon)\partial_i u_\gamma(\varepsilon)\partial_j v_\gamma \,\mathrm{d}\boldsymbol{x} + \varepsilon^2 \int_\Omega \sigma_{ij}(\varepsilon)\partial_i u_3(\varepsilon)\partial_j v_3 \,\mathrm{d}\boldsymbol{x} = \int_\Omega f_i v_i \,\mathrm{d}\boldsymbol{x} + \int_\Gamma g_i v_i \,\mathrm{d}a, \quad \textit{for all } \boldsymbol{v} \in V(\Omega), \tag{41.26}$$

where

$$\sigma_{ij}(\varepsilon) = \hat{\sigma}_{ij}(\varepsilon; \boldsymbol{\nabla u}(\varepsilon)), \tag{41.27}$$

(b) *The functions* $\sigma_{ij}(\varepsilon) \in L^2(\Omega)$ *are related to the components* $\sigma_{ij}^{\varepsilon} \in L^2(\Omega^{\varepsilon})$ *of the second Piola–Kirchhoff stress tensor through the following relations*:

$$\begin{aligned}
&\sigma_{\alpha\beta}^{\varepsilon}(\boldsymbol{x}^{\varepsilon}) = \varepsilon^4 \sigma_{\alpha\beta}(\varepsilon)(\boldsymbol{x}), \\
&\sigma_{3\beta}^{\varepsilon}(\boldsymbol{x}^{\varepsilon}) = \varepsilon^3 \sigma_{3\beta}(\varepsilon)(\boldsymbol{x}), \quad \textit{for all } \boldsymbol{x}^{\varepsilon} = \Pi^{\varepsilon}\boldsymbol{x} \in \overline{\Omega}^{\varepsilon}. \\
&\sigma_{33}^{\varepsilon}(\boldsymbol{x}^{\varepsilon}) = \varepsilon^2 \sigma_{33}(\varepsilon)(\boldsymbol{x}),
\end{aligned} \tag{41.28}$$

PROOF. Part (a) is an equivalent way of stating Theorem 41.1. Formulas in part (b) follow from the definitions of the functions $\sigma_{ij}(\varepsilon)$ given in part (a). □

In other words, part (b) shows that functions $\sigma_{ij}(\varepsilon) = \hat{\sigma}_{ij}(\varepsilon; \nabla \boldsymbol{u}(\varepsilon))$ which thus far simply appeared as convenient "computational intermediaries", can also be seen as the components of a *scaled* second Piola–Kirchhoff stress tensor field $(\sigma_{ij}(\varepsilon)) : \overline{\Omega} \to \boldsymbol{S}^3$. This viewpoint plays a crucial role in the *displacement–stress approach* described in Section 46.

42. Cancellation of the factors of ε^q, $-4 \leqslant q \leqslant 0$, in the scaled three-dimensional problem

The identification of the leading term $\boldsymbol{u}^0$ of the formal expansion of $\boldsymbol{u}(\varepsilon)$ will be carried out in several stages; to begin with, we gather all the information that can be derived from the cancellation of the factors of ε^q, $-4 \leqslant q \leqslant 0$, in the scaled problem (41.8)–(41.13).

More specifically, we first show that the cancellation of the factors of ε^q, $-4 \leqslant q \leqslant -1$, implies that *the formal expansion of the tensor* $\hat{\boldsymbol{\Sigma}}(\varepsilon; \nabla \boldsymbol{u}(\varepsilon))$ *induced by that of the scaled displacement, does not contain any negative powers of* ε.

This is a particularly striking simplification, since an inspection reveals that the expansion of $\hat{\boldsymbol{\Sigma}}(\varepsilon; \nabla \boldsymbol{u}(\varepsilon))$ is a priori of the form $\{\varepsilon^{-4}\boldsymbol{\Sigma}^{-4} + \varepsilon^{-3}\boldsymbol{\Sigma}^{-3} + \varepsilon^{-2}\boldsymbol{\Sigma}^{-2} +$ h.o.t.$\}$; secondly, we show that the cancellation of the factor ε^0 provides *variational equations*, which play a key role in the sequel. The following is an adaptation to the rod case of an equivalent result for plates owed to RAOULT [1988, Chapter 2, Section 2.2].

We need to introduce the following space of Bernoulli–Navier type displacements:

$$\begin{aligned}
\tilde{V}_{\mathrm{BN}}(\Omega) = \{\boldsymbol{v} \in \boldsymbol{W}^{1,4}(\Omega)&: v_\alpha \in W_0^{2,4}(0, L), \\
&v_3 = z_3 - x_\alpha \partial_3 v_\alpha,\ z_3 \in W_0^{1,4}(0, L)\}.
\end{aligned} \tag{42.1}$$

THEOREM 42.1. *Assume that the scaled displacement can be written as*

$$\boldsymbol{u}(\varepsilon) = \boldsymbol{u}^0 + \varepsilon \boldsymbol{u}^1 + \varepsilon^2 \boldsymbol{u}^2 + \varepsilon^3 \boldsymbol{u}^3 + \varepsilon^4 \boldsymbol{u}^4 + \textit{h.o.t.},$$

where the terms satisfy

$$\boldsymbol{u}^p = (u_i^p) \in V(\Omega), \quad p \geqslant 0, \tag{42.2}$$

and

$$\partial_\alpha u_\beta^0 \in C^0(\overline{\Omega}). \tag{42.3}$$

Let

$$\hat{\boldsymbol{\Sigma}}(\varepsilon; \nabla \boldsymbol{u}(\varepsilon)) = \varepsilon^{-4}\boldsymbol{\Sigma}^{-4} + \varepsilon^{-3}\boldsymbol{\Sigma}^{-3} + \varepsilon^{-2}\boldsymbol{\Sigma}^{-2} + \varepsilon^{-1}\boldsymbol{\Sigma}^{-1} + \boldsymbol{\Sigma}^0 + h.o.t., \tag{42.4}$$

with $\boldsymbol{\Sigma}^q = (\sigma_{ij}^q)$, *denoting the induced formal expansion* (41.17) *of the tensor* $\hat{\boldsymbol{\Sigma}}(\varepsilon; \nabla \boldsymbol{u}(\varepsilon))$.

Then the cancellation of the factors of ε^q, $-4 \leqslant q \leqslant 0$, *in problem* (41.8)–(41.13) *is equivalent to the following equations:*

Case $q = -4$.

(i) $\boldsymbol{\Sigma}^{-4} = \mathbf{0}, \qquad E_{\alpha\beta}^0(\boldsymbol{u}^0) = 0, \qquad \sigma_{33}^{-2} = 0,$ (42.5)

(ii) *Displacements* u_β^0 *are a rigid deformation, that is*

$$\begin{aligned} u_1^0(x_1, x_2, x_3) &= (\cos(\theta(x_3)) - 1)x_1 - \sin(\theta(x_3))x_2 + z_1^0(x_3), \\ u_2^0(x_1, x_2, x_3) &= \sin(\theta(x_3))x_1 + (\cos(\theta(x_3)) - 1)x_2 + z_2^0(x_3), \end{aligned} \tag{42.6}$$

where functions θ *and* z_α *are such that*

$$\theta, \; z_\alpha \in W_0^{1,4}(0, L). \tag{42.7}$$

Case $q = -3$.

$$\boldsymbol{\Sigma}^{-3} = \mathbf{0}. \tag{42.8}$$

Case $q = -2$.

$$\boldsymbol{\Sigma}^{-2} = \mathbf{0}, \qquad \boldsymbol{u}^0 \in \tilde{V}_{\mathrm{BN}}(\Omega), \tag{42.9}$$

$$u_3^0 = \underline{u}_3^0 - x_\alpha \partial_3 u_\alpha^0, \tag{42.10}$$

Case $q = -1$.

(i) $\boldsymbol{\Sigma}^{-1} = \mathbf{0}, \qquad \boldsymbol{u}^1 \in \tilde{V}_{\mathrm{BN}}(\Omega), \qquad u_3^1 = \underline{u}_3^1 - x_\alpha \partial_3 u_\alpha^1,$ (42.11)

(ii)
$$\begin{aligned} \sigma_{33}^0 &= \mu \frac{3\lambda + 2\mu}{\lambda + \mu}[\partial_3 \underline{u}_3^0 - x_\alpha \partial_{33} u_\alpha^0 + \Lambda_{\alpha\alpha}] \\ &= E[\partial_3 \underline{u}_3^0 - x_\alpha \partial_{33} u_\alpha^0 + \Lambda_{\alpha\alpha}], \end{aligned} \tag{42.12}$$

where matrix $\boldsymbol{\Lambda} = (\Lambda_{\alpha\beta})$ *is given by*

$$\boldsymbol{\Lambda} = (\Lambda_{\alpha\beta}) = \tfrac{1}{2}\begin{pmatrix} \partial_3 u_1^0 \partial_3 u_1^0 & \partial_3 u_1^0 \partial_3 u_2^0 \\ \partial_3 u_2^0 \partial_3 u_1^0 & \partial_3 u_2^0 \partial_3 u_2^0 \end{pmatrix}. \tag{42.13}$$

$$\text{(iii)}\ \sigma^1_{33} = E[\partial_3 \underline{u}^1_3 - x_\alpha \partial_{33} u^1_\alpha + \partial_3 u^0_\gamma \partial_3 u^1_\gamma]. \tag{42.14}$$

$$\begin{aligned}\text{(iv)}\ \partial_1 u^2_1 &= -\nu[\partial_3 \underline{u}^0_3 - x_\alpha \partial_{33} u^0_\alpha + \Lambda_{\alpha\alpha}] - \Lambda_{11},\\ \partial_1 u^2_2 &= -\nu[\partial_3 \underline{u}^0_3 - x_\alpha \partial_{33} u^0_\alpha + \Lambda_{\alpha\alpha}] - \Lambda_{22}.\end{aligned} \tag{42.15}$$

Case $q = 0$.

$$\begin{aligned}&\int_\Omega \sigma^0_{ij}\partial_j v_i \,\mathrm{d}\boldsymbol{x} + \int_\Omega \sigma^0_{3j}\partial_3 u^0_\gamma \partial_j v_\gamma \,\mathrm{d}\boldsymbol{x}\\ &\quad = \int_\Omega f_i v_i \,\mathrm{d}\boldsymbol{x} + \int_\Gamma g_i v_i \,\mathrm{d}a, \quad \textit{for all } \boldsymbol{v} \in V(\Omega),\end{aligned} \tag{42.16}$$

Consequently, we have the following expressions for the zero- and first order bending moment and axial force components:

$$m^0_\beta = \int_\omega x_\beta \sigma^0_{33} \,\mathrm{d}\omega = -EI_\beta \partial_{33} u^0_\beta \quad (\textit{no sum on } \beta), \tag{42.17}$$

$$n^0_{33} = \int_\omega \sigma^0_{33} \,\mathrm{d}\omega = E[\partial_3 \underline{u}^0_3 - \Lambda_{\alpha\alpha}], \tag{42.18}$$

$$m^1_\beta = \int_\omega x_\beta \sigma^1_{33} \,\mathrm{d}\omega = -EI_\beta \partial_{33} u^1_\beta \quad (\textit{no sum on } \beta), \tag{42.19}$$

$$n^1_{33} = \int_\omega \sigma^1_{33} \,\mathrm{d}\omega = E[\partial_3 \underline{u}^1_3 - \partial_3 u^0_\gamma \partial_3 u^1_\gamma]. \tag{42.20}$$

PROOF. The proof of this result is presented in several steps.

Step 1. The asymptotic expansion in the scaled displacements induces the following expansion in the components of the scaled Green–Saint Venant strain tensor:

$$\begin{aligned}E^0_{ij}(\boldsymbol{u}(\varepsilon)) = {}& e_{ij}(\boldsymbol{u}^0) + \tfrac{1}{2}\partial_i u^0_\gamma \partial_j u^0_\gamma\\ &+ \varepsilon[e_{ij}(\boldsymbol{u}^1) + \tfrac{1}{2}(\partial_i u^0_\gamma \partial_j u^1_\gamma + \partial_i u^1_\gamma \partial_j u^0_\gamma)]\\ &+ \varepsilon^2[e_{ij}(\boldsymbol{u}^2) + \tfrac{1}{2}(\partial_i u^0_\gamma \partial_j u^2_\gamma + \partial_i u^1_\gamma \partial_j u^1_\gamma + \partial_i u^2_\gamma \partial_j u^0_\gamma)]\\ &+ \varepsilon^3[e_{ij}(\boldsymbol{u}^3) + \tfrac{1}{2}(\partial_i u^0_\gamma \partial_j u^3_\gamma + \partial_i u^1_\gamma \partial_j u^2_\gamma + \partial_i u^2_\gamma \partial_j u^1_\gamma + \partial_i u^3_\gamma \partial_j u^0_\gamma)]\\ &+ \varepsilon^4[e_{ij}(\boldsymbol{u}^4) + \tfrac{1}{2}(\partial_i u^0_\gamma \partial_j u^4_\gamma + \partial_i u^1_\gamma \partial_j u^3_\gamma + \partial_i u^2_\gamma \partial_j u^2_\gamma + \partial_i u^3_\gamma \partial_j u^1_\gamma\\ &\quad + \partial_i u^4_\gamma \partial_j u^0_\gamma)] + \text{h.o.t.},\end{aligned} \tag{42.21}$$

$$
\begin{aligned}
E_{ij}^2(\boldsymbol{u}(\varepsilon)) = &\tfrac{1}{2}\partial_i u_3^0 \partial_j u_3^0 \\
&+ \varepsilon \tfrac{1}{2}(\partial_i u_3^0 \partial_j u_3^1 + \partial_i u_3^1 \partial_j u_3^0) \\
&+ \varepsilon^2 \tfrac{1}{2}(\partial_i u_3^0 \partial_j u_3^2 + \partial_i u_3^1 \partial_j u_3^1 + \partial_i u_3^2 \partial_j u_3^0) \\
&+ \varepsilon^3 \tfrac{1}{2}(\partial_i u_3^0 \partial_j u_3^3 + \partial_i u_3^1 \partial_j u_3^2 + \partial_i u_3^2 \partial_j u_3^1 + \partial_i u_3^3 \partial_j u_3^0) \\
&+ \varepsilon^4 \tfrac{1}{2}(\partial_i u_3^0 \partial_j u_3^4 + \partial_i u_3^1 \partial_j u_3^3 + \partial_i u_3^2 \partial_j u_3^2 + \partial_i u_3^3 \partial_j u_3^1 + \partial_i u_3^4 \partial_j u_3^0) \\
&+ \text{h.o.t.}
\end{aligned}
\tag{42.22}
$$

and consequently, the tensor $\hat{\boldsymbol{\Sigma}}(\varepsilon; \boldsymbol{\nabla u}(\varepsilon)) = (\hat{\sigma}_{ij})$ is expanded in the following way:

$$
\begin{aligned}
\hat{\sigma}_{\alpha\beta} &= \varepsilon^{-4}\sigma_{\alpha\beta}^{-4} + \varepsilon^{-3}\sigma_{\alpha\beta}^{-3} + \varepsilon^{-2}\sigma_{\alpha\beta}^{-2} + \text{h.o.t.}, \\
\hat{\sigma}_{3\beta} &= \varepsilon^{-2}\sigma_{3\beta}^{-2} + \varepsilon^{-1}\sigma_{3\beta}^{-1} + \sigma_{3\beta}^{0} + \text{h.o.t.}, \\
\hat{\sigma}_{33} &= \varepsilon^{-2}\sigma_{33}^{-2} + \varepsilon^{-1}\sigma_{33}^{-1} + \sigma_{33}^{0} + \text{h.o.t.},
\end{aligned}
\tag{42.23}
$$

with

$$
\begin{aligned}
\sigma_{\alpha\beta}^{-4} = &\lambda[e_{\rho\rho}(\boldsymbol{u}^0) + \tfrac{1}{2}\partial_\rho u_\gamma^0 \partial_\rho u_\gamma^0]\delta_{\alpha\beta} \\
&+ 2\mu[e_{\alpha\beta}(\boldsymbol{u}^0) + \tfrac{1}{2}\partial_\alpha u_\gamma^0 \partial_\beta u_\gamma^0], \\
\sigma_{\alpha\beta}^{-3} = &\lambda[e_{\rho\rho}(\boldsymbol{u}^1) + \tfrac{1}{2}(\partial_\rho u_\gamma^0 \partial_\rho u_\gamma^1 + \partial_\rho u_\gamma^1 \partial_\rho u_\gamma^0)]\delta_{\alpha\beta} \\
&+ 2\mu[e_{\rho\rho}(\boldsymbol{u}^1) + \tfrac{1}{2}(\partial_\alpha u_\gamma^0 \partial_\beta u_\gamma^1 + \partial_\alpha u_\gamma^1 \partial_\beta u_\gamma^0)], \\
\sigma_{\alpha\beta}^{-2} = &\lambda[e_{\rho\rho}(\boldsymbol{u}^2) + \tfrac{1}{2}(\partial_\rho u_\gamma^0 \partial_\rho u_\gamma^2 + \partial_\rho u_\gamma^1 \partial_\rho u_\gamma^1 + \partial_\rho u_\gamma^2 \partial_\rho u_\gamma^0)]\delta_{\alpha\beta} \\
&+ 2\mu[e_{\rho\rho}(\boldsymbol{u}^2) + \tfrac{1}{2}(\partial_\alpha u_\gamma^0 \partial_\beta u_\gamma^2 + \partial_\alpha u_\gamma^1 \partial_\beta u_\gamma^1 + \partial_\alpha u_\gamma^2 \partial_\beta u_\gamma^0)] \\
&+ \lambda[\tfrac{1}{2}\partial_\rho u_3^0 \partial_\rho u_3^0 + \partial_3 u_3^0 + \tfrac{1}{2}\partial_3 u_\gamma^0 \partial_3 u_\gamma^0]\delta_{\alpha\beta} + 2\mu \tfrac{1}{2}\partial_\alpha u_3^0 \partial_\beta u_3^0,
\end{aligned}
\tag{42.24}
$$

$$
\begin{aligned}
\sigma_{3\alpha}^{-2} = &2\mu[\tfrac{1}{2}(\partial_3 u_\alpha^0 + \partial_\alpha u_3^0) + \tfrac{1}{2}\partial_3 u_\gamma^0 \partial_\alpha u_\gamma^0], \\
\sigma_{3\alpha}^{-1} = &2\mu[\tfrac{1}{2}(\partial_3 u_\alpha^1 + \partial_\alpha u_3^1) + \tfrac{1}{2}(\partial_3 u_\gamma^0 \partial_\alpha u_\gamma^1 + \partial_3 u_\gamma^1 \partial_\alpha u_\gamma^0)], \\
\sigma_{3\alpha}^{0} = &2\mu[\tfrac{1}{2}(\partial_3 u_\alpha^2 + \partial_\alpha u_3^2) + \tfrac{1}{2}(\partial_3 u_\gamma^0 \partial_\alpha u_\gamma^2 + \partial_3 u_\gamma^1 \partial_\alpha u_\gamma^1 + \partial_3 u_\gamma^2 \partial_\alpha u_\gamma^0)] \\
&+ 2\mu \tfrac{1}{2}\partial_3 u_3^0 \partial_\alpha u_3^0,
\end{aligned}
\tag{42.25}
$$

$$
\begin{aligned}
\sigma_{33}^{-2} = &\lambda[\partial_\alpha u_\alpha^0 + \tfrac{1}{2}\partial_\alpha u_\gamma^0 \partial_\alpha u_\gamma^0], \\
\sigma_{33}^{-1} = &\lambda[\partial_\alpha u_\alpha^1 + \tfrac{1}{2}(\partial_\alpha u_\gamma^0 \partial_\alpha u_\gamma^1 + \partial_\alpha u_\gamma^1 \partial_\alpha u_\gamma^0)], \\
\sigma_{33}^{0} = &\lambda[\partial_\alpha u_\alpha^2 + \tfrac{1}{2}(\partial_\alpha u_\gamma^0 \partial_\alpha u_\gamma^2 + \partial_\alpha u_\gamma^1 \partial_\alpha u_\gamma^1 + \partial_\alpha u_\gamma^2 \partial_\alpha u_\gamma^0)] \\
&+ (\lambda + 2\mu)[\partial_3 u_3^0 + \tfrac{1}{2}\partial_3 u_\gamma^0 \partial_3 u_\gamma^0] + \lambda \tfrac{1}{2}\partial_\alpha u_3^0 \partial_\alpha u_3^0.
\end{aligned}
\tag{42.26}
$$

On the other hand, the bilinear and trilinear forms present in the scaled principle of virtual work, or variational formulation, (41.15) are expanded according to

$$
\mathscr{B}(\boldsymbol{\sigma}(\varepsilon), \boldsymbol{v}) = \varepsilon^{-4}\mathscr{B}(\boldsymbol{\sigma}^{-4}, \boldsymbol{v}) + \varepsilon^{-3}\mathscr{B}(\boldsymbol{\sigma}^{-3}, \boldsymbol{v}) + \varepsilon^{-2}\mathscr{B}(\boldsymbol{\sigma}^{-2}, \boldsymbol{v}) + \text{h.o.t.},
\tag{42.27}
$$

$$\begin{aligned}\mathcal{T}^0(\boldsymbol{\sigma}(\varepsilon),\boldsymbol{u}(\varepsilon),\boldsymbol{v}) &= \varepsilon^{-4}\mathcal{T}^0(\boldsymbol{\sigma}^{-4},\boldsymbol{u}^0,\boldsymbol{v})\\ &\quad+\varepsilon^{-3}[\mathcal{T}^0(\boldsymbol{\sigma}^{-3},\boldsymbol{u}^0,\boldsymbol{v})+\mathcal{T}^0(\boldsymbol{\sigma}^{-4},\boldsymbol{u}^1,\boldsymbol{v})]\\ &\quad+\varepsilon^{-2}[\mathcal{T}^0(\boldsymbol{\sigma}^{-2},\boldsymbol{u}^0,\boldsymbol{v})+\mathcal{T}^0(\boldsymbol{\sigma}^{-3},\boldsymbol{u}^1,\boldsymbol{v})+\mathcal{T}^0(\boldsymbol{\sigma}^{-4},\boldsymbol{u}^2,\boldsymbol{v})]\\ &\quad+\varepsilon^{-1}[\mathcal{T}^0(\boldsymbol{\sigma}^{-1},\boldsymbol{u}^0,\boldsymbol{v})+\mathcal{T}^0(\boldsymbol{\sigma}^{-2},\boldsymbol{u}^1,\boldsymbol{v})+\mathcal{T}^0(\boldsymbol{\sigma}^{-3},\boldsymbol{u}^2,\boldsymbol{v})\\ &\qquad+\mathcal{T}^0(\boldsymbol{\sigma}^{-4},\boldsymbol{u}^3,\boldsymbol{v})]\\ &\quad+\varepsilon^{0}[\mathcal{T}^0(\boldsymbol{\sigma}^{0},\boldsymbol{u}^0,\boldsymbol{v})+\mathcal{T}^0(\boldsymbol{\sigma}^{-1},\boldsymbol{u}^1,\boldsymbol{v})+\mathcal{T}^0(\boldsymbol{\sigma}^{-2},\boldsymbol{u}^2,\boldsymbol{v})\\ &\qquad+\mathcal{T}^0(\boldsymbol{\sigma}^{-3},\boldsymbol{u}^3,\boldsymbol{v})+\mathcal{T}^0(\boldsymbol{\sigma}^{-4},\boldsymbol{u}^4,\boldsymbol{v})]+\text{h.o.t.},\end{aligned} \tag{42.28}$$

$$\begin{aligned}\varepsilon^2\mathcal{T}^2(\boldsymbol{\sigma}(\varepsilon),\boldsymbol{u}(\varepsilon),\boldsymbol{v}) &= \varepsilon^{-2}\mathcal{T}^2(\boldsymbol{\sigma}^{-4},\boldsymbol{u}^0,\boldsymbol{v})\\ &\quad+\varepsilon^{-1}[\mathcal{T}^2(\boldsymbol{\sigma}^{-3},\boldsymbol{u}^0,\boldsymbol{v})+\mathcal{T}^2(\boldsymbol{\sigma}^{-4},\boldsymbol{u}^1,\boldsymbol{v})]\\ &\quad+\varepsilon^{0}[\mathcal{T}^2(\boldsymbol{\sigma}^{-2},\boldsymbol{u}^0,\boldsymbol{v})+\mathcal{T}^2(\boldsymbol{\sigma}^{-3},\boldsymbol{u}^1,\boldsymbol{v})+\mathcal{T}^2(\boldsymbol{\sigma}^{-4},\boldsymbol{u}^2,\boldsymbol{v})]\\ &\quad+\varepsilon[\mathcal{T}^2(\boldsymbol{\sigma}^{-1},\boldsymbol{u}^0,\boldsymbol{v})+\mathcal{T}^2(\boldsymbol{\sigma}^{-2},\boldsymbol{u}^1,\boldsymbol{v})+\mathcal{T}^2(\boldsymbol{\sigma}^{-3},\boldsymbol{u}^2,\boldsymbol{v})\\ &\qquad+\mathcal{T}^2(\boldsymbol{\sigma}^{-4},\boldsymbol{u}^3,\boldsymbol{v})]+\text{h.o.t.}\end{aligned} \tag{42.29}$$

Step 2. Cancellation of the factor of ε^{-4}.
The factor of ε^{-4} is

$$\mathcal{B}(\boldsymbol{\sigma}^{-4},\boldsymbol{v})+\mathcal{T}(\boldsymbol{\sigma}^{-4},\boldsymbol{u}^0,\boldsymbol{v})=0,\quad \text{for all } \boldsymbol{v}\in V(\Omega), \tag{42.30}$$

or equivalently

$$\int_\Omega \sigma_{ij}^{-4}\partial_j v_i\,\mathrm{d}\boldsymbol{x}+\int_\Omega \sigma_{ij}^{-4}\partial_i u_\gamma^0\partial_j v_\gamma^0\,\mathrm{d}\boldsymbol{x}=0,\quad \text{for all } \boldsymbol{v}\in V(\Omega), \tag{42.31}$$

from which we conclude that

$$\sigma_{ij}^{-4}=0,\quad \text{or } \partial_1 u_1^0=-1,\qquad \partial_2 u_1^0=0,\qquad \partial_1 u_2^0=0,\qquad \partial_2 u_2^0=-1. \tag{42.32}$$

Since u_α^0 vanishes on $\Gamma_0\cup\Gamma_L$ the second hypothesis is not possible and we have the first part of (42.5). Moreover, from (42.24)

$$\begin{aligned}\sigma_{\alpha\beta}^{-4} &= \lambda[e_{\rho\rho}(\boldsymbol{u}^0)+\tfrac{1}{2}\partial_\rho u_\gamma^0\partial_\rho u_\gamma^0]\delta_{\alpha\beta}+2\mu[e_{\alpha\beta}(\boldsymbol{u}^0)+\tfrac{1}{2}\partial_\alpha u_\gamma^0\partial_\beta u_\gamma^0]\\ &= \lambda E_{\gamma\gamma}^0(\boldsymbol{u}^0)\delta_{\alpha\beta}+2\mu E_{\alpha\beta}(\boldsymbol{u}^0)=0,\end{aligned} \tag{42.33}$$

which is the "planar" component of the Green–Saint Venant strain tensor. Thus u_α^0 must be a planar rigid body displacement in the cross section $\omega(x_3)\times\{x_3\}$, that is there exist functions $\theta(x_3)$ and $z_\alpha^0(x_3)$ belonging to $W_0^{1,4}(0,L)$ and such that

$$u_\alpha^0(x_1,x_2,x_3)=[Q(\theta(x_3))-\mathrm{Id}]\begin{Bmatrix}x_1\\ x_2\end{Bmatrix}+z_\alpha^0(x_3), \tag{42.34}$$

where, $Q(\theta(x_3))$ is the rotation matrix

$$Q(\theta(x_3)) = \begin{bmatrix} \cos(\theta(x_3)) & -\sin(\theta(x_3)) \\ \sin(\theta(x_3)) & \cos(\theta(x_3)) \end{bmatrix}, \quad \text{for all } x_3 \in [0, L]. \tag{42.35}$$

Explicitly, from (42.34) we have expression (42.6) and from (42.26) we obtain $\sigma_{33}^{-2} = 0$.

Step 3. *Cancellation of the factor of* ε^{-3}.

The factor of ε^{-3} is

$$\mathscr{B}(\boldsymbol{\sigma}^{-3}, \boldsymbol{v}) + \mathscr{T}^0(\boldsymbol{\sigma}^{-3}, \boldsymbol{u}^0, \boldsymbol{v}) + \mathscr{T}^0(\boldsymbol{\sigma}^{-4}, \boldsymbol{u}^1, \boldsymbol{v}) = 0, \quad \text{for all } \boldsymbol{v} \in V(\Omega). \tag{42.36}$$

From (42.24) and since $\boldsymbol{\sigma}^{-4} = \mathbf{0}$ we obtain

$$\int_\Omega \sigma_{ij}^{-3} \partial_j v_i \, d\boldsymbol{x} + \int_\Omega \sigma_{ij}^{-3} \partial_i u_\gamma^0 \partial_j v_\gamma^0 \, d\boldsymbol{x} = 0, \quad \text{for all } \boldsymbol{v} \in V(\Omega), \tag{42.37}$$

and, as before, we conclude (42.8) and, then

$$\begin{aligned} \sigma_{\alpha\beta}^{-3} &= \lambda[e_{\rho\rho}(\boldsymbol{u}^1) + \tfrac{1}{2}(\partial_\rho u_\gamma^0 \partial_\rho u_\gamma^1 + \partial_\rho u_\gamma^1 \partial_\rho u_\gamma^0)]\delta_{\alpha\beta} \\ &\quad + 2\mu[e_{\alpha\beta}(\boldsymbol{u}^1) + \tfrac{1}{2}(\partial_\alpha u_\gamma^0 \partial_\beta u_\gamma^1 + \partial_\alpha u_\gamma^1 \partial_\beta u_\gamma^0)] = 0. \end{aligned}$$

Step 4. *Cancellation of the factor of* ε^{-2}.

The factor of ε^{-2} is

$$\begin{aligned} &\mathscr{B}(\boldsymbol{\sigma}^{-2}, \boldsymbol{v}) + \mathscr{T}^0(\boldsymbol{\sigma}^{-2}, \boldsymbol{u}^0, \boldsymbol{v}) + \mathscr{T}^0(\boldsymbol{\sigma}^{-3}, \boldsymbol{u}^1, \boldsymbol{v}) \\ &\quad + \mathscr{T}^0(\boldsymbol{\sigma}^{-4}, \boldsymbol{u}^2, \boldsymbol{v}) + \mathscr{T}^2(\boldsymbol{\sigma}^{-4}, \boldsymbol{u}^0, \boldsymbol{v}) = 0, \quad \text{for all } \boldsymbol{v} \in V(\Omega). \end{aligned}$$

Since $\boldsymbol{\sigma}^{-4} = \mathbf{0}$ and $\boldsymbol{\sigma}^{-3} = \mathbf{0}$ and $\sigma_{33}^{-2} = 0$, this reduces to

$$\int_\Omega (\sigma_{\gamma j}^{-2} \partial_j v_\gamma + \sigma_{3\gamma}^{-2} \partial_\gamma v_3 + \sigma_{ij}^{-2} \partial_i u_\gamma^0 \partial_j v_\gamma) \, d\boldsymbol{x} = 0, \quad \text{for all } \boldsymbol{v} \in V(\Omega).$$

Choosing $v_\gamma = 0$ we obtain $\sigma_{3\alpha}^{-2} = 0$. Consequently, we are left with an equation for $\sigma_{\alpha\beta}^{-2}$ of the type (42.31) or (42.37) and, with a similar reasoning, we obtain $\sigma_{\alpha\beta}^{-2} = 0$.

Now we prove that as a consequence of these results $\boldsymbol{u}^0 \in \tilde{V}_{\mathrm{BN}}(\Omega)$. In fact, from the first equation of (42.25) we obtain

$$\partial_3 u_\alpha^0 + \partial_\alpha u_3^0 + \partial_3 u_\gamma^0 \partial_\alpha u_\gamma^0 = 0. \tag{42.38}$$

Substituting the expression given by (42.6) for u_α^0 into this equation, we obtain

$$\begin{aligned}
&\partial_1 u_3^0 + [\partial_3\theta(-x_1\sin(\theta) - x_2\cos(\theta)) + \partial_3 z_1^0]\cos(\theta)\\
&\quad + [\partial_3\theta(x_1\cos(\theta) - x_2\sin(\theta)) + \partial_3 z_2^0]\sin(\theta) = 0,\\
&\partial_2 u_3^0 + [\partial_3\theta(x_1\cos(\theta) - x_2\sin(\theta)) + \partial_3 z_2^0]\cos(\theta)\\
&\quad + [\partial_3\theta(-x_1\sin(\theta) - x_2\cos(\theta)) + \partial_3 z_2^0](-\sin(\theta)) = 0.
\end{aligned} \tag{42.39}$$

Differentiating the first equation with respect to x_2 and the second one with respect to x_1, one obtains, in the distributional sense

$$\left.\begin{aligned}\partial_{12}u_3^0 &= \partial_3\theta,\\ \partial_{21}u_3^0 &= -\partial_3\theta,\end{aligned}\right\} \quad \text{for all } x_3 \in [0, L]. \tag{42.40}$$

Therefore, since $\theta \in W_0^{1,4}(0, L)$ one must have $\partial_{12}u_3^0 = \partial_3\theta = 0$, which implies that θ is a constant and therefore equal to zero and consequently

$$u_\alpha^0(x_1, x_2, x_3) = z_\alpha^0(x_3) \in W_0^{1,4}(0, L). \tag{42.41}$$

Moreover, (42.39) now reduces to

$$\begin{aligned}\partial_1 u_3^0 + \partial_3 z_1^0 &= 0,\\ \partial_2 u_3^0 + \partial_3 z_2^0 &= 0.\end{aligned} \tag{42.42}$$

From which we obtain (cf. Section 4) the existence of a function $\underline{u}_3^0(x_3)$ such that

$$u_3^0(x_1, x_2, x_3) = \underline{u}_3^0(x_3) - x_\alpha \partial_3 u_\alpha^0, \tag{42.43}$$

with $\underline{u}_3^0(x_3) \in W_0^{1,4}(0, L)$ and $u_\alpha^0 \in W_0^{2,4}(0, L)$. That is to say $\boldsymbol{u}^0 \in \tilde{V}_{\mathrm{BN}}(\Omega)$.

Step 5. Cancellation of the factor of ε^{-1}.

The factor of ε^{-1} is

$$\begin{aligned}
&\mathcal{B}(\boldsymbol{\sigma}^{-1}, \boldsymbol{v}) + \mathcal{T}^0(\boldsymbol{\sigma}^{-1}, \boldsymbol{u}^0, \boldsymbol{v}) + \mathcal{T}^0(\boldsymbol{\sigma}^{-2}, \boldsymbol{u}^1, \boldsymbol{v})\\
&\quad + \mathcal{T}^0(\boldsymbol{\sigma}^{-3}, \boldsymbol{u}^2, \boldsymbol{v}) + \mathcal{T}^0(\boldsymbol{\sigma}^{-4}, \boldsymbol{u}^3, \boldsymbol{v})\\
&\quad + \mathcal{T}^2(\boldsymbol{\sigma}^{-3}, \boldsymbol{u}^0, \boldsymbol{v}) + \mathcal{T}^2(\boldsymbol{\sigma}^{-4}, \boldsymbol{u}^1, \boldsymbol{v}) = 0, \quad \text{for all } \boldsymbol{v} \in V(\Omega).
\end{aligned} \tag{42.44}$$

Using the previous results and with similar arguments as before one obtains properties (42.11).

It is now possible to evaluate σ_{33}^0 and $\partial_\alpha u_\alpha^2$. In fact from (42.26) and the fact that $\boldsymbol{u}^0$ and $\boldsymbol{u}^1$ belong to $\tilde{V}_{\mathrm{BN}}(\Omega)$ gives:

$$\sigma_{33}^0 = \lambda(\partial_\alpha u_\alpha^0 + \tfrac{1}{2}\partial_\alpha u_3^0 \partial_\alpha u_3^0) + (\lambda + 2\mu)[\partial_3 u_3^0 + \tfrac{1}{2}\partial_3 u_\gamma^0 \partial_3 u_\gamma^0]. \tag{42.45}$$

From (42.24) and the fact that $\sigma^{-2}_{\alpha\beta} = 0$ we obtain, after a contraction in the indices,

$$\partial_\alpha u^2_\alpha = -\Lambda_{\alpha\alpha} - \frac{\lambda}{\lambda + \mu}[\partial_3 \underline{u}^0_3 - x_\alpha \partial_{33} u^0_\alpha + \Lambda_{\alpha\alpha}]. \tag{42.46}$$

Substituting back into (42.47) we have (42.12), (42.13).

In a similar way it is also possible to evaluate formula (42.14) for σ^1_{33} and consequently the bending moment and axial force components given by (42.17)–(42.20) are immediately obtained.

Before passing to the cancellation of the factors of ε^0 we remark that with the information obtained so far, besides being possible to evaluate $\partial_\alpha u^2_\alpha$, it is actually possible to calculate $\partial_1 u^2_1$ and $\partial_2 u^2_2$. In fact, using (42.24) and solving $\sigma^{-2}_{11} = \sigma^{-2}_{22} = 0$ in terms of $\partial_1 u^2_1$ and $\partial_2 u^2_2$ we obtain (42.15).

Step 6. *Cancellation of the factor of* ε^0.

The factor of ε^0 is

$$\begin{aligned}
&\mathcal{B}(\boldsymbol{\sigma}^0, \boldsymbol{v}) + \mathcal{T}^0(\boldsymbol{\sigma}^0, \boldsymbol{u}^0, \boldsymbol{v}) + \mathcal{T}^0(\boldsymbol{\sigma}^{-1}, \boldsymbol{u}^1, \boldsymbol{v}) \\
&\quad + \mathcal{T}^0(\boldsymbol{\sigma}^{-2}, \boldsymbol{u}^2, \boldsymbol{v}) + \mathcal{T}^0(\boldsymbol{\sigma}^{-3}, \boldsymbol{u}^3, \boldsymbol{v}) + \mathcal{T}^0(\boldsymbol{\sigma}^{-4}, \boldsymbol{u}^4, \boldsymbol{v}) \\
&\quad + \mathcal{T}^2(\boldsymbol{\sigma}^{-2}, \boldsymbol{u}^0, \boldsymbol{v}) + \mathcal{T}^2(\boldsymbol{\sigma}^{-3}, \boldsymbol{u}^1, \boldsymbol{v}) + \mathcal{T}^2(\boldsymbol{\sigma}^{-4}, \boldsymbol{u}^2, \boldsymbol{v}) \\
&\quad = \mathcal{L}(\boldsymbol{v}) = \int_\Omega f_i v_i \,\mathrm{d}\boldsymbol{x} + \int_\Gamma g_i v_i \,\mathrm{d}a, \quad \text{for all } \boldsymbol{v} \in V(\Omega).
\end{aligned} \tag{42.47}$$

Using the previous results this reduces to (42.16). □

REMARK 42.1. (i) Note the crucial role played (in Step 2) by the regularity assumption "$\partial_\alpha u^0_\beta \in C^0(\overline{\Omega})$".

(ii) In order to avoid cumbersome assumptions we assumed that all the relevant terms of the formal asymptotic expansion belong to the same space $V(\Omega)$. However, an inspection of the proof reveals that the same conclusions could have been drawn under weaker assumptions, since not all partial derivatives $\partial_j u^p_i$ occur for a given exponent p.

(iii) It is not possible to evaluate u^2_α due to a boundary layer phenomena. In order to proceed we must weaken the boundary conditions and consider, for instance, the weakly clamping condition $\int_{\Gamma_0 \cup \Gamma_L} u_\alpha \,\mathrm{d}\omega = \int_{\Gamma_0 \cup \Gamma_L} (x_2 u_1 - x_1 u_2) \,\mathrm{d}\omega = 0$ (see Sections 4 and 5). This does not change the calculations made so far.

(iv) The integral $\int_\Omega \hat{\sigma}_{ij}(\varepsilon; \boldsymbol{\nabla u}(\varepsilon)) \partial_i u_3(\varepsilon) \partial_j v_3 \,\mathrm{d}\boldsymbol{x}$ that factorizes ε^2 in (41.8) plays no part here; it would only if the formal expansion of $\boldsymbol{u}(\varepsilon)$ was pursued until it included the terms $\varepsilon^6 \boldsymbol{u}^6$, and if the factors of ε and ε^2 were also cancelled.

Using the formula of change of variable, once again, and the definition of Q_1, (cf. MIARA and TRABUCHO [1990, 1992] or (33.29), (33.31)), we deduce that

$$\sum_{k=1}^{N} \varepsilon^{2k} \int_{\gamma} Q^k(x_1,x_2)\,\mathrm{d}\gamma \int_a^b |\partial_3^{k-1} G_3^\varepsilon(y_3)|^2\,\mathrm{d}y_3 = \mathrm{O}(\varepsilon^2). \tag{39.35}$$

From (39.32)–(39.35) we conclude that

$$\int_{\Gamma_{\mathrm{sb}}} G_3^\varepsilon \Psi_{\mathrm{sb}3}^{\varepsilon N}\,\mathrm{d}\Gamma_{\mathrm{sb}} = -C_1|C_2 + \mathrm{O}(\varepsilon^m)| + \mathrm{O}(\varepsilon^2), \tag{39.36}$$

where C_1, C_2 are constants independent of ε and N and $C_1 \geqslant 0$. This completes the proof of the theorem. □

THEOREM 39.5. *With the assumptions of Theorem 39.4, there exists a constant C independent of ε and N such that for ε sufficiently small*

$$\|\boldsymbol{\Psi}_{\mathrm{sb}}^N(\varepsilon)\|_{V(\Omega)} \leqslant \varepsilon^{2N-2} C_N^\varepsilon C. \tag{39.37}$$

PROOF. Due to the triangular inequality and to (39.22), we have

$$\begin{aligned}\|\boldsymbol{\Psi}_{\mathrm{sb}}^N(\varepsilon)\|_{V(\Omega)} &\leqslant \|\boldsymbol{\Psi}_{\mathrm{sb}}^N(\varepsilon) - \boldsymbol{u}_{\mathrm{sb}}^N(\varepsilon)\|_{V(\Omega)} + \|\boldsymbol{u}_{\mathrm{sb}}^N(\varepsilon)\|_{V(\Omega)}\\ &\leqslant C_N^\varepsilon \varepsilon^{2N-2} C + \|\boldsymbol{u}_{\mathrm{sb}}^N(\varepsilon)\|_{V(\Omega)}.\end{aligned} \tag{39.38}$$

Consequently, in order to have an estimate for $\|\boldsymbol{\Psi}_{\mathrm{sb}}^N(\varepsilon)\|_{V(\Omega)}$, it is enough to obtain an estimate for $\|\boldsymbol{u}_{\mathrm{sb}}^N(\varepsilon)\|_{V(\Omega)}$. As the bilinear form $C(\varepsilon)(\cdot,\cdot)$ associated to the "straight beam" is $\boldsymbol{V}(\Omega)$-elliptic (cf. (38.38)) we have

$$\|\boldsymbol{u}_{\mathrm{sb}}^N(\varepsilon)\|_{\boldsymbol{V}(\Omega)}^2 \leqslant \varepsilon^{-6} C_1 C(\varepsilon)(\boldsymbol{u}_{\mathrm{sb}}^N(\varepsilon), \boldsymbol{u}_{\mathrm{sb}}^N(\varepsilon)). \tag{39.39}$$

But $C(\varepsilon)(\boldsymbol{u}_{\mathrm{sb}}^N(\varepsilon), \boldsymbol{u}_{\mathrm{sb}}^N(\varepsilon)) = C^\varepsilon(\boldsymbol{u}_{\mathrm{sb}}^{\varepsilon N}, \boldsymbol{u}_{\mathrm{sb}}^{\varepsilon N})$ (cf. (39.20), (39.7), (39.8)), where $C^\varepsilon(\cdot,\cdot)$ is an inner product in the space $\boldsymbol{V}(\Omega_{\mathrm{sb}})$ and $\boldsymbol{u}_{\mathrm{sb}}^{\varepsilon N}$ is the orthogonal projection of $\boldsymbol{u}_{\mathrm{sb}}^\varepsilon$ in $\boldsymbol{V}^N(\Omega_{\mathrm{sb}})$. Let us denote by $(\cdot,\cdot)_{\mathrm{sb}}$ and $\|\cdot\|_{\mathrm{sb}}$ the inner product and the norm, respectively, induced in $\boldsymbol{V}(\Omega_{\mathrm{sb}})$ by the bilinear form $C^\varepsilon(\cdot,\cdot)$. We then have

$$\|\boldsymbol{u}_{\mathrm{sb}}^{\varepsilon N}\|_{\mathrm{sb}}^2 = \tfrac{1}{4}\big(\|\boldsymbol{u}_{\mathrm{sb}}^{\varepsilon N} + \boldsymbol{u}_{\mathrm{sb}}^\varepsilon\|_{\mathrm{sb}}^2 - \|\boldsymbol{u}_{\mathrm{sb}}^{\varepsilon N} - \boldsymbol{u}_{\mathrm{sb}}^\varepsilon\|_{\mathrm{sb}}^2\big). \tag{39.40}$$

On the other hand by the mean value lemma (cf. SCHWARTZ [1979]) one obtains

$$\begin{aligned}\|\boldsymbol{u}_{\mathrm{sb}}^{\varepsilon N} + \boldsymbol{u}_{\mathrm{sb}}^\varepsilon\|_{\mathrm{sb}}^2 &= 2\big(\|\boldsymbol{\Psi}_{\mathrm{sb}}^{\varepsilon N} - \boldsymbol{u}_{\mathrm{sb}}^{\varepsilon N}\|_{\mathrm{sb}}^2 + \|\boldsymbol{\Psi}_{\mathrm{sb}}^{\varepsilon N} + \boldsymbol{u}_{\mathrm{sb}}^\varepsilon\|_{\mathrm{sb}}^2\big)\\ &\quad - 4\Big\|\boldsymbol{\Psi}_{\mathrm{sb}}^{\varepsilon N} - \frac{\boldsymbol{u}_{\mathrm{sb}}^{\varepsilon N} - \boldsymbol{u}_{\mathrm{sb}}^\varepsilon}{2}\Big\|_{\mathrm{sb}}^2,\\ \|\boldsymbol{u}_{\mathrm{sb}}^{\varepsilon N} - \boldsymbol{u}_{\mathrm{sb}}^\varepsilon\|_{\mathrm{sb}}^2 &= 2\big(\|\boldsymbol{\Psi}_{\mathrm{sb}}^{\varepsilon N} - \boldsymbol{u}_{\mathrm{sb}}^{\varepsilon N}\|_{\mathrm{sb}}^2 + \|\boldsymbol{\Psi}_{\mathrm{sb}}^{\varepsilon N} - \boldsymbol{u}_{\mathrm{sb}}^\varepsilon\|_{\mathrm{sb}}^2\big)\\ &\quad - 4\Big\|\boldsymbol{\Psi}_{\mathrm{sb}}^{\varepsilon N} - \frac{\boldsymbol{u}_{\mathrm{sb}}^{\varepsilon N} + \boldsymbol{u}_{\mathrm{sb}}^\varepsilon}{2}\Big\|_{\mathrm{sb}}^2.\end{aligned}$$

43. Identification of the leading term $\boldsymbol{u}^0$ of the formal expansion

We have shown in Theorem 42.1 that the following variational equations are verified:

$$\int_\Omega \sigma_{ij}^0 \partial_j v_i \,\mathrm{d}\boldsymbol{x} + \int_\Omega \sigma_{3j}^0 \partial_3 u_\gamma^0 \partial_j v_\gamma \,\mathrm{d}\boldsymbol{x} = \int_\Omega f_i v_i \,\mathrm{d}\boldsymbol{x} + \int_\Gamma g_i v_i \,\mathrm{d}a, \quad \text{for all } \boldsymbol{v} \in V(\Omega). \tag{43.1}$$

We also established in the same theorem that $\partial_\alpha u_\alpha^2$, found in the expression for σ_{33}^0, is in fact a known function of $\boldsymbol{u}^0$. Hence, *if in the above equations we restrict functions $\boldsymbol{v} = (v_i)$ to the space $\tilde{V}_{\mathrm{BN}}(\Omega)$, defined in* (42.27), *we are left with a variational problem posed over this space and whose only unknown is the function $\boldsymbol{u}^0 \in \tilde{V}_{\mathrm{BN}}(\Omega)$*, since the integrals involving functions σ_{i3}^0 vanish in this case.

Let us state this property as a theorem, where we also show that this variational problem may be in fact posed over a space larger than $\tilde{V}_{\mathrm{BN}}(\Omega)$, namely $V_{\mathrm{BN}}(\Omega)$.

THEOREM 43.1. *Let the assumptions be as in Theorem* 42.1. *Then the leading term $\boldsymbol{u}^0$ of the formal expansion solves the following problem*:

$$\begin{aligned} \boldsymbol{u}^0 &= (u_i^0) \in V_{\mathrm{BN}}(\Omega) \\ &= \{\boldsymbol{v} = (v_i)\colon v_\alpha = z_\alpha,\ v_3 = z_3 - x_\alpha \partial_3 z_\alpha, \text{ with} \\ &\qquad z_\alpha \in H_0^2(0,L),\quad z_3 \in H_0^1(0,L)\}, \\ &\int_\Omega \sigma_{33}^0 \partial_3 v_3 \,\mathrm{d}\boldsymbol{x} + \int_\Omega \sigma_{33}^0 \partial_3 u_\gamma^0 \partial_3 v_\gamma \,\mathrm{d}\boldsymbol{x} \\ &\quad = \int_\Omega f_i v_i \,\mathrm{d}\boldsymbol{x} + \int_\Gamma g_i v_i \,\mathrm{d}a, \quad \text{for all } \boldsymbol{v} \in V_{\mathrm{BN}}(\Omega), \end{aligned} \tag{43.2}$$

where σ_{33}^0 is given by (42.39), (42.40).

PROOF. As was already seen in the previous section, Eq. (43.2) follows from Theorem 41.1 and the definition of the space $\tilde{V}_{\mathrm{BN}}(\Omega)$. It thus remains to verify that the same problem is still well posed if we assume that $\boldsymbol{u}^0$ and the functions $\boldsymbol{v}$ belong to the space $V_{\mathrm{BN}}(\Omega)$, instead of the smaller space $\tilde{V}_{\mathrm{BN}}(\Omega)$. To see this, we observe that the inclusions $H^1(0,L) \hookrightarrow L^q(0,L)$ hold for any $q \geqslant 1$. Hence, $u_\alpha^0 \in H_0^2(0,L)$ implies that $\partial_3 u_\alpha^0 \partial_3 u_\beta^0 \in L^r(0,L)$, for all $r \geqslant 1$. Likewise $\underline{u}_3^0 \in H_0^1(0,L)$ implies that $\sigma_{33}^0 \in L^2(\Omega)$.

Since all the multilinear forms found in (43.2) after replacing σ_{33}^0 by (42.39), (42.40) remain continuous over this larger space and since $W^{m,4}(0,L)$ is a dense subspace of $H^m(0, L)$, any solution of these equations obtained by letting $\boldsymbol{v}$ vary in the space $\tilde{V}_{\mathrm{BN}}(\Omega)$ is also a solution of the same variational equations when $\boldsymbol{v}$ varies in the space $V_{\mathrm{BN}}(\Omega)$. □

Using the fact that both the unknown $\boldsymbol{u}^0$ and the functions $\boldsymbol{v}$ belong to the space $V_{\mathrm{BN}}(\Omega)$, we next show that *problem* (43.2) *is in fact a one dimensional problem,* in the sense that it is equivalent to a variational problem posed over the set $(0, L)$.

THEOREM 43.2. *The function* $\boldsymbol{u}^0 = (u_i^0)$ *solves problem* (43.2) *if and only if*

$$\begin{aligned} u_\alpha^0 &= u_\alpha^0(x_3), \\ u_3^0 &= \underline{u}_3^0 - x_\alpha \partial_3 u_\alpha^0(x_3), \end{aligned} \tag{43.3}$$

with $u_\alpha^0 \in H_0^2(0, L)$ *and* $\underline{u}_3^0 \in H_0^1(0, L)$, *and* u_α^0 *solving the following variational problem (no sum on* β*):*

$$\begin{aligned} &\int_0^L E I_\beta \partial_{33} u_\beta^0 \partial_{33} v_\beta \,\mathrm{d}x_3 + \int_0^L E(\partial_3 \underline{u}_3^0 + \Lambda_{\gamma\gamma}) \partial_3 u_\beta^0 \partial_3 v_\beta \,\mathrm{d}x_3 \\ &\quad = \int_0^L \left[\int_\omega f_\beta \,\mathrm{d}\omega + \int_\gamma g_\beta \,\mathrm{d}\gamma \right] v_\beta \,\mathrm{d}x_3 \\ &\qquad - \int_0^L \left[\int_\omega x_\beta f_3 \,\mathrm{d}\omega + \int_\gamma x_\beta g_3 \,\mathrm{d}\gamma \right] \partial_3 v_\beta \,\mathrm{d}x_3, \quad \textit{for all } v_\beta \in H_0^2(0, L), \end{aligned} \tag{43.4}$$

$$\begin{aligned} &\int_0^L E(\partial_3 \underline{u}_3^0 + \Lambda_{\gamma\gamma}) \partial_3 v_3 \,\mathrm{d}x_3 \\ &\quad = \int_0^L \left[\int_\omega f_3 \,\mathrm{d}\omega + \int_\gamma g_3 \,\mathrm{d}\gamma \right] v_3 \,\mathrm{d}x_3, \quad \textit{for all } v_3 \in H_0^1(0, L), \end{aligned} \tag{43.5}$$

where matrix $\boldsymbol{\Lambda} = (\Lambda_{\alpha\beta})$ *is given by*

$$\boldsymbol{\Lambda} = (\Lambda_{\alpha\beta}) = \frac{1}{2} \begin{pmatrix} \partial_3 u_1^0 \partial_3 u_1^0 & \partial_3 u_1^0 \partial_3 u_2^0 \\ \partial_3 u_2^0 \partial_3 u_1^0 & \partial_3 u_2^0 \partial_3 u_2^0 \end{pmatrix}. \tag{43.6}$$

PROOF. Let functions $\boldsymbol{v} = (v_i) \in V_{\mathrm{BN}}(\Omega)$ in problem (43.2) be of the particular form

$$\boldsymbol{v} = (0, 0, v_3(x_3)), \quad v_3(x_3) \in H_0^1(0, L).$$

We obtain

$$\int_{\Omega} \sigma_{33}^0 \partial_3 v_3 \,\mathrm{d}\boldsymbol{x} = \int_{\Omega} f_3 v_3 \,\mathrm{d}\boldsymbol{x} + \int_{\Gamma} g_3 v_3 \,\mathrm{d}a, \quad \text{for all } v_3 \in H_0^1(0,L). \tag{43.7}$$

Next, let the functions $\boldsymbol{v} = (v_i) \in V_{\mathrm{BN}}(\Omega)$ in problem (43.2) be of the particular form

$$\boldsymbol{v} = (v_1(x_3), v_2(x_3), -x_\beta \partial_3 v_\beta(x_3)), \qquad v_\beta(x_3) \in H_0^2(0,L),$$

which gives

$$\begin{aligned} &\int_{\Omega} x_\beta \sigma_{33}^0 \partial_{33} v_\beta \,\mathrm{d}\boldsymbol{x} + \int_{\Omega} \sigma_{33}^0 \partial_3 u_\gamma^0 \partial_3 v_\gamma \,\mathrm{d}\boldsymbol{x} \\ &\quad = \int_{\Omega} f_\beta v_\beta \,\mathrm{d}\boldsymbol{x} + \int_{\Gamma} g_\beta v_\beta \,\mathrm{d}a - \int_{\Omega} x_\beta f_3 \partial_3 v_\beta \,\mathrm{d}\boldsymbol{x} \\ &\qquad - \int_{\Gamma} x_\beta g_3 \partial_3 v_\beta \,\mathrm{d}a, \quad \text{for all } v_\beta \in H_0^2(0,L). \end{aligned} \tag{43.8}$$

Writing integrals $\int_\Omega \varphi \,\mathrm{d}\boldsymbol{x} = \int_0^L [\int_\omega \varphi \,\mathrm{d}\omega] \,\mathrm{d}x_3$ and $\int_\Gamma \varphi \,\mathrm{d}a = \int_0^L [\int_\gamma \varphi \,\mathrm{d}\gamma] \,\mathrm{d}x_3$, we obtain

$$\begin{aligned} &\int_0^L m_\beta^0 \partial_{33} v_\beta \,\mathrm{d}x_3 + \int_0^L n_{33}^0 \partial_3 u_\beta^0 \partial_3 v_\beta \,\mathrm{d}x_3 \\ &\quad = \int_0^L \left[\int_\omega f_\beta \,\mathrm{d}\omega + \int_\gamma g_\beta \,\mathrm{d}\gamma \right] v_\beta \,\mathrm{d}x_3 - \int_0^L \left[\int_\omega x_\beta f_3 \,\mathrm{d}\omega + \int_\gamma x_\beta g_3 \,\mathrm{d}\gamma \right] \partial_3 v_\beta \,\mathrm{d}x_3, \\ &\quad \text{for all } v_\beta \in H_0^2(0,L), \end{aligned} \tag{43.9}$$

$$\int_0^L n_{33}^0 \partial_3 v_3 \,\mathrm{d}x_3 = \int_0^L \left[\int_\omega f_3 \,\mathrm{d}\omega + \int_\gamma g_3 \,\mathrm{d}\gamma \right] v_3 \,\mathrm{d}x_3, \quad \text{for all } v_3 \in H_0^1(0,L). \tag{43.10}$$

Finally, using definitions (42.39)–(42.42) we see that Eqs. (43.9), (43.10) are equivalent to Eqs. (43.4)–(43.6).

Conversely, if functions ζ_i solve the variational problem (43.4)–(43.6), using definitions (42.39)–(42.42), one sees that $\boldsymbol{u}^0 = (u_i^0)$, with $u_\alpha^0 = \zeta_\alpha$ and $u_3^0 = \zeta_3 - x_\alpha \partial_3 \zeta_\alpha$, solves problem (43.2). □

In order to complete this study, and be able to compare with the classical equations used in engineering literature (cf. for example BRUSH and ALMROTH [1975, Chapter 2]), we next write the one–dimensional boundary value problem, equivalent to the variational problem (43.4)–(43.6) obtained by applying Green's formula.

THEOREM 43.3. *A smooth enough solution* (u_1^0, u_2^0, u_3^0) *of the variational problem* (43.4)–(43.6) *is also a solution of the following classical one-dimensional nonlinear rod problem*:

$$\begin{aligned}
&\partial_{33} m_\beta^0 - \partial_3(n_{33}^0 \partial_3 u_\beta^0) = \tilde{F}_\beta, \quad \textit{in } (0, L),\\
&-\partial_3 n_{33}^0 = F_3, \quad \textit{in } (0, L),\\
&u_i^0 = 0, \quad \textit{on } (0, L),\\
&\partial_3 u_\alpha^0 = 0, \quad \textit{on } (0, L),
\end{aligned} \tag{43.11}$$

where functions m_β^0, n_{33}^0 *are given by* (42.41), (42,42) *and where*

$$\tilde{F}_\beta = \int_\omega f_\beta \, \mathrm{d}\omega + \int_\gamma g_\beta \, \mathrm{d}\gamma + \int_\omega x_\beta \partial_3 f_3 \, \mathrm{d}\omega + \int_\gamma x_\beta \partial_3 g_3 \, \mathrm{d}\gamma, \tag{43.12}$$

$$F_3 = \int_\omega f_3 \, \mathrm{d}\omega + \int_\gamma g_3 \, \mathrm{d}\gamma. \tag{43.13}$$

44. Existence of the leading term $\boldsymbol{u}^0$ in the displacement approach

It has been shown that the leading term $\boldsymbol{u}^0$ of the formal expansion of the scaled displacement $\boldsymbol{u}(\varepsilon)$ should solve the variational problem (43.3)–(43.6) posed in the space $V_{\mathrm{BN}}(\Omega)$. It thus remains to see whether this problem or equivalently the variational problem (43.11)–(43.13), does indeed possess solutions.

THEOREM 44.1. *Assume that the applied force densities are such that*

$$f_i \in L^2(\Omega), \qquad g_i \in L^2(\Gamma). \tag{44.1}$$

Then, the variational problem (43.3)–(43.6) *possesses at least one solution.*

PROOF. This existence result may be established as in LIONS [1969, p. 56] or CIMETIÈRE, GEYMONAT, LE DRET, RAOULT and TUTEK [1986, 1988]. □

For convenience, we now gather in a single statement the properties established thus far, about the leading term $\boldsymbol{u}^0$ (cf. CIMETIÈRE, GEYMONAT, LE DRET, RAOULT and TUTEK [1986, 1988]).

THEOREM 44.2. *In the displacement approach, the leading term $\boldsymbol{u}^0$ of the formal expansion of the scaled displacement $\boldsymbol{u}(\varepsilon)$ solves the following problem:*

$$\begin{aligned}
&\boldsymbol{u}^0 = (u_i^0) \in V_{\mathrm{BN}}(\Omega) \\
&\quad = \{\boldsymbol{v} = (v_i)\colon v_\alpha = z_\alpha,\ v_3 = z_3 - x_\alpha \partial_3 z_\alpha,\ \textit{with} \\
&\qquad z_\alpha \in H_0^2(0,L),\ z_3 \in H_0^1(0,L)\}, \\
&\int_\Omega \sigma_{33}^0 \partial_3 v_3 \,\mathrm{d}\boldsymbol{x} + \int_\Omega \sigma_{33}^0 \partial_3 u_\gamma^0 \partial_3 v_\gamma \,\mathrm{d}\boldsymbol{x} \\
&\quad = \int_\Omega f_i v_i \,\mathrm{d}\boldsymbol{x} + \int_\Gamma g_i v_i \,\mathrm{d}a, \quad \textit{for all } \boldsymbol{v} \in V_{\mathrm{BN}}(\Omega),
\end{aligned} \tag{44.2}$$

where σ_{33}^0 is given by

$$\sigma_{33}^0 = \mu \frac{3\lambda + 2\mu}{\lambda + \mu}[\partial_3 \underline{u}_3^0 - x_\alpha \partial_{33} u_\alpha^0 + \Lambda_{\alpha\alpha}] = E[\partial_3 \underline{u}_3^0 - x_\alpha \partial_{33} u_\alpha^0 + \Lambda_{\alpha\alpha}], \tag{44.3}$$

and where matrix $\boldsymbol{\Lambda} = (\Lambda_{\alpha\beta})$ is given by

$$\boldsymbol{\Lambda} = (\Lambda_{\alpha\beta}) = \frac{1}{2}\begin{pmatrix} \partial_3 u_1^0 \partial_3 u_1^0 & \partial_3 u_1^0 \partial_3 u_2^0 \\ \partial_3 u_2^0 \partial_3 u_1^0 & \partial_3 u_2^0 \partial_3 u_2^0 \end{pmatrix}. \tag{44.4}$$

Any function $\boldsymbol{u}^0$ solves this problem if and only if it is also a solution of

$$\begin{aligned}
&u_\alpha^0 = u_\alpha^0(x_3), \\
&u_3^0 = \underline{u}_3^0 - x_\alpha \partial_3 u_\alpha^0(x_3),
\end{aligned} \tag{44.5}$$

with $u_\alpha^0 \in H_0^2(0,L)$ and $\underline{u}_3^0 \in H_0^1(0,L)$, and u_1^0 solving the following variational problem (no sum on β):

$$\begin{aligned}
&\int_0^L EI_\beta \partial_{33} u_\beta^0 \partial_{33} v_\beta \,\mathrm{d}x_3 + \int_0^L E(\partial_3 \underline{u}_3^0 + \Lambda_{\gamma\gamma}) \partial_3 u_\beta^0 \partial_3 v_\beta \,\mathrm{d}x_3 \\
&\quad = \int_0^L \left[\int_\omega f_\beta \,\mathrm{d}\omega + \int_\gamma g_\beta \,\mathrm{d}\gamma\right] v_\beta \,\mathrm{d}x_3 \\
&\qquad - \int_0^L \left[\int_\omega x_\beta f_3 \,\mathrm{d}\omega + \int_\gamma x_\beta g_3 \,\mathrm{d}\gamma\right] \partial_3 v_\beta \,\mathrm{d}x_3, \quad \textit{for all } v_\beta \in H_0^2(0,L),
\end{aligned} \tag{44.6}$$

$$\int_0^L E(\partial_3 \underline{u}_3^0 + \Lambda_{\gamma\gamma})\partial_3 v_3 \, dx_3$$
$$= \int_0^L \left[\int_\omega f_3 \, d\omega + \int_\gamma g_3 \, d\gamma \right] v_3 \, dx_3, \quad \textit{for all } v_3 \in H_0^1(0, L), \tag{44.7}$$

or equivalently

$$\int_0^L m_\beta^0 \partial_{33} v_\beta \, dx_3 + \int_0^L n_{33}^0 \partial_3 u_\beta^0 \partial_3 v_\beta \, dx_3$$
$$= \int_0^L \left[\int_\omega f_\beta \, d\omega + \int_\gamma g_\beta \, d\gamma \right] v_\beta \, dx_3$$
$$- \int_0^L \left[\int_\omega x_\beta f_3 \, d\omega + \int_\gamma x_\beta g_3 \, d\gamma \right] \partial_3 v_\beta \, dx_3, \quad \textit{for all } v_\beta \in H_0^2(0, L), \tag{44.8}$$

$$\int_0^L n_{33}^0 \partial_3 v_3 \, dx_3 = \int_0^L \left[\int_\omega f_3 \, d\omega + \int_\gamma g_3 \, d\gamma \right] v_3 \, dx_3, \quad \textit{for all } v_3 \in H_0^1(0, L). \tag{44.9}$$

where the zeroth order bending moment components m_β^0 *are*

$$m_\beta^0 = \int_\omega x_\beta \sigma_{33}^0 \, d\omega = -EI_\beta \partial_{33} u_\beta^0, \quad \textit{no sum on } \beta, \tag{44.10}$$

and the zeroth order axial force component n_{33}^0 *is given by*

$$n_{33}^0 = \int_\omega \sigma_{33}^0 \, d\omega = E[\partial_3 \underline{u}_3^0 + \Lambda_{\alpha\alpha}]. \tag{44.11}$$

Moreover, assuming that

$$\begin{aligned} &f_\alpha \in L^2(\Omega), \qquad g_\alpha \in L^2(\Gamma), \\ &f_3 \in H[0, L; L^2(\omega)], \qquad g_3 \in H[0, L; L^2(\gamma)], \end{aligned} \tag{44.12}$$

one has

$$\begin{aligned} &u_\alpha^0 \in H^4(0, L) \cap H_0^2(0, L), \qquad u_3^0 \in H^3(0, L) \cap H_0^1(0, L) \\ &m_\beta^0 \in H^2(0, L), \qquad n_{33}^0 \in H^2(0, L), \end{aligned} \tag{44.13}$$

and consequently,

$$\begin{aligned} &\partial_{33}m^0_\beta - \partial_3(n^0_{33}\partial_3 u^0_\beta) = F_\beta, \ \textit{in}\ (0,L),\\ &-\partial_3 n^0_{33} = F_3, \ \textit{in}\ (0,L),\\ &u^0_i = 0, \ \textit{on}\ (0,L),\\ &\partial_3 u^0_\alpha = 0, \ \textit{on}\ (0,L), \end{aligned} \tag{44.14}$$

where

$$F_\beta = \int_\omega f_\beta \,\mathrm{d}\omega + \int_\gamma g_\beta \,\mathrm{d}\gamma + \int_\omega x_\beta \partial_3 f_3 \,\mathrm{d}\omega + \int_\gamma x_\beta \partial_3 g_3 \,\mathrm{d}\gamma, \tag{44.15}$$

$$F_3 = \int_\omega f_3 \,\mathrm{d}\omega + \int_\gamma g_3 \,\mathrm{d}\gamma \tag{44.16}$$

We remark that except for the last two terms in (44.15), problem (44.14)–(44.16) is the classical stretching–bending problem for a geometrically nonlinear elastic rod (cf. BRUSH and ALMROTH [1975, Chapter 2], LANDAU and LIFCHITZ [1967, Eq. (20.14)]).

We shall see in the next sections that the functions σ^0_{33} introduced in the statement of problem (44.2)–(44.4) are precisely the components with indices 33 of the leading term of the formal expansion of the "scaled stress tensor" in the "displacement–stress approach".

By pursuing this method, one could think of evaluating higher order terms. It is then observed that we are faced with a boundary layer problem related to Saint Venant's principle. However, as already shown for the linearized case (cf. Sections 7, 12, 14 and 15) these higher order terms are extremely important. Consequently, in order to overcome this difficulty, we shall proceed with a weaker boundary condition (weakly clamping) which does not change the results obtained so far, in a similar way to what happens for the linearized case (cf. Sections 5, 6 and 7).

45. The displacement–stress approach for the nonlinear case

In Theorem 42.1, we saw that the formal expansion of the tensor $\hat{\boldsymbol{\Sigma}}(\varepsilon; \boldsymbol{\nabla u}(\varepsilon))$ induced by that of the scaled displacement does not contain any factor of ε^q, $-4 \leqslant q \leqslant -1$. We also showed that the pair $(\boldsymbol{u}^0, \boldsymbol{\Sigma}^0)$, with $\boldsymbol{u}^0 = (u^0_i)$ and $\boldsymbol{\Sigma}^0 = (\sigma^0_{ij})$, satisfies

$$\begin{aligned} &\int_\Omega \sigma^0_{ij}\partial_j v_i \,\mathrm{d}\boldsymbol{x} + \int_\Omega \sigma^0_{ij}\partial_i u^0_\gamma \partial_j v_\gamma \,\mathrm{d}\boldsymbol{x}\\ &\quad = \int_\Omega f_i v_i \,\mathrm{d}\boldsymbol{x} + \int_\Gamma g_i v_i \,\mathrm{d}a, \quad \text{for all } \boldsymbol{v} \in V(\Omega). \end{aligned} \tag{45.1}$$

But then a simple inspection reveals that we would have obtained the same equations if we had required *at the onset* not only that $\boldsymbol{u}(\varepsilon)$ be of the form $\boldsymbol{u}(\varepsilon) = \boldsymbol{u}^0 + \text{h.o.t.}$, but also that the tensor $\boldsymbol{\Sigma}(\varepsilon)$ be of the form $\boldsymbol{\Sigma}(\varepsilon) = \boldsymbol{\Sigma}^0 +$ h.o.t., in the scaled principle of virtual work found in Theorem 41.2. This simple observation is the basis for another way of applying the method of asymptotic expansions, which we shall call the *displacement–stress approach* considered in the first place by CIMETIÈRE, GEYMONAT, LE DRET, RAOULT and TUTEK [1986, 1988]. As for the linearized case the two approaches are closely related, as we shall see later.

To begin with, we consider a weakly clamped boundary condition and reformulate the three-dimensional problem (40.11) of a clamped beam (cf. Section 40) as an equivalent, but different problem: first, the stress tensor field is considered as an unknown per se; secondly, the constitutive equation (40.5) that relates the displacement vector field $\boldsymbol{u}^\varepsilon = (u_i^\varepsilon)$ and the stress tensor field $\boldsymbol{\Sigma}^\varepsilon = (\sigma_{ij}^\varepsilon)$ is *inverted*, i.e., the Green–Saint Venant strain tensor $\boldsymbol{E}^\varepsilon(\boldsymbol{u}^\varepsilon) = (E_{ij}^\varepsilon(\boldsymbol{u}^\varepsilon))$ is expressed as a function of the stress tensor through an *inverted constitutive equation*, given by

$$\begin{aligned} E_{ij}^\varepsilon &= \frac{1}{2}(\partial_i^\varepsilon u_j^\varepsilon + \partial_j^\varepsilon u_i^\varepsilon + \partial_i^\varepsilon u_k^\varepsilon \partial_j^\varepsilon u_k^\varepsilon) \\ &= \frac{1}{2\mu^\varepsilon}\sigma_{ij}^\varepsilon - \frac{\lambda^\varepsilon}{2\mu^\varepsilon(3\lambda^\varepsilon + 2\mu^\varepsilon)}\sigma_{pp}^\varepsilon \delta_{ij} \\ &= \frac{1+\nu^\varepsilon}{E^\varepsilon}\sigma_{ij}^\varepsilon - \frac{\nu^\varepsilon}{E^\varepsilon}\sigma_{pp}^\varepsilon \delta_{ij}. \end{aligned} \tag{45.2}$$

We consider that *the displacement field* $\boldsymbol{u}^\varepsilon = (u_i^\varepsilon)$ *and the second Piola–Kirchhoff tensor field* $\boldsymbol{\Sigma}^\varepsilon = (\sigma_{ij}^\varepsilon)$ *solve the following problem* (compare with (40.9)–(40.11)):

$$\boldsymbol{u}^\varepsilon \in V(\Omega^\varepsilon) = \{\boldsymbol{v}^\varepsilon = (v_i^\varepsilon) \in W^{1,4}(\Omega^\varepsilon): \int_{\Gamma_0^\varepsilon \cup \Gamma_L^\varepsilon} \boldsymbol{v}^\varepsilon \, d\omega^\varepsilon = \int_{\Gamma_0^\varepsilon \cup \Gamma_L^\varepsilon} \boldsymbol{x}^\varepsilon \wedge \boldsymbol{v}^\varepsilon \, d\omega^\varepsilon = \boldsymbol{0}\},$$

$$\boldsymbol{\Sigma}^\varepsilon \in \boldsymbol{L}_s^2(\Omega^\varepsilon) = \{\boldsymbol{\tau}^\varepsilon = (\tau_{ij}^\varepsilon) \in L^2(\Omega^\varepsilon): \tau_{ij}^\varepsilon = \tau_{ji}^\varepsilon\},$$

$$\int_{\Omega^\varepsilon} \left(\sigma_{ij}^\varepsilon \partial_j^\varepsilon v_i^\varepsilon + \sigma_{kj}^\varepsilon \partial_k^\varepsilon u_i^\varepsilon \partial_j^\varepsilon v_i^\varepsilon\right) d\boldsymbol{x}^\varepsilon = \int_{\Omega^\varepsilon} f_i^\varepsilon v_i^\varepsilon \, d\boldsymbol{x}^\varepsilon + \int_{\Gamma^\varepsilon} g_i^\varepsilon v_i^\varepsilon \, da^\varepsilon, \quad \text{for all } \boldsymbol{v}^\varepsilon \in V(\Omega^\varepsilon), \tag{45.3}$$

$$\int_{\Omega^\varepsilon} \left(\frac{1}{2\mu^\varepsilon}\sigma_{ij}^\varepsilon - \frac{\lambda^\varepsilon}{2\mu^\varepsilon(3\lambda^\varepsilon + 2\mu^\varepsilon)}\sigma_{pp}^\varepsilon \delta_{ij}\right) \tau_{ij}^\varepsilon \, d\boldsymbol{x}^\varepsilon - \int_{\Omega^\varepsilon} \tfrac{1}{2}(\partial_i^\varepsilon u_j^\varepsilon + \partial_j^\varepsilon u_i^\varepsilon + \partial_i^\varepsilon u_k^\varepsilon \partial_j^\varepsilon u_k^\varepsilon)\tau_{ij}^\varepsilon \, d\boldsymbol{x}^\varepsilon = 0, \quad \text{for all } \boldsymbol{\tau}^\varepsilon \in \boldsymbol{L}_s^2(\Omega^\varepsilon), \tag{45.4}$$

where the constitutive equation (45.2) is written in a variational form. As in Section 41, with the displacement vector field $\boldsymbol{u}^\varepsilon = (u_i^\varepsilon) : \overline{\Omega}^\varepsilon \to \mathbb{R}^3$, we associate the *scaled displacement vector field* $\boldsymbol{u}(\varepsilon) = (u_i(\varepsilon)) : \overline{\Omega} \to \mathbb{R}^3$, defined by the *scalings*

$$\begin{aligned} u_\alpha^\varepsilon(\boldsymbol{x}^\varepsilon) &= \varepsilon u_\alpha(\varepsilon)(\boldsymbol{x}), \\ u_3^\varepsilon(\boldsymbol{x}^\varepsilon) &= \varepsilon^2 u_3(\varepsilon)(\boldsymbol{x}), \end{aligned} \quad \text{for all } \boldsymbol{x}^\varepsilon = \Pi^\varepsilon \boldsymbol{x} \in \overline{\Omega}^\varepsilon. \tag{45.5}$$

Likewise, we associate with any function $\boldsymbol{v}^\varepsilon = (v_i^\varepsilon) \in V(\Omega^\varepsilon)$ the *scaled function* $\boldsymbol{v} = (v_i) : \overline{\Omega} \to \mathbb{R}^3$ defined by the *scalings*

$$\begin{aligned} v_\alpha^\varepsilon(\boldsymbol{x}^\varepsilon) &= \varepsilon v_\alpha(\boldsymbol{x}), \\ v_3^\varepsilon(\boldsymbol{x}^\varepsilon) &= \varepsilon^2 v_3(\boldsymbol{x}), \end{aligned} \quad \text{for all } \boldsymbol{x}^\varepsilon = \Pi^\varepsilon \boldsymbol{x} \in \overline{\Omega}^\varepsilon. \tag{45.6}$$

In addition, with the stress tensor field $\boldsymbol{\Sigma}^\varepsilon = (\sigma_{ij}^\varepsilon) : \overline{\Omega}^\varepsilon \to \boldsymbol{S}^3$, we associate the *scaled stress field* $\boldsymbol{\Sigma}(\varepsilon) = (\sigma_{ij}(\varepsilon)) : \overline{\Omega} \to \boldsymbol{S}^3$, defined by the scalings

$$\begin{aligned} \sigma_{\alpha\beta}^\varepsilon(\boldsymbol{x}^\varepsilon) &= \varepsilon^4 \sigma_{\alpha\beta}(\varepsilon)(\boldsymbol{x}), \\ \sigma_{3\beta}^\varepsilon(\boldsymbol{x}^\varepsilon) &= \varepsilon^3 \sigma_{3\beta}(\varepsilon)(\boldsymbol{x}), \quad \text{for all } \boldsymbol{x}^\varepsilon = \Pi^\varepsilon \boldsymbol{x} \in \overline{\Omega}^\varepsilon, \\ \sigma_{33}^\varepsilon(\boldsymbol{x}^\varepsilon) &= \varepsilon^2 \sigma_{33}(\varepsilon)(\boldsymbol{x}), \end{aligned} \tag{45.7}$$

and we call *scaled stresses* the functions $\sigma_{ij}(\varepsilon) : \Omega \to \mathbb{R}$. Moreover, we associate with any function $\boldsymbol{\tau}^\varepsilon = (\tau_{ij}^\varepsilon) \in \boldsymbol{L}_s^2(\Omega^\varepsilon)$ the scaled function $\boldsymbol{\tau} = (\tau_{ij}) : \overline{\Omega} \to \mathbb{R}^3$ defined by the *scalings*

$$\begin{aligned} \tau_{\alpha\beta}^\varepsilon(\boldsymbol{x}^\varepsilon) &= \varepsilon^4 \tau_{\alpha\beta}(\varepsilon)(\boldsymbol{x}), \\ \tau_{3\beta}^\varepsilon(\boldsymbol{x}^\varepsilon) &= \varepsilon^3 \tau_{3\beta}(\varepsilon)(\boldsymbol{x}), \quad \text{for all } \boldsymbol{x}^\varepsilon = \Pi^\varepsilon \boldsymbol{x} \in \overline{\Omega}^\varepsilon. \\ \tau_{33}^\varepsilon(\boldsymbol{x}^\varepsilon) &= \varepsilon^2 \tau_{33}(\varepsilon)(\boldsymbol{x}), \end{aligned} \tag{45.8}$$

Finally, we make the same assumptions on the data as in Section 41.

Combining the scalings on the displacements and on the stresses with the assumptions on the data, we then reformulate problem (45.3), (45.4) as a problem posed over the set $\overline{\Omega}$, which we call the *scaled three-dimensional equation of a weakly clamped beam in the displacement–stress approach*. As shown in the following result this will consist of a *scaled principle of virtual work* and of a *scaled inverted constitutive equation written in the variational form*.

THEOREM 45.1. *Assume that* $\boldsymbol{u}^\varepsilon \in \boldsymbol{W}^{1,4}(\Omega^\varepsilon)$. *Then, the scaled displacement* $\boldsymbol{u}(\varepsilon) = (u_i(\varepsilon))$ *and the scaled stress* $\boldsymbol{\Sigma}(\varepsilon) = (\sigma_{ij}(\varepsilon))$ *solve the following problem*:

$$\boldsymbol{u}(\varepsilon) \in V(\Omega = \{\boldsymbol{v} = (v_i) \in \boldsymbol{W}^{1,4}(\Omega) \colon \int_{\Gamma_0 \cup \Gamma_L} \boldsymbol{v}^\varepsilon \, \mathrm{d}\omega = \int_{\Gamma_0 \cup \Gamma_L} \boldsymbol{x} \wedge \boldsymbol{v} \, \mathrm{d}\omega = \boldsymbol{0}\},$$

$$\boldsymbol{\Sigma}(\varepsilon) \in \boldsymbol{L}_s^2(\Omega) = \{\boldsymbol{\tau} = (\tau_{ij}) \in L^2(\Omega):\ \tau_{ij} = \tau_{ji}\},$$

$$\int_\Omega (\sigma_{ij}(\varepsilon)\partial_j v_i + \sigma_{ij}\partial_i u_\alpha \partial_j v_\alpha)\,\mathrm{d}\boldsymbol{x} + \varepsilon^2 \int_\Omega (\sigma_{ij}(\varepsilon)\partial_j u_3(\varepsilon)\partial_j v_3\,\mathrm{d}\boldsymbol{x}$$
$$= \int_\Omega f_i v_i\,\mathrm{d}\boldsymbol{x} + \int_\Gamma g_i v_i\,\mathrm{d}a, \quad \textit{for all } \boldsymbol{v} \in V(\Omega), \tag{45.9}$$

$$\int_\Omega \left[\frac{1}{2\mu}\varepsilon^4\sigma_{\alpha\beta}(\varepsilon) - \frac{\lambda}{2\mu(3\lambda+2\mu)}(\varepsilon^4\sigma_{\alpha\alpha}(\varepsilon) + \varepsilon^3\sigma_{33}(\varepsilon)\delta_{\alpha\beta}\right]\tau_{\alpha\beta}\,\mathrm{d}\boldsymbol{x}$$
$$+ \int_\Omega \frac{1}{\mu}\varepsilon^2\sigma_{3\alpha}(\varepsilon)\tau_{3\alpha}\,\mathrm{d}\boldsymbol{x}$$
$$+ \int_\Omega \left[\frac{1}{2\mu}\sigma_{33}(\varepsilon) - \frac{\lambda}{2\mu(3\lambda+2\mu)}(\varepsilon^2\sigma_{\alpha\alpha}(\varepsilon) + \sigma_{33}(\varepsilon))\right]\tau_{33}\,\mathrm{d}\boldsymbol{x}$$
$$- \int_\Omega [E^0_{\alpha\beta}(\boldsymbol{u}(\varepsilon)) + \varepsilon^2 E^2_{\alpha\beta}(\boldsymbol{u}(\varepsilon))]\tau_{\alpha\beta}\,\mathrm{d}\boldsymbol{x}$$
$$- \int_\Omega 2[E^0_{3\beta}(\boldsymbol{u}(\varepsilon)) + \varepsilon^2 E^2_{3\beta}(\boldsymbol{u}(\varepsilon))]\tau_{3\beta}\,\mathrm{d}\boldsymbol{x}$$
$$- \int_\Omega [E^0_{33}(\boldsymbol{u}(\varepsilon)) + \varepsilon^2 E^2_{33}(\boldsymbol{u}(\varepsilon))]\tau_{33}\,\mathrm{d}\boldsymbol{x} = 0, \quad \textit{for all } \boldsymbol{\tau}^\varepsilon \in \boldsymbol{L}_s^2(\Omega^\varepsilon), \tag{45.10}$$

where

$$\begin{aligned} 2E^0_{ij}(\boldsymbol{u}(\varepsilon)) &= \partial_i u_j(\varepsilon) + \partial_j u_i(\varepsilon) + \partial_i u_\gamma(\varepsilon)\partial_i u_\gamma(\varepsilon), \\ 2E^2_{ij}(\boldsymbol{u}(\varepsilon)) &= \partial_i u_3(\varepsilon)\partial_j u_3(\varepsilon). \end{aligned} \tag{45.11}$$

Alternatively, we could also have used the *scaled inverted constitutive equation* defined by

$$\begin{aligned} &E^0_{\alpha\beta}(\boldsymbol{u}(\varepsilon)) + \varepsilon^2 E^2_{\alpha\beta}(\boldsymbol{u}(\varepsilon)) \\ &\quad = -\frac{\lambda}{2\mu(3\lambda+2\mu)}(\varepsilon^4\sigma_{\alpha\beta}(\varepsilon) + \varepsilon^2\sigma_{33}(\varepsilon))\delta_{\alpha\beta} + \frac{1}{2\mu}\varepsilon^4\sigma_{\alpha\beta}(\varepsilon), \\ &E^0_{3\beta}(\boldsymbol{u}(\varepsilon)) + \varepsilon^2 E^2_{3\beta}(\boldsymbol{u}(\varepsilon)) = \frac{1}{2\mu}\varepsilon^2\sigma_{3\beta}(\varepsilon), \\ &E^0_{33}(\boldsymbol{u}(\varepsilon)) + \varepsilon^2 E^2_{33}(\boldsymbol{u}(\varepsilon)) \\ &\quad = -\frac{\lambda}{2\mu(3\lambda+2\mu)}(\varepsilon^2\sigma_{\alpha\beta}(\varepsilon) + \varepsilon^2\sigma_{33}(\varepsilon))\delta_{\alpha\beta} + \frac{1}{2\mu}\sigma_{33}(\varepsilon). \end{aligned} \tag{45.12}$$

As in the displacement approach, the dependence on parameter ε in now explicit and "polynomial": more specifically, problem (45.9)–(45.11) reads

$$a_0(\boldsymbol{\Sigma}(\varepsilon), \boldsymbol{\tau}) + \varepsilon^2 a_2(\boldsymbol{\Sigma}(\varepsilon), \boldsymbol{\tau}) + \varepsilon^4 a_4(\boldsymbol{\Sigma}(\varepsilon), \boldsymbol{\tau}) + b(\boldsymbol{\tau}, \boldsymbol{u}(\varepsilon))$$

$$+ C_0(\boldsymbol{\tau}, \boldsymbol{u}(\varepsilon), \boldsymbol{u}(\varepsilon)) + \varepsilon^2 C_2(\boldsymbol{\tau}, \boldsymbol{u}(\varepsilon), \boldsymbol{u}(\varepsilon)) = 0, \quad \text{for all } \boldsymbol{\tau} \in \boldsymbol{L}_s^2(\Omega), \tag{45.13}$$

$$b(\boldsymbol{\Sigma}(\varepsilon), \boldsymbol{v}) + 2C_0(\boldsymbol{\Sigma}(\varepsilon), \boldsymbol{u}(\varepsilon), \boldsymbol{v}) + 2C_2(\boldsymbol{\Sigma}(\varepsilon), \boldsymbol{u}(\varepsilon), \boldsymbol{v}) = F(\boldsymbol{v}), \quad \text{for all } \boldsymbol{v} \in V(\Omega), \tag{45.14}$$

where the linear form $F(\cdot)$, the bilinear forms $a_0(\cdot,\cdot)$, $a_2(\cdot,\cdot)$, $a_4(\cdot,\cdot)$, $b(\cdot,\cdot)$, and the trilinear forms $c_0(\cdot,\cdot,\cdot)$ and $c_2(\cdot,\cdot,\cdot)$ are defined by

$$F(\boldsymbol{v}) = -\int_\Omega f_i v_i \,\mathrm{d}\boldsymbol{x} - \int_\Gamma g_i v_i \,\mathrm{d}a, \quad \text{for all } \boldsymbol{v} \in V(\Omega), \tag{45.15}$$

$$a_0(\boldsymbol{\sigma}, \boldsymbol{\tau}) = \int_\Omega \frac{1}{E} \sigma_{33}\tau_{33} \,\mathrm{d}\boldsymbol{x}, \quad \text{for all } \boldsymbol{\sigma}, \boldsymbol{\tau} \in \boldsymbol{L}_s^2(\Omega), \tag{45.16}$$

$$a_2(\boldsymbol{\sigma}, \boldsymbol{\tau}) = \int_\Omega \left[\frac{2(1+\nu)}{E} \sigma_{3\beta}\tau_{3\beta} - \frac{\nu}{E}(\sigma_{\mu\mu}\tau_{33} + \sigma_{33}\tau_{\mu\mu}) \right] \mathrm{d}\boldsymbol{x},$$
$$\text{for all } \boldsymbol{\sigma}, \boldsymbol{\tau} \in \boldsymbol{L}_s^2(\Omega), \tag{45.17}$$

$$a_4(\boldsymbol{\sigma}, \boldsymbol{\tau}) = \int_\Omega \left(\frac{1+\nu}{E} \sigma_{\alpha\beta} - \frac{\nu}{E} \sigma_{\mu\mu}\delta_{\alpha\beta} \right) \tau_{\alpha\beta} \,\mathrm{d}\boldsymbol{x}, \quad \text{for all } \boldsymbol{\sigma}, \boldsymbol{\tau} \in \boldsymbol{L}_s^2(\Omega), \tag{45.18}$$

$$b(\boldsymbol{\tau}, \boldsymbol{v}) = -\int_\Omega \tau_{ij}\partial_i v_j \,\mathrm{d}\boldsymbol{x}, \quad \text{for all } \boldsymbol{\tau} \in \boldsymbol{L}_s^2(\Omega), \text{ and all } \boldsymbol{v} \in V(\Omega), \tag{45.19}$$

$$c_0(\boldsymbol{\tau}, \boldsymbol{u}, \boldsymbol{v}) = -\int_\Omega \tfrac{1}{2}\partial_i u_\alpha \partial_j v_\alpha \tau_{ij} \,\mathrm{d}\boldsymbol{x},$$
$$\text{for all } \boldsymbol{\tau} \in \boldsymbol{L}_s^2(\Omega), \text{ and all } \boldsymbol{u}, \boldsymbol{v} \in V(\Omega), \tag{45.20}$$

$$c_2(\boldsymbol{\tau}, \boldsymbol{u}, \boldsymbol{v}) = -\int_\Omega \tfrac{1}{2}\partial_i u_3 \partial_j v_3 \tau_{ij} \,\mathrm{d}\boldsymbol{x},$$
$$\text{for all } \boldsymbol{\tau} \in \boldsymbol{L}_s^2(\Omega), \text{ and all } \boldsymbol{u}, \boldsymbol{v} \in V(\Omega). \tag{45.21}$$

Alternatively, we could also have written the inverted constitutive equation in the form

$$\boldsymbol{E}^0(\boldsymbol{u}(\varepsilon)) + \varepsilon^2 \boldsymbol{E}^2(\boldsymbol{u}(\varepsilon)) = (\boldsymbol{B}^0 + \varepsilon^2\boldsymbol{B}^2 + \varepsilon^4\boldsymbol{B}^4)\boldsymbol{\Sigma}(\varepsilon), \tag{45.22}$$

where the tensor valued mappings $\boldsymbol{E}^0$, $\boldsymbol{E}^2$ and the fourth order tensors $\boldsymbol{B}^0$, $\boldsymbol{B}^2$, $\boldsymbol{B}^4$, are all independent of ε, where

$$\boldsymbol{E}^0(\boldsymbol{u}(\varepsilon)) = E_{ij}^0(\boldsymbol{u}(\varepsilon)), \qquad \boldsymbol{E}^2(\boldsymbol{u}(\varepsilon)) = E_{ij}^2(\boldsymbol{u}(\varepsilon)), \tag{45.23}$$

$$\boldsymbol{B}^0\boldsymbol{\Sigma} = \begin{bmatrix} 0 & 0 & 0 \\ 0 & 0 & 0 \\ 0 & 0 & \frac{1}{E}\sigma_{33} \end{bmatrix}, \tag{45.24}$$

$$\boldsymbol{B}^2\boldsymbol{\Sigma} = \begin{bmatrix} -\frac{\nu}{E}\sigma_{33} & 0 & \frac{1+\nu}{E}\sigma_{13} \\ 0 & -\frac{\nu}{E}\sigma_{33} & \frac{1+\nu}{E}\sigma_{23} \\ \frac{1+\nu}{E}\sigma_{31} & \frac{1+\nu}{E}\sigma_{32} & -\frac{\nu}{E}\sigma_{33} \end{bmatrix}, \tag{45.25}$$

$$\boldsymbol{B}^4\boldsymbol{\Sigma} = \begin{bmatrix} \frac{1+\nu}{E}\sigma_{11} - \frac{\nu}{E}(\sigma_{11}+\sigma_{22}) & \frac{1+\nu}{E}\sigma_{12} & 0 \\ \frac{1+\nu}{E}\sigma_{21} & \frac{1+\nu}{E}\sigma_{22} - \frac{\nu}{E}(\sigma_{11}+\sigma_{22}) & 0 \\ 0 & 0 & 0 \end{bmatrix} \tag{45.26}$$

for any arbitrary matrix $\boldsymbol{\Sigma} = (\sigma_{ij})$.

The polynomial dependence of these relations with respect to parameter ε again leads us to apply the method of asymptotic expansions in the following way:

(i) Write $\boldsymbol{u}(\varepsilon)$ and $\boldsymbol{\Sigma}(\varepsilon)$ *as formal expansions* a priori:

$$\begin{aligned} \boldsymbol{u}(\varepsilon) &= \boldsymbol{u}^0 + \varepsilon\boldsymbol{u}^1 + \varepsilon^2\boldsymbol{u}^2 + \text{h.o.t.}, \\ \boldsymbol{\Sigma}(\varepsilon) &= \boldsymbol{\Sigma}^0 + \varepsilon\boldsymbol{\Sigma}^1 + \varepsilon^2\boldsymbol{\Sigma}^2 + \text{h.o.t.} \end{aligned} \tag{45.27}$$

(ii) Equate to zero the factor of the successive powers of ε, arranged by increasing values of the exponents, found in the scaled principle of virtual work and in the (variational) scaled inverted constitutive equations (45.13), (45.14), when both $\boldsymbol{u}(\varepsilon)$ and $\boldsymbol{\Sigma}(\varepsilon)$ are replaced by their formal expansions.

(iii) Assuming ad hoc properties on the successive terms found in both formal expansions, pursue this procedure until the physically meaningful terms are fully identified.

While all terms $\boldsymbol{u}^0, \boldsymbol{u}^1, \boldsymbol{u}^2, \boldsymbol{u}^3, \boldsymbol{u}^4$, in the formal expansion of $\boldsymbol{u}^0$ where needed in order that $\boldsymbol{u}^0$ could be identified, the identification of the leading term, in the displacement–stress approach, $\boldsymbol{u}^0$, $\boldsymbol{\Sigma}^0$ requires fewer terms in the formal expansions of $\boldsymbol{u}(\varepsilon)$ and of $\boldsymbol{\Sigma}(\varepsilon)$, as we show in the following result.

THEOREM 45.2. *Assume that the scaled displacement and stress fields can be written as*

$$\begin{aligned} \boldsymbol{u}(\varepsilon) &= \boldsymbol{u}^0 + \varepsilon\boldsymbol{u}^1 + \varepsilon^2\boldsymbol{u}^2 + h.o.t., \\ \boldsymbol{\Sigma}(\varepsilon) &= \boldsymbol{\Sigma}^0 + \varepsilon\boldsymbol{\Sigma}^1 + \varepsilon^2\boldsymbol{\Sigma}^2 + h.o.t., \end{aligned}$$

and that the leading terms of the formal expansions satisfy

$$\begin{aligned} &\boldsymbol{u}^0 = (u_i^0) \in V(\Omega), \qquad \boldsymbol{u}^1 = (u_i^1) \in V(\Omega), \qquad \boldsymbol{u}^2 = (u_i^2) \in V(\Omega), \\ &\partial_\alpha u_\beta^0 \in C^0(\overline{\Omega}), \\ &\boldsymbol{\Sigma}^0 = (\sigma_{ij}^0) \in \boldsymbol{L}_s^2(\Omega), \qquad \boldsymbol{\Sigma}^1 = (\sigma_{ij}^1) \in \boldsymbol{L}_s^2(\Omega), \qquad \boldsymbol{\Sigma}^2 = (\sigma_{ij}^2) \in \boldsymbol{L}_s^2(\Omega), \end{aligned} \tag{45.28}$$

where

$$V(\Omega) = \{\boldsymbol{v} = (v_i) \in W^{1,4}(\Omega) \colon \int_{\Gamma_0\cup\Gamma_L} \boldsymbol{v}^\varepsilon \,\mathrm{d}\omega = \int_{\Gamma_0\cup\Gamma_L} \boldsymbol{x}\wedge\boldsymbol{v}\,\mathrm{d}\omega = \boldsymbol{0}\},$$

$$\boldsymbol{\Sigma}(\varepsilon) \in \boldsymbol{L}_s^2(\Omega) = \{\boldsymbol{\tau} = (\tau_{ij}) \in L^2(\Omega) \colon \tau_{ij} = \tau_{ji}\}.$$

Then, the cancellation of the factor ε^0 in the scaled principle of virtual work (45.14), *and in the variational scaled inverted constitutive equation* (45.13), *implies that the leading terms $\boldsymbol{u}^0$, $\boldsymbol{\Sigma}^0$ solve problem* (44.2)–(44.11), *that is*

$$\begin{aligned}
&\boldsymbol{u}^0 = (u_i^0) \in V_{\mathrm{BN}}(\Omega) \\
&\quad = \{\boldsymbol{v} = (v_i)\colon v_\alpha = z_\alpha,\ v_3 = z_3 - x_\alpha \partial_3 z_\alpha, \text{ with } \\
&\qquad z_\alpha \in H_0^2(0,L),\ z_3 \in H_0^1(0,L)\}, \\
&\int_\Omega \sigma_{33}^0 \partial_3 v_3 \,\mathrm{d}\boldsymbol{x} + \int_\Omega \sigma_{33}^0 \partial_3 u_\gamma^0 \partial_3 v_\gamma \,\mathrm{d}\boldsymbol{x} \\
&\quad = \int_\Omega f_i v_i \,\mathrm{d}\boldsymbol{x} + \int_\Gamma g_i v_i \,\mathrm{d}a, \quad \text{for all } \boldsymbol{v} \in V_{\mathrm{BN}}(\Omega),
\end{aligned} \tag{45.29}$$

where σ_{33}^0 is given by

$$\begin{aligned}
\sigma_{33}^0 &= \mu \frac{3\lambda + 2\mu}{\lambda + \mu} \left[\partial_3 \underline{u}_3^0 - x_\alpha \partial_{33} u_\alpha^0 + \Lambda_{\alpha\alpha}\right] \\
&= E[\partial_3 \underline{u}_3^0 - x_\alpha \partial_{33} u_\alpha^0 + \Lambda_{\alpha\alpha}],
\end{aligned} \tag{45.30}$$

and where matrix $\boldsymbol{\Lambda} = (\Lambda_{\alpha\beta})$ is given by

$$\boldsymbol{\Lambda} = (\Lambda_{\alpha\beta}) = \frac{1}{2} \begin{pmatrix} \partial_3 u_1^0 \partial_3 u_1^0 & \partial_3 u_1^0 \partial_3 u_2^0 \\ \partial_3 u_2^0 \partial_3 u_1^0 & \partial_3 u_2^0 \partial_3 u_2^0 \end{pmatrix}. \tag{45.31}$$

Or, more specifically, any function $\boldsymbol{u}^0$ solves this problem if and only if it is also a solution of

$$\begin{aligned}
u_\alpha^0 &= u_\alpha^0(x_3), \\
u_3^0 &= \underline{u}_3^0 - x_\alpha \partial_3 u_\alpha^0(x_3),
\end{aligned} \tag{45.32}$$

with $u_\alpha^0 \in H_0^2(0,L)$ and $\underline{u}_3^0 \in H_0^1(0,L)$, and u_1^0 solve the following variational problem (no sum on β):

$$\begin{aligned}
&\int_0^L E I_\beta \partial_{33} u_\beta^0 \partial_{33} v_\beta \,\mathrm{d}x_3 + \int_0^L E(\partial_3 \underline{u}_3^0 + \Lambda_{\gamma\gamma}) \partial_3 u_\beta^0 \partial_3 v_\beta \,\mathrm{d}x_3 \\
&\quad = \int_0^L \left[\int_\omega f_\beta \,\mathrm{d}\omega + \int_\gamma g_\beta \,\mathrm{d}\gamma\right] v_\beta \,\mathrm{d}x_3 \\
&\qquad - \int_0^L \left[\int_\omega x_\beta f_3 \,\mathrm{d}\omega + \int_\gamma x_\beta g_3 \,\mathrm{d}\gamma\right] \partial_3 v_\beta \,\mathrm{d}x_3, \quad \text{for all } v_\beta \in H_0^2(0,L),
\end{aligned} \tag{45.33}$$

$$\int_0^L E(\partial_3 \underline{u}_3^0 + \Lambda_{\gamma\gamma})\partial_3 v_3 \,\mathrm{d}x_3$$
$$= \int_0^L \left[\int_\omega f_3 \,\mathrm{d}\omega + \int_\gamma g_3 \,\mathrm{d}\gamma \right] v_3 \,\mathrm{d}x_3, \quad \textit{for all } v_3 \in H_0^1(0, L), \tag{45.34}$$

or equivalently

$$\int_0^L m_\beta^0 \partial_{33} v_\beta \,\mathrm{d}x_3 + \int_0^L n_{33}^0 \partial_3 u_\beta^0 \partial_3 v_\beta \,\mathrm{d}x_3$$
$$= \int_0^L \left[\int_\omega f_\beta \,\mathrm{d}\omega + \int_\gamma g_\beta \,\mathrm{d}\gamma \right] v_\beta \,\mathrm{d}x_3$$
$$- \int_0^L \left[\int_\omega x_\beta f_3 \,\mathrm{d}\omega + \int_\gamma x_\beta g_3 \,\mathrm{d}\gamma \right] \partial_3 v_\beta \,\mathrm{d}x_3,$$
$$\textit{for all } v_\beta \in H_0^2(0, L), \tag{45.35}$$

$$\int_0^L n_{33}^0 \partial_3 v_3 \,\mathrm{d}x_3 = \int_0^L \left[\int_\omega f_3 \,\mathrm{d}\omega + \int_\gamma g_3 \,\mathrm{d}\gamma \right] v_3 \,\mathrm{d}x_3,$$
$$\textit{for all } v_3 \in H_0^1(0, L). \tag{45.36}$$

where the zeroth order bending moment components m_β^0 *are*

$$m_\beta^0 = \int_\omega x_\beta \sigma_{33}^0 \,\mathrm{d}\omega = -EI_\beta \partial_{33} u_\beta^0, \quad \textit{no sum on } \beta, \tag{45.37}$$

and the zeroth order axial force component n_{33}^0 *is given by*

$$n_{33}^0 = \int_\omega \sigma_{33}^0 \,\mathrm{d}\omega = E[\partial_3 \underline{u}_3^0 + \Lambda_{\alpha\alpha}]. \tag{45.38}$$

PROOF. Cancelling the factor of ε^0 in (45.14) immediately gives

$$b(\boldsymbol{\sigma}^0, \boldsymbol{v}) + 2c_0(\boldsymbol{\sigma}^0, \boldsymbol{u}^0, \boldsymbol{v}) = F(\boldsymbol{v}), \quad \text{for all } \boldsymbol{v} \in V(\Omega), \tag{45.39}$$

which by definition of the linear, bilinear and trilinear forms gives

$$\int_\Omega \sigma_{ij}^0 \partial_j v_i \,\mathrm{d}\boldsymbol{x} + \int_\Omega \sigma_{ij}^0 \partial_i u_\gamma^0 \partial_j v_\gamma^0 \,\mathrm{d}\boldsymbol{x} = \int_\Omega f_i v_i \,\mathrm{d}\boldsymbol{x} + \int_\Gamma g_i v_i \,\mathrm{d}a,$$
$$\text{for all } \boldsymbol{v} \in V(\Omega). \tag{45.40}$$

Cancelling the factor of ε^0 in (45.13) gives

$$a_0(\boldsymbol{\sigma}^0, \boldsymbol{\tau}) + b(\boldsymbol{\tau}, \boldsymbol{u}^0) + c_0(\boldsymbol{\tau}, \boldsymbol{u}^0, \boldsymbol{u}^0) = 0, \quad \text{for all } \boldsymbol{\tau} \in \boldsymbol{L}^2_s(\Omega), \tag{45.41}$$

or equivalently

$$\int_\Omega \frac{1}{E}\sigma^0_{33}\tau_{33}\,\mathrm{d}\boldsymbol{x} - \int_\Omega \partial_3 u^0_3 \tau_{33}\,\mathrm{d}\boldsymbol{x} - \int_\Omega \frac{1}{2}\partial_3 u^0_\gamma \partial_3 u^0_\gamma \tau_{33}\,\mathrm{d}\boldsymbol{x} = 0, \quad \text{for all } \tau_{33} \in L^2(\Omega), \tag{45.42}$$

$$- \int_\Omega (\partial_3 u^0_\beta + \partial_\beta u^0_3)\tau_{3\beta}\,\mathrm{d}\boldsymbol{x} - \int_\Omega \partial_3 u^0_\gamma \partial_\beta u^0_\gamma \tau_{3\beta}\,\mathrm{d}\boldsymbol{x} = 0, \quad \text{for all } (\tau_{3\beta} = \tau_{\beta 3}) \in [L^2(\Omega)]^2, \tag{45.43}$$

$$- \int_\Omega (\partial_\alpha u^0_\beta + \partial_\beta u^0_\alpha)\tau_{\alpha\beta}\,\mathrm{d}\boldsymbol{x} - \int_\Omega \partial_\alpha u^0_\gamma \partial_\beta u^0_\gamma \tau_{\alpha\beta}\,\mathrm{d}\boldsymbol{x} = 0, \quad \text{for all } (\tau_{\alpha\beta} = \tau_{\beta\alpha}) \in [L^2(\Omega)]^4. \tag{45.44}$$

From Eq. (45.44) we get $2E^0_{\alpha\beta}(\boldsymbol{u}^0) = \partial_\alpha u^0_\beta + \partial_\beta u^0_\alpha + \partial_\alpha u^0_\gamma \partial_\beta u^0_\gamma = 0$. Thus u^0_α must be a planar rigid body displacement in the cross section $\omega(x_3) \times \{x_3\}$, that is there exist functions $\theta(x_3)$ and $z^0_\alpha(x_3)$ belonging to $W^{1,4}_0(0, L)$ and such that

$$u^0_\alpha(x_1, x_2, x_3) = [Q(\theta(x_3)) - \mathrm{Id}]\begin{Bmatrix} x_1 \\ x_2 \end{Bmatrix} + z^0_\alpha(x_3),$$

where $Q(\theta(x_3))$ is the rotation matrix

$$Q(\theta(x_3)) = \begin{bmatrix} \cos(\theta(x_3)) & -\sin(\theta(x_3)) \\ \sin(\theta(x_3)) & \cos(\theta(x_3)) \end{bmatrix}, \quad \text{for all } x_3 \in [0, L].$$

These equations are exactly (42.17), (42.18) and now the proof follows as in Section 42. □

We remark that the fact that we used a weaker boundary condition does not change the type of boundary conditions associated with the zeroth order term.

The *displacement–stress* approach looks simpler than the *displacement approach*, since Eqs. (45.29) and (45.30) which characterize the zeroth order term are obtained at a lower cost than in Theorem 42.1: it now suffices to cancel the factors of ε^0 in the variational scaled inverted constitutive equation (45.13) and in the scaled principle of virtual work (45.14), while all the factors of ε^q,

$-4 \leqslant q \leqslant 0$ have to be cancelled in problem (41.8), (41.9) in order for the same equations to be obtained in the displacement approach (cf. Theorem 42.1). However, as in the linearized case, this observation must be corrected in two ways:

(i) It must first be shown that all the factors of ε^q, $-4 \leqslant q \leqslant -1$ in the formal expansion of the scaled stress tensor $\boldsymbol{\Sigma}(\varepsilon)$ do indeed vanish. As shown in Theorem 42.1, this is a consequence of the assumption that the scaled displacement vector $\boldsymbol{u}(\varepsilon)$ can be physically expanded as $\varepsilon^0\boldsymbol{u}^0 + \varepsilon^1\boldsymbol{u}^1 + \varepsilon^2\boldsymbol{u}^2 + \varepsilon^3\boldsymbol{u}^3 + \varepsilon^4\boldsymbol{u}^14 + \text{h.o.t.}$ Hence, the displacement approach provides us this invaluable information.

(ii) Once it is known that the formal expansion of $\boldsymbol{\Sigma}(\varepsilon)$ starts with the zeroth order term, it is clear that it is easier to use the variational inverted scaled constitutive equation (45.13), which only involves factors of ε^q, $q \geqslant 0$. If instead the scaled constitutive equation was used, all factors of ε^q, $-4 \leqslant q \leqslant 0$, would have again to be cancelled in order to provide the same information as in Theorem 42.1.

Because of this simplification we could proceed in the evaluation of the higher order terms using the displacement–stress approach but, of course, the same results would have been obtained had we used the displacement approach. We shall not present these results here because they are extremely long and refer to CAMOTIM and TRABUCHO [1989] where some of them are shown.

CHAPTER X

Asymptotic Method for Nonlinear Elastic Rods. Dynamic Case

The main goal of this chapter is the mathematical justification of the well-known semilinear evolution equation modelling the bending of a cylindrical rod submitted to the effects of an axial loading at both ends. This equation takes the following form (see, for example, BALL, MARSDEN and SLEMROD [1982], HUGHES and MARSDEN [1983]):

$$\rho\partial_{tt}u_\alpha + EI\partial_{3333}u_\alpha + p(t)\partial_{33}u_\alpha = 0,$$

where, in addition to the previous notation, we have

$u_\alpha(x_3, t)$: transverse displacement of the rod in the direction Ox_α,
ρ: mass density of the material of the rod,
$p(t)$: axial load applied on each end.

If we suppose that the rod is clamped at both ends then this equation must be completed by the following boundary and initial conditions:

$$u_\alpha(0,t) = u_\alpha(L,t) = 0,$$
$$\partial_3 u_\alpha(0,t) = \partial_3 u_\alpha(L,t) = 0,$$
$$u_\alpha(x_3, 0) = \overline{u}_0(x_3).$$

In order to justify this model ÁLVAREZ-VÁZQUEZ [1991], ÁLVAREZ-VÁZQUEZ and VIAÑO [1992b] use the asymptotic method on an evolution nonlinear three-dimensional elasticity model just as in Chapter IX. Their results are summarized in this chapter.

We start, in Section 46, by formulating the three-dimensional nonlinear dynamic elasticity problem, for a Saint Venant–Kirchhoff material, subjected to axial forces and with appropriate boundary conditions in order to be able to pass to the limit as the diameter of the cross section goes to zero. The by now standard asymptotic analysis is performed in Section 47 and the first term in the asymptotic expansion is calculated in Section 48. The physical interpretation of the results obtained is done in Section 49 and in the last section the results are extended to more general loading conditions.

46. The three-dimensional equations for the nonlinear elastodynamic case

In addition to the notation used so far we also set $\partial_t u$ the partial derivative $\partial u(t,\boldsymbol{x})/\partial t$ of a function $u(t,\boldsymbol{x})$.

In what follows we shall omit the dependence in t of functions, except in those cases where explicit expression is required.

We try to study the dynamic problem corresponding to the mechanical behaviour during the time interval $[0,T]$ of a nonlinearly elastic beam. We suppose the beam to be "weakly" clamped and submitted to axial loads at both ends, in such a way that only the resultant loading H^ε is known. Consequently, H^ε is considered to be defined at the centroid of each end cross-section of the beam.

We assume the constitutive material of the beam to be an homogeneous isotropic elastic material of Saint Venant–Kirchhoff type with Young's modulus E^ε, Poisson's ratio ν^ε and mass density ρ^ε. Then, the displacement field $\boldsymbol{u}^\varepsilon = (u_i^\varepsilon)$ and the second Piola–Kirchhoff stress tensor $\boldsymbol{\sigma}^\varepsilon = (\sigma_{ij}^\varepsilon)$ are the solution of the following problem (see CIARLET [1980, 1988]):

$$\rho^\varepsilon\ \partial_{tt}u_i^\varepsilon - \partial_j^\varepsilon\ (\sigma_{ij}^\varepsilon + \sigma_{kj}^\varepsilon\ \partial_k^\varepsilon u_i^\varepsilon) = 0 \quad \text{in } \Omega^\varepsilon \times (0,T), \tag{46.1}$$

$$\begin{aligned}(A^\varepsilon\ \boldsymbol{\sigma}^\varepsilon)_{ij} &= \frac{1+\nu^\varepsilon}{E^\varepsilon}\ \sigma_{ij}^\varepsilon - \frac{\nu^\varepsilon}{E^\varepsilon}\ \sigma_{kk}^\varepsilon\ \delta_{ij} \\ &= \tfrac{1}{2}(\partial_i^\varepsilon u_j^\varepsilon + \partial_j^\varepsilon u_i^\varepsilon + \partial_i^\varepsilon u_k^\varepsilon\ \partial_j^\varepsilon u_k^\varepsilon) \quad \text{in } \Omega^\varepsilon \times (0,T),\end{aligned} \tag{46.2}$$

with boundary conditions

$$u_\alpha^\varepsilon = 0 \text{ on } (\Gamma_0^\varepsilon \cup \Gamma_L^\varepsilon) \times (0,T), \tag{46.3}$$

$$u_3^\varepsilon \quad \text{independent of } x_\alpha^\varepsilon \text{ on } (\Gamma_0^\varepsilon \cup \Gamma_L^\varepsilon) \times (0,T), \tag{46.4}$$

$$(\sigma_{i\alpha}^\varepsilon + \sigma_{k\alpha}^\varepsilon\ \partial_k^\varepsilon u_i^\varepsilon)\ n_\alpha = 0 \quad \text{on } \Gamma^\varepsilon \times (0,T), \tag{46.5}$$

$$\frac{1}{|\omega^\varepsilon|}\int_{\omega^\varepsilon} (\sigma_{33}^\varepsilon + \sigma_{k3}^\varepsilon\ \partial_k^\varepsilon u_3^\varepsilon)\ n_3 = H^\varepsilon \quad \text{in } \{0,L\} \times (0,T), \tag{46.6}$$

and initial conditions

$$\boldsymbol{u}^\varepsilon(0) = \overline{\boldsymbol{u}}^\varepsilon, \tag{46.7}$$

$$\partial_t \boldsymbol{u}^\varepsilon(0) = \overline{\boldsymbol{v}}^\varepsilon. \tag{46.8}$$

REMARK 46.1. The choice of the boundary conditions (46.3)–(46.6) seems to be essential for the success of the method. In fact, if we consider the boundary condition (46.6) in the way

$$(\sigma_{33}^\varepsilon + \sigma_{k3}^\varepsilon\ \partial_k^\varepsilon u_3^\varepsilon)\ n_3 = \tilde{H}^\varepsilon \quad \text{on } (\Gamma_0^\varepsilon \cup \Gamma_L^\varepsilon) \times (0,T), \tag{46.6'}$$

serious difficulties will appear in later calculations, although it is a perfectly admissible condition for the three-dimensional problem (cf. DUVAUT and LIONS [1972a] for the linearized case). Actually, condition (46.6) is weaker then (46.6′), because it is obtained from the latter just by taking the average in ω^ε, i.e.,

$$H^\varepsilon = \frac{1}{|\omega^\varepsilon|} \int_{\omega^\varepsilon} \tilde{H}^\varepsilon \, \mathrm{d}\omega^\varepsilon,$$

which is considered as a function defined in $\{0, L\} \times (0, T)$.

Following CIARLET [1980] we consider the functional spaces of admissible displacements and stresses

$$\begin{aligned} V(\Omega^\varepsilon) = \{ & \boldsymbol{v}^\varepsilon \in [W^{1,4}(\Omega^\varepsilon)]^3\colon v_\alpha^\varepsilon = 0 \text{ on } (\Gamma_0^\varepsilon \cup \Gamma_L^\varepsilon), \\ & v_3^\varepsilon \text{ independent of } x_\alpha^\varepsilon \text{ on } (\Gamma_0^\varepsilon \cup \Gamma_L^\varepsilon)\}, \end{aligned} \tag{46.9}$$

$$\Sigma(\Omega^\varepsilon) = [L^2(\Omega^\varepsilon)]_s^9 := \{\boldsymbol{\tau}^\varepsilon \in [L^2(\Omega^\varepsilon)]^9\colon \tau_{ij}^\varepsilon = \tau_{ji}^\varepsilon\}. \tag{46.10}$$

Then, we have the following classical result (see LIONS [1969], DUVAUT and LIONS [1972b], CIARLET and RABIER [1980]).

THEOREM 46.1. *If the solution of problem* (46.1)–(46.8) *is regular enough, then* $(\boldsymbol{u}^\varepsilon, \boldsymbol{\sigma}^\varepsilon)$ *is characterized as the solution of the following variational problem* (P^ε):

$$\begin{aligned} & \int_{\Omega^\varepsilon} (A^\varepsilon \boldsymbol{\sigma}^\varepsilon)_{ij} \tau_{ij}^\varepsilon \, \mathrm{d}\boldsymbol{x}^\varepsilon - \int_{\Omega^\varepsilon} e_{ij}^\varepsilon(\boldsymbol{u}^\varepsilon) \, \tau_{ij}^\varepsilon \, \mathrm{d}\boldsymbol{x}^\varepsilon \\ & \quad - \tfrac{1}{2} \int_{\Omega^\varepsilon} \partial_i^\varepsilon u_k^\varepsilon \partial_j^\varepsilon u_k^\varepsilon \, \tau_{ij}^\varepsilon \, \mathrm{d}\boldsymbol{x}^\varepsilon = 0, \quad \textit{for all } \boldsymbol{\tau}^\varepsilon \in \Sigma(\Omega^\varepsilon), \end{aligned} \tag{46.11}$$

$$\begin{aligned} & \langle \rho^\varepsilon \partial_{tt} u_k^\varepsilon, v_k^\varepsilon \rangle_\varepsilon + \int_{\Omega^\varepsilon} \sigma_{ij}^\varepsilon \, e_{ij}^\varepsilon(\boldsymbol{v}^\varepsilon) \, \mathrm{d}\boldsymbol{x}^\varepsilon + \int_{\Omega^\varepsilon} \sigma_{ij}^\varepsilon \, \partial_i^\varepsilon u_k^\varepsilon \partial_j^\varepsilon v_k^\varepsilon \, \mathrm{d}\boldsymbol{x}^\varepsilon \\ & \quad = H^\varepsilon(0) \int_{\omega^\varepsilon \times \{0\}} v_3^\varepsilon \, \mathrm{d}\omega^\varepsilon + H^\varepsilon(L) \int_{\omega^\varepsilon \times \{L\}} v_3^\varepsilon \, \mathrm{d}\omega^\varepsilon, \quad \textit{for all } \boldsymbol{v}^\varepsilon \in V(\Omega^\varepsilon), \end{aligned} \tag{46.12}$$

$$\boldsymbol{u}^\varepsilon(0) = \overline{\boldsymbol{u}}^\varepsilon, \qquad \partial_t \boldsymbol{u}^\varepsilon(0) = \overline{\boldsymbol{v}}^\varepsilon, \tag{46.13}$$

where for an arbitrary vector field $\boldsymbol{v}^\varepsilon = (v_i^\varepsilon)$,

$$e_{ij}^\varepsilon(\boldsymbol{v}^\varepsilon) = \tfrac{1}{2}(\partial_i^\varepsilon v_j^\varepsilon + \partial_j^\varepsilon v_i^\varepsilon),$$

and where $\langle \cdot, \cdot \rangle_\varepsilon$ *stands for the duality on* $W^{1,4}(\Omega^\varepsilon)$.

Very little can be said about existence and regularity of the solution of problem (46.1)–(46.8). Existence results for the nonlinear three-dimensional elastodynamic problem are rather limited. Only for the case of pure displacement problems are existence results for small enough values of $T > 0$ known. These results have been obtained by HUGHES, KATO and MARSDEN [1976] for the case in which the body is all of $\mathbb{R}^3$, and by KATO [1979], DAFERMOS and HRUSA [1985] and others, for domains of $\mathbb{R}^3$.

The common feature of these results is their local character in time, although smooth enough solutions are obtained: displacements are, at least, C^2 in the time variable (cf. HUGHES and MARSDEN [1983, p. 401]).

Global existence results are few, but we must emphasize the ones achieved by GLIMM [1965] and DAFERMOS [1973] for "large" times in the one-dimensional case (see CIARLET [1988] for a more exhaustive bibliography).

47. The asymptotic expansion method for the nonlinear elastodynamic case

As in other cases, in order to study the behaviour of $(\boldsymbol{u}^\varepsilon, \boldsymbol{\sigma}^\varepsilon)$ as ε becomes small we start by a change of variable from Ω^ε to the fixed domain $\Omega = \Omega^1$:

$$\begin{aligned}\Pi^\varepsilon : \boldsymbol{x} = (x_1, x_2, x_3) \in \Omega &\to \boldsymbol{x}^\varepsilon \\ &= \Pi^\varepsilon(\boldsymbol{x}) = (x_1^\varepsilon, x_2^\varepsilon, x_3^\varepsilon) \\ &= (\varepsilon x_1, \varepsilon x_2, x_3) \in \Omega^\varepsilon.\end{aligned}$$

As hypothesis on the data, we assume that there exist ρ, ν, E, independent of ε, such that

$$\rho^\varepsilon = \varepsilon^4 \rho, \quad \rho > 0, \tag{47.1}$$

$$\nu^\varepsilon = \nu, \quad 0 < \nu < \tfrac{1}{2}, \tag{47.2}$$

$$E^\varepsilon = E, \quad E > 0. \tag{47.3}$$

REMARK 47.1. The ε^4 dependence of ρ^ε given by (47.1) is necessary in order to obtain a model sensitive to the inertia effects (see RAOULT [1980] for a complete study in the case of a linearized plate) and it is similar to the dependence assumed in the study of other physical phenomena: for example, flux in porous media (SANCHEZ-PALENCIA [1980]), eigenvalues for linear plates (CIARLET and KESAVAN [1981]).

We now scale the different fields appearing in the variational formulation (46.11)–(46.13). This scaling is the same as for the stationary case (see Section 41). With functions $\boldsymbol{u}^\varepsilon \in \boldsymbol{V}(\Omega^\varepsilon)$ and $\boldsymbol{\sigma}^\varepsilon \in \boldsymbol{\Sigma}(\Omega^\varepsilon)$ we associate the scaled displacement vector $\boldsymbol{u}(\varepsilon) = (u_i(\varepsilon)) \in V(\Omega)$ and the scaled stress tensor

$\boldsymbol{\sigma}(\varepsilon) = (\sigma_{ij}(\varepsilon)) \in \Sigma(\Omega)$ defined by

$$\begin{aligned} &u_\alpha^\varepsilon(\boldsymbol{x}^\varepsilon) = \varepsilon\ u_\alpha(\varepsilon)(\boldsymbol{x}),\\ &u_3^\varepsilon(\boldsymbol{x}^\varepsilon) = \varepsilon^2\ u_3(\varepsilon)(\boldsymbol{x}),\\ &\sigma_{\alpha\beta}^\varepsilon(\boldsymbol{x}^\varepsilon) = \varepsilon^4\ \sigma_{\alpha\beta}(\varepsilon)(\boldsymbol{x}), \quad \text{for all } \boldsymbol{x}^\varepsilon = \pi^\varepsilon,\ \boldsymbol{x} \in \Omega^\varepsilon.\\ &\sigma_{\alpha 3}^\varepsilon(\boldsymbol{x}^\varepsilon) = \varepsilon^3\ \sigma_{\alpha 3}(\varepsilon)(\boldsymbol{x}),\\ &\sigma_{33}^\varepsilon(\boldsymbol{x}^\varepsilon) = \varepsilon^2\ \sigma_{33}(\varepsilon)(\boldsymbol{x}), \end{aligned} \tag{47.4}$$

In a similar way we assume that the axial loading and the initial data are such that there exist functions h, $\overline{\boldsymbol{u}}$, $\overline{\boldsymbol{v}}$, independent of ε, verifying

$$H^\varepsilon(\boldsymbol{x}^\varepsilon) = \varepsilon^2\ h(\boldsymbol{x}) \quad \text{in } \{0, L\} \times (0, T), \tag{47.5}$$

$$\overline{\boldsymbol{u}}_\alpha^\varepsilon(\boldsymbol{x}^\varepsilon) = \varepsilon \overline{\boldsymbol{u}}_\alpha(\boldsymbol{x}), \qquad \overline{\boldsymbol{u}}_3^\varepsilon(\boldsymbol{x}^\varepsilon) = \varepsilon^2 \overline{\boldsymbol{u}}_3(\boldsymbol{x}), \qquad \text{for all } \boldsymbol{x}^\varepsilon = \pi^\varepsilon,\ \boldsymbol{x} \in \Omega^\varepsilon, \tag{47.6}$$

$$\overline{\boldsymbol{v}}_\alpha^\varepsilon(\boldsymbol{x}^\varepsilon) = \varepsilon \overline{\boldsymbol{v}}_\alpha(\boldsymbol{x}), \qquad \overline{\boldsymbol{v}}_3^\varepsilon(\boldsymbol{x}^\varepsilon) = \varepsilon^2 \overline{\boldsymbol{v}}_3(\boldsymbol{x}), \qquad \text{for all } \boldsymbol{x}^\varepsilon = \pi^\varepsilon,\ \boldsymbol{x} \in \Omega^\varepsilon. \tag{47.7}$$

REMARK 47.2. The dependence of the axial force on ε^2 (relation (47.5)) is a consequence of causing the load term in the limit equation to appear (see CIARLET [1980] for analogous conditions in plates).

The scaling of $\overline{\boldsymbol{u}}^\varepsilon$ and $\overline{\boldsymbol{v}}^\varepsilon$ agrees with the scaling (47.4) carried out for the displacement field. It is possible to replace assumptions (47.5)–(47.7) by the more general conditions

$$\begin{aligned} &H^\varepsilon(\boldsymbol{x}^\varepsilon) = \varepsilon^2\ h(\varepsilon)(\boldsymbol{x}) \quad \text{in } \{0, L\} \times (0, T),\\ &\overline{\boldsymbol{u}}_\alpha^\varepsilon(\boldsymbol{x}^\varepsilon) = \varepsilon \overline{\boldsymbol{u}}_\alpha(\varepsilon)(\boldsymbol{x}), \qquad \overline{\boldsymbol{u}}_3^\varepsilon(\boldsymbol{x}^\varepsilon) = \varepsilon^2 \overline{\boldsymbol{u}}_3(\varepsilon)(\boldsymbol{x}), \qquad \text{for all } \boldsymbol{x}^\varepsilon = \pi^\varepsilon,\ \boldsymbol{x} \in \Omega^\varepsilon,\\ &\overline{\boldsymbol{v}}_\alpha^\varepsilon(\boldsymbol{x}^\varepsilon) = \varepsilon \overline{\boldsymbol{v}}_\alpha(\varepsilon)(\boldsymbol{x}), \qquad \overline{\boldsymbol{v}}_3^\varepsilon(\boldsymbol{x}^\varepsilon) = \varepsilon^2 \overline{\boldsymbol{v}}_3(\varepsilon)(\boldsymbol{x}), \qquad \text{for all } \boldsymbol{x}^\varepsilon = \pi^\varepsilon,\ \boldsymbol{x} \in \Omega^\varepsilon, \end{aligned}$$

where we assumed that

$$\begin{aligned} &h(\varepsilon) = h^0 + \varepsilon\ h^1 + \varepsilon^2\ h^2 + \varepsilon^3\ h^3 + \text{h.o.t.},\\ &\overline{\boldsymbol{u}}(\varepsilon) = \overline{\boldsymbol{u}}^0 + \varepsilon\ \overline{\boldsymbol{u}}^1 + \varepsilon^2\ \overline{\boldsymbol{u}}^2 + \varepsilon^3\ \overline{\boldsymbol{u}}^3 + \text{h.o.t.},\\ &\overline{\boldsymbol{v}}(\varepsilon) = \overline{\boldsymbol{v}}^0 + \varepsilon\ \overline{\boldsymbol{v}}^1 + \varepsilon^2\ \overline{\boldsymbol{v}}^2 + \varepsilon^3\ \overline{\boldsymbol{v}}^3 + \text{h.o.t.}, \end{aligned}$$

then, case (47.4)–(47.7) corresponds to having $h^0 = h$, $h^p = 0$, $\overline{\boldsymbol{u}}^0 = \overline{\boldsymbol{u}}$, $\overline{\boldsymbol{u}}^p = 0$, $\overline{\boldsymbol{v}}^0 = \overline{\boldsymbol{v}}$, $\overline{\boldsymbol{v}}^p = 0$, for all $p \geqslant 1$.

Thus, as an immediate consequence of these change of variables, we obtain the following.

THEOREM 47.1. *For a solution $(\boldsymbol{u}^\varepsilon, \boldsymbol{\sigma}^\varepsilon)$ of the variational problem* (46.11)–(46.13), *the scaled displacement–stress field $(\boldsymbol{u}(\varepsilon), \boldsymbol{\sigma}(\varepsilon))$ solves the following problem*

$(P(\varepsilon))$:

$$a_0(\boldsymbol{\sigma}(\varepsilon),\boldsymbol{\tau})+\varepsilon^2\,a_2(\boldsymbol{\sigma}(\varepsilon),\boldsymbol{\tau})+\varepsilon^4\,a_4(\boldsymbol{\sigma}(\varepsilon),\boldsymbol{\tau})+b(\boldsymbol{\tau},\boldsymbol{u}(\varepsilon))$$
$$+\,c_0(\boldsymbol{\tau},\boldsymbol{u}(\varepsilon),\boldsymbol{u}(\varepsilon))+\varepsilon^2\,c_2(\boldsymbol{\tau},\boldsymbol{u}(\varepsilon),\boldsymbol{u}(\varepsilon))=0,\quad \textit{for all } \boldsymbol{\tau}\in\Sigma(\Omega), \tag{47.8}$$
$$b(\boldsymbol{\sigma}(\varepsilon),\boldsymbol{v})+2\,c_0(\boldsymbol{\sigma}(\varepsilon),\boldsymbol{u}(\varepsilon),\boldsymbol{v})+2\,\varepsilon^2\,c_2(\boldsymbol{\sigma}(\varepsilon),\boldsymbol{u}(\varepsilon),\boldsymbol{v})$$
$$=F(\boldsymbol{v})+\rho\langle\partial_{tt}u_\alpha(\varepsilon),v_\alpha\rangle+\varepsilon^2\rho\langle\partial_{tt}u_3(\varepsilon),v_3\rangle,\quad \textit{for all } \boldsymbol{v}\in V(\Omega), \tag{47.9}$$
$$\boldsymbol{u}(\varepsilon)(0)=\overline{\boldsymbol{u}},\qquad \partial_t\boldsymbol{u}(\varepsilon)(0)=\overline{\boldsymbol{v}}, \tag{47.10}$$

where for any $\boldsymbol{u}$, $\boldsymbol{v}\in V(\Omega)$, $\boldsymbol{\sigma}$, $\boldsymbol{\tau}\in\Sigma(\Omega)$ *we define*

$$a_0(\boldsymbol{\sigma},\boldsymbol{\tau})=\int_\Omega \frac{1}{E}\,\sigma_{33}\tau_{33}\,\mathrm{d}\boldsymbol{x},$$
$$a_2(\boldsymbol{\sigma},\boldsymbol{\tau})=\int_\Omega\left[\frac{2(1+\nu)}{E}\sigma_{\alpha3}\tau_{\alpha3}-\frac{\nu}{E}(\sigma_{33}\tau_{\alpha\alpha}+\sigma_{\alpha\alpha}\tau_{33})\right]\mathrm{d}\boldsymbol{x},$$
$$a_4(\boldsymbol{\sigma},\boldsymbol{\tau})=\int_\Omega\left[\frac{(1+\nu)}{E}\sigma_{\alpha\beta}\tau_{\alpha\beta}-\frac{\nu}{E}\sigma_{\alpha\alpha}\tau_{\beta\beta}\right]\mathrm{d}\boldsymbol{x},$$
$$b(\boldsymbol{\tau},\boldsymbol{u})=-\int_\Omega e_{ij}(\boldsymbol{u})\tau_{ij}\,\mathrm{d}\boldsymbol{x},$$
$$c_0(\boldsymbol{\tau},\boldsymbol{u},\boldsymbol{v})=-\frac{1}{2}\int_\Omega \partial_i u_\alpha\partial_j v_\alpha\tau_{ij}\,\mathrm{d}\boldsymbol{x},$$
$$c_2(\boldsymbol{\tau},\boldsymbol{u},\boldsymbol{v})=-\frac{1}{2}\int_\Omega \partial_i u_3\partial_j v_3\tau_{ij}\,\mathrm{d}\boldsymbol{x},$$
$$F(\boldsymbol{v})=-h(0)\int_{\omega\times\{0\}} v_3\,\mathrm{d}\omega-h(L)\int_{\omega\times\{L\}} v_3\,\mathrm{d}\omega.$$

Following the standard asymptotic technique we assume that $(\boldsymbol{u}(\varepsilon),\boldsymbol{\sigma}(\varepsilon))$ admits an asymptotic expansion of the type

$$(\boldsymbol{u}(\varepsilon),\boldsymbol{\sigma}(\varepsilon))=(\boldsymbol{u}^0,\boldsymbol{\sigma}^0)+\varepsilon^1(\boldsymbol{u}^1,\boldsymbol{\sigma}^1)+\varepsilon^2(\boldsymbol{u}^2,\boldsymbol{\sigma}^2)+\text{h.o.t.} \tag{47.11}$$

We substitute this expression in (47.8)–(47.10) and identify the terms corresponding to the same powers of ε. We then obtain, that the first term of the expansion satisfies the problem $P(0)$,

$$a_0(\boldsymbol{\sigma}^0,\boldsymbol{\tau})+b(\boldsymbol{\tau},\boldsymbol{u}^0)+c_0(\boldsymbol{\tau},\boldsymbol{u}^0,\boldsymbol{u}^0)=0,\quad \text{for all } \boldsymbol{\tau}\in\Sigma(\Omega), \tag{47.12}$$
$$b(\boldsymbol{\sigma}^0,\boldsymbol{v})+2c_0(\boldsymbol{\sigma}^0,\boldsymbol{u}^0,\boldsymbol{v})=F(\boldsymbol{v})+\rho\langle\partial_{tt}u^0_\alpha,v_\alpha\rangle,\quad \text{for all } \boldsymbol{v}\in V(\Omega), \tag{47.13}$$
$$\boldsymbol{u}^0(0)=\overline{\boldsymbol{u}},\qquad \partial_t\boldsymbol{u}^0(0)=\overline{\boldsymbol{v}}. \tag{47.14}$$

If we introduce the function spaces

$$\tilde{\Sigma}(\Omega) = \{\boldsymbol{\tau} \in [L^2(\Omega)]^4\colon\ \tau_{\alpha\beta} = \tau_{\beta\alpha}\}, \tag{47.15}$$

$$W_1(\Omega) = W_2(\Omega) = \{v \in W^{1,4}(\Omega)\colon\ v = 0 \text{ on } \Gamma_0 \cup \Gamma_L\}, \tag{47.16}$$

$$W_3(\Omega) = \{v \in W^{1,4}(\Omega)\colon\ v \text{ independent of } x_\alpha \text{ on } \Gamma_0 \cup \Gamma_L\}, \tag{47.17}$$

we have

$$V(\Omega) = W_1(\Omega) \times W_2(\Omega) \times W_3(\Omega),$$

and consequently, we immediately obtain the following result.

THEOREM 47.2. *Problem* (47.12)–(47.14) *admits the following equivalent formulation*:

$$\int_\Omega \frac{1}{E}\sigma_{33}^0 \tau_{33}\,\mathrm{d}\boldsymbol{x} - \int_\Omega \partial_3 u_3^0 \tau_{33}\,\mathrm{d}\boldsymbol{x} - \tfrac{1}{2}\int_\Omega \partial_3 u_\alpha^0 \partial_3 u_\alpha^0 \tau_{33}\,\mathrm{d}\boldsymbol{x} = 0, \quad \textit{for all } \tau_{33} \in L^2(\Omega), \tag{47.18}$$

$$-\int_\Omega (\partial_3 u_\alpha^0 + \partial_\alpha u_3^0)\tau_{\alpha 3}\,\mathrm{d}\boldsymbol{x} - \int_\Omega \partial_\alpha u_\beta^0 \partial_3 u_\beta^0 \tau_{\alpha 3}\,\mathrm{d}\boldsymbol{x} = 0, \quad \textit{for all } (\tau_{\alpha 3}) \in [L^2(\Omega)]^2, \tag{47.19}$$

$$-\int_\Omega \partial_\alpha u_\beta^0 \tau_{\alpha\beta}\,\mathrm{d}\boldsymbol{x} - \tfrac{1}{2}\int_\Omega \partial_\alpha u_\gamma^0 \partial_\beta u_\gamma^0 \tau_{\alpha\beta}\,\mathrm{d}\boldsymbol{x} = 0, \quad \textit{for all } (\tau_{\alpha\beta}) \in \tilde{\Sigma}(\Omega), \tag{47.20}$$

$$-\rho\langle \partial_{tt} u_\alpha^0, v_\alpha\rangle - \int_\Omega \sigma_{i\beta}^0 \partial_i v_\beta\,\mathrm{d}\boldsymbol{x} - \int_\Omega \sigma_{ij}^0 \partial_i u_\alpha^0 \partial_j v_\alpha = 0, \quad \textit{for all } (v_\alpha) \in W_1(\Omega) \times W_2(\Omega), \tag{47.21}$$

$$-\int_\Omega \sigma_{i3}^0 \partial_i v_3\,\mathrm{d}\boldsymbol{x} = -h(0)\int_{\omega\times\{0\}} v_3\,\mathrm{d}\boldsymbol{x} - h(L)\int_{\omega\times\{L\}} v_3\,\mathrm{d}\boldsymbol{x}, \quad \textit{for all } v_3 \in W_3(\Omega), \tag{47.22}$$

$$u_\alpha^0(0) = \bar{u}_\alpha, \qquad \partial_t u_\alpha^0(0) = \bar{v}_\alpha. \tag{47.23}$$

48. Computation of the first term in the expansion for the nonlinear elastodynamic case

In this section we give a partial characterization of the first term $(\boldsymbol{u}^0, \boldsymbol{\sigma}^0)$ of the asymptotic expansion assumed for $(\boldsymbol{u}(\varepsilon), \boldsymbol{\sigma}(\varepsilon))$. The limit problem obtained for $(\boldsymbol{u}^0, \boldsymbol{\sigma}^0)$ will be studied more thoroughly in Section 5. The main result of this section is the following.

THEOREM 48.1. *Assume that a solution* $(\boldsymbol{u}^0, \boldsymbol{\sigma}^0)$ *of Eqs.* (47.18)–(47.23) *satisfies*

$$(\boldsymbol{u}^0, \boldsymbol{\sigma}^0) \in C^0([0,T]; V(\Omega)) \times C^0([0,T]; \Sigma(\Omega)), \tag{48.1}$$

$$\partial_\beta u^0_\alpha \in C^0([0,T]; C^0(\overline{\Omega})). \tag{48.2}$$

then $\boldsymbol{u}^0$ *and* σ^0_{33} *are necessarily of the form*

$$u^0_\alpha(x_1, x_2, x_3) = u^0_\alpha(x_3) \quad \textit{in } [0,T], \tag{48.3}$$

$$u^0_3((x_1, x_2, x_3) = \underline{u}^0_3(x_3) - x_\alpha \partial_3 u^0_\alpha(x_3) \quad \textit{in } [0,T], \tag{48.4}$$

$$\sigma^0_{33} = E(\partial_3 \underline{u}^0_3 - x_\alpha \partial_{33} u^0_\alpha + \tfrac{1}{2}\partial_3 u^0_\alpha \partial_3 u^0_\alpha) \quad \textit{in } [0,T], \tag{48.5}$$

with $(u^0_1, u^0_2, \underline{u}^0_3)$ *a solution of the following one-dimensional problem*:

$$u^0_\alpha \in C^0([0,T]; W^{2,4}_0(0,L)), \qquad u^0_3 \in C^0([0,T]; W^{1,4}(0,L)), \tag{48.6}$$

$$-\rho\langle \partial_{tt} u^0_\alpha, v^0_\alpha\rangle - EI_\alpha \int_0^L \partial_{33} u^0_\alpha \partial_{33} v^0_\alpha \, \mathrm{d}x_3$$

$$-\int_0^L q^0_3 \partial_3 u^0_\alpha \partial_3 v^0_\alpha \, \mathrm{d}x_3 = 0, \quad \textit{for all } (v^0_\alpha) \in [W^{2,4}_0(0,L)]^2, \tag{48.7}$$

$$\int_0^L q^0_3 \partial_3 v^0_3 dx_3 = h(0) v^0_3(0) + h(L) v^0_3(L), \quad \textit{for all } v^0_3 \in W^{1,4}(0,L), \tag{48.8}$$

$$u^0_\alpha(0) = \overline{u}_\alpha, \qquad \partial_t u^0_\alpha(0) = \overline{v}_\alpha, \tag{48.9}$$

where

$$q^0_3 := \int_\omega \sigma^0_{33} \, \mathrm{d}\omega = E(\partial_3 \underline{u}^0_3 + \tfrac{1}{2}\partial_3 u^0_\alpha \partial_3 u^0_\alpha). \tag{48.10}$$

Besides, the bending moment and the shear forces take the form

$$m^0_\alpha := \int_\omega x_\alpha \sigma^0_{33} \, \mathrm{d}\omega = -EI_\alpha \partial_{33} u^0_\alpha \quad (\textit{no sum on } \alpha), \tag{48.11}$$

$$q^0_\alpha := \int_\omega \sigma^0_{\alpha 3} = \partial_3 m^0_\alpha = -EI_\alpha \partial_{333} u^0_\alpha \quad (\textit{no sum on } \alpha). \tag{48.12}$$

Conversely, let $(u_1^0, u_2^0, \underline{u}_3^0)$ *be a solution of problem* (48.6)–(48.9). *Then, the element* $(\boldsymbol{u}^0, \sigma_{33}^0)$ *given by* (48.3)–(48.5) *belongs to* $C^0([0,T]; V(\Omega)) \times C^0([0,T]; L^2(\Omega))$, *and it is a solution of problem* (47.18)–(47.20) *with initial conditions* (47.23).

PROOF. For the sake of simplicity this proof, which is inspired by the results of CIARLET [1980] and CIMITIÈRE, GEYMONAT, LE DRET, RAOULT and TUTEK [1988], will be divided in five steps:

Step 1. As a consequence of Eq. (47.18) we obtain

$$\sigma_{33}^0 = E(\partial_3 u_3^0 + \tfrac{1}{2}\partial_3 u_\alpha^0 \partial_3 u_\alpha^0). \tag{48.13}$$

Then an element $(\boldsymbol{u}^0, \boldsymbol{\sigma}^0) \in C^0([0,T]; V(\Omega)) \times C^0([0,T]; \Sigma(\Omega))$ satisfies Eq. (47.18) if and only if relation (48.13) holds.

Step 2. Equations (47.19)–(47.20) give respectively

$$\partial_3 u_\alpha^0 + \partial_\alpha u_3^0 + \partial_\alpha u_\beta^0 + \partial_3 u_\beta^0 = 0, \tag{48.14}$$

$$\partial_\alpha u_\beta^0 + \partial_\beta u_\alpha^0 + \partial_\alpha u_\gamma^0 \partial_\beta u_\gamma^0 = 0. \tag{48.15}$$

From (48.2) and (48.15) we obtain that $(u_\alpha^0(x_3)) \in C^0([0,T]; \mathcal{R})$ a.e. $x_3 \in [0, L]$, where $\mathcal{R} \subset C^\infty(\omega, \mathbb{R}^2)$ is the set of rigid deformations in $\mathbb{R}^2$, i.e. (cf. CIARLET [1988]),

$$\begin{Bmatrix} u_1^0 \\ u_2^0 \end{Bmatrix} = \begin{Bmatrix} \underline{u}_1^0(x_3) \\ \underline{u}_2^0(x_3) \end{Bmatrix} + \begin{bmatrix} \cos\theta(x_3) & -\sin\theta(x_3) \\ \sin\theta(x_3) & \cos\theta(x_3) \end{bmatrix} \begin{Bmatrix} x_1 \\ x_2 \end{Bmatrix} - \begin{Bmatrix} x_1 \\ x_2 \end{Bmatrix}.$$

Substituting this expression in (48.14) and differentiating with respect to x_1 and x_2 we get

$$\partial_{12} u_3^0 - \partial_3\theta \cos^2\theta - \partial_3\theta \sin^2\theta = 0,$$
$$\partial_{21} u_3^0 + \partial_3\theta \sin^2\theta + \partial_3\theta \cos^2\theta = 0,$$

from which we can conclude that $\partial_{12} u_3^0 = \partial_3\theta = 0$ and then, θ is a constant function on x_3: $\theta(x_3) \equiv \theta$.

Since $\partial_\alpha u_\beta^0 \in C^0([0,T]; C^0(\overline{\Omega}))$ and $u_\alpha^0 = 0$ on $\Gamma_0 \cup \Gamma_L$, then $\partial_\beta u_\alpha^0 = 0$ on $\Gamma_0 \cup \Gamma_L$. Thus

$$\partial_1 u_1^0 = \cos\theta - 1 = 0,$$
$$\partial_2 u_2^0 = \sin\theta = 0,$$

from which we obtain that θ is zero. So

$$u_\alpha^0 \in C^0([0,T]; W_0^{1,4}(0,L)).$$

Then, relation (48.13) yields

$$\partial_\alpha u_3^0 = -\partial_3 u_\alpha^0.$$

Consequently

$$\partial_{\alpha\beta} u_3^0 = -\partial_{3\beta} u_\alpha^0 = 0,$$

which means that

$$u_3^0 = \underline{u}_3^0(x_3) + x_\alpha \overline{\underline{u}}_\alpha^0(x_3),$$

where

$$\underline{u}_3^0,\ \overline{\underline{u}}_\alpha^0 \in C^0([0,T]; W^{1,4}(0,L).$$

Since u_3^0 is independent of x_α on Γ_0, we get

$$\overline{\underline{u}}_\alpha^0 \in C^0([0,T]; W_0^{1,4}(0,L).$$

But, on the other hand

$$\partial_3 u_\alpha^0 = -\partial_\alpha u_3^0 = -\overline{\underline{u}}_\alpha^0.$$

Thus,

$$u_\alpha^0 \in C^0([0,T]; W_0^{2,4}(0,L)).$$

Then we can summarize by saying that an element $\boldsymbol{u}^0 \in C^0([0,T]; V(\Omega))$ with regularity (48.2), satisfies Eqs. (47.19)–(47.20) if and only if $\boldsymbol{u}^0$ is a Bernoulli–Navier displacement field, that is to say

$$u_\alpha^0 \in C^0([0,T]; W_0^{2,4}(0,L), \tag{48.16}$$

$$u_3^0 = \underline{u}_3^0 - x_\alpha \partial_3 u_\alpha^0, \quad \text{where } \underline{u}_3^0 \in C^0([0,T]; W^{1,4}(0,L). \tag{48.17}$$

Step 3. As a consequence of expressions (48.13) and (48.17), the component σ_{33}^0 of the stress tensor is determined by expression (48.5). Therefore, the normal force q_3^0 and the bending moments m_α^0 are given, respectively, by expressions (48.10) and (48.11).

Then we can write

$$\sigma_{33} = q_3^0 + \frac{x_\alpha}{I_\alpha} m_\alpha^0,$$

where

$$q_3^0,\ m_\alpha^0 \in C^0([0,T]; L^2(0,L)).$$

Property (48.12) is obtained from Eq. (47.32) with $v_3 = x_\alpha\, v_3^0$, $v_3^0 \in W_0^{1,4}(0,L)$.

Step 4. If we take $v_3 = v_3^0 \in W^{1,4}(0,L)$ as a test function in Eq. (47.22), we have

$$\int_0^L q_3^0 \partial_3 v_3^0 dx_3 = h(0)v_3^0(0) + h(L)v_3^0(L), \quad \text{for all } v_3^0 \in W^{1,4}(0,L). \tag{48.18}$$

Taking now $v_\beta = v_\beta^0 \in W_0^{2,4}(0,L)$ in relation (47.21),

$$- \rho\langle \partial_{tt} u_\alpha^0, v_\alpha^0\rangle - \int_\Omega \sigma_{3\beta}\partial_3 v_\beta^0 \,\mathrm{d}\boldsymbol{x}$$

$$- \int_0^L q_3^0 \partial_3 u_\alpha^0 \partial_3 v_\alpha^0 \,\mathrm{d}x_3 = 0, \quad \text{for all } (v_\beta^0) \in [W_0^{2,4}(0,L)]^2.$$

Choosing $v_3 = x_\beta \partial_3 v_\beta^0$ as a test function in (47.22) we obtain

$$- \int_\Omega \sigma_{3\beta}^0 \partial_3 v_\beta^0 - \int_0^L m_\beta^0 \partial_{33} v_\beta^0 \,\mathrm{d}x_3 = 0, \quad \text{for all } (v_\beta^0) \in [W_0^{2,4}(0,L)]^2.$$

Thus

$$- \rho\langle \partial_{tt} u_\alpha^0, v_\alpha^0\rangle - \int_0^L m_\beta^0 \partial_{33} v_\beta^0 \,\mathrm{d}\boldsymbol{x}$$

$$- \int_0^L q_3^0 \partial_3 u_\alpha^0 \partial_3 v_\alpha^0 = 0, \quad \text{for all } (v_\beta^0) \in [W_0^{2,4}(0,L)]^2.$$

Substituting (48.11) in this equation we obtain from (48.18) that $(u_\alpha^0, \underline{u}_3^0)$ solves problem (48.6)–(48.9).

Step 5. Summarizing the previous steps, we have proved that a solution of (47.18)–(47.23) $\boldsymbol{u}^0 \in C^0([0,T]; V(\Omega))$ with the additional regularity (48.2) takes necessarily the form (48.3)–(48.4) (Step 2), where $(u_1^0, u_2^0, \underline{u}_3^0)$ are necessarily solutions of (48.6)–(48.9) (Step 4). Besides, the component σ_{33}^0 takes the form (48.5) (Step 3).

Conversely, if $(u_1^0, u_2^0, \underline{u}_3^0)$ is a solution of (48.6)–(48.9), then the displacement field $\boldsymbol{u}^0$ given by (48.3)–(48.4) and the stress tensor component σ_{33}^0 defined by (48.5) verify

$$\boldsymbol{u}^0 \in C^0([0,T]; V(\Omega)), \qquad \sigma_{33}^0 \in C^0([0,T]; L^2(\Omega)),$$

and they are solutions of problem (47.18)–(47.20), (47.23) (Steps 1 and 2). □

REMARK 48.1. As will be shown in the next section, it will be necessary for the axial loading h to satisfy a compatibility condition in order to obtain a solution for problem (48.6)–(48.9).

This necessary condition is of the form

$$h(0) + h(L) = 0,$$

as can be seen just by choosing in Eq. (48.18) $v_3^0 \equiv 1$ as a test function in $H^1(0, L)$.

49. The limit problem for the nonlinear elastodynamic case

This section deals with the study of the properties of the one-dimensional limit problem whose justification as a valid limit behaviour for the three-dimensional beam has been made in the previous section.

We must remember that the first term of the displacement is a Bernoulli–Navier field, that is to say, it takes the form

$$\begin{aligned} &u_\alpha^0(t) \in W_0^{2,4}(0, L), \\ &u_3^0(t) = \underline{u}_3^0(t) - x_\alpha \partial_3 u_\alpha^0(t), \qquad \underline{u}_3^0(t) \in W^{1,4}(0, L), \end{aligned}$$

(u_α^0, u_3^0) being solution of the variational problem (48.6)–(48.9).

Using the strong formulation of this problem we obtain that the first term of the displacement field is solution of the following evolution system coupling an hyperbolic equation with a quasi-static one:

$$-\rho \partial_{tt} u_\alpha^0 - EI_\alpha \partial_{3333} u_\alpha^0 + \partial_3(q_3^0 \partial_3 u_\alpha^0) = 0 \quad \text{in } (0, L) \times (0, T), \tag{49.1}$$

$$\partial_3 q_3^0 = 0 \quad \text{in } (0, L) \times (0, T), \tag{49.2}$$

where q_3^0 is the normal force given by the expression (48.10), with boundary conditions

$$u_\alpha^0(0, t) = u_\alpha^0(L, t) = 0, \tag{49.3}$$

$$\partial_3 u_\alpha^0(0, t) = \partial_3 u_\alpha^0(L, t) = 0, \tag{49.4}$$

$$q_3^0(0, t) = -h(0, t), \tag{49.5}$$

$$q_3^0(L, t) = h(L, t), \tag{49.6}$$

and with initial conditions

$$u_\alpha^0(x_3, 0) = \overline{u}_\alpha, \tag{49.7}$$

$$\partial_3 u_\alpha^0(x_3, 0) = \overline{v}_\alpha. \tag{49.8}$$

THEOREM 49.1. *A necessary condition for problem* (49.1)–(49.8) *to admit a solution is that*

$$-h(0,t) = h(L,t) = p(t). \tag{49.9}$$

If condition (49.9) *is satisfied, we have*

(i) *The normal force* q_3^0 *verifies*

$$q_3^0(t) = -p(t) \tag{49.10}$$

(ii) u_α^0 *is the solution of problem* (P):

$$\begin{aligned}
&\rho\partial_{tt}u_\alpha^0 + EI_\alpha\partial_{3333}u_\alpha^0 + p(t)\partial_{33}u_\alpha^0 = 0 \quad \textit{in } (0,L)\times(0,T),\\
&u_\alpha^0(0,t) = u_\alpha^0(L,t) = 0,\\
&\partial_3 u_\alpha^0(0,t) = \partial_3 u_\alpha^0(L,t) = 0,\\
&u_\alpha^0(x_3,0) = \overline{u}_\alpha,\\
&\partial_3 u_\alpha^0(x_3,0) = \overline{v}_\alpha.
\end{aligned} \tag{49.11}$$

(iii) $\underline{u}_3^0$ *is the solution of the equation*

$$-E\partial_3\underline{u}_3^0 = p(t) + \tfrac{1}{2}E\partial_3 u_\alpha^0\partial_3 u_\alpha^0 \quad \textit{in } (0,L)\times(0,T). \tag{49.12}$$

PROOF. Since Eq. (49.2) implies that q_3^0 is independent of x_3, we conclude from (49.5) and (49.6) that a necessary condition for the existence of solution is (49.9). In this case we get that $q_3^0 = -p(t)$, and Eqs. (49.1)–(49.2) turn into

$$-\rho\partial_{tt}u_\alpha^0 - EI_\alpha\partial_{3333}u_\alpha^0 - p(t)\partial_{33}u_\alpha^0 = 0 \quad \text{in } (0,L)\times(0,T),$$

with boundary conditions (49.3)–(49.4) and initial conditions (49.7)–(49.8), i.e., u_α^0 is the solution of problem (P).

On the other hand, since u_α^0 can be computed from (49.11), we obtain from (48.10) and (49.10) that the displacement $\underline{u}_3^0$ is a solution of Eq. (49.12). □

REMARK 49.1. We remark that in problem (P) obtained in Theorem 49.1 we recover the problem stated in the beginning of the chapter, as was our intention.

REMARK 49.2. Note that condition (49.9) necessary for the existence of the solution for the limit problem is the same condition achieved in the previous section (see Remark 48.1).

Next, we obtain a result of existence and regularity of solution for the limit problem (P).

THEOREM 49.2. (i) *If* $(\overline{u}_\alpha, \overline{v}_\alpha) \in H_0^2(0,L) \times L^2(0,L)$ *and* $p \in L^1((0,T))$, *then problem* (P) *has a solution in the whole interval* $[0,T]$. *Besides, this solution is unique within the class of functions with regularity* $C^0([0,T]; H_0^2(0,L)) \cap C^1([0,T]; L^2(0,L))$.

(ii) *If* $(\overline{u}_\alpha, \overline{v}_\alpha) \in [H^4(0,L) \cap H_0^2(0,L)] \times H_0^2(0,L)$ *and* $p \in C^1([0,T])$, *then the unique solution of problem* (P) *has the following regularity*:

$$u_\alpha^0 \in C^0([0,T]; H^4(0,L) \cap H_0^2(0,L)) \\ \cap C^1([0,T]; H_0^2(0,L)) \cap C^2([0,T]; L^2(0,L)). \tag{49.13}$$

PROOF. The proof of the first part of the theorem, by means of the use of the technique of semigroup theory, can be seen in BALL, MARSDEN and SLEMROD [1982]. The regularity of the solution u_α^0 follows, in an immediate way, from classical results of semigroup theory. See, for instance, SEGAL [1963] or TANABE [1979]. □

50. The general nonlinear elastodynamic case

In this section, we are going to deal with a more general problem in which the nonlinearly elastic beam is "weakly" clamped at both ends and, besides the axial loads H^ε, it is submitted to body forces $\boldsymbol{f}^\varepsilon$ in Ω^ε and surface tractions $\boldsymbol{g}^\varepsilon$ over Γ^ε. Then, the mechanical response of the beam Ω^ε throughout the time interval $[0,T]$ is modelled by the system of equations

$$\rho^\varepsilon \partial_{tt} u_i^\varepsilon - \partial_j^\varepsilon(\sigma_{ij}^\varepsilon + \sigma_{kj}^\varepsilon \partial_k^\varepsilon u_i^\varepsilon) = f_i^\varepsilon \quad \text{in } \Omega^\varepsilon \times (0,T), \tag{50.1}$$

$$(\boldsymbol{A}^\varepsilon \boldsymbol{\sigma}^\varepsilon)_{ij} := \frac{1+\nu^\varepsilon}{E^\varepsilon}\sigma_{ij}^\varepsilon - \frac{\nu^\varepsilon}{E^\varepsilon}\sigma_{kk}^\varepsilon \delta_{ij} \\ = \tfrac{1}{2}(\partial_i^\varepsilon u_j^\varepsilon + \partial_j^\varepsilon u_i^\varepsilon + \partial_i^\varepsilon u_k^\varepsilon \partial_j^\varepsilon u_k^\varepsilon) \quad \text{in } \Omega^\varepsilon \times (0,T), \tag{50.2}$$

with boundary conditions

$$u_\alpha^\varepsilon = 0 \quad \text{on } (\Gamma_0^\varepsilon \cup \Gamma_L^\varepsilon) \times (0,T), \tag{50.3}$$

$$u_3^\varepsilon \quad \text{independent of } x_\alpha \text{ on } (\Gamma_0^\varepsilon \cup \Gamma_L^\varepsilon) \times (0,T), \tag{50.4}$$

$$(\sigma_{i\alpha}^\varepsilon + \sigma_{k\alpha}^\varepsilon \partial_k^\varepsilon u_i^\varepsilon) n_\alpha = g_i^\varepsilon \quad \text{on } \Gamma^\varepsilon \times (0,T), \tag{50.5}$$

$$\frac{1}{|\omega^\varepsilon|}\int_{\omega^\varepsilon} (\sigma_{33}^\varepsilon + \sigma_{k3}^\varepsilon \partial_k^\varepsilon u_3^\varepsilon) n_3 = H^\varepsilon \quad \text{in } \{0,L\} \times (0,T), \tag{50.6}$$

and with initial conditions on the displacement and the velocity fields

$$\boldsymbol{u}^\varepsilon(0) = \overline{\boldsymbol{u}}^\varepsilon, \tag{50.7}$$

$$\partial_t \boldsymbol{u}^\varepsilon(0) = \overline{\boldsymbol{v}}^\varepsilon. \tag{50.8}$$

Obviously, the original problem (46.1)–(46.8) corresponds to taking the body forces $\boldsymbol{f}^{\varepsilon}$ and the surface forces $\boldsymbol{g}^{\varepsilon}$ equal to zero in (50.1)–(50.8).

We consider the change of variable to the fixed domain Ω, as was shown in Section 47. We assume that, besides (47.5)–(47.7), the system of applied loads verifies the relations

$$\begin{aligned} f_{\alpha}^{\varepsilon}(\boldsymbol{x}^{\varepsilon}) &= \varepsilon^{3} f_{\alpha}(\boldsymbol{x}), \\ f_{3}^{\varepsilon}(\boldsymbol{x}^{\varepsilon}) &= \varepsilon^{2} f_{3}(\boldsymbol{x}), \\ g_{\alpha}^{\varepsilon}(\boldsymbol{x}^{\varepsilon}) &= \varepsilon^{4} g_{\alpha}(\boldsymbol{x}), \\ g_{3}^{\varepsilon}(\boldsymbol{x}^{\varepsilon}) &= \varepsilon^{3} g_{3}(\boldsymbol{x}). \end{aligned} \tag{50.9}$$

Then, problem (50.1)–(50.8) admits the equivalent variational formulation (47.8)–(47.10), the only difference being the new definition of $F(\boldsymbol{v})$,

$$F(\boldsymbol{v}) = -\int_{\Omega} f_i v_i \,\mathrm{d}\boldsymbol{x} - \int_{\Gamma} g_i v_i \,\mathrm{d}a - h(0) \int_{\omega\times\{0\}} v_3 \,\mathrm{d}\boldsymbol{x} - h(L) \int_{\omega\times\{L\}} v_3 \,\mathrm{d}\boldsymbol{x}. \tag{50.10}$$

An asymptotic expansion of the form (47.11) for the pair $(\boldsymbol{u}(\varepsilon), \boldsymbol{\sigma}(\varepsilon))$ is still supposed, and then the variational problem corresponding to the first term of the expansion can be formulated as

$$\int_{\Omega} \frac{1}{E}\sigma_{33}^{0}\tau_{33}\,\mathrm{d}\boldsymbol{x} - \int_{\Omega} \partial_3 u_3^0 \tau_{33}\,\mathrm{d}\boldsymbol{x} - \tfrac{1}{2}\int_{\Omega} \partial_3 u_{\alpha}^0 \partial_3 u_{\alpha}^0 \tau_{33}\,\mathrm{d}\boldsymbol{x} = 0, \quad \text{for all } \tau_{33} \in L^2(\Omega), \tag{50.11}$$

$$-\int_{\Omega} (\partial_3 u_{\alpha}^0 + \partial_{\alpha} u_3^0)\tau_{\alpha 3}\,\mathrm{d}\boldsymbol{x} - \int_{\Omega} \partial_{\alpha} u_{\beta}^0 \partial_3 u_{\beta}^0 \tau_{\alpha 3}\,\mathrm{d}\boldsymbol{x} = 0, \quad \text{for all } (\tau_{\alpha 3}) \in [L^2(\Omega)]^2, \tag{50.12}$$

$$-\int_{\Omega} \partial_{\alpha} u_{\beta}^0 \tau_{\alpha\beta}\,\mathrm{d}\boldsymbol{x} - \tfrac{1}{2}\int_{\Omega} \partial_{\alpha} u_{\gamma}^0 \partial_{\beta} u_{\gamma}^0 \tau_{\alpha\beta}\,\mathrm{d}\boldsymbol{x} = 0, \quad \text{for all } (\tau_{\alpha\beta}) \in \tilde{\Sigma}(\Omega), \tag{50.13}$$

$$-\rho\langle \partial_{tt} u_{\alpha}^0, v_{\alpha}\rangle - \int_{\Omega} \sigma_{i\beta}^0 \partial_i v_{\beta}\,\mathrm{d}\boldsymbol{x} - \int_{\Omega} \sigma_{ij}^0 \partial_i u_{\alpha}^0 \partial_j v_{\alpha}\,\mathrm{d}\boldsymbol{x} = -\int_{\Omega} f_{\alpha} v_{\alpha}\,\mathrm{d}\boldsymbol{x} - \int_{\Gamma} g_{\alpha} v_{\alpha}\,\mathrm{d}a, \quad \text{for all } (v_{\alpha}) \in W_1(\Omega) \times W_2(\Omega), \tag{50.14}$$

$$-\int_{\Omega}\sigma_{i3}^0\partial_i v_3\,\mathrm{d}\boldsymbol{x} = -\int_{\Omega} f_3 v_3\,\mathrm{d}\boldsymbol{x} - \int_{\Gamma} g_3 v_3\,\mathrm{d}a$$

$$-h(0)\int_{\omega\times\{0\}} v_3\,\mathrm{d}\boldsymbol{x} - h(L)\int_{\omega\times\{L\}} v_3\,\mathrm{d}\boldsymbol{x}, \quad \text{for all } v_3\in W_3, \tag{50.15}$$

$$u_\alpha^0(0)=\overline{u}_\alpha, \qquad \partial_t u_\alpha^0(0)=\overline{v}_\alpha. \tag{50.16}$$

Following the same technique already employed in Sections 48 and 49 we can give a characterization of the leading term $(\boldsymbol{u}^0,\boldsymbol{\sigma}^0)$ satisfying the variational formulation (50.11)–(50.16).

THEOREM 50.1. *Let* $(\boldsymbol{u}^0,\boldsymbol{\sigma}^0)\in C^0([0,T];V(\Omega))\times C^0([0,T];\Sigma(\Omega))$ *be a solution of Eqs.* (50.11)–(50.16) *that satisfies*

$$\partial_\beta u_\alpha^0\in C^0([0,T];C^0(\overline{\Omega})), \tag{50.17}$$

then $\boldsymbol{u}^0$ *and* σ_{33}^0 *take the form*

$$u_\alpha^0\in C^0([0,T];W_0^{2,4}(0,L)), \tag{50.18}$$

$$u_3^0=\underline{u}_3^0-x_\alpha\partial_3 u_\alpha^0, \qquad \underline{u}_3^0\in C^0([0,T];W^{1,4}(0,L)), \tag{50.19}$$

$$\sigma_{33}^0=E(\partial_3\underline{u}_3^0-x_\alpha\partial_{33}u_\alpha^0+\tfrac{1}{2}\partial_3u_\alpha^0\partial_3u_\alpha^0), \tag{50.20}$$

$(u_1^0,u_2^0,\underline{u}_3^0)$ *being solutions of the evolution problem*

$$-\rho\partial_{tt}u_\alpha^0-EI_\alpha\partial_{3333}u_\alpha^0+\partial_3(q_3^0\partial_3u_\alpha^0)$$

$$=-\int_\omega f_\alpha\,\mathrm{d}\boldsymbol{x}-\int_\gamma g_\alpha\,\mathrm{d}\gamma$$

$$-\partial_3\left[\int_\omega x_\alpha f_3\,\mathrm{d}\boldsymbol{x}+\int_\gamma x_\alpha g_3\,\mathrm{d}\gamma\right] \quad \textit{in } (0,L)\times(0,T), \tag{50.21}$$

$$\partial_3 q_3^0=-\int_\omega f_3\,\mathrm{d}\boldsymbol{x}-\int_\gamma g_3\,\mathrm{d}\gamma \quad \textit{in } (0,L)\times(0,T), \tag{50.22}$$

$$u_\alpha^0(0,t)=u_\alpha^0(L,t)=0, \tag{50.23}$$

$$\partial_3u_\alpha^0(0,t)=\partial_3u_\alpha^0(L,t)=0, \tag{50.24}$$

$$q_3^0(0,t)=-h(0,t), \tag{50.25}$$

$$q_3^0(L,t)=h(L,t), \tag{50.26}$$

$$u_\alpha^0(x_3,0)=\overline{u}_\alpha, \tag{50.27}$$

$$\partial_3u_\alpha^0(x_3,0)=\overline{v}_\alpha, \tag{50.28}$$

where the normal force q_3^0 *is given by the expression*

$$q_3^0=E(\partial_3\underline{u}_3^0+\tfrac{1}{2}\partial_3u_\alpha^0\partial_3u_\alpha^0). \tag{50.29}$$

Besides, the bending moments and shear forces take the form

$$m_\alpha^0 = -EI_\alpha \partial_{33} u_\alpha^0, \quad \textit{no sum on } \alpha, \tag{50.30}$$

$$q_\alpha^0 = \int_\omega x_\alpha f_3 \,\mathrm{d}\boldsymbol{x} + \int_\gamma x_\alpha g_3 \,\mathrm{d}\gamma - EI_\alpha \partial_{333} u_\alpha^0, \quad \textit{no sum on } \alpha. \tag{50.31}$$

Conversely, let $(u_1^0, u_2^0, \underline{u}_3^0)$ *be a solution of problem* (50.21)–(50.28) *with the regularity expressed in* (50.18)–(50.19), *then element* $(\boldsymbol{u}^0, \sigma_{33}^0)$ *given by* (50.18)–(50.20) *belongs to* $C^0([0,T]; V(\Omega)) \times C^0([0,T]; L^2(\Omega))$ *and it is a solution of the problem* (50.11)–(50.13) *with initial conditions* (50.16).

PROOF. The proof of this result is obtained in a similar way as the one in Section 49. □

REMARK 50.1. As before, a necessary condition for system (50.21)–(50.28) to admit a solution is that the following compatibility condition be verified:

$$\int_\Omega f_3(t) \,\mathrm{d}\boldsymbol{x} + \int_\Gamma g_3(t) \,\mathrm{d}a + h(0,t) + h(L,t) = 0. \tag{50.32}$$

From a physical point of view this condition is equivalent to the fact that the final resultant of all the axial loads applied over the beam vanishes.

Finally, we note that condition (50.32) agrees with the compatibility condition (49.9) in the case where the forces f_3 and g_3 are identically zero.

REMARK 50.2. We must recall once more that if in Eqs. (50.21)–(50.28) we neglect the terms corresponding to the derivatives in the time variable, we recognize the model (44.14)–(44.16) for nonlinear elastostatic rods.

References

ADAMS, A. (1975), *Sobolev Spaces* (Academic Press, New York).

ÁLVAREZ-DIOS, J.A. (1986), Obtención y resolución numérica de nuevos modelos unilaterales de vigas en flexión, Tesina de Licenciatura, Dpto. de Ecuaciones Funcionales, Universidad de Santiago de Compostela, Santiago de Compostela.

ÁLVAREZ-DIOS, J.A. (1992), Teoría asintótica de vigas elásticas lineales anisótropas y no homogéneas, Thesis, Dpto. Matemática Aplicada, Universidad de Santiago de Compostela, Santiago de Compostela.

ÁLVAREZ-DIOS, J.A. and J.M. VIAÑO (1988), Asymptotic derivation of bending and torsion models for homogeneous anisotropic elastic beams, in: M.L. MASCARENHAS and L. TRABUCHO eds., *Proceedings II Week on Asymptotic Analysis*. Textos e Notas **43** (C.M.A.F., Lisbon).

ÁLVAREZ-DIOS, J.A. and J.M. VIAÑO (1991a), Una teoría asintótica de flexión y torsión para vigas elásticas anisótropas no homogéneas, in: *Actas XII C.E.D.Y.A. / II Congreso de Matemática Aplicada* (Universidad de Oviedo, Oviedo), 351–355.

ÁLVAREZ-DIOS, J.A. and J.M. VIAÑO (1991b), On a bending and torsion asymptotic theory for linear nonhomogeneous anisotropic elastic rods, *Asymptotic Anal.*, to be published.

ÁLVAREZ-DIOS, J.A. and J.M. VIAÑO (1992a), An asymptotic general theory for linear elastic nonhomogeneous anisotropic rods, in: C. HIRSH et al., eds., *Numerical Methods in Engineering'92* (Elsevier, Amsterdam), 511–518.

ÁLVAREZ-DIOS, J.A. and J.M. VIAÑO (1992b), An asymptotic general model for linear elastic homogeneous anisotropic rods, *Internat. J. Num. Meth. Eng.*, to be published.

ÁLVAREZ-VÁZQUEZ, L. (1991), Tratamiento asintótico de algunos problemas en elasticidad de placas y vigas, Thesis, Dpto. de Matemática Aplicada, Universidad de Santiago de Compostela, Santiago de Compostela.

ÁLVAREZ-VÁZQUEZ, L. and J.M. VIAÑO (1989), Obtención de un modelo de torsión termoelástica lineal por métodos asintóticos, in: *Actas del XI C.E.D.Y.A. / I Congreso de Matemática Aplicada* (Universidad de Málaga, Málaga) 155–159.

ÁLVAREZ-VÁZQUEZ, L. and J.M. VIAÑO (1991a), Justification asymptotique d'un modéle thermoélastique lineaire d'evolution pour les poutres, *C. R. Acad. Sci. Paris Sér. I* **313**, 813–816.

ÁLVAREZ-VÁZQUEZ, L. and J.M. VIAÑO (1991b), Asymptotic derivation of an evolution thermoelastic model for linear rods, in: R. VITHNEVETSKY and J.J.H. MILLER, eds., *13th World Congress on Computation and Applied Mathematics* Vol. **2** (IMACS '91, Dublin), 888–889.

ÁLVAREZ-VÁZQUEZ, L. and J.M. VIAÑO (1991c), Derivación asintótica de un modelo evolutivo termoelástico lineal para vigas, in: *Actas del XII C.E.D.Y.A. / II Congreso de Matemática Aplicada* (Universidad de Oviedo, Oviedo), 357–361.

ÁLVAREZ-VÁZQUEZ, L. and J.M. VIAÑO (1991d), Justificación asintótica de un modelo de evolución para vigas elásticas no lineales, in: *Actas del XII C.E.D.Y.A. / II Congreso de Matemática Aplicada* (Universidad de Oviedo, Oviedo), 363–368.

ÁLVAREZ-VÁZQUEZ, L. and J.M. VIAÑO (1992a), Asymptotic justification of an evolution linear thermoelastic model for rods, *Comp. Meth. Appl. Mech. Eng.*, to be published.

ÁLVAREZ-VÁZQUEZ, L. and J.M. VIAÑO (1992b), Derivation of an evolution model for nonlinearly elastic beams by asymptotic methods, *Comp. Meth. Appl. Mech. Eng.*, to be published.

Antman, S.S. (1972), The theory of rods, *Handbuch der Physik*, Vol. **VIa/2** (Springer, Berlin).

Antman, S.S. (1976), Ordinary differential equations of the one dimensional elasticity, *Arch. Rational Mech. Anal.* **61**, 307–393.

Antman, S.S. and L. Kenney (1981), Large buckled states of nonlinear elastic rods under torsion, thrust and gravity, *Arch. Rational Mech. Anal.* **76**, 2289–338.

Aufranc M. (1990), Sur quelques problèmes de jonctions dans les multistructures elastiques, Thesis, Université Pierre et Marie Curie, Paris.

Axelsson, O. and V. A. Baker (1984), *Finite Element Solution of Boundary Value Problems: Theory and Computation* (Academic Press, New York).

Ball J.M. (1972), Topological methods in the nonlinear analysis of beams, Thesis, University of Sussex, U.K.

Ball, J.M. (1977), Convexity conditions and existence theorems in nonlinear elasticity, *Arch. Rational Mech. Anal.* **63**, 337–403.

Ball, J.M., J. Marsden and M. Slemrod (1982), Controlability of distributed bilinear systems, *SIAM J. Control Optim.* **20**, 579–597.

Ball, J.M. and F. Murat (1984), $W^{1,p}$-Quasiconvexity and variational problems for multiple integrals, *J. Funct. Anal.* **58**, 225–253.

Barros J.C. (1989), Definição e cálculo através do método de expansão assimptótica de constantes de Timoshenko, de constantes de empeno e de centros de corte para vigas de parede fina e aberta, Thesis, Instituto Superior Técnico da Universidade Técnica de Lisboa, Lisbon.

Barros J.C., L. Trabucho and J.M. Viaño (1989), Definições, generalização e cálculo de propiedades geométricas de secções em teoría de vigas, in: *Actas de MECOM 89*, Porto Vol. **2**, A465–A480.

Bathe, K.J. (1982), *Finite Element Procedures in Engineering Analysis* (Prentice-Hall, Englewood Cliffs, NJ).

Bathe, K.J. and E.L. Wilson (1976), *Numerical Methods in Finite Element Analysis* (Prentice-Hall, Englewood Cliffs, NJ).

Benssoussan, A., J.L. Lions, and G. Papanicolaou (1978), *Asymptotic Analysis for Periodic Structures* (North-Holland, Amsterdam).

Bercovier, M. (1982), On C^0 beam elements with shear and their corresponding penalty function, *Comput. Math. Appl.* **8**, 245–256.

Bermúdez, A. and J. Fernández (1986), Solving unilateral problems for beams by finite element methods, *Comp. Meth. Appl. Mech. Eng.* **54**, 67–73.

Bermúdez, A. and J.M. Viaño (1983), Étude de deux schémas numériques pour les équations de la thermoélasticité, *RAIRO Anal. Numér.* **17**, 121–136.

Bermúdez, A. and J.M. Viaño (1984), Une justification des équations de la thermo-élasticité des poutres à section variable par des méthodes asymptotiques, *RAIRO Anal. Numér.* **18**, 347–376.

Bernadou, M. (1987), Formulation variationnelle, approximation et implementation de problèmes de barres et de poutres bi- et tri-dimensionnelles. Partie **A**: Barres et poutres tridimensionnelles, *Rapports de Recherche* **731** (INRIA, France). Partie **B**: Barres et poutres bidimensionnelles, *Rapports Techniques* **86** (INRIA, France).

Bernadou, M. and J.M. Boisserie (1982), *The Finite Element Method in Thin Shell Theory: Application to Arch Dam Simulation* (Birkhäuser, Boston).

Bernadou, M. and Y. Ducatel (1978), Méthodes d'élements finis avec integration numérique pour des problèmes elliptiques du quatriéme ordre, *RAIRO Anal. Numér.* **12**, 3–26.

Bernadou, M. and Y. Ducatel (1982), Approximation of general arch problems by straight beam elements, *Numer. Math.* **40**, 1–29.

Bernadou, M., S. Fayolle and F. Léné (1989), Numerical analysis of junctions between plates, *Comput. Methods Appl. Mech. Engrg.* **74**, 307–326.

Bielak, J. and E. Stephan (1983), A modified Galerkin procedure for bending of beams in elastic foundations, *SIAM J. Sci. Statist. Comput.* **4**, 340–352.

Blanchard, D. (1981), Justification de modèles de plaques correspondant a différentes conditions aux limites, Thesis, Université Pierre et Marie Curie, Paris.

BLANCHARD, D. and P.G. CIARLET (1983), A remark on the von Kărman equations, *Comp. Meth. Appl. Mech. Eng.* **37**, 79–92.

BLANCHARD, D. and G. FRANCFORT (1987), Asymptotic thermoelastic behaviour of flat plates, *Quart. Appl. Math.* **45**, 645–667.

BOURGAT, J.F., J.M. DUMAY and R. GLOWINSKI (1980), Large displacement calculations of flexible pipelines by finite element and nonlinear programming methods, *SIAM J. Statist. Comput.* **1**, 34–81.

BOURGAT, J.F., P. LE TALLEC and S. MANI (1988), Modélisation des grands déplacements de tuyaux élastiques en flexion torsion, *J. Méc. Théor. Appl.* **7**, 1–30.

BOURQUIN, F. (1991), Modeling decomposition and eigenvalue approximation for elastic multi-structures, Thesis. Université Pierre et Marie Curie, Paris.

BOURQUIN, F. and P.G. CIARLET, (1989), Modelling and justification of eigenvalue problems for junctions between elastic structures, *J. Funct. Anal.* **87**, 392–427.

BREBBIA, C.A. and J.J. CONNOR (1974), *Fundamentals of Finite Element Technique for Structural Engineers* (Wiley, New York).

BREZIS, H. (1971), Problèmes unilateraux, *J. Math. Pures Appl. (9)* **72**, 1–68.

BREZIS, H. (1983), *Analyse Fonctionelle: Théorie et Applications* (Masson, Paris).

BREZIS, H., M. CRANDAL and A. PAZY (1970), Perturbation of non linear monotone sets in Banach spaces, *Comm. Pure Appl. Math.* **23**, 123–144.

BREZZI, F. (1974), On the existence, uniqueness and approximation of saddle-point problems arising from Lagrange multipliers, *RAIRO Anal. Numér.* **8**, 129–151.

BREZZI, F., W. HAGER and P.A. RAVIART (1977), Error estimates for the finite element solution of variational inequalities, *Numer. Math.* **28**, 431–443.

BRUSH, D.O. and B.O. ALMROTH (1975), *Buckling of Bars, Plates and Shells* (McGraw-Hill, New York).

BUDIANSKY, B. and J.L. SANDERS (1967), On the best first order linear shell theory, in: *Prog. Appl. Mech.*, W. PRAGER Anniversary Volume (MacMillan, New York), 129–140.

CAILLERIE, D. (1980), The effect of a thin inclusion of high rigidity in an elastic body, *Math. Methods Appl. Sci.* **2**, 251–270.

CAILLERIE, D. (1984), Thin elastic and periodic plates, *Math. Methods Appl. Sci.* **6**, 159–191.

CAILLERIE, D. (1987), Non homogeneous plate theory and conduction in fibered composites, in: E. SANCHEZ-PALENCIA and A. ZAOUI, eds., *Homogenisation Techniques for Composite Media* (Springer, Berlin), 1–62.

CAMOTIM, D. and L. TRABUCHO (1989), A derivation of some nonlinear elastic beam theories, in: C.R. STEELE and R. BEVILACQUA, eds., *Proceedings I Pan American Congress of Applied Mechanics*, Rio de Janeiro, 545–548.

CARLSON, D.E. (1972) Linear thermoelasticity, in: I. FLÜGGE, ed., *Encyclopedia of Physics* **VI/a2** (Springer, Berlin).

CIARLET, P.G. (1974), Quelques méthodes d'éléments fini pour le probléme d'une plaque encastrée, in: R. GLOWINSKI and J.L. LIONS, eds., *Computing Methods in Applied Sciences and Engineering*, Lecture Notes in Mathematiques **363** (Springer, Berlin), 156–176.

CIARLET, P.G. (1978), *The Finite Element Method for Elliptic Problems* (North-Holland, Amsterdam).

CIARLET, P.G. (1980), A justification of the von Kàrmàn equations, *Arch. Rational Mech. Anal.* **73**, 349–389.

CIARLET, P.G. (1982), *Introduction à l'Analyse Numérique Matricielle et à l'Optimisation* (Masson, Paris).

CIARLET, P.G. (1985), *Élasticité Tridimensionnelle* (Masson, Paris).

CIARLET, P.G. (1987), Recent progresses in the two-dimensional approximation of three-dimensional plate models in nonlinear elasticity, in: E. ORTIZ, ed., *Numerical Approximations of Partial Differential Equations* (North-Holland, Amsterdam), 3–19.

CIARLET, P.G. (1988), *Mathematical Elasticity*, Vol. **I**: Three Dimensional Elasticity (North-Holland, Amsterdam).

Ciarlet, P.G. (1989), The method of asymptotic expansions for a nonlinearly elastic clamped plate, in: H. Brezis and J.L. Lions, eds., *Nonlinear Partial Differential Equations and their Applications* (Collège de France Seminar), Vol. **X**.

Ciarlet P.G. (1990), *Plates and Junctions in Elastic Multi-structures. An Asymptotic Analysis* (Masson, Paris).

Ciarlet, P.G. (1992), Basic error estimates for elliptic problems, in: P. G. Ciarlet and J.L. Lions, eds., *Handbook of Numerical Analysis*, Vol. **II** (North-Holland, Amsterdam), 1–351.

Ciarlet, P.G. and P. Destuynder (1977), Une justification du modèle biharmonique en théorie linéaire de plaques, *C. R. Acad. Sci. Paris* **18**, 851–854.

Ciarlet, P.G. and P. Destuynder (1979a), A justification of the two dimensional linear plate model, *J. Mécanique* **18**, 315–344.

Ciarlet, P.G. and P. Destuynder (1979b), A justification of a nonlinear model in plate theory, *Comput. Methods Appl. Mech. Engrg.* **17/18**, 227–258.

Ciarlet P.G. and S. Kesavan (1979), Approximation bidimensionelle du problème de valeurs propres pour une plaque, *C. R. Acad. Sci. Paris* **289**, 579–582.

Ciarlet, P.G. and S. Kesavan (1981), Two-dimensional approximation of three-dimensional eigenvalue problems in plate theory, *Comp. Meth. Appl. Mech. Eng.* **26**, 145–172.

Ciarlet P.G. and H. Le Dret (1989), Justification of the boundary conditions of a clamped plate by an asymptotic analysis, *Asymptotic Anal.* **2**, 257–277.

Ciarlet P.G., H. Le Dret and R. Nzengwa (1987), Modélisation de la jonction entre un corps tridimensionnel et une plaque, *C. R. Acad. Sci. Paris* **305**, Série **I**, 55–58.

Ciarlet P.G., H. Le Dret and R. Nzengwa (1989), Junctions between three dimensional and two-dimensional linearly elastic structures, *J. Math. Pures Appl.* **68**, 261–295.

Ciarlet P.G. and B. Miara (1992), Justification of the two–dimensional equations of a linearly elastic shallow shell, *Comm. Pure Appl. Math.* **XLV**, 327–360.

Ciarlet, P.G. and J.C. Paumier (1985), A justification of the Marguerre–von Kàrmàn equations, *C. R. Acad. Sci.* **301**, 857–860.

Ciarlet, P.G. and J.C. Paumier (1986), A justification of the Marguerre–von Kàrmàn equations, *Comput. Mech.* **1**, 177–202.

Ciarlet, P.G. and P. Rabier (1980), *Les Equation de von Kàrmàn*, Lecture Notes in Mathematics **826** (Springer, Berlin).

Cimatti, G. (1973), The constrained elastic beams, *Meccanica* **8**, 119–124.

Cimetière, A., G. Geymonat, H. Le Dret, A. Raoult and Z. Tutek (1986), Une dérivation d'un modéle non linéaire de poutres a partir de l'élasticité tridimensionelle, *C. R. Acad. Sci.* **302**, 697–700.

Cimetière, A., G. Geymonat, H. Le Dret, A. Raoult and Z. Tutek (1988), Asymptotic theory and analysis for displacement and stress distributions in nonlinear elastic straight slender rods, *J. Elasticity* **19**, 111–161.

Cioranescu, D. and J. Saint Jean Paulin (1979), Homogenization in open sets with holes, *J. Math. Anal. Appl.* **71**, 590–607.

Cioranescu, D. and J. Saint Jean Paulin (1986), Reinforced and honey-comb structures, *J. Math. Pures Appl.* **65**, 403–422.

Cioranescu, D. and J. Saint Jean Paulin (1988), Elastic behaviour of very thin cellular structures, in: J.M. Ball ed., *Material Instabilities in Continuum Mechanics* (Clarendon Press, Oxford), 64–75.

Cioranescu, D. and J. Saint Jean Paulin (1989), Structures très minces en élasticité linearisée: Tours et grillages, *C. R. Acad. Sci. Paris Sér. I*, **308**, 41–46.

Cole, J.D. (1968), *Perturbation Methods in Applied Mathematics* (Ginn Blaisdfell, Boston).

Compe, C., P.L. George, B. Rousselet and M. Vidrascu (1986), Les éléments de poutres en élasticité linéaire de la Biblioteque Modulef, *Rapport de Recherche* **562** (Inria, France).

Crisfield, M.A. (1986), *Finite Elements and Solution Procedures for Structural Analysis* (Pineridge Press, Swansea).

Dafermos, C.M. (1973), Solutions of the Riemann problem for a class of hyperbolic systems of conservation laws by the viscosity method, *Arch. Rational Mech. Anal.* **52**, 1–9.

Dafermos, C.M. and W.J. Hrusa (1985), Energy methods for quasilinear hyperbolic initial-boundary value problems: Applications to elastodynamics, *Arch. Rational Mech. Anal.* **87**, 267–292.

Dautray, R. and J.L. Lions (1984), *Analyse Mathematique et Calcul Numérique pour les Sciences et les Techniques*. Collection du Comissariat à l'Energie Atomique (Masson, Paris).

Davet, J.L. (1986), *Justification de modèles de plaques nonlinéaires pour des lois de comportement generales, Modern Math. Anal. Numer.* **20**, 225–249.

Destuynder, P. (1980), Sur une justification de modèles de plaques et de coques par les methodes asymptotiques, Thesis, Université Pierre et Marie Curie, Paris.

Destuynder, P. (1981), Comparaison entre les modèles tridimensionnels et bidimensionnels de plaques en élasticité, *RAIRO Anal. Numér.* **15**, 331–369.

Destuynder, P. (1982), Sur les modèles de plaques minces en élastoplasticité, *J. Méc. Théor. et Appl.* **1**, 73–80.

Destuynder, P. (1985), A classification of thin shell theory, *Acta Appl. Math.* **4**, 15–63.

Destuynder, P. (1986), *Une théorie asymptotique de plaques minces en élasticité linéaire* (Masson, Paris).

Dhat, G. and G. Touzot (1984), *The Finite Element Method Displayed* (Wiley, New York).

Dikmen, M. (1982), *Theory of Thin Elastic Shells* (Pitman, Boston).

Dökmeci, M.C. (1972), A general theory of elastic beams, *Internat. J. Solids and Structures* **8**, 1205–1222.

Duvaut, G. (1990), *Mecánique des Milieux Continus* (Masson, Paris).

Duvaut, G. and J.L. Lions (1972a), *Les Inéquations en Mécanique et en Physique* (Dunod, Paris).

Duvaut, G. and J.L. Lions (1972b), Inéquations en thermoélasticité et magneto–hydrodynamique, *Arch. Rational Mech. Anal.* **46**, 241–279.

Dym, C.L. and I.H. Shames (1973), *Solid Mechanics. A variational approach* (McGraw-Hill, New York).

Eckhaus, W. (1979), *Asymptotic Analysis of Singular Perturbations* (North-Holland, Amsterdam).

Ekeland, I. and R. Temam (1975), *Convex Analysis and Variational Problems* (North-Holland, Amsterdam).

Fairweather, G. (1978), *Finite Element Galerkin Methods for Differential Equations* (Dekker, New York).

Fayolle, S. (1987), Sur l'analyse numérique de raccords de poutres et de plaques. Thesis 3ème Cycle, Université Pierre et Marie Curie, Paris.

Fernandez, M.T. and J.M. Viaño (1992), Comparación numérica de los modelos reducidos con el modelo tridimensional en vigas elásticas, *Publ. Dep. Mat. Apl.* **1** (Universidad de Santiago de Compostela, Spain).

Fichera, G. (1972), Existence Theorems in Elasticity, in: *Handbuch der Physik* Vol. **VIa/2** (Springer, Berlin), 347–389.

Figueiredo, I.N. (1989), Modèles de coques elastiques non linéaires: Méthode asymptotique et existence de solutions, Thesis, Université Pierre et Marie Curie, Paris.

Figueiredo, I.N. (1990), A justification of the Donnell–Mushtari–Vlasov model by the asymptotic expansion method, *Asymptotic Anal.* **36**, 221–234.

Figueiredo, I.N. and L. Trabucho (1992), A Galerkin approximation for linear elastic shallow shells, *Comput. Mech.* **10**, 107–120.

Figueiredo, I.N. and L. Trabucho (1993), A Galerkin approximation for curved beams, *Comp. Meth. Appl. Mech. Engrg.* **102**, 235–253.

Fraejis de Veubeke, B.M. (1979), *A Course in Elasticity* (Springer, Berlin).

Francfort, G. (1983), Homogenization and linear thermoelasticity, *SIAM J. Math. Anal.* **14**, 696–708.

Friedrichs, K.O. and R.F. Dressler (1961), A boundary layer theory for elastic plates, *Comm. Pure Appl. Math.* **14**, 1–33.

Fujita, H. and T. Suzuki (1991), Evolution problems, in: P. G. Ciarlet and J.L. Lions, eds., *Handbook of Numerical Analysis*, Vol. **II** (North-Holland, Amsterdam).

Fung, Y.C. (1965), *Foundations of Solid Mechanics* (Prentice-Hall, Englewood Cliffs, NJ).

Germain, P. (1962), *Mécanique des Milieux Continus* (Masson, Paris).

GERMAIN, P. (1972), *Mécanique des Milieux Continus, Tome 1* (Masson, Paris).

GEYMONAT, G. , F. KRASUCKI and J.J. MARIGO (1987a), Stress distribution in anisotropic elastic composite beams, in: P.G. CIARLET and E. SANCHEZ-PALENCIA, ed., *Applications of Multiple Scalings in Mechanics*, RMA 4 (Masson, Paris), 118–133.

GEYMONAT, G., F. KRASUCKI and J.J. MARIGO (1987b), Sur la comutativité des passages à la limite en théorie asymptotique des poutres composites, *C. R. Acad. Sci. Paris Sér. II*, **305**, 225–228.

GIRAULT, V. and P.A. RAVIART (1981), *Finite Element Approximation of the Navier–Stokes Equations*, Lecture Notes in Mathematics **749** (Springer, Berlin).

GLIMM, J. (1965), Solutions in the large for nonlinear hyperbolic systems of equations, *Comm. Pure Appl. Math.* **18**, 697–715.

GLOWINSKI, R. (1984), *Numerical Methods for Nonlinear Variational Problems* (Springer, New York).

GOLDENVEIZER, A.L. (1962), Derivation of an approximated theory of bending of a plate by the method of asymptotic integration of the equations of the theory of elasticity, *Plikl. Mat. Mech.* **26**, 668–686.

GOLDENVEIZER, A.L. (1963), Derivation of an approximate theory of bending a plate by a method of asymptotic integration of the equations in the theory of elasticity, *J. Appl. Math.* **19**, 1000–1025.

GOODIER, J.N. (1938), On the problems of the beam and the plate in the theory of elasticity, *Trans. Roy. Soc. Canada* **32**, 65–88.

GREEN, A.E. and P.M. NAGHDI (1990), A direct theory for composite rods, in: EASON G. and R.W. OGDEN, eds., *Elasticity: Mathematical Methods and Applications* (Ellis Horwood, Chichester).

GREEN, A.E. and W. ZERNA (1968), *Theoretical Elasticity* (University Press, Oxford).

GRUAIS, I. (1990), Modélisation de la jonction entre une poutre et une plaque en elasticité, Thesis, Université Pierre et Marie Curie, Paris.

GURTIN, M.E. (1972), The linear theory of elasticity, in: FLUGGE S. and C. TRUESDELL, eds., *Handbuch der Physik* **VI/2** (Springer, Berlin), 1–295.

HASLINGER, J. (1977), Finite element analysis for unilateral problems with obstacles on the boundary, *Apl. Mat.* **22**, 180–188.

HASLINGER, J. (1980), Convergence and dual finite element approximations for unilateral boundary value problems, *Aplik. Matematiky* **25**, 375–386.

HASLINGER, J. (1981), Mixed formulation of elliptic variational inequalities and its approximation, *Apl. Mat.* **26**, 462–475.

HINTON, E. and D.R.J. OWEN (1979), *An Introduction to Finite Element Computations* (Pineridge Press, Swansea).

HLAVÁČEK, I., J. HASLINGER, I. NEČAS and J. LOVÍŠEK (1986), *Solution of Variational Inequalities in Mechanics* (Springer, Berlin).

HOMMAN-CANCELO, R.N. (1980), Une justification des modéles de plaques en élasticité: Conditions aux limites du type Fourier, problèmes unilateraux, Thèse de 3ème Cycle, Université Pierre et Marie Curie, Paris.

HUGHES, T.J.R. (1987), *The Finite Element Method: Linear, Static and Dynamic Finite Element Analysis* (Prentice-Hall, Englewood Cliffs, NJ).

HUGHES, T.J.R., T. KATO and J.E. MARSDEN (1976), Well-posed quasi-linear second-order hyperbolic systems with applications to nonlinear elastodynamics and general relativity, *Arch. Rational Mech. Anal.* **63**, 273–294.

HUGHES, T.J.R., and J.E. MARSDEN (1983), *Mathematical Foundations of Elasticity* (Prentice-Hall, Englewood Cliffs, NJ).

JOHN, F. (1971), Refined interior equations for thin elastic shells, *Comm. Pure Appl. Math.* **24**, 583–615.

JOHNSON, C. (1987), *Numerical Solution of Partial Differential Equations by the Finite Element Method* (Cambridge Univ. Press, Cambridge).

JOHNSON, C., U. NÄVERT and J. PITKÄRANTA (1984) Finite element methods for linear hyperbolic problems, *Comp. Meth. Appl. Mech. Engrg.* **45**, 285–312.

KAMAL, M.M. and J.A.WOLF (1977), *Finite Element Applications in Vibration Problems* (Amer. Soc. of Mech. Engineers, New York)

KAPLUN, S. (1967), *Fluid Mechanics and Singular Perturbations* (Academic Press, New York).

KARWOWSKI, A. (1990), Asymptotic models for a long elastic cylinder, *J. Elasticity* **24**, 229–287.

KATO, T. (1979), Linear and quasi-linear equations of evolution of hyperbolic type, CIME Lectures (Cortona).

KERR, A.D. (1964), Elastic and viscoelastic foundation models, *J. Appl. Mech.* **31**, 491–498.

KIKUCHI, N. (1986), *Finite Element Methods in Mechanics* (Cambridge Univ. Press, Cambridge).

KIKUCHI, N. and J.T. ODEN (1988), *Contact Problems in Elasticity* (SIAM, Philadelphia, PA).

KOHN, R.V. and M. VOGELIUS (1984), A new model for thin plates with rapidly varying thickness. I, *Internat. J. Engrg. Sci.* **20**, 333–350.

KOHN, R.V. and M. VOGELIUS (1985), A new model for thin plates with rapidly varying thickness. II: A convergence proof, *Quart. Appl. Math.* **43**, 1–21.

KOHN, R.V. and M. VOGELIUS (1986), A new model for thin plates with rapidly varying thickness. III: Comparison of different scalings, *Quart. Appl. Math.* **44**, 35–48.

KOITER, W.T. (1970), On the foundations of the linear theory of thin elastic shells, *Proc. Kon. Ned. Akad. Wetensch.* **B 73**, 169–195.

KONDRATIEV, V.A. and O.A. OLEINIK (1989), On Korn's inequalities, *C. R. Acad. Sci. Paris Sér. I* **308**, 483–487.

LAGESTROM, P.A. and R.G. CASTEN (1972), Basic concepts underlying singular perturbation techniques, *SIAM Rev.* **14**, 63–120.

LANDAU, L. and LIFCHITZ (1967), *Théorie de l'Élasticité* (Mir, Moscow).

LAROZE, S. (1988), *Mécanique des Structures. Tome 1: Millieux Continus, Solides, Plaques et Coques. Tome 2: Poutres* (Eyrolles, Masson, Paris).

LEBELTEL C. (1989), Methode des developpements asymptotiques pour un probleme de plaque thermoelastique, Rapport de Recherche **1108**, I.N.R.I.A., Rocquencourt, France.

LE DRET, H. (1989), Modelling of the junction between two rods, *J. Math. Pures Appl.* **68**, 365–397.

LE DRET, H. (1990), Modelling of a folded plate, *Comput. Mech.* **5**, 401–416.

LE DRET, H. (1991), *Problèmes Variationnels dans les Multi-domaines. Modélisation des Jonctions et Applications* (Masson, Paris).

LE DRET, H. (1993), Convergence of displacements and stresses in linearly elastic slender rods as the thickness goes to zero, to appear.

LEGUILLON, D. and E. SANCHEZ-PALENCIA (1990), Approximation of a two dimensional problem of junctions, *Comput. Mech.* **6**, 435–455.

LEKHNITSKII, E.A. (1977), *Theory of Elasticity of an Anisotropic Body* (Mir, Moscow).

LÉNÉ, F. (1984), Contribution à l'étude des matériaux composites et de leur endommagement, Thèse de Doctorat d'État, Université Pierre et Marie Curie, Paris.

LE TALLEC, P., S. MANI and F. A. ROCHINHA (1992), Finite element computation of hyperelastic rods in large displacements, *Mat. Model. Numer. Anal.* **26**, 595–625.

LIONS, J.L. (1969), *Quelques Méthodes de Résolution des Problèmes aux Limites Non Linéaires* (Dunod, Paris).

LIONS, J.L. (1973), *Perturbations Singulières dans les Problèmes aux Limites et en Contrôle Optimal*, Lecture Notes in Mathematics **323** (Springer, Berlin).

LIONS, J.L. and E. MAGENES (1968), *Problèmes aux Limites Non Homogènes et Applications* (Dunod, Paris).

LODS, V. (1992), Formulation mixte d'un problème de jonctions de poutres adaptée à la résolution d'un probèma d'optimisation, *Math. Methods Numer. Anal.* **26**, 523–553.

LOVE, A.E.H. (1929), *A Treatise on the Mathematical Theory of Elasticity* (Cambridge Univ. Press, London).

MAMPASSI, B. (1992), Un type de jonction bidimensionnelle d'une tige et d'un massif élastiques, *C.R. Acad. Sci. Paris* **315**, 261–266.

MAMPASSI, B. and E. SANCHEZ-PALENCIA (1992), Etat local de contraintes à une jonction de plaque avec corps tridimensionnel, *C.R. Acad. Sci. Paris* **315**, 129–135.

MARSDEN, J.E. and T.J.R. HUGHES (1983), *Mathematical Foundations of Elasticity* (Prentice-Hall, Englewood Cliffs, NJ).

MASCARENHAS M.L. and L. TRABUCHO (1990), Homogenized behaviour of a beam with a multicellular cross section, *Appl. Anal.* **38**, 97–119.

MASCARENHAS M.L. and L. TRABUCHO (1991), Homogenization and a Galerkin approximation in three-dimensional beam theory, *Appl. Anal.* **42**, 83–111.

MASCARENHAS M.L. and L. TRABUCHO (1992), Asymptotic, homogenisation and Galerkin methods in three-dimensional beam theory, in: W.F. AMES and P.J. VAN DER HOUWEN, eds., *Computational and Applied Mathematics II-IMACS* (North-Holland, Amsterdam), 85–91.

MEIROVITCH, L. (1980), *Computational Methods in Structural Dynamics* (Sijthoff & Noordhoff, Alphen aan den Rijn).

MIARA, B. (1989), Optimal spectral approximation in linearized plate theory, *Appl. Anal.* **31**, 291–307.

MIARA, B. (1992a), Justification des mises à l'échelle et des hypothéses sur les donnés dans l'analyse asymptotique des modéles bidimensionnelles de plaques minces élastiques. I-Le cas linéaire, *C.R. Acad. Sci. Paris* **314**, 687–690.

MIARA, B. (1992b), Justification des mises à l'échelle et des hypothéses sur les donnés dans l'analyse asymptotique des modéles bidimensionnelles de plaques minces élastiques. II-Le cas nonlinéaire, *C.R. Acad. Sci. Paris* **314**, 965–968.

MIARA, B. and L. TRABUCHO (1990), Approximation spectrale pour une poutre en élasticité linéarisée, *C.R. Acad. Sci. Paris* **311**, 659–662.

MIARA, B. and L. TRABUCHO (1992), A Galerkin spectral approximation in linearized beam theory, *Math. Model. and Numer. Anal.* **26**, 425–446.

MINDLIN, R.D. (1951), Influence of rotatory inertia and shear on flexural motions of isotropic elastic plates, *J. Appl. Mech.* **18**, 31–38.

MITCHELL, A.R. and D.F. GRIFFITHS (1980), *The Finite Difference Method in Partial Differential Equations* (Wiley, Chichester).

MURAT, F. (1977/1978), *H-Convergence*, Séminaire d'Analyse Fonctionnelle et Numérique (Multigraphed, Alger).

MURAT, F. (1978), Compacité par compensation, *Ann. Scuola Norm. Sup. Pisa, Sci. Fis. Mat. (IV)* **V** (3), 489–507.

MURRAY, N.W. (1986), *Introduction to the Theory of Thin-walled Structures* (Clarendon Press, Oxford).

NAGHDI, P.M. (1972), The theory of plates and shells, in: S. FLUGGE and C. TRUESDELL, eds., *Handbuch der Physik*, **VI a/2** (Springer, Berlin).

NAYFEH, A. (1973), *Perturbations Methods* (Wiley, New York).

NEČAS, J. (1967), *Les Méthodes Directes en Théorie des Equations Elliptiques* (Masson, Paris).

NEČAS, J. and I. HLAVÁČEK (1981), *Mathematical Theory of Elastic and Elasto-Plastic Bodies: An Introduction* (Elsevier, New York).

NIORDSON, F.I., (ed.) (1969), *Theory of Thin Shells* (Springer, Berlin).

NOVOZHILOV, V.V. (1959), *The Theory of Thin Shells* (Noordhoff, Groningen).

NOWACKI, W. (1986), *Thermoelasticity* (Pergamon Press, Oxford).

NOWINSKI, J.L. (1978), *Theory of Thermoelasticity with Applications* (Sijthoff and Noordhoff, Alphen aan den Rijn).

ODEN, J.T. (1967), *Mechanics of Elastic Structures* (McGraw-Hill, New York).

ODEN, J.T., and G.F. CAREY (1981–1984), *Finite Elements I: An Introduction* (with E.B. BECKER); *II: A Second Course; III: Computational Aspects; IV: Mathematical Aspects; V: Special Problems in Solid Mechanics* (Prentice Hall, Englewood Cliffs, NJ).

ODEN, J.T. and S.J. KIM (1982), Interior penalty methods for finite element approximations of the Signorini problem in elastostatics, *Comput. Math. Appl.* **8**, 35–56.

ODEN, J.T. and E.B. PIRES (1983), Non local and nonlinear friction laws and variational principles for contact problems in elasticity, *J. Appl. Mech.* **50**.

ODEN, J.T. and J.N. REDDY (1976), *An Introduction to the Mathematical Theory of Finite Elements* (Wiley Interscience, New York).

ODEN, J.T. and E.A. RIPPERGER (1981), *Mechanics of Elastic Structures* (McGraw-Hill, New York).

O'MALLEY, R.E. (1974), *Introduction to Singular Perturbations* (Academic Press, New York).

PANAGIOTOPOULOS, P.D. (1985), *Inequality Problems in Mechanics and Applications* (Birkhäuser, Boston).

QUINTELA-ESTÉVEZ, P. (1989), A new model for nonlinear elastic plates with rapidly varying thickness, *Appl. Anal.* **32**, 107–127.

RABIER, P. and J.M. THOMAS (1985), *Exercices d'Analyse Numérique des Équations aux Derivées Partielles* (Masson, Paris).

RAOULT, A. (1980), Contribution à l'étude des modèles d'évolution de plaques et à l'approximation d'équations d'évolution linéaires de second ordre par des méthodes multipas, Thèse de 3ème Cycle, Université Pierre et Marie Curie, Paris.

RAOULT, A. (1985), Construction d'un modèle d'évolution de plaques avec terme d'inértie de rotation, *Ann. Math. Pura Appl.* **139**, 361–400.

RAOULT, A. (1988), Analyse mathémathique de quelques modèles de plaques et de poutres elastiques ou elastoplastiques, Thesis, Université Pierre et Marie Curie, Paris.

RAOULT, A. (1990a), Asymptotic theory of nonlinear elastic dynamic plates, to appear.

RAOULT, A. (1990b), The dynamics of rods, in: *First Rutgers Conference on Theoretical Mechanics*, (New Brunswick, NJ).

RAOULT, A. (1991), Personal communication.

RAOULT, A. (1992), Asymptotic modeling of the electrodynamics of a multi-structure, *Asymptotic Analysis* **6**, 73–108.

RAVIART, P.A. and J.M. THOMAS (1983), *Introduction a l'Analyse Numérique des Equations aux Derivées Partielles* (Masson, Paris).

REDDY, J.N. (1984), *An Introduction to the Finite Element Method* (McGraw-Hill, New York).

RIGOLOT, A. (1972), Sur une théorie asymptotique des poutres, *J. Mécanique* **11**, 673–703.

RIGOLOT, A. (1976), Sur une théorie asymptotique des poutres, Thesis, Université Pierre et Marie Curie, Paris.

RIGOLOT, A. (1977a), Approximation asymptotique des vibrations de flexion des poutres droites élastiques, *J. Mécanique* **16**, 498–529.

RIGOLOT, A. (1977b), Déplacements et petites deformations des poutres droites: Analyse asymptotique de la solution à grande distance des bases, *J. Méc. Appl.* **1**, 175–206.

RODRIGUES, J.F. (1987), *Obstacle Problems in Mathematical Physics* (North-Holland, Amsterdam).

RODRÍGUEZ, J.M. (1990), Cálculo asintótico y numérico de constantes en vigas elásticas de perfil rectangular fino, Tesina de Licenciatura, Dpto. de Matemática Aplicada, Universidad de Santiago de Compostela, Santiago de Compostela.

RODRÍGUEZ, J.M. (1993), Una teoria asintótica en vigas elásticas de perfil fino y su justificación matemática, Thesis, Dept. Matemática Aplicada, Universidad de Santiago de Compostela, Spain, to appear.

RODRÍGUEZ, J.M. and J.M. VIAÑO (1991), Límite asintótico de la ecuación de Laplace con condiciones Dirichlet o Neumann en un rectángulo de pequeño espesor, in: *Actas XII C.E.D.Y.A. / II Congreso de Matemática Aplicada* (Universidad de Oviedo, Oviedo), 569–572.

RODRÍGUEZ, J.M. and J.M. VIAÑO (1992a), Limit behaviour of Laplace equation in small thickness multi-rectangular domains. Application to thin-walled beams torsion theory, in: C. HIRSH et al., eds., *Numerical Methods in Engineering '92* (Elsevier, Amsterdam), 833–838.

RODRÍGUEZ, J.M. and J.M. VIAÑO (1992b), Asymptotic analysis of Poisson's equation in a thin domain "without" junctions and its application to thin-walled elastic beams theories, to appear.

RODRÍGUEZ, J.M. and J.M. VIAÑO (1992c), Asymptotic derivation of a general linear model for thin-walled elastic rods without junctions, to appear.

RUDIN, W. (1970), *Real and Complex Analysis* (McGraw-Hill, New York).

SAINT VENANT, A.B. (1855), Mémoire sur la torsion des prismes, *Mémoires de l'Academie des Sciences des Savants Étrangers* **14**, 233–560.

SALENÇON, J. (1988), *Mécanique des Milieux Continus* (Ellipses, Paris).

SANCHEZ-HUBERT, J. and E. SANCHEZ-PALENCIA (1991), Couplage fléxion-torsion-traction dans les poutres anisotropes à section hétérogène, *C. R. Acad. Sci. Paris* **312**, 337–344.

SANCHEZ-PALENCIA, E. (1980), *Non Homogeneous Media and Vibration Theory*, Lecture Notes in Physics **127** (Springer, Berlin).

SANCHEZ-PALENCIA, E. (1989a), Statique et dynamique des coques minces. I.-Le cas de fléxion pure non inhibée, *C. R. Acad. Sci. Paris* **309**, 411–417.

SANCHEZ-PALENCIA, E. (1989b), Statique et dynamique des coques minces. I.-Le cas de fléxion pure inhibée, Approximation membranaire, *C. R. Acad. Sci. Paris* **309**, 531–537.

SANCHEZ-PALENCIA, E. (1990), Passage à la limite de l'élasticité tridimensionnelle à la théorie asymptotique de coques minces, *C. R. Acad. Sci. Paris* **311**, 906–916.

SCHWARTZ, L. (1979), *Analyse Hilbertienne* (Hermann, Paris).

SEGAL, I. (1963), Nonlinear Semigroups, *Ann. Math.* **78**, 339–364.

SIGNORINI, A. (1943), Transformazioni termoelastiche finite, Memória 1ª, *Ann. Mat. Pura Appl.* **22**, 33–143.

SIMO, J.C. and L. VU QUOC (1986), A three dimensional finite strain rod model. Part II: Computational Aspects, *Comput. Methods Appl. Mech. Engrg.* **58**, 79–116.

SOKOLNIKOFF, I.S. (1956), *Mathematical Theory of Elasticity* (McGraw-Hill, New York).

TANABE, H. (1979), *Equations of Evolution* (Pitman, London).

TARTAR, L. (1977), *Problèmes d'Homogéneization dans les Equations aux Dérivées Partielles*, Cours Peccot, Collège de France, Paris.

THOMÉE, V. (1984), *Finite Difference Methods for Parabolic Problems*, Lecture Notes in Mathematics **1054** (Springer, Berlin).

THOMÉE, V. (1990), Difference methods for linear parabolic equations, in: P.G. CIARLET and J.L. LIONS, eds., *Handbook of Numerical Analysis* **I** (North-Holland, Amsterdam).

TIMOSHENKO, S. (1921), On the correction for shear of the differential equation for transverse vibration of prismatic bars, *Philos. Mag. Ser. (6)* **41**, 744–746.

TIMOSHENKO, S. and J.N. GOODIER (1951), *Theory of Elasticity* (McGraw-Hill, New York).

TIMOSHENKO, S. and S. WOINOVSKY-KRIEGER (1959), *Theory of Plates and Shells* (McGraw-Hill, New York).

TRABUCHO, L. and J.M. VIAÑO (1987), Derivation of generalized models for linear elastic beams by asymptotic expansion methods, in: P.G. CIARLET and E. SANCHEZ-PALENCIA, eds., *Applications of Multiple Scalings in Mechanics*, RMA 4 (Masson, Paris), 302–315.

TRABUCHO, L. and J.M. VIAÑO (1988), A derivation of generalized Saint Venant's torsion theory from three dimensional elasticity by asymptotic expansion methods, *Appl. Anal.* **31**, 129–148.

TRABUCHO, L. and J.M. VIAÑO (1989), Existence and characterization of higher order terms in an asymptotic expansion method for linearized elastic beams, *J. Asymptotic Anal.* **2**, 223–255.

TRABUCHO, L. and J.M. VIAÑO (1990a), A new approach of Timoshenko's beam theory by the asymptotic expansion method, *Math. Model. and Numer. Anal.* **24**, 651–680.

TRABUCHO, L. and J.M. VIAÑO (1990b), Revisión de la teoría de Vlassov en flexión-torsión de vigas elásticas por métodos asintóticos, in: *Actas del XI C.E.D.Y.A. / I Congreso de Matemática Aplicada* (Universidad de Málaga, Málaga).

TRUESDELL, C. and W. NOLL (1965), The nonlinear field theories of mechanics, in: FLUGGE S. and C. TRUESDELL, eds., *Handbuch der Physik* **III/3** (Springer, Berlin).

TUTEK, Z. (1987), A homogenized model of rod in linear elasticity, in: P.G. CIARLET and E. SANCHEZ-PALENCIA, eds. *Applications of Multiple Scalings in Mechanics*, RMA 4 (Masson, Paris).

TUTEK, Z. and I. AGANOVIČ (1986), A justification of the one-dimensional linear model of elastic beam, *Math. Methods Appl. Sci.* **8**, 502–515.

VALID, R. (1981), *Mechanics of Continuous Media and Analysis of Structures* (North-Holland, Amsterdam).

VAN DYKE, M. (1964), *Perturbation Methods in Fluid Mechanics* (Academic Press, New York).

VEIGA, M.F. (1993a), A Galerkin approximation for homogeneous anisotropic elastic beams, to appear.

VEIGA, M.F. (1993b), Algumas contribuições para as teorias de vigas e placas, Thesis, Dept. Matemática FCUL, Lisbon, to appear.

VEIGA, M.F. (1993c), Asymptotic approximation of an elastic beam with a rotating cross section, to appear.

VIAÑO, J.M. (1981), Existencia y aproximación de soluciones en termoelasticidad y elastoplasticidad, Thesis, Dpto. de Ecuaciones Funcionales, Universidad de Santiago de Compostela, Santiago de Compostela.

VIAÑO, J.M. (1982), Justificacion de modelos en termoelasticidad de placas por métodos asintóticos, in: *Actas del V C.E.D.Y.A.* (Universidad de La Laguna, Tenerife, España), 323–355.

VIAÑO, J.M. (1983), Contribution à l'étude des modèles bidimensionels en thermoélasticité de plaques d'épaisseur non constante, Thesis 3ème Cycle, Université Pierre et Marie Curie, Paris.

VIAÑO, J.M. (1985a), Generalizacion y justificacion de modelos unilaterales en vigas elasticas sobre fundacion, in: *Actas del VIII C.E.D.Y.A.* (Universidad de Santander, España).

VIAÑO, J.M. (1985b), Análisis de un método numérico con elementos finitos para problemas de contacto unilateral sin rozamiento en elasticidad (I), *Rev. Internac. Métod. Numér. Cálc. Diseñ. Ingr.* **1**, 79–93.

VIAÑO, J.M. (1986a), Análisis de un método numérico con elementos finitos para problemas de contacto unilateral sin rozamiento en elasticidad (II), *Rev. Interne. Métod. Numér. Cálc. Diseñ. Ingr.* **2**, 63–86.

VIAÑO, J.M. (1986b), Analyse numerique d'une methode de résolution de problémes de contact unilateral en élasticité et son implementation dans Modulef, Rapport de Recherche **457**, I.N.R.I.A., Rocquencourt, France.

VIAÑO, J.M. (1990), Definición asintótica y cálculo numérico de la constante de Timoshenko en vigas elásticas, in: WINTER, G. and H. GALANTE eds., *Métodos Numéricos en Ingeniería* (Soc. Esp. Métod. Num. Ingr., Las Palmas de Gran Canaria), 386–392.

VIDRASCU, M. (1984), Comparaison numérique entre les solutions bidimensionelles et tridimensionnelles d'un problème de plaque encastré, Rapport de Recherche **309**, I.N.R.I.A., Rocquencourt, France.

VILARES, M. (1988), Cálculo numérico de nuevas constantes de Timoshenko en vigas elásticas, Tesina de Licenciatura, Dpto. de Matemática Aplicada, Universidad de Santiago de Compostela, Santiago de Compostela.

VILLAGIO, P. (1967), Monodimensional solids with constrained solutions, *Meccanica* **2**, 65–68.

VINSON, J.R. (1974), *Structural Mechanics: The Behavior of Plates and Shells* (Wiley-Interscience, New York).

VINSON, J.R. (1989), *The Behavior of Thin-walled Structures: Beams, Plates and Shells* (Kluwer, Dordrecht).

VINSON, J.R. and R.L. SIERAKOWSKI (1986), *The Behavior of Structures Composed of Composite Materials* (Martinus Nijhoff, Dordrecht).

VLASSOV, B.Z. (1961), *Thin-walled Elastic Beams*, translated from Russian (Israel Program for Scientific Translations, Jerusalem).

VLASSOV, B.Z. (1962), *Pièces Longues en Voiles Minces*, translated from Russian by G. SMIRNOFF (Eyrolles, Paris).

VOGELIUS, M. and I. BABUŠKA (1981), On a dimensional reduction method I. The optimal selection of basis functions, *Math. of Comp.* **37**, 31–46.

WANG C.C. and TRUESDELL C. (1973), *Introduction to Rational Elasticity* (Noordhoff, Groningen).

WASHIZU, K. (1982), *Variational Methods in Elasticity and Plasticity* (Pergamon Press, Oxford).

WEMPNER, G. (1981), *Mechanics of Solids with Applications to Thin Bodies* (Sijthoff and Noordhoff, Alphen aan den Rijn).

WIDERA, G.E.O., H. FAN and P. AFSHARI (1989), Applicability of asymptotic beam theories to thin (thick) walled pipes, in: A.K. NOOR, T. BELYTSCHKO, J.C. SIMO, eds., *Analytical and Computational Models of Shells* (ASME, New York), 33–52.

XIANG, Y. (1989), Un modèle de plaque en élasticité linéaire, Convergence du développement asymptotique de la plaque élastique nonlinéaire, Thesis, Université Pierre et Marie Curie, Paris.

YOSIDA, K. (1980), *Functional Analysis* (Springer, Berlin).

ZIENKIEWICZ, O.C. (1971), *The Finite Element Method in Engineering Science* (McGraw-Hill, London).

Printed and bound by CPI Group (UK) Ltd, Croydon, CR0 4YY
13/10/2024
01773498-0001

Subject Index